CONCEPTS OF
GENETICS

CONCEPTS OF
GENETICS

Sixth Edition

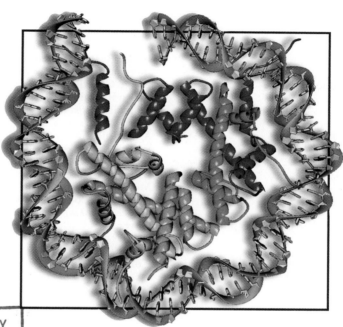

William S. Klug
The College of New Jersey

Michael R. Cummings
University of Illinois, Chicago

With contributions by

Elliott S. Goldstein   *Arizona State University*
Jon C. Herron   *University of Washington*
Marcia O'Connell   *The College of New Jersey*
Charlotte Spencer   *Cross Cancer Institute*

PRENTICE HALL   Upper Saddle River, New Jersey 07458

Library of Congress Cataloging-in-Publication Data
Klug, William S.
    Concepts of genetics / William S. Klug, Michael R.
Cummings with:
  contributions by Elliott S. Goldstein, Jon C. Herron, Marcia
O'Connell, Charlotte Spencer. — 6th ed.
        p.   cm.
    ISBN 0-13-081626-4
    1. Genetics.   I. Cummings, Michael R.   II. Title.
QH430.K574   2000
576.5—dc21   99-32928
                                                      CIP

Executive Editor: *Sheri L. Snavely*
Editor-in-Chief: *Paul F. Corey*
Assistant Vice President of Production and
   Manufacturing: *David W. Riccardi*
Executive Managing Editor: *Kathleen Schiaparelli*
Production Editor: *Alison Lorber*
Director of Creative Services: *Paul Belfanti*
Art Manager: *Gus Vibal*
Art Editors: *Karen Branson* and *Xiaohong Zhu*
Associate Creative Director: *Amy Rosen*
Art Director: *Ann France*
Project Manager: *Karen Horton*
Manufacturing Manager: *Trudy Pisciotti*
Editor-in-Chief of Development: *Ray Mullaney*
Development Editor: *Susan Weisberg*
Director of Marketing: *John Tweeddale*
Executive Marketing Manager: *Jennifer Welchans*
Image Resource Center Director: *Lori Morris-Nantz*

Image Resource Center Associate Director: *Melinda Reo*
Interior Image Specialist: *Beth Boyd*
Image Coordinator: *Debbie Hewitson*
Photo Researcher: *Stuart Kenter*
Editorial Assistant: *Lisa Tarabokjia*
Copy Editor: *Jane Loftus*
Art Studio: *Boston Graphics*
Prentice Hall Artists: *Patricia Gutierrez* and *Charles Pelletreau*
Media Editor: *Andrew Stull*
Assistant Managing Editor, Media: *Amy Reed*
Media Project Manager: *Michael Banino*
Animation Concept & Development: *Mark Shanley* and
Michael Cummings
Cover Art: © *Timothy J. Richmond/ETH Zurich (Swiss Federal
Institute of Technology). Reprinted with permission from Nature,
September 18, 1997 issue.*
Cover Photo: *Normal Chromosomes (human)* © *Dan McCoy/
Rainbow*

Printed in the United States of America

10   9   8   7   6   5   4   3   2

ISBN   0-13-081626-4

Prentice-Hall International (UK) Limited, *London*
Prentice-Hall of Australia Pty. Limited, *Sydney*
Prentice-Hall Canada Inc., *Toronto*
Prentice-Hall Hispanoamericana, S.A., *Mexico*
Prentice-Hall of India Private Limited, *New Delhi*
Prentice-Hall of Japan, Inc., *Tokyo*
Prentice-Hall (Singapore) Pte. Ltd.
Editora Prentice-Hall do Brasil Ltda., *Rio de Janeiro*

There are many references to genetics and heredity in literature that provide a deeper meaning to this discipline than the many facts and technical aspects that collectively help define it. None seems more poignant than the words of the slave Cinque in the Alexs Pate novel *Amistad.** During the trial to determine his freedom, Cinque faced a hopeless situation but spoke with great confidence to John Quincy Adams about where he would turn for help:

> *"I will call into the past,*
> *far back to the beginning*
> *of time, and beg them*
> *(my ancestors) to come help*
> *me at the judgment.*
> *I will reach back and*
> *draw them into me.*
> *And they must come,*
> *for at this moment,*
> *I am the whole reason*
> *they have existed at all."*

About The Authors

WILLIAM S. KLUG is currently Professor of Biology at The College of New Jersey (formerly Trenton State College) in Ewing, New Jersey. He served as Chairman of the Biology Department for 17 years, a position to which he was first elected in 1974. He received his B.A. degree in Biology from Wabash College in Crawfordsville, Indiana and his Ph.D. from Northwestern University in Evanston, Illinois. Prior to coming to Trenton State College, he returned to Wabash College as an Assistant Professor, where he first taught genetics as well as general biology and electron microscopy. His research interests have involved ultrastructural and molecular genetic studies of oogenesis in *Drosophila*. He has taught the genetics course as well as the senior capstone seminar course in human and molecular genetics to undergraduate Biology majors for each of the last 30 years.

MICHAEL R. CUMMINGS is currently Associate Professor in the Department of Biological Sciences and in the Department of Molecular Genetic at the University of Illinois at Chicago. He has also served on the faculty at Northwestern University and Florida State University. He received his B.A. from St. Mary's College in Winona, Minnesota, and his M.S. and Ph.D. from Northwestern University in Evanston, Illinois. He has also written textbooks in human genetics and general biology for non-majors. His research interests center on the molecular organization and physical mapping of human acrocentric chromosomes. At the undergraduate level, he teaches courses in Mendelian genetics, human genetics, and general biology for non-majors. He has received numerous teaching awards given by the university and by student organizations.

Preface

Concepts of Genetics is now completing its second decade of providing support to students as they study one of the most fascinating scientific disciplines. Certainly no subject area has had a more sustained impact on shaping our knowledge of the living condition. As a result of discoveries over the past 50 years, we now understand with reasonable clarity the underlying genetic mechanisms that explain how organisms develop into and then function as adults. We also better understand the basis of biological diversity and have greater insight into the evolutionary process.

Advances in genetic technology are now having a profound impact on our knowledge of human genetics. More than any other event, the launching of the Human Genome Project in 1990 symbolizes our commitment to the pursuit of such knowledge. As we enter the twenty-first century, the application of genetics to the betterment of the human condition will become commonplace. While this era will be filled with the excitement of scientific discovery, many accompanying problems and controversies will also face us. Of these, how we utilize our knowledge of the nucleotide sequence of the human genome promises to be the most significant. This growing body of information has already generated many legal and ethical issues. Currently, the implications of genetic testing and gene therapy are becoming important societal concerns.

As geneticists and students of genetics, the thrill of being part of this era must be balanced by a strong sense of responsibility and careful attention to the many related issues that will undoubtedly arise. The formulation of proper laws and policies will depend on a comprehensive knowledge of genetics and measured responses to these issues. As a result, here has never been a higher premium or greater need for a useful and up-to-date genetics textbook.

The sixth edition of *Concepts of Genetics*, as with all past efforts, has been designed to achieve five major goals:

1. To establish a conceptual framework that represents a sound approach to learning and facilitates the comprehension of the vast amount of information constituting the field of genetics.

2. To emphasize the rich history of scientific discovery and analytical thought, so prevalent in genetics, that provides an opportunity for students to hone their problem-solving abilities and to explore how we know what we know.

3. To cover material with a clear, crisp organizational format, both within each chapter and throughout the chapter sequence, to facilitate effective use of the text.

4. To present carefully designed figures with strong pedagogic value that teach rather than merely illustrate complex topics and analytical experiments.

5. To provide students with clearly written, straightforward explanations that elucidate difficult, complex topics without oversimplifying presentations.

Writing a textbook that achieves these goals and having the opportunity to continually improve on each effort has been a labor of love for us. The creation of each of the six editions is a reflection not only of our passion for teaching genetics, but the constructive feedback and encouragement provided over the past two decades from adopters, reviewers, and our students.

FEATURES OF THIS EDITION

New Conceptual Headings

The organization of each chapter has been revised to include conceptual headings that initiate the discussion of all major topics. These are presented using the narrative style. Collectively, these are presented at the onset of each chapter in a section entitled *Key Concepts*. This organizational feature reinforces our emphasis on conceptual issues in genetics. This is in keeping not only with the title of this textbook, but with our ardent belief that learning depends on the understanding of concepts, which are not easily forgotten, in contrast with countless details that may be memorized but not always retained. We believe that this new approach to the organization of each chapter will pay substantial dividends in improving the pedagogic value of *Concepts of Genetics*.

New Organization of the Table of Contents

As the field of genetics has expanded, many topics have taken on greater importance and stature within the discipline. Once again, as with previous editions, this has necessitated substantial reorganization, which is apparent in the Table of Contents. This edition continues to reflect our belief that coverage of the more classical studies of genetics, which we refer to as *Heredity and the Phenotype*, should initiate the text as Part I. At the urging of many reviewers, we have included coverage of bacterial and phage genetics, as well as extranuclear inheritance in this section. Additionally, this initial part of the text now includes a new chapter, *Sex Determination and Sex Chromosomes* (Chapter 9). Part I then concludes with the chapter on *Chromosome Mutations*.

Part II, entitled *The Molecular Basis of Inheritance*, includes a group of chapters that center around the storage, expression, and regulation of genetic information. These topics collectively form what is classically thought of as molecular genetics. While we have decided to keep many DNA-related topics in Part II, those who wish to begin their courses at this point in the text should not have difficulty doing so. Because the chapters have been written to be independent of the information presented in Part I, many adopters have used this approach successfully in previous editions.

With the strong foundation of DNA structure and function in hand, Part III includes four chapters that address the newest area of study in genetics, entitled *Genomics*. This section opens with coverage of how DNA clones are constructed and analyzed, and follows with chapters that address how DNA sequences and genes are organized within chromosomes. Part III concludes with the consideration of the applications and ethics of genetic technology. Obviously, Part III represents the "cutting edge" of genetics and contains the most exciting, far-reaching findings in this field.

We conclude the text by returning to the consideration of organisms and populations in Part IV. Initial chapters in this section address the role of genetic information in development, cancer, and behavior. We conclude our coverage with a consideration of the genetics of populations and the role of genetics in the process of evolution.

While we realize that no Table of Contents in a diverse field such as genetics can satisfy everyone, the present organization can be used in all first courses in undergraduate genetics, whether offered in a semester, quarter, or two-quarter format. The Parts and Chapters within the text are written so as to be used interchangeably, providing flexibility for the instructor.

Revised Art Program

In this edition, we have substantially refined the entire art program. Every figure has been scrutinized, and most have been either redrawn or revised. A number of the most complex figures have been more carefully and closely linked to their accompanying discussion. In conjunction with many new striking photographs, the text provides an attractive and pedagogically superior illustration program to enhance and support discussions of every major topic.

As in past editions, our goal has been to create figures that facilitate learning. In many cases, "flowchart" figures have been developed to illustrate complex experimental approaches or genetic processes that are significant to our understanding of genetics.

Revisions and Modernization

The entire text has been revised with an eye to providing clarity and coverage of the latest cutting-edge topics in genetics. One of the most significant paradigms in genetic investigation involves the utilization of DNA cloning technology to study many areas of biology. In this edition, this trend is reflected particularly in the revision and modernization of the information presented in Part III—*Genomics*. For example, Chapter 20—*Organization of Genes in Chromosomes*, is a new chapter that includes the most recent findings involving our understanding of genomic organization in a wide range of organisms.

While many areas of genetics have been enhanced by the ability to clone and analyze DNA, human genetics has benefited more than any other field. The expansion of knowledge about our own species represents a second major trend in genetics. As a result, we have increased our coverage of human genetics throughout the text. This is particularly evident in two chapters, *Applications and Ethics of Genetic Technology* (Chapter 21) and *Genetics and Cancer* (Chapter 23), both of which heavily emphasize human genetics. These chapters convey some of the vast amount of information provided by the Human Genome Project. Two other chapters important to the understanding of genetics have been revised substantially: *Population Genetics* (Chapter 25) and *Genetics and Evolution* (Chapter 26). The revisions reflect recent progress in these fields, as well as the influence of molecular genetics on these disciplines. This effort is in keeping with the important role population and evolutionary biology plays in the study of biology.

New Essays

In keeping with the influence of genetics on our everyday lives, we have revised and expanded the essays entitled *Genetics, Technology, and Society*, which appear near the end of most chapters. A number of these are new to this edition, including "Genetics and Society at the Millenium" in Chapter 1 and "Gene Pools and Endangered Species" in Chapter 25. These essays, which contain accompanying references, can readily serve as the basis of class discussions in genetics courses.

New Media Resources

Concepts of Genetics takes a significant step forward with the integration of *interactive media resources* into the text. This is done through the **GenCDX** Student CD-ROM, the text-specific companion website (**http://www.prenhall.com/klug**), and the "Genetics MediaLab" sections at the end of most of the chapters. The integration of these resources provides both instructors and students with new learning tools to help explain difficult concepts, and to explore the applications and implications of genetics in a broader context.

The **GenCDX** Student CD-ROM, designed specifically for this edition, contains over two dozen animations and tutorials, and is the starting point for the web problems in the Genetics MediaLabs. These animations and tutorials are taken directly from the text's art program and cover a wide range of topics, including Mendelian inheritance, gene mutation, transcription, translation, gene cloning and population genetics. The tutorials consist of an introduction, followed by an animation with accompanying text, self-test questions and a conclusion. Each is followed by a set of related questions with hints and explanations accompanying each answer. Icons placed in selected text figures link the tutorials to concepts in the book. The **GenCDX** Student CD-ROM also contains a set of self-grading chapter problems, "Genetics, Technology and Society" exercises, Chapter Search Terms, a Student Bulletin Board and other resources.

The Companion Website for the text provides access to the websites for the MediaLab investigations, linked web destinations in genetics, links to genetics newsgroups, web navigation tips, additional self-grading problems, and other resources.

The "Genetics MediaLab" section at the end of each chapter contains a list of chapter-specific resources available on the **GenCDX** Student CD-ROM and the companion website. This section also contains several interactive web-linked problems designed to enhance the topics presented in the chapter. To complete these problems, students must actively participate in exercises such as virtual experiments in *Drosophila* genetics, bacterial conjugation and the nature of human genetic disorders. Each of these problems has been selected and screened for pedagogical value by genetics instructors. For reference, the estimated time required to solve the problem is noted at the beginning of the exercise.

Coupled to the book, these media resources provide a powerful tool for teaching and learning genetics. It is our hope that both students and instructors will explore these resources and make them an integral part of a genetics course.

INTERACTIVE MEDIA RESOURCES

For Students and Professors

Introducing

GenCDX — an interactive student CD-ROM packaged with the text.

Students can explore the most challenging concepts in genetics using this series of animated tutorials. Each animated tutorial is linked closely with specific concepts in the textbook and utilizes interactive art and multiple-choice questions to foster learning. For additional study, there are self-grading problems linked to each chapter. GenCDX serves as a launching point for access to the Companion Website including the Web Problems listed in the MediaLabs found at the end of every chapter.

The animations include:

Animation title	Chapter	Figure	Animation title	Chapter	Figure
Prophase I of Meiosis	2	Fig. 2.13	Transcription	13	Fig. 13.9
An Overview of Meiosis	2	Fig. 2.14	Eukaryotic mRNA Processing	13	Fig 13.10
Monohybrid Cross	3	Fig. 3.2	Translation	14	Figs. 14.6,
Punnett Square	3	Fig. 3.3			14.7, 14.8
Dihybrid Cross	3	Fig. 3.5	Regulation of Gene Expression	15	Fig. 15.5
Linkage and Single Gene Crossover	6	Fig. 6.5	Gene Mutations	17	Fig. 17.6
Linkage and Double Crossovers	6	Fig. 6.8	DNA Repair	17	Fig. 17.14
Structural Changes in	10	Figs. 10.14	DNA Repair	17	Fig. 17.19
Chromosomes		and 10.15	DNA Repair	17	Fig. 17.20
Structural Changes in Chromosomes	10	Fig. 10.19	Gene Cloning	18	Fig. 18.13
Structural Changes in Chromosomes	10	Fig. 10.24	Polymerase Chain Reaction	18	Fig. 18.22
Hershey-Chase Experiment	11	Fig. 11.6	Chromatin Structure	19	Fig. 19.12
Reassociation Kinetics and C_0t	11	Fig. 11.21	Immunoglobulin Expression	20	Figs. 20.17
DNA Replication	12	Figs. 12.11, 12.12			and 20.18
		and 12.14	DNA Fingerprinting	21	Fig. 21.15
DNA Recombination	12	Fig. 12.20	Population Genetics	25	Fig. 25.3

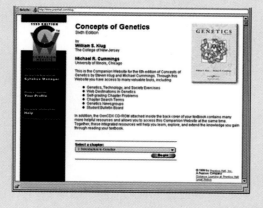

◀ Companion Website — www.prenhall.com/klug

Navigating the Website is as easy as opening the textbook— all of the features are organized by text chapters. The Companion Website provides study aid tools, guided explorations, and updates on current topics relevant to genetics courses. Other features for the student include: Self-grading Chapter Problems; Genetics, Technology and Society Exercises; Chapter Search Terms; and Syllabus Manager.

In addition to providing resources for the student, the Companion Website offers unique tools and support that make it easy to integrate Prentice Hall's on-line resources into the course.

Genetics MediaLab ▶

The Genetics MediaLabs section at the end of each chapter contains a list of chapter-specific resources available on the GenCDX CD-ROM and the Companion Website. This section also contains several web-linked problems designed to enhance the topics presented in the chapter. To complete these problems, students must actively participate in the exercises and virtual experiments. For reference, the estimated time required to solve the problem is noted at the beginning of the exercise.

◄ **Instructor CD-ROM Image Bank: Presentation Manager 3.0 (0-13-84430-6)**

Create multi-media presentations with this text-specific CD-ROM image bank. The image bank includes all illustrations and color images from the textbook. The Presentation Manager provides you with maximum flexibility and ease of use when previewing and selecting images, creating overlays for each image and creating class handouts. Images can also be imported into Powerpoint.

Additional Supplemental Resources for the Student

Student Handbook, Solutions Manual and Art Notebook (0-13-084436-5)
By Harry Nickla, *Creighton University* ►

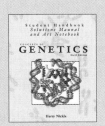

This valuable handbook provides a detailed step-by-step solution or lengthy discussion for every problem in the text. The handbook also features additional study aids including extra study problems, chapter outlines, vocabulary exercises and an overview of how to study genetics. New to this edition is an art notebook containing art reproduced from the textbook. Each figure corresponds to one of the overhead transparencies making it easier for students to take notes on the art during lectures.

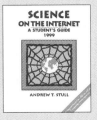

◄ **Science on the Internet: A Student's Guide (0-13-021308-X)**
By Andrew T. Stull

The perfect tool to help your students take advantage of the Concepts of Genetics Companion Website. This informative resource helps students locate and explore the myriad of science resources on the Web. It also provides an overview of the Web, general navigational strategies, and brief student activities. It is available FREE when packaged with the text.

The New York Times Themes of the Times (0-13-014323-5) ►

Compiled by Harry Nickla, Creighton University
This newspaper format supplement brings together recent articles on the latest issues in genetics taken directly from the pages of the New York Times. This free supplement, available in quantity through your local representative, helps students make connections between the classroom and the world around them.

Additional Supplemental Resources for the Instructor

Transparencies (013-084439-X)

170 of the line illustrations from the text are available as overhead transparencies that are reproduced in full color with extra large type for easy classroom viewing.

Instructor's Manual with Testbank
Harry Nickla, *Creighton University*
(0-13-84434-9)

This manual and testbank contains over 1000 questions and problems an instructor can use to prepare exams. The manual also provides optional course sequences, a guide to media supplements, and several "starter references " for term papers and special projects. The test bank portion is also available in Windows and Macintosh formats.

Prentice Hall Custom Test
Windows (0-13-084437-3) and Macintosh (0-13-84438-1)

Powerful and easy to use, this package allows you to edit and add your own questions, manage your students' test records and offer on-line testing.

EXTEND • EXCURSION • EXAMINE • EXCITE

Acknowledgments

No text can be the sole work of its authors. In addition to the input of the many reviewers, we have been blessed with extensive input from several colleagues who are deserving of special recognition for their writing efforts. Elliott Goldstein at Arizona State University is responsible for the creation of Chapter 20—*The Organization of Genes in Chromosomes* as well as the revision of Chapter 16—*Regulation of Gene Expression in Eukaryotes*. Likewise, Jon C. Herron at the University of Washington extensively revised Chapters 25 and 26—*Population Genetics* and *Genetics and Evolution*. At The College of New Jersey, Marcia O'Connell revised Chapter 22—*Developmental Genetics* and assisted with Chapter 21—*Applications and Ethics of DNA Biotechnology*. As in past editions, Charlotte Spencer at Cross Cancer Institute, wrote or revised many of the *Genetics, Society, and Technology* essays. Others were previously contributed by Mark Shotwell at Slippery Rock University.

In addition to the above individuals, we also express our thanks to Harry Nickla at Creighton University. In his role as author of the *Student Handbook* and the *Instructor's Manual*, he has reviewed and edited the problems at the end of each chapter. He also provided the brief answers to selected problems that appear in Appendix C. We appreciate the contributions of all of these geneticists and, particularly, the pleasant manner and dedication to this text displayed during our many interactions. We are forever grateful to them for their efforts.

At Prentice Hall, we express appreciation and high praise for the editorial guidance of Sheri Snavely, whose ideas and efforts have helped to shape and refine the features of this and the previous editions of the text. She has worked tirelessly to provide us with reviews from leading specialists who are also dedicated teachers, and to ensure that the pedagogy and design of the book are at the cutting edge of a rapidly changing discipline. We also appreciate the production efforts of Alison Lorber, whose quest for perfection is reflected throughout the text. Without her work ethic and dedication, the text would never have come to fruition. Marketing is being handled with talent and enthusiasm by Jennifer Welchans. Karen Horton, biology project manager, worked to ensure that the book, supplements, and cutting-edge interactive media all work together seamlessly.

Skillful developmental editing was provided by Susan Weisberg, whose efforts were directed by Ray Mullaney at Prentice Hall. Susan provided assiduous input into both text and art, often refining the pedagogic links between the two. Finally, the beauty and consistent presentation of the art work is the product of Paul Foti and his staff at Boston Graphics. Both Susan and Paul have truly behaved as colleagues during this mammoth project, and we are most pleased to have had the opportunity to work with them.

We also thank several individuals whose efforts have been crucial to the quality of the text. Stuart Kenter did the photo research, resulting in the many striking photographs found throughout the text. We also wish to thank Ann France, art director, for creatively guiding the text and cover design. At The College of New Jersey, Nikhil Bhagat and Archna Prasad provided critical research and proofreading assistance.

We would like to express our sincere appreciation for the team of people responsible for integrating interactive media resources into the new edition. Special thanks to W.R. Pensyl and his talented team of animators for providing the excellent animations on the **GenCDX** Student CD-ROM. Michael R. Cummings and Mark Shanley at the University of North Texas worked tirelessly with the animators to create animations and tutorials that will add great value to students' learning experience. We also are grateful to George Gilchrist at Clarkson University, and Kelly Owens at the University of Washington for thoughtfully creating the "Genetics MediaLabs" found at the end of each chapter. Special thanks to Anthony Maffia for his skillful effort in updating and improving the text specific website. We would like to give special recognition to Andrew Stull, biology media editor at Prentice Hall, for coordinating the complete technology package along with his colleagues Michael Banino and Amy Reed.

While we assume complete responsibility for any errors herein, we gratefully acknowledge the advice, contributions, and suggestions made by reviewers of all editions, and particularly those who were involved in this edition:

Laura Adamkewicz, *George Mason University*
Fred P. Andoli, *California State Polytechnic University*
James B. Courtright, *Marquette University*
Gam Dykstra, *University of Colorado, Boulder*
Johnny El-Rady, *University of South Florida*
Bert Ely, *University of South Carolina*
Thomas J. Glover, *Hobart & William Smith Colleges*
Elliott S. Goldstein, *Arizona State University*
Douglas Harrison, *University of Kentucky*
Tamara Horton, *Princeton University*
David P. Kuhn, *University of Central Florida*
Maria Gallo-Meagher, *University of Florida*
Berl R. Oakley, *Ohio State University*
Karen Rasmussen, *Bates College*
Dennis T. Ray, *University of Arizona*
Richard H. Shippee, *Vincennes University*
Albrecht von Arnim, *University of Tennessee*
David S. Wofford, *University of Florida*
Wade B. Worthen, *Furman University*
Kenneth G. Wunch, *Tulane University*
Robert S. Zemetra, *University of Idaho*

As the above acknowledgments make clear, a text such as this is a collective enterprise. All of the above individuals deserve to share in any success this text enjoys. We want them to know that our gratitude is equaled only by the extreme dedication evident in their many efforts. Many, many thanks to you all.

BRIEF CONTENTS

CONTENTS

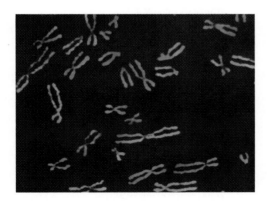

4 Extensions of Mendelian Genetics   77

Alleles alter phenotypes in different ways   78

Geneticists use a variety of symbols for alleles   78

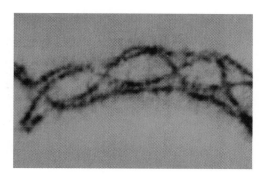

10   Chromosome Mutations: Variations in Chromosome Number and Arrangement   251

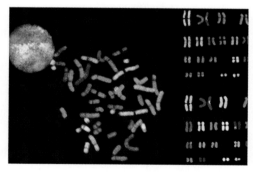

PART TWO
THE MOLECULAR BASIS OF HEREDITY   283

11   DNA Structure and Analysis   283

12   DNA Replication and Recombination   321

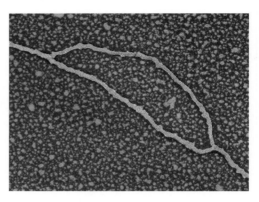

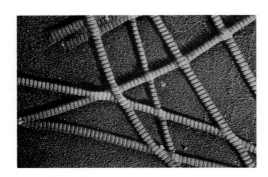

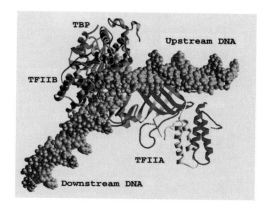

17 Gene Mutation, DNA Repair, and Transposable Elements 455

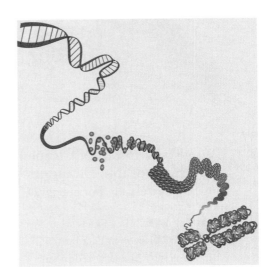

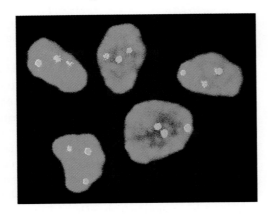

PART FOUR
GENETICS OF ORGANISMS
AND POPULATIONS 607

22 Developmental Genetics 607

23 Genetics and Cancer 635

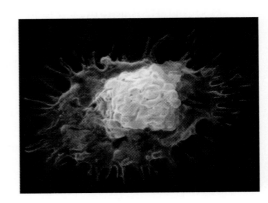

CONCEPTS OF
GENETICS

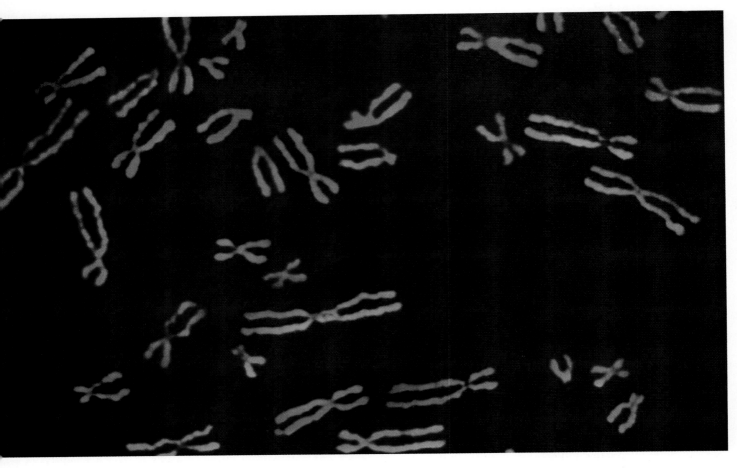

Human metaphase chromosomes, each composed of two sister chromatids joined at a common centromere.

1

Introduction to Genetics

KEY CONCEPTS

- **Genetics has a rich and interesting history**
  Prehistoric Times: Domesticated Animals and Cultivated Plants
  The Greek Influence: Hippocrates and Aristotle
  1600–1850: The Dawn of Modern Biology
  Charles Darwin and Evolution

- **Conceptual issues in genetics provide a useful overview of this discipline**

- **Genetics has been investigated using many different approaches**

- **Genetics has a profound impact on society**
  Eugenics and Euphenics
  Genetic Advances in Agriculture and Medicine

In early 1999, a controversy was growing among the 270,000 residents of the remote island nation of Iceland. Following months of fierce and sometimes heated debate, Iceland's Parliament passed a law giving a local biotechnology company the rights to develop a large database containing detailed DNA profiles of every resident of the country. This genetic information will be correlated with the genealogical and medical records of each entry and marketed to researchers around the world over the next decade.

This is not a passage drawn from Aldous Huxley's *Brave New World* but rather an example of the current interface between genetics and society as we enter the new millennium. There are many interesting reasons why such a possible invasion of genetic privacy is about to occur to the residents of this small outpost country. They represent a nearly unique case of genetic uniformity seldom seen and accessible to scientific investigation. Indeed, for a variety of reasons, most Icelanders share a close genetic resemblance, not only to one another, but to their Viking ancestors who settled this country over 1000 years ago. Because of this, geneticists believe that the Icelandic population is a tremendous asset in studying the link between genetics and disease. Because of the state-supported health-care system, medical records exist for all residents as far back as 1900. Genealogical information is available for every living resident, as well as for over 500,000 of the estimated 750,000 individuals who have ever lived in Iceland.

On the flip side, the debate centers around privacy, consent, and commercialism, the very same issues that will be the focal point of most future dilemmas and controversies stemming from the application of newly acquired genetic technology. The central question currently being asked throughout the worldwide scientific community is what we actually will do with the vast amount of genetic information that we are soon to acquire, and where society at large factors into the decision. For example, how will the knowledge of the complete nucleotide sequence of the human genome be used? More than at any other time in the history of science, addressing the ethical questions surrounding an emerging technology is clearly as important as the information that arises from that technology.

As you launch your study of the discipline of genetics, you should remain sensitive to these issues. As you proceed through this text, the knowledge gained will become a rich resource for achieving a thorough understanding of modern-day genetics. On the one hand, never has there been a more exciting time to be immersed in the study of an area of science. On the other hand, never has such a need for caution been so apparent to society. Along the way, enjoy your studies, but take your responsibility as a novice geneticist most seriously.

Genetics has a rich and interesting history

In the chapters that follow, we will discuss the nature of chromosomes, the way in which genetic information is transmitted from one generation to the next, and the way in which this information is stored, altered, regulated, and expressed. Significant scientific findings that provide the initial foundation for our discussions were obtained in the nineteenth century. As the twentieth century dawned, discoveries were made that began to clarify the understanding of the physical basis of living organisms and their relationship to one another. Several related ideas were gaining acceptance at that time and were particularly significant: (1) matter is composed of atoms; (2) cells are the fundamental units of living organisms; (3) nuclei somehow serve as the "life force" of cells; and (4) chromosomes housed within nuclei somehow play an important role in heredity. When these ideas were correlated with the newly rediscovered genetic findings of Gregor Mendel and integrated with Darwin's theory of natural selection and the origin of species, a more complete picture of life at the level of the individual and of the population emerged. The era of modern-day biology was initiated on this foundation.

But what of the many important ideas and hypotheses that served as forerunners of nineteenth-century thought? In the following short section, we consider some of these, several of which can be traced back well over 1000 years!

Prehistoric Times: Domesticated Animals and Cultivated Plants

While we don't know when people first recognized the existence of heredity, various forms of archeological evidence (e.g., primitive art, preserved bones and skulls, and dried seeds) have provided insights. Such evidence documents the successful domestication of animals and cultivation of plants thousands of years ago. These efforts represent artificial selection of genetic variants within populations. For example, between 8000 and 1000 B.C., horses, camels, oxen, and various breeds of dogs (derived from the wolf family) were domesticated to serve various roles. The cultivation of many plants, including maize, wheat, rice, and the date palm, is thought to have been initiated around 5000 B.C. The remains of maize dating to this period have been recovered in caves in the Tehuacán Valley of Mexico. Assyrian art depicts artificial pollination of the date palm, thought to have originated in Babylonia (Figure 1.1). Such cultivation is believed to have been the source of date

palms found in that region today, where over 400 varieties exist in just four oases in the Sahara Desert. They differ from one another in various traits, such as fruit taste.

Prehistoric evidence of cultivated plants and domesticated animals documents our ancient ancestors' successful attempts to manipulate the genetic composition of useful species. There is little doubt that people soon learned that desirable and undesirable traits are passed to successive generations and more desirable varieties of animals and plants could be selected. Human awareness of heredity was thus apparent during prehistoric times.

The Greek Influence: Hippocrates and Aristotle

Although few, if any, significant ideas were put forward to explain heredity during prehistoric times, philosophers directed much more attention to this subject during the Golden Age of Greek culture. They paid considerable attention to the subjects of reproduction and heredity, particularly as related to the origin of humans. This is particularly evident in the writings of the Hippocratic school of medicine (500–400 B.C.) and subsequently of the philosopher and naturalist Aristotle (384–322 B.C.).

Central to their explanation of the hereditary basis of the reproduction of animals were hypotheses concerning (1) the source of the *physical substance* of the offspring and (2) the nature of the *generative force* that directs the physical substance as it materializes (develops) into an adult organism.

For example, the Hippocratic treatise *On the Seed* argues that male semen is formed in numerous parts of the body and is transported through blood vessels to the testicles. Active "humors" act as the bearers of hereditary traits and are drawn from various parts of the body to the semen. These humors could be healthy or diseased, the latter condition accounting for the appearance of newborns exhibiting congenital disorders or deformities. Furthermore, it was believed that these humors could be altered in individuals and, in their new form, could be passed on to offspring. In this way, newborns could "inherit" traits that their parents had "acquired" because of their environment.

Aristotle (Figure 1.2), who was enormously interested in living organisms, and who had studied under Plato for some 20 years, was more critical and more expansive than Hippocrates in his analysis of human origins and heredity. Aristotle proposed that male semen was formed from blood rather than from each organ and that its generative power resided in a "vital heat" that it contained. This vital heat had the capacity to produce offspring of the same "form" (i.e., basic structure and capacities) as the parent. Aristotle believed that it generated offspring by cooking and shaping the menstrual blood produced by the female, which was the "physical substance" giving rise to an offspring. The embryo developed from the initial "set-

FIGURE 1.1   Relief carving depicting artificial pollination of date palms (800 B.C.). From the Metropolitan Museum of Art, gift of John D. Rockefeller, Jr. (32.143.3). Photo © 1983 The Metropolitan Museum of Art.

FIGURE 1.2 Aristotle describes the animals which Alexander has sent him. (A fresco from a mural in The Main Hall of The Assemblée Nationale in Paris, France. Painting by Eugene Delacroix). © Photo by Erich Lessing/Art Resource, NY.

ting" of the menstrual blood by the semen into a mature offspring, not because it already contained the parts in miniature (as some Hippocratics had thought), but because of the shaping power of the vital heat. These ideas constituted only one part of Aristotelian philosophy of order in the living world.

Although the ideas of Hippocrates and Aristotle may sound primitive and naive today, we should recall that, prior to the 1800s, neither sperm nor eggs had yet been observed in mammals. Thus, the Greek philosophers' ideas were worthy ones in their time and for centuries to come. As we will see, their thinking was not so different from that of Charles Darwin in his formal proposal of the theory of pangenesis put forward during the nineteenth century.

1600–1850: The Dawn of Modern Biology

During the ensuing 1900 years (300 B.C.–A.D. 1600), the theoretical understanding of genetics was not extended by significant new ideas. However, between 1600 and 1850,

major strides were made that provided much greater insights into the biological basis of life, setting the scene for the revolutionary work and principles presented by Charles Darwin and Gregor Mendel. In the 1600s the English anatomist William Harvey (1578–1657), better known for his experiments demonstrating that the blood is pumped by the heart through a circulatory system made up of the arteries and veins, also wrote a treatise on reproduction and development, patterned after Aristotle's work. In it he is credited with the earliest statement of the **theory of epigenesis**—that an organism is derived from substances present in the egg, which differentiate into adult structures during embryonic development. Patterned after Aristotle's ideas, epigenesis holds that new structures, such as body organs, are not present initially but instead are formed *de novo* in the embryo. Indeed, Harvey had studied Aristotle and was well aware of his ideas.

The theory of epigenesis conflicts directly with the theory of preformation, first put forward in the seventeenth century. Stating that sex cells contain a complete miniature adult called the **homunculus** (Figure 1.3), perfect in every form, preformation was popular well into the eighteenth century. However, work by the embryologist Casper Wolff (1733–1794) and others clearly disproved this theory, thus favoring epigenesis. Wolff was convinced

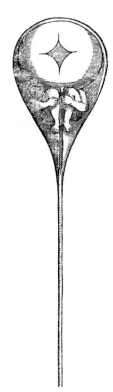

FIGURE 1.3 Depiction of the "homunculus," a sperm containing a miniature adult, perfect in proportion and fully formed.

that several structures, such as the alimentary canal, were not initially present in the earliest embryos he studied, but instead were formed later during development.

During this same period, other significant chemical and biological findings affected future scientific thinking. In 1808, John Dalton expounded his **atomic theory**, which stated that all matter is composed of small, invisible units called atoms. Improved microscopes became available, and around 1830 Matthias Schleiden and Theodor Schwann proposed the **cell theory**: All organisms are composed of basic visible units called cells, which are derived from similar preexisting structures. By this time, the idea of **spontaneous generation**, the creation of living organisms from nonliving components, had clearly been disproved by the experiments of Francesco Redi (1621–1697), Lazzaro Spallanzani (1729–1799), and Louis Pasteur (1822–1895), among others. Thus, living organisms were considered to be derived from preexisting organisms and to consist of cells made up of atoms.

Another prevailing notion had a major influence on nineteenth-century thinking, the **fixity of species**. According to this doctrine, animal and plant groups remain unchanged in form from the moment of their appearance on Earth. Embraced particularly by those who also adhered to a belief in special creation, this doctrine was popularized by several people, including the Swedish physician and plant taxonomist Carolus Linnaeus (1707–1778), who is better known for devising the binomial system of classification.

The influence of this tenet is illustrated by considering the work of the German plant breeder Joseph Gottlieb Kolreuter (1733–1806), who worked with tobacco. He crossbred two groups and derived a new hybrid form, which he then converted back to one of the parental types by repeated backcrosses. In other breeding experiments using carnations, he clearly observed segregation of traits, which was to become one of Mendel's principles of genetics. These results seemed to contradict the idea of "fixed species" that do not change with time. Because of Kolreuter's belief in both special creation and the fixity of species, he was puzzled about these outcomes and failed to recognize the real significance of his findings.

Charles Darwin and Evolution

With this information as background, we conclude our coverage of the historical context of genetics with a brief discussion of the work of Charles Darwin, who in 1859 published the book-length statement of his evolutionary theory, *On the Origin of Species*. Darwin's many geological, geographical, and biological observations convinced him that existing species arose by descent with modification from other ancestral species. Greatly influenced by his now-famous voyage on the H.M.S. *Beagle* (1831–1836),

Darwin's thinking culminated in his formulation of the **theory of natural selection**, which attempted to explain the causes of evolutionary change. Formulated and proposed at the same time, but independently, by Alfred Russel Wallace, natural selection is based on the observation that populations tend to consist of more offspring than the environment can support, leading to a struggle for survival among them. Those organisms with heritable traits that allow them to adapt to their environment are better able to survive and reproduce than those with less-adaptive traits. Over a long period of time, slight but advantageous variations will accumulate. If a population bearing these inherited variations becomes reproductively isolated, a new species may result.

The primary gap in Darwin's theory was a lack of understanding of the genetic basis of variation and inheritance, a gap that left it open to reasonable criticism well into the twentieth century. Aware of this weakness in his theory of evolution, in 1868 Darwin published a second book, *Variations in Animals and Plants under Domestication*, in which he attempted to provide a more definitive explanation of how heritable variation arises gradually over time. Two of his major ideas, pangenesis and the inheritance of acquired characteristics, have their roots in the theories involving "humors," as put forward by Hippocrates and Aristotle.

In his provisional hypothesis of **pangenesis**, Darwin coined the term *gemmules* (rather than humors) to describe the physical units representing each body part that were gathered by the blood into the semen. Darwin felt that these gemmules determined the nature or form of each body part. He further believed that gemmules could respond in an adaptive way to an individual's external environment. Once altered, such changes would be passed on to offspring, allowing for the inheritance of acquired characteristics. Lamarck had previously formalized this idea in his 1809 treatise, *Philosophie Zoologique*. Lamarck's theory, which became known as the **doctrine of use and disuse**, proposed that organisms acquire or lose characteristics that then become heritable.

Even though Darwin never understood the basis for inherited variation, his ideas concerning evolution may be the most influential theory ever put forward in the history of biology. He was able to distill his extensive observations and synthesize his ideas into a cohesive hypothesis describing the origin of diversity of organisms populating Earth.

As Darwin's work ensued, the experiments of Gregor Johann Mendel (Figure 1.4) were performed between 1856 and 1863, forming the basis for his classic 1866 paper. In it, Mendel demonstrated a number of statistical patterns underlying inheritance and developed a theory involving hereditary factors in the germ cells to explain these patterns. His research was virtually ignored until it was partially duplicated and then cited by Carl Correns, Hugo de

FIGURE 1.4 Gregor Johann Mendel, who in 1866 put forward the major postulates of transmission genetics as a result of experiments with the garden pea.

Vries, and Eric Von Tschermak around 1900 and subsequently championed by William Bateson.

By the early part of the twentieth century, chromosomes were discovered and support for the epigenetic interpretation of development had grown considerably. It gradually became clear that heredity and development were dependent on "information" contained in chromosomes, which were contributed by gametes to each individual. The "gap" in Darwin's theory had narrowed considerably.

In Chapter 3, we will return to a thorough analysis of Mendel's findings, which have served to this day as the foundation of genetics. His work was but one important part of the body of knowledge that would initiate the era of modern biological thought in the twentieth century.

Conceptual issues in genetics provide a useful overview of this discipline

We now turn to a review of some of the basic conceptual issues in genetics. They provide an overview of the field of genetics and establish the fundamental vocabulary that will

be useful as we proceed through the text. We shall approach these basic concepts by asking and answering a series of questions. You may wish to write or think through an answer before reading our answer to each question. A number of issues that are addressed here are summarized in Figure 1.5. You should refer to it as you read through this section. Throughout the text, the answers to these questions will be expanded as more detailed information is presented.

What does "genetics" mean?

Genetics is the branch of biology concerned with heredity and variation. This discipline involves the study of cells, individuals, their offspring, and the populations within which organisms live. Geneticists investigate all forms of inherited variation, as well as the molecular basis underlying such characteristics.

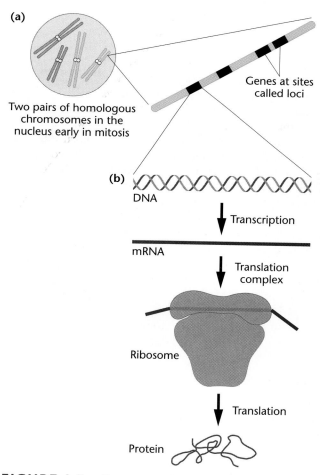

FIGURE 1.5 Depiction of (a) the storage of genetic information in homologous chromosomes, which contain genes made up of DNA; and (b) genetic expression involving the transcription of DNA into mRNA, which can be translated on a ribosome into a protein.

What is the center of heredity in a cell?

In eukaryotic organisms, the **nucleus** contains the genetic material. In prokaryotes, such as bacteria, the genetic material exists in an unenclosed but recognizable area of the cell called the **nucleoid region** (Figure 1.6). In viruses, which are not true cells, the genetic material is ensheathed in the protein coat, together constituting the viral head or capsid.

What is the genetic material?

In eukaryotes and prokaryotes, **DNA** serves as the molecule storing genetic information. In viruses, either DNA or RNA serves this function.

What do DNA and RNA stand for?

DNA and RNA are abbreviations for **deoxyribonucleic acid** and **ribonucleic acid**, respectively. These are the two types of nucleic acids found in organisms. Nucleic acids, along with carbohydrates, lipids, and proteins, compose the four major classes of organic biomolecules found in living things.

How is DNA organized to serve as the genetic material?

DNA, although single-stranded in a few viruses, is usually a double-stranded molecule organized as a double helix (Figure 1.7). Contained within each DNA molecule are hereditary units called **genes**, which are part of larger elements, the **chromosomes**.

What is a gene?

In simplest terms, the gene is the functional unit of heredity. In chemical terms, it is a linear array of nucleotides—the chemical building blocks of DNA and RNA. A more conceptual approach is to consider it to be an informational storage unit capable of undergoing replication, expression, and mutation. As investigations have progressed, the gene has been found to be a very complex element.

What is a chromosome?

In viruses and bacteria, which have only a single chromosome, it is most simply thought of as a long, usually circular DNA molecule organized into genes. Most eukaryotes have many chromosomes (Figure 1.8) that are composed of linear DNA molecules intimately associated with proteins. In addition, eukaryotic chromosomes contain many nongenic regions. It is not yet clear what role, if any, is played by many of these regions. Our knowledge of the chromosome, like that of the gene, is continually expanding.

When and how can chromosomes be visualized?

If the chromosomes are released from the viral head or the bacterial cell, they can be visualized under the electron microscope. In eukaryotes, chromosomes are most easily visualized under the light microscope when they are undergoing **mitosis** or **meiosis**. In these division processes, the material constituting chromosomes is tightly coiled and condensed, giving rise to the characteristic image of chromosomes. Following division, this material, called chromatin, uncoils during interphase, where it can be studied under the electron microscope.

How many chromosomes does an organism have?

Although there are exceptions, members of most eukaryotic species have a specific number of chromosomes, called the **diploid number (2n)**, in each somatic cell. For example, humans have a diploid number of 46 (Figure 1.9). Upon close analysis, these chromosomes are found to occur in pairs, each member of which shares a nearly identical appearance when visible during cell division. Called **homologous chromosomes**, the members of each pair are identical in their length and in the location of the

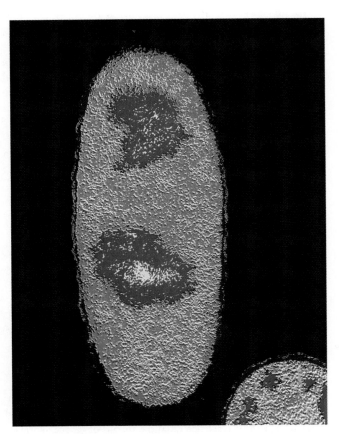

FIGURE I.6 Enhanced electron micrograph of *Escherichia coli*, demonstrating the nucleoid regions (shown in blue). The bacterium has replicated its DNA and is about to begin cell division.

FIGURE 1.7   A bronze sculpture of the DNA double helix.

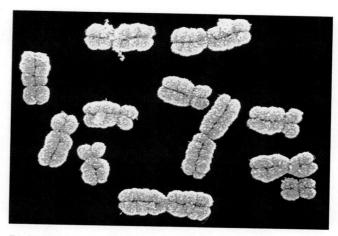

FIGURE 1.8   Human mitotic chromosomes visualized under the scanning electron microscope.

centromere, the point of spindle-fiber attachment during division. They also contain the same sequence of gene sites, or loci, and pair with one another during gamete formation (the process of meiosis).

The number of different *types* of chromosomes in any diploid species is equal to half the diploid number and is called the **haploid number (*n*)**. Some organisms, such as yeasts, are haploid during most of their life cycle and contain only one "set" of chromosomes. Other organisms, especially many plant species, are sometimes characterized by more than two sets of chromosomes and are said to be **polyploid**.

What is accomplished during the processes of mitosis and meiosis?

Mitosis is the process by which the genetic material of eukaryotic cells is duplicated and distributed during cell division. Meiosis is the process whereby cell division produces gametes in animals and spores in most plants. While mitosis occurs in somatic tissue and yields two progeny cells with an amount of genetic material identical to that of the progenitor cell, meiosis creates cells with precisely one-half of the genetic material. Each gamete receives one member of each homologous pair of chromosomes and is haploid. This reduction in chromosome number is essential if the offspring arising from two gametes are to maintain a constant number of chromosomes characteristic of their parents and other members of the species.

What are the sources of genetic variation?

Classically, there are two sources of genetic variation, **chromosomal mutations** and **gene mutations**. The former, also called chromosomal aberrations, include duplication, deletion, or rearrangement of chromosome segments. Gene mutations result from a change in the stored chemical information in DNA, collectively referred to as an organism's **genotype**. Such a change may include substitution, duplication, or deletion of nucleotides, which compose this chemical information. Alternative forms of the gene, which result from mutation, are called **alleles**. Genetic variation frequently, but not always, results in a change in some characteristic of an organism, referred to as its **phenotype**.

How does DNA store genetic information?

There are four forms of chemical building blocks called **nucleotides** in a segment of DNA constituting a gene. The sequence of nucleotides making up a gene encodes the chemical nature (the amino acid composition) of a protein, the end product of genetic expression. Mutations are produced when the nucleotide sequence (the genetic code) is altered, creating alternate forms of the genes, called alleles.

How is the genetic code organized?

There are four nucleotides in DNA, each varying in one of its components, the **nitrogenous base**. The **genetic code** is a triplet; therefore, each combination of three nucleotides constitutes a code word. Almost all the possible codes specify 1 of 20 **amino acids**, the chemical building blocks of proteins.

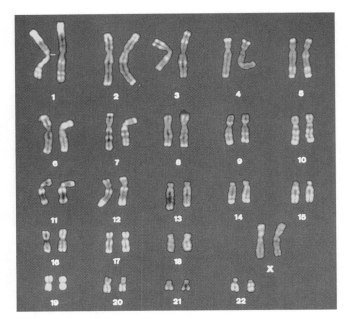

FIGURE 1.9 The human karyotype.

How is the genetic code expressed?

The coded information in DNA is first transferred during a process called **transcription** into a **messenger RNA (mRNA)** molecule. The mRNA subsequently associates with a cellular organelle, the **ribosome**, where it undergoes **translation** into a **protein**, the end product of almost all genes.

Are there exceptions, where proteins are not the end product of a gene?

Yes. For example, genes coding for **ribosomal RNA (rRNA)**, which is part of the ribosome, and for **transfer RNA (tRNA)**, which is involved in the translation process, are transcribed but not translated. Therefore, RNA is sometimes the end product of stored genetic information.

Why are proteins so important to living organisms that they serve as the end product of the vast majority of genes?

Many proteins serve as highly specific biological catalysts, or enzymes. In this role, these proteins control cellular metabolism, determining which carbohydrates, lipids, nucleic acids, and other proteins are present in the cell. Many other proteins perform nonenzymatic roles. For example, hemoglobin, collagen, immunoglobulins, and some hormones are proteins that play diverse roles in living organisms.

Why are enzymes necessary in living organisms?

As biological catalysts, enzymes lower the **energy of activation** required for most biochemical reactions and speed the attainment of equilibrium. Otherwise, these reactions would proceed so slowly as to be ineffectual in organisms living under the physical conditions on Earth. Some genes control the variety of enzymes present in any cell type, dictating its overall biochemical composition.

Genetics has been investigated using many different approaches

The scope of topics encompassed in the field of genetics is enormous. Studies have involved viruses, bacteria, and a wide variety of plants and animals and have spanned all levels of biological organization, from molecules to populations. It is helpful, before we embark on a detailed study of genetics, to categorize the types of investigations that have been used most often in this field. Although some overlap exists, most have used one of four basic approaches.

The most classic investigative approach is the study of **transmission genetics**, in which the patterns of inheritance of traits are examined. Experiments are designed so that the transmission of traits from parents to offspring can be analyzed through several generations. Patterns of inheritance are sought that will provide insights into genetic principles. The first significant experimentation of this kind to have a major impact on the understanding of heredity was performed by Gregor Mendel in the middle of the nineteenth century. The information derived from his work serves today as the foundation of transmission genetics. In human studies, where designed matings are neither possible nor desirable, **pedigree analysis** is often useful. In pedigree analysis, patterns of inheritance are traced through as many generations as possible, leading to inferences concerning the mode of inheritance of the trait under investigation.

The second approach involves **cytogenetics**—the study of chromosomes. The earliest such studies used light

FIGURE 1.10 Visualization of DNA fragments under ultraviolet light. The bands were produced using recombinant DNA technology.

microscopy. The initial discovery of chromosome behavior during mitosis and meiosis, early in the twentieth century, was a critical event in the history of genetics. In addition to playing an important role in the rediscovery and acceptance of Mendelian principles, these observations served as the basis of the chromosomal theory of inheritance. This theory, which viewed the chromosome as the carrier of genes and the functional unit of transmission of genetic information, was the cornerstone for further studies in genetics throughout the first half of this century.

The light microscope continues to be useful in the investigation of chromosome structure and abnormalities and is instrumental in preparing **karyotypes**, which illustrate the chromosomes characteristic of any species arranged in a standard sequence. With the advent of electron microscopy, the repertoire of investigative approaches in genetics has grown. In high-resolution microscopy, genetic molecules and their behavior during gene expression can be visualized directly.

The third general approach involves **molecular genetic analysis**, which has had the greatest impact on the recent growth of genetic knowledge. Molecular studies, initiated in the early 1940s, have consistently expanded our knowledge of the role of genetics in life processes. Although experiments initially relied on bacteria and the viruses that infect them, extensive information is now available concerning the nature, expression, replication, and regulation of the genetic information in eukaryotes as well. The precise nucleotide sequence has been determined for many genes cloned in the laboratory. Recombinant DNA studies (Figure 1.10), in which genes from another organism are literally spliced into bacterial or viral DNA and cloned *en masse*, serve as the basis of a far-reaching research technology used in molecular genetic investigations. Building on this approach, the new field of DNA biotechnology now exists whereby genes are identified, isolated, cloned, and sequenced. Using this new technology, it is now possible to probe gene function in extreme detail. Such molecular and biochemical analysis has created the potential for gene therapy and has profound implications in medicine, agriculture, and bioethics.

Perhaps the most striking achievement in the history of biotechnology occurred in 1996 at the Roslin Institute in Scotland, when the world's most famous lamb, Dolly, was born (Figure 1.11). Representing the first animal ever to be cloned from an adult somatic cell, Dolly was the result of the research of Ian Wilmut, who fused the nucleus of an udder cell taken from a six-year-old sheep with an enucleated oocyte of another sheep. Following implantation into a surrogate mother, complete embryonic and fetal development was achieved under the direction of the genetic material of the udder cell. While the implications of this event are profound and raise numerous ethical con-

FIGURE 1.11   Dolly, a Finn Dorset sheep cloned from the genetic material of an adult mammary cell, shown next to her first-born lamb Bonnie.

cerns, the goal of Wilmut's research is to ultimately use cloned animals as models to study human disease and to produce therapeutic drugs beneficial to humans.

The final approach involves the study of **population genetics**. In these investigations, scientists attempt to define how and why certain genetic variation is maintained in populations, while other variation diminishes or is lost with time (Figure 1.12). Such information is critical to the understanding of evolutionary processes. Population genetics also allows us to predict gene frequencies in future generations.

Together these varied approaches used in investigative genetics have transformed a subject that was only poorly understood in 1900 into one of the most advanced scientific disciplines today. As a result, the impact of genetics on society has been immense. We shall discuss many

FIGURE 1.12   Genetic variation exhibited in the skin of corn snakes. The wild type (normal) variety displays orange and black markings.

examples of the applications of genetics in the following section and throughout the text.

Genetics has a profound impact on society

In addition to acquiring information for the sake of extending knowledge in any discipline of science—an experimental approach called **basic research**—scientists conduct investigations to solve problems facing society or simply to improve the well-being of members of our society—an approach called **applied research**. Together, both types of genetic research have combined to enhance the quality of our existence on this planet and to provide a more thorough understanding of life processes. As we shall see throughout this text, there is very little in our lives that genetics fails to touch.

Eugenics and Euphenics

There is always the danger that scientific findings will be used to formulate policies and/or actions that are unjust or even tragic. This section reviews such a case, which began near the end of the nineteenth century. At that time, Darwin's theory of natural selection provided a major influence on some people's thinking concerning the human condition. Our story recounts the initial attempt to apply genetic knowledge directly for the improvement of human existence. Championed in England by Francis Galton, the general approach is called **eugenics**, a term Galton coined in 1883.

Galton, a cousin of Charles Darwin, believed that many human characteristics were inherited and subject to artificial selection if human matings could be controlled. *Positive eugenics* encouraged parents displaying favorable characteristics to have large families. Superior intelligence, intellectual achievement, and artistic talent are examples. *Negative eugenics*, on the other hand, attempted to restrict the reproduction of parents displaying unfavorable characteristics. Low intelligence, mental retardation, and criminal behavior are examples.

In the United States, the eugenics movement was a significant social force and led to state and federal laws that required the sterilization of those considered "genetically inferior." Over half of the states passed such laws, commencing in 1907 with Indiana. Sterilization was mandated for "imbeciles, idiots, convicted rapists, and habitual criminals." By 1931, involuntary sterilization also applied to "sexual perverts, drug fiends, drunkards, and epileptics." Immigration to the United States from certain areas of Europe and from Asia was also restricted, to prevent the influx of what were regarded as genetically inferior people. In addition to the violation of individual human rights, such policies were seriously flawed by an inadequate understanding of the genetic basis of various characteristics. The formulation of eugenic policies was premised on the mistaken notions that "superior" and "inferior" traits are totally under genetic control and that genes deemed unfavorable could be removed from a population by selecting against (sterilizing) individuals expressing those traits. The potential impact of the environment and the genetic theory underlying population genetics were largely ignored as eugenic policies were developed.

In Nazi Germany in the 1930s, the concept of achieving a superior, racially pure group was an extension of the eugenics movement. Initially applied to those considered socially and physically defective, the underlying rationale of negative eugenics was soon applied to entire ethnic groups, including Jews and Gypsies. Fueled by various forms of racial prejudice, Adolf Hitler and the Nazi regime took eugenics to its extreme by instituting policies aimed at the extinction of these "impure" human populations. This deplorable disregard for human life was preceded by incremental policies involving forced sterilization and mercy killings. This movement, based on scientifically invalid premises, soon led to mass murder.

FIGURE 1.13   *Triticale,* a hybrid grain derived from wheat and rye, produced as a result of applied genetic research.

Even before the Nazi party came to power in 1933, English and American geneticists began separating themselves from the eugenics movement. They were concerned about the validity of the premises underlying the movement and the evidence in support of these premises. Thus, many geneticists chose not to study human genetics for fear of being grouped with those who supported eugenics.

However, since the end of World War II, tremendous strides have been made in human genetics research. Today, a new term, **euphenics**, has replaced eugenics. Euphenics refers to medical and/or genetic intervention designed to reduce the impact of defective genotypes on individuals. The use of insulin by diabetics and the dietary control of newborn phenylketonurics are longstanding examples. Today, "genetic surgery" to replace defective genes is clearly on the horizon. Furthermore, social policies now have a solid genetic foundation on which they may be based. Nevertheless, caution is still required to ensure that our expanded knowledge of human genetics does not obscure the role played by the environment in determining an individual's phenotype.

Genetic Advances in Agriculture and Medicine

As a result of research in genetics, major benefits have accrued to society in the fields of agriculture and medicine. Although the cultivation of plants and domestication of animals had begun long before, the rediscovery of Mendel's work in the early twentieth century spurred scientists to apply genetic principles to these human endeavors. The use of selective breeding and hybridization techniques has had the most significant impact in agriculture.

Plants have been improved in four major ways: (1) enhanced potential for more vigorous growth and increased yields; (2) increased resistance to natural predators and pests, including insects and disease-causing microorganisms; (3) production of hybrids exhibiting a combination of superior traits derived from two different strains or even two different species (Figure 1.13); and (4) selection of genetic variants with desirable qualities such as increased protein value; increased content of limiting amino acids, which are essential in the human diet; or smaller plant size, reducing vulnerability to adverse weather conditions.

Over the past five decades, these improvements have resulted in a tremendous increase in yield and nutrient value in such crops as barley, beans, corn, oats, rice, rye, and wheat. It is estimated that in the United States the use of improved genetic strains has led to a threefold increase in crop yield per acre. In Mexico, where corn is the staple crop, a significant increase in protein content and yield has occurred. A substantial effort has also been made to improve the growth of Mexican wheat. Led by Norman Borlaug, a team of researchers developed varieties of wheat that incorporated favorable genes from other strains found in various parts of the world, revolutionizing wheat production in Mexico and other underdeveloped countries. Because of this effort, which led to the well-publicized "Green Revolution," Borlaug received the Nobel Peace Prize in 1970. There is little question that this application of genetics, which continues even today (see the essay at the end of Chapter 5), has contributed greatly to the well-being of our own species by improving the quality of nutrition worldwide.

Applied research in genetics has also resulted in the development of superior breeds of livestock (Figure 1.14). Selective breeding has produced chickens that grow faster, produce more high-quality meat per chicken, and lay a greater number of larger eggs. In larger animals, including pigs and cows, the use of artificial insemination has been particularly important. Sperm samples derived from a single male with superior genetic traits may now be used to fertilize thousands of females located in all parts of the world.

Equivalent strides have been made in medicine as a result of advances in genetics, particularly since 1950. Numerous disorders in humans have been discovered to result from either a single mutation or a specific chromosomal abnormality. For example, the genetic basis of a large number of diseases such as sickle-cell anemia, erythroblastosis fetalis, cystic fibrosis, hemophilia, muscular dystrophy, Tay-Sachs disease, Down syndrome, and many metabolic disorders is now well documented and often understood at the molecular level. The importance of acquiring knowledge of inherited disorders is underscored by the estimate that more than 10 million children or adults in the United States suffer from some form of genetic affliction and that every child-bearing couple stands an approximately 3 percent risk of having a child with some form of genetic anomaly.

FIGURE 1.14   The effects of breeding and selection, as illustrated by the production of this Vietnamese pot-bellied pig.

GENETICS, TECHNOLOGY, AND SOCIETY

Genetics and Society at the Millennium

Since the beginning of human civilization, we have defined ourselves as the masters of the biological world. Civilization began when humans domesticated plants and animals and settled into societies—as recently as 12 millennia ago. Genetics, in the form of selective breeding, became the foundation of agricultural progress and contributed to the rise and fall of civilizations over thousands of years. Our ability to harness nature is reflected in religions and philosophies that place humans at the center of the universe, at the pinnacle of creation, above all other creatures.

Now, as we enter the 21st century, a revolution of biological thought has begun—a revolution that is changing both our mastery over living things, and our perception of ourselves.

The revolution began in 1953 with Watson and Crick's elucidation of the molecular structure of DNA. The structure of the DNA molecule immediately provided elegant solutions to age-old questions about the mechanisms of heredity, mutation, and evolution. Some of the greatest mysteries of life could suddenly be explained by the beauty and simplicity of a chemical that replicates and shuffles the code of life.

Over the next 30 years, DNA became the focus of laboratories around the world. Geneticists, biochemists, and molecular biologists quickly devised methods to purify, mutate, cut, and paste DNA in the test tube. DNA molecules from one organism were spliced into DNA molecules from another, and the chimeric molecules were introduced into bacteria or cells in culture. The nucleotide sequences of genes were defined and modified *in vitro*. Gene promoters, gene transcription units, and gene activities were assayed. The genetic traits of organisms such as bacteria, fungi, and fruit flies were modified by the removal or addition of genes from similar, or different, species. Genetic engineering had begun.

The DNA revolution has advanced at an explosive rate. In the 27 years since the first gene was cloned, scientists have discovered the genes that control hereditary diseases such as sickle-cell anemia, cystic fibrosis, and Tay-Sachs disease. Prenatal diagnosis for these and other genetic diseases is now available. Biotechnologists have genetically modified bacteria, plants, and animals to express proteins of agricultural and medical importance. Biotechnology now offers DNA forensic tests that have helped convict criminals, exonerate the innocent, and establish paternity. Scientists have even cloned mammals such as sheep and mice from adult somatic cells. DNA manipulation is now a powerful tool of medical research. Scientists are rapidly dissecting the molecular genetic mechanisms that control cell growth, aging, death, and cancer.

The effect that the DNA revolution has had on our views of ourselves and the world is reflected in everyday culture. Although scientists dismiss the idea that humans are simply the products of their genes, popular culture endows DNA with almost magical powers. In television sit-coms, maga-

Additionally, it has gradually become clear over the past decade that most, if not all, forms of cancer have a genetic basis. Although cancer is not usually an inherited disorder, it is now very clear that *cancer is a genetic disorder at the somatic cell level.* That is, most cancers are derived from somatic cells that have undergone some type of genetic change; malignant tumors are then derived from the genetically altered cell. In some cases, such changes are inherited, conferring a genetic predisposition to cancer.

The recognition of the molecular basis of human genetic disorders and cancer has provided the impetus for the development of methods for detection and treatment. Based on advances in molecular genetics, particularly in the manipulation and analysis of DNA, prenatal detection of affected fetuses has become routine. Parents can also learn of their status as "carriers" of a large number of inherited disorders. Genetic counseling provides couples with objective information on which they can base rational decisions about child-bearing. In the case of cancer, recent genetic discoveries have already led to more effective early detection and more efficient approaches to treatment.

Applied research in genetics has also provided other medical benefits. Advances in immunogenetics have made possible compatible blood transfusions as well as organ transplants. In conjunction with immunosuppressive drugs, the number of transplant operations involving human organs, including the heart, liver, pancreas, and kidney, are increasing annually.

The most recent advances in human genetics have been dependent on the application of DNA biotechnology. First developed in the 1970s, recombinant DNA techniques paved the way for manipulating and cloning a

zines, and daily conversation, genes are said to explain personality, career choice, criminality, intelligence—even fashion preferences and political attitudes. Advertisements hijack the language of genetics in order to grant inanimate objects a "genealogy" or "genetic advantage." Popular culture speaks of DNA as an immortal force, with the ability to affect morality and fate. DNA is defined as the "essence of life" and the "immortal text," with the power to shape our future. Biological or genetic explanations for antisocial behavior appear to have more resonance for us than explanations involving social or economic factors.

But what of the future? Can we predict how DNA and genetics will help us shape the world in the next millennium? Although prophecy is certainly a risky business, some developments seem assured. The Human Genome Project promises to decode the DNA sequence of the entire human genome by the year 2003. This will lead to the identification of many genes that control normal and abnormal processes. In turn, this may enhance our ability to diagnose and predict genetic diseases. Scientists predict that the next millennium will bring us biotechnologies as complex as gene therapy and prenatal diagnosis and correction of genetic defects. Undoubtedly, the application of genetic engineering to agriculture will continue, as we manipulate plant genes for disease resistance, productivity, color, and flavor. The genetic engineering of farm animals is also likely to continue.

It is clear that the DNA revolution will have far-reaching practical consequences for humanity. It will also change how we think about ourselves and the world. As human genes are discovered, sequenced, and compared to those of other animals, it will become increasingly evident that we are closely related genetically to the rest of the animate world. The nucleotide sequence of our genome differs only about 1 percent from that of chimpanzees, and some of our genes are virtually identical to homologous genes in plants, animals, and bacteria. Will this knowledge alter our relationships with animals and with each other, as we realize the extent of our genetic kinship? As more genes are discovered that contribute to phenotypic traits as simple as eye color and as complicated as intelligence or sexual orientation, will this lead us to define ourselves more as genetic beings and less as creatures of free will or as the products of our environment?

As the next millennium unfolds, we will inevitably be faced with the practical and philosophical consequences of the DNA revolution. Will society harness DNA for everyone's benefit, or will the new genetic knowledge be used as a vehicle for discrimination? At the same time that modern genetics grants us more dominion over life, will it paradoxically increase our feelings of powerlessness? Will our new genetic view of life increase our compassion for all life forms, or will it increase our perceived separation from the natural world? We will make our choices, and human history will proceed.

REFERENCES:

COLLINS, F.S. et al. 1998. New goals for the U.S. human genome project: 1998–2003. *Science* 282: 682–89.

NELKIN, D., and LINDEE, M.S. 1995. *The DNA mystique, the gene as a cultural icon.* New York: W.H. Freeman.

WILKIE, T. 1993. *Perilous knowledge, the human genome project and its implications.* London: Faber and Faber.

variety of genes, including those that encode many medically important molecules, such as insulin, blood-clotting factors, growth hormone, and interferon. Human genes were isolated and spliced into vectors and transferred to host cells that serve as "production centers" for the synthesis of these proteins.

Recombinant DNA techniques have now been extended considerably. The DNA of any organism of interest is routinely manipulated in the laboratory. The human genes responsible for inherited disorders, such as cystic fibrosis and Huntington disease, have been identified, isolated, cloned, and studied. It is hoped that such research will pave the way for gene therapy, whereby genetic disorders are treated by inserting normal copies of genes into the cells of afflicted individuals.

Perhaps the most far-reaching use of DNA biotechnology involves the Human Genome Project, in which the entire genetic complement (the genome) of several species, including our own, is being sequenced. The genomic sequencing of several bacterial species, as well as yeast, is now complete. The sequencing of the entire 3.3 billion nucleotides constituting the human genome is scheduled for completion in the year 2003.

In later chapters, the applications of DNA biotechnology to agriculture and medicine are discussed in greater detail. Although other scientific disciplines are also expanding in knowledge, none has paralleled the growth of information that is occurring in genetics. As we pointed out at the outset of this chapter, while there never has been a more exciting time to be immersed in the study of genetics, the potential impact of this discipline on society has never been more profound. By the end of this course, we are confident you will agree that the present truly represents the "Age of Genetics."

CHAPTER SUMMARY

1. The history of genetics, which emerged as a fundamental discipline of biology early in the twentieth century, dates back to prehistoric times.

2. The basic concepts and vocabulary useful in the study of genetics have been presented.

3. Four investigative approaches are most often used in the study of genetics: transmission genetic studies, cytogenetic analyses, molecular experimentation, and inquiries into the genetic structure of populations.

4. Genetic research can be either basic or applied. Basic genetic research extends our knowledge of the discipline; the objective of applied genetics research is to solve specific problems affecting the quality of our lives and society in general.

5. Eugenics, the application of the knowledge of genetics for the improvement of human existence, has a long and controversial history. Euphenics, genetic intervention designed to ameliorate the impact of genotypes on individuals, represents the modern eugenic approach.

6. Genetic research has had a highly positive impact on many facets of agriculture and medicine.

7. DNA biotechnology has greatly expanded our research capability and now interfaces directly with society. It has had a profound impact in elucidating the basis of inherited diseases, has made possible the mass production of medically important gene products, and will serve as the foundation on which gene therapy is developed.

GENETICS MediaLab

This is the first in a series of special learning features, which are called *Genetic MediaLabs*. Within the MediaLab at the end of every chapter, you will find a list of media tools on the GenCDX CD-ROM packaged with your book and at its Companion Website **http://www.prenhall.com/klug/**. Each MediaLab will include several Web Problems, designed to help you learn, explore, and extend the knowledge you gain through reading the chapter.

Gen CDX Resources:
The GenCDX CD-ROM is part of your textbook and is attached to the inside back cover. In addition to being the launch point for access to the Web Problems listed in the MediaLabs, you will discover

Animated Tutorials

Self-grading Chapter Problems

If you have not done so, follow the installation instructions and take a moment to explore these resources.

Web Resources:
The Companion Website to your textbook can be reached from the following Web address: **http://www.prenhall.com/klug/**. You may access the Web Problems listed in your textbook's MediaLabs from this site. Additionally, the following valuable tools are also available:

Genetics, Technology, and Society Exercises

Web Destinations in Genetics

Self-grading Chapter Problems

Chapter Search Terms

Genetics Newsgroups

Student Bulletin Board

To become familiar with these resources, take this time to visit and explore the resources on your Companion Website.

Web Problem 1:
Time for completion = 5 minutes
Should you believe everything that you read, see, or hear? At this point, you should be aware that fiction and misinformation are often disguised as truth; therefore, a healthy dose of skepticism is a good thing to have. Being wary of possible misinformation is only half the battle. Skills to critique what you read, see, and hear will prove valuable, especially in using the Internet. To learn more about techniques for critiquing information, visit Web Problem 1 in Chapter 1 of your Companion Website and select the keyword **CRITIQUE.**

Web Problem 2:
Time for completion = 5 minutes
How do you cite and reference the valuable material that you find on the Internet? Communication is a large part of a scientist's career. An important discovery made but never shared is of little benefit to the world. After completing Web Problem 1, you should now be able to identify truthful and valuable information. Next you need to know how to cite this information. To learn more about how to properly cite online information, visit Web Problem 2 in Chapter 1 of your Companion Website and select the keyword **REFERENCE.**

PROBLEMS AND DISCUSSION QUESTIONS

1. Describe and contrast the ideas of Hippocrates and Aristotle relating to the genetic basis of life.
2. Define and contrast pangenesis, epigenesis, and preformationism.
3. Which ideas and doctrines that preceded Darwin were central to his thinking?
4. Describe Darwin's and Wallace's theory of natural selection. What information was lacking from it; that is, what gap remained in it?
5. Contrast chromosomes and genes and describe their role in heredity.
6. Describe the four major investigative approaches used in studying genetics.
7. Contrast basic and applied research.
8. Norman Borlaug received the Nobel Peace Prize for his work in genetics. Why do you think he was awarded this prize?
9. Contrast positive and negative eugenics. Which of these categories includes the approach called euphenics? Define this approach.
10. How has genetic research been applied to agriculture and to medicine?

SELECTED READINGS

ALLEN, G.E. 1996. Science misapplied: The eugenics age revisited. *Technol. Rev.* 99:23–31.

ANDERSON, W.F., and DIRCUMAKOS, E.G. 1981. Genetic engineering in mammalian cells. *Sci. Am.* (July) 245:106–21.

BORLAUG, N.E. 1983. Contributions of conventional plant breeding to food production. *Science* 219:689–93.

BOWLER, P.J. 1989. *The Mendelian revolution: The emergence of hereditarian concepts in modern science and society.* London: Athione.

COCKING, E.C., DAVEY, M.R., PENTAL, D., and POWER, J.B. 1981. Aspects of plant genetic manipulation. *Nature* 293:265–70.

DAY, P.R. 1977. Plant genetics: Increasing crop yield. *Science* 197:1334–39.

DUNN, L.C. 1965. *A short history of genetics.* New York: McGraw-Hill.

GARDNER, E.J. 1972. *History of biology*, 3rd ed. New York: Macmillan.

GARVER, K.L., and GARVER, B. 1991. Eugenics: Past, present, and future. *Am. J. Hum. Genet.* 49:1109–18.

GASSER, C.S., and FRALEY, R.T. 1989. Genetically engineering plants for crop improvement. *Science* 244:1293–99.

———. 1992. Transgenic crops. *Sci. Am.* (June) 266:62–69.

HORGAN, J. 1993. Eugenics revisited. *Sci. Am.* (June) 268:123–31.

KING, R.C., and STANSFIELD, W.D. 1997. *A dictionary of genetics*, 5th ed. New York: Oxford University Press.

KOLATA, G. 1998. *Clone—The road to Dolly, and the path ahead.* New York: William Morrow and Co.

OLBY, R.C. 1985. *Origins of Mendelism*, 2nd ed. London: Constable.

SILVER, L. 1998. *Remaking Eden: Cloning and beyond in a brave new world.* New York: Avon Books.

STUBBE, H. 1972. *History of genetics: From prehistoric times to the rediscovery of Mendel*, trans. by T.R.W. Waters. Cambridge, MA: MIT Press.

TORREY, J.G. 1985. The development of plant biotechnology. *Am. Sci.* 73:354–63.

VASIL, I.K. 1990. The realities and challenges of plant biotechnology. *Bio/Technology* 8:296–301.

WEINBERG, R.A. 1985. The molecules of life. *Sci. Am.* (Oct.) 253:48–57.

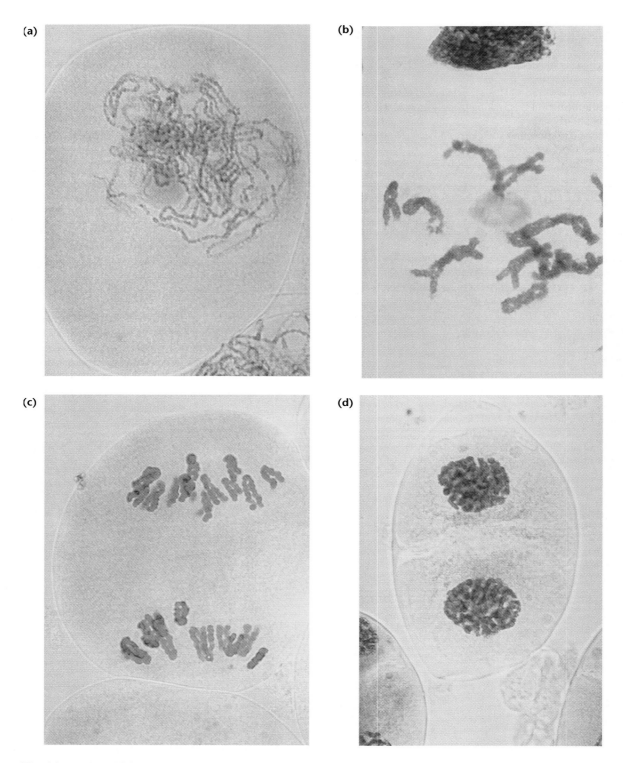

The (a) prophase, (b) metaphase, (c) anaphase, and (d) telophase stages of meiosis I.

2

Mitosis and Meiosis

KEY CONCEPTS

- **Cell structure is closely tied to genetic function**

  Cell Boundaries

  The Nucleus

  The Cytoplasm and Organelles

- **Homologous chromosomes provide the genetic basis of haploid and diploid cells**

- **Mitosis duplicates the genetic material precisely**

  Interphase and the Cell Cycle

  Prophase

  Prometaphase and Metaphase

  Anaphase

  Telophase

- **The cell cycle is genetically regulated**

- **Meiosis converts diploid cells to haploid cells during the formation of gametes and spores**

An Overview of Meiosis

The First Meiotic Division: Prophase I

Metaphase, Anaphase, and Telophase I

The Second Meiotic Division

- **The development of gametes varies during spermatogenesis and oogenesis**

- **Meiosis is critical to the successful sexual reproduction of all diploid organisms**

- **Electron microscopy has revealed the cytological nature of the mitotic and meiotic chromosome**

  Chromatin vs. Chromosomes

  The Synaptonemal Complex

GenCDX

When you see this icon, there are related animations and exercises on the CD accompanying this text.

In every living thing there exists a substance referred to as **genetic material.** Except in certain viruses, this material is composed of the nucleic acid DNA. A molecule of DNA contains many units called **genes,** the products of which direct all metabolic activities of cells. DNA, with its array of genes, is organized into **chromosomes,** structures that serve as the vehicle for transmission of genetic information. The manner in which chromosomes are transmitted from one generation of cells to the next, and from organisms to their descendants, is exceedingly precise. In this chapter, we will consider how such transmission is accomplished as we pursue the topic of genetic continuity between cells and organisms.

Two major processes are involved in eukaryotes: mitosis and meiosis. Although the mechanisms of the two processes are similar in many ways, the outcomes are quite different. **Mitosis** leads to the production of two cells, each with the number of chromosomes identical to the parental cell. **Meiosis,** on the other hand, reduces the amount of genetic material and the number of chromosomes by precisely half. This reduction is essential if sexual reproduction is to occur without doubling the amount of genetic material in each generation. Strictly speaking, mitosis is that portion of the cell cycle during which the hereditary components are precisely and equally divided into daughter cells. Meiosis is part of a special type of cell division leading to the production of sex cells: gametes and spores. This process is an essential step in the transmission of genetic information from an organism to its offspring.

In most cases, chromosomes are visible only when cells are actually dividing, that is, during mitosis or meiosis. When cells are not undergoing division, the genetic material making up chromosomes unfolds and uncoils into a diffuse network within the nucleus. Collectively, this uncoiled genetic material is referred to as **chromatin.**

In this chapter, we will review the structure of cells, emphasizing those cellular components that are of particular significance to genetics. Then we will examine the behavior of chromosomes undergoing division.

Cell structure is closely tied to genetic function

Many components of cells are linked to genetic processes. Before describing mitosis and meiosis, we will briefly review some of them. As we shall see, many cellular components, such as the nucleolus, ribosome, and centriole, are involved directly or indirectly in genetic function. Other components, the mitochondria and chloroplasts, contain their own unique genetic information. It is also useful for us to compare the structural differences of the bacterial prokaryotic cell with the eukaryotic cell to understand how variation in cell structure and function depends on the specific genetic expression by each cell type.

Before 1940, knowledge of cell structure was based on information obtained with the light microscope. About 1940 the transmission electron microscope was in the early stages of development, and by 1950 many details of cell ultrastructure had been unveiled. Under the electron microscope, cells were seen as highly organized, precise structures. A new world of whorled membranes, miniature organelles, microtubules, granules, and filaments was revealed. These discoveries revolutionized thinking in the entire field of biology. Many of the parts of the cell are described in Figure 2.1, which depicts a typical animal cell. Our discussion will concentrate on those aspects of cell structure relating to genetic processes.

Cell Boundaries

The cell is surrounded by a **plasma membrane,** an outer covering that defines the cell boundary and delimits the cell from its immediate external environment. This membrane is not passive; rather, it controls the movement of material such as gases, nutrients, and waste products into and out of the cell. In addition to this membrane, plant cells have an outer covering called the **cell wall.** One of the major components of this rigid structure is a polysaccharide called **cellulose.** Bacterial cells also have a cell wall, but its chemical composition is quite different from that of the plant cell wall, the major component being a complex macromolecule called a **peptidoglycan.** As its name suggests, the molecule consists of peptide and sugar units. Long polysaccharide chains are cross-linked with short peptides, which impart great strength and rigidity to the bacterial cell. Some bacterial cells have still another covering, a **capsule.** This mucus-like material protects these bacteria from phagocytic activity by the host during their pathogenic invasion of eukaryotic organisms. The pres-

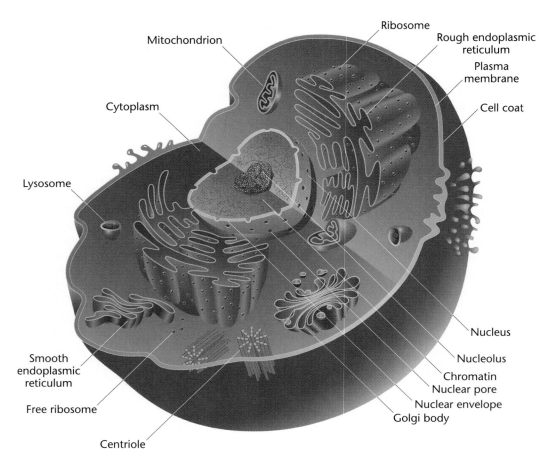

FIGURE 2.1 Drawing of a generalized animal cell. Emphasis has been placed on the cellular components discussed in the text.

ence of the capsule is under genetic control. In fact, as we will see in Chapter 11, its loss due to mutation in the pneumonia-causing bacterium *Diplococcus pneumoniae* provided the underlying basis for a critical experiment proving that DNA is the genetic material.

Activities at cell boundaries are dynamic physiological processes. Both transport into and out of cells and communication between cells are critical to normal function. Because physiological processes are biochemical in nature, we would expect that many genes and their products are essential to these activities. This is indeed the case, and as such, mutations in these genes can alter or interrupt normal physiological functions, often with severe consequences. For example, the inherited disorder **Duchenne muscular dystrophy** is the result of complete loss of function of the gene product **dystrophin,** which is believed to function at the cell membrane of muscle cells. As shown in Figure 2.2, using immunofluorescent localization technology, dystrophin is completely absent from skeletal muscle of an afflicted individual compared to muscle from a control subject.

Many, if not most, animal cells have a covering over the plasma membrane called a **cell coat.** Consisting of glycoproteins (sometimes called the **glycocalyx**) and polysaccharides, the cell coat differs chemically from comparable structures in either plants or bacteria. One function served by the cell coat is to provide biochemical identity at the surface of cells. All forms of biochemical identity at the cell surface are under genetic control, and many have been thoroughly investigated. Among other forms of molecular recognition, various antigens are part of the cell coat. For example, the **AB** and **MN antigens,** which may elicit an immune response during blood transfusions, are found on the surface of red blood cells. In other cells, the **histo-compatibility antigens,** which elicit an immune response during tissue and organ transplants, are part of the cell coat. Further, a variety of highly specific **receptor molecules** are integral components of the cell surface. These constitute recognition sites that receive and transfer chemical signals into the cell. Such signals may initiate a variety of chemical activities and may ultimately signal specific genes to turn on. We may conclude that the cell surface links its host cell to its outside world and is tied in numerous ways to genetic processes.

The Nucleus

The presence of the **nucleus** and other membranous organelles characterizes eukaryotic cells. The nucleus houses the genetic material, **DNA,** which is found in association with large numbers of acidic and basic proteins. During nondivisional phases of the cell cycle this DNA/protein complex exists in an uncoiled, dispersed state called **chromatin.** As you will soon see, during mitosis and meiosis this material coils up and condenses into distinct structures called chromosomes. Also present in the nucleus is the **nucleolus,** an amorphous component where ribosomal RNA (rRNA) is synthesized and where the initial stages of ribosomal assembly occur. One or more nucleoli may be present. Each forms around areas of DNA encoding rRNA referred to as the **nucleolar organizer regions** or the **NORs.**

The lack of a nuclear envelope and membraneous organelles is characteristic of prokaryotes. In bacteria such as *E. coli* the genetic material is present as a long circular DNA molecule that is compacted into an area referred to as the **nucleoid region.** Part of the DNA may be attached to the cell membrane, but in general the nucleoid region

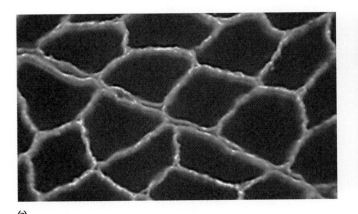

(a)

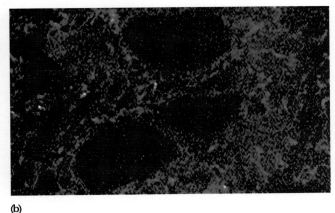

(b)

FIGURE 2.2   (a) Localization of dystrophin in the sarcolemmal areas of normal skeletal muscle using fluorescent immunoperoxidase staining. (b) Complete absence of immunoreactive dystrophin in the skeletal muscle of a patient with Duchenne muscular dystrophy.

constitutes a large area throughout the cell. Although the DNA is compacted, it does not undergo the extensive coiling characteristic of eukaryotic chromosomes. Nor is the DNA in these organisms as extensively associated with proteins as is eukaryotic DNA. Figure 2.3 shows the formation of two bacteria during cell division, where you can see the nucleoid regions. Prokaryotic cells do not have a distinct nucleolus, but they do contain genes that specify rRNA molecules.

The Cytoplasm and Organelles

The remainder of the eukaryotic cell enclosed by the plasma membrane, exclusive of the nucleus, is known as the **cytoplasm.** Cytoplasm consists of a nonparticulate colloidal material referred to as the **cytosol,** which surrounds and encompasses numerous types of cellular **organelles.** The **cytoskeleton,** an extensive system of tubules and filaments, provides a lattice of support structures within the cytoplasm. Consisting primarily of tubulin-derived microtubules and actin-derived microfilaments, this structural framework maintains cell shape, facilitates cell mobility, and anchors the various organelles. Tubulin and actin are both proteins found abundantly in eukaryotic cells.

One organelle, the membranous **endoplasmic reticulum (ER),** compartmentalizes the cytoplasm, greatly increasing the surface area available for biochemical synthesis. The ER may appear smooth, in which case it serves as the site for synthesis of fatty acids and phospholipids. Or the ER may appear rough because it is studded with ribosomes. Ribosomes, which we will discuss in detail in Chapter 14,

serve as sites for the translation of genetic information contained in messenger RNA (mRNA) into proteins.

Three other cytoplasmic structures are very important in the eukaryotic cell's activities: mitochondria, chloroplasts, and centrioles. **Mitochondria,** found in both animal and plant cells, are the sites of the oxidative phases of **cell respiration.** These chemical reactions generate large amounts of adenosine triphosphate (ATP), an energy-rich molecule. The **chloroplast** is one type of plastid found in plants, algae, and some protozoans. This organelle is associated with **photosynthesis,** the major energy-trapping process on Earth. Both mitochondria and chloroplasts contain a type of DNA distinct from that found in the nucleus. Furthermore, these organelles can duplicate themselves and transcribe and translate their genetic information. It is interesting to note that the genetic machinery of mitochondria and chloroplasts closely resembles that of prokaryotic cells. This and other observations have led to the proposal that these organelles were once primitive free-living organisms that established a symbiotic relationship with a primitive eukaryotic cell. This proposal, which describes the evolutionary origin of these organelles, is called the **endosymbiotic hypothesis.**

Animal and some plant cells also contain a pair of complex structures called the **centrioles.** These cytoplasmic bodies, found within a specialized region called the **centrosome,** are associated with the organization of the spindle fibers that function in mitosis and meiosis. In some organisms, the centriole is derived from another structure, the **basal body,** which is associated with the formation of cilia and flagella.

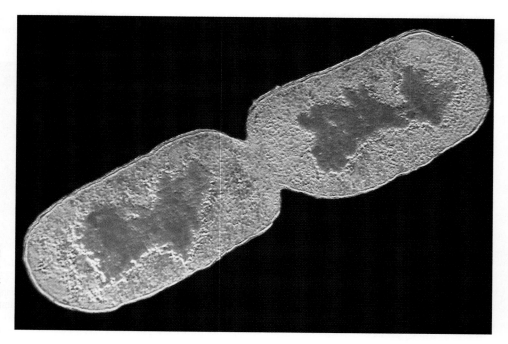

FIGURE 2.3 Color-enhanced electron micrograph of *E. coli* undergoing cell division. Particularly prominent are the two chromosomal areas (shown in red) that have been partitioned into the daughter cells.

The organization of **spindle fibers** by the centrioles occurs during the early phases of mitosis and meiosis. Composed of arrays of microtubules, these fibers play an important role in the movement of chromosomes as they separate during cell division. The microtubules consist of polymers of alpha and beta subunits of the protein tubulin. We will discuss the interaction of the chromosomes and spindle fibers later in this chapter.

Homologous chromosomes provide the genetic basis of haploid and diploid cells

To discuss the processes of mitosis and meiosis, it is important to understand clearly the concept of homologous chromosomes. Such an understanding will also be critical to our future discussions of Mendelian genetics. Thus, before looking more closely at cell division, we will address this topic and introduce other important terminology.

Chromosomes are most easily visualized during mitosis. When they are examined carefully, they are seen to take on distinctive lengths and shapes. Each contains a condensed or constricted region called the **centromere,** which establishes the general appearance of each chro-

mosome. Figure 2.4 illustrates chromosomes with centromere placements at different points along their lengths. Chromosomes are classified as **metacentric, submetacentric, acrocentric,** or **telocentric** on the basis of the centromere location. Extending from either side of the centromere are the arms of the chromosome. Depending on the position of the centromere, different arm ratios are produced. The shorter arm, by convention, is shown above the centromere and is called the **p arm** (*p* stands for "petite"). The longer arm is shown below the centromere and is called the **q arm** (*q* being the next letter in the alphabet). Typically, chromosomes are illustrated as they appear during metaphase. Figure 2.4 also shows the shape during anaphase; you will understand the importance of this when we discuss the phases of cell division.

When studying mitosis, we may make several other important observations. First, each somatic cell within members of the same species contains an identical number of chromosomes. This is called the **diploid number (2*n*).** When the lengths and centromere placements of all such chromosomes are examined, a second general feature is apparent. Nearly all of the chromosomes exist in pairs with regard to these two criteria. The members of each pair are called **homologous chromosomes.** For each chromosome exhibiting a specific length and centromere

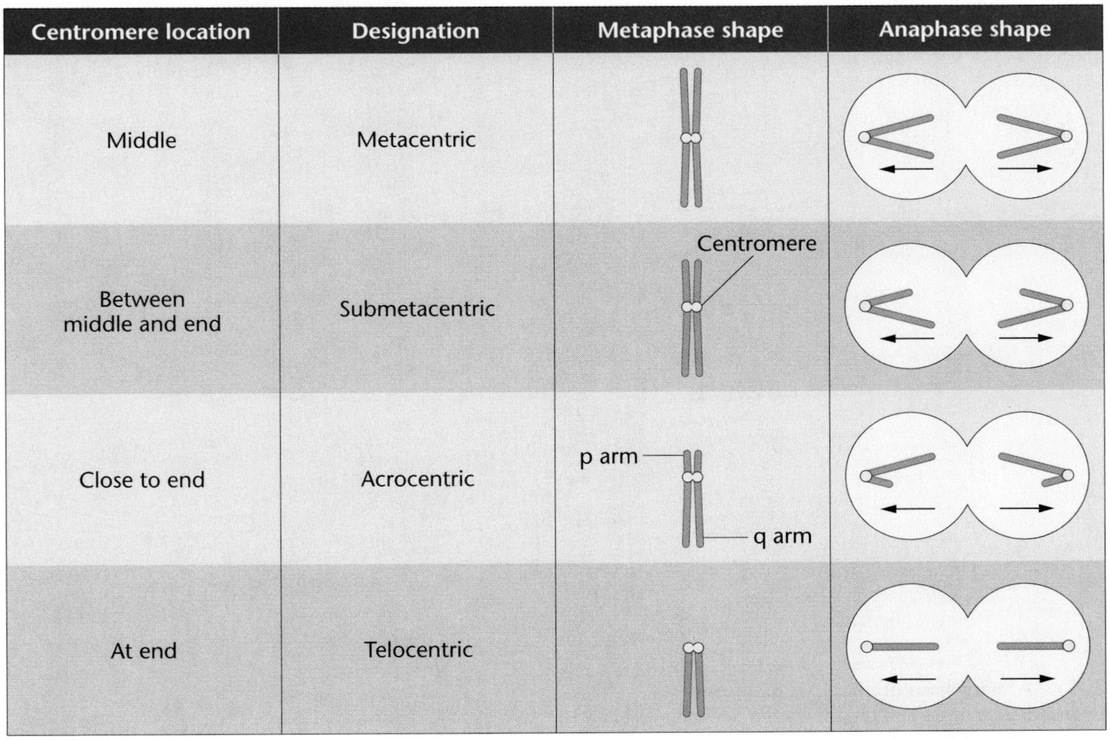

Centromere location	Designation	Metaphase shape	Anaphase shape
Middle	Metacentric		
Between middle and end	Submetacentric	Centromere	
Close to end	Acrocentric	p arm q arm	
At end	Telocentric		

FIGURE 2.4 Centromere locations and designations of chromosomes based on centromere location. Note that the shape of the chromosome during anaphase is determined by the position of the centromere.

placement, another exists with identical features. There are, of course, exceptions to the rule of chromosomes in pairs. Bacteria and viruses have but a single chromosome. In eukaryotes, yeasts, molds, and certain plants such as bryophytes spend most of their life cycle in the haploid stage.

Figure 2.5 illustrates the nearly identical physical appearance of different pairs of homologous chromosomes. There, the human mitotic chromosomes have been photographed (shown at the top of the figure), cut out of the print, and matched up, creating a **karyotype** (shown at the bottom of the figure). As you can see, humans have a *2n* number of 46 and exhibit a diversity of sizes and centromere placements. Note also that each one of the 46 chromosomes is clearly a double structure, consisting of two parallel sister chromatids connected by a common centromere. Had these chromosomes been allowed to continue dividing, each pair of sister chromatids, which are replicas of one another, would have separated into two new cells as division continued.

The **haploid number (*n*)** of chromosomes is one-half of the diploid number. Collectively, the total set of genes contained on one member of each homologous pair of chromosomes constitutes the **haploid genome** of the species. Table 2.1 illustrates the wide range of *n* values found in 32 diverse species of plants and animals.

Homologous pairs of chromosomes have important genetic similarity. They contain identical gene sites along their lengths, each called a **locus** (pl. **loci**). Thus, they have identical genetic potential. In sexually reproducing organisms, one member of each pair is derived from the maternal parent (through the ovum) and one from the paternal parent (through the sperm). Therefore, each diploid organism contains two copies of each gene as a consequence of **biparental inheritance.** As we will see in the following chapters on transmission genetics, the members of each pair of genes, while influencing the same characteristic or trait, need not be identical. In a population of members of the same species, many different alternative forms of the same gene, called **alleles,** may exist.

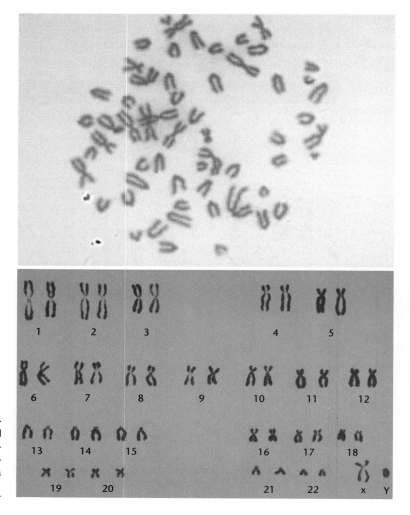

FIGURE 2.5 A metaphase preparation of chromosomes derived from a human male and the karyotype derived from it. All but the X and Y chromosomes are present in homologous pairs. Each chromosome is clearly a double structure, constituting a pair of sister chromatids joined at a common centromere.

TABLE 2.1   The Haploid Number of Chromosomes for Diverse Organisms

Common Name	Scientific Name	Haploid No.	Common Name	Scientific Name	Haploid No.
Black bread mold	*Aspergillus nidulans*	8	House mouse	*Mus musculus*	20
Broad bean	*Vicia faba*	6	Human	*Homo sapiens*	23
Cat	*Felis domesticus*	19	Jimson weed	*Datura stramonium*	12
Cattle	*Bos taurus*	30	Mosquito	*Culex pipiens*	3
Chicken	*Gallus domesticus*	39	Mustard plant	*Arabidopsis thaliana*	5
Chimpanzee	*Pan troglodytes*	24	Pink bread mold	*Neurospora crassa*	7
Corn	*Zea mays*	10	Potato	*Solanum tuberosum*	24
Cotton	*Gossypium hirsutum*	26	Rhesus monkey	*Macaca mulatta*	21
Dog	*Canis familiaris*	39	Rice	*Oryza sativa*	12
Evening primrose	*Oenothera biennis*	7	Roundworm	*Caenorhabditis elegans*	6
Frog	*Rana pipiens*	13	Silkworm	*Bombyx mori*	28
Fruit fly	*Drosophila melanogaster*	4	Slime mold	*Dictyostelium discoidium*	7
Garden onion	*Allium cepa*	8	Snapdragon	*Antirrhinum majus*	8
Garden pea	*Pisum sativum*	7	Tobacco	*Nicotiana tabacum*	24
Grasshopper	*Melanoplus differentialis*	12	Tomato	*Lycopersicon esculentum*	12
Green alga	*Chlamydomonas reinhardi*	18	Water fly	*Nymphaea alba*	80
Horse	*Equus caballus*	32	Wheat	*Triticum aestivum*	21
House fly	*Musca domestica*	6	Yeast	*Saccharomyces cerevisiae*	16

The concepts of haploid number, diploid number, and homologous chromosomes are important in the process of meiosis. During the formation of gametes or spores, meiosis converts the diploid number of chromosomes to the haploid number. As a result, haploid gametes or spores contain precisely one member of each homologous pair of chromosomes, that is, one complete haploid set. Following fusion of two gametes in fertilization, the diploid number is reestablished; that is, the zygote contains two complete haploid sets of chromosomes. The constancy of genetic material is thus maintained from generation to generation.

There is one important exception to the concept of homologous pairs of chromosomes. In many species, one pair, the **sex-determining chromosomes,** is often not homologous in size, centromere placement, arm ratio, or genetic potential. For example, in humans, females carry two homologous X chromosomes, whereas males contain a Y chromosome and only one X chromosome (Figure 2.5). The X and Y chromosomes are not strictly homologous. The Y is considerably smaller and lacks most of the loci contained on the X. Nevertheless, in meiosis the two behave as homologs so that gametes produced by males receive either the X or Y chromosome.

Mitosis duplicates the genetic material precisely

The process of **mitosis** is critical to all eukaryotic organisms. In many single-celled organisms that reproduce by cell division, such as protozoans, algae, and some fungi, mitosis provides the mechanism underlying their asexual reproduction. Multicellular diploid organisms begin life as single-celled fertilized eggs or **zygotes.** The mitotic activity of the zygote and the subsequent daughter cells is the foundation for development and growth of the organism. In adult organisms, mitotic activity associated with cell division is prominent in wound healing and other forms of cell replacement in certain tissues. For example, the epidermal skin cells of humans are continuously being sloughed off and replaced—as many as 100 billion (10^{11}) cells are lost daily according to one estimate! Mitosis is responsible for replacing them. Cell division also results in a continuous production of reticulocytes, cells that eventually shed their nuclei and are necessary for replenishing the supply of red blood cells in vertebrates. In abnormal situations, somatic cells may exhibit uncontrolled cell divisions, resulting in cancer.

The genetic material is partitioned into daughter cells during nuclear division or **karyokinesis.** This process is quite complex and requires great precision. The chromosomes first must be exactly replicated and then accurately partitioned. The end result is the production of two daughter nuclei, each with a chromosome composition identical to that of the parent cell.

Karyokinesis is followed by cytoplasmic division or **cytokinesis.** The less complex division of the cytoplasm requires a mechanism that results in a partitioning of the volume into two parts, followed by the enclosure of both new cells within a distinct plasma membrane. Cytoplasmic organelles either replicate themselves, arise from existing membrane structures, or are synthesized *de novo* (anew) in each cell. The subsequent proliferation of these structures is a reasonable and adequate mechanism for reconstituting the cytoplasm in daughter cells.

Usually following cell division, the initial size of each new daughter cell is approximately one-half the size of the parent cell. The nucleus of each new cell is not appreciably smaller than the nucleus of the original cell, however. Quantitative measurements of DNA confirm that there is an equivalent amount of genetic material in the daughter nuclei as in the parent cell.

Interphase and the Cell Cycle

Many cells undergo a continuous alternation between division and nondivision. The events occurring from the completion of one division until the beginning of the next division constitute the **cell cycle** (Figure 2.6). We will consider the initial phase of the cycle as the interval between divisions. It is called **interphase.** It was once thought that the biochemical activity during interphase was devoted solely to the cell's growth and its normal function. However, we now know that another biochemical step critical to the next mitotic division occurs during interphase: **the replication of the DNA of each chromosome.** Occurring as the cell prepares to enter nuclear division (mitosis), this period during which DNA is synthesized is called the **S phase.** The initiation and completion of synthesis can be detected by monitoring the incorporation of radioactive DNA precursors such as ^{3}H-thymidine. Their incorporation can be monitored using the technique of autoradiography (Appendix A).

Investigations of this nature have demonstrated two periods during interphase, before and after S, when no DNA synthesis occurs. These are designated **G1 (gap1)** and **G2 (gap2),** respectively. During both of these periods, as well as during S phase, intensive metabolic activity, cell growth, and cell differentiation occur. By the end of G2, the volume of the cell has roughly doubled, DNA has been replicated, and mitosis (M) is initiated. Following mitosis, continuously dividing cells repeat this cycle (G1, S, G2, M) over and over, as illustrated in Figure 2.6.

Much is known about the cell cycle based on *in vitro* (test tube) studies. When grown in culture, many cell types in different organisms traverse the complete cell cycle in about 16 hours. The actual process of mitosis occupies only a small part of the cycle, usually about an hour. The lengths of the S and G2 stages of interphase are fairly consistent among different cell types. Most variation is seen in the length of time spent in the G1 stage. Figure 2.7 illustrates the relative length of these periods in a typical human cell.

The G1 period is of great interest in the study of cell proliferation and its control. At a point late in G1, all cells follow one of two paths. They either withdraw from the cycle and enter a resting phase in the **G0 stage** (see Figure 2.6), or they become committed to initiating DNA synthesis and completing the cycle. Cells that enter G0 remain viable and metabolically active but are nonproliferative. Cancer cells apparently avoid entering G0, or else they pass through it very quickly. Other cells enter G0 and never reenter the cell cycle. Still others can remain quiescent in G0, but they may be stimulated to return to G1, reentering the cycle. We will return to this topic when we discuss cell cycle regulation.

Cytologically, interphase is characterized by the absence of visible chromosomes. Instead, a distinct nucleus is evident, filled with chromatin that has formed as the chromosomes have unfolded and uncoiled following the previous mitosis. This is depicted diagrammatically in Figure 2.8(a). When viewed under the light microscope [Figure 2.9(a)], the nucleus appears to be filled with a speckled, granular material. This is the result of having sectioned through the uncoiled chromatin fibers.

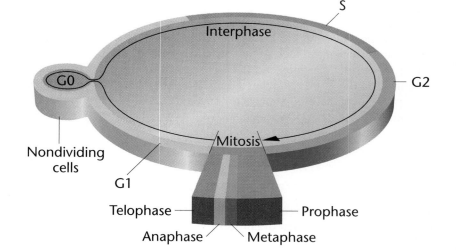

FIGURE 2.6 Diagrammatic representation of the stages comprising an arbitrary cell cycle. Following mitosis (M), cells enter the G1 stage of interphase, initiating a new cycle. Cells may become nondividing (G0) or continue through G1, where they become committed to begin DNA synthesis (S) and complete the cycle (G2 and Mitosis). Following mitosis, two daughter cells are produced.

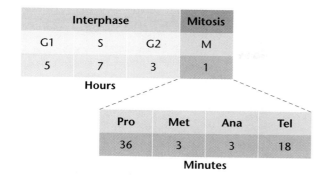

Interphase			Mitosis
G1	S	G2	M
5	7	3	1

Hours

Pro	Met	Ana	Tel
36	3	3	18

Minutes

FIGURE 2.7 The time spent in each phase of one complete cell cycle of a typical cell in culture. Times vary according to cell types and conditions.

Once G1, S, and G2 are completed, mitosis is initiated. Mitosis is a dynamic period of vigorous and continual activity. For ease of discussion, the entire process can be subdivided into discrete phases, each including specific events characteristic of each stage. These stages, in order

of occurrence following interphase, are **prophase, prometaphase, metaphase, anaphase,** and **telophase.** As in interphase, each of these stages is depicted in a drawing in Figure 2.8 and is shown as it actually occurs during plant mitosis in Figure 2.9.

Prophase

A significant portion of mitosis is spent in prophase, a stage characterized by several essential activities. One of the early events in prophase of all animal cells involves the migration of two pairs of centrioles to opposite ends of the cell. These structures are found just outside the nuclear envelope in an area of differentiated cytoplasm called the **centrosome.** It is thought that each pair of centrioles consists of one mature unit and a smaller, newly formed centriole.

The direction of migration of the centrioles is such that two poles are established at opposite ends of the cell. Following their migration, the centrioles are responsible for the organization of cytoplasmic microtubules into a series of **spindle fibers** that are formed and run between these poles. This creates the axis along which chromosomal

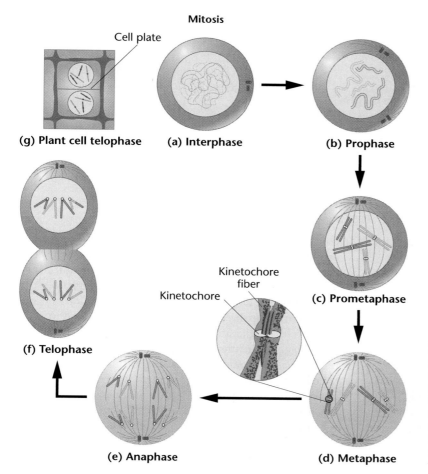

FIGURE 2.8 Mitosis in an animal cell with a diploid number of 4. The events occurring in each stage are described in the text. Of the two homologous pairs of chromosomes, one contains longer metacentric members, and the other, shorter submetacentric members. The maternal and paternal members of each homologous pair of chromosomes are shown in different colors. Part (g), showing the telophase stage in a plant cell, illustrates the formation of the cell plate and lack of centrioles.

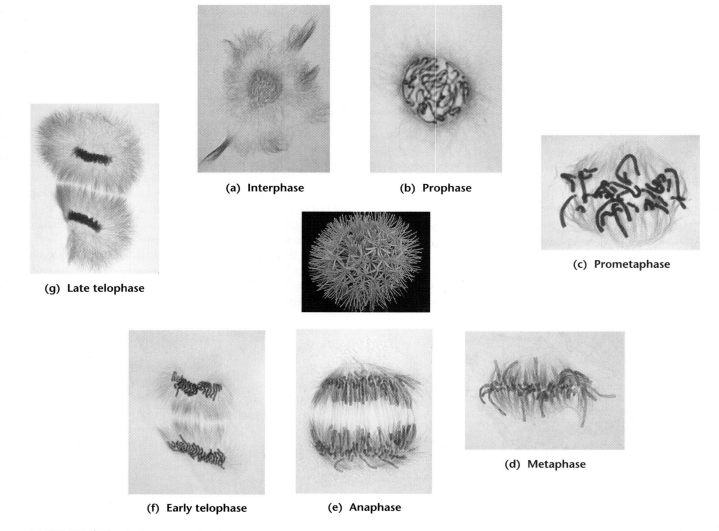

(a) Interphase

(b) Prophase

(c) Prometaphase

(g) Late telophase

(d) Metaphase

(f) Early telophase

(e) Anaphase

FIGURE 2.9 Light micrographs illustrating the stages of mitosis depicted in Figure 2.8. These stages are derived from the flower of *Haemanthus,* shown in the center of the figure.

separation will occur. We will discuss the formation and role of these spindle fibers in greater detail below.

Interestingly, cells of most plants (there are a few exceptions), fungi, and certain algae seem to lack centrioles. Spindle fibers are nevertheless apparent during mitosis. Therefore, centrioles are not universally responsible for the organization of spindle fibers. If some other center that organizes microtubules into spindle fibers exists in the cells of these organisms, it has yet to be discovered.

As the centrioles migrate, the nuclear envelope begins to break down and gradually disappears. In a similar fashion, the nucleolus disintegrates within the nucleus. While these events are taking place, the diffuse chromatin—the characteristic uncoiled form of the genetic material during interphase-begins to condense, a process that continues until distinct threadlike structures, or chromosomes, be-

come visible. Further condensation occurs throughout prophase and by the completion of this stage, it is now apparent that each chromosome is a double structure split longitudinally except at a single point of constriction, the **centromere.** The two parts of each chromosome are called **chromatids.** Because the DNA contained in each pair of chromatids represents the duplication of a single chromosome during the S phase of the previous interphase, these chromatids are genetically identical and are called **sister chromatids.** In humans, with a diploid number of 46, a cytological preparation of late prophase will reveal 46 such chromosomes. At the completion of prophase, the sister chromatids are found randomly distributed in the area formerly occupied by the nucleus.

You will often see the term **kinetochore** used in conjunction with discussions of the centromere. Intimately associated with each of the two sister chromatids, this structure is a multilayered platelike component that forms on opposite sides of the centromere. The external areas of the kinetochore structures ultimately attach to microtubules that make up the spindle fibers, and during the anaphase stage of mitosis the two kinetochores of each pair of sister chromatids are pulled to opposite poles of the cell. The internal areas of the kinetochore are closely aligned with the centromere, which consists of specific DNA regions of each chromosome. The relationship among the kinetochore, the centromere, and attached microtubules is illustrated in the electron micrograph in Figure 2.10.

Prometaphase and Metaphase

The distinguishing event of the next stage of mitosis is the migration of each chromosome, led by its centromeric region, to the equatorial plane of the cell. In some descriptions the term **prometaphase** refers to the period of chromosome movement, and **metaphase** is applied strictly to the chromosome configuration following this movement, as depicted in Figures 2.8(c) and 2.9(c). The equatorial plane, also referred to as the **metaphase plate,** is the midline region of the cell, a plane that lies perpendicular to the axis established by the spindle fibers.

Migration is made possible by the binding of microtubules to the kinetochore regions associated with the centromere of the chromosomes. Spindle fibers actually consist of **microtubules,** which themselves consist of molecular subunits of the protein **tubulin.** Microtubules seem to orig-

inate and "grow" out of the two centrosome regions (containing the centrioles) at opposite poles of the cell. They are dynamic structures that lengthen and shorten as a result of the addition or loss of polarized tubulin subunits. There are two major categories of microtubules. Those most directly responsible for chromosome migration during anaphase make contact with and adhere to kinetochores as they grow from the centrosome region. They are referred to as **kinetochore microtubules** and have one end near the centrosome region (at one of the poles of the cell) and the other anchored to the kinetochore. It is interesting to note that the number of microtubules that bind to the kinetochore varies greatly between organisms. Yeast (*Saccharomyces*) have only a single microtubule bound to each platelike structure of the kinetochore. Mitotic cells of mammals, at the other extreme, reveal 30 to 40 microtubules bound to each portion of the kinetochore.

Those microtubules that do not adhere to kinetochores make contact with growing microtubules from the opposite pole of the cell, often interdigitating with one another. The polarized nature of the tubulin subunits provides the force that joins them together. They are referred to as **polar microtubules** (and sometimes, spindle microtubules) and provide the cytoplasmic framework of spindle fibers that establish and maintain the separation of the two poles during chromosome separation. Kinetochore microtubules are apparent in Figure 2.10.

At the completion of metaphase, the chromosomes are arranged randomly, with each centromere aligned at the metaphase plate and the chromosome arms extending outward. This configuration is shown in Figures 2.8(d) and 2.9(d).

Anaphase

Events critical to chromosome distribution characterize the shortest stage of mitosis, **anaphase.** It is during this phase that sister chromatids of each double chromosomal structure separate from each other and migrate to opposite ends of the cell. The event that marks the end of metaphase and the initiation of anaphase is the division of each centromeric region into two. Once this has occurred, each chromatid is referred to as a **daughter chromosome.**

As discussed above, movement of daughter chromosomes to the opposite poles of the cell is dependent upon the centromere–spindle fiber (kinetochore–microtubulin) attachment. Recent investigations have revealed that chromosome migration results from the activity of a series of specific proteins, generally called motor proteins. These proteins use the energy generated by the hydrolysis of ATP, and their activity is said to constitute **molecular motors** in the cell. These motors act at several positions within the dividing cell, but all are involved in the activity of microtubules and ultimately serve to propel the chromosomes to

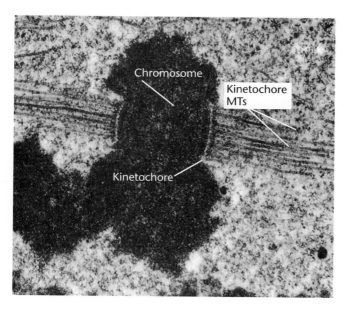

FIGURE 2.10 An electron micrograph showing the association between microtubules constituting the spindle fibers and the kinetochore (in association with a centromere) during mitosis.

opposite ends of the cell. The centromeres of each chromosome *appear* to lead the way during migration, with the chromosome arms trailing behind. The location of the centromere determines the shape of the chromosome during separation, as you can see in Figure 2.4.

The steps occurring during anaphase are critical in providing each subsequent daughter cell with an identical set of chromosomes. In human cells there would now be 46 chromosomes at each pole, one from each original sister pair. Figures 2.8(e) and 2.9(e) illustrate anaphase prior to its completion.

Telophase

Telophase is the final stage of mitosis and is depicted in Figures 2.8(f–g) and 2.9(f–g). At its beginning, there are two complete sets of chromosomes, one at each pole. The most significant event is **cytokinesis,** the division or partitioning of the cytoplasm. Cytokinesis is essential if two new cells are to be produced from one. The mechanism differs greatly in plant and animal cells. In plant cells, a **cell plate** is synthesized and laid down across the dividing cell in the region of the metaphase plate. Animal cells, however, undergo a constriction of the cytoplasm in much the same way a loop might be tightened around the middle of a balloon. The end result is the same: Two distinct cells are formed.

It is not surprising that the process of cytokinesis varies among cells of different organisms. Plant cells, which are more regularly shaped and structurally rigid, require a mechanism for the deposition of new cell wall material around the plasma membrane. The cell plate, laid down during telophase, becomes the **middle lamella.** Subsequently, the primary and secondary layers of the cell wall are deposited between the cell membrane and middle lamella on both sides of the boundary between the two daughter cells. In animals, complete constriction of the cell membrane produces the **cell furrow** characteristic of newly divided cells.

Other events initiated during late telophase represent a general reversal of those that occurred during prophase. In each new cell, the chromosomes begin to uncoil and become diffuse chromatin once again, while the nuclear envelope reforms around them. The nucleolus gradually reforms and is completely visible in the nucleus during early interphase. The spindle fibers also disappear. At the completion of telophase, the cell enters interphase.

The cell cycle is genetically regulated

The cell cycle, including mitosis, is fundamentally the same in all eukaryotic organisms. The similarity of the events leading to cell duplication in many diverse organisms suggests that the cell cycle is governed by a genetic program that has been conserved throughout evolution and is therefore genetically regulated. Elucidation of this genetic program therefore provides information basic to our understanding of the nature of living organisms. Furthermore, because disruption of this regulation may lead to uncontrolled cell division characterizing malignancy, interest in how genes regulate the cell cycle has been great.

A mammoth research effort over the past decade has paid high dividends, and we now have knowledge of many genes involved in the control of the cell cycle. As with other studies of genetic input into essential biological processes, investigation has focused on the discovery of mutations that interrupt the cell cycle and the subsequent study of the effects of these mutations. As we shall return to this subject in even greater detail in Chapter 23 during our consideration of cancer, what follows is a brief overview.

Many mutations are now known that exert their effect at various stages of the cell cycle. First discovered in yeast but now evident in all organisms, including humans, such mutations were originally designated as *cdc* mutations (*cell division cycle* mutations). The study of these mutations has established that during the cell cycle, at least three major **checkpoints** exist, where the cell is monitored or "checked" before it can proceed to the next stage of the cycle.

The products of many of these genes are enzymes called *cdc* **kinases** that can add phosphates to other proteins. They serve as "master control" molecules that work in conjunction with proteins called **cyclins.** These kinases phosphorylate cyclins and influence their activity at the cell-cycle checkpoints. These activities thus regulate the cell cycle. When a *cdc* kinase works in conjunction with a cyclin, it is called a **Cdk protein,** for *Cyclin-dependent kinase protein.*

Figure 2.11 identifies the location of three "checkpoints" within the cell cycle. The first is the **G1/S checkpoint,** which monitors the size that the cell has achieved following the previous mitosis and whether the DNA has been damaged. If the cell has not achieved an adequate size or if the DNA has been damaged, further progress through the cycle is arrested until these conditions are "corrected," so to speak. If both conditions are initially "normal," then the checkpoint is traversed and the cell proceeds to the S phase of the cycle.

The second important checkpoint is the **G2/M checkpoint,** where physiological conditions in the cell are monitored prior to entering mitosis. If DNA replication or repair to any DNA damage has not been completed, the cell cycle is arrested until these processes are completed. The final checkpoint occurs during mitosis and is called the **M checkpoint.** Here, both the successful formation of the spindle fiber system and the attachment of spindle fibers to the kinetochores associated with the centromeres are monitored. If spindle fibers are not properly formed or attachment is inadequate, mitosis is arrested.

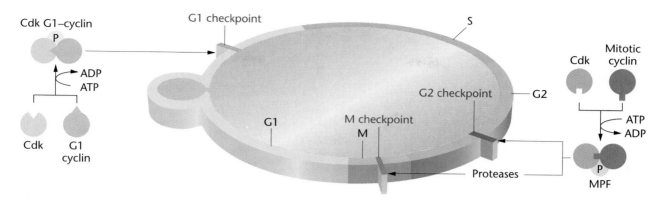

FIGURE 2.11   Illustration of the three major checkpoints in the cell cycle.

The importance of cell-cycle control and these checkpoints can be illustrated by considering what happens when this regulatory system is impaired. If, for example, a cell has incurred damage to its DNA and is allowed to proceed through the cell cycle, it may begin a series of uncontrolled cell divisions—precisely the definition of a cancerous cell. As we saw above, such a damaged cell would normally be arrested at either the G1/S or the G2/M checkpoint.

An interesting related finding involves the protein product of the *p53* **gene** in humans and its involvement during scrutiny at the G1/M checkpoint. This protein functions during the regulation of **apoptosis,** the genetic process whereby programmed cell death occurs. When the normal *p53* gene product is present, a proliferative cell that has incurred severe damage to its DNA will be targeted for programmed cell death at the G1/M checkpoint and, thus, effectively removed from the cell population. However, if the *p53* gene has mutated, resulting in abnormal function of the *p53* gene product, the damaged cell may proceed through the checkpoint and continue to proliferate in an uncontrolled manner.

In fact, a high percentage of human cancers has been found to contain mutations in the *p53* gene. These include a wide variety of cancers, including colon, breast, lung, and bladder malignancies. In the language of cancer genetics, *p53* is referred to as a **tumor-supressor gene.** The role of this and other genes involved in the cell cycle has stirred great interest in cancer research.

Meiosis converts diploid cells to haploid cells during the formation of gametes and spores

The process of meiosis, unlike mitosis, reduces the amount of genetic material by one-half. Whereas in diploids, mitosis produces daughter cells with a full diploid complement, meiosis produces gametes or spores with only one haploid set of chromosomes. During sexual reproduction, gametes then combine in fertilization to reconstitute the diploid complement found in parental cells. Figure 2.12 provides a comparison of the two processes, where two pairs of homologous chromosomes are followed.

Meiosis must be highly specific since, by definition, haploid gametes or spores must contain precisely one member of each homologous pair of chromosomes. Successfully completed, meiosis ensures genetic continuity from generation to generation.

The process of sexual reproduction also ensures genetic variety among members of a species. As you study meiosis, you will see that this process results in gametes with many unique combinations of maternally and paternally derived chromosomes among the haploid complement. With such a tremendous genetic variation among the gametes, a large number of chromosome combinations are possible at fertilization. Furthermore, we will see that the meiotic event referred to as **crossing over** results in genetic exchange between members of each homologous pair of chromosomes. This creates intact chromosomes that are mosaics of the maternal and paternal homologs from which they are derived, further enhancing the potential genetic variation in gametes and the offspring derived from them. Sexual reproduction, therefore, reshuffles the genetic material, producing offspring that often differ greatly from either parent. This process constitutes the major form of genetic recombination within species.

An Overview of Meiosis

Above, we have established what might be considered the goals of meiosis. Before proceeding to systematically consider the stages of this process, we will briefly describe how diploid cells give rise to haploid gametes or spores. You should refer to the meiotic portion of Figure 2.12 during the following discussion.

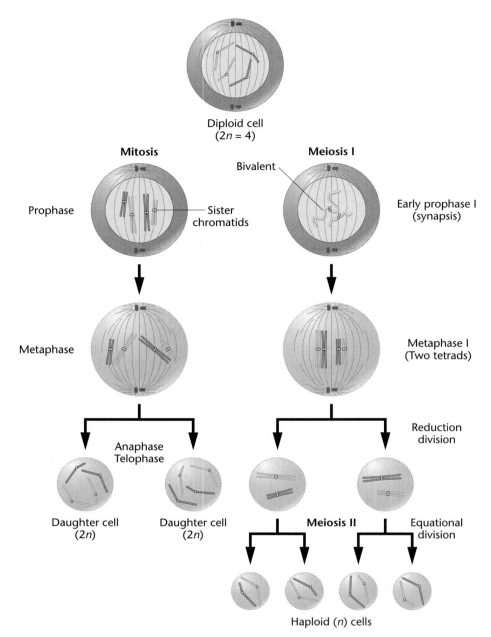

FIGURE 2.12 A comparative overview of the major events and outcomes of mitosis and meiosis. As in Figure 2.8, two pairs of homologous chromosomes are followed.

You have seen that in mitosis each paternally and maternally derived member of any given homologous pair of chromosomes behaves autonomously during division. By contrast, in meiosis, homologous chromosomes pair together; that is, they **synapse.** Each synapsed structure, called a **bivalent,** gives rise to a unit, the **tetrad,** consisting of four chromatids. The presence of four chromatids demonstrates that both chromosomes have duplicated. In order to achieve haploidy, two divisions are necessary. In meiosis I, described as a **reductional division** (because the number of centromeres, each representing one chromosome, is *reduced* by one-half following this division), components of each

tetrad—representing the two homologs—separate, yielding two **dyads.** Each dyad is composed of two sister chromatids joined at a common centromere. During meiosis II, described as **equational** (because the number of centromeres remains *equal* following this division), each dyad splits into two **monads** of one chromosome each. Thus, the two divisions potentially produce four haploid cells.

The First Meiotic Division: Prophase I

We turn now to a detailed account of meiosis. As in mitosis, meiosis is a continuous process. We give names to the parts of each stage of division only for the convenience of discussion. From a genetic standpoint, three events characterize the initial stage, prophase I (Figure 2.13). First, as in mitosis, chromatin present in interphase thickens and coils into visible chromosomes. Second, unlike mitosis, members of each homologous pair of chromosomes undergo synapsis. Third, crossing over, an exchange process, occurs between synapsed homologs. Because of the complexity of these genetic events, this stage of meiosis has been further subdivided into five substages: leptonema,* zygonema,* pachynema,* diplonema,* and diakinesis. As we discuss them, you should be aware that, even though it is not immediately apparent in the earliest phases of meiosis, the DNA of chromosomes has been replicated during the prior interphase stage.

Leptonema

During the **leptotene stage,** the interphase chromatin material begins to condense, and the chromosomes, although still extended, become visible. Along each chromosome are **chromomeres,** localized condensations that resemble beads on a string. Recent evidence suggests that a process called **homology search,** which precedes and is essential to the initial pairing of homologs, begins during leptonema.

Zygonema

The chromosomes continue to shorten and thicken during the **zygotene stage.** During the process of homology search, homologous chromosomes undergo initial alignment with one another. This establishes what is called rough pairing, which is complete by the end of zygonema. In yeast, homologs are separated by about 300 nm, and near the end of zygonema, structures referred to as lateral elements are visible between paired homologs. As meiosis proceeds, the overall length of the lateral elements increases and a more extensive ultrastructural component, the synaptonemal complex, begins to form between the homologs. We will discuss this meiotic component later in the chapter.

At the completion of zygonema, the paired homologs take the form of **bivalents.** Although both members of each bivalent have already replicated their DNA, it is not yet visually apparent that each member is a double structure. The number of bivalents in each species is equal to the haploid (n) number.

Pachynema

In the transition from the zygotene to the **pachytene stage,** coiling and shortening of chromosomes continues, and further development of the synaptonemal complex occurs between the two members of each bivalent. This leads

---

*These are the noun forms of these stages. The adjective forms (leptotene, zygotene, pachytene, and diplotene) are also used.

Meiotic prophase I

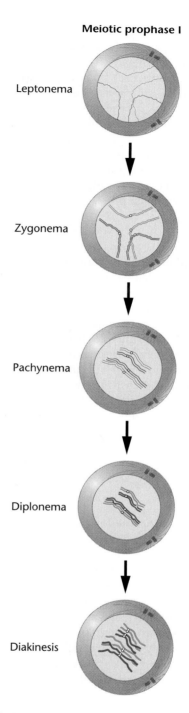

Leptonema

Zygonema

Pachynema

Diplonema

Diakinesis

FIGURE 2.13 Diagrammatic representation of the stages of meiotic prophase I. The same chromosomes depicted in Figure 2.8 and 2.12 are followed.   🌐 GenCDX

to a more intimate pairing, referred to as **synapsis.** Compared to the rough-pairing characteristic of yeast pachynema, homologs are now separated by only 100 nm.

During pachynema, each homolog is first evident as a double structure, providing visual evidence of the earlier replication of the DNA of each chromosome. Thus, each bivalent contains four member chromatids. As in mitosis, replicates are called sister chromatids, while chromatids from maternal vs. paternal members of a homologous pair are called nonsister chromatids. The four-member structure is also referred to as a tetrad, and each tetrad contains two pairs of sister chromatids.

Diplonema

During observation of the ensuing **diplotene stage,** it is even more apparent that each tetrad consists of two pairs of sister chromatids. Within each tetrad, each pair of sister chromatids begins to separate. However, one or more areas remain in contact where chromatids are intertwined. Each such area, called a **chiasma** (pl. **chiasmata**), is thought to represent a point where nonsister chromatids have undergone genetic exchange through the process of crossing over. Although the physical exchange between chromosome areas occurred during the previous pachytene stage, the result of crossing over is visible only when the duplicated chromosomes begin to separate. Crossing over is an important source of genetic variability. As indicated earlier, new combinations of genetic material are formed during this process.

Diakinesis

The final stage of prophase I is **diakinesis.** The chromosomes pull farther apart, but nonsister chromatids remain loosely associated via the chiasmata. As separation proceeds, the chiasmata move toward the ends of the tetrad. This process, called **terminalization,** begins in late diplonema, and is completed during diakinesis. During this final period of prophase I, the nucleolus and nuclear envelope break down, and the two centromeres of each tetrad become attached to the recently formed spindle fibers. By the completion of prophase I, the centromeres of each tetrad structure are present on the equatorial plate of the cell.

Metaphase, Anaphase, and Telophase I

The remainder of the meiotic process is depicted in Figure 2.14. Following the first meiotic prophase stage, steps similar to those of mitosis occur. In the metaphase stage of the first division (**metaphase I**), the chromosomes have maximally shortened and thickened. The terminal chiasmata of each tetrad are visible and appear to be the only factor holding the nonsister chromatids together. Each tetrad interacts with spindle fibers, facilitating movement to the metaphase plate.

The alignment of each tetrad prior to this first anaphase stage is random. One-half of each tetrad will be pulled to one or the other pole at random, and the other half will move to the opposite pole. This random separation of dyads is the basis for the Mendelian postulate of **independent assortment,** which we will discuss in Chapter 3. You may wish to return to this discussion when you study this principle.

During the stages of meiosis I, a single centromere holds each pair of sister chromatids together. It does *not* divide. At **anaphase I,** one-half of each tetrad (one pair of sister chromatids-called a **dyad**) is pulled toward each pole of the dividing cell. This separation process is the physical basis of what we refer to as **disjunction,** the separation of chromosomes from one another. Occasionally, errors in meiosis occur and separation is not achieved, as we will see later in this chapter. The term **nondisjunction** describes such an error. At the completion of the normal anaphase I, a series of dyads equal to the haploid number is present at each pole.

If no crossing over had occurred in the first meiotic prophase, each dyad at each pole would consist solely of either paternal or maternal chromatids. However, the exchanges produced by crossing over create mosaic chromatids of paternal and maternal origin.

In many organisms, **telophase I** reveals a nuclear membrane forming around the dyads. Next, the nucleus enters into a short interphase period. In other cases, the cells go directly from the first anaphase into the second meiotic division. If an interphase period occurs, the chromosomes do not replicate since they already consist of two chromatids. In general, meiotic telophase is much shorter than the corresponding stage in mitosis.

The Second Meiotic Division

A second division, referred to as **meiosis II,** is essential if each gamete or spore is to receive only one chromatid from each original tetrad. The stages characterizing meiosis II are also diagrammed in Figure 2.14. During **prophase II,** each dyad is composed of one pair of sister chromatids attached by a common centromere. During **metaphase II,** the centromeres are positioned on the equatorial plate. When they divide, **anaphase II** is initiated, and the sister chromatids of each dyad are pulled to opposite poles. Because the number of dyads is equal to the haploid number, **telophase II** reveals one member of each pair of homologous chromosomes present at each pole. Each chromosome is referred to as a **monad.** Following cytokinesis in telophase II, four haploid gametes may result from a single meiotic event. At the conclusion of meiosis, not only has the haploid state

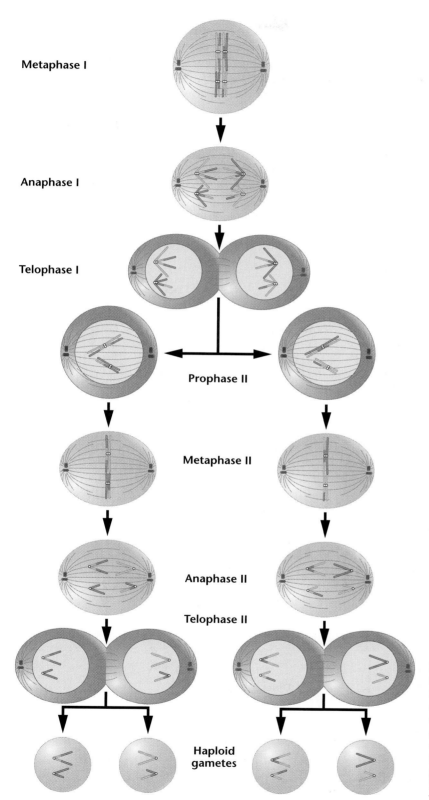

Metaphase I

Anaphase I

Telophase I

Prophase II

Metaphase II

Anaphase II

Telophase II

Haploid gametes

FIGURE 2.14 A continuation of the major events occurring during meiosis in an animal with a diploid number of 4, beginning with metaphase I. Note that the combination of chromosomes contained in the cells produced following telophase II is dependent on the random alignment of each tetrad and dyad on the equatorial plate during metaphase I and metaphase II. Several other combinations (not shown) can be formed. Note that actual photographs of the stages of meiosis I are found on the initial page of this chapter. ⊛ GenCDX

been achieved, but if crossing over has occurred, each monad is a combination of maternal and paternal genetic information. As a result, the offspring produced by any gamete will receive a mixture of genetic information originally present in his or her grandparents.

The development of gametes varies during spermatogenesis and oogenesis

Although events that occur during the meiotic divisions are similar in all cells that participate in gametogenesis in most animal species, there are certain differences between the production of a male gamete (spermatogenesis) and a female gamete (oogenesis). Figure 2.15 summarizes these processes.

Spermatogenesis takes place in the testes, the male reproductive organs. The process begins with the expanded growth of an undifferentiated diploid germ cell called a **spermatogonium.** This cell enlarges to become a **primary spermatocyte,** which undergoes the first meiotic division. The products of this division, called **secondary spermatocytes,** contains a haploid number of dyads. The secondary spermatocytes then undergo the second meiotic division, and each of these cells produces two haploid **spermatids.** Spermatids go through a series of developmental changes, **spermiogenesis,** and become highly specialized, motile **spermatozoa** or **sperm.** All sperm cells produced during spermatogenesis receive equal amounts of genetic material and cytoplasm.

Spermatogenesis may be continuous or occur periodically in mature male animals, with its onset determined by

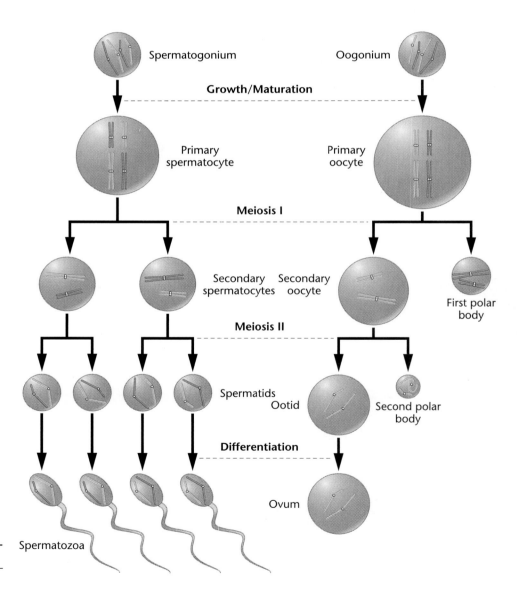

FIGURE 2.15 Spermatogenesis and oogenesis in animal cells.

the nature of the species' reproductive cycle. Animals that reproduce year-round produce sperm continuously, whereas those whose breeding period is confined to a particular season produce sperm only during that time.

In animal **oogenesis,** the formation of **ova** (singular: **ovum**), or eggs, occurs in the ovaries, the female reproductive organs. The daughter cells resulting from the two meiotic divisions receive equal amounts of genetic material, but they do *not* receive equal amounts of cytoplasm. Instead, during each division, almost all the cytoplasm of the **primary oocyte,** itself derived from the **oogonium,** is concentrated in one of the two daughter cells. The concentration of cytoplasm is necessary because a major function of the mature ovum is to nourish the developing embryo following fertilization.

During the first meiotic anaphase in oogenesis, the tetrads of the primary oocyte separate, and the dyads move toward opposite poles. During the first telophase, the dyads present at one pole are pinched off with very little surrounding cytoplasm to form the **first polar body.** The other daughter cell produced by this first meiotic division contains most of the cytoplasm and is called the **secondary oocyte.** The first polar body may or may not divide again to produce two small haploid cells. The mature ovum will be produced from the secondary oocyte during the second meiotic division. During this division, the cytoplasm of the secondary oocyte again divides unequally, producing an **ootid** and a **second polar body.** The ootid then differentiates into the mature ovum.

Unlike the divisions of spermatogenesis, the two meiotic divisions of oogenesis may not be continuous. In some animal species, the two divisions may directly follow each other. In others, including humans, the first division of all oocytes begins in the embryonic ovary but arrests in prophase I. Many years later, the meiosis resumes in each oocyte just prior to its ovulation. The second division is completed only after fertilization.

Meiosis is critical to the successful sexual reproduction of all diploid organisms

The process of meiosis is highly significant during the sexual reproduction of all diploid organisms. It is the means by which the diploid amount of genetic information is reduced to the haploid amount. In animals, meiosis leads to the formation of gametes, whereas in plants haploid spores are produced, which in turn lead to the formation of haploid gametes.

Furthermore, the mechanism of meiosis is the basis for the production of extensive genetic variation among members of a population. As we have learned, each diploid or-

ganism contains its genetic information in the form of homologous pairs of chromosomes, one member of each pair derived from the maternal parent and one member from the paternal parent. Following the reduction to haploidy, gametes or spores contain either the paternal or the maternal representative of every homologous pair of chromosomes. During sexual reproduction, this process has the potential of producing huge quantities of genetically dissimilar gametes. As the number of homologous chromosomes (the haploid number) increases, the possibilities of different combinations of maternal and paternal chromosomes in any given gamete increase. For example, an organism can produce 2^n number of combinations, where n represents the haploid number. For example, an organism with a haploid number of 10 will produce 2^{10} or 1024 combinations. Now calculate the number of different combinations of sperm or eggs in our own species: 2^{23}. When you arrive at the answer, you cannot help but be impressed with the potential for genetic variation resulting from meiosis.

The process of crossing over during meiotic prophase I further reshuffles the genetic information between the maternal and paternal members of each homologous pair. As a result, endless varieties of each homolog may occur in gametes, ranging from either intact maternal or paternal chromosomes, where no exchange occurred, to any mixture of maternal and paternal components, depending on where one or more exchanges occurred during crossing over.

In summary, the two most significant points about meiosis are that the process is responsible for:

1. The maintenance of a constancy of genetic information between generations; and
2. Extensive genetic variation within populations.

Two other aspects of meiosis are also important in the study of genetics, and we will touch on them briefly. First, meiosis plays an important role in the life cycles of fungi and plants. In many fungi, the predominant stage of the life cycle consists of haploid cells. They arise through meiosis and proliferate by mitotic cell division. In multicellular plants, the life cycle alternates between the diploid **sporophyte stage** and the haploid **gametophyte stage.** While one or the other predominates in different plant groups during this "alternation of generations," the processes of meiosis and fertilization constitute the "bridge" between the sporophyte and gametophyte generations (Figure 2.16). Therefore, meiosis is essential in the life cycle of plants.

Finally, it is important to know what happens when meiosis fails to achieve the expected outcome. In rare cases during meiosis I or meiosis II, separation, or disjunction, of the chromatids of a tetrad or dyad fails to occur. Instead, both members move to the same pole during anaphase. Such an event is called **nondisjunction,** because the two members fail to disjoin.

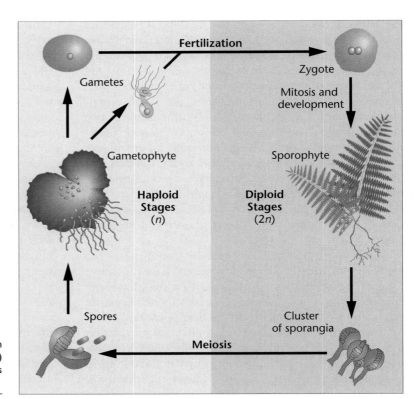

FIGURE 2.16 Alternation of generations between the diploid sporophyte (2*n*) and the haploid gametophyte (*n*) in the fern (a multicellular plant). The process of meiosis bridges the two phases of the life cycle.

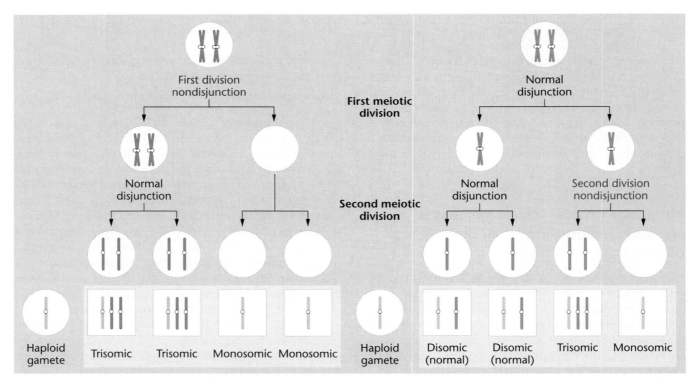

FIGURE 2.17 Diagram illustrating nondisjunction during the first and second meiotic divisions. In both cases, some gametes are formed either containing two members of a specific chromosome or lacking that chromosome altogether. Following fertilization by a normal haploid gamete, monosomic, disomic (normal), or trisomic zygotes result.

The results of nondisjunction during meiosis I and meiosis II for just one chromosome of a diploid genome are shown in Figure 2.17. As you can see, for the affected chromosome, abnormal gametes may be formed that contain either two members or none at all. Fertilization of these with a normal gamete produces a zygote with either three members (trisomy) or only one member (monosomy) of this chromosome. While these conditions are more frequently tolerated in plants, they usually have severe or lethal effects in animals. Trisomy and monosomy will be described in greater detail in Chapter 10.

Electron microscopy has revealed the cytological nature of the mitotic and meiotic chromosome

Thus far in this chapter, we have focused on mitotic and meiotic chromosomes, emphasizing their behavior during cell division and gamete formation. Initially, biologists knew about these chromosomes only from routine observations made with the light microscope. Geneticists were curious about why chromosomes were invisible during interphase but appeared during the prophase stage of mitosis and meiosis. Sub-sequent studies using electron microscopy have clearly shown why chromosomes are visible only during division stages.

Chromatin vs. Chromosomes

During interphase, only dispersed chromatin fibers are present in the nucleus [Figure 2.18(a)]. It is now believed that during interphase, starting in G1, mitotic chromosomes unwind to form these long fibers, which consist of DNA and associated proteins, particularly proteins called histones. It is in this physical arrangement that DNA can most efficiently function during transcription and can be replicated.

Once mitosis begins, however, the fibers coil and fold up, condensing into typical mitotic chromosomes [Figure 2.18(b)]. If the fibers making up the mitotic chromosome are loosened, areas of greatest spreading reveal individual fibers similar to those seen in interphase chromatin [Figure 2.18(c)]. Very few fiber ends seem to be present, and in some cases none can be seen. Instead, individual fibers always seem to loop back into the interior. Such fibers are obviously twisted and coiled around one another, forming the regular pattern of the mitotic chromosome.

Electron microscopic observations of mitotic chromosomes in varying states of coiling led Ernest DuPraw to

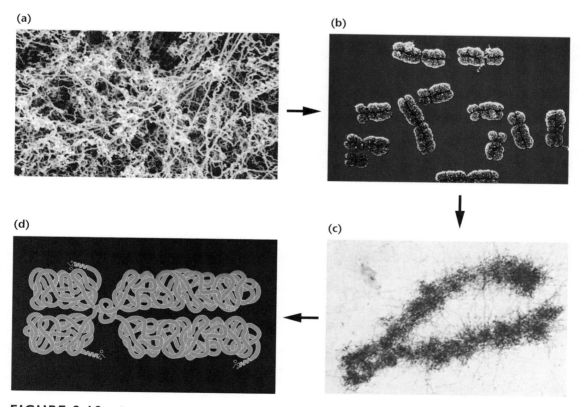

FIGURE 2.18 A comparison of (a) the chromatin fibers characteristic of the interphase nucleus with (b) and (c) metaphase chromosomes that are derived from chromatin during mitosis. Part (d) depicts the folded-fiber model showing how chromatin is condensed into a metaphase chromosome. Parts (a) and (c) are transmission electron micrographs, whereas part (b) is a scanning electron micrograph.

(a)

(b)

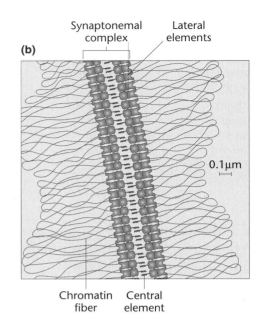

Synaptonemal complex   Lateral elements

0.1μm

Chromatin fiber   Central element

FIGURE 2.19 (a) Electron micrograph of a portion of a synaptonemal complex found between synapsed bivalents of *Neotiella rutilans*. (b) Schematic interpretation of the components making up the synaptonemal complex. The lateral elements, central element, and chromatin are labeled.

postulate the **folded-fiber model,** illustrated in Figure 2.18(d). During metaphase, each chromosome consists of two sister chromatids joined at the centromeric region. Each arm of the chromatid appears to consist of a single fiber wound up much like a skein of yarn. The fiber is composed of double-stranded DNA and protein tightly coiled together. An orderly coiling–twisting–condensing process appears to be involved in the transition of the interphase chromatin to the more condensed, mitotic chromosomes. It is estimated that during the transition from interphase to prophase, a 5000-fold contraction occurs in the length of DNA within the chromatin fiber! This process must indeed be extremely precise, given the highly ordered nature and consistent appearance of mitotic chromosomes in all eukaryotes. Note particularly in the micrographs the clear distinction between the sister chromatids constituting each chromosome. They are joined only by the common centromere that they share prior to anaphase.

After we have provided a more thorough description of DNA structure, we will return to this topic in Chapter 19 to explore the molecular nature of the chromatin fiber.

The Synaptonemal Complex

The electron microscope has also been used to visualize another ultrastructural component of the chromosome found only in cells undergoing meiosis. This structure, first introduced during our earlier discussion of the first meiotic prophase stage, is found between synapsed homologs and is called the **synaptonemal complex.*** In 1956, Montrose Moses observed this complex in spermatocytes of crayfish,

and Don Fawcett saw it in pigeon and human spermatocytes. Because there was not yet any satisfactory explanation of the mechanism of synapsis or of crossing over and chiasmata formation, many researchers became interested in this structure. With few exceptions, the ensuing studies revealed the synaptonemal complex to be present in most plant and animal cells visualized during meiosis.

As you can see in the electron micrograph in Figure 2.19(a) the synaptonemal complex is a tripartite structure. The central element is usually less dense and thinner (100–150 Å) than the two identical outer elements (500 Å). The outer structures, called lateral elements, are intimately associated with the synapsed homologs on either side. Selective staining has revealed that these lateral elements consist primarily of DNA and protein, suggesting that chromatin is an essential part of them. Some DNA fibrils traverse these lateral elements, making connections with the central element, which is composed primarily of protein. Figure 2.19(b) provides a diagrammatic interpretation of the electron micrograph consistent with the above description.

The formation of the synaptonemal complex begins prior to the pachytene stage. As early as leptonema of the first meiotic prophase, lateral elements are seen in association with sister chromatids. Homologs have yet to associate with one another and are randomly dispersed in the nucleus. As we saw earlier, by the next stage, zygonema, homologous chromosomes begin to align with one another in what is called rough pairing, but they remain distinctly apart by some 300 nm. Then, during pachynema, the intimate association referred to as synapsis between homologs occurs as formation of the complex is complet-

---

*An alternative spelling of this term is *synaptinemal complex*.

ed. In some diploid organisms, this occurs in a zipperlike fashion beginning at the ends of the chromosomes, which may be attached to the nuclear envelope.

It is now agreed that the synaptonemal complex is the vehicle for the pairing of homologs and their subsequent segregation during meiosis. However, some degree of synapsis can occur in certain cases where no synaptonemal complexes are formed. Thus, it is possible that the function of this structure may go beyond its involvement in the formation of bivalents.

In certain instances where no synaptonemal complexes are formed during meiosis, synapsis is not complete and crossing over is reduced or eliminated. For example, in male *Drosophila melanogaster*, where synaptonemal complexes are not usually seen, meiotic crossing over rarely, if

ever, occurs. This observation suggests that the synaptonemal complex may be important in order for chiasmata to form and crossing over to occur.

The study of *ZIP1*, a mutation in the yeast *Saccharomyces cerevisiae*, has provided further insights into chromosome pairing. Cells bearing this mutation can undergo the initial alignment stage (rough pairing) and full-length central and lateral element formation, but fail to achieve the intimate pairing characteristic of synapsis. It has been suggested that the gene product of the *ZIP1* locus is a protein component of the central element of the synaptonemal complex since it is absent in mutant cells. This observation further suggests that a complete and intact synaptonemal complex is essential during the transition from the initial rough alignment stage to the intimate pairing characteristic of synapsis.

CHAPTER SUMMARY

1. The structure of cells is elaborate and complex. Many components of cells are involved directly or indirectly with genetic processes.

2. In diploid organisms, chromosomes exist in homologous pairs. Each pair shares the same size, centromere placement, and gene loci. One member of each pair is derived from the maternal parent, and one is derived from the paternal parent.

3. Mitosis and meiosis are mechanisms by which cells distribute genetic information contained in their chromosomes to their descendants in a precise orderly fashion.

4. Mitosis, or nuclear division, is part of the cell cycle and is the basis of cellular reproduction. Daughter cells are produced that are genetically identical to their progenitor cell.

5. Mitosis is subdivided into discrete phases: prophase, prometaphase, metaphase, anaphase, and telophase. Condensation of chromatin into chromosome structures occurs during prophase. During prometaphase, chromosomes take on the appearance of double structures, each represented by a pair of sister chromatids. In metaphase, chromosomes line up on the equatorial plane of the cell. During anaphase, sister chromatids of each chromosome are pulled apart and directed toward opposite poles. Telophase completes daughter cell formation and is characterized by cytokinesis, the division of the cytoplasm.

6. The cell cycle is characteristic of all eukaryotes and is under the control of an elaborate genetic program that regulates its activities. At least three checkpoints are present where cellular activities are monitored and where the cycle may be arrested if further progress jeopardizes the cell.

7. Meiosis, the underlying basis of sexual reproduction, results in the conversion of a diploid cell to a haploid gamete or spore. As a result of chromosome duplication and two subsequent divisions, each haploid cell receives one member of each homologous pair of chromosomes.

8. A major difference exists between animal meiosis in males and females. Spermatogenesis partitions cytoplasmic volume equally and produces four haploid sperm cells. Oogenesis, on the other hand, accumulates the cytoplasm around one egg cell and reduces the other haploid sets of genetic material to polar bodies. The extra cytoplasm contributes to zygote development following fertilization.

9. Meiosis not only maintains the constancy of genetic material from generation to generation, but also results in extensive genetic variation. This is produced by virtue of the random distribution of either a maternal or a paternal member of every homologous pair of chromosomes into each gamete. Variation is further enhanced as a result of the exchange during crossing over between maternal and paternal homologs. In addition, meiosis plays an important role in the life cycle of fungi and plants, serving as the bridge between alternating generations.

10. Mitotic and meiotic chromosomes are produced as a result of the coiling and condensation of chromatin fibers characteristic of interphase. This transition is described by the folded-fiber model.

11. The synaptonemal complex is an ultrastructural component of the cell present during meiotic prophase I. It is important to the process of synapsis of homologs and may play a role in crossing over.

INSIGHTS AND SOLUTIONS

*With this initial appearance of "Insights and Solutions," it is appropriate to provide a brief description of its value to you as a student. This section in each chapter precedes the "Problems and Discussion Questions" and provides sample problems and solutions that will illustrate approaches useful in **genetic analysis**. Insights that you gain will help you arrive at correct solutions to ensuing problems.*

1. In an organism with a haploid number of 3, how many individual chromosomal structures will align on the metaphase plate during (a) mitosis (b) meiosis I and (c) meiosis II? Describe each configuration.

   Solution:

   (a) In mitosis, where homologous chromosomes do not synapse, there will be 6 double structures, each consisting of a pair of sister chromatids. The number of structures is equivalent to the diploid number.

   (b) In meiosis I, the homologs have synapsed, reducing the number of structures to 3. Each is called a *tetrad* and consists of two pairs of sister chromatids.

   (c) In meiosis II, the same number of structures exist (3), but in this case they are called *dyads*. Each consists of a pair of sister chromatids. When crossing over has occurred, each chromatid may contain part of one of its nonsister chromatids obtained during exchange in prophase I.

2. For the chromosomes illustrated in Figure 2.14, draw all possible alignment configurations that may occur during metaphase of meiosis I.

   Solution: As shown in the illustration below, there are four configurations possible when $n = 2$.

3. Assume that there is one gene on both of the larger chromosomes and two alleles, *A* and *a*, as shown. *Also assume* a second gene with two alleles on the smaller chromosomes, using *B* and *b*. Calculate the probability of generating each gene combination (*AB, Ab, aB, ab*) following meiosis I.

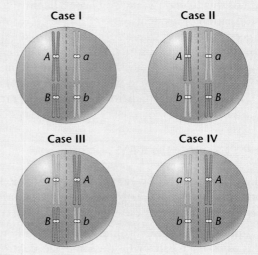

Solution:

Case I	*AB* and *ab*
Case II	*Ab* and *aB*
Case III	*aB* and *Ab*
Case IV	*ab* and *AB*

Total: $AB = 2$ $(p = \frac{1}{4})$

        $Ab = 2$ $(p = \frac{1}{4})$

        $aB = 2$ $(p = \frac{1}{4})$

        $ab = 2$ $(p = \frac{1}{4})$

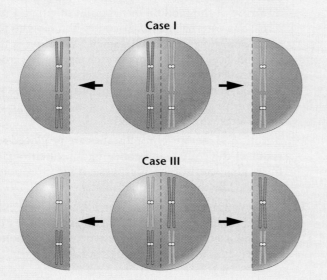

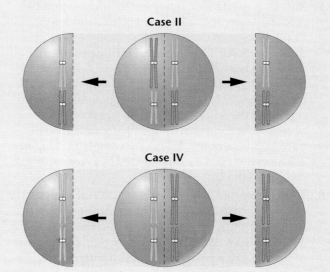

4. How many different chromosome configurations can occur following meiosis I if three different pairs of chromosomes are present ($n = 3$)?

Solution: If $n = 3$, then eight different configurations would be possible. The formula 2^n, where n equals the haploid number, will allow you to calculate the number of potential alignment patterns. As we will see in the next chapter, these patterns are produced as a result of the Mendelian postulate called *segregation* and they serve as the physical basis of the Mendelian postulate of *independent assortment*.

5. Assuming that comparable chromosomes in different individuals are genetically dissimilar because of different alleles, how many unique zygotic combinations are possible following fertilization in an organism where $n = 3$?

Solution: Assuming that no crossing over occurs (and that maternal and paternal chromatids remain intact), then each organism can produce 2^n or 2^3 different chromosome combinations in their gametes. Following random fertilization, $(2^3)(2^3) = 2^6 = 64$ unique combinations are possible in the offspring.

PROBLEMS AND DISCUSSION QUESTIONS

1. What role do the following cellular components play in the storage, expression, or transmission of genetic information: (a) chromatin, (b) nucleolus, (c) ribosome, (d) mitochondrion, (e) centriole, (f) centromere?

2. Discuss the concepts of homologous chromosomes, diploidy, and haploidy. What characteristics are shared between two chromosomes considered to be homologous?

3. If two chromosomes of a species are the same length and have similar centromere placements, yet are *not* homologous, what *is* different about them?

4. Describe the events that characterize each stage of mitosis.

5. If an organism has a diploid number of 16, how many chromatids are visible at the end of mitotic prophase? How many chromosomes are moving to each pole during anaphase of mitosis?

6. How are chromosomes named on the basis of centromere placement?

7. Contrast telophase in plant and animal mitosis.

8. Outline and discuss the events, including regulatory checkpoints, of the cell cycle. What experimental technique was used to demonstrate the existence of the S phase?

9. Examine Figure 2.15 showing oogenesis in animal cells. Will the genetic composition of the second polar bodies (derived from meiosis II) always be identical to that of the ootid? Why or why not?

10. Contrast the end results of meiosis with those of mitosis.

11. Define the following terms and discuss their relevance to meiosis: (a) synapsis, (b) bivalents, (c) chiasmata, (d) crossing over, (e) chromomeres, (f) sister chromatids, (g) tetrads, (h) dyads, (i) monads, and (j) synaptonemal complex.

12. Contrast the genetic content and the origin of sister vs. non-sister chromatids during their earliest appearance in prophase I of meiosis. How might the genetic content of these change by the time tetrads have aligned at the equatorial plate during metaphase I?

13. Given the end results of the two types of division, why is it necessary for homologs to pair during meiosis and not desirable for them to pair during mitosis?

14. If an organism has a diploid number of 16 in an oocyte,
    (a) how many tetrads are present in the first meiotic prophase?

    (b) how many dyads are present in the second meiotic prophase?

    (c) how many monads migrate to each pole during the second meiotic anaphase?

    (d) what is the probability that a gamete will contain only paternal chromosomes?

15. Contrast spermatogenesis and oogenesis. What is the significance of the formation of polar bodies?

16. Explain why meiosis leads to significant genetic variation while mitosis does not.

17. During oogenesis in an animal species with a haploid number of 6, one dyad undergoes second division nondisjunction. Following the second meiotic division, the involved dyad ends up intact in the ovum. How many chromosomes are present in (a) the mature ovum, and (b) the second polar body? (c) Following fertilization by a normal sperm, what chromosome condition is created?

18. What is the probability that in an organism with a haploid number of 10 that a sperm will be formed that contains all 10 chromosomes whose centromeres were derived from maternal homologs?

19. During the first meiotic prophase,
    (a) when does crossing over occur?
    (b) when does synapsis occur?
    (c) during which stage are the chromosomes least condensed?
    (d) when are chiasmata first visible?

20. What is the role of meiosis in the life cycle of a higher plant such as an angiosperm?

21. Describe the transition of a chromatin fiber into a mitotic chromosome. What is the name of the model that depicts this transition? How was this model developed?

22. When during the cell cycle does the transition described in Problem 21 first occur?

23. What are checkpoints within the cell cycle, and why are they important to multicellular organisms?

24. What are *cdc* mutations? How did their study extend our knowledge of the cell cycle?

25. Discuss the role of the *p53* gene in the regulation of the cell cycle. What consequences result when this gene loses normal function as a result of mutation?

GENETICS MediaLab

The following resources will help you achieve a better understanding of the concepts presented in this chapter. These resources can be found on the CD packaged with this textbook and on the Companion Website at **http://www.prenhall.com/klug/**.

CD Resources:

Animated Tutorial: *An Overview of Meiosis*

Animated Tutorial: *Prophase I of Meiosis*

Self-grading Chapter Problems

Web Resources:

Web Destinations in Genetics

Self-grading Chapter Problems

Chapter Search Terms

Genetics Newsgroups

Student Bulletin Board

Web Problem 1:

Time for completion = 10 minutes

What is the relative timing of events in the cell cycle? In this exercise, you will examine micrographs of onion cells during interphase and mitosis and empirically determine the proportion of time these cells spend in each phase of mitosis. Once you reach The Biology Project's tutorial on the cell cycle, follow the instructions and then answer these questions. What part of the cell cycle takes the most time? What are the events during that phase that take so much time? Radiation tends to damage cells undergoing mitosis to a greater degree than cells in interphase; what does this tell you about the use of radiation in treating cancer (see Chapter 23 for information on cancer and genetics)? To complete this exercise, visit Web Problem 1 in Chapter 2 of your Companion Website and select the keyword **CELL CYCLE.**

Web Problem 2:

Time for completion = 10 minutes

What are some key differences between cellular events in mitosis and meiosis? Both processes begin in interphase, but the end products are quite different. Mitosis produces a pair of genetically identical cells, whereas meiosis produces four genetically different cells. You will use micrographs of the two processes to answer the following questions. At what stage in each process do the centromeres joining the sister chromatids separate? In meiosis, at what stage do the cells become haploid? What structures are involved in moving the chromosomes to the poles in mitosis? Are the same structures acting in meiosis? In Down syndrome, one extra copy of chromosome 21 is present in all the cells of the afflicted individual. Do you think the extra copy was added in mitosis or meiosis? At what stage was it added? Where did the extra copy come from? To complete this exercise, visit Web Problem 2 in Chapter 2 of your Companion Website and select the keyword **MITOSIS.**

Web Problem 3:

Time for completion = 10 minutes

How does meiosis generate genetic variation among the gametes? Two processes occur during meiosis that generate this variation. After reading through the short tutorial on recombination during meiosis, identify the two sources of variation. Which of the processes generates variation among the four copies of a given chromosome? What would happen at the level of the genes if the homologs were not perfectly aligned during crossing over, so that unequal parts of the chromosomes were exchanged? Which process accounts for Mendel's postulates of segregation and independent assortment (Chapter 3)? To complete this exercise, visit Web Problem 3 in Chapter 2 of your Companion Website and select the keyword **MEIOSIS.**

EXTRA-SPICY PROBLEMS

As part of the "Problems and Discussion Questions" section in each chapter, we shall present one or more "extra-spicy" genetics problems. We have chosen to set these apart in order to identify problems that are particularly challenging. You may be asked to examine and assess actual data, to design genetics experiments, or to engage in cooperative learning. Like genetic varieties of peppers, some of these experiences are just spicy and some are very hot. Hopefully, all of them will leave an aftertaste that is pleasing to those who indulge themselves.

26. A diploid cell contains three pairs of chromosomes designated *A*, *B*, and *C*. Each pair contains a maternal and a paternal member (e.g., A^m and A^p, etc.). Using these designations, demonstrate your understanding of mitosis and meiosis by drawing chromatid combinations in response to the following questions. Be sure to indicate when chromatids are paired as a result of replication and/or synapsis. You may wish to use a large piece of brown manila wrapping paper or a cut-up paper grocery bag and work with a partner as you deal with this problem. Such cooperative learning may be a useful approach as you solve problems throughout the text.

(a) In mitosis, what chromatid combination(s) will be present during metaphase? What combination(s) will be present at each pole at the completion of anaphase?

(b) During meiosis I, assuming no crossing over, what chromatid combination(s) will be present at the completion of prophase? Draw all possible alignments of chromatids as migration begins during early anaphase.

(c) Are there any possible combinations present during prophase of meiosis II other than those that you drew in (b)? If so, draw them. If not, then proceed to (d).

(d) Draw all possible combinations of chromatids during the early phases of anaphase in meiosis II.

(e) Assume that during meiosis I none of the *C* chromosomes disjoin at metaphase, but they separate into dyads (instead of monads) during meiosis II. How would this change the alignments that you constructed during the anaphase stages in meiosis I and II? Draw them.

(f) Assume that each gamete resulting from (e) participated in fertilization with a normal haploid gamete. What combinations will result? What percentage of zygotes will be diploid, containing one paternal and one maternal member of each chromosome pair?

SELECTED READINGS

ALBERTS, B. et al. 1994. *Molecular biology of the cell,* 3rd ed. New York: Garland Publishing.

BAKER, B.A. et al. 1976. The genetic control of meiosis. *Annu. Rev. Genet.* 10:53–134.

BASERGA, R., and KISIELESKI, W. 1963. Autobiographics of cells. *Sci. Am.* (Aug.) 209:103–10.

BRACHET, J., and MIRSKY, A. E. 1961. *The cell: Meiosis and mitosis,* Vol. 3. Orlando, FL: Academic Press.

CARPENTER, A.T.C. 1994. Chiasma function. *Cell* 77:959–62.

DUPRAW, E.J. 1970. *DNA and chromosomes.* New York: Holt, Rinehart & Winston.

GLOVER, D.M., GONZALEZ, C., and RAFF, J.W. 1993. The centrosome. *Sci. Am.* (June) 268:62–68.

GOLOMB, H.M., and BAHR, G.F. 1971. Scanning electron microscopic observations of surface structures of isolated human chromosomes. *Science* 171:1024–26.

HALL, J.L., RAMANIS, Z., and LUCK, D.J. 1989. Basal body/centriolar DNA: Molecular genetic studies in *Chlamydomonas. Cell* 59:121–32.

HARTWELL, L. H. et al. 1974. Genetic control of the cell division cycle in yeast. *Science* 183:46–51.

HARTWELL, L.H., and KASTAN, M.B. 1994. Cell cycle control and cancer. *Science* 266:1821–28.

HARTWELL, L.H., and WEINERT, T.A. 1989. Checkpoint controls that ensure the order of cell cycle events. *Science* 246:629–34.

HAWLEY, R.S., and ARBEL, T. 1993. Yeast genetics and the fall of the classical view of meiosis. *Cell* 72:301–303.

KLECKNER, N. 1996. Meiosis: How could it work? *Proc. Natl. Acad. Sci.* 93:8167–74.

KOSHLAND, D. 1994. Mitosis: Back to the basics. *Cell* 77:951–54.

MAZIA, D. 1961. How cells divide. *Sci. Am.* (Jan.) 205:101–20.

—— 1974. The cell cycle. *Sci. Am.* (Jan.) 235:54–64.

MCINTOSH, J.R., and MCDONALD, K.L. 1989. The mitotic spindle. *Sci. Am.* (Oct.) 261:48–56.

MCKIM, K. et al. 1998. Meiotic synapsis in the absence of recombination. *Science* 279:876–78.

MOENS, P.B. 1973. Mechanisms of chromosome synapsis at meiotic prophase. *Int. Rev. Cytol.* 35:117–34.

MURRAY, A.W., and KIRSCHNER, M. 1991. What controls the cell cycle? *Sci. Am.* (March) 263:56–63.

—— 1993. *The cell cycle: An introduction.* New York: Oxford University Press.

PRESCOTT, D.M. 1977. *Reproduction of eukaryotic cells.* Orlando, FL: Academic Press.

PRESCOTT, D.M., and FLEXER, A.S. 1986. *Cancer, the misguided cell,* 2nd ed. Sunderland, MA: Sinauer.

SAWIN, K.E. et al. 1992. Mitotic spindle organization by a plus end-directed microtubular motor. *Nature* 359:540–43.

SWANSON, C.P., MERZ, T., and YOUNG, W.J. 1981. *Cytogenetics, the chromosome in division, inheritance, and evolution,* 2nd ed. Englewood Cliffs, NJ: Prentice-Hall.

WADSWORTH, P. 1993. Mitosis: Spindle assembly and chromosome movement. *Curr. Opin. Cell Biol.* 5:93–99.

WESTERGAARD, M., and VON WETTSTEIN, D. 1972. The synaptinemal complex. *Annu. Rev. Genet.* 6:71–110.

WHEATLEY, D.N. 1982. *The centriole: A central enigma of cell biology.* New York: Elsevier/North-Holland Biomedical.

YUNIS, J.J., and CHANDLER, M.E. 1979. Cytogenetics. In *Clinical diagnosis and management by laboratory methods,* ed. J.B. Henry, Vol. 1. Philadelphia: W.B. Saunders.

Mendel's garden, as seen in the 1980s.

Mendelian Genetics

KEY CONCEPTS

- **Mendel used a model experimental approach to study patterns of inheritance**

- **The monohybrid cross reveals how one trait is transmitted from generation to generation**

  Mendel's First Three Postulates
  Modern Genetic Terminology
  Mendel's Analytical Approach
  Punnett Squares
  The Test Cross: One Character

- **Mendel's dihybrid cross revealed his fourth postulate: Independent assortment**

  Independent Assortment
  The Test Cross: Two Characters

- **The trihybrid cross demonstrates that Mendel's principles apply to inheritance of multiple traits**

  The Forked-Line Method, or Branch Diagram

- **Mendel's work was rediscovered in the early twentieth century**

- **The correlation of Mendel's postulates with the behavior of chromosomes formed the foundation of modern transmission genetics**

  Unit Factors, Genes, and Homologous Chromosomes

- **Independent assortment leads to extensive genetic variation**

- **Laws of probability help to explain genetic events**

  The Product Law and the Sum Law
  Conditional Probability
  The Binomial Theorem

- **Chi-square analysis evaluates the influence of chance on genetic data**

- **Pedigrees reveal patterns of inheritance in humans**

GenCDX

When you see this icon, there are related animations and exercises on the CD accompanying this text.

lthough inheritance of biological traits has been recognized for thousands of years, the first significant insights into the mechanisms involved occurred less than a century and a half ago. In 1866, Gregor Mendel published the results of a series of experiments that would lay the foundation for the formal discipline of genetics. By the beginning of the twentieth century, the concept of the gene as a distinct hereditary unit was established, and the ways in which genes are transmitted to offspring and control traits were clarified. Interest in these areas was accelerated in the first half of the twentieth century. The resultant findings served as the foundation for an accelerated research effort in genetics beginning about 1940. It is safe to say that studies in genetics, most recently those at the molecular level, have remained continually at the forefront of biological research since the early 1900s.

When Mendel began his studies of inheritance using *Pisum sativum*, the garden pea, there was no knowledge of chromosomes or of the role and mechanism of meiosis. Nevertheless, he was able to determine that distinct **units of inheritance** exist and to predict their behavior during the formation of gametes. Subsequent investigators, with access to cytological data, were able to relate their observations of chromosome behavior during meiosis to Mendel's principles of inheritance. Once this correlation was made, Mendel's postulates were accepted as the basis for the study of what is known as **Mendelian** or **transmission genetics.** These principles describe how genes are transmitted from parents to offspring and were derived directly from Mendel's experimentation. Even today, they serve as the cornerstone of the study of inheritance. In this chapter we focus on the development of the principles that Mendel established.

Mendel used a model experimental approach to study patterns of inheritance

Johann Mendel was born in 1822 to a peasant family in the central European village of Heinzendorf. An excellent student in high school, he studied philosophy for several years afterward, and in 1843 was admitted to the Augustinian Monastery of St. Thomas in Brno, now part of the Czech Republic, taking the name of Gregor. In 1849, he was relieved of pastoral duties and received a teaching appointment that lasted several years. From 1851 to 1853, he attended the University of Vienna, where he studied physics and botany. In 1854 he returned to Brno, where for the next 16 years he taught physics and natural science. Mendel received support from the monastery for his studies and research throughout his life.

In 1856, Mendel performed his first set of hybridization experiments with the garden pea. The research phase of his career lasted until 1868, when he was elected abbot of the monastery. Although he retained his interest in genetics, his new responsibilities demanded most of his time. In 1884, Mendel died of a kidney disorder. The local newspaper paid him the following tribute: "His death deprives the poor of a benefactor, and mankind at large of a man of the noblest character, one who was a warm friend, a promoter of the natural sciences, and an exemplary priest."

Mendel first reported the results of some simple genetic crosses between certain strains of the garden pea in 1865. Although his was not the first attempt to provide experimental evidence pertaining to inheritance, Mendel's success where others had failed can be attributed, at least in part, to his elegant model of experimental design and analysis.

Mendel showed remarkable insight into the methodology necessary for good experimental biology. First, he chose an organism that was easy to grow and to hybridize artificially. The pea plant is self-fertilizing in nature but is easy to cross-breed experimentally. It reproduces well and grows to maturity in a single season. Mendel then chose to follow seven visible features (unit characters), each represented by two contrasting forms or traits (Figure 3.1). For the character stem height, for example, he experimented with the traits *tall* and *dwarf*. He selected six other contrasting pairs of traits involving seed shape and color, pod shape and color, and pod and flower arrangement. From local seed merchants, Mendel obtained true-breeding strains, those in which each trait appeared unchanged generation after generation in self-fertilizing plants.

Several factors in addition to the choice of a suitable organism led to Mendel's success. He restricted his examination to one or very few pairs of contrasting traits in each experiment. He also kept accurate quantitative records, a necessity in genetic experiments. From the analysis of his data, Mendel derived certain postulates that have become the principles of transmission genetics.

The results of Mendel's experiments went unappreciated until the turn of the century, well after his death. Once Mendel's publications were rediscovered by geneticists in-

vestigating the function and behavior of chromosomes, however, the implications of his postulates were immediately apparent. He had discovered the basis for the transmission of hereditary traits!

The monohybrid cross reveals how one trait is transmitted from generation to generation

The simplest experiments Mendel performed involved only one pair of contrasting traits. In what is called a **monohybrid cross,** he mated individuals from two parent strains, each of which exhibits one of the two contrasting forms of the character under study. The original parents in a genetic cross are called the **P₁** or **parental generation,** and their offspring are the **F₁** or **first filial generation.** If individuals of the F_1 generation undergo self-fertilization, or selfing, their offspring are called the **F₂** or **second filial generation.** We can, of course, continue to follow subsequent generations, if desirable.

The cross between true-breeding peas with tall stems and dwarf stems is representative of Mendel's monohybrid crosses. *Tall* and *dwarf* represent contrasting forms or traits of the character of stem height. Unless tall or dwarf plants are crossed together or with another strain, they will undergo self-fertilization and breed true, producing their respective trait generation after generation. However, when Mendel crossed tall plants with dwarf plants, the resulting F_1 generation consisted of only tall plants. When members of the F_1 generation were selfed, Mendel observed that 787 of 1064 F_2 plants were tall, while 277 of 1064 were dwarf. Note that in this cross (Figure 3.1), the dwarf trait disappeared in the F_1 generation only to reappear in the F_2. Mendel made similar crosses between pea plants exhibiting each of the other pairs of contrasting traits. Results of these crosses are also shown in Figure 3.1. In every case, the outcome was similar to the tall/dwarf cross.

Genetic data are usually expressed and analyzed as ratios. In this particular example, many identical P_1 crosses were made and many F_1 plants—all tall—were produced. Of the 1064 F_2 offspring, 787 were tall and 277 were dwarf—a ratio of approximately 2.8:1.0, or about 3:1. Three-fourths appeared like the F_1 plants, while one-fourth exhibited the contrasting trait, which had disappeared in the F_1 generation.

One further aspect of the monohybrid crosses was important. In each cross, the F_1 and F_2 patterns of inheritance were similar regardless of which P_1 plant served as the source of pollen, or sperm, and which served as the source of the ovum, or egg. The crosses could be made either way—that is, pollen from the tall plant pollinating

Character	Contrasting traits		F$_1$ results	F$_2$ results	F$_2$ ratio
Seeds	round/wrinkled		all round	5474 round 1850 wrinkled	2.96:1
	yellow/green		all yellow	6022 yellow 2001 green	3.01:1
	full/constricted		all full	882 full 299 constricted	2.95:1
Pods	green/yellow		all green	428 green 152 yellow	2.82:1
	axial/terminal		all axial	651 axial 207 terminal	3.14:1
Flowers	violet/white		all violet	705 violet 224 white	3.15:1
Stem	tall/dwarf		all tall	787 tall 277 dwarf	2.84:1

FIGURE 3.1   A summary of the seven pairs of contrasting traits and the results of Mendel's seven monohybrid crosses using the garden pea, *Pisum sativum*. In each case, pollen derived from plants exhibiting one contrasting trait was used to fertilize the ova of plants exhibiting the other contrasting trait. In the F$_1$ generation one of the two traits, referred to as dominant, was exhibited by all plants. The contrasting trait, referred to as recessive, then reappeared in approximately one-fourth of the F$_2$ plants.

dwarf plants, or vice versa. These are called **reciprocal crosses.** Therefore, the results of Mendel's monohybrid crosses were not sex-dependent.

To explain these results, Mendel proposed the existence of what he called particulate **unit factors** for each trait. He suggested that these factors serve as the basic units of heredity and are passed unchanged from generation to generation, determining various traits expressed by each individual plant. Using these general ideas, Mendel proceeded to hypothesize precisely how such factors could account for the results of the monohybrid crosses.

Mendel's First Three Postulates

Using the consistent pattern of results in the monohybrid crosses, Mendel derived the following three postulates or principles of inheritance.

1. *Unit Factors in Pairs*

Genetic characters are controlled by unit factors that exist in pairs in individual organisms.

In the monohybrid cross involving tall and dwarf stems, a specific unit factor exists for each trait. Each diploid individual receives one factor from each parent. Because the factors occur in pairs, three combinations are possible: two factors for tallness, two factors for dwarfness, or one of each factor. Every individual contains one of these three combinations, which determines stem height.

2. *Dominance/Recessiveness*

When two unlike unit factors responsible for a single character are present in a single individual, one unit factor is dominant to the other, which is said to be recessive.

In each monohybrid cross, the trait expressed in the F_1 generation results from the presence of the dominant unit factor. The trait that is not expressed in the F_1, but which reappears in the F_2, is under the genetic influence of the recessive unit factor. Note that this dominance/recessiveness relationship pertains only when unlike unit factors are present together in an individual. The terms **dominant** and **recessive** are also used to designate the traits. In the above case, the trait tall stem is said to be dominant to the recessive trait, dwarf stem.

3. *Segregation*

During the formation of gametes, the paired unit factors separate, or segregate, randomly so that each gamete receives one or the other with equal likelihood.

If an individual contains a pair of like unit factors (for example, both specify tall), then all gametes receive one tall unit factor. If an individual contains unlike unit factors (e.g., one for tall and one for dwarf), then each gamete has a 50 percent probability of receiving either the tall *or* the dwarf unit factor.

These postulates provide a suitable explanation for the results of the monohybrid crosses. For example, consider the tall/dwarf cross. Mendel reasoned that P_1 tall plants contained identical paired unit factors, as did the P_1 dwarf plants. The gametes of tall plants all received one tall unit factor as a result of segregation. Likewise, the gametes of dwarf plants all received one dwarf unit factor. Following fertilization, all F_1 plants received one unit factor from each parent, a tall factor from one and a dwarf factor from the other, reestablishing the paired relationship. Because tall is dominant to dwarf, all F_1 plants were tall.

When F_1 plants form gametes, the postulate of segregation demands that each gamete randomly receive *either* the tall *or* dwarf unit factor. Following random fertilization

events during F_1 selfing, four F_2 combinations will result in equal frequency:

   (1) tall/tall

   (2) tall/dwarf

   (3) dwarf/tall

   (4) dwarf/dwarf

Combinations (1) and (4) will clearly result in tall and dwarf plants, respectively. According to the postulate of dominance/recessiveness, combinations (2) and (3) will both yield tall plants. Therefore, we predict that the F_2 will consist of three-fourths tall and one-fourth dwarf plants, or a ratio of 3:1. This is approximately what Mendel observed in his cross between tall and dwarf plants. A similar pattern was observed in each of the other monohybrid crosses.

Modern Genetic Terminology

To understand the monohybrid cross and Mendel's first three postulates in a modern context, we must introduce several new terms as well as a set of symbols for the unit factors. Traits such as tall or dwarf are physical expressions of the information contained in unit factors. We now call the physical expression of a trait the **phenotype** of the individual.

Mendel's unit factors represent units of inheritance—what modern geneticists now call **genes.** For any given character, such as plant height, the phenotype is determined by different combinations of alternative forms of a single gene called **alleles.** For example, tall and dwarf are alleles determining the height of the pea plant.

Geneticists use several different conventions involving gene symbols to represent genes. In Chapter 4 we will review some of these, but for now we will adopt one that we can use consistently through the next few chapters. According to this convention, the first letter of the recessive trait is chosen to symbolize the character in question. The lowercase form of the letter designates the allele for the recessive trait, and the uppercase letter designates the allele for the dominant trait. Further, these gene symbols are italicized. Therefore, *d* stands for the dwarf allele and *D* represents the tall allele. When alleles are written in pairs to represent the two unit factors present in any individual (*DD*, *Dd*, or *dd*), these symbols are referred to as the **genotype.** This term reflects the genetic makeup of an individual whether it is haploid or diploid. By following the principle of dominance and recessiveness, we can tell the phenotype of the individual from its genotype: *DD* and *Dd* are tall, and *dd* is dwarf. When identical alleles constitute the genotype (*DD* or *dd*), the individual is said to be **homozygous** or a **homozygote;** when alleles are different (*Dd*), we use the term **heterozygous** or **heterozygote.** Figure 3.2 illustrates the complete monohybrid cross using modern terminology.

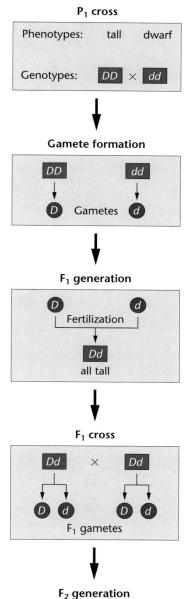

FIGURE 3.2 An explanation of the monohybrid cross between tall and dwarf pea plants. The symbols *D* and *d* are used to designate the tall and dwarf unit factors, respectively, in the genotypes of mature plants and gametes. All individuals are shown in rectangles. All gametes are shown in circles. 🌐GenCDX

Mendel's Analytical Approach

What led Mendel to deduce unit factors in pairs? Because there were two contrasting traits for each character, it seemed logical that two distinct factors must exist. However, why does one of the two traits or phenotypes disappear in the F_1 generation? Observation of the F_2 generation helps to answer this question. The recessive trait and its unit factor do not actually disappear in the F_1; they are merely hidden or masked, only to reappear in one-fourth of the F_2 offspring. Therefore, Mendel concluded that one unit factor for tall and one for dwarf were transmitted to each F_1 individual; but because the tall factor or allele is dominant to the dwarf factor or allele, all F_1 plants are tall. Given the above, we can ask how Mendel explained the 3:1 F_2 ratio. As shown in Figure 3.2, Mendel deduced that the tall and dwarf alleles of the F_1 heterozygote segregate randomly into gametes. If fertilization is random, this ratio is predicted. If a large population of offspring are generated, the outcome of such a cross should reflect the 3:1 ratio.

Because he operated without the hindsight that modern geneticists enjoy, Mendel's analytical reasoning must be considered a truly outstanding scientific achievement. On the basis of rather simple, but precisely executed, breeding experiments, he proposed that discrete **particulate units of heredity** exist, and he explained how they are transmitted from one generation to the next.

Punnett Squares

The genotypes and phenotypes resulting from the recombination of gametes during fertilization can be easily visualized by constructing a **Punnett square,** named after Reginald C. Punnett, who first devised this approach. Figure 3.3 illustrates this method of analysis for the $F_1 \times F_1$ monohybrid cross. Each of the possible gametes is assigned to an individual column or a row, with the vertical column representing those of the female parent and the horizontal row those of the male parent. After we enter the gametes in rows and columns, we can predict the new generation by combining the male and female gametic information for each combination and entering the resulting genotypes in the boxes. This process represents all possible random fertilization events. The genotypes and phenotypes of all potential offspring are ascertained by reading the entries in the boxes.

The Punnett square method is particularly useful when first learning about genetics and how to solve problems. In Figure 3.3, note the ease with which the 3:1 phenotypic ratio and the 1:2:1 genotypic ratio may be derived in the F_2 generation.

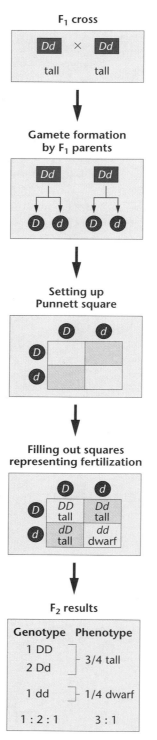

FIGURE 3.3 The use of a Punnett square in generating the F_2 ratio of the $F_1 \times F_1$ cross shown in Figure 3.2. **GenCDX**

The Test Cross: One Character

Tall plants produced in the F_2 generation are predicted to have either the *DD* or the *Dd* genotypes. You might wonder if there is a way to distinguish the genotype of a plant expressing the dominant phenotype. Mendel devised a rather simple method that is still used today in breeding procedures of plants and animals: the **test cross**. The organism of the dominant phenotype but unknown genotype is crossed to a **homozygous recessive individual.** For example, as shown in Figure 3.4(a), if a tall plant of genotype *DD* is test-crossed to a dwarf plant, which must have the *dd* genotype, all offspring will be tall phenotypically and *Dd* genotypically. However, as shown in Figure 3.4(b), if a tall plant is *Dd* and is crossed to a dwarf plant (*dd*), then one-half of the offspring will be tall (*Dd*) and the other half will be dwarf (*dd*). Therefore, a 1:1 ratio of tall/dwarf phenotypes demonstrates the heterozygous na-

ture of the tall plant of unknown genotype. The results of test crosses reinforced Mendel's conclusion that separate unit factors control the tall and dwarf traits.

Mendel's dihybrid cross revealed his fourth postulate: Independent assortment

A natural extension of performing monohybrid crosses was for Mendel to design experiments where two characters were examined simultaneously. Such a cross, involving two pairs of contrasting traits, is called a **dihybrid cross,** or a **two-factor cross.** For example, if pea plants having yellow seeds that are also round were bred with those having green seeds that are also wrinkled, the results shown in Figure 3.5 will occur. The F_1 offspring are all yellow and

Test cross results

(a) 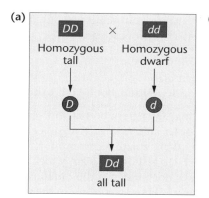 **(b)**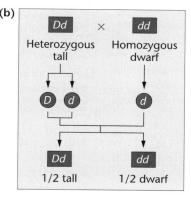

FIGURE 3.4   The test cross illustrated with a single character. In (a), the tall parent is homozygous. In (b), the tall parent is heterozygous. The genotypes of each tall parent may be determined by examining the offspring when each is crossed to the homozygous recessive dwarf plant.

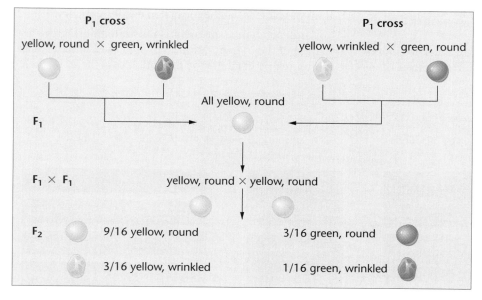

FIGURE 3.5   The F_1 and F_2 results of Mendel's dihybrid crosses between yellow round and green wrinkled pea plants, and between yellow wrinkled and green round pea plants.   ◉ GenCDX

round. It is therefore apparent that yellow is dominant to green, and that round is dominant to wrinkled. In this dihybrid cross, the F_1 individuals are selfed, and approximately 9/16 of the F_2 plants express yellow and round, 3/16 express yellow and wrinkled, 3/16 express green and round, and 1/16 express green and wrinkled.

A variation of this cross is also shown in Figure 3.5. Instead of crossing one P_1 parent with both dominant traits (yellow, round) to one with both recessive traits (green, wrinkled), plants with yellow wrinkled seeds are crossed to those with green round seeds. Despite the change in the P_1 phenotypes, both the F_1 and F_2 results remain unchanged. It will become clear in the next section why this is so.

Independent Assortment

We can most easily understand the results of a dihybrid cross if we consider it theoretically as consisting of two monohybrid crosses conducted separately. Think of the two sets of traits as being inherited independently of each other; that is, the chance of any plant having yellow or green seeds is not at all influenced by the chance that this plant will also have round or wrinkled seeds. Thus, because yellow is dominant to green, all F_1 plants in the first theoretical cross would have yellow seeds. In the second theoretical cross, all F_1 plants would have round seeds because round is dominant to wrinkled. When Mendel examined the F_1 plants of his dihybrid cross, all were yellow and round, as predicted.

The predicted F_2 results of the first cross are 3/4 yellow and 1/4 green. Similarly, the second cross should yield 3/4 round and 1/4 wrinkled. Figure 3.5 shows that in the dihybrid cross, 12/16 of all F_2 plants are yellow while 4/16 are green, exhibiting the 3:1 ratio. Similarly, 12/16 of all F_2 plants have round seeds while 4/16 have wrinkled seeds, again revealing the 3:1 ratio.

Because the two pairs of contrasting traits are inherited independently, we can predict the frequencies of all possible F_2 phenotypes by applying the "product law" of probabilities: **When two independent events occur simultaneously, the combined probability of the two outcomes is equal to the product of their individual probabilities of occurrence.** For example, the probability of an F_2 plant having yellow *and* round seeds is (3/4)(3/4), or 9/16, because 3/4 of all F_2 plants should be yellow and 3/4 of all F_2 plants should be round.

In a like way, we can calculate the probabilities of the other three F_2 phenotypes: Yellow (3/4) *and* wrinkled (1/4) are predicted to be present together 3/16 of the time; green (1/4) *and* round (3/4) are predicted 3/16 of the time; and green (1/4) *and* wrinkled (1/4) are predicted 1/16 of the time. These calculations are illustrated in Figure 3.6. It is now apparent why the F_1 and F_2 results are identical whether the parents of the initial cross are yellow and round bred with green and wrinkled or if they are yellow and wrinkled bred with green and round. In both crosses, the F_1 genotype of all plants is identical. Each plant is heterozygous for both gene pairs. As a result, the F_2 generation is also identical in both crosses.

On the basis of similar results in numerous dihybrid crosses, Mendel proposed a fourth postulate.

4. *Independent Assortment*

During gamete formation, segregating pairs of unit factors assort independently of each other.

This postulate stipulates that any pair of unit factors segregate independently of all other unit factors. Remember that as a result of segregation, each gamete receives one member of every pair of unit factors. For one pair, whichever unit factor is received does not influence the outcome of segregation of any other pair. Thus, according to the postulate of **independent assortment,** all possible combinations of gametes will be formed in equal frequency.

Independent assortment is illustrated during the formation of the F_2 generation, shown in the Punnett square in Figure 3.7. Examine the formation of gametes by the F_1

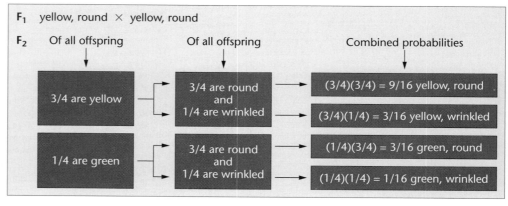

FIGURE 3.6 The determination of the combined probabilities of each F_2 phenotype for two independently inherited characters. The probability of each plant being yellow or green is independent of the probability of it being round or wrinkled.

F_1 yellow, round × yellow, round

F_2 Of all offspring   Of all offspring   Combined probabilities

3/4 are yellow → 3/4 are round and 1/4 are wrinkled → (3/4)(3/4) = 9/16 yellow, round

(3/4)(1/4) = 3/16 yellow, wrinkled

1/4 are green → 3/4 are round and 1/4 are wrinkled → (1/4)(3/4) = 3/16 green, round

(1/4)(1/4) = 1/16 green, wrinkled

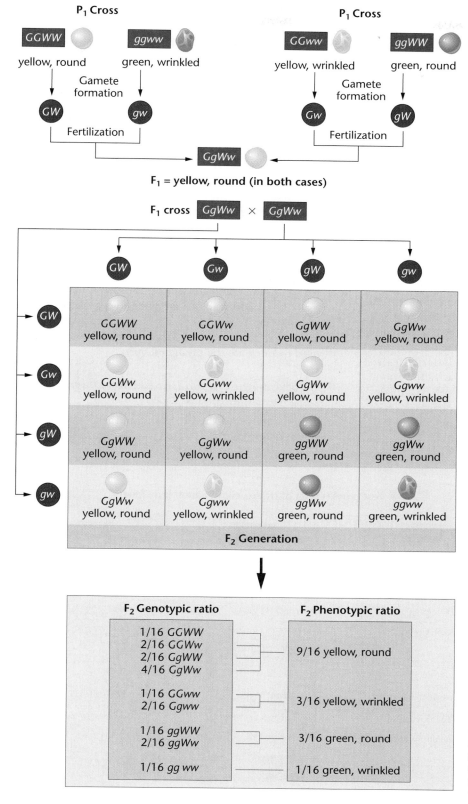

FIGURE 3.7 Diagram of the dihybrid crosses shown in Figure 3.5. The F_1 heterozygous plants are self-fertilized to produce an F_2 generation, which is computed using a Punnett square. Both the phenotypic and genotypic F_2 ratios are shown.

plants. Segregation prescribes that every gamete receives either a *G* or *g* allele and a *W* or *w* allele. Independent assortment stipulates that all four combinations (*GW*, *Gw*, *gW*, and *gw*) will be formed with equal probabilities.

In every F₁ × F₁ fertilization event, each zygote has an equal probability of receiving one of the four combinations from each parent. If a large number of offspring are produced, 9/16 are yellow and round, 3/16 are yellow and wrinkled, 3/16 are green and round, and 1/16 are green and wrinkled, yielding what is designated as **Mendel's 9:3:3:1 dihybrid ratio.** This is an ideal ratio based on probability events involving segregation, independent assortment, and random fertilization. Because of deviation due strictly to chance, particularly if small numbers of offspring are produced, actual results will seldom match the ideal ratio exactly.

The Test Cross: Two Characters

The test cross may also be applied to individuals that express two dominant traits, but whose genotypes are unknown. For example, the expression of the yellow round phenotype in the F₂ generation just described may result from the *GGWW*, *GGWw*, *GgWW*, and *GgWw* genotypes. If an F₂ yellow round plant is crossed with the homozygous recessive green wrinkled plant (*ggww*), analysis of the offspring will indicate the exact genotype of that yellow round plant. Each of the above genotypes will result in a different set of gametes and, in a test cross, a different set of phenotypes in the resulting offspring. Three cases are illustrated in Figure 3.8.

The trihybrid cross demonstrates that Mendel's principles apply to inheritance of multiple traits

So far, we have considered inheritance of up to two pairs of contrasting traits. Mendel demonstrated that the identical processes of segregation and independent assortment apply to three pairs of contrasting traits in what is called a **trihybrid cross,** or a **three-factor cross.**

Although a trihybrid cross is somewhat more complex than a dihybrid cross, its results are easily calculated if the principles of segregation and independent assortment are followed. For example, consider the cross shown in Figure 3.9 where the gene pairs representing theoretical contrasting traits are symbolized *A/a*, *B/b*, and *C/c*. In the cross between *AABBCC* and *aabbcc* individuals, all F₁ individuals are heterozygous for all three gene pairs. Their genotype, *AaBbCc*, results in the phenotypic expression of the dominant *A*, *B*, and *C* traits. When F₁ individuals are parents, each produces eight different gametes in equal frequencies. At this point, we could construct a Punnett square with 64 separate boxes and read out the phenotypes. Because such a method is cumbersome in a cross involving so many factors, another approach, the forked-line method, has been devised to calculate the predicted ratio.

The Forked-Line Method, or Branch Diagram

It is much less difficult to consider each contrasting pair of traits separately and then to combine these results using the **forked-line method,** which was first illustrated in Fig-

Test cross results of three yellow, round individuals

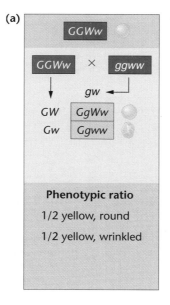
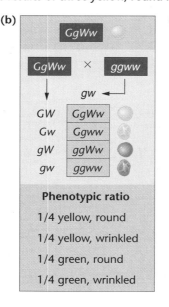
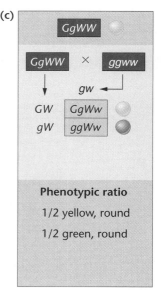

FIGURE 3.8 The test cross illustrated with two independent characters.

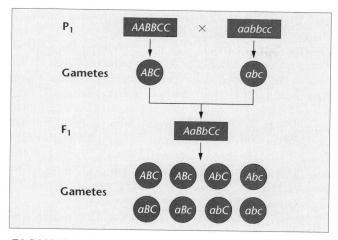

FIGURE 3.9   The formation of P₁ and F₁ gametes in a trihybrid cross.

ure 3.6. This method, also called a **branch diagram**, relies on the simple application of the laws of probability established for the dihybrid cross. Each gene pair is assumed to behave independently during gamete formation.

When the monohybrid cross *AA* × *aa* is made, we know that:

1. All F₁ individuals have the genotype *Aa* and express the phenotype represented by the *A* allele, which is called the *A* phenotype in the following discussion.

2. The F₂ generation consists of individuals with either the *A* phenotype or the *a* phenotype in the ratio of 3:1.

The same generalizations apply to the *BB* × *bb* and *CC* × *cc* crosses. Thus, in the F₂ generation, 3/4 of all organisms will express phenotype *A*, 3/4 will express *B*, and 3/4 will express *C*. Similarly, 1/4 of all organisms will express phenotype *a*, 1/4 will express *b*, and 1/4 will express *c*. The proportions of organisms expressing each phenotypic combination can be predicted by assuming that fertilization, following the independent assortment of these three gene pairs during gamete formation, is a random process. We once again simply apply the product law of probabilities.

The phenotypic proportions of the F₂ generation calculated using the forked-line method are illustrated in Figure 3.10. They fall into the trihybrid ratio of 27:9:9:9:3:3:3:1. The same method can be applied when solving crosses involving any number of gene pairs, *provided* that all gene pairs assort independently from each other. We will see later that this is not always the case, though it appeared to be true for all of Mendel's characters.

Note that in Figure 3.10 only phenotypic ratios of the F₂ generation have been derived. It is possible to generate genotypic ratios as well. To do so, we again consider the *A/a*, *B/b*, and *C/c* gene pairs separately. For example, for the *A/a* pair the F₁ cross is *Aa* × *Aa*. Phenotypically, an F₂ ratio of 3/4 *A*:1/4 *a* is produced. Genotypically, however, the F₂ ratio is different; 1/4 *AA*:1/2 *Aa*:1/4 *aa* will result. Using Figure 3.10 as a model, we would enter these genotypic frequencies on the left side of the calculation. Each would be connected by three lines to 1/4 *BB*, 1/2 *Bb*, and 1/4 *bb*, respectively. From each of these nine designations, three more lines would extend to the 1/4 *CC*, 1/2 *Cc*, and 1/4 *cc* genotypes. On the right side of the completed diagram, 27 genotypes and their frequencies of occurrence would appear. One of the problems at the end of this chapter asks you to use the forked-line or branch diagram method to determine the genotypic ratios generated in a trihybrid cross (see Problem 17).

In crosses involving two or more gene pairs, the calculation of gametes and genotypic and phenotypic results is quite complex. Several simple mathematical rules will enable you to check the accuracy of various steps required in working genetic problems. First, you must determine the number of *heterozygous* gene pairs (*n*) involved in the

Generation of F₂ trihybrid phenotypes

A or a	B or b	C or c	Combined proportion	
3/4 A	3/4 B	3/4 C	(3/4)(3/4)(3/4) *ABC* = 27/64	*ABC*
		1/4 c	(3/4)(3/4)(1/4) *ABc* = 9/64	*ABc*
	1/4 b	3/4 C	(3/4)(1/4)(3/4) *AbC* = 9/64	*AbC*
		1/4 c	(3/4)(1/4)(1/4) *Abc* = 3/64	*Abc*
1/4 a	3/4 B	3/4 C	(1/4)(3/4)(3/4) *aBC* = 9/64	*aBC*
		1/4 c	(1/4)(3/4)(1/4) *aBc* = 3/64	*aBc*
	1/4 b	3/4 C	(1/4)(1/4)(3/4) *abC* = 3/64	*abC*
		1/4 c	(1/4)(1/4)(1/4) *abc* = 1/64	*abc*

FIGURE 3.10   The generation of the F₂ trihybrid ratio using the forked-line, or branch diagram, method, which is based on the expected probability of occurrence of each phenotype.

cross. For example, where $AaBb \times AaBb$ represents the cross, $n = 2$; for $AaBbCc \times AaBbCc$, $n = 3$; for $AaBBCcDd \times AaBBCcDd$, $n = 3$ (because the B genes are not heterozygous). Once n is determined, 2^n is the number of different gametes that can be formed by each parent; 3^n is the number of different genotypes that result following fertilization; and 2^n is the number of different phenotypes that are produced from these genotypes. Table 3.1 summarizes these rules, which may be applied to crosses involving any number of genes, *provided that they assort independently from one another.*

Mendel's work was rediscovered in the early twentieth century

Mendel's work, initiated in 1856, was presented to the Brünn Society of Natural Science in 1865 and published the following year. However, his findings went largely unnoticed for about 35 years. Many reasons have been suggested to explain why the significance of his research was not immediately recognized.

First, Mendel's adherence to mathematical analysis of probability events was quite an unusual approach in biological studies and may have seemed foreign to his contemporaries. More important, his conclusions drawn from such analyses did not fit well with the existing hypotheses involving the source of variation among organisms. Students of evolutionary theory, stimulated by the proposals developed by Charles Darwin and Alfred Russel Wallace, believed in **continuous variation,** where offspring were a blend of their parents' phenotypes. By contrast, Mendel hypothesized that heredity was due to discrete or particulate units, resulting in **discontinuous variation.** The F_2 offspring of a dihybrid cross, for example, were merely expressing traits produced by new combinations of previously existing unit factors. Thus, Mendel's hypotheses did

not fit well with the evolutionists' preconceptions about causes of variation.

It is also likely that Mendel's contemporaries failed to realize that Mendel's postulates explained *how* variation was transmitted to offspring. Instead, they may have attempted to interpret his work in a way that addressed the issue of *why* certain phenotypes survive preferentially. It was this latter question that had been addressed in the theory of natural selection, but it was not addressed by Mendel. It may well be, therefore, that the collective vision of Mendel's scientific colleagues was obscured by the impact of this extraordinary theory of organic evolution.

The correlation of Mendel's postulates with the behavior of chromosomes formed the foundation of modern transmission genetics

Near the end of the nineteenth century, a remarkable observation set the scene for the rebirth of Mendel's work—Walter Flemming's discovery in 1879 of chromosomes. Flemming was able to describe the behavior of these threadlike structures in the nuclei of salamander cells during cell division. As a result of the findings of Flemming and many other cytologists, the presence of a nuclear component soon became an integral part of ideas surrounding inheritance. It was in this setting that scientists were able to reexamine Mendel's findings.

In the early twentieth century, research led to the rebirth of Mendel's work. Hybridization experiments similar to Mendel's were independently performed by three botanists, Hugo DeVries, Karl Correns, and Erich Tschermak. DeVries's work, for example, had focused on unit characters, and he demonstrated the principle of segrega-

TABLE 3.1 Simple Mathematical Rules Useful in Working Genetics Problems

Crosses between Organisms Heterozygous for Genes Exhibiting Independent Assortment			
Number of Heterozygous Gene Pairs	Number of Different Types of Gametes Formed	Number of Different Genotypes Produced	Number of Different Phenotypes Produced*
n	2^n	3^n	2^n
1	2	3	2
2	4	9	4
3	8	27	8
4	16	81	16

*The fourth column assumes that dominance and recessiveness are operational for all gene pairs.

tion in his experiments with several plant species. He had apparently searched the existing literature and found that Mendel's work had anticipated his own conclusions. Correns and Tschermak had independently reached findings similar to those of Mendel.

In 1902, two cytologists, Walter Sutton and Theodor Boveri, independently published papers linking their discoveries of the behavior of chromosomes during meiosis to the Mendelian principles of segregation and independent assortment. They pointed out that the separation of chromosomes during meiosis could serve as the cytological basis of these two postulates. Although they thought that Mendel's unit factors were probably chromosomes rather than genes on chromosomes, their findings reestablished the importance of Mendel's work, which served as the foundation of ensuing genetic investigations.

Based on their studies, Sutton and Boveri are credited with initiating the **chromosomal theory of heredity.** As we will see in subsequent chapters, work by Thomas H. Morgan, Alfred H. Sturtevant, Calvin Bridges, and others using fruit flies established beyond a reasonable doubt that Sutton's and Boveri's hypothesis was correct.

Unit Factors, Genes, and Homologous Chromosomes

Because the correlation between Sutton's and Boveri's observations and Mendelian principles is the foundation for the modern interpretation of transmission genetics, we will examine it in some detail.

As we pointed out in Chapter 2, each species possesses a specific number of chromosomes in each somatic (body) cell nucleus. For diploid organisms, this number is called the **diploid number (2n)** and is characteristic of that species. During the formation of gametes, this number is precisely halved (n), and when two gametes combine during fertilization, the diploid number is reestablished. During meiosis, the chromosome number is not reduced in a random manner, however. It was apparent to early cytologists that the diploid number of chromosomes is composed of homologous pairs identifiable by their morphological appearance and behavior. The gametes contain one member of each pair. The chromosome complement of a gamete is thus quite specific, and the number of chromosomes in each gamete is equal to the haploid number.

With this basic information, we can see the correlation between the behavior of unit factors and chromosomes and genes. Figure 3.11 shows three of Mendel's postulates (in the left column) and the chromosomal explanation of each (in the right column). Unit factors are really genes located on homologous pairs of chromo-

somes [Figure 3.11(a)]. Members of each pair of homologs separate, or segregate, during gamete formation [Figure 3.11(b)]. Two different alignments are possible, both of which are shown.

To illustrate the principle of independent assortment, it is important to distinguish between members of any given homologous pair of chromosomes. One member of each pair is derived from the **maternal parent,** while the other comes from the **paternal parent.** We represent different parental origins by different colors. As shown in Figure 3.11(c), the two pairs of homologs undergo segregation independently of one another during gamete formation. Each gamete receives one chromosome from each pair. All possible combinations are formed. If we add the symbols used in Mendel's dihybrid cross (G, g and W, w) to the diagram, we see why equal numbers of the four types of gametes are formed. The independent behavior of Mendel's pairs of unit factors (G and W in this example) was due to the fact that they were on separate pairs of homologous chromosomes.

From observations of the phenotypic diversity of living organisms, we see that it is logical to assume that there are many more genes than chromosomes. Therefore, each homolog must carry genetic information for more than one trait. The currently accepted concept is that a chromosome is composed of a large number of linearly ordered, information-containing units called **genes.** Mendel's unit factors (which determine tall or dwarf stems, for example) actually constitute a pair of genes located on one pair of homologous chromosomes. The location on a given chromosome where any particular gene occurs is called its **locus** (pl. **loci**). The different forms taken by a given gene, called **alleles** (G or g), contain slightly different genetic information (green or yellow) that determines the same character (seed color). Alleles are alternative forms of the same gene. Although we have only discussed genes with two alternative alleles, most genes have *more* than two allelic forms. We discuss the concept of multiple alleles in Chapter 4.

We conclude this section by reviewing the criteria necessary to classify two chromosomes as a homologous pair:

1. During mitosis and meiosis, when chromosomes are visible as distinct structures, both members of a homologous pair are the same size and exhibit identical centromere locations.

2. During early stages of meiosis, homologous chromosomes pair together, or synapse.

3. Homologs contain identical, linearly ordered, gene loci though these are not generally visible microscopically.

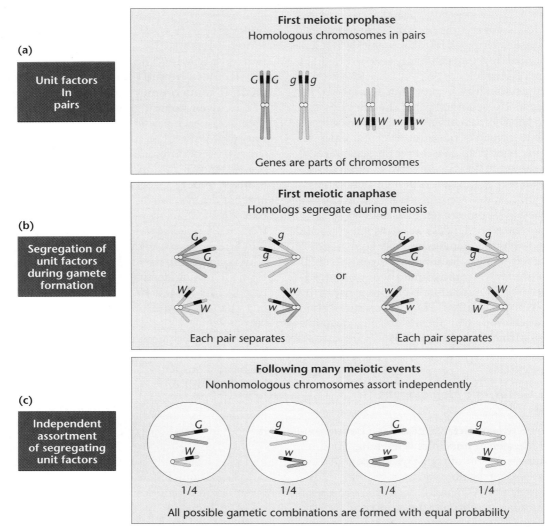

FIGURE 3.11 The correlation between the Mendelian postulates of (a) unit factors in pairs, (b) segregation, and (c) independent assortment, and the presence of genes located on homologous chromosomes and their behavior during meiosis.

Independent assortment leads to extensive genetic variation

One of the major consequences of independent assortment is the production of genetically dissimilar gametes by one individual. Genetic variation results because the two members of any homologous pair of chromosomes are rarely, if ever, genetically identical. Because independent assortment leads to the production of all possible chromosome combinations, extensive genetic diversity results.

We have seen that for any individual, the number of possible gametes, each with different chromosome compositions, is 2^n, where n equals the haploid number. Thus, if a species has a haploid number of 4, then 2^4 or 16 different gamete combinations can be formed as a result of independent assortment. Although this number is not high, consider the human species, where $n = 23$. If we calculate 2^{23}, we find more than 8×10^6, or over 8 million, different types of gametes are possible. Because fertilization represents an event involving only one of approximately 8×10^6 possible gametes from each of two parents, each offspring represents only one of $(8 \times 10^6)^2$ or 64×10^{12} potential genetic combinations! This number of combinations of chromosomes is far greater than the number of humans who have ever lived on Earth! It is no wonder that, except for identical twins, each member of the human species demonstrates such a distinctive appearance and individuality. Genetic variation resulting from independent assortment has been extremely important to the process of evolution in all organisms.

Laws of probability help to explain genetic events

Genetic ratios are most properly expressed as probabilities (e.g., 3/4 tall:1/4 dwarf). These values predict the outcome of each fertilization event, so we say that the probability of each zygote having the genetic potential for becoming tall is 3/4, whereas the potential for becoming dwarf is 1/4. Probabilities range from 0, where an event *is certain not* to occur, to 1.0, where an event *is certain* to occur. In this section, we consider the relation of probability to genetics.

The Product Law and the Sum Law

When two or more events occur independently of one another, but at the same time, we can calculate the probability of possible outcomes when they occur together. To do this, we apply the **product law.** As we mentioned in our earlier discussion of independent assortment, it states that the probability of two or more outcomes occurring simultaneously is equal to the *product* of their individual probabilities. Two or more events are independent of one another if the outcome of each one does not affect the outcome of any of the others under consideration.

To illustrate the use of the product law, consider the possible results of an event where you toss a penny (P) and a nickel (N) at the same time and examine all combinations of heads (H) and tails (T) that can occur. There are four possible outcomes:

$$(P_H{:}N_H) \; = \; (1/2)(1/2) = 1/4$$
$$(P_T{:}N_H) \; = \; (1/2)(1/2) = 1/4$$
$$(P_H{:}N_T) \; = \; (1/2)(1/2) = 1/4$$
$$(P_T{:}N_T) \; = \; (1/2)(1/2) = 1/4$$

The probability of obtaining a head or a tail with either coin is 1/2 and is unrelated to the outcome of the other coin. All four possible combinations are predicted to occur with equal probability.

If we were interested in calculating the probability of a generalized outcome that can be accomplished in more than one way, we would apply the **sum law** to the individual mutually exclusive outcomes. For example, we can ask: What is the probability of tossing our penny and nickel and obtaining one head and one tail? In such a case, we do not care whether it is the penny or the nickel that comes

up heads, provided that the other coin has the alternative outcome. As we can see above, there are two ways in which the desired outcome can be accomplished [($P_H{:}N_T$) and ($P_T{:}N_H$)], each with a probability of 1/4. Thus, according to the sum law, the overall probability is equal to

$$(1/4) + (1/4) = 1/2$$

One-half of all such tosses are predicted to yield the desired outcome.

These simple probability laws will be useful throughout our discussions of transmission genetics, and as you solve genetics problems. In fact, we already applied the product law earlier when we used the forked-line method to calculate the phenotypic results of Mendel's dihybrid and trihybrid crosses. When we wish to know the results of a cross, we need only calculate the probability of each possible outcome. The results of this calculation then allow us to predict the proportion of offspring expressing each phenotype or each genotype.

There is a very important point to remember when dealing with probability. Predictions of possible outcomes are usually realized only with large sample sizes. If we predict that 9/16 of the offspring of a dihybrid cross will express both dominant traits, it is unlikely, in a small sample, that exactly 9 out of every 16 will do so. Instead, our prediction is that of a large number of offspring approximately 9/16 will express this phenotype. The deviation from the predicted ratio in small sample sizes is attributed to chance, a subject we deal with in our discussion of statistics in the next section. As we will see, the impact of deviation due strictly to chance is diminished as the sample size increases.

Conditional Probability

Sometimes we may wish to calculate the probability of an outcome that is dependent on a specific condition related to that outcome. For example, in the F_2 of Mendel's monohybrid cross involving tall and dwarf plants, we might wonder what the probability is that a tall plant is heterozygous (and not homozygous). The "condition" we have set is to consider only tall F_2 offspring since we know that all dwarf plants are homozygous. Of any F_2 tall plant, what is the probability of it being heterozygous?

Because the outcome and specific condition are not independent, we cannot apply the product law of probability. The likelihood of such an outcome is referred to as a **conditional probability.** In its simplest terms, we are asking what is the probability that one outcome will occur, given the specific condition upon which this outcome is dependent. Let us call this probability p_c.

To solve for p_c, we must consider the probability of both the outcome of interest and that of the specific condition that includes the outcome. These are: (a) the probability of an F_2 plant being heterozygous as a result of receiving both a dominant and a recessive allele (p_a) and (b) the probability of the condition under which the event is being assessed, that is, being tall (p_b).

$$p_a = \text{probability of any } F_2 \text{ plant inheriting one dominant and one recessive allele (i.e., being a heterozygote)}$$

$$= 1/2$$

$$p_b = \text{probability of an } F_2 \text{ plant of a monohybrid cross being tall}$$

$$= 3/4$$

To calculate the conditional probability (p_c), we divide p_a by p_b:

$$p_c = p_a / p_b$$
$$= (1/2)/(3/4)$$
$$= (1/2) \cdot (4/3)$$
$$= 4/6$$
$$p_c = 2/3$$

The conditional probability of any tall plant being heterozygous is two-thirds (2/3). Thus, on the average, two-thirds of the F_2 tall plants will be heterozygous. We can confirm this calculation by reexamining Figure 3.3.

Conditional probability has many applications in genetics. During genetic counseling, for example, it is possible to calculate the probability (p_c) that an unaffected sibling of a brother or sister expressing a recessive disorder is a carrier of the disease-causing allele (i.e., a heterozygote). Assuming that both parents are unaffected (and are therefore carriers), the calculation of p_c is identical to the preceding example. The value of $p_c = 2/3$.

The Binomial Theorem

The final example of probability that we shall discuss involves cases where one of two alternative outcomes is possible during each of a number of trials. By applying the **binomial theorem,** we can rather quickly calculate the probability of any specific set of outcomes among a large number of potential events. For example, in families of any size, we can calculate the probability of any combination of male and female children, (e.g., in a family of four, we can calculate the probability of having two children of one sex and two children of the other sex).

The expression of the binomial theorem is

$$(a + b)^n = 1$$

where a and b are the respective probabilities of the two alternative outcomes and n equals the number of trials. For each value of n, the binomial must be expanded,

n	Binomial	Expanded Binomial
1	$(a + b)^1$	$a + b$
2	$(a + b)^2$	$a^2 + ab + b^2$
3	$(a + b)^3$	$a^3 + 3a^2b + 3ab^2 + b^3$
4	$(a + b)^4$	$a^4 + 4a^3b + 6a^2b^2 + 4ab^3 + b^4$
5	$(a + b)^5$	$a^5 + 5a^4b + 10a^3b^2 + 10a^2b^3 + 5ab^4 + b^5$

etc.

To expand any binomial, the various exponents (e.g., a^3b^2) are determined using the pattern

$$(a + b)^n = a^n, a^{n-1}b, a^{n-2}b^2, a^{n-3}b^3, \ldots, b^n$$

The numerical coefficients preceding each expression can be most easily determined using Pascal's triangle,

n															
								1							
1								1		1					
2							1		2		1				
3						1		3		3		1			
4					1		4		6		4		1		
5				1		5		10		10		5		1	
6			1		6		15		20		15		6		1
7		1		7		21		35		35		21		7	1

etc.

Notice that all numbers other than the 1s are equal to the sum of the two numbers directly above them.

Using the above methods, the initial expansion of $(a + b)^7$ is

$$a^7 + 7a^6b + 21a^5b^2 + 35a^4b^3 + \cdots + b^7$$

Applying the binomial theorem, we can return to our original question: *What is the probability that in a family of four children, two are male and two are female?*

First, assign initial probabilities to each outcome,

$$a = \text{male} = 1/2$$
$$b = \text{female} = 1/2$$

Then locate the appropriate term in the expanded binomial, where $n = 4$,

$$(a + b)^4 = a^4 + 4a^3b + 6a^2b^2 + 4ab^3 + b^4$$

In each term, the exponent of *a* represents the number of males and the exponent of *b* represents the number of females. Therefore, the correct expression of *p* is

$$p = 6a^2b^2$$
$$= 6(1/2)^2(1/2)^2$$
$$= 6(1/2)^4$$
$$= 6(1/16)$$
$$= 6/16$$
$$p = 3/8$$

Thus, the probability of families of four children having two boys and two girls is 3/8. Of all families with four children, 3 out of 8 are predicted to have two boys and two girls.

Before examining one other example, we should note that a single formula can be applied in determining the numerical coefficient for any set of exponents,

$$n!/(s!t!)$$

where

n	=	the total number of events
s	=	the number of times outcome *a* occurs
t	=	the number of times outcome *b* occurs

Therefore, $n = s + t$.

The symbol ! means "factorial." For example,

$$5! = (5)(4)(3)(2)(1) = 120$$

Note that when using factorials, $0! = 1$.

Using the formula, let's determine the probability, in a family of seven, that five males and two females will occur. Thus, $n = 7$, $s = 5$, and $t = 2$. We first extend our equation to include five events of outcome *a* and two events of outcome *b*. The appropriate term is

$$p = \frac{n!}{s!t!}a^sb^t$$
$$= \frac{7!}{5!2!}(1/2)^5(1/2)^2$$
$$= \frac{(7)\cdot(6)\cdot(5)\cdot(4)\cdot(3)\cdot(2)\cdot(1)}{(5)\cdot(4)\cdot(3)\cdot(2)\cdot(1)\cdot(2)\cdot(1)}(1/2)^7$$
$$= \frac{(7)\cdot(6)}{(2)\cdot(1)}(1/2)^7$$
$$= \frac{42}{2}(1/2)^7$$
$$= 21(1/2)^7$$
$$= 21(1/128)$$
$$p = 21/128$$

Of families with seven children, on the average, 21/128 are predicted to have five males and two females.

Calculations using the binomial theorem have various applications in genetics, including the analysis of polygenic traits (see Chapter 5) and in population equilibrium studies (see Chapter 25).

Chi-square analysis evaluates the influence of chance on genetic data

Mendel's 3:1 monohybrid and 9:3:3:1 dihybrid ratios are hypothetical predictions based on the following assumptions: (1) each allele is dominant or recessive, (2) segregation is operative, (3) independent assortment occurs, and (4) fertilization is random. The last three assumptions are influenced by chance events and therefore are subject to random fluctuation. This concept, called **chance deviation,** is most easily illustrated by tossing a single coin numerous times and recording the number of heads and tails observed. In each toss, there is a probability of 1/2 that a head will occur and a probability of 1/2 that a tail will occur. Therefore, the expected ratio of many tosses is 1:1. If a coin were tossed 1000 times, we would usually expect *about* 500 heads and 500 tails to be observed. Any reasonable fluctuation from this hypothetical ratio (e.g., 486 heads and 514 tails) would be attributed to chance.

As the total number of tosses is reduced, the impact of chance deviation increases. For example, if a coin were tossed only four times, you wouldn't be too surprised if all four tosses resulted in only heads or only tails. But, for 1000 tosses, 1000 heads or 1000 tails would be most unexpected. In fact, you might believe that such a result would be impossible. Actually, the probability of all heads or all tails in 1000 tosses can be predicted to occur with a probability of only $(1/2)^{1000}$. Because $(1/2)^{20}$ is equivalent to less than 1 in 1 million times, an event occurring with a probability as small as $(1/2)^{1000}$ would be virtually impossible.

Two major points are significant here:

1. The outcomes of segregation, independent assortment, and fertilization, like coin tossing, are subject to random fluctuations from their predicted occurrences as a result of chance deviation.

2. As the sample size increases, the average deviation from the expected fraction or ratio is expected to decrease. Therefore, a larger sample size diminishes the impact of chance deviation on the final outcome.

It is important in genetics to be able to evaluate observed deviation. When we assume that data will fit a given ratio such as 1:1, 3:1, or 9:3:3:1, we establish what is called the **null hypothesis.** It is so named because the hypothesis assumes that there is no *real difference* between the **measured values** (or ratio) and the **predicted values** (or ratio). The *apparent* difference can be attributed purely to chance. To evaluate the validity of the null hypothesis in a given case, we use statistical analysis. On this basis, the null hypothesis may either: (1) be rejected, or (2) fail to be rejected. If it is rejected, the observed deviation from the expected is *not* attributed to chance alone; the null hypothesis and the underlying assumptions leading to it must be reexamined. If the null hypothesis fails to be rejected, any observed deviations *can* be attributed to chance.

Thus, statistical analysis provides a mathematical basis for examining how well observed data fit or differ from predicted or expected occurrences, testing what is called the **goodness of fit.** Assuming that the data do not "fit" exactly, just how much deviation can be allowed before the null hypothesis is rejected?

One of the simplest statistical tests devised to assess the goodness of fit of the null hypothesis is **chi-square (χ^2) analysis.** This test takes into account the observed deviation in each component of an expected ratio as well as the sample size and reduces them to a single numerical value. This χ^2 value is then used to estimate how frequently the observed deviation, or even more deviation than was observed, can be expected to occur strictly as a result of chance. The formula used in chi-square analysis is

$$\chi^2 = \sum \frac{(o - e)^2}{e}$$

In this equation,

o = the observed value for a given category

e = the expected value for that category

Σ (sigma) = the sum of the calculated values for each category of the ratio

Because $(o - e)$ is the deviation (d) in each case, the equation can be reduced to

$$\chi^2 = \sum \frac{d^2}{e}$$

Table 3.2(a) illustrates the step-by-step procedure necessary to calculate χ^2 for the F_2 results of a monohybrid cross. If you were analyzing these data, you would work from left to right, calculating and entering the appropriate numbers in each column. Regardless of whether the calculated deviation ($o - e$) is initially positive or negative, it becomes positive after the number is squared. Table 3.2(b) illustrates the analysis of the F_2 results of a hypothetical dihybrid cross. Based on your study of the calculations involved in the monohybrid cross, check to make certain that you understand how each number was calculated in the dihybrid example.

The final step in the analysis is to interpret the χ^2 value. As the number of categories that an outcome may fall into increases, so does the *expected* amount of deviation due strictly to chance. Because of this, we must take into ac-

TABLE 3.2 Chi-Square Analysis

			(a) Monohybrid Cross		
Expected Ratio	Observed (o)	Expected (e)	Deviation ($o - e$)	Deviation^2 (d^2)	Deviation^2/ Expected (d^2/e)
3/4	740	3/4 (1000) = 750	740 − 750 = −10	$(-10)^2 = 100$	100/750 = 0.13
1/4	260	1/4 (1000) = 250	260 − 250 = +10	$(+10)^2 = 100$	100/250 = 0.40
	Total = 1000				$\chi^2 = 0.53$
					$p = 0.48$

			(b) Dihybrid Cross		
Expected Ratio	o	e	$o - e$	d^2	d^2/e
9/16	587	567	+20	400	0.71
3/16	197	189	+ 8	64	0.34
3/16	168	189	−21	441	2.33
1/16	56	63	− 7	49	0.78
	Total = 1008				$\chi^2 = 4.16$
					$p = 0.26$

count the value called the **degrees of freedom (df),** which in χ^2 analyses is equal to $n - 1$ where n is the number of different categories into which any outcome may fall. For the 3:1 ratio, plants can have one of *two* phenotypes. Thus, $n = 2$, so $df = 2 - 1 = 1$. For the 9:3:3:1 ratio, $df = 3$.

Once we have determined the degrees of freedom, we can interpret the χ^2 value in terms of a corresponding **probability value (p).** Because this calculation is complex, we usually take the p value from a standard table or graph. Figure 3.12 shows the wide range of χ^2 and p values for numerous degrees of freedom in both graphed and tabular form. We will use the graph to explain how to determine the p value. The caption for Figure 3.12(b) explains how to use the table.

To determine p, execute the following steps:

1. Locate the χ^2 value on the abscissa (the horizontal x axis).
2. Draw a vertical line from this point up to the line on the graph representing the appropriate df.
3. Extend a horizontal line from this point to the left until it intersects the ordinate (the vertical y axis).
4. Estimate, by interpolation, the corresponding p value.

For our first example (the monohybrid cross) in Table 3.2, the p value of 0.48 may be estimated in this way and is illustrated in Figure 3.12(a). For the dihybrid cross, use this method to see if you can determine the p value. χ^2 is 4.16 and df equals 3. A p value of 0.26 is the approximate value. Use of the table rather than the graph confirms that both p values are between 0.20 and 0.50. Examine Table 3.2 to confirm this.

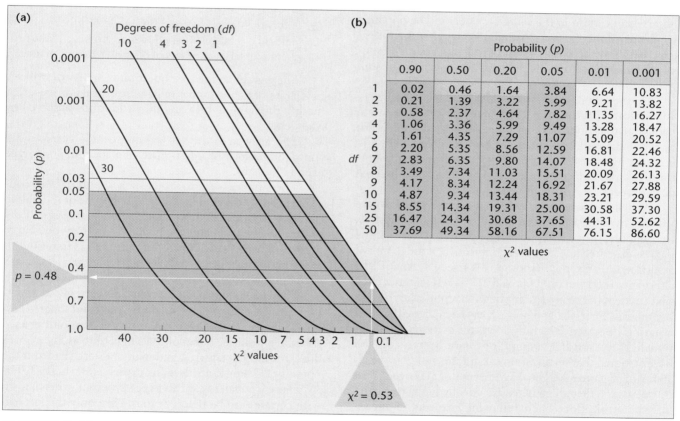

FIGURE 3.12 (a) A graph used to convert χ^2 values to p values. The interpolation of a χ^2 value of 0.53 with 1 degree of freedom (*df*) to an estimated probability value of 0.48 is illustrated. (b) A table showing χ^2 values for a variety of combinations of *df* and *p*. To use the table, locate the line with the appropriate *df* and along that line, determine the two values between which the calculated χ^2 value resides. These columns represent the correct range of the p value (shown along the top). In our example, a χ^2 value of 0.53 for a *df* of 1 is converted to a probability value between 0.20 and 0.50. From our graph in (a), the more precise value ($p = 0.48$) was estimated by interpolation. All values that serve to fail to reject the null hypothesis (0.05) are more darkly shaded in both the graph and the chart.

Thus far we have been concerned only with the determination of p. The most important aspect of χ^2 analysis is understanding what the p value actually means. We will use the example of the dihybrid cross ($p = 0.26$) to illustrate. In these discussions, it is simplest to think of the p value as a percentage (e.g., $0.26 = 26$ percent).

In our example, the p value indicates that, were the same experiment repeated many times, 26 percent of the trials would be expected to exhibit chance deviation as great as or greater than that seen in the initial trial. Conversely, 74 percent of the repeats would show less deviation as a result of chance than initially observed.

These interpretations of the p value reveal that a hypothesis (a 9:3:3:1 ratio in this case) is never proved or disproved absolutely. Instead, a relative standard is set that serves as the basis for either rejecting or failing to reject the hypothesis. This standard is most often a probability value of 0.05. When applied to chi-square analysis, a p value less than 0.05 means the probability is less than 5% that the observed deviation in the set of results could be obtained by chance alone. Such a p value indicates that the difference between the observed and predicted results is substantial and thus serves as the basis for rejecting the null hypothesis.

On the other hand, p values of 0.05 or greater (0.05 to 1.0) indicate that the probability of the observed deviation being due to chance is 5% or more. The conclusion is not to reject the null hypothesis. In our example where $p = 0.26$, the hypothesis that independent assortment accounts for the results fails to be rejected. Therefore, the observed deviation can be reasonably attributed to chance.

A final note is relevant here concerning the case where the null hypothesis is rejected, that is, $p < 0.05$. Suppose, for example, the null hypothesis being tested was that the data represented a 9:3:3:1 ratio, indicative of independent assortment. If the null hypothesis is rejected, what is the alternative interpretation of the data? The researcher must first reassess the many assumptions that underlie the null hypothesis. First, the assumption has been made that segregation operated faithfully for both gene pairs. Additionally, several other assumptions have been made. We have assumed that fertilization is random, and that the viability of all gametes is equal irrespective of genotype. That is, we assume that all gametes are equally likely to participate in fertilization. Then, following fertilization, the assumption is that all preadult stages and adult offspring are equally viable, regardless of their genotype.

A specific example will help to clarify this further. If we were to predict that a dihybrid cross between fruit flies would result in 3/16 mutant wingless flies, that proportion of mutant zygotes may, in fact, occur at fertilization. However, these mutant embryos may not survive as well during their preadult development or as young adults, compared to flies whose genotype give rise to wings. As a result, when the data are gathered, there will be fewer than 3/16 wingless flies. Rejection of the null hypothesis would then not be cause for us to disregard the validity of the postulates of segregation and independent assortment.

The above discussion serves to point out that statistical information must be assessed carefully on a case-by-case basis. When we reject a null hypothesis, we must examine all underlying assumptions. If there is no concern about their validity, then we must consider other alternative hypotheses to explain the results.

Pedigrees reveal patterns of inheritance in humans

In all the crosses we have discussed so far, one of the two traits for each character has been dominant to the other. This observation leads to two significant questions:

1. Does the expression of all genes occur in this fashion?
2. Is it possible to ascertain the mode of inheritance of genes in organisms where designed crosses and the production of large numbers of offspring are not practical?

The answer to the first question is no. As we will see in Chapter 4, many modes of genetic expression exist that modify the monohybrid and dihybrid ratios observed by Mendel.

The answer to the second question is yes. We can study the pattern of inheritance of a specific phenotype even in humans.

The simplest way to study such a pattern is to construct a family tree indicating the phenotype of the trait in question for each member. Such a family tree is called a **pedigree.** By analyzing the pedigree, we may be able to predict how the gene controlling the trait is inherited. If many similar pedigrees for the same trait are found, the prediction is strengthened.

Figure 3.13 shows the conventions used in constructing a pedigree. Circles represent females, and squares designate males. If the sex is unknown, a diamond is used (II-2). If a pedigree traces only a single trait, as Figure 3.13 does, the circles, squares, and diamonds are shaded if the phenotype being considered is expressed (I-1, III-3, III-4). Those who fail to express a recessive trait, when known with certainty to be heterozygous, have only the left half of their square or circle shaded (II-3 and II-4).

Parents are connected by a horizontal line, and vertical lines lead to their offspring. The offspring of one set of parents are called **sibs** and are connected by a horizontal **sibship line.** Sibs are placed from left to right according to birth order and are labeled with arabic numerals. Each generation is indicated by a roman numeral.

Twins are indicated by diagonal lines stemming from a vertical line connected to the sibship line. For **monozygot-**

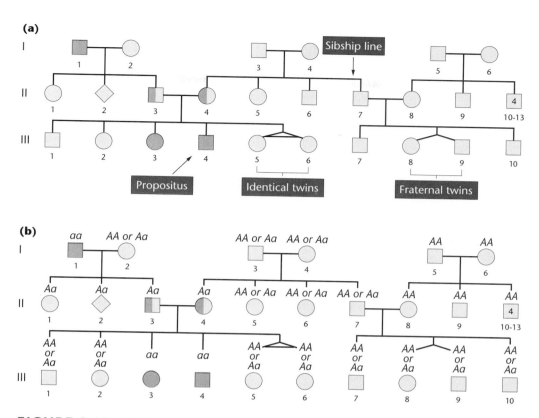

FIGURE 3.13   (a) A representative pedigree for a single character through three generations. (b) The most probable genotypes of each individual in the pedigree.

ic or **identical twins,** the diagonal lines are linked by a horizontal line (III-5,6). **Dizygotic** or **fraternal twins,** lack this connecting line (III-8,9). A number within one of the symbols (II-10–13) represents numerous sibs of the same or unknown phenotypes. A male whose phenotype drew the attention of a physician or geneticist is called the **propositus,** while a female in the same circumstance is called a **proposita.** Such an individual is indicated by an arrow (III-4).

The pedigree shown in Figure 3.13 traces the pattern of inheritance of the human trait albinism. By analyzing the pedigree, we can see that albinism is inherited as a recessive trait.

One of the parents of the first generation (I-1), is affected. Because none of his offspring show the disorder, we might conclude that the unaffected female parent (I-2) is a homozygous normal individual. Had she been heterozygous, one-half of the offspring would be expected to exhibit albinism. However, such a small sample (three offspring) prevents any certainty in the matter.

An unaffected second generation is characteristic of a rare recessive trait. If albinism were inherited as a dominant trait, individual II-3 would have to express the disorder in order to pass it to his offspring (III-3 and III-4). He does not. Inspection of the offspring constituting the third generation (row III) further supports the hypothesis that albinism is a recessive trait. If it is so, parents II-3 and II-4 are both heterozygous, and approximately one-fourth of their offspring should be affected. Two of the six offspring do show albinism. This deviation from the expected ratio is not unexpected in crosses with few offspring.

Based on this pedigree analysis and the conclusion that albinism is a recessive trait, we can predict the most probable genotypes of all the individuals in the pedigree. For both the first and second generations, this can be done with some certainty in only a few cases. In the third generation, we cannot determine whether most phenotypically normal individuals are homozygous or heterozygous. Figure 3.13(b) shows the most probable genotypes for each individual.

Pedigree analysis of many traits has been an extremely valuable research technique in human genetic studies. However, this approach does not usually provide the certainty in drawing conclusions that is afforded by designing crosses between experimental organisms that yield large numbers of offspring. Nevertheless, analysis of many independent pedigrees of the same trait or disorder often leads to consistent conclusions. Table 3.3 lists numerous human traits and classifies them according to their recessive or dominant expression. As we will see in Chapter 4, the genes controlling some of these traits are located on the sex-determining chromosomes.

TABLE 3.3 Representative Recessive and Dominant Human Traits

Recessive Traits	Dominant Traits
Albinism	Achondroplasia
Alkaptonuria	Brachydactyly
Ataxia telangiectasia	Congenital stationary night blindness
Color blindness	Ehlers–Danlos syndrome
Cystic fibrosis	Facioscapulohumeral muscular dystrophy
Duchenne muscular dystrophy	Huntington disease
Galactosemia	Hypercholesterolemia
Hemophilia	Marfan syndrome
Lesch–Nyhan syndrome	Neurofibromatosis
Phenylketonuria	Phenylthiocarbamide tasting (PTC)
Sickle-cell anemia	Porphyria
Tay–Sachs disease	Widow's peak

CHAPTER SUMMARY

1. Over a century ago, Gregor Mendel studied inheritance patterns in the garden pea, establishing the principles of transmission genetics.

2. Mendel's postulates help describe the basis for the inheritance of phenotypic expression. He showed that unit factors, later called alleles, exist in pairs and exhibit a dominant/recessive relationship in determining the expression of traits.

3. Mendel postulated that unit factors must segregate during gamete formation, such that each gamete receives only one of the two factors with equal probability.

4. Mendel's postulate of independent assortment states that each pair of unit factors segregates independently of other such pairs. As a result, all possible combinations of gametes will be formed with equal probability.

5. Mendel designed the test cross to determine the exact genotype of peas expressing dominant traits.

6. The discovery of chromosomes in the late 1800s and subsequent studies of their behavior during meiosis linked the behavior of Mendel's unit factors with that of chromosomes during meiosis.

7. The Punnett square and the forked-line methods are used to predict the probabilities of phenotypes (and genotypes) from crosses involving two or more gene pairs.

8. Genetic ratios are expressed as probabilities. Thus, deriving outcomes of genetic crosses relies on an understanding of the laws of probability, particularly the sum law, the product law, conditional probability, and the binomial theorem.

9. Statistical analysis is used to test the validity of experimental outcomes. In genetics, variations from the expected ratios are anticipated owing to chance deviation.

10. Chi-square analysis allows us to predict the probability that variation observed in an experiment, such as a genetic cross, is being generated by chance alone. Calculating χ^2 provides the basis for assessing the null hypothesis, namely that there is no real difference between the expected and observed values.

11. Pedigree analysis provides a method for studying the inheritance pattern of human traits over several generations. This often provides the basis for determining the mode of inheritance of human characteristics and disorders.

INSIGHTS AND SOLUTIONS

Students demonstrate their knowledge of transmission genetics by solving genetics problems. Success at this task represents not only comprehension of theory but also its application to more practical genetic situations. Most students find problem solving in genetics to be challenging but rewarding. This section is designed to provide basic insights into the reasoning essential to this process.

Genetics problems are in many ways similar to algebraic word problems. The approach taken should be identical: (1) analyze the problem carefully; (2) translate words into symbols, defining each one; and (3) choose and apply a specific technique to solve the problem. The first two steps are the most critical. The third step is largely mechanical.

The simplest problems are those that state all necessary information about the P_1 generation and ask you to find the expected ratios of the F_1 and F_2 genotypes and/or phenotypes. Always follow these steps when you encounter this type of problem:

1) Determine to the extent possible the genotypes of the individuals in the P_1 generation.

2) Determine what gametes may be formed by the P_1 parents.

3) Recombine gametes either by the Punnett square method, by the forked-line method, or, when you become more accomplished, by inspection. From the genotypes of the F_1 generation, determine the phenotypes.

4) Repeat the process to obtain information about the F_2 generation.

Performing the above steps requires an understanding of the basic theory of transmission genetics. For example, consider the following problem:

A recessive mutant allele, *black*, causes a very dark body in *Drosophila* when homozygous. The normal wild-type color is described as gray. What F_1 phenotypic ratio is predicted when a black female is crossed to a gray male whose father was black?

To work this problem, you must understand dominance and recessiveness as well as the principle of segregation. Further, you must use the information about the male parent's father. You can work out the problem as follows.

1) Because the female parent is black, she must be homozygous for the mutant allele (*bb*).

2) The male parent is gray; therefore, he must have at least one dominant allele (*B*). Because his father was black (*bb*) and he received one of the chromosomes bearing these alleles, the male parent must be heterozygous (*Bb*).

From here, the problem is simple.

Apply this approach to the following problems.

1. In Mendel's work using peas, he found that full pods are dominant to constricted pods, whereas round seeds are dominant to wrinkled seeds. One of his crosses was between full, round plants and constricted, wrinkled plants. From this cross, he obtained an F_1 that was all full and round. In the F_2, Mendel obtained his classic 9:3:3:1 ratio. Using the above information, determine the expected F_1 and F_2 results of a cross between homozygous constricted, round and full, wrinkled plants.

Solution: First, define gene symbols for each pair of contrasting traits. Select the lower-case forms of the first letter of the recessive traits to designate those traits, and use the upper-case forms to designate the dominant traits. For example, use C and c to indicate full and constricted, and use W and w to indicate the round and wrinkled phenotypes, respectively.

Now, determine the genotypes of the P_1 generation, form gametes, reconstitute the F_1 generation, and determine the F_1 phenotype(s):

P_1:	*ccWW*	×	*CCww*
	constricted, round		full, wrinkled
	↓		↓
Gametes:	*cW*		*Cw*
F_1:		*CcWw*	
		full, round	

You can see immediately that the F_1 generation expresses both dominant phenotypes and is heterozygous for both gene pairs. We can expect that the F_2 generation will yield the classic Mendelian ratio of 9:3:3:1. Let's work it out anyway, just to confirm this, using the forked-line method. Because both gene pairs are heterozygous and can be expected to assort independently, we can predict the F_2 outcomes from each gene pair separately and then proceed with the forked-line method.

bb	×	**Bb**	
Homozygous black female		Heterozygous gray male	

B *b*

b | *Bb* | *bb* | → F_1

1/2 Heterozygous gray males and females, *Bb*

1/2 Homozygous black males and females, *bb*

Every F_2 offspring is subject to the following probabilities:

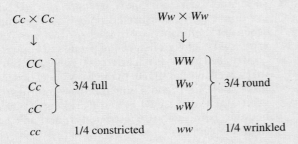

The forked-line method then allows us to confirm the 9:3:3:1 phenotypic ratio. Remember that this represents proportions of 9/16:3/16:3/16:1/16. Note that we are applying the product law as we compute the final probabilities:

$$3/4 \text{ full} \begin{cases} 3/4 \text{ round} \xrightarrow{(3/4)(3/4)} 9/16 \text{ full, round} \\ 1/4 \text{ wrinkled} \xrightarrow{(3/4)(1/4)} 3/16 \text{ full, wrinkled} \end{cases}$$

$$1/4 \text{ constricted} \begin{cases} 3/4 \text{ round} \xrightarrow{(1/4)(3/4)} 3/16 \text{ constricted, round} \\ 1/4 \text{ wrinkled} \xrightarrow{(1/4)(1/4)} 1/16 \text{ constricted, wrinkled} \end{cases}$$

2. Determine the probability that a plant of genotype $CcWw$ will be produced from parental plants of the genotypes $CcWw$ and $Ccww$.

   Solution: Because the two gene pairs independently assort during gamete formation, we need only calculate the individual probabilities of the two separate events (Cc and Ww) and apply the product law to calculate the final probability.

   $$Cc \times Cc \rightarrow 1/4 \, CC : 1/2 \, Cc : 1/4 \, cc$$
   $$Ww \times ww \rightarrow 1/2 \, Ww : 1/2 \, ww$$
   $$p = (1/2 \, Cc)(1/2 \, Ww) = 1/4 \, CcWw$$

3. In another cross, involving parent plants of unknown genotype and phenotype, the offspring shown below were obtained. Determine the genotypes and phenotypes of the parents.

   Offspring: 3/8 full, round

   3/8 full, wrinkled

   1/8 constricted, round

   1/8 constricted, wrinkled

Solution: This problem is more difficult and requires keener insights since you must work backwards. The best approach is to consider the outcomes of pod shape separately from those of seed texture.

Of all plants, 6/8 (=3/4) are full and 2/8 (=1/4) are constricted. Of the various genotypic combinations that can serve as parents, which will give rise to a ratio of 3/4:1/4? Because this ratio is identical to Mendel's monohybrid F_2 results, we can propose that both unknown parents share the same genetic characteristic as the monohybrid F_1 parents: They must both be heterozygous for the genes controlling pod shape and thus are

$$Cc$$

Before accepting this hypothesis, let's consider the possible genotypic combinations that control seed texture. If we consider this characteristic alone, we can see that the traits are expressed in a ratio of 4/8 (=1/2) round:4/8 (=1/2) wrinkled. In order to generate such a ratio, the parents *cannot* both be heterozygous or their offspring would yield a 3/4:1/4 phenotypic ratio. They *cannot* both be homozygous or all offspring would express a single phenotype. Thus, we are left with testing the hypothesis that one parent is homozygous and one is heterozygous for the alleles controlling texture. The potential case of $WW \times Ww$ will not work, for it also would yield only a single phenotype. This leaves us with the potential case of $Ww \times ww$. Offspring in such a mating will yield 1/2 Ww (round):1/2 ww (wrinkled), exactly the outcome we are seeking.

Now, let's combine our hypotheses and predict the outcome of crossing. In our solution, we will use the symbol "—" to indicate that the second allele may be either dominant or recessive, since we are only predicting phenotypes.

$$3/4 \, C \begin{cases} 1/2 \, Ww \xrightarrow{(3/4)(1/2)} 3/8 \, C{-}Ww \text{ full, round} \\ 1/2 \, ww \xrightarrow{(3/4)(1/2)} 3/8 \, C{-}ww \text{ full, wrinkled} \end{cases}$$

$$1/4 \, cc \begin{cases} 1/2 \, Ww \xrightarrow{(1/4)(1/2)} 1/8 \, ccWw \text{ constricted, round} \\ 1/2 \, ww \xrightarrow{(1/4)(1/2)} 1/8 \, ccww \text{ constricted, wrinkled} \end{cases}$$

As we can see, this cross produces offspring in accordance with the information provided, solving the problem. Note that in this solution we have used *genotypes* in the forked-line method, in contrast to the use of *phenotypes* in the solution to the first problem.

4. In the laboratory, a genetics student crossed flies with normal long wings to flies with mutant *dumpy* wings, which she believed was a recessive trait. In the F_1, all flies had long wings. In the F_2, the following results were obtained:

792 long-winged flies

208 dumpy-winged flies

The student tested the hypothesis that the *dumpy* wing is inherited as a recessive trait by performing χ^2 analysis of the F_2 data.

(a) What ratio did the student hypothesize?

(b) Did the χ^2 analysis support the hypothesis?

(c) What do the data suggest about the *dumpy* mutation?

Solution:

(a) The student hypothesized that the F_2 data (792:208) fit Mendel's 3:1 monohybrid ratio for recessive genes.

(b) The initial step in χ^2 analysis is to calculate the expected results (e) assuming a ratio of 3:1. Then calculate the deviations (d) between the expected values and observed values (the actual data),

Ratio	o	e	d	d^2	d^2/e
3/4	792	750	42	1764	2.35
1/4	208	250	−42	1764	7.06

Total = 1000

$$\chi^2 = \sum \frac{d^2}{e}$$
$$= 2.35 + 7.06$$
$$= 9.41$$

Consulting Figure 3.12 allows us to determine the probability (p). This value will let us determine whether the deviations from the null hypothesis can be attributed to chance. There are two possible outcomes (n), so the degrees of freedom (df) = $n - 1$ or 1. The table in Figure 3.12 shows that $p = 0.01$ to 0.001. The graph gives an estimate of about 0.001. That p is less than 0.05 causes us to reject the null hypothesis. The data do not statistically fit a 3:1 ratio.

(c) When we accept Mendel's 3:1 ratio as a valid expression of the monohybrid cross, we make numerous assumptions. One of these may explain why the null hypothesis was rejected. We must assume *that all genotypes are equally viable*. That is, at the time the data are collected, genotypes yielding long wings are equally likely to survive from fertilization through adulthood as the genotype yielding *dumpy* wings. Further study would reveal that *dumpy* flies are somewhat less viable than normal flies. As a result, we would expect *less* than 1/4 of the total offspring to express dumpy. This observation is borne out in the data, although we have not proved this.

5. If two parents, both heterozygous carriers of the autosomal recessive gene causing cystic fibrosis, have five children, what is the probability that three will be normal?

Solution: First, the probability of having a normal child during each pregnancy is

$$p_a = \text{normal} = 3/4$$

while the probability of having an afflicted offspring is

$$p_b = \text{afflicted} = 1/4$$

Then apply the formula

$$\frac{n!}{s!t!} a^s b^t$$

where $n = 5$, $s = 3$, and $t = 2$.

$$p = \frac{(5) \cdot (4) \cdot (3) \cdot (2) \cdot (1)}{(3) \cdot (2) \cdot (1) \cdot (2) \cdot (1)} (3/4)^3 (1/4)^2$$

$$= \frac{(5) \cdot (4)}{(2) \cdot (1)} (3/4)^3 (1/4)^2$$

$$= 10(27/64) \cdot (1/16)$$

$$= 10(27/1024)$$

$$= 270/1024$$

$$p = \sim 0.26$$

PROBLEMS AND DISCUSSION QUESTIONS

When working genetics problems in this and succeeding chapters, always assume that members of the P_1 generation are homozygous, unless the information given indicates or requires otherwise.

1. In a cross between a black and a white guinea pig, all members of the F_1 generation are black. The F_2 generation is made up of approximately 3/4 black and 1/4 white guinea pigs.
   (a) Diagram this cross, showing the genotypes and phenotypes.
   (b) What will the offspring be like if two F_2 white guinea pigs are mated?
   (c) Two different matings were made between black members of the F_2 generation with the results shown below. Diagram each of the crosses.

Cross	Offspring
Cross 1	All black
Cross 2	3/4 black, 1/4 white

2. Albinism in humans is inherited as a simple recessive trait. For the following families, determine the genotypes of the parents and offspring. When two alternative genotypes are possible, list both.
   (a) Two normal parents have five children, four normal and one albino.
   (b) A normal male and an albino female have six children, all normal.
   (c) A normal male and an albino female have six children, three normal and three albino.
   (d) Construct a pedigree of the families in (b) and (c). Assume that one of the normal children in (b) marries one of the albino children in (c) and that they have eight children.

3. Which of Mendel's postulates are illustrated by the pedigree in Problem 2? List and define these postulates.

4. Discuss the rationale relating Mendel's monohybrid results to his postulates.

5. What advantages were provided by Mendel's choice of the garden pea in his experiments?

6. Pigeons may exhibit a checkered or plain pattern. In a series of controlled matings, the following data were obtained:

P_1 Cross	F_1 Progeny	
	Checkered	Plain
(a) checkered × checkered	36	0
(b) checkered × plain	38	0
(c) plain × plain	0	35

Then, F_1 offspring were selectively mated with the following results. The P_1 cross giving rise to each F_1 pigeon is indicated in parentheses.

$F_1 \times F_1$ Crosses	Progeny	
	Checkered	Plain
(d) checkered (a) × plain (c)	34	0
(e) checkered (b) × plain (c)	17	14
(f) checkered (b) × checkered (b)	28	9
(g) checkered (a) × checkered (b)	39	0

How are the checkered and plain patterns inherited? Select and define symbols for the genes involved and determine the genotypes of the parents and offspring in each cross.

7. Mendel crossed peas having round seeds and yellow cotyledons (seed leaves) with peas having wrinkled seeds and green cotyledons. All the F_1 plants had round seeds with yellow cotyledons. Diagram this cross through the F_2 generation using both the Punnett square and forked-line, or branch diagram, methods.

8. Based on the above cross, in the F_2 generation, what is the probability that an organism will have round seeds and green cotyledons *and* be true-breeding?

9. Based on the same characters and traits as in Problem 7, determine the genotypes of the parental plants involved in the crosses shown below by analyzing the phenotypes of their offspring.

Parental Plants	Offspring
(a) round, yellow × round, yellow	3/4 round, yellow 1/4 wrinkled, yellow
(b) wrinkled, yellow × round, yellow	6/16 wrinkled, yellow 2/16 wrinkled, green 6/16 round, yellow 2/16 round, green
(c) round, yellow × round, yellow	9/16 round, yellow 3/16 round, green 3/16 wrinkled, yellow 1/16 wrinkled, green
(d) round, yellow × wrinkled, green	1/4 round, yellow 1/4 round, green 1/4 wrinkled, yellow 1/4 wrinkled, green

10. Which of the crosses in Problem 9 is a test cross?

11. Which of Mendel's postulates can only be demonstrated in crosses involving at least two pairs of traits? State the postulate.

12. Correlate Mendel's four postulates with what is now known about homologous chromosomes, genes, alleles, and the process of meiosis.

13. What is the basis for homology among chromosomes?

14. Distinguish between homozygosity and heterozygosity.

15. In *Drosophila, gray* body color is dominant to *ebony* body color, while *long* wings are dominant to *vestigial* wings. Work

the following crosses through the F_2 generation and determine the genotypic and phenotypic ratios for each generation. Assume the P_1 individuals are homozygous.

(a) gray, long × ebony, vestigial

(b) gray, vestigial × ebony, long

(c) gray, long × gray, vestigial

16. How many different types of gametes can be formed by individuals of the following genotypes: (a) *AaBb*, (b) *AaBB*, (c) *AaBbCc*, (d) *AaBBcc*, (e) *AaBbcc*, and (f) *AaBbCcDdEe?* What are the gametes in each case?

17. Using the forked-line, or branch diagram, method, determine the genotypic and phenotypic ratios of the trihybrid crosses: (a) *AaBbCc × AaBBCC*, (b) *AaBBCc × aaBBCc*, and (c) *AaBbCc × AaBbCc*.

18. Mendel crossed peas with green seeds to those with yellow seeds. The F_1 generation produced only yellow seeds. In the F_2, the progeny consisted of 6022 plants with yellow seeds and 2001 plants with green seeds. Of the F_2 yellow-seeded plants, 519 were self-fertilized with the following results: 166 bred true for yellow and 353 produced a 3:1 ratio of yellow:green. Explain these results by diagramming the crosses.

19. In a study of black and white guinea pigs, 100 black animals were crossed individually to white animals and each cross was carried to an F_2 generation. In 94 of the cases, the F_1 individuals were all black, and an F_2 ratio of 3 black:1 white was obtained. In the other 6 cases, half of the F_1 animals were black and the other half were white. Why? Predict the results of crossing the black and white F_1 guinea pigs from the 6 exceptional cases.

20. Mendel crossed peas with round green seeds to ones with wrinkled yellow seeds. All F_1 plants had seeds that were round and yellow. Predict the results of test-crossing these F_1 plants.

21. Thalassemia is an inherited anemic disorder in humans. Individuals can be completely normal, they can exhibit a "minor" anemia, or they can exhibit a "major" anemia. Assuming that only a single gene pair and two alleles are involved in the inheritance of these conditions, which phenotype is recessive?

22. Below are shown F_2 results of two of Mendel's monohybrid crosses. State a null hypothesis to be tested using X^2 analysis. Calculate the X^2 value and determine the p value for both. Interpret the p values. Can the deviation in each case be attributed to chance or not? Which of the two crosses shows a greater amount of deviation?

(a) Full pods   882
    Constricted pods   299
(b) Violet flowers   705
    White flowers   224

23. In one of Mendel's dihybrid crosses, he observed 315 round yellow, 108 round green, 101 wrinkled yellow, and 32 wrinkled green F_2 plants. Analyze these data using the X^2 test to see if:

(a) They fit a 9:3:3:1 ratio.

(b) The round:wrinkled data fit a 3:1 ratio.

(c) The yellow:green data fit a 3:1 ratio.

24. In assessing data that fell into two phenotypic classes, a geneticist observed values of 250:150. She decided to perform a X^2 analysis using two different null hypotheses: (a) the data fit a 3:1 ratio; and (b) the data fit a 1:1 ratio. Calculate the X^2 values for each hypothesis. What can be concluded about each hypothesis?

25. The basis for rejection of any null hypothesis is arbitrary. The researcher can set more or less stringent standards by deciding to raise or lower the p value used to reject or fail to reject the hypothesis. In the case of chi-square analysis of genetic crosses, would the use of a standard of $p = 0.10$ be more or less stringent in failing to reject the null hypothesis? Explain.

26. For the following pedigree, predict the mode of inheritance and the most probable genotypes of each individual. Assume that the alleles A and a control the expression of the trait.

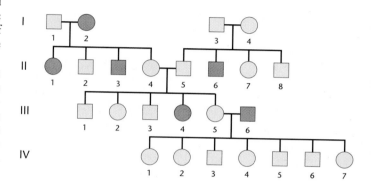

27. The following pedigree is for myopia (near-sightedness) in humans. Predict whether the disorder is inherited as the result of a dominant or recessive trait. Determine the most probable genotype for each individual based on your prediction.

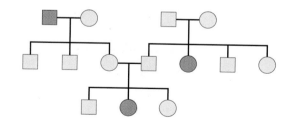

GENETICS MediaLab

The following resources will help you achieve a better understanding of the concepts presented in this chapter. These resources can be found on the CD packaged with this textbook and on the Companion Website at **http://www.prenhall.com/klug/**.

CD Resources:
  Animated Tutorial: *Monohybrid Cross*
  Animated Tutorial: *Punnett Square*
  Animated Tutorial: *Dihybrid Cross*
  Self-grading Chapter Problems

Web Resources:
  Web Destinations in Genetics
  Self-grading Chapter Problems
  Chapter Search Terms
  Genetics Newsgroups
  Student Bulletin Board

Web Problem 1:
Time for completion = 10 minutes
In this exercise, you will practice making predictions about the genetic basis of flower color in peas, one of the traits studied by Mendel. Peas have either white or purple flowers. Mendel postulated that some factor present in the gametes affected flower color. He was confident that when the sperm and the ova united, two copies of the white gamete would make a white flower and two copies of the purple gamete would make a purple flower, but what if the zygote received one white and one purple? Clearly such flowers were either white or purple, not intermediate, so one factor must be dominant over the other one. Read about dominance in Chapter 3. How would you set up a test to see if the white factor was dominant over purple or vice versa? State your hypothesis regarding dominance and outline the predictions you would make if a true-breeding white stock was crossed with a true-breeding purple stock. What genotype(s) would be present in each stock? What would be the genotype

28. Draw all possible conclusions concerning the mode of inheritance of the trait denoted in each of the following limited pedigrees. (Each case is based on a different trait.)

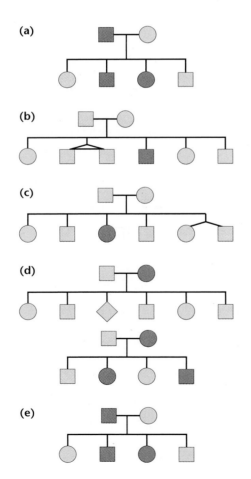

(a)

(b)

(c)

(d)

(e)

29. Consider three independently assorting gene pairs, *A/a*, *B/b*, and *C/c*, where each demonstrates typical dominance (*A–*, *B–*, *C–*, and recessiveness (*aa*, *bb*, *cc*). What is the probability of obtaining an offspring that is *AABbCc* from parents that are *AaBbCC* and *AABbCc?*

30. What is the probability of obtaining a triply recessive individual from the parents shown in Problem 29?

31. Of all offspring of the parents in Problem 29, what proportion will express all three dominant traits?

32. When a die (one of a pair of dice) is rolled, it has an equal probability of landing on any of its six sides.
    (a) What is the probability of rolling a 3 with a single throw?
    (b) When a die is rolled twice, what is the probability that the first throw will be a 3 and the second will be a 6?
    (c) If two dice are rolled together, what is the probability that one will be a 3 and the other will be a 6?
    (d) If one die is rolled and it comes up as an odd number, what is the probability that it is a 5?

33. Consider the F_2 offspring of Mendel's dihybrid cross. Determine the conditional probability that F_2 plants expressing both dominant traits are heterozygous at both loci.

34. Cystic fibrosis is an autosomal recessive disorder. A male whose brother has the disease marries a female whose sister has the disease. It is not known if either the male or the female is a carrier. If the male and female have one child, what is the probability that the child will have cystic fibrosis?

35. In a family of five children, what is the probability that
    (a) All are males?
    (b) Three are males and two are females?
    (c) Two are males and three are females?
    (d) All are the same sex?
    Assume that the probability of a male child is equal to the probability of a female child (*p* = 1/2).

36. In a family of eight children, where both parents are heterozygous for albinism, what mathematical expression predicts the probability that six are normal and two are albinos?

of their progeny? Interpret the results of the crosses. Which hypothesis was supported? To complete this exercise, visit Web Problem 1 in Chapter 3 of your Companion Website and select the keyword **MENDEL**.

Web Problem 2:
Time for completion = 10 minutes
In the Web Problem 1, you determined dominance by crossing pure-breeding lines with white and purple flowers. Suppose you were asked to determine the genotype of a purple-flowering plant that a fellow student found growing in a field of white-flowering plants. How would you proceed? Assume that you have access to pure-breeding white- and purple-flowering peas. Read about Punnett squares and test crosses in Chapter 3. When two or more outcomes are possible, scientists often specify multiple hypotheses and devise tests that will eliminate one or more hypotheses. State two hypotheses regarding the genetic makeup of your purple-flowered plant. Devise crosses using your purple plant and/or your true-breeding stocks to test your hypotheses. Use the monohybrid Punnett squares module to predict the outcomes of each cross under each of the two scenarios. How many crosses would you need to carry out to determine the genotype of the purple plant? Can you think of an alternative way to test your hypotheses? To complete this exercise, visit Web Problem 2 in Chapter 3 of your Companion Website and select the keyword **TESTCROSS**.

Web Problem 3:
Time for completion = 15 minutes
Now that you have practiced formulating hypotheses and tests for single traits, let's try a dihybrid cross. Wild-type *Drosophila melanogaster* have oval eyes and brown bodies. Suppose you are given a fly with *lobe* eyes and an *ebony* body. Assume that these traits are not sex-linked (Chapter 4) and are on separate chromosomes (Chapter 5). You should read the introduction to using Virtual Fly Lab once you enter the linked website. Set up a cross with a wild-type female and a *lobe-eyed, ebony-bodied* male. Would it matter if one fly were *lobe-eyed* and one were *ebony-bodied*? Write down your hypotheses regarding the dominance and recessiveness of these two traits and derive a set of predictions using Punnett squares. Make the cross and observe the phenotype of the F_1. Does this provide enough information to test your predictions or do you need to cross the F_1 and observe the F_2? If you think you have enough information, elect to test your hypothesis and enter your predictions in the box, ignoring the information regarding sex. Does the χ^2 test allow you to reject your hypothesis or not? To complete this exercise, visit Web Problem 1 in Chapter 3 of your Companion Website and select the keyword **DIHYBRID**.

EXTRA-SPICY PROBLEMS

37. Two true-breeding pea plants were crossed. One parent is round, terminal, violet, constricted, while the other expresses the respective contrasting phenotypes of wrinkled, axial, white, full. The four pairs of contrasting traits are controlled by four genes, each located on a separate chromosome. In the F_1, only round, axial, violet, and full were expressed. In the F_2, all possible combinations of these traits were expressed in ratios consistent with Mendelian inheritance.
   (a) What conclusion about the inheritance of the traits can be drawn based on the F_1 results?
   (b) In the F_2 results, which phenotype appeared most frequently? Write a mathematical expression that predicts the probability of occurrence of this phenotype.
   (c) Which F_2 phenotype is expected to occur least frequently? Write a mathematical expression that predicts this probability.
   (d) In the F_2 generation, how often is either of the P_1 phenotypes likely to occur?
   (e) If the F_1 plants were test-crossed, how many different phenotypes would be produced? How does this number compare to the number of different phenotypes in the F_2 generation discussed above?

38. Tay–Sachs disease (TSD) is an inborn error of metabolism that results in death by the age of two. You are a genetic counselor and one day you interview a phenotypically normal couple, where the male had a female first cousin (on his father's side) who died from TSD, and where the female had a maternal uncle with TSD. There are no other known cases in either of the families and none of the matings were/are between related individuals. Assume that this trait is very rare.
   (a) Draw a pedigree of the families of this couple, showing the relevant individuals.
   (b) Calculate the probability that both the male and female are carriers for TSD.
   (c) What is the probability that neither of them are carriers?
   (d) What is the probability that one of them is a carrier and the other is not? (*Hint:* The p values in (b), (c), and (d) should equal 1.)

39. The wild-type (normal) fruit fly, *Drosophila melanogaster*, has straight wings and long bristles. Mutant strains have been isolated that have either curled wings or shaven bristles. The

genes representing these two mutant traits are located on separate autosomes. Carefully examine the data from the five crosses below. (a) For each mutation, determine whether it is dominant or recessive. In each case, identify which crosses support your answer; and (b) define gene symbols, and for each cross, determine the genotypes of the parents.

40. What do you suppose the following mathematical expression applies to? Can you think of a genetic example where it might have application?

$$\frac{n!}{s!t!u!}\, a^s b^t c^u$$

	Cross		straight wings, long bristles	straight wings, short bristles	curled wings, long bristles	curled wings, short bristles
				Number of Progeny		
1	straight, short	straight, short	30	90	10	30
2	straight, long	straight, long	120	0	40	0
3	curled, long	straight, short	40	40	40	40
4	straight, short	straight, short	40	120	0	0
5	curled, short	straight, short	20	60	20	60

SELECTED READINGS

CARLSON, E.A. 1987. *The gene: A critical history*, 2nd ed. Philadelphia, PA: Saunders.

CUMMINGS, M.R. 2000. *Human heredity: Principles and issues*, 5th ed., Belmont, CA: Wadsworth Publ. Co.

DUNN, L.C. 1965. *A short history of genetics*. New York: McGraw-Hill.

MILLER, J.A. 1984. Mendel's peas: A matter of genius or of guile? *Sci. News* 125:108–109.

OLBY, R.C. 1985. *Origins of Mendelism*, 2nd ed. London: Constable.

OREL, V. 1996. *Gregor Mendel: The first geneticist*. Oxford: Oxford University Press.

PETERS, J., ed. 1959. *Classic papers in genetics*. Englewood Cliffs, NJ: Prentice-Hall.

SNEDECOR, G.W., and COCHRAN, W.G. 1980. *Statistical methods*, 7th ed. Ames, IA: Iowa State University Press.

SOKAL, R.R., and ROHLF, F.J. 1987. *Introduction to biostatistics*, 2nd ed. New York: W.H. Freeman.

SOUDEK, D. 1984. Gregor Mendel and the people around him. *Am. J. Hum. Genet.* 36:495–98.

STERN, C., and SHERWOOD, E. 1966. *The origins of genetics: A Mendel source book*. San Francisco: W.H. Freeman.

STUBBE, H. 1972. *History of genetics: From prehistoric times to the rediscovery of Mendel's laws*. Cambridge, MA: MIT Press.

STURTEVANT, A. H. 1965. *A history of genetics*. New York: Harper & Row.

TSCHERMAK-SEYSENEGG, E. 1951. The rediscovery of Mendel's work. *J. Hered.* 42:163–72.

VOELLER, B. R., ed. 1968. *The chromosome theory of inheritance: Classical papers in development and heredity*. New York: Appleton-Century-Crofts.

Mice expressing the agouti, yellow, and black phenotypes.

Extensions of Mendelian Genetics

KEY CONCEPTS

- **Alleles alter phenotypes in different ways**

- **Geneticists use a variety of symbols for alleles**

- **In incomplete, or partial, dominance, neither allele is dominant**

- **In codominance, both alleles in a heterozygote are jointly expressed**

- **Multiple alleles of a gene may exist in a population**
  The ABO Blood Groups
  The A and B Antigens
  The Bombay Phenotype
  The Secretor Locus
  The Rh Antigens
  The *white* Locus in *Drosophila*

- **Lethal alleles may be recessive or dominant**

- **Combination of two gene pairs involving two modes of inheritance modify the 9:3:3:1 ratio**

- **Phenotypes are often controlled by more than one gene**
  Epistasis
  Novel Phenotypes
  Other Modified Dihybrid Ratios

- **Complementation analysis can determine whether or not two similar mutations are alleles**

- **X-linkage describes genes on the X chromosome**
  X-Linkage in *Drosophila*
  X-Linkage in Humans

- **In sex-limited and sex-influenced inheritance, an individual's sex influences the phenotype**

- **Phenotypic expression is not always a direct reflection of the genotype**
  Penetrance and Expressivity
  Genetic Background: Suppression and Position Effects
  Temperature Effects
  Nutritional Effects
  Onset of Genetic Expression
  Genetic Anticipation
  Genomic (Parental) Imprinting

In Chapter 3, we discussed the fundamental principles of transmission genetics. We saw that genes are present on homologous chromosomes and that during gamete formation these chromosomes segregate from each other and independently assort with other segregating chromosomes. These two postulates are the basic principles of gene transmission from parent to offspring. Once an offspring has received the total set of genes, however, it is the expression of genes that determines the organism's phenotype. When gene expression does not adhere to a simple dominant/recessive mode, or when more than one pair of genes influences the expression of a single character, the classic 3:1 and 9:3:3:1 F_2 ratios are usually modified. In this and the next several chapters we consider more complex modes of inheritance. Nevertheless fundamental principles of segregation and independent assortment still hold true in these situations.

We will begin by discussing the inheritance of traits that are controlled by only one set of genes. In diploid organisms, where homologous pairs of chromosomes exist, two copies of each gene influence such traits. The copies need not be identical since alternative forms of genes, or **alleles,** occur within populations. How alleles act to influence a given phenotype will be our major consideration. We will then turn to **gene interaction,** a situation in which a single phenotype is controlled by more than one gene. Numerous examples will be presented to illustrate a variety of heritable patterns observed in such situations. Thus far we have restricted our discussion to chromosomes other than the X and Y pair. By examining cases where genes are present on the X chromosome, illustrating **X-linkage,** we will see yet another modification of Mendelian ratios. Our discussion of modified ratios concludes with the consideration of sex-limited and sex-influenced inheritance, cases where the sex of the individual, but not necessarily the X chromosome, influences the phenotype. We conclude the chapter by entertaining the idea that the expression of a given phenotype often varies depending on the overall environment in which a cell or an organism finds itself. This discussion points out that phenotypic expression depends on more that just the genotype of an organism.

Alleles alter phenotypes in different ways

Following the rediscovery of Mendel's work in the early 1900s, research focused on the many ways in which genes can influence an individual's phenotype. This course of investigation, stemming from Mendel's findings, is called neo-Mendelian genetics (*neo* from the Greek word meaning "since" or "new").

Each type of inheritance described in this chapter was investigated when observations of genetic data did not precisely conform to the expected Mendelian ratios. Hypotheses that modified and extended the Mendelian principles were proposed and tested with specifically designed crosses. Explanations for these observations were in accord with the principle that a phenotype is under the control of one or more genes located at specific loci on one or more pairs of homologous chromosomes.

To understand the various modes of inheritance, we must first examine the potential function of an allele. Alleles are alternative forms of the same gene. The allele that occurs most frequently in a population, the one that is arbitrarily designated as normal, is often referred to as the **wild-type allele.** This common allele is usually dominant—the allele for tall plants in the garden pea, for example—and its product is therefore functional in the cell. Wild-type alleles are responsible, of course, for the corresponding wild-type phenotype and serve as standards for comparison against all mutations occurring at a particular locus.

A mutant allele contains modified genetic information and often specifies an altered gene product. For example, in human populations there are many known alleles of the gene that encodes the β-chain of human hemoglobin. All such alleles store information necessary for the synthesis of the β-chain polypeptide, but each allele specifies a slightly different form of the same molecule. Once manufactured, the product of an allele may or may not have its function altered.

The process of mutation is the source of new alleles. Each new allele leads to a change in the phenotype. A new phenotype results from a change in functional activity of the cellular product controlled by that gene. Usually, the alteration or mutation is expressed as a loss of the specific wild-type function. For example, if a gene is responsible for the synthesis of a specific enzyme, a mutation in the gene may change the conformation of this enzyme and eliminate its affinity for the substrate. Such a mutation will result in a total loss of function. Conversely, another organism may have a different mutation in the same gene, representing a different allele. The resulting enzyme may demonstrate a reduced or increased affinity for binding the substrate, or it may not have its affinity altered at all.

Correspondingly, the functional capacity of the enzyme may be reduced, enhanced, or be unchanged.

Although phenotypic traits may be affected by a single mutation, traits are often influenced by many gene products. In the case of enzymatic reactions, most are part of complex metabolic pathways. Therefore, phenotypic traits may be under the control of more than one gene and the allelic forms of each gene involved. In each of the many crosses discussed in the next few chapters, only one or relatively few gene pairs are involved. Keep in mind that in each cross all genes that are not under consideration are assumed to have no effect on the inheritance patterns described.

Geneticists use a variety of symbols for alleles

In Chapter 3, we learned to symbolize alleles for very simple Mendelian traits. The lowercase form of the initial letter of a recessive trait's name denotes the recessive allele, and the same letter in uppercase refers to the dominant allele. Thus, for *tall* and *dwarf*, where *dwarf* is recessive, D and d represent the alleles responsible for these respective traits. Mendel used upper- and lowercase letters such as these to symbolize his unit factors.

Another useful system was developed in genetic studies of the fruit fly *Drosophila melanogaster* to discriminate between wild-type and mutant traits. This system uses the initial letter, or a combination of two or three letters, of the name of the mutant trait. If the trait is recessive, the lowercase form is used; if it is dominant, the uppercase form is used. The contrasting wild-type trait is denoted by the same letter, but with a superscript $+$.

For example, *ebony* is a recessive body-color mutation in *Drosophila*. The normal wild-type body color is gray. Using the above system, *ebony* is denoted by the symbol e, while gray is denoted by e^+. The responsible locus may be occupied by either the wild-type allele (e^+) or the mutant allele (e). A diploid fly may thus exhibit one of three possible genotypes:

e^+/e^+:	gray homozygote (wild type)
e^+/e:	gray heterozygote (wild type)
e/e:	ebony homozygote (mutant)

The slash between the letters indicates that the two allele designations represent the same locus on two homologous chromosomes. If we were instead considering a dominant wing mutation in *Drosophila*, such as *Wrinkled* (*Wr*), the three possible designations would be Wr^+/Wr^+, Wr^+/Wr, and Wr/Wr. The latter two genotypes express the wrinkled-wing phenotype.

One advantage of this *Drosophila* system is that the symbols may be abbreviated further if necessary: The wild-type allele may simply be denoted by the + symbol. For example, in an *ebony/gray* cross, the designations of the three possible genotypes become:

+/+ :	gray homozygote (wild type)
+/ *e* :	gray heterozygote (wild type)
e / *e* :	ebony homozygote (mutant)

This system works nicely for other organisms as well, provided there is a distinct wild-type phenotype for the character under consideration. As we will see in Chapter 6, it is particularly useful when two or three genes linked together on the same chromosome are considered simultaneously. If no dominance exists, we may simply use uppercase letters and superscripts to denote alternative alleles (e.g., R^1 and R^2, L^M and L^N, I^A and I^B). Their use will become apparent in ensuing sections of this chapter.

Two other points are important. First, although we have adopted a standard convention for assigning genetic symbols, there are many diverse systems of genetic nomenclature used to identify genes in various organisms. Usually, the symbol selected reflects the function of the gene, or occasionally a disorder caused by a mutant gene. For example, the *cdc* gene discussed in Chapter 2 refers to *cell division cycle* genes discovered in yeast. In bacteria, *leu⁻* refers to a mutation that interrupts the biosynthesis of the amino acid leucine, where the wild-type gene is designated *leu⁺*. The symbol *dnaA* represents a bacterial gene involved in DNA replication (and DnaA designates the protein made by that gene). In humans, capital letters are used to name genes: *BRCA1* represents a gene associated with susceptibility to *br*east *ca*ncer. Although these different systems may sometimes be confusing, they all represent different ways to symbolize genes.

In incomplete, or partial, dominance, neither allele is dominant

A cross between parents with contrasting traits may generate offspring with an intermediate phenotype. For example, if plants such as four-o'clocks or snapdragons with red flowers are crossed with white-flowered plants, the offspring have pink flowers. Because some red pigment is produced in the F_1 intermediate-colored pink flowers, neither red nor white flower color is dominant. Such a situation is known as **incomplete** or **partial dominance**.

If this phenotype is under the control of a single gene and two alleles, where neither is dominant, the results of the F_1 (pink) × F_1 (pink) cross can be predicted. The resulting F_2 generation shown in Figure 4.1 confirms the

hypothesis that only one pair of alleles determines these phenotypes. The genotypic ratio (1:2:1) of the F_2 generation is identical to that of Mendel's monohybrid cross. However, because there is no dominance, the phenotypic ratio is identical to the genotypic ratio. Note here that because neither of the alleles is recessive, we have chosen not to use upper- and lowercase letters as symbols. Instead, we have chosen R^1 and R^2 to denote the red and white alleles. We could have chosen W^1 and W^2 or still other designations such as C^W and C^R, where C indicates "color" and W and R indicate "white" and "red."

Clear-cut cases of incomplete dominance, which result in intermediate expression of the overt phenotype, are relatively rare. However, even when complete dominance seems apparent, careful examination of the gene product,

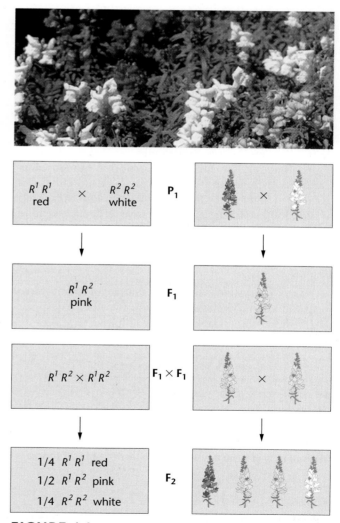

FIGURE 4.1 Incomplete dominance illustrated by flower color. The photograph illustrates red, white, and pink snapdragons.

rather than the phenotype, often reveals an intermediate level of gene expression. An example is the human biochemical disorder **Tay-Sachs disease,** in which homozygous recessive individuals are severely affected with a fatal lipid-storage disorder and neonates die during their first one to three years of life. In afflicted individuals, there is almost no activity of the responsible enzyme **hexosaminidase,** which is normally involved in lipid metabolism. Heterozygotes, with only a single copy of the mutant gene, are phenotypically normal, but express only about 50 percent of the enzyme activity found in homozygous normal individuals. Fortunately, this level of enzyme activity is adequate to achieve normal biochemical function. This situation is not uncommon in enzyme disorders.

In codominance, the expression of both alleles in a heterozygote is clearly evident

If two alleles of a single gene are responsible for the production of two distinct and detectable gene products, a situation different from incomplete dominance or dominance/recessiveness arises. When the distinct expression of both alleles in a heterozygote is clearly evident, the mode of inheritance is referred to as **codominance.** The **MN blood group** in humans illustrates this phenomenon. Karl Landsteiner and Philip Levin discovered a glycoprotein molecule found on the surface of red blood cells that acts as a native antigen, providing biochemical and immunological identity to individuals. In the human population, two forms of this glycoprotein exist, designated M and N. An individual may exhibit either one or both of them.

The MN system is under the control of an autosomal locus found on chromosome 4, with two alleles designated L^M and L^N. Because humans are diploid, three combinations are possible, each resulting in a distinct blood type.

Genotype	Phenotype
$L^M L^M$	M
$L^M L^N$	MN
$L^N L^N$	N

As predicted, a mating between two heterozygous MN parents may produce children of all three blood types.

$$L^M L^N \times L^M L^N$$
$$\downarrow$$

1/4	$L^M L^M$
1/2	$L^M L^N$
1/4	$L^N L^N$

The above example illustrates that codominant inheritance is characterized by the production of distinct gene products by both alleles. For codominance to be studied, both products must be phenotypically detectable. This characteristic distinguishes codominance from other modes of inheritance, such as incomplete dominance, where heterozygotes express an intermediate, or blended, phenotype.

Multiple alleles of a gene may exist in a population

Because the information stored in any gene is extensive, mutations can modify the gene in many ways. Each change has the potential for producing a different allele. Therefore, for any gene, the number of alleles within members of a population of individuals is not necessarily restricted to two. When three or more alleles of the same gene are found, **multiple alleles** are said to be present, creating a unique mode of inheritance. It is important to realize that *multiple alleles can be studied only in populations.* Any individual diploid organism has, at most, two homologous gene loci that may be occupied by different alleles of the same gene. However, many alternative forms of the same gene may exist among members of a species.

The ABO Blood Groups

The simplest possible case of multiple alleles is that in which there are three alternative alleles of one gene. This situation is illustrated in the inheritance of the **ABO blood groups** in humans, discovered by Karl Landsteiner in the early 1900s. Like the antigens responsible for the MN blood types, the A and B antigens are also present on the surface of red blood cells. However, the A and B antigens are distinct from the MN antigens and are under the control of a different gene, located on chromosome 9. As in the MN system, one combination of alleles in the ABO system exhibits a codominant mode of inheritance.

The ABO phenotype of any individual is ascertained by mixing a blood sample with antiserum containing type A or type B antibodies. If the antigen is present on the surface of the person's red blood cells, it will react with the corresponding antibody and cause clumping (or agglutination) of the red blood cells. When an individual is tested in this way, one of four phenotypes may be revealed. Each individual has either the A antigen (A phenotype), the B antigen (B phenotype), the A and B antigens (AB phenotype), or neither antigen (O phenotype). Based on studies of the blood types of many different families, it was hypothesized in 1924 that these phenotypes were inherited as the result of three alleles of a single gene. This hypothesis was based on studies of the blood types of many different families.

Although different designations can be used, we will use the symbols I^A, I^B, and I^O for the three alleles. The I designation stands for **isoagglutinogen,** another term for antigen. If we assume that the I^A and I^B alleles are responsible for the production of their respective A and B antigens and that I^O is an allele that does not produce any detectable A or B antigens, we can list the various genotypic possibilities and the corresponding phenotype assigned to each.

Genotype	Antigen	Phenotype
$I^A I^A$	A	A
$I^A I^O$	A	
$I^B I^B$	B	B
$I^B I^O$	B	
$I^A I^B$	A, B	AB
$I^O I^O$	Neither	O

Note that in these assignments, the I^A and I^B alleles behave dominantly to the I^O allele, but codominantly to each other.

We can test the hypothesis that three alleles control ABO blood groups by examining potential offspring from the various combinations of matings, as shown in Table 4.1. If we assume heterozygosity wherever possible, we can predict which phenotypes can occur. These theoretical predictions have been upheld in numerous studies examining the blood types of children of parents with all possible phe-

notypic combinations. The hypothesis that three alleles control ABO blood types in the human population is now universally accepted.

Our knowledge of human blood types has several practical applications. One of the most important is testing the compatibility of blood transfusions. In cases of disputed parentage, as when newborns are inadvertently mixed up in the hospital, or when it is uncertain whether a specific male is the father of a child, an examination of the ABO blood groups as well as other inherited antigens of the possible parents and the child may help to resolve the situation. For example, of all the matings shown in Table 4.1, the only one that can result in offspring with all four phenotypes is that between two heterozygous individuals, one showing the A phenotype and the other showing the B phenotype. On genetic grounds alone, then, a male or female may be unequivocally ruled out as the parent of a certain child. This type of genetic evidence never proves parenthood, however.

The A and B Antigens

The biochemical basis of the ABO blood type system has now been carefully worked out. The A and B antigens are actually carbohydrate groups (sugars) that are bound to lipid molecules (fatty acids) protruding from the membrane of the red blood cell. The specificity of the A and B antigens is based on the terminal sugar of the carbohydrate group.

Almost all individuals possess what is called the **H substance,** to which one or two terminal sugars are added. As

TABLE 4.1 Potential Phenotypes in the Offspring of Parents with All Possible ABO Blood Group Combinations, Assuming Heterozygosity When Applicable

Parents		Potential Offspring			
Phenotypes	Genotypes	A	B	AB	O
A × A	$I^A I^O \times I^A I^O$	3/4	—	—	1/4
B × B	$I^B I^O \times I^B I^O$	—	3/4	—	1/4
O × O	$I^O I^O \times I^O I^O$	—	—	—	all
A × B	$I^A I^O \times I^B I^O$	1/4	1/4	1/4	1/4
A × AB	$I^A I^O \times I^A I^B$	1/2	1/4	1/4	—
A × O	$I^A I^O \times I^O I^O$	1/2	—	—	1/2
B × AB	$I^B I^O \times I^A I^B$	1/4	1/2	1/4	—
B × O	$I^B I^O \times I^O I^O$	—	1/2	—	1/2
AB × O	$I^A I^B \times I^O I^O$	1/2	1/2	—	—
AB × AB	$I^A I^B \times I^A I^B$	1/4	1/4	1/2	—

shown in Figure 4.2, the H substance itself consists of three sugar molecules, galactose (gal), N-acetylglucosamine (AcGluNH), and fucose, chemically linked together. The I^A allele is responsible for an enzyme that can add the terminal sugar N-acetylgalactosamine (AcGalNH) to the H substance. The I^B allele is responsible for a modified en-zyme that cannot add N-acetylgalactosamine but instead can add a terminal galactose. Heterozygotes ($I^A I^B$) add either one or the other sugar at the many sites (substrates) available on the surface of the red blood cell, illustrating the biochemical basis of codominance in individuals of the AB blood type. Finally, persons of type O ($I^O I^O$) cannot

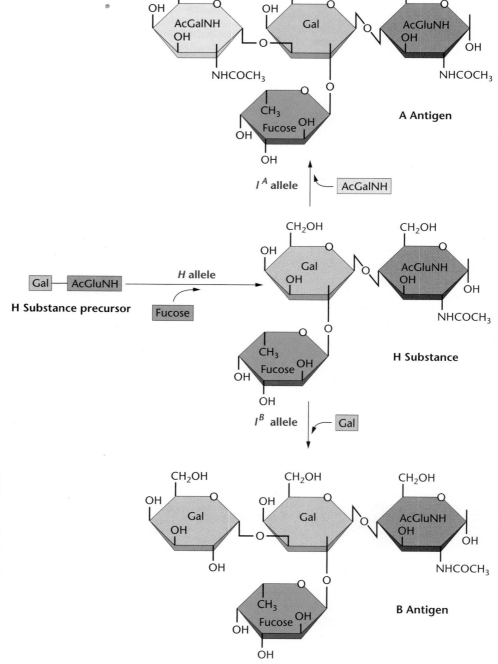

FIGURE 4.2 The biochemical basis of the ABO blood groups. The *H* allele, present in almost all humans, directs the conversion of a precursor molecule to the H substance by adding a molecule of fucose. Failure to do so results in the Bombay phenotype. The I^A and I^B alleles are then able to direct the addition of terminal sugar residues to the H substance. The I^O allele is unable to direct either of these terminal additions. Gal: galactose; AcGluNH: N-acetyl-glucosamine; AcGalNH: N-acetylgalactosamine.

add either terminal sugar, possessing only the H substance protruding from the surface of their red blood cells.

The molecular genetic basis of the mutations leading to the I^A, I^B, and I^O alleles has been elucidated. We shall return to this topic in Chapter 17 when we discuss mutation and mutagenesis.

The Bombay Phenotype

In 1952, a very unusual situation provided information concerning the genetic basis of the H substance. A woman in Bombay displayed a unique genetic history that was inconsistent with her blood type. In need of a transfusion, she was found to lack both the A and B antigens and was thus typed as O. However, as shown in the partial pedigree in Figure 4.3, one of her parents was type AB and she was the obvious donor of an I^B allele to two of her offspring. Thus, she was genetically type B but functionally type O!

The Bombay woman was subsequently shown to be homozygous for a rare recessive mutation, h, which prevented her from synthesizing the complete H substance. In this mutation, the terminal portion of the carbohydrate chain protruding from the red cell membrane was shown to lack fucose. In the absence of fucose, the enzymes specified by the I^A and I^B alleles are apparently unable to recognize the incomplete H substance as a proper substrate. Thus, neither the terminal galactose nor N-acetylgalactosamine can be added, even though the appropriate enzymes capable of doing so are present and functional. As a result, the ABO system genotype cannot be expressed in individuals of genotype hh, and they are functionally type O. To distinguish them from the rest of the population, they are said to demonstrate the Bombay phenotype. The frequency of the h allele is exceedingly low. Hence, the vast majority of the human population is of the HH or Hh genotype (almost all HH) and can synthesize the H substance.

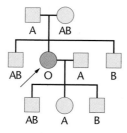

FIGURE 4.3 A partial pedigree of a woman displaying the Bombay phenotype. Functionally, her ABO blood group behaves as type O. Genetically, she is type B.

The Secretor Locus

Expression of the ABO blood type system is affected by still a third gene, found at what is called the **secretor locus.** In about 80 percent of the human population, the A and B antigens are present in various body secretions as well as on the membrane of red blood cells. The ability to secrete these antigens in body fluids such as saliva, gastric juice, semen, and vaginal fluids is under the influence of the dominant allele, Se (Se/Se or Se/se). Nonsecretors (se/se) make the antigens as specified by the ABO loci but do not secrete them. Secretion of these antigens has great significance in forensic science (the application of scientific knowledge to legal proceedings) since ABO typing can be performed on tissue samples other than blood.

The Rh Antigens

Another set of antigens thought by some geneticists to illustrate multiple allelism includes those designated Rh. Discovered by Landsteiner, Levine, and others about 1940, the **Rh antigens** have received a great deal of attention because of their direct involvement in the disorder **erythroblastosis fetalis.** Also referred to as **hemolytic disease of the newborn (HDN),** erythroblastosis fetalis is a form of anemia. It occurs in an Rh-positive fetus whose mother is Rh-negative and whose father is Rh-positive, contributing that allele to the fetus. Such a genetic combination results in a potential immunological incompatibility between the mother and fetus. If fetal blood passes through the ruptured placenta at birth and enters the maternal circulation, the mother's immune system recognizes the Rh antigen as foreign and builds antibodies against it. During a second pregnancy, the antibody concentration becomes high enough that when maternal antibodies pass across the placenta, they enter the fetus's circulation and begin to destroy the fetus's red blood cells. This causes the hemolytic anemia.

About 10 percent of all human pregnancies demonstrate Rh incompatibility. However, for numerous reasons, less than 0.5 percent actually result in anemia. Currently, incompatible mothers are given anti-Rh antibodies immediately after giving birth to all Rh-positive babies. This destroys any Rh-positive cells that have entered the mother's circulation so that she does not produce her own anti-Rh antibodies. Before this treatment was developed, many fetuses failed to survive to term. For those that did, complete blood transfusions were often necessary, but even then many newborns failed to survive.

Initial genetic investigations led to the belief that in the human population only two alleles controlled the presence or absence of the antigen. It was thought that the Rh allele determined the presence of the antigen and behaved as a dominant gene. The rh allele resulted in the absence of the antigen.

With the development of more refined tests determining the presence of the antigen, it became apparent that the genetic control of Rh antigens was much more complex than originally thought. For example, two closely related antigens are sometimes detected in individuals who test negatively for the original Rh antigen. Alexander Wiener subsequently proposed that a series of multiple alleles at a single Rh locus accounted for these variants. Other investigators, including Ronald A. Fisher and Robert R. Race, proposed an alternative type of inheritance involving three closely linked genes, each with two alleles, involved in the inheritance of Rh factors. The term linkage is used to describe genes located on the same chromosome. We will discuss this concept in detail in Chapter 6.

Table 4.2 presents the two systems and the nomenclature each uses to designate the alleles. In the Wiener system, the presence of at least one of the four dominant alleles is sufficient to yield the Rh-positive blood type. In the Fisher–Race system, the genetic locus bearing the set of *D* and *d* alleles is most critical. The presence of at least one dominant *D* allele results in the Rh-positive phenotype, whereas the *dd* genotype ensures the Rh-negative blood type. While the *C*, *c*, *E*, and *e* alleles specify related antigens, they are less immunologically significant. Because of the complexity of the antigenic patterns, even modern molecular analysis has yet to provide strong favor of one system over the other.

The white *Locus in* Drosophila

Many other phenotypes in both plants and animals are controlled by multiple allelic inheritance. In *Drosophila*, for example, where the induction of mutations has been used extensively as a method of investigation, many alleles are present at practically every locus. The recessive eye mutation, *white*, discovered by Thomas H. Morgan and Calvin Bridges in 1912, represents only one of over 100 alleles that may occupy this locus. In this allelic series, eye colors range from complete absence of pigment in the *white* allele, to deep ruby in the *white-satsuma* allele, to orange in the *white-apricot* allele, to a buff color in the *white-buff* allele. These alleles are designated *w*, w^{sat}, w^a, and w^{bf}, respectively (Table 4.3). In each of these cases, the total amount of pigment in these mutant eyes is reduced to less than 20 percent of that found in the brick-red wild-type eye.

Lethal alleles may be recessive or dominant

Many gene products are essential to an organism's survival. Mutations resulting in a gene product that is nonfunctional can sometimes be tolerated in the heterozygous state; that is, one wild-type allele may produce enough of the essential product for the organism to survive. However, such a mutation behaves as a **recessive lethal allele;** homozygous recessive individuals will not survive. The time of death will usually depend upon when the product is essential to development but may occur during adulthood.

In some cases, the allele responsible for a lethal effect in homozygotes may result in a distinctive mutant phenotype in heterozygotes. Such an allele is behaving as a recessive lethal but is dominant with respect to the phenotype. For example, in the early part of this century, researchers discovered a mutation causing a yellow coat in mice that varied from the normal agouti coat phenotype, as illustrated in Figure 4.4. Crosses between the various combinations of the two strains yielded unusual results:

Crosses
A: agouti × agouti → all agouti
B: yellow × yellow → 2/3 yellow: 1/3 agouti
C: agouti × yellow → 1/2 yellow: 1/2 agouti

TABLE 4.2 A Comparison of the Alleles in the Wiener System with Those of the Fisher–Race System Used to Explain the Genetic Basis of the Rh Blood Groups

Wiener Nomenclature	Fisher–Race Nomenclature	Phenotype
R^1	C D E	
R^2	C D e	
R^0	c D E	
R^z	c D e	Rh$^+$
r	C d E	
r'	C d e	
r''	c d E	
r^y	c d e	Rh$^-$

TABLE 4.3 Some of the Alleles Present at the *white* Locus of *Drosophila melanogaster* and Their Eye Color Phenotypes

Allele	Name	Eye Color
w	*white*	pure white
w^a	*white-apricot*	yellowish orange
w^{bf}	*white-buff*	light buff
w^{bl}	*white-blood*	yellowish ruby
w^{cf}	*white-coffee*	deep ruby
w^e	*white-eosin*	yellowish pink
w^{mo}	*white-mottled*	orange light mottled orange
w^{sat}	*white-satsuma*	deep ruby
w^{sp}	*white-spotted*	fine grain, yellow mottling
w^t	*white-tinged*	light pink

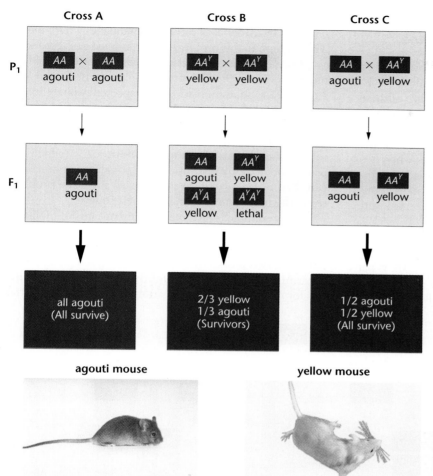

agouti mouse

yellow mouse

FIGURE 4.4 Inheritance patterns in three crosses involving the wild-type agouti allele (*A*) and the mutant yellow allele (*A*^{Y}) in the mouse. Note that the mutant allele behaves dominantly to the normal allele (*A*) in controlling coat color, but it also behaves as a homozygous lethal allele. The genotype *A*^{Y}*A*^{Y} does not survive.

These results are explained on the basis of a single pair of alleles. The mutant yellow allele A^Y is dominant to the wild-type agouti allele A, so heterozygous mice will have yellow coats. However, the yellow allele also behaves as a homozygous lethal. Homozygous mice of the genotype $A^Y A^Y$ die before birth, so no homozygous yellow mice are ever recovered. The genetic basis for these three crosses is provided in Figure 4.4.

Many genes are known to behave similarly in other organisms. In *Drosophila*, *Curly* wing (*Cy*), *Plum* eye (*Pm*), *Dichaete* wing (D), *Stubble* bristle (*Sb*), and *Lyra* wing (*Ly*) behave as homozygous lethals but are dominant with respect to the expression of the mutant phenotype when heterozygous.

In instances where one copy of the wild-type gene is not sufficient for normal development, even the heterozygote will not survive. In this case, the mutation is behaving as a **dominant lethal allele** because its presence somehow overrides the expression of the wild-type product, or the amount of wild-type product is simply insuffi-

cient to support its essential function. For example, in humans, a disorder called **Huntington disease** (also referred to as Huntington's chorea) is due to a dominant allele, *H*. The onset of the disease in heterozygotes (*Hh*) is delayed, usually well into adulthood, typically at about age 40. Affected individuals then undergo gradual neural degeneration until they die. This lethal disorder is particularly tragic because it has such a late onset and an affected individual may have produced a family before discovering the condition. Each child has a 50 percent probability of inheriting the lethal allele and transmitting it to his or her offspring. The American folk singer and composer Woody Guthrie, father of Arlo Guthrie, died from this disease.

Dominant lethal alleles are rarely observed. For them to exist in a population, the affected individual must reproduce before the allele's lethality is expressed, as can occur in Huntington disease. If all affected individuals die before reaching the reproductive age, the mutant gene will not be passed to future generations and will disappear from the population unless it recurs again as a result of a new mutation.

Combinations of two gene pairs involving two modes of inheritance modify the 9:3:3:1 ratio

Each example discussed so far modifies Mendel's 3:1 F_2 monohybrid ratio. Therefore, combining any two of these modes of inheritance in a dihybrid cross will also modify the classical 9:3:3:1 ratio. Having established the foundation for incomplete dominance, codominance, multiple alleles, and lethal alleles, let us see what happens when two modes of inheritance occur simultaneously. Mendel's principle of independent assortment applies to these situations when the genes controlling each character are not located on the same chromosome.

Consider, for example, a mating between two humans who are both heterozygous for the autosomal recessive gene that causes albinism and who are both of blood type AB. What is the probability of any particular phenotypic combination occurring in each of their children? Albinism is inherited in

the simple Mendelian fashion, and the blood types are determined by the series of three multiple alleles, I^A, I^B, and I^O. The solution to this problem is diagrammed in Figure 4.5.

Instead of this dihybrid cross yielding the classical four phenotypes in a 9:3:3:1 ratio, six phenotypes occur in a 3:6:3:1:2:1 ratio, establishing the expected probability for each phenotype. Figure 4.5 solves the problem using the forked-line method first described in Chapter 3. Recall that the forked-line method requires that the phenotypic ratios for each trait be computed individually (which can usually be done by inspection); all possible combinations can then be calculated. We could also solve the problem using the more conventional Punnett square.

This example is just one of many variants of modified ratios possible when different modes of inheritance are combined. We can deal in a similar way with any combination of two modes of inheritance. You will be asked to determine the phenotypes and their expected probabilities for many of these combinations in the problems at the

$$A\,a\,I^A\,I^B \times A\,a\,I^A\,I^B$$

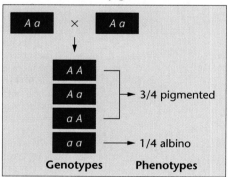

Consideration of pigmentation alone

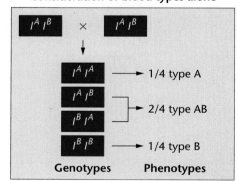

Consideration of blood types alone

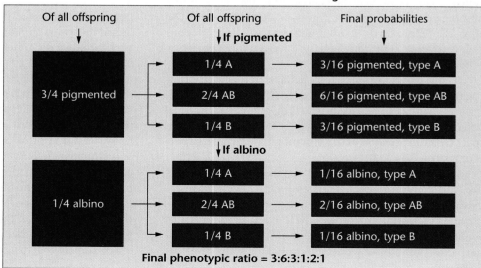

FIGURE 4.5 The calculation of the probabilities in a mating involving the ABO blood type and albinism in humans using the forked-line method.

end of the chapter. In each case, the final phenotypic ratio is a modification of the 9:3:3:1 dihybrid ratio.

Phenotypes are often controlled by more than one gene

Soon after the rediscovery of Mendel's work, experimentation revealed that individual characteristics displaying discrete phenotypes were often under the control of more than one gene. This discovery was significant because it revealed that genetic influence on the phenotype is much more complex than that which Mendel encountered in his crosses with the garden pea. Instead of single genes controlling the development of individual parts of a plant or animal body, it soon became clear that phenotypic characteristics, such as eye color, hair color, or fruit shape, are influenced by many genes and their resultant products.

The term **gene interaction** is often used to describe the idea that several genes influence a particular characteristic. This does not mean, however, that two or more genes or their products necessarily interact directly with one another to influence a particular phenotype. Rather, the cellular function of numerous gene products contributes to the development of a common phenotype. For example, the development of an organ like the eye of an insect is exceedingly complex and leads to a structure with multiple phenotypic manifestations—most simply described as a specific size, shape, texture, and color. The development of the eye can best be envisioned as a complex cascade of developmental events. This process illustrates the developmental concept of **epigenesis** (originally introduced in Chapter 1), whereby each ensuing step of development increases the complexity of this sensory organ and is under the control and influence of one or more genes. To clarify, we will present several examples that illustrate gene interaction at the biochemical level.

Epistasis

Some of the best examples of gene interaction are those illustrating the phenomenon of **epistasis**. Derived from the Greek word meaning "stoppage," epistasis occurs when genes control the expression of the same general phenotypic character. Sometimes epistasis is antagonistic, as when the expression of one gene pair masks or modifies the expression of another gene pair. In other cases, the genes may exert their influence on one another in a complementary, or cooperative, fashion.

For example, the homozygous presence of a recessive allele may prevent or override the expression of other alleles at a second locus (or several other loci). In this case, the alleles at the initial locus are said to be **epistatic** to those at the second locus. The alleles at the second locus, which are masked, are described as being **hypostatic** to those at the first locus. In another example, a single dominant allele at the first locus may influence the expression of the alleles at a second gene locus. In a third case, two gene pairs may complement one another such that at least one dominant allele at each locus is required to express a particular phenotype.

The Bombay phenotype discussed earlier is an example of the homozygous recessive condition at one locus masking the expression of a second locus. There, the homozygous condition (hh) masked the expression of the I^A and I^B alleles. Only individuals with the H–genotype can form the A or B antigens. As a result, individuals whose genotypes include the I^A or I^B allele and who are also hh express the type O phenotype. An example of the outcome of matings between individuals heterozygous at both loci is illustrated in Figure 4.6. If many individuals of the genotype $I^A I^B Hh$ have children, the phenotypic ratio of 3 A: 6 AB: 3 B: 4 O is expected in their offspring.

It is important to note two things when examining this cross and the predicted phenotypic ratio:

1. In illustrating gene interaction, an important distinction exists in this cross compared to the modified dihybrid cross illustrated in Figure 4.5: *only one characteristic—blood type—is being followed.* In the modified dihybrid cross in Figure 4.5, blood type *and* skin pigmentation are followed as separate phenotypic characteristics.

2. Even though only a single character was followed, the phenotypic ratio was expressed in sixteenths. If we knew nothing about the H substance and the genes controlling it, we could still be confident that a second gene pair, other than that controlling the A and B antigens, was involved in the phenotypic expression. *When studying a single character, a ratio that is expressed in 16 parts (e.g., 3:6:3:4) suggests that two gene pairs are "interacting" during the expression of the phenotype under consideration.*

The study of gene interaction has revealed a number of inheritance patterns that modify the classical Mendelian dihybrid F_2 ratio (9:3:3:1) in other ways. In several of the subsequent examples, epistasis has the effect of combining one or more of the four phenotypic categories in various ways. The generation of these four groups is reviewed in Figure 4.7, along with several modified ratios.

As we discuss these and other examples, we will make several assumptions and adopt certain conventions.

1. In each case, distinct phenotypic classes are produced, each clearly discernible from all others. Such traits il-

$$I^A I^B \; H h \; \times \; I^A I^B \; H h$$

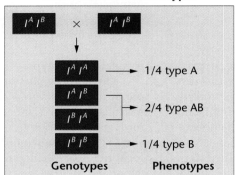

Consideration of blood types

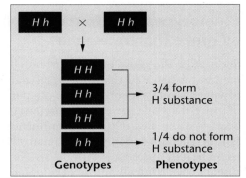

Consideration of H substance

Consideration of both gene pairs together

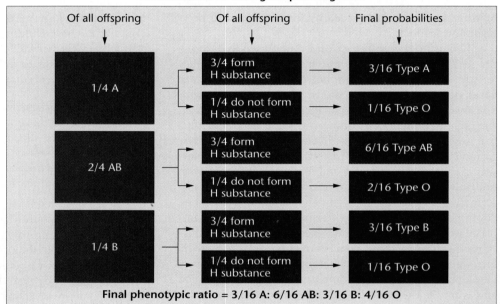

Final phenotypic ratio = 3/16 A : 6/16 AB : 3/16 B : 4/16 O

FIGURE 4.6 The outcome of a mating between individuals heterozygous at two genes determining their ABO blood type. Final phenotypes are calculated by considering both genes separately and then combining the results using the forked-line method.

lustrate discontinuous variation, where phenotypic categories are discrete and qualitatively different from one another.

2. The genes considered in each cross are not linked and therefore assort independently of one another during gamete formation. So that you may easily compare the results of different crosses, we designate alleles as *A*, *a* and *B*, *b* in each case.

3. When we assume that complete dominance exists between the alleles of any gene pair, such that *AA* and *Aa* or *BB* and *Bb* are equivalent in their genetic effects, we use the designations *A–* or *B–* for both combinations. Therefore, the dash (–) indicates that either allele may be present, without consequence to the phenotype.

4. All P_1 crosses involve homozygous individuals (e.g., *AABB* × *aabb*, *AAbb* × *aaBB*, or *aaBB* × *AAbb*). Therefore, each F_1 generation will consist of only heterozygotes of genotype *AaBb*.

5. In each example, the F_2 generation produced from these heterozygous parents will be the main focus of analysis. When two genes are involved (Figure 4.7), the F_2 genotypes fall into four categories: 9/16 *A–B–*, 3/16 *A–bb*, 3/16 *aaB–*, and 1/16 *aabb*. Because of dominance, all genotypes in each category are equivalent in their effect on the phenotype.

The first case to be discussed is the inheritance of coat color in mice (Case 1 of Figure 4.8). As we saw in our discussion of lethal alleles, wild-type coat color is agouti, a

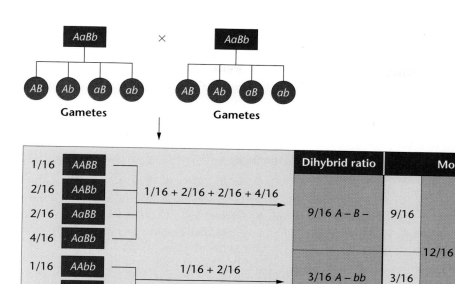

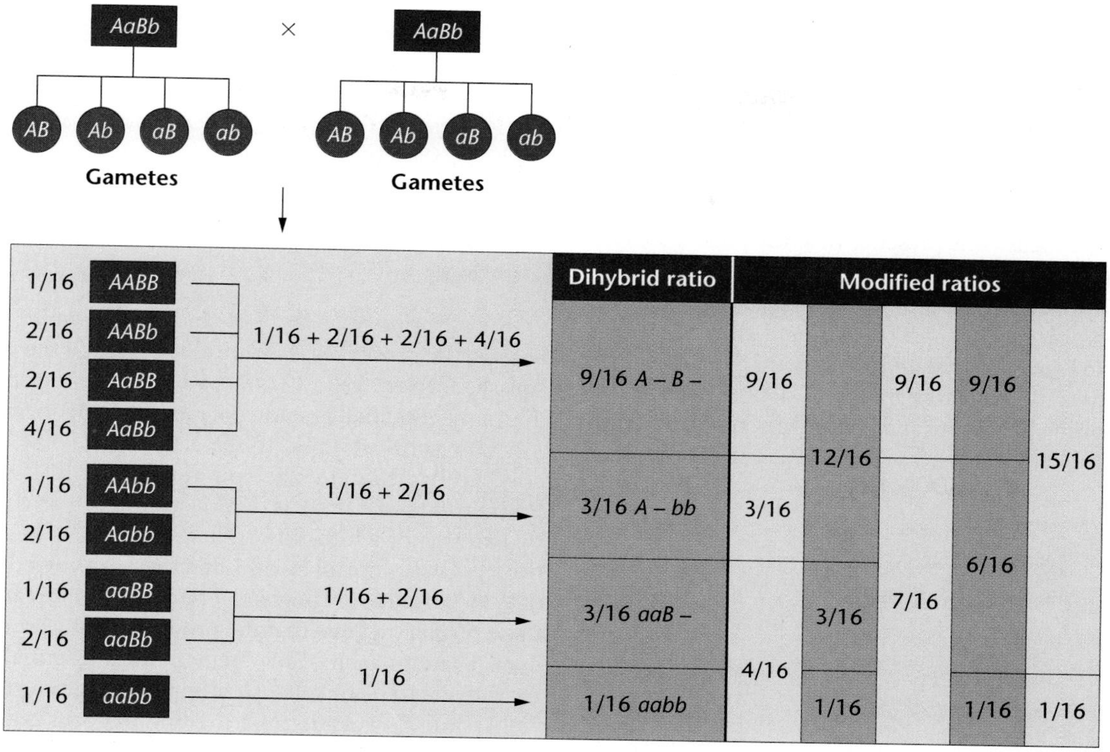

FIGURE 4.7   Generation of the various modified dihybrid ratios from the nine unique genotypes produced in a cross between individuals heterozygous at two genes.

Case	Organism	Character	F₂ Phenotypes				Modified ratio
			9/16	3/16	3/16	1/16	
1	Mouse	Coat color	agouti	albino	black	albino	9:3:4
2	Squash	Color	white		yellow	green	12:3:1
3	Pea	Flower color	purple		white		9:7
4	Squash	Fruit shape	disc		sphere	long	9:6:1
5	Chicken	Color	white		colored	white	13:3
6	Mouse	Color	white-spotted	white	colored	white-spotted	10:3:3
7	Shepherd's purse	Seed capsule	triangular			ovoid	15:1
8	Flour beetle	Color	red | sooty | red | sooty	black	jet	black	6:3:3:4

FIGURE 4.8   The basis of modified dihybrid F₂ phenotypic ratios, resulting from crosses between doubly heterozygous F₁ individuals. The four groupings of the F₂ phenotypes shown across the top of this figure are combined in various ways to produce these ratios.

grayish pattern formed by alternating bands of pigment on each hair (see Figure 4.4). Agouti is dominant to black (nonagouti) hair, which results from the homozygous expression of a recessive allele, a. Thus, $A-$ results in agouti, whereas aa yields black coat color. When homozygous, a recessive mutation, b, at a separate locus, eliminates pigmentation altogether, yielding albino mice, regardless of the genotype at the a locus. The presence of at least one B allele allows pigmentation to occur in much the same way that the H allele in humans allows the production of the A and B antigens. In a cross between agouti ($AABB$) and albino ($aabb$), members of the F$_1$ are all $AaBb$ and have agouti coat color. In the F$_2$ progeny of a cross between two F$_1$ double heterozygotes, the following genotypes and phenotypes are observed:

$$\text{F}_1: AaBb \times AaBb$$
$$\downarrow$$

F$_2$ Ratio	Genotype	Phenotype	Final Phenotypic Ratio
9/16	$A-B-$	agouti	9/16 agouti
3/16	$A-bb$	albino	4/16 albino
3/16	$aa\,B-$	black	3/16 black
1/16	$aa\,bb$	albino	

Gene interaction yielding the observed 9:3:4 F$_2$ ratio might be envisioned hypothetically as a two-step process.

	Gene B		Gene A	
Precursor	$\downarrow$	Black	$\downarrow$	agouti
Molecule	$\longrightarrow$	Pigment	$\longrightarrow$	Pattern
(colorless)	$B-$		$A-$	

In the presence of a B allele, black pigment can be made from a colorless substance. In the presence of an A allele, the black pigment is deposited during the development of hair in a pattern producing the agouti phenotype. If the aa genotype occurs, all of the hair remains black. If the bb genotype occurs, no black pigment is produced, regardless of the presence of the A or a alleles, and the mouse is albino. Therefore, the bb genotype masks or suppresses the expression of the A allele, illustrating epistasis.

A second type of epistasis occurs when a dominant allele at one genetic locus masks the expression of the alleles of a second locus. For instance, Case 2 of Figure 4.8 deals with the inheritance of fruit color in summer squash. Here, the dominant allele A results in white fruit color regardless of the genotype at a second locus, B. In the absence of a dominant A allele (the aa genotype), BB or Bb results in yellow color while bb results in green color. Therefore, if two white-colored double heterozygotes ($AaBb$) are crossed, an interesting genetic ratio occurs because of this type of epistasis.

$$\text{F}_1: AaBb \times AaBb$$
$$\downarrow$$

F$_2$ Ratio	Genotype	Phenotype	Final Phenotypic Ratio
9/16	$A-B-$	white	12/16 white
3/16	$A-bb$	white	
3/16	$aa\,B-$	yellow	3/16 yellow
1/16	$aa\,bb$	green	1/16 green

Of the offspring, 9/16 are $A-B-$ and are thus white. The 3/16 bearing the genotypes $A-bb$ are also white. Of the remaining squash, 3/16 are yellow ($aaB-$), while 1/16 are green ($aabb$). Thus, the modified phenotypic ratio of 12:3:1 occurs.

Our third example (Case 3 of Figure 4.8), first discovered by William Bateson and Reginald Punnett (of Punnett-square fame), is demonstrated in a cross between two strains of white-flowered sweet peas. Unexpectedly, the F$_1$ plants were all purple, and the F$_2$ occurred in a ratio of 9/16 purple to 7/16 white. The proposed explanation for these results suggests that the presence of at least one dominant allele of each of two gene pairs is essential in order for flowers to be purple. All other genotype combinations yield white flowers because the homozygous condition of either recessive allele masks the expression of the dominant allele at the other locus.

The cross is shown as follows:

$$\text{P}_1: AAbb \times aaBB$$
$$\text{white} \quad \text{white}$$
$$\downarrow$$
$$\text{F}_1: \text{All } AaBb \text{ (purple)}$$
$$\downarrow$$

F$_2$ Ratio	Genotype	Phenotype	Final Phenotypic Ratio
9/16	$A-B-$	purple	9/16 purple
3/16	$A-bb$	white	
3/16	$aa\,B-$	white	7/16 white
1/16	$aa\,bb$	white	

We can envision the way in which two gene pairs might yield such results.

	Gene A		Gene B	
Precursor	$\downarrow$	Intermediate	$\downarrow$	Final
Substance	$\longrightarrow$	Product	$\longrightarrow$	Product
(colorless)	$A-$	(colorless)	$B-$	(purple)

At least one dominant allele from each pair of genes is necessary to ensure both biochemical conversions to the final product, yielding purple flowers. In the cross above, this will occur in 9/16 of the F$_2$ offspring. All other plants have flowers that remain white.

These three examples illustrate in a simple way how the products of two genes "interact" to influence the development of a common phenotype. In other instances, more than two genes and their products are involved in controlling phenotypic expression.

Novel Phenotypes

Other cases of gene interaction yield novel, or new, phenotypes in the F_2 generation, in addition to producing modified dihybrid ratios. Case 4 in Figure 4.8 depicts the inheritance of fruit shape in the summer squash *Cucurbita pepo*. When plants with disc-shaped fruit (*AABB*) are crossed to plants with long fruit (*aabb*), the F_1 generation all have disc fruit. However, in the F_2 progeny, fruit with a novel shape—sphere—appear, as well as fruit exhibiting the parental phenotypes. These phenotypes are shown in Figure 4.9.

The F_2 generation, with a modified 9:6:1 ratio, is as follows:

$$F_1: AaBb \times AaBb$$
$$\text{disc} \downarrow \text{disc}$$

F_2 Ratio	Genotype	Phenotype	Final Phenotypic Ratio
9/16	*A– B–*	disc	9/16 disc
3/16	*A– bb*	sphere	
3/16	*aa B–*	sphere	6/16 sphere
1/16	*aa bb*	long	1/16 long

In this example of gene interaction, both gene pairs influence fruit shape equivalently. A dominant allele at either locus ensures a sphere-shaped fruit. In the absence of dom-

inant alleles, the fruit is long. However, if both dominant alleles (*A* and *B*) are present, the fruit is flattened into a disc shape.

Another interesting example of an unexpected phenotype arising in the F_2 generation is the inheritance of eye color in *Drosophila melanogaster*. The wild-type eye color is brick red. When two autosomal recessive mutants, *brown* and *scarlet*, are crossed, the F_1 generation consists of flies with wild-type eye color. In the F_2 generation wild, scarlet, brown, and white-eyed flies are found in a 9:3:3:1 ratio. While this ratio is numerically the same as Mendel's dihybrid ratio, the *Drosophila* cross involves only one character, eye color. This is an important distinction to make as modified dihybrid ratios resulting from "gene interaction" are studied.

This cross is an excellent example of gene interaction because the biochemical basis of eye color in this organism has been determined (Figure 4.10). *Drosophila*, as a typical arthropod, has compound eyes made up of hundreds of individual visual units called ommatidia. The wild-type eye color is due to the deposition and mixing of two separate pigments in each ommatidium—the bright-red pigment **drosopterin** and the brown pigment **xanthommatin**. Each pigment is produced by a separate biosynthetic pathway. Each step of each pathway is catalyzed by a separate enzyme and is thus under the control of a separate gene. As shown in Figure 4.10, the *brown* mutation, when homozygous, interrupts the pathway leading to the synthesis of the bright-red pigment. Because only xanthommatin pigments are present, the eye is brown. The *scarlet* mutation, affecting a gene located on a separate autosome, interrupts the pathway leading to the synthesis of the brown xanthommatins and renders the eye color bright red in homozygous mutant flies. Each mutation apparently causes the production of a nonfunctional enzyme. Flies that are double mutants and thus homozygous for both *brown* and *scarlet* lack both functional enzymes and can make neither of the pigments; they represent the novel white-eyed flies appearing in 1/16 of the F_2 generation.

Other Modified Dihybrid Ratios

Cases (5–8) in Figure 4.8 illustrate additional modifications of the dihybrid ratio and provide still other examples of gene interactions. All eight cases have two things in common. First, in arriving at a suitable explanation of the inheritance pattern of each one, we have not violated the principles of segregation and independent assortment. Therefore, the added complexity of inheritance in these examples does not detract from the validity of Mendel's conclusions. Second, the F_2 phenotypic ratio in each

FIGURE 4.9 Summer squash exhibiting various fruit-shape phenotypes, where disc (white), long (orange gooseneck), and sphere (bottom left) are apparent.

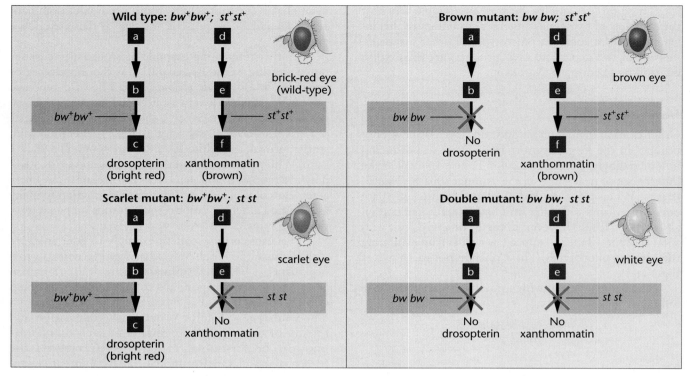

FIGURE 4.10 A theoretical explanation of the biochemical basis of the four eye-color phenotypes produced in a cross between *Drosophila* with brown eyes and scarlet eyes. In the presence of at least one wild-type bw^+ allele, an enzyme is produced that converts substance b to c, and the pigment drosopterin is synthesized. In the presence of at least one wild-type st^+ allele, substance e is converted to f, and the pigment xanthommatin is synthesized. The homozygous presence of the recessive bw and st mutant alleles blocks the synthesis of these respective pigment molecules. Either one, both, or neither of these pathways can be blocked, depending on the genotype.

example has been expressed in sixteenths. Therefore, when geneticists observe a similar ratio, in crosses where the inheritance pattern is unknown, it is likely that two gene pairs are controlling the observed phenotypes. You should make the same inference in the analysis of genetics problems. Other insights into solving genetics problems are provided in the "Insights and Solutions" section at the conclusion of this chapter.

Complementation analysis can determine whether or not two similar mutations are alleles

An interesting situation arises when two mutations, both of which produce a similar phenotype, are isolated independently. For example, suppose that two investigators, one in a genetics laboratory in the United States and the other in a genetics laboratory in Canada, independently isolated and established a true-breeding strain of wingless *Drosophila* and demonstrated that each was due to a recessive mutation. We might assume that both strains contained mutations in the same gene. However, since we

know that many genes are involved in the formation of wings, mutations in any one of them might inhibit wing formation during development. The experimental approach called **complementation analysis** allows us to determine whether two such mutations are in the same gene—that is, to determine if they are alleles—or if they represent mutations in separate genes.

The analysis seeks to answer this simple question: *Are two mutations that yield similar phenotypes present in the same gene or in two different genes?* To find the answer, we cross the two mutant strains together and analyze the F₁ generation. There are two alternative outcomes and interpretations of such a cross, as illustrated in Figure 4.11 and discussed below. The mutation isolated in the U.S. is designated m^{usa}, while the mutation isolated in Canada is designated m^{can}.

Case 1. *All offspring develop normal wings.*

Interpretation: The two recessive mutations are in separate genes and are not alleles of one another. Following the cross, all F₁ flies are heterozygous for both genes. **Complementation** occurs. Because each mutation is in a separate gene and each F₁ fly is heterozygous at both loci, the normal products of both

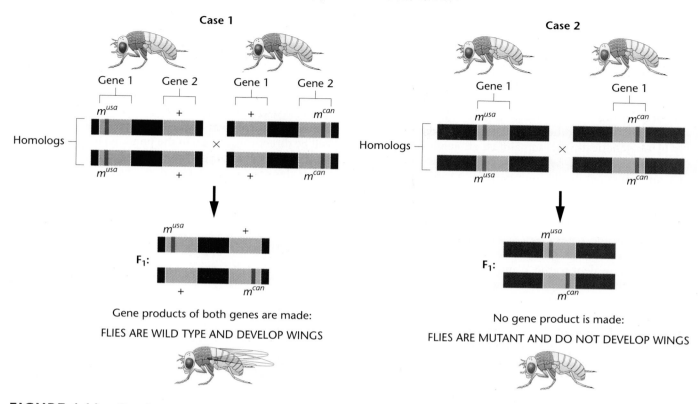

FIGURE 4.11   Complementation analysis where two wingless mutants are investigated to determine if the responsible mutations are in separate genes (Case 1) or if they are alleles of the same gene (Case 2).

genes are produced (by the one normal copy of each gene). Wings develop.

Case 2. *All offspring fail to develop wings.*

Interpretation: The two mutations affect the same gene and are alleles of one another. *No complementation* occurs. Because the two mutations affect the same gene, the F_1 flies are homozygous for the two mutant alleles (one is the m^{usa} allele and the other is the m^{can} allele). No normal product of the gene is produced. In the absence of this essential product, wings do not form.

Complementation analysis, as originally devised by the *Drosophila* geneticist Edward B. Lewis, is often called the ***cis-trans* test.** *Cis* (along side of one another) refers to the case where the two mutations are on the same homolog; *trans* (opposite one another) refers to the case where the mutations are on separate homologs. These designations are borrowed from nomenclature used in organic chemistry. As shown in Figure 4.11, the *trans* configuration is critical in determing whether or not the two mutations are alleles of the same gene. If we determine that the two mutations are indeed alleles, their presence as heterozygotes in the *cis* configuration serves as an important control in complementation analyses. In this configuration, flies will develop wings.

Complementation analysis may be used to screen any number of individual mutations that result in the same phenotype. Such an analysis may reveal that only a single gene is involved, or that two or more genes are involved. All mutations determined to be present in any single gene are said to fall into the same **complementation group** and to *complement* mutations in all other groups.

If large numbers of mutations affecting the same trait are available and studied using complementation analysis, it is possible to predict the total number of genes involved in the determination of that trait. We will return to the use of complementation analysis in Chapter 17, where we discuss the inherited human disorder xeroderma pigmentosum. There, complementation analysis reveals that mutations in any of seven complementation groups (genes) lead to the disorder.

X-linkage describes genes on the X chromosome

In many animal and some plant species, one of the sexes contains a pair of unlike chromosomes that are involved in sex determination. In many cases, these are designated as the X and Y. For example, in both *Drosophila* and humans, males contain an X and a Y chromosome, whereas females contain two X chromosomes. The Y chromosome must contain a region of pairing homology with the X chromosome if the two are to synapse and segregate during meiosis. For example, on the human X and Y, two such areas have been identified, one on each end of both chromosomes, referred to as **pseudoautosomal regions (PARs).** However, the remainder of the Y chromosome in humans as well as other species is considered to be largely blank genetically, lacking almost all genes present on the X chromosome. As a result, genes present on the X chromosome exhibit unique patterns of inheritance in comparison to autosomal genes. The term **X-linkage** is used to describe such situations.

We shall return to the consideration of the role of the X and Y chromosomes in sex determination in Chapter 9. There, we will also learn that some organisms lack a chromosome equivalent to the Y and that it is not always the case that the male has unlike sex-determining chromosomes. We will also see that modern molecular techniques have allowed the identification of a few human Y-linked genes that have counterparts on the X chromosome. Below, however, we will focus on genes present on the X but absent from the Y chromosome, causing a modification of Mendelian ratios, the central theme of this chapter.

X-Linkage in Drosophila

One of the first cases of X-linkage was documented in 1910 by Thomas H. Morgan during his studies of the *white* eye mutation in *Drosophila*. We will use this case to illustrate X-linkage. The normal wild-type red eye color is dominant to white eye color.

Morgan's work established that the inheritance pattern of the white-eye trait was clearly related to the sex of the parent carrying the mutant allele. Unlike the outcome of the typical Mendelian monohybrid cross where F_1 and F_2 data were very similar regardless of which P_1 parent exhibited the recessive mutant trait, reciprocal crosses between white-eyed and red-eyed flies did not yield identical results. Morgan's analysis led to the conclusion that the *white* locus is present on the X chromosome rather than on one of the autosomes; and it is therefore said to be X-linked.

Results of reciprocal crosses between white-eyed and red-eyed flies are shown in Figure 4.12. The obvious differences in phenotypic ratios in both the F_1 and F_2 generations are dependent on whether or not the P_1 white-eyed parent was male or female.

Morgan was able to correlate these observations with the difference found in the sex chromosome composition between male and female *Drosophila*. He hypothesized that in males with white eyes the recessive allele for white eye is found on the X chromosome, but its corresponding locus is absent from the Y chromosome. Females thus have two

available gene sites, one on each X chromosome, while males have only one available gene site on their single X chromosome.

Morgan's interpretation of X-linked inheritance, shown in Figure 4.13, provides a suitable theoretical explanation for his results. Because the Y chromosome lacks homology with most genes on the X chromosome, whatever alleles are present on the X chromosome of the males will be directly expressed in the phenotype. As males cannot be either homozygous or heterozygous for X-linked genes, this condition is referred to as **hemizygous.** In such

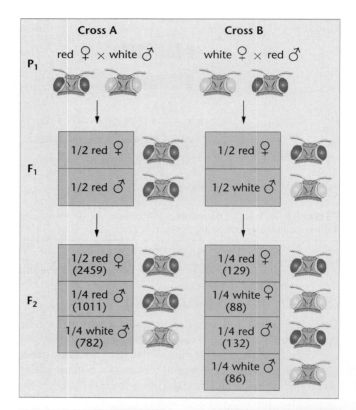

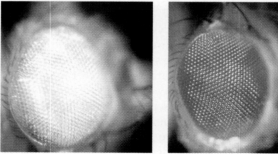

FIGURE 4.12 The F_1 and F_2 results of T. H. Morgan's reciprocal crosses involving the X-linked *white* mutation in *Drosophila melanogaster*. The actual F_2 results are shown in parentheses. The photograph contrasts white eyes with the brick-red wild-type eye color.

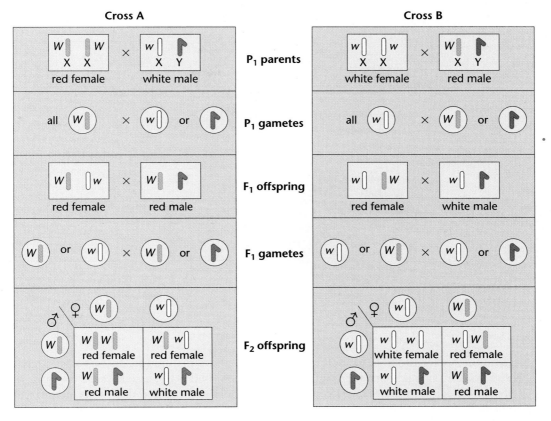

FIGURE 4.13   The chromosomal explanation of the results of the X-linked crosses shown in Figure 4.12.

cases, no alternative alleles are present, and the concept of dominance and recessiveness is irrelevant.

One result of X-linkage is the **crisscross pattern of inheritance,** whereby phenotypic traits controlled by recessive X-linked genes are passed from homozygous mothers to all sons. This pattern occurs because females exhibiting a recessive trait must contain the mutant allele on both X chromosomes. Because male offspring receive one of their mother's two X chromosomes and are hemizygous for all alleles present on that X, all sons will express the same recessive X-linked traits as their mother. This pattern of inheritance is apparent in Figure 4.14(b).

In addition to documenting the phenomenon of X-linkage, Morgan's work has taken on great historical significance. By 1910, the correlation between Mendel's work and the behavior of chromosomes during meiosis had provided the basis for the **chromosome theory of inheritance,** as postulated by Sutton and Boveri (see Chapter 3). Morgan's work, and subsequently that of his student Calvin Bridges, providing direct evidence that genes are transmitted on specific chromosomes, is considered the first solid experimental evidence in support of this theory. In

the following two decades, further research provided indisputable evidence in support of the chromosome theory of inheritance.

X-Linkage in Humans

In humans, many genes and the traits controlled by them are recognized as being linked to the X chromosome. These X-linked traits can be easily identified in pedigrees because of the crisscross pattern of inheritance. A pedigree for one form of human color blindness is shown in Figure 4.14. The mother in generation I passes the trait to all her sons but to none of her daughters. If the offspring in generation II marry normal individuals, the color-blind sons will produce all normal male and female offspring (III-1, 2, and 3); the normal-vision daughters will produce normal-vision female offspring (III-4, 6, and 7), as well as color-blind (III-8) and normal-vision (III-1 and 5) male offspring.

Many X-linked human genes have now been identified, as shown in Table 4.4. For example, the genes controlling two forms of hemophilia and two forms of muscular

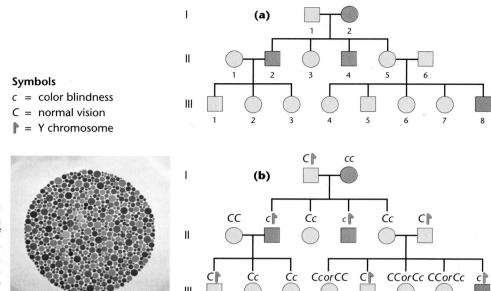

Symbols

c = color blindness
C = normal vision
⚑ = Y chromosome

FIGURE 4.14 (a) A human pedigree of the X-linked color blindness trait. (b) The most probable genotypes of each individual in the pedigree. The photograph is of an Ishihara color blindness chart. Red-green color blind individuals see a 3 rather than the 8 visualized by those with normal color vision.

dystrophy are located on the X chromosome. In addition, numerous genes whose expression yields enzymes are X-linked. Glucose-6-phosphate dehydrogenase and hypoxanthine-guanine-phosphoribosyl transferase are two examples. In the latter case, the severe Lesch–Nyhan syndrome results from the mutant form of the X-linked gene product.

Because of the way in which X-linked genes are transmitted, unusual circumstances may be associated with recessive X-linked disorders in comparison to recessive autosomal disorders. For example, if an X-linked disorder debilitates or is lethal to the affected individual prior to

reproductive maturation, the disorder occurs exclusively in males. This is the case because the only sources of the lethal allele in the population are heterozygous females who are "carriers" and do not express the disorder. They pass the allele to one-half of their sons, who develop the disorder because they are hemizygous but who rarely, if ever, reproduce. Heterozygous females also pass the allele to one-half of their daughters, who become carriers but do not develop the disorder. An example of such an X-linked disorder is the Duchenne form of muscular dystrophy. The disease has an onset prior to age 6 and is often lethal prior to age 20. It normally occurs only in males.

TABLE 4.4 Human X-linked Traits

Condition	Characteristics
Color blindness, deutan type	Insensitivity to green light.
Color blindness, protan type	Insensitivity to red light.
Fabry's disease	Deficiency of galactosidase A; heart and kidney defects, early death.
G-6-PD deficiency	Deficiency of glucose-6-phosphate dehydrogenase, severe anemic reaction following intake of primaquines in drugs and certain foods, including fava beans.
Hemophilia A	Classical form of clotting deficiency; lack of clotting factor VIII.
Hemophilia B	Christmas disease; deficiency of clotting factor IX.
Hunter syndrome	Mucopolysaccharide storage disease resulting from iduronate sulfatase enzyme deficiency; short stature, clawlike fingers, coarse facial features, slow mental deterioration, and deafness.
Ichthyosis	Deficiency of steroid sulfatase enzyme; scaly dry skin, particularly on extremities.
Lesch–Nyhan syndrome (HGPRT)	Deficiency of hypoxanthine-guanine phosphoribosyl transferase enzyme leading to motor and mental retardation, self-mutilation, and early death.
Muscular dystrophy (Duchenne type)	Progressive, life-shortening disorder characterized by muscle degeneration and weakness; sometimes associated with mental retardation; deficiency of the protein dystrophin.

In sex-limited and sex-influenced inheritance, an individual's sex influences the phenotype

In still other instances, inheritance is affected by the sex of an individual, though not necessarily by genes on the X chromosome. There are numerous examples in different organisms where the sex of the individual plays a determining role in the expression of certain phenotypes. In **sex-limited inheritance**, the expression of a specific phenotype is absolutely limited to one sex. In **sex-influenced inheritance**, the sex of an individual influences the way a phenotype is expressed.

In both types of inheritance, autosomal genes are responsible for the existence of contrasting phenotypes, but the expression of these genes is dependent on the hormone constitution of the individual. Thus, the heterozygous genotype may exhibit one phenotype in males and the contrasting one in females. In domestic fowl, for example, tail and neck plumage is often distinctly different in males and females (Figure 4.15), demonstrating sex-limited inheritance. Cock-feathering is longer, more curved, and pointed, whereas hen-feathering is shorter and more rounded. The inheritance of feather type is due to a single pair of autosomal alleles. Expression of the alleles is modified by the individual's sex hormones to produce different phenotypes.

As shown below, hen-feathering is due to a dominant allele, *H*. Even females homozygous for the recessive *h* allele, however, remain hen-feathered. Only in males does the *hh* genotype result in cock-feathering.

Genotype	Phenotype ♀	♂
HH	Hen-feathered	Hen-feathered
Hh	Hen-feathered	Hen-feathered
hh	Hen-feathered	Cock-feathered

In certain breeds of fowl, the hen-feathering or cock-feathering allele has become fixed in the population. In the Leghorn breed, all individuals are of the *hh* genotype; as a result, male plumage differs from female plumage. Sebright bantams are all *HH*, showing no sexual distinction in feathering.

Cases of sex-influenced inheritance include horn formation in sheep, certain coat patterns in cattle, and **pattern baldness** in humans, where the hair is very thin on the top of the head (Figure 4.16). Pattern baldness is inherited in the following way:

Genotype	Phenotype ♀	♂
BB	Bald	Bald
Bb	Not bald	Bald
bb	Not bald	Not bald

FIGURE 4.15   Hen-feathering (left) versus cock-feathering (right) in domestic fowl. The feathers in the hen are shorter and less curved.

Even though females may display pattern baldness, this phenotype is much more prevalent in males. When females do inherit the *BB* genotype, the phenotype is much less pronounced than in males and is expressed later in life.

Phenotypic expression is not always a direct reflection of the genotype

Thus far in the text, we have assumed that the phenotype expressed by an organism directly reflects the genotype. For example, peas homozygous for the recessive *w* allele

FIGURE 4.16   Pattern baldness, a sex-influenced autosomal trait in humans.

(*ww*) will always be wrinkled. We have discussed gene expression as though the genes operate in a closed "black box" system in which the presence or absence of functional products directly determines the collective phenotype of an individual. The situation is actually much more complex. Most gene products function within the internal milieu of the cell, and cells interact with one another in various ways. Further, the organism must survive under diverse environmental influences. Thus, gene expression and the resultant phenotype are often modified through the interaction between an individual's particular genotype and the internal and external environment.

The degree of environmental influence can vary from inconsequential to subtle to very strong. Subtle interactions are the most difficult to detect and document, and they have led to "nature vs. nurture" questions in which scientists debate the relative importance of genes versus environment—usually inconclusively. In this final section of this chapter, we will deal with some of the variables known to modify gene expression.

Penetrance and Expressivity

Some mutant genotypes are always expressed as a distinct phenotype, whereas others produce a proportion of individuals whose phenotypes cannot be distinguished from normal (wild type). The degree of expression of a particular trait may be studied quantitatively by determining the penetrance and expressivity of the genotype under investigation. The percentage of individuals that show at least some degree of expression of a mutant genotype defines the **penetrance** of the mutation. For example, the phenotypic expression of many mutant alleles in *Drosophila* may overlap with wild type. If 15 percent of mutant flies show the wild-type appearance, the mutant gene is said to have a penetrance of 85 percent.

By contrast, **expressivity** reflects the *range of expression* of the mutant genotype. Flies homozygous for the recessive mutant gene *eyeless* yield phenotypes that range from the presence of normal eyes to a partial reduction in size to the complete absence of one or both eyes (Figure 4.17). Although the average reduction of eye size is one-fourth to one-half, expressivity ranges from complete loss of both eyes to completely normal eyes.

Examples such as the expression of the *eyeless* gene have provided the basis for experiments to determine the causes of phenotypic variation. If the laboratory environment is held constant and extensive variation is still observed, other genes may be influencing or modifying the eyeless phenotype. On the other hand, if the genetic background is not the cause of the phenotypic variation, environmental factors such as temperature, humidity, and nutrition may be involved. In the case of the eyeless phe-

notype, experiments have shown that both genetic background and environmental factors influence its expression.

Genetic Background: Suppression and Position Effects

Though it is difficult to assess the specific effect of the **genetic background** and the expression of a gene responsible for determining a potential phenotype, two effects of genetic background have been well characterized.

First, the expression of other genes throughout the genome may have an effect on the phenotype produced by the gene in question. The phenomenon of **genetic suppression** is an example. Mutant genes such as *suppressor of sable* (*su-s*), *suppressor of forked* (*su-f*), and *suppressor of Hairy-wing* (*su-Hw*) in *Drosophila* completely or partially restore the normal phenotype in an organism that is homozygous (or hemizygous) for the *sable*, *forked*, and *Hairy-wing* mutations, respectively. For example, flies hemizygous for both *forked* (a bristle mutation) and *su-f* have normal bristles. In each case, the suppressor gene causes the complete reversal of the expected phenotypic expression of the original mutation. Suppressor genes are excellent examples of the genetic background modifying primary gene effects.

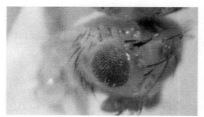

FIGURE 4.17 Variable penetrance as illustratrated by the expression of the *eyeless* mutation in *Drosophila*. Gradations in phenotype range from wild type to partial reduction to eyeless.

Second, the physical location of a gene in relation to other genetic material may influence its expression. Such a situation is called a **position effect.** For example, if a region of a chromosome is relocated or rearranged (called a translocation or inversion event), normal expression of genes in that chromosomal region may be modified. This is particularly true if the gene is relocated to or near certain areas of the chromosome that are prematurely condensed and genetically inert, referred to as **heterochromatin.**

An example of a position effect involves female *Drosophila* heterozygous for the X-linked recessive eye-color mutant *white* (*w*). The w^+/w genotype normally results in a wild-type brick-red eye color. However, if the region of the X chromosome containing the wild-type w^+ allele is translocated so that it is close to a heterochromatic region, expression of the w^+ allele is modified. Instead of having a red color, the eyes are variegated, or mottled with red and white patches (Figure 4.18). Therefore, following translocation, the dominant effect of the normal w^+ allele is reduced. A similar position effect is produced if a heterochromatic region is relocated next to the *white* locus on the X chromosome. Apparently, heterochromatic regions inhibit the expression of adjacent genes. Loci in many other organisms also exhibit position effects, providing proof that alteration of the normal arrangement of genetic information can modify its expression.

Temperature Effects

Because chemical activity depends on the kinetic energy of the reacting substances, which in turn depends on the surrounding temperature, we can expect temperature to influence phenotypes. One example is the evening primrose, which produces red flowers when grown at 23°C and white flowers when grown at 18°C. An even more striking example is seen in Siamese cats and Himalayan rabbits, which

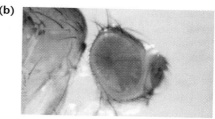

FIGURE 4.18   Position effect, as illustrated in the eye phenotype in two female *Drosophila* heterozygous for the gene *white*. (a) Normal dominant phenotype showing brick-red eye color. (b) Variegated color of an eye caused by rearrangement of the *white* gene to another location in the genome.

exhibit dark fur in certain body regions where the body temperature is slightly cooler, particularly the nose, ears, and paws (Figure 4.19). In these cases, it appears that the wild-type enzyme responsible for pigment production is functional at the lower temperatures present in the extremities, but it loses its catalytic function at the slightly higher temperatures found throughout the rest of the body.

Mutations that are affected by temperature are said to be **conditional** and are called **temperature sensitive.** Examples are known in viruses and a variety of organisms, including bacteria, fungi, and *Drosophila*. In extreme cases, an organism carrying a mutant allele may express a mutant phenotype

FIGURE 4.19   (a) A Himalayan rabbit. (b) A Siamese cat. Both show dark fur color on the the snout, ears, and paws. These patches are due to expression of a temperature-sensitive allele responsible for pigment production at the lower temperatures of the extremities but which is inactive at slightly higher temperatures.

when grown at one temperature, but express the wild-type phenotype when reared at another temperature. This type of temperature effect is useful in studying mutations that interrupt essential processes during development and are thus normally detrimental to the organism. For example, if bacterial viruses carrying certain temperature-sensitive mutations infect bacteria cultured at 42°C, known as the *restrictive condition*, infection progresses up to the point where the essential gene product is required (e.g., for viral assembly) and then arrests. If cultured under *permissive conditions* of 25°C, the gene product is functional, infection proceeds normally, and new viruses are produced. The use of temperature-sensitive mutations, which may be induced and isolated, has added immensely to the study of viral genetics.

Nutritional Effects

Another example of conditional mutations involves nutrition. In microorganisms, mutations that prevent synthesis of nutrient molecules are quite common. These **nutritional mutants** arise when an enzyme essential to a biosynthetic pathway becomes inactive. A microorganism bearing such a mutation is called an **auxotroph.** If the end product of a biochemical pathway can no longer be synthesized, and if that molecule is essential to normal growth and development, the mutation prevents growth and may be lethal. For example, if the bread mold *Neurospora* can no longer synthesize the amino acid leucine, proteins cannot be synthesized. If leucine is present in the growth medium, the detrimental effect is overcome. Nutritional mutants have been crucial to genetic studies in bacteria and also served as the basis for George Beadle and Edward Tatum's proposal, in the early 1940s, that one gene functions to produce one enzyme (see Chapter 14).

A slightly different set of circumstances exists in humans. The presence or absence of certain dietary substances, which normal individuals may consume without harm, can adversely affect individuals with abnormal genetic constitutions. Often, a mutation may prevent an individual from metabolizing some substance commonly found in normal diets. For example, those afflicted with the genetic disorder **phenylketonuria** cannot metabolize the amino acid phenylalanine. Those with **galactosemia** cannot metabolize galactose. However, if the dietary intake of the molecule is drastically reduced or eliminated, the associated phenotype may be reversed.

The fairly common case of **lactose intolerance,** in which individuals are intolerant of the milk sugar lactose, illustrates the general principles involved. Lactose is a disaccharide consisting of a molecule of glucose linked to a molecule of galactose and makes up 7 percent of human milk and 4 percent of cow's milk. To metabolize lactose, humans require the enzyme **lactase,** which cleaves the disaccharide. Adequate amounts of lactase are produced during the first few years after birth.

However, in many people, the level of this enzyme soon drops drastically, and as adults they become intolerant of milk. The major phenotypic effect involves severe intestinal diarrhea, flatulence, and abdominal cramps. This condition (while not limited to these groups), is particularly prevalent in Eskimos, Africans, Asians, and Americans with these heritages. In these cultures, milk is often converted to cheese, butter, and yogurt where the amount of lactose is reduced significantly and adverse effects can be nearly eliminated. In the United States, milk low in lactose is commercially available, and to aid in the digestion of other lactose-containing foods, lactase is now a commercial product that can be ingested.

Onset of Genetic Expression

Not all genetic traits are expressed at the same time during an organism's life span. In most cases, the age at which a gene is expressed corresponds to the normal sequence of growth and development. In humans, the prenatal, infant, preadult, and adult phases require different genetic information. Similarly, many genetic disorders can be expected to manifest themselves at different stages of life.

In humans, many severe inherited disorders are often not manifested until after birth. For example, **Tay–Sachs disease,** inherited as an autosomal recessive, is a lethal lipid-metabolism disease involving an abnormal enzyme, hexosaminidase A. Newborns appear phenotypically normal for the first few months. Then, developmental retardation, paralysis, and blindness ensue, and most affected children die by the age of three.

The **Lesch–Nyhan syndrome,** inherited as an X-linked recessive disease, is characterized by abnormal nucleic acid metabolism (biochemical salvage of nitrogenous purine bases), leading to the accumulation of uric acid in blood and tissues, mental retardation, palsy, and self-mutilation of the lips and fingers. The disorder is due to a mutation in the gene encoding hypoxanthine-guanine phosphoribosyl transferase (HPRT). Newborns are normal for six to eight months prior to the onset of the first symptoms. Still another example involves **Duchenne muscular dystrophy (DMD),** an X-linked recessive disorder associated with progressive muscular wasting. It is not usually diagnosed until the age of 3 to 5 years. Even with modern medical intervention, the disease is often fatal in the early 20s.

Perhaps the most variable of all inherited human disorders regarding age of onset is **Huntington disease.** Inherited as an autosomal dominant, Huntington disease affects the frontal lobes of the cerebral cortex, where progressive cell death occurs over a period of more than a decade. Brain deterioration is accompanied by spastic uncontrolled movements, intellectual and emotional deterioration, and ultimately death. While onset has been reported at all ages, it most frequently occurs between ages 30 and 50, with a mean onset of 38 years.

Conditions such as these support the concept that the critical expression of normal genes varies throughout the life cycle of organisms, including humans. Gene products may play more essential roles at certain times. Further, it is likely that the internal physiological environment of an organism changes with age.

Genetic Anticipation

Interest in studying the genetic onset of phenotypic expression has intensified with the discovery of heritable disorders that exhibit a progressively earlier age of onset and an increased severity of the disorder in each successive generation. This general phenomenon is referred to as **genetic anticipation.**

Myotonic dystrophy (DM), the most common type of adult muscular dystrophy, clearly illustrates genetic anticipation. Individuals afflicted with this autosomal dominant disorder exhibit extreme variation in the severity of symptoms. Mildly affected individuals develop cataracts as adults but have little or no muscular weakness. Severely affected individuals demonstrate more extensive myopathy and may be mentally retarded. In its most extreme form, the disease is fatal just after birth. A great deal of excitement was generated in 1989 when C. J. Howeler and colleagues confirmed the correlation of increased severity with earlier onset. They studied 61 parent–child pairs that expressed the disorder and, in 60 of the cases, age of onset was earlier in the child than in his or her parent.

In 1992 an explanation was put forward to explain both the molecular cause of the mutation responsible for DM as well as the basis of genetic anticipation. As we will see later, a particular region of the DM gene is repeated a variable number of times and is unstable. Normal individuals average about five copies of this region, minimally affected individuals reveal about 50 copies, and severely affected individuals demonstrate over 1000 copies. The most remarkable observation was that, in successive generations, the size of the repeated segment increases. Although it is not yet clear how the expansion in size affects onset and phenotypic expression, the correlation is extremely strong. Several other inherited human disorders, including the fragile-X syndrome, Kennedy disease, and Huntington disease also reveal an association between the size of specific regions of the responsible gene and disease severity. We will return to this general topic when we discuss the molecular explanation in detail in Chapter 17.

Genomic (Parental) Imprinting

Our final example of modification of the laws of Mendelian inheritance involves the variation of phenotypic expression depending strictly on the parental origin of the chromosome carrying a particular gene, a phenomenon called **genomic** (or **parental**) **imprinting.** In some species, certain chromosomal regions and the genes contained within them somehow retain a memory, or an "imprint," of their parental origin that influences whether specific genes either are expressed or remain genetically silent (i.e., are not expressed).

The imprinting step is thought to occur before or during gamete formation, leading to differentially marked genes (or chromosome regions) in sperm-forming versus egg-forming tissues. The process is clearly different from mutation because the imprint can be reversed in succeeding generations as genes pass from mother to son to granddaughter, and so on.

An example of imprinting involves the inactivation of one of the X chromosomes in mammalian females. In mice, prior to development of the embryo, imprinting occurs in tissues such that the X chromosome of paternal origin is genetically inactivated in all cells while the genes on the maternal X chromosome remain genetically active. As embryonic development is subsequently initiated, the imprint is "released" and random inactivation of either the paternal *or* the maternal X chromosome occurs.

In 1991, more specific information established that three specific mouse genes undergo imprinting. One is the gene encoding insulinlike growth factor II (*Igf2*). A mouse that carries two nonmutant alleles of this gene is normal in size, whereas a mouse that carries two mutant alleles lacks a growth factor and is a dwarf. The size of a heterozygous mouse (one allele normal and one mutant; see Figure 4.20) depends on the parental origin of the normal

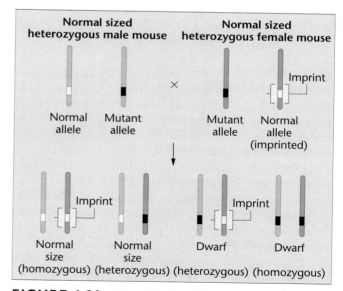

FIGURE 4.20   The effect of imprinting on the mouse *Igf2* gene, which produces dwarf mice in the homozygous condition. Heterozygous offspring that receive the normal allele from their father are normal in size. Heterozygotes that receive the normal allele from their mother, which has been imprinted, are dwarf.

Improving the Genetic Fate of Purebred Dogs

For a dog-lover, there is nothing quite so heart-breaking as watching a dog slowly go blind, standing by helplessly as he struggles to adapt to a life of perpetual darkness. That's what happens in progressive retinal atrophy (PRA), an inherited disorder first described in Gordon setters in 1911. Since then, PRA has been found in many other breeds of dogs, including Irish setters, border collies, Norwegian elkhounds, toy poodles, miniature schnauzers, cocker spaniels, and Siberian huskies.

The products of many genes are required for the development and maintenance of a healthy retina, and a defect in any one of them has the potential to cause retinal dysfunction. Decades of research have led to the identification of four such genes (*rcd1, rcd2, erd,* and *prcd*), and more are likely to be discovered. Different genes are mutated in different breeds (e.g., *rcd1* in Irish setters and *prcd* in Labrador retrievers). In all dogs examined so far, PRA shows a recessive pattern of inheritance.

Whichever mutation is responsible, PRA is more common in certain purebred dogs than in mixed breeds. The development of distinct breeds of dogs has involved the intensive selection for desirable attributes, whether a particular size, shape, color, or behavior. Since most desired characteristics are determined by recessive alleles, the fastest way to increase the homozygosity of these alleles and establish the characteristics in the population is to mate close relatives, which are likely to carry the same alleles. In the practice known as line breeding, for example, dogs may be mated to a cousin or a grandparent.

Unfortunately, the generations of inbreeding that have established favorable characteristics in pure breeds have also increased the homozygosity of certain harmful recessive alleles such as in PRA, resulting in other inherited diseases. Many breeds are plagued with inherited hip dysplasia, which is particularly prevalent in German shepherds. Deafness and kidney disorders are common genetic maladies in dalmatians. Over 300 such diseases have been characterized in purebred dogs, and most breeds of dogs are affected by one or more of them. By some estimates, fully 25 percent of purebred dogs are affected with one genetic ailment or another. Inbreeding is not the cause of these genetic diseases, but there is general agreement that breeding practices, if not executed properly, can increase the frequency of disease.

Inbreeding is the likely explanation for the elevated frequency of PRA in certain breeds. Fortunately, advances in canine genetics are beginning to provide new tools for the breeding of

allele. The mouse is normal in size if the normal allele came from the father, but dwarf if the normal allele came from the mother. From this we can deduce that the normal *Igf 2* gene is imprinted to function poorly during the course of egg production in females but functions normally when it has passed through sperm-producing tissue in males.

Imprinting continues to depend on whether the gene passes through sperm-producing or egg-forming tissue leading to the next generation. For example, a heterozygous normal-sized male will donate to half his offspring a "normal-functioning" wild-type allele that will counteract a mutant allele received from the mother.

In humans, two distinct genetic disorders are thought to be caused by differential imprinting of the same region of chromosome 15 (15q1). In both cases, the disorders *appear* to be due to an identical deletion of this region in one member of the chromosome 15 pair. The first disorder, **Prader–Willi syndrome (PWS),** results when only an undeleted maternal chromosome remains. If only an undeleted paternal chromosome remains, an entirely different disorder, **Angelman syndrome (AS),** results.

The two conditions are clearly different phenotypically. PWS entails mental retardation as well as a severe eating disorder marked by an uncontrollable appetite, obesity, and diabetes. AS involves distinct behavioral manifestations as well as mental retardation. We can conclude that region 15q1 is imprinted differently in male versus female gametes and that both a maternal and paternal region are required for normal development.

Many questions remain unanswered in the area of genomic imprinting. It is not known how many genes are subject to imprinting, nor its developmental role. While it appears that regions of chromosomes rather than specific genes are imprinted, the molecular mechanism of imprinting is still a matter for conjecture. It has been hypothesized that DNA methylation may be involved. In

healthy dogs. Since 1995, a genetic test has been available to identify mutations in the *rcd1* gene, which is responsible for the form of PRA that occurs in Irish setters. This test is now being used to identify heterozygous carriers of *rcd1* mutations, dogs that show no symptoms of PRA but, if mated with other carriers, pass the trait on to about 25 percent of their offspring. Eliminating PRA carriers from breeding programs could, in theory, eradicate this condition from Irish setters in just a few generations.

It took many years to identify *rcd1* and other genes responsible for PRA. In the future, the isolation of genes underlying canine inherited disease should be faster, thanks to the Dog Genome Project, a collaborative effort involving scientists at the University of California, the University of Oregon, the Fred Hutchinson Cancer Research Center in Seattle, and other research centers. Its goal is the creation of a complete genetic map of the 39 chromosomes in the dog. In late 1997, preliminary stages were completed, paving the way for the development of a more comprehensive map. The identification of genes that cause inherited disease will allow genetic tests for the diagnosis of disease before symptoms develop, as well as for ensuring that breeding animals are free of harmful recessive alleles.

The Dog Genome Project may have benefits for humans beyond the reduction of disease in their canine companions. This is because about 85 percent of the genes in the dog genome have equivalents in humans, so the identification of a disease-causing dog gene may be a short-cut to the isolation of the corresponding gene in humans. For example, some forms of PRA in dogs appear to be equivalent to retinitis pigmentosa (RP) in humans, which afflicts about 1.5 million people worldwide. Despite much research, RP remains poorly understood, and current treatments only slow its progress. Understanding the genetic basis of PRA could lead to breakthroughs in the diagnosis and treatment of RP, potentially saving the sight of thousands of people every year. By contributing to the cure of human diseases, dogs may prove to be "man's best friend" in an entirely new way.

REFERENCES

BERSON, E.L. 1996. Retinitis pigmentosum: Unfolding its mystery. *Proc. Natl. Acad. Sci. USA* 93:4526–28.

RAY, K., BALDWIN, V., ACLAND, G., and AQUIRRE, G. 1995. Molecular diagnostic tests for ascertainment of genotype at the rod cone dysplasia (*rcd1* locus) in Irish setters. *Curr. Eye Res.* 14:243.

SMITH, C.A. 1994. New hope for overcoming canine inherited disease. *Am. J. Vet. Med. Assoc.* 204:41–46.

vertebrates, methyl groups can be added to the carbon atom at position 5 in cytosine (see Chapter 11) as a result of the activity of the enzyme DNA methyl transferase. Methyl groups are added when the dinucleotide CpG or groups of CpG units (called CpG islands) are present along a DNA chain. DNA methylation is a reasonable mechanism for establishing a molecular imprint, since there is some evidence that a high level of methylation can inhibit gene activity and that active genes (or their regulatory sequences) are often undermethylated. Additionally, variation in methylation has been noted in the mouse genes that undergo imprinting. Whatever the cause of this phenomenon, it is a fascinating topic and one that will receive significant attention in future research studies.

CHAPTER SUMMARY

1. Ever since Mendel's work was rediscovered, the study of transmission genetics has been expanded to include many alternative modes of inheritance. In many cases, phenotypes may be influenced by two or more genes in a variety of ways.

2. Incomplete, or partial, dominance is exhibited when intermediate phenotypic expression of a trait occurs in a heterozygote.

3. Codominance is exhibited when distinctive expression of two alleles occurs in a heterozygous organism.

4. Since a diploid organism may host only two alternative alleles at any given locus, multiple allelism applies only to populations, where many alternative alleles of the same gene can occur.

5. Lethal mutations usually result in the inactivation or the lack of synthesis of gene products essential during development. Such mutations may be recessive or dominant. Some lethal genes, such as the gene causing Huntington disease, are not expressed until adulthood.

6. Phenotypes are often controlled by more than one gene, leading to a variety of modifications of the Mendelian dihybrid and trihybrid ratios. Such cases are often classified under the general heading of gene interaction.

7. Epistasis occurs when two or more genes influence a single characteristic and one of the genes masks or modifies the expression of the other gene or genes.

8. Complementation analysis allows the determination whether or not two mutations producing similar phenotypes are alleles.

9. Genes located on the X chromosome display a unique mode of inheritance referred to as X-linkage. Unique X-linked ratios result because hemizygous individuals (those with an X and a Y chromosome) express all alleles present on their X chromosome.

10. Both sex-limited and sex-influenced inheritance occur when the sex of the organism affects the phenotype under the control of an autosomal gene.

11. Phenotypic expression is not always the direct reflection of the genotype. Penetrance measures the percentage of organisms in a given population that exhibits evidence of the corresponding mutant phenotype. Expressivity, on the other hand, measures the range of phenotypic expression of a given genotype.

12. Phenotypic expression can be modified by genetic background, temperature, and nutrition. Genetic expression and position effects illustrate the existence of genetic background that affects phenotypic expression.

13. The time of onset of gene expression in organisms varies as the need for certain gene products occurs at different periods during development, growth, and aging.

14. Genetic anticipation refers to the phenomenon where the onset of phenotypic expression occurs earlier and becomes more severe in each ensuing generation.

15. Genomic imprinting is a process whereby a region of either the paternal or maternal chromosome is modified (marked or imprinted), thereby affecting phenotypic expression. Expression therefore depends upon which parent contributes a mutant allele.

INSIGHTS AND SOLUTIONS

Genetic problems take on an added complexity if they involve two independent characters and multiple alleles, incomplete dominance, or epistasis. The most difficult types of problems were those faced by pioneering geneticists in the laboratory. In these problems they had to determine the mode of inheritance by working backwards from the observations of offspring to parents of unknown genotype.

1. Consider the problem of comb shape inheritance in chickens, where walnut, rose, pea, and single are observed as distinct phenotypes. How is comb shape inherited, and what are the genotypes of the P_1 generation of each cross? Use the following data to answer these questions.

Cross 1:	single × single	→ all single
Cross 2:	walnut × walnut	→ all walnut
Cross 3:	rose × pea	→ all walnut
Cross 4:	F_1 × F_1 of Cross 3	
	walnut × walnut	→ 93 walnut
		28 rose
		32 pea
		10 single

Walnut

Pea

Rose

Single

Solution: This problem may initially appear quite difficult. To solve it, you must use a multistep process. First, carefully scrutinize the information about the parents and offspring. Look for a pattern of inheritance that is consistent with at least some of the data. Then, follow an empirical approach by formulating a hypothesis about the mode of inheritance and testing it against all of the data. It will become apparent if you have arrived at the correct solution when it is consistent with all outcomes. If there is an inconsistency between your solution (hypothesis) and the data, look through the information again and modify your hypothesis.

In this problem there are two immediately useful facts. First, in Cross 1, P_1 singles breed true. Second, while P_1 walnut breeds true (Cross 2), a walnut phenotype is also produced in crosses between rose and pea (Cross 3). When these F_1 walnuts are mated (Cross 4), all four comb shapes are produced in a ratio that approximates 9:3:3:1. This observation should immediately suggest a cross involving two gene pairs because the resulting data closely resemble the ratio of Mendel's dihybrid crosses. Because only one character is involved and discontinuous phenotypes occur (comb shape), perhaps some form of epistasis is occurring. This may serve as a working hypothesis, and we must now propose how the two gene pairs interact to produce each phenotype.

If we call the allele pairs A, a and B, b, we might predict that, since walnut represents 9/16 in Cross 4, A–B– will produce walnut. We might also hypothesize that in the case of Cross 2, the genotypes were $AABB \times AABB$ where walnut was seen to breed true. (Recall that A– and B– mean AA or Aa and BB or Bb, respectively.) Since single is the phenotype representing 1/16 of the offspring of Cross 4, we could predict that this phenotype is the result of the $aabb$ genotype. This is consistent with Cross 1.

Now we have only to determine the genotypes for rose and pea. A logical prediction would be that at least one dominant A or B allele combined with the double recessive condition of the other allele pair can account for these phenotypes. For example,

$$A\text{–}bb \rightarrow \text{rose}$$
$$aa\,B\text{–} \rightarrow \text{pea}$$

If, in Cross 3, $AAbb$ (rose) were crossed with aaBB (pea), all offspring would be AaBb (walnut). This is consistent with the data, and we must now look at Cross 4. We predict these walnut genotypes to be $AaBb$ (as before), and from the cross

$$AaBb \times AaBb$$
$$\text{(walnut)} \quad \text{(walnut)}$$

We expect

9/16 A–B– (walnut)
3/16 A–bb (rose)
3/16 $aa\,B$– (pea)
1/16 $aa\,bb$ (single)

Our prediction is consistent with the data. The initial hypothesis of the epistatic interaction of two gene pairs proves consistent throughout, and the problem has been solved.

This example illustrates the need to have a basic theoretical knowledge of transmission genetics. Then, you must search for the appropriate clues so that you can proceed in a stepwise fashion toward a solution.

2. In radishes, flower color may be red, purple, or white. The edible portion of the radish may be long or oval. When only flower color is studied, red $\times$ white yields all purple. If these F_1 purples are interbred, no dominance is evident, and the F_2 generation consists of 1/4 red:1/2 purple:1/4 white. Regarding radish shape, long is dominant to oval in a normal Mendelian fashion.

(a) Determine the F_1 and F_2 phenotypes from a cross between a true-breeding red long radish and one that is white oval. Be sure to define all gene symbols initially.

Solution: This is a modified dihybrid cross where the gene pair controlling color exhibits incomplete dominance. Shape is controlled conventionally. First, establish gene symbols:

$$RR = \text{red} \quad Rr = \text{purple} \quad rr = \text{white}$$
$$O\text{–} = \text{long} \quad oo = \text{oval}$$

P_1: $RROO$ $\times$ $rroo$
(red long) (white oval)

F_1: all $RrOo$ (purple long)
$F_1 \times F_1$: $RrOo \times RrOo$

F_2:	1/4 RR	3/4 O–	3/16 $RR\ O$–	red long
		1/4 oo	1/16 $RR\ oo$	red oval
	2/4 Rr	3/4 O–	6/16 $Rr\ O$–	purple long
		1/4 oo	2/16 $Rr\ oo$	purple oval
	1/4 rr	3/4 O–	3/16 $rr\ O$–	white long
		1/4 oo	1/16 $rr\ oo$	white oval

Note above that to generate the F_2 results, we have used the forked-line method. First, the outcome of crossing F_1 parents for the color genes is considered ($Rr \times Rr$). Then the outcome of shape is considered ($Oo \times Oo$).

(b) A red oval plant was crossed with a plant of unknown genotype and phenotype, yielding the data shown below. Determine the genotype and phenotype of the unknown plant.

Offspring: 103 red long:101 red oval:
98 purple long:100 purple oval

Solution: Since the two characters are inherited independently, consider them separately. The data indicate a 1/4:1/4:1/4:1/4 proportion. First, consider color:

P_1: red × ??? (unknown)
F_1: 204 red (1/2)
198 purple (1/2)

Because the red parent must be RR, the unknown must have a genotype of Rr to produce these results. It is thus purple. Now, consider shape:

P_1: oval × ??? (unknown)
F_1: 201 long (1/2)
201 oval (1/2)

Because the oval plant must be oo, the unknown plant must have a genotype of Oo to produce these results. It is thus long. The unknown plant is thus

$RrOo$ purple long

3. In humans, red–green color blindness is inherited as an X-linked recessive trait. A woman with normal vision whose father is color-blind marries a male who has normal vision. Predict the color vision of their male and female offspring.

Solution: The female is heterozygous since she inherited an X chromosome with the mutant allele from her father. Her husband is normal. Therefore, the parental genotypes are

$Cc \times C\Upsilon$ (↑ is the Y chromosome)

All female offspring are normal (CC or Cc). One-half of the male children will be color-blind ($c\Upsilon$), and the other half will have normal vision ($C\Upsilon$).

PROBLEMS AND DISCUSSION QUESTIONS

1. In shorthorn cattle, coat color may be red, white, or roan. Roan is an intermediate phenotype expressed as a mixture of red and white hairs. The following data were obtained from various crosses:

red × red ⟶ all red
white × white ⟶ all white
red × white ⟶ all roan
roan × roan ⟶ 1/4 red: 1/2 roan: 1/4 white

How is coat color inherited? What are the genotypes of parents and offspring for each cross?
2. Contrast incomplete dominance and codominance.
3. In foxes, two alleles of a single gene, P and p, may result in lethality (PP), platinum coat (Pp), or silver coat (pp). What ratio is obtained when platinum foxes are interbred? Is the P allele behaving dominantly or recessively in causing lethality? In causing platinum coat color?
4. In mice, a short-tailed mutant was discovered. When it was crossed to a normal long-tailed mouse, 4 offspring were short-tailed and 3 were long-tailed. Two short-tailed mice from the F_1 generation were selected and crossed. They produced 6 short-tailed and 3 long-tailed mice. These genetic experiments were repeated three times with approximately the same results. What genetic ratios are illustrated? Hypothesize the mode of inheritance and diagram the crosses.
5. List all possible genotypes for the A, B, AB, and O phenotypes. Is the mode of inheritance of the ABO blood types representative of dominance? Of recessiveness? Of codominance?
6. With regard to the ABO blood types in humans, determine the genotype of the male parent and female parent below:

Male Parent: Blood type B; mother type O
Female Parent: Blood type A; father type B

Predict the blood types of the offspring that this couple may have and the expected proportion of each.
7. In a disputed parentage case, the child is blood type O while the mother is blood type A. What blood type would exclude a male from being the father? Would the other blood types prove that a particular male was the father?
8. The A and B antigens in humans may be found in water-soluble form in secretions, including saliva, of some individuals (Se/Se and Se/se) but not in others (se/se). The population thus contains "secretors" and "nonsecretors."
   (a) Determine the proportion of various phenotypes (blood type and ability to secrete) in matings between individuals that are blood type AB and type O, both of whom are Se/se.
   (b) How will the results of such matings change if both parents are heterozygous for the gene controlling the synthesis of the H substance (Hh)?
9. Distinguish between epistasis and polygenic inheritance, and between discontinuous and continuous variation.

10. In rabbits, a series of multiple alleles controls coat color in the following way. C is dominant to all other alleles and causes full color. The chinchilla phenotype is due to the c^{ch} allele, which is dominant to all alleles other than C. The c^h allele, dominant only to c^a (albino), results in the Himalayan coat color. Thus, the order of dominance is $C > c^{ch} > c^h > c^a$. For each of the three cases below, the phenotypes of the P_1 generations of two crosses are shown, as well as the phenotype of one member of the F_1 generation. For each case, determine the genotypes of the P_1 generation and the F_1 offspring and predict the results of making each cross between F_1 individuals as shown.

Chinchilla rabbit

Himalayan rabbit

	P_1 Phenotypes		F_1 Phenotypes
(a)	Himalayan × Himalayan	⟶	albino
		× ⟶ ??	
	full color × albino	⟶	chinchilla
(b)	albino × chinchilla	⟶	albino
		× ⟶ ??	
	full color × albino	⟶	full color
(c)	chinchilla × albino	⟶	Himalayan
		× ⟶ ??	
	full color × albino	⟶	Himalayan

11. In the guinea pig, one locus involved in the control of coat color may be occupied by any of four alleles: C (black), c^k (sepia), c^d (cream), or c^a (albino). Like coat color in rabbits (Problem 10), an order of dominance exists: $C > c^k > c^d > c^a$. In the following crosses, write the parental genotypes and predict the phenotypic ratios that would result.

(a) sepia × cream, where both guinea pigs had an albino parent
(b) sepia × cream, where the sepia guinea pig had an albino parent and the cream guinea pig had two sepia parents
(c) sepia × cream, where the sepia guinea pig had two full color parents and the cream guinea pig had two sepia parents
(d) sepia × cream, where the sepia guinea pig had a full color parent and an albino parent and the cream guinea pig had two full color parents.

12. Three gene pairs located on separate autosomes determine flower color and shape as well as plant height. The first pair exhibits incomplete dominance where color can be red, pink (the heterozygote), or white. The second pair leads to per- sonate (dominant) or peloric (recessive) flower shape, while the third gene pair produces either the dominant tall trait or the recessive dwarf trait. Homozygous plants that are red, personate, and tall are crossed to those that are white, peloric, and dwarf. Determine the F_1 genotype(s) and phenotype(s). If the F_1 plants are interbred, what proportion of the off- spring will exhibit the same phenotype as the F_1 plants?

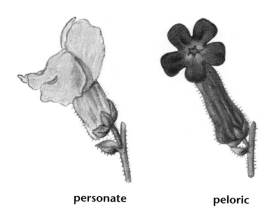

personate   **peloric**

13. As in Problem 12, color may be red, white, or pink, and flower shape may be personate or peloric. For the following cross- es, determine the P_1 and F_1 genotypes. What phenotypic ra- tios would result from crossing the F_1 of (a) to the F_1 of (b)?

(a)	red peloric × white personate	⟶	F_1: all pink personate
(b)	red personate × white peloric	⟶	F_1: all pink personate
(c)	pink personate × red peloric	⟶	F_1: 1/4 red personate
			1/4 red peloric
			1/4 pink personate
			1/4 pink peloric
(d)	pink personate × white peloric	⟶	F_1: 1/4 white personate
			1/4 white peloric
			1/4 pink personate
			1/4 pink peloric

14. Horses can be cremello (a light cream color), chestnut (a brownish color), or palomino (a golden color with white in the horse's tail and mane). Of these phenotypes, only palominos never breed true.
(a) From the results shown below, determine the mode of inheritance by assigning gene symbols and indicating which genotypes yield which phenotypes.

cremello × palomino	⟶	1/2 cremello
		1/2 palomino
chestnut × palomino	⟶	1/2 chestnut
		1/2 palomino
palomino × palomino	⟶	1/4 chestnut
		1/2 palomino
		1/4 cremello

(b) Predict the F_1 and F_2 results of many initial matings be- tween cremello and chestnut horses.

15. With reference to the eye color phenotypes produced by the recessive, autosomal, unlinked *brown* and *scarlet* loci in *Drosophila* (see Figure 4.10), predict the F_1 and F_2 results of the following P_1 crosses. Recall that when both the *brown* and *scarlet* alleles are homozygous, no pigment is produced, and the eyes are white.
    (a) wild type × white
    (b) wild type × scarlet
    (c) brown   × white

16. Pigment in the mouse is only produced when the *C* allele is present. Individuals of the *cc* genotype have no color. If color is present, it may be determined by the *A*, *a* alleles. *AA* or *Aa* results in agouti color, while *aa* results in black coats.
    (a) What F_1 and F_2 genotypic and phenotypic ratios are obtained from a cross between *AACC* and *aacc* mice?
    (b) In three crosses between agouti females whose genotypes were unknown and males of the *aacc* genotype, the following phenotypic ratios were obtained:

(1) 8 agouti	(2) 9 agouti	(3) 4 agouti
8 colorless	10 black	5 black
		10 colorless

    What are the genotypes of these female parents?

17. In some plants a red pigment, cyanidin, is synthesized from a colorless precursor. The addition of a hydroxyl group (OH^-) to the cyanidin molecule causes it to become purple. In a cross between two randomly selected purple plants, the following results were obtained:

94 purple
31 red
43 colorless

    How many genes are involved in the determination of these flower colors? Which genotypic combinations produce which phenotypes? Diagram the purple × purple cross.

18. In rats, the following genotypes of two independently assorting autosomal genes determine coat color:

A–	*B–*	(gray)
A–	*bb*	(yellow)
aa	*B–*	(black)
aa	*bb*	(cream)

    A third gene pair on a separate autosome determines whether or not any color will be produced. The *CC* and *Cc* genotypes allow color according to the expression of the *A* and *B* alleles. However, the *cc* genotype results in albino rats regardless of the *A* and *B* alleles present. Determine the F_1 phenotypic ratio of the following crosses:
    (a) *AA bb CC* × *aa BB cc*
    (b) *Aa BB CC* × *AA Bb cc*
    (c) *Aa Bb Cc* × *Aa Bb cc*
    (d) *Aa BB Cc* × *Aa BB Cc*
    (e) *AA Bb Cc* × *AA Bb Cc*

19. Given the inheritance pattern of coat color in rats, as described in Problem 18, predict the genotype and phenotype of the parents who produced the following F_1 offspring:

    (a) 9/16 gray:3/16 yellow:3/16 black:1/16 cream
    (b) 9/16 gray:3/16 yellow:4/16 albino
    (c) 27/64 gray:16/64 albino:9/64 yellow:9/64 black: 3/6 cream
    (d) 3/8 black:3/8 cream:2/8 albino
    (e) 3/8 black:4/8 albino:1/8 cream

20. In a species of the cat family, eye color can be gray, blue, green, or brown, and each trait is true-breeding. In separate crosses involving homozygous parents, the following data were obtained:

Cross	P_1	F_1	F_2
A	green × gray	all green	3/4 green: 1/4 gray
B	green × brown	all green	3/4 green: 1/4 brown
C	gray × brown	all green	9/16 green: 3/16 brown: 3/16 gray: 1/16 blue

    (a) Analyze the data: How many genes are involved? Define gene symbols and indicate which genotypes yield each phenotype.
    (b) In a cross between a gray-eyed cat and one of unknown genotype and phenotype, the F_1 generation was not observed. However, the F_2 resulted in the same F_2 ratio as in cross C. Determine the genotypes and phenotypes of the unknown P_1 and F_1 cats.

21. In a plant, a tall variety was crossed with a dwarf variety. All F_1 plants were tall. When F_1 × F_1 plants were interbred, 9/16 of the F_2 were tall and 7/16 were dwarf.
    (a) Explain the inheritance of height by indicating the number of gene pairs involved and by designating which genotypes yield tall and which yield dwarf (use dashes where appropriate).
    (b) Of the F_2 plants, what proportion of them will be true-breeding if self-fertilized? List these genotypes.

22. In a unique species of plants, flowers may be yellow, blue, red, or mauve. All colors may be true-breeding. If plants with blue flowers are crossed to red-flowered plants, all F_1 plants have yellow flowers. When carried to an F_2 generation, the following ratio was observed:

    9/16 yellow:3/16 blue:3/16 red:1/16 mauve

    In still another cross using true-breeding parents, yellow-flowered plants are crossed with mauve-flowered plants. Again, all F_1 plants had yellow flowers and the F_2 showed a 9:3:3:1 ratio, as shown here.
    (a) Describe the inheritance of flower color by defining gene symbols and designating which genotypes give rise to each of the four phenotypes.
    (b) Determine the F_1 and F_2 results of a cross between true-breeding red and true-breeding mauve-flowered plants.

23. Shown below are five human matings (1–5), including both maternal and paternal phenotypes for ABO, MN, and Rh blood-group antigen status. Each mating resulted in one of the five offspring shown to the right (a–e). Match each offspring with one correct set of parents using each parental set only once. Is there more than one set of correct answers?

Parental Phenotypes	Offspring
(1) A, M, Rh⁻ × A, N, Rh⁻	(a) A, N, Rh⁻
(2) B, M, Rh⁻ × B, M, Rh⁺	(b) O, N, Rh⁺
(3) O, N, Rh⁺ × B, N, Rh⁺	(c) O, MN, Rh⁻
(4) AB, M, Rh⁻ × O, N, Rh⁺	(d) B, M, Rh⁺
(5) AB, MN, Rh⁻ × AB, MN, Rh⁻	(e) B, MN, Rh⁺

24. A husband and wife have normal vision, although both of their fathers are red–green color-blind, which is inherited as an X-linked recessive condition. What is the probability that their first child will be (a) a normal son? (b) a normal daughter? (c) a color-blind son? (d) a color-blind daughter?

25. In humans, the ABO blood type is under the control of autosomal multiple alleles. Color blindness is a recessive X-linked trait. If two parents who are both type *A* and have normal vision produce a son who is color-blind and is type O, what is the probability that their next child will be a female who has normal vision and is type O?

26. In *Drosophila*, an X-linked recessive mutation, scalloped (*sd*), causes irregular wing margins. Diagram the F_1 and F_2 results if: (a) a scalloped female is crossed with a normal male; (b) a scalloped male is crossed with a normal female. Compare these results to those that would be obtained if scalloped were not X-linked.

27. Another recessive mutation in *Drosophila*, ebony (*e*), is on an autosome (chromosome 3) and causes darkening of the body compared with wild-type flies. What phenotypic F_1 and F_2 male and female ratios will result if a scalloped-winged female with normal body color is crossed with a normal-winged ebony male? Work this problem by both the Punnett square method and the forked-line method.

28. In *Drosophila*, the X-linked recessive mutation vermilion (*v*) causes bright-red eyes, which is in contrast to brick-red eyes of wild type. A separate autosomal recessive mutation, suppressor of vermilion (*su-v*), causes flies homozygous or hemizygous for *v* to have wild-type eyes. In the absence of vermilion alleles, *su-v* has no effect on eye color. Determine the F_1 and F_2 phenotypic ratios from a cross between a female with wild-type alleles at the vermilion locus, but who is homozygous for *su-v*, with a vermilion male who has wild-type alleles at the *su-v* locus.

29. While vermilion is X-linked and brightens the eye color, brown is an autosomal recessive mutation that darkens the eye. Flies carrying both mutations lose all pigmentation and are white eyed. Predict the F_1 and F_2 results of the following crosses:
(a) vermilion females × brown males
(b) brown females × vermilion males
(c) white females × wild-type males

30. In a cross in *Drosophila* involving the X-linked recessive eye mutation *white* and the autosomally linked recessive eye mutation *sepia* (resulting in a dark eye), predict the F_1 and F_2 results of crossing true-breeding parents of the following phenotypes:
(a) white females × sepia males
(b) sepia females × white males
Note that white is epistatic to the expression of sepia.

31. Consider the following three pedigrees all involving a single human trait.

(a) Which conditions, if any, can be excluded?

Conditions: dominant and X-linked
dominant and autosomal
recessive and X-linked
recessive and autosomal

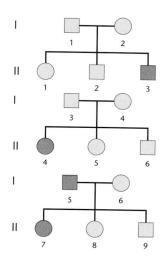

(b) For any condition that you excluded, indicate the single individual in generation II (e.g., II-1, II-2) that was most instrumental in your decision to exclude that condition. If none were excluded, answer "none apply."
(c) Given your conclusions above, indicate the genotype of the individuals listed below. If more than one possibility applies, list all possibilities. Use the symbols *A* and *a* for the genotypes.

II-1;   II-6;   II-9

32. In spotted cattle, the colored regions may be mahogany or red. If a red female and a mahogany male, both derived from separate true-breeding lines, are mated and the cross carried to an F_2 generation, the following results are obtained:

F_1: 1/2 mahogany males
1/2 red females

F_2: 3/8 mahogany males
1/8 red males
1/8 mahogany females
3/8 red females

When the reciprocal of the initial cross is performed (mahogany female and red male), identical results are obtained. Explain these results by postulating how the color is genetically determined. Diagram the crosses.

33. Predict the F_1 and F_2 results of crossing a male fowl that is cock-feathered with a true-breeding hen-feathered female fowl. Recall that these traits are sex-limited.

34. Two mothers give birth to sons at the same time at a busy urban hospital. The son of couple 1 is afflicted with hemophilia, a disease caused by an X-linked recessive allele. Neither parent has the disease. Couple 2 have a normal son, despite the fact that the father has hemophilia. Several years later, couple 1 sues the hospital, claiming that these two newborns were swapped in the nursery following their birth. You are called, as a genetic counselor, to testify. What information can you provide the jury concerning the allegation?

35. Discuss the topic of phenotypic expression and the many factors that impinge on it.

36. Contrast penetrance and expressivity as the terms relate to phenotypic expression.

37. Contrast the phenomena of genetic anticipation and genomic imprinting.

EXTRA-SPICY PROBLEMS

38. Labrador retrievers may be black, brown, or golden in color. While each color may breed true, many different outcomes occur if many litters are examined from a variety of matings, where the parents are not necessarily true-breeding. Shown below are just some of the many possibilities. Propose a mode of inheritance that is consistent with these data, and indicate the corresponding genotypes of the parents in each mating. Indicate as well the genotypes of dogs that breed true for each color.

(a) black × brown	⟶	all black
(b) black × brown	⟶	1/2 black
		1/2 brown
(c) black × brown	⟶	3/4 black
		1/4 golden
(d) black × golden	⟶	all black
(e) black × golden	⟶	4/8 golden
		3/8 black
		1/8 brown
(f) black × golden	⟶	2/4 golden
		1/4 black
		1/4 brown
(g) brown × brown	⟶	3/4 brown
		1/4 golden
(h) black × black	⟶	9/16 black
		4/16 golden
		3/16 brown

39. Members of a strain of inbred skunks whose stripes are uniformly 20 cm long are crossed to those of another strain whose stripes are uniformly 24 cm. The F_1 hybrids all have stripes that are 22 cm. When the F_1 skunks were crossed, the F_2 generation produced animals with stripes of 16, 18, 20, 22, 24, 26, and 28 cm. The 22-cm variety was most frequent and the 16- and 28-cm variety least frequent. Of 1000 F_2 skunks, 15 were 16 cm and 16 were 28 cm.
  (a) What mode of inheritance is illustrated?
  (b) How many gene pairs are involved?
  (c) What is the contribution of each additive allele to stripe length?
  (d) What are the genotypes of the P_1 parents and F_1 offspring?
  (e) List four genotypes that give rise to skunks whose stripes are 20 cm long.

40. A true-breeding purple-leafed plant isolated from one side of the rain forest in Puerto Rico (El Yunque) was crossed to a true-breeding white variety found on the other side of the rain forest. The F_1 offspring were all purple. A large number of crosses resulted in the following results:

  purple: 4219   white: 5781   (Total = 10,000)

Propose an explanation for the inheritance of leaf color. As a geneticist, how might you go about testing your hypothesis? Describe the genetic experiments that you would conduct.

41. In Dexter and Kerry cattle, animals may be polled (hornless) or horned. The Dexter animals have short legs, whereas the Kerry animals have long legs. When many offspring were obtained from matings between polled Kerrys and horned Dexters, one-half were found to be polled Dexters and one-half polled Kerrys. When these F_1 cattle were interbred, the following F_2 data were obtained:

3/8 polled Dexters
3/8 polled Kerrys
1/8 horned Dexters
1/8 horned Kerrys

A geneticist was puzzled by these data and interviewed farmers who had bred these cattle for decades. She learned that Kerrys were true-breeding. Dexters, on the other hand, were not true-breeding and never produced as many offspring as Kerrys. Provide a genetic explanation for these observations.

42. An alien geneticist escaping from a planet where genetic research had recently been prohibited brought with him to Earth two pure-breeding lines of pet frogs. One line croaked by uttering *rib-it rib-it* and had purple eyes. The other line croaked more softly by muttering *knee-deep knee-deep* and had green eyes. With a new-found freedom of inquiry, he mated the two types of frogs. In the F_1, all frogs had blue eyes and uttered *rib-it rib-it*. He proceeded to make many $F_1 \times F_1$ crosses, and when he fully analyzed the F_2 data, he realized they could be reduced to the following ratio:

27/64 blue-eyed, rib-it utterer
12/64 green-eyed, rib-it utterer
9/64 blue-eyed, knee-deep mutterer
9/64 purple-eyed, rib-it utterer
4/64 green-eyed, knee-deep mutterer
3/64 purple-eyed, knee-deep mutterer

(a) How many total gene pairs are involved in the inheritance of both traits? Support your answer.
(b) Of these, how many are controlling eye color? How can you tell? How many are controlling croaking?
(c) Assign gene symbols for all phenotypes and indicate the genotypes of the P_1 and F_1 frogs.
(d) Indicate the genotypes of the six F_2 phenotypes.
(e) After years of experiments, the geneticist isolated pure-breeding strains of all six F_2 phenotypes. Indicate the F_1 and F_2 phenotypic ratios of the following cross using these pure-breeding strains:

blue-eyed, knee-deep mutterer
×
purple-eyed, rib-it utterer

(f) One set of crosses with his true-breeding lines initially caused the geneticist some confusion. When he crossed true-breeding purple-eyed, knee-deep mutterers with true-breeding green-eyed, knee-deep mutterers, he often got different results. In some matings, all offspring were blue-eyed, knee-deep mutterers, but in other matings all offspring were purple-eyed, knee-deep mutterers. In still a third mating, 1/2 blue-eyed, knee-deep mutterers and 1/2 purple-eyed, knee-deep mutterers were observed. Explain why the results differed.
(g) In another experiment, the geneticist crossed two purple-eyed, rib-it utterers together with the results shown below. What were the genotypes of the two parents?

9/16 purple-eyed, rib-it utterer
3/16 purple-eyed, knee-deep mutterer
3/16 green-eyed, rib-it utterer
1/16 green-eyed, knee-deep mutterer

43. The following pedigree is characteristic of an inherited condition known as male precocious puberty, where affected males show signs of puberty by age 4. Propose a genetic explanation of this phenotype. [*Reference:* A. Shenker (1993).]

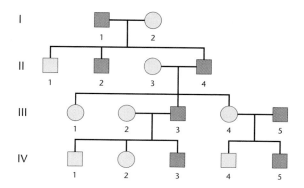

44. In birds, the male is the homogametic sex and the female is the heterogametic sex. The sex chromosome compositions are designated ZZ and ZW, respectively. In parakeets, two genes control feather color. The presence of the dominant Y allele at the first gene results in the production of a yellow pigment. The dominant B allele at the second gene controls melanin production. When both genes are active, a green pigment results. If only the Y gene is active, the feathers are yellow. If only the B gene is active, a blue color is exhibited. If neither gene is active, the birds are albinos. Therefore, with our conventional designations, phenotypes are produced as follows.

$Y-$	$B-$	green
$Y-$	bb	yellow
yy	$B-$	blue
yy	bb	albino

(a) A series of crosses established that one of the genes is autosomal and one is Z-linked. Based on the results of the following cross shown below, where both parents are true-breeding, determine which gene is Z-linked. Support your answer by establishing the genotypes of the P_1 parents and working the cross through the F_2 generation.

P_1: green male × albino female
F_1: 1/2 green males;   1/2 green females
F_2: 6/16 green males
2/16 yellow males
3/16 green females
1/16 yellow females
3/16 blue females
1/16 albino females

GENETICS MediaLab

The following resources will help you achieve a better understanding of the concepts presented in this chapter. These resources can be found on the CD packaged with this textbook and on the Companion Website at **http://www.prenhall.com/klug/**.

CD Resources:
Self-grading Chapter Problems

Web Resources:
Genetics and Society Issue: *Improving the Fate of Purebred Dogs*

Web Destinations in Genetics

Self-grading Chapter Problems

Chapter Search Terms

Genetics Newsgroups

Student Bulletin Board

Web Problem 1:
Time for completion = 10 minutes
Is dominance always complete? You will study flower color in the four-o'clock plant, *Mirabilis*. As in peas, flower color is determined by a single locus with two alleles. In this case, however, dominance is incomplete. Read about incomplete dominance in Chapter 4. Once you complete the F_1 cross in the tutorial, go on to the next module to study the F_2. Suppose you begin by crossing true-breeding stocks with red and white flowers. What genotypic ratio would you predict in the F_1? What is the expected phenotypic ratio in the F_1? How do these results differ from the crosses of the peas you did in the Chapter 3 MediaLab? Suppose you cross the F_1; what ratios of phenotypes and genotypes would you expect in the F_2 generation? Do the ratios of genotypes differ from a cross of heterozygotes where one allele is dominant? To complete this exercise, visit Web Problem 1 in Chapter 4 of your Companion Website and select the keyword **INCOMPLETE DOMINANCE.**

Web Problem 2:
Time for completion = 10 minutes
Epistasis occurs when the genotype at one locus modifies the phenotypic expression of another locus. Epistatic effects are very common among the genes affecting mammalian fur color. You will examine a table of loci that influence the coloration of horses and search for cases of epistatic interaction. In horses, the main gene affecting color is the E locus. The black allele (E) is dominant over the red allele (e). One example of epistasis is the A locus; if a horse is black, then the presence of the dominant A allele will restrict the black coloring to the "points" (i.e., the legs, nose, ears, and tail), whereas a recessive homozygote will be black all over. Suppose you encounter an all-black horse. Use the table to determine what the genotype might be at the C and D loci. What might the genotype of an all-white horse be at the E locus? Is the W locus epistatic? Why might you not want to breed two all-white horses? To complete this exercise, visit Web Problem 2 in Chapter 4 of your Companion Website and select the keyword **EPISTASIS.**

Web Problem 3:
Time for completion = 15 minutes
X-linked genes (Chapter 4) reside on the X chromosome. In many organisms, such as humans and fruit flies, males are usually XY, while females are XX. Since the Y chromosome carries relatively little genetic information, males are effectively haploid for most of the genes on X. Both dominant and recessive traits are expressed in such males and they may exhibit different phenotypic ratios than females. We are going to conduct reciprocal dihybrid crosses with two traits in *Drosophila*, at least one of which is sex linked. First cross a white-eyed, curve-winged male with a wild-type female. Record the phenotypic ratios (including sex) in the F_1, then cross the F_1 and record the phenotypic ratios in the F_2. Repeat the exercise using the reciprocal cross (white-eyed, curved-winged female × wild-type male). Which of the traits is sex linked? How could you tell? Examine your data and propose a hypothesis regarding the dominance of each of the two mutations. If you had only done the reciprocal F_1 crosses, could you determine the dominance of the sex-linked trait(s)? Test your hypothesis using the X-square test. To complete this exercise, visit Web Problem 3 in Chapter 4 of your Companion Website and select the keyword **SEX LINKAGE.**

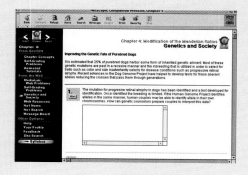

(b) In a cross where the parental genotypes were unknown, the offspring from repeated matings were recorded, as shown below. Further, it was not known if the parents were true breeding. Based on the results, determine the phenotypes and genotypes of the parents.

13 green males
3 yellow males
11 blue females
5 albino females

SELECTED READINGS

BRINK, R.A., ed. 1967. *Heritage from Mendel.* Madison: University of Wisconsin Press.

BULTMAN, S.J., MICHAUD, E.J., and WOYCHIK, R.P. 1992. Molecular characterization of the mouse *agouti* locus. *Cell* 71:1195–1204.

CATTANACH, B.M., and JONES, J. 1994. Genetic imprinting in the mouse: Implications for gene regulation. *J. Inherit. Metab. Dis.* 17:403–20.

CHAPMAN, A.B. 1985. *General and quantitative genetics.* Amsterdam: Elsevier.

CLARKE, C.A. 1968. The prevention of "Rhesus" babies. *Sci. Am.* (Nov.) 219:46–52.

CORWIN, H.O., and JENKINS, J.B. 1976. *Conceptual foundations of genetics: Selected readings.* Boston: Houghton-Mifflin.

CROW, J.F. 1983. *Genetics notes,* 8th ed. New York: Macmillan.

DRAYNA, D., and WHITE, R. 1985. The genetic linkage map of the human X chromosome. *Science* 230:753–58.

DUNN, L.C. 1966. *A short history of genetics.* New York: McGraw-Hill.

EAST, E.M. 1910. A Mendelian interpretation of variation that is apparently continuous. *Am. Naturalist* 44:65–82.

———. 1916. Studies on size inheritance in *Nicotiana. Genetics* 1:164–76.

FALCONER, D.S. 1981. *Introduction to quantitative genetics,* 2nd ed. New York: Longman.

FEIL, R., and KELSEY, G. 1997. Genomic imprinting: A chromatin connection. *Am. J. Hum. Genet.* 61:1213–19.

FOSTER, H.L. et al., eds. 1981. *The mouse in biomedical research, Vol. 1. History, genetics, and wild mice.* Orlando, FL: Academic Press.

FOSTER, M. 1965. Mammalian pigment genetics. *Adv. Genet.* 13:311–39.

GRANT, V. 1975. *Genetics of flowering plants.* New York: Columbia University Press.

HARPER, P.S. et al. 1992. Anticipation in myotonic dystrophy: New light on an old problem. *Am. J. Hum. Genet.* 51:10–16.

HOWELER, C.J. et al. 1989. Anticipation in myotonic dystrophy: Fact or fiction? *Brain* 112:779–97.

LINDSLEY, D.C., and GRELL, E.H. 1967. *Genetic variations of Drosophila melanogaster.* Washington, DC: Carnegie Institute of Washington.

MAHEDEVAN, M. et al. 1992. Myotonic dystrophy mutation: An unstable CTG repeat in the 3′ untranslated region of the gene. *Science* 255:1253–58.

MCKUSICK, V.A. 1962. On the X chromosome of man. *Quart. Rev. Biol.* 37:69–175.

MORGAN, T.H. 1910. Sex-limited inheritance in *Drosophila. Science* 32:120–22.

NOLTE, D.J. 1959. The eye-pigmentary system of *Drosophila. Heredity* 13:233–41.

PAWELEK, J.M., and KÖRNER, A.M. 1982. The biosynthesis of mammalian melanin. *Am. Sci.* 70:136–45.

PETERS, J.A., ed. 1959. *Classic papers in genetics.* Englewood Cliffs, NJ: Prentice-Hall.

RACE, R.R., and SANGER, R. 1975. *Blood groups in man,* 6th ed. Oxford: Blackwell Scientific Publishers.

SAPIENZA, C. 1990. Parental imprinting of genes. *Sci. Am.* (Oct.) 363:52–60.

SHENKER, A. 1993. A constitutively activating mutation of the luteinizing hormone receptor in familial male precocious puberty. *Nature* 365:652–54.

SIRACUSA, L.D. 1994. The *agouti* gene: Turned on to yellow. *Cell* 10:423–28.

SMITH, C.A. 1995. New hope for overcoming canine inherited disease. *J. Am. Veter. Med. Assoc.* 204:41–46.

STERN, C. 1973. *Principles of human genetics,* 3rd ed. New York: W.H. Freeman.

TEARLE, R.G. et al. 1989. Cloning and characterization of the scarlet gene of *Drosophila melanogaster. Genetics* 122:595–606.

VOELLER, B.R., ed. 1968. *The chromosome theory of inheritance—Classic papers in development and heredity.* New York: Appleton-Century-Crofts.

VOGEL, F., and MOTULSKY, A.G. 1997. *Human genetics: Problems and approaches,* 3rd ed. New York: Springer-Verlag.

WATKINS, M.W. 1966. Blood group substances. *Science* 152:172–81.

WIENER, A.S., ed. 1970. *Advances in blood groupings,* Vol. 3. New York: Grune and Stratton.

YOSHIDA, A. 1982. Biochemical genetics of the human blood group ABO system. *Am. J. Hum. Genet.* 34:1–14.

ZIEGLER, I. 1961. Genetic aspects of ommochrome and pterin pigments. *Adv. Genet.* 10:349–403.

Ripe Gaines red wheat, growing in the state of Washington.

5

Quantitative Genetics

KEY CONCEPTS

- **Continuous variation characterizes the inheritance of quantitative traits**

- **Quantitative traits can be explained in Mendelian terms**
  Additive Alleles: The Basis of Continuous Variation
  Calculating the Number of Polygenes
  The Significance of Polygenic Control

- **The study of polygenic traits relies on statistical analysis**
  The Mean
  Variance
  Standard Deviation

Standard Error of the Mean
Analysis of a Quantitative Character

- **Heritability is a measure of the genetic contribution to phenotypic variability**
  Broad-Sense Heritability
  Narrow-Sense Heritability
  Artificial Selection

- **Twin studies allow an estimation of heritability in humans**

- **Quantitative trait loci can be mapped**

In the preceding chapter, numerous examples were discussed that illustrated gene interaction. In each case, the resultant phenotypic variation was classified into distinct traits. Pea plants were tall or dwarf; squash shape was spherical, disc-shaped, or elongated; and fruit fly eye color was red or white. These phenotypes are examples of discontinuous variation, in which discrete phenotypic categories exist. Many other traits in a population demonstrate considerably more variation and are not as easily categorized into distinct classes. Such phenotypes are thus said to demonstrate continuous variation.

Traits exhibiting continuous variation are most often controlled by two or more genes that provide an additive component to the phenotype that can be quantified. In this chapter, we will examine cases that illustrate such patterns of inheritance and outline some statistical techniques used to study such traits. They illustrate what is described as **quantitative**, or **polygenic, inheritance.** In addition, we will consider how geneticists assess the relative importance of genetic versus environmental factors as they contribute to phenotypic variation, and we will discuss an approach used to map these "quantitative" genes.

large peas were crossed to those with small peas, the F_1 plants contained peas that were all of an intermediate diameter. In the F_2 generation, peas were of many sizes, as large or as small as the original parents, and many sizes in between.

This example illustrates a pattern of inheritance encountered by other investigators, including at least one cross made by Mendel. The F_1 generations were an intermediate blend of the parental phenotypes, and the F_2 generation exhibited a more or less continuous phenotypic variation expressed, for example, as size, height, weight, and color.

Not surprisingly, these traits were difficult to study, and their mode of inheritance was not clarified until early in the twentieth century. Until then, they posed apparent exceptions to the patterns observed by Mendel in most of his crosses and clouded the initial understanding of transmission genetics. This no doubt delayed the acceptance of Mendel's work. Subsequently, the genetic explanation of continuous variation provided the foundation for our current understanding of the field of genetics now called quantitative inheritance.

Continuous variation characterizes the inheritance of quantitative traits

Throughout the eighteenth and nineteenth centuries, geneticists studied traits that exhibited a continuous gradation of phenotypes. For example, Sir Francis Galton investigated the diameter of sweet peas. When plants with

Quantitative traits can be explained in Mendelian terms

One of the first cases of continuous phenotypic variation was encountered by Josef Gottlieb Kölreuter when he crossed tall and dwarf varieties of the tobacco plant *Nicotiana longiflora*. The plants of the F_1 generation were all intermediate in height. When the F$_2$ generation was examined, individuals showed continuous variation in height, ranging from tall to dwarf, like the original parents, including many heights in between. A critical observation involved the distribution of phenotypes in the second generation: The majority of the F$_2$ plants were intermediate like the F$_1$, but only a few were as tall or dwarf as the P$_1$ parents. These distributions are depicted in histograms in Figure 5.1. Note that the F$_2$ data demonstrate a normal distribution, as evidenced by the bell-shaped curve in the histogram.

At the beginning of the twentieth century, geneticists noted that many characters in different species had similar patterns of inheritance, such as height and stature in humans, seed size in the broad bean, grain color in wheat, and kernel number and ear length in corn. In each case, offspring in the succeeding generation seemed to be a blend of their parents' characteristics.

The issue of whether continuous variation could be accounted for in Mendelian terms caused considerable controversy in the early 1900s. William Bateson and Gudny Yule, who adhered to the Mendelian explanation of inheritance, suggested that a large number of factors or genes could account for the observed patterns. This proposal, called the **multiple-factor** or **multiple-gene hypothesis,** implied that many factors or genes contribute to the phenotype in a *cumulative* or *quantitative* way. However, other geneticists argued that Mendel's unit factors could not account for the blending of parental phenotypes characteristic of these patterns of inheritance and were thus skeptical of this hypothesis.

By 1920, the conclusions of several critical sets of experiments largely resolved the controversy and demonstrated that Mendelian factors could account for continuous variation. In one experiment, Edward M. East performed crosses between two strains of tobacco plant. The fused inner petals of the flower, or corollas, of strain A were decidedly shorter than the corollas of strain B. With only minor variation, each strain was true breeding. Thus, the differences between them were clearly under genetic control.

When plants from the two strains were crossed, examination of the F$_1$, F$_2$, and selected F$_3$ data revealed a very distinct pattern (Figure 5.2). The F$_1$ generation displayed corollas that were intermediate in length compared with the P$_1$ varieties and showed only minor variation among individuals. While corolla lengths of the P$_1$ plants were about 40 mm and 94 mm, the F$_1$ generation contained plants with corollas that were all about 64 mm. In the F$_2$ generation, lengths varied much more, ranging from 52 mm to 82 mm. The majority of individuals were similar to their F$_1$ parents, and as the deviation from this average increased, fewer and fewer plants were observed. When the data are plotted graphically (frequency vs. length), a bell-shaped curve results.

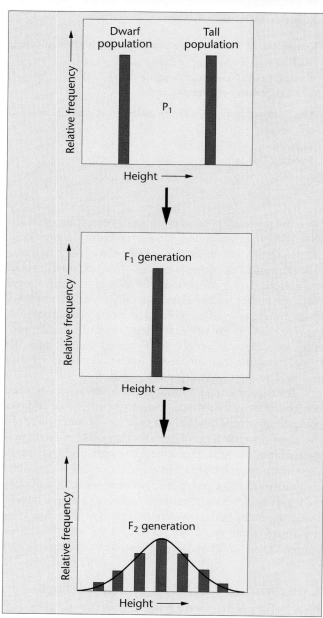

FIGURE 5.1   Histograms showing the relative frequency of individuals expressing various height phenotypes derived from Kölreuter's cross between dwarf and tall tobacco plants carried to the F$_2$ generation. The photograph shows a field of tobacco plants.

— Corolla

FIGURE 5.2   The F_1, F_2, and selected F_3 results of East's cross between two strains of *Nicotiana* with different corolla lengths. Corollas of strain A plants vary from 37 to 43 mm, while corollas of strain B plants vary from 91 to 97 mm. The photograph and drawing illustrate the flower and corolla of a tobacco plant.

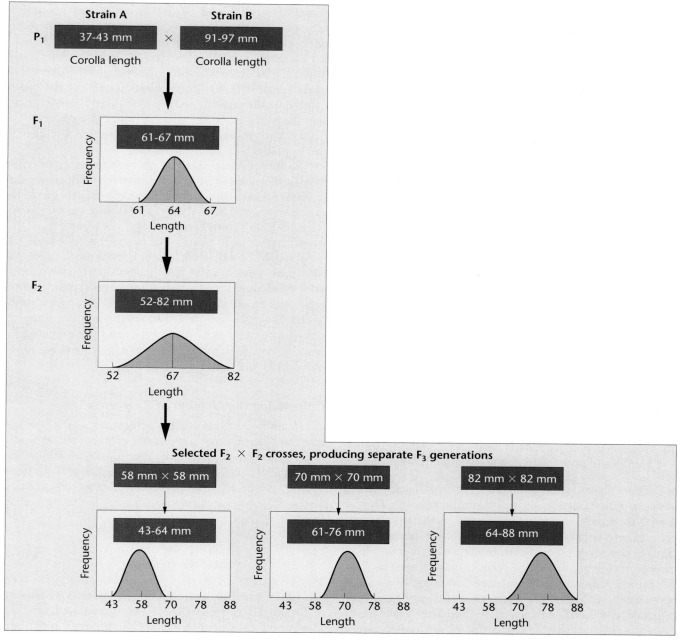

East further experimented with this population by selecting F_2 plants of various corolla lengths and allowing them to produce separate F_3 generations. Several are illustrated in Figure 5.2. In each case, a bell-shaped distribution was observed, with most individuals similar in height to the selected F_2 parents, but with considerable variation around this value.

East's experiments demonstrated that although the variation in corolla length, at first glance, seemed continuous, experimental crosses nevertheless resulted in the segregation of distinct phenotypic classes, as observed in the three independent F_3 categories. As we will see below, this key finding was the basis for his explanation of how the multiple-factor hypothesis could account for traits that deviate considerably in their expression.

Additive Alleles: The Basis of Continuous Variation

The multiple-factor hypothesis, suggested by the observations of East and others, embodies the following major points:

1. Characters that exhibit continuous variation can usually be quantified by measuring, weighing, counting, and so on.

2. Two or more pairs of genes, located throughout the genome, account for the hereditary influence on the phenotype in an *additive way*. Because many genes may be involved, inheritance of this type is often called *polygenic*.

3. Each gene locus may be occupied by either an **additive allele,** which contributes a set amount to the phenotype, or by a **nonadditive allele,** which does not contribute quantitatively to the phenotype.

4. The total effect on the phenotype of each additive allele, while small, is approximately equivalent to all other additive alleles at other gene sites.

5. Together, the genes controlling a single character produce substantial phenotypic variation.

6. Analysis of polygenic traits requires the study of large numbers of progeny from a population of organisms.

These points center around the concept that additive alleles at numerous loci control quantitative traits. Herman Nilsson-Ehle's experiments early in the twentieth century involving grain color in wheat illustrate the concept. In one set of experiments, wheat with red grain was crossed to wheat with white grain (Figure 5.3). The F_1 generation demonstrated an intermediate color. In the F_2, approximately 15/16 of the plants showed some degree of red grain, while 1/16 of the plants showed white grain. Because the ratio occurred in sixteenths, we can hypothesize

that two gene pairs control the phenotype and, if so, that they segregate independently from one another in a Mendelian fashion.

Careful examination of the F_2 revealed that grain with color could be classified into four different shades of red. If two gene pairs were operating, each with one potential additive allele and one potential nonadditive allele, we can envision how the multiple-factor hypothesis could account for this variation. In the P_1, both parents were homozygous; the red parent contains only additive alleles (uppercase letters in Figure 5.3), while the white parent contains only nonadditive alleles (lowercase letters). The F_1, being heterozygous, contains only two additive alleles and expresses an intermediate phenotype. In the F_2, each offspring has either 4, 3, 2, 1, or 0 additive alleles (Figure 5.3). Wheat with no additive alleles (1/16) is white like one of the P_1 parents, while wheat with 4 additive alleles is red like the other P_1 parent. Plants with 3, 2, or 1 additive alleles constitute the other three categories of red color observed in the F_2, with most (6/16) having 2 additive alleles like the F_1 plants.

Multiple-factor inheritance, where additive alleles influence the phenotype in a quantitative manner, results in continuous variation. Therefore, continuous variation can be explained in a Mendelian fashion. As we saw in Nilsson-Ehle's initial cross, if two gene pairs were involved, only five F_2 phenotypic categories, in a 1:4:6:4:1 ratio, would be expected. There is no reason, however, why three, four, or more gene pairs cannot function in a similar fashion in controlling various phenotypes. As more gene pairs become involved, greater and greater numbers of classes would be expected to appear in more complex ratios. The number of phenotypes and the expected F_2 ratios of crosses involving up to five gene pairs are illustrated in Figure 5.4.

Calculating the Number of Polygenes

When additive effects control polygenic traits, it is of interest to determine the number of genes that are involved. If the ratio (proportion) of F_2 individuals resembling *either* of the two most extreme phenotypes (the parental phenotypes) can be determined, then the number of gene pairs involved (n) may be calculated using the following simple formula:

$$\frac{1}{4^n} = \text{ratio of } F_2 \text{ individuals expressing either extreme phenotype}$$

In our previous example, the P_1 phenotypes shown in Figure 5.3 represent these two extremes. In the F_2 generation, for example, 1/16 of the progeny are either

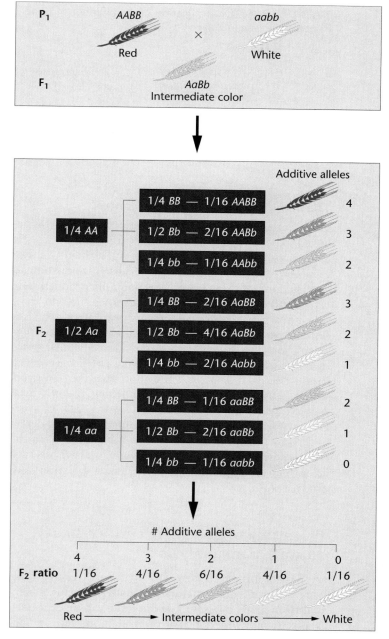

FIGURE 5.3 An illustration of how the multiple-factor hypothesis can account for the 1:4:6:4:1 phenotypic ratio of grain color. All alleles designated by an uppercase letter are additive and contribute an equal amount of pigment to the phenotype.

red *or* white like the P_1 classes; this ratio can be substituted on the right side of the equation before solving for *n*:

$$\frac{1}{4^n} = \frac{1}{16}$$

$$\frac{1}{4^2} = \frac{1}{16}$$

$$n = 2$$

Table 5.1 lists the ratio and the number of F_2 phenotypic classes produced in crosses involving up to five gene pairs.

For low numbers of gene pairs, it is sometimes easier to use the $(2n + 1)$ rule. If *n* equals the number of gene pairs, $2n + 1$ will determine the total number of categories of possible phenotypes. When $n = 2$, $2n + 1$ = 5. Each phenotypic category can have 4, 3, 2, 1, or 0 additive alleles. When $n = 3$, $2n + 1 = 7$, each phenotypic category could have 6, 5, 4, 3, 2, 1, or 0 additive alleles, and so on.

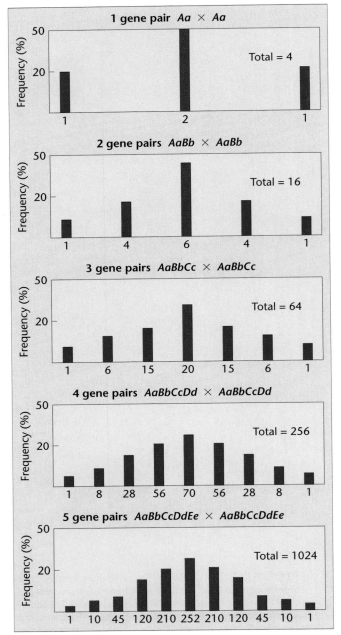

FIGURE 5.4 The results of crossing two heterozygotes when polygenic inheritance is in operation with one to five gene pairs. Each histogram bar indicates a distinct F₂ phenotypic class from one extreme (left end) to the other extreme (right end). Each phenotype results from a different number of additive alleles.

TABLE 5.1 Determination of the Number of Gene Pairs (n) Involved in Polygenic Crosses

n	Ratio of Individuals Expressing an Extreme Phenotype	Number of Distinct F₂ Phenotypic Classes
1	1/4	3
2	1/16	5
3	1/64	7
4	1/256	9
5	1/1024	11

The Significance of Polygenic Control

Polygenic control is a significant concept because it is believed to be the mode of inheritance for a vast number of traits involved in animal breeding and agriculture. For example, height, weight, and physical stature in animals, size and grain yield in crops, beef and milk production in cattle, and egg production in chickens are thought to be under polygenic control. Polygenic traits are also an important part of human genetics; traits such as skin pigmentation, intelligence, obesity, and predisposition to certain diseases are thought to be under polygenic control. In most cases, it is important to note that the genotype, which is fixed at fertilization, establishes the potential range in which a particular phenotype may fall. However, as we saw in Chapter 4, environmental factors determine how much of the potential will be realized. In the crosses described thus far in this chapter, we have assumed an optimal environment, which minimizes variation from external sources.

The study of polygenic traits relies on statistical analysis

Analysis of any given polygenic trait involves quantitative measurements, usually from many offspring generated from many crosses. The outcome can be expressed as a frequency diagram that often demonstrates a normal (bell-shaped) distribution (Figure 5.5). While it is hoped that each series of crosses is representative of the population at large, variation in samples due strictly to chance may influence the data gathered. To assess the experimental validity of the data, geneticists use statistical techniques, which were first devised by Galton early in the twentieth century. Galton's efforts to assess the inheritance of traits exhibiting continuous variation served as the initial basis of the field of study called **biometry**, a quantitative, statistics-based approach to the study of biology.

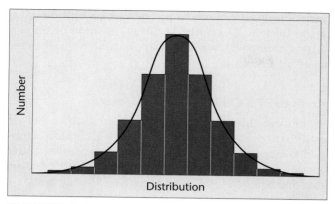

FIGURE 5.5 A normal frequency distribution characterized by a bell-shaped curve.

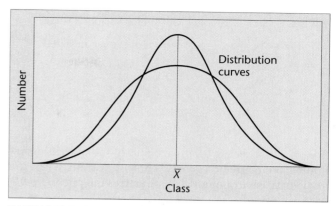

FIGURE 5.6 Two normal frequency distributions with the same mean but different amounts of variation.

Statistical analysis serves three purposes:

1. Data can be mathematically reduced to provide a descriptive summary of the sample.

2. Data from a small but random sample can be used to infer information about groups larger than those from which the original data were obtained (statistical inference).

3. Two or more sets of experimental data may be compared to determine whether they represent significantly different populations of measurements.

Several statistical methods are useful in the analysis of traits that exhibit a normal distribution, including the mean, variance, standard deviation, and standard error of the mean.

The Mean

The distribution of two sets of phenotypic measurements that is graphed in Figure 5.6 tends to cluster around a central value. This clustering is called a **central tendency,** one measurement of which is the **mean** $(\overline{X})$.

The mean is simply the arithmetic average of a set of measurements or data and is calculated as:

$$\overline{X} = \frac{\Sigma X_i}{n}$$

where $\overline{X}$ is the mean, ΣX_i represents the sum of all individual values in the sample, and n is the number of individual values.

Although the mean provides a descriptive summary of the sample, it is of itself of limited value. As illustrated in Figure 5.6, a symmetrical distribution of values in the sample may, in one case, be clustered near the mean. Or,

a set of values may have the same mean, but be distributed widely around it. These contrasting conditions represent different types of variation within each sample called the **frequency distribution.** Whether due to chance or to one or more experimental variables, such variation creates the need for methods to describe sample measurements statistically.

Variance

As shown in Figure 5.6, the range and distribution of values on either side of the mean determines the shape of the distribution curve. The degree to which values within this distribution diverge from the mean is called the sample **variance** (s^2) and is used as an estimate of the variation present in an infinitely large population. The variance for a sample is calculated as

$$s^2 = \frac{\Sigma(X_i - \overline{X})^2}{n - 1}$$

where the sum (Σ) of the squared differences between each measured value (X_i) and the mean ($\overline{X}$) is divided by one less than the total sample size ($n - 1$). To avoid the numerous subtraction functions necessary in calculating s^2 for a large sample, we can convert the equation to its algebraic equivalent:

$$s^2 = \frac{\Sigma X_i{}^2 - n\overline{X}^2}{n - 1}$$

The variance is a valuable measure of sample variability. Two distributions may have identical means ($\overline{X}$), yet vary considerably in their frequency distribution around the mean. The variance represents the average squared deviation of the measurements from the mean. The estima-

TABLE 5.2 Sample inclusion for various s values

Multiples of s	Percent of Sample Included
$\overline{X} \pm 1s$	68.3%
$\overline{X} \pm 1.96s$	95.0
$\overline{X} \pm 2s$	95.5
$\overline{X} \pm 3s$	99.7

tion of variance has been particularly valuable in determining the degree of genetic control of traits when the immediate environment also influences the phenotype.

Standard Deviation

Because the variance is a squared value, its unit of measurement is also squared (*e.g.*, cm^2). To express variation around the mean in the original units of measurement, it is necessary to calculate the square root of the variance, a term called the **standard deviation** (*s*):

$$s = \sqrt{s^2}$$

Table 5.2 shows the percentage of the individual values within a normal distribution that is included with different multiples of the standard deviation. Sixty-eight percent of all values in the sample fall within the mean plus or minus one standard deviation ($\overline{X} \pm 1s$). Over 95 percent of all values are found within two standard deviations ($\overline{X} \pm 2s$). As such, the standard deviation provides an important descriptive summary of a set of data. Furthermore, *s* can be interpreted as a probability. The mean plus or minus one standard deviation ($\overline{X} \pm 1s$) indicates that there is a 68 percent probability that a measured value picked at random will fall within that range.

Standard Error of the Mean

To estimate how much the means of other similar samples drawn from the same population might vary, we can calculate the **standard error of the mean** ($S_{\overline{X}}$):

$$S_{\overline{X}} = \frac{s}{\sqrt{n}}$$

where *s* is the standard deviation, and $\sqrt{n}$ is the square root of the sample size. The standard error of the mean is a measure of the accuracy of the sample mean—that is, the

variation of sample means in replications of the experiment. Because the standard error of the mean is computed by dividing *s* by $\sqrt{n}$, it is always a smaller value than the standard deviation.

Analysis of a Quantitative Character

To illustrate how biometric methods are used to analyze quantitative characters statistically, we will consider a simplified example involving fruit weight in tomatoes. Let us assume that fruit weight is a quantitative character and that one highly inbred strain (one that is highly homozygous) produces tomatoes averaging 18 oz. in weight, and another highly inbred strain produces fruit averaging 6 oz. in weight. These two varieties are crossed and produce an F_1 generation with weights ranging from 10 oz. to 14 oz. The F_2 population contains individuals that produce fruit ranging from 6 oz. to 18 oz. The results characterizing both generations are shown in Table 5.3.

The mean value for the fruit weight in the F_1 generation can be calculated as:

$$\overline{X} = \frac{\sum X_i}{n} = \frac{626}{52} = 12.04$$

Similarly, the mean value for fruit weight in the F_2 generation is calculated as:

$$\overline{X} = \frac{\sum X_i}{n} = \frac{872}{72} = 12.11$$

Average fruit weight is 12.04 oz. in the F_1 generation and 12.11 oz. in the F_2 generation. Although these mean values are similar, it is apparent from the frequency distributions (Table 5.3) that there is more variation present in the F_2 generation. Fruit weight ranges from 6 oz. to 18 oz. in the F_2 generation, but only from 10 oz. to 14 oz. in the F_1 generation.

To assist in quantifying the amount of variation present in each generation, we can calculate the variance (Table 5.4). As noted above, the sample variance can be calculated as the sum of the squared differences between each value and the mean, divided by one less than the total number of observations. However, in the case where a number of observations (f) have been grouped into representative classes (x), the variance can be calculated according to the formula:

TABLE 5.3 Distribution of F_1 and F_2 progeny

														Weight in oz.		
		6	7	8	9	10	11	12	13	14	15	16	17	18
Number of	F_1:					4	14	16	12	6				
Individuals	F_2:	1	1	2	0	9	13	17	14	7	4	3	0	1

TABLE 5.4 Calculation of variance

	F_1				F_2		
x	f	$f(x)$	$f(x)^2$	x	f	$f(x)$	$f(x)^2$
6				6	1	6	36
7				7	1	7	48
8				8	2	16	128
9				9	0	0	0
10	4	40	400	10	9	90	900
11	14	154	1694	11	13	143	1573
12	16	192	2304	12	17	204	2448
13	12	156	2028	13	14	182	2366
14	6	84	1176	14	7	98	1372
15				15	4	60	900
16				16	3	48	768
17				17	0	0	0
18				18	1	18	324
	$n = \overline{52}$	$\sum fx = \overline{626}$	$\sum fx^2 = \overline{7602}$		$n = \overline{72}$	$\sum fx = \overline{872}$	$\sum fx^2 = \overline{10{,}864}$

For F_1:

$$s^2 = \frac{52 \times 7602 - (626)^2}{52(52 - 1)}$$

$$= \frac{395{,}304 - 391{,}876}{2652}$$

$$= 1.29$$

For F_2:

$$s^2 = \frac{72 \times 10{,}864 - (872)^2}{72(72 - 1)}$$

$$= \frac{782{,}208 - 760{,}384}{5112}$$

$$= 4.27$$

$$s^2 = \frac{n \sum f(x^2) - (\sum fx)^2}{n(n - 1)}$$

As shown in Table 5.4, the variance for the F_1 generation is 1.29, and for the F_2 generation it is 4.27. When converted to the standard deviation $s = \sqrt{s^2}$ the values become 1.13 and 2.06, respectively. Therefore, the distribution of tomato weight in the F_1 generation can be described as 12.04 ± 1.13, and that in the F_2 generation can be described as 12.11 ± 2.06. This analysis indicates that the mean fruit weight of the F_1 is identical to that of the F_2, but the F_2 generation shows greater variability in the distribution of weights than does the F_1.

Observations about the inheritance of fruit weight in crosses between these two strains of tomatoes meet the expectations for polygenic traits. For the sake of this example, if we assume that each parental strain is homozygous for the additive or nonadditive alleles that control fruit weight, we can estimate the number of gene pairs involved in controlling this trait in these two strains of tomatoes. Since 1/72 of the F_2 offspring have a phenotype that overlaps one of the parental strains (72 total F_2 offspring: one weighs 6 oz., one weighs 18 oz.; see Table 5.3), the use of the formula $1/4^n = 1/72$ indicates that n is between 3 and 4, indicative of the number of genes that control fruit weight in these tomato strains. If this experiment were repeated many times, with similar results, our confidence in this conclusion would be bolstered.

Heritability is a measure of the genetic contribution to phenotypic variability

Having just introduced several ways in which quantitative or continuous variation can be measured and characterized in populations, we now consider how we assess the extent to which genetic factors contribute to such phenotypic variation. Often, much of the variation can be attributed to genetic factors, with the total environment having less impact. In other cases, the environment may have a greater impact on phenotypic variation within a population. The following discussion considers how geneticists attempt to define the impact of heredity versus environment on phenotypic variation.

Broad-Sense Heritability

Provided that a trait can be measured quantitatively, experiments on many plants and animals can test the origin of variation. One approach is to use inbred strains containing individuals of a relatively homogeneous (highly homozygous) genetic background. Experiments are then designed to test the effects of the range of prevailing environmental conditions on phenotypic variability. Variation observed *between* different inbred strains reared in a constant environment is due predominantly to genetic factors. Variation observed *among* members of the same inbred strain reared under different environmental conditions is due to nongenetic factors, which are generally categorized as "environmental."

The relative importance of genetic versus environmental factors may be formally assessed by examining the **heritability index** (H^2), which can be calculated using an analysis of variance among individuals of a known genetic relationship. (In the ensuing discussion, we will assign the term V to designate variance.) This is an important approach when investigating organisms with long generation times. Also called **broad-sense heritability,** H^2 measures the degree to which **phenotypic variance** (V_P) is due to variation in genetic factors for a single population under the limits of environmental variation during the study. It is important to note that calculation of H^2 *does not* determine the proportion of the total phenotype attributable to genetic factors. It *does* estimate the proportion of observed variation in the phenotype attributable to genetic factors compared to environmental factors.

Phenotypic variance is due to the sum of three components: **environmental variance** (V_E), **genetic variance** (V_G), and variance resulting from the **interaction of genetics and environment** (V_{GE}). Therefore, phenotypic variance (V_P) is theoretically expressed as:

$$V_P = V_E + V_G + V_{GE}$$

Because V_{GE} is often negligible, it is usually omitted. Therefore, the simpler equation is generally used:

$$V_P = V_E + V_G$$

Broad heritability expresses that proportion of variance due to the genetic component:

$$H^2 = \frac{V_G}{V_P}$$

An H^2 value that approaches 1.0 indicates that the environmental conditions have had little impact on phenotypic variance in the population studied. An H^2 value close to 0.0 indicates that variation in the environment has been almost solely responsible for the observed phenotypic variation within the population studied. Rarely do phenotypic characteristics demonstrate H^2 values that are close to either extreme, indicative of the joint role that both genetics and environment play in expression of most phenotypic characteristics.

It is not possible to obtain an absolute H^2 value for any given character. If measured in a different population under a greater or lesser degree of environmental variability, H^2 might well change for that character. For that reason, broad-sense heritability estimates are most useful in highly inbred strains or genetic clones such as artificially selected plant populations that are asexually propagated by cuttings. Furthermore, broad-sense heritability estimates are not very accurate in estimating the selection potential of quantitative traits since H^2 calculations take into account all forms of genetic variation in general, not specifically additive genetic effects. Therefore, another type of calculation, narrow-sense heritability, has been devised that is of more practical use.

Narrow-Sense Heritability

Information regarding heritability is most useful in animal and plant breeding as a measure of potential response to selection. In this case, a different estimate of heritability must be used, based on a subcomponent of V_G:

$$V_G = V_A + V_D + V_I$$

V_A represents **additive variance** that results from the average effect of additive components of genes. V_D represents **dominance variance,** deviation from the additive components that results when phenotypic expression in heterozygotes is not precisely intermediate between the two homozygotes. V_I reflects **interactive variance,** deviation from the additive components that occurs when two or more loci behave epistatically. V_I reflects the variance that is not associated with the average effect of V_A and is often negligible. Thus, it is often excluded from calculations.

When V_G is partitioned into V_A and V_D, a new assessment of heritability, h^2, or **narrow-sense heritability,** may be calculated. It is h^2 that is useful in assessing selection potential in randomly breeding animal and plant populations:

$$h^2 = \frac{V_A}{V_P}$$

Because $V_P = V_E + V_G$ and $V_G = V_A + V_D$, we obtain:

$$h^2 = \frac{V_A}{V_E + V_A + V_D}$$

Artificial Selection

The process of selecting a specific group of organisms from an initially heterogeneous population for future breeding purposes is referred to as **artificial selection.** A relatively high h^2 value is a prediction of the impact that selection will have in altering a population. As you can imagine based on our preceding discussion, measuring the components necessary to calculate h^2 is a complex task. A more simplified approach involves measurement of the central tendencies (the means) of a trait from (1) a parental population exhibiting a bell-shaped distribution (M), (2) a "selected" segment of the parental population that expresses the most desirable quantitative phenotypes (M1), and (3) the offspring (M2) resulting from interbreeding the selected M1 group. When this is accomplished, the following relationship of the three means and h^2 exists:

$$M2 = M + h^2(M1 - M)$$

Solving this equation for h^2 leads to the formula:

$$h^2 = \frac{M2 - M}{M1 - M}$$

Once the above relationships are established, the equation can be further simplified by defining $M2 - M$ as the **response** (R) and $M1 - M$ as the **selection differential** (S), where h^2 reflects the ratio of the response observed to the total response possible. Thus,

$$h^2 = \frac{R}{S}$$

To illustrate the assessment of selection, assume that we measure the diameter of corn kernels in a population where the mean diameter (M) was much larger than desirable (20 mm), and from that population we select a group with the smallest diameters, for which the mean (M1) equals 10 mm. If plants that yielded this selected population are then interbred and the progeny kernels yield a mean (M2) of 13 mm, then we can calculate h^2 in order to estimate the potential for artificial selection on kernel size:

$$\begin{aligned} h^2 &= \frac{13 - 20}{10 - 20} \\ &= \frac{-7}{-10} \\ &= 0.70 \end{aligned}$$

On the basis of this calculation, we may conclude that the selection potential for kernel size is relatively high.

Corn is involved in the longest-running artificial selection experiment known, which is still being conducted at the State Agricultural Laboratory in Illinois. Since 1896, corn has been selected for both high and low oil content. After 76 generations, selection continues to result in increased oil content (Figure 5.7). Predictably, as selection has progressed successfully, a greater number of plants possess a higher percentage of additive alleles involved in oil production. Thus, heritability (h^2) of increased oil content in succeeding generations has declined (see parenthetical values at generations 9, 25, 52, and 76 in Figure 5.7) as artificial selection comes closer and closer to optimizing the genetic potential for oil production. Theoretically, the process will continue until all individuals in the

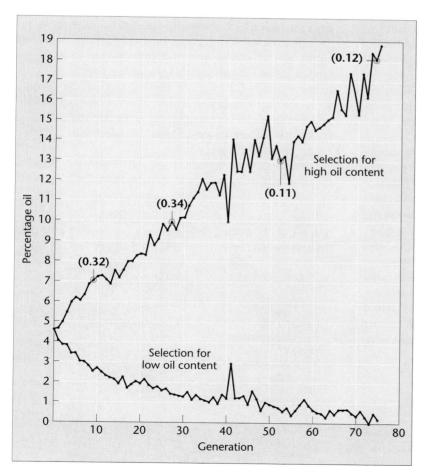

FIGURE 5.7 Response of corn selected for high and low oil content over 76 generations. The numbers in parentheses at generations 9, 25, 52, and 76 for the "high oil" line indicate the calculation of heritability at these points in the continuing experiment.

TABLE 5.5 Estimates of Heritability for Traits in Different Organisms

Trait	Heritability (h^2)
Mice	
Tail length	60%
Body weight	37
Litter size	15
Drosophila	
Abdominal bristle number	52
Wing length	45
Egg production	18
Chickens	
Body weight	50
Egg production	20
Egg hatchability	15
Cattle	
Birth weight	51
Milk yield	44
Conception rate	3

population contain a uniform genotype for the additive alleles responsible for oil content. At that point, heritability will be reduced to zero, and artificial selection will cease. Examination of the selection for low oil content shows that heritability is approaching this point.

Similar calculations involving artificial selection for other traits in a variety of organisms have established whether artificial selection will be an effective approach in obtaining populations exhibiting desirable phenotypes. Table 5.5 provides estimates of narrow heritability for a variety of traits in various organisms. These proportions are expressed as percentage values. As you can see, heritability varies considerably among traits.

In general, heritability is low for traits that are essential to an organism's survival, primarily because the genetic component must be largely optimized during evolution. Egg production, litter size, and conception rate are examples where such physiological limitations on selection have already been established. Traits that are less critical to survival, such as body weight, tail length, and wing length, show higher heritability values. Narrow-sense heritability estimates are more valuable when they have been based on data collected in many populations and environments and where a clear trend is established. Based on such estimates, selection techniques have led to vast improvements in the quality of animal and plant products.

Twin studies allow an estimation of heritability in humans

Traditional heritability studies are not possible in humans, for obvious reasons. However, human twins are very useful subjects for studying the heredity versus environment question. **Monozygotic (MZ)** or **identical twins,** derived from the division and splitting of a single egg following fertilization, are identical in their genetic compositions. Although most identical twins are reared together and are exposed to very similar environments, some pairs are separated and raised in different settings. For any particular trait, average similarities or differences can be investigated. Such an analysis is particularly useful because characteristics that remain similar in different environments are believed to have a strong genetic component. These data can then be compared with a similar analysis of **dizygotic (DZ)** or **fraternal twins,** who originate from two separate fertilization events. Dizygotic twins are thus no more genetically similar than any two siblings, sharing (on average) one-half of their genes.

Another approach involves the measuremet of **concordance** values of phenotypic expression in twin pairs raised together in a similar environment. Twins are said to be concordant for a given trait if both express it or neither expresses it. If one expresses the trait and the other does not, the pair is said to be **discordant.** Comparison of concordance values of MZ vs. DZ twins reared together (Table 5.6) illustrates the potential value for heritability assessment.

These data must be examined very carefully before any conclusions are drawn. If the concordance value approaches 90 to 100 percent in monozygotic twins, we might be inclined to interpret this value as indicating a large genetic contribution to the expression of the trait. In some cases—for example, blood types and eye color—we know that this is indeed true. In the case of measles, however, a high concordance value merely indicates that the trait is almost always induced by a factor in the environment—in this case, a virus.

It is more meaningful to compare the *difference* between the concordance values of monozygotic and dizygotic twins.

TABLE 5.6 A Comparison of Concordance of Various Traits Between Monozygotic (MZ) and Dizygotic (DZ) Twins

Trait	Concordance	
	MZ	DZ
Blood types	100%	66%
Eye color	99	28
Mental retardation	97	37
Measles	95	87
Idiopathic epilepsy	72	15
Schizophrenia	69	10
Diabetes	65	18
Identical allergy	59	5
Tuberculosis	57	23
Cleft lip	42	5
Club foot	32	3
Mammary cancer	6	3

If these values are significantly higher for monozygotic twins than for dizygotic twins, we suspect that there is a strong genetic component involved in the determination of the trait. We reach this conclusion because monozygotic twins, with identical genotypes, would be expected to show a greater concordance than genetically related, but not genetically identical, dizygotic twins. In the case of measles, where concordance is high in both types of twins, the environment is assumed to contribute significantly.

Even though a particular trait may demonstrate considerable genetically based variation, it is often difficult to formulate a precise mode of inheritance based on available data. In many cases, the trait is considered to be controlled by multiple-factor inheritance. However, when the environment is also exerting a partial influence, such a conclusion is particularly difficult to prove.

Quantitative trait loci can be mapped

Because quantitative traits are influenced by numerous genes, geneticists would like to know where these genes are located in the genome. Are they linked on a single chromosome, or scattered throughout the genome among many chromosomes? As we will see in the next two chapters, locating, or "mapping," genes within the genome represents an important step in establishing the genetic identity of organisms. The initial approach is to localize genes controlling quantitative traits on a particular chromosome or chromosomes. In the context of this discussion, these genes are called **quantitative trait loci (QTLs)**. As we will see, some rather ingenious methods have been devised to map these genes.

As an example of this type of analysis, we will consider the phenotypic trait in *Drosophila* of resistance to the insecticide DDT, which has been shown to be under polygenic control. To find the loci responsible, strains selected for resistance to DDT and strains selected for sensitivity to DDT were crossed to flies carrying dominant genes that serve to "mark" each of *Drosophila*'s four chromosomes. These markers, known to be located on particular chromosomes, serve as reference points during mapping experiments. Following a variety of crosses, offspring were produced that contained many different combinations of marker chromosomes and chromosomes from either resistant or sensitive strains, as shown in Figure 5.8. Flies with various chromosome combinations were then tested for resistance (their survival) when exposed to DDT. As shown in Figure 5.8, results indicate that each of the chromosomes in *Drosophila* contains genes that contribute to resistance. In other words, the loci bearing the genes that control DDT resistance are scattered throughout the genome.

It is possible to map the position of these genes more specifically along each chromosome in *Drosophila* because

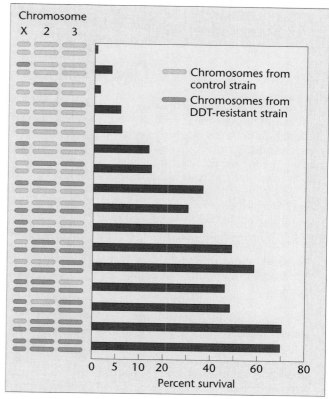

FIGURE 5.8   The differential survival of *Drosophila* carrying different combinations of chromosomes from DDT-resistant and DDT-sensitive (control) strains when exposed to DDT. The results indicate that DDT resistance is polygenic, with genes on each of the chromosomes making a major contribution. The *Drosophila* chromosome 4 carries only a few genes and was omitted from this analysis.

of the presence of molecular markers on each chromosome. The locations of QTLs are determined relative to the known positions of these markers. Called **restriction fragment length polymorphisms (RFLPs),** these markers represent sites along the chromosome where a specific nucleotide sequence exists that can be cleaved by a nuclease (called a *restriction enzyme*) that recognizes that sequence (see Chapter 21 for a detailed discussion of RFLPs). They provide a new approach to enumerating and mapping the loci responsible for quantitative traits. RFLP markers are now available for many organisms of agricultural importance, making possible systematic mapping of QTLs.

For example, in the tomato, hundreds of RFLP markers have been located. They are spaced along all 12 chromosomes of this organism. Analysis is performed by crossing plants with extreme, but opposite phenotypes, and following the crosses through several generations. When both a marker and a phenotypic trait of interest are expressed together, they are said to *cosegregate*. Consistent cosegregation establishes the presence of a QTL at or near the RFLP marker along the chromosome. Whenever both

The Green Revolution Revisited

Of the 5.9 billion people now living on Earth, about 750 million don't have enough to eat. And despite efforts to limit population growth, an additional 1 million people are expected to go hungry each year for the next several decades.

Will we be able to solve this problem? The past gives us some reasons to be optimistic. In the 1950s and 1960s, in the face of looming population increases, plant scientists around the world set about to increase the production of crop plants, including the three most important grains, wheat, rice, and maize. These efforts became known as the Green Revolution. The approach was three-pronged: (1) to increase the use of fertilizers, pesticides, and irrigation water, (2) to bring more land under cultivation, and (3) to develop improved varieties of crop plants by intensive plant breeding. While highly successful, in recent years the rate of increase in grain yields has slowed. If food production is to keep pace with the projected increase in the world's population, plant breeders will

have to depend more and more on the genetic improvement of crop plants to provide higher yields. But is this possible? Are we approaching the theoretical limits of yield in important crop plants? Recent work with rice suggests that the answer to this question is a resounding no.

Rice ranks third in worldwide production, just behind wheat and maize. About 2 billion people, fully one-third of Earth's population, depend on rice for their basic nourishment. The majority of the world's rice crop is grown and consumed in Asia, but it is also a dietary staple in Africa and Central America. The Green Revolution for rice began in 1960 with the establishment of the International Rice Research Institute (IRRI), headquartered at Los Baños, Philippines. The goal was to breed rice with improved disease resistance and higher yield. Breeders were almost too successful: The first high-yield varieties were so top-heavy with grain that they tended to fall over. (Plant breeders call this "lodging.") To

reduce lodging, IRRI breeders crossed a high-yield line with a dwarf native variety to create semi-dwarf lines, which were introduced to farmers in 1966. Due in large part to the adoption of the semi-dwarf lines, the world production of rice doubled in the next 25 years.

Rice breeders cannot afford to rest on their laurels, however, as the yield of modern rice varieties has not improved much in recent years. Predictions suggest that a 70 percent increase in the annual rice harvest may be necessary to keep pace with anticipated population growth during the next 30 years. Breeders are now looking to wild rice varieties for further crop improvement. Leading the way are Susan McCouch, Steven Tanksley, and their co-workers at Cornell University. To test the hypothesis that wild rice species carry genes that will improve the yield of cultivated varieties, they crossed cultivated rice (*Oryza sativa*) with a low-yield wild ancestor species (*Oryza rufipogon*) and then successively backcrossed the interspecific hybrid to cultivated rice for

an RFLP marker and a QTL responsible for the trait under investigation are closely linked along a single chromosome, they are more likely to demonstrate an association together throughout a pedigree than if they are not closely linked. When numerous QTLs are located, a genetic map is created for the involved genes.

RFLP analysis has resulted in mapping of QTLs for fruit weight, soluble solids, and acidity in the tomato. Six loci responsible for fruit weight were found on six different chromosomes, four loci for soluble solids were identified on five chromosomes, and five loci for acidity were found on four different chromosomes. Several chromosomes contained loci for all three traits. Further investigation has revealed that the genes present at these loci

account for approximately 50 percent of the phenotypic variation expressed in these traits. Note that the RFLP method simply allows the identification of chromosomal regions, not individual genes, although each locus may indeed house only a single gene.

The determination of the locations of QTLs for agriculturally important characteristics has permitted increased efficiency in selection and will open the door to their future genetic manipulation and transfer between organisms. As these mapping techniques became more routine in the 1990s, they became an important tool in the repertoire of genetic engineering. Eventually, such methods will be able to be used to isolate and characterize the individual genes that control quantitative traits.

three generations. In theory, this would create lines whose genomes were about 95 percent from *O. sativa* and 5 percent from *O. rufipogon*. When testing these backcrossed lines for grain yield, they found that several of them outproduced cultivated rice by as much as 30 percent. These results demonstrated strikingly that even though wild rice relatives have low yields and appear to be inferior to cultivated rice, they still carry genes that will increase the yield of elite rice varieties. It will now be up to breeders to exploit the wild rice relatives.

But introducing favorable genes from a wild relative into a cultivated variety by conventional breeding is a long and involved process, often requiring a decade or more of crossing, selection, backcrossing, and more selection. Future improvements in cultivated species must be quicker if crop yields are to keep up with population growth. Fortunately, modern gene-mapping techniques are now leading to the identification of *quantitative trait loci* (QTL) that control complex traits such as yield and disease resistance. This makes possible a more direct approach to crop improvement, which

has been termed the *advanced backcross QTL method*.

First, a cultivated variety is crossed with a wild relative, just as *O. sativa* was interbred with *O. rufipogon*. The hybrid is then backcrossed to the cultivated variety to generate lines that contain only a small fraction of the "wild" genome. The backcross lines with the best qualities (e.g., highest yield and most disease resistance) are selected, and the "wild QTL" responsible for the superior performance are identified using a detailed molecular linkage map. Once beneficial QTL are discovered by this method, they may be introduced into other cultivated varieties.

In order for this strategy to succeed, it is essential that wild crop relatives be preserved as a storehouse of potentially useful genes. Efforts were begun in the 1970s to protect the existing crop relatives of many plants in their natural habitats and to preserve them in seed banks. As the work of McCough and Tanksley and others has shown, it is not possible to predict which wild varieties may be needed decades or even centuries from now to

contribute their beneficial alleles to cultivated varieties. To prevent the loss of superior genes, the widest possible spectrum of wild species must be preserved, even those that have no obvious favorable characteristics.

Almost 60 years ago, the great Russian plant geneticist N. I. Vavilov suggested that wild relatives of crop plants could be the source of genes to improve agriculture. In the coming century, Vavilov's vision may finally be realized, as genes from long-neglected wild crop relatives, identified by new molecular methods, spark a revitalized Green Revolution.

REFERENCES

MANN, C. 1997. Reseeding the Green Revolution. *Science* 277:1038–43.

RONALD, P.C. 1997. Making rice disease-resistant. *Sci. Am.* (Nov.) 277:98–105.

TANKSLEY, S.D., and McCOUCH, S.R. 1997. Seed banks and molecular maps: Unlocking genetic potential from the wild. *Science* 277:1063–66.

XIAO, J. et al. 1996. Genes from wild rice improve yield. *Nature* 384:223–24.

CHAPTER SUMMARY

1. Continuous variation—variation that is not easily categorized into distinct phenotypic classes—is exhibited in crosses involving traits under polygenic control. Such traits are quantitative in nature and are inherited as a result of the cumulative impact of additive alleles.

2. Polygenic characteristics can be analyzed using statistical methods, which include the mean, the variance, the standard deviation, and the standard error of the mean. Such statistical analysis can be descriptive, can be used to make inferences about a population, or can be used to compare and contrast sets of data.

3. For many phenotypic characteristics, it is difficult to ascertain when variation is due to genetic or to environmental factors. Heritability, the estimate of the rel-

ative importance of genetic versus nongenetic factors in determining genetic variation in populations, can be calculated for many characters and is especially useful in selective breeding of commercially valuable plants and animals.

4. Studies involving twins are aimed at resolving the question of heredity versus environment in human traits. The degree of concordance of a trait may be compared in monozygotic (identical) and dizygotic (fraternal) twins raised together or apart.

5. Loci-bearing genes that control a quantitative trait are called quantitative trait loci (QTLs). Using either genetic or molecular markers, QTLs may be mapped in order to ascertain their location and distribution within the genome.

INSIGHTS AND SOLUTIONS

1. In a plant, height varies from 6 to 36 cm. When 6-cm and 36-cm plants were crossed, all F_1 plants were 21 cm. In the F_2 generation, a continuous range of heights was observed. Most were around 21 cm, and 3 of 200 were as short as the 6-cm P_1 parent.

   (a) What mode of inheritance is illustrated, and how many gene pairs are involved?

   Solution: Polygenic inheritance is illustrated where a continuous trait is involved and where alleles contribute additively to the phenotype. The 3/200 ratio of F_2 plants is the key to determining the number of gene pairs. This reduces to a ratio of 1/66.7, very close to 1/64. Using the formula $1/4^n = 1/64$ (where 1/64 is equal to the proportion of F_2 phenotypes as extreme as either P_1 parent), $n = 3$. Therefore, three gene pairs are involved.

   (b) How much does each additive allele contribute to height?

   Solution: The variation between the two extreme phenotypes is:

   $$36 - 6 = 30 \text{ cm}$$

   Because there are six potential additive alleles (*AABBCC*), each contributes

   $$30 / 6 = 5 \text{ cm}$$

   to the base height of 6 cm, which results when no additive alleles (*aabbcc*) are part of the genotype.

   (c) List all genotypes that give rise to plants that are 31 cm.

   Solution: All genotypes that include 5 additive alleles will be 31 cm (5 alleles × 5 cm + 6 cm base height = 31 cm):

   Therefore, *AABBCc*, *AABbCC*, and *AaBBCC* are the genotypes that will result in plants that are 31 cm.

   (d) In a cross separate from the F_1 above, a plant of unknown phenotype and genotype was test-crossed with the following results:

   1/4 11 cm
   2/4 16 cm
   1/4 21 cm

An astute genetics student realized that the unknown plant could be only one phenotype, but could be any of three genotypes. What were they?

Solution: When test-crossed (with *aabbcc*), the unknown plant must be able to contribute either one, two, or three additive alleles in its gametes in order to yield the three phenotypes in the offspring. Since no 6-cm offspring are observed, the unknown plant never contributes all nonadditive alleles (*abc*). Only plants that are homozygous at one locus and heterozygous at the other two loci will meet these criteria. Therefore, the unknown parent can be any of three genotypes, all of which have a phenotype of 26 cm:

> *AABbCc*
> *AaBbCC*
> *AaBBCc*

For example, in the first genotype (*AABbCc*):

> *AABbCc* × *aabbcc*

> 1/4 *AaBbCc* ⟶ 21 cm
> 1/4 *AaBbcc* ⟶ 16 cm
> 1/4 *AabbCc* ⟶ 16 cm
> 1/4 *Aabbcc* ⟶ 11 cm

which is the ratio of phenotypes observed.

2. The results shown in the table along the bottom of this page were recorded for ear length in corn:

   For each of the parental strains and the F_1, calculate the mean values for ear length.

   Solution: The mean values can be calculated as follows:

   $$\overline{X} = \frac{\sum X_i}{n} \qquad P_A: \overline{X} = \frac{\sum X_i}{n} = \frac{378}{57} = 6.63$$

   $$P_B: \overline{X} = \frac{\sum X_i}{n} = \frac{1697}{101} = 16.80$$

   $$F_1: \overline{X} = \frac{\sum X_i}{n} = \frac{836}{69} = 12.11$$

(Problem 2)—Length of Ear in cm

	5	6	7	8	9	10	11	12	13	14	15	16	17	18	19	20	21
Parent A	4	21	24	8													
Parent B									3	11	12	15	26	15	10	7	2
F_1					1	12	12	14	17	9	4						

3. For the corn plant described in Problem 2, compare the mean of the F_1 with that of each parental strain. What does this tell you about the type of gene action involved?

Solution: The F_1 mean (12.11) is almost midway between the parental means of 6.63 and 16.80. This indicates that the genes in question may be additive in effect.

4. The mean and variance of corolla length in two highly inbred strains of *Nicotiana* and their progeny are shown below. One parent (P_1) has a short corolla and the other parent (P_2) has a long corolla.

Strain	Mean (mm)	Variance
P_1 short	40.47	3.12
P_2 long	93.75	3.87
F_1 ($P_1 \times P_2$)	63.90	4.74
F_2 ($F_1 \times F_1$)	68.72	47.70

Calculate the heritability (H^2) of corolla length in this plant.

Solution: The formula for estimating heritability is $H^2 = V_G/V_P$, where V_G and V_P are the genetic and phenotypic components of variation, respectively. The main issue in this problem is obtaining some estimate of two components of phenotypic variation: genetic and environmental factors. V_P is the combination of genetic and environmental variance. Because the two parental strains are true breeding, they are assumed to be homozygous, and the variance of 3.12 and 3.87 is considered to be the result of environmental influences. The average of these two values is 3.50. The F_1 is also genetically homogeneous and gives us an additional estimate of the environmental factors. By averaging with the parents,

$$\frac{3.50 + 4.74}{2} = 4.12$$

we obtain a relatively good idea of environmental impact on the phenotype. The phenotypic variance in the F_2 is the sum of the genetic (V_G) and environmental (V_E) components. We have estimated the environmental input as 4.12, so 47.70 minus 4.12 gives us an estimate of V_G, which is 43.58. Heritability then becomes 43.58/47.70 or 0.91. This value, when viewed in percentage form, indicates that about 91 percent of the variation in corolla length is due to genetic influences.

PROBLEMS AND DISCUSSION QUESTIONS

1. Distinguish between discontinuous and continuous variation. Which type relates to inheritance of a quantitative nature?

2. Define the following: (a) polygenes, (b) additive alleles, (c) multiple-factor hypothesis, (d) monozygotic and dizygotic twins, (e) concordance and discordance, and (f) heritability.

3. A dark-red strain and a white strain of wheat are crossed and produce an intermediate, medium red F_1. When the F_1 plants are interbred, an F_2 generation is produced in a ratio of 1 dark red:4 medium-dark red:6 medium red:4 light red:1 white. Further crosses reveal that the dark red and white F_2 plants are true breeding.
  (a) Based on the ratio of offspring in the F_2, how many genes are involved in the production of color?
  (b) How many additive alleles are needed to produce each possible phenotype?
  (c) Assign symbols to these alleles and list possible genotypes that give rise to the medium-red and the light-red phenotypes.
  (d) Predict the outcome of the F_1 and F_2 generations in a cross between a true-breeding medium-red plant and a white plant.

4. Height in humans depends on the additive action of genes. Assume that this trait is controlled by the four loci R, S, T, and U and that environmental effects are negligible. Instead of additive versus nonadditive alleles, assume that additive and partially additive alleles exist. Additive alleles contribute two units, and partially additive alleles contribute one unit to height.
  (a) Can two individuals of moderate height produce offspring that are much taller or shorter than either parent? If so, how?
  (b) If an individual with the minimum height specified by these genes marries an individual of intermediate or moderate height, will any of their children be taller than the tall parent? Why or why not?

5. An inbred strain of plants has a mean height of 24 cm. A second strain of the same species from a different geographical region also has a mean height of 24 cm. When plants from the two strains are crossed together, the F_1 plants are the same height as the parent plants. However, the F_2 generation shows a wide range of heights; the majority are like the P_1 and F_1 plants, but approximately 4 of 1000 are only 12 cm high, and about 4 of 1000 are 36 cm high.
  (a) What mode of inheritance is occurring here?
  (b) How many gene pairs are involved?
  (c) How much does each gene contribute to plant height?
  (d) Indicate one possible set of genotypes for the original P_1 parents and the F_1 plants that could account for these results.
  (e) Indicate three possible genotypes that could account for F_2 plants that are 18 cm high and three that account for F_2 plants that are 33 cm high.

6. Erma and Harvey were a compatible barnyard pair, but a curious sight. Harvey's tail was only 6 cm, while Erma's was 30 cm. Their F_1 piglet offspring all grew tails that were 18 cm.

When inbred, an F_2 generation resulted in many piglets (Erma and Harvey's grandpigs), whose tails ranged in 4-cm intervals from 6 to 30 cm (6, 10, 14, 18, 22, 26, 30). Most had 18-cm tails, while 1/64 had 6-cm tails and 1/64 had 30-cm tails.

(a) Explain how tail length is inherited by describing the mode of inheritance, indicating how many gene pairs are at work, and designating the genotypes of Harvey, Erma, and their 18-cm offspring.

(b) If one of the 18-cm F_1 pigs is mated with the 6-cm F_2 pigs, what phenotypic ratio would be predicted if many offspring resulted? Diagram the cross.

7. In the following table, average differences of height and weight between monozygotic twins (reared together and apart), dizygotic twins, and non-twin siblings are compared. Draw as many conclusions as you can concerning the effects of genetics and the environment in influencing these human traits.

Trait	MZ Reared Together	MZ Reared Apart	DZ Reared Together	Sibs Reared Together
Height (cm)	1.7	1.8	4.4	4.5
Weight (kg)	1.9	4.5	4.5	4.7

8. Discuss how monozygotic and dizygotic twins reared together and apart are useful in assessing the genetic component responsible for phenotypic variation in humans.

9. List as many human traits as you can that are likely to be under the control of a polygenic mode of inheritance.

10. Corn plants from a test plot are measured, and the distribution of heights at 10-cm intervals is recorded in the following table:

Height (cm)	Plants (no.)
100	20
110	60
120	90
130	130
140	180
150	120
160	70
170	50
180	40

Calculate (a) the mean height, (b) the variance, (c) the standard deviation, and (d) the standard error of the mean. Plot a rough graph of plant height versus frequency. Do the values represent a normal distribution? Based on your calculations, how would you assess the variation within this population?

11. Contrast broad-sense heritability (H^2) and narrow-sense heritability (h^2). To what type of population is each calculation applicable? Which is useful in artificial selection procedures? Why?

12. The mean and variance of plant height of two highly inbred strains (P_1 and P_2) and their progeny (F_1 and F_2) are shown below. Calculate the broad-sense heritability (H^2) of plant height in this species.

Strain	Mean (cm)	Variance
P_1	34.2	4.2
P_2	55.3	3.8
F_1	44.2	5.6
F_2	46.3	10.3

13. A hypothetical study investigated the vitamin A content and the cholesterol content of eggs from a large population of chickens. The variances (V) were calculated, as shown below:

	Trait	
Variance	Vitamin A	Cholesterol
V_P	123.5	862.0
V_E	96.2	484.6
V_A	12.0	192.1
V_D	15.3	185.3

(a) Calculate the narrow heritability (h^2) for both traits.

(b) Which trait, if either, is likely to respond to selection?

14. In an assessment of learning in *Drosophila*, flies may be trained to avoid certain olfactory cues. In one population, a mean of 8.5 trials was required. A subgroup of this parental population that was trained most quickly (mean = 6.0) was interbred and their progeny examined. These flies demonstrated a mean training value of 7.5. Calculate and interpret narrow-sense heritability for olfactory learning in *Drosophila*.

15. In a population of tomato plants mean fruit weight is 60 g and h^2 is 0.3. Predict the results of artificial selection (the mean weight of the progeny) if tomato plants whose fruit averaged 80 g were selected from the original population and interbred.

EXTRA-SPICY PROBLEMS

16. A mutant strain of *Drosophila* was isolated and shown to be resistant to an experimental insecticide, whereas normal (wild type) flies were sensitive to the chemical. Following a cross between resistant flies and sensitive flies, isolated populations were derived that had various combinations of chromosomes from the two strains. Each was tested for resistance, as shown at the top of the next page. Analyze the data and draw any appropriate conclusion about which chromosome(s) contain a gene responsible for inheritance of resistance to the insecticide.

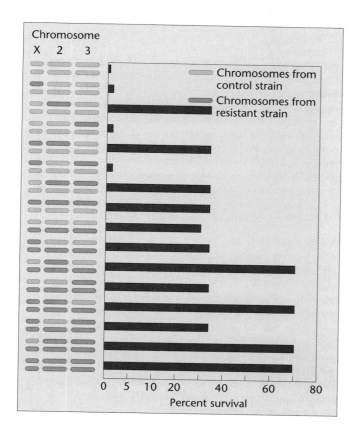

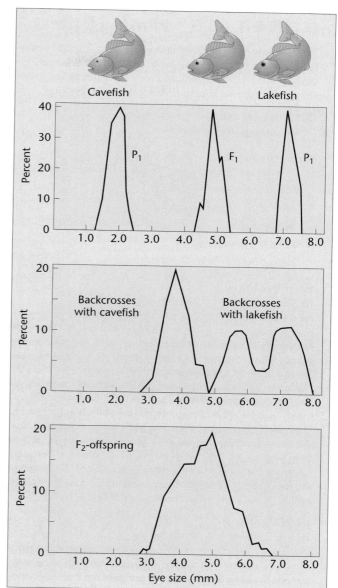

17. In 1988, Horst Wilkens investigated blind cavefish, comparing them to members of a sibling species with normal vision that are found in a lake (we will call them cavefish and lakefish). He found that cavefish eyes are about seven times smaller than lakefish eyes. F_1 hybrids have eyes of intermediate size. These data as well as the $F_1 \times F_1$ cross and those from backcrosses ($F_1 \times$ cavefish and $F_1 \times$ lakefish) are depicted below. Examine Wilkens' results and respond to the following questions.

(a) Based strictly on the F_1 and F_2 results of Wilkens' initial crosses, what possible explanation concerning the inheritance of eye size seems most feasible?

(b) Based on the results of the F_1 backcross with cavefish, is your explanation supported? Explain.

(c) Based on the results of the F_1 backcross with lakefish, is your explanation supported? Explain.

(d) Wilkens examined about 1000 F_2 progeny and estimated that 6–7 genes are involved in determining eye size. Is this sample size adequate to justify this conclusion? (You may want to refer to Chapter 4.) Propose an experimental protocol to test this hypothesis.

(e) A comparison of the embryonic eye in cavefish and lakefish revealed that both reach approximately 4 mm in diameter. However, lakefish continue to grow, while cavefish eye size is greatly reduced. Speculate on the role of the genes involved in this problem.

[*Reference:* Wilkens, H. (1988). *Ecol. Biol.* 23:271–367.]

GENETICS MediaLab

The following resources will help you achieve a better understanding of the concepts presented in this chapter. These resources can be found on the CD packaged with this textbook and on the Companion Website at **http://www.prenhall.com/klug/**.

CD Resources:

Self-grading Chapter Problems

Web Resources:

Genetics and Society Issue: *The Green Revolution Revisited*

Web Destinations in Genetics

Self-grading Chapter Problems

Chapter Search Terms

Genetics Newsgroups

Student Bulletin Board

Web Problem 1:

Time for completion = 10 minutes

How do we study polygenic traits? The genes underlying quantitative traits are assumed to behave as Mendelian factors, following the same rules of segregation and assortment as the genes we have studied in Chapters 3 and 4. Quantitative traits are also assumed to be influenced by the environment. Read Dr. Phillip McClean's explanation of genetic and environmental effects on quantitative traits. What does "additive genetic effects" mean? Examine the data presented on the yields of winter wheat. Does one genotype produce a higher yield in most years than the other two? Can you tell if any one genotype is less variable from year to year? Which genotype would offer the best performance over the broadest range of conditions? To complete this exercise, visit Web Problem 1 in Chapter 5 of your Companion Website and select the keyword **WHEAT.**

Web Problem 2:

Time for completion = 10 minutes

What does heritability measure? Genetic and environmental variances can be statistically estimated through a carefully planned breeding design. The proportion of the total phenotypic variance that can be associated with genetic effects is termed the heritability. Read the article about estimating heritability in New Zealand Angus cattle. [Note: EBV stands for "expected breeding value." It is the difference between the mean of the trait within a family and the mean within the entire herd. Thus, a bull with a large EBV for yearling weight comes from a family with a higher mean weight than most of the families in the herd.] Consider the broad-sense heritability described in Chapter 5. Would a herd with a high genetic variance have a higher or lower heritability than one with a low genetic variance, assuming the environmental variances are equal? If the heritability of body size was high within a family, would the calves be expected to be more or less similar to calves from a population where the heritability was low? Would such a comparison, where the populations are drawn from different environments, be valid? (*Hint:* Think about the components of variance.) To complete this exercise, visit Web Problem 2 in Chapter 5 of your Companion Website and select the keyword **HERITABILITY.**

Web Problem 3:

Time for completion = 10 minutes

How can we study quantitative genetic traits in humans? With experimental organisms, we can set up families and breeding experiments to estimate the genetic influences on quantitative traits, but it would be unethical to perform experiments of this sort on people. Many researchers studying physical and mental disease, personality traits, and estimates of intelligence use data on monozygotic (MZ) and dizygotic (DZ) twins to infer what portion of such traits might have a genetic basis. Read the summary of twin research on left-handedness. Why does the similarity in concordance of handedness in MZ and DZ twins imply an environmental influence? The summary suggests a possible genetic effect on handedness due to correlations between handedness in mothers and both sons and daughters, and between handedness in fathers and sons. Can you think of an environmental explanation for these correlations? To complete this exercise, visit Web Problem 3 in Chapter 5 of your Companion Website and select the keyword **TWINS.**

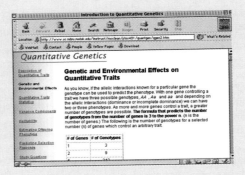

SELECTED READINGS

BRINK, R., ed. 1967. *Heritage from Mendel.* Madison: University of Wisconsin Press.

CROW, J.F. 1993. Francis Galton: Count and measure, measure and count. *Genetics* 135:1.

DUDLEY, J.W. 1977. 76 generations of selection for oil and protein percentage in maize. In: *Proc. Intern. Conf. on Quant. Genet.* E. Pollack, O. Kempthorne, and T. Bailey eds. pp. 459–73. Ames: Iowa State University Press.

FALCONER, D.S., and MACKAY, F. C. 1996. *Introduction to quantitative genetics,* 4th ed. Essex, England: Longman.

FARBER, S. 1980. *Identical twins reared apart.* New York: Basic Books.

FELDMAN, M.W., and LEWONTIN, R.C. 1975. The heritability hangup. *Science* 190:1163–66.

FOWLER, C., and MOONEY, P. 1990. *Shattering: Food, politics, and the loss of genetic diversity.* Tucson: University of Arizona Press.

HALEY, C. 1991. Use of DNA fingerprints for the detection of major genes for quantitative traits in domestic species. *Anim. Genet.* 22:259–77.

———1996. Livestock QTLs: Bringing home the bacon. *Trends Genet.* 11:488–90.

LANDER, E., and BOTSTEIN, D. 1989. Mapping mendelian factors underlying quantitative traits using RFLP linkage maps. *Genetics* 121:185–99.

LANDER, E., and SCHORK, N. 1994. Genetic dissection of complex traits. *Science* 265:2037–48.

LEWONTIN, R.C. 1974. The analysis of variance and the analysis of causes. *Am. J. Hum. Genet.* 26:400–11.

NEWMAN, H.H., FREEMAN, F.N., and HOLZINGER, K.T. 1937. *Twins: A study of heredity and environment.* Chicago: University of Chicago Press.

PATERSON, A., DEVERNA, J., LANINI, B., and TANKSLEY, S. 1990. Fine mapping of quantitative traits loci using selected overlapping recombinant chromosomes in an interspecific cross of tomato. *Genetics* 124:735–42.

PLOMIN, R., MCCLEARN, G., GORA-MASLAK, G., and NEIDERHISER, J. 1991. Use of recombinant inbred strains to detect quantitative trait loci associated with behavior. *Behav. Genet.* 21:99–116.

STUBER, C.W. 1996. Mapping and manipulating quantitative traits in maize. *Trends Genet.* 11:477–87.

ZAR, J.H. 1996. *Biostatistical analysis,* 3rd ed. Upper Saddle River, NJ: Prentice Hall.

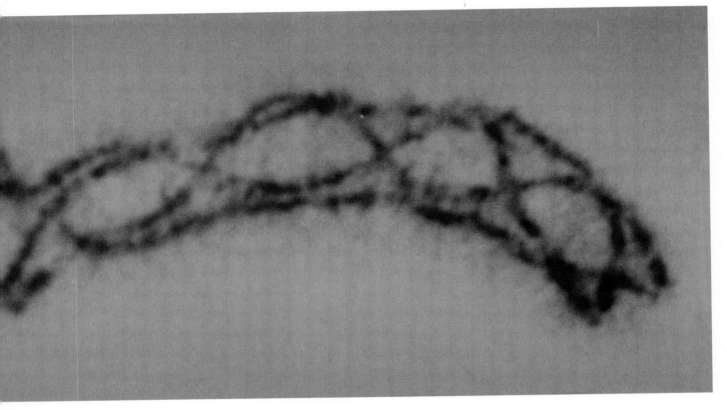

Chiasmata present between synapsed homologues during the first meiotic prophase.

Linkage, Crossing Over, and Mapping in Eukaryotes

KEY CONCEPTS

- **Genes linked on the same chromosome segregate together**

  The Linkage Ratio

- **Crossing over serves as the basis of determining the distance between genes during mapping**

  Morgan and Crossing Over
  Sturtevant and Mapping
  Single Crossovers

- **Determining the gene sequence during mapping relies on the analysis of multiple crossovers**

  Multiple Exchanges
  Three-Point Mapping in *Drosophila*
  Determining the Gene Sequence
  A Mapping Problem in Maize

- **Interference affects the recovery of multiple exchanges**

- **As the distance between two genes increases, mapping experiments become more inaccurate**

- ***Drosophila* genes have been extensively mapped**

- **Crossing over involves a physical exchange between chromatids**

- **Recombination occurs between mitotic chromosomes**

- **Exchanges also occur between sister chromatids**

- **Linkage analysis and mapping can be performed in haploid organisms**

  Gene-to-Centromere Mapping
  Ordered versus Unordered Tetrad Analysis
  Linkage and Mapping

- **Somatic cell hybridization is an important technique in creating human chromosome maps**

- **Did Mendel encounter linkage?**

GenCDX

When you see this icon, there are related animations and exercises on the CD accompanying this text.

As early as 1903, Walter Sutton, who was instrumental in uniting the fields of cytology and genetics, pointed out the likelihood that organisms contain many more "unit factors" than chromosomes. Soon thereafter, genetic studies with several organisms revealed that certain genes were not transmitted according to the law of independent assortment, rather these genes seemed to segregate as if they were somehow joined or linked together. Further investigations showed that such genes were part of the same chromosome, and they were indeed transmitted as a single unit.

We now know that most chromosomes consist of very large numbers of genes. Those that are part of the same chromosome are said to be **linked** and to demonstrate **linkage** in genetic crosses.

Because the chromosome, not the gene, is the unit of transmission during meiosis, linked genes are not free to undergo independent assortment. In theory, the alleles at all loci of one chromosome should be transmitted as a unit during gamete formation. However, in many instances this does not occur. As we saw in Chapter 2, during the first meiotic prophase, when homologs are paired (or synapsed), a reciprocal exchange of chromosome segments may take place. This event, called **crossing over,** results in the reshuffling or **recombination** of the alleles between homologs.

Crossing over is currently viewed as an actual physical breaking and rejoining process that occurs during meiosis. You can see an example in the micrograph that opens this chapter. This exchange of chromosome segments provides an enormous potential for genetic variation in the gametes formed by any individual. This type of variation, in combination with that resulting from independent assortment, ensures that all offspring will contain a diverse mixture of maternal and paternal alleles.

The degree of crossing over between any two loci on a single chromosome is proportional to the distance between them, known as the **interlocus distance.** Thus, the percentage of recombinant gametes varies, depending on which loci are being considered. This correlation serves as the basis for the construction of **chromosome maps,** which indicate the relative locations of genes on the chromosomes.

In this chapter, we will discuss linkage, crossing over, and chromosome mapping. We will also consider a variety of other topics involving the exchange of genetic information, concluding the chapter with the rather intriguing question of why Mendel, who studied seven genes, did not encounter linkage. Or did he?

Genes linked on the same chromosome segregate together

To provide a simplified overview of the major theme of this chapter, Figure 6.1 illustrates and contrasts the meiotic consequences of (a) independent assortment, (b) linkage *without* crossing over, and (c) linkage *with* crossing over. Figure 6.1(a) illustrates the results of independent assortment of two pairs of chromosomes, each containing one heterozygous gene pair. No linkage is exhibited. When a large number of meiotic events are observed, four genetically different gametes are formed in equal proportions. Each contains a different combination of alleles of the two genes.

We can compare these results with those that occur if the same genes are linked on the same chromosome. If no crossing over occurs between the two genes [Figure 6.1(b)], only two genetically different gametes are formed. Each gamete receives the alleles present on one homolog or the other, which has been transmitted intact as the result of segregation. This case illustrates **complete linkage,** which results in the production of only **parental** or **noncrossover gametes.** The two parental gametes are formed in equal proportions. Though complete linkage between two genes seldom occurs, it is useful to consider the theoretical consequences of this concept when learning about crossing over.

Figure 6.1(c) illustrates the results of crossing over between two linked genes. As you can see, this crossover involves only two nonsister chromatids of the four chromatids present in the tetrad. This exchange generates two new allele combinations, called **recombinant** or **crossover gametes.** The two chromatids not involved in the exchange result in noncrossover gametes, like those in Figure 6.1(b).

The frequency with which crossing over occurs between any two linked genes is generally proportional to the distance separating the respective loci along the chromosome. In theory, two randomly selected genes can be so close to each other that crossover events are too infrequent to be easily detected. As shown in Figure 6.1(b), this circumstance, complete linkage, produces only parental gametes. On the other hand, if a small, but distinct distance separates two genes, few recombinant and many parental gametes will be formed. As the distance between the two genes increases, the proportion of recombinant gametes increases and that of the parental gametes decreases. Thus, as shown in Figure 6.1(c), the proportion of crossover gametes varies depending on the distance between the genes studied.

FIGURE 6.1 A comparison of the results of gamete formation where two heterozygous genes are (a) on two different pairs of chromosomes; (b) on the same pair of homologs, but where no exchange occurs between them; and (c) on the same pair of homologs, where an exchange occurs between two nonsister chromatids.

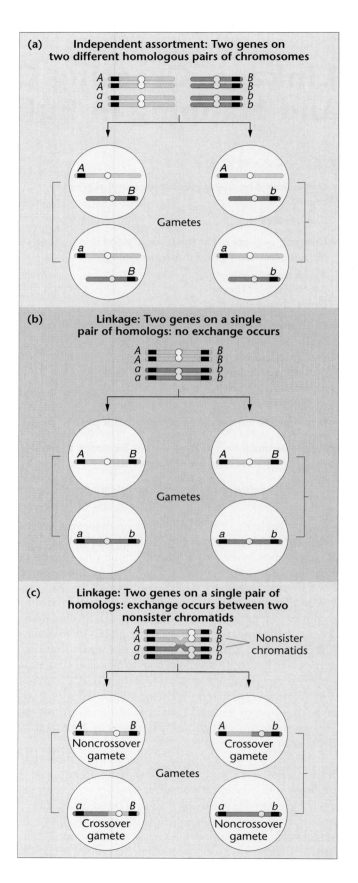

As we will discuss again later in this chapter, when the loci of two linked genes are far apart, the number of recombinant gametes approaches, but does not exceed, 50 percent. If 50 percent recombinants occurred, a 1:1:1:1 ratio of the four types (two parental and two recombinant gametes) would result. In such a case, transmission of two linked genes would be indistinguishable from that of two unlinked, independently assorting genes. That is, the proportion of the four possible genotypes would be identical, as shown in Figure 6.1(a,c).

The Linkage Ratio

If complete linkage exists between two genes because of their close proximity, and organisms heterozygous at both loci are mated, an F_2 phenotypic ratio results that is unique, which we shall designate the **linkage ratio**. To illustrate this ratio, we will consider a cross involving the closely linked recessive mutant genes *brown* (*bw*) eye and *heavy* (*hv*) wing vein in *Drosophila melanogaster* (Figure 6.2). The normal wild-type alleles bw^+ and hv^+ are

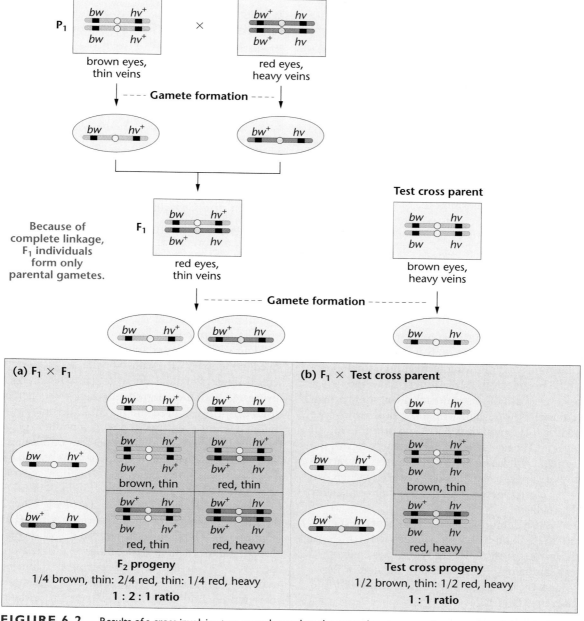

FIGURE 6.2   Results of a cross involving two genes located on the same chromosome where complete linkage is demonstrated. Part (a) generates the F_2 results of the cross; (b) generates the results of a test cross involving the F_1 progeny.

both dominant and result in red eyes and thin wing veins, respectively.

In this cross, flies with mutant brown eyes and normal thin veins are mated to flies with normal red eyes and mutant heavy veins. To be more concise, we will discuss the flies simply in terms of their mutant phenotypes and say that brown-eyed flies are crossed with heavy-veined flies. Using the genetic symbols established in Chapter 4, we can represent linked genes by placing their allele designations above and below a single or double horizontal line. Those placed above the line are located at loci on one homolog and those below the line are at the homologous loci on the other homolog. Thus, we can represent the P_1 generation as follows:

$$P_1: \quad \frac{bw \; hv^+}{bw \; hv^+} \times \frac{bw^+ \; hv}{bw^+ \; hv}$$

$$\text{brown, thin} \quad \text{red, heavy}$$

Because the genes are located on an autosome, and not the X chromosome, no designation between males and females is necessary.

In the F_1 generation each fly receives one chromosome of each pair from each parent; all flies are heterozygous for both gene pairs and exhibit the dominant traits of red eyes and thin veins:

$$F_1: \quad \frac{bw \quad hv^+}{bw^+ \quad hv}$$

$$\text{red, thin}$$

As shown in Figure 6.2, because of complete linkage, when the F_1 generation is interbred each F_1 individual forms only parental gametes. Following fertilization, the F_2 generation will be produced in a 1:2:1 phenotypic and genotypic ratio. One-fourth of this generation will show brown eyes and thin veins; one-half will show both wild-type traits, namely red eyes and thin veins; and one-fourth will show red eyes and heavy veins. In more concise terms, the ratio is 1 brown:2 wild:1 heavy. Such a ratio is characteristic of complete linkage, which is observed only when two genes are very close together and the number of progeny is relatively small.

Figure 6.2 also demonstrates the results of a test cross with the F_1 flies. Such a cross produces a 1:1 ratio of brown, thin and red, heavy flies. Had the genes controlling these traits been incompletely linked or located on separate autosomes, the test cross would have produced four phenotypes rather than two.

When we consider large numbers of mutant genes in any given species, genes located on the same chromosome will show evidence of linkage to one another. As a result, we can establish a **linkage group** for each chromosome. In theory, the number of linkage groups should correspond to the haploid number of chromosomes. In organisms with large numbers of mutant genes available for genetic study, this correlation has been confirmed.

Crossing over serves as the basis of determining the distance between genes during chromosome mapping

It is highly improbable that two randomly selected genes that are linked on the same chromosome will be so close to one another along the chromosome that they demonstrate complete linkage. Instead, crosses involving two such genes will almost always produce a percentage of offspring resulting from recombinant gametes. This percentage is variable and depends upon the distance between the two genes along the chromosome. This phenomenon was first explained in 1911 by two *Drosophila* geneticists, Thomas H. Morgan and his undergraduate student, Alfred H. Sturtevant.

Morgan and Crossing Over

As you may recall from our discussion in Chapter 4, Morgan first discovered the phenomenon of X-linkage. In his studies, he investigated numerous *Drosophila* mutations located on the X chromosome. When he analyzed crosses involving only one trait, he was able to deduce the mode of X-linked inheritance. However, when he made crosses that simultaneously involved two X-linked genes, his results were at first puzzling. For example, as shown in Cross A of Figure 6.3, he crossed mutant *yellow*-bodied (*y*) and *white*-eyed (*w*) females with wild-type males (gray body and red eyes). The F_1 females were wild type, whereas the F_1 males expressed both mutant traits. In the F_2, 98.7 percent of the total offspring showed the parental phenotypes—yellow-bodied, white-eyed flies and wild-type flies (gray-bodied, red-eyed). The remaining 1.3 percent of the flies were either yellow-bodied with red eyes or gray-bodied with white eyes. It was as if the genes had somehow separated from each other during gamete formation in the F_1 flies.

When Morgan made crosses involving other X-linked genes, the results were even more puzzling (Cross B of Figure 6.3). The same basic pattern was observed, but the proportion of F_2 phenotypes differed. For example, when he crossed *white*-eye, *miniature*-wing mutants with wild-type flies, only 62.8 percent of all the F_2 flies showed the parental phenotypes, while 37.2 percent of the offspring appeared as if the mutant genes had been separated during gamete formation.

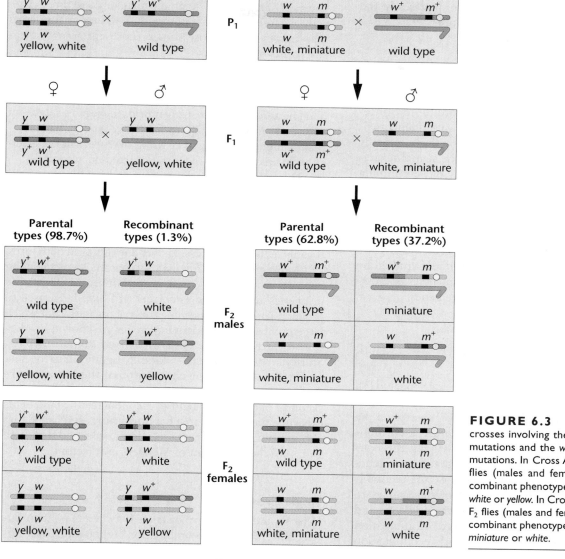

FIGURE 6.3 The F$_1$ and F$_2$ results of crosses involving the *yellow*-body, *white*-eye mutations and the *white*-eye, *miniature*-wing mutations. In Cross A, 1.3 percent of the F$_2$ flies (males and females) demonstrate recombinant phenotypes, which express either *white* or *yellow*. In Cross B, 37.2 percent of the F$_2$ flies (males and females) demonstrate recombinant phenotypes, which express either *miniature* or *white*.

As a result of these observations, Morgan was faced with two questions: (1) What was the source of gene separation? and (2) Why did the frequency of the apparent separation vary depending on the genes being studied? The answer Morgan proposed to the first question was based on his knowledge of earlier cytological observations. F. Janssens and others had observed that synapsed homologous chromosomes in meiosis wrapped around each other, creating **chiasmata** (singular, **chiasma**) where points of overlap are evident. Morgan proposed that these chiasmata could represent points of genetic exchange.

In the crosses shown in Figure 6.3, Morgan postulated that if an exchange occurred between the mutant genes on the two X chromosomes of the F$_1$ females, it would lead to the percentages in the observed results. He suggested that such exchanges led to 1.3 percent recombinant gametes in the *yellow–white* cross and 37.2 percent in the *white–miniature* cross. On the basis of this and other experimentation, Morgan concluded that if linked genes exist in a linear order along the chromosome, then a variable amount of exchange occurs between any two genes.

As an answer to the second question, Morgan proposed that two genes located relatively close to each other along a chromosome are less likely to have a chiasma form between them than if the two genes are farther apart on the chromosome. Therefore, the closer two genes are, the less likely it is that a genetic exchange will occur between them. Morgan proposed the term **crossing over** to describe the physical exchange leading to recombination.

Sturtevant and Mapping

Morgan's student, Alfred H. Sturtevant, was the first to realize that his mentor's proposal could also be used to map the sequence of linked genes. According to Sturtevant, "In a conversation with Morgan . . . I suddenly realized that the variations in strength of linkage, already attributed by Morgan to differences in the spatial separation of the genes, offered the possibility of determining sequences in the linear dimension of a chromosome. I went home and spent most of the night (to the neglect of my undergraduate homework) in producing the first chromosomal map. . . ." For example, Sturtevant compiled data on recombination between the genes represented by the *yellow*, *white*, and *miniature* mutants initially studied by Morgan and observed the following frequencies of crossing over between each pair of these three genes:

(1)	*yellow, white*	0.5%
(2)	*white, miniature*	34.5%
(3)	*yellow, miniature*	35.4%

Because the sum of (1) and (2) is approximately equal to (3), Sturtevant argued that the recombination frequencies between linked genes are additive. On this assumption, he predicted that the order of the genes on the X chromosome was *yellow–white–miniature*. In arriving at this conclusion, he reasoned as follows. The *yellow* and *white* genes are apparently close to each other because the recombination frequency is low. However, both of these genes are quite far from *miniature* because the *white, miniature* and *yellow, miniature* combinations show large recombination frequencies. Because *miniature* shows more recombination with *yellow* than with *white* (35.4 vs. 34.5), it follows that *white* is between the other two genes, not outside of them.

Sturtevant knew from Morgan's work that the frequency of exchange could be taken as an estimate of the relative distance between two genes or loci along the chromosome. He constructed a map of the three genes on the X chromosome, with one map unit equated with 1 percent recombination between two genes.* In the preceding example, the distance between *yellow* and *white* would be 0.5 map unit, and between *yellow* and *miniature* 35.4 map units. It follows that the distance between *white* and *minia-*

---

*In honor of Morgan's work, map units are often referred to as centimorgans (cM).

---

ture should be (35.4 − 0.5) or 34.9. This estimate is close to the actual frequency of recombination between *white* and *miniature* (34.5). The simple map for these three genes is shown in Figure 6.4. It is important to keep in mind that map units such as those shown represent *relative* measurements of distance.

In addition to these three genes, Sturtevant considered two other genes on the X chromosome and produced a more extensive map including all five genes. Soon thereafter, he and a colleague, Calvin Bridges, began a search for autosomal linkage in *Drosophila*. By 1923, they had clearly shown that linkage and crossing over were not restricted to X-linked genes, but also occurred on autosomes.

During their work, Sturtevant and Bridges made another interesting observation. In *Drosophila*, crossing over was shown to occur only in females. The fact that no crossing over occurs in males made the analysis of genetic mapping in *Drosophila* much less complex. In most other organisms, however, crossing over does occur in both sexes.

Although many refinements in chromosome mapping have developed since Sturtevant's initial work, his basic principles are accepted as correct and have been used to produce detailed chromosome maps of organisms for which large numbers of linked mutant genes are known. In addition to providing the basis for chromosome mapping, Sturtevant's findings were historically significant to the field of genetics. In 1910, the **chromosomal theory of inheritance** was still being widely disputed. Even Morgan was skeptical of this theory before he conducted the bulk of his experimentation. Research has now firmly established that chromosomes contain genes in a linear order and that these genes are the equivalent of Mendel's unit factors.

Single Crossovers

Why should the relative distance between two loci influence the amount of recombination and crossing over observed between them? The basis for this variation is explained in the following analysis.

During meiosis, a limited number of crossover events occur in each tetrad. These recombinant events occur randomly along the length of the tetrad. Therefore, the clos-

FIGURE 6.4 A map of the *yellow* (*y*), *white* (*w*), and *miniature* (*m*) genes on the X chromosome of *Drosophila melanogaster*. Each number represents the percentage of recombinant offspring produced in one of three crosses, each involving two different genes.

er that two loci reside along the axis of the chromosome, the less likely it is that any **single crossover** event will occur in between them. The same reasoning suggests that the farther apart two linked loci are, the more likely it is that a random crossover event will occur in between them.

In Figure 6.5(a), a single crossover occurs between two nonsister chromatids, but in an area of the chromosome that is not in between the two loci. Therefore, the crossover goes undetected because no recombinant gametes are produced. In Figure 6.5(b), where two loci are quite far apart, the crossover occurs in an area of the chromosome between them. This separates them, yielding recombinant gametes.

When a single crossover occurs between two nonsister chromatids, the other two chromatids of the tetrad will not be involved in this exchange. Following the completion of meiosis, they represent noncrossover gametes. In this case, even if a single crossover occurs 100 percent of the time between two linked genes, recombination will subsequently be observed in only 50 percent of the potential gametes formed. This concept is diagrammed in Figure 6.6.

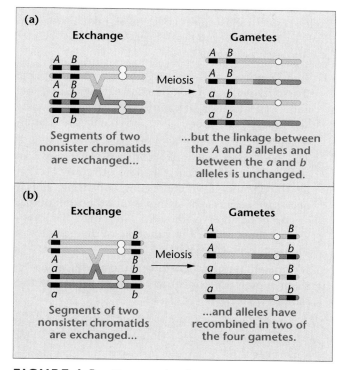

FIGURE 6.5 Two examples of a single crossover between two nonsister chromatids and the gametes subsequently produced. In (a) the exchange does not alter the linkage arrangement between the alleles of the two genes, only parental gametes are formed, and the exchange goes undetected. In (b) the exchange separates the alleles, resulting in recombinant gametes, which are detectable. ⬤GenCDX

Determining the gene sequence during mapping relies on the analysis of multiple crossovers

The study of single crossovers between two linked genes provides the basis of determining the *distance* between them. However, when many linked genes are studied, their *sequence* along the chromosome is more difficult to determine. Fortunately, the discovery that multiple exchanges occur between the chromatids of a tetrad has facilitated the process of producing more extensive chromosome maps. As we shall see below, when three or more linked genes are investigated simultaneously, it is possible to determine first the sequence of and then the distances between genes.

Multiple Exchanges

It is possible that in a single tetrad, two, three, or more exchanges will occur between nonsister chromatids as a result of several crossing over events. Double exchanges of genetic material result from **double crossovers,** as shown

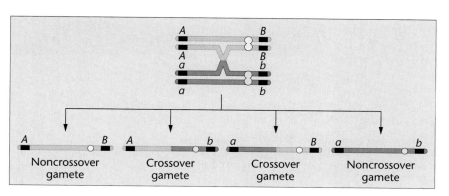

Noncrossover gamete

Crossover gamete

Crossover gamete

Noncrossover gamete

FIGURE 6.6 The consequences of a single exchange between two nonsister chromatids occurring in the tetrad stage. Two noncrossover (parental) and two crossover (recombinant) gametes are produced.

in Figure 6.7. To study a double exchange, three gene pairs must be investigated, each heterozygous for two alleles. Before we can determine the frequency of recombination among all three loci, we must review some simple probability calculations.

As we have seen, the probability of a single exchange occurring in between the *A* and *B* or the *B* and *C* genes is directly related to the physical distance separating the loci. The closer *A* is to *B* and *B* is to *C*, the less likely it is that a single exchange will occur in between either of the two sets of loci. In the case of a double crossover, two separate and independent events or exchanges must occur simultaneously. The mathematical probability of two independent events occurring simultaneously is equal to the product of the individual probabilities.

Suppose that crossover gametes result from single exchanges between *A* and *B* 20 percent of the time ($p = 0.20$) and between *B* and *C* 30 percent of the time ($p = 0.30$). The probability of recovering a double-crossover gamete arising from two exchanges, between *A* and *B* and between *B* and *C*, is predicted to be $(0.20)(0.30) = 0.06$, or 6 percent. This calculation makes apparent that the expected frequency of double-crossover gametes is always much lower than that of either single-crossover class of gametes alone.

If three genes are relatively close together along one chromosome, the expected frequency of double-crossover gametes is extremely low. For example, assume the *A–B* distance in Figure 6.7 to be 3 map units and the *B–C* distance to be 2 map units. The expected double-crossover frequency would be $(0.03)(0.02) = 0.0006$, or 0.06 percent. This translates to only 6 events in 10,000. In such a mapping experiment, where closely linked genes are involved, very large numbers of offspring are required to detect double-crossover events. In this example, it would be unlikely that a double crossover would be observed even if 1000 offspring were examined. If we extend these probability considerations, it is evident that if four or five genes were being mapped, even fewer triple and quadruple crossovers could be expected to occur.

Three-Point Mapping in Drosophila

The information presented in the previous section serves as the basis for mapping three or more linked genes in a single cross. To illustrate this, let us examine a situation involving three linked genes. Three criteria must be met for a successful mapping cross:

1. The genotype of the organism producing the crossover gametes must be heterozygous at all loci under consideration. If homozygosity occurred at any locus, all gametes produced would contain the same allele, precluding mapping analysis.

2. The cross must be constructed so that genotypes of all gametes can be accurately determined by observing the phenotypes of the resulting offspring. This is necessary because the gametes and their genotypes can never be observed directly. Thus, each phenotypic class must reflect the genotype of the gametes of the parents producing it.

3. A sufficient number of offspring must be produced in the mapping experiment to recover a representative sample of all crossover classes.

These criteria are met in the three-point mapping cross from *Drosophila melanogaster* shown in Figure 6.8. This cross involves three X-linked recessive mutant genes—

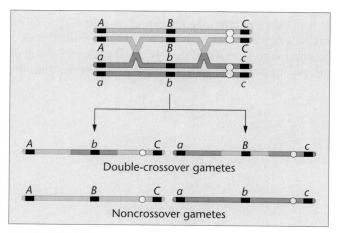

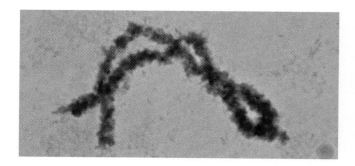

FIGURE 6.7 Consequences of a double exchange occurring between two nonsister chromatids. Because the exchanges involve only two strands, two noncrossover gametes and two double-crossover gametes are produced. The photograph illustrates several chiasmata found in a tetrad isolated during the first meiotic prophase stage.

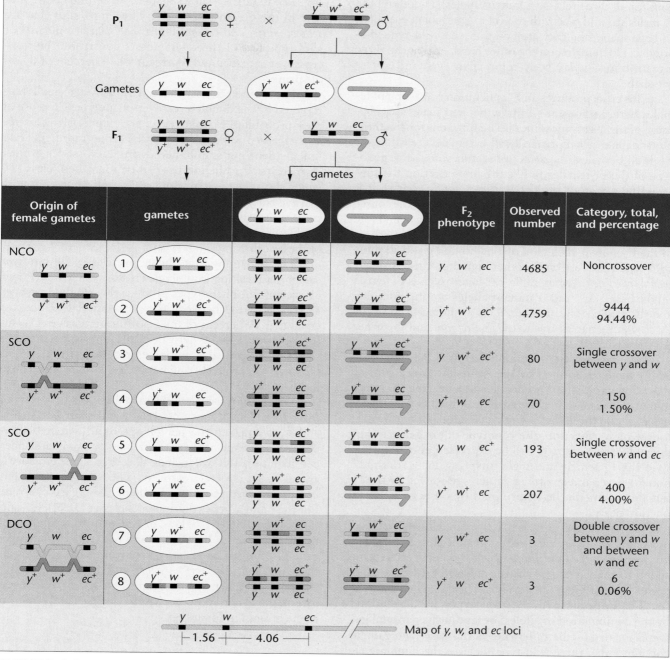

FIGURE 6.8 A three-point mapping cross involving the *yellow* (*y* or *y*⁺), *white* (*w* or *w*⁺), and *echinus* (*ec* or *ec*⁺) genes in *Drosophila melanogaster.* NCO, SCO, and DCO refer to noncrossover, single-crossover, and double-crossover groups, respectively. Because of the complexity of this and several of the ensuing figures, centromeres have not been included on the chromosomes, and only two nonsister chromatids are initially shown. Crossing over always occurs in the four-strand tetrad stage. 💿GenCDX

yellow body color, *white* eye color, and *echinus* eye shape. To diagram the cross, we must assume some theoretical sequence, even though we do not yet know if it is correct. In Figure 6.8, we initially assume the sequence of the three

genes to be *y–w–ec.* If this is incorrect, our analysis will reveal the correct sequence.

In the P₁ generation, males hemizygous for all three wild-type alleles are crossed to females that are homozygous for all three recessive mutant alleles. Therefore, the P_1 males are wild type with respect to body color, eye color, and eye shape, and they are said to have a wild-type phenotype. The females, on the other hand, exhibit the three mutant traits—*yellow* body color, *white* eyes, and *echinus* eye shape.

This cross produces an F_1 generation consisting of females heterozygous at all three loci and males that, because of the Y chromosome, are hemizygous for the three mutant alleles. Phenotypically, all F_1 females are wild type, while all F_1 males are *yellow*, *white*, and *echinus*. The genotype of the F_1 females fulfills the first criterion for constructing a map of the three linked genes; that is, it is heterozygous at the three loci and may serve as the source of recombinant gametes generated by crossing over. Note that because of the genotypes of the P_1 parents, all three of the F_1 mutant alleles are on one homolog and all three wild-type alleles are on the other homolog. *Other arrangements are possible.* For example, the heterozygous F_1 female might have the *y* and *ec* mutant alleles on one homolog and the *w* allele on the other. This would occur if, in the P_1 cross, one parent was *yellow* and *echinus* and the other parent was *white*.

In our cross, the second criterion is met by virtue of the gametes formed by the F_1 males. Every gamete will contain either an X chromosome bearing the three mutant alleles or a Y chromosome, which is genetically inert for the three loci being considered. In both cases, following fertilization, the genotype of the gamete produced by the F_1 female will be expressed phenotypically in the F_2 female and male offspring. As a result, all noncrossover and crossover gametes produced by the F_1 female parent can be determined by observing the F_2 phenotypes.

With these two criteria met, we can construct a chromosome map from the crosses illustrated in Figure 6.8, but first, we must determine which F_2 phenotypes correspond to the various noncrossovers and crossover categories. Two of these categories can be easily determined.

The **noncrossover** F_2 phenotypes are determined by the combination of alleles present in the parental gametes formed by the F_1 female. In this case each gamete contains either three wild-type alleles *or* three mutant alleles, depending on which of the X chromosomes is unaffected by crossing over. As a result of segregation, approximately equal proportions of the two types of gametes, and subsequently the F_2 phenotypes, are produced. Because the F_2 phenotypes complement one another (i.e., one is wild type and the other is mutant for all three genes), they are called **reciprocal classes** of phenotypes.

The two noncrossover phenotypes are most easily recognized because *they occur in the greatest proportion of offspring.* Figure 6.8 shows that classes (1) and (2) are present in the greatest numbers. Therefore, flies that are *yellow*, *white*, and *echinus* and those that are normal or wild type for all three characters constitute the noncrossover category and represent 94.44 percent of the F_2 offspring.

The second category that can be easily detected is represented by the double-crossover phenotypes. Because of their probability of occurrence, *they must be present in the least numbers.* Remember that this group represents two independent but simultaneous single-crossover events. Two reciprocal phenotypes can be identified: class (7), which shows the mutant traits *yellow* and *echinus* but normal eye color; and class (8), which shows the mutant trait *white* but normal body color and eye shape. Together these double-crossover phenotypes constitute only 0.06 percent of the F_2 offspring.

The remaining four phenotypic classes fall into two categories resulting from single crossovers. Classes (3) and (4), reciprocal phenotypes produced by single-crossover events occurring between the *yellow* and *white* loci, are equal to 1.50 percent of the F_2 offspring. Classes (5) and (6), constituting 4.00 percent of the F_2 offspring, represent the reciprocal phenotypes resulting from single-crossover events occurring between the *white* and *echinus* loci.

We can now calculate the map distances between the three loci. The distance between *y* and *w*, or between *w* and *ec*, is equal to the percentage of all detectable exchanges occurring between them. For any two genes under consideration, this will include all related single crossovers as well as all double crossovers. The latter are included because they represent two simultaneous single crossovers. For the *y* and *w* genes this includes classes (3), (4), (7), and (8), totaling 1.50% + 0.06%, or 1.56 map units (mu). Similarly, the distance between *w* and *ec* is equal to the percentage of offspring resulting from an exchange between these two loci, classes (5), (6), (7), and (8), totaling 4.00% + 0.06%, or 4.06 map units (mu). The map of these three loci on the X chromosome, based on these data, is shown at the bottom of Figure 6.8.

Determining the Gene Sequence

In the preceding example, we assumed the order, or sequence, of the three genes along the chromosome to be *y–w–ec*. Our analysis established that this sequence is consistent with the data. In many mapping experiments, the gene sequence is not known, and this constitutes another variable in the analysis. Had the gene order been unknown in this example, we could have used one of two methods to determine it. In your own work, you should select one of these methods and use it consistently.

Method I

This method is based on the fact that there are only three possible orders, each containing one of the three genes in between the other two.

(I) $w–y–ec$ (y is in the middle)
(II) $y–ec–w$ (ec is in the middle)
(III) $y–w–ec$ (w is in the middle)

The steps below will allow you to determine which gene order is correct.

1. Assuming any of the three orders, first determine the *arrangement of alleles* along each homolog of the heterozygous parent giving rise to noncrossover and crossover gametes (the F_1 female in our example).

2. Determine whether a double-crossover event occurring within that arrangement will produce the *observed double-crossover phenotypes*. Remember that you can easily identify these phenotypes because they occur least frequently.

3. If this order does not produce the correct phenotypes, try each of the other two orders. One of the three must work!

Figure 6.9 illustrates these steps for the cross shown in Figure 6.8. The three possible orders are labeled (I), (II), and (III). Either y, ec, or w must be in the middle.

1. If we assume order I with y between w and ec, the arrangement of alleles along the homologs of the F_1 heterozygote is:

$$(I) \quad \frac{w \quad y \quad ec}{w^+ \quad y^+ \quad ec^+}$$

We know this because of the way in which the P_1 generation was crossed. The P_1 female contributed an X chromosome bearing the w, y, and ec alleles, while the P_1 male contributed an X chromosome bearing the w^+, y^+, and ec^+ alleles.

2. A double crossover within the above arrangement would yield the following gametes:

$$\underline{w \quad y^+ \quad ec} \quad \text{and} \quad \underline{w^+ \quad y \quad ec^+}$$

If y is in the middle, following fertilization, the F_2 double-crossover phenotypes will correspond to the above gametic genotypes, yielding offspring that are white, echinus and offspring that are yellow. Determination of the actual double crossovers, however, reveals them to be yellow, echinus flies and white flies. Therefore, our assumed order is incorrect.

3. If we consider the other orders, one with ec/ec^+ alleles in the middle (II) or one with the w/w^+ alleles in the middle (III),

$$(II) \quad \frac{y \quad ec \quad w}{y^+ \quad ec^+ \quad w^+} \quad \text{or} \quad (III) \quad \frac{w \quad y \quad ec}{w^+ \quad y^+ \quad ec^+}$$

we see that arrangement II again provides *predicted* double-crossover phenotypes that do not correspond to the *actual* double-crossover phenotypes. The predicted phenotypes are yellow, white flies and echinus flies in the F_2 generation. Therefore, this order is also incorrect. However, arrangement III *will* produce the *observed* phenotypes, yellow, echinus flies and white flies. Therefore, this order, where the w gene is in the middle, is correct.

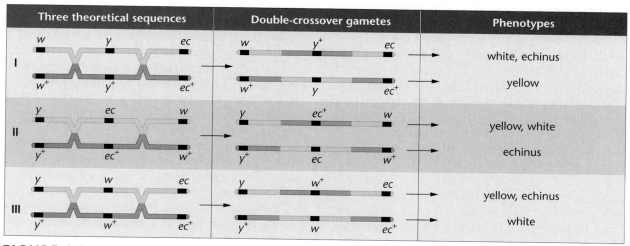

FIGURE 6.9 A summary of the three possible sequences of the *white*, *yellow*, and *echinus* genes, the results of a double crossover in each case, and the resulting phenotypes produced in a test cross. The two noncrossover chromatids of each tetrad have been omitted to simplify the illustration.

To summarize method I: Determine the arrangement of alleles on the homologs of the heterozygote yielding the crossover gametes. Then test each of three possible orders to determine which one yields the observed (actual) double-crossover phenotypes. Whichever of the three does so represents the correct order. You can review this method, which is summarized in Figure 6.9.

Method II

This method also begins by assuming the arrangement of alleles along each homolog of the heterozygous parent. In addition, it requires one further assumption:

Following a double-crossover event, the allele in the middle position will fall between the outside, or flanking, alleles that were present on the opposite parental homolog.

To illustrate, assume order (I), *w–y–ec*, in the arrangement:

$$\text{(I)} \quad \frac{w \quad y \quad ec}{w^+ \quad y^+ \quad ec^+}$$

Following a double-crossover event, the *y* and y^+ alleles would be switched to the arrangement:

$$\frac{w \quad y^+ \quad ec}{w^+ \quad y \quad ec^+}$$

Following segregation, two gametes would be formed:

$$\underline{w \quad y^+ \quad ec} \quad \text{and} \quad \underline{w^+ \quad y \quad ec^+}$$

Because the genotype of the gamete will be expressed directly in the phenotype following fertilization, the double-crossover phenotypes will be:

white, echinus flies   and   yellow flies

Note that the *yellow* allele, assumed to be in the middle, is now associated with the two outside markers of the other homolog, w^+ and ec^+. However, these predicted phenotypes do not coincide with the observed double-crossover phenotypes. The *yellow* gene, therefore, is not in the middle.

This same reasoning can be applied to the assumption that the *echinus* gene or the *white* gene is in the middle. In the former case, we will reach a negative conclusion. If we assume that the *white* gene is in the middle, the *predicted* and *actual* double crossovers coincide. Therefore, we conclude that the *white* gene is located between the *yellow* and *echinus* genes.

To summarize method II: Determine the arrangement of alleles on the homologs of the heterozygote yielding

crossover gametes. Then determine the actual double-crossover phenotypes. Simply select the single allele that has been switched so that it is now no longer associated with its original neighboring alleles.

In our example *y*, *ec*, and *w* are together in the F_1 heterozygote, as are y^+, ec^+, and w^+. In the F_2 double-crossover classes, it is *w* and w^+ that have been switched. The *w* allele is now associated with y^+ and ec^+, while the w^+ allele is now associated with the *y* and *ec* alleles. Therefore, the *white* gene is in the middle, and the *yellow* and *echinus* genes are the flanking markers.

A Mapping Problem in Maize

Having established the basic principles of chromosome mapping, we will now consider a problem in maize (corn) where the gene sequence and interlocus distances are unknown. This analysis differs from the preceding discussion in several ways and therefore will expand your knowledge of mapping procedures:

1. The previous mapping cross involved X-linked genes. Here, autosomal genes are considered.

2. In the previous discussion we initially assumed a particular gene sequence, and we then introduced methods to determine whether it or a different sequence was correct. In this analysis of maize, the sequence is *initially* unknown.

3. In the discussion of this cross we will use different symbols for alleles. Instead of using the symbols bm^+, v^+, and pr^+ we will simply use + to denote each wild-type allele. As first described in Chapter 4, this annotation of symbols is easier to manipulate, but it requires a strong understanding of mapping procedures.

When we consider three autosomally linked genes in maize, the experimental cross must still meet the same three criteria established for the X-linked genes in *Drosophila*: (1) one parent must be heterozygous for all traits under consideration, (2) the gametic genotypes produced by the heterozygote must be apparent from observing the phenotypes of the offspring, and (3) a sufficient sample size must be available.

In maize, the recessive mutant genes *bm* (brown midrib), *v* (virescent seedling), and *pr* (purple aleurone) are linked on chromosome 5. Assume that a female plant is known to be heterozygous for all three traits. Nothing is known about the arrangement of the mutant alleles on the maternal and paternal homologs of this heterozygote, the sequence of genes, or the map distances between the genes. What genotype must the male plant have to allow successful mapping? To meet the second criterion, the male must be homozygous for all three recessive mutant alleles. Otherwise, offspring of this cross showing a given phenotype might represent more than one genotype, making accurate mapping impossible. Note that this is equivalent to performing a test cross.

Figure 6.10 diagrams this cross. As shown, we do not know either the arrangement of alleles or the sequence of loci in the heterozygous female. Some of the possibilities are shown, but we have yet to determine which is correct. In the test-cross male parent and the offspring, *we do not know the sequence*, but must write one down initially as a random selection. Note that we have chosen to place *v* in the middle. This may or may not be correct.

The offspring have been arranged in groups of two for each pair of reciprocal phenotypic classes. The two members of each reciprocal class are derived from either no

crossing over (NCO), one of two possible single-crossover events (SCO), or a double-crossover event (DCO).

To solve this problem, consider the following questions. As you do, it will be helpful to refer to Figures 6.10 and 6.11.

1. *What is the correct heterozygous arrangement of alleles in the female parent?*

   Identify the two noncrossover classes, those that occur in the highest frequency. In this case, they are <u>+ *v* *bm*</u> and <u>*pr* + +</u>. Therefore, the arrangement of alleles on the homologs of the female parent must be as shown in Figure 6.11(a). These homologs segregate into gametes, unaffected by any recombination event. Any other arrangement of alleles could not yield the observed noncrossover classes. (Remember that

(a) Possible allele arrangements and gene sequences in a heterozygous female

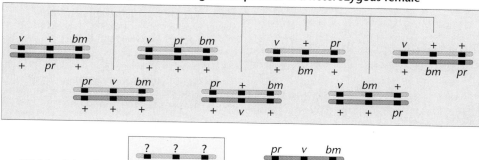

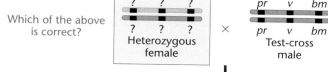

(b) Actual results of mapping cross

Phenotypes of offspring			Number	Total and percentage	Exchange classification
+	*v*	*bm*	230	467 42.1%	Noncrossover (NCO)
pr	+	+	237		
+	+	*bm*	82	161 14.5%	Single crossover (SCO)
pr	*v*	+	79		
+	*v*	+	200	395 35.6%	Single crossover (SCO)
pr	+	*bm*	195		
pr	*v*	*bm*	44	86 7.8%	Double crossover (DCO)
+	+	+	42		

FIGURE 6.10 Part (a) shows some possible allele arrangements and gene sequences in a heterozygous female. The data from a three-point mapping cross, depicted in (b), where the female is test-crossed, provide the basis for determining which combination of arrangement and sequence is correct [see Figure 6.11 (d)].

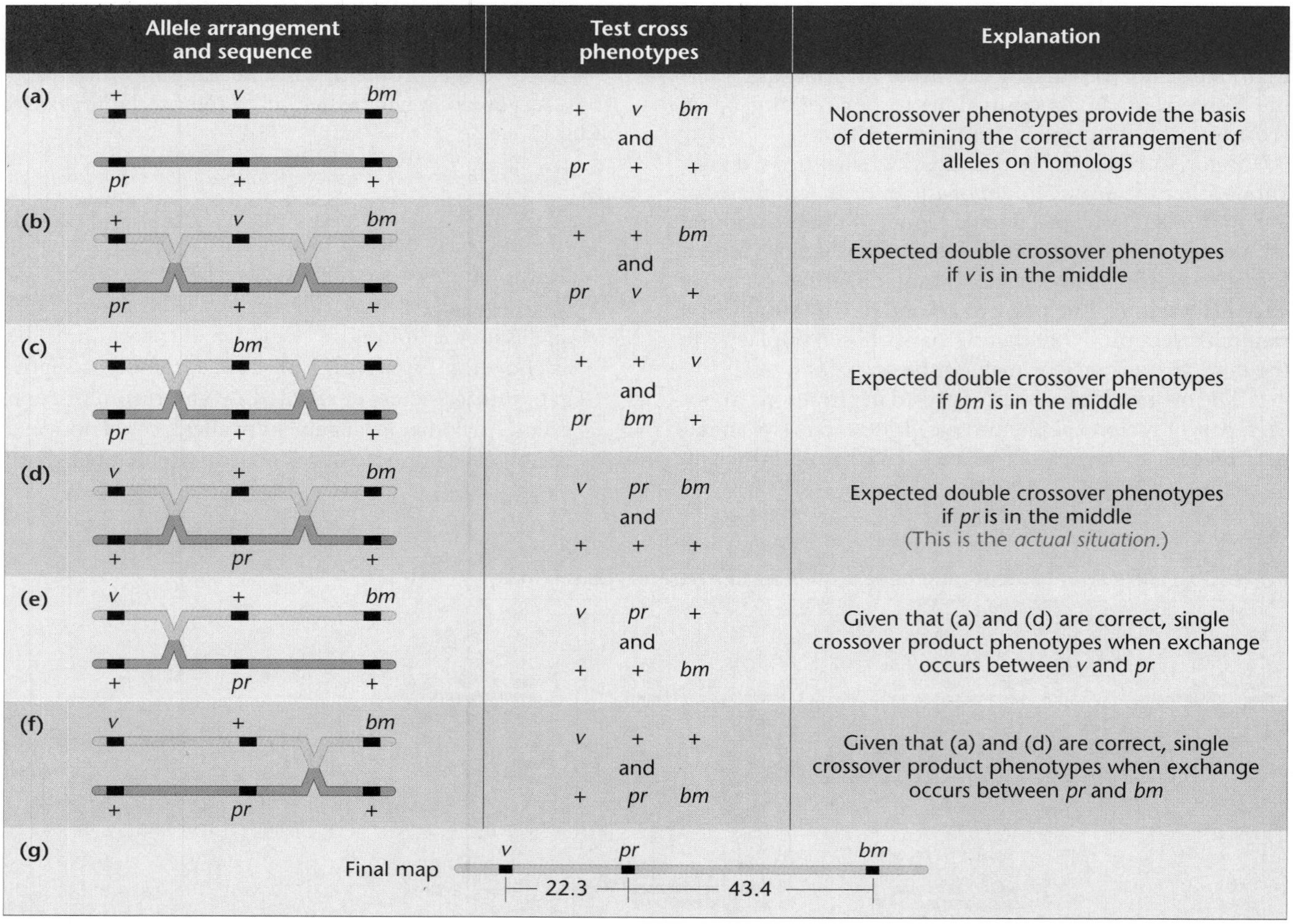

Allele arrangement and sequence	Test cross phenotypes	Explanation
(a)	+ v bm and pr + +	Noncrossover phenotypes provide the basis of determining the correct arrangement of alleles on homologs
(b)	+ + bm and pr v +	Expected double crossover phenotypes if *v* is in the middle
(c)	+ + v and pr bm +	Expected double crossover phenotypes if *bm* is in the middle
(d)	v pr bm and + + +	Expected double crossover phenotypes if *pr* is in the middle (This is the *actual situation*.)
(e)	v pr + and + + bm	Given that (a) and (d) are correct, single crossover product phenotypes when exchange occurs between *v* and *pr*
(f)	v + + and + pr bm	Given that (a) and (d) are correct, single crossover product phenotypes when exchange occurs between *pr* and *bm*
(g) Final map		

FIGURE 6.11 Steps used in producing a map of the three genes involved in the cross shown in Figure 6.10, where neither the arrangement of alleles nor the sequence of genes is initially known in the heterozygous female parent.

$\underline{+ \quad v \quad bm}$ is equivalent to $pr^+ \ v \ bm$, and that $\overline{pr \ + \ +}$ is equivalent to $pr \ v^+ \ bm^+$.)

2. *What is the correct sequence of genes?*

To answer this question we will first use the approach described in Method I. We know that the arrangement of alleles is

$$\frac{+ \quad v \quad bm}{pr \quad + \quad +}$$

But, is the assumed sequence correct? That is, will a double-crossover event yield the observed double-crossover phenotypes following fertilization? Simple observation shows that it will not [Figure 6.11(b)]. Try the other two orders [Figure 6.11(c), (d)], keeping the same arrangement:

$$\frac{+ \quad bm \quad v}{pr \quad + \quad +} \quad \text{and} \quad \frac{v \quad + \quad bm}{+ \quad pr \quad +}$$

Only the latter case will yield the observed double-crossover classes [Figure 6.11(d)]. Therefore, the pr gene is in the middle.

The same conclusion is reached if we used Method II to analyze the problem. In this case, no assumption of gene sequence is necessary. The arrangement of alleles in the heterozygous parent is

$$\frac{+ \quad v \quad bm}{pr \quad + \quad +}$$

The double-crossover gametes are also known:

$$\underline{pr \quad v \quad bm} \qquad \text{and} \qquad \underline{+ \quad + \quad +}$$

We can see that the *pr* allele has shifted so as to be associated with *v* and *bm* following a double crossover. The latter two alleles were present together on one homolog, and they stayed together. Therefore, *pr* is the odd gene, so to speak, and it is in the middle.

3. *What is the distance between each pair of genes?*

Having established the correct sequence of loci as *v–pr–bm*, we can now determine the distance between *v* and *pr* and between *pr* and *bm*. Remember that the map distance between two genes is calculated on the basis of all detectable recombinational events occurring between them. This includes both the single- and double-crossover events involving the two genes being considered.

Figure 6.11(e) shows that the phenotypes $\underline{v \; pr \; +}$ and $\underline{+ \; + \; bm}$ result from single crossovers between *v* and *pr*; and Figure 6.10 shows that single crossovers account for 14.5 percent of the offspring. By adding the percentage of double crossovers (7.8 percent) to the number obtained for single crossovers, we calculate the total distance between *v* and *pr* to be 22.3 map units.

Figure 6.11(f) shows that the phenotypes $\underline{v \; + \; +}$ and $\underline{+ \; pr \; bm}$ result from single crossovers between *pr* and *bm*, totaling 35.6 percent according to Figure 6.10. With the addition of the double-crossover classes (7.8 percent) the distance between *pr* and *bm* is 43.4 map units. The final map for all three genes in this example is shown in Figure 6.11(g).

Interference affects the recovery of multiple exchanges

Based on the above discussion the expected frequency of multiple exchanges, such as double crossovers, can be predicted once the distance between genes is established. For example, in the above cross in maize, the distance between *v* and *pr* is 22.3 map units and the distance between *pr* and *bm* is 43.4 map units. If the two single crossovers that make up a double crossover occur independently of one another, we can calculate the expected frequency of double crossovers (DCO_{exp}):

$$DCO_{exp} = (0.223) \times (0.434) = 0.097 = 9.7 \text{ percent}$$

Most often in mapping experiments, the observed DCO frequency is less than the expected number of DCOs. In the maize cross, for example, only 7.8 percent DCOs were observed when 9.7 percent were expected. The phenomenon called **interference** is used to explain this reduction; this is when a crossover event in one region of the chromosome inhibits a second event in nearby regions.

To quantify these disparities that result from interference, we can calculate the **coefficient of coincidence (*C*)**:

$$C = \frac{\text{Observed DCO}}{\text{Expected DCO}}$$

In the maize cross, we have:

$$C = \frac{0.078}{0.097} = 0.804$$

Once we have found *C*, we can quantify interference (*I*) by using the simple equation:

$$I = 1 - C$$

In the maize cross, we have:

$$I = 1.000 - 0.804 = 0.196$$

If interference is complete and no double crossovers occur, then $I = 1.0$. If fewer DCOs than expected occur, *I* is a positive number and **positive interference** has occurred. If more DCOs than expected occur, *I* is a negative number and **negative interference** has occurred. In the above example, *I* is a positive number (0.196), indicating that 19.6 percent fewer double crossovers occurred than expected.

Positive interference is most often observed in eukaryotic systems. In general, the closer genes are to one another along the chromosome, the more positive interference occurs. In *Drosophila*, within a distance of 10 map units, interference is often complete and no multiple crossovers are recovered. This observation suggests that interference may be explained by physical constraints that prevent the formation of closely aligned chiasmata. This interpretation is consistent with the finding that interference decreases as the genes in question are located farther apart. In the maize cross illustrated in Figures 6.10 and 6.11, the three genes are relatively far apart, and 80 percent of the expected double crossovers are observed.

As the distance between two genes increases, mapping experiments become more inaccurate

In theory, the frequency of crossing over between any two genes in a mapping experiment is expected to be directly proportional to the actual distance between two genes. However, in most cases, the experimentally derived mapping distance between two genes is an underestimate, and the farther apart the two genes are, the greater is the inaccuracy. The discrepancy is due primarily to multiple exchanges that are predicted to occur between the two genes, but which are not recovered during experimental mapping. As we will explain below, this inaccuracy is the result of probability events that can be described using the **Poisson distribution.**

First, let us examine a mapping experiment that involves two exchanges between two genes that are far apart on a chromosome. As shown in Figure 6.12, there are three possible ways that two exchanges (equivalent to a double-crossover event) can occur between nonsister chromatids within a tetrad. A **two-strand double exchange** yields no recombinant chromatids, a **three-strand double exchange** yields 50 percent recombinant chromatids, and a **four-strand double exchange** yields 100 percent recombinant chromatids. In the aggregate, therefore, these multiple events "even out" and two genes that are far apart on the chromosome theoretically yield the maximum of 50 percent recombination, essential for *accurate* gene mapping.

In such a mapping experiment, double exchanges that occur between two genes (and, for that matter, all multiple exchanges) are relatively infrequent in comparison to the total number of single crossovers. As such, the *actual* occurrence of the infrequent events is subject to probability considerations based on the Poisson distribution. In our case, this distribution allows us to predict mathematically the frequency of samples that will *actually* undergo double exchanges. It is the failure of such exchanges to occur that leads to the underestimate of mapping distance.

The Poisson distribution is a mathematical function that assigns probabilities of observing various numbers of a specific event in a sample. To illustrate the effect of Poisson distribution, consider the analogy of an Easter egg hunt where 1000 children, randomly search a large area for 1000 randomly hidden eggs. In one hour, all eggs are recovered. If all children are equally adept in the search, we can safely predict that many children will have one egg, but also that many will have either no eggs or more than one egg. The Poisson distribution allows us to mathematically predict the frequency (probability) of each outcome, that is, the freqency of children within the sample that recovered 0, 1, 2, 3, 4, . . . eggs. Poisson distribution applies when the average number of events is small (a child finds an egg), while the total number of times the event that can occur within the sample is relatively large (1000 eggs can be found).

The Poisson terms used to calculate predicted distributions of events are:

Distribution of Events	Probability
0	e^{-m}
1	me^{-m}
2	$(m^2/2)(e^{-m})$
3	$(m^3/6)(e^{-m})$
etc.	

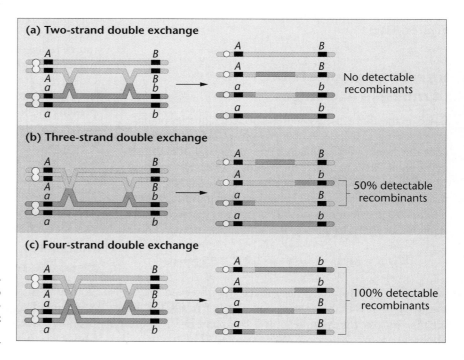

FIGURE 6.12 Three types of double exchanges that may occur between two genes. Two of them, (b) and (c), involve more than two chromatids. In each case, the detectable recombinant chromatids are bracketed.

where the mean number of independently occurring events is m, and e represents the base of natural logarithms (e = about 2.7). For the Easter egg analogy, a calculation will reveal that over 300 children will fail to find an egg. Had we attempted to estimate the total number of youngsters in the hunt by tabulating the number who found at least one egg, we would have seriously underestimated the number of participants in the hunt.

In relation to chromosome mapping, we are interested in the cases where double exchanges may potentially occur between two genes, but because of the predictions based on Poisson distribution, no such exchanges actually occur within the data sample. Such an analysis creates what is called a **mapping function** that relates recombination (crossover) frequency (RF) to map distance.

We can now apply the Poisson distribution, assuming no interference occurs (see the previous section). Any class where m is one or more (one or more random crossovers) will yield, on average, 50 percent recombinant chromatids. Thus, we are interested in the zero term, which effectively reduces the number of recombinant chromatids. The proportion of meioses with one or more crossovers is equal to 1 − [fraction of zero crossovers], whereby 50 percent recombinant chromatids will occur. Therefore,

$$\text{percent observed recombination (RF)} = 0.5(1 - e^{-m}) \times 100$$

Solving this equation generates the curve (mapping function) shown in Figure 6.13(a) is generated. This is compared to the hypothetical case [Figure 6.13(b)] where recombination is directly proportional to mapping distance—that is, where interference is complete and no multiple exchanges occur.

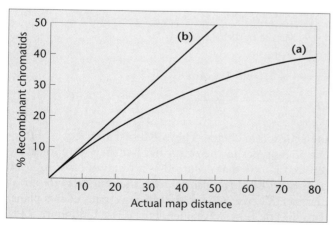

FIGURE 6.13   The relationship between the percentage of recombinant chromatids that occur and actual map distance when (a) Poisson distribution is used to predict the frequency of recombination in relation to map distance; and (b) there is a direct relationship between recombination and map distance.

Careful examination of this graph reveals two important observations. When the actual map distance is low (i.e. 0–7 mu), the two lines coincide. *When two genes are close together, the accuracy of a mapping experiment is very high! However, as the distance between two genes increases, the accuracy of the experiment diminishes.* As predicted by the Poisson distribution, the absence of multiple exchanges has a very significant impact. For example, when 25 percent recombinant chromatids are detected, actual map distance is almost 35 percent! When just over 30 percent recombinants are detected, the true distance, discounting any interference, is 50 mu! Such inaccuracy has been well documented in a number of studies involving various organisms, including maize, *Drosophila*, and *Neurospora*.

Drosophila genes have been extensively mapped

In organisms such as *Drosophila*, maize, and the mouse, where large numbers of mutants have been discovered and experimental crosses are easy to perform, extensive maps of each chromosome have been made. Figure 6.14 presents partial maps of the four chromosomes of *Drosophila*. Virtually every morphological feature of the fruit fly has been altered by mutation. Each locus affected by mutation is first localized to one of the four chromosomes, or linkage groups, and then mapped in relation to other linked genes of that group. As you can see, the genetic map of the X chromosome is somewhat less extensive than that of autosome 2 or 3. In comparison to these three, autosome 4 is minuscule. Cytological evidence has shown that the relative lengths of the genetic maps correlate roughly with the relative physical lengths of these chromosomes.

Crossing over involves a physical exchange between chromatids

Once the process of genetic mapping was established, it was of great interest to investigate the relationship between chiasmata observed in meiotic prophase I and crossing over. For example, are chiasmata visible manifestations of crossing over events? If so, then crossing over in higher organisms appears to be the result of an actual physical exchange between homologous chromosomes. That this is the case was demonstrated independently in the 1930s by Harriet Creighton and Barbara McClintock in *Zea mays* and by Curt Stern in *Drosophila*.

Because the experiments are similar, we will consider only one of them, the work with maize. Creighton and McClintock studied two linked genes on chromosome 9. At one locus, the alleles *colorless* (*c*) and *colored* (*C*) control

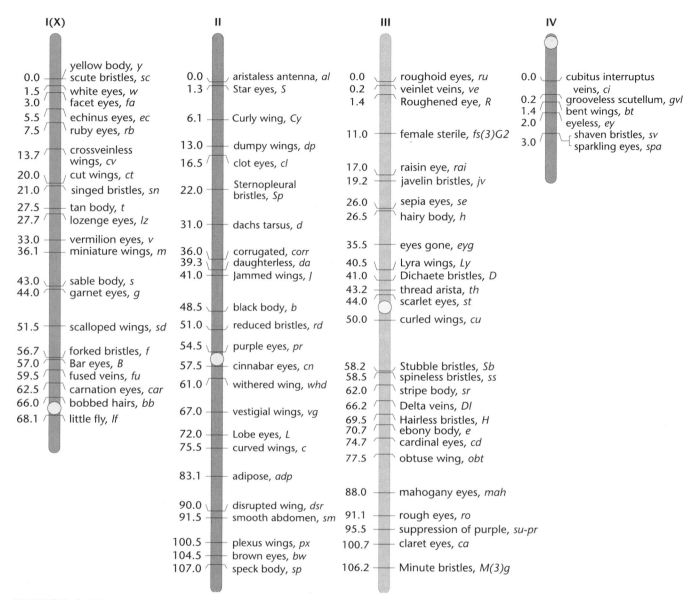

FIGURE 6.14 A partial genetic map of the four chromosomes of *Drosophila melanogaster*. The circle on each chromosome represents the position of the centromere.

endosperm coloration. At the other locus, the alleles *starchy* (*Wx*) and *waxy* (*wx*) control the carbohydrate characteristics of the endosperm. The maize plant studied was heterozygous at both loci. The key to the experiment was that one of the homologs contained two unique cytological markers. The markers consisted of a densely stained knob at one end of the chromosome and a translocated piece of another chromosome (8) at the other end. The arrangements of alleles and cytological markers could be detected cytologically and are shown in Figure 6.15.

Creighton and McClintock crossed this plant to one homozygous for the color allele (*c*) and heterozygous for

the endosperm alleles. They obtained a variety of different phenotypes in the offspring, but they were most interested in one that occurred as a result of crossing over involving the chromosome with the unique cytological markers. They examined the chromosomes of this plant, with the colorless, waxy phenotype (Case I in Figure 6.15) for the presence of the cytological markers. If genetic crossing over is accompanied by a physical exchange between homologs, the translocated chromosome would still be present, but the knob would not. This was the case! In a second plant (Case II), the phenotype colored, starchy could result from either nonrecombinant gametes or from

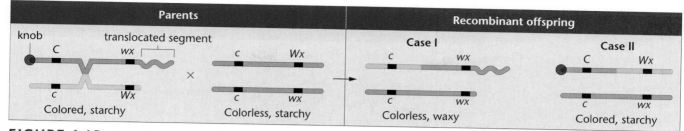

FIGURE 6.15   The phenotypes and chromosome compositions of parents and recombinant offspring in Creighton and McClintock's experiment in maize. The knob and translocated segment served as cytological markers that established that crossing over involves an actual exchange of chromosome arms.

crossing over. Some of the cases then ought to contain chromosomes with the dense knob but not the translocated chromosome. This condition was also found, and the conclusion that a physical exchange had taken place was again supported. Along with Stern's findings with *Drosophila*, this work clearly established that crossing over has a cytological basis.

Once we have introduced the topics of the chemical structure and replication of DNA (Chapters 11 and 12), we will return to the topic of crossing over and look at how breakage and reunion occur between the strands of DNA making up chromatids. This discussion will provide a better understanding of genetic recombination.

Recombination occurs between mitotic chromosomes

In 1936, Curt Stern considered the question as to whether exchanges similar to crossing over occur during mitosis. He was able to demonstrate that this indeed is the case in *Drosophila*. This finding, the first to demonstrate **mitotic recombination,** was considered unusual because homologs do not normally pair up during mitosis in most organisms. However, such synapsis appears to be the rule in *Drosophila*. Since Stern's discovery, genetic exchange during mitosis has also been shown to be a general event in certain fungi.

Stern observed small patches of mutant tissue in females heterozygous for the X-linked recessive mutations *yellow* body and *singed* bristles. Under normal circumstances, a heterozygous female is completely wild type (gray-bodied with straight, long bristles). He explained the appearance of the mutant patches by postulating that, during mitosis in certain cells during development, homologous exchanges could occur between the loci for *yellow* (*y*) and *singed* (*sn*) or between *singed* and the centromere. Figure 6.16 diagrams the nature of the proposed exchanges, in contrast to the case where no exchange occurs. The mutant patches that will occur are also depicted. When no exchange occurs, all tissue is wild type (gray body and gray, straight bristles). After the two types of exchanges, tissues

are produced with either a *yellow* patch *or* with an adjacent *yellow* and *singed* patch (called a **twin spot**).

In 1958 George Pontecorvo and others described a similar phenomenon in the fungus *Aspergillus*. Although the vegetative stage is normally haploid, some cells fuse. The resultant diploid cells then divide mitotically. As in *Drosophila*, crossing over occasionally occurs between linked genes during mitosis in this diploid stage, resulting in recombinant cells. Pontecorvo referred to these events that produce genetic variability as the **parasexual cycle.** On the basis of such exchanges, genes can be mapped by estimating the frequency of recombinant classes.

As a rule, if mitotic recombination occurs at all in an organism, it does so at a much lower frequency than meiotic crossing over. We assume that there is always at least one exchange per meiotic tetrad. By contrast, in organisms that demonstrate mitotic exchange, it occurs in 1 percent or less of mitotic divisions.

Exchanges also occur between sister chromatids

Because homologous chromosomes do not usually pair up or synapse in somatic cells of diploid organisms (*Drosophila* is an exception), each individual chromosome in prophase and metaphase of mitosis consists of two identical sister chromatids, joined at a common centromere. Surprisingly, a number of experiments have demonstrated that reciprocal exchanges similar to crossing over occur even between sister chromatids. While these **sister chromatid exchanges (SCEs)** do not produce new allelic combinations, there is increasing evidence that these events are significant.

Identification and study of sister chromatid exchanges are facilitated by several modern staining techniques. In one approach, cells are allowed to replicate for two generations in the presence of the thymidine analog bromodeoxyuridine (BUdR).* Following two rounds of

---

*The abbreviation BrdU is also used to denote bromodeoxyuridine.

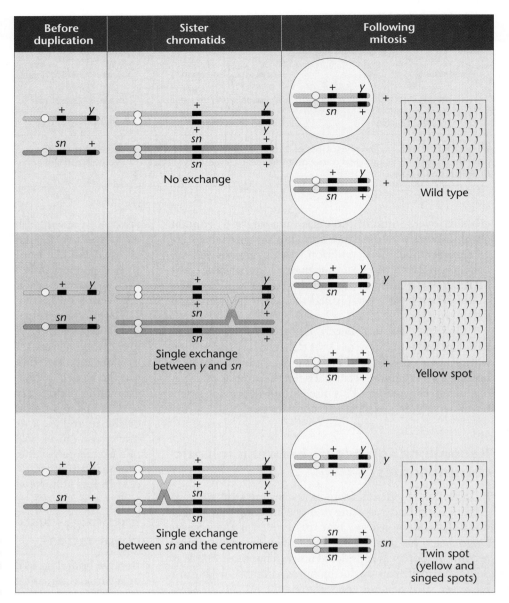

FIGURE 6.16 The production of mutant tissue in a female *Drosophila* heterozygous for the recessive *yellow* (*y*) and *singed* (*sn*) alleles as a result of mitotic recombination.

semiconservative replication, chromatids are found with either one or two DNA strands containing BUdR. Chromatids with both strands containing BUdR stain less brightly than do chromatids with only one BUdR strand (Figure 6.17). Thus, for each pair of sister chromatids, one chromatid contains both DNA strands "labeled" and one chromatid has only one DNA strand "labeled." In Figure 6.17, numerous instances of SCE events can be detected.

Although the significance of sister chromatid exchanges is still uncertain, several observations have led to great interest in this phenomenon. It is known, for example, that agents that induce chromosome damage (such as viruses, X rays, ultraviolet light, and certain chemical mutagens) also increase the frequency of sister chromatid exchanges. Further, an elevated frequency of SCEs is characteristic of **Bloom syndrome,** a human disorder caused by a mutation in the chromosome 15 BLM gene. This rare, recessively inherited disease is characterized by prenatal and postnatal retardation of growth, a great sensitivity of the facial skin to the sun, immune deficiency, a predisposition to malignant and benign tumors, and abnormal behavior patterns. The chromosomes from cultured leukocytes, bone marrow cells, and fibroblasts derived from homozygotes are very fragile and unstable when compared to those derived from homozygous and heterozygous normal individuals. Increased breaks and rearrangements between nonhomologous chro-

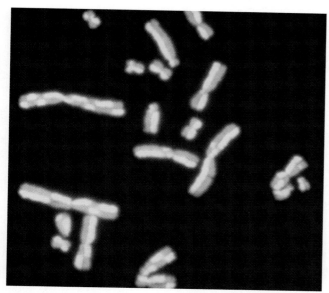

FIGURE 6.17 Demonstration of sister chromatid exchanges (SCEs) in mitotic chromosomes. Sometimes called **harlequin chromosomes** because of their patchlike appearance, chromatids containing the thymidine analog BUdR in both DNA strands fluoresce *less* brightly than do those with the analog in only one strand. These chromosomes were stained with 33258-Hoechst reagent and acridine orange and then viewed under fluorescence microscopy.

mosomes are observed in addition to excessive amounts of sister chromatid exchanges. Recent work by James German and colleagues suggest that the BLM gene encodes an enzyme called **DNA helicase,** which is best known for its role in DNA replication (see Chapter 12).

The mechanisms of exchange between nonhomologous chromosomes and between sister chromatids may prove to share common features because the frequency of both events increases substantially in individuals with genetic disorders. These findings suggest that further study of sister chromatid exchange may contribute to an increased understanding of recombination mechanisms and the relative stability of normal and genetically abnormal chromosomes. We shall encounter still another demonstration of SCEs in Chapter 12 when we consider replication of DNA (see Figure 12.5).

Linkage analysis and mapping can be performed in haploid organisms

We turn now to still another topic that is an extension of our study of transmission genetics: linkage analysis and chromosome mapping in haploid eukaryotes. As we shall see, even though analysis of the location of genes relative to one another throughout the genome of haploid organisms may *seem* a bit more complex than in diploid organisms, the basic underlying principles are the same. In fact,

many basic principles of inheritance were established during the study of haploid fungi.

While many single-celled eukaryotes are haploid during the vegetative stages of their life cycle, they also form reproductive cells that fuse during fertilization, forming a diploid zygote. The zygote then undergoes meiosis and reestablishes haploidy. The haploid meiotic products are the progenitors of the subsequent members of the vegetative phase of the life cycle. Figure 6.18 illustrates this type of cycle in the green alga *Chlamydomonas.* Even though the haploid cells that fuse during fertilization *look* identical (and are thus called **isogametes**), a chemical identity that distinguishes two distinct types exists on their surface. As a result, all strains are either "+" or "−" and fertilization occurs only between unlike cells.

To perform genetic experiments with haploid organisms, genetic strains of different genotypes are isolated and crossed to one another. Following fertilization and meiosis, the meiotic products are retained together and can be analyzed. Such is the case in *Chlamydomonas* as well as in the fungus *Neurospora*, which we shall use as an example in the ensuing discussion. Following fertilization in *Neurospora* (Figure 6.19), meiosis occurs in a saclike structure called the **ascus** and the initial set of haploid products, called a **tetrad,** are retained within it. This term has a quite different meaning here from its use to describe the four-stranded chromosome configuration characteristic of meiotic prophase I in diploids.

Following meiosis in *Neurospora*, each cell in the ascus divides mitotically, producing eight haploid **ascospores.** These can be dissected and examined morphologically or tested to determine their genotypes and phenotypes. Because the eight cells reflect the *sequence* of their formation following meiosis, the tetrad is "ordered" and we can do **ordered tetrad analysis.** This process is critical to our subsequent discussion.

Gene-to-Centromere Mapping

When a single gene (*a*) is analyzed in *Neurospora*, as diagrammed in Figure 6.20, the data can be used to calculate the map distance between that gene and the centromere. This process is sometimes referred to as **mapping the centromere.** It can be accomplished by experimentally determining the frequency of recombination using tetrad data.

When no crossover event occurs between the gene under study and the centromere, the pattern of ascospores in the ascus appears as shown in Figure 6.20(a) (*aaaa++++*). This pattern represents **first-division segregation** because the two alleles are separated during the first meiotic division. However, crossover events will alter this pattern, as shown in Figure 6.20(b) (*aa++aa++*) and 6.20(c) (*++aaaa++*). Two other recombinant patterns can also occur, depending on which chromatids are involved in the crossover event,

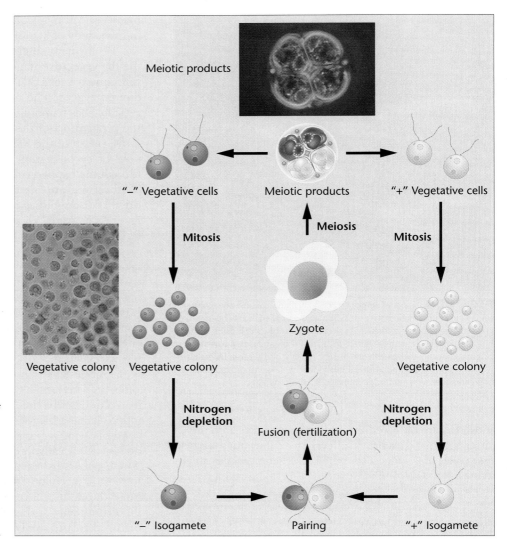

FIGURE 6.18 The life cycle of *Chlamydomonas*. The diploid zygote (in the center) undergoes meiosis, producing "+" or "−" haploid cells that undergo mitosis, yielding vegetative colonies. Unfavorable conditions stimulate them to form isogametes, which fuse in fertilization, producing a zygote, and repeating the cycle. Two of these stages are illustrated photographically.

(*++aa++aa*) and (*aa++++aa*). All four of these patterns resulting from a crossover event between the *a* gene and the centromere reflect **second-division segregation** because the two alleles are not separated until the second meiotic division. Since the mitotic division simply replicates the patterns (from 4 to 8 ascospores), ordered tetrad data are usually condensed to reflect the genotypes reflected in ascospore pairs that may be distinguished from one another. Five unique combinations are possible.

First-Division Segregation

a a + +

Second-Division Segregation

a + a +
+ a + a
+ a a +
a + + a

To calculate the distance between the gene and the centromere, data must be tabulated from a large number of asci resulting from a controlled cross. Using these data, the distance is calculated as

$$\frac{1/2(\text{second-division segregant asci})}{\text{total asci scored}}$$

The recombination percentage is only one-half the number of second-division segregants because crossing over in each of them has occurred in only two of the four strands during meiosis.

To illustrate, assume that *a* represents albino and + represents wild type in *Neurospora*. In crosses between the two genetic types, suppose the following data were observed.

65 first-division segregants
70 second-division segregants

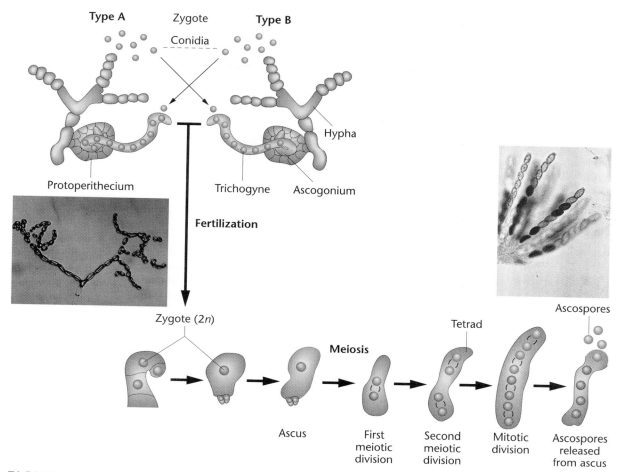

FIGURE 6.19 Sexual reproduction during the life cycle of *Neurospora* is initiated following fusion of conidia of opposite mating types. Following fertilization, each diploid zygote becomes enclosed in an ascus where meiosis occurs, leading to four haploid cells, two of each mating type. A mitotic division then occurs and the eight haploid ascospores are released. Upon germination, the cycle may be repeated. The photographs show the vegetative stage of the organism and several asci that may form in a single structure, even though we have illustrated the events occurring in only one ascus.

The distance between *a* and the centromere is thus

$$\frac{(1/2)(70)}{135} = 0.259$$

or about 26 map units.

As the distance increases up to 50 units, in theory all asci should reflect second-division segregation. However, numerous factors prevent this. As in diploid organisms, accuracy is greatest when the gene and the centromere are relatively close together.

Ordered versus Unordered Tetrad Analysis

In our previous discussion we assumed that the genotype of each ascospore and its position in the tetrad can be determined. To perform such an **ordered tetrad analysis,** individual asci must be dissected and each ascospore must

be tracked as it germinates. This is a tedious process, but it is essential for two types of analysis:

1. To distinguish between first-division segregation and second-division segregation of alleles in meiosis.
2. To determine whether recombinational events are reciprocal or not.

In the first case, such information is essential to "map the centromere," as we have just discussed. Thus, ordered tetrad analysis must be performed in order to map the distance between a gene and the centromere.

In the second case, ordered tetrad analysis has revealed that recombinational events are not always reciprocal, particularly when closely linked genes are studied in Ascomycetes. This observation has led to the investigation of the phenomenon called **gene conversion.** Because its discussion requires a background in DNA structure and analysis, we will return to this topic in Chapter 12.

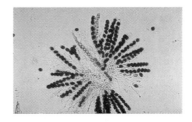

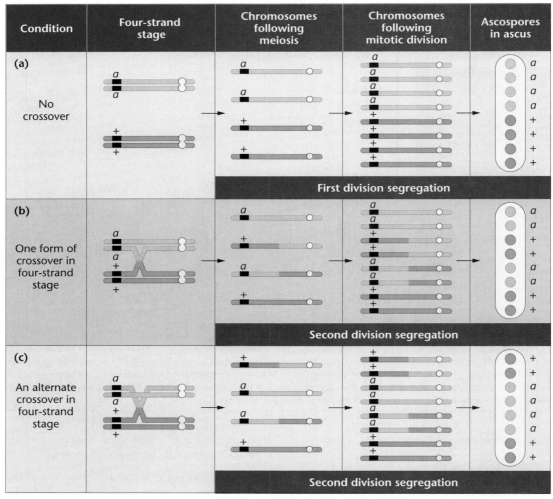

FIGURE 6.20 Three ways in which different ascospore patterns can be generated in *Neurospora*. Analysis of these patterns can serve as the basis of gene-to-centromere mapping.

It is much less tedious to isolate individual asci, allow them to mature, and then determine the genotypes of each ascospore, but not in any particular order. This approach is referred to as **unordered tetrad analysis.** As we shall see in the next section, this type of analysis can be used to determine whether or not two genes are linked on the same chromosome, and if so, to determine the map distance between them.

Linkage and Mapping

Analysis of genetic data derived from haploid organisms can be used to distinguish between linkage and independent assortment of two genes; it further allows mapping distances to be calculated between gene loci once linkage is established. In the following discussion we shall consider tetrad analysis in the alga *Chlamydomonas*. With the exception that the four meiotic products are not ordered and *do not* undergo a mitotic division following the completion of meiosis, the general principles discussed for *Neurospora* also apply to *Chlamydomonas*.

To compare independent assortment and linkage, we will consider two theoretical mutant alleles, *a* and *b*, representing two distinct loci in *Chlamydomonas*. Suppose that 100 tetrads derived from the cross *ab* × ++ yield the tetrad data shown in Table 6.1. As you can see, all tetrads produce one of three patterns. For example, all tetrads in category I produce two ++ cells and two *ab* cells and are designated as **parental ditypes (P).** Category II tetrads produce two *a*+ cells and two +*b* cells and are called **nonparental ditypes (NP).** Category III tetrads produce one cell of each of the four possible genotypes and are thus termed **tetratypes (T).**

These data support the hypothesis that the genes represented by the *a* and *b* alleles are located on separate chromosomes. To understand why, you should refer to Figure 6.21. In parts (a) and (b) of that figure, the origin of parental (P) and nonparental (NP) ditypes is demonstrated for two unlinked genes. According to the Mendelian principle of independent assortment of unlinked genes, approximately equal propor-

TABLE 6.1   Tetrad Analysis in *Chlamydomonas*

Category	I	II	III
Tetrad type	Parental (P)	Nonparental (NP)	Tetratypes (T)
Genotypes present	+ + + + *a b* *a b*	*a* + *a* + + *b* + *b*	+ + *a* + + *b* *a b*
Number of tetrads	43	43	14

tions of these tetrad types are predicted. Thus, when the parental ditypes are equal to the nonparental ditypes, then the two genes are not linked. The data in Table 6.1 confirm this prediction. Because independent assortment has occurred, it can be concluded that the two genes are located on separate chromosomes.

The origin of category III, the **tetratypes,** is diagrammed in Figure 6.21(c,d). The genotypes of tetrads in

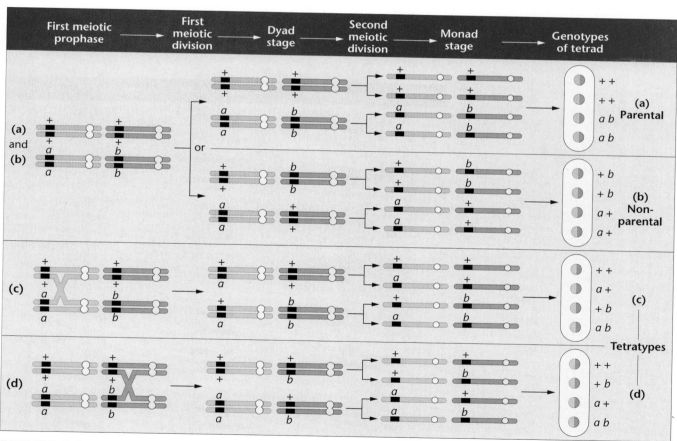

FIGURE 6.21   The origin of various genotypes found in tetrads in *Chlamydomonas* when two genes located on separate chromosomes are considered.

this category can be generated in two possible ways. Both involve a crossover event between one of the genes and the centromere. In Figure 6.21(c) the exchange involves one of the two chromosomes and occurs between gene *a* and the centromere; in Figure 6.21(d), the other chromosome is involved and the exchange occurs between gene *b* and the centromere.

Production of tetratype tetrads does not alter the final ratio of the four genotypes present in all meiotic products. If the genotypes from 100 tetrads (which yield 400 cells) are computed, 100 of each genotype are found. This 1:1:1:1 ratio is predicted according to independent assortment.

Now consider the case where the genes *a* and *b* are linked (Figure 6.22). The same categories of tetrads will be produced. However, parental and nonparental ditypes will not necessarily occur in equal proportions; nor will the four genotypic combinations be found in equal numbers if the genotypes of all meiotic products are computed. For example, the following data might be encountered.

Category I	Category II	Category III
P	NP	T
64	6	30

Since the parental and nonparental categories are not produced in equal proportions, we can conclude that independent assortment is not in operation and that the two genes are linked. We can then proceed to determine the map distance between them.

In the analysis of these data, we are concerned with the determination of which tetrad types represent genetic exchanges between the two genes. The **parental ditype tetrads (P)** arise only when no crossing over occurs between the two genes. The **nonparental ditype tetrads (NP)** arise only when a double exchange involving all four chromatids occurs between two genes. The **tetratype tetrads (T)** arise when either a single crossover occurs or when an alternative type of double exchange occurs between the two genes. The various types of exchanges described here are diagrammed in Figure 6.22.

When the proportion of the three tetrad types has been determined, it is possible to calculate the map distance between the two linked genes. The following formula computes the exchange frequency, which is proportional to the map distance between the two genes.

$$\text{exchange frequency } (\%) = \frac{\text{NP} + 1/2(\text{T})}{\substack{\text{total number} \\ \text{of tetrads}}} \times 100$$

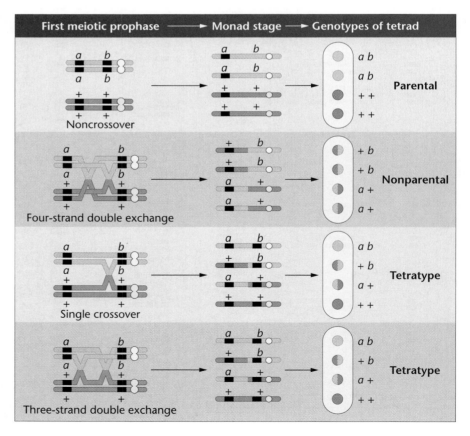

FIGURE 6.22 The various types of exchanges leading to the genotypes found in tetrads in *Chlamydomonas* when two genes located on the same chromosome are considered.

In this formula NP represents the nonparental tetrads; all meiotic products represent an exchange. The tetratype tetrads are represented by T; assuming only single exchanges, one-half of the meiotic products represents exchanges. The sum of the scored tetrads that fall into these categories is then divided by the total number of tetrads examined (P + NP + T). If this calculated number is multiplied by 100, it is converted to a percentage, which is directly equivalent to the map distance between the genes.

In our example, the calculation reveals that genes *a* and *b* are separated by 21 map units.

$$\frac{6 + 1/2(30)}{100} = \frac{6 + 15}{100} = \frac{21}{100} = 0.21 \times 100 = 21\,\%$$

Although we have considered linkage analysis and mapping of only two genes at a time, such studies often involve three or more genes. In these cases, both gene sequence and map distances can be determined.

Somatic-cell hybridization is an important technique in creating human chromosome maps

For obvious reasons, our own species is not a good source of data for the types of extensive linkage analysis performed with experimental organisms. Thus, in humans, the earliest linkage studies had to rely on pedigree analysis. Attempts were made to establish whether a trait was X-linked or autosomal. As we established in Chapter 4, traits determined by genes located on the X chromosome result in characteristic pedigrees, and thus such genes were easier to identify. For autosomal traits, geneticists tried to distinguish clearly whether pairs of traits demonstrated linkage or independent assortment. When extensive pedigrees are available, it is possible to ascertain that the genes under consideration are closely linked (i.e., rarely separated by crossing over) from the fact that the two traits segregate together. For example, this approach established linkage between the genes encoding the **Rh antigens** and the gene responsible for the phenotype referred to as **elliptocytosis** (where the shape of erythrocytes is oval).

The difficulty arises, however, when two genes of interest are separated on a chromosome such that recombinant gametes are formed, obscuring linkage in a pedigree. In such cases, an approach relying on probability calculations, called the **lod score method,** helps to demonstrate linkage. First devised by J. B. S. Haldane and C. A. Smith in 1947 and refined by Newton Morton in 1955, the lod score (standing for *l*og of the *od*ds favoring linkage) assesses the probability that a particular pedigree involving two traits reflects linkage. First, the probability is calculated that family data (pedigrees) concerning two or more traits conform to the transmission of traits without linkage. Then the probability is calculated that the identical family data following these same traits result from linkage with a specified recombination frequency. The ratio of these probability values expresses the "odds" for and against linkage. The lod score method represented an important advance in assigning human genes to specific chromosomes and constructing preliminary human chromosome maps. Its accuracy is limited by the extent of the family data, however, and the initial results were discouraging because of these limitations and because of the relatively high haploid number of human chromosomes (23). By 1960, short of the assignment of some genes to the X chromosome, very little autosomal linkage or mapping information had become available.

In the 1960s, a new technique, **somatic-cell hybridization,** was developed that aided immensely in the initial step of human gene mapping, assigning genes to their respective chromosomes. This technique, first discovered by Georges Barsky, relies on the fact that two cells in culture can be induced to fuse into a single hybrid cell. Barsky used two mouse cell lines, but it soon became evident that cells from different organisms will also fuse together. This event produces an initial cell type called a **heterokaryon,** which contains two nuclei in a common cytoplasm. With somatic-cell hybridization techniques, it is possible to fuse human and mouse cells, for example, and isolate the hybrids from the parental cells.

As the heterokaryons are cultured *in vitro*, two interesting changes occur. Eventually, the nuclei fuse together, creating what is termed a **synkaryon.** Then, as culturing is continued for many generations, chromosomes from one of the two parental species are gradually lost. In the case of the human–mouse hybrid, human chromosomes are lost randomly until, eventually, the synkaryon has a full complement of mouse chromosomes and only a few human chromosomes. As we will see, it is the preferential loss of human chromosomes rather than mouse chromosomes that makes possible the assignment of human genes to the chromosomes upon which they reside.

The experimental rationale is straightforward. For example, if a specific human gene product is synthesized in a synkaryon containing three human chromosomes, then the gene responsible for that product must reside on one of the three human chromosomes remaining in the hybrid cell. Or, if the human gene product is absent, the responsible gene cannot be present on any of the remaining three human chromosomes. Ideally, a panel of 23 hybrid cell lines, each with only one unique human chromosome, would allow the immediate assignment to a particular chromosome of any human gene for which the product could be characterized.

Hybrid cell lines	Human chromosomes present								Gene products expressed			
	1	2	3	4	5	6	7	8	*A*	*B*	*C*	*D*
23	present	present	present	present					–	+	–	+
34	present	present			present	present			+	–	–	+
41	present		present		present		present		+	+	–	+

FIGURE 6.23 A hypothetical grid of data used in synteny testing to assign genes to their appropriate human chromosomes. Three somatic hybrid cell lines, designated 23, 34, and 41, have each been scored for the presence or absence of human chromosomes 1 through 8, as well as for their ability to produce the hypothetical human gene products A through D.

In practice, a panel of cell lines, each with several remaining human chromosomes, is most often used. The correlation of the presence or absence of each chromosome with the presence or absence of each gene product is called **synteny testing.** Consider, for example, the hypothetical data provided in Figure 6.23, where four gene products (*A*, *B*, *C*, and *D*) are tested in relationship to eight human chromosomes. Let us carefully analyze the gene that produces product *A*.

1. Product *A* is not produced by cell line 23, but chromosomes 1, 2, 3, and 4 are present in cell line 23. Therefore, we can rule out the presence of gene *A* on those four chromosomes and conclude that it must be on chromosome 5, 6, 7, or 8.
2. Product *A* is produced by cell line 34, which contains chromosomes 5 and 6, but not 7 and 8. Therefore, gene *A* is on chromosome 5 or 6.
3. Product *A* is also produced by cell line 41, which contains chromosome 5 but not chromosome 6. Therefore, gene *A* is on chromosome 5, according to this analysis.

Using a similar approach, we can assign gene *B* to chromosome 3. You should perform this analysis to demonstrate for yourself that this is correct.

Gene *C* presents a unique situation. The data indicate that it is not present on any of the first seven chromosomes (1–7) because it is not produced by any of the three cell lines that collectively contain those chromosomes. While it might be on chromosome 8, no direct evidence supports this conclusion. Other panels are needed. We leave gene *D* for you to analyze. Upon what chromosome does it reside?

Using the approach described above, literally hundreds of human genes have been assigned to specific chromosomes. Figure 6.23 illustrates some gene locations on two human chromosomes—X and 1. Some of the assign-

Key		
	AMY	Amylase (Salivary and Pancreatic)
	AT3	Antithrombin (Clotting Factor IV)
	CB	Colorblindness
	DMD	Duchenne Muscular Dystrophy
	FHM	Fumarate Hydratase (Mitochondrial)
	GDH	Glucose Dehydrogenase
	G6PD	Glucose-6-Phosphate Dehydrogenase
	HEMA	Hemophilia A (Classic)
	HGPRT	Hypoxanthine-Guanine-Phosphoribosyl Transferase (Lesch–Nyhan Syndrome)
	PEPC	Peptidase C
	PGK	Phosphoglycerate Kinase
	PGM	Phosphoglucomutase
	Rh	Rhesus Blood Group (Erythroblastosis Fetalis)

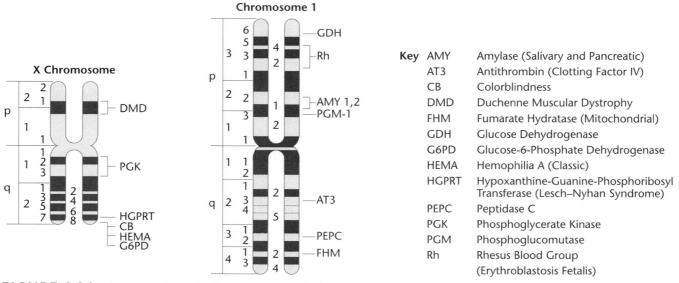

FIGURE 6.24 Representative regional gene assignments for human chromosome 1 and the X chromosome. Many assignments were initially derived using somatic-cell hybridization techniques.

ments were derived in this way. For mapping still other genes, where the products have yet to be discovered, researchers have had to rely on other linkage techniques. For example, by combining a rather sophisticated approach using recombinant DNA technology with pedigree analysis, it has been possible to assign the genes responsible for **Huntington disease, cystic fibrosis,** and **neurofibromatosis** to their respective chromosomes, 4, 7, and 17. We will discuss this approach in Chapter 21.

We conclude this discussion by addressing the next step in human gene mapping, assigning genes to different regions of a given chromosome. Sometimes in hybrid cell lines, fragments of a particular chromosome become transferred to another chromosome, resulting in a **translocation.** It is possible using chromosome banding techniques to identify the exact origin of the translocation and correlate the presence of a chromosomal segment in hybrid cells with specific gene expression. In this way, gene maps of human chromosomes may be compiled. The partial maps shown in Figure 6.24 illustrate this point. Although human chromosome maps are not as specific as the genetic map of *Drosophila*, we are beginning to learn a great deal about the chromosome locations of a multitude of human genes. As we will see in Chapters 18 and 21, modern technology involving recombinant DNA and the Human Genome Project have greatly extended our knowledge of gene locations within the human genome.

Did Mendel encounter linkage?

It has been said often that Mendel had extremely good fortune in his classical experiments with the garden pea. In none of his crosses did he encounter apparent linkage relationships between any of the seven mutant characters. Had Mendel obtained highly variable data characteristic of linkage and crossing over, these unorthodox observations might have hindered his successful analysis and interpretation.

The accompanying article by Stig Blixt, reprinted in its entirety, presents a modern-day interpretation of Mendel's experiments, demonstrating the inadequacy of this hypothesis. As we shall see, some of Mendel's genes were indeed linked. We shall leave it to Stig Blixt to enlighten you as to why Mendel did not detect linkage.

Why Didn't Gregor Mendel Find Linkage?

It is quite often said that Mendel was very fortunate not to run into the complication of linkage during his experiments. He used seven genes, and the pea has only seven chromosomes. Some have said that had he taken just one more, he would have had problems. This, however, is a gross oversimplification. The actual situation, most probably, is shown in Table 1. This shows that Mendel worked with three genes in chromosome 4, two genes in chromosome 1, and one gene in each of chromosome(s) 5 and 7. It seems at first glance that, out of the 21 dihybrid combinations Mendel theoretically could have studied, no less than four (that is, *a-i, v-fa, v-le, fa-le*) ought to have resulted in linkages. As found, however, in hundreds of crosses and shown by the genetic map of the pea,* *a* and *i* in chromosome 1 are so distantly located on the chromosome that no linkage is normally detected. The same is true for *v* and *le* on the one hand, and *fa* on the other, in chromosome 4. This leaves *v-le*, which ought to have shown linkage.

Mendel, however, seems not to have published this particular combination and thus, presumably, never made the appropriate cross to obtain both genes segregating simultaneously. It is therefore not so astonishing that Mendel did not run into the complication of linkage, although he did not avoid it by choosing one gene from each chromosome.

Weibullsholm Plant Breeding Institute, Landskrona, Sweden and Centro Energia Nucleare na Agricultura, Piraciaba, SP, Brazil

STIG BLIXT

Received March 5; accepted June 4, 1975.

*Blixt, S. 1974. In *Handbook of genetics*, ed. R. C. King. New York: Plenum Press.

TABLE 1 Relationship between Modern Genetic Terminology and Character Pairs Used by Mendel

Character Pair Used by Mendel	Alleles in Modern Terminology	Located in Chromosome
Seed color, yellow-green	*l-i*	1
Seed coat and flowers, colored-white	*A-a*	1
Mature pods, smooth expanded-wrinkled indented	*V-v*	4
Inflorescences, from leaf axis-umbellate in top of plant	*Fa-fa*	4
Plant height, 1 m-around 0.5 m	*Le-le*	4
Unripe pods, green-yellow	*Gp-gp*	5
Mature seeds, smooth-wrinkled	*R-r*	7

CHAPTER SUMMARY

1. Genes located on the same chromosome are said to be linked. Alleles located on the same homolog, therefore, can be transmitted together during gamete formation. However, the mechanism of crossing over between homologs during meiosis results in the reshuffling of alleles, thereby contributing to genetic variability within gametes.

2. Early in this century, geneticists realized that crossing over could provide an experimental basis for mapping the location of linked genes relative to one another along the chromosome.

3. Interference is a phenomenon describing the extent to which a crossover in one region of a chromosome influence the occurrence of a crossover in an adjacent region of the chromosome. The coefficient of coincidence (C) is a quantitative estimate of interference calculated by dividing the observed double crossovers by the expected double crosssovers.

4. Due to statistical considerations described by the Poisson distribution, as the actual distance between two genes increases, experimentally determined mapping distances become more and more inaccurate (underestimated).

5. Extensive genetic maps have been created in organisms such as corn, mice, and *Drosophila*.

6. Cytological investigations of both maize and *Drosophila* have revealed that crossing over involves a physical exchange of segments between nonsister chromatids.

7. In a few organisms, including *Drosophila* and *Aspergillus*, homologs pair during mitosis and crossing over occurs between them at a frequency far less than during meiosis.

8. An exchange of genetic material between sister chromatids may occur during mitosis as well. These events are referred to as sister chromatid exchanges (SCEs).

9. Linkage analysis and chromosome mapping are possible in haploid eukaryotes. Our discussion has included gene-to-centromere and gene-to-gene mapping as well as the consideration of how to distinguish between linkage and independent assortment.

10. Somatic-cell hybridization techniques have made possible linkage and mapping analysis of human genes.

11. Evidence now suggests that several of Mendel's seven genetic characters are linked. However, in the cases where they were studied the genes were sufficiently far apart along the chromosome to prevent linkage from being detected.

INSIGHTS AND SOLUTIONS

1. In a series of two-point map crosses involving three genes linked on chromosome 3 in *Drosophila*, the following distances were calculated.

    cd–sr   13 mu
    cd–ro   16 mu

    (a) Determine the sequence and construct a map of these three genes.

    Solution: It is impossible to do so. There are two possibilities based on these limited data,

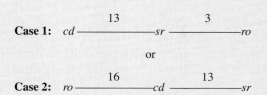

    (b) What mapping data will resolve this?

    Solution: The map distance determined by crossing over between *ro* and *sr*. If Case 1 is correct, it should be 3 map units (mu), and if Case 2 is correct, it should be 29 mu. In fact, this distance is 29 mu, demonstrating that Case 2 is correct.

(c) Can we tell which of the sequences shown below is correct?

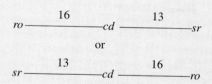

Solution: No; based on the mapping data, they are equivalent.

2. In rabbits, *black* (B) is dominant to *brown* (b), while *full color* (C) is dominant to *chinchilla* (c^{ch}). The genes controlling these traits are linked. Rabbits that are heterozygous for both traits and express *black, full color* were crossed to rabbits that express *brown, chinchilla*, with the following results:

> 31 *brown, chinchilla*
> 35 *black, full color*
> 16 *brown, full color*
> 19 *black, chinchilla*

Determine the arrangement of alleles in the heterozygous parents and the map distance between the two genes.

Solution: This is a two-point map problem, where the two reciprocal noncrossover phenotypes are recognized as those present in the highest numbers (*brown, chinchilla* and *black, full*). The less frequent reciprocal phenotypes (*brown, full* and *black, chinchilla*) arise from a single crossover. The arrangement of alleles is derived from the noncrossover phenotypes because they enter gametes intact. The cross is shown as follows.

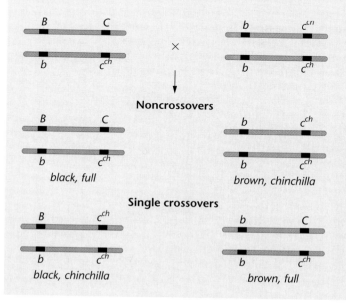

The single crossovers give rise to 35/100 offspring (35 percent). Therefore, the distance between the two genes is 35 map units (mu).

3. In *Drosophila*, *Lyra (Ly)* and *Stubble (Sb)* are dominant mutations located at locus 40 and 58, respectively, on chromosome 3. A recessive mutation with bright-red eyes was discovered and shown also to be on chromosome 3. A map was obtained by crossing a female who was heterozygous for all three mutations to a male homozygous for the bright-red mutation (which we refer to here as *br*). The following data were obtained. Determine the location of the bright-red mutation on chromosome 3. By referring to Figure 6.14, predict what mutation has been discovered. How could you be sure?

Phenotype			Number
(1) Ly	Sb	br	404
(2) +	+	+	422
(3) Ly	+	+	18
(4) +	Sb	br	16
(5) Ly	+	br	75
(6) +	Sb	+	59
(7) Ly	Sb	+	4
(8) +	+	br	2
			1000

Solution: First determine the *arrangement* of the alleles on the homologs of the heterozygous crossover parent (the female in this case). This is done by locating the most frequent reciprocal phenotypes, which arise from the noncrossover gametes. These are phenotypes (1) and (2). Each one represents the arrangement of alleles on one of the homologs. Therefore, the arrangement is

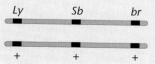

Second, determine the correct *sequence* of the three loci along the chromosome. This is done by determining which sequence will yield the *observed* double-crossover phenotypes, and which are the least frequent reciprocal phenotypes (7 and 8).

If the sequence is correct as written, then a double crossover depicted here,

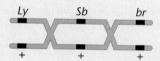

will yield *Ly + br* and *+ Sb +* as phenotypes. Inspection shows that these categories (5 and 6) are actually single crossovers, not double crossovers. Therefore, the sequence, as written, is incorrect. There are only two other possible sequences. The *br* gene is either to the left of *Ly*, or it is between *Ly* and *Sb*.

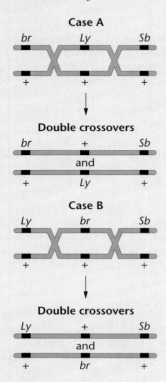

Case A

Double crossovers

Case B

Double crossovers

Comparison with the actual data shows that Case B is correct. The double-crossover gametes (7) and (8) yield flies that express *Ly* and *Sb*, but not *br*, or express *br*, but not *Ly* and *Sb*. Therefore, the correct arrangement *and* sequence are:

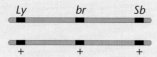

Once this has been determined, it is possible to determine the location of *br* relative to *Ly* and *Sb*. A single crossover between *Ly* and *br*, as shown here yields flies that are *Ly + +* and *+ br Sb* (categories 3 and 4).

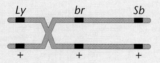

Therefore, the distance between the *Ly* and *br* loci is equal to

18 + 16 + 4 + 2 / 1000 = 40/1000 = 0.04 = 4 map units

Remember that because we need to know the frequency of all crossovers between *Ly* and *br*, we must add in the double crossovers, since they represent two single crossovers occurring simultaneously.

Similarly, the distance between the *br* and *Sb* loci is derived mainly from single crossovers between them:

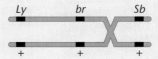

This event yields *Ly br +* and *+ + Sb* phenotypes (categories 5 and 6). Therefore, the distance equals

75 + 59 + 4 + 2 / 1000 = 140 / 1000 = 0.14 = 14 map units

The final map shows that *br* is located at locus 44, since *Lyra* and *Stubble* are known.

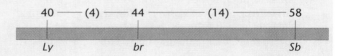

Inspection of Figure 6.14 reveals that the mutation *scarlet*, which has bright-red eyes, is known to exist at locus 44, so it is reasonable to hypothesize that the bright-red eye mutation is an allele of *scarlet*. To test this hypothesis, we could cross females of our bright-red mutant with known *scarlet* males. If they are alleles, all progeny will reveal bright-red mutant eyes. If not, all progeny will show normal brick-red wild-type eyes, indicating that our mutant and scarlet are actually at slightly different loci (very near 44.0). In such a case, all progeny will be heterozygous and will not show bright-red eyes. This cross represents what is called an **allelism test**.

4. In a cross in *Neurospora* where one parent expresses the mutant allele *a* and the other expresses a wild-type phenotype (+), the following data were obtained in the analysis of ascospores. Calculate the gene-to-centromere distance.

		Asci Types			
1	2	3	4	5	6
+	*a*	*a*	+	*a*	+
+	*a*	*a*	+	*a*	+
+	*a*	+	*a*	+	*a*
+	*a*	+	*a*	+	*a*
a	+	*a*	+	+	*a*
a	+	*a*	+	+	*a*
a	+	+	*a*	*a*	+
a	+	+	*a*	*a*	+
39	33	5	4	9	10

Sequence of ascospores in ascus

Total = 100

Solution: Ascus types 1 and 2 represent first-division segregants (fds) where no crossing over occurred between the *a* locus and the centromere. All others (3–6) represent second-division segregation (sds). By applying the formula

$$\text{distance} = 1/2 \text{ sds} / \text{total asci}$$

we obtain the following result.

$$d = 1/2(5 + 4 + 9 + 10) / 100$$
$$= 1/2(28) / 100$$
$$= 0.14$$
$$= 14 \text{ map units}$$

PROBLEMS AND DISCUSSION QUESTIONS

1. What is the significance of genetic recombination to the process of evolution?
2. Describe the cytological observation that suggests that crossing over occurs during the first meiotic prophase.
3. Why does more crossing over occur between two distantly linked genes than between two genes that are very close together on the same chromosome?
4. Why is a 50 percent recovery of single-crossover products the upper limit, even when crossing over *always* occurs between two linked genes?
5. Why are double-crossover events expected in lower frequency than single-crossover events?
6. What is the proposed basis for positive interference?
7. What two essential criteria must be met in order to execute a successful mapping cross?
8. The genes *dumpy wing* (*dp*), *clot eye* (*cl*), and *apterous wing* (*ap*) are linked on chromosome 2 of *Drosophila*. In a series of two-point mapping crosses, the following genetic distances were determined.

dp–ap	42
dp–cl	3
ap–cl	39

What is the sequence of the three genes?

9. Consider two hypothetical recessive autosomal genes *a* and *b*. Where a heterozygote is test-crossed to a double-homozygous mutant, predict the phenotypic ratios under the following conditions:
   (a) *a* and *b* are located on separate autosomes.
   (b) *a* and *b* are linked on the same autosome but are so far apart that a crossover always occurs between them.
   (c) *a* and *b* are linked on the same autosome but are so close together that a crossover almost never occurs.
   (d) *a* and *b* are linked on the same autosome about 10 map units apart.

10. In corn, colored aleurone (in the kernels) is due to the dominant allele *R*. The recessive allele *r*, when homozygous, produces colorless aleurone. The plant color (not the kernel color) is controlled by another gene with two alleles, *Y* and *y*. The dominant *Y* allele results in green color, whereas the homozygous presence of the recessive *y* allele causes the plant to appear yellow. In a test cross between a plant of unknown genotype and phenotype and a plant that is homozygous recessive for both traits, the following progeny were obtained:

Colored green	88
Colored yellow	12
Colorless green	8
Colorless yellow	92

Explain how these results were obtained by determining the exact genotype and phenotype of the unknown plant, including the precise association of the two genes on the homologs (i.e., the arrangement).

11. In the cross shown involving two linked genes, *ebony* (*e*) and *claret* (*ca*), in *Drosophila*, where crossing over does not occur in males, offspring were produced in a 2 +:1 ca:1 e phenotypic ratio.

♀		♂
$\dfrac{e \quad ca^+}{e^+ \quad ca}$	×	$\dfrac{e \quad ca^+}{e^+ \quad ca}$

These genes are 30 units apart on chromosome 3. What contribution did crossing over in the female make to these phenotypes?

12. With two pairs of genes involved (*P/p* and *Z/z*), a test cross (*ppzz*) with an organism of unknown genotype indicated that the gametes produced were in the following proportions:

PZ, 42.4 percent; *Pz,* 6.9 percent; *pZ,* 7.1 percent; and *pz,* 43.6 percent

Draw all possible conclusions from these data.

13. In a series of two-point map crosses involving five genes located on chromosome II in *Drosophila*, the following recombinant (single-crossover) frequencies were observed.

pr–adp	29
pr–vg	13
pr–c	21
pr–b	6
adp–b	35
adp–c	8
adp–vg	16
vg–b	19
vg–c	8
c–b	27

(a) If *adp* gene is present near the end of chromosome II (locus 83), construct a map of these genes.
(b) In another set of experiments, a sixth gene, *d*, was tested against *b* and *pr:*

d–b	17%
d–pr	23%

Predict the results of two-point maps between *d* and *c*, *d* and *vg*, and *d* and *adp*.

14. Two different female *Drosophila* were isolated, each heterozygous for the autosomally linked genes *b* (*black body*), *d* (*dachs tarsus*), and *c* (*curved wings*). These genes are in the order *d–b–c*, with *b* being closer to *d* than to *c*. Shown is the genotypic arrangement for each female along with the various gametes formed by both. Identify which categories are noncrossovers (NCO), single crossovers (SCO), and double crossovers (DCO) in each case. Then, indicate the relative frequency in which each will be produced.

Female A		Female B	
d b +		*d + +*	
+ + c		*+ b c*	
↓ Gamete formation ↓			
(1) *d b c*	(5) *d + +*	(1) *d b +*	(5) *d b c*
(2) *+ + +*	(6) *+ b c*	(2) *+ + c*	(6) *+ + +*
(3) *+ + c*	(7) *d + c*	(3) *d + c*	(7) *d + +*
(4) *d b +*	(8) *+ b +*	(4) *+ b +*	(8) *+ b c*

15. In *Drosophila*, a cross was made between females expressing the three X-linked recessive traits, *scute* (*sc*) *bristles*, *sable body* (*s*), and *vermilion eyes* (*v*), and wild-type males. In the F_1, all females were wild type, while all males expressed all three mutant traits. The cross was carried to the F_2 generation and 1000 offspring were counted, with the results shown here. No determination of sex was made in the F_2 data.

Phenotype			Offspring
sc	*s*	*v*	314
+	+	+	280
+	*s*	*v*	150
sc	+	+	156
sc	+	*v*	46
+	*s*	+	30
sc	*s*	+	10
+	+	*v*	14

(a) Determine the genotypes of the P_1 and F_1 parents, using proper nomenclature.
(b) Determine the sequence of the three genes and the map distance between them.
(c) Are there more or fewer double crossovers than expected? Calculate the coefficient of coincidence. Does this represent positive or negative interference?

16. Another cross in *Drosophila* involved the recessive, X-linked genes *yellow* (*y*), *white* (*w*), and *cut* (*ct*). A female that was yellow-bodied and white-eyed with normal wings was crossed to a male whose eyes and body were normal but whose wings were cut. The F_1 females were wild type for all three traits, while the F_1 males expressed the yellow-body, white-eye traits. The cross was carried to an F_2, and only male offspring were tallied. On the basis of the data shown below, a genetic map was constructed.

Phenotype			Male Offspring
y	+	*ct*	9
+	*w*	+	6
y	*w*	*ct*	90
+	+	+	95
+	+	*ct*	424
y	*w*	+	376
y	+	+	0
+	*w*	*ct*	0

(a) Diagram the genotypes of the F_1 parents.
(b) Construct a map, assuming that *white* is at locus 1.5 on the X chromosome.
(c) Were any double-crossover offspring expected?
(d) Could the F_2 female offspring be used to construct the map? Why or why not?

17. In *Drosophila*, *Dichaete* (*D*) is a mutation on chromosome 3 mutation with a dominant effect on wing shape. It is lethal when homozygous. The genes *ebony* (*e*) and *pink* (*p*) are recessive mutations on chromosome 3 affecting the body and eye color, respectively. Flies from a *Dichaete* stock were crossed to homozygous *ebony, pink* flies, and the F_1 progeny,

with a *Dichaete* phenotype, were backcrossed to the *ebony, pink* homozygotes. The results of this backcross were

Phenotype	Number
Dichaete	401
ebony, pink	389
Dichaete, ebony	84
pink	96
Dichaete, pink	2
ebony	3
Dichaete, ebony, pink	12
wild type	13

(a) Diagram this cross, showing the genotypes of the parents and offspring of both crosses.

(b) What is the sequence and interlocus distance between these three genes?

18. *Drosophila* females homozygous for the third chromosomal genes *pink* and *ebony* (the same genes from Problem 17) were crossed with males homozygous for the second chromosomal gene *dumpy*. Because these genes are recessive, all offspring were wild type (normal). F_1 females were test-crossed to triply recessive males. If we assume that the two linked genes, *pink* and *ebony*, are 20 map units apart, predict the results of this cross. If the reciprocal cross were made (F_1 males—where no crossing over occurs—with triply recessive females), how would the results vary, if at all?

19. In *Drosophila*, two mutations, *Stubble* (*Sb*) and *curled* (*cu*), are linked on chromosome 3. *Stubble* is a dominant gene that is lethal in a homozygous state, and *curled* is a recessive gene. If a female of the genotype

$$\frac{Sb \quad cu}{+ \quad +}$$

is to be mated to detect recombinants among her offspring, what male genotype would you choose as a mate?

20. In *Drosophila*, a heterozygous female for the X-linked recessive traits *a*, *b*, and *c* was crossed to a male that was phenotypically *a b c*. The offspring occurred in the following phenotypic ratios:

+	*b*	*c*	460
a	+	+	450
a	*b*	*c*	32
+	+	+	38
a	+	*c*	11
+	*b*	+	9

No other phenotypes were observed.

(a) What is the genotypic arrangement of the alleles of these genes on the X chromosomes of the female?

(b) Determine the correct sequence and construct a map of these genes on the X chromosome.

(c) What progeny phenotypes are missing? Why?

21. Why did Stern observe more "twin spots" than *singed* spots in his study of somatic crossing over? If he had been studying *tan* body color (locus 27.5) and *forked* bristles (locus 56.7) on the X chromosome of heterozygous females, what relative frequencies of tan spots, forked spots, and "twin spots" would you predict might occur?

22. Are mitotic recombinations and sister chromatid exchanges effective in producing genetic variability in an individual? In the offspring of individuals?

23. What possible conclusions can be drawn from the observations that no synaptonemal complexes are observed in male *Drosophila* and female *Bombyx* and that no crossing over occurs in these organisms?

24. An organism of the genotype *AaBbCc* was test-crossed to a triply recessive organism (*aabbcc*). The genotypes of the progeny were as follows:

20	*Aa Bb Cc*	20	*Aa Bb cc*
20	*aa bb Cc*	20	*aa bb cc*
5	*Aa bb Cc*	5	*Aa bb cc*
5	*aa Bb Cc*	5	*aa Bb cc*

(a) If these three genes were all assorting independently, how many genotypic and phenotypic classes would result in the offspring, and in what proportion, assuming simple dominance and recessiveness in each gene pair?

(b) Answer the same question assuming the three genes are so tightly linked on a single chromosome that no crossover gametes were recovered in the sample of offspring.

(c) What can you conclude from the *actual* data about the location of the three genes in relation to one another?

25. Based on our discussion of the potential inaccuracy of mapping (see Figure 6.13), would you revise your answer to Problem 24? If so, how?

26. In a plant, fruit was either red or yellow and was either oval or long, where *red* and *oval* are the dominant traits. Two plants, both heterozygous for these traits, were test-crossed, with the following results:

Phenotype	Progeny Plant A	Progeny Plant B
red, long	46	4
yellow, oval	44	6
red, oval	5	43
yellow, long	5	47
	100	100

Determine the location of the genes relative to one another and the genotypes of the two parental plants.

27. In a plant heterozygous for two gene pairs (*Ab/aB*), where the two loci are linked and 25 map units apart, two such individuals were crossed together. Assuming that crossing over occurs during the formation of both male and female gametes, and that the *A* and *B* alleles are dominant, determine the phenotypic ratio of the offspring.

28. In a cross in *Neurospora* involving two alleles *B* and *b*, the following tetrad patterns were observed. Calculate the distance between the gene and the centromere.

Tetrad Pattern	Number
BBbb	36
bbBB	44
BbBb	4
bBbB	6
BbbB	3
bBBb	7

29. In *Neurospora*, the cross a + × + b yielded only two types of ordered tetrads in approximately equal numbers.

	Spore Pair			
	1–2	3–4	5–6	7–8
Tetrad Type 1	a +	a +	+ b	+ b
Tetrad Type 2	+ +	+ +	a b	a b

What can be concluded?

30. Below are two sets of data derived from crosses in *Chlamydomonas*, involving three genes represented by the mutant alleles *a*, *b*, and *c*. Determine as much as you can concerning genetic arrangement of these three genes relative to one another. Describe the expected results of Cross 3, assuming that *a* and *c* are linked and are 38 map units apart, and that 100 tetrads are produced.

Cross	Genes Involved	Tetrads P	NP	T
1	a and b	36	36	28
2	b and c	79	3	18
3	a and c	?	?	?

31. In *Chlamydomonas*, a cross ab × ++ yielded the following unordered tetrad data where *a* and *b* are linked.

(1)	+ + + + a b a b	38	(3)	a + a + + b + b	6	(5)	a b + + + b a +	2
(2)	+ + a b + + a b	5	(4)	a b a + + b + +	17	(6)	a b + b a + + +	3

(a) Identify the categories representing parental ditypes (P), nonparental ditypes (NP), and tetratypes (T).
(b) Explain the origin of category (2).
(c) Determine the map distance between *a* and *b*.

32. The following results are ordered tetrad pairs from a cross between strain *cd* and strain ++. They are summarized by tetrad classes.

	Tetrad Class						
	1	2	3	4	5	6	7
	c + c +	c + c d	c d c d	+ d c +	c + + +	c d + +	c + + d
	+ d + d	+ + + d	+ + + +	c + + d	c d + d	c d + +	c d + +
	1	17	41	1	5	3	1

(a) Name the ascus type of each class from 1 to 7 (P, NP, or T).
(b) The data support the conclusion that the *c* and *d* loci are linked. State the evidence in support of this conclusion.
(c) Calculate the gene–centromere distance for each locus.
(d) Calculate the distance between the two linked loci.
(e) Draw a linkage map including the centromere and explain the discrepancy between the distances determined by the two different methods in parts (c) and (d).
(f) Describe the arrangement of crossovers needed to produce the ascus class 6 above.

33. In a cross in *Chlamydomonas*, AB × ab, 211 unordered asci were recovered:

10	AB, Ab, aB, ab
102	Ab, aB, Ab, aB
99	AB, AB, ab, ab

(a) Correlate each of the three tetrad types in the problem with their appropriate tetrad designations (names).
(b) Are genes *A* and *B* linked?
(c) If they are linked, determine the map distance between the two genes. If they are unlinked, provide the maximum information you can about why you drew this conclusion.

34. A number of human-mouse somatic-cell hybrid clones were examined for the expression of specific human genes and the presence of human chromosomes. The table here summarizes the results. Assign each gene to the chromosome upon which it is located.

	Hybrid cell clone					
	A	B	C	D	E	F
Genes expressed						
ENO1 (enolase-1)	−	+	−	+	+	−
MDH1 (malate dehydrogenase-1)	+	+	−	+	−	+
PEPS (peptidase S)	+	−	+	−	−	−
PGM1 (phosphoglucomutase-1)	−	+	−	+	+	−
Chromosomes (present or absent)						
1	−	+	−	+	+	−
2	+	+	−	+	−	+
3	+	+	−	−	+	−
4	+	−	+	−	−	−
5	−	+	+	+	+	+

GENETICS MediaLab

The following resources will help you achieve a better understanding of the concepts presented in this chapter. These resources can be found on the CD packaged with this textbook and on the Companion Website at **http://www.prenhall.com/klug/**.

CD Resources:

Animated Tutorial: *Linkage and Single Gene Crossover*

Animated Tutorial: *Linkage and Double Gene Crossover*

Self-grading Chapter Problems

Web Resources:

Web Destinations in Genetics

Self-grading Chapter Problems

Chapter Search Terms

Genetics Newsgroups

Student Bulletin Board

Web Problem 1:

Time for completion = 10 minutes

What are the consequences of crossing over, which occurs during prophase of meiosis I when the homologous chromosomes are paired? Read this article from NIH about the physical events associated with crossing over. How do we define linkage between genes? How does crossing over affect variation among gametes? What other process causes variation among gametes? After examining the figure under "Crossing over and recombination in meiosis," you will note that equal portions of the chromosomes have been exchanged. What might happen if unequal portions were exchanged? To complete this exercise, visit Web Problem 1 in Chapter 6 of your Companion Website and select the keyword **CROSSING OVER**.

Web Problem 2:

Time for completion = 10 minutes

How can we determine the order of and distance between genes? The analysis of recombination between linked marker loci forms the basis for constructing chromosome maps. In the first part of the exercise you will review material on linkage (Chapter 6). The second part gives an excellent overview of how to analyze a suite of mutations using the techniques of classical genetics. How do you distinguish between parental and recombinant chromosomes? What is the largest distance, in centimorgans, that you can measure between two linked loci using nothing but mutant and wild-type stocks for those two loci? Why is the recombination frequency between unlinked loci not 100 percent? How do double crossovers affect your estimate of distance between loci? To complete this exercise, visit Web Problem 2 in Chapter 6 of your Companion Website and select the keyword **LINKAGE**.

Web Problem 3:

Time for completion = 20 minutes

In Web Problem 2, you studied how linkage analysis works. The final example showed how a geneticist studying three mutations in *Drosophila* might go about determining their genetic basis (Chapters 3 and 4) and linkage (Chapter 6). In this exercise you will use The Virtual Flylab to try this on your own. The mutations you will study are *spineless* (bristles) *ebony* (body) and *sepia* (eyes). Using the methods outlined in your textbook and in the final example of Web Problem 2, determine the genetic basis (dominant vs. recessive, sex-linked vs. autosomal, viable vs. lethal) of these traits, determine if they are linked or not, and construct a genetic map of the loci. Determine the genotypes of the parental and F_1 flies, using standard *Drosophila* notation. To complete this exercise, visit Web Problem 3 in Chapter 6 of your Companion Website and select the keyword **MAPPING**.

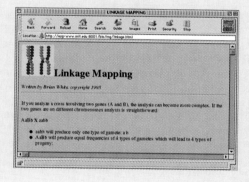

EXTRA-SPICY PROBLEMS

35. A female of genotype

$$\frac{a \quad b \quad c}{+ \; + \; +}$$

produces 100 meiotic tetrads. Of these, 68 show no crossover events. Of the remaining 32, 20 show a crossover between *a* and *b*, 10 show a crossover between *b* and *c*, and 2 show a double crossover between *a* and *b* and between *b* and *c*. Of the 400 gametes produced, how many of each of the 8 different genotypes will be produced? Assuming the order *a–b–c* and the allele arrangement shown here, what is the map distance between these loci?

36. In laboratory, a genetics student was assigned an unknown mutation in *Drosophila* that had a whitish eye. He crossed females from his true-breeding mutant stock to wild-type (brick-red-eyed) males, recovering all wild-type F_1 flies. In the F_2 generation, the following offspring were recovered in the following proportions:

wild type	5/8
bright red	1/8
brown eye	1/8
white eye	1/8

The student was stumped until the instructor suggested that perhaps the whitish eye in the original stock was the result of homozygosity for a mutation causing brown eyes *and* a mutation causing bright-red eyes, illustrating gene interaction (see Chapter 4). After much thought, the student was able to analyze the data, explain the results, and learn several things about the location of the two genes relative to one another. One key to his understanding was that crossing over occurs in *Drosophila* females but not in males. What did the student learn about the two genes based on his analysis?

37. *Drosophila melanogaster* has one pair of sex chromosomes (XX or XY) and three autosomes, referred to as chromosomes 2, 3, and 4. A genetics student discovered a male fly with very short legs. Using this male, the student was able to establish a pure breeding stock of this mutant and found that it was recessive. She then incorporated this mutant into a stock containing the recessive gene *black* (body color located on chromosome 2) and the recessive gene *pink* (eye color located on chromosome 3). A female from the ho-

mozygous *black, pink, short* stock was then mated to a wild-type male. The F_1 males of this cross were all wild type and were then backcrossed to the homozygous *b p sh* females. The F_2 results appeared as shown in the table. No other phenotypes were observed.

	Wild	Pink*	Black, Short	Black, Pink, Short
Females	63	58	55	69
Males	59	65	51	60

*Pink indicates that the other two traits are wild type, and so on.

(a) Based on these results, the student was able to assign *short* to a linkage group (a chromosome). Which one was it? Include a step-by-step reasoning.

(b) The student repeated the experiment, making the reciprocal cross, F_1 females backcrossed to homozygous *b p sh* males. She observed that 85 percent of the offspring fell into the above classes, but that 15 percent of the offspring were equally divided among *b + p*, *b + +*, *+ sh p*, and *+ sh +* phenotypic males and females. How can these results be explained and what information can be derived from the data?

SELECTED READINGS

ALLEN, G.E. 1978. *Thomas Hunt Morgan: The man and his science.* Princeton, NJ: Princeton University Press.

BLIXT, S. 1975. Why didn't Gregor Mendel find linkage? *Nature* 256:206.

CATCHESIDE, D.G. 1977. *The genetics of recombination.* Baltimore, MD: University Park Press.

CHAGANTI, R., SCHONBERG, S., and GERMAN, J. 1974. A manyfold increase in sister chromatid exchange in Bloom's syndrome lymphocytes. *Proc. Natl. Acad. Sci. USA* 71:4508–12.

CREIGHTON, H.S., and McCLINTOCK, B. 1931. A correlation of cytological and genetical crossing over in *Zea mays. Proc. Natl. Acad. Sci. USA* 17:492–97.

DOUGLAS, L., and NOVITSKI, E. 1977. What chance did Mendel's experiments give him of noticing linkage? *Heredity* 38:253–57.

ELLIS, N.A. et al. 1995. The Bloom's syndrome gene product is homologous to RecQ helicases. *Cell* 83:655–66.

EPHRUSSI, B., and WEISS, M.C. 1969. Hybrid somatic cells. *Sci. Am.* (April) 220:26–35.

GARCIA-BELLIDO, A. 1972. Some parameters of mitotic recombination in *Drosophila melanogaster. Molec. Genet.* 115:54–72.

HOTTA, Y., TABATA, S., and STERN, H. 1984. Replication and nicking of zygotene DNA sequences: Control by a meiosis-specific protein. *Chromosoma* 90:243–53.

KING, R.C. 1970. The meiotic behavior of the *Drosophila* oocyte. *Int. Rev. Cytol.* 28:125–68.

LATT, S.A. 1981. Sister chromatid exchange formation. *Annu. Rev. Genet.* 15:11–56.

LINDSLEY, D.L., and GRELL, E.H. 1972. *Genetic variations of Drosophila melanogaster.* Washington, DC: Carnegie Institute of Washington.

MOENS, P.B. 1977. The onset of meiosis. In *Cell biology—A comprehensive treatise, Vol. 1. Genetic mechanisms of cells,* ed. L. Goldstein and D.M. Prescott, pp. 93–109. Orlando, FL: Academic Press.

MORGAN, T.H. 1911. An attempt to analyze the constitution of the chromosomes on the basis of sex-linked inheritance in *Drosophila. J. Exp. Zool.* 11:365–414.

MORTON, N.E. 1955. Sequential test for the detection of linkage. *Am. J. Hum. Genet.* 7:277–318.

———. 1995. LODs—past and present. *Genetics* 140:7–12.

NEUFFER, M.G., JONES, L., and ZOBER, M. 1968. *The mutants of maize.* Madison, WI: Crop Science Society of America.

PERKINS, D. 1962. Crossing-over and interference in a multiply marked chromosome arm of *Neurospora. Genetics* 47:1253–74.

RUDDLE, F.H., and KUCHERLAPATI, R.S. 1974. Hybrid cells and human genes. *Sci. Am.* (July) 231:36–49.

STAHL, F.W. 1979. *Genetic recombination.* New York: W.H. Freeman.

STERN, C. 1936. Somatic crossing over and segregation in *Drosophila melanogaster. Genetics* 21:625–31.

STERN, H., and HOTTA, Y. 1973. Biochemical controls in meiosis. *Annu. Rev. Genet.* 7:37–66.

————. 1974. DNA metabolism during pachytene in relation to crossing over. *Genetics* 78:227–35.

STURTEVANT, A.H. 1913. The linear arrangement of six sex-linked factors in *Drosophila*, as shown by their mode of association. *J. Exp. Zool.* 14:43–59.

————. 1965. *A history of genetics*. New York: Harper & Row.

TAYLOR, J.H., ed. 1965. *Selected papers on molecular genetics*. Orlando, FL: Academic Press.

VOELLER, B.R., ed. 1968. *The chromosome theory of inheritance: Classical papers in development and heredity*. New York: Appleton-Century-Crofts.

VON WETTSTEIN, D., RASMUSSEN, S.W., and HOLM, P.B. 1984. The synaptonemal complex in genetic segregation. *Annu. Rev. Genet.* 18:331–414.

WOLFF, S., ed. 1982. *Sister chromatid exchange*. New York: Wiley–Interscience.

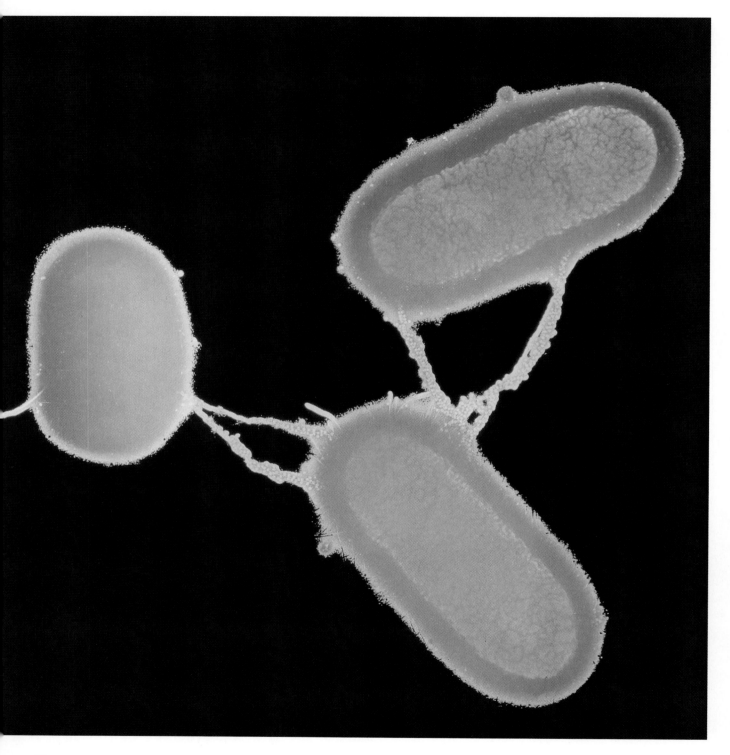

Transmission electron micrograph of conjugating *E. coli.*

Genetic Recombination and Mapping in Bacteria and Bacteriophages

KEY CONCEPTS

- **Bacteria mutate spontaneously and grow at an exponential rate**

- **Conjugation is one means of genetic recombination in bacteria**

  F^+ and F^- Bacteria
  Hfr Bacteria and Chromosome Mapping
  Recombination in $F^+ \times F^-$ Matings: A Reexamination
  The F' State and Merozygotes

- **Rec proteins are essential to bacterial recombination**

- **F factors are plasmids**

- **Transformation is another process leading to genetic recombination in bacteria**

  The Transformation Process
  Transformation and Mapping

- **Bacteriophages are bacterial viruses**

  Phage T4: Structure and Life Cycle
  The Plaque Assay
  Lysogeny

- **Transduction is virus-mediated bacterial DNA transfer**

  The Lederberg-Zinder Experiment
  The Nature of Transduction
  Transduction and Mapping
  Specialized Transduction

- **Bacteriophages undergo intergenic recombination**

  Intergenic Recombination and Mapping in Bacterial Viruses

- **Intragenic recombination also occurs in phage T4**

  The *rII* Locus of Phage T4
  Complementation by *rII* Mutations
  Recombinational Analysis
  Deletion Testing of the *rII* Locus
  The *rII* Gene Map

I n this chapter we shift from the consideration of transmission of genetic information in eukaryotes to a discussion of various genetic phenomena in **bacteria** (prokaryotes) and **bacteriophages,** viruses that have bacteria as their host. The study of bacteria and bacteriophages has been essential to the accumulation of knowledge in many areas of genetic study. For example, much of what is known about molecular genetics, recombinational phenomena, and gene structure initially was derived from experimental work with these organisms. Furthermore, as we shall see in Chapters 18 and 21, our knowledge of bacteria and their resident plasmids has served as the basis for their widespread use in DNA cloning and other recombinant DNA studies.

Bacteria and their viruses have been especially useful research organisms in genetics for a number of reasons. First, they have extremely short reproductive cycles. Literally hundreds of generations, giving rise to billions of genetically identical bacteria or phages, can be produced in short periods of time. Furthermore, they can be studied in pure cultures. That is, a single species or mutant strain of bacteria or one type of virus can be isolated and investigated independently of other similar organisms.

Our major emphasis in this chapter will be on genetic recombination in both bacteria and bacteriophages. Complex processes have evolved in these microorganisms and viruses that facilitate genetic recombination within populations. As we will see, these processes were the basis for performing chromosome mapping analysis that became the cornerstone of molecular genetic investigations.

Bacteria mutate spontaneously and grow at an exponential rate

Genetic studies using bacteria depend upon our ability to isolate mutations in these organisms. Although it was known well before 1943 that pure cultures of bacteria could give rise to small numbers of cells exhibiting heritable variation, particularly with respect to survival under different environmental conditions, the source of the variation was hotly debated. The majority of bacteriologists believed that environmental factors induced changes in certain bacteria that led to their survival or adaptation to the new conditions. For example, strains of *E. coli* are known to be *sensitive* to infection by the bacteriophage T1.

Infection by the bacteriophage leads to the reproduction of the virus at the expense of the bacterial cell, which is lysed or destroyed. If a plate of *E. coli* is homogeneously sprayed with T1, almost all cells are lysed. Rare *E. coli* cells, however, survive infection and are not lysed. If these cells are isolated and established in pure culture, all their descendants are *resistant* to T1 infection. It might be argued that the mutations responsible for T1 resistance were "induced" by the presence of the T1 viruses, and that, in the absence of the T1 viruses, the mutations would not have occurred. However, in 1943 Salvador Luria and Max Delbruck elegantly proved that such T1-resistant cells result from **spontaneous mutation.** We will explore this discovery further in Chapter 17.

Bacterial cells that bear spontaneous mutations, such as T1 resistance, can be isolated and established independently from the parent strain by means of various selection techniques. As a result, we can now induce and isolate mutations for almost any desired characteristic. Because bacteria and the viruses that infect them are haploid, all mutations are expressed directly in the descendants of mutant cells, adding to the ease with which these microorganisms can be studied.

Bacteria are grown in either a liquid culture medium or in a petri dish on a semisolid agar surface. If the nutrient components of the growth medium are very simple and consist only of an organic carbon source (such as a glucose or lactose) and a variety of inorganic ions, including Na^+, K^+, Mg^{++}, Ca^{++}, and NH_4^+, present as inorganic salts, it is called **minimal medium.** To grow on such a medium, a bacterium must be able to synthesize all essential organic compounds (e.g., amino acids, purines, pyrimidines, sugars, vitamins, and fatty acids). A bacterium that can accomplish this remarkable biosynthetic feat—one that we ourselves cannot duplicate—is termed a **prototroph.** It is said to be wild type for all growth requirements and can grow on minimal medium. On the other hand, if a bacterium loses, through mutation, the ability to synthesize one or more organic components, it is said to be an **auxotroph.** For example, a bacterium that loses the ability to make histidine is designated as a *his⁻* auxotroph, as opposed to its prototrophic *his⁺* counterpart. For the *his⁻* bacterium to grow, this amino acid must be added as a supplement to the minimal medium. Note that medium that has been extensively supplemented is referred to as complete medium.

To study mutant bacteria in a quantitative fashion, an inoculum (e.g., 0.1 ml, 1.0 ml) of bacteria is placed in liquid culture medium. The bacteria exhibit a characteristic growth pattern, as illustrated in Figure 7.1. Initially, during the **lag phase,** growth is slow. Then, a period of rapid growth follows called the **log phase,** during which cells divide many times with a fixed time interval between cell divisions, resulting in logarithmic growth. When the bacteria reach a cell density of about 10^9 cells per milliliter, nu-

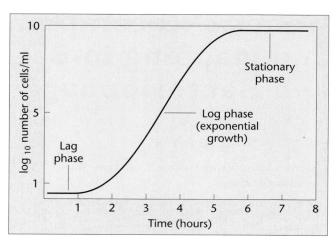

FIGURE 7.1 A typical bacterial population growth curve illustrating the initial lag phase, the subsequent log phase where exponential growth occurs, and the stationary phase that occurs when nutrients are exhausted.

trients and oxygen become limiting and cells enter the **stationary phase.** As the doubling time during the log phase may be as short as 20 minutes, an initial inoculum of a few thousand cells added to the culture can easily achieve a maximum cell density overnight.

Once cells are grown in liquid medium they can be quantitated. First, the bacteria are plated on (transferred to) semisolid medium in a petri dish where, following incubation and many divisions, each cell gives rise to a colony visible on the surface of the medium. From the number of colonies that subsequently grow, it is possible to estimate the number of bacteria present in the original culture. If the number of colonies is too great to count, then serial dilutions of the original liquid culture can be made and plated, until the colony number is reduced to the point where it can be counted (Figure 7.2).

For example, assume that the three petri dishes in Figure 7.2 represent dilutions of 10^{-3}, 10^{-4}, and 10^{-5}, respectively (left to right). We select the dish where there are few enough colonies to be accurately counted. Because each colony presumably arose from a single bacterium, the number of colonies times the dilution factor represents the number of bacteria in the initial milliliter. In this case, the dish farthest to the right contains 15 colonies. Since it represents a dilution of 10^{-5}, we can estimate the initial number of bacteria to be 15×10^5 per milliliter. Calculations such as these are useful in a number of studies.

Conjugation is one means of genetic recombination in bacteria

Development of techniques that allowed the identification and study of bacterial mutations led to detailed investigations of the arrangement of genes on the bacterial chro-

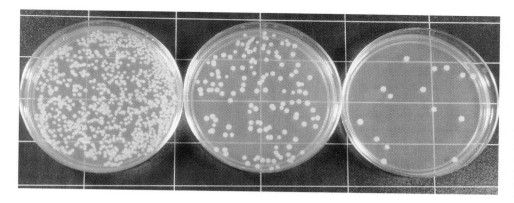

FIGURE 7.2   Results of the serial dilution technique and subsequent culture of bacteria. Each of the dilutions varies by a factor of 10. Each colony was derived from a single bacterial cell.

mosome. Such studies began in 1946 when Joshua Lederberg and Edward Tatum showed that bacteria undergo **conjugation,** a parasexual process in which the genetic information from one bacterium is transferred to and recombined with that of another bacterium (see the photograph that opens this chapter). Like meiotic crossing over in eukaryotes, genetic recombination in bacteria provided the basis for the development of methodology for chromosome mapping. Note that the term **genetic recombination,** as applied to bacteria and bacteriophages, leads to the *replacement* of one or more genes present in one strain with those from a genetically distinct strain. While this is somewhat different from our use of genetic recombination in eukaryotes, where the term describes crossing over that results in *reciprocal exchange events,* the overall effect is the same: Genetic information is transferred from one chromosome to another, resulting in an altered genotype. Two other phenomena that result in the transfer of genetic information from one bacterium to another, **transformation** and **transduction,** have also served as a basis for determining the arrangement of genes on the bacterial chromosome. We will return to a discussion of these processes in later sections of this chapter.

Lederberg and Tatum's initial experiments were performed with two multiple auxotroph strains (nutritional mutants) of *E. coli* K12. As shown in Figure 7.3, Strain A required methionine (met) and biotin (bio) in order to grow, whereas strain B required threonine (thr), leucine (leu), and thiamine (thi). Neither strain would grow on minimal medium. The two strains were first grown separately in supplemented media, and then cells from both were mixed and grown together for several more generations. They were then plated on minimal medium. Any bacterial cells that grew on minimal medium are prototrophs (wild-type bacteria that did not need nutritional supplements). It was highly improbable that any of the cells that contained two or three mutant genes would undergo spontaneous mutation simultaneously at two or three gene sites. Therefore, any prototrophs recovered must have arisen as a result of some form of genetic exchange and recombination.

In this experiment, prototrophs were recovered at a rate of $1/10^7$ (10^{-7}) cells plated. The controls for this experiment involved separate plating of cells from strains A and B on minimal medium. No prototrophs were recovered. On the basis of these observations, Lederberg and Tatum proposed that, while the events were indeed quite rare, genetic recombination had occurred.

F^+ and F^- Bacteria

Lederberg and Tatum's findings were soon followed by numerous experiments designed to elucidate the genetic basis of conjugation. It quickly became evident that different strains of bacteria were involved in a unidirectional transfer of genetic material. When cells serve as donors of parts of their chromosomes, they are designated as F^+ **cells** (F for "fertility"). Recipient bacteria receive the donor chromosome material (now known to be DNA), and recombine it with part of their own chromosome. They are designated as F^- **cells.**

It was established subsequently that cell contact is essential to chromosome transfer. Support for this concept was provided by Bernard Davis, who designed a U-tube in which to grow F^+ and F^- cells (Figure 7.4). At the base of the tube was a glass filter with pores large enough to allow passage of the liquid medium, but too small to allow the passage of bacteria. The F^+ cells were placed on one side of the filter and F^- cells on the other side. The medium was moved back and forth across the filter so that the bacterial cells essentially shared a common medium during incubation. Samples from both sides of the tube were then plated independently on minimal medium, but no prototrophs were found. Davis concluded that *physical contact is essential to genetic recombination.* Such physical interaction is the initial stage of the process of conjugation and is mediated through a conjugation tube called the **F,** or **sex, pilus.** Bacteria often have many pili, which are microscopic tube-like extensions of the cell. Different types of pili perform different cellular functions, but all pili are involved in some way with adhesion. After contact has been initiated between mating pairs via the F pili (Figure 7.5), transfer of DNA begins.

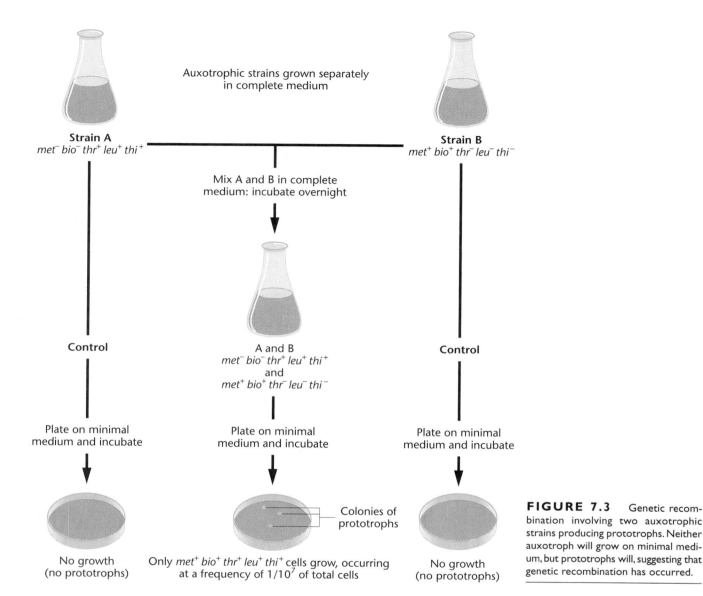

Auxotrophic strains grown separately
in complete medium

Strain A
met⁻ bio⁻ thr⁺ leu⁺ thi⁺

Strain B
met⁺ bio⁺ thr⁻ leu⁻ thi⁻

Mix A and B in complete
medium: incubate overnight

Control

A and B
met⁻ bio⁻ thr⁺ leu⁺ thi⁺
and
met⁺ bio⁺ thr⁻ leu⁻ thi⁻

Control

Plate on minimal
medium and incubate

Plate on minimal
medium and incubate

Plate on minimal
medium and incubate

Colonies of
prototrophs

No growth
(no prototrophs)

Only *met⁺ bio⁺ thr⁺ leu⁺ thi⁺* cells grow, occurring
at a frequency of 1/10⁷ of total cells

No growth
(no prototrophs)

FIGURE 7.3 Genetic recombination involving two auxotrophic strains producing prototrophs. Neither auxotroph will grow on minimal medium, but prototrophs will, suggesting that genetic recombination has occurred.

Later evidence established that F⁺ cells contained a **fertility factor** (called the **F factor**) that confers the ability to donate part of their chromosome during conjugation. Experiments by Joshua and Esther Lederberg and by William Hayes and Luca Cavalli-Sforza showed that certain conditions could eliminate the F factor in otherwise fertile cells. However, if these "infertile" cells were then grown with fertile donor cells, the F factor was regained.

The conclusion that the F factor is a mobile element was further supported by the observation that, following conjugation and genetic recombination, recipient cells always become F⁺. Thus, in addition to the *rare* cases of transfer of genes from the bacterial chromosome (genetic recombination), the F factor itself is passed to *all* recipient cells. On this basis, the initial crosses of Lederberg and Tatum (Figure 7.3) may be designated

STRAIN A		STRAIN B
F⁺	×	F⁻
DONOR		RECIPIENT

Isolation of the F factor confirmed these conclusions. Like the bacterial chromosome, though distinct from it, the F factor has been shown to consist of a circular, double-stranded DNA molecule, equivalent to about 2 percent of the bacterial chromosome (about 100,000 nucleotide pairs). Contained in the F factor, among others, are 19 genes, the products of which are involved in the transfer of genetic information (*tra* genes). These include those essential to the formation of the sex pilus.

As we soon shall see, the F factor is in reality an autonomous genetic unit referred to as a **plasmid.** However, in our historical coverage of its discovery, we will continue in this chapter to refer to it as a "factor."

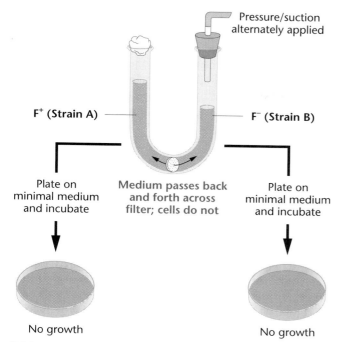

Pressure/suction alternately applied

F⁺ (Strain A)   F⁻ (Strain B)

Plate on minimal medium and incubate

Medium passes back and forth across filter; cells do not

Plate on minimal medium and incubate

No growth   No growth

FIGURE 7.4   When strain A and B auxotrophs are grown in a common medium but separated by a filter, no genetic recombination occurs and no prototrophs are produced. This apparatus is called a Davis U-tube.

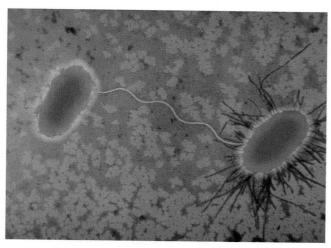

FIGURE 7.5   An electron micrograph of conjugation between an F⁺ *E. coli* cell and an F⁻ cell. The sex pilus linking them is clearly visible.

It is believed that the transfer of the F factor during conjugation involves separation of two strands of the F factor double helical DNA and the movement of one of the two strands into the recipient. The other strand remains in the donor cell. Both of these parental strands serve as templates for DNA replication, resulting in two intact F factors, one in each of the two cells. Both cells are now F⁺. This process is diagrammed in Figure 7.6.

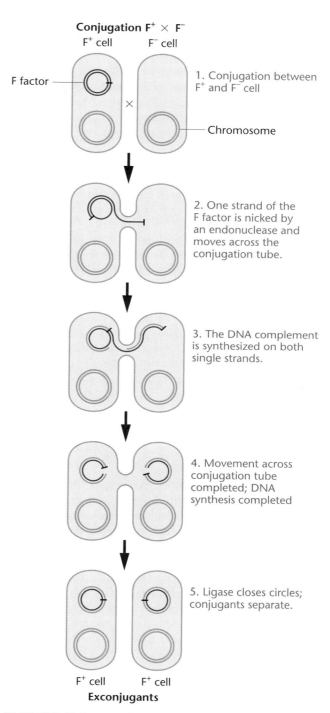

Conjugation F⁺ × F⁻

F⁺ cell   F⁻ cell

F factor

1. Conjugation between F⁺ and F⁻ cell

Chromosome

2. One strand of the F factor is nicked by an endonuclease and moves across the conjugation tube.

3. The DNA complement is synthesized on both single strands.

4. Movement across conjugation tube completed; DNA synthesis completed

5. Ligase closes circles; conjugants separate.

F⁺ cell   F⁺ cell

Exconjugants

FIGURE 7.6   An F⁺ × F⁻ mating demonstrating how the recipient F⁻ cell is converted to F⁺. During conjugation, the DNA of the F factor is replicated with one new copy entering the recipient cell, converting it to F⁺. The black bar has been added to the F factors to follow their clockwise rotation during replication.

To summarize, *E. coli* cells may or may not contain the F factor. When it is present, the cell is able to form a sex pilus and potentially serve as a donor of genetic information. During conjugation, a copy of the F factor is almost always transferred from the F$^+$ cell to the F$^-$ recipient, converting it to the F$^+$ state. The question remains as to exactly how a very low percentage of F$^-$ cells undergo genetic recombination. The answer awaited further experimentation. Subsequent discoveries not only clarified how genetic recombination occurs but also defined a mechanism by which the *E. coli* chromosome could be mapped. We shall first address chromosome mapping.

Hfr Bacteria and Chromosome Mapping

In 1950, Cavalli-Sforza treated an F$^+$ strain of *E. coli* K12 with nitrogen mustard, a potent chemical known to induce mutations. From these treated cells, he recovered a strain of donor bacteria that underwent recombination at a rate of $1/10^4$ (10^{-4}), 1000 times more frequently than the original F$^+$ strains. In 1953, William Hayes isolated another strain demonstrating a similar elevated frequency. Both strains were designated **Hfr,** or **high-frequency recombination.** Because Hfr cells behave as chromosome donors, they are a special class of F$^+$ cells.

Another important difference was noted between Hfr strains and the original F$^+$ strains. If the donor is an Hfr strain, recipient cells, while sometimes displaying genetic recombination, never become Hfr; that is, they remain F$^-$. In comparison, then,

$$\text{F}^+ \times \text{F}^- \rightarrow \text{F}^+ \quad \text{(low rate of recombination)}$$
$$\text{Hfr} \times \text{F}^- \rightarrow \text{F}^- \quad \text{(higher rate of recombination)}$$

Perhaps the most significant characteristic of Hfr strains is the nature of recombination. In any given strain, certain genes are more frequently recombined than others, and some do not recombine at all. This *nonrandom pattern of gene transfer* was shown to vary from Hfr strain to Hfr strain. While these results were puzzling, Hayes interpreted them to mean that some physiological alteration of the F factor had occurred, resulting in the production of Hfr strains of *E. coli*.

In the mid-1950s, experimentation by Ellie Wollman and François Jacob explained the differences between cells that are Hfr and those that are F$^+$ and showed how Hfr strains allow genetic mapping of the *E. coli* chromosome. Wollman and Jacob first incubated a culture containing a mixture of an Hfr strain and an F$^-$ strain. To facilitate the recovery of only recombinants, the Hfr strain was sensitive to an antibiotic while the recipient strain was resistant. At various intervals, the researchers removed samples and placed them in a blender. The shear forces created in the blender separated conjugating bacteria so that the transfer

of the chromosome was effectively terminated. To assay the cells for genetic recombination following the blender treatment, they were grown on medium *containing* the antibiotic in order to ensure the recovery of only recipient cells.

This process, called the **interrupted mating technique,** demonstrated that specific genes of a given Hfr strain were transferred and recombined sooner than others. Figure 7.7 illustrates this point. During the first 8 minutes after the two strains were initially mixed, no genetic recombination could be detected. In cells assayed at about 10 minutes, recombination of the *azi*^{R} gene could be detected, but no transfer of the *ton*^{S}, *lac*$^+$, or *gal*$^+$ genes was noted. By 15 minutes, 70 percent of the recombinants were *azi*^{R}; 30 percent were now also *ton*^{S}; but none was *lac*$^+$ or *gal*$^+$. Within 20 minutes, the *lac*$^+$ gene was found among the recombinants; and within 30 minutes, *gal*$^+$ was also being transferred. Wollman and Jacob had demonstrated an *oriented transfer of genes* that was correlated with the length of time conjugation was allowed to proceed.

It appeared that the chromosome of the Hfr bacterium was transferred linearly and that the gene order and distance between genes, as measured in minutes, could be predicted from such experiments (Figure 7.8). This information served as the basis for the first genetic map of the *E. coli* chromosome. "Minutes" in bacterial mapping are equivalent to "map units" in eukaryotes.

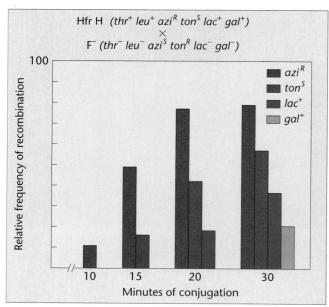

FIGURE 7.7 The progressive transfer during conjugation of various genes from a specific Hfr strain of *E. coli* to an F$^-$ strain. Certain genes (*azi* and *ton*) are transferred sooner than others and recombine more frequently. Others (*lac* and *gal*) take longer to be transferred and recombine with a lower frequency. Others (*thr* and *leu*) are always transferred and are used in the initial screen for recombinants.

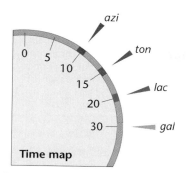

FIGURE 7.8   A time map of the genes studied in the experiment depicted in Figure 7.7.

Wollman and Jacob repeated the same type of experimentation with other Hfr strains, obtaining similar results with one important difference. Although genes were always transferred linearly with time, as in their original experiment, which genes entered first and which followed later seemed to vary from Hfr strain to Hfr strain [Figure 7.9(a)]. When they reexamined the rate of entry of genes, and thus the different genetic maps for each strain, a definite pattern emerged. The major difference between each strain was simply the point of the

origin and the direction in which entry proceeded from that point [Figure 7.9(b)].

To explain these results, Wollman and Jacob postulated that the *E. coli* chromosome is circular. If the point of origin (*O*) varied from strain to strain, a different sequence of genes would be transferred in each case. But what determines *O*? They proposed that in various Hfr strains, the F factor integrates into the chromosome at different points and its position determines the *O* site. One such case of integration is shown in Figure 7.10 (Step 1). During conjugation between this Hfr and an F⁻ cell, the position of the F factor determines the initial point of transfer (Step 2 and 3). Those genes adjacent to *O* are transferred first. *The F factor becomes the last part to be transferred* (Step 4). Apparently, conjugation rarely, if ever, lasts long enough to allow the entire chromosome to pass across the conjugation tube (Step 5). This proposal explains why recipient cells, when mated with Hfr cells, remain F⁻.

Figure 7.10 also depicts the way in which the two strands making up a DNA molecule unwind during transfer, allowing for the entry of one of the strands of DNA into the recipient (Step 3). Following replication, the entering DNA now has the potential to recombine with its homologous region of the host chromosome. The DNA strand that remains in the donor also undergoes replication.

(a)

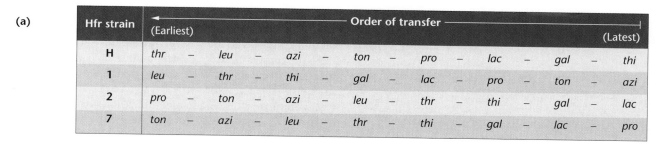

Hfr strain	Order of transfer							
	(Earliest)							(Latest)
H	*thr* –	*leu* –	*azi* –	*ton* –	*pro* –	*lac* –	*gal* –	*thi*
1	*leu* –	*thr* –	*thi* –	*gal* –	*lac* –	*pro* –	*ton* –	*azi*
2	*pro* –	*ton* –	*azi* –	*leu* –	*thr* –	*thi* –	*gal* –	*lac*
7	*ton* –	*azi* –	*leu* –	*thr* –	*thi* –	*gal* –	*lac* –	*pro*

(b)

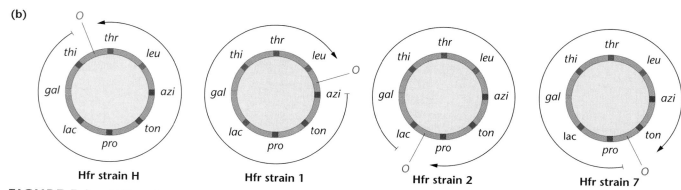

FIGURE 7.9   (a) The order of gene transfer in four Hfr strains, suggesting that the *E. coli* chromosome is circular. (b) The point where transfer originates (*O*) is identified in each strain. Note that transfer can proceed in either direction, depending on the strain. The origin is determined by the point of integration into the chromosome of the F factor, and the direction of transfer is determined by the orientation of the F factor as it integrates.

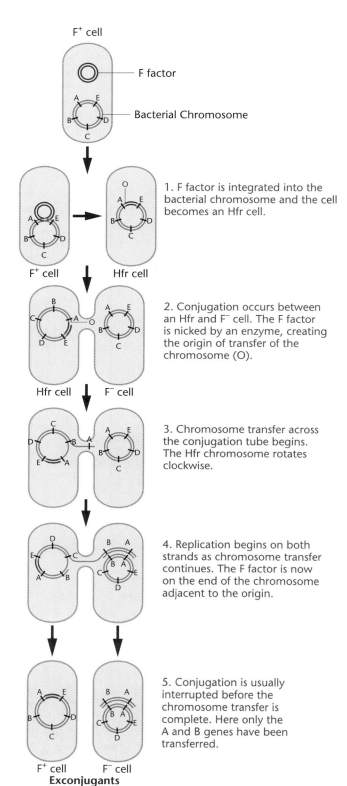

F⁺ cell

F factor

Bacterial Chromosome

F⁺ cell   Hfr cell

1. F factor is integrated into the bacterial chromosome and the cell becomes an Hfr cell.

Hfr cell   F⁻ cell

2. Conjugation occurs between an Hfr and F⁻ cell. The F factor is nicked by an enzyme, creating the origin of transfer of the chromosome (O).

3. Chromosome transfer across the conjugation tube begins. The Hfr chromosome rotates clockwise.

4. Replication begins on both strands as chromosome transfer continues. The F factor is now on the end of the chromosome adjacent to the origin.

5. Conjugation is usually interrupted before the chromosome transfer is complete. Here only the A and B genes have been transferred.

F⁺ cell   F⁻ cell
Exconjugants

FIGURE 7.10   Conversion of F⁺ to an Hfr state occurs by the integration of the F factor into the bacterial chromosome. The point of integration determines the origin (*O*) of transfer. During conjugation, the F factor, now integrated into the host chromosome, is nicked by an enzyme, initiating transfer of the chromosome at that point. Conjugation is usually interrupted prior to complete transfer. Above, only the *A* and *B* genes are transferred to the F⁻ cell, which may recombine with the host chromosome.

The use of the interrupted mating technique with different Hfr strains has provided the basis for mapping the entire *E. coli* chromosome. Mapped in time units, strain K12 (or *E. coli* K12) is 100 minutes long. Over 900 genes have now been placed on the map. In most instances, only a single copy of each gene exists.

Recombination in F⁺ × F⁻ Matings: A Reexamination

The above model has helped geneticists to better understand how genetic recombination occurs during the F⁺ × F⁻ matings. Recall that recombination occurs much less frequently in them than in Hfr × F⁻ matings, and that random gene transfer is involved. The current belief is that when F⁺ and F⁻ cells are mixed, conjugation occurs readily and that each F⁻ cell involved in conjugation with an F⁺ cell receives a copy of the F factor, *but that no genetic recombination occurs*. However, at an extremely low frequency in a population of F⁺ cells, the F factor integrates spontaneously from the cytoplasm to a random point in the bacterial chromosome, converting the F⁺ cell to the Hfr state as we saw in Figure 7.10 Therefore, in F⁺ × F⁻ crosses, the extremely low frequency of genetic recombination (10^{-7}) is attributed to the rare, newly formed Hfr cells, which then undergo conjugation with F⁻ cells. Because the point of integration of the F factor is random, a nonspecific gene transfer ensues, leading to the low-frequency, random genetic recombination observed in the F⁺ × F⁻ experiment. Unless the recipient cell simultaneously or subsequently undergoes conjugation with a separate F⁺ cell, it will remain F⁻. Most often, the recombinants become F⁺.

The F′ State and Merozygotes

In 1959, during experiments with Hfr strains of *E. coli*, Edward Adelberg discovered that the F factor could lose its integrated status, causing the cell to revert to the F⁺ state (Figure 7.11 Step 1). When this occurs, the F factor frequently carries several adjacent bacterial genes along with it (Step 2). Adelberg labeled this condition **F′** to distinguish it from F⁺ and Hfr. F′, like Hfr, is thus another special case of F⁺. This conversion is described as one from Hfr to F′.

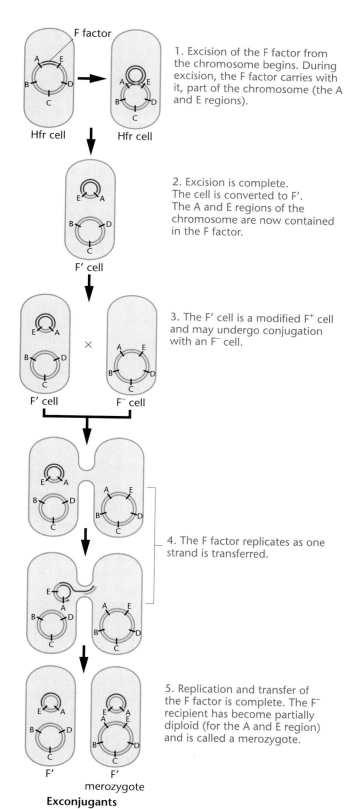

1. Excision of the F factor from the chromosome begins. During excision, the F factor carries with it, part of the chromosome (the A and E regions).

2. Excision is complete. The cell is converted to F′. The A and E regions of the chromosome are now contained in the F factor.

3. The F′ cell is a modified F⁺ cell and may undergo conjugation with an F⁻ cell.

4. The F factor replicates as one strand is transferred.

5. Replication and transfer of the F factor is complete. The F⁻ recipient has become partially diploid (for the A and E region) and is called a merozygote.

FIGURE 7.11 Conversion of an Hfr bacterium to F′ and its subsequent mating with an F⁻ cell. The conversion occurs when the F factor loses its integrated status. During excision from the chromosome, it carries with it one or more chromosomal genes (A and E). Following conjugation with an F⁻ cell, the recipient cell becomes partially diploid and is called a merozygote. It also behaves as an F⁺ donor cell.

The presence of bacterial genes within a cytoplasmic F factor creates an interesting situation. An F′ bacterium behaves like an F⁺ cell, initiating conjugation with F⁻ cells (Figure 7.11-Step 3). When this occurs, the F factor, containing chromosomal genes, is transferred to the F⁻ cell (Step 4). As a result, whatever chromosomal genes are part of the F factor are now present in duplicate in the recipient cell (Step 5), because the recipient still has a complete chromosome. This creates a partially diploid cell called a **merozygote.** Pure cultures of F′ merozygotes can be established. They have been extremely useful in the study of bacterial genetics, particularly in genetic regulation.

Rec proteins are essential to bacterial recombination

Once researchers established that a unidirectional transfer of DNA occurs between bacteria, they were interested in determining how the actual recombination event occurs in the recipient cell. Just how does the donor DNA replace the comparable region in the recipient chromosome? As with many systems, the biochemical mechanism by which recombination occurs was deciphered through genetic studies. Major insights were gained as a result of the isolation of a group of mutations representing genes called *rec*.

The first relevant observation in this case involved a series of mutant genes labeled *recA*, *recB*, *recC*, and *recD*. The first mutant gene, *recA*, was found to diminish genetic recombination in bacteria 1000-fold, nearly eliminating it altogether. The other *rec* mutations reduced recombination by about 100 times. Clearly, the normal wild-type products of these genes play some essential role in the process of recombination.

By looking for a functional gene product present in normal cells but missing in mutant cells, researchers subsequently isolated several gene products and showed that they played a role in genetic recombination. The first is called the **RecA protein.*** The second is a more complex

---

*Note that the names of bacterial genes begin with lowercase letters and are italicized. The names of the corresponding gene products (proteins) begin with an uppercase letter and are not italicized. For example, the *recA* gene encodes the RecA protein.

protein called the **RecBCD product,** an enzyme consisting of polypeptide subunits encoded by three other *rec* **genes.** The roles of these proteins have now been elucidated *in vitro*. As a result of this genetic research, our knowledge of the process of recombination has been extended considerably. These discoveries underscore the value of isolating mutations, establishing their phenotypes, and determining the biological role of the normal, wild-type gene as a result of subsequent investigation. In Chapter 12, once the topic of DNA structure has been thoroughly explored, we will examine the molecular role of the Rec proteins in recombination.

F factors are plasmids

In the preceding sections we have examined the extrachromosomal heredity unit called the F factor. When it exists autonomously in the bacterial cytoplasm, the F factor is composed of a double-stranded closed circle of DNA [Figure 7.12(a)]. These characteristics place the F factor in the more general category of genetic structures called **plasmids.** These structures contain one or more genes—often, quite a few. Their replication depends on the same enzymes that replicate the chromosome of the host cell, and they are distributed to daughter cells along with the host chromosome during cell division.

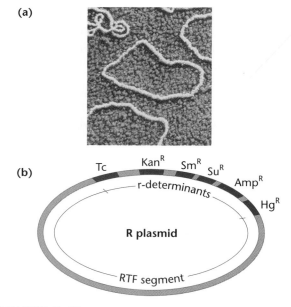

FIGURE 7.12 (a) Electron micrograph of a plasmid isolated from *E. coli*; (b) diagramatic representation of an R plasmid containing resistance transfer factors (RTFs) and multiple r-determinants (Tc, tetracycline; Kan, kanamycin; Sm, streptomycin; Su, sulfonamide; Amp, ampicillin; and Hg, mercury).

Plasmids are generally classified according to the genetic information specified by their DNA. The F factor confers fertility and contains genes essential for sex pilus formation, upon which genetic recombination depends. Other examples of plasmids include the R and the Col plasmids.

Most **R plasmids** consist of two components: the **RTF (resistance transfer factor)** and one or more **r-determinants** [Figure 7.12(b)]. The RTF encodes genetic information essential to transfer of the plasmid between bacteria, and the r-determinants are genes conferring resistance to antibiotics. While RTFs are quite similar in a variety of plasmids from different bacterial species, there is a wide variation in r-determinants, each of which is specific for resistance to one class of antibiotic. Sometimes, a bacterial cell contains r-determinant plasmids but no RTF is present. Such a cell is resistant but cannot transfer the genetic information for resistance to recipient cells. The most commonly studied plasmids, however, contain the RTF as well as one or more r-determinants. Resistances to tetracycline, streptomycin, ampicillin, sulfonamide, kanamycin, and chloramphenicol are most frequently encountered. Sometimes these occur in a single plasmid, conferring multiple resistance to several antibiotics [Figure 7.12(b)]. Bacteria bearing such plasmids are of great medical significance, not only because of their multiple resistance, but because of the ease with which the plasmids may be transferred to other bacteria.

The **Col plasmid,** ColE1, derived from *E. coli*, is clearly distinct from R plasmids. It encodes one or more proteins that are highly toxic to bacterial strains that do not harbor the same plasmid. These proteins, called **colicins,** may kill neighboring bacteria. Bacteria that carry the plasmid are said to be colicinogenic. Present in 10 to 20 copies per cell, the plasmid also contains a gene encoding an immunity protein that protects the host cell from the toxin. Unlike an R plasmid, the Col plasmid is not usually transmissible to other cells.

Interest in plasmids has increased dramatically because of their role in the genetic technology known as recombinant DNA research, which we discuss in Chapters 18 and 21. Specific genes from any source can be inserted into a plasmid, which may then be artificially inserted into a bacterial cell. As the altered cell replicates its DNA and undergoes division, the foreign gene is also replicated, thus cloning the genes.

Transformation is another process leading to genetic recombination in bacteria

Transformation is another process that provides a mechanism for the recombination of genetic information in some bacteria. In transformation, small pieces of extracellular DNA are taken up by a living bacterium, ultimately leading to a stable genetic change in the recipient cell. We are interested in transformation in this chapter because, in those bacterial species where it occurs, the process can be used to map bacterial genes, although in a more limited way than conjugation. We will return to the topic of transformation in Chapter 11 because of the central role the process played in experiments providing evidence that DNA is the genetic material.

The Transformation Process

Transformation (Figure 7.13) consists of numerous steps that can be divided into two main categories: (1) entry of DNA into a recipient cell, and (2) recombination of the donor DNA with its homologous region in the recipient chromosome. In a population of cells, only those in a particular physiological state, referred to as **competence,** take up DNA. Entry is thought to occur at a limited number of receptor sites on the surface of the bacterial cell. Passage across the cell wall and membrane is an active process requiring energy and specific transport molecules. This concept is supported by the fact that substances that inhibit energy production or protein synthesis in the recipient cell also inhibit the transformation process.

During the process of entry, one of the two strands of the invading DNA molecule is digested by nucleases, leaving only a single strand to participate in transformation (Step 2 and 3). The surviving DNA strand aligns with its complementary region of the bacterial chromosome. In a process involving several enzymes, this segment of DNA replaces its counterpart in the chromosome, which is excised and degraded (Step 4).

For recombination to be detected, the transforming DNA must be derived from a different strain of bacteria, bearing some genetic variation. Once integrated into the chromosome, the recombinant region contains one DNA strand from the bacterial chromosome and one from the transforming DNA. Because these strands are not genetically identical, this helical region is referred to as a **heteroduplex.** Following one round of replication, one chromosome is restored to its original configuration, identical to that of the recipient cell, and the other contains the transformed gene. Cell division produces one host cell and one transformed cell (Step 5).

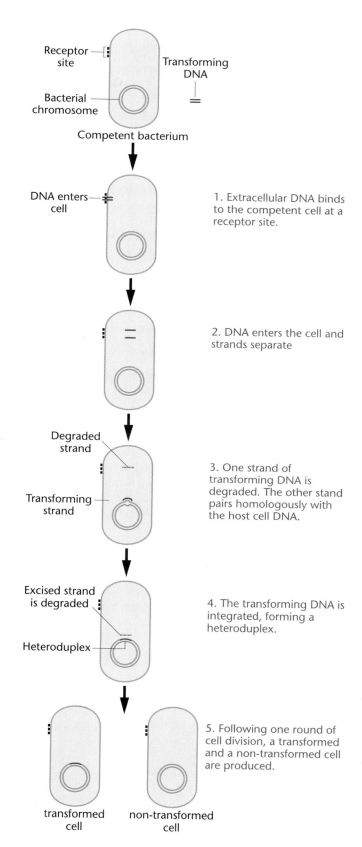

1. Extracellular DNA binds to the competent cell at a receptor site.

2. DNA enters the cell and strands separate

3. One strand of transforming DNA is degraded. The other stand pairs homologously with the host cell DNA.

4. The transforming DNA is integrated, forming a heteroduplex.

5. Following one round of cell division, a transformed and a non-transformed cell are produced.

FIGURE 7.13   Proposed steps leading to transformation of a bacterial cell by exogenous DNA. Only one of the two strands of the entering DNA is involved in the transformation event, which is completed following cell division.

Transformation and Mapping

The size of DNA that is effective in transformation is between 10,000 to 20,000 nucleotide pairs, about 1/200 of the *E. coli* chromosome. The DNA thus contains sufficient genetic information to encode several genes. Genes that are adjacent or very close to one another on the bacterial chromosome may be carried on a single piece of DNA of this size, so a single transfer event may result in the **cotransformation** of several genes simultaneously. Genes that are close enough to each other to be cotransformed are said to be linked. Note that here *linkage* refers to the proximity of genes, in contrast to the use of this term in eukaryotes to indicate that genes are on a single chromosome but not necessarily close to one another.

If two genes are not linked, simultaneous transformation can occur only as a result of two independent events involving two distinct segments of DNA. The probability of two independent events occurring simultaneously is equal to the product of the individual probabilities. Thus, the frequency of two unlinked genes being transformed simultaneously is much lower than if they are linked.

Linked bacterial genes were first demonstrated in 1954 during studies of *Pneumococcus* by Rollin Hotchkiss and Julius Marmur. They were examining transformation at the *streptomycin* and *mannitol* loci. Recipient cells were *str^s^* and *mtl^−^*, meaning that they were sensitive to streptomycin and could not ferment mannitol. Cells that were *str^r^ mtl^+^*, which are resistant to streptomycin and able to ferment mannitol, were used to derive the transforming DNA. If the *str* and *mtl* genes are not linked, simultaneous transformation for both genes will occur with such a minimal probability as to be nearly undetectable. However, as shown in Table 7.1, a low but detectable rate (0.17 percent) of cotransformation did occur.

A second experiment confirmed that the genes were indeed linked. Instead of *str^r^ mtl^+^* DNA, a mixture of *str^r^ mtl^−^* and *str^s^ mtl^+^* DNA served as the donor DNA. Thus, both the *str^r^* and *mtl^+^* genes were available to recipient cells, but on separate segments of DNA. If the observed rate of double transformants in the previous experiment had been due to two independent events and the genes were not linked, then a similar rate would also be observed under these conditions. Instead, the number of double transformants was 25 times fewer, confirming linkage. Careful examination of the data in Table 7.1 confirms the linkage of these loci.

In addition to establishing linkage relationships, relative mapping distances between linked genes can also be determined from the recombination data provided by transformation experiments. Though the analysis is more complex, such data are interpreted in a manner analogous to chromosome mapping in eukaryotes. In the Problems and Discussion Questions section at the end of the chapter we consider data that provides mapping information.

Subsequent studies of transformation have shown that various bacteria readily undergo this process of recombination (e.g., *Bacillus subtilis* and *Shigella paradysenteriae*). Other investigations have established that culture conditions can be adjusted to "induce" artificial transformation, an important component of recombinant DNA technology.

Bacteriophages are bacterial viruses

Bacteriophages, or phages as they are commonly known, are viruses that have bacteria as their hosts. During their reproduction, phages can be involved in still another mode of bacterial genetic recombination called transduction. In order to understand this process, we first must consider the genetics of bacteriophages, which themselves also undergo recombination. In this section, we will first examine the structure and life cycle of one type of bacteriophage. We will then discuss how these phages are studied during their infection of bacteria. Finally, we will contrast two possible modes of behavior once initial phage infection occurs. This information will serve as background for our subsequent discussion of transduction and bacteriophage recombination. Furthermore, a great deal of genetic research has been done using bacteriophages as a model system, making them a worthy subject of discussion.

TABLE 7.1   Results of Several Transformation Experiments That Establish Linkage between the *str* and *mtl* Loci in *Pneumococcus*

Donor DNA	Recipient Cell Genotype	Transformed Genotypes (%)		
		*str^r^ mtl^−^*	*str^s^ mtl^+^*	*str^r^ mtl^+^*
*str^r^ mtl^+^*	*str^s^ mtl^−^*	4.3	0.40	0.17
*str^r^ mtl^−^* and *str^s^ mtl^+^*	*str^s^ mtl^−^*	2.8	0.85	0.0066

Source: Data from Hotchkiss and Marmur, 1954, p. 55.

Phage T4: Structure and Life Cycle

Bacteriophage T4 is one of a group of related bacterial viruses referred to as T-even phages. It exhibits an intricate structure, as illustrated in Figure 7.14. Its genetic material, DNA, is contained within an icosahedral (a polyhedron with 20 faces) protein coat, together making up the head of the virus. The DNA is sufficient in quantity to encode more than 150 average-size genes. The head is connected to a tail that contains a collar and a contractile sheath surrounding a central core. Tail fibers, which protrude from the tail, contain binding sites in their tips that specifically recognize unique areas of the outer surface of the cell wall of the bacterial host, *E. coli*.

The life cycle of phage T4 (Figure 7.15) is initiated when the virus binds by adsorption to the bacterial host cell. Then, an ATP-driven contraction of the tail sheath causes the central core to penetrate the cell wall. The DNA in the head is extruded, where it then moves across the cell membrane into the bacterial cytoplasm. Within minutes, all bacterial DNA, RNA, and protein synthesis is inhibited, and synthesis of viral molecules begins. At the same time, degradation of the host DNA is initiated.

A period of intensive viral gene activity characterizes infection. Initially, phage DNA replication ensues, leading to a pool of viral DNA molecules. Then, the components of the head, tail, and tail fibers are synthesized. The assembly of mature viruses is a complex process that has been well studied by William Wood, Robert Edgar, and others. Three sequential pathways occur: (1) DNA packaging as the viral heads are assembled, (2) tail assembly, and (3) tail fiber asssembly. Once DNA is packaged into the head, it combines with the tail components, to which tail fibers are added. Total construction is a combination of self-assembly and enzyme-directed processes.

When approximately 200 viruses are constructed, the bacterial cell is ruptured by the action of lysozyme (a phage gene product) and the mature phages are released from the host cell. The 200 new phages will infect other available bacterial cells, and the process will be repeated over and over again.

The Plaque Assay

The experimental study of bacteriophages and other viruses has played a critical role in our understanding of molecular genetics. During infection of bacteria, enormous quantities of bacteriophages may be obtained for investigation. Often, over 10^{10} viruses per milliliter of culture medium are produced. Many genetic studies have relied on our ability to quantitate the number of phages produced following infection under specific culture conditions using a technique called the **plaque assay.**

This assay is illustrated in Figure 7.16, where actual plaques are also shown. A serial dilution of the original virally infected bacterial culture is first performed. Then, a 0.1-ml aliquot from one or more dilutions is added to a small volume of melted nutrient agar (about 3 ml) into which a few drops of a healthy bacterial culture have been mixed. The solution is then poured evenly over a base of solid nutrient agar in a petri dish and allowed to solidify prior to incubation. A clear area develops at each place where a single virus has initially infected one bacterium in the lawn that has grown up during incubation. This area, where all bacteria have been lysed, is called a **plaque.** It represents multiple clones of the single infecting T4 bacteriophage. If the dilution factor is too low, the plaques are plentiful and will fuse, lysing the entire lawn. This has occurred in the 10^{-3} dilution in Figure 7.16. On the other hand, if the dilution factor is increased, fewer phages are present in a given aliquot and plaques can be counted. From such data, the density of viruses in the initial culture can be estimated. The calculation is similar to that used for determining bacterial density by counting colonies following serial dilution of an initial culture:

$$(\text{plaque number/ml}) \times (\text{dilution factor})$$

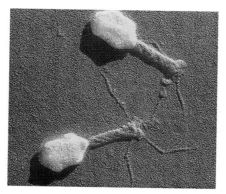

Mature T4 phage

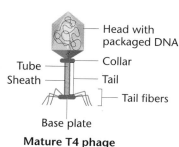

FIGURE 7.14 The structure of bacteriophage T4 including an icosahedral head filled with DNA, a tail consisting of a collar, tube, sheath, and base plate, and tail fibers. During assembly, the tail components are added to the head and then tail fibers are added.

Using the results shown in Figure 7.16, it is observed that there are 23 phage plaques derived from the 0.1 ml aliquot of the 10^{-5} dilution. Therefore, we can estimate that there are 230 phages per milliliter at this dilution. This initial viral density in the undiluted sample, where 23 plaques are observed from 0.1 ml of the 10^{-5} dilution, is calculated as

$$(230/ml) \times (10^5) = 230 \times 10^5/ml = 2.3 \times 10^7/ml$$

Because this figure is derived from the 10^{-5} dilution, we can estimate that there will be only 0.23 phage per 0.1 ml in the 10^{-7} dilution. As a result, when 0.1 ml from this tube is assayed, we can expect that no phage particles will be present. This prediction is borne out in Figure 7.16, where an intact lawn of bacteria exists. The dilution factor is simply too great.

The use of the plaque assay has been invaluable in mutational and recombinational studies of bacteriophages. We will apply this technique more directly later in this chapter, when we discuss Seymour Benzer's elegant genetic analysis of a single gene in phage T4.

Lysogeny

The relationship between virus and bacterium does not always result in viral reproduction and lysis. As early as the 1920s, it was known that some bacteriophages could enter a bacterial cell and establish a symbiotic relationship with it. The precise molecular basis of this symbiosis is now well understood. Upon entry, the viral DNA, instead of replicating in the bacterial cytoplasm, is integrated into the bacterial chromosome, a step that characterizes the developmental stage referred to as **lysogeny.** Subsequently, each time the bacterial chromosome is replicated, the viral DNA is also replicated and passed to daughter bacterial cells following division. No new viruses are produced and no lysis of the bacterial cell occurs. However, in response to certain stimuli, such as chemical or ultraviolet-light treatment, the viral DNA may lose its integrated status and initiate replication, phage reproduction, and lysis of the bacterium.

Several terms are used to describe this relationship. The viral DNA integrated into the bacterial chromosome is called a **prophage.** Viruses that can either lyse the cell or behave as a prophage are called **temperate.** Those that can only lyse the cell are referred to as **virulent.** A bacterium harboring a prophage has been **lysogenized** and is said to be **lysogenic;** that is, it is capable of being lysed as a result of induced viral reproduction. The viral DNA, which can replicate either in the bacterial cytoplasm or as part of the bacterial chromosome, is sometimes classified as an **episome.**

Transduction is virus-mediated bacterial DNA transfer

In 1952, Norton Zinder and Joshua Lederberg were investigating possible recombination in the bacterium *Salmonella typhimurium*. Although they recovered prototrophs

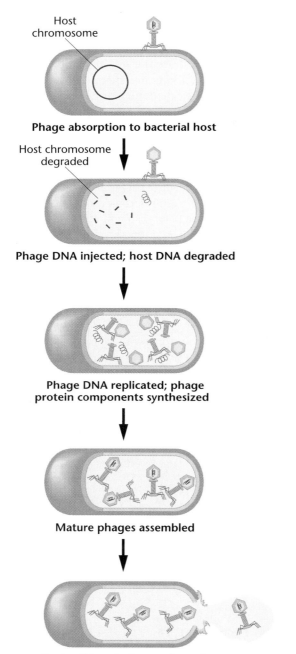

Host chromosome

Phage absorption to bacterial host

Host chromosome degraded

Phage DNA injected; host DNA degraded

Phage DNA replicated; phage protein components synthesized

Mature phages assembled

Host cell lysed; phage released

FIGURE 7.15 Life cycle of bacteriophage T4.

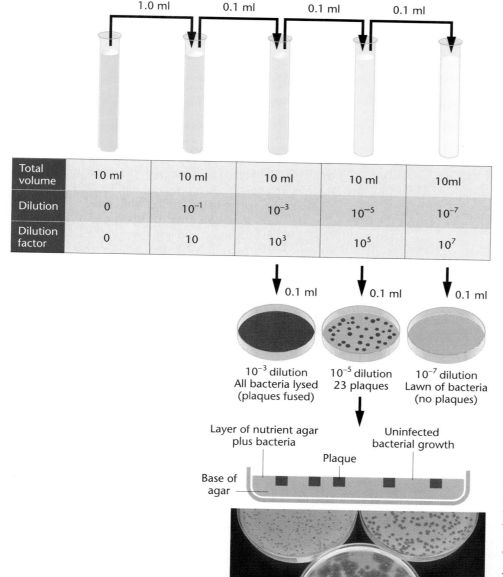

FIGURE 7.16 Diagrammatic illustration of the plaque assay for bacteriophage analysis. Serial dilutions of a bacterial culture infected with bacteriophages are first made. Then three of the dilutions (10^{-3}, 10^{-5}, and 10^{-7}) are analyzed using the plaque assay technique. In each case, 0.1 ml of the diluted culture is used. Each plaque represents the initial infection of one bacterial cell by one bacteriophage. In the 10^{-3} dilution, so many phages are present that all bacteria are lysed. In the 10^{-5} dilution, 23 plaques are produced. In the 10^{-7} dilution, the dilution factor is so great that no phages are present in the 0.1 ml sample, and thus no plaques form. From the 0.1 ml sample of the 10^{-5} dilution, the original bacteriophage density can be calculated as $23 \times 10 \times 10^5$ phages/ml (23×10^6 or 2.3×10^7). The photograph illustrates phage T2 plaques on lawns of *E. coli*.

from mixed cultures of two different auxotrophic strains, subsequent investigations revealed that recombination was occurring in a manner different from that attributable to the presence of an F factor, as in *E. coli*. What they discovered was a process of bacterial recombination mediated by bacteriophages and now called **transduction**.

The Lederberg–Zinder Experiment

Lederberg and Zinder mixed the *Salmonella* auxotrophic strains LA-22 and LA-2 together and, when the mixture was plated on minimal medium, they recovered prototroph cells. LA-22 was unable to synthesize the amino acids phenylalanine and tryptophan (phe^- trp^-), and LA-2 could not synthesize the amino acids methionine and histidine (met^- his^-). Prototrophs (phe^+ trp^+ met^+ his^+) were recovered at a rate of about $1/10^5$ (10^{-5}) cells.

Although these observations at first suggested that the recombination involved was the type observed earlier in conjugative strains of *E. coli*, experiments using the Davis U-tube soon showed otherwise (Figure 7.17). The two auxotrophic strains were separated by a glass-sintered fil-

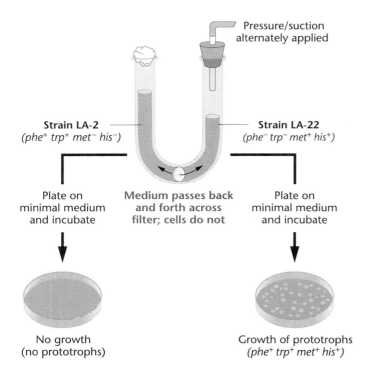

Strain LA-2
(*phe⁺ trp⁺ met⁻ his⁻*)

Strain LA-22
(*phe⁻ trp⁻ met⁺ his⁺*)

Pressure/suction
alternately applied

Plate on
minimal medium
and incubate

Medium passes back
and forth across
filter; cells do not

Plate on
minimal medium
and incubate

No growth
(no prototrophs)

Growth of prototrophs
(*phe⁺ trp⁺ met⁺ his⁺*)

FIGURE 7.17 The Lederberg–Zinder experiment using *Salmonella*. After placing two auxotrophic strains on opposite sides of a Davis U-tube, Lederberg and Zinder recovered prototrophs from the side containing the LA-22 strain but not from the side containing the LA-2 strain. These initial observations led to the discovery of the phenomenon called transduction.

ter, thus preventing cell contact but allowing growth to occur in a common medium. Surprisingly, when samples were removed from both sides of the filter and plated independently on minimal medium, prototrophs were recovered only from the side of the tube containing LA-22 bacteria. Recall that if conjugation were responsible, the conditions in the Davis U-tube would be expected to prevent recombination altogether (see Figure 7.4).

Since LA-2 cells appeared to be the source of the new genetic information (*phe⁺* and *trp⁺*), how that information crossed the filter from the LA-2 cells to the LA-22 cells, allowing recombination to occur, was a mystery. The unknown source was designated simply as a **filterable agent (FA).**

Three subsequent observations were useful in identifying the FA:

1. The FA was produced by the LA-2 cells only when they were grown in association with LA-22 cells. If LA-2 cells were grown independently and that culture medium was then added to LA-22 cells, recombination did not occur. Therefore, LA-22 cells play some role in the production of FA by LA-2 cells and do so only when the two share common growth medium.

2. The presence of DNase, which enzymatically digests DNA, did not render the FA ineffective. Therefore, the FA is not naked DNA, ruling out transformation.

3. The FA could not pass across the filter of the Davis U-tube when the pore size was reduced below the size of bacteriophages.

Aided by these observations and aware of temperate phages that could lysogenize *Salmonella*, researchers proposed that the genetic recombination event was mediated by bacteriophage P22, present initially as a prophage in the chromosome of the LA-22 *Salmonella* cells. It was hypothesized that rarely P22 prophages might enter the vegetative or lytic phase, reproduce, and be released by the LA-22 cells. Such phages, being much smaller than a bacterium, were then able to cross the filter of the U-tube and subsequently infect and lyse some of the LA-2 cells. In the process of lysis of LA-2, the P22 phages occasionally packaged in their heads a region of the LA-2 chromosome. If this region contained the *phe⁺* and *trp⁺* genes, and if the phages subsequently passed back across the filter and infected LA-22 cells, these newly lysogenized cells would behave as prototrophs. This process of transduction, whereby bacterial recombination is mediated by bacteriophage P22, is diagrammed in Figure 7.18.

The Nature of Transduction

Further studies revealed the existence of transducing phages in other species of bacteria. For example, *E. coli* can be transduced by phages P1 and λ. *Bacillus subtilis* and *Pseudomonas aeruginosa* can be transduced by the phages SPO1 and F116, respectively. The details of several different modes of transduction have also been established. Even though the initial discovery of transduction involved a temperate phage and a lysogenized bacterium, the same process can occur during the normal lytic cycle. Sometimes a small piece of bacterial DNA is packaged *along with* the viral chromosome so that the transducing phage contains both viral and bacterial DNA. In such cases, only a few bacterial genes are present in the transducing phage. However, when *only* bacterial DNA is packaged, regions as large as 1 percent of the bacterial chromosome may become enclosed in the viral head. In either case, the ability to infect is unrelated to the type of DNA in the phage head, making transduction possible.

When bacterial rather than viral DNA is injected into the bacterium, it can either remain in the cytoplasm or recombine with the homologous region of the bacterial chromosome. If the bacterial DNA remains in the cytoplasm, it does not replicate but is transmitted to one of the progeny cells following each division. When this happens, only a single cell, partially diploid for the transduced genes, is produced—a phenomenon called **abortive transduction.** If

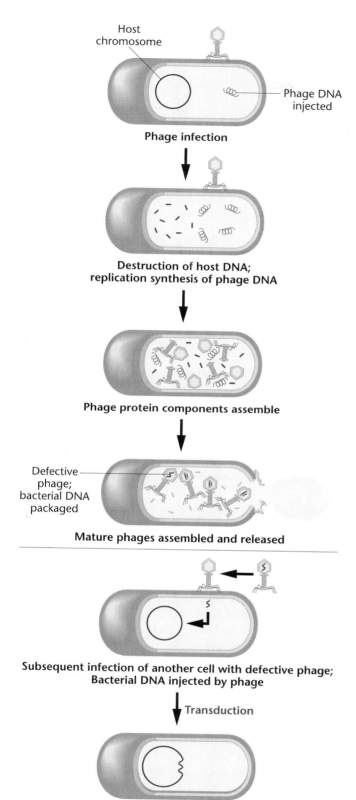

FIGURE 7.18   Generalized transduction.

the bacterial DNA recombines with its homologous region of the bacterial chromosome, the transduced genes are replicated as part of the chromosome and passed to all daughter cells. This process is called **complete transduction.**

Both abortive and complete transduction are subclasses of the broader category of **generalized transduction,** which is characterized by the random nature of DNA fragments and genes transduced. Each fragment of the bacterial chromosome has a finite but small chance of being packaged in the phage head. Most cases of generalized transduction are of the abortive type; some data suggest that complete transduction occurs 10 to 20 times less frequently.

Transduction and Mapping

Like transformation, generalized transduction has been used in linkage and mapping studies of the bacterial chromosome. The fragment of bacterial DNA involved in a transduction event is large enough to include numerous genes. As a result, two genes that are closely aligned (linked) on the bacterial chromosome may be simultaneously transduced, a process called **cotransduction.** Two genes that are not close enough to one another along the chromosome to be included on a single DNA fragment require two independent events in order to be transduced into a single cell. Since this occurs with a much lower probability than cotransduction, linkage can be determined.

By concentrating on two or three linked genes, transduction studies can also determine the precise order of these genes. The closer to each other linked genes are, the greater the frequency of cotransduction. Mapping studies involving three closely aligned genes can be executed. The analysis of such an experiment is predicated on the same rationale underlying other mapping techniques.

Specialized Transduction

In some instances, only certain genes are recombined, a situation called **specialized transduction.** This is in contrast to generalized transduction described above, where all genes have an equal probability of being recombined. One of the best examples involves transduction of *E. coli* by the temperate phage λ.

In this case, transduction is restricted to the *gal* (galactose) or *bio* (biotin) genes. The reason why transduction involves only these genes became clear when it was learned that the λ DNA always integrates into the region of the *E. coli* chromosome between these two genes at a site called *att* [Figure 7.19(a)]. Phage λ DNA that is integrated in the bacterial chromosome can subsequently detach from it, reproduce, and lyse the host cell [Figure 7.19(b)]. Sometimes the excision process occurs incorrectly and carries either the *gal* or *bio E. coli* genes in place of part of the viral

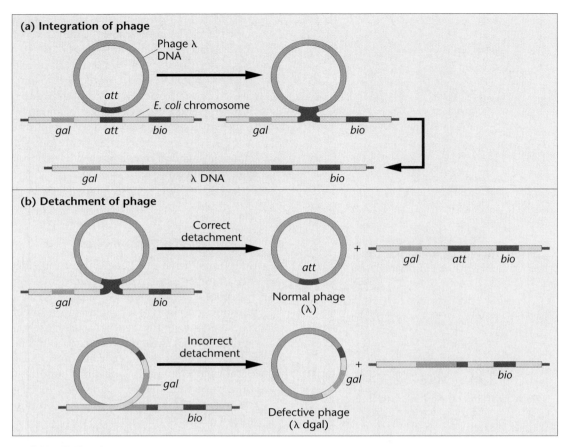

FIGURE 7.19   The production of defective phage λ*dgal*, which can result in specialized transduction following another round of infection of *E. coli*. If detachment occurs correctly, no transduction will result.

DNA [Figure 7.19(b)]. The resulting phage chromosome is defective because it has lost some of its own genetic information, but it is nevertheless replicated and packaged during the formation of mature phage particles. The virus can subsequently inject the defective chromosome into another bacterial cell.

In this case of specialized transduction, the defective phage chromosome is integrated into the bacterial chromosome and is replicated along with it. Such bacterial cells contain the defective phage DNA , making them diploid for the *gal* or *bio* genes. The presence of this transducing *gal*+ or *bio*+ DNA causes these auxotrophs to revert to a *gal*+ or *bio*+ phenotype.

Bacteriophages undergo intergenic recombination

Around 1947, several research teams demonstrated that genetic recombination also occurs in bacteriophages. These studies relied on the discovery of numerous phage muta-tions that could be visualized or assayed. Before considering recombination in these bacterial viruses, we will briefly introduce several of the mutations that were studied.

Phage mutations often affect the morphology of the plaques formed following lysis of bacterial cells. For example, in 1946 Alfred Hershey observed unusual T2 plaques on plates of *E. coli* strain B. Where the normal T2 plaques are small and have a clear center surrounded by a diffuse (nearly invisible) halo, the unusual plaques were larger and possessed a more distinctive outer perimeter (Figure 7.20). When the viruses were isolated from these plaques and replated on *E. coli* B cells, the resulting plaque appearance was identical. Thus, the plaque phenotype was an inherited trait resulting from the reproduction of mutant phages. Hershey named the mutant *rapid lysis* (*r*) because the plaques were larger, apparently resulting from a more rapid or more efficient life cycle of the phage. It is now known that in wild-type phages, reproduction is inhibited once a particular-sized plaque has been formed. The *r* mutant T2 phages are able to overcome this inhibition, producing larger plaques.

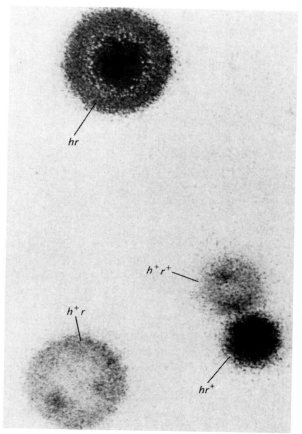

FIGURE 7.20 Plaque morphology phenotypes observed following simultaneous infection of *E. coli* by two strains of phage T2, h^+r and hr^+. In addition to the parental genotypes, recombinant plaques hr and h^+r^+ were also recovered.

Another bacteriophage mutation, *host range (h)*, was discovered by Luria. This mutation extends the range of bacterial hosts that the phage can infect. Although wild-type T2 phages can infect *E. coli* B, they cannot normally attach or be adsorbed to the surface of *E. coli* B-2. The *h* mutation, however, provides the basis for adsorption and subsequent infection of *E. coli* B-2. When grown on a mixture of *E. coli* B and B-2, the center of the *h* plaque appears much darker than the h^+ plaque (Figure 7.20).

Table 7.2 lists other types of mutations that have been isolated and studied in the T-even series of bacteriophages (e.g., T2, T4, T6). These mutations are important to the study of genetic phenomena in bacteriophages.

Intergenic Recombination and Mapping in Bacterial Viruses

The discovery of genetic recombination in bacteriophages was made during **mixed infection experiments** in which two distinct mutant strains were allowed to *simultaneously*

TABLE 7.2 Some Mutant Types of T-Even Phages

Name	Description
minute	Small plaques
turbid	Turbid plaques on *E. coli* B
star	Irregular plaques
uv-sensitive	Alters UV sensitivity
acriflavin-resistant	Forms plaques on acriflavin agar
osmotic shock	Withstands rapid dilution into distilled water
lysozyme	Does not produce lysozyme
amber	Grows in *E. coli* K12, but not B
temperature-sensitive	Grows at 25°C, but not at 42°C

infect the same bacterial culture. These studies were designed so that the number of viral particles sufficiently exceeded the number of bacterial cells so as to ensure simultaneous infection of most cells by both viral strains. Because two loci are involved, recombination is referred to as intergenic.

For example, in one study using the T2/*E. coli* system, the parental viruses were of either the h^+r (wild-type host range, rapid lysis) or hr^+ (extended host range, normal lysis) genotype. If no recombination occurred, these two parental genotypes would be the only expected phage progeny. However, the recombinants h^+r^+ and hr were detected in addition to the parental genotypes (see Figure 7.20). As with eukaryotes, the percentage of recombinant plaques divided by the total number of plaques reflects the relative distance between the genes.

$$(h^+r^+) + (hr) \text{ / total plaques} \times 100 = \text{recombinational frequency}$$

Sample data for the *h* and *r* loci are shown in Table 7.3.

Similar recombinational studies have been performed with a large number of mutant genes in a variety of bacteriophages. Data are analyzed in much the same way as they are in eukaryotic mapping experiments. Two- and three-point mapping crosses are possible, and the percentage of recombinants in the total number of phage

TABLE 7.3 Results of a Cross Involving the *h* and *r* Genes in Phage T2 ($hr^+ \times h^+r$)

Genotype	Plaques	Designation
h r^+	42	Parental progeny 76%
h^+ r	34	
h^+ r^+	12	Recombinants 24%
h r	12	

Source: Data derived from Hershey and Rotman, 1949.

progress is calculated. This value is proportional to the relative distance between two genes along the DNA molecule constituting the chromosome.

An interesting phenomenon in phage crosses is **negative interference.** Recall that in eukaryotic mapping crosses, positive interference leads to the recovery of fewer than expected double-crossover events. In phage crosses, often just the opposite occurs. In three-point analysis, a greater than expected frequency of double exchanges is often observed.

Investigations into phage recombination support a model similar to that of eukaryotic crossing over—a breakage and reunion process between the viral chromosomes. A fairly clear picture of the dynamics of viral recombination has emerged. Following the early phase of infection, the chromosomes of the phages begin replication. As this stage progresses, a pool of chromosomes accumulates in the bacterial cytoplasm. If double infection by phages of two genotypes has occurred, then the pool of chromosomes initially consists of the two parental types. Genetic exchange between these two types will occur before, during, and after replication, producing recombinant chromosomes.

In the case of the h^+r and hr^+ example discussed here, recombinant h^+r^+ and hr chromosomes are produced. Each of these chromosomes may undergo replication, with new replicates undergoing exchange with each other and with parental chromosomes. Furthermore, recombination is not restricted to exchange between two chromosomes— three or more may be involved simultaneously. As phage development progresses, chromosomes are randomly removed from the pool and packed into the phage head, forming mature phage particles. Thus, a variety of parental and recombinant genotypes are represented in progeny phages.

As we will see in the following section, powerful selection systems have made it possible to detect *intragenic* recombination in viruses, where exchanges occur at points within a single gene. Such studies have led to the fine-structure analysis of the gene, as discussed in an ensuing section.

Intragenic recombination occurs in phage T4

We conclude this chapter with an account of an ingenious example of genetic analysis. In the early 1950s, Seymour Benzer undertook a detailed examination of a single locus, *rII*, in phage T4. Benzer successfully designed experiments to recover the extremely rare genetic recombinants arising as a result of **intragenic exchange.** Such recombination is

equivalent to eukaryotic crossing over, but in this case, *within a gene rather than between two genes.* Benzer demonstrated that such recombination occurs between the DNA of individual bacteriophages during simultaneous infection of the host bacterium *E. coli.*

The end result of Benzer's work was the production of a detailed map of the *rII* locus. Because of the extremely detailed information provided from his analysis, Benzer's work was described as the "fine-structure analysis" of the gene. Because these experiments occurred decades before DNA-sequencing techniques were developed, the insights concerning the internal structure of the gene were particularly noteworthy.

The rII Locus of Phage T4

The primary requirement in genetic analysis is the isolation of a large number of mutations in the gene being investigated. Mutants at the *rII* locus produce distinctive plaques when plated on *E. coli* strain B, allowing their easy identification. Figure 7.20 illustrates mutant *r* plaques compared to their wild type r^+ counterparts in the related T2 phage. Benzer's approach was to isolate many independent *rII* mutants—he eventually obtained about 20,000—and to perform recombinational studies so as to produce a genetic map of this locus.

The key to Benzer's analysis was that *rII* mutant phages, though capable of infecting and lysing *E. coli* B, could not successfully lyse a second related strain, *E. coli* K12(λ)*. Wild-type phages, by contrast, could lyse both the B and the K12 strains. Benzer reasoned that these conditions provided the potential for a highly sensitive screening system. If phages from any two different mutant strains were allowed to simultaneously infect *E. coli* B, exchanges between the two mutant sites within the locus would produce rare wild-type recombinants (Figure 7.21). If the phage population, which contained over 99.9 percent *rII* phages and less than 0.1 percent wild-type phages, were then allowed to infect strain K12, the wild-type recombinants would successfully reproduce and produce wild-type plaques. *This is the critical step in recovering and quantifying rare recombinants.*

By using serial dilution techniques, Benzer was able to determine the total number of mutant *rII* phages produced

---

*The inclusion of (λ) in the designation of K12 indicates that this bacterial strain is lysogenized by phage λ. This, in fact, is the reason that *rII* mutants cannot lyse such bacteria. Note that as we refer to this strain in future discussions, we will abbreviate it simply as *E. coli* K12.

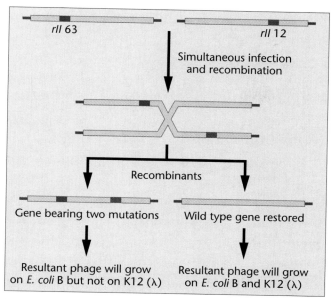

FIGURE 7.21 Illustration of intragenic recombination between two mutations in the *rII* locus of phage T4. The result is the production of a wild-type phage that will grow on *E. coli* B and K12 and a phage that has incorporated both mutations into the *rII* locus. It will grow on *E. coli* B, but not on *E. coli* K12.

on *E. coli* B and the total number of recombinant wild-type phages that would lyse *E. coli* K12. These data provided the basis for calculating the frequency of recombination, a value proportional to the distance within the gene between the two mutations being studied. As we will see, this experimental design was extraordinarily sensitive. Remarkably, it was possible for Benzer to detect as few as one recombinant wild-type phage among 100 million mutant phages. When information from many such experiments is combined, a detailed map of the locus is possible.

Before we discuss this mapping we need to describe an important discovery Benzer made during the early development of his screen, for this discovery led to the development of a technique used widely in genetics labs today, the **complementation assay**. Recall that we first introduced this topic in Chapter 4. You may wish to review this topic before proceeding.

Complementation by rII Mutations

Before Benzer was able to initiate intragenic recombinational studies, he had to resolve a problem encountered during the early stages of his experimentation. While doing a control for his experiment in which K12 bacteria were simultaneously infected with pairs of different *rII*

mutant strains, Benzer sometimes found that the K12 bacteria were lysed. This was initially quite puzzling since only the wild-type *rII* was supposed to be capable of lysing K12 bacteria. How could two mutant strains of *rII*, each of which was thought to contain a defect in the same gene, show a wild-type function?

Benzer reasoned that, during simultaneous infection, each mutant strain provided something that the other lacked, thus restoring wild-type function. This phenomenon, which he called **complementation**, is illustrated in Figure 7.22(a). When many pairs of mutations were tested, each mutation fell into one of two possible **complementation groups**, A or B. Those that failed to complement one another were placed in the same complementation group, while those that did complement one another were assigned to different complementation groups. Benzer coined the term **cistron** to describe each complementation group, which he defined as the smallest functional genetic unit.

We now know that Benzer's A and B cistrons represent the two genes in what we originally referred to as the *rII* locus. Complementation results when K12 bacteria are infected with two *rII* mutants, one with a mutation in the A gene, and one with a mutation in the B gene. Therefore, there is a source of both wild-type gene products, since the A mutant provides wild-type B, and the B mutant provides wild-type A. We can also explain why two strains that fail to complement, say two A cistron mutants, are actually mutations in the same gene. In this case, if two A cistron mutants are combined, there will be a source of the wild-type B product, but no source of wild-type A product [Figure 7.22(b)].

Once Benzer was able to place all *rII* mutations in either the A or the B cistron, he was set to return to his intragenic recombination studies, testing mutations in the A cistron against each other and testing mutations in the B cistron against each other.

Recombinational Analysis

Of the approximately 20,000 *rII* mutations, roughly half fell into each cistron. Benzer set about to map the mutations within each one. For example, focusing initially on the A cistron, if two *rIIA* mutants were first allowed to infect *E. coli* B in a liquid culture, and if a recombination event occurred between the mutational sites in the A cistron, then wild-type progeny viruses would be produced at low frequency. If samples of the progeny viruses from such an experiment were then plated on *E. coli* K12, only the wild type recombinants would lyse the bacteria and produce plaques. The total number of nonrecombinant progeny viruses was also determined by plating samples on *E. coli* B.

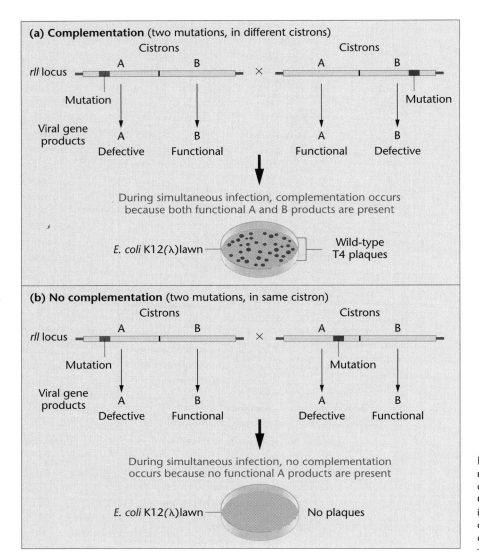

(a) Complementation (two mutations, in different cistrons)

During simultaneous infection, complementation occurs because both functional A and B products are present

(b) No complementation (two mutations, in same cistron)

During simultaneous infection, no complementation occurs because no functional A products are present

FIGURE 7.22 Comparison of two *rII* mutations that either (a) complement one another; or (b) do not complement one another. Complementation occurs when each mutation is in a separate cistron. Failure to complement occurs when the two mutations are in the same cistron.

This experimental protocol is illustrated in Figure 7.23. The percentage of recombinants can be determined by counting the plaques at the appropriate dilution in each case. As in eukaryotic mapping experiments, the frequency of recombination is an estimate of the distance between the two mutations within the cistron. For example, if the number of recombinants is equal to 4×10^3/ml, and the total number of progeny is 8×10^9/ml, then the frequency of recombination between the two mutants is:

$$2\left(\frac{4 \times 10^3}{8 \times 10^9}\right) = 2(0.5 \times 10^{-6})$$

$$= 10^{-6}$$

$$= 0.000001$$

Multiplying by 2 is necessary because each recombinant event yields two reciprocal products. Only one of them—the wild type—is detected.

Deletion Testing of the rII Locus

Although the system for assessing recombination frequencies described above allowed for mapping mutations within each cistron, testing 1000 mutants two at a time in all combinations would have required millions of experiments. Fortunately, Benzer was able to overcome this obstacle when he discovered that some of the *rII* mutations were in reality deletions of small parts of each cistron. That is, the genetic changes giving rise to the *rII* properties were not **point mutations** involving a single nucleotide change, but

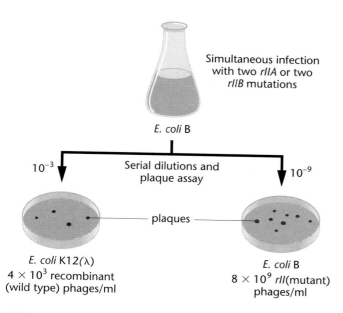

Simultaneous infection with two *rIIA* or two *rIIB* mutations

E. coli B

10^{-3} Serial dilutions and plaque assay 10^{-9}

plaques

E. coli K12(λ)
4×10^3 recombinant (wild type) phages/ml

E. coli B
8×10^9 *rII*(mutant) phages/ml

FIGURE 7.23 The experimental protocol for recombinational studies between pairs of *rIIA* or *rIIB* mutations.

instead were due to the loss or deletion of a variable number of nucleotides within the cistron. These mutations, which we shall call **deletion mutations,** could be identified by their failure to ever revert to wild type; reverting to wild type is a characteristic of point mutations. Most important, when a deletion mutation was tested during simultaneous infection with a point mutation located in the deleted part of the same cistron, it *never* yielded wild-type recombinants. The basis for the failure to do so is illustrated in Figure 7.24. Thus, a method was available that could roughly but quickly localize any mutation, provided it was contained within a region covered by a deletion.

Deletion testing could serve as the basis for the initial localization of each mutation. For example, as shown in Figure 7.25, seven overlapping deletions spanning various regions of the A and B cistrons were used for initial screening of the point mutations. Depending on whether the viral chromosome bearing a point mutation does or does not undergo recombination with the chromosome bearing a deletion, each point mutation can be assigned to a specific area of the cistron. Further deletions within each of the seven areas can be used to localize, or map, each *rII* point mutation more specifically. Remember that, in each case, a point mutation is localized in the area of a deletion when it fails to give rise to any wild-type recombinants.

The rII Gene Map

After several years of work, Benzer produced a genetic map of the two cistrons composing the *rII* locus of phage T4 (Figure 7.26). Of the 20,000 mutations analyzed, 307 distinct sites within this locus were mapped in relation to one another. Areas containing many mutations, designated as **hot spots,** were apparently more susceptible to mutation than areas where only one or a few mutations were found. Additionally, Benzer discovered areas within the cistrons where no mutations were localized. He estimated that as many as 200 recombinational units had not been localized by his studies.

The significance of Benzer's work is his application of genetic analysis to what had previously been considered an abstract unit—the gene. Benzer had demonstrated in 1955 that a gene is not an indivisible particle but instead consists of mutational and recombinational units that are arranged in a specific order—today, we know these are nucleotides composing DNA. His analysis, performed prior to the detailed molecular studies of the gene in the 1960s, is considered one of the classic examples of genetic experimentation.

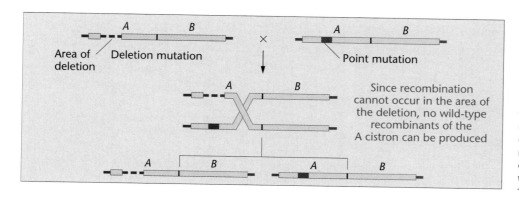

Area of deletion   Deletion mutation   ×   Point mutation

Since recombination cannot occur in the area of the deletion, no wild-type recombinants of the A cistron can be produced

FIGURE 7.24 Demonstration that recombination between a phage chromosome with a deletion in the A cistron and another phage with a point mutation overlapped by that deletion cannot yield a chromosome with wild-type A and B cistrons.

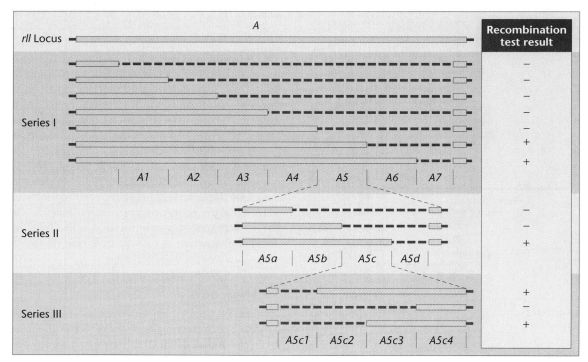

FIGURE 7.25   Three series of overlapping deletions in the A cistron of the *rII* locus used to localize the position of an unknown *r* mutation. For example, if a mutant strain tested against each deletion (dashed areas) in Series I for the production of recombinant wild-type progeny shows the results at the right (+ or −), the mutation must be in segment A5. In Series II, the mutation is further narrowed to segment A5c, and in Series III to segment A5c3.

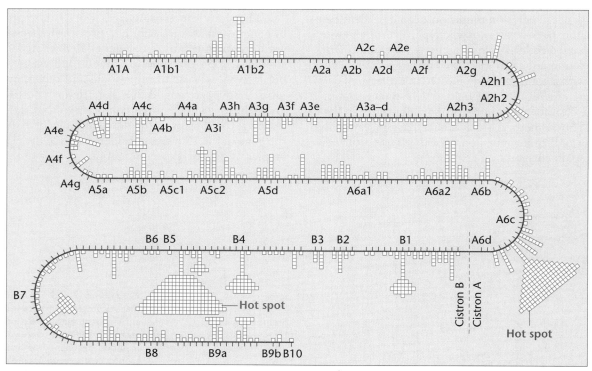

FIGURE 7.26   A partial map of mutations in the A and B cistrons of the *rII* locus of phage T4. Each square represents an independently isolated mutation. Note the two areas where the largest number of mutations are present, referred to as "hot spots" (A6cd and B5).

CHAPTER SUMMARY

1. Phenotypic variation in bacteria results from spontaneous mutation.

2. Genetic recombination in bacteria may result from three different modes: conjugation, transformation, and transduction.

3. Conjugation is initiated by a bacterium housing a plasmid called the F factor. If the F factor is in the cytoplasm of a donor cell (F$^+$), the recipient cell receives a copy of the F factor, converting it to the F$^+$ status.

4. If the F factor is integrated into the donor cell chromosome (Hfr), recombination is initiated with the recipient cell, with genetic information flowing unidirectionally to it. Time mapping of the bacterial chromosome is based on the orientation and position of the F factor in the donor chromosome.

5. The products of a group designated as the *rec* genes are directly involved in the process of recombination between the invading DNA and the recipient bacterial chromosome.

6. Plasmids, such as the F factor, are autonomously replicating DNA molecules found in the bacterial cytoplasm. Plasmids contain unique genes, such as those conferring antibiotic resistance, as well as those necessary for their transfer during conjugation.

7. The phenomenon of transformation, which does not require cell contact, involves the entry of exogenous DNA into the host chromosome of a recipient bacterial cell. Linkage mapping of closely aligned genes may be performed using this process.

8. Bacteriophages, which are viruses that infect bacteria, demonstrate a defined life cycle during which they reproduce within the host cell. They can be studied using the plaque assay.

9. Bacteriophages may be lytic, where they infect, reproduce, and lyse the host cell; or they may lysogenize the host cell, where they infect it, integrate their DNA into the host chromosome, and do not reproduce.

10. Transduction is virus-mediated bacterial DNA recombination. When a lysogenized bacterium subsequently reenters the lytic cycle, the new bacteriophages serve as the vehicle for the transfer of host (bacterial) DNA. In the process of generalized transduction, a random part of the bacterial chromosome is transferred. In specialized transduction, only specific genes adjacent to the point of insertion of the prophage are transferred.

11. Transduction may also be used for bacterial linkage and mapping studies.

12. Various mutant phenotypes, including plaque morphology and host range, have been studied in bacteriophages. These have served as the basis for investigating genetic exchange and mapping between these viruses.

13. Genetic analysis of the *rII* locus in bacteriophage T4 allowed Seymour Benzer to study intragenic recombination. By isolating *rII* mutants, performing complementation analysis, recombinational studies, and deletion mapping, Benzer was able to locate and map over 300 distinct sites within the two cistrons of the *rII* locus.

INSIGHTS AND SOLUTIONS

1. Time mapping was performed in a cross involving the genes *his, leu, mal,* and *xyl.* The recipient cells were auxotrophic for all four genes. After 25 minutes, mating was interrupted with the following results in recipient cells.

90 percent were *xyl*$^+$
80 percent were *mal*$^+$
20 percent were *his*$^+$
none were *leu*$^+$

What are the positions of these genes relative to the origin (*O*) of the F factor and to one another?

Solution: Because the *xyl* gene was transferred most frequently, it is closest to *O* (very close). The *mal* gene is next and reasonably close to *xyl*, followed by the *his*

gene. The *leu* gene is well beyond these three, since no recombinants are recovered that include it. The diagram here illustrates these relative locations along a piece of the circular chromosome.

2. Three strains of bacteria, each bearing a separate mutation, $a^-, b^-,$ or c^- were used as the sources of donor DNA in a transformation experiment. Recipient cells were wild type for those genes, but expressed the mutant d^-.

 (a) Based on the data, and assuming that the location of the *d* gene precedes the *a, b,* and *c* genes, propose a linkage map for the four genes.

DNA Donor	Recipient	Trans-formants	Frequency of Trans-formants
$a^- \, d^+$	$a^+ \, d^-$	$a^+ \, d^+$	0.21
$b^- \, d^+$	$b^+ \, d^-$	$b^+ \, d^+$	0.18
$c^- \, d^+$	$c^+ \, d^-$	$c^+ \, d^+$	0.63

Solution: These data reflect the relative distances between each of the a, b, c genes and the d gene. The a and b genes are about the same distance away from the d gene and are thus tightly linked to one another. The c gene is more distant. Assuming that the d gene precedes the others, the map looks like this:

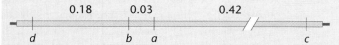

(b) If the donor DNA were wild type and the recipient cells were either $a^- b^-$, $a^- c^-$, or $b^- c^-$, in which case would wild-type transformants be expected most frequently?

Solution: Because the a and b genes are closely linked, they are most likely to be cotransformed in a single event. Thus, recipient cells of $a^- b^-$ are most likely to be converted to wild type.

3. In four Hfr strains of bacteria, all derived from an original F^+ culture grown over several months, a group of hypothetical genes was studied and shown to be transferred in the following orders:

Hfr Strain			Order of Transfer			
1	E	R	I	U	M	B
2	U	M	B	A	C	T
3	C	T	E	R	I	U
4	R	E	T	C	A	B

Assuming that B is the first gene along the chromosome, determine the sequence of all genes shown. One strain creates an apparent dilemma. Which one is it? Explain why the dilemma is only apparent and not real.

Solution: The solution is arrived at by overlapping the genes in each strain in the sequence in which they were transferred:

Strain: 2   U M B A C T
3   C T E R I U
1   E R I U M B

Starting with B, the sequence of genes is: *BACTERIUM*.

Strain 4 creates an apparent dilemma, which is resolved by realizing that the F factor is integrated in the opposite orientation; thus, the genes enter in the opposite sequence, starting with gene R.

MUIRETCAB →

4. In Benzer's fine-structure analysis of the *rII* locus in phage T4, he was able to perform complementation testing of any pair of mutations once it was clear that the locus contained two cistrons. Complementation was assayed by simultaneously infecting *E. coli* K12 with two phage strains, each with an independent mutation, neither of which could alone lyse K12. From the following data, determine which mutations are in which cistron, assuming that mutation 1 (M-1) is in the A cistron and mutation 2 (M-2) is in the B cistron. Are there any cases where the mutation cannot be properly assigned?

Test Pair	Results
1, 2	+
1, 3	−
1, 4	−
1, 5	+
2, 3	−
2, 4	+
2, 5	−

Solution: M-1 and M-5 complement one another and are therefore not in the same cistron. Thus, M-5 must be in the B cistron. M-2 and M-4 complement one another. Using the same reasoning, M-4 is not with M-2 and, therefore, is in the A cistron.

M-3 fails to complement either M-1 or M-2 and so would seem to be in both cistrons. One explanation is that the physical basis of M-3 somehow overlaps both the A and B cistrons. It might be a double mutation with one in each cistron. It might also be a deletion that overlaps both cistrons, making it impossible for it to complement either M-1 or M-2.

5. Another mutation, M-6, was tested with the results shown below. Draw all possible conclusions about M-6.

Test Pair	Results
1, 6	+
2, 6	−
3, 6	−
4, 6	+
5, 6	−

Solution: These results are consistent with assigning M-6 to the B cistron.

6. Recombination testing was then performed for M-2, M-5, and M-6 so as to map the B cistron. Recombination analysis using both *E. coli* B and K12 showed that recombination occurred between M-2 and M-5 and between M-5 and M-6, but not between M-2 and M-6. Why not?

Solution: Either M-2 and M-6 represent identical mutations or one of them may be a deletion that overlaps the other, but not M-5. Furthermore, the data cannot rule out the possibility that both are deletions.

7. In recombination studies, what is the significance of the value determined by calculating growth on the K12

versus the B strains of *E. coli* following simultaneous infection in *E. coli*? Which value is always greater?

Solution: By performing plaque analysis on *E. coli* B, where wild-type and mutant phages are both lytic, the total number of phages per milliliter can be determined. Because almost all cells are *rII* mutants of one type or another, this value is much larger. To avoid total lysis of the plate, extensive dilutions are necessary. On K12, *rII* mutations will not grow, but wild-type phages will. Since wild-type phages are the rare recombinants, there are relatively few of them and extensive dilution is not required.

PROBLEMS AND DISCUSSION QUESTIONS

1. Distinguish among the three modes of recombination in bacteria.
2. With respect to F^+ and F^- bacterial matings, answer the following questions.
   (a) How was it established that physical contact was necessary?
   (b) How was it established that chromosome transfer was unidirectional?
   (c) What is the physical–chemical basis of a bacterium being F^+?
3. List all major differences (a) between the $F^+ \times F^-$ and the $Hfr \times F^-$ bacterial crosses; and (b) between F^+, F^-, Hfr, and F' bacteria.
4. Describe the basis for chromosome mapping in the Hfr $\times$ F^- crosses.
5. Why are the recombinants produced from an Hfr $\times$ F^- cross never F^+?
6. Describe the origin of F' bacteria and merozygotes.
7. Describe what is known about the mechanisms of the transformation process.
8. In a transformation experiment involving a recipient bacterial strain of genotype a^-b^-, the following results were obtained.

Transforming DNA	Transformants (%)		
	a^+b^-	a^-b^+	a^+b^+
a^+b^+	3.1	1.2	0.04
a^+b^- and a^-b^+	2.4	1.4	0.03

What can you conclude about the location of the *a* and *b* genes relative to each other?

9. In a transformation experiment, donor DNA was obtained from a prototroph bacterial strain ($a^+b^+c^+$), and the recipient was a triple auxotroph ($a^-b^-c^-$). The following transformant classes were recovered.

$a^+ \; b^- \; c^-$	180
$a^- \; b^+ \; c^-$	150
$a^+ \; b^+ \; c^-$	210
$a^- \; b^- \; c^+$	179
$a^+ \; b^- \; c^+$	2
$a^- \; b^+ \; c^+$	1
$a^+ \; b^+ \; c^+$	3

What general conclusions can you draw about the linkage relationships among the three genes?

10. Explain the observations that led Zinder and Lederberg to conclude that the prototrophs recovered in their transduction experiments were not the result of F^- mediated conjugation.
11. Define plaque, lysogeny, and prophage.
12. Differentiate between generalized and restricted transduction. Which can be used in mapping and why?
13. Two theoretical genetic strains of a virus ($a^-b^-c^-$ and $a^+b^+c^+$) are used to simultaneously infect a culture of host bacteria. Of 10,000 plaques scored, the following genotypes were observed.

$a^+ \; b^+ \; c^+$	4100	$a^- \; b^+ \; c^-$	160
$a^- \; b^- \; c^-$	3990	$a^+ \; b^- \; c^+$	140
$a^+ \; b^- \; c^-$	740	$a^- \; b^- \; c^+$	90
$a^- \; b^+ \; c^+$	670	$a^+ \; b^+ \; c^-$	110

Determine the genetic map of these three genes on the viral chromosome. Determine whether interference was positive or negative.

14. Describe the conditions under which genetic recombination may occur in bacteriophages.
15. If a single bacteriophage infects one *E. coli* cell present on a lawn of bacteria, and upon lysis yields 200 viable viruses, how many phages will exist in a single plaque if only three more lytic cycles occur?

16. A phage-infected bacterial culture was subjected to a series of dilutions, and a plaque assay was performed in each case, with the results shown here. What conclusion can be drawn in the case of each dilution?

	Dilution Factor	Assay Results
(a)	10^4	All bacteria lysed
(b)	10^5	14 plaques
(c)	10^6	0 plaques

17. In complementation studies of the *rII* locus of phage T4, three groups of three different mutations were tested. For each group, only two combinations were tested. On the basis of each set of data, predict the results of the third experiment.

Group A	Group B	Group C
$d \times e$—lysis	$g \times h$—no lysis	$j \times k$—lysis
$d \times f$—no lysis	$g \times i$—no lysis	$j \times l$—lysis
$e \times f$—?	$h \times i$—?	$k \times l$—?

18. In an analysis of other *rII* mutants, complementation testing yielded the following results:

Mutants	Results (lysis)
1, 2	+
1, 3	+
1, 4	−
1, 5	−

(a) Predict the results of testing 2 and 3, 2 and 4, and 3 and 4 together.

(b) If further testing yielded the following results, what would you conclude about mutant 5?

Mutants	Results
2, 5	−
3, 5	−
4, 5	−

(c) Following mixed infection of mutants 2 and 3 on *E. coli* B, progeny viruses were plated in a series of dilutions on both *E. coli* B and K12 with the following results. What is the recombination frequency between the two mutants?

Strain Plated	Dilution	Plaques
E. coli B	10^{-5}	2
E. coli K12	10^{-1}	5

(d) Another mutation, 6, was then tested in relation to mutations 1 through 5. In initial testing, mutant 6 complemented mutants 2 and 3. In recombination testing with 1, 4, and 5, mutant 6 yielded recombinants with 1 and 5, but not with 4. What can you conclude about mutation 6?

19. When the interrupted mating technique was used with five different strains of Hfr bacteria, the following order of gene entry and recombination was observed. On the basis of these data, draw a map of the bacterial chromosome. Do the data support the concept of circularity?

Hfr Strain	Order				
1	T	C	H	R	O
2	H	R	O	M	B
3	M	O	R	H	C
4	M	B	A	K	T
5	C	T	K	A	B

EXTRA-SPICY PROBLEMS

20. During the analysis of seven *rII* mutations in phage T4, mutants 1, 2, and 6 were in cistron A, while mutants 3, 4, and 5 were in cistron B. Of these, mutant 4 was a deletion overlapping mutant 5. The remainder were point mutations. Nothing was known about mutant 7.
    (a) Predict the results of complementation (+ or −): 1 and 2; 1 and 3; 2 and 4; and 4 and 5.
    (b) In recombination studies between 1 and 2, the following results were obtained. Calculate the recombination frequency.

Strain	Dilution	Plaques	Phenotypes
E. coli B	10^{-7}	4	r
E. coli K12	10^{-2}	8	+

(c) When mutant 6 was tested for recombination with mutant 1, the data were the same for strain B as shown in part (b), but not for K12. The researcher lost the K12 data but remembered that recombination was 10 times more frequent than when mutants 1 and 2 were tested. What were the lost values (dilution and colony numbers)?
(d) Mutant 7 failed to complement any of the other mutants (1–6). Define the nature of mutant 7.

21. In *Bacillus subtilis*, linkage analysis of two mutant genes affecting the synthesis of the two amino acids tryptophan (trp_2^-) and tyrosine (tyr_1^-) was performed using transformation. Examine the data below and draw all possible conclusions regarding linkage. What is the role of Part B of the experiment? [*Reference:* E. Nester, M. Schafer, and J. Lederberg (1963).]

	Donor DNA	Recipient Cell	Transformants	No.
A.	$trp_2^+ tyr_1^+$	$trp_2^- tyr_1^-$	$trp^+ tyr^-$	196
			$trp^- tyr^+$	328
			$trp^+ tyr^+$	367
B.	$trp_2^+ tyr_1^-$ and $trp_2^- tyr_1^+$	$trp_2^- tyr_1^-$	$trp^+ tyr^-$	190
			$trp^- tyr^+$	256
			$trp^+ tyr^+$	2

22. An Hfr strain is used to map three genes in an interrupted mating experiment. The cross is $Hfr/a^+ b^+ c^+ rif^s \times F^-/a^- b^- c^- rif^r$ (no map order is implied in the listing of the alleles; *rif* = the antibiotic rifampicin). The a^+ gene is required for the biosynthesis of nutrient A, the b^+ gene for nutrient B, and c^+ for nutrient C. The minus alleles are auxotrophs for these nutrients. The cross is initiated

GENETICS MediaLab

The following resources will help you achieve a better under-standing of the concepts presented in this chapter. These resources can be found on the CD packaged with this textbook and on the Companion Website at **http://www.prenhall.com/klug/**.

CD Resources:

Self-grading Chapter Problems

Web Resources:

Web Destinations in Genetics

Self-grading Chapter Problems

Chapter Search Terms

Genetics Newsgroups

Student Bulletin Board

Web Problem 1:

Time for completion = 10 minutes
Do bacterial genomes recombine? This exercise uses an anima-tion of bacterial conjugation to explore some of the ways that genetic material is transferred from one cell to another. What are the differences between an F$^+$ cell, an Hfr cell, and an F$^-$ cell? Which strand of DNA in the rolling circle model of gene transfer and replication is used as the template for replicating the donor genome? Which strand is replicated in the recipient's genome? Once the recipient cell has replicated the donor's DNA, how is the genetic information incorporated into the genome? To complete this exercise, visit Web Problem 1 in Chap-ter 7 of your Companion Website and select the keyword **CONJUGATION.**

Web Problem 2:

Time for completion = 10 minutes
How do viruses participate in bacterial recombination? Trans-duction refers to the transfer of genetic material between two bacterial cells using a bacteriophage as a vector. Follow the tu-torial on the linked Website and use the material in Chapter 7 to answer the following questions. What is the key difference between the genes transferred in generalized transduction and those transferred through specialized transduction? Must the viral DNA incorporate into the host's genome (as a prophage) in order to recombine with the bacterial chromosome? What is the role of the *rec* proteins in transduction? Do they play any role in either conjugation or transformation? Would specialized trans-duction or generalized transduction be more useful in linkage and mapping studies? Why? To complete this exercise, visit Web Problem 2 in Chapter 7 of your Companion Website and select the keyword **TRANSDUCTION.**

Web Problem 3:

Time for completion = 10 minutes
Can viruses themselves recombine? If two phages infect a single cell, is it possible that, within that cell, the phage DNA may re-combine to create new bacteriophage genotypes? Seymour Ben-zer conducted an amazing set of experiments using a single locus in the bacteriophage T4. Using the system of *rII* mutants, Benz-er constructed the first map of a gene. Why did Benzer use two strains of *E. coli* (the B and the K12 strains)? Why, when Benzer mixed the mutant and wild-type bacterial strains together and plated them on K12, did he sometimes get no plaques and other times only a few plaques? Were Benzer's recombinant phage de-tected in cells that were in the lysogenic or the lytic phase of the phage life cycle? To complete this exercise, visit Web Problem 3 in Chapter 7 of your Companion Website and select the key-word **VIRAL RECOMBINATION.**

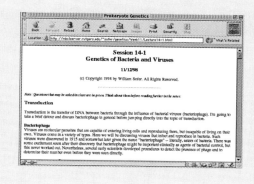

---

at time = 0, and at various times the mating mixture is plated on three types of medium. Each plate contains minimal medium (MM) *plus* rifampicin *plus* specific sup-plements that are indicated in the table. The results for each time point are shown as number of colonies grow-ing on each plate.

Supplements Added to MM	Time of Interruption			
	5 min	10 min	15 min	20 min
Nutrients A and B	0	0	4	21
Nutrients B and C	0	5	23	40
Nutrients A and C	4	25	60	82

(a) What is the purpose of rifampicin in the experiment?
(b) Based on these data, determine the approximate loca-tion on the chromosome of the *a*, *b*, and *c* genes relative to one another and relative to the F factor.
(c) Can the location of the *rif* gene be determined in this experiment? If not, design an experiment to determine the location of *rif* relative to the F factor and gene *B*.

SELECTED READINGS

ADELBERG, E.A. 1960. *Papers on bacterial genetics.* Boston: Little, Brown.

BENZER, S. 1955. Fine structure of a genetic region in bacteriophage. *Proc. Natl. Acad. Sci. USA* 42:344–54.

———. 1961. On the topography of the genetic fine structure. *Proc. Natl. Acad. Sci. USA* 47:403–15.

———. 1962. The fine structure of the gene. *Sci. Am.* (Jan.) 206:70–87.

BIRGE, E.A. 1988. *Bacterial and bacteriophage genetics—An introduction.* New York: Springer-Verlag.

BROCK, T. 1990. *The emergence of bacterial genetics.* Cold Spring Harbor, NY: Cold Spring Harbor Laboratory Press.

BRODA, P. 1979. *Plasmids.* New York: W.H. Freeman.

CAIRNS, J., STENT, G.S., and WATSON, J.D., eds. 1966. *Phage and the origins of molecular biology.* Cold Spring Harbor, NY: Cold Spring Harbor Laboratory Press.

CAMPBELL, A.M. 1976. How viruses insert their DNA into the DNA of the host cell. *Sci. Am.* (Dec.) 235:102–13.

FOX, M.S. 1966. On the mechanism of integration of transforming deoxyribonucleate. *J. Gen. Physiol.* 49:183–96.

HAYES, W. 1953. The mechanisms of genetic recombination in *Escherichia coli. Cold Spring Harbor Symp. Quant. Biol.* 18:75–93.

———. 1968. *The genetics of bacteria and their viruses,* 2nd ed. New York: Wiley.

HERSHEY, A.D. 1946. Spontaneous mutations in a bacterial virus. *Cold Spring Harbor Symp. Quant. Biol.* 11:67–76.

HERSHEY, A.D., and CHASE, M. 1951. Genetic recombination and heterozygosis in bacteriophage. *Cold Spring Harbor Symp. Quant. Biol.* 16:471–79.

HERSHEY, A.D., and ROTMAN, R. 1949. Genetic recombination between host range and plaque-type mutants of bacteriophage in single cells. *Genetics* 34:44–71.

HOTCHKISS, R.D., and MARMUR, J. 1954. Double marker transformations as evidence of linked factors in deoxyribonucleate transforming agents. *Proc. Natl. Acad. Sci. USA* 40:55–60.

JACOB, F., and WOLLMAN, E.L. 1961a. *Sexuality and the genetics of bacteria*. Orlando, FL: Academic Press.

———. 1961b. Viruses and genes. *Sci. Am.* (June) 204:92–106.

LEDERBERG, J. 1986. Forty years of genetic recombination in bacteria: A fortieth anniversary reminiscence. *Nature* 324:627–28.

LWOFF, A. 1953. Lysogeny. *Bacteriol. Rev.* 17:269–337.

MESELSON, M., and WEIGLE, J.J. 1961. Chromosome breakage accompanying genetic recombination in bacteriophage. *Proc. Natl. Acad. Sci. USA* 47:857–68.

MILLER, J.H. 1992. *A short course in bacterial genetics*. Cold Spring Harbor, NY: Cold Spring Harbor Press.

MORSE, M.L., LEDERBERG, E.M., and LEDERBERG, J. 1956. Transduction in *Escherichia coli* K12. *Genetics* 41:141–56.

NESTER, E., SCHAFER, M., and LEDERBERG, J. 1963. Gene linkage in DNA transfer: A cluster of genes in *Bacillus subtilis*. *Genetics* 48:529–51.

NOVICK, R.P. 1980. Plasmids. *Sci. Am.* (Dec.) 243:102–27.

OZEKI, H., and IKEDA, H. 1968. Transduction mechanisms. *Annu. Rev. Genet.* 2:245–78.

SMITH, H.O., DANNER, D.B., and DEICH, R.A. 1981. Genetic transformation. *Annu. Rev. Biochem.* 50:41–68.

SMITH-KEARY, P.F. 1989. *Molecular genetics of Escherichia coli*. New York: Guilford Press.

STAHL, F.W. 1979. *Genetic recombination: Thinking about it in phage and fungi*. New York: W.H. Freeman.

———. 1987. Genetic recombination. *Sci. Am.* (Nov.) 256:91–101.

STENT, G.S. 1963. *Molecular biology of bacterial viruses*. New York: W.H. Freeman.

———. 1966. *Papers on bacterial viruses*, 2nd ed. Boston: Little, Brown.

TESSMAN, I. 1965. Genetic ultrafine structure in the T4 *rII* region. *Genetics* 51:63–75.

VISCONTI, N., and DELBRUCK, M. 1953. The mechanism of genetic recombination in phage. *Genetics* 38:5–33.

WOLLMAN, E.L., JACOB, F., and HAYES, W. 1956. Conjugation and genetic recombination in *Escherichia coli* K-12. *Cold Spring Harbor Symp. Quant. Biol.* 21:141–62.

ZINDER, N.D. 1958. Transduction in bacteria. *Sci. Am.* (Nov.) 199:38–47.

———. 1992. Forty years ago: The discovery of bacterial transduction. *Genetics* 132:291–94.

ZINDER, N.D., and LEDERBERG, J. 1952. Genetic exchange in *Salmonella*. *J. Bacteriol.* 64:679–99.

Variegation illustrated in *Coleus*.

Extrachromosomal Inheritance

KEY CONCEPTS

- **Organelle heredity involves DNA in chloroplasts and mitochondria**

    Chloroplasts: Variegation in Four O'Clock Plants
    Iojap in Maize
    Chloroplast Mutations in *Chlamydomonas*
    Mitochondrial Mutations: The case of *poky* in *Neurospora*
    Petite in *Saccharomyces*

- **Mutations in mitochondrial DNA cause human disorders**

- **Infectious heredity is based on a symbiotic relationship between host organism and invader**

    Kappa in *Paramecium*
    Infective Particles in *Drosophila*

- **In maternal effect, the maternal genotype has a strong influence during early development**

    Ephestia Pigmentation
    Limnaea Coiling
    Embryonic Development in *Drosophila*

- **Genomic imprinting cause genes to exhibit uniparental inheritance**

Throughout the history of genetics, occasional reports have challenged the basic tenet of Mendelian transmission genetics—that the phenotype is transmitted by nuclear genes located on chromosomes of both parents. Observations have revealed inheritance patterns that reflect neither Mendelian nor neo-Mendelian principles and some indicate an apparent extranuclear or extrachromosomal influence on the phenotype. Such observations were commonly regarded with skepticism, but, with the increasing knowledge of molecular genetics and the discovery of DNA in mitochondria and chloroplasts, **extranuclear inheritance** is now recognized as an important aspect of genetics.

There are several varieties of extranuclear inheritance. One major type is described as maternal inheritance, whereby DNA contained in mitochondria or chloroplasts determines the phenotype of the offspring. Examples are recognized on the basis of the uniparental transmission of these organelles to progeny. A second type involves infectious heredity, resulting from the symbiotic or parasitic association of a microorganism. In such cases, an inherited phenotype is affected by the presence of the microorganism in the cytoplasm of the host cells. The third variety involves the maternal effect on the phenotype whereby nuclear gene products are stored and then transmitted through the cytoplasm to offspring. Finally, the phenomenon of genomic imprinting in mammals leads to the differential expression of the same gene, depending on its parental origin. Since both maternal effect and genomic imprinting result in forms of uniparental inheritance, we include both topics in our discussion of extranuclear inheritance.

Organelle heredity involves DNA in chloroplasts and mitochondria

In this section we will examine examples of inheritance patterns related to chloroplast and mitochondrial function. Before DNA was discovered in these organelles, the genetic processes occurring within them were categorized as *cytoplasmic inheritance*. That is, certain mutant phenotypes seemed to be inherited through the cytoplasm rather than through the genetic information of chromosomes. Transmission was most often from the maternal parent through the ooplasm, causing the results of reciprocal crosses to vary. Such a pattern is now more appropriately called **organelle heredity.**

Analysis of the hereditary transmission of mutant alleles of chloroplast and mitochondrial DNA has been difficult. First, the function of these organelles is dependent upon gene products of both nuclear and organelle DNA. Second, more than one organelle is contributed to each progeny. If many chloroplasts and/or mitochondria are contributed and only one or a few contain a mutant gene, the corresponding mutant phenotype may not be revealed. Analysis is thus much more involved than for Mendelian characters.

In this section we discuss examples of inheritance patterns related to these organelles. We will return to this topic and provide a more detailed analysis of the molecular organization of the DNA in Chapter 19.

Chloroplasts: Variegation in Four O'Clock Plants

In 1908, Carl Correns (one of the rediscoverers of Mendel's work) provided the earliest example of inheritance linked to chloroplast transmission. Correns discovered a variant of the four o'clock plant, *Mirabilis jalapa*, that had branches with either white, green, or variegated leaves. As shown in Table 8.1, inheritance in all possible combinations of crosses is strictly determined by the *phenotype* of the ovule source. For example, if the seeds (representing the progeny) were derived from ovules on branches with green leaves, all progeny plants bore only green leaves, regardless of the phenotype of the source of pollen.

Correns concluded that inheritance was through the cytoplasm of the maternal parent because the pollen, which contributes little or no cytoplasm to the zygote, had no apparent influence on the progeny phenotypes. Since leaf coloration involves the chloroplast, genetic information either in that organelle, or in the cytoplasm and influencing the chloroplast, could be responsible for this inheritance pattern.

Iojap in Maize

A phenotype similar to *Mirabilis* but with a different pattern of inheritance has been analyzed in maize by Marcus M. Rhoades. In this case, the color of the leaves is influenced not only by the cytoplasm but also by a nuclear gene located on chromosome 7. This locus is called **iojap,** after the recessive mutation *ij* located there. The wild-type allele, designated *Ij*, produces green leaves. Plants homozygous for the mutation (*ij/ij*) may have green-and-white striped leaves. Reciprocal crosses between plants with striped leaves (*ij/ij*) and green leaves (*Ij/Ij*) produce varied results, depending on which parent is mutant (Figure 8.1). If the female is striped (*ij/ij*) and the male is green (*Ij/Ij*), the progeny have colorless, striped, or green leaves. If the male parent is striped (*ij/ij*) and the female is green (*Ij/Ij*), only plants with green leaves are produced! In both types of crosses, all offspring have identical genotypes (*Ij/ij*). We may conclude that, although a nuclear gene seems to be involved, the inheritance pattern is influenced strictly maternally.

We can understand this pattern better by examining the offspring resulting from self-fertilization of heterozygous plants. As Figure 8.1 shows, the striped plant gives rise to progeny with colorless, striped, or green leaves, regardless of the genotype of the progeny. Green plants give rise to both green and striped progeny in a 3:1 ratio. Results of these self-fertilizations substantiate that the mutant chloroplasts are transmitted solely through the female cytoplasm, regardless of the plant's nuclear genotype.

Apparently, the nuclear genotype *ij/ij* somehow alters chloroplasts, which are then transmitted maternally. The colorless areas of the leaf are due to the lack of the green chlorophyll pigment in the chloroplasts, originally caused by the *ij* allele. Once acquired, chlorophyll-deficient (colorless) chloroplasts are transmitted through the maternal cytoplasm, establishing the leaf phenotype of progeny plants.

Chloroplast Mutations in Chlamydomonas

The unicellular green alga *Chlamydomonas reinhardi* has provided an excellent system for the investigation of plastid inheritance. This haploid eukaryotic organism (Figure 8.2) has a single large chloroplast containing over 50 copies of a circular double-stranded DNA molecule. Matings that reestablish diploidy are immediately followed by meiosis, and the various stages of the life cycle are easily studied in culture in the laboratory. The first cytoplasmic mutant, *streptomycin resistance* (*str^R*), was reported in 1954 by Ruth Sager. Although *Chlamydomonas*' two mating types—*mt$^+$* and *mt$^-$*—appear to make equal cytoplasmic contributions to the zygote, Sager determined that the *str* phenotype is transmitted only through the *mt$^+$* parent (Figure 8.2). Reciprocal crosses between sensitive and resistant strains yield different results depending on the genotype of the *mt$^+$* parent, which is expressed in all offspring. As shown, 1/2 of the offspring are *mt$^+$* and 1/2 of

TABLE 8.1 Offspring from Crosses Between Flowers from Various Branches of Variegated Four O'Clock Plants

	Location of Ovule		
Source of Pollen	*White branch*	*Green branch*	*Variegated branch*
White branch	White	Green	White, green, or variegated
Green branch	White	Green	White, green, or variegated
Variegated branch	White	Green	White, green, or variegated

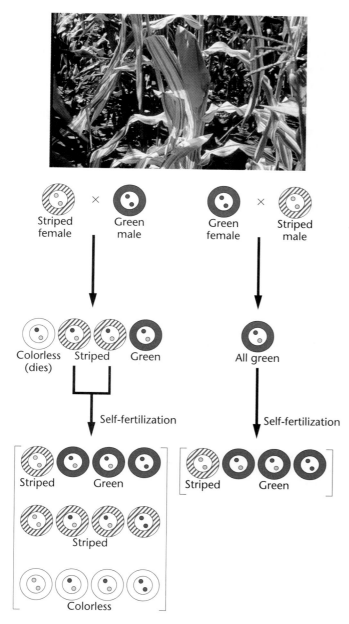

Key

○ iojap (ij)

● Wild type (Ij)

FIGURE 8.1 Maternal inheritance of striping in maize (illustrated at top). Regardless of their *iojap* genotype, offspring reflect the maternal phenotypes in their own appearance. Because color is due to chloroplasts in the leaves, inheritance is controlled by these organelles as they are passed through the maternal cytoplasm.

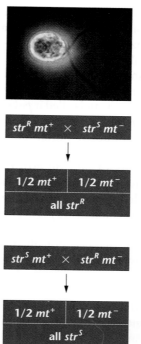

$str^R \ mt^+ \ \times \ str^S \ mt^-$

1/2 mt^+	1/2 mt^-
all str^R	

$str^S \ mt^+ \ \times \ str^R \ mt^-$

1/2 mt^+	1/2 mt^-
all str^S	

FIGURE 8.2 The results of reciprocal crosses between streptomycin-resistant (str^R) and streptomycin-sensitive (str^S) strains in the green alga *Chlamydomonas*.

uniparental inheritance pattern have been discovered. These mutations have all been linked to the transmission of the chloroplast, and their study has extended our knowledge of chloroplast inheritance.

Following fertilization, which involves the fusion of two cells of opposite mating type, the single chloroplasts of the two mating types fuse. After the resulting zygote has undergone meiosis and haploid cells are produced, it is apparent that the genetic information of the chloroplasts of progeny cells is derived only from the mt^+ parent. The genetic information present within the mt^- chloroplast has degenerated.

It is interesting to note that the inheritance of phenotypes affected by DNA present in mitochondria is also uniparental. However, studies of the transmission of several cases of antibiotic resistance have shown that it is the mt^- parent that transmits the genetic information to progeny cells. This is just the opposite of what occurs with chloroplast-derived phenotypes such as str^R. The significance of inheriting one organelle from one parent and the other organelle from the other parent is not yet established.

Mitochondrial Mutations: The Case of poky *in* Neurospora

Mutations affecting mitochondrial function have been discovered and studied, revealing that mitochondria also contain a distinctive genetic system. As with chloroplasts, mitochondrial mutations are transmitted through the cy-

them are mt^-, indicating that mating type is controlled by a nuclear gene that segregates in a Mendelian fashion.

Since Sager's discovery, a number of other *Chlamydomonas* mutations (including resistance to or dependence on a variety of bacterial antibiotics) that show a similar

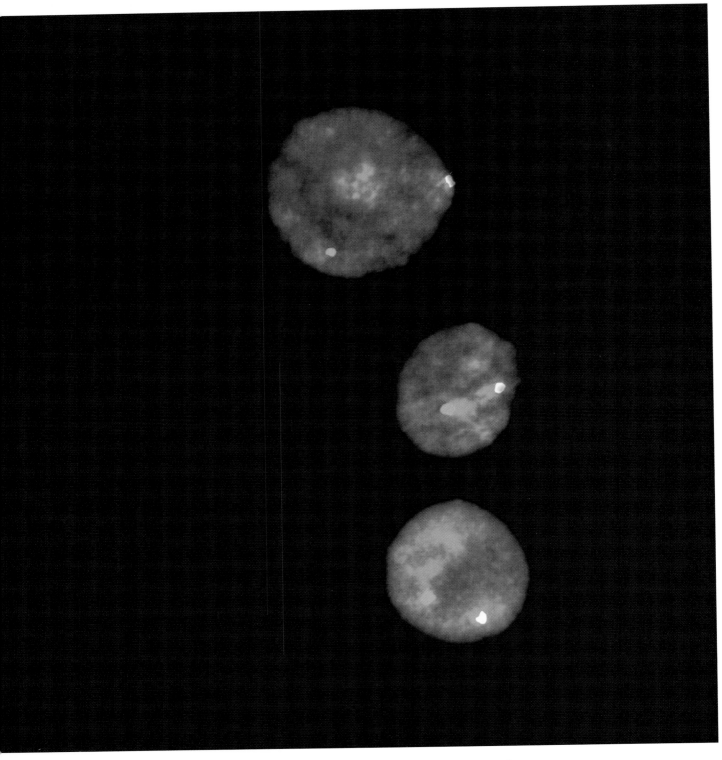

Demonstration of the X and Y chromosomes (the blue and the pink dots, respectively,) in mammalian fetal cells using fluorescent *in situ* hybridization (FISH).

Sex Determination and Sex Chromosomes

KEY CONCEPTS

- **Sexual differentiation plays a varying role in the life cycle of organisms**

  Chlamydomonas
  Zea mays
  C. elegans

- **X and Y chromosomes were first linked to sex determination early in the twentieth century**

- **The Y chromosome determines maleness in humans**

  Klinefelter and Turner Syndromes
  47,XXX Syndrome
  47,XYY Condition
  Sexual Differentiation in Humans
  The Y Chromosome and Male Development

- **The ratio of males to females in humans is not 1.0**

- **Dosage compensation prevents excessive expression of X-linked genes in humans and other mammals**

  Barr Bodies
  The Lyon Hypothesis
  The Mechanism of Inactivation

- **The ratio of X chromosomes to sets of autosomes determines sex in *Drosophila***

  Dosage Compensation in *Drosophila*
  Drosophila Mosaics

- **Temperature variation controls sex determination in reptiles**

In the biological world, a wide range of reproductive modes and life cycles are recognized. Asexual organisms are those for which no evidence of sexual reproduction is known. Other species alternate between short periods of sexual reproduction and prolonged periods of asexual reproduction. In most diploid eukaryotes, however, sexual reproduction is the only natural mechanism resulting in new members of a species. The orderly transmission of genetic units from parents to offspring, and thus any phenotypic variability, relies on the processes of segregation and independent assortment that occur during meiosis. The meiotic process also results in the production of haploid gametes so that, following fertilization, the resulting offspring maintain the diploid number of chromosomes characteristic of their species. Thus, meiosis ensures genetic constancy within members of the same species.

These events, which are involved in the perpetuation of all sexually reproducing organisms, depend ultimately on an efficient union of gametes during fertilization. In turn, successful mating between organisms, the basis for fertilization, depends on some form of sexual differentiation in organisms. Although not overtly apparent, this differentiation occurs as low on the evolutionary scale as bacteria and single-celled eukaryotic algae. In evolutionarily higher forms of life, the differentiation of the sexes is more evident as phenotypic dimorphism in the male and female members of each species. The shield and spear ♂, the ancient symbols of iron and Mars, and the mirror ♀, the symbol of copper and Venus, represent the maleness and femaleness acquired by individuals.

While dissimilar, or **heteromorphic,** chromosomes, such as the X-Y pair, often characterize one sex or the other, resulting in their label as **sex chromosomes,** genes rather than chromosomes serve as the underlying basis of sex determination. As we will see, some of these genes are present on sex chromosomes, but others are autosomal. Extensive investigation has revealed a wide variation in sex chromosome systems, even in closely related organisms, suggesting that mechanisms controlling sex determination have undergone rapid evolution in many instances.

In this chapter we will first review several representative modes of sexual differentiation by examining the life cycle of three organisms often studied in genetics: the green alga *Chlamydomonas;* the maize plant, *Zea mays;* and the nematode (roundworm), *Caenorhabditis elegans* (most often referred to as *C. elegans*). These will serve to contrast the different roles that sexual differentiation plays in the lives of diverse organisms. Then, we will delve more deeply into what is known about the genetic basis for the determination of sexual differences, with a particular emphasis on two other organisms: our own species, representative of mammals; and *Drosophila*, where pioneering sex-determining studies were performed.

Sexual differentiation plays a varying role in the life cycle of organisms

In multicellular organisms, it is important to distinguish between **primary sexual differentiation,** which involves only the gonads where gametes are produced, and **secondary sexual differentiation,** which involves the overall appearance of the organism, including clear differences in such organs as mammary glands and external genitalia. In plants and animals, the terms **unisexual, dioecious,** and **gonochoric** are equivalent; they all refer to an individual containing only male or only female reproductive organs. Conversely, the terms **bisexual, monoecious,** and **hermaphroditic** refer to individuals containing both male and female reproductive organs, a common occurrence in both the plant and animal kingdoms. These organisms can produce fertile gametes of both sexes. The term **intersex** is usually reserved for individuals of intermediate sexual differentiation, who are most often sterile.

Chlamydomonas

The life cycle of the green alga *Chlamydomonas,* shown in Figure 9.1, is representative of organisms that exhibit only infrequent periods of sexual reproduction. Such organisms spend most of their life cycle in the haploid phase, asexually producing daughter cells by mitotic divisions. However, under unfavorable nutrient conditions, such as nitrogen depletion, certain daughter cells function as gametes. Following fertilization, a diploid zygote, which can withstand the unfavorable environment, is formed. When conditions become more suitable, meiosis ensues and haploid vegetative cells are produced.

In such species, there is little visible difference between the haploid vegetative cells that reproduce asexually and the haploid gametes that are involved in sexual reproduction. The two gametes that fuse together during mating are not usually morphologically distinguishable. Such gametes are called **isogametes,** and species producing them are said to be **isogamous.**

In 1954, Ruth Sager and Sam Granik demonstrated that gametes in *Chlamydomonas* could be subdivided into two mating types. Working with clones derived from single haploid cells, they showed that cells from a given clone would mate with cells from some but not all other clones. When they tested mating abilities of large numbers of clones, all could be placed into one of two mating categories, either mt^+ or mt^-. "Plus" cells would only mate with "minus" cells, and vice versa, as represented in Figure 9.2. Following fertilization and meiosis, the four hap-

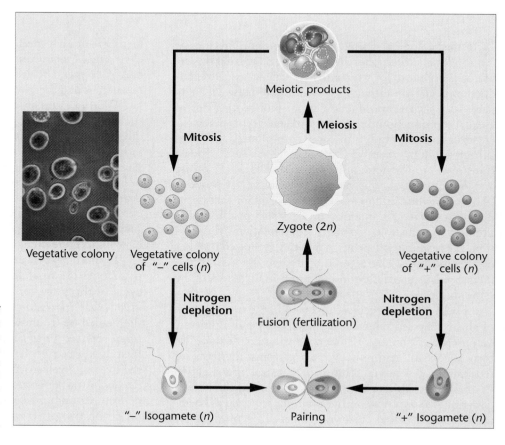

FIGURE 9.1 The life cycle of *Chlamydomonas.* Unfavorable conditions stimulate the formation of isogametes of opposite mating type that may fuse in fertilization. The resulting zygote undergoes meiosis, producing two haploid cells of each mating type. The photograph shows vegetative cells of this green alga.

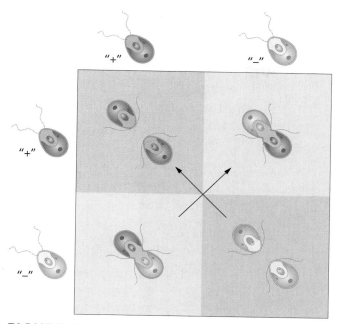

FIGURE 9.2   Illustration of mating types during fertilization in *Chlamydomonas*. Only when plus (+) and minus (−) cells are together will mating occur.

loid cells (**zoospores**) produced were found to consist of two plus types and two minus types.

Further experimentation established that there is a chemical difference between plus and minus cells. When extracts were prepared from cloned *Chlamydomonas* cells (or their flagella), and then added to cells of the opposite mating type, clumping or agglutination occurred. No such agglutination occurred if the extract were added to cells of the mating type from which it was derived. These observations suggest that despite the morphological similarities between isogametes, a chemical differentiation has occurred between them. Therefore, in this alga, a primitive means of sex differentiation exists even though there is no morphological indication that such differentiation has occurred.

Zea mays

In many plants, life cycles alternate between the haploid **gametophyte stage** and the diploid **sporophyte stage.** The processes of meiosis and fertilization link the two phases during the life cycle. The relative amount of time spent in the two phases varies between the major plant groups. In some nonseed plants, such as mosses, the haploid gametophyte phase and the morphological structures representing this stage predominate. The reverse is true in seed plants.

Maize (*Zea mays*) is an example of a monoecious seed plant that bears both male and female structures on a sin-

gle diploid **sporophyte.** Thus, sex determination must occur differently in different tissues of the same organism, as illustrated in the life cycle of this familiar plant (Figure 9.3). The **stamen,** or **tassels,** produce diploid microspore mother cells, each of which undergoes meiosis and gives rise to four haploid microspores. Each haploid microspore in turn develops into a mature male **microgametophyte**—the pollen grain—which contains two sperm nuclei.

Equivalent female diploid cells, known as megaspore mother cells, exist in the **pistil** of the sporophyte. Following meiosis, only one of the four haploid megaspores survives. It usually divides mitotically three times, producing a total of eight haploid nuclei enclosed in the embryo sac. Two of these nuclei unite near the center of the embryo sac, becoming the **endosperm nuclei.** At the micropyle end of the sac where the sperm enters, three nuclei remain: the **oocyte nucleus** and two **synergids.** The other three **antipodal nuclei** are clustered at the opposite end of the embryo sac.

Pollination occurs when pollen grains make contact with the silks (or stigma) of the pistil and develop extensive pollen tubes that grow toward the embryo sac. When contact is made at the micropyle, the two sperm nuclei enter the embryo sac. One sperm nucleus unites with the haploid oocyte nucleus and the other sperm nucleus unites with two endosperm nuclei. This process, known as **double fertilization,** results in the diploid zygote nucleus and the triploid endosperm nucleus, respectively. Each ear of corn may contain as many as 1000 of these structures, each of which develops into a single kernel. Each kernel, if allowed to germinate, will give rise to a new plant, the sporophyte.

The mechanism of sex determination and differentiation in a monoecious plant like *Zea mays*, where the tissues that form both male and female gametes are of the same genetic constitution, was difficult to comprehend at first. However, the discovery of a large number of mutant genes that disrupt normal tassel and pistil formation supports the concept that normal products of these genes play an important role in sex determination by affecting the differentiation of male or female tissue in several ways.

For example, mutant genes that cause sex reversal provide valuable information. When homozygous, all mutations classified as *tassel seed* (*ts*) interfere with tassel production and induce the formation of female structures. Thus, it is possible for a single gene to cause a normally monoecious plant to become functionally female. On the other hand, the recessive mutations *silkless* (*sk*) and *barren stalk* (*ba*) interfere with the development of the pistil, resulting in plants with only functional male reproductive organs.

Data gathered from studies of these and other mutants suggest that the products of many wild-type alleles of these genes interact in controlling sex determination. During development, certain cells are "determined" to become male or female structures. Following sexual differentia-

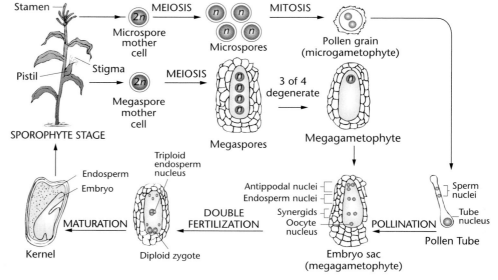

FIGURE 9.3 The life cycle of maize (*Zea mays*). The diploid sporophyte bears stamens and pistils that give rise to haploid microspores and megaspores, which develop into the pollen grain and the embryo sac that ultimately house the sperm and oocyte, respectively. Following fertilization, the embryo develops within the kernel and is nourished by the endosperm. Germination of the kernel gives rise to a new sporophyte (the mature corn plant), and the cycle repeats itself.

tion into either male or female structures, male or female gametes are produced.

C. elegans

The roundworm *C. elegans* [Figure 9.4(a)] has become a popular organism in genetic studies, particularly during the investigation of the genetic control of development. Its usefulness is based on the fact that the adult consists of only about 1000 cells, the precise lineage of which can be traced back to specific embryonic origins. Among many interesting mutant phenotypes that have been stud-

ied, behavioral modifications have also been a favorite topic of inquiry.

There are two sexual phenotypes in these worms: males, which have only testes, and hermaphrodites that contain both testes and ovaries. During larval development of hermaphrodites, testes form that produce sperm, which is stored. Ovaries are also produced, but oogenesis does not occur until the adult stage is reached several days later. The eggs that are then produced are fertilized by the stored sperm in the process of self-fertilization.

The outcome of this process is quite interesting [Figure 9.4(b)]. The vast majority of organisms that result, like

(a)

(b)

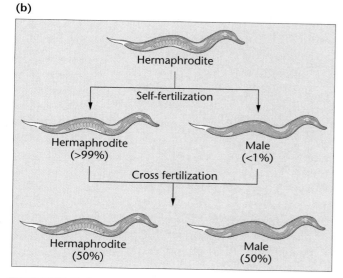

FIGURE 9.4 (a) Photomicrograph of an hermaphroditic nematode, *C. elegans*; (b) The outcomes of self-fertilization in an hermaphrodite and a mating of a hermaphrodite and a male worm.

the parental worm, are hermaphrodites; less than 1 percent of the offspring are males. As adults, they can mate with hermaphrodites, producing about half male and half hermaphrodite offspring.

The genetic signal that determines maleness in contrast to hermaphroditic development is provided by genes located on both the X chromosome and autosomes. *C. elegans* lacks a Y chromosome altogether. Hermaphrodites have two X chromosomes while males have only one X chromosome. It is believed that the ratio of X chromosomes to the number of sets of autosomes ultimately determines the sex of these worms. A ratio of 1.0 results in hermaphrodites and a ratio of 0.5 results in males. The absence of a heteromorphic Y chromosome is not uncommon in organisms.

X and Y chromosomes were first linked to sex determination early in the twentieth century

How sex is determined has long intrigued geneticists. In 1891, H. Henking identified a nuclear structure in the sperm of certain insects, which he labeled the X-body. Several years later, Clarance McClung showed that some

grasshopper sperm contain an unusual genetic structure, which he called a **heterochromosome**, but the remainder lack such a structure. He mistakenly associated the presence of the heterochromosome with the production of male progeny. In 1906, Edmund B. Wilson clarified the findings of Henking and McClung when he demonstrated that female somatic cells in the insect *Protenor* contain 14 chromosomes, including 2 X chromosomes. During oogenesis, an even reduction occurs, producing gametes with 7 chromosomes, including one X. Male somatic cells, on the other hand, contain only 13 chromosomes, including a single X chromosome. During spermatogenesis, gametes are produced containing either 6 chromosomes, without an X, or 7 chromosomes, one of which is an X. Fertilization by X-bearing sperm results in female offspring, and fertilization by X-deficient sperm result in male offspring [Figure 9.5(a)].

The presence or absence of the X chromosome in male gametes provides an efficient mechanism for sex determination in this species and also produces a 1:1 sex ratio in the resulting offspring. This mechanism, now called the **XX/XO** or **Protenor mode of sex determination**, depends on the random distribution of the X chromosome into one-half of the male gametes during segregation. As we saw above, *C. elegans* exhibits this system of sex determination.

Wilson also experimented with the hemipteran insect *Lygaeus turicus*, in which both sexes have 14 chromosomes. Twelve of these are autosomes. In addition, the females have 2 X chromosomes, while the males have only a single X and a smaller heterochromosome labeled the **Y chromosome**. Females in this species produce only gametes of the (6A + X) constitution, but males produce two types of gametes in equal proportions: (6A + X) and (6A + Y). Therefore, following random fertilization, equal numbers of male and female progeny will be produced with distinct chromosome complements. This mode of sex determination is called the **Lygaeus** or **XX/XY** type [Figure 9.5(b)].

The *Protenor* and *Lygaeus* examples presented above are ones where males produce unlike gametes. As a result, they are described as the **heterogametic sex,** and in effect, their gametes ultimately determine the sex of the progeny in those species. In such cases, the female, who has like sex chromosomes, is the **homogametic sex,** producing uniform gametes with regard to chromosome numbers and types.

The male is not always the heterogametic sex. In other organisms, the female produces unlike gametes, exhibiting either the *Protenor* (XX/XO) or *Lygaeus* (XX/XY) mode of sex determination. Examples include moths and butterflies, most birds, some fish, reptiles, amphibians, and at least one species of plants (*Fragaria orientalis*). To immediately distinguish situations in which the female is the

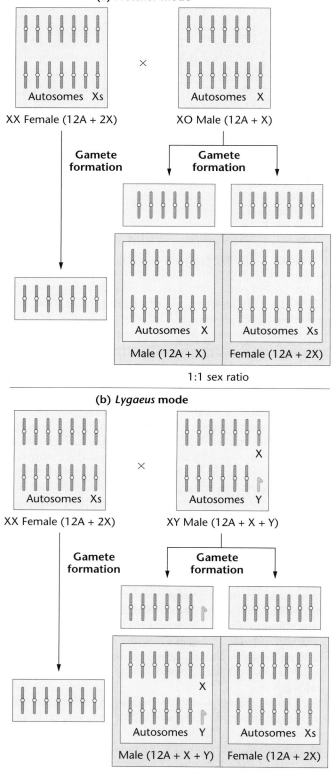

(a) Protenor mode

Autosomes Xs

Autosomes X

XX Female (12A + 2X)   XO Male (12A + X)

Gamete formation   Gamete formation

Autosomes X   Autosomes Xs

Male (12A + X)   Female (12A + 2X)

1:1 sex ratio

(b) Lygaeus mode

Autosomes Xs

Autosomes Y

XX Female (12A + 2X)   XY Male (12A + X + Y)

Gamete formation   Gamete formation

Autosomes Y   Autosomes Xs

Male (12A + X + Y)   Female (12A + 2X)

1:1 sex ratio

heterogametic sex, some workers use the notation **ZZ/ZW**, where ZW is the heterogamous female, instead of the XX/XY notation.

The situation with fowl (chickens) illustrates the difficulty in establishing which sex is heterogametic and whether the *Protenor* or *Lygaeus* mode is operable. While genetic evidence supported the hypothesis that the female is the heterogametic sex, the cytological identification of the sex chromosome was not accomplished until 1961 because of the large number of chromosomes (78) characteristic of chickens. When the sex chromosomes were finally identified, the female was shown to contain an unlike chromosome pair, including a heteromorphic chromosome (the W chromosome). Thus, in fowl, the female is indeed heterogametic and is characterized by the *Lygaeus* type of sex determination.

The Y chromosome determines maleness in humans

The first attempt to understand sex determination in our own species occurred almost 100 years ago and involved the examination of chromosomes present in dividing cells. Efforts were made to accurately determine the diploid chromosome number of humans, but because of the relatively large number of chromosomes, this proved to be quite difficult. In 1912, H. von Winiwarter counted 47 chromosomes in a spermatogonial metaphase preparation. It was believed that the sex-determining mechanism in humans was based on the presence of an extra chromosome in females, who were thought to have 48 chromosomes. However, in the 1920s, Theophilus Painter observed between 45 and 48 chromosomes in cells of testicular tissue and also discovered the small Y chromosome, which is now known to occur only in males. In his original paper, Painter favored 46 as the diploid number in humans, but he later concluded incorrectly that 48 was the chromosome number in both males and females.

For 30 years, this number was accepted. Then, in 1956, Joe Hin Tjio and Albert Levan discovered a better way to prepare chromosomes. This improved technique led to a strikingly clear demonstration of metaphase stages showing that 46 was indeed the human diploid number. Later that same year, C. E. Ford and John L. Hamerton,

FIGURE 9.5   (a) The *Protenor* mode of sex determination where the heterogametic sex (the male in this example) is XO and produces gametes with or without the X chromosome; (b) The *Lygaeus* mode of sex determination, where the heterogametic sex (again, the male in this example) is XY and produces gametes with either an X or a Y chromosome. In both cases, the chromosome composition of the offspring determines its sex.

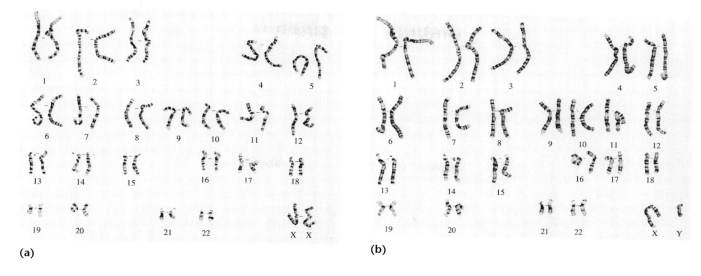

FIGURE 9.6 The traditional human karyotypes derived from a normal female and a normal male. Each contains 22 pairs of autosomes and two sex chromosomes. The female (a) contains two identical X chromosomes, while the male (b) contains one X and one Y chromosome (the Y is often referred to as a heterochromosome).

also working with testicular tissue, confirmed this finding. The familiar karyotype of humans (Figure 9.6) is based on Tjio and Levan's technique.

Within the normal 23 pairs of human chromosomes, one pair was shown to vary in configuration in males and females. These two chromosomes were designated the X and Y sex chromosomes. The human female has two X chromosomes, and the human male has one X and one Y chromosome.

We might believe that this observation is sufficient to conclude that the Y chromosome determines maleness. However, several other interpretations are possible. The Y could play no role in sex determination; the presence of two X chromosomes could cause femaleness; or maleness could result from the lack of a second X chromosome. The evidence that clarified which explanation was correct awaited the study of variations in the human sex chromosome composition. As such investigations revealed, the Y chromosome does indeed determine maleness in humans.

Klinefelter and Turner Syndromes

About 1940, scientists identified two human abnormalities characterized by aberrant sexual development, **Klinefelter syndrome** and **Turner syndrome.*** Individuals

with Klinefelter syndrome have genitalia and internal ducts that are usually male, but their testes are underdeveloped and fail to produce sperm.

Although some masculine development does occur, feminine sexual development is not entirely suppressed. Slight enlargement of the breasts is common, for example. This ambiguous sexual development, referred to as **intersexuality**, may lead to abnormal social development.

In Turner syndrome, the affected individual has female external genitalia and internal ducts, but the ovaries are rudimentary. Other characteristic abnormalities include short stature (usually under five feet), a webbed neck, and a broad, shieldlike chest.

In 1959, the karyotypes of individuals with these syndromes were determined to be abnormal with respect to the sex chromosomes. Individuals with Klinefelter syndrome have more than one X chromosome. Most often they have an XXY complement in addition to 44 autosomes [Figure 9.7(a)]. People with this karyotype are designated **47,XXY**. Individuals with Turner syndrome are most often monosomic and have only 45 chromosomes, including just a single X chromosome. They are designated **45,X** [Figure 9.7(b)]. Note the convention used in designating the above chromosome compositions. The number indicates the total number of chromosomes present, and the information after the comma designates the relevant deviation from the normal diploid content. Both conditions result from nondisjunction of the X chromosomes during meiosis (see Figures 2.17 and 10.1).

---

*Although the possessive form of the names of most syndromes (eponyms) is sometimes used (e.g., Klinefelter's), the current preference is to use the nonpossessive form, which we have adopted for all human syndromes.

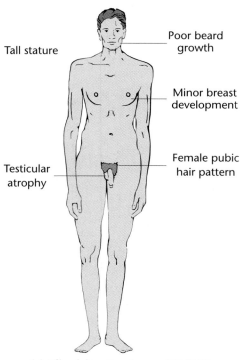

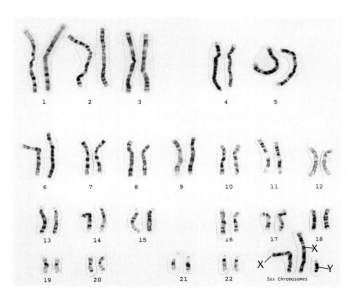

Tall stature

Poor beard growth

Minor breast development

Testicular atrophy

Female pubic hair pattern

(a) Klinefelter Syndrome (47,XXY)

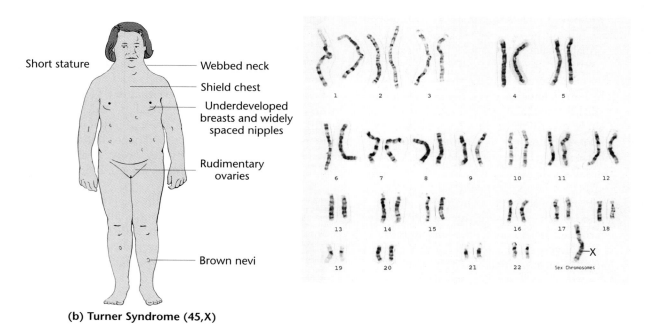

Short stature

Webbed neck

Shield chest

Underdeveloped breasts and widely spaced nipples

Rudimentary ovaries

Brown nevi

(b) Turner Syndrome (45,X)

FIGURE 9.7   The karyotypes and phenotypic depictions of individuals with (a) Klinefelter syndrome (47,XXY) and (b) Turner syndrome (45,X).

These Klinefelter and Turner karyotypes and their corresponding sexual phenotypes allow us to conclude that the Y chromosome determines maleness in humans. In its absence, the sex of the individual is female, even if only a single X chromosome is present. The presence of the Y chromosome in the individual with Klinefelter syndrome is sufficient to determine maleness, even though its expression is not complete. Similarly, in the absence of a Y chromosome, as in the case of individuals with Turner syndrome, no masculinization occurs.

Klinefelter syndrome occurs in about 2 of every 1000 male births. The karyotypes **48,XXXY, 48,XXYY, 49,XXXXY,** and **49,XXXYY** are similar phenotypically to **47,XXY,** but manifestations are often more severe in individuals with a greater number of X chromosomes.

Turner syndrome can also result from karyotypes other than 45,X, including individuals called **mosaics** whose somatic cells display two different genetic cell lines, each exhibiting a different karyotype. Such cell lines result from a mitotic error during early development, the most common chromosome combinations being **45,X/46,XY** and **45,X/46,XX.** Thus, an embryo that began life with a normal karyotype can give rise to an individual whose cells show a mixture of karyotypes and who expresses this syndrome.

Turner syndrome is observed in about 1 in 3000 female births, a frequency much lower than that for Klinefelter syndrome. One explanation for this difference is the observation that a substantial majority of 45,X fetuses die *in utero* and are aborted spontaneously. Thus, a similar frequency of the two syndromes may occur at conception.

47,XXX Syndrome

The presence of three X chromosomes along with a normal set of autosomes **(47,XXX)** results in female differentiation. This syndrome, which is estimated to occur in about 1 of 1200 female births, is highly variable in expression. Frequently, 47,XXX women are perfectly normal. In other cases, underdeveloped secondary sex characteristics, sterility, and mental retardation may occur. In rare instances, **48,XXXX** and **49,XXXXX** karyotypes have been reported. The syndromes associated with these karyotypes are similar to, but more pronounced than, the 47,XXX. Thus in many cases, the presence of additional X chromosomes appears to disrupt the delicate balance of genetic information essential to normal female development.

47,XYY Condition

Another human condition involving the sex chromosomes, **47,XYY,** has been discovered and intensively investigated. Studies of this condition, where the only deviation from diploidy is the presence of an additional Y chromosome in an otherwise normal male karyotype, have led to an interesting controversy.

In 1965, Patricia Jacobs discovered 9 of 315 males in a Scottish maximum security prison to have the 47,XYY karyotype. These males were significantly above average in height and had been involved in criminal acts of serious social consequence. Of the nine males studied, seven were of subnormal intelligence, and all suffered personality disorders. Several other studies produced similar findings.

The possible correlation between this chromosome composition and antisocial and criminal behavior was of considerable interest and led to extensive investigations of the phenotype and frequency of the 47,XYY condition in both criminal and noncriminal populations. Above-average height (usually over six feet tall) and subnormal intelligence have been generally substantiated, and the frequency of males displaying this karyotype is indeed higher in penal and mental institutions compared with unincarcerated males (see Table 9.1). A particularly relevant question involves the characteristics displayed by XYY males who are not incarcerated. The only nearly constant association is that such individuals are over 6 feet tall!

A study that addressed this issue was initiated to identify 47,XYY individuals at birth and to follow their be-

TABLE 9.1 Frequency of XYY Individuals in Various Settings

Setting	Restriction	Number Studied	Number XYY	Frequency XYY
Control population	Newborns	28,366	29	0.10%
Mental–penal	No height restriction	4,239	82	1.93
Penal	No height restriction	5,805	26	0.44
Mental	No height restriction	2,562	8	0.31
Mental-penal	Height restriction	1,048	48	4.61
Penal	Height restriction	1,683	31	1.84
Mental	Height restriction	649	9	1.38

Source: Compiled from data presented in Hook, 1973, Tables 1–8. Copyright 1973 by the American Association for the Advancement of Science.

offspring has been investigated and is referred to as the **sex ratio.** It can be assessed in two ways. The **primary sex ratio** reflects the proportion of males to females conceived in a population. The **secondary sex ratio** reflects the proportion of each sex that is born. The secondary sex ratio is much easier to determine, but has the disadvantage of not accounting for any disproportionate embryonic or fetal mortality.

When the secondary sex ratio in the human population was determined in 1969 using worldwide census data, it was found not to equal 1.0. For example, in the Caucasian population in the United States, the secondary ratio was a little less than 1.06, indicating that about 106 males are born for each 100 females. (In 1995, this ratio dropped to slightly less than 1.05). In the African-American population in the United States, the ratio was 1.025. In other countries the excess of male births is even greater than reflected in these values. For example, in Korea, the secondary sex ratio was 1.15.

Despite these ratios, it is possible that the *primary sex ratio* is 1.0, and it is altered between conception and birth. For the secondary ratio to exceed 1.0, then, prenatal female mortality would have to be greater than prenatal male mortality. However, this hypothesis has been examined and shown to be false. In fact, just the opposite occurs. In a Carnegie Institute study, reported in 1948, the sex of approximately 6000 embryos and fetuses recovered from miscarriages and abortions was determined, and fetal mortality was actually higher in males. On the basis of the data derived from that study, the primary sex ratio in U.S. Caucasians was estimated to be 1.079. More recent data have estimated that this figure is much higher—between 1.20 and 1.60, suggesting that many more males than females are conceived in the human population.

It is not clear why such a radical departure from the expected primary sex ratio of 1.0 occurs. To come up with a suitable explanation, we should examine the assumptions upon which the theoretical ratio is based:

1. Because of segregation, males produce equal numbers of X- and Y-bearing sperm.

2. Each type of sperm has equivalent viability and motility in the female reproductive tract.

3. The egg surface is equally receptive to both X- and Y-bearing sperm.

While there is no direct experimental evidence that contradicts any of these assumptions, the human Y chromosome is smaller than the X chromosome and therefore of less mass. Thus, it has been speculated that Y-bearing sperm are more motile than X-bearing sperm. If this is true, then the probability of a fertilization event leading to

a male zygote is increased, providing one possible explanation for the observed primary ratio.

Dosage compensation prevents excessive expression of X-lined genes in humans and other mammals

The presence of two X chromosomes in normal human females and only one X in normal human males is unique compared with the equal numbers of autosomes present in the cells of both sexes. On theoretical grounds alone, it is possible to speculate that this disparity should create a "genetic dosage" problem between males and females for all X-linked genes. Recall that in Chapter 4 we have previously discussed the topic of X-linkage, the inheritance of traits under the control of genes located on one of the sex chromosomes. There, we saw that during meiosis, sex chromosomes, like autosomes, are subject to the laws of segregation and independent assortment during their distribution into gametes. Since females have two copies of the X chromosome and males only one, there is the potential for females to produce twice as much of each gene product for all X-linked genes. The additional X chromosomes in both males and females exhibiting the various syndromes discussed earlier in this chapter should compound this dosage problem even more. In this section, we will describe certain research findings regarding X-linked gene expression that demonstrate a genetic mechanism allowing for **dosage compensation.**

Barr Bodies

Murray L. Barr and Ewart G. Bertram's experiments with female cats, and Keith Moore and Barr's subsequent study with humans, demonstrate a genetic mechanism in mammals that compensates for X chromosome dosage disparities. Barr and Bertram observed a darkly staining body in interphase nerve cells of female cats that was absent in similar cells of males. In humans this body can be easily demonstrated in female cells derived from the buccal mucosa or in fibroblasts but not in similar male cells (Figure 9.9). This highly condensed structure, about 1 μm in diameter, lies against the nuclear envelope of interphase cells. It stains positively in the Feulgen reaction for DNA.

Current experimental evidence demonstrates that this body, called a **sex chromatin body** or simply a **Barr body,** is an inactivated X chromosome. Susumo Ohno was the first to suggest that the Barr body arises from one of the two X chromosomes. This hypothesis is attractive because it provides a mechanism for dosage compensation. If one

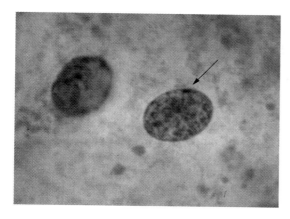

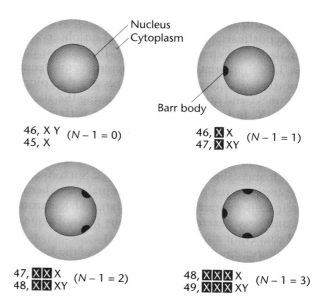

FIGURE 9.10 Diagrammatic representation of Barr body occurrence in various human karyotypes, where all X chromosomes except one (N − 1) are inactivated.

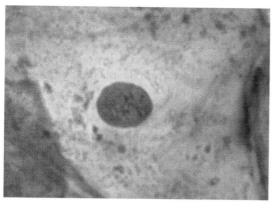

FIGURE 9.9 Photomicrographs comparing cheek epithelial cell nuclei from a male that fails to reveal Barr bodies (bottom) with a female that demonstrates Barr bodies (indicated by an arrow in top image). This structure, also called a sex chromatin body, represents an inactivated X chromosome.

of the two X chromosomes is inactive in the cells of females, the dosage of genetic information that can be expressed in males and females is equivalent. Convincing but indirect evidence for this hypothesis comes from the study of the sex chromosome syndromes described earlier in this chapter. Regardless of how many X chromosomes exist, all but one of them appear to be inactivated and can be seen as Barr bodies. For example, no Barr body is seen in Turner 45,X females; one is seen in Klinefelter 47,XXY males; two in 47,XXX females; three in 48,XXXX females; and so on (Figure 9.10). Therefore, the number of Barr bodies follows an N − 1 rule, where N is the total number of X chromosomes present.

Although this mechanism of inactivation of all but one X chromosome increases our understanding of dosage compensation, it further complicates our perception of other matters. Because one of the two X chromosomes is inactivated in normal human females, why then is the

Turner 45,X individual not entirely normal? Why aren't females with the triplo-X and tetra-X karyotypes (47,XXX and 48,XXXX) normal? Further, in Klinefelter syndrome (47,XXY), X chromosome inactivation effectively renders such individuals 46,XY. Why aren't they unaffected by the additional X chromosome in their nuclei?

One possible explanation is that chromosome inactivation does not normally occur in the very early stages of development of those cells destined to form gonadal tissues. Another possible explanation is that not all of each X chromosome forming a Barr body is inactivated. If either hypothesis is correct, excessive expression of certain X-linked genes might still occur despite apparent inactivation of additional X chromosomes.

The Lyon Hypothesis

In mammalian females one X chromosome is of maternal origin, and the other is of paternal origin. Which one is inactivated? Is the inactivation random? Is the same chromosome inactive in all somatic cells? In 1961, Mary Lyon and Liane Russell independently proposed a hypothesis that answers these questions. They postulated that the inactivation of X chromosomes occurs randomly in somatic cells at a point early in embryonic development and that once inactivation has occurred, all progeny cells have the same X chromosome inactivated.

Figure 2.17), **nondisjunction** is the failure of paired chromosomes to segregate or separate during the anaphase stage of the first or second meiotic divisions. The result is the production of two abnormal gametes, one of which contains an extra chromosome ($n + 1$) and the other which lacks a chromosome ($n - 1$). Fertilization of such gametes with a haploid gamete produces ($2n + 1$) or ($2n - 1$) zygotes. As in humans, if nondisjunction involves the X chromosome, in addition to the normal complement of autosomes, either an XXY or an X0 sex chromosome composition results. (The zero signifies that the second chromosome is absent.) Contrary to what was later discovered in humans, Bridges found that the XXY flies were normal females, and the X0 flies were sterile males. The presence of the Y chromosome in the XXY flies did not cause maleness, and its absence in the X0 flies did not produce femaleness. From these data he concluded that the Y chromosome in *Drosophila* lacks male-determining factors but, since the X0 males were sterile, it does contain genetic information essential to male fertility.

Bridges was able to clarify the mode of sex determination in *Drosophila* by studying the progeny of triploid females ($3n$), which have three copies each of the haploid complement of chromosomes. *Drosophila* has a haploid number of four, thereby displaying three pairs of autosomes in addition to its pair of sex chromosomes. Triploid females apparently originate from rare diploid eggs fertilized by normal haploid sperm. Triploid females have heavy-set bodies, coarse bristles, and coarse eyes, and they may be fertile. Because of the odd number of each chromosome (three), during meiosis a wide range of chromosome complements is distributed into gametes that give rise to offspring with a variety of abnormal chromosome constitutions. A correlation among the sexual morphology, chromosome composition, and Bridges' interpretation is shown in Figure 9.13.

Bridges realized that the critical factor in determining sex is the **ratio of X chromosomes to the number of haploid sets of autosomes (A) present.** Normal (2X:2A) and triploid (3X:3A) females each have a ratio equal to 1.0, and both are fertile. As the ratio exceeds unity (3X:2A, or 1.5, for example), what was originally called a superfemale is produced. Because this female is rather weak and infertile and has lowered viability, this type is now more appropriately called a **metafemale.**

Normal (XY:2A) and sterile (X0:2A) males each have a ratio of 1:2, or 0.5. When the ratio decreases to 1:3, or 0.33, as in the case of an XY:3A male, infertile **metamales** result. Other flies recovered by Bridges in these studies contained an X:A ratio intermediate between 0.5 and 1.0. These flies were generally larger, and they exhibited a variety of morphological abnormalities and rudimentary bisexual gonads and genitalia. They were invariably sterile

and expressed both male and female morphology, thus being designated as **intersexes.**

Bridges' results indicate that in *Drosophila*, factors that cause a fly to develop into a male are not localized on the sex chromosomes, but are instead found on the autosomes. Some female-determining factors, however, are localized on the X chromosomes. Thus, with respect to primary sex determination, male gametes containing one of each autosome plus a Y chromosome result in male offspring, not because of the presence of the Y but because of the lack of a second X chromosome. This mode of sex determination is explained by the **genic balance theory.** Bridges proposed that a threshold for maleness is reached when the X:A ratio is 1:2 (X:2A), but that the presence of an additional X (XX:2A) alters this balance and results in female differentiation.

Numerous mutant genes have been identified that are involved in sex determination in *Drosophila*. The recessive autosomal gene *transformer* (*tra*), discovered over 50 years ago by Alfred H. Sturtevant, clearly demonstrated that a single autosomal gene could have a profound impact on sex determination. Females homozygous for *tra* are transformed into sterile males, but homozygous males are unaffected.

More recently, another gene, *Sex-lethal* (*Sxl*), has been shown to play a critical role, serving as a "master switch" in sex determination. Activation of the X-linked *Sxl* gene, which relies on a ratio of X chromosomes to sets of autosomes that equals 1.0, is essential to female development. In the absence of activation, resulting, for example, from an X:A ratio of 0.5, male development occurs. It is interesting to note that mutations that inactivate the *Sxl* gene, as originally studied in 1960 by Hermann J. Muller, kill female embryos, but have no effect on male embryos, consistent with the role of the gene, as described above.

While it is not yet exactly clear how this ratio influences the *Sxl* locus, we do have some insights into the question. The *Sxl* locus is part of a hierarchy of gene expression and exerts control over still other genes, including *tra* (discussed above), and *dsx* (*doublesex*), as well as others. The wild-type allele of *tra* is activated by the product of *Sxl* only in females, which in turn influences the expression of *dsx*. Depending on how the initial RNA transcript of *dsx* is processed (spliced), the resultant dsx protein activates either male- or female-specific genes required for sexual differentiation. Each step in this regulatory cascade requires a form of processing called **RNA splicing,** in which portions of the RNA are removed and the remaining fragments "spliced" back together prior to translation into a protein. In the case of the *Sxl* gene, its transcript may be spliced in several different ways, a phenomenon called **alternative splicing.** Two different RNA transcripts are produced in females and males, respectively. In potential females, the transcript is active and initiates a cascade of regulatory gene expression, ultimately leading to female differentiation. In

Chromosome composition	Chromosome formulation	Ratio of X chromosomes to autosome sets	Sexual morphology
	3X/2A	1.5	Metafemale
	3X/3A	1.0	Female
	2X/2A	1.0	Female
	3X/4A	0.75	Intersex
	2X/3A	0.67	Intersex
	X/2A	0.50	Male
	XY/2A	0.50	Male
	XY/3A	0.33	Metamale

Normal diploid male

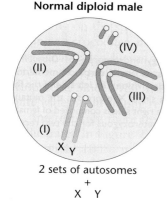

(II) (IV) (III) (I)

X Y

2 sets of autosomes
+
X   Y

FIGURE 9.13   Chromosome compositions, the ratios of X chromosomes to sets of autosomes, and the resultant sexual morphology in *Drosophila melanogaster*. The normal diploid male chromosome composition is shown as a reference on the left (XY/2A).

potential males, the transcript is inactive, leading to different gene activity, whereby male differentiation occurs. We will return to this topic in Chapter 16, when alternative splicing is again addressed as one of the mechanisms involved in the regulation of genetic expression in eukaryotes.

Dosage Compensation in Drosophila

Since *Drosophila* females contain two copies of X-linked genes whereas males contain only one copy, a dosage problem exists, as it does in mammals such as humans and mice. However, the mechanism of dosage compensation in *Drosophila* differs considerably from that in

mammals, since X chromosome inactivation is not observed. Instead, male X-linked genes are transcribed at twice the level of the comparable genes in females. Interestingly, if groups of X-linked genes are moved (translocated) to autosomes, dosage compensation also affects them even when they are no longer part of the X chromosome.

As in mammals, considerable gains have been made recently in understanding the process of dosage compensation in *Drosophila*. At least four autosomal genes are known to be involved, under the same master-switch gene, *Sxl*, that induces female differentiation during sex determination. Mutations in any of these genes severely reduce

CHAPTER SUMMARY

1. In sexually reproducing organisms, meiosis, which both creates genetic variability and ensures genetic constancy, depends on fertilization. Fertilization ultimately relies on some form of sexual differentiation, which is achieved by a variety of sex-determining mechanisms.

2. The genetic basis of sexual differentiation is often related to different chromosome compositions in the two sexes. The heterogametic sex either lacks one chromosome or contains a unique heteromorphic chromosome, usually referred to as the Y chromosome.

3. In humans, the study of individuals with altered sex chromosome compositions has established that the Y chromosome is responsible for male differentiation. The absence of the Y leads to female differentiation. Similar studies in *Drosophila* have excluded the Y in such a role, instead demonstrating that a balance between the number of X chromosomes and sets of autosomes is the critical factor.

4. The primary sex ratio in humans substantially favors males at conception. During embryonic and fetal development, male mortality is higher than that of females. The secondary sex ratio at birth still favors males by a small margin.

5. Dosage compensation mechanisms limit the expression of X-linked genes in females, who have two X chromosomes, as compared to males who have only one X. In mammals, compensation is achieved by the inactivation of either the maternal or paternal X early in development. This process results in the formation of Barr bodies in female somatic cells. In *Drosophila*, compensation is achieved by stimulation of genes located on the single X chromosome to double their genetic expression.

6. The Lyon hypothesis states that, early in development, inactivation is random between the maternal and paternal X. All subsequent progeny cells inactivate the same X as their progenitor cell. Mammalian females thus develop as genetic mosaics with respect to their expression of heterozygous X-linked alleles.

7. In many reptiles, the incubation temperature at a critical time during embryogenesis is often responsible for sex determination. Temperature influences the activity of enzymes involved in the metabolism of steroids related to sexual differentiation.

INSIGHTS AND SOLUTIONS

1. In *Drosophila*, the X chromosomes may become attached to one another ($\widehat{XX}$) such that they always segregate together. Some flies contain both an attached X chromosome and a Y chromosome.

   (a) What sex would such a fly be? Explain why this is so.

   Solution: The fly would be a female. The ratio of X chromosomes to sets of autosomes would be 1.0, leading to normal female development. The Y chromosome has no influence on sex determination in *Drosophila*.

   (b) Given this answer, predict the sex of the offspring that would occur in a cross between this fly and a normal one of the opposite sex.

   Solution: All flies would have two sets of autosomes, but each would have one of the following sex chromosome compositions:

   (1) $\widehat{XX}X \rightarrow$ a metafemale with 3 Xs (called a trisomic)

   (2) $\widehat{XX}Y \rightarrow$ a female like her mother

   (3) X Y $\rightarrow$ a normal male

   (4) Y Y $\rightarrow$ no development occurs

   (c) If the above offspring were allowed to interbreed, what would be the outcome?

   Solution: A true-breeding stock would be created that would maintain the attached X females generation after generation.

2. The X_g cell-surface antigen is coded for by a gene that is located on the X chromosome. There is no equivalent gene on the Y chromosome. Two codominant alleles of this gene have been identified: *Xg1* and *Xg2*. A woman of genotype *Xg2/Xg2* marries a man of genotype *Xg1/Y* and they produce a son with Klinefelter syndrome of genotype *Xg1/Xg2/Y*. Using proper genetic terminology, briefly explain how this individual was generated. In which parent and in which meiotic division did the mistake occur?

   Solution: Because the son with Klinefelter syndromome is *Xg1 Xg2/Y*, he must have received both the *Xg1* allele and the Y chromosome from his father. Therefore, nondisjunction must have occurred during meiosis I in the father.

PROBLEMS AND DISCUSSION QUESTIONS

1. Define the terms heteromorphic and heterogamy as they are related to sex determination.
2. Contrast the *Protenor* and *Lygaeus* modes of sex determination.
3. Contrast the evidence explaining the different modes of sex determination in *Drosophila* and humans.
4. Devise a method of nondisjunction in human female gametes that would give rise to Klinefelter and Turner syndrome offspring following fertilization by a normal male gamete.
5. An insect species is discovered in which the heterogametic sex is unknown. An X-linked recessive mutation for *reduced wing* (*rw*) is discovered. Contrast the F_1 and F_2 generations from a cross between a female with reduced wings and a male with normal-sized wings when:
   (a) the female is the heterogametic sex.
   (b) the male is the heterogametic sex.
   (c) Is it possible to distinguish between the *Protenor* and *Lygaeus* mode of sex determination based on the outcome of these crosses?
6. When cows have twin calves of unlike sex (fraternal twins), the female twin is usually sterile and has masculinized reproductive organs. This calf is referred to as a freemartin. In cows, twins may share a common placenta and thus fetal circulation. Predict why a freemartin develops.
7. An attached X female fly, as described in the Insights and Solutions section of this chapter, expresses the recessive X-linked *white* eye phenotype. It is crossed to a male fly that expresses the X-linked recessive miniature wing phenotype. Determine the outcome of this cross regarding the sex, eye color, and wing size of the offspring.
8. It has been suggested that any male-determining genes contained on the Y chromosome in humans cannot be located in the limited region that synapses with the X chromosome during meiosis. What might be the outcome if such genes were located in this region?
9. What is a Barr body?
10. Indicate the expected number of Barr bodies in interphase cells of the following individuals: Klinefelter syndrome; Turner syndrome; and karyotypes 47,XYY, 47,XXX, and 48,XXXX.
11. Define the Lyon hypothesis.
12. Relate the potential effect of the Lyon hypothesis on the retina of a human female heterozygous for the X-linked red-green color-blindness trait.
13. Cat breeders are aware that kittens expressing the X-linked calico coat pattern are almost invariably females. Why?
14. What does the apparent need for dosage compensation mechanisms suggest about the expression of genetic information in normal diploid individuals?
15. The marine echiurid worm *Bonellia viridis* is an extreme example of the environment's influence on sex determination. Undifferentiated larvae either remain free-swimming and differentiate into females, or they settle on the proboscis of an adult female and become males. If larvae that have been on a female proboscis for a short period are removed and placed in seawater, they develop as intersexes. If larvae are forced to develop in an aquarium where pieces of proboscises have been placed, they develop into males. Contrast this mode of sexual differentiation with that of mammals. Suggest further experimentation to elucidate the mechanism of sex determination in *Bonellia*.
16. Discuss the possible reasons why the primary sex ratio in humans is as high as 1.40 to 1.60.
17. In mice, the *Sry* gene (described in this chapter) is located on the Y chromosome very close to the pseudoautosomal region that pairs with the X chromosome during male meiosis. Given this information, propose a model to explain the generation of unusual males who have two X chromosomes (with an *Sry*-containing piece of the Y attached to one X chromosome).

EXTRA-SPICY PROBLEMS

18. The genes encoding the red and green color-detecting proteins of the human eye are located next to one another on the X chromosome and probably arose during evolution from a common ancestral pigment gene. The two proteins demonstrate 96 percent homology in their amino acid sequences. A normal woman with one copy of each gene on each of her two X chromosomes has a red–color-blind son who was shown to contain one copy of the green gene and no copies of the red gene. Devise an explanation at the chromosomal level (during meiosis) that explains these observations.
19. The X-linked dominant mutation in the mouse, *Testicular feminization* (*Tfm*), eliminates the normal response to the testicular hormone testosterone during sexual differentiation. An XY animal bearing the *Tfm* allele on the X chromosome develops testes, but no further male differentiation occurs. The external genitalia of such an animal are female. From this information, what might you conclude about the role of the *Tfm* gene product and the X and Y chromosomes in sex determination and sexual differentiation in mammals? Can you devise an experiment, assuming you can "genetically engineer" the chromosomes of mice, to test and confirm your explanation?
20. Campomelic dysplasia (CMD1) is a congenital human syndrome, featuring malformation of bone and cartilage. It is caused by an autosomal dominant mutation of a gene located on chromosome 17. Consider the following observations in sequence, and in each case, draw whatever appropriate conclusions are warranted.
    (a) Of those with the syndrome who are karyotypically 46,XY, approximately 75 percent are sex-reversed, exhibiting a wide range of female characteristics.
    (b) The nonmutant form of the gene, called *SOX9* is expressed in the developing gonad of the XY male, but not the XX female.
    (c) The *SOX9* gene shares 71 percent amino acid coding-sequence homology with the Y-linked *SRY* gene (see p. 237).

GENETICS MediaLab

The following resources will help you achieve a better understanding of the concepts presented in this chapter. These resources can be found on the CD packaged with this textbook and on the Companion Website at **http://www.prenhall.com/klug/**.

CD Resources:
   Self-grading Chapter Problems

Web Resources:
   Web Destinations in Genetics

   Self-grading Chapter Problems

   Chapter Search Terms

   Genetics Newsgroups

   Student Bulletin Board

Web Problem 1:
Time for completion = 15 minutes
Why is the Y chromosome so small? In many cases of genetic sex determination, females are XX and males are XY. As you know from our discussion of sex linkage in Chapter 4, the X chromosome carries a number of loci; however, the Y carries only a few loci in humans. The difference in size has important consequences for genetics, as explored in the linked website article by Ken Miller. Read the article and then think about the following questions. In humans, is the Y chromosome necessary for male development? Why are errors on the X chromosome repaired through sexual reproduction but errors on the Y are not? How does this relate to the size of the chromosomes? Bill Rice set up an experiment in which specific autosomes were not allowed to recombine. Did his results support the idea that mistakes accumulating on the Y might be responsible for its reduction in size? To complete this exercise, visit Web Problem 1 in Chapter 9 of your Companion Website and select the keyword **Y CHROMOSOME.**

Web Problem 2:
Time for completion = 10 minutes
Human females have twice as many copies of the genes on the X chromosome as do males; however, it is unlikely that they need twice as many of the proteins encoded by those genes. Dosage compensation refers to the mechanism by which the transcription of the genetic material on the X chromosome is altered in one or the other sex. Read the article about dosage compensation in female mammals and the related material in Chapter 9. What did Mary Lyon's hypothesis regarding dosage compensation in female mammals predict about the genetic constitution of male vs. female cells? Is X inactivation random in all mammals? Is there evidence that X inactivation is not complete? Are Barr bodies ever found in the cells of males? Is dosage compensation implemented in female *Drosophila* as it is in female mammals? Explain your answer. To complete this exercise, visit Web Problem 2 in Chapter 9 of your Companion Website and select the keyword **DOSAGE COMPENSATION.**

Web Problem 3:
Time for completion = 10 minutes
The Y chromosome appears to be necessary for the development of "maleness" in mammals, but is it also sufficient? Recent research suggests that some autosomal genes are necessary for complete male development. Read the linked website article on sex determination in mice and use the information there and in Chapter 9 to answer the following questions. Dr. Eicher and her colleagues observed an excess of female mice in their colony; what was the genotype of the excess females? Does the *Sry* gene on the Y chromosome appear to regulate the expression of any autosomal genes? What are some hypotheses for the function of those genes? Why would information about sex determination in mice be important for understanding human sex-reversal disorders? To complete this exercise, visit Web Problem 3 in Chapter 9 of your Companion Website and select the keyword **GENDER GENES.**

(d) CMD1 patients who exhibit a 46,XX karyotype develop as females, with no gonadal abnormalities.

(e) The product of the *SRY* gene is believed to be a transcription factor, which by definition (see Chapter 16) activates the expression of other genes. Can you hypothesize how *SRY* and *SOX9* might be related?

21. In the wasp, *Bracon hebetor*, a form of parthenogenesis (where unfertilized eggs initiate development) resulting in haploid organisms is not uncommon. All haploids are males. When offspring arise from fertilization, females almost invariably result. P. W. Whiting has shown that an X-linked gene with 9 multiple alleles (X_a, X_b, etc.) controls sex determination. Any homozygous or hemizygous condition results in males and any heterozygous condition results in females. If an X_a/X_b female mates with an X_a male and lays 50 percent fertilized and 50 percent unfertilized eggs, what proportion of male and female offspring will result?

SELECTED READINGS

BARR, M.L. 1966. The significance of sex chromatin. *Int. Rev. Cytol.* 19:35–39.

BURGOYNE, P.S. 1998. The mammalian Y chromosome: a new perspective. *Bioessays* 20: 363–66.

CARREL, L., and WILLARD, H.F. 1998. Counting on Xist. *Nature Genetics* 19: 211–12.

COURT-BROWN, W.M. 1968. Males with an XYY sex chromosome complement. *J. Med. Genet.* 5:341–59.

DAVIDSON, R., NITOWSKI, H., and CHILDS, B. 1963. Demonstration of two populations of cells in human females heterozygous for glucose-6-phosphate dehydrogenase variants. *Proc. Natl. Acad. Sci. USA* 50:481–85.

DELLAPORTA, S.L., and CALDERON-URREA, A. 1994. The sex determination process in maize. *Science* 266: 1501–05.

ERICKSON, J.D. 1976. The secondary sex ratio of the United States, 1969–71: Association with race, parental ages, birth order, paternal education and legitimacy. *Ann. Hum. Genet.* (London) 40:205–12.

FREIJE, D., et al. 1992. Identification of a second pseudoautosomal region near the Xq and Yq telomeres. *Science* 258: 1784–87.

GORMAN, M., KURODA, M., and BAKER, B.S. 1993. Regulation of sex-specific binding of maleness dosage compensation protein to the male X chromosome in *Drosophila*. Cell 72:39–49.

HASELTINE, F.P., and OHNO, S. 1981. Mechanisms of gonadal differentiation. *Science* 211:1272–78.

HODGKIN, J. 1990. Sex determination compared in *Drosophila* and *Caenorhabditis*. *Nature* 344:721–28.

HOOK, E.B. 1973. Behavioral implications of the humans XYY genotype. *Science* 179:139–50.

IRISH, E.E. 1996. Regulation of sex determination in maize. *BioEssays* 18:363–69.

JACOBS, P.A., et al. 1974. A cytogenetic survey of 11,680 newborn infants. *Ann. Hum. Genet.* 37:359–76.

KOOPMAN, P., et al. 1991. Male development of chromosomally female mice transgenic for *Sry*. *Nature* 351:117–21.

LAHN, B.T., and PAGE, D.C. 1997. Functional coherence of the human Y chromosome. *Science* 278:675–80.

LUCCHESI, J. 1983. The relationship between gene dosage, gene expression, and sex in *Drosophila*. *Dev. Genet.* 3:275–82.

LYON, M.F. 1961. Gene action in the X-chromosome of the mouse (*Mus musculus L.*). *Nature* 190:372–73.

———. 1972. X-chromosome inactivation and developmental patterns in mammals. *Biol. Rev.* 47:1–35.

———. 1988. X-chromosome inactivation and the location and expression of X-linked genes. *Am. J. Hum. Genet.* 42:8–16.

———. 1998. X-Chromosome inactivation spreads itself: effects in autosomes. *Am. J. Hum. Genet.* 63: 17–19.

MARIN, I., and BAKER, S.S. 1998. The evolutionary dynamics of sex determination. *Science* 281: 1990–94.

MARSHALL GRAVES, J.A. 1998. Interaction between *SRY* and *SOX* genes in mammalian sex determination. *BioEssays* 20:264–69.

MCMILLEN, M.M. 1979. Differential mortality by sex in fetal and neonatal deaths. *Science* 204:89–91.

PAGE, D.C., et al. 1987. The sex-determining region of the human Y chromosome encodes a finger protein. *Cell* 51:1091–1104.

PENNY, G.D., et al. 1996. Requirement for Xist in X chromosome inactivation. *Nature* 379:131–37.

PIEAU, C. 1996. Temperature variation and sex determination in reptiles. *BioEssays* 18:19–26.

REDDY, K.S., and SULCOVA, V. 1998. Pathogenetics of 45,X/46,XY gonadal mosaicism. *Cytogenet. Cell Genet.* 82:52–57.

SCHAFER, A.J. 1996. Sex determination in humans. *BioEssays* 18:955–63.

VAINIO. S., et al. 1999. Female Development in mammals is regulated by Wnt-4 signalling. *Nature* 397:405–09.

WESTERGAARD, M. 1958. The mechanism of sex determination in dioecious flowering plants. *Adv. Genet.* 9:217–81.

WHITING, P.W. 1939. Multiple alleles in sex determination in *Habrobracon*. *J. Morphology* 66:323–55.

WITKIN, H.A., et al. 1976. Criminality in XYY and XXY men. *Science* 193:547–55.

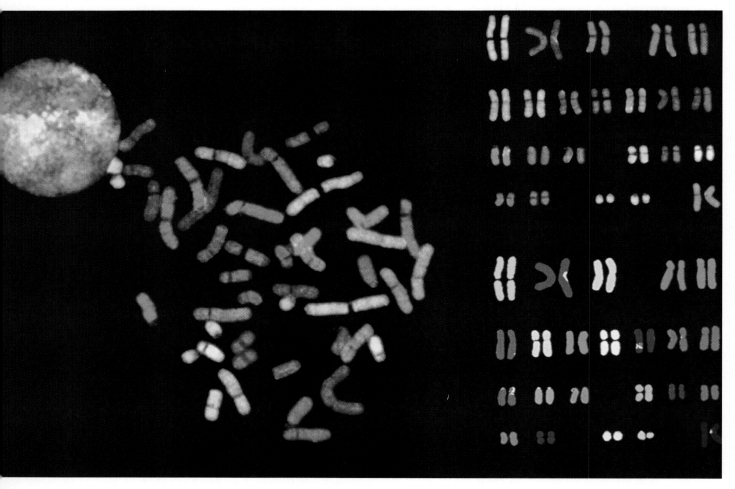

Spectral karyotyping of human chromosomes utilizing differentially labeled "painting" probes.

Chromosome Mutations: Variation in Chromosome Number and Arrangement

KEY CONCEPTS

- **Specific terminology describes variations in chromosome number**

- **Variation in the number of chromosomes results from nondisjunction**

- **Monosomy, the loss of a single chromosome, may have severe phenotypic effects**

  Partial Monosomy in Humans: The Cri-du-Chat Syndrome

- **Trisomy involves the addition of a chromosome to a diploid genome**

  Down Syndrome
  Patau Syndrome
  Edwards Syndrome
  Viability in Human Aneuploidy

- **Polyploidy, in which more than two haploid sets of chromosomes are present, is prevalent in plants**

  Autopolyploidy
  Allopolyploidy
  Endopolyploidy

- **Variation occurs in the structure and arrangement of chromosomes**

- **A deletion is a lost portion of a chromosome**

- **A duplication is a repeated part of the genetic material**

  Gene Redundancy and Amplification: Ribosomal RNA Genes
  The *Bar* Eye Mutation in *Drosophila*
  The Role of Gene Duplication in Evolution

- **Inversions rearrange the linear gene sequence**

  Consequences of Inversions during Gamete Formation
  Position Effects of Inversions
  Evolutionary Advantages of Inversions

- **Translocations alter the location of chromosomal segments in the genome**

  Translocations in Humans: Familial Down Syndrome

- **Fragile sites in humans are susceptible to chromosome breakage**

  Fragile X Syndrome (Martin-Bell Syndrome)
  Fragile Sites and Cancer

GenCDX

When you see this icon, there are related animations and exercises on the CD accompanying this text.

W e have previously emphasized how mutations and the resulting alleles affect an organism's phenotype and how traits are passed from parents to offspring according to Mendelian principles. In this chapter, we shall look at phenotypic variation that occurs as a result of changes that are more substantial than alterations of individual genes—modifications at the level of the chromosome.

Although most members of diploid species normally contain precisely two haploid chromosome sets, there are many known cases of variations from this pattern. Modifications include a change in the total number of chromosomes, the deletion or duplication of genes or segments of a chromosome, and rearrangements of the genetic material either within or among chromosomes. Taken together, such changes are called **chromosome mutations** or **chromosome aberrations,** to distinguish them from gene mutations. Because, according to Mendelian laws, the chromosome is the unit of genetic transmission, chromosome aberrations are passed on to offspring in a predictable manner, resulting in many heritable phenotypic variations.

The genetic component of an organism is delicately balanced. Even minor alterations of either the content or location of genetic information within the genome may result in some form of phenotypic variation. More substantial changes may be lethal, particularly in animals. Throughout the chapter, we shall consider the many types of chromosomal aberrations, the phenotypic consequences for the organism that harbors an aberration, and the impact of the aberration on offspring of the affected individual. We will also discuss the role of chromosome aberrations in the evolutionary process.

Specific terminology describes variations in chromosome number

Variation in chromosome number ranges from the addition or loss of one or more chromosomes to the addition of one or more haploid sets of chromosomes. Before we embark on our discussion, it is useful to establish the terminology that describes such changes. In the condition known as **aneuploidy,** an organism gains or loses one or more chromosomes, but not a complete set. The loss of a single chromosome from an otherwise diploid genome is called monosomy. The gain of one chromosome results in trisomy. These are contrasted with the condition of **euploidy,** where complete haploid sets of chromosomes are present. If there are three or more sets, the general term polyploidy is applicable. Organisms with three sets are specifically triploid; those with four sets are tetraploid, and so on. Table 10.1 provides an organizational framework for you to follow as we discuss each of these categories of aneuploid and euploid variation and the subsets within them.

Variation in the number of chromosomes results from nondisjunction

It is useful, as we consider cases that include the gain or loss of chromosomes, to discuss how such aberrations originate. For instance, how do the syndromes arise where the number of sex-determining chromosomes in humans is altered, as first described in Chapter 9? As you may recall, in spite of a mechanism in somatic cells that inactivates all X chromosomes in excess of one, the gain (47,XXY), or the loss (45,X) of a sex-determining chromosome from an otherwise diploid genome alters the normal phenotype, resulting in **Klinefelter syndrome** and **Turner syndrome,** respectively (see Figure 9.7). Human females are also known who have extra X chromosomes (e.g., 47,XXX, 48,XXXX), and males may have an extra Y chromosome (47,XYY).

Such chromosomal variation originates as the result of an error during meiosis. As first introduced in Chapter 2, **nondisjunction,** the failure of paired homologs to disjoin during segregation, is a process where normal distribution of chromosomes into gametes is disrupted. Figure 10.1 illustrates nondisjunction of the X chromosome during human oogenesis. As shown, nondisjunction may occur during either the first or second meiotic division. Provided that the remainder of meiosis occurs normally, gametes are produced with either zero, one, or two X chromosomes. Following fertilization by a normal Y-bearing or X-bearing sperm, zygotes are produced that are either lethal, normal, or that express Klinefelter or Turner syndrome. You may wish to review the various sex chromosome aberrations discussed in greater detail in Chapter 9. Other chromosome nondisjunctions lead to a variety of autosomal aneuploid conditons in humans and other organisms.

Monosomy, the loss of a single chromosome, may have severe phenotypic effects

We turn now to a consideration of variations in the number of autosomes and the genetic consequence of such changes. The most common examples of aneuploidy, where an organism has a chromosome number other than an exact multiple of the haploid set, are cases in which a single chromosome is either added to or lost from a normal diploid set. The loss of one chromosome produces a $2n - 1$ complement and is called **monosomy.**

Although monosomy for the X chromosome occurs in humans, as we have seen in 45,X Turner syndrome, monosomy for any of the autosomes is not usually tolerated in humans or other animals. In *Drosophila*, flies monosomic for the very small chromosome 4—a condition referred to as **Haplo-IV**—survive, but they develop more slowly, exhibit a reduced body size, and have impaired viability. Monosomy for the larger chromosomes 2 and 3 is apparently lethal because such flies have never been recovered.

The failure of monosomic individuals to survive in many animal species is at first quite puzzling, since at least a single copy of every gene is present in the remaining homolog. However, if an organism is heterozygous for just one recessive lethal allele, and loses the homologous chromosome bearing the normal allele (which prevents lethality), the unpaired chromosome condition leads to the death of the organism. Thus, recessive lethals that are tolerated in heterozygotes may be "unmasked" as a result of monosomy. Another possible explanation is that the expression of genetic information during early development is carefully regulated such that a delicate equilibrium of gene products is required to ensure normal development. While this is thought to be true during animal development, such a requirement does not appear to be so stringent in the plant kingdom, where aneuploidy is tolerated. Monosomy for autosomal chromosomes has been observed in maize, tobac-

TABLE 10.1 Terminology for Variation in Chromosome Numbers

Term	Explanation
Aneuploidy	$2n$ plus or minus chromosomes
Monosomy	$2n - 1$
Trisomy	$2n + 1$
Tetrasomy, pentasomy, etc.	$2n + 2, 2n + 3$, etc.
Euploidy	Multiples of n
Diploid	$2n$
Polyploidy	$3n, 4n, 5n, \ldots$
Triploid	$3n$
Tetraploidy, pentaploidy, etc.	$4n, 5n$, etc.
Autopolyploidy	Multiples of the same genome
Allopolyploidy (Amphidiploidy)	Multiples of different genomes

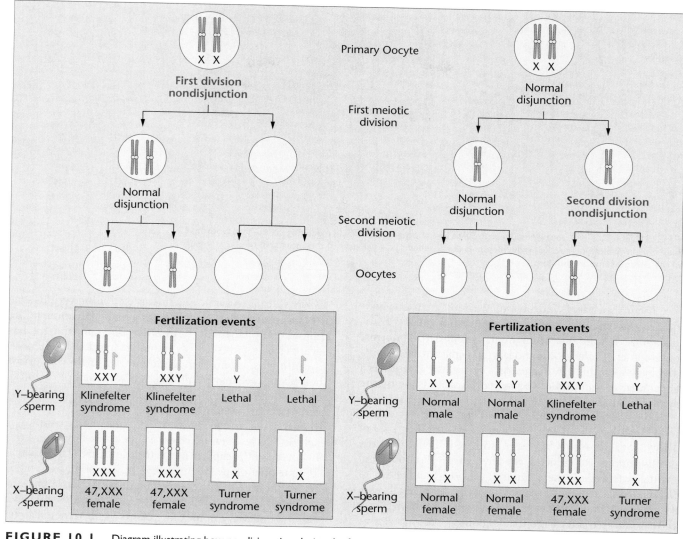

FIGURE 10.1   Diagram illustrating how nondisjunction during the first or second meiotic division of human oogenesis can lead to the various X and Y chromosome compositions. Oocytes are produced that contain zero, one, or two X chromosomes. Following fertilization by a Y-bearing or X-bearing sperm, zygotes are produced that are either lethal, normal, or that express Klinefelter or Turner syndrome.

co, the evening primrose *Oenothera*, and the jimsonweed *Datura*, among other plants. Nevertheless, such monosomic plants are almost always less viable than their diploid derivatives. Since both pollen grains and ovules must undergo extensive development after undergoing meiosis, but before participating in fertilization, they are particularly sensitive to the lack of one chromosome and are often inviable.

Partial Monosomy in Humans: The Cri-du-Chat Syndrome

In humans, autosomal monosomy has not been reported beyond birth. Individuals with such chromosome com-

plements are undoubtedly conceived, but none apparently survive embryonic and fetal development. There are, however, examples of survivors with **partial monosomy,** where only part of one chromosome is lost. These cases are also referred to as **segmental deletions.** One such case was first reported by Jerome LeJeune in 1963 when he described the clinical symptoms of the **cri-du-chat syndrome** ("cry of the cat"). This syndrome is associated with the loss of about one-half of the short arm of chromosome 5 (Figure 10.2). Thus, the genetic constitution may be designated as **46,−5p,** meaning that such an individual has all 46 chromosomes but that some or all of the p arm (the petite or short arm) of chromosome 5 is missing.

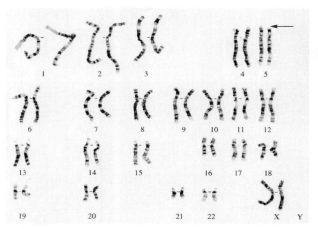

FIGURE 10.2 A representative karyotype and photograph of a child exhibiting cri-du-chat syndrome (46,−5p). In the karyotype, the arrow identifies the absence of a small part of the short arm of one member of the chromosome 5 homologs.

Infants with this syndrome may exhibit anatomic malformations, including gastrointestinal and cardiac complications, and they are often mentally retarded. Abnormal development of the glottis and larynx is characteristic of individuals with this syndrome. As a result, the infant has a cry similar to that of the meowing of a cat, thus giving the syndrome its name.

Since 1963, hundreds of cases of cri-du-chat syndrome individuals have been reported worldwide. An incidence of 1 in 50,000 live births has been estimated. The size of the deletion appears to influence the physical, psychomotor, and mental skill levels of these children. Although the effects of the syndrome are severe, many individuals achieve a level of social development in the trainable range. Those who receive home care and early special schooling are ambulatory, develop self-care skills, and learn to communicate verbally.

Trisomy involves the addition of a chromosome to a diploid genome

In general, the effects of trisomy ($2n + 1$) parallel those of monosomy. However, the addition of an extra chromosome produces somewhat more viable individuals in both animal and plant species than does the loss of a chromosome.

As in monosomy, the sex chromosome variation of the trisomic type has a less dramatic effect on the phenotype than does autosomal variation. Recall from our previous discussion that *Drosophila* females with three X chromosomes and a normal complement of two sets of autosomes (3X:2A) survive and reproduce, but are less viable than normal 2X:2A females. In humans, the addition of an extra X or Y chromosome to an otherwise normal male or female chromosome constitution (47,XXY, 47,XYY, and

47,XXX) leads to viable individuals exhibiting various syndromes. However, the addition of a large autosome to the diploid complement in both *Drosophila* and humans has severe effects and is usually lethal during development.

In plants, trisomic individuals are usually viable, but their phenotype may be altered. A classic example involves the jimsonweed *Datura*, a plant long known for its narcotic effect, whose diploid number is 24. Twelve different primary trisomic conditions are possible, and examples of each one have been recovered. Each trisomy alters the phenotype of the capsule of the fruit sufficiently to produce a unique phenotype (Figure 10.3). Still another example is seen in the rice plant (*Oryza sativa*), which has a haploid number of 12. Trisomic strains for each chromosome have been isolated and studied. The plants of 11 of them can be distinguished from one another and from wild type. Trisomics for the longer chromosomes are the most distinctive and grow more slowly than the rest. This is in keeping with the belief that larger chromosomes cause greater genetic imbalance than smaller ones. In addition to growth rate, leaf structure, foilage, stems, grain morphology, and plant height vary between the various trisomies.

In plants as well as animals, trisomy may be detected during cytological observations of meiotic divisions. Since three copies of one of the chromosomes are present, pairing configurations are usually irregular. At any particular region along the chromosome length, only two of the three homologs may synapse, though different regions of the trio may be paired. When three copies of a chromosome are paired, the configuration is called a **trivalent,** which may be arranged on the spindle so that during anaphase one member moves to one pole, and two go to the opposite pole (Figure 10.4). In some cases, one bivalent and one univalent (an unpaired chromosome) may be present in-

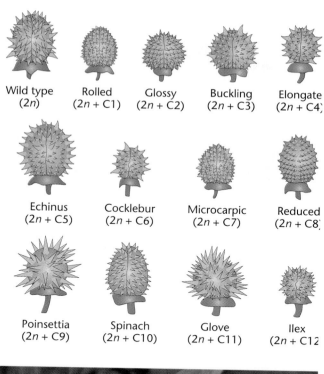

Wild type
(2n)

Rolled
(2n + C1)

Glossy
(2n + C2)

Buckling
(2n + C3)

Elongate
(2n + C4)

Echinus
(2n + C5)

Cocklebur
(2n + C6)

Microcarpic
(2n + C7)

Reduced
(2n + C8)

Poinsettia
(2n + C9)

Spinach
(2n + C10)

Glove
(2n + C11)

Ilex
(2n + C12)

FIGURE 10.3   Drawings of capsule phenotypes of the fruits of the jimsonweed *Datura stramonium*. In comparison with wild type, each phenotype is the result of trisomy of 1 of the 12 chromosomes characteristic of the haploid genome. The photograph illustrates the plant itself.

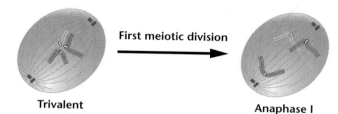

First meiotic division

Trivalent   Anaphase I

FIGURE 10.4   Diagrammatic representation of one possible pairing arrangement during meiosis I of three copies of a single chromosome, forming a trivalent configuration. During anaphase I, two chromosomes move toward one pole and one chromosome toward the other pole.

Down Syndrome

The only human autosomal trisomy in which a significant number of individuals survive longer than a year past birth was discovered in 1866 by Langdon Down. The condition is now known to result from trisomy of chromosome 21, one of the G group* (Figure 10.5), and is called **Down syndrome** or simply **trisomy 21** (designated **47,+21**). This trisomy is found in approximately one infant in every 800 live births.

The external phenotype of these individuals is similar so that they bear a striking resemblance to one another. They display a prominent epicanthic fold in the corner of each eye and are characteristically short. They may have flat, round heads; protruding, furrowed tongues, which cause the mouth to remain partially open; and short, broad hands with fingers showing characteristic palm and fingerprint patterns. Physical, psychomotor, and mental development is retarded, and poor muscle tone is characteristic. Their life expectancy is shortened, although individuals are known to survive into their 50s.

Down children are prone to respiratory disease and heart malformations and show an incidence of leukemia approximately 15 times higher than that of the normal population. However, careful medical scrutiny and treatment throughout their lives has extended their survival significantly. A striking observation is that death of older Down syndrome adults is frequently due to Alzheimer's disease.

The most characteristic origin of this trisomic condition is through nondisjunction of chromosome 21 during meiosis. Failure of paired homologs to disjoin during either anaphase I or II can result in male or female gametes

stead of a trivalent prior to the first meiotic division. Meiosis thus produces gametes with a chromosome composition of (*n* + 1), which can perpetuate the trisomic condition.

---

*On the basis of size and centromere placement, human autosomal chromosomes are divided into seven groups: A (1–3); B (4–5); C (6–12); D (13–15); E (16–18); F (19–20); and G (21–22).

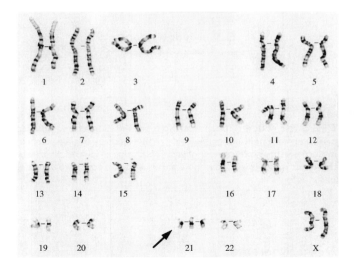

FIGURE 10.5 The karyotype and a photograph of a child with Down syndrome. In the karyotype, three members of the G-group chromosome 21 are present, creating the 47,+21 condition.

with the $n + 1$ chromosome composition. Following fertilization with a normal gamete, the trisomic condition is created. Chromosome analysis has shown that while the additional chromosome may be derived from either the mother or father, the ovum is the source in about 95% of these trisomies.

Before the development of techniques that distinguish paternal from maternal homologs, this conclusion was supported by other indirect evidence derived from studies of the age of mothers giving birth to Down infants. Figure 10.6 shows the relationship between the incidence of Down births and maternal age, illustrating the dramatic increase as the age of the mother increases. While the frequency is about 1 in 1000 at maternal age 30, a 10-fold increase to a frequency of 1 in 100 is noted at age 40. The frequency increases still further to about 1 in 50 at age 45. In spite of these statistics, it is important to point out that, overall, more than half of affected births occur to women who are under 35 years old, primarily because there are so many more pregnancies in this group of women.

While the nondisjunctional event that produces Down syndrome seems more likely to occur during oogenesis in women between the ages of 35 and 45, we do not know with certainty why this is so. However, one observation may be relevant. In human females, all primary oocytes have been formed by birth. Therefore, once ovulation begins, each succeeding ovum has been arrested in meiosis for about a month longer than the one preceding it. As a result, women 30 or 40 years old produce ova that are sig-

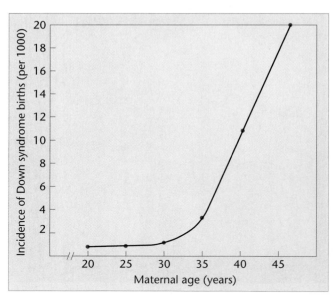

FIGURE 10.6 Incidence of Down syndrome births contrasted with maternal age.

nificantly older and arrested longer than those they ovulated 10 or 20 years previously. However, it is not yet known whether ovum age is the cause of the increased incidence of nondisjunction leading to Down syndrome.

These statistics are the basis of a serious issue facing parents when pregnancy occurs late in a woman's reproductive years. Genetic counseling early in such pregnancies is highly recommended. It informs the parents about the probability that their child will be affected and educates them about Down syndrome. Although some affected individuals may be institutionalized, this is not necessarily the case. Down syndrome children benefit greatly from special-education programs and a stimulating home environment. Further, they are noted as being affectionate, loving children. Intelligence varies greatly, and many have graduated from high school and even college.

A genetic counselor may recommend a prenatal diagnostic technique, where fetal cells are isolated and cultured. **Amniocentesis** and **chorionic villus sampling (CVS)** are the most familiar approaches, whereby fetal cells are obtained from the amniotic fluid or the chorion of the placenta, respectively. In a more current approach, fetal cells are derived directly from the maternal circulation. Once perfected, this approach will be preferable because it is noninvasive, posing no risk to the fetus. Once fetal cells are obtained, the karyotype can then be determined by cytogenetic analysis. If the fetus is diagnosed as having Down syndrome, a therapeutic abortion is one option currently available to parents. Obviously, this is a difficult decision involving a number of religious and ethical issues.

Because Down syndrome appears to be caused by a random error—nondisjunction of chromosome 21 during maternal or paternal meiosis—the occurrence of the disorder is not expected to run in families. Nevertheless, in rare cases, it does. These instances, referred to as **familial Down syndrome,** involve a **translocation** of chromosome 21, another type of chromosomal aberration, which we will discuss later in this chapter.

Patau Syndrome

In 1960, Klaus Patau and his associates observed an infant with severe developmental malformations with a karyotype of 47 chromosomes (Figure 10.7). The additional chromosome was medium-sized, one of the acrocentric D group. It is now designated as chromosome 13. The trisomy 13 condition has since been described in many newborns and is called **Patau syndrome (47,+13).** Affected infants are not mentally alert, are thought to be deaf, and characteristically have a harelip, cleft palate, and demonstrate polydactyly. Autopsies have revealed congenital malformation of most organ systems, a condition indicative of abnormal developmental events occurring as early as five to six weeks of gestation. The average survival of these infants is about three months.

The average maternal and paternal ages of parents of Patau infants are higher than the ages of parents of normal children, but they are not as high as the average maternal age in cases of Down syndrome. Both male and female parents

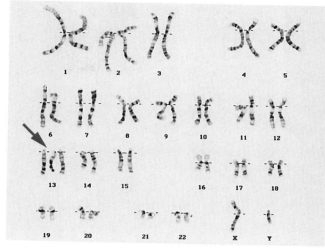

Mental retardation
Growth failure
Low set
deformed ears
Deafness
Atrial septal defect
Ventricular septal
defect
Abnormal
polymorphonuclear
granulocytes
Microcephaly
Cleft lip and palate
Polydactyly
Deformed finger nails
Kidney cysts
Double ureter
Umbilical hernia
Developmental uterine
abnormalities
Cryptorchidism

FIGURE 10.7   The karyotype and phenotypic characteristics of Patau syndrome, where three members of the D-group chromosome 13 are present, creating the 47,+13 condition.

average about 32 years of age when the affected child is born. Because the condition is so rare, occurring as infrequently as 1 in 19,000 live births, it is not known whether the origin of the extra chromosome is more often maternal or paternal or whether it arises equally from either parent.

Edwards Syndrome

In 1960, John H. Edwards and his colleagues reported on an infant trisomic for a chromosome in the E group, now known to be chromosome 18 (Figure 10.8). This aberration has been named **Edwards syndrome (47,+18).** The

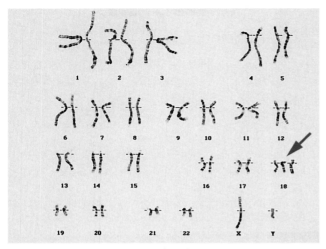

Growth failure

Mental retardation

Open skull sutures
at birth

High arched eyebrows

Low set deformed ears

Short sternum

Ventricular septal defect

Flexion deformities
of fingers

Abnormal kidneys

Persistent ductus
arteriosus

Deformity of hips

Prominent external
genitalia

Muscular hypertonus

Prominent heel

Dorsal flexion of big toes

FIGURE 10.8 The karyotype and phenotypic characteristics of Edwards syndrome. Three members of the E-group chromosome 18 are present, creating the 47,+18 condition.

phenotype of this child, like that of individuals with Down and Patau syndromes, illustrates that the presence of an extra autosome produces congenital malformations and reduced life expectancy. These infants are smaller than the average newborn. Their skulls are elongated in an anterior–posterior direction, and their ears are set low and malformed. A webbed neck, congenital dislocation of the hips, and a receding chin are often characteristic of such individuals. Although the frequency of trisomy 18 is somewhat greater than that of trisomy 13, the average survival time is about the same, less than four months. Death is usually caused by pneumonia or heart failure.

Again, the average maternal age is high—34.7 years by one calculation. In contrast to Patau syndrome, the prepon-

derance of Edwards syndrome infants are females. In one set of observations based on 143 cases, 80 percent were female. Overall, about 1 in 8000 live births exhibits this malady.

Viability in Human Aneuploidy

The reduced viability of individuals with recognized monosomic and trisomic conditions leads us to believe that many other aneuploid conditions may arise, but the affected fetuses do not survive to term. This observation has been confirmed by karyotypic analysis of spontaneously aborted fetuses.

These studies have revealed some striking statistics. At least 15 to 20 percent of all conceptions are terminated in spontaneous abortion (some estimates are considerably higher). About 30 percent of all spontaneous abortuses demonstrate some form of chromosomal anomaly, and approximately 90 percent of all chromosomal anomalies are terminated prior to birth as a result of spontaneous abortion.

A large percentage of spontaneous abortuses demonstrating chromosomal abnormalities are aneuploids. The aneuploid with highest incidence among abortuses is the 45,X condition, which produces an infant with Turner syndrome if the fetus survives to term.

An extensive review of this subject by David H. Carr has also revealed that a significant percentage of abortuses are trisomic for one of the autosomal chromosome groups. Trisomies for every human chromosome have been recovered. Monosomies were almost never found, however, even though nondisjunction should produce $n - 1$ gametes with a frequency equal to $n + 1$ gametes. This finding suggests that either gametes lacking a single chromosome are so functionally impaired that they never participate in fertilization, or that the embryo dies so early in its development that the pregnancy goes undetected. Various forms of polyploidy (see below) and other miscellaneous chromosomal anomalies were also found in Carr's study.

These observations support the hypothesis that normal embryonic development requires a precise diploid complement of chromosomes that maintains a delicate equilibrium of expression of genetic information. The prenatal mortality of most aneuploids provides a barrier against the introduction of a variety of chromosome-based genetic anomalies into the human population.

Polyploidy, in which more than two haploid sets of chromosomes are present, is prevalent in plants

The term **polyploidy** describes instances where more than two multiples of the haploid chromosome set are found. The naming of polyploids is based on the number of sets of chromosomes found: a **triploid** has $3n$ chromosomes; a **tetraploid** has $4n$; a **pentaploid**, $5n$; and so forth. Several general statements may be made about polyploidy. This condition is relatively infrequent in most animal species, but is well known in lizards, amphibians, and fish. It is much more common in plant species. Odd numbers of chromosome sets are not usually maintained reliably from generation to generation because a polyploid organism with an uneven number of homologs usually does not produce genetically balanced gametes. For this reason, triploids, pentaploids, and so on are not usually found in species that depend solely upon sexual reproduction for propagation.

Polyploidy can originate in two ways: (1) The addition of one or more extra sets of chromosomes, identical to the normal haploid complement of the same species, results in **autopolyploidy;** and (2) the combination of chromosome sets from different species may occur as a consequence of interspecific matings resulting in **allopolyploidy** (from the Greek word *allo*, meaning other or different). The distinction between auto- and allopolyploidy is based on the genetic origin of the extra chromosome sets, as illustrated in Figure 10.9.

In our discussion of polyploidy, we will use the following symbols to clarify the origin of additional chromosome sets. For example, if A represents the haploid set of chromosomes of any organism, then

$$A = a_1 + a_2 + a_3 + a_4 + \cdots + a_n$$

where a_1, a_2, and so on represent individual chromosomes, and where n is the haploid number. Using this nomenclature, a normal diploid organism would be represented simply as AA.

Autopolyploidy

In **autopolyploidy,** each additional set of chromosomes is identical to the parent species. Therefore, **triploids** are represented as AAA, **tetraploids** are $AAAA$, and so forth.

Autotriploids may arise in several ways. A failure of all chromosomes to segregate during meiotic divisions (first-division or second-division nondisjunction) may lead to the production of a diploid gamete. If such a gamete survives and is fertilized by a haploid gamete, a zygote with three sets of chromosomes is produced. Or, occasionally two sperm may fertilize an ovum, resulting in a triploid zygote. Triploids can also be produced under experimental conditions by crossing diploids with tetraploids. Diploid organisms produce gametes with n chromosomes, whereas tetraploids produce $2n$ gametes. Upon fertilization, the desired triploid is produced.

Because they have an even number of chromosomes, **autotetraploids** ($4n$) are theoretically more likely to be found in nature than are autotriploids. Unlike triploids, which often produce genetically unbalanced gametes with odd numbers of chromosomes, tetraploids are more likely to produce balanced gametes when involved in sexual reproduction.

How polyploidy arises naturally is of great interest. In theory, if chromosomes have replicated, but the parent cell never divides and reenters interphase, the chromosome number will be doubled. That this very likely occurs is supported by the observation that tetraploid cells can be produced experimentally from diploid cells. This can be accomplished by applying cold or heat shock to meiotic cells or by applying colchicine to somatic cells undergoing mitosis. **Colchicine,** an alkaloid derived from the autumn crocus, interferes with spindle formation, and thus replicated chromosomes cannot be separated at anaphase and

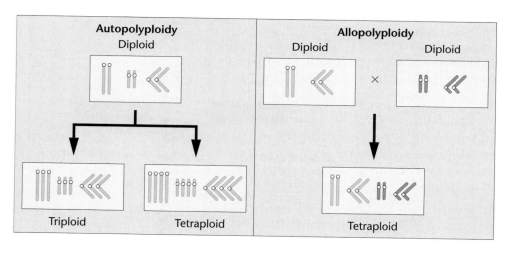

FIGURE 10.9 Contrasting chromosome origins of an autopolyploid vs. an allopolyploid karyotype.

FIGURE 10.10 The potential involvement of colchicine in doubling the chromosome number, as occurs during the production of an autotetraploid. Two pairs of homologous chromosomes are followed. While each chromosome has replicated its DNA earlier during interphase, the chromosomes do not appear as double structures until late prophase. When anaphase fails to occur normally, the chromosome number doubles if the cell reenters interphase.

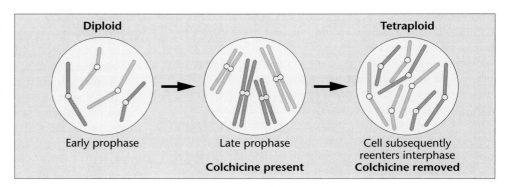

migrate to the poles. When colchicine is removed, the cell may reenter interphase. When the paired sister chromatids separate and uncoil, the nucleus will contain twice the diploid number of chromosomes and is therefore 4*n*. This process is illustrated in Figure 10.10.

In general, autopolyploids are larger than their diploid relatives. This increase seems to be due to larger cell size rather than greater cell number. Although autopolyploids do not contain new or unique information compared with the diploid relative, the flower and fruit of plants are often increased in size, making such varieties of greater horticultural or commercial value. Economically important triploid plants include several potato species of the genus *Solanum*, Winesap apples, commercial bananas, seedless watermelons, and the cultivated tiger lily *Lilium tigrinum*. These plants are propagated asexually. Diploid bananas contain hard seeds, but the commercial, triploid, "seedless" variety has edible seeds. Tetraploid alfalfa, coffee, peanuts, and McIntosh apples are also of economic value because they are either larger or grow more vigorously than do their diploid or triploid counterparts. The commercial strawberry is an octoploid.

Allopolyploidy

Polyploidy can also result from hybridization of two closely related species. If a haploid ovum from a species with chromosome sets *AA* is fertilized by a haploid sperm from a species with sets *BB*, the resulting hybrid is *AB*, where $A = a_1, a_2, a_3, \ldots, a_n$ and $B = b_1, b_2, b_3, \ldots, b_n$. The hybrid plant may be sterile because of its inability to produce viable gametes. Most often, this occurs when some or all of the *a* and *b* chromosomes are not homologous and therefore cannot synapse in meiosis. As a result, unbalanced genetic conditions result. If, however, the new *AB* genetic combination undergoes a natural or induced chromosomal doubling, two copies of all *a* chromosomes and two copies of all *b* chromosomes are now present, and they will pair during meiosis. As a result, a fertile *AABB* tetraploid is produced. These events are illustrated in Fig-

ure 10.11. Since this polyploid contains the equivalent of four haploid genomes derived from separate species, such an organism is called an **allotetraploid**. In such cases, when both original species are known, an equivalent term, **amphidiploid**, is preferred to describe the allotetraploid.

Amphidiploid plants are often found in nature. Their reproductive success is based on their potential for form-

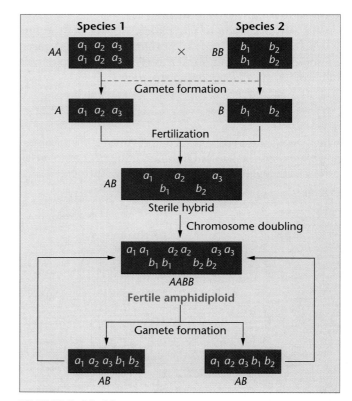

FIGURE 10.11 The origin and propagation of an amphidiploid. Species 1 contains genome A consisting of three distinct chromosomes, a_1, a_2, and a_3. Species 2 contains genome B consisting of two distinct chromosomes, b_1 and b_2. Following fertilization between members of the two species and chromosome doubling, a fertile amphidiploid containing two complete diploid genomes (*AABB*) is formed.

ing balanced gametes. Since two homologs of each specific chromosome are present, meiosis can occur normally (Figure 10.11), and fertilization can successfully propagate the plant sexually. This discussion assumes the simplest situation, where none of the chromosomes in set *A* are homologous to those in set *B*. In amphidiploids formed from closely related species, some homology between *a* and *b* chromosomes is likely. In this case, meiotic pairing is more complex. Multivalents, which are more complex than the trivalent shown in Figure 10.4, may be formed, resulting in the production of unbalanced gametes. In such cases, aneuploid varieties of amphidiploids may arise. Allopolyploids are rare in most animals because mating behavior is most often species-specific, and thus the initial step in hybridization is unlikely to occur.

A classic example of amphidiploidy in plants is the cultivated species of American cotton, *Gossypium* (Figure 10.12). This species has 26 pairs of chromosomes: 13 are large and 13 are much smaller. When it was discovered that Old World cotton had only 13 pairs of large chromosomes, allopolyploidy was suspected. After an examination of wild American cotton revealed 13 pairs of small chromosomes, this speculation was strengthened. J. O. Beasley was able to reconstruct the origin of cultivated cotton experimentally. He crossed the Old World strain with the wild American strain and then treated the hybrid with colchicine to double the chromosome number. The result of these treatments was a fertile amphidiploid variety of cotton. It contained 26 pairs of chromosomes and characteristics similar to the cultivated variety.

Amphidiploids often exhibit traits of both parental species. An interesting example, but one with no practical economic importance, is that of the hybrid formed between the radish *Raphanus sativus* and the cabbage *Brassica oleracea*. Both species have a haploid number of 9. The initial hybrid consists of 9 *Raphanus* and 9 *Brassica* chromosomes (9R +9B). While hybrids are almost always sterile, some fertile amphidiploids (18R + 18B) have been produced. Unfortunately, the root of this plant is more like the cabbage and its shoot more like the radish. Had the converse occurred, the hybrid might have been of economic importance.

A much more successful commercial hybridization has been performed using the grasses wheat and rye. Wheat (genus *Triticum*) has a basic haploid genome of 7 chromosomes. In addition to normal diploids ($2n = 14$), cultivated allopolyploids exist, including tetraploid ($4n = 28$) and hexaploid ($6n = 42$) species. Rye (genus *Secale*) also has a genome consisting of 7 chromosomes. The only cultivated species is the diploid plant ($2n = 14$).

Using the technique outlined in Figure 10.11, geneticists have produced various hybrids. When tetraploid wheat is crossed with diploid rye, and the F_1 treated with colchicine, a hexaploid variety ($6n = 42$) is derived. The hybrid, designated *Triticale* (see Figure 1.13), represents a new genus. Fertile hybrid varieties derived from various wheat and rye species can be crossed together or backcrossed. These crosses have created many variations of the genus *Triticale*. The hybrid plants demonstrate characteristics of both wheat and rye. For example, certain hybrids combine the high protein content of wheat with the high content of the amino acid lysine of rye. The lysine content is low in wheat and thus is a limiting nutritional factor. Wheat is considered a high-yielding grain, whereas rye is noted for its versatility of growth in unfavorable environments. *Triticale* species, combining both traits, have the potential of significantly increasing grain production. Programs designed to improve crops through hybridization have long been underway in several underdeveloped countries of the world, as discussed in Chapter 1.

Recall that in a previous chapter, we discussed the use of **somatic cell hybridization** to map human genes (see Chapter 6). This technique has also been applied to the production of amphidiploid plants (Figure 10.13). Cells from the developing leaves of plants can be treated to remove their cell wall, resulting in **protoplasts.** These altered cells can be maintained in culture and stimulated to fuse with other protoplasts, producing somatic cell hybrids. If cells from different plant species are fused in this way, hybrid amphidiploid cells can be produced. Since protoplasts can be induced to divide and differentiate into stems that develop leaves, the potential for producing allopolyploids is available in the research laboratory. In some cases, entire plants can be derived from cultured protoplasts. If only stems and leaves are produced, these can be grafted onto the stem of another plant. If flowers are

FIGURE 10.12   The pods of the amphidiploid form of *Gossypium,* the cultivated cotton plant.

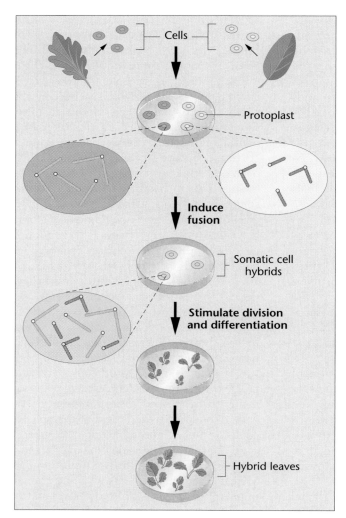

FIGURE 10.13 Application of the somatic cell hybridization technique in the production of an amphidiploid. Cells from the leaves of two species of plants are removed and cultured. The cell walls are digested away and the resultant protoplasts are induced to undergo cell fusion. The hybrid cell is selected and stimulated to divide and differentiate, as illustrated in the photograph. An amphidiploid has a complete set of chromosomes from each parental cell type and displays phenotypic characteristics of each. Two pairs of chromosomes from each species are depicted.

formed, fertilization may yield mature seeds, which, upon germination, yield an allopolyploid plant.

There are now many examples of allopolyploids that have been created commercially using the above approach. Although most of them are not true amphidiploids (one or more chromosomes are missing), precise allotetraploids are sometimes produced.

Endopolyploidy

Endopolyploidy is the condition in which certain cells in an otherwise diploid organism are polyploid. In such cells, replication and separation of chromosomes occur without

nuclear division. Endopolyploidy is the result of a process called **endomitosis.**

Numerous examples of naturally occurring endopolyploidy have been observed. For example, vertebrate liver cell nuclei, including human ones, often contain $4n$, $8n$, or $16n$ chromosome sets. The stem and parenchymal tissue of apical regions of flowering plants are also often endopolyploid. Cells lining the gut of mosquito larvae attain a $16n$ ploidy, but during the pupal stages such cells undergo very quick reduction divisions, giving rise to smaller diploid cells. In the water strider *Gerris*, wide variations in chromosome numbers are found in different tissues, with as many as 1024 to 2048 copies of each chromosome in the salivary gland cells. Since the diploid number in this organism is 22, the nuclei of these cells may contain over 40,000 chromosomes!

Although the role of endopolyploidy is not clear, the proliferation of chromosome copies often occurs in cells where high levels of certain gene products are required. In fact, it is well established that certain genes whose product is in high demand in *every* cell exist naturally in multiple copies in the genome. Ribosomal and transfer RNA genes are examples of multiple-copy genes. In certain cells of organisms, where even this condition may not allow for a sufficient amount of a particular gene product, it may be necessary to replicate the entire genome, allowing an even greater rate of expression of that gene.

Variation occurs in the structure and arrangement of chromosomes

The second general class of chromosome aberrations includes structural changes that delete, add, or rearrange substantial portions of one or more chromosomes. Included in this broad category are deletions and duplications of genes or part of a chromosome and rearrangements of genetic material in which a chromosome segment is inverted, exchanged with a segment of a nonhomologous chromosome, or merely transferred to another chromosome. Exchanges and transfers are called translocations, in which the loca-

tion of a gene is altered within the genome. These types of chromosome alterations are illustrated in Figure 10.14.

In most instances, these structural changes are due to one or more breaks along the axis of a chromosome, followed by either the loss or rearrangement of genetic material. Chromosomes can break spontaneously, but the rate of breakage may increase in cells exposed to chemicals or radiation. Although the actual ends of chromosomes, known as telomeres, do not readily fuse with newly created ends of "broken" chromosomes or with other telomeres, the ends produced at points of breakage are "sticky" and can rejoin other broken ends. If breakage and rejoining does not reestablish the original relationship, and if the alteration occurs in germ plasm, the gametes will contain a structural rearrangement that will be heritable.

When the aberration is found in one homolog but not the other, the individual is said to be heterozygous for the aberration. In such cases, unusual but characteristic pairing configurations are formed during meiotic synapsis. These patterns are useful in identifying the type of change that has occurred. If no loss or gain of genetic material occurs, individuals bearing the aberration "heterozygously" will likely be unaffected phenotypically. However, the unusual pairing arrangements often lead to gametes that are duplicated or deficient for chromosomal regions. When this occurs, the offspring of "carriers" of certain aberrations often have an increased probability of demonstrating phenotypic manifestations.

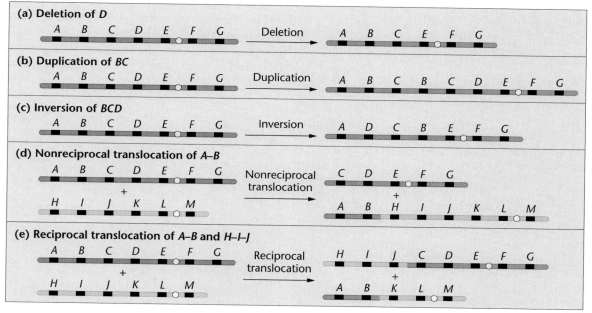

FIGURE 10.14   Overview of the five different types of rearrangement of chromosome segments.   ⬤GenCDX

A deletion is a lost portion of a chromosome

When a chromosome breaks in one or more places, and a portion of it is lost, the missing piece is referred to as a **deletion** (or a **deficiency**). The deletion can occur either near one end of the chromosome or from its interior. These are called **terminal** or **intercalary deletions,** respectively [Figure 10.15(a) and (b)]. The portion of the chromosome retaining the centromere region will usually be maintained when the cell divides, whereas the segment without the centromere will eventually be lost in progeny cells following mitosis or meiosis. For synapsis to occur between a chromosome with a large intercalary deficiency and a normal complete homolog, the unpaired region of the normal homolog must "buckle out" into a **deficiency** or **compensation loop** [Figure 10.15(c)].

As seen earlier in our discussion of the cri-du-chat syndrome, where only part of the short arm of chromosome 5 is lost, a deficiency need not be very great before the effects become severe. If even more genetic information is lost as a result of a deletion, the aberration could be

lethal. Such chromosome mutations never become available for study.

A final consequence of deletions can be noted in organisms heterozygous for a deficiency. Consider the mutant *Notch* phenotype in *Drosophila*. In these flies, the wings are notched on the posterior and lateral margins. Data from breeding studies indicate that the phenotype is controlled by an X-linked dominant mutation because heterozygous females have notched wings and transmit this allele to one-half of their female progeny. The mutation also appears to behave as a homozygous and hemizygous lethal because such females and males are never recovered. It has also been noted that if notched-winged females are also heterozygous for the closely linked recessive mutations *white*-eye, *facet*-eye, or *split*-bristle, they express these mutant phenotypes as well as *Notch*. Because these mutations are recessive, heterozygotes should express the normal, wild-type phenotype. These genotypes and phenotypes are summarized in Table 10.2.

These observations have been explained through a cytological examination of the particularly large chromosomes that characterize certain larval cells of many insects. Called polytene chromosomes (see Chapter 19), they demonstrate a banding pattern specific to each region of each chromosome. In such cells of heterozygous *Notch* females a deficiency loop was found along the X chromosome from the band designated 3C2 through band 3C11, as shown in Figure 10.16. These bands had previously been shown to include the loci for the *white, facet,* and *split* genes, among others. This region's deficiency in one of the two homologous X chromosomes has two distinct effects. First, it results in the *Notch* phenotype. Second, by deleting the loci for genes whose mutant alleles are present on the other X chromosome, the deficiency creates a partially hemizygous condition so that the recessive *white, facet,* or *split* alleles are expressed. This type of phenotypic expression of

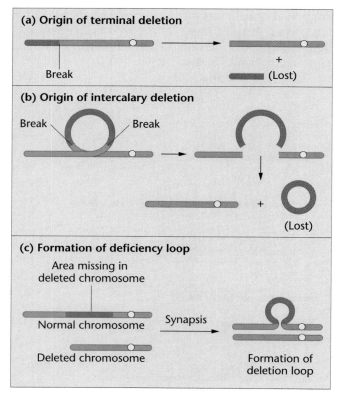

FIGURE 10.15 Origins of (a) a terminal and (b) intercalary deletion. In part (c), pairing occurs between a normal chromosome and one with an intercalary deletion by looping out the undeleted portion to form a deficiency (or a compensation) loop. *GenCDX*

TABLE 10.2 *Notch* Genotypes and Phenotypes

Genotype	Phenotype
$\frac{N^+}{N}$	*Notch* female
$\frac{N}{N}$	Lethal
$\frac{N}{\Rightarrow}$	Lethal
$\frac{N^+ w}{N (w^+)^*}$	*Notch, white* female
$\frac{N^+ fa\ spl}{N (fa^+ spl^+)^*}$	*Notch, facet, split* female

*(deleted)

Bands
3C2–3C11

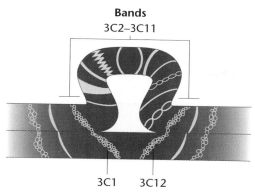

FIGURE 10.16 Deficiency loop formed in salivary chromosomes of *Drosophila melanogaster* where the fly is heterozygous for a deletion. The deletion encompasses bands 3C2 through 3C11, corresponding to the region associated with the *Notch* phenotype.

recessive genes in association with a deletion is an example of the phenomenon called **pseudodominance.**

Many independently arising *Notch* phenotypes have been investigated. The common deficient band for all *Notch* phenotypes has now been designated as 3C7. In every case that *white* was also expressed pseudodominantly, the band 3C2 was also missing. The bands that cytologically distinguish the *Notch* locus from the *white* locus have been confirmed in this manner.

A duplication is a repeated part of the genetic material

When any part of the genetic material—a single locus or a large piece of a chromosome—is present more than once in the genome, it is called a **duplication.** As in deletions, pairing in heterozygotes may produce a compensation loop. Duplications can arise as the result of unequal crossing over between synapsed chromosomes during meiosis (Figure 10.17) or through a replication error prior to meiosis. In the former case, both a duplication and a deficiency are produced.

Three interesting aspects of duplications will be considered. First, they may result in gene redundancy. Sec-

ond, as with deletions, duplications can produce phenotypic variation. Third, according to one convincing theory, duplications have also been an important source of genetic variability during evolution.

Gene Redundancy and Amplification: Ribosomal RNA Genes

Although many gene products are not needed in every cell of an organism, other gene products are known to be essential components of all cells. For example, ribosomal RNA must be present in abundance to support protein synthesis. The more metabolically active a cell is, the higher the demand for this molecule, which becomes a part of each ribosome. In theory, a single copy of the gene encoding rRNA may be inadequate in many cells. Studies using the technique of molecular hybridization, which can determine the percentage of the genome coding for specific RNA sequences, show that in most organisms there are multiple copies of the genes coding for rRNA. Such DNA is called **rDNA,** and the general phenomenon is called **gene redundancy.** For example, in the common intestinal bacterium *Escherichia coli* (*E. coli*) about 0.4 percent of the haploid genome consists of rDNA. This is equivalent to 5 to 10 copies of the gene. In *Drosophila melanogaster,* 0.3 percent of the haploid genome, equivalent to 130 copies, consists of rDNA. Although the presence of multiple copies of the same gene is not restricted to those coding for rRNA, we will focus on them in this section.

Studies of *Drosophila* have documented the critical importance of the extensive amounts of rRNA and ribosomes that is made possible by multiple copies of these genes. In this organism, the X-linked mutation *bobbed* has, in fact, been shown to be due to a deletion of a variable number of the 130 genes coding for rRNA. When the number of these genes is reduced and inadequate rRNA is produced, the result is mutant flies that have low viability, are underdeveloped, and have bristles reduced in size. Both general development and bristle formation, which occur very rapidly during normal pupal development, apparently depend on a great number of ribosomes to support protein synthesis. Many *bobbed* alleles have been studied, and each has been shown to involve a deletion, often of a unique

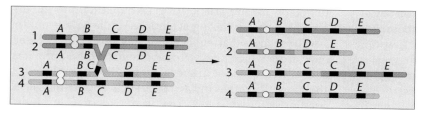

FIGURE 10.17 The origin of duplicated and deficient regions of chromosomes as a result of unequal crossing over. The tetrad at the left is mispaired during synapsis. A single crossover between chromatids 2 and 3 results in deficient and duplicated chromosomal regions (see chromosomes 2 and 3, respectively, on the right). The two chromosomes uninvolved in the crossover event remain normal in their gene sequence and content.

size. The extent to which both viability and bristle length are decreased correlates well with the relative number of rRNA genes deleted. We may conclude that the normal rDNA redundancy observed in wild-type *Drosophila* is near the minimum required for adequate ribosome production during normal development.

In some cells, particularly oocytes, even the normal redundancy of rDNA may be insufficient to provide adequate amounts of rRNA and ribosomes. Oocytes store abundant nutrients in the ooplasm for use by the embryo during early development. In fact, more ribosomes are included in the oocytes than in any other cell type. By considering how the amphibian *Xenopus laevis* (the South African clawed frog) acquires this abundance of ribosomes, we will see a second way in which the amount of rRNA is increased. This phenomenon is referred to as **gene amplification.**

The genes that code for rRNA are located in an area of the chromosome known as the **nucleolar organizer region (NOR).** The NOR is intimately associated with the nucleolus, which is a processing center for ribosome production. Molecular hybridization analysis has shown that each NOR in *Xenopus* contains the equivalent of 400 redundant gene copies coding for rRNA. Even this number of genes is apparently inadequate to synthesize the vast amount of ribosomes that must accumulate in the amphibian oocyte to support development following fertilization. To amplify further the number of rRNA genes, the rDNA is selectively replicated, and each new set of genes is released from its template. Because each new copy is equivalent to the NOR, multiple small nucleoli are formed around each NOR in the oocyte. As many as 1500 of these "micronucleoli" have been observed in a single oocyte. If we multiply the number of micronucleoli (1500) by the number of gene copies in each NOR (400), we see that amplification in *Xenopus* oocytes can result in over half a million gene copies! If each copy is transcribed only 20 times during the maturation of a single oocyte, in theory, sufficient copies of rRNA will be produced to result in well over one million ribosomes.

The Bar Eye Mutation in Drosophila

Duplications can cause phenotypic variations that at first might appear to be caused by a simple gene mutation. The *Bar* eye phenotype (Figure 10.18) in *Drosophila* is a classic example. Instead of the normal oval eye shape, *Bar*-eyed flies have narrow, slitlike eyes. This phenotype appears to be inherited as a dominant X-linked mutation. However, because both heterozygous females and hemizygous males exhibit the trait, but homozygous females show a more pronounced phenotype than either of these two cases, the inheritance is more accurately described as **semidominant.**

In the early 1920s Alfred H. Sturtevant and Thomas H. Morgan discovered and investigated this "mutation." As illustrated in Figure 10.18(a), normal wild-type females (B^+/B^+) have almost 800 facets in each eye. Heterozygous females (B^+/B) have about 350 facets, whereas homozygous females (B/B) average fewer than 50 facets. Females are occasionally recovered with even fewer facets and are designated as *double Bar* (B^D/B^+).

Some 10 years later, Calvin Bridges and Herman J. Muller compared the polytene X chromosome-banding pattern of the *Bar* fly with that of the wild-type fly. Recall from our discussion of deletions that such chromosomes contain specific banding patterns that characterize specific chromosomal regions. Their studies revealed that one copy of region 16A of the X chromosome was present in wild-type flies and that this region was duplicated in *Bar* flies and triplicated in *double Bar* flies. These observations provided evidence that the *Bar* phenotype is not the result of a simple chemical change in the gene, but is instead a duplication. The *double Bar* condition originates as a result of unequal crossing over, which produces the triplicated 16A region [Figure 10.18(b)].

Figure 10.18 also illustrates what is referred to as a **position effect.** You may recall from our introduction of this term in Chapter 4 that a position effect refers to altered gene expression resulting from new "positioning" of a gene within the genome. In this case, when the eye facet phenotypes of B/B and B^D/B^+ flies are compared, an average of 68 and 45 facets are found, respectively. In both cases, there are two extra 16A regions. However, when the repeated segments are distributed on the same homolog instead of being positioned on two homologs, the phenotype is more pronounced. Thus, the same amount of genetic information produces an altered *phenotype* depending on the *position* of the genes.

The Role of Gene Duplication in Evolution

One of the most intriguing aspects of the study of evolution is the consideration of the mechanisms for genetic variation. The origin of unique gene products present in phylogenetically advanced organisms but absent in less advanced, ancestral forms is a topic of particular interest. In other words, how do "new" genes arise?

In 1970, Susumo Ohno published the provocative monograph *Evolution by Gene Duplication* in which he elaborated on the importance of gene duplication to the origin of new genes during evolution. While he was not the first to suggest that duplications might provide a "reservoir" from which new genes might arise, Ohno provided a detailed account of this idea. Ohno's thesis was based on the supposition that the gene products of essential genes, present as only a single copy in the genome, are indis-

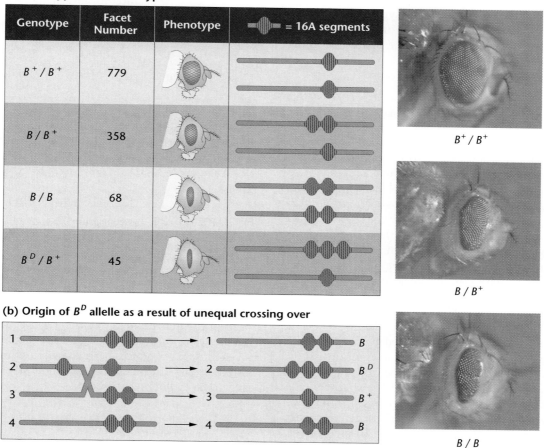

(a) Genotypes and Phenotypes

(b) Origin of B^D allele as a result of unequal crossing over

FIGURE 10.18   (a) Summary of the duplication genotypes and resultant *Bar* eye phenotypes in *Drosophila*. (b) The origin of the B^D (*double Bar*) allele. (c) Photographs illustrate the actual *Bar* eye phenotypes compared to wild type (B^+/B^+).

---

pensable to the survival of members of any species during evolution. Therefore, these genes are not free to accumulate mutations that alter their primary function and potentially give rise to new genes.

However, if an essential gene were to become duplicated in a germ cell, major mutational changes in this extra copy would be tolerated because the original gene still provides the genetic information for its essential function. The duplicated copy would be free to acquire large numbers of mutational changes over extended periods of time. Over short intervals, the new genetic information may be of no practical advantage. However, over long evolutionary periods, the duplicated gene may change sufficiently so that its product assumes a divergent role in the cell. The new function may impart an "adaptive" advantage to organisms carrying such unique genetic information, enhancing their fitness. Ohno has outlined a mechanism through which sustained genetic variability may have originated.

Ohno's thesis is supported by the discovery of genes that have a substantial amount of their DNA sequence in common, but whose gene products are distinct. For example, the vertebrate digestive enzymes, trypsin and chymotrypsin, fit this description, as do the respiratory proteins, myoglobin and hemoglobin. The DNA sequence homology is great enough in these cases to conclude that members of each gene pair arose from a common ancestral gene through duplication. During evolution, the related genes diverged sufficiently so that their products became unique.

Other support includes the presence of **gene families,** regional groups of genes whose products perform the same general function. Again, members of a family show DNA sequence homology sufficient to conclude that they share a common origin. The various types of globin chains that are part of hemoglobin is one example. As we will see in Chapter 14, various globin chains function at different times during development, but they are all a part of hemoglobin, functioning to transport oxygen. Other examples include the immunologically important T-cell receptors and antigens encoded by the major histocompatibility complex (MHC).

Inversions rearrange the linear gene sequence

The **inversion,** another class of structural variation, is a type of chromosomal aberration in which a segment of a chromosome is turned around 180° within a chromosome.

An inversion does not involve a loss of genetic information, but simply rearranges the linear gene sequence. An inversion requires two breaks along the length of the chromosome and subsequent reinsertion of the inverted segment. Figure 10.19 illustrates how one type of inversion might arise. By forming a chromosomal loop prior to breakage, the newly created "sticky ends" are brought close together and rejoined. The inverted segment may be short or quite long and may or may not include the centromere. If the centromere *is* included in the inverted segment, the term **pericentric** describes the inversion, as in Figure 10.19.

If the centromere is *not* part of the rearranged chromosome segment, the inversion is said to be **paracentric.** Although the gene sequence has been reversed in the paracentric inversion, the ratio of arm lengths extending from the centromere is unchanged (Figure 10.20). In contrast, some pericentric inversions create chromosomes with arms of different lengths from those of the noninverted chromosome, thereby changing the **arm ratio.** The change in arm lengths may sometimes be detected during the metaphase stage of mitotic or meiotic divisions.

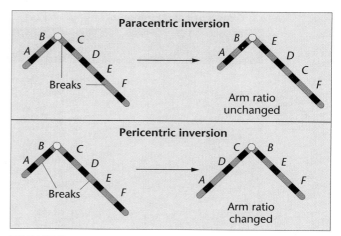

FIGURE 10.20 A comparison of the arm ratios of a submetacentric chromosome before and after the occurrence of a paracentric and pericentric inversion. Only the pericentric inversion results in an alteration of the original ratio.

Although inversions may seem to have a minimal impact on the genetic material, their consequences are interesting to geneticists. Organisms heterozygous for inversions may produce aberrant gametes. Inversions may also result in position effects and might play an important role in the evolutionary process.

Consequences of Inversions during Gamete Formation

If only one member of a homologous pair of chromosomes has an inverted segment, normal linear synapsis during meiosis is not possible. Organisms with one inverted chromosome and one noninverted homolog are called **inversion**

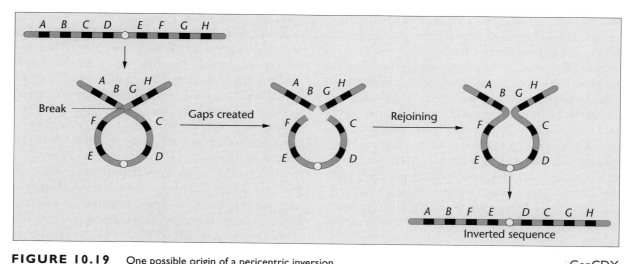

FIGURE 10.19 One possible origin of a pericentric inversion.

GenCDX

heterozygotes. Pairing between two such chromosomes in meiosis may be accomplished only if they form an **inversion loop** (Figure 10.21). In other cases, if no loop is formed, the homologs are seen to synapse everywhere but along the length of the inversion, where they appear separated.

Provided that crossing over does not occur within the inverted segment of the inversion heterozygote, the homologs will segregate and result in two normal and two inverted chromatids that are distributed into gametes, and the inversion will be passed on to one-half of the offspring. However, if crossing over occurs within the inversion loop, abnormal chromatids are produced. For example, the effect of a single exchange event within a paracentric inversion is diagrammed in Figure 10.22(a). As in any meiotic tetrad, a single crossover between nonsister chromatids produces two parental chromatids and two recombinant chromatids. As shown, one recombinant chromatid is **dicentric** (two centromeres), and one recombinant chromatid is **acentric** (lacking a centromere). Both contain duplications and deletions of chromosome segments, as well. During anaphase, an acentric chromatid moves randomly to one pole or the other, or it may be lost, where-

as a dicentric chromatid is pulled in two directions. This polarized movement produces a **dicentric bridge** that is cytologically recognizable. A dicentric chromatid will usually break at some point so that part of the chromatid goes into one gamete and part into another gamete during the reduction divisions. Therefore, gametes containing either broken chromatid are deficient in genetic material.

A similar chromosomal imbalance is produced as a result of a crossover event between a chromatid bearing a pericentric inversion and its noninverted homolog [Figure 10.22(b)]. The recombinant chromatids that are directly involved in the exchange have duplications and deletions. However, no acentric or dicentric chromatids are produced.

In plants, gametes receiving such aberrant chromatids usually fail to develop normally, leading to aborted pollen or ovules. Thus, lethality occurs prior to fertilization, and inviable seeds result. In animals, the gametes have developed prior to the meiotic error, so fertilization is more likely to occur in spite of the chromosome error. However, the end result is the production of inviable embryos following fertilization. In both cases, viability is reduced.

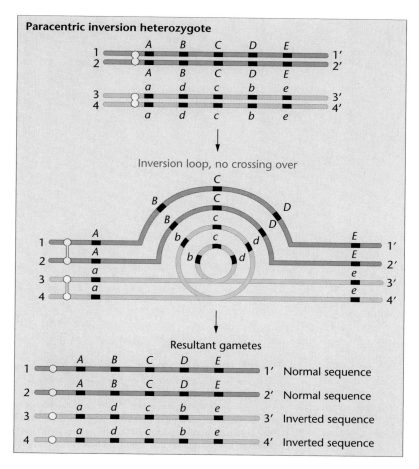

FIGURE 10.21   Illustration of how synapsis occurs in a paracentric inversion heterozygote by virtue of the formation of an inversion loop.

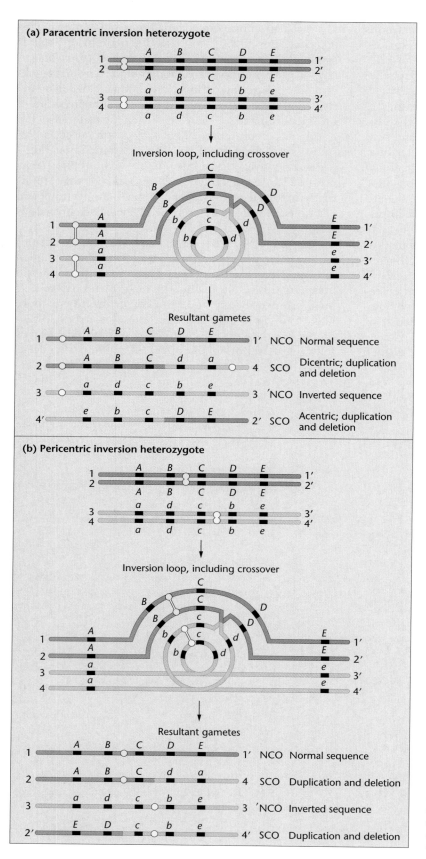

FIGURE 10.22 The effects of a single crossover with an inversion loop in cases involving (a) a paracentric inversion, and (b) a pericentric inversion. In (a), two altered chromosomes are produced, one that is acentric and one that is dicentric. Both chromosomes also contain duplicated and deficient regions. In (b), two altered chromosomes are produced, both with duplicated and deficient regions.

Since in either plants or animals, viable offspring do not result, it *appears* as if the inversion suppresses crossing over since offspring bearing crossover gametes are not recovered. *Actually,* in inversion heterozygotes, the inversion has the effect of suppressing the *recovery* of crossover products when chromosome exchange occurs within the inverted region. If crossing over always occurred within a paracentric or pericentric inversion, 50 percent of the gametes would be ineffective. The viability of the resulting zygotes is therefore greatly diminished. Furthermore, up to one-half of the viable gametes have the inverted chromosome, and the inversion will be perpetuated within the species. The cycle will be repeated continuously during meiosis in future generations.

Position Effects of Inversions

Another consequence of inversions involves the new positioning of genes relative to other genes and particularly to areas of the chromosome that do not contain genes, such as the centromere. If the expression of the gene is altered as a result of its relocation, a change in phenotype may result. Such a change is an example of what is called a **position effect,** as introduced in our earlier discussion of the *Bar* duplication.

We introduced the general topic of position effect in Chapter 4 during our discussion of phenotypic expression. As we saw there, in *Drosophila* females heterozygous for the X-linked recessive mutation *white* eye (w^+/w), the X chromosome bearing the wild-type allele (w^+) may be inverted such that the *white* locus is moved to a point adjacent to the centromere. If the inversion is not present, a heterozygous female has wild-type red eyes, since the *white* allele is recessive. Females with the X chromosome inversion described above have eyes that are mottled or variegated, that is, with red and white patches (see Figure 4.18). Placement of the w^+ allele next to the centromere apparently inhibits wild-type gene expression, resulting in the loss of complete dominance over the w allele. Other genes, also located on the X chromosome, behave in the same manner when they are similarly relocated. Reversion to wild-type expression has sometimes been noted. When this has occurred, cytological examination has shown that the inversion has reestablished the normal sequence along the chromosome.

Evolutionary Advantages of Inversions

One major effect of an inversion is the maintenance of a set of specific alleles at a series of adjacent loci, provided that they are contained within the inversion. Because the recovery of crossover products is suppressed in inversion heterozygotes, a particular combination of alleles is preserved intact in the viable gametes. If the alleles of the in-volved genes provide a survival advantage to organisms maintaining them, the inversion is beneficial to the evolutionary survival of the species.

For example, if the set of alleles *ABcDef* is more adaptive than the sets *AbCdeF* or *abcdEF*, the favorable set of genes will not be disrupted by crossing over if it is maintained within a heterozygous inversion. As we will see in Chapter 26, there are documented examples where inversions are adaptive in this way. Specifically, Theodosius Dobzhansky has shown that the maintenance of different inversions on chromosome 3 of *Drosophila pseudoobscura* through many generations has been highly adaptive to this species. Certain inversions seem to be characteristic of enhanced survival under specific environmental conditions.

Translocations alter the location of chromosomal segments in the genome

Translocation, as the name implies, is the movement of a chromosomal segment to a new location in the genome. Reciprocal translocation, for example, involves the exchange of segments between two nonhomologous chromosomes. The least complex way for this event to occur is for two nonhomologous chromosome arms to come close to each other so that an exchange is facilitated. Figure 10.23(a) illustrates a simple reciprocal translocation in which only two breaks are required. If the exchange includes internal chromosome segments, four breaks are required, two on each chromosome.

The genetic consequences of reciprocal translocations are, in several instances, similar to those of inversions. For example, genetic information is not lost or gained. Rather, there is only a rearrangement of genetic material. The presence of a translocation does not, therefore, directly alter viability of individuals bearing it. Like an inversion, a translocation may also produce a position effect, because it may realign certain genes in relation to other genes. This exchange may create a new genetic linkage relationship that can be detected experimentally.

Homologs heterozygous for a reciprocal translocation undergo unorthodox synapsis during meiosis. As shown in Figure 10.23(b), pairing results in a crosslike configuration. As with inversions, genetically unbalanced gametes are also produced as a result of this unusual alignment during meiosis. In the case of translocations, however, aberrant gametes are not necessarily the result of crossing over. To see how unbalanced gametes are produced, focus on the homologous centromeres in Figure 10.23(b) and (c). According to the principle of independent assortment, the chromosome containing centromere 1 will migrate randomly toward one pole of the spindle during the first meiotic anaphase; it will travel with the chromosome *either* having centromere 3 *or* centromere 4. The chromosome

exhibit Down syndrome. Other potential surviving off-spring contain either the standard diploid genome (without a translocation) or the balanced translocation like the parent. Both cases result in phenotypically normal individuals. Knowledge of translocations has allowed geneticists to resolve the seeming paradox of an inherited trisomic phenotype in an individual with an apparent diploid number of chromosomes.

It is interesting to note that the "carrier," who has 45 chromosomes and exhibits a normal phenotype, does not contain the *complete* diploid amount of genetic material. A small region is lost from both chromosomes 14 and 21 during the translocation event. This occurs because the ends of both chromosomes have broken off prior to their fusion. These specific regions are known to be two of many chromosomal locations housing multiple copies of the genes encoding rRNA, the major component of ribosomes. Despite the loss of up to 20 percent of these genes, the carrier individual is unaffected.

Fragile sites in humans are susceptible to chromosome breakage

We conclude this chapter by briefly discussing the results of an interesting discovery made about 1970 during observations of metaphase chromosomes prepared from human cell cultures derived from various individuals. In some cells, a specific area along one of the chromosomes failed to stain, giving the appearance of a gap. In other cells, derived from different individuals, chromosomes displayed the gap in different positions within the set of chromosomes. Such areas eventually became known as **fragile sites** because they were susceptible to chromosome breakage when cultured in the absence of certain chemicals such as folic acid, which is normally present in the culture medium.

The cause of the fragility at these sites is of great interest. Because they appear to represent points along the chromosome susceptible to breakage, they may indicate regions where chromatin is not tightly coiled or compacted. Note that even though almost all studies of fragile sites have been carried out *in vitro* using mitotically dividing cells, clear associations have been established between several of these sites and altered phenotypes, including mental retardation and cancer.

Fragile X Syndrome (Martin-Bell Syndrome)

Many fragile sites do not appear to be associated with any clinical syndrome. However, individuals bearing a *folate-sensitive site* on the X chromosome (Figure 10.26) exhibit the **fragile X syndrome** (or **Martin-Bell syndrome**), the most common form of inherited mental retardation. This syndrome affects about 1 in 1250 males

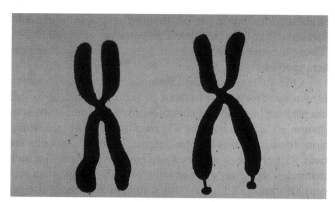

FIGURE 10.26   A normal human X chromosome (left) contrasted with a fragile X chromosome (right). The "gap" region (near the bottom of the chromosome) is associated with the fragile X syndrome.

and 1 in 2500 females. Because it is a dominant trait, females carrying only one fragile X chromosome can be mentally retarded. Fortunately, the trait is not fully expressed, as only about 30 percent of fragile X-bearing females are retarded, whereas about 80 percent of fragile X-bearing males are mentally retarded. In addition to mental retardation, affected males have characteristic long, narrow faces with protruding chins, enlarged ears, and increased testicular size.

A gene that is found within the fragile site is responsible for this syndrome. Known as *FMR-1*, it is one of a growing number of genes where a sequence of three nucleotides is repeated many times, expanding the size of one end of the gene. This phenomenon, called **trinucleotide repeats,** has been recognized in other human disorders, including Huntington disease. In *FMR-1*, the repeated trinucleotide sequence CGG is present in an untranslated area prior to the coding sequence of the gene (called the "upstream" region). The number of repeats varies immensely within the human population, and a high number correlates directly with expression of fragile X syndrome. Normal individuals have 6 to 54 repeats, whereas those with 55 to 200 repeats are considered to be "carriers" of the disorder. Above 200 repeats leads to expression of the syndrome.

It is thought that when the number of repeats reaches this level, the CGG regions of the gene become chemically modified whereby the bases within and around the repeat are methylated, causing inactivation of the gene. The normal product of the gene is an RNA-binding protein that is known to be expressed in the brain. However, the relationship between the absence of this protein and fragile X syndrome is not yet clear.

From a genetic standpoint, perhaps the most interesting aspect of fragile X syndrome is the instability of the CGG repeats. An individual with 6 to 54 repeats transmits a gene containing the same number to his or her offspring. However, those with 55 to 200 repeats, while not at risk to develop the syndrome, may transmit to their offspring a gene where the number of repeats has increased. The number of repeats then continues to increase in future generations, illustrating the phenomenon known as **genetic anticipation,** first introduced in Chapter 4. Once the threshold of 200 is exceeded, expression of the malady becomes more severe in each successive generation as the number of trinucleotide repeats increases.

While the mechanism that leads to the trinucleotide expansion has not yet been established, several factors are known that influence the instability. Most significant is the observation that expansion from the carrier status (55–200 repeats) to the syndrome status (over 200 repeats) occurs during the transmission of the gene by the maternal parent, but not by the paternal parent. Furthermore, several reports suggest that male offspring are more likely to receive the increased repeat size leading to the syndrome than are female offspring. Obviously, there is much yet to be learned about the genetic basis of instability and expansion of DNA sequences.

Fragile Sites and Cancer

A second link between a fragile site and a human condition was reported in 1996 by Carlo Croce, Kay Huebner, and their colleagues, who demonstrated an association between an autosomal site and cancer. They showed that the gene *FHIT* (standing for *f*ragile *hi*stidine *t*riad), located within a well-defined fragile site on chromosome 3, is often altered or missing in cells taken from tumors recovered from individuals with lung cancer. A variety of mutations were found in cells derived from the tumors where the DNA had apparently been broken and incorrectly fused back together, leading to deletions within the gene. In most cases, these mutations caused the *FHIT* gene to become genetically inactivated.

This gene is part of the fragile area of the autosome designated *FRA3B*, and has also been linked to other cancers, including the esophagus, breast, cervix, kidney, colon, and stomach. The nature of the genetic alterations found in cancer cells suggests that the *FHIT* gene, because it is within a fragile region, may be highly susceptible to induced breaks in DNA. If they are incorrectly repaired, cancer-specific chromosome alterations may occur. Thus, this region of the chromosome appears to be particularly sensitive to carcinogen-induced damage, creating a susceptibility to cancer.

Still in question, however, is whether the alteration in the fragile site, leading to inactivation of what appears to be a tumor-suppressor gene, causes cancer, *or* whether the behavior of malignant cells somehow induces breaks at fragile sites, subsequently inactivating this gene. Another important question under investigation is whether molecular polymorphism at this fragile site exists within the human population, causing some individuals to be more susceptible to the effects of carcinogens than others. Whatever the answers, these are significant and exciting studies.

CHAPTER SUMMARY

1. Investigations into the uniqueness of each organism's chromosomal constitution have further enhanced our understanding of genetic variation. Alterations of the precise diploid content of chromosomes are referred to as chromosomal aberrations or chromosomal mutations.

2. Deviations from the expected chromosomal number, or mutations in the structure of the chromosome, are inherited in predictable Mendelian fashion; they often result in inviable organisms or substantial changes in the phenotype.

3. Aneuploidy is the gain or loss of one or more chromosomes from the diploid content, resulting in conditions of monosomy, trisomy, tetrasomy, etc. Studies of monosomic and trisomic disorders have increased our understanding of the delicate genetic balance that must exist for normal development to occur.

4. When complete sets of chromosomes are added to the diploid genome, polyploidy is created. These sets can have identical or diverse genetic origin, creating either autopolyploidy or allopolyploidy, respectively.

5. Large segments of the chromosome can be modified by deletions or duplications. Deletions can produce serious conditions such as the cri-du-chat syndrome in humans, whereas duplications can be particularly important as a source of redundant or new genes.

6. Inversions and translocations, while altering the gene order along chromosomes, initially cause little or no loss of genetic information or deleterious effects. However, heterozygous combinations may cause genetically abnormal gametes following meiosis, often causing lethality.

7. Fragile sites in human mitotic chromosomes have sparked research interest because one such site on the X chromosome is associated with the most common form of inherited mental retardation. Another fragile site, located on chromosome 3, has been linked to lung cancer.

INSIGHTS AND SOLUTIONS

1. In a cross involving three linked genes, *a*, *b*, and *c* in maize, a heterozygote (*abc*/+++) was test-crossed (to *abc*/*abc*). Even though the three genes were separated by at least 5 map units, only two phenotypes were recovered: *abc* and +++. Additionally, the cross produced significantly fewer viable seeds (and thus plants) than expected. Can you propose why no other phenotypes were recovered and why viability was reduced?

   Solution: One of the two chromosomes contains an inversion that overlaps all three genes, effectively precluding the recovery of any "crossover" offspring. If this is a paracentric inversion and the genes are clearly separated (assuring that a significant number of crossovers will occur between them), then numerous acentric and dicentric chromosomes will be formed, resulting in the observed reduction in viability.

2. A male *Drosophila* from a wild-type stock was discovered with only seven chromosomes, whereas the normal 2*n* number is eight. Close examination revealed that one member of chromosome IV (the smallest chromosome) was attached to (translocated to) the distal end of chromosome II and was missing its centromere, thus accounting for the reduction in chromosome number.

   (a) Diagram all members of chromosomes II and IV during synapsis in Meiosis I.

   Solution:

   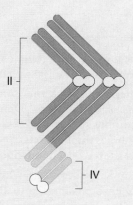

   (b) If this male mates with a female with a normal chromosome composition who is homozygous for the recessive chromosome IV mutation *eyeless* (*ey*), what chromosome compositions will occur in the offspring regarding chromosomes II and IV?

   Solution:

   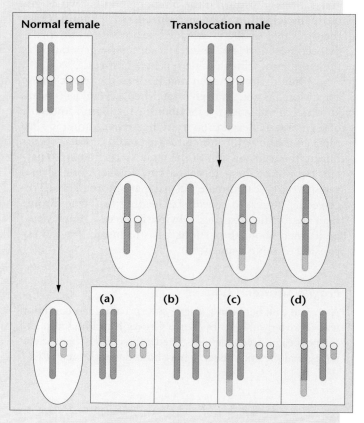

   (c) What phenotypic ratio will result regarding the presence of eyes, assuming all abnormal chromosome compositions survive?

   Solution:
   (a) normal (heterozygous)

   (b) eyeless (monosomic, contains chromosome IV from mother)

   (c) normal (heterozygous)

   (d) normal (heterozygous)

   The final ratio is 3/4 normal:1/4 eyeless.

3. If a Haplo-IV female *Drosophila* (containing only one chromosome 4, but an otherwise normal set of chromosomes) that has white eyes (an X-linked trait) and normal bristles is crossed with a male with a diploid set of chromosomes and normal red eye color, but who is homozygous for the recessive chromosome 4 bristle mutation, *shaven* (*sv*), what F$_1$ phenotypic ratio might be expected?

Solution: Let us first consider only the eye color phenotypes. This is a straightforward X-linked cross. Offspring will appear as 1/2 red females:1/2 white males as shown below:

P$_1$: ww × $w^+/$ ↑
        white female   red male

F$_1$:   1/2 ww^+ :   1/2 $w/$ ↑
        red female   white male

The bristle phenotypes will be governed by the fact that the normal-bristled P$_1$ female produces gametes, one-half of which contain a chromosome 4 (sv^+) and one-half that have no chromosome 4 (− designates no chromosome). Following fertilization by sperm from the *shaven* male, one-half of the offspring will receive

two members of chromosome 4 and be heterozygous for *sv*, expressing normal bristles. The other half will have only one copy of chromosome 4. Because its origin is from the male parent, where the chromosome bears the *sv* allele, these flies will express *shaven* because there is no wild-type allele present to mask this recessive allele.

P$_1$: $sv^+/-$ × sv/sv
        normal female   shaven male

F$_1$: 1/2 sv^+/sv males and females = 1/2 wild type
      1/2 $sv^+/-$ males and females = 1/2 shaven

Using the forked-line method, we can consider both eye color and bristle phenotypes together:

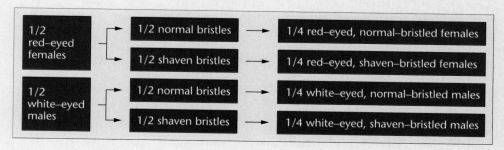

PROBLEMS AND DISCUSSION QUESTIONS

1. For a species with a diploid number of 18, indicate how many chromosomes will be present in the somatic nuclei of individuals who are haploid, triploid, tetraploid, trisomic, and monosomic.

2. Define and distinguish between the following pairs of terms:
   aneuploidy/euploidy
   monosomy/trisomy
   Patau syndrome/Edwards syndrome
   autopolyploidy/allopolyploidy
   autotetraploid/amphidiploid
   polyteny/endopolyploidy
   paracentric inversion/pericentric inversion

3. Contrast the relative survival times of individuals with Down syndrome, Patau syndrome, and Edwards syndrome. Speculate as to why such differences exist.

4. What evidence suggests that Down syndrome is more often the result of nondisjunction during oogenesis rather than during spermatogenesis?

5. What evidence exists that humans with aneuploid karyotypes occur at conception, but are usually inviable?

6. Contrast the fertility of an allotetraploid with an autotriploid and an autotetraploid.

7. When two plants belonging to the same genus but different species were crossed together, the F$_1$ hybrid was more viable and had more ornate flowers. Unfortunately, this hybrid was sterile and could only be propagated by vegetative cuttings. Explain the sterility of the hybrid. How might a horticulturist attempt to reverse its sterility?

8. Describe the origin of cultivated American cotton.

9. Predict how the synaptic configurations of homologous pairs of chromosomes might appear when one member is normal and the other member has sustained a deletion or duplication.

10. Inversions are said to "suppress crossing over." Is this terminology technically correct? If not, restate the description accurately.

11. Contrast the genetic composition of gametes derived from tetrads of inversion heterozygotes where crossing over occurs within a paracentric and pericentric inversion.

12. Contrast the *Notch* locus with the *Bar* locus in *Drosophila*. What phenotypic ratios would be produced in a cross between *Notch* females and *Bar* males?

13. Discuss Ohno's hypothesis on the role of gene duplication in the process of evolution.

14. What roles have inversions and translocations played in the evolutionary process?

15. A human female with Turner syndrome also expresses the X-linked trait hemophilia, as did her father. Which of her parents underwent nondisjunction during meiosis, giving rise to the gamete responsible for the syndrome?

16. The primrose, *Primula kewensis*, has 36 chromosomes that are similar in appearance to the chromosomes in two related species, *Primula floribunda* ($2n = 18$) and *Primula verticillata* ($2n = 18$). How could *P. kewensis* arise from these species? How would you describe *P. kewensis* in genetic terms?

17. Varieties of chrysanthemums are known that contain 18, 36, 54, 72, and 90 chromosomes; all are multiples of a basic set of 9 chromosomes. How would you describe these varieties genetically? What feature is shared by the karyotypes of each variety? A variety with 27 chromosomes was discovered, but it was sterile. Why?

18. What is the effect of a rare double crossover within a pericentric inversion present heterozygously? Within a paracentric inversion present heterozygously?

19. *Drosophila* may be monosomic for chromosome 4 and remain fertile. Contrast the F$_1$ and F$_2$ results of the following crosses involving the recessive chromosome 4 trait, *bent bristles*:

    (a) monosomic IV, bent bristles × diploid, normal bristles
    (b) monosomic IV, normal bristles × diploid, bent bristles

20. *Drosophila* may also be trisomic for chromosome 4 and remain fertile. Predict the F$_1$ and F$_2$ results of crossing:

    trisomic, bent bristles ($b/b/b$) × diploid, normal brisles (b^+/b^+)

21. Mendelian ratios are modified in crosses involving autotetraploids. Assume that one plant expresses the dominant trait green (seeds) and is homozygous ($WWWW$). This plant is crossed to one with white seeds that is also homozygous ($wwww$). If only one dominant allele is sufficient to produce green seeds, predict the F$_1$ and F$_2$ results of such a cross. Assume that synapsis between chromosome pairs is random during meiosis.

22. Having correctly established the F$_2$ ratio above, predict the F$_2$ ratio of a "dihybrid" cross involving two independently assorting characteristics (e.g., P$_1$ = $WWWWAAAA$ X $wwwwaaaa$).

23. A couple, in looking ahead to planning a family, were aware that through the past three generations on the male's side a substantial number of stillbirths had occurred and several malformed babies were born who died early in childhood. The female had studied genetics and urged her husband to visit a genetic counseling clinic, where a complete karyotype-banding analysis was subsequently performed. Although it was found that he had a normal complement of 46 chromosomes, banding analysis revealed that one member of the chromosome 1 pair (in Group A) contained an in-

version covering 70 percent of its length. The homologue of chromosome 1 and all other chromosomes showed the normal banding sequence.

(a) How would you explain the high incidence of past stillbirths?

(b) What would you predict about the probability of abnormality/normality of their future children?

(c) Would you advise the female that she would have to "wait out" each pregnancy to term so as to determine whether the fetus was normal? If not, what would you suggest she do?

EXTRA-SPICY PROBLEMS

24. In a cross in *Drosophila*, a female heterozygous for the autosomally linked genes *a*, *b*, *c*, *d*, and *e* ($abcde/+++++$) was test-crossed to a male homozygous for all recessive alleles. Even though the distance between each of the above loci was at least 3 map units, only four phenotypes were recovered, yielding the following data:

Phenotype	No. of Flies
+ + + + +	440
a b c d e	460
+ + + + *e*	48
a b c d +	52
	Total 1000

Why are many expected crossover phenotypes missing? Can any of these loci be mapped from the data given here? If so, determine map distances.

25. A woman, seeking genetic counseling, was found to be heterozygous for a chromosomal rearrangement between the second and third chromosomes. Her chromosomes, compared to those in a normal karyotype, are diagrammed below:

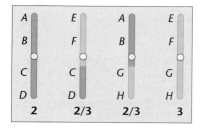

(a) What kind of chromosomal aberration is shown here?

(b) Using a drawing, demonstrate how these chromosomes would pair during meiosis. Be sure to label the different segments of the chromosomes.

(c) This woman is phenotypically normal. Does this surprise you? Why or why not? Under what circumstances might you expect a phenotypic effect of such a rearrangement?

(d) This woman has had two miscarriages. She has come to you, an established genetic counselor, for advice. She

GENETICS MediaLab

The following resources will help you achieve a better understanding of the concepts presented in this chapter. These resources can be found on the CD packaged with this textbook and on the Companion Website at **http://www.prenhall.com/klug/**.

CD Resources:

Animated Tutorial: *Structural Changes in Chromosomes*

Self-grading Chapter Problems

Web Resources:

Web Destinations in Genetics

Self-grading Chapter Problems

Chapter Search Terms

Genetics Newsgroups

Student Bulletin Board

Web Problem 1:

Time for completion = 15 minutes

How do changes in chromosome number (aneuploidy and polyploidy) result in new species or varieties of plants and agricultural cultivars? You will learn about the genetics of polyploids and aneuploids, as well as some of the morphological and physiological changes that occur because of extra copies of the chromosomes. Read through the linked website article on "Chromosome number, translocations, and their consequences." Additional information can be found in Chapter 10. Characterize the difference between autopolyploids and allopolyploids. What are the potential meiotic problems with triploid autopolyploids? Why do you think so many agricultural cultivars are polyploids? Why are trisomics more common among aneuploids than monosomics? Timothy grass, *Phleum pratense,* at first appeared to be an allopolyploid (from crossing *Phleum nodosum* and *Phleum alpinum*), but later was shown to be an autopolyploid; how was this possible? To complete this exercise, visit Web Problem 1 in Chapter 10 of your Companion Website and select the keyword **POLYPLOIDY.**

Web Problem 2:

Time for completion = 10 minutes

How do chromosomal mutations affect human biology? Down syndrome is a relatively common birth defect, occurring at a rate of about 1 in 800 births, that results from extra chromosome 21 material. Read the linked website article on the genetics and risks of Down syndrome. The most common form is a simple trisomic 21, with an entire extra copy of the chromosome. What are the other genetic mechanisms associated with Down syndrome? How many chromosomes are present in a phenotypically normal carrier of a Robertsonian translocation of chromosomes 14 and 21? Robertsonian translocations are inherited from a carrier parent in only about 25 percent of translocation Down syndrome patients; how did the translocation arise in these cases? What would be the genotypes of the gametes produced by a translocation carrier? Are there testing methods with a lower risk factor than amniocentesis that could detect a fetus with Down syndrome? To complete this exercise, visit Web Problem 2 in Chapter 10 of your Companion Website and select the keyword **DOWN SYNDROME.**

Web Problem 3:

Time for completion = 10 minutes

How are some genetic diseases caused by repeated trinucleotide DNA sequences that increase in copy number from generation to generation? Fragile X syndrome and myotonic distrophy (introduced in Chapter 17) are two examples of such diseases. Read the three linked website pages about the genetics of myotonic dystrophy. Additional information can be found in Chapter 10. What is the genetic basis of myotonic dystrophy? The disease is associated with CTG tandem repeats within the myotonin protein kinase gene; what is the difference between normal and afflicted individuals? The number of repeats varies from generation to generation; what mechanism could be responsible for this kind of variation? What is the relationship between the number of repeats and the onset and severity of the disease? What particular problems might a genetic counselor want to consider when discussing whether or not a person with a very mild form of the disease should have children? To complete this exercise, visit Web Problem 3 in Chapter 10 of your Companion Website and select the keyword **MYOTONIC DYSTROPHY.**

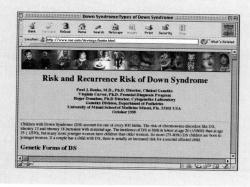

raises the following questions; provide an informed response to her concerns:

Is there a genetic explanation of her frequent miscarriages? Should she abandon her attempts to have a child of her own? If not, what is the chance that she could have a normal child?

26. A son with Klinefelter syndrome is born to a mother who is phenotypically normal and a father who has the X-linked skin condition called anhidrotic ectodermal dysplasia. The mother's skin is completely normal with no signs of the skin abnormality. In contrast, her son has patches of normal skin and patches of abnormal skin.

(a) Which parent contributed the abnormal gamete?

(b) Using the appropriate genetic terminology, describe the meiotic mistake that occurred. Be sure to indicate which division the mistake occurred in.

(c) Using the appropriate genetic terminology, explain the son's skin phenotype.

27. 200 human oocytes that had failed to be fertilized during *in vitro* fertilization procedures were subsequently examined in order to investigate the origin of nondisjunction (Angel, R. 1997. *Am. J. Hum. Genet.* 61:23–32.). These oocytes had completed meiosis I and were arrested in metaphase II (MII). The majority (67 percent) had a normal MII-metaphase complement, showing 23 chromosomes, each consisting of two sister chromatids joined at a common centromere. The remaining oocytes all had abnormal chromosome compositions. Surprisingly, when trisomy is considered, none of the abnormal oocytes had 24 chromosomes.

(a) Interpret these results in regard to the origin of trisomy, as it relates to nondisjunction and when it occurs. Why are the results surprising?

A large number of the abnormal oocytes contained $22\frac{1}{2}$ chromosomes, that is, 22 chromosomes plus a single chromatid representing the $\frac{1}{2}$ chromosome.

(b) What chomosome compositions will result in the zygote if such oocytes proceed through meiosis and are fertilized by normal sperm?

(c) How could the complement of $22\frac{1}{2}$ chromosomes arise? Answer this question with a drawing that includes several pairs of MII chromosomes.

(d) Do these findings support or dispute the generally accepted theory regarding nondisjunction and trisomy, as outlined in Figure 10.1?

SELECTED READINGS

ANTONARAKIS, S.E. 1998. Ten years of *Genomics*, chromosome 21, and Down syndrome. *Genomics* 51:1–16.

ASHLEY-KOCH, A.E., et al. 1997. Examination of factors associated with instability of the *FMR1* CGG repeat. *Am. J. Hum. Genet.* 63:776–85.

BEASLEY, J.O. 1942. Meiotic chromosome behavior in species, species hybrids, haploids, and induced polyploids of *Gossypium*. *Genetics*. 27:25–54.

BLAKESLEE, A.F. 1934. New jimson weeds from old chromosomes. *J. Hered.* 25:80–108.

BORGAONKER, D.S. 1989. *Chromosome variation in man: A catalogue of chromosomal variants and anomalies*, 5th ed. New York: Alan R. Liss.

BOUE, A. 1985. Cytogenetics of pregnancy wastage. *Adv. Hum. Genet.* 14:1–58.

BURGIO, G.R., et al., eds. 1981. *Trisomy 21*. New York: Springer-Verlag.

CARR, D.H. 1971. Genetic basis of abortion. *Annu. Rev. Genet.* 5:65–80.

CROCE, C.M., et al. 1996. The *FHIT* gene at 3p14.2 is abnormal in lung cancer. *Cell*. 85:17–26.

CUMMINGS, M.R. 2000. *Human heredity: Principles and issues*. 5th ed. Pacific Grove, CA: Brooks/Cole.

DEARCE, M.A., and KEARNS, A. 1984. The fragile X syndrome: The patients and their chromosomes. *J. Med. Genet.* 21:84–91.

FELDMAN, M., and SEARS, E.R. 1981. The wild gene resources of wheat. *Sci. Am.* (Jan.) 244:102–12.

GERSH, M., et al. 1995. Evidence for a distinct region causing a cat-like cry in patients with 5p deletions. *Am. J. Hum. Genet.* 56:1404–10.

GUPTA, P.K., and PRIYADARSHAN, P.M. 1982. *Triticale:* Present status and future prospects. *Adv. Genet.* 21:256–346.

HASSOLD, T., et al. 1980. Effect of maternal age on autosomal trisomies. *Ann. Hum. Genet.* (London). 44:29–36.

HASSOLD, T.J., and JACOBS, P.A. 1984. Trisomy in man. *Annu. Rev. Genet.* 18:69–98.

HECHT, F. 1988. Enigmatic fragile sites on human chromosomes. *Trends Genet.* 4:121–22.

HENIKOFF, S. 1994. A reconsideration of the mechanism of position effect. *Genetics*. 138:1–5.

HULSE, J.H., and SPURGEON, D. 1974. Triticale. *Sci. Am.* (Aug.) 231:72–81.

KAISER, P. 1984. Pericentric inversions: Problems and significance for clinical genetics. *Hum. Genet.* 68:1–47.

KHUSH, G.S. 1973. *Cytogenetics of aneuploids*. Orlando, FL: Academic Press.

KHUSH, G.S., et al. 1984. Primary trisomics of rice: Origin, morphology, cytology and use in linkage mapping. *Genetics*. 107:141–63.

LEWIS, E.B. 1950. The phenomenon of position effect. *Adv. Genet.* 3:73–115.

LEWIS, W.H., ed. 1980. *Polyploidy: Biological relevance*. New York: Plenum Press.

LUPSKI, J.R., ROTH, J.R., and WEINSTOCK, G.M. 1996. Chromosomal duplications in bacteria, fruit flies, and humans. *Am. J. Hum. Genet.* 58: 21–26.

MADAN, K. 1995. Paracentric inversions: a review. *Hum. Genet.* 96:503–515.

OBE, G., and BASLER, A. 1987. *Cytogenetics: Basic and applied aspects*. New York: Springer-Verlag.

OHNO, S. 1970. *Evolution by gene duplication*. New York: Springer-Verlag.

OOSTRA, B.A., and VERKERK, A.J. 1992. The fragile X syndrome: Isolation of the *FMR-1* gene and characterization of the fragile X mutation. *Chromosoma*. 101:381–87.

PAGE, S.L., and SHAFFER, L.G. 1997. Nonhomologous Robertsonian translocations form predominantly during female meiosis. *Nature Genetics* 15:231–32.

PATTERSON, D. 1987. The causes of Down syndrome. *Sci. Am.* (Aug.) 257:52–61.

SCHIMKE, R.T., ed. 1982. *Gene amplification*. Cold Spring Harbor, NY: Cold Spring Harbor Laboratory Press.

SHEPARD, J.F. 1982. The regeneration of potato plants from protoplasts. *Sci. Am.* (May) 246:154–166.

SHEPARD, J.F., et al. 1983. Genetic transfer in plants through interspecific protoplast fusion. *Science* 219:683–88.

SIMMONDS, N.W., ed. 1976. *Evolution of crop plants*. London: Longman.

SMITH, G.F., ed. 1984. *Molecular structure of the number 21 chromosome and Down syndrome*. New York: New York Academy of Sciences.

STEBBINS, G.L. 1966. Chromosome variation and evolution. *Science* 152:1463–69.

STRICKBERGER, M.W. 1996. *Evolution*. 2nd ed. Boston: Jones and Bartlett.

SUTHERLAND, G. 1984. The fragile X chromosome. *Int. Rev. Cytol.* 81:107–43.

———. 1985. The enigma of the fragile X chromosome. *Trends Genet.* 1:108–11.

SWANSON, C.P., MERZ, T., and YOUNG, W.J. 1981. *Cytogenetics: The chromosome in division, inheritance, and evolution*. 2nd ed. Englewood Cliffs, NJ: Prentice-Hall.

TAYLOR, A.I. 1968. Autosomal trisomy syndromes: A detailed study of 27 cases of Edwards syndrome and 27 cases of Patau syndrome. *J. Med. Genet.* 5:227–52.

THERMAN, E. and SUSMAN, B. 1995. *Human chromosomes*. 3rd ed.. New York: Springer-Verlag.

TJIO, J.H., and LEVAN, A. 1956. The chromosome number of man. *Hereditas*. 42:1–6.

TURPIN, R., and LEJEUNE, J. 1969. *Human afflictions and chromosomal aberrations*. Oxford: Pergamon Press.

WHARTON, K.A., et al. 1985. *opa:* A novel family of transcribed repeats shared by the *Notch* locus and other developmentally regulated loci in *D. melanogaster. Cell* 40:55–62.

WILKINS, L.E., BROWN, J.A., and WOLF, B. 1980. Psychomotor development in 65 home-reared children with cri-du-chat syndrome. *J. Pediatr.* 97:401–5.

YUNIS, J.J., ed. 1977. *New chromosomal syndromes*. Orlando, FL: Academic Press.

Bronze sculpture of double-helical DNA.

DNA Structure and Analysis

KEY CONCEPTS

- **The genetic material must exhibit four characteristics**

- **Until 1944, observations favored protein as the genetic material**

- **Evidence favoring DNA as the genetic material was first obtained during the study of bacteria and bacteriophages**

  Transformation Studies
  The Hershey–Chase Experiment
  Transfection experiments

- **Indirect and direct evidence supports the concept that DNA is the genetic material in eukaryotes**

  Indirect Evidence: Distribution of DNA
  Indirect Evidence: Mutagenesis
  Direct Evidence: Recombinant DNA Studies

- **RNA serves as the genetic material in some viruses**

- **Knowledge of nucleic acid chemistry is essential to the understanding of DNA structure**

  Nucleotides: Building Blocks of Nucleic Acids
  Nucleoside Diphosphates and Triphosphates
  Polynucleotides

- **The structure of DNA holds the key to understanding its function**

  Base Composition Studies
  X-Ray Diffraction Analysis
  The Watson–Crick Model

- **Alternative forms of DNA exist**

- **The structure of RNA is chemically similar to DNA, but single-stranded**

- **Many analytical techniques have been useful during the investigation of DNA and RNA**

  Absorption of Ultraviolet Light (UV)
  Sedimentation Behavior
  Denaturation and Renaturation of Nucleic Acids
  Molecular Hybridization
  Reassociation Kinetics and Repetitive DNA
  Electrophoresis of Nucleic Acids

GenCDX

When you see this icon, there are related animations and exercises on the CD accompanying this text.

In the first part of the text, we discussed the presence of genes on chromosomes that control phenotypic traits and the way in which the chromosomes are transmitted through gametes to future offspring. Logically, some form of information must be contained in genes, which, when passed to a new generation, influences the form and characteristics of the offspring; this is called the **genetic information.** We might also conclude that this same information in some way directs the many complex processes leading to the adult form.

Until 1944 it was not clear what chemical component of the chromosome makes up genes and constitutes the genetic material. Because chromosomes were known to have both a nucleic acid and a protein component, both were considered candidates. In 1944, however, there emerged direct experimental evidence that the nucleic acid, DNA, serves as the informational basis for the process of heredity.

Once the importance of DNA in genetic processes was realized, work was intensified with the hope of discerning not only the structural basis of this molecule but also the relationship of its structure to its function. Between 1944 and 1953, many scientists sought information that might answer one of the most significant and intriguing questions in the history of biology: How does DNA serve as the genetic basis for living processes? The answer was believed to depend strongly on the chemical structure of the DNA molecule, given the complex but orderly functions ascribed to it.

These efforts were rewarded in 1953 when James Watson and Francis Crick set forth their hypothesis for the double-helical nature of DNA. The assumption that the molecule's functions would be clarified more easily once its general structure was determined proved to be correct. This chapter initially reviews the evidence that DNA is the genetic material and then discusses the elucidation of its structure.

The genetic material must exhibit four characteristics

For a molecule to serve the role of the genetic material, it must possess four major characteristics: **replication, storage of information, expression of that information,** and **variation by mutation.** "Replication" of the genetic material is one facet of the cell cycle, a fundamental property of all living organisms. Once the genetic material of cells has been replicated, it must then be partitioned equally into daughter cells. During the formation of gametes, the genetic material is also replicated but is partitioned so that each cell gets only one-half of the original amount of genetic material. This process is called *meiosis*. Although the products of mitosis and meiosis are different, these processes are both part of the more general phenomenon of cellular reproduction.

The characteristic of "storage" may be viewed as genetic information that is present as a repository of all hereditary characteristics of an organism. However, that information may or may not be expressed. It is clear that, whereas most cells contain a complete complement of DNA, at any given point they express only a part of this genetic potential. For example, bacteria turn many genes on only in response to specific environmental conditions, only to turn them off when such conditions change. In vertebrates, skin cells may display active melanin genes, but never activate their hemoglobin genes; digestive cells activate many genes specific to their function, but do not activate their melanin genes.

Inherent in the concept of storage is the need for the genetic material to be able to encode the nearly infinite variety of gene products found among the countless forms of life present on our planet. The chemical language of the genetic material must be capable of this potential task as it stores information and as it is transmitted to progeny cells and organisms.

"Expression" of the stored genetic information is a complex process and is the basis for the concept of **information flow** within the cell. Figure 11.1 shows a simplified illustration of this concept. The initial event is the **transcription** of DNA, resulting in the synthesis of three types of RNA molecules: **messenger RNA (mRNA),** **transfer RNA (tRNA),** and **ribosomal RNA (rRNA).** Of these, mRNAs are translated into proteins. Each type of mRNA is the product of a specific gene and leads to the synthesis of a different protein. **Translation** occurs in conjunction with rRNA-containing ribosomes and involves tRNA, which acts as an adaptor to convert the chemical information in mRNA to the amino acids that make up proteins. Collectively, these processes serve as the foundation for the **central dogma of molecular genetics:** "DNA makes RNA, which makes proteins."

The genetic material is also the source of newly arising "variability" among organisms through the process of mutation. If a mutation—a change in the chemical composition of DNA—occurs, the alteration will be reflected during transcription and translation, often affecting the specified protein. If a mutation is present in gametes, it will be passed to future generations and, with time, may

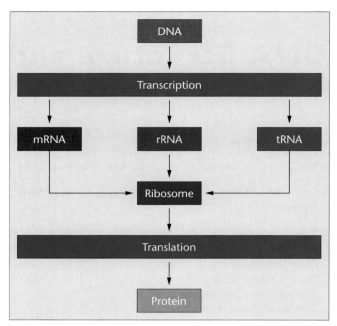

FIGURE 11.1 A simplified view of information flow involving DNA, RNA, and proteins within cells.

become distributed in the population. Genetic variation, which also includes rearrangements within and between chromosomes (see Chapter 10), provides the raw material for the process of evolution.

Until 1944, observations favored protein as the genetic material

The idea that genetic material is physically transmitted from parent to offspring has been accepted for as long as the concept of inheritance has existed. Beginning in the late nineteenth century, research into the structure of biomolecules progressed considerably, setting the stage for the description of the genetic material in chemical terms. Although proteins and nucleic acid were both considered major candidates for the role of the genetic material, many geneticists, until the 1940s, favored proteins. Three factors contributed to this belief.

First, proteins are abundant in cells. Although the protein content may vary considerably, these molecules account for over 50 percent of the dry weight of cells. Because cells contain such a large amount and variety of proteins, it is not surprising that early geneticists believed that some of this protein could function as the genetic material.

The second factor was the accepted proposal for the chemical structure of nucleic acids during the early to mid-1900s. DNA was first studied in 1868 by Friedrick

Miescher, a Swiss chemist. He was able to separate nuclei from the cytoplasm of cells and then isolate from these nuclei an acidic substance that he called **nuclein**. Miescher showed that nuclein contained large amounts of phosphorus and no sulfur, characteristics that differentiate it from proteins.

As analytical techniques were improved, nucleic acids, including DNA, were shown to be composed of four similar molecular building blocks called nucleotides. About 1910, Phoebus A. Levene proposed the **tetranucleotide hypothesis** to explain the chemical arrangement of these nucleotides in nucleic acids. He proposed that a very simple four-nucleotide unit, as shown in Figure 11.2, was repeated over and over in DNA. Levene based his proposal on studies of the composition of the four types of nucleotides. Although his actual data revealed proportions of the four nucleotides that varied considerably, he assumed a 1:1:1:1 ratio. The discrepancy was ascribed to inadequate analytical technique.

Because a single covalently bonded tetranucleotide structure was relatively simple, geneticists believed nucleic acids could not provide the large amount of chemical variation expected for the genetic material. Proteins, on the other hand, contain 20 different amino acids, thus providing the basis for substantial variation. As a result, attention was directed away from nucleic acids, strengthening the speculation that proteins served as the genetic material.

The third contributing factor simply concerned the areas of most active research in genetics. Before 1940, most geneticists were engaged in the study of transmission genetics and mutation. The excitement generated in these areas undoubtedly diluted the concern for finding the precise molecule that serves as the genetic material. Thus, proteins were the most promising candidate and were accepted rather passively.

Between 1910 and 1930, other proposals for the structure of nucleic acids were advanced, but they were generally overturned in favor of the tetranucleotide hypothesis. It was not until the 1940s that the work of Erwin Chargaff led to the realization that Levene's hypothesis was incorrect. Chargaff showed that, for most organisms, the 1:1:1:1 ratio was inaccurate, disproving Levene's hypothesis.

Evidence favoring DNA as the genetic material was first obtained during the study of bacteria and bacteriophages

The 1944 publication by Oswald Avery, Colin MacLeod, and Maclyn McCarty concerning the chemical nature of a "transforming principle" in bacteria was the initial event leading to the acceptance of DNA as the genetic material. Their work, along with the subsequent findings of other research teams, constituted direct experimental proof that DNA, and not protein, is the biomolecule responsible for heredity. It marked the beginning of the era of molecular genetics, a period of discovery in biology that has served as the foundation for current studies in biotechnology and has moved us closer to an understanding of the basis of life. The impact of the initial findings on future research and thinking paralleled that of the publication of Darwin's theory of evolution and the subsequent rediscovery of Mendel's postulates of transmission genetics. Together, these events constituted the three great revolutions in biology.

Transformation Studies

The research that provided the foundation for Avery, MacLeod, and McCarty's work was initiated in 1927 by Frederick Griffith, a medical officer in the British Ministry of Health. He performed experiments with several different strains of the bacterium *Diplococcus pneumoniae.** Some were **virulent** strains, which cause pneumonia in certain vertebrates (notably humans and mice), whereas others were **avirulent** strains, which do not cause illness.

The difference in virulence is related to the polysaccharide capsule of the bacterium. Virulent strains have this capsule, whereas avirulent strains do not. The nonencapsulated bacteria are readily engulfed and destroyed by

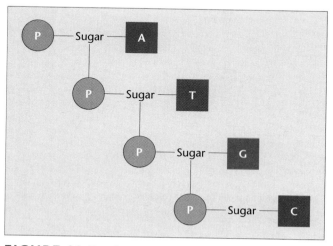

FIGURE 11.2 Diagrammatic depiction of Levene's proposed tetranucleotide containing one molecule each of the four nitrogenous bases: adenine (A), thymine (T), guanine (G), and cytosine (C). Each block represents a nucleotide. Other potential sequences of four nucleotides are possible.

*This organism is now named *Streptococcus pneumoniae*.

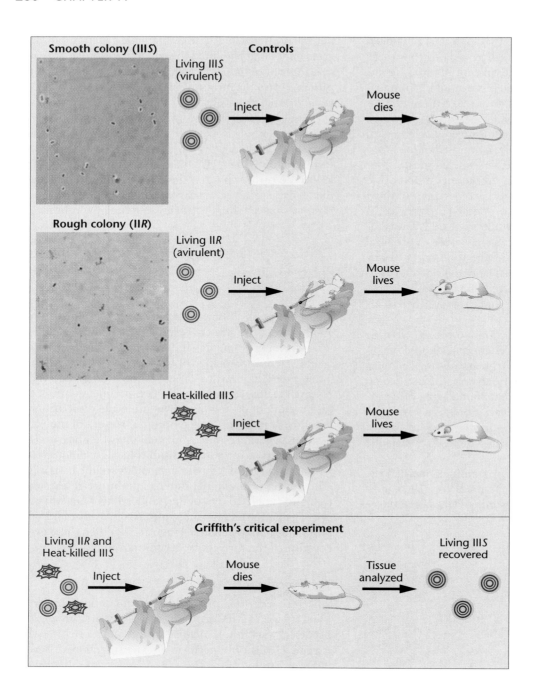

FIGURE 11.3 Summary of Griffith's transformation experiment. The photographs show bacterial colonies containing cells with capsules (type IIIS) and without capsules (type IIR).

phagocytic cells in the animal's circulatory system. Virulent bacteria, which possess the polysaccharide coat, are not easily engulfed; they are able to multiply and cause pneumonia.

The presence or absence of the capsule is the basis for another characteristic difference between virulent and avirulent strains. Encapsulated bacteria form a **smooth,** shiny-surfaced colony (**S**) when grown on an agar culture plate; nonencapsulated strains produce **rough** colonies (**R**) (Figure 11.3). This allows virulent and avirulent strains to be distinguished easily by standard microbiological culture techniques.

Each strain of *Diplococcus* may be one of dozens of different types called **serotypes.** The specificity of the serotype is due to the detailed chemical structure of the polysaccharide constituent of the thick, slimy capsule. Serotypes are identified by immunological techniques and are usually designated by Roman numerals. In the United States, types I and II are most common in causing pneumonia. Griffith used types II and III in the critical experiments that led to new concepts about the genetic material. Table 11.1 summarizes the characteristics of Griffith's two strains.

Griffith knew from the work of others that only living virulent cells would produce pneumonia in mice. If heat-killed virulent bacteria are injected into mice, no pneumonia results, just as living avirulent bacteria fail to produce the disease. Griffith's critical experiment (Figure 11.3) involved an injection into mice of living II*R* (avirulent) cells combined with heat-killed III*S* (virulent) cells. Since neither cell type caused death in mice when injected alone, Griffith expected that the double injection would not kill the mice. But, after five days, all mice receiving double injections were dead. The analysis of the blood of the dead mice revealed a large number of living type III*S* (virulent) bacteria!

As far as could be determined regarding the capsular polysaccharide, these III*S* bacteria were identical to the III*S* strain from which the heat-killed cell preparation had been made. The control mice, injected only with living avirulent II*R* bacteria for this set of experiments, did not develop pneumonia and remained healthy. This finding ruled out the possibility that the avirulent II*R* cells had simply changed (or mutated) to virulent III*S* cells in the absence of the heat-killed III*S* fraction. Instead, some type of interaction was required between living II*R* and heat-killed III*S* cells.

Griffith concluded that the heat-killed III*S* bacteria were somehow responsible for converting live avirulent II*R* cells into virulent III*S* ones. Calling the phenomenon **transformation,** he suggested that the **transforming principle** might be some part of the polysaccharide capsule *or* some compound required for capsule synthesis, although the capsule alone did not cause pneumonia. To use Griffith's term, the transforming principle from the dead III*S* cells served as a "pabulum" for the II*R* cells.

Griffith's work led other physicians and bacteriologists to research the phenomenon of transformation. By 1931, Henry Dawson at the Rockefeller Institute had confirmed Griffith's observations and extended his work one step further. Dawson and his co-workers showed that transformation could occur *in vitro* (in a test tube). When heat-killed III*S* cells were incubated with living II*R* cells, living III*S* cells were recovered. Therefore, injection into mice was not necessary for transformation to occur. By 1933, Lionel J. Alloway had refined the *in vitro* system by using crude extracts of *S* cells and living *R* cells. The soluble filtrate from the heat-killed *S* cells was as effective in inducing transformation as were the intact cells! Alloway and others did not view transformation as a genetic event but rather as a physiological modification of some sort. Nevertheless, the experimental evidence that a chemical substance was responsible for transformation was quite convincing.

Then, in 1944, after 10 years of work, Avery, MacLeod, and McCarty published their results in what is now regarded as a classic paper in the field of molecular genetics. They reported that they had obtained the transforming principle in a purified state and, that beyond reasonable doubt, the molecule responsible for transformation was DNA.

The details of the work, sometimes called the Avery, MacLeod, and McCarty experiment, are outlined in Figure 11.4. These researchers began their isolation procedure with large quantities (50 to 75 liters) of liquid cultures of type III*S* virulent cells. The cells were centrifuged, collected, and heat-killed. Following homogenization and several extractions with the detergent deoxycholate (DOC), they obtained a soluble filtrate that, when tested, still contained the transforming principle. Protein was removed from the active filtrate by several chloroform extractions, and polysaccharides were enzymatically digested and removed. Finally, precipitation with ethanol yielded a fibrous mass that still retained the ability to induce transformation of type II*R* avirulent cells. From the original 75-liter sample, the procedure yielded 10 to 25 mg of the "active factor."

T A B L E   I I . I   Strains of *Diplococcus pneumoniae* Used by Frederick Griffith in His Original Transformation Experiments

Serotype	Colony Morphology	Capsule	Virulence
II*R*	Rough	Absent	Avirulent
III*S*	Smooth	Present	Virulent

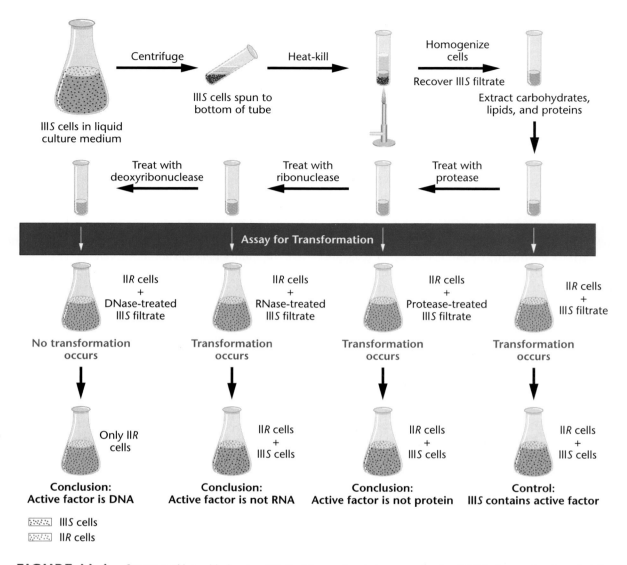

FIGURE 11.4 Summary of Avery, MacLeod, and McCarty's experiment demonstrating that DNA is the transforming principle.

Further testing established beyond a reasonable doubt that the transforming principle was DNA. The fibrous mass was first analyzed for its nitrogen/phosphorus ratio, which was shown to coincide with the ratio of "sodium desoxyribonucleate," the chemical name then used to describe DNA. To solidify their findings, they sought to eliminate, to the greatest extent possible, all probable contaminants from their final product. Thus, it was treated with the proteolytic enzymes trypsin and chymotrypsin and then with an RNA-digesting enzyme, called ribonuclease. Such treatments destroyed any remaining activity of proteins and RNA. Nevertheless, transforming activity still remained. Chemical testing of the final product gave

strong positive reactions for DNA. The final confirmation came with experiments using crude samples of the DNA-digesting enzyme **deoxyribonuclease,** which was isolated from dog and rabbit sera. Digestion with this enzyme was shown to destroy transforming activity. There could be little doubt that the active transforming principle was DNA!

The great amount of work, the confirmation and reconfirmation of the conclusions drawn, and the unambiguous logic of the experimental design involved in the research of these three scientists are truly impressive. The conclusion to the 1944 publication was, however, very simply stated: "The evidence presented supports the belief that a nucleic acid of the desoxyribose* type is the fundamental unit of the transforming principle of *Pneumococcus* Type III."

Avery and his co-workers recognized the genetic and biochemical implications of their work when they observed that "nucleic acids of this type must be regarded not merely as structurally important but as functionally active in determining the biochemical activities and specific characteristics of pneumococcal cells." This suggested that the transforming principle interacts with the II*R* cell and gives rise to a coordinated series of enzymatic reactions culminating in the synthesis of the type III*S* capsular polysaccharide. They emphasized that, once transformation occurs, the capsular polysaccharide is produced in successive generations. Transformation is therefore heritable, and the process affects the genetic material.

Immediately after the publication of the report, several investigators turned to or intensified their studies of transformation in order to clarify the role of DNA in genetic mechanisms. In particular, the work of Rollin Hotchkiss was instrumental in confirming that the critical factor in transformation was DNA and not protein. In 1949, in a separate study, Harriet Taylor isolated an **extremely rough (*ER*)** mutant strain from a rough (*R*) strain. This *ER* strain produced colonies that were more irregular than the *R* strain. The DNA from *R* accomplished the transformation of *ER* to *R*. Thus, the *R* strain, which served as the recipient in the Avery experiments, was shown to be able to serve as the DNA donor in transformation also.

Transformation, which was introduced previously in Chapter 7, has now been shown to occur in *Haemophilus influenzae*, *Bacillus subtilis*, *Shigella paradysenteriae*, and *Escherichia coli*, among many other microorganisms. Transformation of numerous genetic traits other than colony morphology has also been demonstrated, including ones involving resistance to antibiotics and the ability to metabolize various nutrients. These observations further strengthened the belief that transformation by DNA is primarily a genetic event, rather than simply a physiological change. This idea is pursued again in the "Insights and Solutions" section at the end of this chapter.

The Hershey–Chase Experiment

The second major piece of evidence supporting DNA as the genetic material was provided by the study of the infection of the bacterium *Escherichia coli* by one of the viruses to which it serves as a host, **bacteriophage T2.** Often referred to simply as a **phage,** the virus consists of a pro-

*Desoxyribose is now spelled deoxyribose.

tein coat surrounding a core of DNA. Electron micrographs have revealed the phage's external structure to be composed of a hexagonal head plus a tail. The general aspects of the life cycle of a T-even bacteriophage such as T2, as known in 1952, is shown in Figure 11.5. Briefly, the phage adsorbs to the bacterial cell and some component of the phage enters the bacterial cell. Following this infection step, the viral information "commandeers" the cellular machinery of the host and proceeds to undergo viral reproduction. In a reasonably short time, many new phages are constructed and the bacterial cell is lysed, releasing the progeny viruses.

In 1952, Alfred Hershey and Martha Chase published the results of experiments designed to clarify the events leading to phage reproduction, as described above. Several of the experiments clearly established the independent functions of phage protein and nucleic acid in the reproduction process associated with the bacterial cell. Hershey and Chase knew from existing data that:

1. T2 phages consist of approximately 50 percent protein and 50 percent DNA.

2. Infection is initiated by adsorption of the phage by its tail fibers to the bacterial cell.

3. The production of new viruses occurs within the bacterial cell.

It appeared that some molecular component of the phage, DNA and/or protein, enters the bacterial cell and directs viral reproduction. Which was it?

Hershey and Chase used the radioisotopes ^{32}P and ^{35}S to follow the molecular components of phages during infection. Because DNA contains phosphorus (P) but not sulfur, ^{32}P effectively labels DNA, and because proteins contain sulfur (S) but not phosphorus, ^{35}S labels protein. *This was a key point in the experiment.* If *E. coli* cells are first grown in the presence of ^{32}P or ^{35}S and then infected with T2 viruses, the progeny phages will have either a radioactively labeled DNA core or a radioactively labeled protein coat, respectively. These labeled phages can be isolated and used to infect unlabeled bacteria (Figure 11.6).

When labeled phages and unlabeled bacteria are mixed, an adsorption complex is formed as the phages attach their tail fibers to the bacterial wall. These complexes were isolated and subjected to a high shear force by placing them in a blender. This force stripped off the attached phages. When the mixture was centrifuged, the lighter phage particles separated from the heavier bacterial cells, allowing the isolation of both components (Figure 11.6). By tracing the radioisotopes, Hershey and Chase were able to demonstrate that most of the ^{32}P-labeled DNA had been transferred into the bacterial cell following adsorption; on the other hand, most of the ^{35}S-labeled

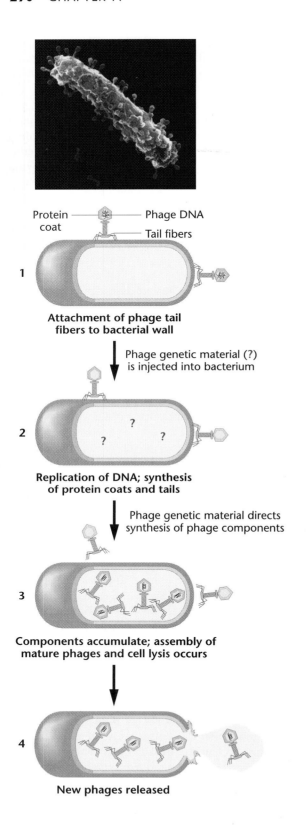

1

**Attachment of phage tail
fibers to bacterial wall**

Phage genetic material (?)
is injected into bacterium

2

**Replication of DNA; synthesis
of protein coats and tails**

Phage genetic material directs
synthesis of phage components

3

**Components accumulate; assembly of
mature phages and cell lysis occurs**

4

New phages released

FIGURE II.5   The life cycle of a T-even bacteriophage. The electron micrograph shows *E. coli* during infection by phage T2.

protein remained outside the bacterial cell and was recovered in the phage "ghosts" (empty phage coats) after the blender treatment. Following this separation, the bacterial cells, which now contained viral DNA, were eventually lysed as new phages were produced. These progeny phages contained ^{32}P but not ^{35}S.

Hershey and Chase interpreted these results to indicate that the protein of the phage coat remains outside the host cell and is not involved in directing the production of new phages. On the other hand, and most important, phage DNA enters the host cell and directs phage multiplication. They had demonstrated that in phage T2, DNA and not protein is the genetic material.

This experimental work, along with that of Avery and his colleagues, provided convincing evidence to most geneticists that DNA was the molecule responsible for heredity. Since then, many significant findings have been based on this supposition. These many findings, constituting the field of molecular genetics, are discussed in detail in subsequent chapters.

Transfection Experiments

During the eight years following the publication of the Hershey–Chase experiment, additional research using bacterial viruses provided even more solid proof that DNA is the genetic material. In 1957, several reports demonstrated that if *E. coli* was treated with the enzyme **lysozyme,** the outer wall of the cell could be removed without destroying the bacterium. Enzymatically treated cells are naked, so to speak, and contain only the cell membrane as their outer boundary. Such structures are called **spheroplasts** (or **protoplasts**). John Spizizen and Dean Fraser independently reported that by using spheroplasts, they were able to initiate phage reproduction with disrupted T2 particles. That is, provided that the cell wall is absent, it is not necessary for a virus to be intact in order for infection to occur. Thus, the outer protein coat structure may be essential to the movement of DNA through the intact cell wall, but it is not essential for infection when spheroplasts are used.

Similar but more refined experiments were reported in 1960 by George Guthrie and Robert Sinsheimer. DNA was purified from bacteriophage φX174, a small phage that contains a single-stranded circular DNA molecule of some 5386 nucleotides. When added to *E. coli* protoplasts, the purified DNA resulted in the production of complete φX174 bacteriophages. This process of infection by only the viral nucleic acid, called **transfection,** proved conclusively that φX174 DNA alone contains all the necessary information for production of mature viruses. Thus, the evidence supporting the conclusion that DNA serves as the genetic

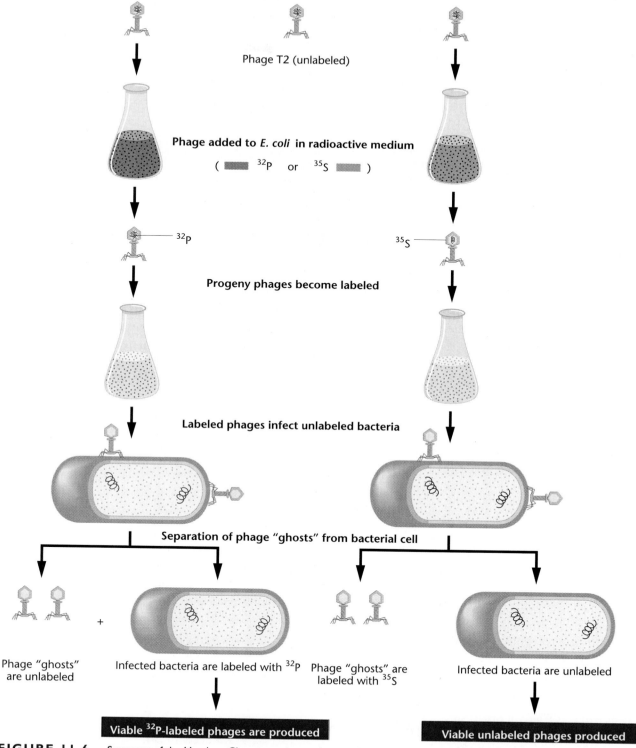

Phage T2 (unlabeled)

Phage added to *E. coli* in radioactive medium

(^{32}P or ^{35}S)

^{32}P ^{35}S

Progeny phages become labeled

Labeled phages infect unlabeled bacteria

Separation of phage "ghosts" from bacterial cell

Phage "ghosts"
are unlabeled

Infected bacteria are labeled with ^{32}P

Phage "ghosts" are
labeled with ^{35}S

Infected bacteria are unlabeled

Viable ^{32}P-labeled phages are produced

Viable unlabeled phages produced

FIGURE 11.6   Summary of the Hershey–Chase experiment demonstrating that DNA, and not protein, is responsible for directing the reproduction of phage T2 during the infection of *E. coli*.   GenCDX

material was further strengthened, even though all direct evidence thus far had been obtained from bacterial and viral studies.

Indirect and direct evidence supports the concept that DNA is the genetic material in eukaryotes

In 1950, eukaryotic organisms were not amenable to the types of experiments performed to demonstrate that DNA is the genetic material in bacteria and viruses. Nevertheless, it was generally assumed that the genetic material would be a universal substance and also serve this role in eukaryotes. Initially, support for this assumption relied on several circumstantial (indirect) observations that, taken together, provided support that DNA serves as the genetic material in eukaryotes. Subsequently, direct evidence established unequivocally the central role of DNA in genetic processes.

Indirect Evidence: Distribution of DNA

The genetic material should be found where it functions—in the nucleus as part of chromosomes. Both DNA and protein fit this criterion. However, protein is also abundant in the cytoplasm, whereas DNA is not. Both mitochondria and chloroplasts are known to perform genetic functions, and DNA is also present in these organelles. Thus, DNA is found only where primary genetic function occurs. Protein, on the other hand, is found everywhere in the cell. These observations are consistent with the interpretation favoring DNA over proteins as the genetic material.

Because it had earlier been established that chromosomes within the nucleus contain the genetic material, a correlation was expected to exist between the ploidy of a cell (e.g., *n*, or 2*n*) and the quantitative amount of the molecule serving as the genetic material. Meaningful comparisons can be made between the amount of DNA and protein in gametes (sperm and eggs) and somatic (or body) cells. The latter are recognized as being **diploid** (2*n*) and containing twice the number of chromosomes as gametes, which are **haploid** (*n*).

Table 11.2 compares the amount of DNA found in haploid sperm and diploid nucleated precursors of red blood cells from a variety of organisms. A close correlation exists between the amount of DNA and the number of sets of chromosomes. No such correlation can be observed between gametes and diploid cells for proteins. These data thus provide further circumstantial evidence favoring DNA over proteins as the genetic material of eukaryotes.

TABLE 11.2 DNA Content of Haploid Versus Diploid Cells of Various Species (in picograms)

Organism	*n*	2*n*
Human	3.25	7.30
Chicken	1.26	2.49
Trout	2.67	5.79
Carp	1.65	3.49
Shad	0.91	1.97

Note: Sperm (*n*) and nucleated precursors to red blood cells (2*n*) were used to contrast ploidy levels.

Indirect Evidence: Mutagenesis

Ultraviolet (UV) light is one of a number of agents capable of inducing mutations in the genetic material. Bacteria and other organisms can be irradiated with various wavelengths of ultraviolet light and the effectiveness of each wavelength measured by the number of mutations it induces. When the data are plotted, an **action spectrum** of UV light as a mutagenic agent is obtained. This action spectrum can then be compared with the **absorption spectrum** of any molecule suspected to be the genetic material (Figure 11.7). The molecule serving as the genetic material is expected to absorb UV light at the wavelengths found to be mutagenic.

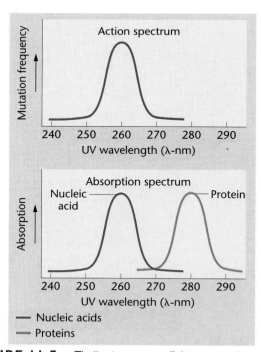

FIGURE 11.7 The "action spectrum" determining the most effective mutagenic wavelength compared with the "absorption spectrum" of nucleic acids vs. proteins in the UV range of the electromagnetic spectrum.

UV light is most mutagenic at the wavelength (λ) of 260 nanometers (nm). Both DNA and RNA absorb UV light most strongly at 260 nm. On the other hand, protein absorbs most strongly at 280 nm, yet no significant mutagenic effects are observed at this wavelength. This indirect evidence supports the idea that a nucleic acid is the genetic material and tends to exclude protein.

Direct Evidence: Recombinant DNA Studies

Although the circumstantial evidence just described does not constitute direct proof that DNA is the genetic material in eukaryotes, these observations spurred researchers to forge ahead, basing their work on this hypothesis. Today, there is no doubt of the validity of this conclusion; DNA *is* the genetic material in all eukaryotes.

The strongest direct evidence has been provided by the application of molecular analysis referred to as **recombinant DNA technology.** In such research, segments of eukaryotic DNA corresponding to specific genes are isolated and literally spliced into bacterial DNA. Such a complex can be inserted into a bacterial cell and its **genetic expression** monitored. If a eukaryotic gene is introduced, the presence of the corresponding eukaryotic protein product demonstrates directly that this DNA is not only present, but functional in the bacterial cell. This has been shown to be the case in countless instances. For example, the human gene products insulin and interferon are produced by bacteria following the insertion of human genes that encode these proteins. As the bacterium divides, the eukaryotic DNA is replicated along with the host DNA and is distributed to the daughter cells, which also express the human genes and synthesize the corresponding proteins.

The availability of vast amounts of DNA coding for specific genes, available as a result of recombinant DNA research, has led to other direct evidence that DNA serves as the genetic material. Work in the laboratory of Beatrice Mintz has demonstrated that DNA encoding the human β-globin gene, when microinjected into a fertilized mouse egg, is later found to be present and expressed in adult mouse tissue and transmitted to and expressed in that mouse's progeny. These mice are examples of what are called **transgenic animals.**

More recent work has introduced *rat* DNA encoding a growth hormone into fertilized *mouse* eggs. About one-third of the resultant mice grew to twice their normal size. This indicated that foreign DNA was present *and* functional in the experimental mice. Subsequent generations of mice inherited this genetic information and also grew to a large size.

We will pursue the topic of recombinant DNA again later (see Chapters 18 and 21). The point to be made here is that in eukaryotes, DNA has been shown directly to meet the requirement of expression of genetic information. Later we will see exactly how DNA is stored, replicated, expressed, and mutated.

RNA serves as the genetic material in some viruses

Some viruses contain an RNA core rather than one composed of DNA. In these viruses, it would thus appear that RNA might serve as the genetic material—an exception to the general rule that DNA performs this function. In 1956, it was demonstrated that when purified RNA from **tobacco mosaic virus (TMV)** was spread on tobacco leaves, the characteristic lesions caused by TMV appeared later on the leaves. It was concluded that RNA is the genetic material of this virus.

Soon afterwards, another type of experiment with TMV was reported by Heinz Fraenkel-Conrat and B. Singer, as illustrated in Figure 11.8. These scientists discovered that the RNA core and the protein coat from wild-type TMV and other viral strains could be isolated separately. In their work, RNA and coat proteins were separated and isolated from TMV and a second viral strain, **Holmes ribgrass (HR).** Then, mixed viruses were reconstituted from the RNA of one strain and the protein of the other. When this "hybrid" virus was spread on tobacco leaves, the lesions that developed corresponded to the type of RNA in the reconstituted virus; that is, viruses with wild-type TMV RNA and HR protein coats produced TMV lesions and vice versa. Again, it was concluded that RNA serves as the genetic material in these viruses.

In 1965 and 1966, Norman R. Pace and Sol Spiegelman further demonstrated that RNA from the phage Qβ could be isolated and replicated *in vitro*. Replication was dependent on an enzyme, **RNA replicase,** which was isolated from host *E. coli* cells following normal infection. When the RNA replicated *in vitro* was added to *E. coli* protoplasts, infection and viral multiplication occurred. Thus, RNA synthesized in a test tube can amply serve as the genetic material in these phages by directing the production of all the components necessary for viral replication.

One other group of RNA-containing viruses bears mentioning. These are the **retroviruses,** which replicate in an unusual way. Following infection of the host cell, their RNA serves as a template for the synthesis of the complementary DNA molecule. The process, designated as **reverse transcription,** occurs under the direction of an RNA-dependent DNA polymerase enzyme called reverse transcriptase. The viral genetic material is "represented" by this DNA intermediate, which may be incorporated into the genome of the host cell. Once present, if the DNA is expressed, transcription produces

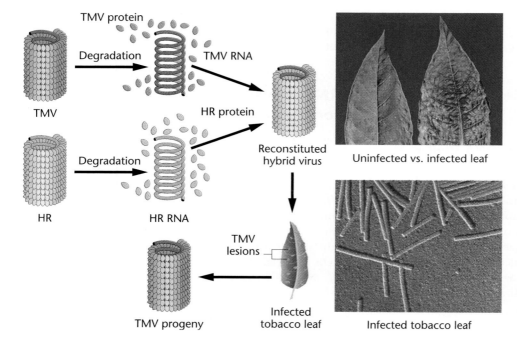

Uninfected vs. infected leaf

Infected tobacco leaf

FIGURE 11.8 Reconstitution of hybrid tobacco mosaic viruses. In the hybrid, RNA is derived from the wild-type TMV virus, while the protein subunits are derived from the HR strain. Following infection, viruses are produced with protein subunits characteristic of the wild-type TMV strain and not those of the HR strain. The top photograph shows TMV lesions on a tobacco leaf compared to an uninfected leaf. At the bottom is an electron micrograph of mature viruses.

copies of the retroviral RNA genome. Among others, the polio virus and the human immunodeficiency virus (HIV), which causes AIDS, are examples of retroviruses.

Knowledge of nucleic acid chemistry is essential to the understanding of DNA structure

Having established thus far the critical importance of DNA and RNA in genetic processes, we shall now provide a brief introduction to the chemical basis of these molecules. As we shall see, the structural components of DNA and RNA are very similar. This chemical similarity is important in the coordinated functions played by these molecules during gene expression. Like the other major groups of organic biomolecules (proteins, carbohydrates, and lipids), nucleic acid chemistry is based on a variety of similar building blocks that are polymerized into chains of varying lengths.

Nucleotides: Building Blocks of Nucleic Acids

Nucleotides are the building blocks of all nucleic acid molecules. These structural units consist of three essential components: a **nitrogenous base,** a **pentose sugar** (a 5-carbon sugar), and a **phosphate group.** There are two kinds of nitrogenous bases: the nine-member double-ring **purines** and the six-member single-ring **pyrimidines.** Two types of purines and three types of pyrimidines are found commonly in nucleic acids. The two purines are **adenine** and **guanine,** abbreviated **A** and **G.** The three pyrimidines

are **cytosine, thymine,** and **uracil,** abbreviated **C, T,** and **U.** The chemical structures of A, G, C, T, and U are shown in Figure 11.9(a). Both DNA and RNA contain A, C, and G; only DNA contains the base T, whereas only RNA contains the base U. Each nitrogen or carbon atom of the ring structures of purines and pyrimidines is designated by an unprimed number. Note that corresponding atoms in the two rings are numbered differently in most cases.

The pentose sugars found in nucleic acids give them their names. Ribonucleic acids (RNA) contain **ribose,** while deoxyribonucleic acids (DNA) contain **deoxyribose.** Figure 10.9(b) shows the ring structures for these two pentose sugars. Each carbon atom is distinguished by a number with a prime sign (e.g., C-1′, C-2′). As you can see, deoxyribose is missing one hydroxyl group at the C-2′ position, compared with ribose. The presence of a hydroxyl group at the C-2′ position distinguishes RNA from DNA.

If a molecule is composed of a purine or pyrimidine base and a ribose or deoxyribose sugar, the chemical unit is called a **nucleoside.** If a phosphate group is added to the nucleoside, the molecule is now called a **nucleotide.** Nucleosides and nucleotides are named according to the specific nitrogenous base (A, T, G, C, or U) that is part of the building block. The nomenclature and general structure of nucleosides and nucleotides are given in Figure 11.10.

The bonding among the components of a nucleotide is highly specific. The C-1′ atom of the sugar is involved in the chemical linkage to the nitrogenous base. If the base is a purine, the N-9 atom is covalently bonded to the sugar. If the base is a pyrimidine, the bonding involves the N-1

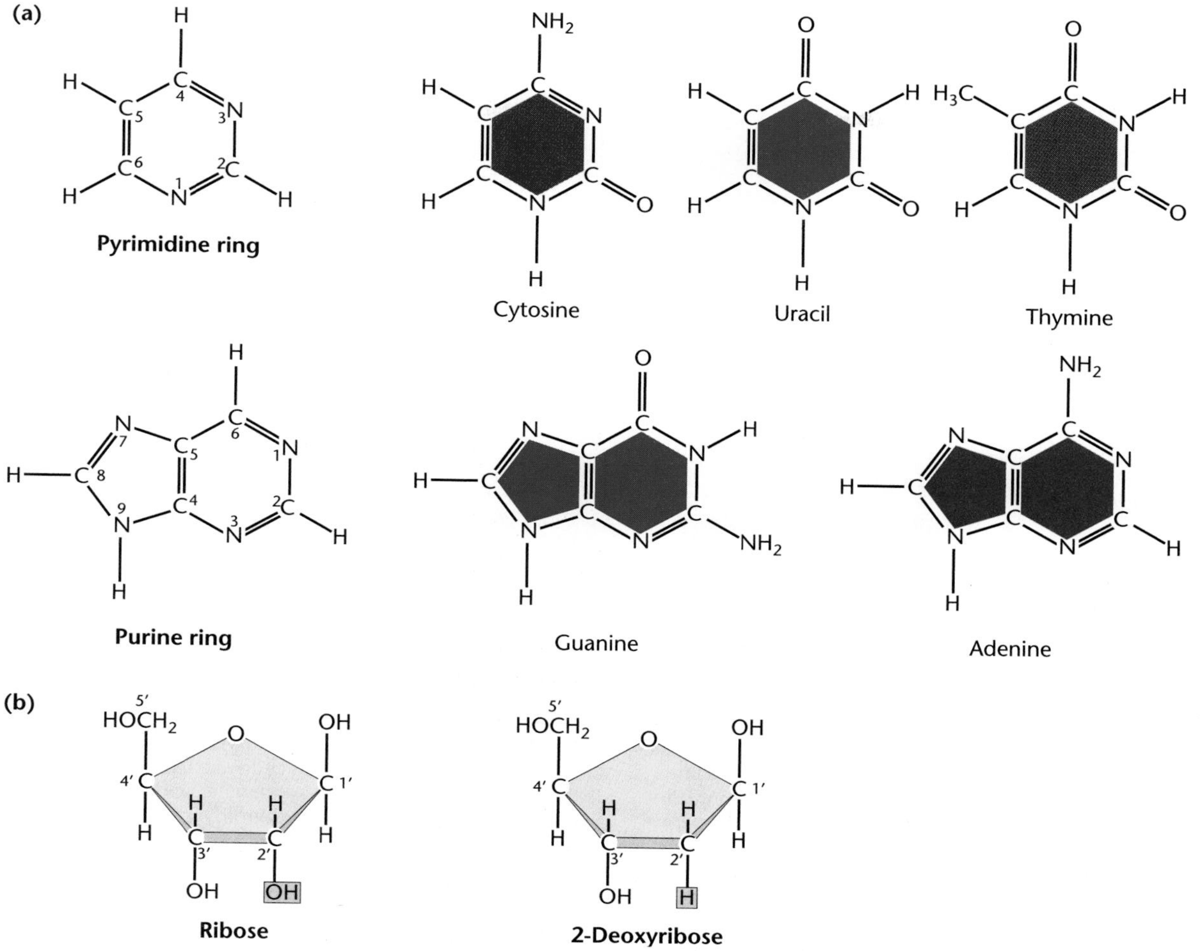

(a)

Pyrimidine ring

Cytosine

Uracil

Thymine

Purine ring

Guanine

Adenine

(b)

Ribose

2-Deoxyribose

FIGURE 11.9 (a) Chemical structures of the pyrimidines and purines that serve as the nitrogenous bases in RNA and DNA. (b) Chemical ring structures of ribose and 2-deoxyribose, which serve as the pentose sugars in RNA and DNA, respectively.

atom. In a nucleotide, the phosphate group may be bonded to the C-2′, C-3′, or C-5′ atom of the sugar. The C-5′–phosphate configuration is shown in Figure 11.10. It is by far the most prevalent one in biological systems and the one found in DNA and RNA.

Nucleoside Diphosphates and Triphosphates

Nucleotides are also described by the term **nucleoside monophosphate (NMP)**. The addition of one or two phosphate groups results in **nucleoside diphosphates (NDPs)** and **triphosphates (NTPs),** as illustrated in Figure 11.11. The triphosphate form is significant because it serves as the precursor molecule during nucleic acid synthesis within the cell (see Chapter 12). Additionally, the triphosphates **adenosine triphosphate (ATP)** and

guanosine triphosphate (GTP) are important in the cell's bioenergetics because of the large amount of energy involved in the addition or removal of the terminal phosphate group. The hydrolysis of ATP or GTP to ADP or GDP and inorganic phosphate (P_i) is accompanied by the release of a large amount of energy in the cell. When the chemical conversion of ATP or GTP is coupled to other reactions, the energy produced may be used to drive the reactions. As a result, both ATP and GTP are involved in many cellular activities, including numerous genetic events.

Polynucleotides

The linkage between two mononucleotides consists of a phosphate group linked to two sugars. A **phosphodiester**

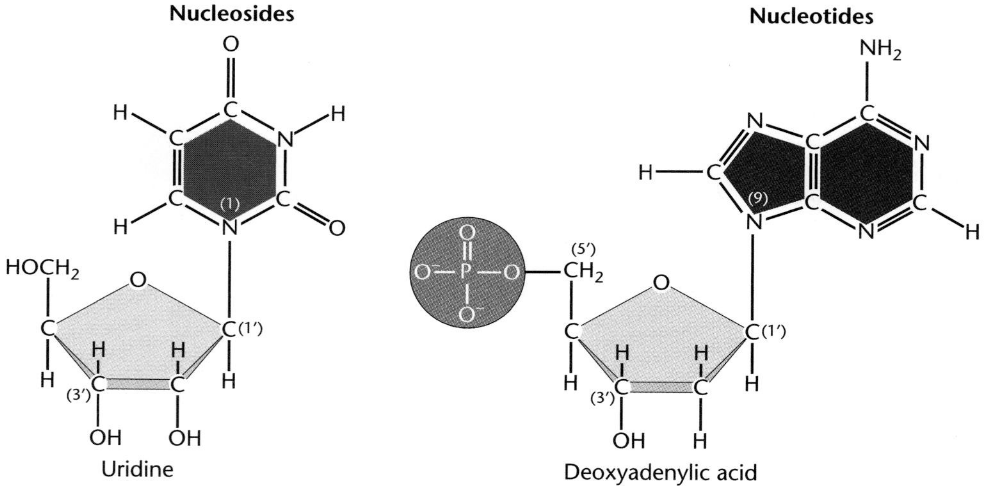

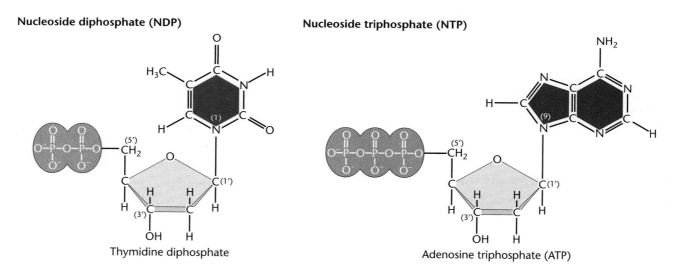

Uridine

Deoxyadenylic acid

Ribonucleosides	Ribonucleotides
Adenosine	Adenylic acid
Cytidine	Cytidylic acid
Guanosine	Guanylic acid
Uridine	Uridylic acid
Deoxyribonucleosides	**Deoxyribonucleotides**
Deoxyadenosine	Deoxyadenylic acid
Deoxycytidine	Deoxycytidylic acid
Deoxyguanosine	Deoxyguanylic acid
Deoxythymidine	Deoxythymidylic acid

FIGURE 11.10 The structures and names of the nucleosides and nucleotides of RNA and DNA.

Nucleoside diphosphate (NDP)

Nucleoside triphosphate (NTP)

Thymidine diphosphate

Adenosine triphosphate (ATP)

FIGURE 11.11 The basic structures of nucleoside diphosphates and triphosphates, as illustrated by thymidine diphosphate and deoxyadenosine triphosphate.

bond is formed, because phosphoric acid has been joined to two alcohols (the hydroxyl groups on the two sugars) by an ester linkage on both sides. Figure 11.12(a) shows the resultant phosphodiester bond in DNA. The same bond is found in RNA. Each structure has a **C-5′ end** and a **C-3′ end.** The joining of two nucleotides forms a **dinucleotide;** of three nucleotides, a **trinucleotide;** and so forth. Short chains consisting of fewer than 20 nucleotides linked together are called **oligonucleotides.** Still longer chains are referred to as **polynucleotides.**

Because drawing the structures in Figure 11.12(a) is time-consuming and complex, a schematic shorthand method has been devised [Figure 11.12(b)]. The nearly vertical lines represent the pentose sugar; the nitrogenous base is attached at the top, or the C-1′ position. The diagonal line, with the P in the middle of it, is attached to the C-3′ position of one sugar and the C-5′ position of the neighboring sugar; it represents the phosphodiester bond. Several modifications of this shorthand method are in use, and they can be understood in terms of these guidelines.

Although Levene's tetranucleotide hypothesis (described earlier in this chapter) was generally accepted before 1940, research in subsequent decades revealed it to be incorrect. It was shown that DNA does not necessarily contain equimolar quantities of the four bases. Additionally, the molecular weight of DNA molecules was determined to be in the range of 10^6 to 10^9 daltons, far in excess of that of a tetranucleotide. The current view of DNA is that it consists of exceedingly long polynucleotide chains.

Long polynucleotide chains would account for the observed molecular weight and provide the basis for the most important property of DNA—storage of vast quantities of genetic information. If each nucleotide position in this long chain is occupied by any one of four nucleotides, extraordinary variation is possible. For example, a polynucleotide that is only 1000 nucleotides in length can be arranged 4^{1000} different ways, each one different from all other possible sequences. This potential variation in molecular structure is essential if DNA is to serve the function of storing the vast amounts of chemical information necessary to direct cellular activities.

The structure of DNA holds the key to understanding its function

The previous sections in this chapter have established that DNA is the genetic material in all organisms (with certain viruses being the exception) and have provided details as to the basic chemical components making up nucleic acids. What remained to be deciphered was the precise structure of DNA. That is, how are polynucleotide chains organized into DNA, which serves as the genetic material? Is DNA composed of a single chain, or more than one? If the latter is the case, how do the chains relate chemically to one another? Do the chains branch? And more important, how does the structure of this molecule relate to the various genetic functions served by DNA (i.e., storage, expression, replication, and mutation)?

From 1940 to 1953, many scientists were interested in solving the structure of DNA. Among others, Erwin Chargaff, Maurice Wilkins, Rosalind Franklin, Linus Pauling, Francis Crick, and James Watson sought information that might answer what many consider to be the most significant and intriguing question in the history of biology: "How does DNA serve as the genetic basis for the living process?" The answer was believed to depend strongly on the chemical structure and organization of the DNA molecule, given the complex but orderly functions ascribed to it.

In 1953, two young scientists, James Watson and Francis Crick, proposed that the structure of DNA is in the form of a double helix. Their proposal was published in a short paper in the journal *Nature*, which is reprinted in its entirety on pages 302–303. In a sense, this publication con-

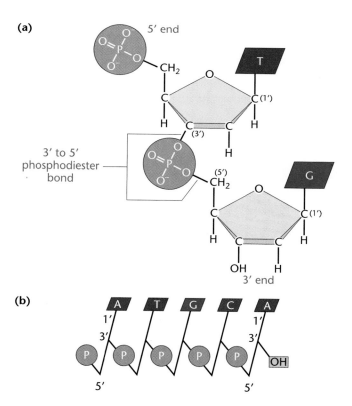

(a)

5′ end

3′ to 5′ phosphodiester bond

3′ end

(b)

A   T   G   C   A

OH

FIGURE 11.12   (a) The linkage of two nucleotides by the formation of a C-3′–C-5′ (3′–5′) phosphodiester bond, producing a dinucleotide. (b) A shorthand notation for a polynucleotide chain.

stituted the finish line in a highly competitive scientific race. This "race," as recounted in Watson's book *The Double Helix* (see Selected Readings), demonstrates the human interaction, genius, frailty, and intensity involved in the scientific effort that eventually led to the elucidation of DNA structure.

The data available to Watson and Crick, which was crucial to the development of their proposal, came primarily from two sources: base composition analysis of hydrolyzed samples of DNA and X-ray diffraction studies of DNA. The analytical success of Watson and Crick can be attributed to model building that conformed to the existing data. If the structure of DNA can be analogized by a puzzle, Watson and Crick, working at the Cavendish Laboratory in Cambridge, England, were the first to put together successfully all of the pieces. Given the far-reaching significance of this discovery, you may be interested to know that Watson, who entered college (University of Chicago) at age 15, was a 24-year-old postdoctoral fellow in 1953. Crick, now considered one of the great theoretical biologists of our time, was still a graduate student.

Base Composition Studies

Between 1949 and 1953, Erwin Chargaff and his colleagues used chromatographic methods to separate the four bases in DNA samples from various organisms. Quantitative methods were then used to determine the amounts of the four bases from each source. Table 11.3(a) provides some of Chargaff's original data. Parts (b) and (c) of this table show more recently derived base-composition information from various organisms that reinforce Chargaff's findings.

On the basis of these data, which you should examine, the following conclusions may be drawn:

1. The amount of adenine residues is proportional to the amount of thymine residues in the DNA of any species (columns 1, 2, and 5). Also, the amount of guanine residues is proportional to the amount of cytosine residues (columns 3, 4, and 6).

2. Based on the above proportionality, the sum of the purines (A + G) equals the sum of the pyrimidines (C + T), as shown in column 7.

3. The percentage of C + G does not necessarily equal the percentage of A + T. As we can see, the ratio between the two values varies greatly among species, as shown in column 8 and as is apparent in Table 11.3(c).

These conclusions indicate definite patterns of base composition of DNA molecules. These data also served as the initial clue to "the puzzle." Additionally, they directly re-

---

T A B L E I I . 3   DNA Base Composition Data

*(a) Chargaff's data**

| Source | Molar proportions[a] | | | | |
	1 A	2 T	3 G	4 C	
Ox thymus	26	25	21	16	
Ox spleen	25	24	20	15	
Yeast	24	25	14	13	
Avian tubercle bacilli	12	11	28	26	
Human sperm	29	31	18	18	

(c) G + C content in several organisms

Organism	%G + C
Phage T2	36.0
Drosophila	45.0
Maize	49.1
Euglena	53.5
Neurospora	53.7

(b) Base compositions of DNAs from various sources

Source	Base composition				Base ratio		A + T/G + C ratio	
	1 A	2 T	3 G	4 C	5 A/T	6 G/C	7 (A + G)/(C + T)	8 (A+ T)/(C + G)
Human	30.9	29.4	19.9	19.8	1.05	1.00	1.04	1.52
Sea urchin	32.8	32.1	17.7	17.3	1.02	1.02	1.02	1.58
E. coli	24.7	23.6	26.0	25.7	1.04	1.01	1.03	0.93
Sarcina lutea	13.4	12.4	37.1	37.1	1.08	1.00	1.04	0.35
T7 bacteriophage	26.0	26.0	24.0	24.0	1.00	1.00	1.00	1.08

Source: From Chargaff, 1950.
[a]Moles of nitrogenous constituent per mole of P (often, the recovery was less than 100 percent).

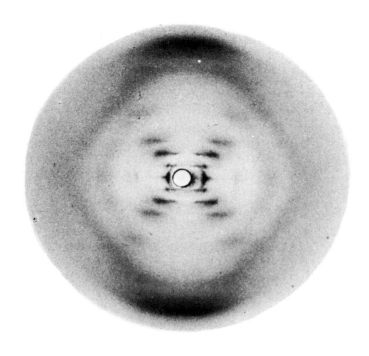

FIGURE 11.13 An X-ray diffraction photograph of the B form of purified DNA fibers. The strong arcs on the periphery reflect information about closely spaced aspects of the molecule, providing an estimate of the periodicity of nitrogenous bases, which are 3.4 Å apart. The inner cross pattern of spots is indicative of the grosser aspect of the molecule, indicating its helical nature.

fute the tetranucleotide hypothesis, which stated that all four bases are present in equal amounts.

X-Ray Diffraction Analysis

When fibers of a DNA molecule are subjected to X-ray bombardment, these rays are scattered according to the molecule's atomic structure. The pattern of scatter (diffraction) can be captured as spots on photographic film and analyzed, particularly for the overall shape of and regularities within the molecule. This process, **X-ray diffraction** analysis, was successfully applied to the study of protein structure by Linus Pauling and other chemists. The technique had been attempted on DNA as early as 1938 by William Astbury. By 1947, he had detected a periodicity (a repeating unit) of 3.4 Å* that suggested to him that the bases were stacked like pennies on top of one another.

Between 1950 and 1953, Rosalind Franklin, working in the laboratory of Maurice Wilkins, obtained improved X-ray data from more purified samples of DNA (Figure 11.13). Her work confirmed the 3.4 Å periodicity seen by Astbury and suggested that the structure of DNA was some sort of helix. However, she did not propose a definitive model. Paul-

*Today, measurement in nanometers (nm) is favored (1 nm = 10 Å).

ing had analyzed the work of Astbury and others and incorrectly proposed that DNA was a triple helix.

The Watson–Crick Model

Watson and Crick published their analysis of DNA structure in 1953 (their original paper is reproduced on pp. 302–303). By building models under the constraints of the information just discussed, they proposed the double-helical form of DNA, as shown in Figure 11.14(a). This model has the following major features:

1. Two long polynucleotide chains are coiled around a central axis, forming a right-handed double helix.

2. The two chains are **antiparallel;** that is, their C-5′-to-C-3′ orientations run in opposite directions.

3. The bases of both chains are flat structures, lying perpendicular to the axis; they are "stacked" on one another, 3.4 Å (0.34 nm) apart, and are located on the inside of the helix.

4. The nitrogenous bases of opposite chains are paired to one another as the result of the formation of **hydrogen bonds** (described in the following discussion); in DNA, only A═T and G≡C pairs are allowed.

5. Each complete turn of the helix is 34 Å (3.4 nm) long; thus, 10 bases exist in each chain per turn.

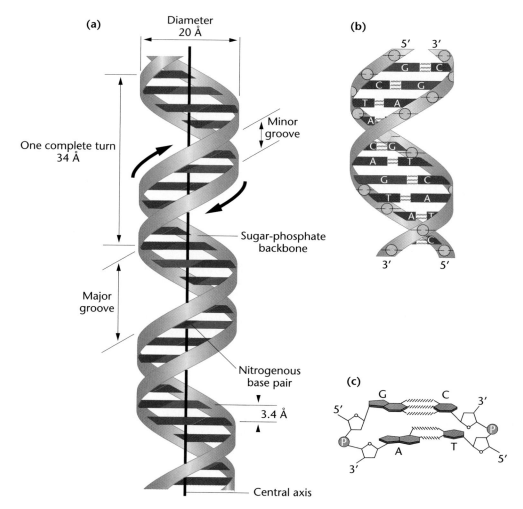

(a)

Diameter
20 Å

One complete turn
34 Å

Minor
groove

Sugar-phosphate
backbone

Major
groove

Nitrogenous
base pair

3.4 Å

Central axis

(b)

5'   3'

G   C
C   G
T   A
A   T
C   G
A   T
G   C
T   A
A   T
C

3'   5'

(c)

G   C
5'   3'
P   P
A   T
P   5'
3'

FIGURE 11.14 (a) A schematic representation of the DNA double helix as proposed by Watson and Crick. The ribbonlike strands constitute the sugar-phosphate backbones, and the horizontal rungs constitute the nitrogenous base pairs, of which there are 10 per complete turn. The major and minor grooves are apparent. A solid vertical bar representing the central axis has been placed through the center of the helix; (b) A more detailed view of the model, depicting the bases, sugars, phophates, and hydrogen bonds of the helix. (c) A demonstration of the antiparallel nature of the helix and the horizontal stacking of the bases.

6. In any segment of the molecule, alternating larger **major grooves** and smaller **minor grooves** are apparent along the axis.

7. The double helix measures 20 Å (2.0 nm) in diameter.

The nature of base-pairing (Point 4) is the most genetically significant feature of the model. Before discussing it in detail, several other important features warrant emphasis. First, the antiparallel nature of the two chains is a key part of the double-helix model. While one chain runs in the 5'-to-3' orientation (what seems right-side up to us), the other chain is in the 3'-to-5' orientation (and thus appears upside down). This is illustrated in Figure 11.14(b). Given the constraints of the bond angles of the various nu-

cleotide components, the double helix could not be constructed easily if both chains ran parallel to one another.

Second, the right-handed nature of the helix is best appreciated by comparing such a structure to its left-handed counterpart, which is a mirror image, as shown in Figure 11.15. The conformation in space of the right-handed helix is most consistent with the data that were available to Watson and Crick. As we shall see momentarily, an alternative form of DNA (Z-DNA) does exist as a left-handed helix.

The key to the model proposed by Watson and Crick is the specificity of base pairing. Chargaff's data had suggested that the amounts of A equaled T and that G equaled C. Watson and Crick realized that if A pairs with T and C pairs with G, thus accounting for these proportions, the

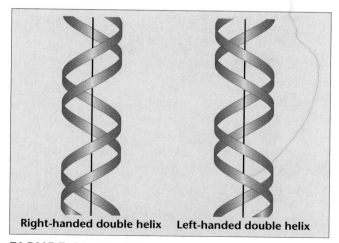

Right-handed double helix **Left-handed double helix**

FIGURE 11.15 The right- and left-handed helical forms of DNA. Note that they are mirror images of one another.

members of each such base pair formed hydrogen bonds [Figure 11.14(b)], providing the chemical stability necessary to hold the two chains together. Arranged in this way, both **major** and **minor grooves** become apparent along the axis. Further, a purine (A or G) opposite a pyrimidine (T or C) on each "rung of the spiral staircase" of the proposed helix accounts for the 20Å (2 nm) diameter suggested by X-ray diffraction studies.

The specific A=T and G≡C base pairing is the basis for the concept of **complementarity**. This term describes the chemical affinity provided by the hydrogen bonds between the bases. As we will see, this concept is very important in the processes of DNA replication and gene expression.

Two questions are particularly worthy of discussion. First, why aren't other base pairs possible? Watson and Crick discounted the A=G and C=T pairs because these represent purine-purine and pyrimidine-pyrimidine pairings, respectively. These pairings would lead to alternating diameters of more than and less than 20 Å because of the respective sizes of the purine and pyrimidine rings; in addition, the three-dimensional configurations formed by such pairings do not produce the proper alignment leading to sufficient hydrogen-bond formations. It is for this reason that A=C and G=T pairings were also discounted, even though these pairs each consist of one purine and one pyrimidine.

The second question concerns hydrogen bonds. Just what is the nature of such a bond, and is it strong enough to stabilize the helix? A **hydrogen bond** is a very weak electrostatic attraction between a covalently bonded hydrogen atom and an atom with an unshared electron pair. The hydrogen atom assumes a partial positive charge, while the unshared electron pair—characteristic of covalently bonded oxygen and nitrogen atoms—assumes a par-

tial negative charge. These opposite charges are responsible for the weak chemical attraction. As oriented in the double helix (Figure 11.16), adenine forms two hydrogen bonds with thymine, and guanine forms three hydrogen bonds with cytosine. Although two or three hydrogen bonds taken alone are very weak, two or three thousand bonds in tandem (which would be found in two long polynucleotide chains) are capable of providing great stability to the helix.

Still another stabilizing factor is the arrangement of sugars and bases along the axis. In the Watson–Crick model, the hydrophobic or "water-fearing" nitrogenous bases are "stacked" almost horizontally on the interior of the axis, shielded from water. The hydrophilic sugar-phosphate backbone is on the outside of the axis, where both components may interact with water. These molecular arrangements provide significant chemical stabilization to the helix.

A more recent and accurate analysis of the form of DNA that served as the basis for the Watson–Crick model has revealed a minor structural difference. A precise measurement of the number of base pairs (bp) per turn has demonstrated a value of 10.4 rather than the 10.0 predicted by Watson and Crick. Where in the classic model each base pair is rotated around the helical axis 36°, relative to the adjacent base pair, the new finding requires a rotation of 34.6°. This results in slightly more than 10 base pairs per 360° turn.

The Watson–Crick model had an immediate effect on the emerging discipline of molecular biology. Even in their initial 1953 article, the authors observed, "It has not escaped our notice that the specific pairing we have postulated immediately suggests a possible copying mechanism for the genetic material." Two months later, in a second article in *Nature*, Watson and Crick pursued this idea, suggesting a specific mode of replication of DNA—the semiconservative model. The second article also alluded to two new concepts: the storage of genetic information in the sequence of the bases and the mutation or genetic change that would result from alteration of the bases. These ideas have received vast amounts of experimental support since 1953 and are now universally accepted.

The "synthesis" of ideas by Watson and Crick was a remarkable feat and was highly significant to subsequent studies in genetics and biology. The nature of the gene and its role in genetic mechanisms could now be viewed and studied in biochemical terms. Recognition of the work leading to the double-helical model subsequently came in the form of the Nobel Prize in Physiology or Medicine awarded in 1962 to Watson, Crick, and Wilkins. Unfortunately, Rosalind Franklin had died in 1958 at the age of 37, making her contributions ineligible for consideration, since the award is not given posthumously. The Nobel Prize was to be one of many such awards bestowed for work in the field of molecular genetics.

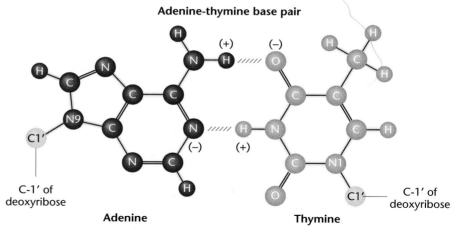

Adenine-thymine base pair

C-1' of
deoxyribose

Adenine

Thymine

C-1' of
deoxyribose

Guanine-cytosine base pair

C-1' of
deoxyribose

C-1' of
deoxyribose

Guanine

Cytosine

//////// Hydrogen bond

FIGURE 11.16 Ball-and-stick models of A $=$ T and G $\equiv$ C base pairs. The cross-hatched strips (//////) represent the hydrogen bonds that form between bases.

Molecular Structure of Nucleic Acids: A Structure for Deoxyribose Nucleic Acid

We wish to suggest a structure for the salt of deoxyribose nucleic acid (D. N. A.). This structure has novel features which are of considerable biological interest. A structure for nucleic acid has already been proposed by Pauling and Corey.[1] They kindly made their manuscript available to us in advance of publication. Their model consists of three intertwined chains, with the phosphates near the fibre axis, and the bases on the outside. In our opinion, this structure is unsatisfactory for two reasons: (1) We believe that the material which gives the X-ray diagrams is the salt, not the free acid. Without the acidic hydrogen atoms it is not clear what forces would hold the structure together, especially as the negatively charged phosphates near the axis will repel each other. (2) Some of the van der Waals distances appear to be too small.

Another three-chain structure has also been suggested by Fraser (in the press). In his model the phosphates are on the outside and the bases on the inside, linked together by hydrogen bonds. This structure as described is rather ill-defined, and for this reason we shall not comment on it.

We wish to put forward a radically different structure for the salt of deoxyribose nucleic acid. This structure has two helical chains each coiled round the same axis. We have made the usual chemical assumptions, namely, that each chain consists of phosphate diester groups joining β-D-deoxyribofuranose residues with 3′,5′ linkages. The two chains (but not their bases) are related by a dyad perpendicular to the fibre axis. Both chains follow right-handed helices, but owing to the dyad the sequences of the atoms in the two chains run in opposite directions. Each chain loosely resembles Furberg's[2] model No. 1; that is, the bases are on the inside of the helix and the phosphates on the outside. The configuration of the sugar and the atoms near it is close to Furberg's "standard configuration," the sugar being roughly perpendicular to the attached base. There is a residue on each chain every 3.4 Å in the z-direction. We have assumed an angle of 36° between adjacent residues in the same chain, so that the structure repeats after 10 residues on each chain, that is, after 34 Å. The distance of a phosphorus atom from the fibre axis is 10 Å. As the phosphates are on the outside, cations have easy access to them.

The structure is an open one, and its water content is rather high. At lower water contents we would expect the bases to tilt so that the structure could become more compact.

The novel feature of the structure is the manner in which the two chains are held together by the purine and pyrimidine bases. The planes of the bases are perpendicular to the fibre axis. They are joined together in pairs, a single base from one chain being hydrogen-bonded to a single base from the other chain, so that the two lie side by side with identical z-co-ordinates. One of the pair must be a purine and the other a pyrimidine for bonding to occur. The hydrogen bonds are made as follows: purine position 1 to pyrimidine position 1; purine position 6 to pyrimidine position 6.

If it is assumed that the bases only occur in the structure in the most plausible tautomeric forms (that is, with the keto rather than the enol configurations) it is found that only specific pairs of bases can bond together. These pairs are: adenine (purine) with thymine (pyrimidine), and guanine (purine) with cytosine (pyrimidine).

In other words, if an adenine forms one member of a pair, on either chain, then on these assumptions the other member must be thymine; similarly for guanine and cytosine. The sequence of bases on a single chain does not appear to be restricted in any way. However, if only specific pairs of bases can be formed, it follows that if the sequence of bases on one chain is given, then the sequence on the other chain is automatically determined.

It has been found experimentally[3,4] that the ratio of the amounts of adenine to thymine, and the ratio of guanine to cytosine, are always very close to unity for deoxyribose nucleic acid.

It is probably impossible to build this structure with a ribose sugar in place of the deoxyribose, as the extra oxygen atom would make too close a van der Waals contact.

The previously published X-ray data[5,6] on deoxyribose nucleic acid are insufficient for a rigorous test of our structure. So far as we can tell, it is roughly compatible with the experimental data, but it must be regarded as unproved until it has been checked against more exact results. Some of these are given in the following communications. We were not aware of the details of the results presented there when we devised our structure, which rests mainly though not entirely on published experimental data and stereochemical arguments.

It has not escaped our notice that the specific pairing we have postulated immediately suggests a possible copying mechanism for the genetic material.

Full details of the structure, including the conditions assumed in building it, together with a set of co-ordinates for the atoms, will be published elsewhere.

We are much indebted to Dr. Jerry Donohue for constant advice and criticism, especially on interatomic distances. We have also been stimulated by a knowledge of the general nature of the unpublished experimental results and ideas of Dr. M. H. F. Wilkins, Dr. R. E. Franklin and their co-workers at King's College, London. One of us (J. D. W.) has been aided by a fellowship from the National Foundation for Infantile Paralysis.

Medical Research Council Unit for the Study of the Molecular Structure of Biological Systems, Cavendish Laboratory, Cambridge, England.

J. D. WATSON
F. H. C. CRICK

[1]Pauling, L., and Corey, R. B., *Nature*, 171, 346 (1953); *Proc. U.S. Nat. Acad. Sci.*, 39, 84 (1953).

[2]Furberg, S., *Acta Chem. Scand.*, 6, 634 (1952).

[3]Chargaff, E., for references see Zamenhof, S., Brawerman, G., and Chargaff, E., *Biochim. et Biophys. Acta*, 9, 402 (1952).

[4]Wyatt, G. R., *J. Gen. Physiol*, 36, 201 (1952).

[5]Astbury, W. T., *Symp. Soc. Exp. Biol. 1, Nucleic Acid*, 66 (Camb. Univ. Press, 1947).

[6]Wilkins, M. H. F., and Randall, J. T., *Biochim. et Biophys. Acta*, 10, 192 (1953).

Alternative forms of DNA exist

Under different conditions of isolation, several conformational forms of DNA have been recognized. At the time Watson and Crick performed their analysis, two forms—**A-DNA** and **B-DNA**—were known. Watson and Crick's analysis was based on X-ray studies of the B form by Rosalind Franklin, which is present under aqueous, low-salt conditions and is believed to be the biologically significant conformation.

Although DNA studies around 1950 relied on the use of X-ray diffraction, more recent investigations have been performed using **single-crystal X-ray analysis.** The earlier studies achieved resolution of about 5 Å, but single crystals diffract X rays at about 1 Å, near atomic resolution. As a result, every atom is "visible," and much greater structural detail is available during analysis.

Using these modern techniques, A-DNA has now been scrutinized. It is prevalent under high-salt or dehydration conditions. In comparison to B-DNA (Figure 11.17), A-DNA is slightly more compact, with 11 base pairs in each complete turn of the helix, which is 23 Å in diameter. While it is also a right-handed helix, the orientation of the bases is somewhat different. They are tilted and displaced laterally in relation to the axis of the helix. As a result of these differences, the appearance of the major and minor grooves is modified compared with those in B-DNA. It seems doubtful that A-DNA occurs under biological conditions.

Three other right-handed forms of DNA helices have been discovered when investigated under laboratory conditions. These have been designated C-, D-, and E-DNA. **C-DNA** is found under even greater dehydration conditions than those observed during the isolation of A- and B-DNA. It has only 9.3 base pairs per turn and is, thus, less compact. Its helical diameter is 19 Å. Like A-DNA, C-DNA does not have its base pairs lying flat; rather, they are tilted relative to the axis of the helix. Two other forms, **D-DNA** and **E-DNA,** occur in helices lacking guanine in

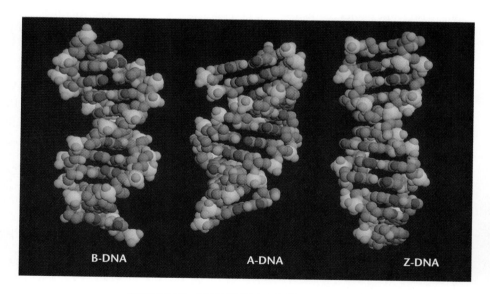

B-DNA   A-DNA   Z-DNA

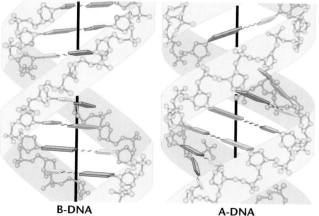

B-DNA   A-DNA

FIGURE II.17   The top half of the figure shows computer-generated space-filling models of B-DNA (left), A-DNA (center), and Z-DNA (right). Below is an artist's depiction illustrating the orientation of the base pairs of B-DNA and A-DNA. (Note that in B-DNA the base pairs are perpendicular to the helix, while they are tilted and pulled away from the helix in A-DNA.)

their base composition. They have even fewer base pairs per turn: 8 and $7\frac{1}{2}$, respectively.

Still another form of DNA, called **Z-DNA,** was discovered by Andrew Wang, Alexander Rich, and their colleagues in 1979 when they examined a synthetic DNA oligonucleotide containing only C-G base pairs. Z-DNA takes on the rather remarkable configuration of a left-handed double helix (Figure 11.17). Like A- and B-DNA, Z-DNA consists of two antiparallel chains held together by Watson–Crick base pairs. Beyond these characteristics, Z-DNA is quite different. The left-handed helix is 18 Å (1.8 nm) in diameter, contains 12 base pairs per turn, and assumes a zigzag conformation (hence its name). The major groove present in B-DNA is nearly eliminated in Z-DNA.

More recent research by Jean-François Allemand and colleagues has shown that if DNA is artificially stretched, still another form of the molecule is assumed, which they have called **P-DNA.** A model of P-DNA contrasted with the B form of DNA is quite interesting since it is longer, more narrow, and the phosphate groups, found on the outside of B-DNA, are present inside the molecule. The nitrogenous bases, present inside the helix in B-DNA, are found closer to the external surface of the helix in P-DNA, and there are 2.62 bases per turn, in contrast to the 10.4 per turn in B-DNA.

The interest in alternative forms of DNA, such as Z and P, stems from the belief that DNA might have to assume a structure other that the B-form as it functions as the genetic material. During both replication and transcription (when its RNA complement is synthesized during gene expression), the strands of the helix must separate and become accessible to large enzymes, as well as a variety of other proteins involved in these processes. It is possible that changes in the shapes of the DNA facilitate these functions. Unique conformations might serve as points of molecular recognition for proteins.

However, verification of the biological significance of alternative forms awaits the clear demonstration that they exist *in vivo*.

The structure of RNA is chemically similar to DNA, but single-stranded

The second category of nucleic acids is the ribonucleic acids, or RNA. The structure of these molecules is similar to DNA, with several important exceptions. Although RNA also has as its building blocks nucleotides linked into polynucleotide chains, the sugar ribose replaces deoxyribose and the nitrogenous base uracil replaces thymine. Another important difference is that most RNA is usually considered to be single-stranded. However, RNA molecules sometimes fold back on themselves to form double-stranded regions following their synthesis; such a configuration results when regions of complementarity occur in positions that allow base pairs to form. Furthermore, some animal viruses that have RNA as their genetic material contain it in the form of double-stranded helices. Thus, there are several instances where RNA does not exist strictly as a linear single-stranded molecule.

At least three major classes of cellular RNA molecules function during the expression of genetic information: **ribosomal RNA (rRNA), messenger RNA (mRNA),** and **transfer RNA (tRNA).** These molecules all originate as complementary copies of one of the two strands of DNA segments during the process of transcription. That is, their nucleotide sequence is complementary to the deoxyribonucleotide sequence of DNA, which served as the template for their synthesis. Because uracil replaces thymine in RNA, uracil is complementary to adenine during transcription and during RNA base pairing.

Table 11.4 characterizes the major forms of RNA found in prokaryotic and eukaryotic cells. Different RNAs

TABLE 11.4   RNA Characterization

RNA Class	% Total RNA*	Components (Svedberg Coefficient)	Eukaryotic (E) or Prokaryotic (P)	Number of Nucleotides
Ribosomal (rRNA)	80	5*S*	P and E	120
		5.8*S*	E	160
		16*S*	P	1542
		18*S*	E	1874
		23*S*	P	2904
		28*S*	E	4718
Transfer (tRNA)	15	4*S*	P and E	75–90
Messenger (mRNA)	5	varies	P and E	100–10,000

*In *E. coli*.

are distinguished according to their sedimentation behavior in a centrifugal field and their size, as measured by the number of nucleotides each contains. Sedimentation behavior depends on a molecule's density, mass, and shape, and its measure is called the **Svedberg coefficient (S).** While higher S values almost always designate molecules of greater molecular weight, the correlation is not direct; that is, a two-fold increase in molecular weight does not lead to a two-fold increase in S. This is because that, in addition to a molecule's mass, the size and the shape of the molecule also impact on its rate of sedimentation (S). As you can see, a wide variation exists in the size of the three classes of RNA.

Ribosomal RNA is generally the largest of these molecules (as is generally reflected in its S values) and usually constitutes about 80 percent of all RNA in the cell. Ribosomal RNAs are important structural components of **ribosomes,** which function as a nonspecific workbench during the synthesis of proteins during the process of translation. The various forms of rRNA found in prokaryotes and eukaryotes differ distinctly in size.

Messenger RNA molecules carry genetic information from the DNA of the gene to the ribosome, where translation occurs. They vary considerably in size, which is a reflection of the variation in the size of the protein encoded by the mRNA as well as the gene serving as the template for transcription of mRNA.

Transfer RNA, the smallest class of RNA molecules, carries amino acids to the ribosome during translation. Because more than one tRNA molecule interacts simultaneously with the ribosome, the molecule's smaller size facilitates these interactions.

We will discuss the functions of the three classes of RNA in much greater detail in Chapter 13. In addition, as we proceed through the text, we will encounter other unique RNAs that perform various roles. For example, **small nuclear RNA (snRNA)** participates in processing mRNAs (Chapter 13). **Telomerase RNA** is involved in DNA replication at the ends of chromosomes (Chapter 12). And **antisense RNA** is involved in gene regulation (Chapter 16). Our purpose in this section has been to contrast the structure of DNA, which stores genetic information, with that of RNA, which most often functions in the expression of that information.

Many analytical techniques have been useful during the investigation of DNA and RNA

Since 1953, the role of DNA as the genetic material and the role of RNA in transcription and translation have been clarified through detailed analysis of nucleic acids. We will consider several methods of analysis of these molecules in this chapter. Some of these, as well as other research procedures, are presented in greater detail in Appendix A.

Absorption of Ultraviolet Light (UV)

Nucleic acids absorb ultraviolet (UV) light most strongly at wavelengths of 254 to 260 nm (see Figure 11.7) due to the interaction between UV light and the ring systems of the purines and pyrimidines. Thus, any molecule containing nitrogenous bases (i.e., nucleosides, nucleotides, and polynucleotides) can be analyzed using UV light. This technique is especially important in the localization, isolation, and characterization of nucleic acids.

Ultraviolet analysis is used in conjunction with many standard procedures that separate molecules. As we shall see in the next section, the use of UV absorption is critical to the isolation of nucleic acids following their separation.

Sedimentation Behavior

Nucleic acid mixtures can be separated by subjecting them to one of several possible **gradient centrifugation** procedures (Figure 11.18). The mixture can be loaded on top of a solution prepared so that a concentration gradient has been formed from top to bottom. Then the entire mixture is centrifuged at high speeds in an ultracentrifuge. The mixture of molecules will migrate downward, with each component moving at a different rate. Centrifugation is stopped and the gradient eluted from the tube. Each fraction can then be measured spectrophotometrically for absorption at 260 nm. In this way, the previous position of a nucleic acid fraction along the gradient can be predicted and the fraction isolated and studied further.

The gradient centrifugations described rely on the sedimentation behavior of molecules in solution. Two major types of gradient centrifugation techniques are employed in the analysis of nucleic acids: sedimentation equilibrium and sedimentation velocity. Both require the use of high-speed centrifugation to create large centrifugal forces upon molecules in a gradient solution.

In **sedimentation equilibrium centrifugation** (sometimes called **density gradient centrifugation**), a density gradient is created that overlaps the densities of the individual components of a mixture of molecules. Usually, the gradient is made of a heavy metal salt, such as cesium chloride (CsCl). During centrifugation, the molecules migrate until they reach a point of neutral buoyant density. At this point, the centrifugal force on them is equal and opposite to the upward diffusion force, and no further migration occurs. If DNAs of different densities are used, they will separate as the molecules of each density reach equilibrium with the corresponding density of CsCl. The gradient may be fractionated and the components isolated (Figure 11.18). When properly executed, this technique

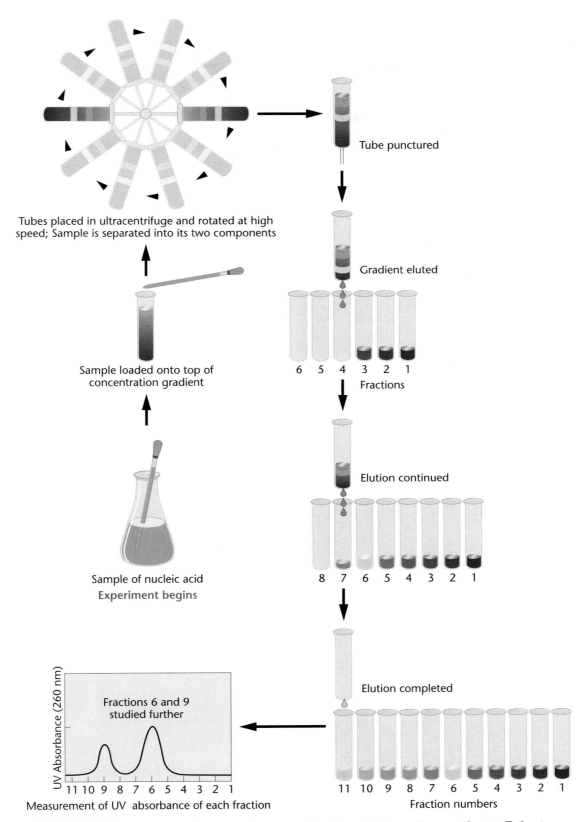

Tubes placed in ultracentrifuge and rotated at high speed; Sample is separated into its two components

Tube punctured

Gradient eluted

6 5 4 3 2 1
Fractions

Sample loaded onto top of concentration gradient

Elution continued

8 7 6 5 4 3 2 1

Sample of nucleic acid
Experiment begins

Elution completed

Fractions 6 and 9 studied further

UV Absorbance (260 nm)

11 10 9 8 7 6 5 4 3 2 1
Measurement of UV absorbance of each fraction

11 10 9 8 7 6 5 4 3 2 1
Fraction numbers

FIGURE 11.18   Separation of a mixture of two types of nucleic acid using gradient centrifugation. To fractionate the gradient, successive samples are eluted from the bottom of the tube. Each is measured for absorbance of ultraviolet light at 260 nm, producing a profile of the sample in graphic form.

provides high resolution in separating mixtures of molecules varying only slightly in density.

Sedimentation equilibrium centrifugation studies can also be used to generate data on the base composition of double-stranded DNA. The G≡C base pairs, compared with A═T pairs, are more compact and dense. As shown in Figure 11.19, the percentage of G≡C pairs in DNA is directly proportional to the molecule's buoyant density. By using this technique, we can make a useful molecular characterization of DNAs from different sources.

The second technique, **sedimentation velocity centrifugation,** employs an analytical centrifuge, which enables the migration of the molecules during centrifugation to be monitored with ultraviolet absorption optics. Thus, the "velocity of sedimentation" can be determined. This velocity has been standardized in units called Svedberg coefficients (*S*), as mentioned earlier.

In this technique, the molecules are loaded on top of the gradient, and the gravitational forces created by centrifugation drive them toward the bottom of the tube. Two forces work against this downward movement: (1) The viscosity of the solution creates a frictional resistance, and (2) part of the force of diffusion is directed upward. Under these conditions, the key variables are the mass and shape of the molecules being examined. In general, the greater the mass, the greater the sedimentation velocity. However, the molecule's shape affects the frictional resistance. Therefore, two molecules of equal mass but different shape will sediment at different rates.

One use of the sedimentation velocity technique is the determination of **molecular weight (MW).** If certain physical–chemical properties of a molecule under study are also known, the MW can be calculated based on the sedimentation velocity. The *S* values increase with molecular weight, but they are not directly proportional to it.

Denaturation and Renaturation of Nucleic Acids

When **denaturation** of double-stranded DNA occurs, the hydrogen bonds of the duplex structure break, the duplex unwinds, and the strands separate. However, no covalent bonds break. During strand separation, which can be induced by heat or chemical treatment, the viscosity of DNA decreases, and both the UV absorption and the buoyant density increase. Denaturation as a result of heating is sometimes referred to as **melting.** The increase in UV absorption of heated DNA in solution, called the **hyperchromic shift,** is easiest to measure. This effect is illustrated in Figure 11.20.

Because G≡C base pairs have one more hydrogen bond than do A═T pairs, they are more stable to heat treatment. Thus, DNA with a greater proportion of G≡C pairs than A═T pairs requires higher temperatures to denature completely. When absorption at 260 nm (OD_{260}) is monitored and plotted against temperature during heating, a **melting profile** of DNA is obtained. The midpoint of this profile, or curve, is called the **melting temperature (T_m)** and represents the point at which 50 percent of the strands are unwound or denatured (Figure 11.20). When the curve

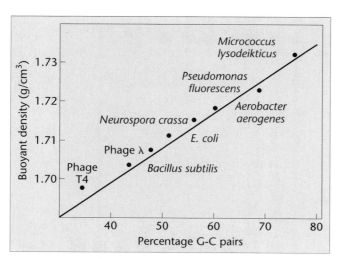

FIGURE 11.19 Percentage of guanine-cytosine (G≡C) base pairs in DNA plotted against buoyant density for a variety of microorganisms.

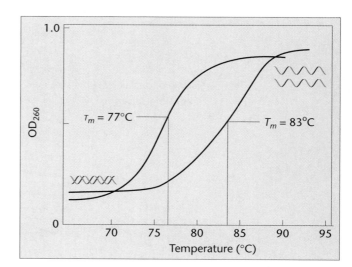

FIGURE 11.20 Comparison of the increase in UV absorbance with an increase in temperature (the hyperchromic effect) for two DNA molecules with different G≡C contents. The molecule with a melting point (T_m) of 83°C has a greater G≡C content than the molecule with a T_m of 77°C.

plateaus at its maximum optical density, denaturation is complete and only single strands exist. Analysis of melting profiles provides a characterization of DNA and an alternative method of estimating the base composition of DNA.

One might ask whether the denaturation process can be reversed; that is, can single strands of nucleic acids reform a double helix, provided that each strand's complement is present? Not only is the answer yes, but such reassociation provides the basis for several important analytical techniques that have provided much valuable information during genetic experimentation.

If DNA that has been denatured thermally is cooled slowly, random collisions between complementary strands will result in their reassociation. At the proper temperature, hydrogen bonds will re-form, securing pairs of strands into duplex structures. With time during cooling, more and more duplexes will form. Depending on the conditions, a complete match is not essential for duplex formation, provided there are at least stretches of base-pairing on two reassociating strands.

Molecular Hybridization

The property of renaturation of complementary single strands of nucleic acids is the basis for a powerful analytical technique in molecular genetics—**molecular hybridization.** This technique derives its name from the fact that renaturing single strands need not originate from the same nucleic acid source. For example, if DNA strands are isolated from different organisms and some degree of base complementarity exists between them, **molecular hybrids** will form during renaturation. Furthermore, when mixtures of DNA and RNA strands are cooled slowly, hybridization may also occur. A case in point is when RNA and the DNA from which it has been transcribed are present together (Figure 11.21). The RNA will find its single-stranded DNA complement and renature. In this example, the DNA strands are heated, causing strand separation, and then slowly cooled in the presence of single-stranded RNA. If the RNA has been transcribed on the DNA used in the experiment, and is therefore complementary to it, molecular hybridization will occur, creating a DNA:RNA duplex. Several methods are available for monitoring the amount of double-stranded molecules produced following strand separation.

In the 1960s molecular hybridization techniques contributed to our increased understanding of transcriptional events occurring at the gene level. Refinements of this process have occurred continually and have been the forerunners of work in studies of molecular evolution as well as the organization of DNA in chromosomes. Hybridization can occur in solution or when DNA is bound either to a gel or to a special type of filter, facilitating recovery of the newly formed hybrids.

The technique can even be performed using cytological preparations, which is called ***in situ* molecular hybridization.** In this procedure mitotic or interphase cells are fixed to slides and subjected to hybridization conditions. Radioactive single-stranded DNA or RNA is added, and hybridization is monitored. The nucleic acid serves as a "probe," since it will hybridize only with the specific chromosomal areas for which it is complementary. It can be either radioactive or involve fluorescence to allow its detection on the slide. In the former case, the technique of **autoradiography** can be used (see Appendix A).

Fluorescent probes are prepared in a different way. When DNA is used, it is first coupled to the small organic molecule biotin (creating biotinylated DNA). Once *in situ* hybridization is completed, another molecule (avidin or streptavidin) that has a high binding affinity for biotin is used. A fluorescent molecule such as fluorescein is linked to avidin (or streptavidin) and the complex is reacted with the cytological preparation. The above procedure represents an extremely sensitive method for localizing the hybridized DNA. Because fluorescence is used, the technique is known by the acronym **FISH (fluorescent *in situ* hybridization).**

Figure 11.22 illustrates the use of FISH in identifying the DNA specific to the centromeres of human chromosomes. The resolution of FISH is great enough to detect just a single gene within an entire set of chromosomes (see Appendix A). The use of this technique in identifying chromosomal locations housing specific genetic information has been a valuable addition to the repertoire of experimental geneticists.

Reassociation Kinetics and Repetitive DNA

One extension of molecular hybridization procedures is the technique that measures the *rate* of reassociation of complementary strands of DNA derived from a single source. This technique, called **reassociation kinetics,** was first refined and studied by Roy Britten and David Kohne.

The DNA used in such studies is first fragmented into small pieces as a result of shearing forces introduced during isolation. The resultant DNA fragments cluster around a uniform average size of several hundred base pairs. These fragments of DNA are then dissociated into single strands by heating. Next, the temperature is lowered and reassociation is monitored. During reassociation, pieces of single-stranded DNA collide randomly. If they are complementary, a stable double strand is formed; if not, they separate and are free to encounter other DNA fragments. The process continues until all matches are made.

The results of such an experiment are presented in Figure 11.23. The percentage of reassociation of DNA fragments is plotted against a logarithmic scale of the prod-

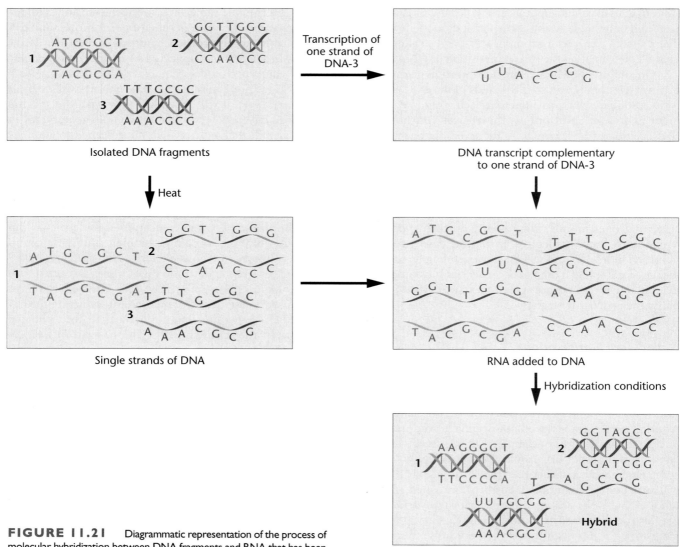

FIGURE II.21 Diagrammatic representation of the process of molecular hybridization between DNA fragments and RNA that has been transcribed on one of the single-stranded fragments. ●GenCDX

uct of C_0 (the initial concentration of DNA single strands in moles per liter of nucleotides), and t (the time, usually measured in minutes). The process of renaturation follows second-order rate kinetics according to the equation

$$\frac{C}{C_0} = \frac{1}{1 + kC_0t}$$

where C is the single-stranded DNA concentration remaining after time t has elapsed and k is the second-order rate constant. Initially, C equals C_0, and the fraction remaining single-stranded is 100 percent.

The initial shape of the curve reflects the fact that in a mixture of unique sequence fragments, each with one complement, initial matches take more time to make. Then, as many single strands are converted to duplexes, matches are made more quickly, reflecting an increase in the "slope" of the curve. Near the end of the reaction, the few remaining single strands require relatively greater time to make the final matches.

A great deal of information can be obtained from studies comparing the reassociation of DNA of different organisms. For example, we may compare the point in the reaction when one-half of the DNA is present as double-stranded fragments. This point is called the $C_0t_{1/2}$, or *half-reaction time*. Provided that all pairs of single-stranded DNA complements consist of unique nucleotide sequences and all are about the same size, $C_0t_{1/2}$ varies directly with

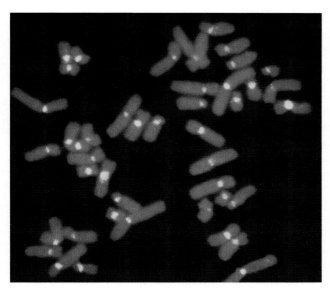

FIGURE 11.22 *In situ* hybridization of human metaphase chromosomes using a fluorescent technique (FISH). The probe, specific to centromeric DNA, produces a yellow fluorescence signal indicating hybridization. The red fluorescence is produced by propidium iodide counterstaining of chromosomal DNA.

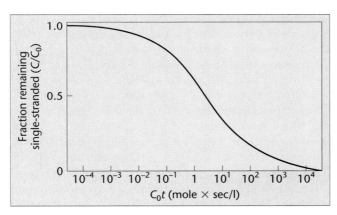

FIGURE 11.23 The ideal time course for reassociation of DNA (C/C_0) when, at time zero, all DNA consists of unique fragments of single-stranded complements. Note that the abscissa (C_0t) is plotted logarithmically.

the complexity of the DNA. Designated as X, complexity represents the length in nucleotide pairs of all unique DNA fragments laid end to end. If the DNA used in an experiment represents the entire genome, and if all DNA sequences are different from one another, then X is equal to the size of the haploid genome.

Figure 11.24 illustrates what is found when DNA from various sources are compared. As can be seen, as genome size increases, the curves obtained are shifted farther and farther to the right, indicative of an increased reassociation time.

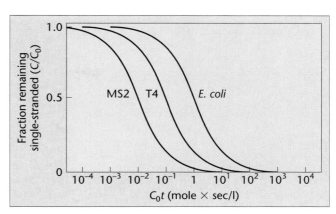

FIGURE 11.24 Comparison of the reassociation rate (C/C_0) of DNA derived from phage MS-2, phage T4, and *E. coli*. The genome of T4 is larger than MS2, and that of *E. coli* is larger than T4.

As shown in Figure 11.25, $C_0t_{1/2}$ is directly proportional to the size of the genome. If nearly the entire genome consists of unique DNA sequences, reassociation experiments can be used to determine the genome size of organisms. This method has been useful in assessing genome size in viruses and bacteria.

However, when reassociation kinetics of DNA from eukaryotic organisms were first studied, a surprising observation was made. The data showed that some of the DNA segments reassociated even more rapidly than did those derived from *E. coli*! The remainder, as expected because of its greater complexity, took longer to reassociate. For example, Britten and Kohne (cited earlier) examined DNA derived from calf thymus tissue (Figure 11.26). Based on these observations, they hypothesized that the rapidly reassociating fraction must represent **repetitive DNA sequences** present many times in the calf genome. This interpretation would explain why these DNA seg-

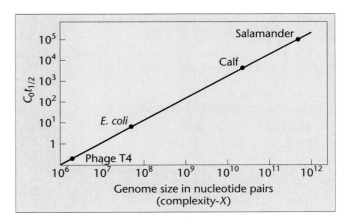

FIGURE 11.25 Comparison of $C_0t_{1/2}$ and genome size for phage T4, *E. coli*, calf, and salamander.

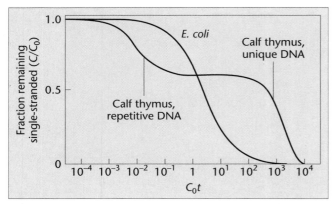

FIGURE II.26 The C_0t curve of calf thymus DNA compared with *E. coli*. The repetitive fraction of calf DNA reassociates more quickly than that of *E. coli*, while the more complex unique calf DNA takes longer to reassociate than that of *E. coli*.

ments reassociate so rapidly. On the other hand, they hypothesized that the remaining DNA segments consist of unique nucleotide sequences present only once in the genome; because there are more of these unique sequences, increasing the DNA complexity in calf thymus (compared with *E. coli*), their reassociation takes longer. The *E. coli* curve has been added to Figure 11.26 for the sake of comparison.

It is now clear that repetitive DNA sequences are prevalent in the genome of eukaryotes. When analyzed carefully, it is apparent that there are various levels of repetition. Cases are known where short DNA sequences are repeated over a million times, where longer DNA sequences are repeated only a few times, and where intermediate levels of sequence redundancy are present. We will return to this topic in Chapter 19, where we will discuss the organization of DNA in genes and chromosomes. For now, we conclude this chapter by pointing out that the discovery of repetitive DNA was one of the first clues that much of the DNA in eukaryotes is not contained in genes that encode proteins. This concept will be developed and elaborated on as we expand our coverage of the molecular basis of heredity.

Electrophoresis of Nucleic Acids

We conclude our discussion by considering one of the most essential techniques involved in the analysis of nucleic

acids, **electrophoresis.** This technique allows for the separation of different-sized fragments of DNA and RNA chains and is invaluable in current day-to-day research investigations in molecular genetics.

In general, electrophoresis separates, *or resolves*, molecules in a mixture by causing them to migrate under the influence of an electric field. A sample is placed on a porous substance (a piece of filter paper or a semisolid gel), which is placed in a solution that conducts electricity. If two molecules have approximately the same shape *and* mass, the one with the greatest net charge will migrate more rapidly toward the electrode of opposite polarity.

As electrophoretic technology was developed, from how it was initially applied to the separation of proteins, it was discovered that the use of gels of varying pore sizes significantly improved the resolution of this research technique. This advance was particularly useful for mixtures of molecules with a similar charge:mass ratio but different sizes. For example, two polynucleotide chains of different *lengths* (e.g., 10 vs. 20 nucleotides) are both negatively charged based on the phosphate group of each nucleotide. While they both move to the positively charged pole (the anode), the charge:mass ratio is the same for both chains, and separation based strictly on the electric field is minimal. However, the use of a porous medium such as **polyacrylamide gels** or **agarose gels,** which can be prepared with various pore sizes, provides the basis for the separation of these two molecules.

In such cases, *the smaller molecules migrate at a faster rate through the gel than the larger molecules* (Figure 11.27). The key to separation is based on the matrix (pores) of the gel, which *restricts migration of a larger molecule more than it restricts a smaller molecule.* The resolving power is so great that polynucleotides that vary by even one nucleotide in length are clearly separated. Once electrophoresis is complete, bands representing the variously sized molecules are identified either by autoradiography (if a component of the molecule is radioactive) or by the use of a fluorescent dye that binds to nucleic acids.

Electrophoretic separation of nucleic acids is at the heart of a variety of commonly used research techniques discussed later in the text (Chapters 18 and 21). Of particular note are the various "blotting" techniques (e.g., Southern blots and Northern blots), as well as DNA sequencing methods. We introduce electrophoresis here to complete our coverage of the analysis of nucleic acids.

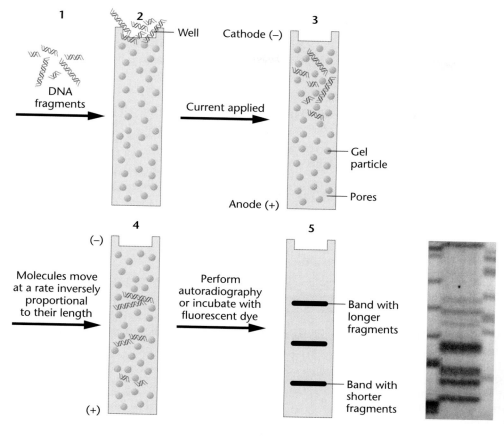

FIGURE 11.27 Diagrammatic illustration of electrophoretic separation of a mixture of DNA fragments that vary in length.

CHAPTER SUMMARY

1. The existence of a genetic material capable of replication, storage, expression, and mutation is deducible from the observed patterns of inheritance in organisms.

2. Both proteins and nucleic acids were initially considered as the possible candidates for genetic material. Proteins are more diverse than nucleic acids, a requirement for the genetic material, and were favored owing to the advances being made in protein chemistry at the time. Additionally, Levene's tetranucleotide hypothesis had underestimated the magnitude of chemical diversity inherent in nucleic acids.

3. By 1952, transformation studies and experiments using bacteria infected with bacteriophages strongly suggested that DNA is the genetic material in bacteria and most viruses.

4. Initially, only circumstantial evidence supported the concept of DNA controlling inheritance in eukaryotes. This included the distribution of DNA in the cell, quantitative analysis of DNA, and studies of UV-induced mutagenesis. Recent recombinant DNA techniques, as well as experiments with transgenic mice, have provided direct experimental evidence that the eukaryotic genetic material is DNA.

5. Numerous viruses provide an important exception to this general rule because many of them use RNA as their genetic material. These include some bacteriophages and some plant and animal viruses, as well as retroviruses.

6. Establishment of DNA as the genetic material paved the way for the expansion of molecular genetics research and has served as the cornerstone for further important studies for nearly half a century.

7. During the late 1940s and early 1950s, a considerable effort was made to integrate accumulated information on the chemical structure of nucleic acids into a model of the molecular structure of DNA. The X-ray crystallography data of Franklin and Wilkins suggested that DNA was some sort of helix. In 1953, Watson and Crick were able to assemble a model of the double-helical DNA

structure based on these X-ray diffraction studies as well as on Chargaff's analysis of base composition of DNA.

8. The DNA molecule exhibits antiparallel orientation and adenine-thymine and guanine-cytosine base-pairing complementarity along the polynucleotide chains. This model of DNA presents an obvious straightforward mechanism for its replication. The structure assumed by the helix appears to be a function of the nucleotide sequence and its chemical environment. Several alternative forms of the DNA helical structure exist. Watson and Crick described a B configuration, one of several right-handed helices. Wang and Rich discovered the left-handed Z-DNA currently being investigated for its physiological and genetic significance.

9. The second category of nucleic acids important in genetic function is RNA, which is similar to DNA with the exceptions that it is usually single-stranded, the sugar ribose replaces the deoxyribose, and the pyrimidine uracil replaces thymine. Classes of RNA—ribosomal, transfer, and messenger—facilitate the flow of information from DNA to RNA to proteins, which are the end products of most genes.

10. The structure of DNA lends itself to various forms of analyses, which have in turn led to studies of the functional aspects of the genetic machinery. Absorption of UV light, sedimentation properties, denaturation–reassociation, and electrophoresis procedures are among the important tools for the study of nucleic acids. Reassociation kinetics analysis enabled geneticists to postulate the existence of repetitive DNA in eukaryotes, where certain nucleotide sequences are present many times in the genome.

INSIGHTS AND SOLUTIONS

In contrast to the preceding chapters, this one does not emphasize genetic problem solving. Instead, it recounts some of the initial experimental analyses that served as the cornerstone of modern genetics. Quite fittingly, then, our Insights and Solutions section shifts its emphasis to experimental rationale and analytical thinking, an approach that will continue through the remainder of the text whenever appropriate.

1. (a) Based strictly on the analysis of the transformation data of Avery, MacLeod, and McCarty, what objection might be made to the conclusion that DNA is the genetic material? What other conclusion might be considered?

    Solution: Based solely on their results, it may be concluded that DNA is essential for transformation. However, DNA might have been a substance that caused capsular formation by *directly* converting nonencapsulated cells to ones with a capsule. That is, DNA may simply have played a catalytic role in capsular synthesis, leading to cells displaying smooth type III colonies.

   (b) What observations argue against this objection?

    Solution: First, transformed cells pass the trait onto their progeny cells, thus supporting the conclusion that DNA is responsible for heredity, not for the direct production of polysaccharide coats. Second, subsequent transformation studies over a period of five years showed that other traits, such as antibiotic resistance, could be transformed. Therefore, the transforming factor has a broad general effect, not one specific to polysaccharide synthesis. This observation is more in keeping with the conclusion that DNA is the genetic material.

2. If RNA were the universal genetic material, how would this have affected the Avery experiment and the Hershey–Chase experiment?

    Solution: In the Avery experiment, digestion of the soluble filtrate with RNase, rather than DNase would have eliminated transformation. Had this occurred, Avery and his colleagues would have concluded that RNA was the transforming factor. The Hershey and Chase results would not have changed, since ^{32}P would also label RNA, but not protein. Had they been using a bacteriophage with RNA as its nucleic acid, and had they known this, they would have concluded that RNA was responsible for directing the reproduction of their bacteriophage.

3. Sea urchin DNA, which is double-stranded, was shown to contain 17.5 percent of its bases in the form of cytosine (C). What percentages of the other three bases are present in this DNA?

    Solution: The amount of C = G, so guanine is also present as 17.5 percent. The remaining bases, A and T, are present in equal amounts and together they represent the rest of the bases $(100 - 35)$. Therefore, A = T = 65/2 = 32.5 percent.

4. The quest to isolate an important disease-causing organism was successful and the molecular biologists were hard at work. The organism contained as its genetic material a remarkable nucleic acid with a base composition of A = 21 percent, C = 29 percent, G = 29 percent, U = 21 percent. When heated, it showed a major hyperchromic effect, and when kinetics were studied, the nucleic acid of this organism provided the C_0t curve shown below, in contrast to that of phage

T4 and *E. coli*. T4 contains 10^5 nucleotide pairs and exhibits a $C_0t_{1/2}$ of 0.5. The unknown organism produced a $C_0t_{1/2}$ of 20. Analyze this information carefully and draw *all* possible conclusions about the genetic material of this organism, based strictly on the above observations. What important, straightforward information is missing and needed to confirm your hypothesis about the nature of this molecule?

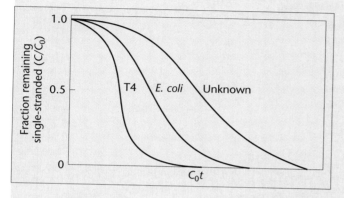

Solution: First of all, because of the presence of uracil (U), the molecule appears to be RNA. As A/U = G/C = 1, the molecule may be a double helix. The hyperchromic shift and reassociation kinetics support this hypothesis. The kinetic study demonstrates several things. First, the shape of the C_0t curve reveals that there is no repetitive sequence DNA. Further, the complexity (X), or total length of unique sequence DNA, is greater than that of either phage T4 or *E. coli*. X can, in fact, be calculated, since a direct proportionality between $C_0t_{1/2}$ and number of base pairs exists when there are only unique sequences present:

$$\frac{0.5}{10^5} = \frac{20}{X}$$

$$0.5X = 20(10^5)$$

$$X = 40(10^5)$$

$$= 4 \times 10^6 \text{ base pairs}$$

There *may* be a greater number of genes present compared to T4 or *E. coli*, but the excessive unique-sequence DNA may serve some other role, or simply play no genetic role. The missing information concerns the sugars. Our model predicts that ribose rather than d-ribose should be present. If not, the organism contains a very unusual molecule as its genetic material.

PROBLEMS AND DISCUSSION QUESTIONS

1. The functions ascribed to the genetic material are replication, expression, storage, and mutation. What does each of these terms mean?

2. Discuss the reasons why proteins were generally favored over DNA as the genetic material before 1940. What was the role of the tetranucleotide hypothesis in this controversy?

3. Contrast the various contributions made to an understanding of transformation by Griffith, by Avery and his co-workers, and by Taylor.

4. When Avery and his colleagues had obtained what was concluded to be purified DNA from the III*S* virulent cells, they treated the fraction with proteases, RNase, and DNase, followed by the assay for retention or loss of transforming ability. What were the purpose and results of these experiments? What conclusions were drawn?

5. Why were ^{32}P and ^{35}S chosen for use in the Hershey–Chase experiment? Discuss the rationale and conclusions of this experiment.

6. Does the design of the Hershey–Chase experiment distinguish between DNA and RNA as the molecule serving as the genetic material? Why or why not?

7. Would an experiment similar to that performed by Hershey and Chase work if the basic design were applied to the phenomenon of transformation? Explain why or why not.

8. What observations are consistent with the conclusion that DNA serves as the genetic material in eukaryotes? List and discuss them.

9. What are the exceptions to the general rule that DNA is the genetic material in all organisms? What evidence supports these exceptions?

10. Draw the chemical structure of the three components of a nucleotide and then link the three together. What atoms are removed from the structures when the linkages are formed?

11. How are the carbon and nitrogen atoms of the sugars, purines, and pyrimidines numbered?

12. Adenine may also be named 6-amino purine. How would you name the other four nitrogenous bases using this alternative system? ($=O$ is oxy, and $—CH_3$ is methyl.)

13. Draw the chemical structure of a dinucleotide composed of A and G. Opposite this structure, draw the dinucleotide TC in an antiparallel (or upside-down) fashion. Form the possible hydrogen bonds.

14. Describe the various characteristics of the Watson–Crick double-helix model for DNA.
15. What evidence did Watson and Crick have at their disposal in 1953? What was their approach in arriving at the structure of DNA?
16. Had Chargaff's data from a single source indicated the following, what might Watson and Crick have concluded?

	A	T	G	C
%	29	19	21	31

Why would this conclusion be contradictory to Wilkins and Franklin's data?
17. How do covalent bonds differ from hydrogen bonds? Define base complementarity.
18. List three main differences between DNA and RNA.
19. What are the three types of RNA molecules? How is each related to the concept of information flow?
20. What component of the nucleotide is responsible for the absorption of ultraviolet light? How is this technique important in the analysis of nucleic acids?
21. Distinguish between sedimentation velocity and sedimentation equilibrium centrifugation.
22. What is the basis for determining base composition using density gradient centrifugation?
23. What is the physical state of DNA following denaturation?
24. Compare the following curves representing reassociation kinetics. What can be said about the DNAs represented by each set of data compared with *E. coli*?

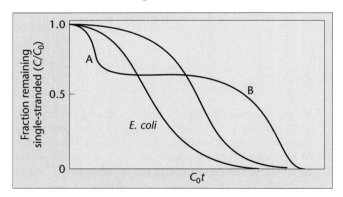

25. What is the hyperchromic effect? How is it measured? What does T_m imply?
26. Why is T_m related to base composition?
27. What is the chemical basis of molecular hybridization?
28. What did the Watson–Crick model suggest about the replication of DNA?
29. A genetics student was asked to draw the chemical structure of an adenine- and thymine-containing dinucleotide derived from DNA. His answer is shown here. The student made more than six major errors. One of them is circled, numbered 1, and explained. Find five others. Circle them, number them 2–6, and briefly explain each, following the example.

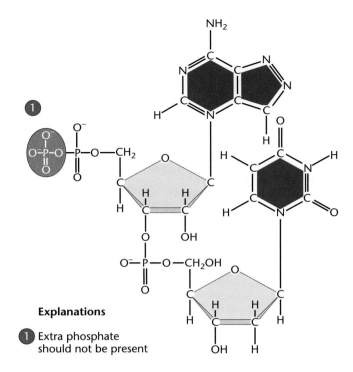

Explanations

1 Extra phosphate should not be present

30. The DNA of the bacterial virus T4 produces a $C_0t_{1/2}$ of about 0.5 and contains 10^5 nucleotide pairs in its genome. How many nucleotide pairs are present in the genome of the virus MS-2 and the bacterium *E. coli*, whose respective DNAs produce $C_0t_{1/2}$ values of 0.001 and 10.0?
31. A primitive eukaryote was discovered that displayed a unique nucleic acid as its genetic material. Analysis revealed the following observations:
    (i) X-ray diffraction studies display a general pattern similar to DNA, but with somewhat different dimensions and more irregularity.
    (ii) A major hyperchromic shift is evident upon heating and monitoring UV absorption at 260 nm.
    (iii) Base composition analysis reveals four bases in the following proportions.

Adenine	=	8%
Guanine	=	37%
Xanthine	=	37%
Hypoxanthine	=	18%

    (iv) About 75 percent of the sugars are d-ribose, while 25 percent are ribose. Attempt to solve the structure of this molecule by postulating a model that is consistent with the above observations.
32. Considering the information in this chapter on B- and Z-DNA and right- and left-handed helices, carefully analyze the structures here and draw conclusions about the helical nature of areas (a) and (b). Which is right-handed and which is left-handed?

GENETICS MediaLab

The following resources will help you achieve a better understanding of the concepts presented in this chapter. These resources can be found on the CD packaged with this textbook and on the Companion Website at **http://www.prenhall.com/klug/**.

CD Resources:
Animated Tutorial: *Hershey–Chase Experiment*

Animated Tutorial: *Reassociation Kinetics and C_0t*

Self-grading Chapter Problems

Web Resources:
Web Destinations in Genetics

Self-grading Chapter Problems

Chapter Search Terms

Genetics Newsgroups

Student Bulletin Board

Web Problem 1:
Time for completion = 10 minutes

How is the DNA molecule suited for its function of encoding genetic information? The DNA molecule is simply a polymer of four kinds of nucleotides, and yet this molecule encodes information for diverse processes in a myriad of distinct organisms. In this exercise you will examine the structure of DNA by viewing four interactive modules illustrating different aspects of DNA structure. You can manipulate each three-dimensional model with buttons that rotate, zoom in on and out from, highlight, and erase structural elements. Discuss how DNA is both uniform and variable. Include how the backbone, bases, major and minor grooves, and 5'- and 3'-ends of the strands affect these qualities. To complete this exercise, visit Web Problem 1 in Chapter 11 of your Companion Website and select the keyword **DNA**.

Web Problem 2:
Time for completion = 10 minutes

In what ways do forms of DNA differ? Data from X-ray crystallography has contributed to an understanding of the three-dimensional properties of DNA. In this exercise, you will examine molecular models of nucleic acids for the A, B, and Z forms of DNA. Read the brief introduction to nucleic acids. Click on the hypertext links for A, B, and Z DNA to view the images. List the differences among these three major forms of DNA. Describe the effect of a base-pair mismatch. To complete this exercise, visit Web Problem 2 in Chapter 11 of your Companion Website and select the keywords **FORMS** and **DNA MODEL**. For a brief overview of the X-ray crystallography process select the keyword **X RAY**.

Web Problem 3:
Time for completion = 10 minutes

Many molecular biology methods are dependent on complementarity between strands of DNA or between DNA and RNA, but what is the relationship between sequences and denaturation? Melting temperature, T_m, of a sequence provides a measure of how easily a double-stranded sequence will denature. In this exercise you will calculate the estimated T_m and other parameters for a given sequence. Enter a sequence in the query box using the letters A, G, C, and T to represent the nucleotides. Click on the calculate button to perform your analysis. Repeating this process, design a set of input sequences to answer the following questions. What is the relation between T_m and the length of a sequence? Between T_m and GC content? What can you conclude about optical density (OD) and molecular weight (MW) of different nucleotides? To complete this exercise, visit Web Problem 3 in Chapter 11 of your Companion Website and select the keyword **TM** for the Mac version or **PRIMER** for the PC version.

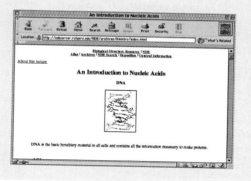

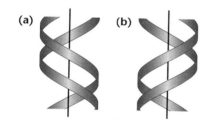

(a)   **(b)**

33. One of the most common spontaneous lesions that occurs in DNA under physiological conditions is the hydrolysis of the amino group of cytosine, converting it to uracil. What would be the effect on DNA structure of a uracil group replacing cytosine?

34. In some organisms, cytosine is methylated at carbon 5 of the pyrimidine ring after it is incorporated into DNA. If a 5-methyl cytosine is then hydrolyzed, as described in Problem 33, what base will be generated?

EXTRA-SPICY PROBLEMS

35. *Newsdate: March 1, 2015.* A unique creature has been discovered during exploration of outer space. Recently, its genetic material has been isolated and analyzed. This material is similar in some ways to DNA in its chemical makeup. It contains in abundance the 4-carbon sugar erythrose and a molar equivalent of phosphate groups. Additionally, it contains six nitrogenous bases: adenine (A), guanine (G), thymine (T), cytosine (C), hypoxanthine (H), and xanthine (X). These bases exist in the following relative proportions:

$$A = T = H \text{ and } C = G = X$$

X-ray diffraction studies have established a regularity in the molecule and a constant diameter of about 30 Å.

Together, these data have suggested a model for the structure of this molecule.
(a) Propose a general model of this molecule. Describe it briefly.
(b) What base-pairing properties must exist for H and for X in the model?
(c) Given the constant diameter of 30 Å, do you think that *either* (i) both H and X are purines or both pyrimidines, *or* (ii) one is a purine and one is a pyrimidine?

36. You are provided with DNA samples from two newly discovered bacterial viruses. Based on the various analytical techniques discussed in this chapter, construct a research protocol that would be useful in characterizing and contrasting the DNA of both viruses. For each technique that you include in the protocol, indicate the type of information you hope to obtain.

SELECTED READINGS

ADLEMAN, L.M. 1998. Computing with DNA. *Sci. Am.* (Aug.)279: 54–61.

ALLOWAY, J.L. 1933. Further observations on the use of pneumococcus extracts in effecting transformation of type *in vitro. J. Exp. Med.* 57:265–78.

AVERY, O.T., MACLEOD, C.M., and MCCARTY, M. 1944. Studies on the chemical nature of the substance inducing transformation of pneumococcal types: Induction of transformation by a desoxyribonucleic acid fraction isolated from pneumococcus type III. *J. Exp. Med.* 79:137–58. (Reprinted in Taylor, J.H. 1965. *Selected papers in molecular genetics.* Orlando, FL: Academic Press.)

BRITTEN, R.J., and KOHNE, D.E. 1968. Repeated sequences in DNA. *Science* 161:529–40.

CAIRNS, J., STENT, G.S., and WATSON, J.D. 1966. *Phage and the origins of molecular biology.* Cold Spring Harbor, NY: Cold Spring Harbor Laboratory Press.

CHARGAFF, E. 1950. Chemical specificity of nucleic acids and mechanism for their enzymatic degradation. *Experientia* 6:201–9.

CRICK, F.H.C., WANG, J.C., and BAUER, W.R. 1979. Is DNA really a double helix? *J. Mol. Biol.* 129:449–61.

DAVIDSON, J.N. 1976. *The biochemistry of the nucleic acids,* 8th ed. Orlando, FL: Academic Press.

DAWSON, M.H. 1930. The transformation of pneumococcal types: I. The interconvertibility of type-specific *S. pneumococci. J. Exp. Med.* 51:123–47.

DEROBERTIS, E.M., and GURDON, J.B. 1979. Gene transplantation and the analysis of development. *Sci. Am.* (Dec.) 241:74–82.

DICKERSON, R.E. 1983. The DNA helix and how it is read. *Sci. Am.* (June) 249:94–111.

DICKERSON, R.E. et al. 1982. The anatomy of A-, B-, and Z-DNA. *Science* 216:475–85.

DUBOS, R.J. 1976. *The professor, the Institute and DNA: Oswald T. Avery, his life and scientific achievements.* New York: Rockefeller University Press.

FELSENFELD, G. 1985. DNA. *Sci. Am.* (Oct.) 253:58–78.

FRAENKEL-CONRAT, H., and SINGER, B. 1957. Virus reconstruction: II. Combination of protein and nucleic acid from different strains.

Biochem. Biophys. Acta 24:530–48. (Reprinted in Taylor, J.H. 1965. *Selected papers in molecular genetics,* Orlando, FL: Academic Press.)

FRANKLIN, R.E., and GOSLING, R.G. 1953. Molecular configuration in sodium thymonucleate. *Nature* 171:740–41.

GEIS, I. 1983. Visualizing the anatomy of A, B and Z-DNAs. *J. Biomol. Struc. Dynam.* 1: 581–91.

GRIFFITH, F. 1928. The significance of pneumococcal types. *J. Hyg.* 27:113–59.

GUTHRIE, G.D., and SINSHEIMER, R.L. 1960. Infection of protoplasts of *Escherichia coli* by subviral particles. *J. Mol. Biol.* 2:297–305.

HERSHEY, A.D., and CHASE, M. 1952. Independent functions of viral protein and nucleic acid in growth of bacteriophage. *J. Gen. Phys.* 36:39–56. (Reprinted in Taylor, J.H. 1965. *Selected papers in molecular genetics.* Orlando, FL: Academic Press.)

HOTCHKISS, R.D. 1951. Transfer of penicillin resistance in pneumococci by the desoxyribonucleate derived from resistant cultures. *Cold Spring Harbor Symp. Quant. Biol.* 16:457–61. (Reprinted in Adelberg, E.A. 1960. *Papers on bacterial genetics.* Boston: Little, Brown.)

JUDSON, H. 1979. *The eighth day of creation: Makers of the revolution in biology.* New York: Simon & Schuster.

LEVENE, P.A., and SIMMS, H.S. 1926. Nucleic acid structure as determined by electrometric titration data. *J. Biol. Chem.* 70:327–41.

MCCARTY, M. 1980. Reminiscences of the early days of transformation. *Annu. Rev. Genet.* 14:1–16.

———. 1985. *The transforming principle: Discovering that genes are made of DNA.* New York: W. W. Norton.

MCCONKEY, E.H. 1993. *Human genetics—The molecular revolution.* Boston: Jones and Bartlett.

OLBY, R. 1974. *The path to the double helix.* Seattle: University of Washington Press.

PALMITER, R.D., and BRINSTER, R.L. 1985. Transgenic mice. *Cell* 41:343–45.

PAULING, L., and COREY, R.B. 1953. A proposed structure for the nucleic acids. *Proc. Natl. Acad. Sci. USA* 39:84–97.

RICH, A., NORDHEIM, A., and WANG, A.H.-J. 1984. The chemistry and biology of left-handed Z-DNA. *Annu. Rev. Biochem.* 53:791–846.

SCHILDKRAUT, C.L., MARMUR, J., and DOTY, P. 1962. Determination of the base composition of deoxyribonucleic acid from its buoyant density in CsCl. *J. Mol. Biol.* 4:430–43.

SPIZIZEN, J. 1957. Infection of protoplasts by disrupted T2 viruses. *Proc. Natl. Acad. Sci. USA* 43:694–701.

STENT, G.S., ed. 1981. *The double helix: Text, commentary, review, and original papers.* New York: W. W. Norton.

STEWART, T.A., WAGNER, E.F., and MINTZ, B. 1982. Human β-globin gene sequences injected into mouse eggs, retained in adults, and transmitted to progeny. *Science* 217:1046–48.

VARMUS, H. 1988. Retroviruses. *Science* 240:1427–35.

WATSON, J.D. 1968. *The double helix.* New York: Atheneum.

WATSON, J.D., and CRICK, F.C. 1953a. Molecular structure of nucleic acids: A structure for deoxyribose nucleic acids. Nature 171:737–38.

———. 1953b. Genetic implications of the structure of deoxyribose nucleic acid. *Nature* 171:964.

WEINBERG, R.A. 1985. The molecules of life. *Sci. Am.* (Oct.) 253:48–57.

WILKINS, M.H F., STOKES, A.R., and WILSON, H.R. 1953. Molecular structure of desoxypentose nucleic acids. *Nature* 171:738–40.

YUNG, J. 1996. New FISH probes—The end in sight. *Nature Genetics* 14:10–12.

ZIMMERMAN, B. 1982. The three-dimensional structure of DNA. *Annu. Rev. Biochem.* 51:395–428.

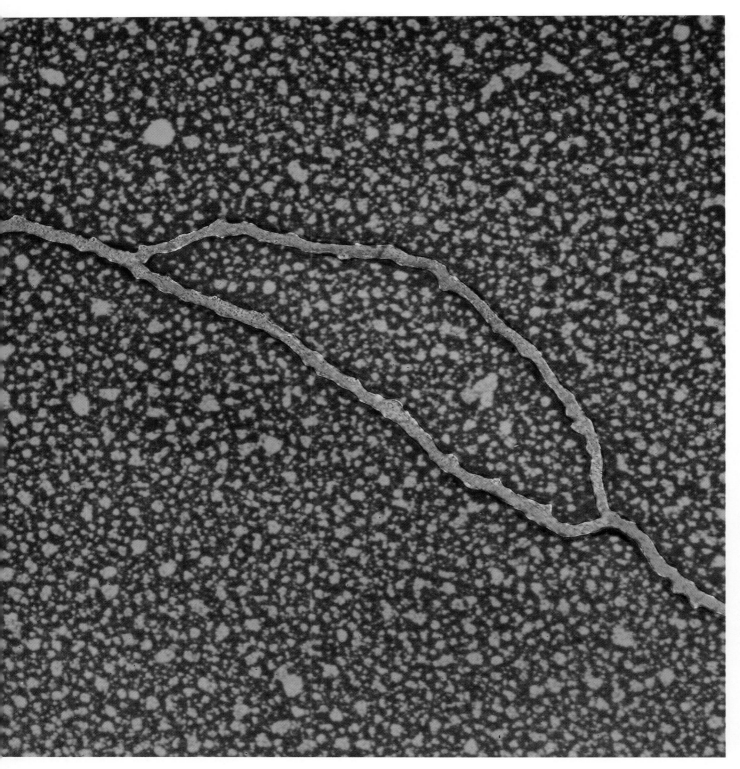

Transmission electron micrograph of human DNA from a HeLa cell, illustrating the replication bubble that characterizes DNA replication within a single replicon.

DNA Replication and Recombination

KEY CONCEPTS

- **DNA is reproduced by semiconservative replication**

  The Meselson–Stahl Experiment
  Semiconservative Replication in Eukaryotes
  Origins, Forks, and Units of Replication

- **DNA synthesis in bacteria involves three polymerases as well as other enzymes**

  DNA Polymerase I
  Synthesis of Biologically Active DNA
  DNA Polymerase II and III

- **Many complex issues must be resolved during DNA replication**

- **The DNA helix must be unwound**

- **Initiation of DNA synthesis requires an RNA primer**

- **Antiparallel strands require continuous and discontinuous DNA synthesis**

- **Concurrent synthesis occurs on the leading and lagging strands**

- **Proofreading and error correction are an integral part of DNA replication**

- **A coherent model summarizes DNA replication**

- **Replication is controlled by a variety of genes**

- **Eukaryotic DNA synthesis is similar to synthesis in prokaryotes but more complex**

  Multiple Replication Origins
  Eukaryotic DNA Polymerases

- **The ends of linear chromosomes are problematic during replication**

- **DNA recombination, like DNA replication, is directed by specific enzymes**

- **Gene conversion is a consequence of DNA recombination**

🔵 GenCDX

When you see this icon, there are related animations and exercises on the CD accompanying this text.

Following Watson and Crick's proposal for the structure of DNA, scientists focused their attention on how this molecule is replicated. This essential function of the genetic material must be executed precisely if genetic continuity between cells is to be maintained following cell division. It is an enormous, complex task. Consider for a second that about 10^9 (3 billion) base pairs exist within the 23 chromosomes of the human genome. To duplicate faithfully the DNA of just one of these chromosomes requires a mechanism of extreme precision. An error rate of only 10^{-6} (one in a million) will still create 3000 errors, obviously an excessive number, during each replication cycle of the genome. While not error free, a much more accurate system of DNA replication has evolved that functions in all organisms.

As Watson and Crick noted in their 1953 paper, the structure of the double helix itself suggests how replication could occur. This mode, called *semiconservative replication*, has since received strong experimental support from studies of viruses, prokaryotes, and eukaryotes.

Once the general mode of replication was made clear, research was intensified to determine the precise details of DNA synthesis. So far, we know that numerous enzymes and other proteins are needed to copy a DNA helix. Because of the complexity of the chemical events during synthesis, this subject remains an extremely active area of research.

This chapter will examine the general mode of replication as well as the specific details of DNA synthesis. The research leading to this knowledge is another link in our understanding of life processes at the molecular level.

DNA is reproduced by semiconservative replication

It was apparent to Watson and Crick that because of the arrangement and nature of the nitrogenous bases, each strand of a DNA double helix could serve as a template for the synthesis of its complementary strand (Figure 12.1). They proposed that if the helix were unwound, each nu-

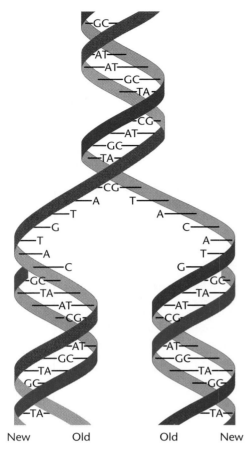

FIGURE 12.1 Generalized model of semiconservative replication of DNA. New synthesis is shown in blue.

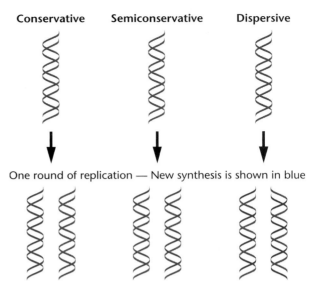

FIGURE 12.2 Results of one round of replication of DNA for each of the three possible modes by which replication could be accomplished.

cleotide along the two parent strands would have an affinity for its complementary nucleotide. As we learned in Chapter 11, the complementarity is due to the potential hydrogen bonds that can be formed. If thymidylic acid (T) were present, it would "attract" adenylic acid (A); if guanidylic acid (G) were present, it would "attract" cytidylic acid (C); and so on. If these nucleotides were then covalently linked into polynucleotide chains along both templates, the result would be the production of two identical double strands of DNA. Each replicated DNA molecule would consist of one "old" and one "new" strand; hence, the process is known as **semiconservative replication.**

Two other modes that also rely on the parental strands as a template were considered as possible ways in which DNA could replicate (Figure 12.2). In **conservative replication,** complementary polynucleotide chains are synthesized as described before, but the two newly created strands then come together and the parental strands reassociate. The original helix is thus "conserved."

In the second alternative mode, called **dispersive replication,** the parental strands cleave during replication

and the DNA is then dispersed into two new double helices. Thus, each strand consists of both old and new DNA. This mode is the most complex of the three possibilities and is therefore considered to be least likely to occur. It could not, however, be ruled out as an experimental model. Figure 12.2 shows the theoretical results of a single round of replication by each of the three different modes.

The Meselson–Stahl Experiment

In 1958, Matthew Meselson and Franklin Stahl published the results of an experiment providing strong evidence that semiconservative replication is the mode used by bacterial cells to produce new DNA molecules. They grew *E. coli* cells for many generations in a medium where $^{15}NH_4Cl$ (ammonium chloride) was the only nitrogen source. A "heavy" isotope of nitrogen, ^{15}N contains one more neutron than the naturally occurring ^{14}N isotope. Unlike radioactive isotopes, ^{15}N is stable. DNA that contains ^{15}N may be distinguished from ^{14}N-containing DNA by the use of **sedimentation equilibrium centrifugation,** where samples are "forced" by centrifugation through a density gradient of a heavy metal salt, such as cesium chloride (see Chapter 11). The more dense ^{15}N-containing DNA will reach equilibrium in the gradient at a point closer to the bottom (where the density is greater) than will ^{14}N-containing DNA.

In Meselson and Stahl's experiment, after many generations, all nitrogen-containing molecules in the *E. coli* cells, including the nitrogenous bases of DNA, contained ^{15}N. The cells were then transferred to a medium containing only $^{14}NH_4Cl$. All subsequent synthesis of DNA

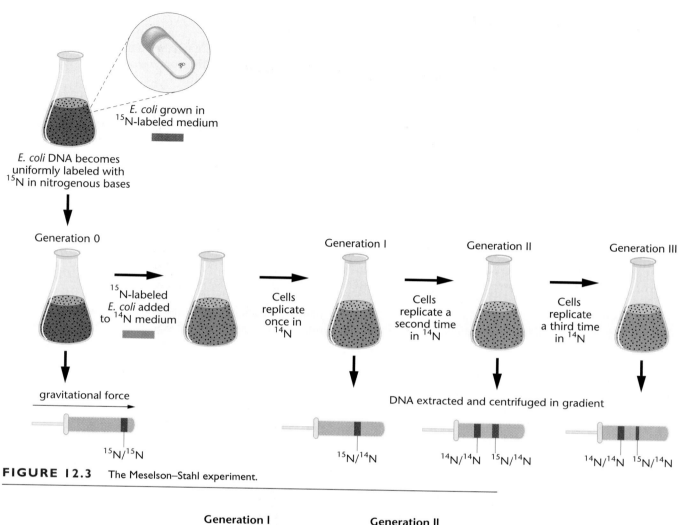

FIGURE 12.3 The Meselson–Stahl experiment.

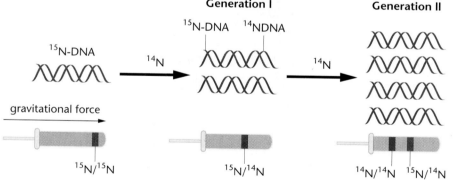

FIGURE 12.4 The expected results of two generations of semiconservative replication in the Meselson–Stahl experiment.

during replication would therefore contain the "lighter" isotope of nitrogen. The *E. coli* cells were allowed to replicate during several generations with cell samples removed at various intervals, with the time of transfer taken as time zero (*t* = 0). From each sample, DNA was isolated and subjected to sedimentation equilibrium centrifugation. The results are depicted in Figure 12.3.

After one generation, the isolated DNA was all present in a single band of intermediate density—the expected result for semiconservative replication, where each replicated molecule was composed of one new ^{14}N-strand and one old ^{15}N-strand (Figure 12.4). This result was not consistent with conservative replication, in which two distinct bands would have been predicted to occur.

After two cell divisions, DNA samples showed two density bands, an intermediate one and a lighter one that corresponded to the ^{14}N position in the gradient. Similar results occurred after a third generation, except that the proportion of the ^{14}N band increased. This was again consistent with the interpretation that replication is semiconservative.

You may have realized that a molecule exhibiting intermediate density is also consistent with dispersive replication. However, Meselson and Stahl ruled out this mode of replication on the basis of two observations. First, after the first generation of replication in an ^{14}N-containing medium, they isolated the hybrid molecule and heat-denatured it. Recall from Chapter 11 that heating will separate a duplex into single strands. When the densities of the single strands of the hybrid were determined, they exhibited *either* an ^{15}N-profile *or* an ^{14}N-profile, but *not* an intermediate density. This observation is consistent with the semiconservative mode but inconsistent with the dispersive mode.

Furthermore, if replication were dispersive, *all* generations after $t = 0$ would demonstrate DNA of an intermediate density. In each generation after the first, the ratio of ^{15}N/^{14}N would decrease and the hybrid band would become lighter and lighter, eventually approaching the ^{14}N band. This result was not observed. The Meselson–Stahl experiment provided conclusive support for semiconservative replication in bacteria and tended to rule out the conservative and dispersive modes.

Semiconservative Replication in Eukaryotes

In 1957, the year before the work of Meselson and his colleagues was published, J. Herbert Taylor, Philip Woods, and Walter Hughes presented evidence that semiconservative replication also occurs in eukaryotes. They experimented with root tips of the broad bean *Vicia faba*, which are an excellent source of dividing cells. These researchers were able to monitor the process of replication by labeling DNA with ^{3}H-thymidine, a radioactive precursor of DNA, and performing autoradiography.

Autoradiography, discussed in detail in Appendix A, is a cytological procedure that allows an isotope to be localized within the cell. In this procedure, a photographic emulsion is placed over a section of cellular material (root tips in this experiment), and the preparation is stored in the dark. The slide is then developed, much as photographic film is processed. Because the radioisotope emits energy, the emulsion turns black at the approximate point of emission following development. The end result is the presence of dark spots or "grains" on the surface of the section, identifying the location of newly synthesized DNA within the cell.

Taylor and his colleagues grew root tips for approximately one generation in the presence of the radioisotope and then placed them in unlabeled medium, where cell division continued. At the conclusion of each generation, they arrested the cultures at metaphase by adding colchicine (a chemical derived from the crocus plant that "poisons" the spindle fibers), and then examined the chromosomes by autoradiography. They found labeled thymidine only in chromosomes that contained newly synthesized DNA. Figure 12.5 illustrates replication of a single chromosome over two division cycles, including the distribution of grains.

These results are compatible with the semiconservative mode of replication. After the first replication cycle in the presence of the isotope, both sister chromatids show radioactivity, indicating that each chromatid contains one "new" radioactive DNA strand and one "old" unlabeled strand. After the second replication cycle, which takes place in unlabeled medium, only one of the two sister chromatids of each chromosome should be radioactive. With only minor exceptions of **sister chromatid exchanges** (see Chapter 6), this result was observed.

Together, the Meselson–Stahl experiment and the experiment by Taylor, Woods, and Hughes soon led to the general acceptance of the semiconservative mode of replication. Later studies with other organisms reached the same conclusion and also strongly supported Watson and Crick's proposal for the double-helix model of DNA.

Origins, Forks, and Units of Replication

Semiconservative replication is the general mode by which DNA is duplicated. Let us flesh out our understanding of this pattern by considering briefly a number of relevant issues. The first concerns the **origin of replication.** Where along the chromosome is DNA replication initiated? Is there only a single origin, or does DNA synthesis begin at more than one point? Is any given point of origin random or is it located at a specific region along the chromosome? Second, once replication begins, does it proceed in a single direction or in both directions away from the origin? In other words, is replication **unidirectional** or **bidirectional?**

To address these issues, we need to introduce two terms. First, at the actual point along the chromosome where replication is occurring, the strands of the helix are unwound, creating what is called a **replication fork.** Such a fork will initially appear at the point of origin of synthesis and then move along the DNA duplex as replication proceeds. If replication is bidirectional, two such forks will be present, migrating in opposite directions away from the origin. Second, the length of DNA that is replicated following one initiation event at a single origin is a unit referred to as the **replicon.**

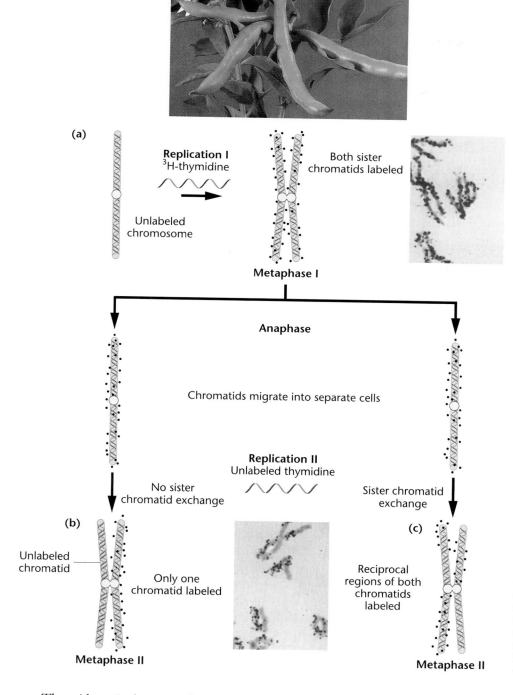

(a)

Replication I
^{3}H-thymidine

Unlabeled chromosome

Both sister chromatids labeled

Metaphase I

Anaphase

Chromatids migrate into separate cells

Replication II
Unlabeled thymidine

No sister chromatid exchange

Sister chromatid exchange

(b)

Unlabeled chromatid

Only one chromatid labeled

Reciprocal regions of both chromatids labeled

(c)

Metaphase II

Metaphase II

FIGURE 12.5 Depiction of the experiment by Taylor, Woods, and Hughes, demonstrating the semiconservative mode of replication of DNA in root tips of *Vicia faba*. A portion of the plant is shown in the top photograph. (a) An unlabeled chromosome proceeds through the cell cycle in the presence of ^{3}H-thymidine. As it enters mitosis, both sister chromatids of the chromosome are labeled, as shown by autoradiography. After a second round of replication, this time in the absence of ^{3}H-thymidine, only one chromatid of each chromosome is expected to be surrounded by grains (b). Except where a reciprocal exchange has occurred between sister chromatids (c), the expectation was upheld. The micrographs are of the actual autoradiograms obtained in the experiment.

The evidence is clear regarding the origin and direction of replication. John Cairns tracked replication in *E. coli* using both radioisotopes and autoradiography. He was able to demonstrate that there is only a single region where replication is initiated. In *E. coli* this specific region, called *oriC*, has been mapped along the chromosome. It consists of 245 base pairs, though only a small number are actually essen-

tial to the initiation of DNA synthesis. Since there is but a single point of origin of DNA synthesis in bacteriophages and bacteria, the entire chromosome constitutes one replicon. The presence of only a single origin is characteristic of bacteria, which have only one circular chromosome.

Results put forward by other researchers, again relying on autoradiography, demonstrated that replication is bidi-

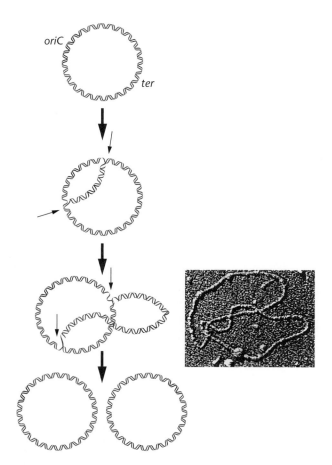

FIGURE 12.6 Bidirectional replication of the *E. coli* chromosome. The thin black arrows identify the advancing replication forks. The micrograph is of a bacterial chromosome in the process of replication, comparable to the figure next to it.

rectional, moving away from *oriC* in both directions (Figure 12.6). This creates two replication forks that migrate farther and farther apart as replication proceeds. These forks eventually merge as semiconservative replication of the entire chromosome is completed at a termination region, called *ter*.

DNA synthesis in bacteria involves three polymerases as well as other enzymes

While replication is semiconservative and bidirectional, this information indicates only the pattern of DNA duplication and the association of finished strands with one another once synthesis is completed. A more complex issue is how the actual *synthesis* of long complementary polynucleotide chains occurs on a DNA template. As in most molecular biological studies, this question was first approached by using microorganisms. Research began about

the same time as the Meselson–Stahl work, and the topic is still an active area of investigation. What is most apparent in this research is the tremendous chemical complexity of the biological synthesis of DNA.

DNA Polymerase I

In 1957 Arthur Kornberg and his colleagues isolated an enzyme from *E. coli* that was able to direct DNA synthesis in a cell-free (*in vitro*) system. The enzyme is now called **DNA polymerase I,** as it was the first of several similar enzymes to be isolated.

Kornberg determined that there were two major requirements for *in vitro* DNA synthesis under the direction of DNA polymerase I:

1. All four deoxyribonucleoside triphosphates (dATP, dCTP, dGTP, dTTP = dNTP).*
2. Template DNA.

If any one of the four deoxyribonucleoside triphosphates was omitted from the reaction, no measurable synthesis occurred. If derivatives of these precursor molecules other than the nucleoside triphosphate were used (nucleotides or nucleoside diphosphates), synthesis also did not occur. If no template DNA was added, synthesis of DNA occurred, but was reduced greatly.

Most of the synthesis directed by Kornberg's enzyme appeared to be exactly the type required for semiconservative replication. The reaction is summarized in Figure 12.7, which depicts the addition of a single nucleotide. The enzyme has since been shown to consist of a single polypeptide containing 928 amino acids.

The way in which each nucleotide is added to the growing chain is a function of the specificity of DNA polymerase I. As shown in Figure 12.8, the precursor dNTP contains the three phosphate groups attached to the 5′-carbon of d-ribose. As the two terminal phosphates are cleaved during synthesis, the remaining phosphate attached to the 5′-carbon is covalently linked to the 3′-OH group of the d-ribose to which it is added. Thus, **chain elongation** occurs in the **5′-to-3′ direction** by the addition of one nucleotide at a time to the growing 3′ end. Each step provides a newly exposed 3′-OH group that can participate in the next addition of a nucleotide as DNA synthesis proceeds.

Having shown how DNA was synthesized, Kornberg sought to demonstrate the accuracy, or fidelity, with which the enzyme replicated the DNA template. Because the

_____

*dNTP designates the deoxyribose forms of the four nucleoside triphosphates; in a similar way, dNMP refers to the monophosphate forms.

FIGURE 12.7 The chemical reaction catalyzed by DNA polymerase I. During each step, a single nucleotide is added to the growing complement of the DNA template using a nucleoside triphosphate as the substrate. The release of inorganic pyrophosphate drives the reaction energetically.

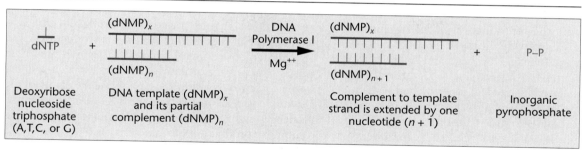

| $dNTP$ | + | $(dNMP)_x$ / $(dNMP)_n$ | DNA Polymerase I / Mg^{++} | $(dNMP)_x$ / $(dNMP)_{n+1}$ | + | $P–P$ |

Deoxyribose nucleoside triphosphate (A,T,C, or G)

DNA template $(dNMP)_x$ and its partial complement $(dNMP)_n$

Complement to template strand is extended by one nucleotide $(n + 1)$

Inorganic pyrophosphate

technology to determine the nucleotide sequences of the template and the product was not yet available in 1957, he had to rely initially on several indirect methods.

One of Kornberg's approaches was to compare the nitrogenous base compositions of the DNA template with those of the recovered DNA product. Table 12.1 shows Kornberg's base composition analysis of three DNA templates. Within experimental error, the base composition of each product agreed with the template DNAs used. These data, along with other types of comparisons of template and product (see Problem 29 at the end of this chapter), suggested that the templates were replicated faithfully.

Synthesis of Biologically Active DNA

Despite Kornberg's extensive work, not all researchers were convinced that DNA polymerase I was the enzyme that replicates DNA within cells (*in vivo*). The primary reservations were that *in vitro* synthesis was much slower than that *in vivo*, that the enzyme was much more effective

replicating single-stranded DNA than double-stranded DNA, and that the enzyme appeared to be able to *degrade* DNA as well as to *synthesize* it.

Uncertain of the true cellular function of DNA polymerase I, Kornberg pursued another approach. He reasoned that if the enzyme could be used to synthesize

TABLE 12.1 Base Composition of the DNA Template and the Product of Replication in Kornberg's Early Work

Organism	Template or Product	%A	%T	%G	%C
T2	Template	32.7	33.0	16.8	17.5
	Product	33.2	32.1	17.2	17.5
E. coli	Template	25.0	24.3	24.5	26.2
	Product	26.1	25.1	24.3	24.5
Calf	Template	28.9	26.7	22.8	21.6
	Product	28.7	27.7	21.8	21.8

Source: Kornberg (1960).

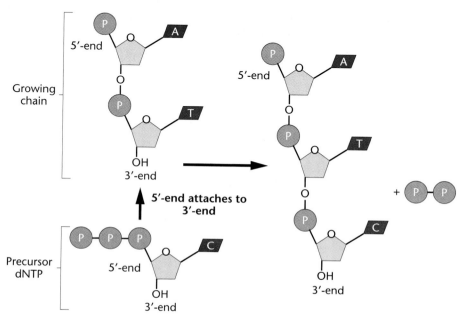

Growing chain

5'-end

A

T

OH 3'-end

5'-end attaches to 3'-end

Precursor dNTP

5'-end

C

OH 3'-end

5'-end

A

T

C

OH 3'-end

+ P P

FIGURE 12.8 Demonstration of 5' to 3' synthesis of DNA.

biologically active DNA *in vitro*, then DNA polymerase I must be the major catalyzing force for DNA synthesis within the cell. The term *biologically active* means that the DNA synthesized is capable of supporting metabolic activities and directing the reproduction of the organism from which it was originally duplicated.

In 1967, Mehran Goulian, Kornberg, and Robert Sinsheimer experimented with the small bacteriophage φX174. This phage provides an ideal experimental system because its genetic material is a very small (5386 nucleotides), circular single-stranded DNA molecule. Because the molecule is a closed circle, the experiment depended on a second enzyme, **DNA ligase** (also called the **polynucleotide joining enzyme**), to join the two ends of the linear molecule following replication.

In the normal course of φX174 infection, the circular single-stranded DNA, referred to as the **(+) strand**, enters the *E. coli* cell and serves as a template for the synthesis of the complementary **(−) strand.** The two strands (+ and −) remain together in a circular double helix, or duplex, called the **replicative form (RF).** The RF serves as the template for its own replication, and subsequently only (+) strands are produced. These strands are then packaged into viral coat proteins to form mature virus particles.

As shown in Figure 12.9, the experiment was carefully designed so that each newly synthesized strand could be distinguished and isolated from the template strand. Initially (step 1), the (+) strand was labeled with tritium (^{3}H)—a radioactive form of hydrogen. During synthesis, two "tags" were used to identify the new (−) strands. First, radioactive ^{32}P was present in the precursor nucleotides. Second, the base analogue 5-bromouracil (BU) was used in place of thymine. In the chemical structure of this analogue, a bromine atom is substituted for a carbon atom in the methyl group at the C-5 position of the pyrimidine ring, increasing the mass and therefore the density. As a result, newly synthesized (−) strands (step 2) were radioactive, and because their mass was greater, they could be isolated from the (+) strands using sedimentation equilibrium centrifugation.

The duplex was "nicked" by the enzyme DNAase to open one strand and heat denatured to separate the strands. The BU-containing strands, representing newly synthesized DNA, were then isolated by centrifugation. The process was repeated in the absence of any tags, using the BU-containing strands as templates. Strands synthesized after this point were detectable because they were "lighter" and nonradioactive. Eventually, newly synthesized infectious (+) strands were isolated. The protocol of the experiment dictated that these (+) strands must have been synthesized *in vitro* under the direction of Kornberg's enzyme.

The critical test of biological activity was the process of transfection (see Chapter 11), in which newly synthesized (+) strands were added to bacterial protoplasts (bacterial cells minus their cell wall). Following infection by the synthetic DNA, mature phages were produced—the synthetic DNA had successfully directed reproduction.

This demonstration of biological activity represented a precise assessment of faithful copying. If even a single error had occurred to alter the base sequence of any of the 5386 nucleotides constituting the φX174 chromosome, the change might easily have caused a mutation that would prohibit the production of viable phages.

DNA Polymerase II and III

Although DNA synthesized under the direction of polymerase I demonstrated biological activity, a more serious reservation about the enzyme's true biological role was raised in 1969. Peter DeLucia and John Cairns discovered a mutant strain of *E. coli* that was deficient in polymerase I activity. The mutation was designated *polA1*. In the absence of the functional enzyme, this mutant strain of *E. coli* still duplicated its DNA and successfully reproduced. However, the cells were highly deficient in their ability to "repair" DNA. For example, the mutant strain is highly sensitive to ultraviolet light (UV) and radiation, both of which damage DNA and are mutagenic. Nonmutant bacteria are able to repair a great deal of UV-induced damage.

These observations led to two conclusions:

1. There must be at least one other enzyme present in *E. coli* cells that is capable of replicating DNA *in vivo*.

2. DNA polymerase I may serve only a secondary function *in vivo*. Kornberg and others now believe that DNA polymerase I is critical to the *fidelity* of DNA synthesis, but is not the enzyme that actually synthesizes the complementary strand.

To date, two other unique DNA polymerases have been isolated from cells lacking polymerase I activity and from normal cells that do contain polymerase I. Table 12.2 shows that the two enzymes, called **DNA polymerase II** and **III**, share several characteristics with DNA polymerase

T A B L E 1 2 . 2 Comparative Properties of the Three Bacterial DNA Polymerases

Properties	I	II	III
Initiation of chain synthesis	−	−	−
5′–3′ polymerization	+	+	+
3′–5′ exonuclease activity	+	+	+
5′–3′ exonuclease activity	+	−	−
Molecules of polymerase/cell	400	?	15

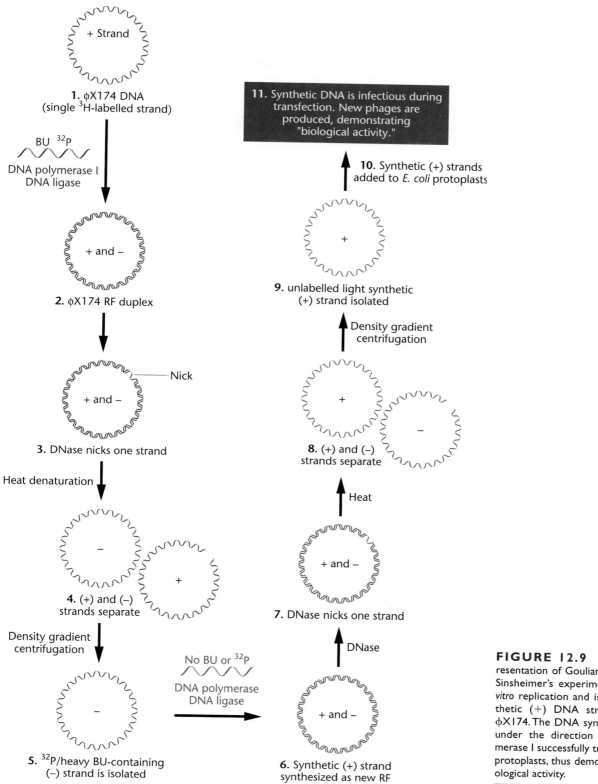

FIGURE 12.9 Schematic representation of Goulian, Kornberg, and Sinsheimer's experiment involving *in vitro* replication and isolation of synthetic (+) DNA strands of phage φX174. The DNA synthesized *in vitro* under the direction of DNA polymerase I successfully transfected *E. coli* protoplasts, thus demonstrating its biological activity.

1. φX174 DNA
(single ³H-labelled strand)

+ Strand

BU   ³²p
DNA polymerase I
DNA ligase

2. φX174 RF duplex

+ and −

3. DNase nicks one strand

+ and −

Nick

Heat denaturation

4. (+) and (−)
strands separate

−   +

Density gradient
centrifugation

5. ³²P/heavy BU-containing
(−) strand is isolated

−

No BU or ³²p
DNA polymerase
DNA ligase

6. Synthetic (+) strand
synthesized as new RF

+ and −

7. DNase nicks one strand

+ and −

DNase

8. (+) and (−)
strands separate

+   −

Heat

9. unlabelled light synthetic
(+) strand isolated

+

Density gradient
centrifugation

+

10. Synthetic (+) strands
added to *E. coli* protoplasts

11. Synthetic DNA is infectious during transfection. New phages are produced, demonstrating "biological activity."

I. While none of the three can *initiate* DNA synthesis on a template, all three can *elongate* an existing DNA strand, called a **primer.** As we shall see, RNA is also an adequate primer and is, in fact, used initially.

The DNA polymerase enzymes are all large complex proteins exhibiting a molecular weight in excess of 100,000 daltons. All three possess 3′–5′ **exonuclease activity,** which means that they have the potential to polymerize in one direction and then pause and excise nucleotides just added. As we will discuss later in this chapter, this activity provides a capacity to proofread newly synthesized DNA and to remove incorrect nucleotides, which may then be replaced.

DNA polymerase I also demonstrates 5′–3′ exonuclease activity. This potentially allows the enzyme to excise nucleotides starting at the end where synthesis begins and proceeding in the direction of synthesis. Thus, DNA polymerase I has the ability to remove the RNA primer. Two final observations probably explain why Kornberg isolated polymerase I and not polymerase III: Polymerase I is present in much greater amounts than is polymerase III and it is also much more stable.

What then are the roles of the three polymerases *in vivo*? Polymerase I is believed to be responsible for removing the primer as well as for the synthesis that fills the gaps that naturally occur as primers are removed. Its exonuclease activity also allows for proofreading during this process. Polymerase II appears to be involved in repair of DNA that has been damaged by external forces, such as ultraviolet light. It is encoded by a gene that may be activated by disruption of DNA synthesis at the replication fork. Polymerase III is considered to be the enzyme responsible for the polymerization essential to replication. Its 3′–5′ exonuclease activity provides its proofreading function, as described.

We end this section by emphasizing the complexity of the DNA polymerase III molecule. Its active form, called a **holoenzyme,** consists of two sets (a dimer) of 10 separate polypeptide subunits (Table 12.3) and has a molecular weight in excess of 600,000 daltons. The largest subunit, α, has a molecular weight of 140,000 daltons and, along with subunits ε and θ, constitutes the "core" enzyme responsible for the polymerization activity of the holoenzyme. The α subunit is responsible for nucleotide polymerization on the template strands, whereas the ε subunit of the core enzyme possesses the 3′–5′ exonuclease activity.

A second group of five subunits (γ, δ, δ′, χ, and ψ) forms what is called the γ complex, which is involved in "loading" the enzyme onto the template at the replication fork. This enzymatic function requires energy and is dependent on the hydrolysis of ATP. The β subunit prevents the core enzyme from falling off the template during polymerization. Finally, the τ subunit functions to hold together the two core polymerases at the replication fork.

TABLE 12.3 Subunits Constituting DNA Polymerase III Holoenzyme

Subunit	Function	Groupings
α	5′–3′ polymerization	"Core" enzyme:
ε	3′–5′ exonuclease	Elongates polynucleotide
θ	??	chain and proofreads.
γ	Loads enzyme	
δ	on template	
δ′	(Serves as	γ complex
χ	clamp loader)	
ψ		
β	Sliding clamp structure (Processivity Factor)	
τ	Dimerizes Core Complex	

The holoenzyme and several other proteins at the replication fork together form a complex nearly as large as a ribosome known as a **replisome.** We consider the function of DNA polymerase III in more detail later.

Many complex issues must be resolved during DNA replication

We have established that in bacteria and viruses replication is semiconservative and bidirectional along a single replicon. We also know that synthesis is in the 5′-to-3′ mode under the direction of DNA polymerase III, creating two replication forks. These move in opposite directions away from the origin of synthesis. As we can see in the points listed here, many issues must be resolved in order to provide a comprehensive understanding of DNA replication.

1. A mechanism must exist by which the helix undergoes localized unwinding and is stabilized in this "open" configuration so that synthesis may proceed along both strands.

2. There must also be a mechanism for reducing the tension that develops as the helix unwinds and coils farther down the strand.

3. A primer of some sort must be synthesized so that polymerization can commence under the direction of DNA polymerase III. Surprisingly, RNA, not DNA, serves as the primer.

4. Once the RNA primers have been synthesized, DNA polymerase III begins to synthesize the DNA complement of both strands of the parent molecule. Because the two strands are antiparallel to one another, continuous synthesis in the direction that the replication fork moves is possible along only one of the two strands. On the other strand, synthesis is discontinuous in the opposite direction.

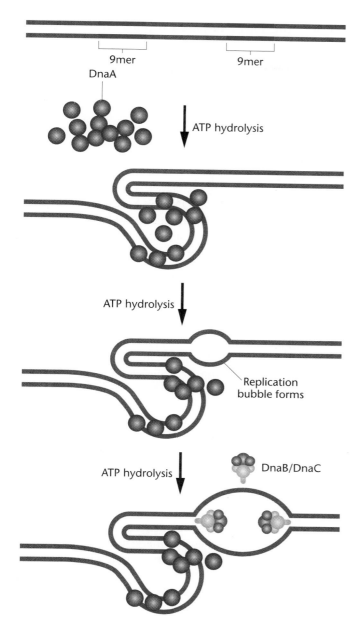

FIGURE 12.10 Helical unwinding of DNA during replication as accomplished by DnaA, DnaB, and DnaC proteins. Initial binding of many monomers of DnaA occurs at DNA sites containing repeating sequences of 9 nucleotides, called *9mers*. Not illustrated are 13mers, which are also involved.

7. While DNA polymerases accurately insert complementary bases during replication, they are not perfect and, occasionally, incorrect bases are added to the growing strand. A proofreading mechanism that also corrects errors is an integral process during DNA synthesis.

As we consider these points, Figures 12.10, 12.11, and 12.12 will be used to illustrate each of the above issues. Figure 12.13 will summarize the model of DNA synthesis.

The DNA helix must be unwound

As discussed earlier, there is a single point of origin along the circular chromosome of most bacteria and viruses where DNA synthesis is initiated. This region of the *E. coli* chromosome has been particularly well studied. Called *oriC,* the origin of replication consists of 245 base pairs characterized by repeating sequences of 9 and 13 bases (called **9mers** and **13mers**). As shown in Figure 12.10, one particular protein, called **DnaA** (because it is encoded by the gene called *dnaA*), is responsible for the initial step in unwinding the helix. A number of subunits of the DnaA protein bind to each of several 9mers. This step facilitates the subsequent binding of **DnaB** and **DnaC** proteins that further open and destabilize the helix. Proteins such as these, which require the energy normally supplied by the hydrolysis of ATP in order to break hydrogen bonds and denature the double helix, are called **helicases.** Other proteins, called **single-stranded binding proteins (SSBPs),** stabilize this conformation.

As unwinding proceeds, a coiling tension is created ahead of the replication fork, often producing **supercoiling.** In circular molecules, supercoiling may take the form of added twists and turns of the DNA, much like the coiling you can create in a rubberband by stretching it out and then twisting one end. Such supercoiling can be relaxed by the **DNA gyrase,** a member of a larger group referred to as **DNA topoisomerases.** The gyrase makes either single- or double-stranded "cuts" and also catalyzes localized movements that have the effect of "undoing" the twists and knots created during supercoiling. The strands are then resealed. These various reactions are driven by the energy released during ATP hydrolysis.

Together, the DNA, the polymerase complex, and associated enzymes make up an array of molecules that participate in DNA synthesis and are part of what we have previously called the replisome.

Initiation of DNA synthesis requires an RNA primer

Once a small portion of the helix is unwound, the initiation of synthesis may occur. As we have seen, DNA polymerase III requires a primer with a free 3′ end in order

5. The RNA primers must be removed prior to completion of replication. The gaps that are temporarily created must be filled with DNA complementary to the template at each location.

6. The newly synthesized DNA strand that fills each temporary gap must be joined to the adjacent strand of DNA.

to elongate a polynucleotide chain. This prompted researchers to investigate how the first nucleotide could be added. Though no free 3'-hydroxyl group is initially present, there is now evidence that, as pointed out above, RNA is the primer that initiates DNA synthesis.

A short segment of RNA, (about 5 to 15 nucleotides long), complementary to DNA, is first synthesized on the DNA template. Synthesis of the RNA is directed by a form of RNA polymerase called **primase,** which does not require a free 3' end to initiate synthesis. It is to this short segment of RNA that DNA polymerase III begins to add 5'-deoxyribonucleotides, initiating DNA synthesis. A conceptual diagram of initiation on a DNA template is shown in Figure 12.11. At a later point, the RNA primer must be clipped out and replaced with DNA. This occurs under the direction of DNA polymerase I. Recognized in viruses, bacteria, and several eukaryotic organisms, RNA priming is a universal phenomenon during the initiation of DNA synthesis.

Antiparallel strands require continuous and discontinuous DNA synthesis

We must now reconsider the fact that the two strands of a double helix are **antiparallel** to each other; that is, one runs in the 5'–3' direction, while the other has the opposite 3'–5' polarity. Because DNA polymerase III synthesizes DNA in only the 5'–3' direction, synthesis along an advancing replication fork occurs in one direction on one strand and in the opposite direction on the other, simultaneously.

As illustrated in Figure 12.12, as the strands unwind and the replication fork progresses down the helix, only one strand can serve as a template for **continuous DNA synthesis.** This strand is called the **leading DNA strand.** As the fork progresses, many points of initiation are necessary on the opposite, or **lagging DNA strand,** resulting in **discontinuous DNA synthesis.**

Evidence in support of discontinuous DNA synthesis was first provided by Reiji Okazaki, Tuneko Okazaki, and their colleagues. They discovered that when bacteriophage DNA is replicated in *E. coli,* some of the newly formed DNA that is hydrogen-bonded to the template strand is present as small fragments containing 1000 to 2000 nucleotides. RNA primers are part of each such fragment. These pieces, called **Okazaki fragments,** are converted into longer and longer DNA strands of higher molecular weight as synthesis proceeds.

Discontinuous synthesis of DNA requires enzymes that remove the RNA primer and that unite the Okazaki fragments into the continuous lagging strand. As we have noted, DNA polymerase I removes the primer and replaces the missing nucleotides. Joining the fragments appears to be the work of **DNA ligase,** which is capable of catalyz-

ing the formation of the phosphodiester bond that closes the gap between the discontinuously synthesized strands. The evidence that DNA ligase performs this function during DNA synthesis is strengthened by the observation of a ligase-deficient mutant strain (*lig*) of *E. coli* in which a large number of unjoined Okazaki fragments accumulate.

Concurrent synthesis occurs on the leading and lagging strands

Given the model just discussed, we might ask how DNA polymerase III synthesizes DNA on both the leading and lagging strands. Can both strands be replicated simultaneously at the same replication fork, or are the events distinct, involving two separate copies of the enzyme? Evidence suggests that both strands can be replicated simultaneously. As Figure 12.13 illustrates, if the lagging strand forms a loop, nucleotide polymerization can occur on both template strands under the direction of the dimer of the enzyme. After the synthesis of 100 to 200 base pairs, the monomer of the enzyme on the lagging strand will encounter a completed Okazaki fragment, at which point it releases the lagging strand. A new loop is then formed with the lagging template strand, and the process is repeated. Looping inverts the orientation of the template, but not the direction of actual synthesis on the lagging strand, which is always in the 5' to 3' direction.

Another important feature of the holoenzyme that facilitates synthesis at the replication fork is a dimer of the β subunit that forms a clamp-like structure around the newly formed DNA duplex. This β-subunit clamp prevents the **core enzyme** (the α, ε, and θ subunits that are responsible for catalysis of nucleotide addition) from falling off the template as polymerization proceeds. Because the entire holoenzyme moves along the parent duplex, advancing the replication fork, the β-subunit dimer is often referred to as a sliding clamp.

Proofreading and error correction are an integral part of DNA replication

The underpinning of DNA replication is the synthesis of a new strand that is precisely complementary to the template strand at each nucleotide position. Although the action of DNA polymerases is very accurate, synthesis is not perfect and a noncomplementary nucleotide is occasionally inserted erroneously. To compensate for such inaccuracies, polymerases I and III both possess **3'–5' exonuclease activity,** which enables them to detect and excise a mismatched nucleotide (in the 3'–5' direction). Once the mismatched nucleotide is removed, 5'–3' synthesis can again proceed. This process, **ex-**

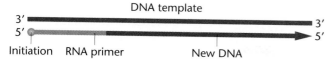

DNA template

3′ ——————————————————————— 3′
5′ ——————————————————————→ 5′

Initiation   RNA primer   New DNA

FIGURE 12.11   A conceptual diagram of the initiation of DNA synthesis. A complementary RNA primer is first synthesized, to which DNA is added. All synthesis is in the 5′ to 3′ direction. Eventually, the RNA primer is replaced with DNA under the direction of DNA polymerase I.   ✐GenCDX

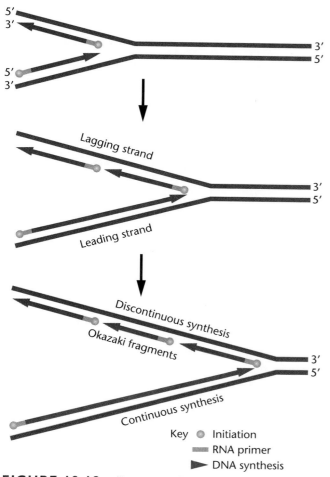

FIGURE 12.12   Illustration of the opposite polarity of DNA synthesis along the two strands, necessary because the two strands of DNA run antiparallel to one another and DNA polymerase III synthesizes only in one direction (5′ to 3′). On the lagging strand, synthesis must be discontinuous, resulting in the production of Okazaki fragments. On the leading strand, synthesis is continuous. RNA primers are used to initiate synthesis on both strands.   ✐GenCDX

onuclease proofreading, increases the fidelity of synthesis. In the case of the holoenzyme form of DNA polymerase III, the epsilon (ϵ) subunit is directly involved in the proofreading step. In strains of *E. coli* where a mutation has rendered the ϵ subunit nonfunc-

tional, the error rate (the mutation rate) during DNA synthesis is increased substantially.

A coherent model summarizes DNA replication

We can now combine the various aspects of DNA replication occurring at a single replication fork into a coherent model, as shown in Figure 12.14. At the advancing fork, a helicase is unwinding the double helix. Once unwound, single-stranded binding proteins associate with the strands, preventing the reformation of the helix. In advance of the replication fork, DNA gyrase functions to diminish the tension created as the helix supercoils. Each half of the dimeric polymerase is a core enzyme bound to one of the template strands by a β-subunit sliding clamp. Continuous synthesis occurs on the leading strand, while the lagging strand must "loop" around in order for simultaneous synthesis to occur on both strands. Not shown in the figure, but essential to replication on the lagging strand, is the action of DNA polymerase I and DNA ligase, which function to replace the RNA primer with DNA and join the Okazaki fragments.

Because the investigation of DNA synthesis is still an extremely active area of research, this model will no doubt be extended in the future. In the meantime, it provides a summary of DNA synthesis against which genetic phenomena can be interpreted.

Replication is controlled by a variety of genes

Much of what we know about DNA replication in viruses and bacteria is based on genetic analysis of the process. For example, we have already discussed studies involving the *polA1* mutation, which revealed that DNA polymerase I is not the major enzyme responsible for replication. Many other mutations interrupt or seriously impair some aspect of replication, such as the ligase-deficient and the proofreading-deficient mutations mentioned previously. Genetic analysis frequently uses **conditional mutations**, which are expressed under one condition but not under a different condition. For example, a **temperature-sensitive mutation** may not be expressed at a particular *permissive* temperature. When mutant cells are grown at a *restrictive* temperature, the mutation is expressed. The investigation of such temperature-sensitive mutants can provide insights into the product and the associated function of the normal, nonmutant gene.

As shown in Table 12.4, a variety of genes in *E. coli* specify the subunits of polymerases I, II, and III and en-

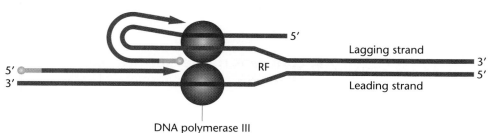

FIGURE 12.13 Illustration of how concurrent DNA synthesis may be achieved on both the leading and lagging strands at a single replication fork. The lagging template strand is "looped" in order to invert the physical direction of synthesis, but not the biochemical direction. The enzyme functions as a dimer, with each core enzyme achieving synthesis on one or the other strand.

code products involved in specification of the origin of synthesis, helix-unwinding and stabilization, initiation and priming, relaxation of supercoiling, repair, and ligation. The discovery of such a large group of genes attests to the complexity of the process of replication, even in the relatively simple prokaryote. Given the enormous quantity of DNA that must be unerringly replicated in a very brief time, this level of complexity is not unexpected. As we see next, the process is even more involved and therefore more difficult to investigate in eukaryotes.

Eukaryotic DNA synthesis is similar to synthesis in prokaryotes but more complex

Research has shown that eukaryotic DNA is replicated in a manner similar to that of bacteria. In both systems, double-stranded DNA is unwound at a replication origin, two replication forks are formed, and bidirectional synthesis of DNA occurs on the leading and lagging strand templates under the direction of DNA polymerase. Eukaryotic polymerases have the same fundamental requirements for DNA synthesis as do bacterial systems: four deoxyribonucleoside triphosphates, a template, and a primer.

TABLE 12.4 A List of Some of the Various *E. coli* Mutant Genes and Their Products or Role in Replication

Mutant Gene	Enzyme or Role
polA	DNA polymerase I
polB	DNA polymerase II
dnaE, N, Q, X, Z	DNA polymerase III subunits
dnaG	Primase
dnaA, I, P	Initiation
dnaB, C	Helicase at *oriC*
oriC	Origin of replication
gyrA, B	Gyrase subunits
lig	Ligase
rep	Helicase
ssb	Single-stranded binding proteins
rpoB	RNA polymerase subunit

However, because eukaryotic cells contain much more DNA per cell and because this DNA is complexed with proteins, eukaryotes face many problems not encountered by bacteria. As we might expect, these complications make the process of DNA synthesis much more complex in eukaryotes and more difficult to study. However, a great deal is now known about the process.

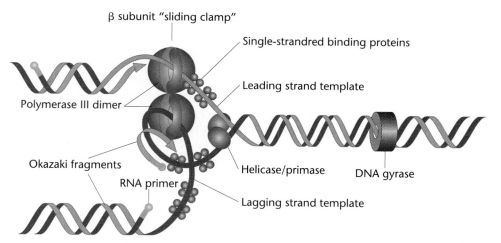

FIGURE 12.14 A summary of DNA synthesis at a single replication fork. Various enzymes and proteins essential to the process are shown. GenCDX

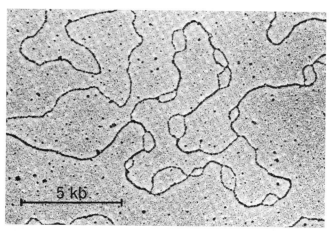

FIGURE 12.15 A demonstration of the multiple origins of replication along a eukaryotic chromosome. Each origin is apparent as a replication bubble along the axis of the chromosome.

Multiple Replication Origins

The most *obvious* difference between eukaryotic and prokaryotic DNA replication is that eukaryotic chromosomes contain multiple replication origins, in contrast to the single site that is part of the *E. coli* chromosome. Multiple origins, visible under the electron microscope (Figure 12.15), are essential if replication of the entire genome of a typical eukaryote is to be completed in a reasonable time, since: 1) eukaryotes have much greater amounts of DNA than bacteria (e.g., yeast has 4 times as much and *Drosophila* has 100 times as much DNA as *E. coli*); and 2) the rate of synthesis by eukaryotic DNA polymerase is only about 50 nucleotides per second, 20 times slower than the comparable bacterial enzyme. Under these conditions, replication from a single origin of a typical eukaryotic genome would take up to 1 month to complete! However, replication is accomplished in as little as 3 minutes in some eukaryotic organisms.

Many insights are now available concerning the molecular nature of the multiple origins and the initiation of DNA synthesis at these sites. Most information was originally derived from the study of yeast (e.g., *Saccharomyces cerevisiae*), which have between 250–400 replicons, with subsequent studies performed using mammalian cells,

which have as many as 25,000 replicons. The origins in yeast have been isolated and are called **autonomously replicating sequences (ARSs).** They consist of a unit of 11 base pairs, flanked by other short sequences involved in initiation efficient. As we know from Chapter 2, DNA synthesis is restricted to the S phase of the eukaryotic cell cycle. Research has shown that the many origins are not all activated at once; instead clusters of 20–80 adjacent replicons are activated sequentially throughout the S phase until all DNA is replicated.

How the polymerase finds the ARS sequences among so much DNA is an obvious recognition problem. The solution involves a mechanism that is initiated prior to the S phase (Figure 12.16). During the G1 phase of the cell cycle, all ARS sequences are bound by a group of proteins (six in yeast), forming what is called the **origin recognition complex (ORC).** Mutations in either the ARS sequence or in any of the genes encoding the proteins of the ORC in yeast abolish initiation of DNA synthesis. Since these recognition complexes are formed in G1, but synthesis is not initiated at these sites until S, there must be still other proteins involved in the actual initiation signal. The most important of these proteins are specific kinases, key enzymes involved in phosphorylation, that are integral parts of cell cycle control. When bound along with ORC, a **prereplication complex (pre-RC)** is formed that is accessible to DNA polymerase. When these kinases are activated, they not only serve to complete the initiation complex, triggering DNA synthesis, but they inhibit reformation of the pre-RC complexes once DNA synthesis has been completed at each replicon. This is an important mechanism since it distinguishes segments of DNA that have completed replication from unreplicated DNA, thus maintaining orderly and efficient replication.

Eukaryotic DNA Polymerases

The most complex aspect of eukaryotic replication is the array of polymerases involved in directing DNA synthesis. As we will see, a total of six different forms of the enzyme have been isolated and studied. In order for the polymerases to have access to DNA, the topology of the helix must first be modified. As synthesis is triggered at each

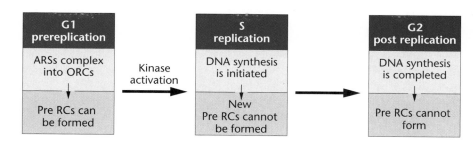

FIGURE 12.16 Summary of the events that occur during the cell cycle that control the activation of the multiple replication origins that characterize eukaryotic DNA.

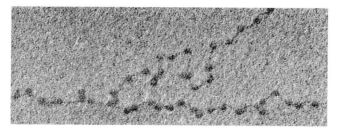

FIGURE 12.17 An electron micrograph of a eukaryotic replicating fork demonstrating the presence of histone protein-containing nucleosomes on both branches.

origin site, the double strands are opened up within an A═T-rich region allowing the entry of a helicase enzyme that proceeds to further unwind the double-stranded DNA. Before polymerases can begin synthesis, histone proteins complexed to the DNA (which form the characteristic nucleosomes of chromatin—see Chapter 19) also must be stripped away or otherwise modified. As DNA synthesis then proceeds, histones reassociate with the newly formed duplexes, reestablishing the characteristic nucleosome pattern (Figure 12.17). In eukaryotes, the synthesis of new histone proteins is tightly coupled to DNA synthesis during the S phase of the cell cycle.

The nomenclature and characteristics of the six polymerases are summarized in Table 12.5. Of these, three (Pol α, δ, and ε) are now considered to be essential to nuclear DNA replication in eukaryotic cells. Two others (Pol β and ζ) are thought to be involved in DNA repair. The sixth form (Pol γ) is involved in the synthesis of mitochondrial DNA. Presumably, its replication function is limited to that organelle even though it is encoded by a nuclear gene. As shown in Table 12.5, all but one of the six forms of the enzyme (the β form) consist of multiple subunits. Different subunits perform different functions during replication.

Pol α appears to be the major form of the enzyme involved in the initiation of synthesis of nuclear DNA. Two of the four subunits of the enzyme function as a primase in the synthesis of the RNA primers on both the leading and lagging template strands. Then, another subunit functions to elongate the RNA primer by adding complementary deoxyribonucleotides, constituting the initial phase of DNA synthesis. Pol α is said to possess low **processivity,** a term that essentially reflects the length of DNA that is synthesized by an enzyme before it dissociates from the template. Thus, after a short DNA sequence is added to the RNA primer, an event known as **polymerase switching** occurs. Pol α dissociates from the template and is replaced by Pol δ. This form of the enzyme possesses high processivity as well as 3′-to-5′ exonuclease activity, which provides it with the potential for proofreading. It is also capable of a 100-fold increase in the rate of synthesis in comparison to Pol α. Thus, under the direction of Pol δ, elongation and proofreading of the growing DNA strand occurs, as synthesis continues. Pol ε, the third essential form, possesses the same general characteristics as Pol δ, but it is believed to operate under different cellular conditions. In yeast, mutations that render Pol ε inactive are lethal, attesting to its essential function during replication.

The process described applies to both leading and lagging strand synthesis. On both strands, RNA primers must be replaced with DNA. On the lagging strand, the Okazaki fragments, which are about 10 times smaller (100–150 nucleotides) in eukaryotes than in prokaryotes, must be ligated.

To accommodate the increased number of replicons, eukaryotic cells contain many more DNA polymerase molecules than do bacteria. While *E. coli* has about 15 copies of DNA polymerase III per cell, there may be up to 50,000 copies of the α form of DNA polymerase in animal cells. As has been pointed out, the presence of greater numbers of smaller replicons in eukaryotes compared to bacteria compensates for the slower rate of DNA synthesis in eukaryotes. *E. coli* requires 20 to 40 minutes to replicate its chromosome, while *Drosophila*, with 40 times more DNA, accomplishes the same task in only 3 minutes during embryonic cell divisions.

TABLE 12.5 Properties of Eukaryotic DNA Polymerases

	Polymerase α	Polymerase β	Polymerase δ	Polymerase ε	Polymerase γ	Polymerase ζ
Location	Nucleus	Nucleus	Nucleus	Nucleus	Mitochondria	Nucleus
Mass of subunits (kDa)	180, 70, 58, 48	39	125, 53, 50	250, 55	125, 35	173, 29
3′–5′ Exonuclease Activity	No	No	Yes	Yes	Yes	No
Essential to replication?	Yes	No	Yes	Yes	No	No

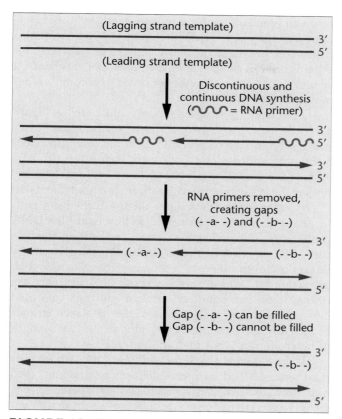

FIGURE 12.18 Diagram illustrating the difficulty encountered during the replication of the ends of linear chromosomes. A gap (- -b- -) is left following synthesis on the lagging strand.

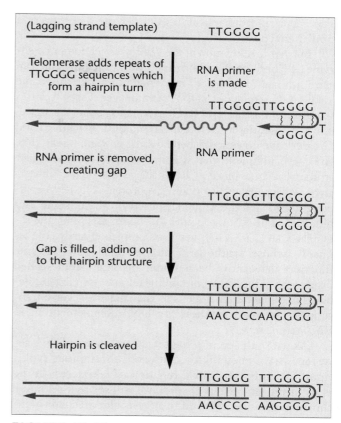

FIGURE 12.19 Diagram of the predicted solution to the problem posed in Figure 12.18. The enzyme telomerase directs synthesis of the TTGGGG sequences resulting in the formation of a hairpin structure. The gap can now be filled, and following cleavage of the hairpin structure, this process averts the creation of a gap during replication of the ends of linear chromosomes.

The ends of linear chromosomes are problematic during replication

A final difference that exists between prokaryotic and eukaryotic DNA synthesis involves the nature of the chromosomes. Unlike the closed, circular DNA of bacteria and most bacteriophages, eukaryotic chromosomes are linear. During replication they face a special problem at the "ends" of these linear molecules, which are part of each telomeric region of the chromosome. While synthesis can proceed normally to the end of the leading strand, a difficulty arises on the **lagging strand** as the RNA primer is removed (Figure 12.18). Normally, the newly created gap would be filled by addition of a nucleotide to the existing 3'-OH group provided during discontinuous synthesis (to the right of gap b in Figure 12.18). However, there is no strand present to provide the 3'-OH group because this is the end of the chromosome. As a result, each successive round of synthesis will theoretically shorten the chromosome by the length of the RNA primer. Because this is such a significant problem, we can suppose, at least for some cells, that a molecular solution would have devel-

oped early in evolution and be shared by all eukaryotes, and this is indeed the case.

Discovery of a unique eukaryotic enzyme, **telomerase,** has helped us understand how more complex organisms solve this problem. In the ciliated protozoan *Tetrahymena*, the many telomeres all terminate in the sequence 5'-TTGGGG-3'. Telomerase adds repeats of TTGGGG to the ends of molecules already containing this sequence, preventing the telomeric ends from shortening after each replication. As illustrated in Figure 12.19, telomerase adds several copies of the six-nucleotide repeat to the 3' end of the lagging strand (using 5'-3' synthesis). These repeats form a "hairpin loop," which is stabilized by unorthodox hydrogen bonding between opposite guanine residues (G ≡ G). This creates a free 3'-OH end that, following removal of the RNA primer, can serve as a substrate for DNA polymerase I to fill the gap. The hairpin loop is then cleaved off, and the potential loss of DNA in each subsequent replication cycle is averted. This process also occurs in other eukaryotes under the direction of a similar enzyme.

Further investigation of the *Tetrahymena* telomerase enzyme by Elizabeth Blackburn and Carol Greider yielded an extraordinary finding. This enzyme adds the same TTGGGG sequence to DNA termini even if they lack this sequence. Thus the TTGGGG sequence is not the signal, nor is it essential for telomerase function. Blackburn and Greider have now established how the enzyme works. The enzyme is unique in that it contains within its molecular structure a short piece of RNA that is essential to its catalytic activity, making it a ribonucleoprotein. The RNA component encodes the sequences used by the enzyme as a template. The RNA contains 159 bases, including the sequence 5′-AACCCC-3′, which is complementary to the sequence whose synthesis it directs. Analogous enzyme functions have now been found in other single-celled organisms. The RNA-containing telomerase enzyme behaves in a manner analogous to the enzyme, reverse transcriptase, in that it synthesizes a DNA complement on an RNA template.

As we shall see in Chapter 19, telomeric DNA sequences have been highly conserved throughout evolution, reflecting the critical function of telomeres. In the essay at the end of this chapter, we will see that telomere shortening has been linked to a molecular mechanism involved in the aging process of cells. In most eukaryotic somatic cells, telomerase is, in fact, not active, and thus with each cell division, the telomeres of each chromosome shorten. After many divisions, the telomere is seriously eroded and the cell loses the capacity for further division. Malignant cells, on the other hand, maintain telomerase activity and are immortalized.

DNA recombination, like DNA replication, is directed by specific enzymes

We conclude this chapter by returning to a topic discussed in Chapter 6–genetic recombination. We saw there that the process of crossing over depends on breakage and rejoining of the DNA strands between homologs. Now that we have discussed the chemistry and replication of DNA, we can consider how recombination occurs at the molecular level. In general, the following information pertains to genetic exchange between any two homologous double-stranded DNA molecules, whether they be viral or bacterial chromosomes or eukaryotic homologs during meiosis. Genetic exchange at equivalent positions along two chromosomes with substantial DNA sequence homology is referred to as **general** or **homologous recombination.**

Several models attempt to explain homologous recombination, all sharing certain features. First, all are based on proposals put forth independently by Robin Holliday and Harold L. K. Whitehouse in 1964. They also depend on the complementarity between DNA strands for the precision of exchange. Finally, each model relies on a series of enzymatic processes in order to accomplish genetic recombination.

The generally accepted model is illustrated in Figure 12.20. It begins with two paired DNA duplexes or homologs (a), in each of which an endonuclease introduces a single-stranded nick at an identical position (b). The ends of the strands produced by these cuts are then displaced and subsequently pair with their complements on the opposite duplex (c). A ligase then seals the loose ends (d), creating hybrid duplexes called **heteroduplex DNA molecules.** The exchange creates a cross-bridged structure. The position of this cross-bridge can then move down the chromosomes by a process referred to as branch migration (e). This occurs as a result of a zipperlike action as hydrogen bonds are broken and then reformed between complementary bases of the displaced strands of each duplex. This migration yields an increased length of heteroduplex DNA on both homologs.

If the duplexes now separate (f), and the bottom portions rotate 180° (g), an intermediate planar structure called a chi form, the characteristic **Holliday structure,** is created. If the two strands on opposite homologs previously uninvolved in the exchange are now nicked by an endonuclease (h), and ligation occurs (i), recombinant duplexes are created. Note in our illustration that the arrangement of alleles is altered as a result of the recombination event.

Evidence supporting this model includes the electron microscopic visualization of chi-form planar molecules from bacteria where four duplex arms are joined at a single point of exchange [Figure 12.20(g)]. Further important evidence comes from the discovery in *E. coli* of the **RecA protein.** This molecule promotes the exchange of reciprocal single-stranded DNA molecules as occurs in step (c) of the model. RecA also enhances the hydrogen-bond formation during strand displacement, thus initiating heteroduplex formation. Finally, many other enzymes essential to the nicking and ligation process have also been discovered and investigated. The products of the *recB*, *recC*, and *recD* genes are thought to be involved in the nicking and unwinding of DNA. Numerous mutations that prevent genetic recombination have been found in viruses and bacteria. These mutations are thought to identify genes, whose products play an essential role in this process.

Gene conversion is a consequence of DNA recombination

A modification of the above model has helped us to better understand a unique genetic phenomenon known as

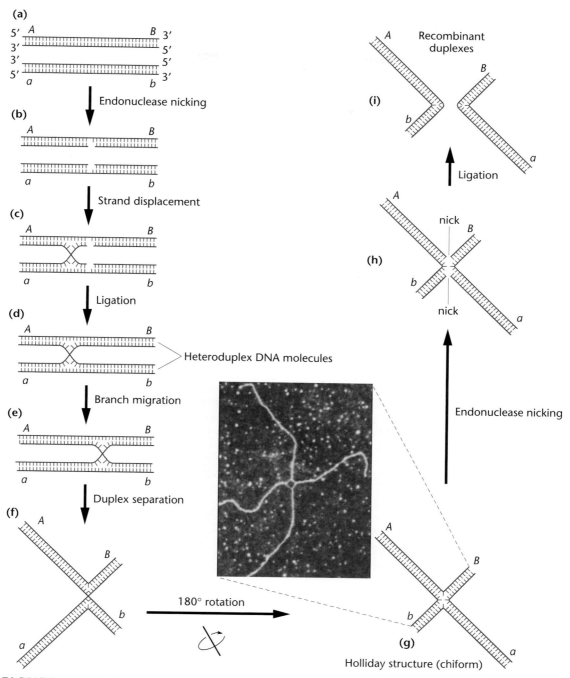

FIGURE 12.20 A model depicting how genetic recombination may occur as a result of the breakage and rejoining of heterologous DNA strands. Each stage is described in the text. The electron micrograph shows DNA in a chi-form structure similar to the diagram in (g). This micrograph shows DNA in an extended Holliday structure, derived from the *Col*E1 plasmid of *E. coli.*

🌐GenCDX

GENETICS, TECHNOLOGY, AND SOCIETY

Telomerase: The Key to Immortality?

Humans, as do all multicellular organisms, grow old and die. As we age, our immune systems become less efficient, wound healing is impaired, and tissues and organs lose resilience. It has always been a mystery why we go through these age-related declines, and why each species has a characteristic finite lifespan. Why do we grow old? Can we reverse this march to mortality? Some recent discoveries suggest that the answers to these questions may lie at the ends of our chromosomes.

The study of human aging begins with a study of human cells growing in culture dishes. Like the organisms from which the cells are taken, cells in culture have a finite life span. This "replicative senescence" was noted over 30 years ago by Hayflick. He reported that normal human fibroblasts lose their ability to grow and divide after about 50 cell divisions. These senescent cells remain metabolically active, but can no longer proliferate. Eventually, they die. Although it is not known whether cellular senescence directly causes organismal aging, the evidence is suggestive. For example, cells from young people go through more divisions in culture than cells from

older people; human fetal cells divide 60–80 times before undergoing senescence, whereas cells from older adults divide only 10–20 times. In addition, cells from species with short life spans stop growing after fewer divisions than cells from species with longer life spans; mouse cells divide 10–15 times in culture, but tortoise cells undergo over 100 divisions. Moreover, cells from patients with genetic premature aging syndromes (such as Werner syndrome) undergo fewer divisions in culture than cells from normal patients.

Another characteristic of aging cells is that their telomeres become shorter. *Telomeres* are the tips of linear chromosomes and consist of several thousand repeats of a short DNA sequence (TTAGGG in humans). Telomeres help preserve the structural integrity of chromosomes by protecting their ends from degrading or from fusing to other chromosomes. Telomeres are created and maintained by *telomerase*—a remarkable RNA-containing enzyme that adds telomeric DNA sequences onto the ends of linear chromosomes. Telomerase also solves the "end-replication" problem—which asserts that linear DNA mole-

cules will become shorter at each replication because DNA polymerase cannot synthesize new DNA at the 3′-ends of each parent strand. By adding numerous telomeric repeat sequences onto the 3′-ends of chromosomes, telomerase prevents the chromosomes from shrinking into oblivion. Unfortunately, normal somatic cells contain little if any telomerase. As a result, telomere length decreases by about 100 base pairs every time a normal cell divides. Telomere shortening may act as a clock that counts cell division and instructs the cell to stop dividing.

Could we gain perpetual youth and vitality by increasing our telomere lengths? A recent study suggests that it may be possible to reverse senescence by artificially increasing the amount of telomerase in our cells. When the investigators introduced cloned telomerase genes into normal human cells in culture, telomeres lengthened by thousands of base pairs and the cells continued to grow long past their senescence point. These observations confirm that telomere length can act as a cellular clock. In addition, they suggest that some of the atrophy of tissues that accompanies old age may some-

---

gene conversion. Initially found in yeast by Carl Lindegren and in *Neurospora* by Mary Mitchell, **gene conversion** is characterized by a *nonreciprocal* genetic exchange between two closely linked genes. For example, if we were to cross two *Neurospora* strains, each bearing a separate mutation ($a + \times + b$), reciprocal recombination between the genes would yield spore pairs of the $++$ *and* the *ab* genotypes. A nonreciprocal exchange, by contrast, yields one pair without the other. Working with pyridoxine mutants, Mitchell observed several asci containing spore pairs with the $++$ genotype, but not the reciprocal product (*ab*). The frequency of these events was higher than the

predicted mutation rate and so could not be accounted for by mutation. They were called *gene conversions* because it appeared that one allele had somehow been "converted" to another where genetic exchange had also occurred. Similar findings come from studies of other fungi.

Gene conversion is now considered to be a consequence of the process of DNA recombination. One possible explanation interprets conversion as a mismatch of base pairs during heteroduplex formation, as shown in Figure 12.21. Mismatched regions of hybrid strands can be repaired by the excision of one of the strands and the synthesis of the complement using the remaining strand as a template. Ex-day be reversed by activating telomerase genes. However, before we rush out to buy telomerase pills, we need to consider a possible consequence of cellular immortality—cancer.

Although normal cells undergo senescence after a specific number of cell divisions, cancer cells do not. It is thought that cancers arise after several genetic mutations accumulate in a cell. These mutations disrupt the normal checks and balances that control cell growth and division. It seems logical that cancer cells would also stop the normal aging clock. If their telomeres became shorter after each cell division, tumor cells would eventually succumb to aging and cease growth. However, if they synthesized telomerase, they would arrest the ticking of the senescence clock and become immortal. In keeping with this idea, over 90% of human tumor cells contain telomerase activity, and have stable telomeres. The correlation between uncontrolled tumor cell growth and the presence of telomerase activity is so good that telomerase assays are being developed as diagnostic markers for cancer. Although there is currently some debate about whether the presence of telomerase is a prerequisite for, or simply a consequence of, cell

transformation, it is possible that the acquisition of telomerase activity may be an important step in the development of a cancer cell. In support of this hypothesis, recent studies show that the induction of telomerase activity, when combined with the inactivation of a tumor-suppressor gene (p16^{INK4a}), results in cell immortalization—an essential step toward tumor development. Therefore, any attempt to increase telomerase activity in normal cells carries the risk of enhancing the development of tumors.

An attractive possibility is that telomerase may be an ideal target for anticancer drugs. Drugs that inhibit telomerase might destroy cancer cells by allowing their telomeres to shorten, thereby forcing the cells into senescence. Because most normal human cells do not express telomerase, such a therapy might be specific for tumor cells and hence less toxic than most current anticancer drugs. Such antitelomerase anticancer drugs are currently under development by Geron Corporation. Although we do not know yet whether this approach will work in animals, it appears to work in cultured tumor cells. Tumor cells that are treated with an antitelomerase agent lose telomeric sequences and die after about 25 cell divisions.

Before antitelomerase drugs can be developed and used on humans, several questions must be answered. Is telomerase required by some normal human cells (such as lymphocytes and germ cells)? If so, antitelomerase drugs may be unacceptably toxic. Could some cancer cells compensate for the loss of telomerase by using other telomere-lengthening mechanisms (such as recombination)? If so, antitelomerase drugs may be doomed to failure. Even if we inhibit telomerase activity in tumor cells, could they undergo multiple divisions before reaching senescence and still damage the host?

Will telomerase allow us to arrest cancers and to reverse the descent into old age? Time will tell.

REFERENCES:

BODNAR, A.G. et al. 1998. Extension of life-span by introduction of telomerase into normal human cells. *Science* 279:349–52.

DELANGE, T. 1998. Telomeres and senescence: Ending the debate. *Science* 279:334–35.

KIYONO, T. et al. 1998. Both Rb/p16^{INK4a} inactivation and telomerase activity are required to immortalize human epithelial cells. *Nature* 396:84–88.

cision may occur in either one of the strands, yielding two possible "corrections." One repairs the mismatched base pair and "converts" it to restore the original sequence. The other also corrects the mismatch, but does so by copying the altered strand, creating a base-pair substitution. Conversion may have the effect of creating identical alleles on the two homologs that were different initially.

In our example in Figure 12.21, suppose the $G \equiv C$ pair on one of the two homologs was responsible for the mutant allele, while the $A = T$ pair was part of the wild-type gene sequence on the other homolog. Conversion of the $G \equiv C$ pair to $A = T$ would have the effect of chang-

ing the mutant allele to wild type, just as Mitchell originally observed.

Gene-conversion events have helped to explain other puzzling genetic phenomena in fungi. For example, when mutant and wild-type alleles of a single gene are crossed, asci should yield equal numbers of mutant and wild-type spores. However, exceptional asci with 3:1 or 1:3 ratios are sometimes observed. These ratios can be understood in terms of gene conversion. The phenomenon has also been detected during mitotic events in fungi, as well as during the study of unique compound chromosomes in *Drosophila*.

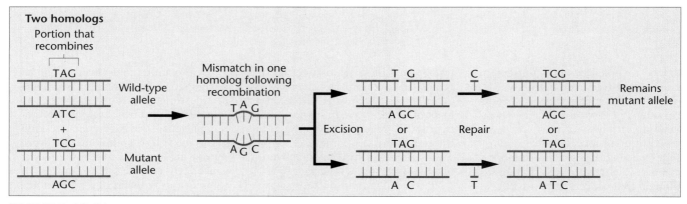

FIGURE 12.21 Illustration of a proposed mechanism that accounts for the phenomenon of gene conversion. A base-pair mismatch occurs in one of the two homologs (bearing the mutant allele) during heteroduplex formation, which accompanies recombination in meiosis. During excision repair, one of the two mismatches is removed and the complement is synthesized. In one case, the mutant base pair is preserved. When it is subsequently included in a recombinant spore, the mutant genotype will be maintained. In the other case, the mutant base pair is converted to the wild-type sequence. When included in a recombinant spore, the wild-type genotype will be expressed, leading to a nonreciprocal exchange ratio.

CHAPTER SUMMARY

1. In theory, three modes of DNA replication are possible: semiconservative, conservative, and dispersive. Although all three rely on base complementarity, semiconservative replication is the most straightforward and was predicted by Watson and Crick.

2. In 1958, Meselson and Stahl resolved this question in favor of semiconservative replication in *E. coli*, showing that newly synthesized DNA consists of one old strand and one new strand. Taylor, Woods, and Hughes used root tips of the broad bean to demonstrate semiconservative replication in eukaryotes.

3. During the same period, Kornberg isolated DNA polymerase I from *E. coli* and demonstrated that it could synthesize DNA *in vitro*, provided that a template and precursor nucleoside triphosphates were supplied.

4. The subsequent discovery of the *polA1* mutant strain of *E. coli*, capable of DNA replication despite its lack of polymerase I activity, cast doubt on the replicative function of DNA polymerase I *in vivo*. DNA polymerases II and III were then isolated. Polymerase III has been identified as the enzyme responsible for DNA replication *in vivo*.

5. During the process of DNA synthesis, the double helix unwinds, forming a replication fork where synthesis begins. Proteins stabilize the unwound helix and assist in relaxing the coiling tension created ahead of the replication.

6. Synthesis is initiated at specific sites along each template strand by RNA primase, which results in a short segment of RNA that provides a suitable 3' end upon which DNA polymerase III can begin polymerization.

7. Because of the antiparallel nature of the double helix, polymerase III synthesizes DNA continuously on the leading strand in a 5'–3' direction. On the opposite strand, called the lagging strand, synthesis results in short Okazaki fragments that are later joined by DNA ligase.

8. DNA polymerase I removes and replaces the RNA primer with DNA, which is joined to the adjacent polynucleotide by DNA ligase.

9. The isolation of numerous phage and bacterial mutant genes affecting many of the molecules involved in the replication of DNA has helped to define the complex genetic control of the entire process.

10. DNA replication in eukaryotes is similar to, but more complex than, replication in prokaryotes. Multiple replication origins exist and multiple forms of DNA polymerase direct DNA synthesis.

11. Replication at the ends (telomeres) of linear molecules poses a special problem in eukaryotes that can be solved by a unique RNA-containing enzyme called telomerase.

12. Homologous recombination between genetic molecules relies on a series of enzymes that can cut, realign, and reseal DNA strands. The phenomenon of gene conversion may be best explained in terms of mismatch repair synthesis during these exchanges.

INSIGHTS AND SOLUTIONS

1. Predict the theoretical results of conservative and dispersive models of DNA synthesis using the conditions of the Meselson–Stahl experiment. Follow the results through two generations of replication after cells have been shifted to an ^{14}N-containing medium, using the following migration standards:

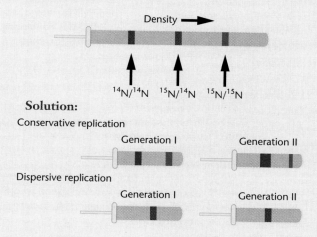

Solution:

2. Mutations in the *dnaA* gene of *E. coli* are lethal and can only be studied following the isolation of conditional, temperature-sensitive mutations. Such mutant strains grow nicely and replicate their DNA at the permissive temperature of 18°C, but they do not grow or replicate their DNA at the restrictive temperature of 37°C. Two observations were useful in determining the function of the DnaA protein product. First, *in vitro* studies using DNA templates that have been nicked (opened) do not require the DnaA protein. Second, if intact cells are grown at 18°C and then shifted to 37°C, DNA synthesis continues at this temperature until one round of replication is completed and then stops. What do these observations suggest about the role of the *dnaA* gene product?

Solution: At 18°C (the permissive temperature) the mutation is not expressed and DNA synthesis begins. Following the shift to the restrictive temperature, the DNA synthesis already initiated continues, but no new synthesis can begin. Because the DnaA protein is not required for synthesis of "nicked" DNA, these observations suggest that *in vivo* the DnaA protein plays an essential role in DNA synthesis by interacting with the intact helix and somehow facilitating the localized denaturing necessary for synthesis to proceed.

3. DNA is allowed to replicate in moderately radioactive ^{3}H-thymidine for several minutes and then switched to a highly radioactive medium for several more minutes. Synthesis is stopped and the DNA is subjected to autoradiography and electron microscopy. Interpret as much as you can regarding DNA replication from the drawing of the micrograph presented here.

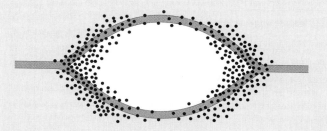

Solution: One interpretation is that there are two advancing replication forks proceeding in opposite directions and that replication is therefore bidirectional. The low density of grains represents synthesis that occurred during the first few minutes, beginning at the origin and proceeding outward in both directions. The higher density of grains represents the later minutes of synthesis when the level of radioactivity was increased.

PROBLEMS AND DISCUSSION QUESTIONS

1. Compare conservative, semiconservative, and dispersive modes of DNA replication.
2. Describe the role of ^{15}N in the Meselson–Stahl experiment.
3. In the Meselson–Stahl experiment, which of the three modes of replication could be ruled out after one round of replication? After two rounds?
4. Predict the results of the experiment by Taylor, Woods, and Hughes if replication were (a) conservative and (b) dispersive.
5. Reconsider Problem 35 in Chapter 11. In the model you proposed, could the molecule be replicated semiconservatively? Why? Would other modes of replication work?
6. What are the requirements for *in vitro* synthesis of DNA under the direction of DNA polymerase I?
7. In Kornberg's initial experiments, he actually grew *E. coli* in Anheuser-Busch beer vats (he was working at Washington University in St. Louis). Why do you think this was helpful to the experiment?

GENETICS MediaLab

The following resources will help you achieve a better understanding of the concepts presented in this chapter. These resources can be found on the CD packaged with this textbook and on the Companion Website at **http://www.prenhall.com/klug/**.

CD Resources:
Animated Tutorial: *DNA Replication*

Animated Tutorial: *DNA Recombination*

Self-grading Chapter Problems

Web Resources:
Genetics and Society Issue: *Telomerase: The Key to Immortality?*

Web Destinations in Genetics

Self-grading Chapter Problems

Chapter Search Terms

Genetics Newsgroups

Student Bulletin Board

Web Problem 1:
Time for completion = 15 minutes

How are the roles of the molecules involved in DNA replication coordinated? Researchers have successfully identified many of the components of the replication machinery and now understand the major steps in replication (local unwinding, leading and lagging strand synthesis using RNA primers, and proofreading). The coordination of these functions is still under investigation. At the linked website you can manipulate models of replication proteins that are derived from X-ray crystallography data. Directions on how to download the CHIME browser plug-in needed to view this module are found at the top of the website. Scroll to the Hall of DNA polymerization and click on the T7 DNA replication complex. Examine how replication proteins direct the addition of free nucleotides to the newly synthesized strand of DNA. Write a brief discussion about how physical attribute(s) of these replication proteins contribute to their function. To complete this exercise, visit Web Problem 1 in Chapter 12 of your Companion Website and select the keyword **REPLICATION.**

Web Problem 2:
Time for completion = 10 minutes

How do scientists construct models of complex molecular processes such as recombination? Our understanding of this physical exchange of DNA strands is derived in part from experiments in which intermediate stages are isolated and identified by microscopy or biochemical analysis. Mutations in recombination proteins have been useful in teasing apart the roles of these molecules. The linked website contains electron micrographs of DNA bound by recombination proteins. Read the figure legend and click on the links to view the images. Describe one of the images you view. Include what stage of recombination you observe and any experimental conditions contributing to the "capture" of this intermediate. To complete this exercise, visit Web Problem 2 in Chapter 12 of your Companion Website and select the keyword **RECOMBINATION.**

Web Problem 3:
Time for completion = 10 minutes

What does a recombination event look like? A crucial step during recombination is the resolution of the intermediate form called a Holliday structure. These structures form when the two homologous DNA molecules are joined in a cross-bridge. In this exercise, you will view a short animation of the rotation of the Holliday junction. Describe what this rotation event accomplishes. The resolution of this structure requires the nicking of two strands and religation to form two separate DNA molecules. There are two possible pairs of DNA strands that can be nicked and religated and therefore two distinct outcomes. Draw a diagram indicating the two pairs of DNA strands and explain briefly what distinguishes one pair of nicked and religated DNA strands of the resulting DNA molecules from the other. To complete this exercise, visit Web Problem 3 in Chapter 12 of your Companion Website and select the keyword **HOLLIDAY.**

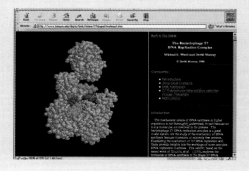

8. How did Kornberg test the fidelity of copying DNA by polymerase I?

9. Which of Kornberg's tests is the more stringent assay? Why?

10. Which characteristics of DNA polymerase I raised doubts that its *in vivo* function is the synthesis of DNA leading to complete replication?

11. Explain the theory of nearest-neighbor frequency.

12. What is meant by "biologically active" DNA?

13. Why was the phage φX174 chosen for the experiment demonstrating biological activity?

14. Outline the experimental design of Kornberg's biological activity demonstration.

15. What was the significance of the *polA1* mutation?

16. Summarize and compare the properties of DNA polymerase I, II, and III.

17. List and describe the function of the 10 subunits constituting DNA polymerase III. Distinguish between the holoenzyme and the core enzyme.

18. Distinguish between (a) unidirectional and bidirectional synthesis and (b) continuous and discontinuous synthesis of DNA.

19. List the proteins that unwind DNA during *in vivo* DNA synthesis. How do they function?

20. Define and indicate the significance of (a) Okazaki fragments, (b) DNA ligase, and (c) primer RNA during DNA replication.

21. Outline the current model for DNA synthesis.

22. Why is DNA synthesis expected to be more complex in eukaryotes than in bacteria? How is DNA synthesis similar in the two types of organisms?

23. If the analysis of DNA from two different microorganisms demonstrated very similar nearest-neighbor frequencies, is the DNA of the two organisms identical in (a) amount, (b) base composition, and/or (c) nucleotide sequences?

24. Suppose that *E. coli* synthesizes DNA at a rate of 100,000 nucleotides per minute and takes 40 minutes to replicate its chromosome.
    (a) How many base pairs are present in the entire *E. coli* chromosomes?
    (b) What is the physical length of the chromosome in its helical configuration; that is, what is the circumference of the circular chromosome?

25. Several temperature-sensitive mutant strains of *E. coli* display various characteristics. Predict what enzyme or function is being affected by each mutation.
    (a) Newly synthesized DNA contains many mismatched base pairs.
    (b) Okazaki fragments accumulate, and DNA synthesis is never completed.
    (c) No initiation occurs.
    (d) Synthesis is very slow.
    (e) Supercoiled strands are found to remain following replication, which is never completed.

26. Define gene conversion and describe how this phenomenon is related to genetic recombination.

27. Many of the gene products involved in DNA synthesis were initially defined by studying mutant *E. coli* strains that could not synthesize DNA.
    (a) The *dnaE* gene encodes the α subunit of DNA polymerase III. What effect is expected from a mutation in this gene? How could the mutant strain be maintained?
    (b) The *dnaQ* gene encodes the ε subunit of DNA polymerase. What effect is expected from a mutation in this gene?

EXTRA-SPICY PROBLEMS

28. The genome of the fruit fly *Drosophila melanogaster* consists of approximately 1.6×10^8 base pairs. DNA synthesis occurs at a rate of 30 base pairs per second. In the early embryo, the entire genome is replicated in 5 minutes. How many *bidirectional origins of synthesis* are required to accomplish this feat?

29. To gauge the fidelity of DNA synthesis, Kornberg used the nearest-neighbor frequency test, which determines the frequency with which any two bases occur adjacent to each other along the polynucleotide chain. This test relies on the enzyme spleen phosphodiesterase. As we saw in Figure 12.8, during synthesis of DNA, 5′-nucleotides are inserted; that is, each nucleotide is added with the phosphate on the C-5′ of deoxyribose. As shown here, however, the phosphodiesterase enzyme cleaves between the phosphate and the C-5′ atom, thereby producing 3′-nucleotides. If the phosphates on only one of the four nucleotides (cytidylic acid, for example) are made radioactive with ^{32}P during DNA synthesis, then after enzymatic cleavage, a radioactive phosphate will be transferred to the base that is the "nearest neighbor" on the 5′ side of all cytidylic acid nucleotides.

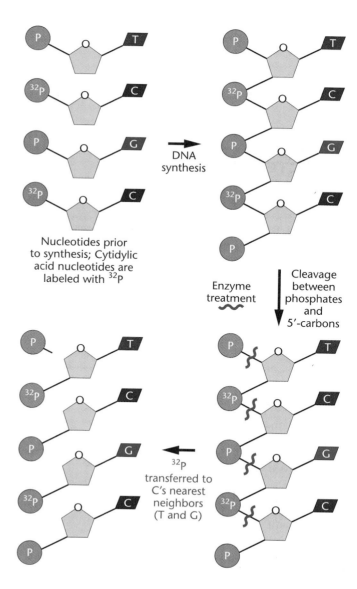

Nucleotides prior to synthesis; Cytidylic acid nucleotides are labeled with ^{32}P

DNA synthesis

Cleavage between phosphates and 5′-carbons

Enzyme treatment

^{32}P transferred to C's nearest neighbors (T and G)

Following four separate experiments, in each of which only one of the four nucleotide types is made radioactive, the frequency of all 16 possible nearest neighbors can be calculated. When Kornberg applied the nearest-neighbor frequency test to the DNA template and the resultant product from a variety of experiments, he found general agreement between the nearest-neighbor frequencies of the two.

Analysis of nearest-neighbor data led Josse, Kaiser, and Kornberg (1961) to conclude that the two strands of the double helix are in opposite polarity to one another. Demonstrate their approach by determining the outcome of such an analysis if the strands of DNA shown here are (a) antiparallel versus (b) parallel.

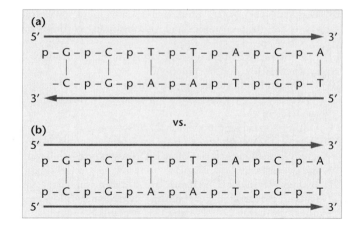

30. Assume a hypothetical organism in which DNA replication is conservative. Design an experiment similar to that of Taylor, Woods, and Hughes that will unequivocally establish this. Using the format established in Figure 12.5, draw sister chromatids and illustrate the expected results establishing this mode of replication.

SELECTED READINGS

BODNAR, A.G. et al. 1998. Extension of life-span by introduction of telomerase into normal human cells. *Science* 279:349–52.

BLACKBURN, E.H. 1991. Structure and function of telomeres. *Nature* 350:569–72.

BRAMHILL, D., and KORNBERG, A. 1988. A model for initiation at origins of DNA replication. *Cell* 54:915–18.

CAMERINI-OTERO, R.D., and HSIEH, P. 1993. Parallel DNA triplexes, homologous recombination, and other homology-dependent DNA interactions. *Cell* 73:217–23.

DARNELL, J., LODISH, H., and BALTIMORE, D. 1990. *Molecular cell biology*, 2nd ed. New York: Scientific American Books.

DAVIDSON, J.N. 1976. *The biochemistry of nucleic acids*, 8th ed. Orlando, FL: Academic Press.

DELUCIA, P., and CAIRNS, J. 1969. Isolation of an *E. coli* strain with a mutation affecting DNA polymerase. *Nature* 224:1164–66.

DENHARDT, D.T., and FAUST, E.A. 1985. Eukaryotic DNA replication. *Bioessays* 2:148–53.

DIFFLEY, J.F.X. 1996. Once and only once upon a time: Specifying and regulating origins of DNA replication in eukaryotic cells. *Genes and Development* 10:2819–30.

DRESSLER, D., and POTTER, H. 1982. Molecular mechanisms in genetic recombination. *Annu. Rev. Biochem.* 51:727–61.

GELLERT, M. 1981. DNA topoisomerases. *Annu. Rev. Biochem.* 50:879–910.

GREIDER, C.W., and BLACKBURN, E.H. 1989. A telomeric sequence in the RNA of *Tetrahymena* telomerase required for telomere repeat synthesis. *Nature* 337:331–36.

GREIDER, C.W. 1996. Telomeres, telomerase, and cancer. *Sci. Am.* (Feb.) 274:92–97.

———. 1998. Telomerase activity, cell proliferation, and cancer. *Proc. Natl. Acad. Sci. USA* 95:90–92.

HERENDEEN, D.R., and KELLY, T.J. 1996. DNA polymerase III: Running rings around the fork. *Cell* 84:5–8.

HINDGES, R., and HÜBSCHER, U. 1997. DNA polymerase essential for DNA transactions *Biol. Chem.* 378: 345–62.

HOLLIDAY, R. 1964. A mechanism for gene conversion in fungi. *Genet. Res.* 5:282–304.

HUBERMAN, J.C. 1987. Eukaryotic DNA replication: A complex picture partially clarified. *Cell* 48:7–8.

JOSSE, J., KAISER, A.D., and KORNBERG, A. 1961. Enzymatic synthesis of deoxyribonucleic acid: VIII. Frequencies of nearest-neighbor base sequences in deoxyribonucleic acid. *J. Biol. Chem.* 236:864–75.

KORNBERG, A. 1960. Biological synthesis of DNA. *Science* 131:1503–8.

———. 1969. Active center of DNA polymerase. *Science* 163:1410–18.

———. 1974. *DNA synthesis.* New York: W. H. Freeman.

———. 1979. Aspects of DNA replication. *Cold Spring Harbor Symp. Quant. Biol.* 43:1–10.

KORNBERG, A., and BAKER, T. 1992. *DNA replication*, 2nd ed. New York: W. H. Freeman.

LEHMAN, I.R. 1974. DNA ligase: Structure, mechanism, and function. *Science* 186:790–97.

LINDEGREN, C.C. 1953. Gene conversion in *Saccharomyces. J. Genet.* 51:625–37.

LODISH, H. et al. 1995. *Molecular cell biology,* 3rd ed. New York: Scientific American Books.

MESELSON, M., and STAHL, F.W. 1958. The replication of DNA in *Escherichia coli. Proc. Natl. Acad. Sci. USA* 44:671–82.

MITCHELL, M.B. 1955. Aberrant recombination of pyridoxine mutants of *Neurospora. Proc. Natl. Acad. Sci. USA* 41:215–20.

OGAWA, T., and OKAZAKI, T. 1980. Discontinuous DNA synthesis. *Annu. Rev. Biochem.* 49:421–57.

OKAZAKI, T. et al. 1979. Structure and metabolism of the RNA primer in the discontinuous replication of prokaryotic DNA. *Cold Spring Harbor Symp. Quant. Biol.* 43:203–22.

RADDING, C.M. 1978. Genetic recombination: Strand transfer and mismatch repair. *Annu. Rev. Biochem.* 47:847–80.

RADMAN, M., and WAGNER, R. 1988. The high fidelity of DNA duplication. *Sci. Am.* (Aug.) 259:40–46.

SCHEKMAN, R., WEINER, A., and KORNBERG, A. 1974. Multienzyme systems of DNA replication. *Science* 186:987–93.

STAHL, F.W. 1979. *Genetic recombination: Thinking about it in phage and fungi.* New York: W. H. Freeman.

———. 1987. Genetic recombination. *Sci. Am.* (Feb.) 256:90–101.

SZOSTAK, J.W., ORR-WEAVER, T.L., and ROTHSTEIN, R.J. 1983. The double-strand-break repair model for recombination. *Cell* 33:25–35.

TAYLOR, J.H., WOODS, P.S., and HUGHES, W.C. 1957. The organization and duplication of chromosomes revealed by autoradiographic studies using tritium-labeled thymidine. *Proc. Natl. Acad. Sci. USA* 48:122–28.

THÖMMES, P., and HÜBSCHER, U. 1992. Eukaryotic DNA helicases: Essential enzymes for DNA transactions. *Chromosoma* 101:467–73.

TOMIZAWA, J., and SELZER, G. 1979. Initiation of DNA synthesis in *E. coli.* Annu. Rev. Biochem. 48:999–1034.

WANG, J.C. 1982. DNA topoisomerases. *Sci. Am.* (July) 247:94–108.

———. 1987. Recent studies of DNA topoisomerases. *Biochim. Biophys. Acta* 909:1–9.

WATSON, J.D. et al. 1987. *Molecular biology of the gene, Vol. 1. General principles,* 4th ed. Menlo Park, CA: Benjamin/Cummings.

WHITEHOUSE, H.L.K. 1982. *Genetic recombination: Understanding the mechanisms.* New York: Wiley.

ZUBAY, G.L., and MARMUR, J., eds. 1973. *Papers in biochemical genetics,* 2nd ed. New York: Holt, Rinehart & Winston.

ZYSKIND, J.W., and SMITH, D.W. 1986. The bacterial origin of replication, *oriC. Cell* 46:489–90.

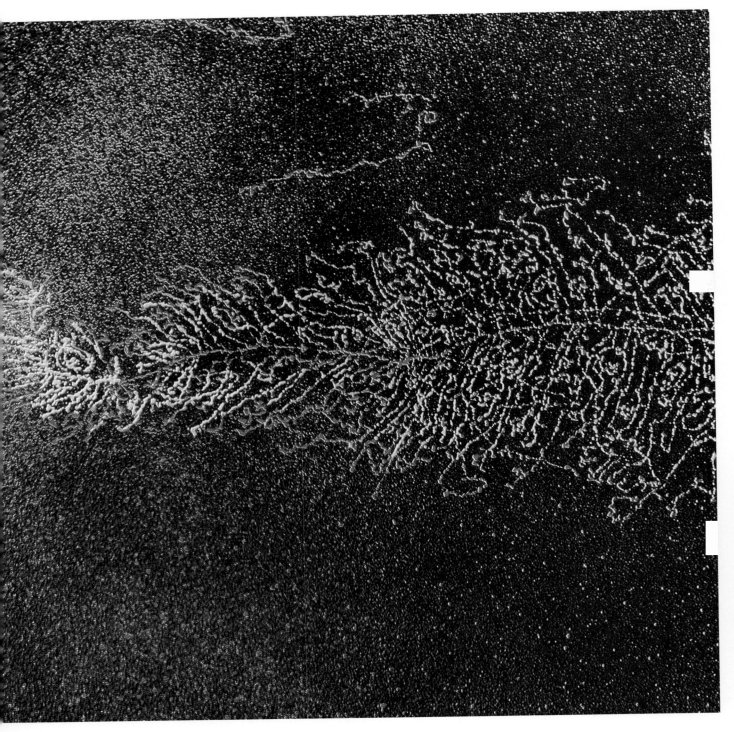

Electron micrograph visualizing the process of transcription.

The Genetic Code and Transcription

KEY CONCEPTS

- **The genetic code exhibits a number of characteristics**

- **Early studies established the basic operational patterns of the code**

  The Triplet Nature of the Code
  The Nonoverlapping Nature of the Code
  The Commaless and Degenerate Nature of the Code

- **Studies by Niremberg, Matthaei, and others led to deciphering of the code**

  Synthesizing Polypeptides in a Cell-Free System
  Homopolymer Codes
  Mixed Copolymers
  The Triplet Binding Assay
  Repeating Copolymers

- **The coding dictionary reveals the function of the 64 triplets**

  Degeneracy and the Wobble Hypothesis
  Initiation, Termination, and Suppression

- **The genetic code has been confirmed in studies of phage MS2**

- **The genetic code is nearly universal**

- **Different initiation points create overlapping genes**

- **Transcription synthesizes RNA on a DNA template**

- **Studies with bacteria and phages provided evidence for the existence of mRNA**

- **RNA polymerase directs RNA synthesis**

  Promoters, Template Binding, and the Sigma Subunit
  Initiation and Elongation of RNA

- **Transcription in eukaryotes differs from prokaryotic transcription in several ways**

  Initiation of Transcription in Eukaryotes
  Heterogeneous Nuclear RNA and Its Processing: Caps and Tails

- **The coding regions of eukaryotic genes are interrupted by intervening sequences**

  Splicing Mechanisms: Autocatalytic RNAs
  Splicing Mechanisms: The Spliceosome
  RNA Editing

- **Transcription has been visualized by electron microscopy**

GenCDX

When you see this icon, there are related animations and exercises on the CD accompanying this text.

As we saw in Chapter 11, the structure of DNA consists of a linear sequence of deoxyribonucleotides. This sequence ultimately dictates the components constituting proteins, the end product of most genes. The central question is how such information stored as a nucleic acid can be decoded into a protein. Figure 13.1 provides a simple overview of how this transfer of information occurs. The first step in gene expression involves the transfer of information present on one of the two strands of DNA (the template strand) into an RNA complement through the process of transcription. Once synthesized, this RNA acts as a "messenger" molecule, bearing the coded information—thus its name, messenger RNA (mRNA). Such RNAs then associate with ribosomes, where decoding into proteins occurs.

In this chapter, we will focus on the initial phases of gene expression by addressing two major questions. First, how is genetic information encoded? Second, how does the transfer from DNA to RNA occur, thus defining the process of transcription? As we will see, ingenious analytical research established that the genetic code is written in units of three letters—ribonucleotides present in mRNA that reflect the stored information in genes. Each triplet code word directs the incorporation of a specific amino acid into a protein as it is synthesized. As we can predict based on our prior discussion of the replication of DNA, transcription is also a complex process dependent on a major polymerase enzyme and a cast of supporting proteins. We will explore what is known about transcription in bacteria and then contrast this prokaryotic model with the differences found in eukaryotes.

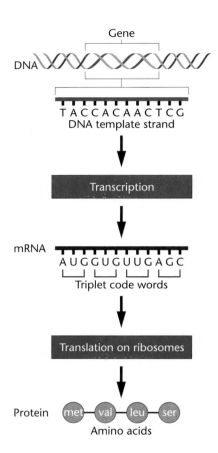

FIGURE 13.1 An overview of the concept of the flow of genetic information encoded in DNA to messenger RNA to protein.

In Chapter 14, we will continue our discussion of gene expression by addressing how translation occurs and then discussing the structure and function of proteins. Together, the information in this and the next chapter provides a comprehensive picture of molecular genetics, which serves as the most basic foundation for the understanding of living organisms.

The genetic code exhibits a number of characteristics

Before we consider the various analytical approaches that led to our current understanding of the genetic code, let us summarize the general features that characterize it.

1. The genetic code is written in linear form, using the ribonucleotide bases that compose mRNA molecules as "letters." The ribonucleotide sequence is derived from the complementary nucleotide bases in DNA.

2. Each "word" within the mRNA contains three ribonucleotide letters. Each group of *three* ribonucleotides, called a **codon**, specifies *one* amino acid; thus, the code is a **triplet.**

3. The code is **unambiguous**, meaning that each triplet specifies only a single amino acid.

4. The code is **degenerate**, meaning that a given amino acid can be specified by more than one triplet codon. This is the case for 18 of the 20 amino acids.

5. The code contains "start" and "stop" signals, certain triplets that are necessary to **initiate** and to **terminate** translation.

6. No internal punctuation ("commas") is used in the code. Thus, the code is said to be **commaless**. Once translation of mRNA begins, the codons are read one after the other with no breaks between them.

7. The code is **nonoverlapping**. Once translation commences, any single ribonucleotide at a specific location within the mRNA is part of only one triplet.

8. The code is nearly **universal**. With only minor exceptions, a single coding dictionary is used by almost all viruses, prokaryotes, archaea, and eukaryotes.

Early studies established the basic operational patterns of the code

In the late 1950s, before it became clear that mRNA serves as an intermediate in transferring genetic information from DNA to proteins, it was thought that DNA itself might directly encode proteins during their synthesis. Because ribosomes had already been identified, the initial thinking was that information in DNA was transferred in the nucleus to the RNA of the ribosome, which served as the template for protein synthesis in the cytoplasm. This concept soon became untenable as accumulating evidence demonstrated that there was an unstable template intermediate. The RNA of ribosomes, on the other hand, was extremely stable. As a result, in 1961 François Jacob and Jacques Monod postulated the existence of **messenger RNA (mRNA)**. Once mRNA was discovered, it was clear that even though genetic information is stored in DNA, the code that is translated into proteins resides in RNA. The central question then was how only four letters—the four nucleotides—could specify 20 words—the amino acids.

The Triplet Nature of the Code

In the early 1960s, Sidney Brenner argued on theoretical grounds that the code must be a triplet since three-letter words represent the minimal use of four letters to specify 20 amino acids. A code of four nucleotides, taken two at a time, for example, would provide only 16 unique code words (4^2). A triplet code provides 64 words (4^3)—clearly more than the 20 needed—and it is much simpler than a four-letter code (4^4), which would specify 256 words.

The ingenious experimental work of Francis Crick, Leslie Barnett, Brenner, and R. J. Watts-Tobin represented the first solid evidence for the triplet nature of the code. These researchers induced insertion and deletion mutations in the B cistron of the *rII* locus of phage T4 (see Chapter 7). The B cistron is one of two functional sites in this locus that, in mutant form, causes rapid lysis and distinctive plaques. Such mutants will successfully infect strain B of *E. coli* but cannot reproduce on a separate strain of *E. coli*, designated K12. Crick and his colleagues used the acridine dye proflavin to induce mutations. This mutagenic agent intercalates within the double helix of DNA, often causing the insertion or the deletion of one or more nucleotides during replication. As shown in Figure 13.2(a), an insertion of a single nucleotide causes the reading frame to shift, changing the specific sequence of all subsequent triplets to the right of the insertion. Upon translation, the amino acid sequence of the encoded pro-

tein will be altered. Such mutations are called **frameshifts**. When they are present at the *rII* locus, T4 will not reproduce on *E. coli* K12.

Crick and his colleagues reasoned that if phages with these induced mutations were treated again with proflavin, still other insertions or deletions would occur. A second change might result in a revertant phage, which would behave like wild type and successfully infect *E. coli* K12. For example, if the original mutant contained an insertion (+), a second event causing a deletion (−) close to the insertion would restore the original reading frame. In the same way, an event resulting in an insertion (+) might correct an original deletion (−).

In studying many mutations of this type, these researchers were able to compare various mutant combinations together within the same DNA sequence. They found that various combinations of one (+) and one (−) indeed caused reversion to wild-type behavior. Still other observations shed light on the number of nucleotides constituting the genetic code. When two (+)s were together or when two (−)s were together, the correct reading frame *was not* reestablished. This argued against a doublet (two-letter) code. However, when three (+)s [Figure 13.2(b)] or three (−)s were present together, the original frame *was* reestablished. These observations strongly supported the triplet nature of the code.

The Nonoverlapping Nature of the Code

Sydney Brenner also argued that the code was nonoverlapping. Assuming a triplet code, Brenner considered restrictions that might be placed on it if it were overlapping. He considered theoretical nucleotide sequences encoding a protein consisting of three amino acids. In the nucleotide sequence GTACA, for example, parts of the central triplet, TAC, are shared by the outer triplets, GTA and ACA. Brenner reasoned that if this were the case, then only certain amino acids should be found adjacent to the one encoded by the central triplet.

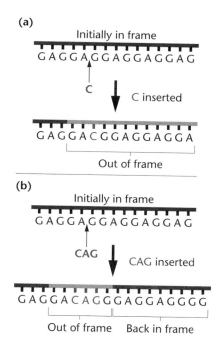

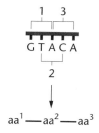

FIGURE 13.2 The effect of frameshift mutations on a DNA sequence repeating the triplet sequence GAG. In part (a) the insertion of a single nucleotide shifts all subsequent triplet reading frames. In part (b) the insertion of three nucleotides changes only two triplets, but the frame of reading is then reestablished to the original sequence.

For any given central amino acid, only 16 combinations (2^4) of "three amino acid" sequences (a tripeptide) are theoretically possible. Brenner concluded that if the code were overlapping, tripeptide sequences within proteins should be somewhat limited. Looking at the available amino acid sequences of proteins that had been studied, he failed to find such restrictions in tripeptide sequences. For any central amino acid, he found many more than 16 different tripeptides.

A second major argument against an overlapping code involved the effect of a single nucleotide change. With an overlapping code, two adjacent amino acids would be affected by such a point mutation. However, mutations in the genes coding for the protein coat of tobacco mosaic virus (TMV), human hemoglobin, and the bacterial enzyme tryptophan synthetase invariably revealed only single amino acid changes.

The third argument against an overlapping code was presented by Francis Crick in 1957, when he predicted that DNA does not serve as a direct template for the formation of proteins. Crick reasoned that any affinity between nucleotides and an amino acid would require hydrogen bonding. Chemically, however, such specific affinities seemed unlikely. Instead, Crick proposed that there must be an **adaptor molecule** that could covalently bind to the amino acid, yet be capable of hydrogen bonding to a nucleotide sequence. Because various adaptors would somehow have to overlap one another at nucleotide sites during translation, Crick reasoned that physical constraints would make the process overly complex and, perhaps, even inefficient during translation. As we will see later in this chapter, Crick's prediction was correct; transfer RNA (tRNA) serves as the adaptor in protein synthesis. And, the ribosome accommodates two tRNA molecules at a time during translation.

Crick's and Brenner's arguments, taken together, strongly suggested that during translation, the genetic code is **nonoverlapping**. Without exception, this concept has been upheld.

The Commaless and Degenerate Nature of the Code

Between 1958 and 1960, information related to the genetic code continued to accumulate. In addition to his adaptor proposal, Crick hypothesized, on the basis of genetic evidence, that the code is **commaless**; that is, he believed no internal punctuation occurs along the reading frame. Crick also speculated that only 20 of the 64 possible triplets specify an amino acid and that the remaining 44 carry no coding assignment.

Was Crick wrong with respect to the 44 "blank" codes? That is, is the code **degenerate**, with more than one triplet coding for the same amino acid? Crick's frameshift studies suggested that, contrary to his earlier proposal, the code is degenerate. The reasoning leading

to this conclusion is as follows. In the cases where wild-type function is restored, that is, (+) and (−), (+ + +) and (− − −), the original frame of reading is also restored. However, there may be numerous triplets between the various additions and deletions that would still be out of frame. If 44 of the 64 possible triplets were blank and did not specify an amino acid, one of these **nonsense triplets** would very likely occur in the length of nucleotides still out of frame. If a nonsense code were encountered during protein synthesis, it was reasoned that the process would stop or be terminated at that point. If so, the product of the *rII*-B locus would not be made, and restoration would not occur. Because the various mutant combinations were able to reproduce on *E. coli* K12, Crick and his colleagues concluded that, in all likelihood, most if not all of the remaining 44 codes were not blank. It follows that the genetic code is **degenerate**. As we shall see, this reasoning proved to be correct.

Studies by Nirenberg, Matthaei, and others led to deciphering of the code

In 1961 Marshall Nirenberg and J. Heinrich Matthaei characterized the first specific coding sequences, which served as a cornerstone for the complete analysis of the genetic code. Their success, as well as that of others who made important contributions in deciphering the code, was dependent on the use of two experimental tools, an *in vitro* **(cell-free) protein-synthesizing system** and an enzyme, **polynucleotide phosphorylase**, which allowed the production of synthetic mRNAs. These mRNAs served as templates for polypeptide synthesis in the cell-free system.

Synthesizing Polypeptides in a Cell-Free System

In the cell-free system, amino acids can be incorporated into polypeptide chains. This *in vitro* mixture, as might be expected, must contain the essential factors for protein synthesis in the cell: ribosomes, tRNAs, amino acids, and other molecules essential to translation. In order to follow (or trace) protein synthesis, one or more of the amino acids must be radioactive. Finally, an mRNA must be added, which serves as the template to be translated.

In 1961 mRNA had yet to be isolated. However, the use of the enzyme polynucleotide phosphorylase allowed the artificial synthesis of RNA templates, which could be added to the cell-free system. This enzyme, isolated from bacteria, catalyzes the reaction shown in Figure 13.3. Discovered in 1955 by Marianne Grunberg-Manago and Severo Ochoa, the enzyme functions metabolically in bacterial cells to degrade RNA. However, *in vitro*, with high

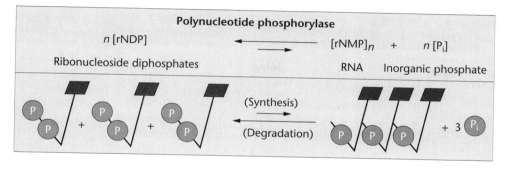

FIGURE 13.3 The reaction catalyzed by the enzyme polynucleotide phosphorylase. Note that the equilibrium of the reaction favors the degradation of RNA but can be "forced" in the direction favoring synthesis.

concentrations of ribonucleoside diphosphates, the reaction can be "forced" in the opposite direction to synthesize RNA, as illustrated.

In contrast to RNA polymerase, polynucleotide phosphorylase requires no DNA template. As a result, each addition of a ribonucleotide is random, based on the relative concentration of the four ribonucleoside diphosphates added to the reaction mixtures. The probability of the insertion of a specific ribonucleotide is proportional to the availability of that molecule, relative to other available ribonucleotides. *This point is absolutely critical to understanding the work of Nirenberg and others in the ensuing discussion.*

Taken together, the cell-free system for protein synthesis and the availability of synthetic mRNAs provided a means of deciphering the ribonucleotide composition of various triplets encoding specific amino acids.

Homopolymer Codes

In their initial experiments, Nirenberg and Matthaei synthesized **RNA homopolymers**, each consisting of only one type of ribonucleotide. Therefore, the mRNA added to the *in vitro* system was either UUUUUU..., AAAAAA..., CCCCCC..., or GGGGGG.... In testing each mRNA, they were able to determine which, if any, amino acids were incorporated into newly synthesized proteins. The researchers determined this by labeling one of the 20 amino acids added to the *in vitro* system and conducting a series of experiments, each with a different amino acid made radioactive. For example, consider the experiments using ^{14}C-phenylalanine (Table 13.1).

From these and related experiments, Nirenberg and Matthaei concluded that the message poly U (polyuridylic acid) directs the incorporation of only phenylalanine into the homopolymer polyphenylalanine. Assuming a triplet code, they had determined the first specific codon assignment: UUU codes for phenylalanine. In the same way, they quickly found that AAA codes for lysine and CCC codes for proline. Poly G did not serve as an adequate template, probably because the molecule folds back on itself. Thus, the assignment for GGG had to await other approaches.

Note that the specific triplet codon assignments were possible only because homopolymers were used. The method yields only the composition of triplets, but since three Us, Cs, or As can have only one possible sequence (i.e., UUU, CCC, and AAA), the actual codon was identified.

Mixed Copolymers

With these techniques in hand, Nirenberg and Matthaei, and Ochoa and co-workers turned to the use of **RNA heteropolymers**. In this technique, two or more different ribonucleoside diphosphates are added in combination to form the artificial message. The researchers reasoned that if they knew the relative proportion of each type of ribonucleoside diphosphate, they could predict the frequency of any particular triplet codon occurring in the synthetic mRNA. If they then added the mRNA to the cell-free system and ascertained the percentage of any particular amino acid present in the new protein, they could analyze the results and predict the *composition* of triplets specifying specific amino acids.

This approach is illustrated in Figure 13.4. Suppose that A and C are added in a ratio of 1A:5C. Now, the insertion of a ribonucleotide at any position along the RNA molecule during its synthesis is determined by the ratio of A:C. Therefore, there is a 1/6 possibility for an A and a 5/6 chance for a C to occupy each position. On this basis, we can calculate the frequency of any given triplet appearing in the message.

TABLE 13.1 Incorporation of ^{14}C-phenylalanine into Protein

Artificial mRNA	Radioactivity (counts/min)
None	44
Poly U	39,800
Poly A	50
Poly C	38

Source: After Nirenberg and Matthaei (1961).

Possible compositions	Probability of occurrence of any triplet	Possible triplets	Final %
3A	$(1/6)^3 = 1/216 = 0.4\%$	AAA	0.4
1C:2A	$(1/6)^2(5/6) = 5/216 = 2.3\%$	AAC ACA CAA	$3 \times 2.3 = 6.9$
2C:1A	$(1/6)(5/6)^2 = 25/216 = 11.6\%$	ACC CAC CCA	$3 \times 11.6 = 34.8$
3C	$(5/6)^3 = 125/216 = 57.9\%$	CCC	57.9
			100.0%

Chemical synthesis of message ↓

———————————————————————————— RNA
C C C C C C C C A C C C C C C A A C C A C C C C C A C C C C C A C C A A

Translation of message ↓

Percentage of amino acids in protein		Probable base composition assignments
Lysine	<1%	AAA
Glutamine	2%	1C:2A
Asparagine	2%	1C:2A
Threonine	12%	2C:1A
Histidine	14%	2C:1A, 1C:2A
Proline	69%	CCC, 2C:1A

FIGURE 13.4 Results and interpretation of a mixed copolymer experiment where a ratio of 1A:5C is used (1/6A:5/6C).

For AAA, the frequency is $(1/6)^3$, or about 0.4 percent. For AAC, ACA, and CAA, the frequencies are identical—that is, $(1/6)^2(5/6)$, or about 2.3 percent for each. Together, all three 2A:1C triplets account for 6.9 percent of the total three-letter sequences. In the same way, each of three 1A:2C triplets accounts for $(1/6)(5/6)^2$, or 11.6 percent (or a total of 34.8 percent). CCC is represented by $(5/6)^3$, or 57.9 percent of the triplets.

By examining the percentages of any given amino acid incorporated into the protein synthesized under the direction of this message, it is possible to propose probable base composition (Figure 13.4). Because proline appears 69 percent of the time and because 69 percent is close to 57.9 percent + 11.6 percent, we can deduce that proline is coded by CCC and by one triplet of the 2C:1A variety. Histidine, at 14 percent, is probably coded by one 2C:1A (11.6 percent) and one 1C:2A (2.3 percent). Threonine, at 12 percent, is likely coded by only one 2C:1A. Asparagine and glutamine each appear to be coded by one of the 1C:2A triplets, and lysine appears to be coded by AAA.

Using as many as all four ribonucleotides to construct the mRNA, the researchers conducted many similar experiments. Although determination of the *composition* of

triplet code words corresponding to all 20 amino acids represented a very significant breakthrough, the *specific sequences* of triplets were still unknown. Their determination awaited still other approaches.

The Triplet Binding Assay

It was not long before more advanced techniques were developed. In 1964 Nirenberg and Philip Leder developed the **triplet binding assay**, which led to specific assignments of triplets. The technique took advantage of the observation that ribosomes, when presented with an RNA sequence as short as three ribonucleotides, will bind to it and form a complex similar to what is found *in vivo*. The triplet acts like a codon in mRNA, attracting the complementary sequence within tRNA (Figure 13.5). Such a triplet sequence in tRNA that is complementary to a codon of mRNA is known as an **anticodon**.

Although it was not yet feasible to chemically synthesize long stretches of RNA, triplets of known sequence could be synthesized in the laboratory to serve as templates. All that was needed was a method to determine which tRNA-amino acid was bound to the triplet RNA-

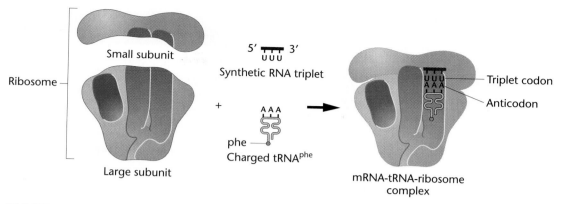

FIGURE 13.5 An example of the triplet binding assay. The UUU triplet acts as a codon, attracting the complementary tRNA^Phe anticodon AAA.

ribosome complex. The test system devised was quite simple. The amino acid to be tested was made radioactive, and a charged tRNA was produced. Because code compositions were known, it was possible to narrow the decision as to which amino acids should be tested for each specific triplet.

The radioactively charged tRNA, the RNA triplet, and ribosomes are incubated together on a nitrocellulose filter, which will retain the larger ribosomes but not the other smaller components, such as charged tRNA. If radioactivity is not retained on the filter, an incorrect amino acid has been tested. If radioactivity remains on the filter, it is retained because the charged tRNA has bound to the triplet associated with the ribosome. In such a case, a specific codon assignment can be made.

Work proceeded in several laboratories, and in many cases clear-cut, unambiguous results were obtained. Table 13.2, for example, shows 26 triplets assigned to 9 amino acids. However, in some cases the degree of triplet

binding was inefficient, and assignments were not possible. Eventually, about 50 of the 64 triplets were assigned. These specific assignments of triplets to amino acids led to two major conclusions. First, the genetic code is degenerate; that is, one amino acid may be specified by more than one triplet. Second, the code is also unambiguous; that is, a single triplet specifies only one amino acid. As we shall see later in this chapter, these conclusions have been upheld with only minor exceptions. The triplet binding technique was a major innovation in deciphering the genetic code.

Repeating Copolymers

Still another innovative technique used to decipher the genetic code was developed in the early 1960s by Gobind Khorana, who was able to chemically synthesize long RNA molecules consisting of short sequences repeated many times. First, he created shorter sequences (e.g., di-, tri-, and tetranucleotides), which were then replicated many times and finally joined enzymatically to form the long polynucleotides. As illustrated in Figure 13.6, a dinucleotide made in this way is converted to a message with two repeating triplets. A trinucleotide is converted to a message with three potential triplets, depending on the point at which initiation occurs, and a tetranucleotide creates four repeating triplets.

When these synthetic mRNAs were added to a cell-free system, the predicted number of amino acids incorporated was upheld. Several examples are shown in Table 13.3. When such data were combined with conclusions drawn from composition assignment and triplet binding, specific assignments were possible.

One example of specific assignments made from such data will illustrate the value of Khorana's approach. Consider the following three experiments in concert with one

TABLE 13.2 Amino Acid Assignments to Specific Trinucleotides Derived from the Triplet Binding Assay

Trinucleotides	Amino Acid
UGU UGC	Cysteine
GAA GAG	Glutamic acid
AUU AUG AUA	Isoleucine
UUA UUG CUU	Leucine
CUC CUA CUG	Leucine
AAA AAG	Lysine
AUG	Methionine
UUU UUC	Phenylalanine
CCU CCC	Proline
CCG CCA	Proline
UCU UCC	Serine
UCA UCG	Serine

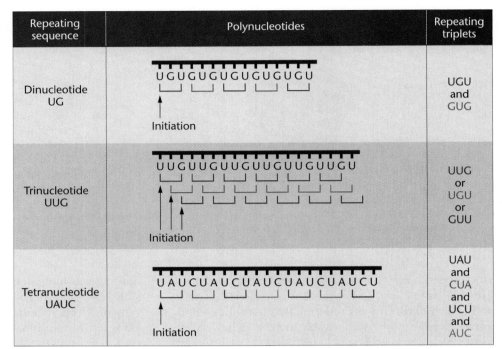

Repeating sequence	Polynucleotides	Repeating triplets
Dinucleotide UG	U G U G U G U G U G U G U G U Initiation	UGU and GUG
Trinucleotide UUG	U U G U U G U U G U U G U U G U Initiation	UUG or UGU or GUU
Tetranucleotide UAUC	U A U C U A U C U A U C U A U C U Initiation	UAU and CUA and UCU and AUC

FIGURE 13.6 The conversion of di-, tri-, and tetranucleotides into repeating copolymers. The triplet codons produced in each case are shown.

another. The repeating trinucleotide sequence UUCUU-CUUC . . . produces three possible triplets: UUC, UCU, and CUU, depending on the initiation point. When placed in a cell-free translation system, the polypeptides containing phenylalanine (phe), serine (ser), and leucine (leu)

TABLE 13.3 Amino Acids Incorporated Using Repeated Synthetic Copolymers of RNA

Repeating Copolymer	Codons Produced	Amino Acids in Polypeptides
UG	UGU GUG	Cysteine Valine
AC	ACA CAC	Threonine Histidine
UUC	UUC UCU CUU	Phenylalanine Serine Leucine
AUC	AUC UCA CAU	Isoleucine Serine Histidine
UAUC	UAU CUA UCU AUC	Tyrosine Leucine Serine Isoleucine
GAUA	GAU AGA UAG AUA	None

are produced. On the other hand, the repeating dinucleotide sequence UCUCUCUC . . . produces the triplets UCU and CUC with the incorporation of leucine and serine into the polypeptide. The overlapping results indicate that the triplets UCU and CUC specify leucine and serine but not which triplet specifies which amino acid. One can further conclude that *either* the CUU *or* the UUC triplet also encodes leucine *or* serine, while the other encodes phenylalanine.

To derive more specific information, we can examine the results of using the repeating tetranucleotide sequence UUAC, which produces the triplets UUA, UAC, ACU, and CUU. The CUU triplet is one of the two in which we are interested. Three amino acids are incorporated: leucine, threonine, and tyrosine. Because CUU must specify only serine or leucine, and because, of these two, only leucine appears, we may conclude that CUU specifies leucine.

Once this is established, we can logically determine all other specific assignments. Of the two triplet pairs remaining (UUC and UCU from the first experiment *and* UCU and CUC from the second experiment), whichever triplet is common to both must encode serine. This is UCU. By elimination, UUC is determined to encode phenylalanine and CUC is determined to encode leucine. While the logic must be carefully followed, four specific triplets encoding three different amino acids have been assigned from these experiments.

From such interpretations, Khorana reaffirmed triplets that were already deciphered and filled in gaps left from other approaches. For example, the use of two tetranucleotide sequences, GAUA and GUAA, suggested that at least two triplets were termination signals. He reached this conclusion because neither of these sequences directed the incorporation of any amino acids into a polypeptide. Because there are no triplets common to both messages, he predicted that each repeating sequence contained at least one triplet that terminates protein synthesis. Table 13.3 lists the possible triplets for the poly-(GAUA) sequence, of which UAG is a termination codon.

The coding dictionary reveals the function of the 64 triplets

The various techniques applied to decipher the genetic code have yielded a dictionary of 61 triplet codon–amino acid assignments. The remaining three triplets are termination signals, not specifying any amino acid. Figure 13.7 designates the assignments in a particularly illustrative form first suggested by Francis Crick.

First position (5′-end)	Second position: U	Second position: C	Second position: A	Second position: G	Third position (3′-end)
U	UUU *phe*	UCU *ser*	UAU *tyr*	UGU *cys*	U
U	UUC *phe*	UCC *ser*	UAC *tyr*	UGC *cys*	C
U	UUA *leu*	UCA *ser*	UAA Stop	UGA Stop	A
U	UUG *leu*	UCG *ser*	UAG Stop	UGG *trp*	G
C	CUU *leu*	CCU *pro*	CAU *his*	CGU *arg*	U
C	CUC *leu*	CCC *pro*	CAC *his*	CGC *arg*	C
C	CUA *leu*	CCA *pro*	CAA *gln*	CGA *arg*	A
C	CUG *leu*	CCG *pro*	CAG *gln*	CGG *arg*	G
A	AUU *ile*	ACU *thr*	AAU *asn*	AGU *ser*	U
A	AUC *ile*	ACC *thr*	AAC *asn*	AGC *ser*	C
A	AUA *ile*	ACA *thr*	AAA *lys*	AGA *arg*	A
A	AUG *met*	ACG *thr*	AAG *lys*	AGG *arg*	G
G	GUU *val*	GCU *ala*	GAU *asp*	GGU *gly*	U
G	GUC *val*	GCC *ala*	GAC *asp*	GGC *gly*	C
G	GUA *val*	GCA *ala*	GAA *glu*	GGA *gly*	A
G	GUG *val*	GCG *ala*	GAG *glu*	GGG *gly*	G

■ Initiation ■ Termination

FIGURE 13.7 The coding dictionary. AUG encodes methionine, which initiates most polypeptide chains. All other amino acids except tryptophan, which is encoded only by UGG, are represented by two to six triplets. The triplets UAA, UAG, and UGA are termination signals and do not encode any amino acids.

Degeneracy and the Wobble Hypothesis

A general pattern of triplet codon assignments becomes apparent when we look at the genetic coding dictionary. Most evident is that the code is degenerate, as the early researchers predicted. That is, almost all amino acids are specified by two, three, or four different codons. Three amino acids (serine, arginine, and leucine) are each encoded by six different codons. Only tryptophan and methionine are encoded by single codons.

Also evident is the pattern of degeneracy. Most often in a set of codons specifying the same amino acid, the first two letters are the same, with only the third differing. Crick observed a pattern in the degeneracy at the third position, and in 1966, he postulated the **wobble hypothesis**.

Crick's hypothesis first predicted that the initial two ribonucleotides of triplet codes are more critical than is the third member in attracting the correct tRNA. He postulated that hydrogen bonding at the third position of the codon–anticodon interactions is *less* constrained and need not adhere as specifically to the established base-pairing rules. The wobble hypothesis thus proposes a set of base-pairing rules at the third position of the codon (Table 13.4).

This relaxed base-pairing requirement, or "wobble," allows the anticodon of a single form of tRNA to pair with more than one triplet in mRNA. Consistent with the wobble hypothesis and the degeneracy of the code, U at the first position (the 5′-end) of the tRNA anticodon may pair with A or G at the third position (the 3′-end) of the mRNA codon, and that G may likewise pair with U or C. Inosine, one of the modified bases found in tRNA, may pair with C, U, or A. Applying these wobble rules, a minimum of

TABLE 13.4 Anticodon–Codon Base-Pairing Rules

Base at 1st position (5′-end) of tRNA	Base at 3rd position (3′-end) of mRNA
A	U
C	G
G	C or U
U	A or G
I	A, U or C

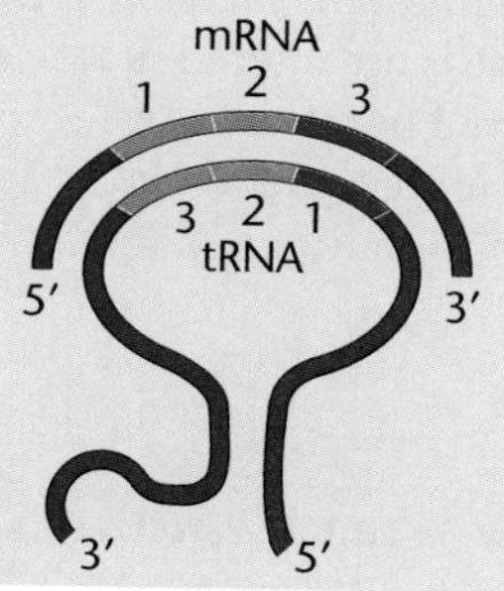

about 30 different tRNA species is necessary to accommodate the 61 triplets specifying an amino acid. If nothing more, wobble can be considered a potential economy measure, provided that the fidelity of translation is not compromised. Current estimates are that 30–40 tRNA species are present in bacteria and up to 50 tRNA species in animal and plant cells.

Initiation, Termination, and Suppression

Initiation of protein synthesis is a highly specific process. In bacteria (in contrast to the *in vitro* experiments discussed earlier), the initial amino acid inserted into all polypeptide chains is a modified form of methionine—**N-formylmethionine (fmet)**. Only one codon, AUG, codes for methionine, and it is sometimes called the **initiator codon**. However, when AUG appears internally in mRNA rather than at an initiating position, unformylated methionine is inserted into the polypeptide chain. Rarely, another triplet, GUG, specifies methionine during initiation, though it is not clear why this happens, since GUG normally encodes valine.

In bacteria, either the formyl group is removed from the initial methionine upon the completion of synthesis of a protein, or the entire formylmethionine residue is removed. In eukaryotes, methionine is also the initial amino acid during polypeptide synthesis. However, it is not formylated.

As mentioned in the preceding section, three other triplets (UAG, UAA, and UGA)* serve as **termination codons**, punctuation signals that do not code for any amino acid. They are not recognized by a tRNA molecule, and translation terminates when they are encountered. Mutations that produce any of the three triplets internally in a gene also result in termination. Consequently, only a partial polypeptide is synthesized, since it is prematurely released from the ribosome. When such a change occurs, it is called a **nonsense mutation**.

Interestingly, a distinct mutation in a second gene may trigger **suppression** of premature termination. These mutations cause the chain-termination signal to be read as a "sense" codon. The "correction" usually inserts an amino acid other than that found in the wild-type protein, but if the protein's structure is not altered drastically, it may function almost normally. Therefore, this second mutation has "suppressed" the mutant character resulting from the initial change to the termination codon.

Some suppressor mutations occur in genes specifying tRNAs. If the mutation results in a change in the anticodon such that it becomes complementary to a termination code, there is the potential for insertion of an amino acid and suppression.

The genetic code has been confirmed in studies of phage MS2

All aspects of the genetic code discussed so far yield a fairly complete picture. The code is triplet in nature, degenerate, unambiguous, and commaless, but it contains punctuation with respect to start and stop signals. These individual principles have been confirmed by the detailed analysis of the RNA-containing bacteriophage MS2 by Walter Fiers and his co-workers.

MS2 is a bacteriophage that infects *E. coli*. Its nucleic acid (RNA) contains only about 3500 ribonucleotides, making up only three genes. These genes specify a coat protein, an RNA-directed replicase, and a maturation protein (the A protein). This simple system of a small genome and few gene products allowed Fiers and his colleagues to sequence the genes and their products. The amino acid sequence of the coat protein was completed in 1970, and the nucleotide sequence of the gene and a number of nucleotides on each end of it were reported in 1972.

When the chemical nature of the gene and protein are compared, a colinear relationship is evident. That is, based on the coding dictionary, the linear sequence of nucleotides (and thus the sequence of triplet codons) corresponds precisely with the linear sequence of amino acids in the protein. Furthermore, the codon for the first amino acid is AUG, the common initiator codon; and the codon for the last amino acid is followed by two consecutive termination codons, UAA and UAG.

By 1976, the other two genes and their protein products were sequenced, providing similar confirmation. The analysis clearly shows that the genetic code, as established in bacterial systems, is identical in this virus. Other evidence suggests that the code is also identical in eukaryotes.

The genetic code is nearly universal

Between 1960 and 1978, it was generally assumed that the genetic code would be found to be universal, applying equally to viruses, bacteria, archaea, and eukaryotes. Certainly, the nature of mRNA and the translation machinery seemed to be very similar in these organisms. For example, cell-free systems derived from bacteria could translate eukaryotic mRNAs. Poly U was shown to stimulate translation of polyphenylalanine in cell-free systems when

---

*Historically, the terms *amber* (UAG), *ochre* (UAA), and *opal* (UGA) have been used to distinguish the three mutation possibilities.

TABLE 13.5 Exceptions to the Universal Code

Triplet	Normal Code Word	Altered Code Word	Source
UGA	termination	trp	Human and yeast mitochondria *Mycoplasma*
CUA	leu	thr	Yeast mitochondria
AUA	ile	met	Human mitochondria
AGA AGG	arg	termination	Human mitochondria
UAA	termination	gln	*Paramecium* *Tetrahymena* *Stylonychia*

the components were derived from eukaryotes. Many recent studies involving recombinant DNA technology (see Chapter 18) have revealed that eukaryotic genes can be inserted into bacterial cells and transcribed and translated. Within eukaryotes, mRNAs from mice and rabbits have been injected into amphibian eggs and efficiently translated. For the many eukaryotic genes that have been sequenced, notably those for hemoglobin molecules, the amino acid sequence of the encoded proteins adheres to the coding dictionary established from bacterial studies.

However, several 1979 reports on the coding properties of DNA derived from mitochondria of yeast and humans (mtDNA) altered the principle of the universality of the genetic language. Since then, mtDNA has been examined in many other organisms.

Cloned mtDNA fragments were sequenced and compared with the amino acid sequences of various mitochondrial proteins, revealing several exceptions to the coding dictionary (Table 13.5). Most surprising was that the codon UGA, normally specifying termination, specifies the insertion of tryptophan during translation in yeast and human mitochondria. In human mitochondria, AUA, which normally specifies isoleucine, directs the internal insertion of methionine. In yeast mitochondria, threonine is inserted instead of leucine when CUA is encountered in mRNA.

In 1985, several other exceptions to the standard coding dictionary were discovered in the bacterium *Mycoplasma capricolum*, and in the nuclear genes of the protozoan ciliates *Paramecium*, *Tetrahymena*, and *Stylonychia*. For example, as shown in Table 13.5, one alteration converts the termination codons (UGA) to tryptophan. Several others convert the normal termination codon to glutamine (gln). These changes are significant because both a prokaryote and several eukaryotes are involved, representing distinct species that have evolved over a long period of time.

Note the apparent pattern in several of the altered codon assignments. The change in coding capacity involves only a shift in recognition of the third, or wobble, position. For example, AUA specifies isoleucine in the cytoplasm and methionine in the mitochondrion. In the cytoplasm, methionine is specified by AUG. In a similar way, UGA calls for termination in the cytoplasm but for tryptophan in the mitochondrion. In the cytoplasm, tryptophan is specified by UGG. It has been suggested that such changes in codon recognition may represent an evolutionary trend toward reducing the number of tRNAs needed in mitochondria; only 22 tRNA species are encoded in human mitochondria, for example. However, until more examples are revealed, the differences must be considered to be exceptions to the previously established general coding rules.

Different initiation points create overlapping genes

Earlier we stated that the genetic code is nonoverlapping—each ribonucleotide in an mRNA is part of only one triplet. However, this characteristic of the code does not rule out the possibility that a single mRNA may have multiple initiation points for translation. If so, these points could theoretically create several different reading frames within the same mRNA, thus specifying more than one polypeptide. This concept of **overlapping genes** is illustrated in Figure 13.8(a).

That this might actually occur in some viruses was suspected when phage φX174 was carefully investigated. The circular DNA chromosome consists of 5386 nucleotides, which should encode a maximum of 1795 amino acids, sufficient for five or six proteins. However, this small virus in fact synthesizes 11 proteins consisting of more than 2300 amino acids. A comparison of the nucleotide sequence of the DNA and the amino acid sequences of the polypeptides synthesized has clarified the

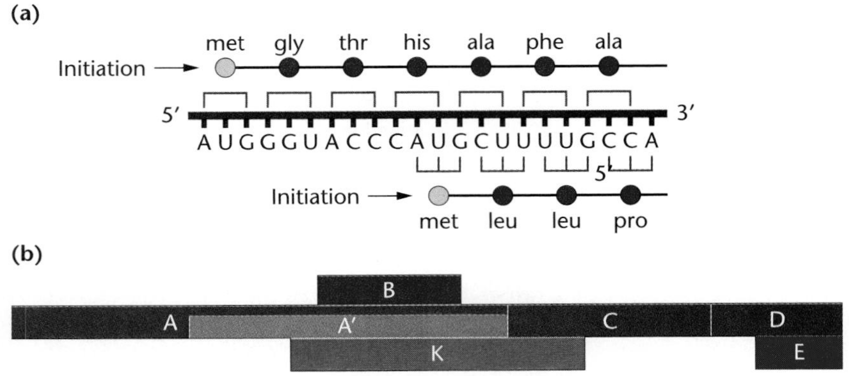

(a)

(b)

FIGURE 13.8 Illustration of the concept of overlapping genes. (a) An mRNA sequence initiated at two different AUG positions out of frame with one another will give rise to two distinct amino acid sequences. (b) The relative positions of the sequences encoding seven polypeptides of the phage φX174.

apparent paradox. At least four cases of multiple initiation have been discovered, creating overlapping genes [Figure 13.8(b)].

The sequences specifying the K and B polypeptides are initiated with separate reading frames within the sequence specifying the A polypeptide. The K sequence overlaps into the adjacent sequence specifying the C polypeptide. The E sequence is out of frame with, but initiated in, that of the D polypeptide. Finally, the A′ sequence, while in frame, begins in the middle of the A sequence. They both terminate at the identical point. In all, seven different polypeptides are created from a DNA sequence that might otherwise have specified only three (A, C, and D).

A similar situation has been observed in other viruses, including phage G4 and the animal virus SV40. Like φX174, phage G4 contains a circular single-stranded DNA molecule. The use of overlapping reading frames optimizes the use of a limited amount of DNA present in these small viruses. However, such an approach to storing information has a distinct disadvantage in that a single mutation may affect more than one protein and thus increase the chances that the change will be deleterious or lethal. In the case we just discussed, a single mutation at the junction of genes *A* and *C* could affect three proteins (the A, C, and K proteins). It may be for this reason that overlapping genes are not common in other organisms.

Transcription synthesizes RNA on a DNA template

Even while the genetic code was being studied, it was quite clear that proteins were the end products of many genes. Thus, while some geneticists were attempting to elucidate

the code, other research efforts were directed toward the nature of genetic expression. The central question was how DNA, a nucleic acid, is able to specify a protein composed of amino acids.

The complex, multistep process begins with the transfer of genetic information stored in DNA to RNA. The process by which RNA molecules are synthesized on a DNA template is called **transcription**. It results in an mRNA molecule complementary to the gene sequence of one of the two strands of the double helix. Each triplet codon in the mRNA is, in turn, complementary to the anticodon region of its corresponding tRNA as the amino acid is correctly inserted into the polypeptide chain during translation. The significance of the process of transcription is enormous, for it is the initial step in the process of information flow within the cell. The idea that RNA is involved as an intermediate molecule in the process of information flow between DNA and protein is suggested by the following observations.

1. DNA is, for the most part, associated with chromosomes in the nucleus of the eukaryotic cell. However, protein synthesis occurs in association with ribosomes located outside the nucleus in the cytoplasm. Therefore, DNA does not appear to participate directly in protein synthesis.

2. RNA is synthesized in the nucleus of eukaryotic cells, where DNA is found, and is chemically similar to DNA.

3. Following its synthesis, most RNA migrates to the cytoplasm, where protein synthesis (translation) occurs.

4. The amount of RNA is generally proportional to the amount of protein in a cell.

Like most new ideas in molecular genetics, the initial supporting experimental evidence for an RNA intermediate was based on studies of bacteria and their phages.

Studies with bacteria and phages provided evidence for the existence of mRNA

In two papers published in 1956 and 1958, Elliot Volkin and his colleagues reported their analysis of RNA produced immediately after bacteriophage infection of *E. coli*. Using the isotope ^{32}P to follow newly synthesized RNA, they found that its base composition closely resembled that of the phage DNA but was different from that of bacterial RNA (Table 13.6). Although this newly synthesized RNA was unstable, or short-lived, its production was shown to precede the synthesis of new phage proteins. Thus, Volkin and his co-workers considered the possibility that synthesis of RNA is a preliminary step in the process of protein synthesis.

Although ribosomes were known to participate in protein synthesis, their role in this process was not clear. As we noted earlier, one possibility was that each ribosome is specific for the protein synthesized in association with it. That is, perhaps genetic information in DNA is transferred to the RNA of a ribosome during its synthesis so that different groups of ribosomes are restricted to the translation of particular proteins. The alternative hypothesis was that ribosomes are nonspecific "workbenches" for protein synthesis and that specific genetic information rests with a "messenger" RNA.

In an elegant experiment using the *E. coli*–phage system, the results of which were reported in 1961, Sidney Brenner, François Jacob, and Matthew Meselson clarified this question. They labeled uninfected *E. coli* ribosomes with heavy isotopes and then allowed phage infection to occur in the presence of radioactive RNA precursors. By following these components during translation, the researchers demonstrated that the synthesis of phage proteins (under the direction of newly synthesized RNA) occurred on bacterial ribosomes that were present prior to infection. The ribosomes appeared to be nonspecific, strengthening the case that another type of RNA serves as an intermediary in the process of protein synthesis.

That same year, Sol Spiegelman and his colleagues reached the same conclusion when they isolated ^{32}P-labeled RNA following the infection of bacteria and used it in molecular hybridization studies. They tried hybridizing this RNA to the DNA of both phages and bacteria in separate experiments. The RNA hybridized only with the phage DNA, showing that it was complementary in base sequence to the viral genetic information.

Results of these experiments agree with the concept of a **messenger RNA (mRNA)** being made on a DNA template and then directing the synthesis of specific proteins in association with ribosomes. This concept was formally proposed by François Jacob and Jacques Monod in 1961 as part of a model for gene regulation in bacteria. Since then, mRNA has been isolated and thoroughly studied. There is no longer any question about its role in genetic processes.

RNA polymerase directs RNA synthesis

To prove that RNA can be synthesized on a DNA template, it was necessary to demonstrate that there is an enzyme capable of directing this synthesis. By 1959, several investigators, including Samuel Weiss, had independently discovered such a molecule from rat liver. Called **RNA polymerase**, it has the same general substrate requirements as does DNA polymerase, the major exception being that the substrate nucleotides contain the ribose rather than the deoxyribose form of the sugar. Unlike DNA polymerase, no primer is required to initiate synthesis. The initial base remains as a nucleoside triphosphate (NTP). The overall reaction summarizing the synthesis of RNA on a DNA template can be expressed as:

$$n(\text{NTP}) \xrightarrow[\text{enzyme}]{\text{DNA}} (\text{NMP})_n + n(\text{PP}_i)$$

As the equation reveals, nucleoside triphosphates (NTPs) serve as substrates for the enzyme, which catalyzes

TABLE 13.6 Base Compositions (in mole percents) of RNA Produced Immediately Following Infection of *E. coli* by the Bacteriophages T2 and T7 in Contrast to the Composition of RNA of Uninfected *E. coli*.

	Adenine	Thymine	Uracil	Cytosine	Guanine
Postinfection RNA in T2-infected cells	33	—	32	18	18
T2 DNA	32	32	—	17*	18
Postinfection RNA in T7-infected cells	27	—	28	24	22
T7 DNA	26	26	—	24	22
E. coli RNA	23	—	22	18	17

*5-hydroxymethyl cytosine.
Source: From Volkin and Astrachan (1956); and Volkin, Astrachan, and Countryman (1958).

the polymerization of nucleoside monophosphates (NMPs), or nucleotides, into a polynucleotide chain (NMP)$_n$. Nucleotides are linked during synthesis by 5'-to-3' phosphodiester bonds (see Figure 11.12). The energy created by cleaving the triphosphate precursor into the monophosphate form drives the reaction, and inorganic phosphates (PP$_i$) are produced.

A second equation summarizes the sequential addition of each ribonucleotide as the process of transcription progresses:

$$(NMP)_n + NTP \xrightarrow[\text{enzyme}]{\text{DNA}} (NMP)_{n+1} + PP_i$$

As this equation shows, each step of transcription involves the addition of one ribonucleotide (NMP) to the growing polyribonucleotide chain (NMP)$_{n+1}$, using a nucleoside triphosphate (NTP) as the precursor.

RNA polymerase from *E. coli* has been extensively characterized and shown to consist of subunits designated α, β, β', and σ. The active form of the enzyme, the **holoenzyme**, contains the subunits $\alpha_2\beta\beta'\sigma$ and has a molecular weight of almost 500,000 daltons. Of these sub-

units, it is the β and β' polypeptides that provide the catalytic basis and active site for transcription. As we will see, the **σ (sigma) subunit** plays a regulatory function involving the initiation of RNA transcription.

While there is but a single form of the enzyme in *E. coli*, there are several different sigma factors, creating variations of the polymerase holoenzyme. On the other hand, eukaryotes display three distinct forms of RNA polymerase, each consisting of a greater number of polypeptide subunits than does the bacterial form of the enzyme.

Promoters, Template Binding, and the Sigma Subunit

Transcription results in the synthesis of a single-stranded RNA molecule complementary to a region along one of the two strands of the DNA double helix. For the purpose of discussion, we will call the DNA strand that is transcribed the **template strand** and its complement the **partner strand** [Figure 13.9(a)].

The initial step is referred to as **template binding** [Figure 13.9(b)]. In bacteria, the site of this initial bind-

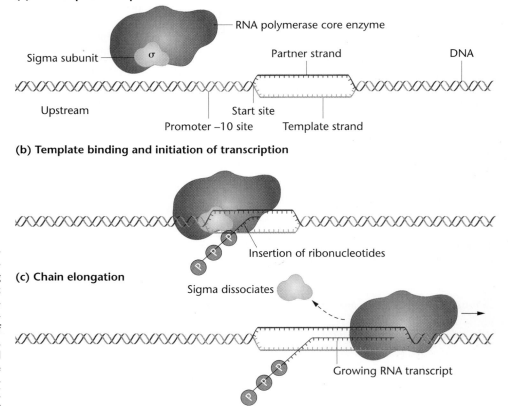

(a) Transcription components

RNA polymerase core enzyme

Sigma subunit — σ

Partner strand

DNA

Upstream

Start site

Promoter −10 site

Template strand

(b) Template binding and initiation of transcription

Insertion of ribonucleotides

FIGURE 13.9 Schematic representation of the early stages of transcription in prokaryotes, showing (a) the components of the process; (b) template binding at the −10 site involving the sigma subunit of RNA polymerase and subsequent initiation of RNA synthesis; and (c) chain elongation, after the sigma subunit has dissociated from the transcription complex and the enzyme moves along the DNA template. ⬤GenCDX

(c) Chain elongation

Sigma dissociates

Growing RNA transcript

ing is established when the RNA polymerase sigma subunit (σ) recognizes specific DNA sequences called **promoters**. These regions are located in the 5′-region upstream from the point of initial transcription of a gene. It is believed that the enzyme "explores" a length of DNA until it recognizes the promoter region and binds to about 60 nucleotide pairs of the helix, 40 of which are upstream from the point of initial transcription. Once this occurs, the helix is denatured or unwound locally, making the DNA template accessible to the action of the enzyme.

The importance of promoter sequences cannot be overemphasized. They govern the efficiency of initiation of transcription. In bacteria, both strong promoters and weak promoters have been discovered, leading to a variation of initiation from once every 1 to 2 seconds to only once every 10 to 20 minutes. Mutations in promoter sequences may have the effect of severely reducing the initiation of gene expression. Because the interaction of promoters with RNA polymerase governs transcription, the nature of the binding between them is at the heart of discussions concerning genetic regulation, the subject of Chapters 15 and 16. While we will pursue information involving promoter–enzyme interactions in greater detail in those chapters, two points are appropriate here.

The first is the concept of **consensus sequences** of DNA. These are sequences that are similar (homologous) in different genes of the same organism or in one or more genes of related organisms. Their conservation throughout evolution attests to the critical nature of their role in biological processes. Two such sequences have been found in bacterial promoters. One, TATAAT, is located 10 nucleotides upstream from the site of initial transcription (the **−10 region**, or **Pribnow box**). The other, TTGACA, is located 35 nucleotides upstream (the **−35 region**). Mutations in both regions diminish transcription, often severely. In most eukaryotic genes studied, a consensus sequence comparable to that in the −10 region has been recognized. Because it is rich in adenine and thymine residues, it is called the **TATA box**.

The second is that RNA polymerase binding to different promoters varies greatly, leading to a substantial variation in the level of gene expression. Currently, this variation, referred to as *promoter strength*, is attributed to sequence variation in the promoters.

A final general point to be made involves the sigma subunit in bacteria. The major form is designated as σ^{70}, based on its molecular weight of 70 kilodaltons (kDa). While the promoters of most bacterial genes recognize this form, there are several alternative forms of RNA polymerase in *E. coli* that have unique σ subunits associated with them (e.g., σ^{28}, σ^{32}, σ^{38}, and σ^{54}). These recognize different promoter sequences and provide specificity to the initiation of transcription.

Initiation and Elongation of RNA

Once it has recognized and bound to the promoter, RNA polymerase catalyzes **initiation**, the insertion of the first 5′-ribonucleoside triphosphate, which is complementary to the first nucleotide at the start site of the DNA template strand [Figure 13.9(b)]. As we noted, no primer is required. Subsequent ribonucleotide complements are inserted and linked together by phosphodiester bonds as RNA polymerization proceeds. This process, called **chain elongation** [Figure 13.9(c)], continues in the 5′-to-3′ direction, creating a temporary DNA/RNA duplex whose chains run antiparallel to one another.

After a few ribonucleotides have been added to the growing RNA chain, the σ subunit dissociates from the holoenzyme, and elongation proceeds under the direction of the core enzyme. In *E. coli* this process proceeds at the rate of about 50 nucleotides/second at 37°C.

Eventually, the enzyme traverses the entire gene until it encounters a specific nucleotide sequence that acts as a termination signal. Such termination sequences, about 40 base pairs in length, are extremely important in prokaryotes because of the close proximity of the end of one gene and the upstream sequences of the adjacent gene. In some cases, the termination of synthesis is dependent upon the **termination factor, rho (ρ)**. Rho is a large hexameric protein that physically interacts with the growing RNA transcript. At the point of termination, the transcribed RNA molecule is released from the DNA template, and the core polymerase enzyme dissociates. The synthesized RNA molecule is precisely complementary to a DNA sequence representing the template strand of a gene. Wherever an A, T, C, or G residue existed, a corresponding U, A, G, or C residue, respectively, is incorporated into the RNA molecule. This RNA molecule ultimately provides the information leading to the synthesis of all proteins present in the cell.

It is important to note that groups of genes whose products are related are often clustered together along the chromosome. In many such cases, they are contiguous and all but the last gene lack the encoded signals for termination of transcription. The result is that during transcription a large mRNA is produced, encoding more than one gene. Since genes in bacteria are sometimes called cistrons (see Chapter 7), the RNA is called a **polycistronic mRNA**. Since the products of genes transcribed in this fashion are usually all needed at the same time, this is an efficient way to transcribe and, subsequently, to translate the needed genetic information. In eukaryotes, **monocistronic** mRNAs are the rule.

Transcription in eukaryotes differs from prokaryotic transcription in several ways

Much of our knowledge of transcription has been derived from studies of prokaryotes. Most of the general aspects of the mechanics of these processes are similar in eukaryotes, though there are several notable differences. We will first summarize some of the major differences and then expand on a number of them in the following sections.

1. Transcription in eukaryotes occurs within the nucleus under the direction of three separate forms of RNA polymerase. Unlike prokaryotes, in eukaryotes the RNA transcript is not free to associate with ribosomes prior to the completion of transcription. For the mRNA to be translated, it must move out of the nucleus into the cytoplasm.

2. Initiation and regulation of transcription involve a more extensive interaction between upstream DNA sequences and protein factors involved in stimulating and initiating transcription. In addition to promoters, other control units called *enhancers* may be located in the 5′-regulatory region upstream from the initiation point, but they have also been found within the gene or even in the 3′ downstream region beyond the coding sequence.

3. Maturation of eukaryotic mRNA from the primary RNA transcript involves many complex stages referred to generally as "processing." An initial processing step involves the addition of a 5′-cap and a 3′-tail to most transcripts destined to become mRNAs. Other extensive modifications occur to the internal nucleotide sequence of eukaryotic RNA transcripts that eventually serve as mRNAs. The initial (or primary) transcripts are most often much larger than those that are eventually translated. Sometimes called **pre-mRNAs**, they are part of a group of molecules found only in the nucleus—a group referred to collectively as **heterogeneous nuclear RNA (hnRNA)**. Such RNA molecules are of variable but large size (up to 10^7 daltons) and are complexed with proteins, forming **heterogeneous nuclear ribonucleoprotein particles (hnRNPs)**. Only about 25 percent of hnRNA molecules are converted to mRNA. In those that are converted, substantial amounts of the ribonucleotide sequence are excised and the remaining segments are spliced back together prior to translation. This phenomenon has given rise to the concepts of **split genes** and **splicing** in eukaryotes.

In the remainder of this chapter, we will elaborate on these differences. We will also return to them and other topics directly related to regulation of eukaryotic transcription in Chapter 16.

Initiation of Transcription in Eukaryotes

The recognition of certain highly specific DNA regions by RNA polymerase is the basis of orderly genetic function in all cells. Eukaryotic RNA polymerase exists in three unique forms, each of which transcribes different types of genes, as indicated in Table 13.7. Each enzyme is larger and more complex than the single prokaryotic polymerase, consisting of two large subunits and 10 to 15 smaller subunits. In regard to the initial template binding step and promoter regions, most is known about polymerase II, which transcribes all mRNAs in eukaryotes.

At least three *cis-acting elements* of a eukaryotic gene aid the efficient initiation of transcription by polymerase II. Recall from our discussion of the *cis-trans* test in Chapter 4 that the use of the term *cis* is drawn from organic chemistry nomenclature, meaning "next to" or on the same side as, in contrast to being "across from" or **trans**, to other functional groups. In molecular genetics, then, *cis*-elements are adjacent parts of the same DNA molecule.

One such element found within the promoter region is called the **Goldberg–Hogness** or **TATA box**, located about 30 nucleotide pairs upstream (−30) from the start point of transcription. The consensus sequence is a heptanucleotide consisting solely of A and T residues (TATAAAA). The sequence and function are analogous to that found in the −10 promoter region of prokaryotic genes. Because such a region is common to most eukaryotic genes, the TATA box is thought to be nonspecific and simply have the responsibility for fixing the site of initiation of transcription by facilitating denaturation of the helix. Such a conclusion is supported by the fact that A $=$ T base pairs are less stable than G $\equiv$ C pairs.

A second common *cis*-acting element is called the **CAAT box**, located upstream within the promoter of many genes at about 80 nucleotides from the start of transcription (−80). It contains the consensus sequence GGCCAATCT. Still other upstream regulatory regions are

TABLE 13.7 RNA Polymerases in Eukaryotes

Form	Product	Location
I	rRNA	Nucleolus
II	mRNA, snRNA	Nucleoplasm
III	5S rRNA	Nucleoplasm
	tRNA	Nucleoplasm

found, and most genes contain one or more of them. Along with the TATA and CAAT box, they influence the efficiency of the promoter. The location of each element is based on studies of deletions of particular regions of the promoter, each of which reduces the efficiency of transcription.

Still another *cis*-acting element is represented by DNA regions called **enhancers**. Though their locations may vary, enhancers are often found even farther upstream than the regions already mentioned, or even downstream or within the gene. Thus they have the potential to modulate transcription from a distance. Although they may not participate directly in template binding, they are essential to highly efficient initiation of transcription. We will return to a discussion of these elements in Chapter 16.

Complementing the *cis*-acting regulatory sequences are various ***trans*-acting factors** that facilitate template binding and, therefore, the initiation of transcription. These are proteins referred to as **transcription factors**. They are essential because RNA polymerase II cannot bind directly to eukaryotic promoter sites and initiate transcription without their presence. The transcription factors involved with human RNA polymerase II-binding are well characterized and designated **TFIIA, TFIIB**, and so on. One of these, TFIID, binds directly to the TATA-box sequence and is sometimes called the **TATA-binding protein (TBP)**. TFIID consists of about 10 polypeptide subunits. Once initial binding to DNA occurs, at least seven other transcription factors bind sequentially to TFIID, forming an extensive pre-initiation complex, which is then bound by RNA polymerase II.

Transcription factors with similar activity have been discovered in a variety of eukaryotes, including *Drosophila* and yeast. The nucleotide sequences in all organisms studied demonstrate a high degree of conservation. Transcription factors appear to supplant the role of the sigma factor in the prokaryotic enzyme. In Chapter 16 we will consider the role of transcription factors in eukaryotic gene regulation, as well as the various DNA-binding domains that characterize them.

Heterogeneous Nuclear RNA and Its Processing: Caps and Tails

The genetic code is written in the ribonucleotide sequence of mRNA. This information originated, of course, in the template strand of DNA, where complementary sequences of deoxyribonucleotides exist. In bacteria, the relationship between DNA and RNA appears to be quite direct. The DNA base sequence is transcribed into an mRNA sequence, which is then translated into an amino acid sequence according to the genetic code. In eukary-

otes, by contrast, complex processing of mRNA occurs before it is transported to the cytoplasm to participate in translation.

By 1970, accumulating evidence showed that eukaryotic mRNA is transcribed initially as a precursor molecule much larger than that which is translated. This notion was based on the observation by James Darnell and his coworkers of heterogeneous nuclear RNA (hnRNA) in mammalian nuclei that contained nucleotide sequences common to the smaller mRNA molecules present in the cytoplasm. They proposed that the initial transcript of a gene results in a large RNA molecule that must first be processed in the nucleus before it appears in the cytoplasm as a mature mRNA molecule. The various processing steps, discussed in the following sections, are summarized in Figure 13.10.

The initial **posttranscriptional modification** of eukaryotic RNA transcripts destined to become mRNAs involves the 5′-end of these molecules, where a **7-methylguanosine (7mG) cap** is added (Figure 13.10, Step 2). The cap, which is added even before the initial transcript is complete, appears to be important to the subsequent processing within the nucleus, perhaps by protecting the 5′-end of the molecule from nuclease attack. Subsequently, this cap may be involved in the transport of mature mRNAs across the nuclear membrane into the cytoplasm. The cap is fairly complex and distinguished by a unique 5′–5′ bonding between the cap and the initial ribonucleotide of the RNA. Some eukaryotes also contain a methyl group ($—CH_3$) on the 2′-carbon of the ribose sugars of the first two ribonucleotides of the RNA.

Further insights into the processing of RNA transcripts during the maturation of mRNA came from the discovery that both hnRNAs and mRNAs contain at their 3′-end a stretch of as many as 250 adenylic acid residues. Such **poly-A sequences** are added after the 5′-7mG cap has been added. First the 3′-end of the initial transcript is cleaved enzymatically at a point some 10 to 35 ribonucleotides from a highly conserved AAUAAA sequence (Step 3). Then polyadenylation occurs by the sequential addition of adenylic acid residues (Step 4). Poly A has now been found at the 3′-end of almost all mRNAs studied in a variety of eukaryotic organisms. The exceptions seem to be the products of histone genes.

While the AAUAAA sequence is not found on all eukaryotic transcripts, mutations in that sequence of those transcripts that do have it cannot add the poly A tail. In the absence of this tail, the RNA transcripts are rapidly degraded. Therefore, both the 5′-cap and the 3′-poly A tail is critical if an RNA transcript is to be further processed and transported to the cytoplasm.

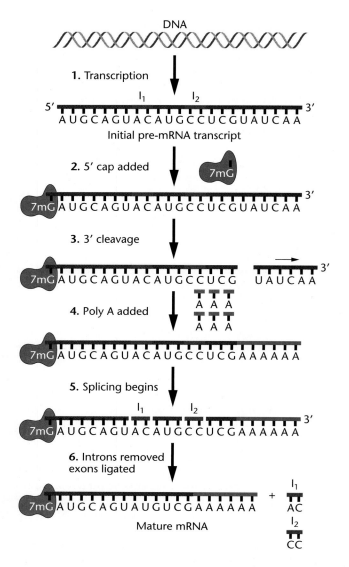

FIGURE 13.10 A summary of posttranscriptional RNA processing in eukaryotes. Heterogeneous nuclear RNA (hnRNA) is converted to messenger (mRNA), which contains a 5'-cap and a 3'-poly-A tail, which then has introns spliced out. ⊜GenCDX

The coding regions of eukaryotic genes are interrupted by intervening sequences

One of the most exciting discoveries in the history of molecular genetics occurred in 1977 when Susan Berget, Philip Sharp, and Richard Roberts presented direct evidence that the genes of animal viruses contain *internal* nucleotide sequences that are not expressed in the amino acid sequence of the proteins they encode. These internal DNA sequences are present in initial RNA transcripts, but they are removed before the mature mRNA is translated (Figure 13.10 Steps 5 and 6). Such nucleotide segments are called **intervening sequences**, and the genes containing them are known as **split genes**. Those DNA sequences that are not represented in the final mRNA product are also called **introns** ("int" for intervening), and those retained and expressed are called **exons** ("ex" for expressed). Splicing involves the removal of the ribonucleotide sequences present in introns as a result of an excision process and the rejoining of exons.

Similar discoveries were soon made in a variety of eukaryotes. Two approaches have been most fruitful. The first involves the molecular hybridization of purified, functionally mature mRNAs with DNA containing the genes specifying that message. Hybridization between nucleic

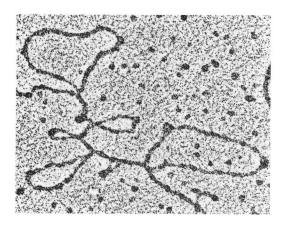

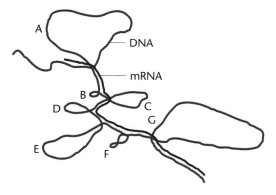

FIGURE 13.11 An electron micrograph and an interpretive drawing of the hybrid molecule (heteroduplex) formed between the template DNA strand of the chicken ovalbumin gene and the mature ovalbumin mRNA. Seven DNA introns, A–G, produce unpaired loops.

acids that are not perfectly complementary results in **heteroduplexes**, in which introns present in the DNA but absent in the mRNA loop out and remain unpaired. Such structures can be visualized with the electron microscope, as shown in Figure 13.11. That heteroduplex contains seven loops (A–G) representing seven introns whose sequences are present in DNA but not in the final mRNA.

The second approach provides more specific information (Figure 13.12). It involves a comparison of nucleotide sequences of DNA with those of mRNA and the correlation with amino acid sequences. Such an approach allows the precise identification of all intervening sequences.

Thus far, most eukaryotic genes have been shown to contain introns. One of the first so identified was the **beta-globin gene** in mice and rabbits, studied independently by Philip Leder and Richard Flavell. The mouse gene contains an intron 550 nucleotides long, beginning immediately after the codon specifying the 104th amino acid. In the rabbit, there is an intron of 580 base pairs near the codon for the 110th amino acid. Additionally, a second intron of about 120 nucleotides exists earlier in both genes. Similar introns have been found in the beta-globin gene in all mammals examined.

The **ovalbumin gene** of chickens has been extensively characterized by Bert O'Malley in the United States and Pierre Chambon in France. As shown in Figure 13.12, the gene contains seven introns. In fact, as you can see, the majority of the gene's DNA sequence is "silent," being composed of introns. The initial RNA transcript is nearly three times the length of the mature mRNA. Compare the ovalbumin gene in Figures 13.11 and 13.12. Can you match the unpaired loops in Figure 13.11 with the sequence of introns specified in Figure 13.12?

The list of genes containing intervening sequences is long. In fact, few eukaryotic genes seem to be without introns. An extreme example of the number of introns in a single gene is that found in the gene coding for one of the subunits of collagen, the major connective tissue protein in vertebrates. The *pro-α-2(1) collagen* gene contains 50 introns. The precision of cutting and splicing that occurs must be extraordinary if errors are not to be introduced into the mature mRNA. What is equally noteworthy is a comparison of the size of genes with the size of the final mRNA once introns are removed. As shown in Table 13.8, only about 15% of the collagen gene consists of exons that finally appear in mRNA. For other proteins, an even more extreme picture emerges. Only about 8% of the albumin gene remains to be translated, and in the largest human gene known, dystrophin (which is the protein product absent in Duchenne muscular dystrophy), less than 1% of the gene sequence is retained in the mRNA. Several other human genes are also contrasted in Table 13.8.

While the vast majority of eukaryotic genes examined thus far contain introns, there are several exceptions. Notably, those coding for histones and for interferon appear to contain no introns. It is not clear why or how the genes encoding these molecules have been maintained throughout evolution without acquiring the extraneous information characteristic of almost all other genes.

Splicing Mechanisms: Autocatalytic RNAs

The discovery of split genes led to intensive attempts to elucidate the mechanism by which introns of RNA are excised and exons are spliced back together. A great deal of progress has already been made. Interestingly, it appears that somewhat different mechanisms exist for different types of RNA, as well as for RNAs produced in mitochondria and chloroplasts.

Introns can be categorized into several groups based on their splicing mechanisms. In group I, represented by introns that are part of the primary transcript of rRNAs, no additional components are required for intron excision; the intron itself is the source of the enzymatic activity nec-

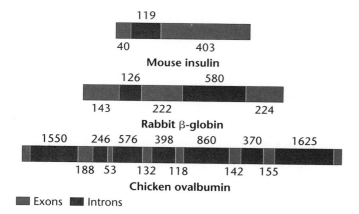

FIGURE 13.12 Intervening sequences in various eukaryotic genes. The numbers indicate the number of nucleotides present in various intron and exon regions.

TABLE 13.8 Contrasting Human Gene Size, mRNA Size, and the Number of Introns

Gene	Gene Size (Kb)	mRNA size (Kb)	Number of Introns
Insulin	1.7	0.4	2
Collagen [*pro-α-(1)*]	38.0	5.0	50
Albumin	25.0	2.1	14
Phenylalanine hydroxylase	90.0	2.4	12
Dystrophin	2000.0	17.0	50

essary for its own removal. This amazing discovery, which contradicted expectations, was made in 1982 by Thomas Cech and his colleagues during a study of the ciliate protozoan *Tetrahymena*. Reflecting their autocatalytic properties, RNAs that are capable of splicing themselves are sometimes called **ribozymes**.

The **self-excision process** is illustrated in Figure 13.13. Chemically, two nucleophilic reactions (called transesterification reactions) occur, the first involving an interaction between guanosine, which acts as a cofactor in the reaction, and the primary transcript [Figure 13.13(a)]. The 3'-OH group of guanosine is transferred to the nucleotide adjacent to the 5'-end of the intron. The second reaction involves the interaction of the newly acquired 3'-OH group on the left hand exon and the phosphate on the 3'-end of the right intron [Figure 13.13(b)]. The intron is spliced out and the two exon regions are ligated, leading to the mature RNA [Figure 13.13(c)].

Self-excision of group I introns, as described, is now known to apply to pre-rRNAs from other protozoans. Self-excision also seems to govern the removal of introns present in the primary mRNA and tRNA transcripts produced in the mitochondria and chloroplasts. These are referred to as group II introns. Like group I molecules, splicing involves two autocatalytic reactions leading to the excision of introns. However, guanosine is not involved as a cofactor.

Splicing Mechanisms: The Spliceosome

Another major group of introns is found in nuclear-derived pre-mRNA transcripts. Compared to the other RNAs we have been discussing, introns in nuclear-derived mRNA can be much larger—up to 20,000 nucleotides—and they are more plentiful. Their removal appears to require a much more complex mechanism, which has been more difficult to define.

Nevertheless, many clues have now emerged, and the model diagrammed in Figure 13.14 illustrates the removal of one intron. You should refer to this figure during this discussion. First, the nucleotide sequences near the perimeters of this type of intron are often similar. Many begin at the 5'-end with a GU dinucleotide sequence and terminate at the 3'-end with an AG dinucleotide sequence. These, as well as other consensus sequences shared by introns, attract specific molecules that form a molecular complex essential to splicing. Such a complex, called a **spliceosome**, has been identified in extracts of yeast as well as mammalian cells. It is very large, being $40S$ in yeast and $60S$ in mammals. Perhaps the most essential component of spliceosomes is the unique set of **small nuclear RNAs (snRNAs)**. These RNAs are usually 100 to 200 nucleotides or less and are often complexed with proteins to form **small nuclear ribonucleoproteins (snRNPs,** or **snurps)**. These are found only in the nucleus. Because they are rich in uridine residues, the snRNAs have been arbitrarily designated U1, U2, . . . , U6.

The snRNA of U1 bears a nucleotide sequence that is homologous to the 5'-end of the intron. Base pairing resulting from this homology promotes binding that represents the initial step in the formation of the spliceosome. Following the addition of the other snurps (U2, U4, U5, and U6), splicing commences. As with group I splicing, two transesterification reactions are involved. The first involves the interaction of the 2'-OH group from an adenine (A) residue present within the region of the intron called the **branch point**. The A residue attacks the 5'-splice site, cutting the RNA chain. In a subsequent step involving several other snurps, an intermediate structure is formed and the second reaction ensues, linking the cut 5'-end of the intron to the A. This results in the formation of a characteristic loop structure referred to as a lariat, which contains the excised intron. The exons are then ligated and the snurps are released.

The processing involved in splicing represents a potential regulatory step during gene expression. For exam-

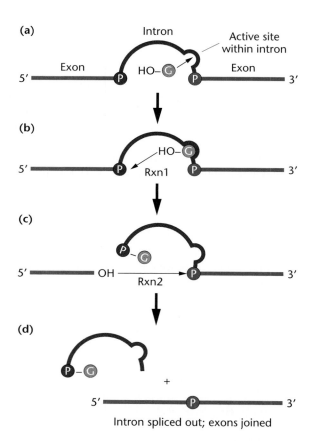

FIGURE 13.13 Splicing mechanism involved with group I introns removed from the primary transcript leading to rRNA. The process is one of self-excision involving two transesterification reactions.

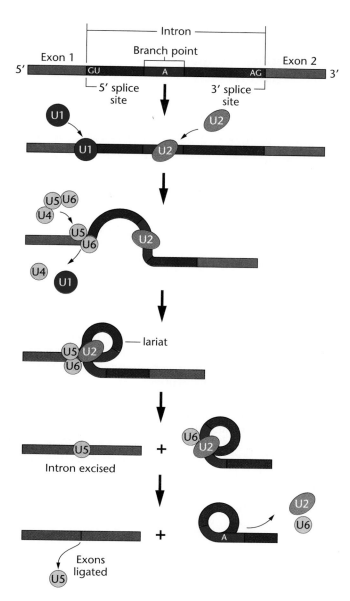

FIGURE 13.14 A model of the splicing mechanism involved with the removal of an intron from a pre-mRNA. Excision is dependent on various snRNAs that combine with proteins to form snurps (U1, U2, . . . , U6), which function as part of a large structure referred to as the spliceosome. The lariat structure in the intermediate stage is characteristic of this mechanism.

ple, several cases are known where introns present in pre-mRNAs *derived from the same gene* are spliced *in more than one way*, thereby yielding different collections of exons in the mature mRNA. This process, referred to as **alternative splicing**, yields a group of mRNAs that, upon translation, result in a series of related proteins called **isoforms**. A growing number of examples are now found in organisms ranging from viruses to *Drosophila* to humans. Alternative splicing of pre-mRNAs provides the basis for producing related proteins from a single gene. We shall return to this topic in our discussion of the regulation of gene expression in eukaryotes (Chapter 16).

RNA Editing

In the late 1980s, still another quite unexpected form of posttranscriptional RNA processing was discovered. In this case, referred to as **RNA editing**, the nucleotide sequence of a pre-mRNA is actually changed prior to translation. As a result, the ribonucleotide sequence of the mature RNA differs from the sequence encoded in the exons of the DNA from which the RNA was transcribed.

There are two main types of RNA editing—**substitution editing**, in which the identities of individual nucleotide bases are altered; and **insertion/deletion editing**, where nucleotides are added to or subtracted from the total number of bases. Substitution editing is used in some nuclear-derived eukaryotic RNAs and is very prevalent in mitochondrial and chloroplast RNAs transcribed in plants. *Physarum polycephalum*, a slime mold, uses both substitution and insertion/deletion editing for its mitochondrial mRNAs.

Trypanosoma, a parasite that causes African sleeping sickness, and its relatives use extensive insertion/deletion editing in mitochondrial RNAs. The number of uridines added to an individual transcript can make up more than 60 percent of the coding sequence, usually forming the initiation codon and placing the rest of the sequence into the proper reading frame. Insertion/deletion editing in trypanosomes is directed by "**gRNA**" (**guide RNA**) templates, which are also transcribed from the mitochondrial genome. These small RNAs are complementary to the edited region of the final, edited mRNAs. They base-pair with the pre-edited mRNAs to direct the editing machinery to make the correct changes.

The most well-studied examples of substitutional editing occur in mammalian nuclear-encoded mRNA transcripts. Apolipoprotein B (apo B) exists in both a long and a short form, though a single gene encodes both proteins. In human intestinal cells, apo B mRNA is edited by a single C to U change, which converts a CAA glutamine codon into a UAA stop codon and terminates the polypeptide at approximately half its genomically encoded length. The editing is performed by a complex of proteins that bind to a "mooring sequence" on the mRNA transcript just downstream of the editing site. In a second system, the synthesis of subunits constituting the glutamate receptor channels (GluR) in mammalian brain tissue is also affected by RNA editing. In this case, adenosine (A) to inosine (I) editing occurs in pre-mRNAs prior to their translation, where I is read as guanosine (G) during translation. A family of three ADAR (adenosine deaminase acting on RNA) enzymes are believed to be responsible for the editing of various sites within the glutamate channel subunits. The double-stranded RNAs required for editing by ADARs are provided by intron/exon pairing of the GluR mRNA transcripts. The editing changes alter the physiological parameters (solute permeability and desensitization response time) of the receptors containing the subunits.

Findings such as these in mammals have established that RNA editing provides still another important mechanism of posttranscriptional modification, and that this process is not restricted to small or asexually reproducing genomes such as mitochondria. Several new examples of RNA editing have been found each year since its discovery, and this trend is likely to continue. Further, the process has important implications for the regulation of genetic expression.

Transcription has been visualized by electron microscopy

We conclude this chapter by presenting a striking visual demonstration of the transcription process based on the electron microscope studies of Oscar Miller, Jr., Barbara Hamkalo, and Charles Thomas. Their combined work has captured the transcription process in both prokaryotes and eukaryotes. Figure 13.15 shows micrographs and interpretive drawings from two organisms, the bacterium *E. coli* and the newt *Notophthalmus viridescens*. In both cases, multiple strands of RNA are seen to emanate from different points along a central DNA template. Many RNA strands result because numerous transcription events are occurring simultaneously along each gene. Progressively longer RNA strands are found farther downstream from the point of initiation of transcription, whereas the shortest strands are closest to the point of initiation.

An interesting picture emerges from the study of *E. coli* [Figure 13.15(a)]. Because prokaryotes lack nuclei, cytoplasmic ribosomes are not separated physically from the chromosome. As a result, ribosomes are free to attach to *partially* transcribed mRNA molecules and initiate translation. The longer RNA strands demonstrate the greatest number of ribosomes. In the case of the newt [Figure 13.15(b)], the segment of DNA is derived from oocytes that are known to produce an enormous amount of ribosomal (rRNA), one of the major components of the ribosome. To accomplish this synthesis, the genes specific for ribosomal RNA (**rDNA**) are replicated many times in the oocyte, a process called **gene amplification**. The micrograph shows several of these genes, each undergoing simultaneous transcription events. For each rRNA gene, longer and longer strands of incomplete rRNA molecules are produced as the enzymes move along the DNA strand. Visualization of transcription confirms our expectations based on the biochemical analysis of this process.

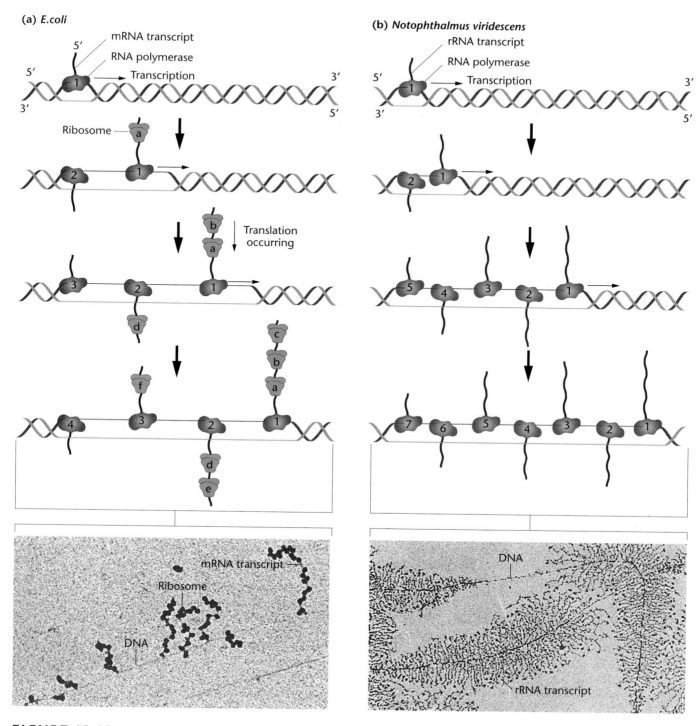

FIGURE 13.15 Electron micrographs and interpretive drawings of simultaneous transcription of genes in (a) *E. coli* and (b) *Notophthalmus (Triturus) viridescens*.

GENETICS, TECHNOLOGY, AND SOCIETY

Antisense Oligonucleotides: Attacking the Messenger

Standard chemotherapies for diseases such as cancer and AIDS are often accompanied by toxic side effects. Conventional therapeutic drugs target both normal and diseased cells, with diseased or infected cells being only slightly more susceptible than the patient's normal cells. Scientists have long wished for a magic bullet that could seek out and destroy the virus or cancer cell, leaving normal cells alive and healthy. Over the last decade, one particularly promising candidate for magic-bullet status has emerged—the *antisense oligonucleotide*.

Antisense therapies have arisen from an understanding of the molecular biology of gene expression. Gene expression is a two-step process. First, a single-stranded messenger RNA (mRNA) is copied from one strand of the duplex DNA molecule. Second, the mRNA is transported to the cytoplasm and complexed with ribosomes, and its genetic information is translated into the amino acid sequence of a polypeptide.

Normally, a gene is transcribed into RNA from only one strand of the DNA duplex. The resulting RNA is known as sense RNA. However, it is sometimes possible for the other DNA strand to be copied into RNA. The RNA produced by the transcription of the "wrong" strand of DNA is called antisense RNA. As with complementary strands of DNA molecules, complementary strands of RNA can form double-stranded molecules.

The formation of duplex structures between a sense and an antisense RNA may affect the sense RNA in at least two ways. The binding of antisense RNA to sense RNA may physically block ribosome binding or elongation, and hence may inhibit translation. Second, the binding of antisense RNA to sense RNA may trigger the degradation of the sense RNA because double-stranded RNA molecules are attacked by intracellular ribonucleases. In both cases, gene expression is blocked.

What makes the antisense approach so exciting is its potential specificity. Scientists can design antisense RNA (or DNA) molecules of known nucleotide sequence and can then synthesize large amounts of these an-

tisense nucleic acids *in vitro*. Usually, oligonucleotides up to 20 nucleotides are used. It is theoretically possible to treat cells with these synthetic antisense oligonucleotides, to have the oligonucleotides enter the cell and bind precise target mRNAs, and to turn off the synthesis of one specific protein. If that protein is necessary for virus reproduction or cancer cell growth (but is not necessary in normal cells), the antisense oligonucleotide should have only therapeutic effects.

In the last few years, laboratory tests of antisense drugs have been so promising that several clinical trials are now in progress. In addition, one antisense drug (Vitravene™, ISIS Pharmaceuticals and CIBA Vision Corp.) has now been approved for sale in the U.S. Vitravene is an antisense oligonucleotide used to treat cytomegalovirus-induced retinitis in AIDS patients. Cytomegalovirus (CMV) is a common virus that infects most people. Although it causes few problems in people with normal immune systems, it can cause serious symptoms in people with conditions

CHAPTER SUMMARY

1. The genetic code, stored in DNA, is copied to RNA, where it is used to direct the synthesis of polypeptide chains. It is degenerate, unambiguous, nonoverlapping, and commaless.

2. The complete coding dictionary, determined using various experimental approaches, reveals that of the 64 possible codons, 61 encode the 20 amino acids found in proteins, while three triplets terminate trans-

lation. One of these 61 is the initiation codon and specifies methionine.

3. The observed pattern of degeneracy often involves only the third letter of a triplet series and led Francis Crick to propose the "wobble hypothesis."

4. Confirmation for the coding dictionary, including codons for initiation and termination, was obtained by comparing the complete nucleotide sequence of

(such as AIDS) that impair the immune system. Up to 40 percent of AIDS patients develop retinitis or blindness as a result of CMV infections of the eye. Clinical trials indicate that the introduction of antisense CMV oligonucleotides into the eyes of AIDS patients with CMV-induced retinitis significantly delays the disease progression compared to untreated groups. Although it is not known how the oligonucleotides operate, laboratory studies suggest that these antisense oligonucleotides may trigger the destruction of CMV mRNA and interfere with the absorption of the virus to the surface of host cells.

Antisense oligonucleotides may also act as anti-inflammatory drugs. Clinical trials are in progress to test antisense compounds in the treatment of asthma, rheumatoid arthritis, psoriasis, ulcerative colitis, and transplant rejection. One interesting antisense oligonucleotide (now in clinical trials conducted by ISIS and Boehringer Ingelheim) binds to the mRNA that encodes ICAM-1. ICAM-1 is a cell surface glycoprotein that helps activate immune system and inflammatory cells. It is often overexpressed in tissues that suffer extreme inflammatory responses. It is hoped that antisense oligonucleotides might temper inflammatory responses by reducing the expression of ICAM-1. The results of a recent clinical trial suggest that antisense ICAM-1 may be an effective treatment for Crohn's disease. Crohn's disease is a form of inflammatory bowel disease that affects about 200,000 people in the U.S. It is a debilitating, chronic disease, and treatments such as steroids and immunosuppressive drugs are often ineffective and have toxic side effects. In one clinical trial, almost 50 percent of Crohn's patients went into remission after treatment with antisense ICAM-1, whereas none of the placebo group did so. In addition, one-third of the treated patients were well enough to stop steroid treatments by the end of the trial period. In this trial, the antisense ICAM-1 drugs appeared to be well-tolerated and safe.

Some interesting clinical trials are in progress to test antisense oligonucleotides as treatments for some types of cancer. These oligonucleotides are designed to reduce the synthesis of proteins that are either overexpressed in cancer cells or are present as mutant forms. These antisense drugs will be evaluated as treatments for tumors of the ovary, prostate, breast, brain, colon, and lung. Other antisense drugs in the pipeline are designed to attack the hepatitis B, hepatitis C, human papilloma, and AIDS viruses.

If antisense therapeutics pass the scrutiny of these scientific and clinical trials, we may indeed have acquired a magic molecular bullet to use in the battle against a variety of diseases.

REFERENCES:

EDGINGTON, S.M. 1997. Antisense '97: A roundtable on the state of the industry. *Nature Biotechnology* 15:519–24.

NYCE, J.W. and METZGER, W.J. 1997. DNA antisense therapy for asthma in an animal model. *Nature* 385:721–25.

ROUSH, W. 1997. Antisense aims for a renaissance. *Science* 276:1192–93.

WEBSITE:
ISIS Pharmaceuticals
http://www.isip.com

phage MS2 with the amino acid sequence of the corresponding proteins. Other findings support the belief that, with only minor exceptions, the code is universal for all organisms.

5. In some bacteriophages, multiple initiation points may occur during the transcription of RNA, resulting in multiple reading frames and overlapping genes.

6. Transcription, the initial step in gene expression, describes the synthesis, under the direction of RNA polymerase, of a strand of RNA complementary to a DNA template.

7. The processes of transcription, like DNA replication, can be subdivided into the stages of initiation, elongation, and termination. Also like DNA replication,

the process relies on base-pairing affinities between complementary nucleotides.

8. Initiation of transcription is dependent on an upstream (5′) DNA region called the promoter that represents the initial binding site for RNA polymerase. Promoters contain specific DNA sequences such as the TATA box that are essential to polymerase binding.

9. Transcription is more complex in eukaryotes than in prokaryotes. The primary transcript is a pre-mRNA that must be modified in various ways before it can be efficiently translated. Processing, which produces a mature mRNA, includes the addition of a 7mG cap and a poly-A tail, and the removal, through splicing, of intervening sequences, or introns. RNA editing of pre-mRNA prior to its translation also occurs in some systems.

INSIGHTS AND SOLUTIONS

1. Had evolution seized on six bases (three complementary base pairs) rather than four bases within the structure of DNA, calculate how many triplet codons would be possible. Would six bases accommodate a two-letter code, assuming 20 amino acids and start-and-stop codons?

   Solution: Six things taken three at a time will produce $(6)^3$ or 216 triplet codes. If the code was a doublet, there would be $(6)^2$ or 36 two-letter codes, more than enough to accommodate 20 amino acids and start–stop punctuation.

2. In a heteropolymer experiment using 1/2C:1/4A:1/4G, how many different triplets will occur in the synthetic RNA molecule? How frequently will the most frequent triplet occur?

   Solution: There will be $(3)^3$ or 27 triplets produced. The most frequent will be CCC, present $1/2^3$ or 1/8 of the time.

3. In a regular copolymer experiment, where UUAC is repeated over and over, how many different triplets will occur in the synthetic RNA, and how many amino acids will occur in the polypeptide when this RNA is translated? Be sure to consult Figure 13.7.

   Solution: The synthetic RNA will repeat four triplets—UUA, UAC, ACU, and CUU—over and over.

   Because both UUA and CUU encode leucine, while ACU and UAC encode threonine and tyrosine, respectively, the polypeptides synthesized under the directions of such an RNA contain three amino acids in the repeating sequence leu-leu-thr-tyr.

4. Actinomycin D inhibits DNA-dependent RNA synthesis. This antibiotic is added to a bacterial culture where a specific protein is being monitored. Compared to a control culture, where no antibiotic is added, translation of the protein declines over a period of 20 minutes, until no further protein is made. Explain these results.

   Solution: The mRNA, which is the basis for the translation of the protein, has a lifetime of about 20 minutes. When actinomycin D is added, transcription is inhibited and no new mRNAs are made. Those already present support the translation of the protein for up to 20 minutes.

5. DNA and RNA base compositions were analyzed from a hypothetical bacterial species with the following results:

	(A + G)(T + C)	(A + T)(C + G)	(A + G)(U + C)	(A + U)(C + G)
DNA	1.0	1.2		
RNA			1.3	1.2

On the basis of these data, what can you conclude about the DNA and RNA of the organism? Are the data consistent with the Watson–Crick model of DNA? Is the RNA single-stranded or double-stranded, or can't we tell? If we assume that the entire length of DNA has been transcribed, do the data suggest that RNA has been derived from the transcription of one or both DNA strands, or can't we tell from these data?

Solution: This problem is a theoretical exercise designed to get you to look at the consequences of base complementarity as it affects the base composition of DNA and RNA. The base composition of DNA is consistent with the Watson–Crick double helix. In a double helix, we expect A + G to equal T + C (the number of purines should equal the number of pyrimidines). In this case, there is a preponderance of A/T base pairs (120 A and T pairs to every 100 G and C pairs).

Given what we know about RNA, there is no reason to expect the RNA to be double-stranded, but if it were double-stranded, then we would expect that A = U and C = G. If so, then (A + G)/(U + C) = 1. Since it doesn't equal unity, we can conclude that the RNA is not double-stranded.

If all of the DNA is transcribed, from either one or both strands, the ratio of (A + U)/(C + G) in RNA should be 1.2, and, as predicted, it is. Note that this ratio will not change, whether only one or both of the strands are transcribed. This is the case because for every A = T pair in DNA, for example, transcription of RNA will yield one A *and* one U if both strands are transcribed. If just one strand is transcribed, transcription will yield one A *or* one U. In either case, the (A + U)/(C + G) ratio in RNA will reflect the (A + T)/(C + G) ratio in the DNA from which it was transcribed.

To prove this to yourself, draw out a DNA molecule with 12 A = T pairs and 10 C ≡ G pairs and transcribe *both* strands and then transcribe *either* strand. Count the bases in the RNAs produced in both cases and calculate the ratios. Thus, we cannot determine whether just one or both strands are transcribed from the (A + U)/(C + G) ratio.

However, if both strands are transcribed, then the ratio of (A + G)/(U + C) should equal 1.0, and it doesn't. It equals 1.3. To verify this conclusion, examine the theoretical data you drew out on paper. One explanation for the observed ratio of 1.3 is that only one of the two strands is transcribed. If this is the case, then the (A + G)/(U + C) will reflect the proportion of A/T pairs that are A and the proportion of the G/C pairs that are G *on the DNA strand that is transcribed*. Another explanation is that transcription occurs only on one strand at any given point (e.g., for one gene), but on the other strand at other points (for other genes).

PROBLEMS AND DISCUSSION QUESTIONS

1. Early proposals regarding the genetic code considered the possibility that DNA served directly as the template for polypeptide synthesis (see Gamow, 1954, in Selected Readings). In eukaryotes, what difficulties would such a system pose? What observations and theoretical considerations argue against such a proposal?

2. Crick, Barnett, Brenner, and Watts-Tobin, in their studies of frameshift mutations, found that either 3 (+)s or 3 (−)s restored the correct reading frame. If the code were a sextuplet (consisting of six nucleotides), would the reading frame be restored by either of the above combinations?

3. In a mixed copolymer experiment using polynucleotide phosphorylase, 3/4G:1/4C was added to form the synthetic message. The resulting amino acid composition of the ensuing protein was determined:

Glycine	36/64	(56%)
Alanine	12/64	(19%)
Arginine	12/64	(19%)
Proline	4/64	(6%)

From this information:
(a) Indicate the percentage (or fraction) of the time each possible triplet will occur in the message.
(b) Determine one consistent base-composition assignment for the amino acids present.
(c) Considering the wobble hypothesis, predict as many specific triplet assignments as possible.

4. When repeating copolymers are used to form synthetic mRNAs, dinucleotides produce a single type of polypeptide that contains only two different amino acids. On the other hand, using a trinucleotide sequence produces three different polypeptides, each consisting of only a single amino acid. Why? What will be produced when a repeating tetranucleotide is used?

5. The mRNA formed from the repeating tetranucleotide UUAC incorporates only three amino acids, but the use of UAUC incorporates four amino acids. Why?

6. In studies using repeating copolymers, ACA . . . incorporates threonine and histidine, and CAACAA . . . incorporates glutamine, asparagine, and threonine. What triplet code can definitely be assigned to threonine?

7. In a coding experiment using repeating copolymers (as shown in Table 13.3), the following data were obtained:

Copolymer	Codons Produced	Amino Acids in Polypeptide
AG	AGA, GAG	Arg, Glu
AAG	AGA, AAG, GAA	Lys, Arg, Glu

AGG is known to code for arginine. Taking into account the wobble hypothesis, assign each of the four remaining different triplet codes to its correct amino acid.

8. In the triplet binding technique, radioactivity remains on the filter when the amino acid corresponding to the triplet is labeled. Explain the basis of this technique.

9. When the amino acid sequences of insulin isolated from different organisms were determined, some differences were noted. For example, alanine was substituted for threonine, serine was substituted for glycine, and valine was substituted

for isoleucine at corresponding positions in the protein. List the single-base changes that could occur in triplets of the genetic code to produce these amino acid changes.

10. In studies of the amino acid sequence of wild-type and mutant forms of tryptophan synthetase in *E. coli*, the following changes have been observed:

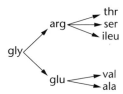

Determine a set of triplet codes in which only a single nucleotide change produces each amino acid change.

11. Why doesn't polynucleotide phosphorylase synthesize RNA *in vivo*?

12. Refer to Table 13.1. Can you hypothesize why a mixture of Poly U + Poly A would not stimulate incorporation of ^{14}C-phenylalanine into protein?

13. Predict the amino acid sequence produced during translation by the following short theoretical mRNA sequences. Note that the second sequence was formed from the first by a deletion of only one nucleotide.

Sequence 1:   AUGCCGGAUUAUAGUUGA
Sequence 2:   AUGCCGGAUUAAGUUGA

What type of mutation gave rise to Sequence 2?

14. A short RNA molecule was isolated that demonstrated a hyperchromic shift indicating secondary structure. Its sequence was determined to be

AGGCGCCGACUCUACU

(a) Propose a two-dimensional model for this molecule.
(b) What DNA sequence would give rise to this RNA molecule through transcription?
(c) If the molecule were a tRNA fragment containing a CGA anticodon, what would the corresponding codon be?
(d) If the molecule were an internal part of a message, what amino acid sequence would result from it following translation? (Refer to the code chart in Figure 13.7.)

15. A glycine residue exists at position 210 of the tryptophan synthetase enzyme of wild-type *E. coli*. If the codon specifying glycine is GGA, how many single-base substitutions will result in an amino acid substitution at position 210? What are they? How many will result if the wild-type codon is GGU?

16. (a) Shown here is a theoretical viral mRNA sequence. Assuming that it could arise from overlapping genes, how many different polypeptide sequences can be produced? Using Figure 13.7, what are the sequences?

5′-AUGCAUACCUAUGAGACCCUUGGGA-3′

(b) A base substitution mutation that altered the sequence in (a) eliminated the synthesis of all but one polypeptide. The altered sequence is shown here. Using Figure 13.7, determine why.

5′-AUGCAUACCUAUGUGACCCUUGGA-3′

17. Most proteins have more leucine than histidine residues, but more histidine than tryptophan residues. Correlate the number of codons for these three amino acids with this information.

18. Define the process of transcription. Where does this process fit into the central dogma of molecular genetics?

19. What was the initial evidence for the existence of mRNA?

20. Describe the structure of RNA polymerase in bacteria. What is the core enzyme? What is the role of the sigma subunit?

21. In a written paragraph, describe the abbreviated chemical reactions that summarize RNA polymerase-directed transcription. What differences exist between the "visualization of transcription" studies of *E. coli* and *Notophthalmus*? Why?

22. Messenger RNA molecules are very difficult to isolate in prokaryotes because they are rather quickly degraded in the cell. Can you suggest a reason why this occurs? Eukaryotic mRNAs are more stable and exist longer in the cell than do prokaryotic mRNAs. Is this an advantage or disadvantage for a pancreatic cell making large quantities of insulin?

23. The following represent deoxyribonucleotide sequences derived from the template strand of DNA:

Sequence 1:   CTTTTTTGCCAT
Sequence 2:   ACATCAATAACT
Sequence 3:   TACAAGGGTTCT

(a) For each strand, determine the mRNA sequence that would be derived from transcription.
(b) Using Figure 13.7, determine the amino acid sequence that is encoded by these mRNAs.
(c) For Sequence 1, what is the sequence of the partner strand?

EXTRA-SPICY PROBLEMS

24. In a mixed copolymer experiment, messengers were created with either 4/5C:1/5A or 4/5A:1/5C. These messages yielded proteins with the following amino acid compositions. Using these data, predict the most specific coding composition for each amino acid.

4/5C:1/5A		4/5A:1/5C	
Proline	63.0%	Proline	3.5%
Histidine	13.0%	Histidine	3.0%
Threonine	16.0%	Threonine	16.6%
Glutamine	3.0%	Glutamine	13.0%
Asparagine	3.0%	Asparagine	13.0%
Lysine	0.5%	Lysine	50.0%
	98.5%		98.5%

GENETICS MediaLab

The following resources will help you achieve a better understanding of the concepts presented in this chapter. These resources can be found on the CD packaged with this textbook and on the Companion Website at **http://www.prenhall.com/klug/**.

CD Resources:

Animated Tutorial: *Transcription*

Animated Tutorial: *mRNA Processing*

Self-grading Chapter Problems

Web Resources:

Genetics and Society Issue: *Antisense Oligonucleotides: Attacking the Messenger*

Web Destinations in Genetics

Self-grading Chapter Problems

Chapter Search Terms

Genetics Newsgroups

Student Bulletin Board

Web Problem 1:

Time for completion = 5 minutes

How does the transcription machinery select where to begin transcription? In eukaryotic cells, RNA polymerase II transcribes genes to produce mRNA. For transcription to begin, the RNA polymerase must be directed to the appropriate region of DNA at the 5′-end of the gene by transcription activators. This is a complex process involving many proteins. Some of the proteins regulating this process are found bound to DNA while others are preassembled with the polymerase. This linked website presents the research of one laboratory in defining how the transcription process occurs. Describe one question the researchers are pursuing and what approach they are taking to answer that question. To complete this exercise, visit Web Problem 1 in Chapter 13 of your Companion Website and select the keyword **POLYMERASE.**

Web Problem 2:

Time for completion = 10 minutes

How universal is the universal genetic code? Not all organisms use this code at all times. A prominent exception is the mitochondrion. In this exercise, you can view variations in the genetic code. In addition, codon use can vary; sometimes one codon is used more or less frequently than another codon that encodes the same amino acid. The linked website has a hypertext link to a searchable database for codon use in many organisms. Choose an example of a genetic code differing from the universal code and list the distinctions. Examine the codon use of one organism from the codon-use database. Using the standard (universal) genetic code, discuss any difference in the use of Ser, Leu, and stop codons, including which codons are used most and least often. To complete this exercise, visit Web Problem 2 in Chapter 13 of your Companion Website and select the keyword **CODON.**

Web Problem 3:

Time for completion = 10 minutes

How are introns removed from a transcript? The spliceosome consists of a complex of proteins and small nuclear RNA (together called snRNP, small nuclear ribonucleoprotein particles) that removes introns and splices exons together. The linked website describes the components of the spliceosome and their functions. Select the hypertext links to view the two splicing animations. The synthesis animation provides an overview of the process and snRNPs involved, while the mechanism animation details the molecular events. Describe the steps involved in splicing a pre-mRNA transcript and which components interact with which portions of the transcript. To complete this exercise, visit Web Problem 3 in Chapter 13 of your Companion Website and select the keyword **SPLICEOSOME.**

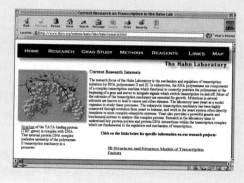

25. Shown below are the amino acid sequences of the wild-type and three mutant forms of a short protein. Use this information to answer the following questions.

Wild type: met-trp-tyr-arg-gly-ser-pro-thr
Mutant 1:   met-trp
Mutant 2:   met-trp-his-arg-gly-ser-pro-thr
Mutant 3:   met-cys-ile-val-val-val-gln-his

(a) Using Figure 13.7, predict the type of mutation that occurred leading to each altered protein.

(b) For each mutant protein, determine the specific ribonucleotide change that led to its synthesis.

(c) The wild-type RNA consists of nine triplets. What is the role of the ninth triplet?

(d) For the first eight wild-type triplets, which, if any, can you determine specifically from an analysis of the mutant proteins? Explain why or why not in each case.

(e) Another mutation (Mutant 4) is isolated. Its amino acid sequence is unchanged, but mutant cells produce abnormally low amounts of the wild-type proteins. As specifically as you can, predict where in the gene this mutation exists.

SELECTED READINGS

ALBERTS, B. et al. 1998. *Essential cell biology*, 3rd ed. New York: Garland Publishing.

BARRALLE, F.E. 1983. The functional significance of leader and trailer sequences in eukaryotic mRNAs. *Int. Rev. Cytol.* 81:71–106.

BARRELL, B.G., AIR, G., and HUTCHINSON, C. 1976. Overlapping genes in bacteriophage φX174. *Nature* 264:34–40.

BARRELL, B.G., BANKER, A.T., and DROUIN, J. 1979. A different genetic code in human mitochondria. *Nature* 282:189–94.

BEARDSLEY, T. 1996. Vital data. *Sci. Am.* (Mar.) 274:100–105.

BIRNSTIEL, M., BUSSLINGER, M., and STRUB, K. 1985. Transcription termination and 3′ processing: The end is in site. *Cell* 41:349–59.

BONITZ, S.G. et al. 1980. Codon recognition rules in yeast mitochondria. *Proc. Natl. Acad. Sci. USA* 77:3167–70.

BREITBART, R.E., and NADAL-GINARD, B. 1987. Developmentally induced, muscle-specific *trans* factors control the differential splicing of alternative and constitutive troponin T-exons. *Cell* 49:793–803.

BRENNER, S. 1989. *Molecular biology: A selection of papers.* Orlando, FL: Academic Press.

BRENNER, S., JACOB, F., and MESELSON, M. 1961. An unstable intermediate carrying information from genes to ribosomes for protein synthesis. *Nature* 190:575–80.

BRENNER, S., STRETTON, A.O.W., and KAPLAN, D. 1965. Genetic code: The nonsense triplets for chain termination and their suppression. *Nature* 206:994–98.

CATTANEO, R. 1991. Different types of messenger RNA editing. *Annu. Rev. Genet.* 25:71–88.

CECH, T.R. 1986. RNA as an enzyme. *Sci. Am.* (Nov.) 255(5):64–75.

———. 1987. The chemistry of self-splicing RNA and RNA enzymes. *Science* 236:1532–39.

CECH, T.R., ZAUG, A.J., and GRABOWSKI, P.J. 1981. *In vitro* splicing of the ribosomal RNA precursor of *Tetrahymena.* Involvement of a guanosine nucleotide in the excision of the intervening sequence. *Cell* 27:487–96.

CHAMBON, P. 1975. Eucaryotic nuclear RNA polymerases. *Annu. Rev. Biochem.* 44:613–38.

———. 1981. Split genes. *Sci. Am.* (May) 244:60–71.

COLD SPRING HARBOR LABORATORY. 1966. The genetic code. *Cold Spring Harbor Symp. Quant. Biol.* Vol. 31.

CRICK, F.H.C. 1962. The genetic code. *Sci. Am.* (Oct.) 207:66–77.

———. 1966a. The genetic code: III. *Sci. Am.* (Oct.) 215:55–63.

———. 1966b. Codon-anticodon pairing: The wobble hypothesis. *J. Mol. Biol.* 19:548–55.

———. 1979. Split genes and RNA splicing. *Science* 204:264–71.

CRICK, F.H.C., BARNETT, L., BRENNER, S., and WATTS-TOBIN, R.J. 1961. General nature of the genetic code for proteins. *Nature* 192:1227–32.

DARNELL, J.E. 1983. The processing of RNA. *Sci. Am.* (Oct.) 249:90–100.

———. 1985. RNA. *Sci. Am.* (Oct.) 253:68–87.

DICKERSON, R.E. 1983. The DNA helix and how it is read. *Sci. Am.* (Dec.) 249:94–111.

DUGAICZK, A. et al. 1978. The natural ovalbumin gene contains seven intervening sequences. *Nature* 274:328–33.

FIERS, W. et al. 1976. Complete nucleotide sequence of bacteriophage MS2 RNA: Primary and secondary structure of the replicase gene. *Nature* 260:500–507.

GAMOW, G. 1954. Possible relation between DNA and protein structures. *Nature* 173:318.

GUTHRIE, C., and PATTERSON, B. 1988. Spliceosomal snRNAs. *Annu. Rev. Genet.* 22:387–419.

HALL, B.D., and SPIEGELMAN, S. 1961. Sequence complementarity of T2-DNA and T2-specific RNA. *Proc. Natl. Acad. Sci. USA* 47:137–46.

HAMKALO, B. 1985. Visualizing transcription in chromosomes. *Trends Genet.* 1:255–60.

HELMAN, J.D., and CHAMBERLIN, M.J. 1988. Structure and function of bacterial sigma factors. *Annu. Rev. Biochem.* 57:839–72.

HODGES, P., and SCOTT, J. 1992. Apolipoprotein B mRNA editing: A new tier for the control of gene expression. *Trends Biochem. Sci.* 17:77–81.

HORTON, H.R. et al. 1996. *Principles of biochemistry*, 2nd ed. Upper Saddle River, NJ: Prentice Hall.

HUMPHREY, T., and PROUDFOOT, N.J. 1988. A beginning to the biochemistry of polyadenylation. *Trends Genet.* 4:243–45.

JUDSON, H.F. 1979. *The eighth day of creation.* New York: Simon & Schuster.

JUKES, T.H. 1963. The genetic code. *Am. Sci.* 51:227–45.

KABLE, M.L. et al. 1996. RNA editing: A mechanism for gRNA-specified uridylate insertion into precursor mRNA. *Science* 273:1189–95.

KHORANA, H.G. 1967. Polynucleotide synthesis and the genetic code. *Harvey Lectures* 62:79–105.

MANIATIS, T., and REED, R. 1987. The role of small nuclear ribonucleoprotein particles in pre-mRNA splicing. *Nature* 325:673–78.

MILLER, O.L., and BEATTY, B.R. 1969. Portrait of a gene. *J. Cell Physiol.* 74 (Suppl. 1): 225–32.

MILLER, O.L., HAMKALO, B., and THOMAS, C. 1970. Visualization of bacterial genes in action. *Science* 169:392–95.

MIN JOU, W., HAGEMAN, G., YSEBART, M., and FIERS, W. 1972. Nucleotide sequence of the gene coding for bacteriophage MS2 coat protein. *Nature* 237:82–88.

NIKOLOV, D.B. and BURLEY, S.K. 1997. RNA polymerase II transcription initiation: A structural view. *Proc. Natl. Acad. Sci. USA* 94:15–22.

NILSEN, T.W. 1994. RNA-RNA interactions in the spliceosome: Unraveling the ties that bind. *Cell* 78:1–4.

NIRENBERG, M.W. 1963. The genetic code: II. *Sci. Am.* (March) 190:80–94.

NIRENBERG, M.W., and LEDER, P. 1964. RNA codewords and protein synthesis. *Science* 145:1399–1407.

NIRENBERG, M.W., and MATTHAEI, H. 1961. The dependence of cell-free protein synthesis in *E. coli* upon naturally occurring or synthetic polyribosomes. *Proc. Natl. Acad. Sci. USA* 47:1588–1602.

O'MALLEY, B. et al. 1979. A comparison of the sequence organization of the chicken ovalbumin and ovomucoid genes. In *Eucaryotic gene regulation*, ed. R. Axel et al., pp. 281–99. Orlando, FL: Academic Press.

PADGETT, R.A., GRABOWSKI, P.J., KONARSKA, M.M., SEILER, S., and SHARP, P.A. 1986. Splicing of messenger RNA precursors. *Annu. Rev. Biochem.* 55:1119–50.

REED, R., and MANIATIS, T. 1985. Intron sequences involved in lariat formation during pre-mRNA splicing. *Cell* 41:95–105.

ROSS, J. 1989. The turnover of mRNA. *Sci. Am.* (April) 260:48–55.

SHARP, P.A. 1987. Splicing of messenger RNA precursors. *Science* 235:766–71.

———— 1994. Nobel Lecture: Split genes and RNA splicing. *Cell* 77:805–15.

SHARP, P.A., and EISENBERG, D. 1987. The evolution of catalytic function. *Science* 238:729–30.

SHATKIN, A.J. 1985. mRNA cap binding proteins: Essential factors for initiating translation. *Cell* 40:223–24.

STEITZ, J.A. 1988. Snurps. *Sci. Am.* (June) 258(6):56–63.

VOLKIN, E., and ASTRACHAN, L. 1956. Phosphorus incorporation in *E. coli* ribonucleic acids after infection with bacteriophage T2. *Virology* 2:149–61.

VOLKIN, E., ASTRACHAN, L., and COUNTRYMAN, J.L. 1958. Metabolism of RNA phosphorus in *E. coli* infected with bacteriophage T7. *Virology* 6:545–55.

WATSON, J.D. 1963. Involvement of RNA in the synthesis of proteins. *Science* 140:17–26.

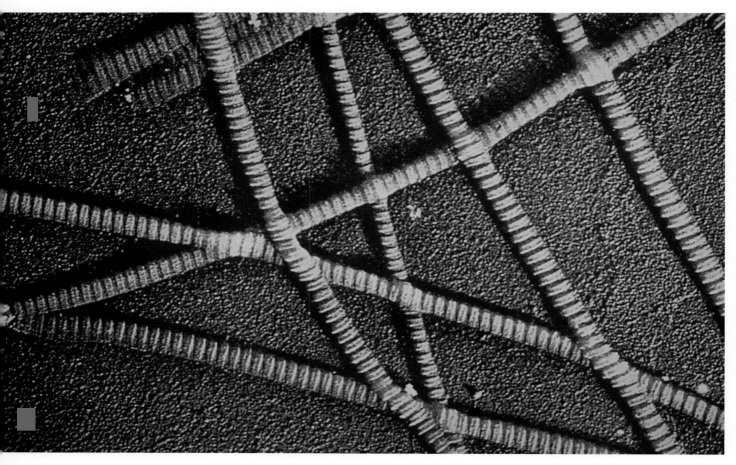

An electron micrograph of collagen fibers, the most abundant protein found in vertebrates.

Translation and Proteins

KEY CONCEPTS

- **Translation of mRNA depends on ribosomes and transfer RNAs**

  Ribosomal Structure
  tRNA Structure
  Charging tRNA

- **Translation of mRNA can be divided into three steps**

  Initiation
  Elongation
  Termination
  Polyribosomes

- **Translation is more complex in eukaryotes**

- **The initial insight that proteins are important in heredity was provided by the study of inborn errors of metabolism**

  Phenylketonuria

- **Studies of *Neurospora* led to the one-gene:one-enzyme hypothesis**

  Beadle and Tatum: *Neurospora* Mutants
  Genes and Enzymes: Analysis of Biochemical Pathways

- **Studies of human hemoglobin established that one gene encodes one polypeptide**

  Sickle-Cell Anemia
  Human Hemoglobins

- **The nucleotide sequence of a gene and the amino acid sequence of the corresponding protein exhibit colinearity**

- **Protein structure is the basis of biological diversity**

  Posttranslational Modification

- **Protein function is directly related to the structure of the molecule**

- **Proteins are made up of one or more functional domains**

  Exon Shuffling and the Origin of Protein Domains

🌐 GenCDX

When you see this icon, there are related animations and exercises on the CD accompanying this text.

I n Chapter 13, we established that there is a genetic code that stores information in the form of triplet nucleotides in DNA and that this information is initially expressed through the process of transcription into a messenger RNA that is complementary to one of the two strands of the DNA helix. However, the final product of gene expression, in almost all instances, is a polypeptide chain consisting of a linear series of amino acids whose sequence has been prescribed by the genetic code. In this chapter, we will examine how the information present in mRNA is processed in order to create polypeptides, which then fold into protein molecules. We will also review the evidence that confirmed that proteins are the end products of genes and discuss briefly the various levels of protein structure, diversity, and function. This information extends our understanding of gene expression and provides an important foundation for interpreting how the mutations that arise in DNA can result in the diverse phenotypic effects observed in organisms.

Translation of mRNA depends on ribosomes and transfer RNAs

Translation of mRNA is the biological polymerization of amino acids into polypeptide chains. This process, alluded to in our discussion in Chapter 13 of the genetic code, occurs only in association with ribosomes, which serve as nonspecific workbenches. The central question in translation is how triplet ribonucleotides of mRNA direct specific amino acids into their correct position in the polypeptide. This question was answered once **transfer RNA (tRNA)** was discovered. This class of molecules adapts specific triplet codons in mRNA to their correct amino acids. This adaptor role of tRNA was postulated by Francis Crick in 1957.

In association with a ribosome, mRNA presents a triplet codon that calls for a specific amino acid. A specific tRNA molecule contains within its nucleotide sequence three consecutive ribonucleotides complementary to the codon, called the **anticodon,** which can base-pair with the codon. Another region of this tRNA is covalently bonded to its corresponding amino acid.

Inside the ribosome, hydrogen bonding of tRNAs to mRNA holds the amino acid in proximity so that a peptide bond can be formed. As this process occurs over and over as mRNA runs through the ribosome, amino acids are polymerized into a polypeptide. Before we discuss the actual process of translation, we will first consider the structures of the ribosome and transfer RNA.

Ribosomal Structure

Because of its essential role in the expression of genetic information, the **ribosome** has been extensively analyzed. One bacterial cell contains about 10,000 of these structures, and a eukaryotic cell contains many times more. Electron microscopy has revealed that the bacterial ribosome is about 250 μm at its largest diameter and consists of two subunits, one large and one small. Both subunits consist of one or more molecules of rRNA and an array of **ribosomal proteins.** When the two subunits are associated with each other in a single ribosome, the structure is sometimes called a **monosome.**

The specific differences between prokaryotic and eukaryotic ribosomes are summarized in Figure 14.1. The subunit and rRNA components are most easily isolated and characterized on the basis of their sedimentation behavior in sucrose gradients (their rate of migration, abbreviated as S, as introduced in Chapter 11). In prokaryotes the monosome is a $70S$ particle, and in eukaryotes it is approximately $80S$. Sedimentation coefficients, which reflect the variable rate of migration of different-size particles and molecules, are not additive. For example, the prokaryotic $70S$ monosome consists of a $50S$ and a $30S$ subunit, and the eukaryotic $80S$ monosome consists of a $60S$ and a $40S$ subunit.

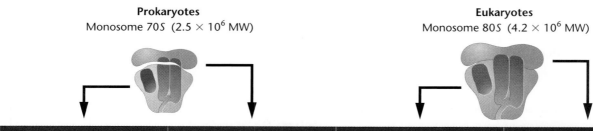

Prokaryotes
Monosome $70S$ (2.5×10^6 MW)

Eukaryotes
Monosome $80S$ (4.2×10^6 MW)

Large subunit	Small subunit	Large subunit	Small subunit
$50S$ 1.6×10^6 MW	$30S$ 0.9×10^6 MW	$60S$ 2.8×10^6 MW	$40S$ 1.4×10^6 MW
$23S$ rRNA (2904 nucleotides)	$16S$ rRNA (1541 nucleotides)	$28S$ rRNA (4718 nucleotides)	$18S$ rRNA (1874 nucleotides)
+ 31 proteins	+ 21 proteins	+ 49 proteins	+ 33 proteins
+ $5S$ rRNA (120 nucleotides)		$5S$ rRNA (120 nucleotides) + $5.8S$ rRNA (160 nucleotides)	

FIGURE 14.1 A comparison of the components in the prokaryotic and eukaryotic ribosome.

The larger subunit in prokaryotes consists of a 23S RNA molecule, a 5S rRNA molecule, and 31 ribosomal proteins. In the eukaryotic equivalent, a 28S rRNA molecule is accompanied by a 5.8S and 5S rRNA molecule and about 50 proteins. The smaller prokaryotic subunits consist of a 16S rRNA component and 21 proteins. In the eukaryotic equivalent, an 18S rRNA component and about 33 proteins are found. The approximate molecular weights and number of nucleotides of these components are shown in Figure 14.1.

Regarding the components of the ribosome, it is now clear that the RNA molecules perform the all important catalytic functions associated with translation. The many proteins, whose functions were long a mystery, are thought to promote the binding of the various molecules involved in translation, and in general, to fine-tune the process. This generalization is based on the observation that some of the catalytic functions in ribosomes still occur in experiments involving "ribosomal protein-depleted" ribosomes.

Molecular hybridization studies have established the degree of redundancy of the genes coding for the rRNA components. The *E. coli* genome contains seven copies of a single sequence that encodes all three components—23S, 16S, and 5S. The initial transcript of these genes produces a 30S RNA molecule that is enzymatically cleaved into these smaller components. Coupling of the genetic information encoding these three rRNA components ensures that, following multiple transcription events, equal quantities of all three will be present as ribosomes are assembled.

In eukaryotes, many more copies of a sequence encoding the 28S and 18S components are present. In *Drosophila*, approximately 120 copies per haploid genome are each transcribed into a molecule of about 34S. This is processed into the 28S, 18S, and 5.8S rRNA species. These are homologous to the three rRNA components of *E. coli*. In *X. laevis*, over 500 copies per haploid genome are present. In mammalian cells, the initial transcript is 45S. The rRNA genes, called **rDNA,** are part of the moderately repetitive DNA fraction and are present in clusters at various chromosomal sites.

Each cluster in eukaryotes consists of **tandem repeats,** with each unit separated by a noncoding **spacer DNA** sequence [see the micrograph in Figure 13.15(b)]. In humans, these gene clusters have been localized near the ends of chromosomes 13, 14, 15, 21, and 22. The unique 5S rRNA component of eukaryotes is not part of this larger transcript. Instead, genes coding for this ribosomal component are distinct and located separately. In humans, a gene cluster encoding them has been located on chromosome 1.

Despite the detailed knowledge available on the structure and genetic origin of the ribosomal components, a complete understanding of the function of these components has eluded geneticists. This is not surprising; the ribosome is perhaps the most intricate of all cellular structures. In bacteria, the monosome has a combined molecular weight of 2.5 million daltons!

tRNA Structure

Because of their small size and stability in the cell, transfer RNAs (tRNAs) have been investigated extensively and are the best characterized RNA molecules. They are composed of only 75 to 90 nucleotides, displaying a nearly identical structure in bacteria and eukaryotes. In both types of organisms, tRNAs are transcribed as larger precursors, which are cleaved into mature 4S tRNA molecules. In *E. coli*, for example, tRNA^{tyr} (the superscript identifies the specific tRNA and the cognate amino acid that binds to it) is composed of 77 nucleotides, yet its precursor contains 126 nucleotides.

In 1965, Robert Holley and his colleagues reported the complete sequence of tRNA^{ala} isolated from yeast. Of great interest was the finding that a number of nucleotides are unique to tRNA. As shown in Figure 14.2, each is a modification of one of the four nitrogenous bases expected in RNA (G, C, A, and U). These include inosinic acid, which contains the purine hypoxanthine, ribothymidylic acid, and pseudouridine, among others. These modified structures, variously referred to as *unusual, rare,* or *odd bases,* are created following transcription, illustrating the more general concept of **posttranscriptional modification.** In this case, the unmodified base is inserted during transcription and, subsequently, enzymatic reactions catalyze the chemical modifications to the base.

Holley's sequence analysis led him to propose the two-dimensional **cloverleaf model of tRNA.** It had been known that tRNA demonstrates a secondary structure due to base pairing. Holley discovered that he could arrange the linear model in such a way that several stretches of base pairing would result. This arrangement created a series of paired stems and unpaired loops resembling the shape of a cloverleaf. Loops consistently contained modified bases that do not generally form base pairs. Holley's model is shown in Figure 14.3.

Because the triplets GCU, GCC, and GCA specify alanine, Holley looked for an anticodon sequence complementary to one of these codons in his tRNA^{ala} molecule. He found it in the form of CGI (the 3'-to-5' direction), in one of the loops of the cloverleaf. The nitrogenous base I (inosinic acid) can form hydrogen bonds with U, C, or A, the third members of the triplets. Thus, the **anticodon loop** was established.

Studies of other tRNA species revealed many constant features. First, at the 3'-end, all tRNAs contain the sequence ...**pCpCpA-3'.** It is at this end of the molecule where the amino acid is covalently joined to the terminal

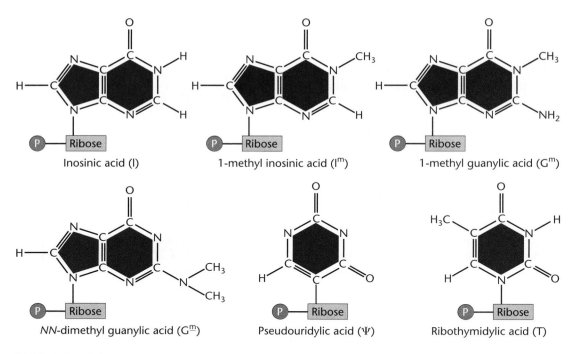

FIGURE 14.2   Unusual nitrogenous bases found in transfer RNA.

adenosine residue. All tRNAs contain **5′-Gp...** at the other end of the molecule.

In addition, the lengths of various stems and loops are very similar. Each tRNA examined also contains an anti-

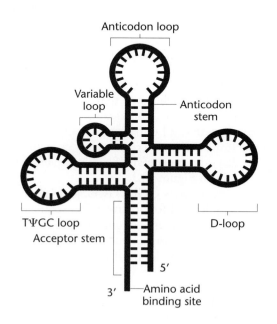

FIGURE 14.3   The two-dimensional cloverleaf model of transfer RNA.

codon complementary to the known amino acid code for which it is specific, and all anticodon loops are present in the same position of the cloverleaf.

Because the cloverleaf model was predicted strictly on the basis of nucleotide sequence, there was great interest in the X-ray crystallographic examination of tRNA, which reveals a three-dimensional structure. By 1974, Alexander Rich and his colleagues in the United States, and J. Roberts, B. Clark, Aaron Klug, and their colleagues in England had been successful in crystallizing tRNA and performing X-ray crystallography at a resolution of 3 Å. At such resolution, the pattern formed by individual nucleotides is discernible.

As a result of these studies, a complete three-dimensional model of tRNA is now available (Figure 14.4). At one end of the molecule is the anticodon loop and stem, and at the other end is the 3′-acceptor region where the amino acid is bound. It has been speculated that the shapes of the intervening loops may be recognized by the specific enzymes responsible for adding the amino acid to tRNA, a subject to which we now turn our attention.

Charging tRNA

Before translation can proceed, the tRNA molecules must be chemically linked to their respective amino acids. This activation process, called **charging,** occurs under the direction of enzymes called **aminoacyl tRNA synthetases.**

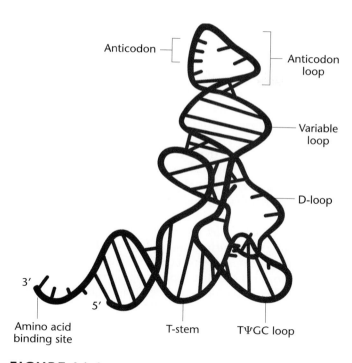

FIGURE 14.4   A three-dimensional model of transfer RNA.

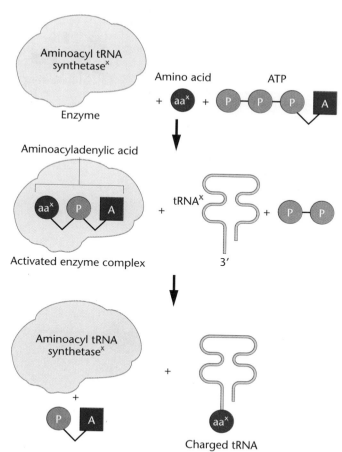

FIGURE 14.5   Steps involved in charging tRNA. The "x" denotes that for each amino acid only the corresponding specific tRNA and specific aminoacyl tRNA synthetase enzyme are involved in the charging process.

Because there are 20 different amino acids, there must be at least 20 different tRNA molecules and as many different enzymes. In theory, because there are 61 triplet codes, there could be the same number of specific tRNAs and enzymes. However, because of the ability of the third member of a triplet code to "wobble," it is now thought that there are at least 32 different tRNAs; it is also believed that there are only 20 synthetases, one for each amino acid, regardless of the greater number of corresponding tRNAs.

The charging process is outlined in Figure 14.5. In the initial step, the amino acid is converted to an activated form, reacting with ATP to create an **aminoacyladenylic acid.** A covalent linkage is formed between the 5′-phosphate group of ATP and the carboxyl end of the amino acid. This molecule remains associated with the synthetase enzyme, forming a complex that then reacts with a specific tRNA molecule. In this next step, the amino acid is transferred to the appropriate tRNA and bonded covalently to the adenine residue at the 3′-end. The charged tRNA may participate directly in protein synthesis. Aminoacyl tRNA synthetases are highly specific enzymes because they recognize only one amino acid and only a subset of corresponding tRNAs called **isoaccepting tRNAs.** This is a crucial point if fidelity of translation is to be maintained.

Translation of mRNA can be divided into three steps

In a way similar to transcription, the process of translation can be best described by breaking it into discrete steps. We will break the process into three steps, each with its own set of illustrations (Figures 14.6, 14.7, and 14.8), but keep in mind that translation is a dynamic, ongoing process. Correlate the following discussion with the step-by-step characterization of the process in these figures. Many of the protein factors and their roles in translation are summarized in Table 14.1.

Initiation

Initiation of translation is depicted in Figure 14.6. Recall that the ribosome serves as a nonspecific workbench for the translation process. Most ribosomes, when they are

Translation components

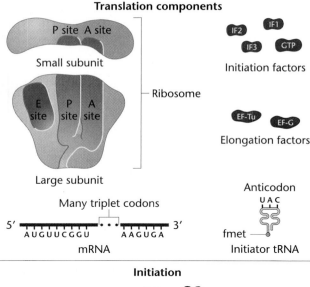

Initiation

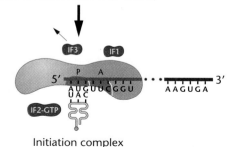

1. mRNA binds to small subunit along with initiation factors (IF1, 2, 3)

Initiation complex

2. Initiator tRNA^fmet binds to mRNA codon in P site; IF3 released

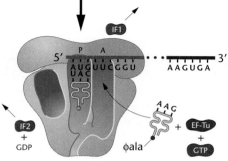

3. Large subunit binds to complex; IF1 and IF2 released; EF-Tu binds to tRNA, facilitating entry into A site

FIGURE 14.6 Schematic representation of initiation of translation. The components are depicted at the top of the figure. ⊙GenCDX

not involved in translation, are dissociated into their large and small subunits. Initiation of translation in *E. coli* involves the small ribosome subunit, an mRNA molecule, a specific charged initiator tRNA, GTP, Mg^{2+}, and at least three proteinaceous initiation factors (IFs) that are required to enhance the binding affinity of the various translational components. In prokaryotes, the initiation codon of mRNA—AUG—calls for the modified amino acid **formylmethionine (f-met).**

The small ribosomal subunit binds to several initiation factors, and this complex in turn binds to mRNA (Step 1). In bacteria, this binding involves a sequence of up to six ribonucleotides (AGGAGG, not shown), which *precedes* the initial AUG start codon of mRNA. This sequence (containing only purines and called the **Shine-Dalgarno sequence)** base-pairs with a region of the 16S rRNA of the small ribosomal subunit, facilitating initiation.

Another initiation protein then enhances the binding of charged formylmethionyl-tRNA to the small subunit in response to the AUG triplet (Step 2). This step "sets" the reading frame so that all subsequent groups of three ribonucleotides are translated accurately. This aggregate represents the **initiation complex,** which then combines with the large ribosomal subunit. In this process, a molecule of GTP is hydrolyzed, providing the required energy, and the initiation factors are released (Step 3).

Elongation

Elongation during translation is depicted in Figure 14.7. Once both subunits of the ribosome are assembled with the mRNA, binding sites for two charged tRNA molecules are formed. These are designated as the **P, or peptidyl,** and the **A, or aminoacyl, sites.** The initiation tRNA binds to the P site, provided that the AUG triplet of mRNA is in the corresponding position of the small subunit.

The increase of the growing polypeptide chain by one amino acid is called **elongation.** The sequence of the second triplet in mRNA dictates which charged tRNA molecule will become positioned at the A site (Step 1). Once it is present, **peptidyl transferase** catalyzes the formation of the peptide bond, which links the two amino acids together (Step 2). This enzyme is part of the large subunit of the ribosome. At the same time, the covalent bond between the amino acid and the tRNA occupying the P site is hydrolyzed (broken). The product of this reaction is a dipeptide, which is attached to the 3'-end of tRNA still residing in the A site.

Before elongation can be repeated, the tRNA attached to the P site, which is now uncharged, must be released from the large subunit. The uncharged tRNA moves transiently through a third site on the ribosome called the **E site** (E stands for exit). The entire **mRNA–tRNA–aa$_2$–aa$_1$** complex then shifts in the direction of the P site by a dis-

Elongation

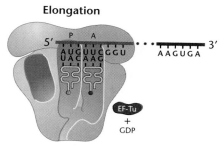

1. Second charged tRNA has entered A site, facilitated by EF-Tu; first elongation step commences

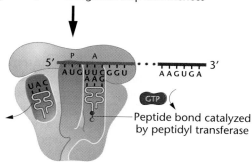

Peptide bond catalyzed by peptidyl transferase

2. Dipeptide bond forms; uncharged tRNA moves to E-site and then out of ribosome

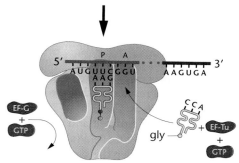

3. mRNA has shifted by 3 bases; EF-G facilitates the translocation step; first elongation step completed

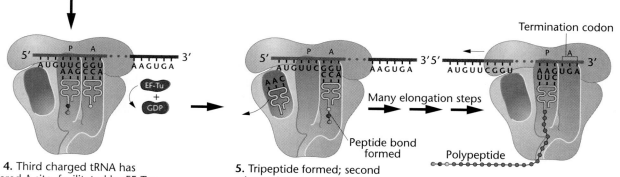

4. Third charged tRNA has entered A site, facilitated by EF-Tu; second elongation step begins

5. Tripeptide formed; second elongation step completed; uncharged tRNA moves to E site

Peptide bond formed

Many elongation steps

Termination codon

Polypeptide

6. Polypeptide chain synthesized and exiting ribosome

FIGURE 14.7 Schematic representation of elongation of the growing polypeptide chain during translation.

GenCDX

eukaryotes. Many of these eukaryotic factors are clearly homologous to their counterparts in prokaryotes. However, in some cases, there are a greater number of factors required during each of these steps, and some are more complex than in prokaryotes.

Finally, it should be recalled that in eukaryotes a large proportion of the ribosomes are found in association with the membranes that make up the endoplasmic reticulum (forming what is referred to as rough ER). Such membranes are absent from the cytoplasm of prokaryotic cells. This association in eukaryotes facilitates the secretion of newly synthesized proteins from the ribosomes directly into the channels of the endoplasmic reticulum. Recent studies using cryo-electron microscopy have established how this occurs. There is a tunnel in the large subunit of ribosomes that begins near the point where the two subunits interface and exits near the back of the large subunit. The location of this tunnel within the large subunit is the basis for the belief that it provides the conduit for the movement of the newly synthesized polypeptide chain out of the ribosome. In studies in yeast, newly synthesized polypeptides enter the ER through a membrane channel formed by a specific protein, Sec61. This channel is perfectly aligned with the exit point of the ribosomal tunnel. In prokaryotes, the polypeptides are released by the ribosome directly into the cytoplasm.

The initial insight that proteins are important in heredity was provided by the study of inborn errors of metabolism

We turn now to the consideration of how we know that proteins are the end products of genetic expression. The first insight into the role of proteins in genetic processes was provided by observations made by Sir Archibald Garrod and William Bateson early in the twentieth century. Garrod was born into an English family of medical scientists. His father was a physician with a strong interest in the chemical basis of rheumatoid arthritis, and his eldest brother was a leading zoologist in London. It is not surprising, then, that as a practicing physician, Garrod became interested in several human disorders that seemed to be inherited. Although he also studied albinism and cystinuria, we will describe his investigation of the disorder **alkaptonuria.** Individuals afflicted with this disorder cannot metabolize the alkapton 2,5-dihydroxyphenylacetic acid, also known as **homogentisic acid.** As a result, an important metabolic pathway (Figure 14.10) is blocked. Homogentisic acid accumulates in cells and tissues and is excreted in the urine. The molecule's oxidation products are black and easily detectable in the diapers of newborns.

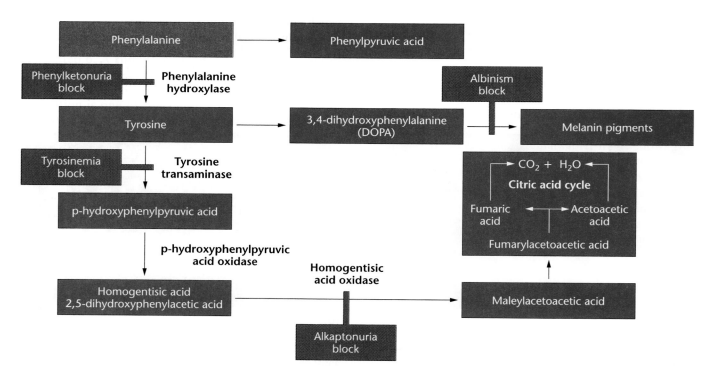

FIGURE 14.10   Metabolic pathway involving phenylalanine and tyrosine. Various metabolic blocks resulting from mutations lead to the disorders phenylketonuria, alkaptonuria, albinism, and tyrosinemia.

The products tend to accumulate in cartilaginous areas, causing a darkening of the ears and nose. In joints, this deposition leads to a benign arthritic condition. This rare disease is not serious, but it persists throughout an individual's life.

Garrod studied alkaptonuria by increasing dietary protein or adding to the diet the amino acids phenylalanine or tyrosine, both of which are chemically related to homogentisic acid. Under such conditions. homogentisic acid levels increase in the urine of alkaptonurics but not in unaffected individuals. Garrod concluded that normal individuals can break down, or catabolize, this alkapton, but that afflicted individuals cannot. By studying the patterns of inheritance of the disorder, Garrod further concluded that alkaptonuria was inherited as a simple recessive trait.

On the basis of these conclusions, Garrod hypothesized that hereditary information controls chemical reactions in the body and that the inherited disorders he studied are the result of alternative modes of metabolism. While *genes* and *enzymes* were not familiar terms during Garrod's time, he used the corresponding concepts of *unit factors* and *ferments*. Garrod published his initial observations in 1902.

Only a few geneticists, including Bateson, were familiar with or referred to Garrod's work. Garrod's ideas fit nicely with Bateson's belief that inherited conditions were caused by the lack of some critical substance. In 1909, Bateson published *Mendel's Principles of Heredity*, in which he linked ferments with heredity. However, for almost 30 years, most geneticists failed to see the relationship between genes and enzymes. Garrod and Bateson, like Mendel, were ahead of their time.

Phenylketonuria

The inherited human metabolic disorder, **phenylketonuria (PKU),** results when another reaction in the pathway shown in Figure 14.10 is blocked. Described first in 1934, this disorder can result in mental retardation and is transmitted as an autosomal recessive disease. Afflicted individuals are unable to convert the amino acid phenylalanine to the amino acid tyrosine. These molecules differ by only a single hydroxyl group ($-$OH), present in tyrosine, but absent in phenylalanine. The reaction is catalyzed by the enzyme **phenylalanine hydroxylase,** which is inactive in affected individuals and active at about a 30-percent level in heterozygotes. The enzyme functions in the liver. While the normal blood level of phenylalanine is about 1 mg/100 ml, phenylketonurics show a level as high as 50 mg/100 ml.

As phenylalanine accumulates, it may be converted to phenylpyruvic acid and, subsequently, to other derivatives. These are less efficiently resorbed by the kidney and tend to spill into the urine more quickly than phenylalanine. Both phenylalanine and its derivatives enter the cerebrospinal fluid, resulting in elevated levels in the brain. The presence of these substances during early development is thought to cause mental retardation.

As a result of early detection based on PKU screening of newborns, retardation can be prevented. When the condition is detected in the analysis of an infant's blood, a strict dietary regimen is instituted. A low-phenylalanine diet can reduce by-products such as phenylpyruvic acid, and the abnormalities characterizing the disease can be diminished. The screening of newborns occurs routinely in all states of the United States. Phenylketonuria occurs in approximately 1 in 11,000 births.

Knowledge of inherited metabolic disorders such as alkaptonuria and phenylketonuria has caused a revolution in medical thinking and practice. No longer is human disease attributed solely to the action of invading microorganisms, viruses, or parasites. We know now that literally thousands of abnormal physiological conditions are caused by errors in metabolism that are the result of mutant genes. These human biochemical disorders are far-ranging and include all classes of organic biomolecules.

Studies of *Neurospora* led to the one-gene: one-enzyme hypothesis

In two separate investigations beginning in 1933, George Beadle was to provide the first convincing experimental evidence that genes are directly responsible for the synthesis of enzymes. The first investigation, conducted in collaboration with Boris Ephrussi, involved *Drosophila* eye pigments. Together, they confirmed that mutant genes that altered the eye color of fruit flies could be linked to biochemical errors that, in all likelihood, involved the loss of enzyme function. Encouraged by these findings, Beadle then joined with Edward Tatum to investigate nutritional mutations in the pink bread mold *Neurospora crassa*. This investigation led to the **one-gene:one-enzyme hypothesis.**

Beadle and Tatum: Neurospora Mutants

In the early 1940s, Beadle and Tatum chose to work with the fungus *Neurospora crassa* because much was known about its biochemistry and mutations could be induced and isolated with relative ease. By inducing mutations, they produced strains that had genetic blocks of reactions essential to the growth of the organism.

Beadle and Tatum knew that *Neurospora* could manufacture nearly everything necessary for normal development. For example, using rudimentary carbon and nitrogen sources, this organism can synthesize 9 water-soluble vitamins, 20 amino acids, numerous carotenoid pigments, purines, and pyrimidines. Beadle and Tatum irradiated asexual conidia (spores) with X rays to increase the frequency of mutations and allowed them to be grown on "complete" medium containing all the necessary growth factors (e.g., vitamins and amino acids). Under such growth conditions, a mutant strain that would be unable to grow on minimal medium was able to grow by virtue of supplements present in the enriched complete medium. All the cultures were then transferred to minimal medium. If growth occurred on the minimal medium, the organisms were able to synthesize all the necessary growth factors themselves, and it was concluded that the culture did not contain a mutation. If no growth occurred, then it was concluded that the culture contained a nutritional mutation, and the only task remaining was to determine its type. Both cases are illustrated in Figure 14.11(a).

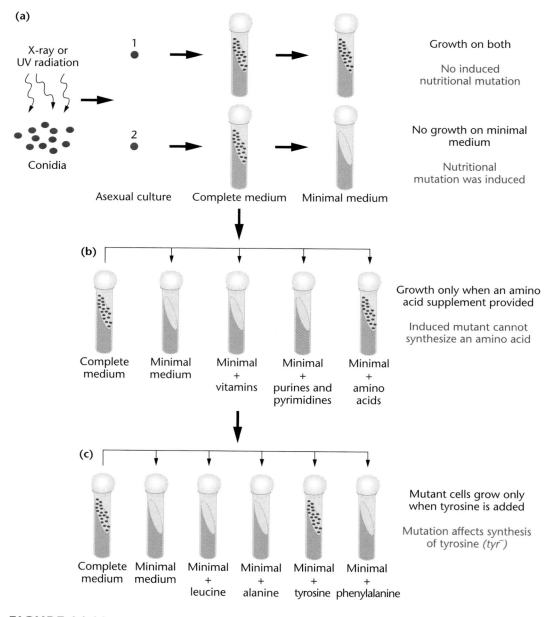

FIGURE 14.11   Induction, isolation, and characterization of a nutritional auxotrophic mutation in *Neurospora*. In (a), most conidia are not affected, but one conidium (shown in red) contains such a mutation. In (b) and (c), the precise nature of the mutation is determined to involve the biosynthesis of tyrosine.

Many thousands of individual spores derived by this procedure were isolated and grown on complete medium. In subsequent tests on minimal medium, many cultures failed to grow, indicating that a nutritional mutation had been induced. To identify the mutant type, the mutant strains were tested on a series of different minimal media [Figure 14.11(b) and (c)], each containing groups of supplements, and subsequently on media containing single vitamins, amino acids, purines, or pyrimidines until one specific supplement that permitted growth was found. Beadle and Tatum reasoned that the supplement that restores growth is the molecule that the mutant strain could not synthesize.

The first mutant strain isolated required vitamin B-6 (pyridoxine) in the medium, and the second one required vitamin B-1 (thiamine). Using the same procedure, Beadle and Tatum eventually isolated and studied hundreds of mutants deficient in the ability to synthesize other vitamins, amino acids, or other substances.

The findings derived from testing over 80,000 spores convinced Beadle and Tatum that genetics and biochemistry have much in common. It seemed likely that each nutritional mutation caused the loss of the enzymatic activity that facilitates an essential reaction in wild-type organisms. It also appeared that a mutation could be found for nearly any enzymatically controlled reaction. Beadle and Tatum had thus provided sound experimental evidence for the hypothesis that **one gene specifies one enzyme,** an idea alluded to over 30 years earlier by Garrod and Bateson. With modifications, this concept was to become a major principle of genetics.

Genes and Enzymes: Analysis of Biochemical Pathways

The one-gene:one-enzyme concept and its attendant methods have been used over the years to work out many details of metabolism in *Neurospora, Escherichia coli,* and a number of other microorganisms. One of the first metabolic pathways to be investigated in detail was that leading to the synthesis of the amino acid arginine in *Neurospora*. By studying seven mutant strains, each requiring arginine for growth (*arg⁻*), Adrian Srb and Norman Horowitz were able to ascertain a partial biochemical pathway that leads to the synthesis of this molecule. Their work illustrates how genetic analysis can be used to establish biochemical information.

Srb and Horowitz tested each mutant strain's ability to reestablish growth if either **citrulline** or **ornithine,** two compounds with close chemical similarity to arginine, was used as a supplement to minimal medium. If either was able to substitute for arginine, they reasoned that it must be involved in the biosynthetic pathway of arginine. They found that both molecules could be substituted in one or more strains.

Of the seven mutant strains, four of them (*arg 4–7*) grew if supplied with either citrulline, ornithine, or arginine. Two of them (*arg 2* and *arg 3*) grew if supplied with citrulline or

arginine. One strain (*arg 1*) would grow only if arginine were supplied; neither citrulline nor ornithine could substitute for it. From these experimental observations, the following pathway and metabolic blocks for each mutation were deduced:

$$\text{Precursor} \xrightarrow[\text{Enzyme A}]{arg\,4\text{–}7} \text{Ornithine} \xrightarrow[\text{Enzyme B}]{arg\,2\text{–}3} \text{Citrulline} \xrightarrow[\text{Enzyme C}]{arg\,1} \text{Arginine}$$

The reasoning supporting these conclusions is based on the following logic. If mutants *arg 4* through *arg 7* can grow regardless of which of the three molecules is supplied as a supplement to minimal medium, the mutations preventing growth must cause a metabolic block that occurs *prior to* the involvement of ornithine, citrulline, or arginine in the pathway. When any one of these three molecules is added, its presence bypasses the block. As a result, it can be concluded that both citrulline and ornithine are involved in the biosynthesis of arginine. However, the sequence of their participation in the pathway cannot be determined on the basis of these data.

On the other had, the *arg 2* and *3* mutations grow if supplied citrulline but not if they are supplied with only ornithine. Therefore, ornithine must occur in the pathway *prior* to the block. Its presence will not overcome the block. Citrulline, however, does overcome the block, so it must be involved beyond the point of blockage. Therefore, the conversion of ornithine to citrulline represents the correct sequence in the pathway.

Finally, we can conclude that *arg 1* represents a mutation preventing the conversion of citrulline to arginine. Neither ornithine nor citrulline can overcome the metabolic block because both participate earlier in the pathway.

Taken together, these reasons support the sequence of biosynthesis outlined here. Since Srb and Horowitz's work in 1944, the detailed pathway has been worked out and the enzymes controlling each step characterized. The abbreviated metabolic pathway is shown in Figure 14.12.

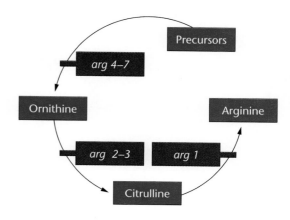

FIGURE 14.12   Abbreviated pathway resulting in the biosynthesis of arginine in *Neurospora*.

Studies of human hemoglobin established that one gene encodes one polypeptide

The concept of one-gene:one-enzyme developed in the early 1940s was not accepted immediately by all geneticists. This is not surprising because it was not yet clear how mutant enzymes could cause variation in many phenotypic traits. For example, *Drosophila* mutants demonstrated altered eye size, wing shape, wing vein pattern, and so on. Plants exhibited mutant varieties of seed texture, height, and fruit size. How an inactive, mutant enzyme could result in such phenotypes was puzzling to many geneticists.

Two factors soon modified the one-gene:one-enzyme hypothesis. First, while nearly all enzymes are proteins, not all proteins are enzymes. As the study of biochemical genetics proceeded, it became clear that all proteins are specified by the information stored in genes, leading to the more accurate phraseology **one-gene:one-protein.** Second, proteins were often shown to have a subunit structure consisting of two or more **polypeptide chains.** This is the basis of the **quaternary structure** of proteins, which we will discuss later in this chapter. Because each distinct polypeptide chain is encoded by a separate gene, a more modern statement of Beadle and Tatum's basic principle is **one-gene:one-polypeptide chain.** These modifications of the original hypothesis became apparent during the analysis of hemoglobin structure in individuals afflicted with sickle-cell anemia.

Sickle-Cell Anemia

The first direct evidence that genes specify proteins other than enzymes came from the work on mutant hemoglobin molecules derived from humans afflicted with the disorder **sickle-cell anemia.** Affected individuals contain erythrocytes that, under low oxygen tension, become elongated and curved because of the polymerization of hemoglobin. The "sickle" shape of these erythrocytes is in contrast to the biconcave disc shape characteristic of normal individuals (Figure 14.13). Individuals with the disease suffer attacks when red blood cells aggregate in the venous side of capillary systems, where oxygen tension is very low. As a result, a variety of tissues may be deprived of oxygen and suffer severe damage. When this occurs, an individual is said to experience a sickle-cell crisis. If untreated, a crisis may be fatal. The kidneys, muscles, joints, brain, gastrointestinal tract, and lungs may be affected.

In addition to suffering crises, these individuals are anemic because their erythrocytes are destroyed more rapidly than are normal red blood cells. Compensatory physiological mechanisms include increased red-cell production by bone marrow and accentuated heart action. These mechanisms lead to abnormal bone size and shape as well as dilation of the heart.

In 1949, James Neel and E. A. Beet demonstrated that the disease is inherited as a Mendelian trait. Pedigree analysis revealed three genotypes and phenotypes controlled by a single pair of alleles, Hb^A and Hb^S. Normal and affected individuals result from the homozygous genotypes $Hb^A Hb^A$ and $Hb^S Hb^S$, respectively. The red blood cells of heterozy-

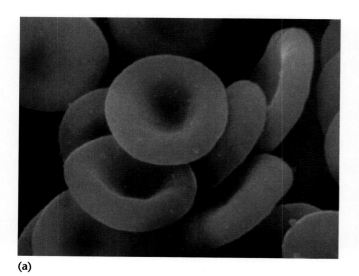

(a)

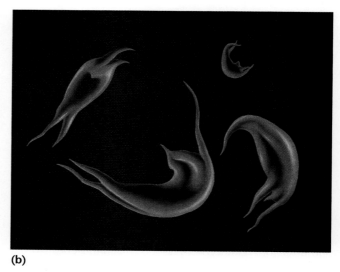

(b)

FIGURE 14.13 A comparison of erythrocytes from normal individuals (left) and from individuals with sickle-cell anemia (right).

gotes, who exhibit the **sickle-cell trait** but not the disease, undergo much less sickling because over half of their hemoglobin is normal. Although largely unaffected, such heterozygotes are "carriers" of the defective gene, which is transmitted, on average, to 50 percent of their offspring.

In the same year, Linus Pauling and his co-workers provided the first insight into the molecular basis of the disease. They showed that hemoglobins isolated from diseased and normal individuals differ in their rates of electrophoretic migration. In this technique (see Chapter 11 and Appendix A), charged molecules migrate in an electric field. If the net charge of two molecules is different, their rates of migration will be different. On this basis, Pauling and his colleagues concluded that a chemical difference exists between normal and sickle-cell hemoglobin. The two molecules are now designated **HbA** and **HbS,** respectively.

Figure 14.14(a) illustrates the migration pattern of hemoglobin derived from individuals of all three possible genotypes when subjected to **starch gel electrophoresis.** The gel provides the supporting medium for the molecules during migration. In this experiment, samples are placed at a point of origin between the cathode (−) and the anode (+), and an electric field is applied. The migration pattern reveals that all molecules move toward the anode, indicating a net negative charge. However, HbA migrates farther than HbS, suggesting that its net negative charge is greater. The electrophoretic pattern of hemoglobin derived from carriers reveals the presence of both HbA and HbS, confirming their heterozygous genotype.

Pauling's findings suggested two possibilities. It was known that hemoglobin consists of four nonproteinaceous, iron-containing **heme groups** and a **globin portion** con-

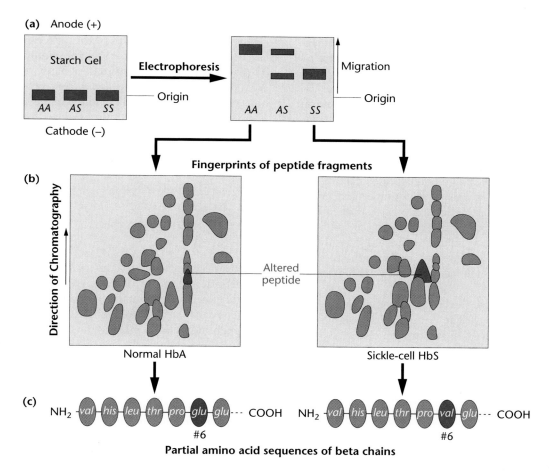

FIGURE 14.14   Investigation of hemoglobin derived from $Hb^A Hb^A$ and $Hb^S Hb^S$ individuals using electrophoresis, fingerprinting, and amino acid analysis. Hemoglobin from individuals with sickle-cell anemia ($Hb^S Hb^S$) (a) migrates differently in an electrophoretic field, (b) shows an altered peptide in fingerprint analysis, and (c) shows an altered amino acid, valine, at the sixth position in the β chain. During electrophoresis, heterozygotes ($Hb^A Hb^S$) reveal both forms of hemoglobin.

taining four polypeptide chains. The elongation in net charge in HbS could be due, theoretically, to a chemical change in either component.

Work carried out between 1954 and 1957 by Vernon Ingram resolved this question. He demonstrated that the chemical change occurs in the primary structure of the globin portion of the hemoglobin molecule. Using the **fingerprinting technique,** Ingram showed that HbS differs in amino acid composition compared to HbA. Human adult hemoglobin contains two identical alpha (α) chains of 141 amino acids and two identical beta (β) chains of 146 amino acids in its quaternary structure.

The fingerprinting technique involves the enzymatic digestion of the protein into peptide fragments. The mixture is then placed on absorbent paper and exposed to an electric field, where migration occurs according to net charge. The paper is then turned at a right angle and placed in a solvent, where chromatographic action causes the migration of the peptides in the second direction. The end result is a two-dimensional separation of the peptide fragments into a distinctive pattern of spots or a "fingerprint." Ingram's work revealed that HbS and HbA differed by only a single peptide fragment [Figure 14.14(b)]. Further analysis then revealed a single amino acid change: Valine was substituted for glutamic acid at the sixth position of the β chain, accounting for the peptide difference [Figure 14.14(c)].

The significance of this discovery has been multifaceted. It clearly establishes that a single gene provides the genetic information for a single polypeptide chain. Studies of HbS also demonstrate that a mutation can affect the phenotype by directing a single amino acid substitution. Also, by providing the explanation for sickle-cell anemia, the concept of inherited **molecular disease** was firmly established. Finally, this work led to a thorough study of human hemoglobins, which has provided valuable genetic insights.

In the United States, sickle-cell anemia is found almost exclusively in the African-American population. It affects about one in every 625 African-American infants. Currently, about 50,000 to 75,000 individuals are afflicted. In about one of every 145 African-American married couples, both partners are heterozygous carriers. In these cases, each of their children has a 25 percent chance of having the disease.

Human Hemoglobins

Having introduced human hemoglobins in an historical context, it may be useful to extend our discussion and provide an update of what is currently known about these molecules in our species. Molecular analysis reveals that a variety of hemoglobin molecules are produced in humans at different stages of the life cycle. All are tetramers consisting of numerous combinations of seven distinct polypeptide chains, each encoded by a separate gene. The expression of these various genes is developmentally regulated.

Almost all adult hemoglobin consists of **HbA,** which contains two α and two β chains. Recall that the mutation in sickle-cell anemia involves the β chain. HbA represents about 98 percent of all hemoglobin found in an individual's erythrocytes after the age of 6 months. The remaining 2 percent consists of **HbA$_2$,** a minor adult component. This molecule contains two α chains and two **delta (δ) chains.** The δ chain is very similar to the β chain, consisting of 146 amino acids.

During embryonic and fetal development, quite a different set of hemoglobins is found. The earliest set to develop is called **Gower 1,** containing two **zeta (ζ) chains,** which are most similar to α chains, and two **epsilon (ϵ) chains,** which are most similar to β chains. By eight weeks of gestation, this embryonic form is gradually replaced by still another hemoglobin molecule with still different chains. This molecule is called **HbF,** or **fetal hemoglobin,** and consists of two α chains and two **gamma (γ) chains.** There are two types of γ chains designated $^G\gamma$ and $^A\gamma$. Both are most similar to β chains and differ from each other by only a single amino acid. The nomenclature and sequence of appearance of the five tetramers described are summarized in Table 14.2.

The nucleotide sequence of a gene and the amino acid sequence of the corresponding protein exhibit colinearity

Once it was established that genes specify the synthesis of polypeptide chains, the next logical question was how the genetic information contained in the nucleotide sequence of a gene can be transferred to the amino acid sequence of a polypeptide chain. It seemed most likely that a **colinear relationship** would exist between the two molecules. That is, the order of nucleotides in the DNA of a gene would

TABLE 14.2 Chain Compositions of Human Hemoglobins from Conception to Adulthood

Hemoglobin Type	Chain Composition
Embryonic-Gower 1	$\zeta_{2G}\epsilon_2$
Fetal-HbF	$\alpha_2\gamma_2$
	$\alpha_2{}^A\gamma_2$
Adult-HbA	$\alpha_2\beta_2$
Minor adult-HbA$_2$	$\alpha_2\delta_2$

correlate directly with the order of amino acids in the corresponding polypeptide.

The initial experimental evidence in support of this concept was derived from studies by Charles Yanofsky of the *trpA* gene that encodes the A subunit of the enzyme **tryptophan synthetase** in *E. coli*. Yanofsky isolated many independent mutants that had lost the activity of the enzyme. He was able to map these mutations and establish their location with respect to one another within the gene. He then determined where the amino acid substitution had occurred in each mutant protein. When the two sets of data were compared, the colinear relationship was apparent. The location of each mutation in the *trpA* gene correlates with the position of the altered amino acid in the A polypeptide of tryptophan synthetase. This comparison is illustrated in Figure 14.15.

Protein structure is the basis of biological diversity

Having established that the genetic information is stored in DNA and influences cellular activities through the proteins it encodes, we turn now to a brief discussion of protein structure. How can these molecules play such a critical role in determining the complexity of cellular activities? As we will see, the fundamental aspects of the structure of proteins provide the basis for incredible complexity and diversity. At the outset, we should differentiate between the terms **polypeptides** and **proteins.** Both describe molecules composed of amino acids. The molecules differ, however, in their state of assembly and functional capacity. Polypeptides are the precursors of proteins. As assembled on the ribosome

during translation, the molecule is called a *polypeptide*. When released from the ribosome following translation, a polypeptide folds up and assumes a higher order of structure. When this occurs, a three-dimensional conformation in space emerges. In many cases, several polypeptides interact to produce this conformation. When the final conformation is achieved, the molecule, now fully functional, is appropriately called a *protein*. It is the three-dimensional conformation that is essential to the function of the molecule.

The polypeptide chains of proteins, like nucleic acids, are linear nonbranched polymers. There are 20 amino acids that serve as the subunits (the building blocks) of proteins. Each amino acid has a **carboxyl group,** an **amino group,** and an **R (radical) group** (a side chain) bound covalently to a **central carbon atom.** The R group gives each amino acid its chemical identity. Figure 14.16 illustrates the 20 R groups, which show a variety of configurations and can be divided into four main classes: (1) **nonpolar (hydrophobic),** (2) **polar (hydrophilic),** (3) **negatively charged,** and (4) **positively charged.** Because polypeptides are often long polymers, and because each position may be occupied by any one of 20 amino acids with unique chemical properties, enormous variation in chemical conformation and activity is possible. For example, if an average polypeptide is composed of 200 amino acids (molecular weight of about 20,000 daltons), 20^{200} different molecules, each with a unique sequence, can be created using 20 different building blocks.

Around 1900, German chemist Emil Fischer determined the manner in which the amino acids are bonded together. He showed that the amino group of one amino acid can react with the carboxyl group of another amino acid during a dehydration reaction, releasing a molecule of

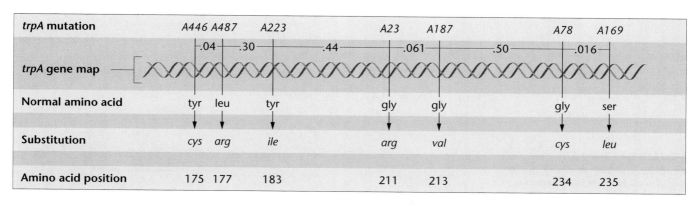

FIGURE 14.15 Demonstration of colinearity between the genetic map of various *trpA* mutations in *E. coli* and the affected amino acids in the protein product. The numbers shown between mutations represent linkage distances.

1. Nonpolar: Hydrophobic

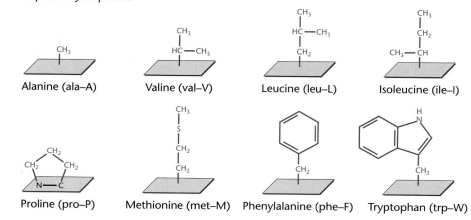

Alanine (ala–A)   Valine (val–V)   Leucine (leu–L)   Isoleucine (ile–I)

Proline (pro–P)   Methionine (met–M)   Phenylalanine (phe–F)   Tryptophan (trp–W)

2. Polar: Hydrophilic

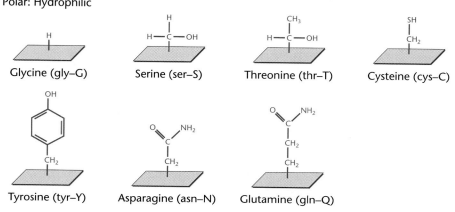

Glycine (gly–G)   Serine (ser–S)   Threonine (thr–T)   Cysteine (cys–C)

Tyrosine (tyr–Y)   Asparagine (asn–N)   Glutamine (gln–Q)

3. Polar: positively charged (basic)

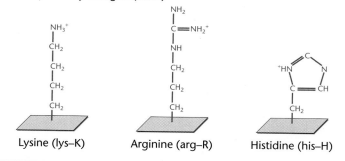

Lysine (lys–K)   Arginine (arg–R)   Histidine (his–H)

4. Polar: negatively charged (acidic)

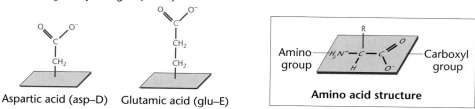

FIGURE 14.16 Chemical structures and designations of the 20 amino acids found in living organisms, divided into four major categories.

Aspartic acid (asp–D)   Glutamic acid (glu–E)

Amino group — Carboxyl group

Amino acid structure

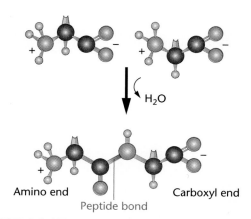

FIGURE 14.17 Peptide bond formation between two amino acids, resulting from a dehydration reaction.

Amino end

Carboxyl end

Peptide bond

H_2O. The resulting covalent bond is known as a peptide bond (Figure 14.17). **Two amino** acids linked together constitute a **dipeptide,** three a **tripeptide,** and so on. Once 10 or more amino acids are linked by peptide bonds, the chain may be referred to as a polypeptide. Generally, no matter how long a polypeptide is, it will contain a free amino group at one end (the **N-terminus**) and a free carboxyl group at the other end (the **C-terminus**).

Four levels of protein structure are recognized: primary, secondary, tertiary, and quaternary. The sequence of amino acids in the linear backbone of the polypeptide constitutes its **primary structure.** This sequence is specified by the sequence of deoxyribonucleotides in DNA via

an mRNA intermediate. The primary structure of a polypeptide helps determine the specific characteristics of the higher orders of organization as a protein is formed.

The **secondary structure** refers to a regular or repeating configuration in space assumed by amino acids aligned closely to one another in the polypeptide chain. In 1951, Linus Pauling and Robert Corey predicted, on theoretical grounds, an **α helix** as one type of secondary structure. The α-helix model [Figure 14.18(a)] has since been confirmed by X-ray crystallographic studies. It is rodlike and has the greatest possible theoretical stability. The helix is composed of a spiral chain of amino acids stabilized by hydrogen bonds.

The side chains (the R groups) of amino acids extend outward from the helix, and each amino acid residue occupies a vertical distance of 1.5 Å in the helix. There are 3.6 residues per turn. While left-handed helices are theoretically possible, all proteins demonstrating an α helix are right-handed.

Also in 1951, Pauling and Corey proposed a second structure, **the β-pleated-sheet** configuration. In this model, a single polypeptide chain folds back on itself, or several chains run in either parallel or antiparallel fashion next to one another. Each such structure is stabilized by hydrogen bonds formed between atoms present on adjacent chains [Figure 14.18(b)]. A single zigzagging plane is formed in space with adjacent amino acids 3.5 Å apart.

As a general rule, most proteins demonstrate a mixture of α-helix and β-pleated-sheet structures. Globular

(a) Alpha helix

(b) Beta-pleated sheet

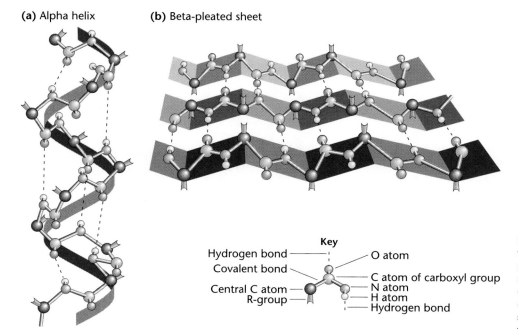

Key

Hydrogen bond

Covalent bond

Central C atom

R-group

O atom

C atom of carboxyl group

N atom

H atom

Hydrogen bond

FIGURE 14.18 (a) The right-handed α helix, which represents one form of secondary structure of a polypeptide chain. (b) The β-pleated-sheet configuration, an alternative form of secondary structure of polypeptide chains. For the sake of clarity, some atoms are not shown.

proteins, most of which are round in shape and water soluble, usually contain a core of β-pleated-sheet structure as well as many areas demonstrating helical structures. The more rigid structural proteins, many of which are water insoluble, rely on more extensive β-pleated-sheet regions for their rigidity. For example, **fibroin,** the protein made by the silk moth, depends extensively on this form of secondary structure.

While the secondary structure describes the arrangement of amino acids within certain areas of a polypeptide chain, **tertiary protein structure** defines the three-dimensional conformation of the entire chain in space. Each protein twists and turns and loops around itself in a very specific fashion, characteristic of the specific protein. Three aspects of this level of structure are most important in determining this conformation and in stabilizing the molecule.

1. Covalent disulfide bonds form between closely aligned cysteine residues to form the unique amino acid cystine.

2. Nearly all of the polar hydrophilic R groups are located on the surface, where they can interact with water.

3. The nonpolar hydrophobic R groups are usually located on the inside of the molecule, where they interact with one another, avoiding interaction with water.

It is important to emphasize that the three-dimensional conformation achieved by any protein is a product of the primary structure of the polypeptide. Thus, the genetic code need only specify the sequence of amino acids in order to encode information that leads ultimately to the final assembly of proteins. The three stabilizing factors depend on the location of each amino acid relative to all others in the chain. As folding occurs, the most thermodynamically stable conformation possible results.

A model of the three-dimensional tertiary structure of the respiratory pigment **myoglobin** is shown in Figure 14.19. This level of organization is extremely important because the specific function of any protein is directly related to its three-dimensional conformation.

The **quaternary level of organization** applies only to proteins composed of more than one polypeptide chain, indicating the conformation of the various chains in relation to one another. This type of protein is called **oligomeric,** and each chain is called a **promoter,** or, less formally, a **subunit.** The individual promoters have conformations that fit together with other subunits in a specific complementary fashion. Hemoglobin, an oligomeric protein consisting of four polypeptide chains, has been studied in great detail. Its quaternary structure is shown in Figure 14.20.

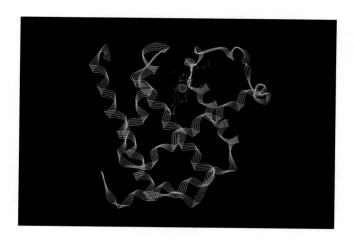

FIGURE 14.19 A ribbon drawing illustrating the tertiary level of protein structure in respiratory pigment myoglobin. The bound oxygen atom is shown in red.

Most enzymes, including DNA and RNA polymerase, demonstrate quaternary structure.

Posttranslational Modification

Before turning to a discussion of protein function, it is important to point out that polypeptide chains, like RNA transcripts, are often modified once they have been synthesized. This additional processing is broadly described as **posttranslational modification.** Although many of these alterations are detailed biochemical transformations and beyond the scope of this discussion, you should be

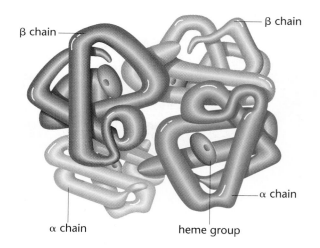

FIGURE 14.20 Schematic drawing illustrating the quaternary level of protein structure as seen in hemoglobin. Four chains (2 α and 2 β) interact with four heme groups to form the functional molecule.

aware that they occur and that they are critical to the functional capability of the final protein product. Several examples of posttranslational modification are presented below.

1. **The N-terminus and C-terminus amino acids are usually removed or modified.** For example, the initial N-terminal formylmethionine residue in bacterial polypeptides is usually removed enzymatically. Often, the amino group of the initial methionine residue is removed, and the amino group of the N-terminal residue is chemically modified (acetylated) in eukaryotic polypeptide chains.

2. **Individual amino acid residues are sometimes modified.** For example, phosphates may be added to the hydroxyl groups of certain amino acids, such as tyrosine. Modifications such as these create negatively charged residues that may ionically bond with other molecules. The process of phosphorylation is extremely important in regulating many cellular activities and is a result of the action of enzymes called **kinases.** In other proteins, methyl groups may be added enzymatically.

3. **Carbohydrate side chains are sometimes attached.** These are added covalently, producing **glycoproteins,** an important category of molecules that includes many antigenic determinants, such as those specifying the antigens in the ABO blood-type system in humans.

4. **Polypeptide chains may be trimmed.** For example, insulin is first translated into a longer molecule that is enzymatically trimmed to its final 51-amino-acid form.

5. **Signal sequences are removed.** At the N-terminal end of some proteins, a sequence of up to 30 amino acids is found that plays an important role in directing the protein to the location in the cell where it functions. This is called a **signal sequence,** and it determines the final destination of a protein in the cell. This process is called **protein targeting.** For example, proteins whose fate involves secretion or that are to become part of the plasma membrane are dependent on specific sequences for their initial transport into the lumen of the endoplasmic reticulum. While the signal sequence of various proteins with a common destination might differ in their primary amino acid sequence, they share many chemical properties. For example, those destined for secretion all contain a string of up to 15 hydrophobic amino acids preceded by a positively charged amino acid at the N-terminus of the signal sequence. Once the polypeptides are transported, but prior to achieving their functional status as proteins, the sig-

nal sequence is enzymatically removed from these polypeptides.

6. **Polypeptide chains are often complexed with metals.** The tertiary and quaternary levels of protein structure often include and are dependent on metal atoms. The function of the protein is thus dependent on the molecular complex that includes both polypeptide chains and metal atoms. Hemoglobin, containing four iron atoms along with four polypeptide chains, is a good example.

These types of posttranslational modifications are obviously important in achieving the functional status specific to any given protein. Because the final three-dimensional structure of the molecule is intimately related to its specific function, how polypeptide chains ultimately fold into their final conformations is also an important topic. For many years, it was thought that protein folding was a spontaneous process whereby the molecule achieved maximum thermodynamic stability, based largely on the combined chemical properties inherent in the amino acid sequence of the polypeptide chain(s) composing the protein. However, numerous studies have shown that, for many proteins, folding is dependent upon members of a family of still other, ubiquitous proteins called **chaperones.** Chaperone proteins (sometimes called *molecular chaperones* or *chaperonins*) function to facilitate the folding of other proteins. While the mechanism by which chaperones function is not yet clear, like enzymes, they do not become part of the final product. Initially discovered in *Drosophila*, where they are called heat-shock proteins, chaperones have been discovered in a variety of organisms, including bacteria, animals, and plants.

Protein function is directly related to the structure of the molecule

The essence of life on Earth rests at the level of diverse cellular function. One can argue that DNA and RNA simply serve as vehicles to store and express genetic information. However, proteins are at the heart of cellular function. And it is the capability of cells to assume diverse structures and functions that distinguishes most eukaryotes from less evolutionarily advanced organisms such as bacteria. Therefore, an introductory understanding of protein function is critical to a complete view of genetic processes.

Proteins are the most abundant macromolecules found in cells. As the end products of genes, they play many diverse roles. For example, the respiratory pigments **hemoglobin** and **myoglobin** transport oxygen, which is essential for cellular metabolism. **Collagen** and **keratin** are structural proteins associated with the skin, connective tissue, and hair of organisms. **Actin** and **myosin** are contractile proteins, found in abundance in muscle tissue. Still other examples are the **immunoglobins,** which function in the immune system of vertebrates; **transport proteins,** involved in movement of molecules across membranes; some of the **hormones** and their **receptors,** which regulate various types of chemical activity; and **histones,** which bind to DNA in eukaryotic organisms.

The largest group of proteins with a related function are the **enzymes.** Since we have been referring to these molecules throughout this chapter, it may be useful to extend our discussion and include a more detailed description of their biological role. These molecules specialize in catalyzing chemical reactions within living cells. Enzymes increase the rate at which a chemical reaction reaches equilibrium, but they do not alter the end point of the chemical equilibrium. Their remarkable, highly specific catalytic properties largely determine the metabolic capacity of any cell type. The specific functions of many enzymes involved in the genetic and cellular processes of cells are described throughout the textbook.

Biological catalysis is a process whereby the **energy of activation** for a given reaction is lowered (Figure 14.21). The energy of activation is the increased kinetic energy state that molecules must usually reach before they react with one another. While this state can be attained as a result of elevated temperatures, enzymes allow biological reactions to occur at lower physiological temperatures. In this way, enzymes make possible life as we know it.

The catalytic properties and specificity of an enzyme are determined by the chemical configuration of the molecule's **active site.** This site is associated with a crevice, a cleft, or a pit on the surface of the enzyme, which binds the reactants, or substrates, facilitating their interaction. Enzymatically catalyzed reactions control metabolic activities in the cell. Each reaction is either **catabolic** or **anabolic.** Catabolism is the degradation of large molecules into smaller, simpler ones with the release of chemical energy. Anabolism is the synthetic phase of metabolism, yielding the various components that make up nucleic acids, proteins, lipids, and carbohydrates.

Proteins are made up of one or more functional domains

We conclude this chapter by briefly discussing the finding that regions made up of specific amino acid sequences are associated with specific functions in protein molecules. Such sequences, usually between 50 and 300 amino acids, constitute what are called **protein domains** and are represented by modular portions of the protein that fold independently of the rest of the molecule into stable, unique conformations. Different domains impart different functional capabilities. Some proteins contain only a single domain, while others contain two or more.

Exon Shuffling and the Origin of Protein Domains

An interesting proposal to explain the genetic origin of protein domains was put forward by Walter Gilbert in 1977. Gilbert suggested that the functional regions of genes in higher organisms consist of collections of exons originally present in ancestral genes that are brought together through recombination during the course of evo-

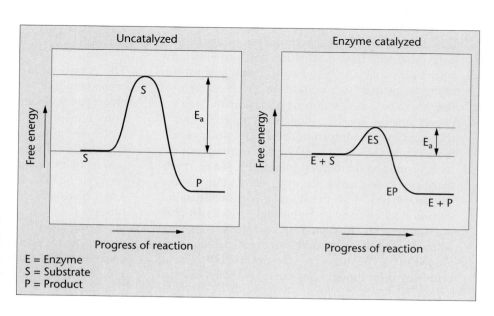

FIGURE 14.21 Energy requirements of an uncatalyzed vs. an enzymatically catalyzed chemical reaction. The energy of activation (E_a) necessary to initiate the reaction is substantially lower as a result of catalysis.

E = Enzyme
S = Substrate
P = Product

lution. Referring to this process as **exon shuffling,** Gilbert proposed that exons are modular in the sense that each might encode a single **protein domain.** Serving as the basis of useful part of a protein, a single type of domain might function in a variety of proteins. In Gilbert's proposal, during evolution many exons could be "mixed and matched" to form unique genes in eukaryotes.

Several observations lend support to this proposal. Most exons are fairly small, averaging about 150 nucleotides and encoding about 50 amino acids. This size is consistent with the production of many functional domains in proteins. Second, recombinational events that lead to exon shuffling would be expected to occur within areas of genes represented by introns. Because introns are free to accumulate mutations without harm to the organism, recombinational events would tend to further randomize their nucleotide sequences. Over extended evolutionary period, diverse sequences would tend to accumulate. This is, in fact, what is observed; introns range from 50 to 20,000 bases and exhibit fairly random base sequences.

Since 1977, a serious research effort has been directed toward the analysis of gene structure. In 1985, more direct evidence in favor of Gilbert's proposal of exon modules was presented. For example, the human gene encoding the membrane receptor for low-density lipoproteins (LDL) was isolated and sequenced. The **LDL receptor protein** is essential to the transport of plasma cholesterol into the cell. It mediates **endocytosis** and is expected to have numerous functional domains. These include the capability of this protein to bind specifically to the LDL substrates and to interact with other proteins at different levels of the membrane during transport across it. In addition, this receptor molecule is modified posttranslationally by the addition of a carbohydrate; a domain must exist that links to this carbohydrate.

Detailed analysis of the gene encoding this protein supports the concept of exon modules and their shuffling during evolution. The gene is quite large—45,000 nucleotides—and contains 18 exons. These represent only slightly less than 2600 nucleotides. These exons are related to the functional domains of the protein *and* appear to have been recruited from other genes during evolution.

Figure 14.22 illustrates these relationships. The first exon encodes a signal sequence that is removed from the protein before the LDL receptor becomes part of the membrane. The next five exons represent the domain specifying the binding site for cholesterol. This domain is made up of a 40-amino-acid sequence repeated seven times. The next domain consists of a sequence of 400 amino acids bearing a striking homology to the mouse peptide hormone **epidermal growth factor (EGF).** This region is encoded by eight exons and contains three repetitive sequences of 40 amino acids. A similar sequence is also found in three blood-clotting proteins. The 15th exon specifies the domain for the posttranslational addition of the carbohydrate, while the remaining two exons specify regions of the protein that are part of the membrane, anchoring the receptor to specific sites called *coated pits* on the cell surface.

These observations concerning the LDL exons are fairly compelling in support of the theory of exon shuffling during evolution. Certainly, there is no disagreement concerning the concept of protein domains being responsible for specific molecular interactions.

What remains controversial and evocative in the exon shuffling theory is the question of when introns first appeared on the evolutionary scene. In 1978, W. Ford Doolittle proposed that these intervening sequences were part of the genome of the most primitive ancestors of modern-day eukaryotes. In support of this idea, Gilbert has argued that if intron similarities in DNA sequence are found in identical positions within genes shared by totally unrelated eukaryotes (such as humans, chickens, and corn), they must have also been present in primitive ancestral genomes.

If Doolittle's proposal is correct, why are introns absent in most prokaryotes? Gilbert argues that they were present at one point during evolution, but as the genome of these primitive organisms evolved, they were lost. This occurred as a result of strong selection pressure to streamline their chromosomes to minimize the energy expenditure supporting replication and gene expression. Further, streamlining led to more error-free mRNA production. However, supporters of the opposing "intron-late" school, including Jeffrey Palmer, argue that introns first appeared much later during evolution, when they became a part of a single group of eukaryotes that are ancestral to modern-day members but not to prokaryotes. Although it may be difficult to resolve, it seems certain that this controversy will persist for many years to come.

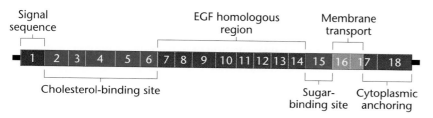

FIGURE 14.22 A comparison of the 18 exons making up the gene encoding the LDL receptor protein. The exons are organized into five functional domains and one signal sequence.

GENETICS, TECHNOLOGY, AND SOCIETY

Mad Cows and Heresies: The Prion Story

On March 20, 1996, the British government announced that a new deadly brain disease had killed 10 young Britons and that the victims might have caught the disease by eating infected beef. The disease, known as "mad cow disease" or bovine spongiform encephalopathy (BSE), slowly destroys brain cells and is always fatal. The ensuing panic triggered political turmoil in Europe, a worldwide ban on British beef, and the near collapse of the $8.9 billion British beef industry. The European Union demanded the slaughter and incineration of 4.7 million British cattle, a campaign that would cost the government over $12 billion to compensate farmers, import milk, and buy cattle for new herds. Although some scientists felt that it was unlikely that humans could catch BSE from eating beef, recent studies confirm that BSE and the human disease, known as variant Creutzfeldt-Jakob disease (vCJD), are so similar at the molecular and pathological levels that it is very likely they are the same disease. At least 24 Europeans have now died of vCJD, and it is still not known how many more

people are infected. Recently, Britain banned the sale of certain cuts of beef and decided to destroy the plasma from all blood donations, due to fears that BSE could be transmitted through blood products. The U.S. Department of Agriculture has banned the importation of cattle or meat products from countries that have BSE. No cases of BSE have been reported in the U.S.

BSE and CJD are members of a group of neurological diseases known as spongiform encephalopathies. These lethal diseases have long incubation times (up to 20 or 30 years), and lead to progressive neurodegeneration. The infected brain tissue resembles a sponge (hence "spongiform") and is riddled with proteinaceous deposits. The victims of the disease lose motor function, become demented, and eventually die. CJD cases arise spontaneously and randomly, at a rate of 1 per million per year worldwide. A less common form of CJD is inherited as an autosomal dominant condition. But the transmitted forms are the strangest. CJD can be transmitted through corneal or nervous tissue grafts or by injection of growth

hormone derived from human pituitary glands. Kuru, a CJD-like disease of the Fore people of New Guinea, was transmitted from person to person via ritualistic cannibalism. Transmissible spongiform encephalopathies in animals include scrapie (sheep and goats) and BSE. Like Kuru, BSE and scrapie are passed from animal to animal by the ingestion of diseased animal remains, particularly neural tissue. The epidemic of BSE in Britain occurred because diseased cows and sheep were processed and fed to cattle as a protein supplement. In 1998, the British government banned the use of cows and sheep as feed for other cows and sheep, and the epidemic of BSE is subsiding.

For many years, the spongiform encephalopathies defied analysis. The diseases are difficult to study, as they require the injection of infected brain material into the brains of experimental animals and the diseases take months or years to develop. In addition, the infectious agent is apparently not a virus or bacterium, and the infected animals do not develop antibodies against these mysterious agents. There are no treat-

CHAPTER SUMMARY

1. Translation describes the synthesis in cells of polypeptide chains, under the direction of mRNA and in association with ribosomes. This process ultimately converts the information stored in the genetic code of the DNA making up a gene into the corresponding sequence of amino acids making up the polypeptide.

2. Translation is a complex energy-requiring process that also depends on charged tRNA molecules and numerous protein factors. Transfer RNA (tRNA) serves as the adaptor molecule between an mRNA triplet and the appropriate amino acid.

3. The processes of translation, like transcription, can be subdivided into the stages of initiation, elonga-

tion, and termination. The process relies on base-pairing affinities between complementary nucleotides and is more complex in eukaryotes than in prokaryotes.

4. The first insight that proteins are the end products of genes was provided by the study of inherited metabolic disorders in humans early in the twentieth century by Garrod. Basic to his studies were inborn errors of metabolism leading to cystinuria, albinism, and alkaptonuria.

5. The investigation of nutritional requirements in *Neurospora* by Beadle and colleagues made it clear that mutations cause the loss of enzyme activity. Their work led to the concept of one-gene:one-enzyme.

ments for the disease, and the only way to make a firm diagnosis is to examine brain tissue, after death. Scrapie infectious material is unaffected by radiation or nucleases that damage nucleic acids; however, it is destroyed by some reagents that hydrolyze or modify proteins. In the early 1980s, American scientist Stanley Prusiner purified the infectious agent and concluded that it consisted only of protein. He proposed that scrapie was spread by an infectious protein particle that he called a "prion." His hypothesis was immediately dismissed by most scientists, as the idea of an infectious agent that contained no DNA or RNA as genetic material was heretical. However, Prusiner and others have presented evidence supporting the prion hypothesis, and the notion that disease can be transmitted by an infectious particle that contains no genetic material has gained acceptance. The prion hypothesis is now widely accepted, and Prusiner was awarded the 1997 Nobel Prize in Physiology or Medicine for his pioneering work on spongiform encephalopathies. Although a few scientists still disagree with the protein-only hypothesis, no one has yet shown that the infectious agent contains genetic material.

If prions are comprised only of protein, how do they cause disease? The answer may be as strange as the disease itself. The protein that makes up a prion (PrP) is also a normal protein that is synthesized in neurons and found in the brains of all adult animals. The difference between normal PrP in a normal brain and prion PrP lies in their protein secondary structures. Normal noninfectious PrP folds into α-helices whereas infectious prion PrP folds into β-sheets. When a normal PrP molecule contacts a prion PrP molecule, the normal protein is somehow unfolded and refolded into the abnormal PrP conformation. Once the normal PrP molecule has been transformed into an abnormal PrP molecule, it spreads its lethal conformation to neighboring normal PrP molecules, and the process takes off in a chain reaction. Normal PrP is a soluble protein that is easily destroyed by heat or enzymes that digest proteins. However, abnormal infectious PrP is insoluble in detergents and resists both heat and protease digestion. It forms deposits in the brain, causes the death of neurons, and is nearly indestructible. Hence, spongiform encephalopathies can be considered diseases of protein secondary structure.

Many urgent questions need to be addressed. Is BSE now out of the food chain? How many humans are infected with BSE but don't yet show symptoms? Can humans or animals act as asymptomatic carriers of prion diseases? Do some people have a genetic susceptibility to spongiform encephalopathies? Can prions exist in other parts of the body besides the brain and spinal cord? If so, can CJD be spread via blood transfusions, from mother to fetus, or by surgical instruments? Can we develop diagnostic tests and therapies for BSE and vCJD? Have we seen the end of the mad cow disease story, or is this just the beginning?

REFERENCES:
AGUZZI, A., and WEISSMANN, C. 1997. Prion research: The next frontiers. *Nature* 389:795–98.
ALMOND, J., and PATTISON, J. 1997. Human BSE. *Nature* 389:437–38.
PRUSINER, S.B. 1998. Prions. *Proc. Natl. Acad. Sci. USA.* 95:15363–83.

WEBSITE:
PBS *Nova* website—The Brain Eater
http://www.pbs.org/wgbh/nova/mad cow

6. The one-gene:one-enzyme hypothesis was later revised. Pauling and Ingram's investigations of hemoglobins from patients with sickle-cell anemia led to the discovery that one gene directs the synthesis of only one polypeptide chain.

7. Thorough investigations have revealed the existence of several major types of human hemoglobin molecules found in the embryo, fetus, and adult. Specific genes control each polypeptide chain constituting these various hemoglobin molecules.

8. The proposal suggesting that a gene's nucleotide specifies in a colinear way the sequence of amino acids in a polypeptide chain was confirmed by experiments involving mutations in the tryptophan synthetase gene in *E. coli.*

9. Proteins, the end products of genes, demonstrate four levels of structural organization that together provide the chemical basis for their three-dimensional conformation, which is the basis of the molecule's function.

10. Of the myriad functions performed by proteins, the most influential role is assumed by enzymes. These highly specific, cellular catalysts play a central role in the production of all classes of molecules in living systems.

11. Proteins consist of one or more functional domains, which are shared by different molecules. The origin of these domains may be the result of exon shuffling during evolution.

INSIGHTS AND SOLUTIONS

1. The following growth responses were obtained using four mutant strains of *Neurospora* and the related compounds A, B, C, and D. None of the mutations grow on minimal medium. Draw all possible conclusions.

		Growth Product			
		C	D	B	A
	1	−	−	−	−
	2	−	+	+	+
Mutation	3	−	−	+	+
	4	−	−	+	−

Solution: First, nothing can be concluded about mutation 1, except that it is lacking some essential growth factor, perhaps even unrelated to the biochemical pathway represented by mutations 2–4; nor can anything be concluded about compound C. If it is involved in the pathway, it is a product synthesized prior to the synthesis of A, B, and D.

We must now analyze these three compounds and the control of their synthesis by the enzymes encoded by genes 2, 3, and 4. Because product B allows growth in all three cases, it may be considered the "end product." It bypasses the block in all three instances. Using similar reasoning, product A precedes B in the pathway, since it allows a bypass in two of the three steps. Product D precedes B, yielding the more complete solution:

$$C\ (?) \longrightarrow D \longrightarrow A \longrightarrow B$$

Now determine which mutations control which steps. Since mutation 2 can be alleviated by products D, B, and A, it must control a step prior to all three products, perhaps the direct conversion to D, although we cannot be certain. Mutation 3 is alleviated by B and A, so its effect must precede them in the pathway. Thus, we will assign it as controlling the conversion of D to A. Likewise, we can assign mutation 4 to the conversion of A to B, leading to the more complete solution:

$$C\ (?) \xrightarrow{\ 2\ (?)\ } D \xrightarrow{\ 3\ } A \xrightarrow{\ 4\ } B$$

PROBLEMS AND DISCUSSION QUESTIONS

1. List and describe the role of all of the molecular constituents present in a functional polyribosome.

2. Contrast the roles of tRNA and mRNA during translation and list all enzymes that participate in the transcription and translation process.

3. Francis Crick proposed the "adaptor hypothesis" for the function of tRNA. Why did he choose that description?

4. During translation, what molecule bears the codon? The anticodon?

5. The α chain of eukaryotic hemoglobin is composed of 141 amino acids. What is the minimum number of nucleotides in an mRNA coding for this polypeptide chain? Assuming that each nucleotide is 0.34 nm long in the mRNA, how many triplet codes can, at one time, occupy space in a ribosome that is 20 nm in diameter?

6. Summarize the steps involved in charging tRNAs with their appropriate amino acids.

7. For it to carry out its role, each transfer RNA requires at least four specific recognition sites that must be inherent in its tertiary structure. What are they?

8. Discuss the potential difficulties involved in designing a diet to alleviate the symptoms of phenylketonuria.

9. Phenylketonurics cannot convert phenylalanine to tyrosine. Why don't these individuals exhibit a deficiency of tyrosine?

10. Phenylketonurics are often more lightly pigmented than are normal individuals. Can you suggest a reason why this is so?

11. The synthesis of flower pigments is known to be dependent on enzymatically controlled biosynthetic pathways. In the crosses shown here, postulate the role of mutant genes and their products in producing the observed phenotypes.

(a) P₁: white strain A × white strain B
   F₁: all purple
   F₂: 9/16 purple: 7/16 white
(b) P₁: white × pink
   F₁: all purple
   F₂: 9/16 purple: 3/16 pink: 4/16 white

12. A series of mutations in the bacterium *Salmonella typhimurium* results in the requirement of either tryptophan or some related molecule in order for growth to occur. From the data shown here, suggest a biosynthetic pathway for tryptophan.

	Growth Supplement				
Mutation	Minimal Medium	Anthra-nilic Acid	Indole Glycerol Phosphate	Indole	Tryptophan
trp-8	−	+	+	+	+
trp-2	−	−	+	+	+
trp-3	−	−	−	+	+
trp-1	−	−	−	−	+

13. The study of biochemical mutants in organisms such as *Neurospora* has demonstrated that some pathways are branched. The data shown here illustrate the branched nature of the pathway resulting in the synthesis of thiamine. Why don't the data support a linear pathway? Can you postulate a pathway for the synthesis of thiamine in *Neurospora*?

		Growth Supplement		
Mutation	Minimal Medium	Pyrimidine	Thiazole	Thiamine
thi-1	–	–	+	+
thi-2	–	+	–	+
thi-3	–	–	–	+

14. Explain why the one-gene:one-enzyme concept is not considered totally accurate today.

15. Why is an alteration of electrophoretic mobility interpreted as a change in the primary structure of the protein under study?

16. Contrast the polypeptide-chain components of each of the hemoglobin molecules found in humans.

17. Using sickle-cell anemia as a basis, describe what is meant by a molecular or genetic disease. What are the similarities and dissimilarities between this type of a disorder and a disease caused by an invading microorganism?

18. Contrast the contributions of Pauling and Ingram to our understanding of the genetic basis for sickle-cell anemia.

19. Hemoglobins from two individuals are compared by electrophoresis and by fingerprinting. Electrophoresis reveals no difference in migration, but fingerprinting shows an amino acid difference. How is this possible?

20. Describe what colinearity means. Of what significance is the concept of colinearity in the study of genetics?

21. Certain mutations called *amber* in bacteria and viruses result in premature termination of polypeptide chains during translation. Many *amber* mutations have been detected at different points along the gene coding for a head protein in phage T4. How might this system be further investigated to demonstrate and support the concept of colinearity?

22. Define and compare the four levels of protein organization.

23. List as many different categories of protein functions as you can. Wherever possible, give an example of each category.

24. How does an enzyme function? Why are enzymes essential for living organisms on Earth?

25. Does Fiers' work with phage MS2, discussed in Chapter 13, constitute more direct evidence in support of colinearity than Yanofsky's work with the *trpA* locus in *E. coli* (discussed in this chapter)? Explain.

26. Shown here are several amino acid substitutions in the α and β chains of human hemoglobin. Using the code table (Figure 13.7), determine how many of them can occur as a result of a single nucleotide change.

Hb Type	Normal Amino Acid	Substituted Amino Acid
HbJ Toronto	ala	asp (α-5)
HbJ Oxford	gly	asp (α-15)
Hb Mexico	gln	glu (α-54)
Hb Bethesda	tyr	his (β-145)
Hb Sydney	val	ala (β-67)
HbM Saskatoon	his	tyr (β-63)

EXTRA-SPICY PROBLEMS

27. In 1962, F. Chapeville and others reported an experiment in which they isolated radioactive ^{14}C-cysteinyl-tRNA^{cys} (charged tRNA^{cys} + cysteine). They then removed the sulfur group from the cysteine, creating alanyl-tRNA^{cys} (charged tRNA^{cys} + alanine). When alanyl-tRNA^{cys} was added to a synthetic mRNA calling for cysteine but not alanine, a polypeptide chain was synthesized containing alanine. What can you conclude from this experiment? [*Reference:* Chapeville et al., *Proc. Natl. Acad. Sci. USA* 48:1086–93 (1962).]

28. Three independently assorting genes are known to control the biochemical pathway below that provides the basis for flower color in a hypothetical plant. Homozygous recessive mutations, which interrupt each step, are known.

$$\text{colorless} \xrightarrow{A-} \text{yellow} \xrightarrow{B-} \text{green} \xrightarrow{C-} \text{speckled}$$

Determine the phenotypic results in the F_1 and F_2 generations resulting from the P_1 crosses given, involving true-breeding plants.

(a) speckled	(*AABBCC*)	×	yellow	(*AAbbCC*)
(b) yellow	(*AAbbCC*)	×	green	(*AABBcc*)
(c) colorless	(*aaBBCC*)	×	green	(*AABBcc*)

29. How would the results vary in cross (a) of Problem 28 if genes A and B were linked with no crossing over between them? How would the results of cross (a) vary if genes A and B were linked and 20 map units apart?

30. Most wood lily plants have orange flowers and are true breeding. A genetics student discovered both red and yellow true-breeding variants and proceeded to hybridize the three strains. Being knowledgeable about transmission genetics, the student surmised that two gene pairs were involved. On this basis, she proposed these genes account for the various colors as shown here.

	P_1	F_1	F_2
(1)	orange × red	all orange	3/4 orange 1/4 red
(2)	orange × yellow	all orange	3/4 orange 1/4 yellow
(3)	red × yellow	all orange	9/16 orange 4/16 yellow 3/16 red

(a) What outcome(s) of the crosses suggested that two gene pairs were at work?

(b) Assuming that she was correct, and using the mutant gene symbols *y* for yellow and *r* for red, propose which genotypes give rise to which phenotypes.

(c) The student then devised several possible biochemical pathways for the production of these pigments, as well as which steps are enzymatically controlled by which genes. These are listed here. Which of these (one or more) is/are consistent with your proposal in part (b); which was based on the original data? Assume that a mixture of red and yellow pigment results in orange flowers. Defend your answer.

I. white precursor I $\xrightarrow{\text{y gene}}$ red pigment

   white precursor II $\xrightarrow{\text{r gene}}$ yellow pigment

II. white precursor I $\xrightarrow{\text{r gene}}$ red pigment

   white precursor II $\xrightarrow{\text{y gene}}$ yellow pigment

III. white precursor $\longrightarrow$ red $\xrightarrow{\text{y gene}}$ yellow $\xrightarrow{\text{r gene}}$ orange

IV. white precursor $\longrightarrow$ red $\xrightarrow{\text{r gene}}$ yellow $\xrightarrow{\text{y gene}}$ orange

V. white precursor $\longrightarrow$ yellow $\xrightarrow{\text{y gene}}$ red $\xrightarrow{\text{r gene}}$ orange

VI. white precursor $\longrightarrow$ yellow $\xrightarrow{\text{r gene}}$ red $\xrightarrow{\text{y gene}}$ orange

31. Deep in a previously unexplored South American rain forest, a species of plants was discovered with true-breeding varieties whose flowers were either pink, rose, orange, or purple. A very astute plant geneticist made a single cross, carried to the F_2 generation, as shown here. Based solely on these data, he was able to propose both a mode of inheritance for flower pigmentation and a biochemical pathway for the synthesis of these pigments.

Carefully study the data. Create a hypothesis of your own to explain the mode of inheritance. Then propose a biochemical pathway consistent with your hypothesis. How could you test the hypothesis by making other crosses?

P_1: purple × pink
F_1: all purple
F_2: 27/64 purple
16/64 pink
12/64 rose
9/64 orange

GENETICS MediaLab

The following resources will help you achieve a better understanding of the concepts presented in this chapter. These resources can be found on the CD packaged with this textbook and on the Companion Website at **http://www.prenhall.com/klug/**.

CD Resources:

Animated Tutorial: *Translation*

Self-grading Chapter Problems

Web Resources:

Genetics and Society Issue: *Mad Cows and Heresies: The Prion Story*

Web Destinations in Genetics

Self-grading Chapter Problems

Chapter Search Terms

Genetics Newsgroups

Student Bulletin Board

Web Problem 1:

Time for completion = 5 minutes

What are the steps in protein synthesis? As with DNA transcription, the process requires three steps. The translational machinery must initiate translation, elongate the polypeptide, and terminate the process. These stages of translation require the coordination of RNA and many protein molecules. The website provides an overview of this process and illustrates the dynamics with a short animation. After viewing the text and animation, describe the physical relationship between the tRNA, mRNA, and the A and P sites of the ribosome. Discuss the role of the endoplasmic reticulum in translation. To complete this exercise, visit Web Problem 1 in Chapter 14 of your Companion Website and select the keyword **TRANSLATION**.

Web Problem 2:

Time for completion = 15 minutes

How is protein synthesis regulated during early development? Many transcripts are produced during oogenesis for use at a later time in development. In this exercise you will read about various mechanisms for controlling the use of transcripts in the oocytes of *Xenopus*, a genus of frog. Create a list of the different mechanisms controlling translation, indicate what portion of the transcript provides the regulatory information, and provide a specific example of each. To complete this exercise, visit Web Problem 2 in Chapter 14 of your Companion Website and select the keyword **TRANSLATIONAL REGULATION**.

Web Problem 3:

Time for completion = 5 minutes

How is genetic information transferred in a cell? DNA replication, transcription, and translation are the events determining flow of genetic information. While each process is distinct in many aspects, there are similarities in the pattern of information processing. The examination of many cellular processes suggests that a general evolutionary strategy has been to repeat successful strategies in new contexts. In this exercise, you will view short animations of DNA transcription and protein translation. After watching these side by side, list as many common features of these processes as you can. To complete this exercise, visit Web Problem 3 in Chapter 14 of your Companion Website and select the keyword **INFORMATION**.

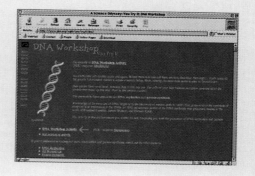

SELECTED READINGS

ANFINSEN, C.B. 1973. Principles that govern the folding of protein chains. *Science* 181:223–30.

BARTHOLOME, K. 1979. Genetics and biochemistry of phenylketonuria—Present state. *Hum. Genet.* 51:241–45.

BATESON, W. 1909. *Mendel's principles of heredity.* Cambridge: Cambridge University Press.

BEADLE, G.W. 1945. Genetics and metabolism in *Neurospora. Physiol. Rev.* 25:643.

BEADLE, G.W., and TATUM, E.L. 1941. Genetic control of biochemical reactions in *Neurospora. Proc. Natl. Acad. Sci. USA* 27:499–506.

BECKMANN, R. et al. 1997. Alignment of conduits for the nascent polypeptide chain in the ribosome-Sec61 complex. *Science* 278:2123–26.

BEET, E.A. 1949. The genetics of the sickle-cell trait in a Bantu tribe. *Ann. Eugenics* 14:279–84.

BOYER, P.D., ed. 1974. *The enzymes,* Vol 10, 3rd ed. Orlando, FL: Academic Press.

BRAY, D. 1995. Protein molecules as computational elements in living cells. *Nature* 376:307–12.

BRENNER, S. 1955. Tryptophan biosynthesis in *Salmonella typhimurium*. *Proc. Natl. Acad. Sci. USA* 41:862–63.

BYERS, P.H. 1989. Inherited disorders of collagen gene structure and expression. *Am. J. Med. Genet.* 34:72–80.

CHAPEVILLE, F. et al. 1962. On the role of soluble ribonucleic acid in coding for amino acids. *Proc. Natl. Acad. Sci. USA* 48:1086–93.

CIGAN, A.M., FENG, L., and DONAHUE, T.F. 1988. tRNA^{met} functions in directing the scanning ribosome to the start site of translation. *Science* 242:93–98.

DAHLBERG, A.E. 1989. The functional role of ribosomal RNA in protein synthesis. *Cell* 57:525–29.

DICKERSON, R.E. 1964. X-ray analysis and protein structure. In *The proteins*, ed. H. Neurath, Vol. 2, 2nd ed. Orlando, FL: Academic Press.

DICKERSON, R.E., and GEIS, I. 1983. *Hemoglobin: Structure, function, evolution, and pathology*. Menlo Park, CA: Benjamin/Cummings.

DOOLITTLE, R.F. 1985. Proteins. *Sci. Am.* (Oct.) 253:88–99.

EZZELL, C. 1994. Evolutions: Molecular chaperones and protein folding. *J. NIH Res.* 6:103.

EPHRUSSI, B. 1942. Chemistry of eye color hormones of *Drosophila*. *Quart. Rev. Biol.* 17:327–38.

FRANK, J. 1998. How the ribosome works. *Amer. Scient.* 86:428–39.

GARROD, A.E. 1902. The incidence of alkaptonuria: A study in chemical individuality. *Lancet* 2:1616–20.

———. 1909. *Inborn errors of metabolism*. London: Oxford University Press. (Reprinted 1963, Oxford University Press, London.)

GARROD, S.C. 1989. Family influences on A. E. Garrod's thinking. *J. Inher. Metab. Dis.* 12:2–8.

HORTON, H.R. et al. 1996. *Principles of biochemistry*, 2nd ed. Upper Saddle River, NJ: Prentice-Hall.

INGRAM, V.M. 1957. Gene mutations in human hemoglobin: The chemical difference between normal and sickle cell hemoglobin. *Nature* 180:326–28.

———. 1963. *The hemoglobins in genetics and evolution*. New York: Columbia University Press.

KOSHLAND, D.E. 1973. Protein shape and control. *Sci. Am.* (Oct.) 229:52–64.

LaDU, B.N., ZANNONI, V.G., LASTER, L., and SEEGMILLER, J.E. 1958. The nature of the defect in tyrosine metabolism in alkaptonuria. *J. Biol. Chem.* 230:251.

LAKE, J.A. 1981. The ribosome. *Sci. Am.* (Aug.) 245:84–97.

LEHNINGER, A.L., NELSON, D.L., and COX, M.M. 1993. *Principles of biochemistry*, 2nd ed. New York: Worth Publishers.

LODMELL, J.S., and DAHLBERG, A.E. 1997. A conformational switch in *Eschericia coli* 16S ribosomal RNA during decoding of messenger RNA. *Science* 277:1262–65.

MANIATIS, T. et al. 1980. The molecular genetics of human hemoglobins. *Annu. Rev. Genet.* 14:145–78.

MOORE, P.B. 1988. The ribosome returns. *Nature* 331:223–27.

MURAYAMA, M. 1966. Molecular mechanism of red cell sickling. *Science* 153:145–49.

NEEL, J.V. 1949. The inheritance of sickle-cell anemia. *Science* 110:64–66.

NIRENBERG, M.W., and LEDER, P. 1964. RNA codewords and protein synthesis. *Science* 145:1399–1407.

NOLLER, H.F. 1973. Assembly of bacterial ribosomes. *Science* 179:864–73.

———. 1984. Structure of ribosomal RNA. *Annu. Rev. Biochem.* 53:119–62.

NOMURA, M. 1984. The control of ribosome synthesis. *Sci. Am.* (Jan.) 250:102–14.

PAULING, L., ITANO, H.A., SINGER, S.J., and WELLS, I.C. 1949. Sickle cell anemia: A molecular disease. *Science* 110:543–48.

PFEFFER, S.R., and ROTHMAN, J.E. 1987. Biosynthetic protein transport and sorting by the endoplasmic reticulum and golgi. *Annu. Rev. Biochem.* 56:829–52.

PORCE, B.T., and GARRETT, R.A. 1999. Ribosomal mechanics, antibiotics, and GTP hydrolysis. *Cell* 97:423–26.

PROCKOP, D.J., and KIVIRIKKO, K.I. 1995. Collagens: Molecular biology, diseases, and potentials for therapy. *Annu. Rev. Biochem.* 64:403–34.

RICH, A., and HOUKIM, S. 1978. The three-dimensional structure of transfer RNA. *Sci. Am.* (Jan.) 238:52–62.

RICH, A., WARNER, J.R., and GOODMAN, H.M. 1963. The structure and function of polyribosomes. *Cold Spring Harbor Symp. Quant. Biol.* 28:269–85.

RICHARDS, F.M. 1991. The protein folding problem. *Sci. Am.* (Jan.) 264:54–63.

ROSS, J. 1989. The turnover of mRNA. *Sci. Am.* (Apr.) 260:48–55.

ROULD, M.A. et al. 1989. Structure of *E. coli* glutaminyl-tRNA synthetase complexed with tRNA^gln and ATP at 2.8 Å resolution. *Science* 246:1135–42.

SAKS, M.E., SAMPSON, J.R., and ABELSON, J.N. 1994. The transfer RNA identity problem: A search for rules. *Science* 263:191–97.

———. 1998. Evolution of a transfer RNA gene through a point mutation in the anticodon. *Science* 279:1665–70.

SCOTT-MONCRIEFF, R. 1936. A biochemical survey of some Mendelian factors for flower colour. *J. Genet.* 32:117–70.

SCRIVER, C.R., and CLOW, C.L. 1980. Phenylketonuria and other phenylalanine hydroxylation mutants in man. *Annu. Rev. Genet.* 14:179–202.

SHARP, P.A., and EISENBERG, D. 1987. The evolution of catalytic function. *Science* 238:729–30.

SMITH, J.D. 1972. Genetics of tRNA. *Annu. Rev. Genet.* 6:235–56.

SRB, A.M., and HOROWITZ, N.H. 1944. The ornithine cycle in *Neurospora* and its genetic control. *J. Biol. Chem.* 154:129–39.

UCHINO, T. et al. 1995. Molecular basis of phenotypic variation in patients with argininemia. *Hum. Genet.* 96:255–60.

WAGNER, R.P., and MITCHELL, H.K. 1964. *Genetics and metabolism*, 2nd ed. New York: John Wiley.

WARNER, J., and RICH, A. 1964. The number of soluble RNA molecules on reticulocyte polyribosomes. *Proc. Natl. Acad. Sci. USA* 51:1134–41.

WITTMAN, H.G. 1983. Architecture of prokaryotic ribosomes. *Annu. Rev. Biochem.* 52:35–65.

YANOFSKY, C., DRAPEAU, G., GUEST, J., and CARLTON, B. 1967. The complete amino acid sequence of the tryptophan synthetase A protein and its colinear relationship with the genetic map of the *A* gene. *Proc. Natl. Acad. Sci. USA* 57:296–98.

ZIEGLER, I. 1961. Genetic aspects of ommochrome and pterin pigments. *Adv. Genet.* 10:349–403.

ZUBAY, G.L., and MARMUR, J., eds. 1973. *Papers in biochemical genetics*, 2nd ed. New York: Holt, Rinehart & Winston.

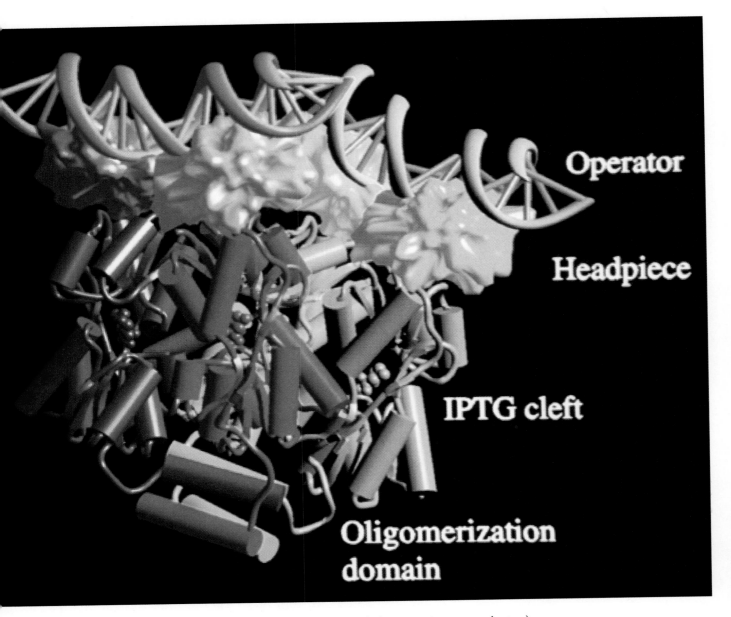

A model of the operator DNA sequences (the red and blue helices running across the top) in close contact with the lac repressor subunits.

CHAPTER 15

Regulation of Gene Expression in Prokaryotes

KEY CONCEPTS

- **Prokaryotes exhibit efficient genetic mechanisms to respond to environmental conditions**

- **Lactose metabolism in *E. coli* is regulated by an inducible system**

  Structural Genes
  The Discovery of Regulatory Mutations
  The Operon Model: Negative Control
  Genetic Proof of the Operon Model
  Isolation of the Repressor

- **The catabolite-activating protein (CAP) exerts positive control over the *lac* operon**

- **Crystal structure analysis of repressor complexes has confirmed the operon model**

- **The *tryptophan* (*trp*) operon in *E. coli* is a repressible gene system**

  Evidence for the *trp* Operon

- **Attenuation is a critical process during the regulation of the *trp* operon**

- **The *ara* operon is controlled by a regulator protein that exerts both positive and negative control**

GenCDX

When you see this icon, there are related animations and exercises on the CD accompanying this text.

I n earlier chapters we established how DNA is organized into genes, how genes store genetic information, and how this information is expressed. We now consider one of the most fundamental issues in molecular genetics: *How is genetic expression regulated?* Evidence in support of the idea that genes can be turned on and off is very convincing. Detailed analysis of proteins in *Escherichia coli*, for example, has shown that concentrations of the 4000 or so polypeptide chains encoded by the genome vary widely. Some proteins may be present in as few as 5 to 10 molecules per cell, whereas others, such as ribosomal proteins and the many proteins involved in the glycolytic pathway, are present in as many as 100,000 copies per cell. Though in prokaryotes most gene products exist continuously at a basal level (a few copies), the level can be increased dramatically. Clearly, fundamental regulatory mechanisms must exist to control the expression of the genetic information.

In this chapter, we will explore some of what is known about the regulation of genetic expression in bacteria. These organisms served as excellent research models during the seminal investigations involving the induction of genetic transcription in response to changes in environmental conditions. Our focus will be on regulation at the level of the gene. Keep in mind that posttranscriptional

regulation also occurs in bacteria. However, we will defer discussion of this level of regulation to the following chapter when we consider eukaryotes.

Prokaryotes exhibit efficient genetic mechanisms to respond to environmental conditions

The regulation of gene expression has been extensively studied in prokaryotes, particularly in *E. coli*. We have learned that highly efficient genetic mechanisms have evolved to turn genes on and off, depending on the cell's metabolic need for the respective gene products. Not only do bacteria respond to changes in their environment, but they also regulate gene activity involved in a variety of normal cellular responses, including the replication, recombination, and repair of their DNA, and in cell division and development.

The idea that microorganisms regulate the synthesis of gene products is not a new one. As early as 1900 it was shown that when lactose (a galactose-glucose–containing disaccharide) is present in the growth medium of yeast, the organisms produce enzymes specific to lactose metabolism. When lactose is absent, the enzymes are not manufactured.

Soon thereafter, investigators were able to generalize that bacteria "adapt" to their chemical environment, producing certain enzymes only when specific substrates are present. Such enzymes were thus referred to as **adaptive** (sometimes also called facultative). In contrast, enzymes that are produced continuously, regardless of the chemical makeup of the environment, were called **constitutive.** Since then, the term *adaptive* has been replaced with the more accurate term **inducible,** reflecting the role of the substrate, which serves as the **inducer** in enzyme production.

More recent investigation has revealed a contrasting system, whereby the presence of a specific molecule inhibits genetic expression. This is usually true for molecules that are end products of anabolic biosynthetic pathways. For example, the amino acid tryptophan can be synthesized by bacterial cells. If a sufficient supply of tryptophan is present in the environment or culture medium, it is energetically inefficient for the organism to synthesize the enzymes necessary for tryptophan production. A mechanism has evolved whereby tryptophan plays a role in repressing transcription of RNA essential to the production of the appropriate biosynthetic enzymes. In contrast to the inducible system controlling lactose metabolism, the system governing tryptophan expression is said to be **repressible,** and the substrate of the enzyme is the **repressor.**

As we will soon see, regulation, whether it is inducible or repressible, may be under either **negative** or **positive control.** Under negative control, genetic expression occurs unless it is shut off by some form of a regulator. In contrast, under positive control, transcription occurs only if a regulator molecule directly stimulates RNA production. In theory, each type of control can govern inducible or repressible systems. Our discussion in the ensuing sections of this chapter will help clarify these contrasting systems of regulation.

Lactose metabolism in *E. coli* is regulated by an inducible system

In prokaryotes, structural genes tend to be organized in clusters controlled from a single regulatory site. The cluster of genes, collectively known as an **operon,** usually produces products with related functions—for example, enzymes involved in a single metabolic pathway. The regulatory site, linked to the gene cluster it controls, usually located upstream (at the 5′-end of the gene), is known as a *cis*-acting element. Interactions at such a regulatory site involve the binding of molecules that control transcription of the gene cluster. Such molecules are called ***trans*-acting elements**, because they are not physically linked to the genes they regulate. In this section, we will discuss how such a bacterial gene cluster is coordinately regulated. The operon involved in the metabolism of lactose in *E. coli* has been extensively investigated and today serves as the paradigm for the study of genetic regulation.

Beginning in 1946 with the studies of Jacques Monod and continuing through the next decade with significant contributions by Joshua Lederberg, François Jacob, and Andre L'woff, genetic and biochemical evidence was amassed. Insights were provided into the way in which the gene activity responsible for lactose metabolism is repressed when lactose is absent but induced when it is available. In the presence of lactose, the concentration of the enzymes responsible for its metabolism increases rapidly from a few molecules to thousands per cell. The enzymes responsible for lactose metabolism are thus *inducible*, and lactose serves as the *inducer.*

Paramount to the understanding of how gene expression is controlled in this system was the discovery of a regulatory gene and a regulatory site that are part of the gene cluster. Neither of these elements encodes enzymes necessary for lactose metabolism; that is the function of three other genes in the cluster. As illustrated in Figure 15.1, the three structural genes and the adjacent regulatory site constitute the **lactose,** or **lac, operon.** Together, the entire gene cluster functions in an integrated fashion to provide a rapid response to the presence or absence of lactose.

Structural Genes

Genes coding for the primary structure of the enzymes are called **structural genes.** In the lactose operon, the so-called *lacZ* gene specifies the amino acid sequence of the enzyme β-**galactosidase,** which converts lactose to glucose and

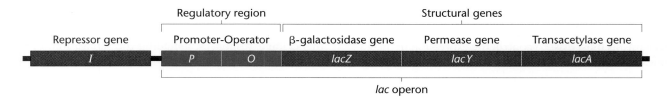

FIGURE 15.1 A simplified overview of the genes and regulatory units involved in the control of lactose metabolism (this region of DNA is not drawn to scale). A more detailed model will be developed later in this chapter (see Figure 15.10).

galactose (Figure 15.2). This conversion is essential if lactose is to serve as the primary energy source in glycolysis. The second gene, *lacY*, specifies the primary structure of **permease,** an enzyme that facilitates the entry of lactose into the bacterial cell. The third gene, *lacA*, codes for the enzyme **transacetylase.** While its physiological role is still not completely clear, there is some thought that it may be involved in the removal of toxic by-products of lactose digestion from the cell.

Studies of the genes coding for these three enzymes relied on the isolation of numerous mutations that eliminated the function of one or the other enzyme. Such *lac⁻* mutants were first isolated and studied by Joshua Lederberg. Mutant cells that fail to produce active β-galactosidase (*lacZ⁻*) or permease (*lacY⁻*) are unable to use lactose as an energy source. Mutations also were found in the transacetylase gene. Mapping studies by Lederberg established that all three genes are closely linked or contiguous to one another in the order Z–Y–A (see Figure 15.1).

Another observation is relevant to what became known about the structural genes. Knowledge of their close linkage led to the discovery that all three genes are transcribed as a single unit, resulting in a polycistronic mRNA (Figure 15.3). This results in the coordinate regulation of all three genes since a single message serves as the basis for translation of all three gene products.

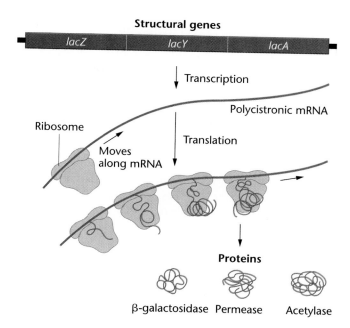

FIGURE 15.3 The structural genes of the *lac* operon are transcribed into a single polycistronic mRNA, which is translated simultaneously by several ribosomes into the three enzymes encoded by the operon.

The Discovery of Regulatory Mutations

How does lactose activate structural genes and induce the synthesis of the related enzymes? A partial answer comes from the discovery and study of **gratuitous inducers,** chemical analogues of lactose such as the sulfur analogue **isopropylthiogalactoside (IPTG),** shown in Figure 15.4. Gratuitous inducers behave like natural inducers, but they do not serve as substrates for the enzymes that are subsequently synthesized. Their discovery provides strong evidence that the primary induction event does *not* depend on the interaction between the inducer and the enzyme.

What, then, is the role of lactose in induction? The answer to this question required the study of another class of mutations called **constitutive mutants.** In this type of

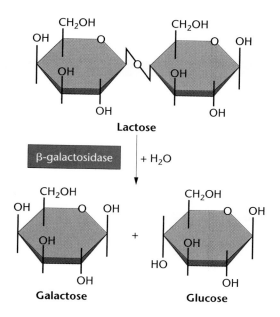

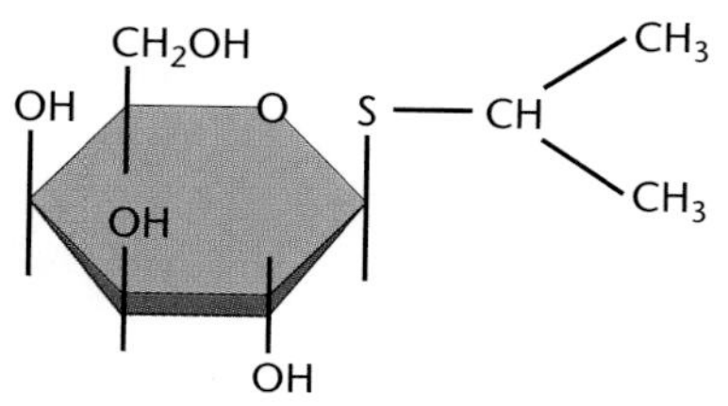

FIGURE 15.2 The catabolic conversion of the disaccharide lactose into its monosaccharide units, galactose and glucose.

FIGURE 15.4 The gratuitous inducer isopropylthiogalactoside (IPTG).

mutation, the enzymes are produced regardless of the presence or absence of lactose. Maps of the first type of constitutive mutation, $lacI^-$, showed that it is located at a site on the DNA close to, but distinct from, the structural genes. As we will soon see, the *lacI* gene is appropriately called a **repressor gene.** We will soon see why this name is appropriate. A second set of constitutive mutations producing identical effects was found in a region immediately adjacent to the structural genes. This class of mutations, designated $lacO^C$, identifies the **operator region** of the operon. Because inducibility has been eliminated in both types of constitutive mutation (the enzymes are continually produced), clearly regulation has been disrupted by genetic changes.

The Operon Model: Negative Control

Around 1960, Jacob and Monod proposed a scheme involving negative control called the **operon model,** where-by a group of genes is regulated and expressed together as a unit. As we saw in Figure 15.1, the *lac* operon they proposed consists of the *Z, Y,* and *A* structural genes as well as the adjacent sequences of DNA referred to as the operator region. They argued that the *lacI* gene regulates the transcription of the structural genes by producing a **repressor molecule,** and that the repressor is **allosteric,** meaning that the molecule reversibly interacts with another molecule, causing both a conformational change in three-dimensional shape and a change in chemical activity. Figure 15.5 illustrates the components of the *lac* operon as well as the action of the *lac* repressor in the presence and absence of lactose. You should refer to it as you read the following section.

Jacob and Monod suggested that the repressor normally interacts with the DNA sequence of the operator region. When it does so, it inhibits the action of RNA polymerase, effectively repressing the transcription of the

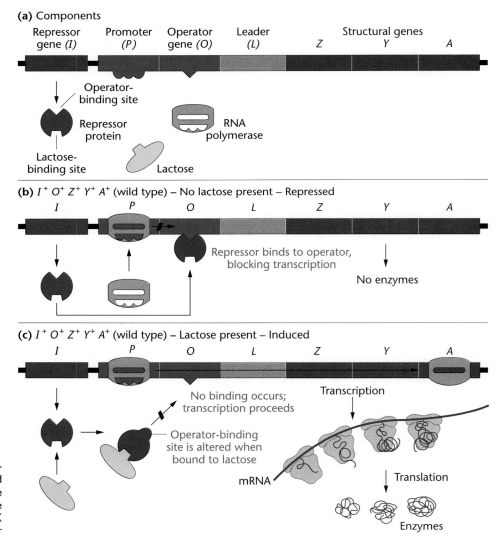

FIGURE 15.5 The components of the wild-type *lac* operon and the response in the absence and the presence of lactose, as described in the text. ◉GenCDX

(a) Components

Repressor gene *(I)* — Promoter *(P)* — Operator gene *(O)* — Leader *(L)* — Structural genes *Z   Y   A*

Operator-binding site

Repressor protein

RNA polymerase

Lactose-binding site

Lactose

(b) $I^+ O^+ Z^+ Y^+ A^+$ (wild type) – No lactose present – Repressed

I   P   O   L   Z   Y   A

Repressor binds to operator, blocking transcription

No enzymes

(c) $I^+ O^+ Z^+ Y^+ A^+$ (wild type) – Lactose present – Induced

I   P   O   L   Z   Y   A

No binding occurs; transcription proceeds

Operator-binding site is altered when bound to lactose

Transcription

mRNA

Translation

Enzymes

structural genes [Figure 15.5(b)]. However, when lactose is present, it binds to the repressor and causes an allosteric conformational change. This change alters the binding site of the repressor, rendering it incapable of interacting with operator DNA [Figure 15.5(c)]. In the absence of the repressor–operator interaction, RNA polymerase transcribes the structural genes, and the enzymes necessary for lactose metabolism are produced. Since transcription occurs only when the repressor *fails* to bind to the operator region, *negative control* is exerted.

The operon model uses these potential molecular interactions to explain the efficient regulation of the structural genes. In the absence of lactose, the enzymes encoded by the genes are not needed and are repressed. When lactose is present, it indirectly induces the activation of the genes by binding with the repressor.* If all lactose is metabolized, none is available to bind to the repressor, which is again free to bind to operator DNA and repress transcription.

---

*Technically, allolactose, an isomer of lactose, is the inducer. Allolactose is produced during the initial step in the metabolism of lactose by β-galactosidase.

Both the I^- and O^C constitutive mutations interfere with these molecular interactions, allowing continuous transcription of the structural genes. In the case of the I^- mutant, seen in Figure 15.6(a), the repressor protein is altered and cannot bind to the operator region, so the structural genes are always turned on. In the case of the O^C mutant [Figure 15.6(b)], the nucleotide sequence of the operator DNA is altered and will not bind with a normal repressor molecule. The result is the same: Structural genes are always transcribed.

Genetic Proof of the Operon Model

The operon model is a good one because it leads to three major predictions that can be tested to determine its validity. The major predictions to be tested are that (1) the I gene produces a diffusible cellular product, (2) the O region is involved in regulation but does not produce a product, and (3) the O region must be adjacent to the structural genes in order to regulate transcription.

The construction of partially diploid bacteria allows us to assess these assumptions, particularly those that predict *trans*-acting regulatory elements. For example, as introduced in Chapter 7, the F plasmid may contain chromo-

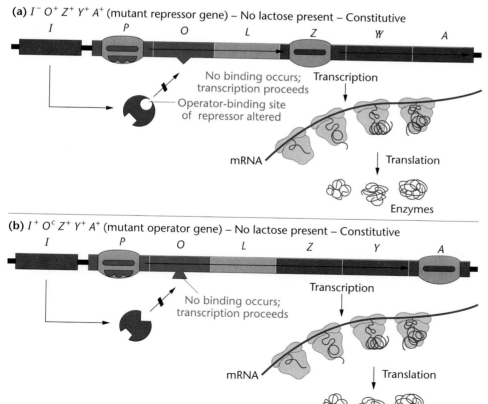

(a) $I^- O^+ Z^+ Y^+ A^+$ (mutant repressor gene) – No lactose present – Constitutive

No binding occurs; transcription proceeds
Operator-binding site of repressor altered
Transcription
mRNA
Translation
Enzymes

(b) $I^+ O^C Z^+ Y^+ A^+$ (mutant operator gene) – No lactose present – Constitutive

No binding occurs; transcription proceeds
Transcription
mRNA
Translation
Enzymes

FIGURE 15.6 The response of the *lac* operon in the absence of lactose when a cell bears either the I^- or the O^C mutation.

somal genes, in which case it is designated F′. When an F⁻ cell acquires such a plasmid, it now contains its own chromosome plus one or more additional genes present in the plasmid, creating a host cell, called a **merozygote,** that is diploid for those genes. The use of such a plasmid makes it possible, for example, to introduce an I^+ gene into a host cell whose gentoype is I^-, or to introduce an O^+ gene into a host cell of genotype O^C. The Jacob-Monod operon model predicts how regulation should be affected in such cells. Adding an I^+ gene to an I^- cell should restore inducibility, because a normal repressor, which is a *trans*-acting factor, would again be produced. Adding an O^+ gene to an O^C cell should have no effect on constitutive enzyme production, since regulation depends on an O^+ gene immediately adjacent to the structural genes, that is, O^+ is a *cis*-acting regulator.

Results of these experiments are shown in Table 15.1, where Z represents the structural genes. The inserted genes are listed after the designation F′. In both cases described above, the Jacob-Monod model is upheld (part B of Table 15.1). Part C shows the reverse experiments, where either an I^- gene or an O^C region is added to cells of normal inducible genotypes. As the model predicts, inducibility is maintained in these partial diploids.

Another prediction of the operon model is that certain mutations in the I gene should have the opposite effect of I^-. That is, instead of being constitutive by failing to interact with the operator, mutant repressor molecules should be produced that cannot interact with the inducer, lactose. As a result, the repressor would always bind to the operator sequence, and the structural genes would be permanently repressed (Figure 15.7). If this were the case, the presence of an additional I^+ gene would have little or no effect on repression.

In fact, such a mutation, I^S, was discovered where the operon is "superrepressed," as shown in part D of Table 15.1. An additional I^+ gene does not effectively relieve repression of gene activity. These observations again support the operon model for gene regulation.

TABLE 15.1 A Comparison of Gene Activity (+ or −) in the Presence or Absence of Lactose for Various *E. coli* Genotypes

	Genotype	Presence of β-Galactosidase Activity	
		Lactose Present	Lactose Absent
A.	$I^+O^+Z^+$	+	−
	$I^+O^+Z^-$	−	−
	$I^-O^+Z^+$	+	+
	$I^+O^CZ^+$	+	+
B.	$I^-O^+Z^+/F'I^+$	+	−
	$I^+O^CZ^+/F'O^+$	+	+
C.	$I^+O^{Z+}/F'I^-$	+	−
	$I^+O^+Z^+/F'O^C$	+	−
D.	$I^SO^+Z^+$	−	−
	$I^SO^+Z^+/F'I^+$	−	−

Note: In parts B to D, most genotypes are partially diploid, containing an F factor plus attached genes (F′).

Isolation of the Repressor

Although the operon theory of Jacob and Monod succeeded in explaining many aspects of genetic regulation in prokaryotes, the nature of the repressor molecule was not known when their landmark paper was published in 1961. While they had assumed that the allosteric repressor was a protein, RNA was also a candidate because activity of the molecule required the ability to bind to DNA. Despite many attempts to isolate and characterize the hypothetical repressor molecule, no direct chemical evidence was immediately forthcoming. A single *E. coli* cell contains no more than 10 or so copies of the *lac* repressor; direct chemical identification of 10 molecules in a population of millions of proteins and RNAs in a single cell presented a tremendous challenge.

In 1966, Walter Gilbert and Benno Müller-Hill reported the isolation of the *lac* repressor in partially puri-

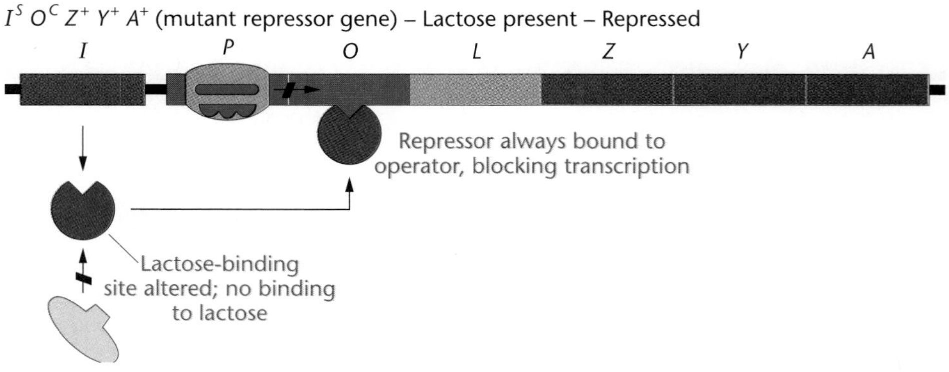

$I^S O^C Z^+ Y^+ A^+$ (mutant repressor gene) – Lactose present – Repressed

Repressor always bound to operator, blocking transcription

Lactose-binding site altered; no binding to lactose

FIGURE 15.7 The response of the *lac* operon in the presence of lactose in a cell bearing the I^S mutation.

fied form. To achieve the isolation, they used a *regulator quantity* (I^q) mutant strain that contains about 10 times as much repressor as do wild-type *E. coli* cells. Also instrumental in their success were the use of the gratuitous inducer, IPTG, which binds to the repressor, and the technique of **equilibrium dialysis.** In this technique, extracts of I^q cells were placed in a dialysis bag and allowed to attain equilibrium with an external solution of radioactive IPTG, which is small enough to diffuse freely in and out of the bag. At equilibrium, the concentration of IPTG was higher inside the bag than in the external solution, indicating that an IPTG-binding material was present in the cell extract and that this material was too large to diffuse across the wall of the bag.

Ultimately the IPTG-binding material was purified and shown to have various characteristics of a protein. In contrast, extracts of I^- constitutive cells having no *lac* repressor activity did not exhibit IPTG-binding activity, strongly suggesting that the isolated protein was the repressor molecule.

To confirm this thinking, Gilbert and Müller-Hill grew *E. coli* cells in a medium containing radioactive sulfur and then isolated the IPTG-binding protein, which was labeled in its sulfur-containing amino acids. This protein was mixed with DNA from a strain of phage lambda (λ) which carries the *lacO*$^+$ gene. The DNA sediments at 40*S*, while the IPTG-binding protein sediments at 7*S*. The DNA and protein were mixed and sedimented in a gradient using ultracentrifugation. The radioactive protein sedimented at the same rate as DNA, indicating that the protein binds to the DNA. Further experiments showed that the IPTG-binding or repressor protein binds only to DNA containing the *lac* region and does not bind to *lac* DNA containing an operator-constitutive (O^C) mutation.

The catabolite-activating protein (CAP) exerts positive control over the *lac* operon

As is apparent from the above discussion of the *lac* operon, the role of β-galactosidase is to cleave lactose into its components glucose and galactose. Then, for galactose to be used by the cell, it is converted to glucose. What if the cell found itself in an environment that contained an ample amount of lactose *and* glucose? It would not be energetically efficient for cell to be "induced" by lactose to make β-galactosidase, since what it really needs, glucose, is already present. As we shall see below, still another molecular component, called the **catabolite-activating protein (CAP),** is involved in effectively repressing the expression of the *lac* operon when glucose is present. This inhibition,

called **catabolite repression,** reflects the greater simplicity with which glucose may be metabolized in comparison to lactose. The cell "prefers" glucose, and if it is available, the *lac* operon is not activated, even when lactose is present.

To understand CAP and its role in regulation, let's backtrack for a moment. When the *lac* repressor is bound to the inducer, the *lac* operon is activated and RNA polymerase transcribes the structural genes. As we have learned in Chapter 13, transcription is initiated as a result of the binding that occurs between RNA polymerase and the nucleotide sequence of the **promoter region,** found upstream (5') from the initial coding sequences. Within the *lac* operon, the promoter is found between the *I* gene and the operator region (O^C) (see Figure 15.1). Careful examination has revealed that polymerase binding is never very efficient unless CAP is also present to facilitate the process.

The mechanism is summarized in Figure 15.8. In the absence of glucose, and under inducible conditions, CAP exerts **positive control** by binding to the CAP site, facilitating RNA polymerase binding at the promoter, and thus transcription. Therefore, for maximal transcription, the repressor must be bound by lactose (so as not to repress operon expression), *and* CAP must be bound to the CAP-binding site.

This leads to the central question about CAP. What role does glucose play in inhibiting CAP binding when it is present? The answer involves still another molecule, **cyclic adenosine monophosphate (cAMP),** upon which CAP binding is dependent. In order to bind to the promoter, CAP must be linked to cAMP. The level of cAMP is itself dependent on an enzyme **adenyl cyclase,** which catalyzes the conversion of ATP to cAMP* (Figure 15.9).

The role of glucose in catabolite repression is now clear. It inhibits the activity of adenyl cyclase, causing a decline in the level of cAMP in the cell. Under this condition, CAP cannot form the CAP–cAMP complex essential to the positive control of transcription.

Like the *lac* repressor, CAP and cAMP–CAP have been examined using X-ray crystallography. CAP is a dimer that inserts into adjacent regions of a specific nucleotide sequence of the DNA making up the promoter. The cAMP–CAP complex, when bound to DNA, bends it, causing it to assume a new conformation.

Binding studies in solution further clarify the mechanism of gene activation. Alone, neither cAMP–CAP nor RNA polymerase has a strong affinity to bind to *lac* pro-

---

*Because of its involvement with cAMP, CAP is also called cyclic AMP receptor protein (CRP), and the gene encoding the protein is named *crp.* Because the protein was first named CAP, we will adhere to the initial nomenclature.

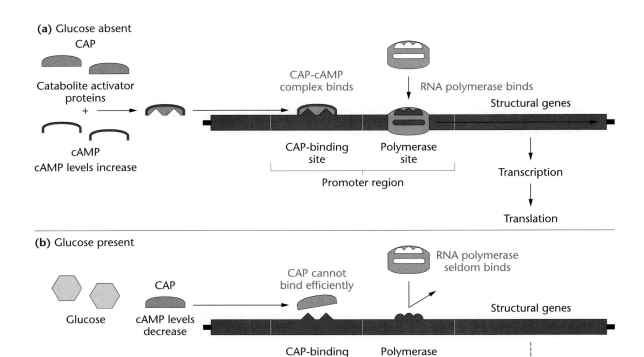

(a) Glucose absent

CAP

Catabolite activator proteins
+
cAMP
cAMP levels increase

CAP-cAMP complex binds

RNA polymerase binds

CAP-binding site
Polymerase site
Promoter region

Structural genes

Transcription

Translation

(b) Glucose present

Glucose

CAP
cAMP levels decrease

CAP cannot bind efficiently

RNA polymerase seldom binds

CAP-binding site
Polymerase site
Promoter region

Structural genes

Transcription diminished

Translation diminished

FIGURE 15.8 Catabolite repression. (a) In the absence of glucose, cAMP levels increase, resulting in the formation of a cAMP–CAP complex, which binds to the CAP site of the promoter, stimulating transcription. (b) In the presence of glucose, cAMP levels decrease, cAMP–CAP complexes are not formed, and transcription is not stimulated.

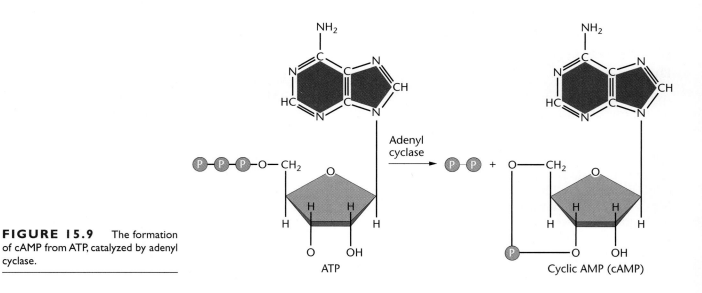

FIGURE 15.9 The formation of cAMP from ATP, catalyzed by adenyl cyclase.

Adenyl cyclase

ATP

Cyclic AMP (cAMP)

moter DNA. Nor does either molecule have a strong affinity to bind to the other. However, when both are together in the presence of the *lac* promoter DNA, a tightly bound complex is formed, an example of what, in biochemical terms, is called **cooperative binding.** In the case of cAMP–CAP and the *lac* operon, the phenomenon illustrates the high degree of specificity that is involved in the genetic regulation of just one small group of genes.

Regulation of the *lac* operon by catabolite repression results in efficient energy use because the presence of glucose will override the need for the metabolism of lactose, should it also be available to the cell. Catabolite repression involving CAP has also been observed for other inducible operons, including those controlling the metabolism of galactose and arabinose.

Crystal structure analysis of repressor complexes has confirmed the operon model

We now have available thorough knowledge of the biochemical nature of the regulatory region of the *lac* operon, identifying the precise locations of its various components relative to one another (Figure 15.10). Just a few years ago, in 1996, Mitchell Lewis, Ponzy Lu, and their colleagues succeeded in determining the crystal structure of the *lac* repressor as well as the structure of the repressor bound to the inducer and to operator DNA. As a result, previous information that was based on genetic and biochemical data has now been complemented with the missing structural interpretation. Together, a nearly complete picture of the regulation of the operon has emerged.

The repressor, as the gene product of the *I* gene, is a monomer consisting of 360 amino acids. Within the monomer, the region of inducer binding has been identified [Figure 15.11(a)]. The functional repressor contains four such subunits, creating a homotetramer. The tetramer can be cleaved with a protease under controlled conditions and yields five fragments. Four are derived from the N-terminal ends of the tetramer, and they bind

to operator DNA. The fifth fragment is the remaining core of the tetramer, derived from the COOH-terminus ends; it binds to lactose and gratuitous inducers such as IPTG. Analysis has revealed that, at any single time, each tetramer can bind to two symmetrical operator DNA helices [Figure 15.11(b)].

The operator DNA that was previously defined by mutational studies (*lacO^C*) and confirmed by DNA-sequencing analysis is located just downstream from the start of the *lacZ* gene, but upstream from the beginning of the actual coding sequence. The crystallographic studies show that the actual region of repressor binding of this primary operator (O_1) consists of 21 base pairs. Two other auxiliary operator regions have been identified, as shown in Figure 15.10. One, O_2, is 401 base pairs downstream from the primary operator, within the *lacZ* gene. The other, O_3 is 93 base pairs upstream from O_1, just above the CAP site. *In vivo*, all three operators must be bound for maximum repression.

Binding by the repressor at two operator sites distorts the conformation of DNA, causing it to bend away from the repressor. When a model is created involving dual binding of operators O_1 and O_3 [Figure 15.11(c)], the 93 base pairs of DNA that intervene must jut out, forming what is called a **repression loop.** This model positions the promoter region that binds RNA polymerase on the inside of the loop, which prevents access during repression. In addition, the repression loop positions the CAP-binding site in a way to facilitate CAP interaction with RNA polymerase upon subsequent induction. The findings of DNA looping during repression is similar to transitions that are predicted to occur in eukaryotic systems (see Chapter 16).

Studies have also defined the three-dimensional conformational changes that accompany the allosteric transitions that occur during the interactions with the inducer molecules. Taken together, the crystallographic studies bring us to a new level of understanding of the regulatory process occurring within the *lac* operon, confirming the findings and predictions of Jacob and Monod in their model set forth over 40 years ago based strictly on genetic grounds.

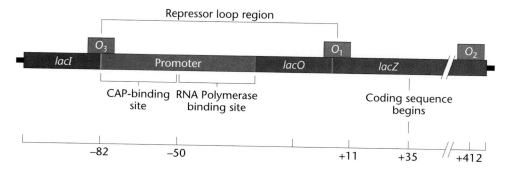

FIGURE 15.10 A detailed depiction of various regulatory regions involved in the control of genetic expression of the *lac* operon, as described in the text. The numbers on the bottom scale represent nucleotide sites upstream and downstream from the initiation of transcription.

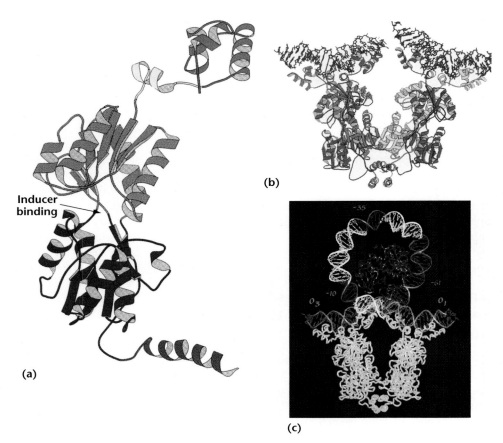

Inducer
binding

FIGURE 15.11 Models of the *lac* repressor and its binding to operator sites with DNA, as generated from crystal structure analysis: (a) the repressor monomer. The arrow points to the inducer binding site. The DNA-binding region is shown in red. (b) The repressor dimer bound to two 21 base pair segments of operator DNA (shown in blue). (c) The repressor and CAP (shown in dark blue) bound to the *lac* DNA. Binding to operator regions O_1 and O_3 creates a 93 base-pair repression loop of promoter DNA.

(a)

(b)

(c)

The *tryptophan* (*trp*) operon in *E. coli* is a repressible gene system

Although the process of induction had been known for some time, it was not until 1953 that Monod and colleagues discovered a repressible operon. Wild-type *E. coli* are capable of producing the enzymes necessary for the biosynthesis of amino acids as well as other essential macromolecules. Focusing his studies on the amino acid tryptophan and the enzyme **tryptophan synthetase,** Monod discovered that if tryptophan is present in sufficient quantity in the growth medium, the enzymes necessary for its synthesis are not produced. Energetically, repression of the genes involved in the production of these enzymes is highly economical for the cell when ample tryptophan is present.

Further investigation showed that a series of enzymes encoded by five contiguous genes on the *E. coli* chromosome is involved in tryptophan synthesis. These genes are part of an operon, and in the presence of tryptophan, all are coordinately repressed and none of the enzymes are produced. Because of the great similarity between this repression and the induction of enzymes for lactose metab-

olism, Jacob and Monod proposed a model of gene regulation analogous to the *lac* system (Figure 15.12).

To account for repression, they suggested the presence of a *normally inactive repressor* that alone cannot interact with the operator region of the operon. However, the repressor is an allosteric molecule that can bind to tryptophan. When this amino acid is present, the resultant complex of repressor and tryptophan attains a new conformation that binds to the operator, repressing transcription. Thus, when tryptophan, the end product of this anabolic pathway, is present, the system is repressed and enzymes are not made. Because the regulatory complex inhibits transcription of the operon, this repressible system is under negative control. And, as tryptophan participates in repression, it is referred to as a **co-repressor** in this regulatory scheme.

Evidence for the trp Operon

Support for the concept of a repressible operon was soon forthcoming, based primarily on the isolation of two distinct categories of constitutive mutations. The first class,

(a) Components

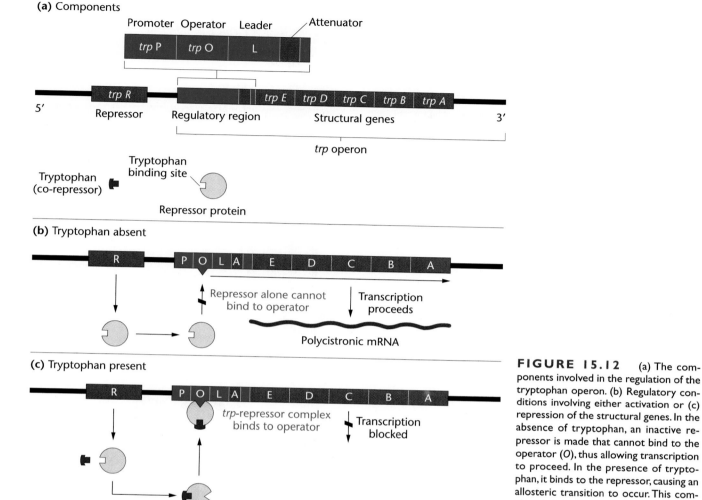

FIGURE 15.12 (a) The components involved in the regulation of the tryptophan operon. (b) Regulatory conditions involving either activation or (c) repression of the structural genes. In the absence of tryptophan, an inactive repressor is made that cannot bind to the operator (*O*), thus allowing transcription to proceed. In the presence of tryptophan, it binds to the repressor, causing an allosteric transition to occur. This complex binds to the operator region, leading to repression of the operon.

trpR⁻, maps at a considerable distance from the structural genes. This locus represents the gene coding for the repressor. Presumably, the mutation either inhibits the interaction of the repressor with tryptophan or inhibits repressor formation entirely. Whichever the case, no repression ever occurs in cells with the *trpR*⁻ mutation. As expected, if the *trpR* gene encodes a repressor molecule, the presence of an additional *trpR*⁺ gene restores repressibility.

The second constitutive mutant is analogous to that of the operator of the lactose operon because it maps immediately adjacent to the structural genes. Furthermore, the addition of a wild-type operator gene into mutant cells (as a *trans*-acting element) does not restore enzyme repression. This is predictable if the mutant operator can no longer interact with the repressor–tryptophan complex.

The entire *trp* operon has now been well defined, as shown in Figure 15.12. Five contiguous structural genes (*trp E, D, C, B,* and *A*) are transcribed as a polycistronic message that directs translation of the enzymes that catalyze the biosynthesis of tryptophan. As in the *lac* operon, a promoter region (*trpP*) represents the binding site for RNA polymerase, and an operator region (*trpO*) binds the repressor. In the absence of binding, transcription is initiated within the overlapping *trpP–trpO* region and proceeds along a **leader sequence** 162 nucleotides prior to the first structural gene (*trpE*). Within this leader sequence, still another regulatory site has been demonstrated, called an attenuator, the subject of the next section of this chapter. As we shall see, this regulatory unit is an integral part of the control mechanism of this operon.

(a) Selected regions of leader transcript and potential base pairing

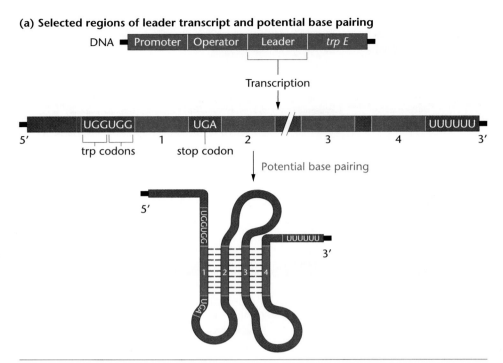

(b) High concentrations of tryptophan: Attenuation occurs

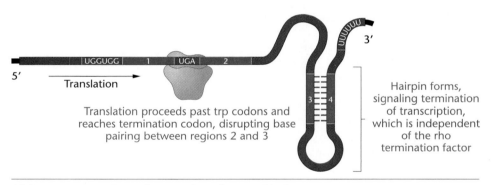

Translation proceeds past trp codons and reaches termination codon, disrupting base pairing between regions 2 and 3

Hairpin forms, signaling termination of transcription, which is independent of the rho termination factor

(c) Low concentrations of tryptophan: Attenuation is overcome

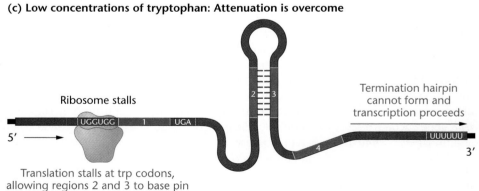

Ribosome stalls

Translation stalls at trp codons, allowing regions 2 and 3 to base pin

Termination hairpin cannot form and transcription proceeds

FIGURE 15.13 Diagram of the involvement of the leader sequence of the mRNA transcript of the *trp* operon of *E. coli* during attenuation. As is common in bacteria, translation by the ribosome proceeds prior to completion of transcription. When tryptophan is abundant, translation proceeds through region 1, whereby attenuation occurs as a result of the formation of a hairpin transcription termination loop. When tryptophan levels are low, translation stalls, the termination loop is not formed, and transcription proceeds, bypassing attenuation.

Attenuation is a critical process during the regulation of the *trp* operon

Charles Yanofsky, his coworker Kevin Bertrand, and their colleagues observed that, even when tryptophan is present and the *trp* operon is repressed, initiation of transcription of the leader sequence usually still occurs. Thus, while the activated repressor binds to the operator region, it does not strongly inhibit the *initial expression* of the operon, suggesting that there must be a subsequent mechanism by which tryptophan inhibits enzyme synthesis. Yanofsky discovered that, in the presence of high concentrations of tryptophan, transcription of the leader sequence of the operon proceeds, but that mRNA synthesis is usually terminated at a point about 140 nucleotides along the transcript. This process is called **attenuation***, since the effect is to severely diminish the genetic expression of the operon. However, when tryptophan is absent, or present in very low concentrations, the repressor is inactive and does not bind to the promoter. Transcription is initiated, but *not* subsequently terminated, instead proceeding through the leader sequence and into the structural genes. As a result, a polycistronic mRNA is transcribed and the enzymes essential to the biosynthesis of tryptophan are subsequently translated. The genetic components involved in the attenuation process are diagrammed in Figure 15.13(a).

Identification of the site involved in attenuation was made possible by the isolation of deletion mutations in the region 123 to 150 nucleotides into the leader sequence. Such mutations abolish attenuation. Thus the site is referred to as the **attenuator**. An explanation of how attenuation occurs and how it is overcome, put forward by Yanofsky and colleagues, is summarized in Figure 15.13(b). Present in the leader is a DNA sequence that, when transcribed, gives rise to an RNA molecule that immediately folds back on itself and forms a **hairpin loop**. The loop region is followed by a polyU sequence, typical of the sequence found in the 3′ termination region of many prokaryotic RNA transcripts. Thus, regardless of whether the repressor is bound, transcription still begins. However, in the presence of tryptophan, the process is almost always terminated prematurely. It should be noted that in contrast to termination of most regular transcripts, attenuation does not require the rho (ρ) termination factor. The attenuator is thus described as being rho-independent.

The question remains as to how the absence (or a low concentration) of tryptophan allows attenuation to be bypassed somehow. The key to Yanofsky's model is that the leader transcript must be translated in order for the hair-

pin to form. He discovered that the leader transcript includes two triplets (UGG) encoding tryptophan. Even though this leader sequence is not part of the structural genes in the operon, following its transcription, an initial AUG sequence prompts translation by ribosomes. When adequate tryptophan is present, charged tRNA^{trp} is also present. As a result, translation proceeds, and the termination hairpin is formed [Figure 15.13(b)]. If cells are starved of tryptophan, charged tRNA^{trp} is unavailable, and the ribosome "stalls" during the translation of triplets calling for charged tRNA^{trp} [Figure 15.13(c)]. This event influences the secondary structure of the transcript such that the termination hairpin does not form. As a result, attenuation is overcome, and transcription and translation of the entire set of structural genes then proceed.

Details of the secondary structures of the transcript have now been worked out and described. They satisfy conditions leading either to continued transcription or termination (attenuation). Additionally, other mutations in the leader sequence have been isolated that are predicted to alter the secondary structure of the transcript and its impact on attenuation. In each case, the predicted result has been upheld. Yanofsky's model, while complex, has significantly extended our knowledge of genetic regulation. Furthermore, the experimental evidence supporting it represents an integrated approach involving genetic and biochemical analysis, which is becoming commonplace in molecular genetic research.

The phenomenon of attenuation appears to be a mechanism common to other bacterial operons that regulate the enzymes essential to the biosynthesis of amino acids. In addition to tryptophan, operons involved in threonine, histidine, leucine, and phenylalanine display attenuators in their leader sequences. As with the *trp* operon, each is known to contain multiple codons calling for the amino acid being regulated, which prompt "stalling" of translation if the appropriate amino acid is missing. For example, the leader sequence in the histidine operon codes for seven histidine residues in a row. The threonine operon leader sequence calls for eight threonine residues. As with tryptophan, when the amino acid is present, stalling does not occur, a hairpin structure is formed, and attenuation occurs.

The *ara* operon is controlled by a regulator protein that exerts both positive and negative control

We conclude this chapter with a brief discussion of the **arabinose (*ara*) operon** as studied in *E. coli*. This inducible operon is unique because the same regulatory protein is capable of exerting both positive or negative control, and as a result, either induces or represses gene expression.

---

*Attenuation is derived from the verb attenuate, meaning "to reduce in strength, weaken, or impair."

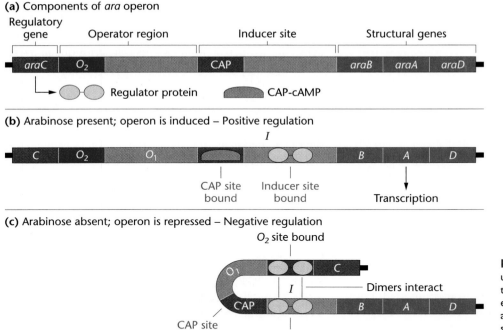

(a) Components of *ara* operon

Regulatory gene | Operator region | Inducer site | Structural genes

araC | O_2 | CAP | araB | araA | araD

Regulator protein   CAP-cAMP

(b) Arabinose present; operon is induced – Positive regulation

I

C | O_2 | O_1 | | | B | A | D

CAP site bound   Inducer site bound   Transcription

(c) Arabinose absent; operon is repressed – Negative regulation

O_2 site bound

O_1 C

I — Dimers interact

CAP   B | A | D

CAP site unbound   Inducer site bound   No transcription

FIGURE 15.14 Genetic regulation of the *ara* operon. The regulatory protein of the *araC* gene acts as either an inducer (in the presence of arabinose) or a repressor (in the absence of arabinose).

The various panels in Figure 15.14 accompany the following description of the conditions where the operon is defined and shown to be either active or inactive.

The metabolism of the sugar arabinose is under the direction of the enzymatic products of three structural genes, *ara B*, *A*, and *D*. Their transcription is controlled by the regulatory protein AraC encoded by the *ara C* gene that interacts with two regulatory regions, *araI* and *araO₂** [Figure 15.14(a)]. These sites can be bound individually or coordinately by the AraC protein.

The *I* region bears that designation since, when only it is bound by AraC, the system is induced [Figure 15.14(b)]. For this to occur, both arabinose and cAMP must be present. Thus, like induction of the *lac* operon, a CAP-binding site is present in the promoter region that modulates catabolite repression in the presence of glucose.

In the absence of both arabinose and cAMP, the AraC protein *also* binds to the O_2 site (so named because it was the second operator region to be discovered in this operon). When both *I* and O_2 are bound by AraC, a conformational change occurs in the DNA whereby a tight loop is formed [Figure 14.11(c)] and the structural genes are repressed.

The O_2 region is found about 200 nucleotides upstream from the *I* region. Binding at either regulatory region involves a dimer of AraC, and when both regions are bound, the dimers interact leading to the loop that causes repression. Presumably, this loop inhibits the access of RNA polymerase to the promoter region. The region of DNA located between *I* and O_2, which loops out, is critical to the formation of the repression complex. Genetic alteration through insertion or deletion of even a few nucleotides is sufficient to interfere with loop formation and therefore repression.

The above findings involving the *ara* operon serve to illustrate the degree of complexity that exists in the regulation of a group of related genes. As we alluded to at the beginning of this chapter, the development of regulatory mechanisms such as these has provided evolutionary advantages to bacterial systems that allow them to adjust to a variety of natural environments. Without question, they are keenly equipped genetically to not only survive under varying physiological conditions, but they do so with great biochemical efficiency.

*An additional operator region (O_1) is also present, but is not involved in the regulation of the *ara* structural genes.

CHAPTER SUMMARY

1. A system of genetic regulation must exist if the complete genome is not to be continuously active in transcription throughout the life of every cell. Highly refined mechanisms have evolved that regulate transcription, maintaining genetic efficiency.

2. Genetic analysis of bacteria has been pursued successfully since the 1940s. The ease of obtaining large quantities of pure cultures of mutant strains of bacteria as experimental material has made bacteria an organism of choice in numerous types of genetic studies.

3. The discovery and study of the *lac* operon in *E. coli* pioneered the study of gene regulation in bacteria. Genes involved in the metabolism of lactose are coordinately regulated by a negative control system, whereby gene expression is induced by lactose. In its absence, the system is shut down.

4. The catabolite-activating protein (CAP) is essential to the binding of RNA polymerase to the promoter and subsequent transcription in the *lac* operon (as well as other operons). Therefore, CAP exerts positive control over gene expression. For CAP to stimulate the *lac* operon, it must be bound to cyclic AMP.

5. When glucose is present, cyclic AMP levels are low, CAP binding does not occur, and the structural genes are not expressed. This phenomenon is called catabolite repression. This mechanism is efficient energetically since glucose is preferred, even in the presence of lactose.

6. The *lac* repressor has been isolated and studied. Crystal structure analysis has demonstrated how it interacts with the DNA of the operon as well as with inducers. These studies have demonstrated conformational changes in DNA leading to the formation of a repression loop that inhibits binding between RNA polymerase and the promoter region of the operon.

7. The biosynthesis of tryptophan in *E. coli* involves a number of enzymes whose presence or absence is controlled by another distinct operon. In contrast to the inducible operon controlling lactose metabolism, the *trp* operon is repressible. In the presence of tryptophan, an active repressor is made that shuts off the expression of the structural genes. Like the *lac* operon, the *trp* operon functions under negative control.

8. An additional regulatory step, referred to as attenuation, has been studied in the *trp* operon. Essential to repression of structural gene expression, transcription is initiated, but interrupted in the presence of tryptophan.

9. The *ara* operon is unique in that the regulator protein exerts both positive and negative control over expression of the genes specifying the enzymes that metabolize the sugar arabinose.

INSIGHTS AND SOLUTIONS

1. A theoretical operon (*theo*) in *E. coli* contains several structural genes encoding enzymes that are involved sequentially in the biosynthesis of an amino acid. Unlike the *lac* operon, where the repressor gene is separate from the operon, the gene encoding the regulator molecule is contained within the *theo* operon. When the end product (the amino acid) is present, it combines with the regulator molecule, and this complex binds to the operator, repressing the operon. In the absence of the amino acid, the regulatory molecule fails to bind to the operator, and transcription proceeds.

   Characterize this operon; then consider the following mutations as well as the situation in which the wild-type gene is present along with the mutant gene in partially diploid cells (F′). In each case, will the operon be active or inactive in transcription, assuming that the mutation affects the regulation of the *theo* operon? Compare each response to the equivalent situation of the *lac* operon.

   (a) Mutation in the operator gene.

   (b) Mutation in the promoter region.

   (c) Mutation in the regulator gene.

   Solution: The *theo* operon is repressible and under negative control. When there is no amino acid present in the medium (or the environment), the product of the regulatory gene cannot bind to the operator region, and transcription proceeds under the direction of RNA polymerase. The enzymes necessary for the synthesis of the amino acid are produced, as is the regulator molecule. If the amino acid *is* present, or after sufficient synthesis occurs, the amino acid binds to the regulator, forming a complex that interacts with the operator region, causing repression of transcription of the genes within the operon.

   The *theo* operon is similar to the tryptophan system with the exception that the regulator gene is within the operon rather than being located separate from it. Therefore, in the *theo* operon, the regulator gene is itself regulated by the presence or absence of the amino acid.

(a) As in the *lac* operon, a mutation in the *theo* operator gene inhibits binding with the repressor complex, and transcription occurs constitutively. The presence of an F' plasmid bearing the wild-type allele would have no effect since it is not adjacent to the structural genes.

(b) A mutation in the *theo* promoter region would no doubt inhibit binding to RNA polymerase and therefore inhibit transcription. This would also happen in the *lac* operon. A wild-type allele present in an F' plasmid would have no effect.

(c) A mutation in the *theo* regulator gene, as in the *lac* system, may inhibit either its binding to the repressor or its binding to the operator gene. In both cases, transcription will be constitutive because the *theo* system is repressible. Both cases result in the failure of the regulator to bind to the operator, allowing transcription to proceed. In the *lac* system, failure to bind the co-repressor lactose would permanently repress the system. The addition of a wild-type allele would restore repressibility, provided that this gene was transcribed constitutively.

PROBLEMS AND DISCUSSION QUESTIONS

1. Contrast the need for the enzymes involved in lactose and tryptophan metabolism in bacteria when lactose and tryptophan, respectively, are: (a) present and (b) absent.
2. Contrast positive vs. negative control of gene expression.
3. Contrast the role of the repressor in an inducible system and in a repressible system.
4. Describe how the *lac* repressor was isolated. What properties demonstrate it to be a protein? Describe the evidence that it indeed serves as a repressor within the operon scheme.
5. Even though the *lac Z, Y,* and *A* structural genes are transcribed as a single polycistronic mRNA, each gene contains the appropriate initiation and termination signals essential for translation. Predict what will happen when a cell growing in the presence of lactose contains a deletion of one nucleotide early in the *Z* gene. Early in the *A* gene?
6. For the *lac* genotypes shown in the accompanying table, predict whether the structural genes (Z) are constitutive,

permanently repressed, or inducible in the presence of lactose.

Genotype	Constitutive	Repressed	Inducible
$I^+O^+Z^+$			X
$I^-O^+Z^+$			
$I^+O^CZ^+$			
$I^-O^+Z^+/F'I^+$			
$I^+O^CZ^+/F'O^+$			
$I^SO^+Z^+$			
$I^SO^+Z^+/F'I^+$			

7. For the genotypes and condition (lactose present or absent) shown in the accompanying table, predict whether functional enzymes are made, nonfunctional enzymes are made, or no enzymes are made.

Genotype	Condition	Functional Enzyme Made	Nonfunctional Enzyme Made	No Enzyme Made
$I^+O^+Z^+$	No lactose			X
$I^+O^CZ^+$	Lactose			
$I^-O^+Z^-$	No lactose			
$I^-O^+Z^-$	Lactose			
$I^-O^+Z^+/F'I^+$	No lactose			
$I^+O^CZ^+/F'O^+$	Lactose			
$I^+O^+Z^-/F'I^+O^+Z^+$	Lactose			
$I^-O^+Z^-/F'I^+O^+Z^+$	No lactose			
$I^SO^+Z^+/F'O^+$	No lactose			
$I^+O^CZ^+/F'O^+Z^+$	Lactose			

8. Predict the level of genetic activity of the *lac* operon as well as the status of the *lac* repressor and the CAP protein under the cellular conditions listed in the accompanying table.

	Lactose	Glucose
(a)	−	−
(b)	+	−
(c)	−	+
(d)	+	+

9. Predict the effect on the induciblity of the *lac* operon of a mutation that disrupts the function of:
(a) the *crp* gene, which encodes the CAP protein.
(b) the CAP-binding site within the promoter.

 ## EXTRA-SPICY PROBLEMS

10. In a theoretical operon, genes *A*, *B*, *C*, and *D* represent the repressor gene, the promoter sequence, the operator gene, and the structural gene, *but not necessarily in that order*. This operon is concerned with the metabolism of a theoretic molecule (tm). From the data provided, first decide if the operon is inducible or repressible. Then assign *A*, *B*, *C*, and *D* to the four parts of the operon. Explain your rationale. (AE = active enzyme; IE = inactive enzyme; NE = no enzyme).

Genotype	tm Present	tm Absent
$A^+B^+C^+D^+$	AE	NE
$A^-B^+C^+D^+$	AE	AE
$A^+B^-C^+D^+$	NE	NE
$A^+B^+C^-D^+$	IE	NE
$A^+B^+C^+D^-$	AE	AE
$A^-B^+C^+D^+/F'A^+B^+C^+D^+$	AE	AE
$A^+B^-C^+D^+/F'A^+B^+C^+D^+$	AE	NE
$A^+B^+C^-D^+/F'A^+B^+C^+D^+$	AE + IE	NE
$A^+B^+C^+D^-/F'A^+B^+C^+D^+$	AE	NE

11. A bacterial operon is responsible for the production of the biosynthetic enzymes needed to make the theoretical amino acid tisophane (tis). The operon is regulated by a separate gene, *R*, deletion of which causes the loss of enzyme synthesis. In the wild-type condition, when tis is present, no enzymes are made. In the absence of tis, the enzymes are made. Mutations in the operator gene (O^-) result in repression regardless of the presence of tis.

Is the operon under positive or negative control? Propose a model for (a) repression of the genes in the presence of tis in wild-type cells, and (b) the O^- mutations.

12. A marine bacterium is isolated and shown to contain an inducible operon whose genetic products metabolize oil when it is encountered in the environment. Investigation demonstrates that the operon is under positive control and that there is a *reg* gene whose product interacts with an operator region (*o*) to regulate the structural genes designated *sg*.

In the attempt to understand how the operon functions, a constitutive mutant strain and several partial diploid strains were isolated and tested with the results shown below. Draw all possible conclusions about the mutation as well as the nature of regulation of the operon. Is the constitutive mutation in the *trans*-acting *reg* element or in the *cis*-acting *o* (operator) element?

Host Chromosome	F' Factor	Phenotype
wild type	none	inducible
wild type	*reg* gene from mutant strain	inducible
wild type	operon from mutant strain	constitutive
mutant strain	*reg* gene from wild type	constitutive

13. The SOS repair genes in *E. coli* (see Chapter 17) are negatively regulated by the *lexA* gene product, called the LexA repressor. When a cell sustains extensive damage to its DNA, the LexA repressor is inactivated by the *recA* gene product (RecA), and transcription of the SOS genes is increased dramatically.

One of the SOS genes is the *uvrA* gene. You are studying the function of the *uvrA* gene product in DNA repair. You isolate a mutant strain that shows constitutive expression of the UvrA protein. You name this mutant strain *uvrA^C*. Shown below is a simple diagram of the *lexA* and *uvrA* operons.
(a) Describe two different mutations that would result in a *uvrA* constitutive phenotype. Indicate the actual genotypes involved.
(b) Outline a series of genetic experiments using partial diploid strains that would allow you to determine which of the two possible mutations you have isolated.
(c) A fellow student considers this problem and argues that there is a more straightforward nongenetic experiment that could differentiate between the two types of mutations. While this experiment requires no fancy genetics, you must be able to easily assay the products of the other SOS genes. Propose such an experiment.

GENETICS MediaLab

The following resources will help you achieve a better understanding of the concepts presented in this chapter. These resources can be found on the CD packaged with this textbook and on the Companion Website at **http://www.prenhall.com/klug/**.

CD Resources:

Animated Tutorial: *Gene Regulation*

Self-grading Chapter Problems

Web Resources:

Web Destinations in Genetics

Self-grading Chapter Problems

Chapter Search Terms

Genetics Newsgroups

Student Bulletin Board

Web Problem 1:

Time for completion = 10 minutes

How are decisions made at the cellular level? To choose among the alternatives, the cell must be able to sense particular conditions and respond differentially. A classic example of such a decision is observed when a bacteriophage infects a bacterial cell. The bacteriophage has two choices, lysis or lysogeny. Ultimately this choice reflects differences in gene regulation that are established immediately after infection. Read about the life cycle of the bacteriophage depicted on the linked website and view the animation of cell lysis. Discuss the key features determining whether a phage enters the lytic or lysogenic pathway. To complete this exercise, visit Web Problem 1 in Chapter 15 of your Companion Website and select the keyword **PHAGE**.

Web Problem 2:

Time for completion = 20 minutes

How does a cell respond to changes in environmental conditions? One of the best examples is the description of the lactose

operon by Jacob, Monod, and Wollman. Test your understanding of *lac* operon regulation with a series of interactive questions at the linked website. Feedback will indicate whether your answers are correct and tutorial information is provided to clarify your understanding. Construct a chart listing the various genotypes and conditions (with and without lactose) and the resulting effect. To complete this exercise, visit Web Problem 2 in Chapter 15 of your Companion Website and select the keyword **OPERON**.

Web Problem 3:

Time for completion = 5 minutes

What mechanisms are used to regulate the expression of bacterial genes? Different operons use positive and/or negative regulation through various mechanisms. Regulatory control reflects the basal state of the cell and how it responds to changing conditions. The linked website describes current research involving the *pyrR* operon of *Bacillus subtilis*. Compare and contrast the proposed model of regulation at this operon to those discussed in your textbook. Discuss how you would test the proposed model. To complete this exercise, visit Web Problem 3 in Chapter 15 of your Companion Website and select the keyword **MECHANISM**.

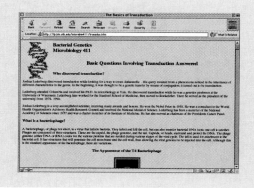

14. In Figure 15.13, numerous base-pairing regions of mRNA are depicted (1–4) that play important roles during the process of attenuation in the trp operon. Shown at the right, beginning with the 5′-end, is the sequence of 80 ribonucleotides that make up part of the mRNA whose transcription is affected by attenuation. Within this molecule are the sequences that serve as the basis of the formation of the secondary structures that are critical in the attenuation process. Obtain a large piece of paper (such as manila wrapping paper) and, along with several colleagues from your genetics class, work through this base sequence, attempting to identify which parts of the mRNA molecule represent the base-pairing regions (1–4) shown in Figure 15.13. Draw the configuration of the molecule under high tryptophan levels, as well as under low tryptophan levels.

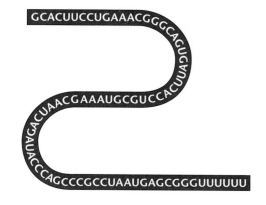

SELECTED READINGS

BECKWITH, J.R., and ZIPSER, D., eds. 1970. *The lactose operon.* Cold Spring Harbor, NY: Cold Spring Harbor Laboratory Press.

BERTRAND, K., et al. 1975. New features of the regulation of the tryptophan operon. *Science* 189:22–26.

COHEN, J.S., and HOGAN, M.E. 1994. New genetic medicines. *Sci. Am.* (Dec.) 271:76–82.

ENGLESBERG, E., and WILCOX, G. 1974. Regulation: Positive control. *Annu. Rev. Genet.* 8:219–42.

GILBERT, W., and MÜLLER-HILL, B. 1966. Isolation of the *lac* repressor. *Proc. Natl. Acad. Sci. USA* 56:1891–98.

——— 1967. The *lac* operator is DNA. *Proc. Natl. Acad. Sci. USA* 58:2415–21.

JACOB, F., and MONOD, J. 1961. Genetic regulatory mechanisms in the synthesis of proteins. *J. Mol. Biol.* 3:318–56.

KELLER, E. B., and CALVO, J.M. 1979. Alternative secondary structures of leader RNAs and the regulation of the *trp, phe, his, thr,* and *leu* operons. *Proc. Natl. Acad. Sci. USA* 76:6186–90.

LEE, D. and SCHLEIF, R. 1989. *In vivo* DNA loops in *araCBAD*: Size limits and helical repeats. *Proc. Natl. Acad. Sci. USA* 86:476-80.

LEWIN, B. 1974. Interaction of regulator proteins with recognition sequences of DNA. *Cell* 2:1–7.

LEWIS, M., et al. 1996. Crystal structure of the lactose operon repressor and its complexes with DNA and inducer. *Science* 271:1247–54.

MILLER, J.H., and REZNIKOFF, W.S. 1978. *The operon.* Cold Spring Harbor, NY: Cold Spring Harbor Laboratory Press.

OLSON, K.E., et al. 1996. Genetically engineered resistance to dengue-2 virus transmission in mosquitoes. *Science* 272:884–86.

PLATT, T. 1981. Termination of transcription and its regulation in the tryptophan operon of *E. coli. Cell* 24:10–23.

STROYNOWSKI, I., and YANOFSKY, C. 1982. Transcript secondary structures regulate transcription termination at the attenuator of *S. marcescens* tryptophan operon. *Nature* 298:34–38.

STUDIER, F.W. 1972. Bacteriophage T7. *Science* 176:367–76.

UMBARGER, H.E. 1978. Amino acid biosynthesis and its regulation. *Annu. Rev. Biochem.* 47:533–606.

WAGNER, R.W. 1994. Gene inhibition using antisense oligodeoxynucleotides. *Nature* 372:333–35.

YANOFSKY, C. 1981. Attenuation in the control of expression of bacterial operons. *Nature* 289:751–58.

YANOFSKY, C., and KOLTER, R. 1982. Attenuation in amino acid biosynthetic operons. *Annu. Rev. Genet.* 16:113–134.

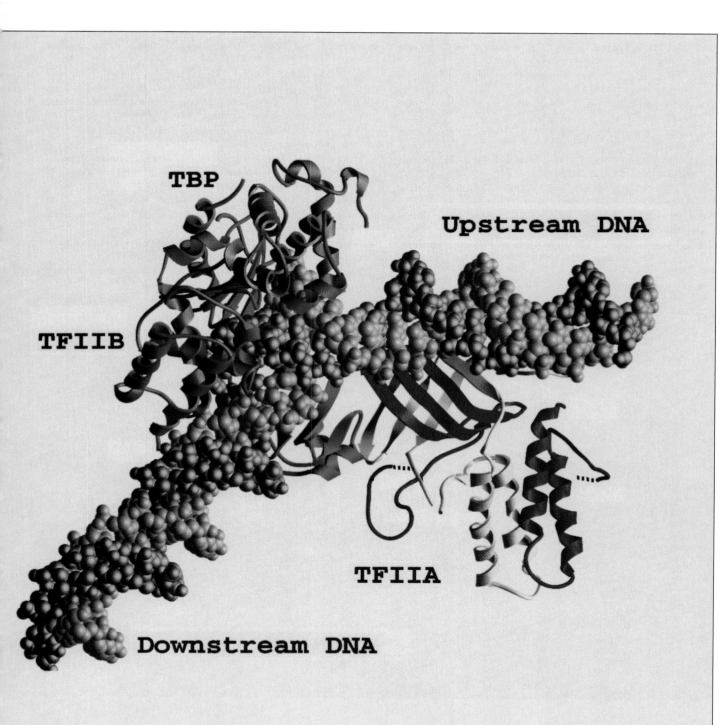

Molecular complex formed between the TATA box-binding protein (TBP), transcription factors TFIIA and TFIIB, and TATA box DNA during the initiation of transcription.

Regulation of Gene Expression in Eukaryotes

KEY CONCEPTS

- Eukaryotic gene regulation is very different from prokaryotic gene regulation

- The promoter is the site of assembly of the basal transcription complex

- Enhancers control chromatin structure and the rate of transcription

  True Activator Transcription Factors
  Antirepressor Transcription Factors

- Yeast *gal* genes are an example of positive inducibility and catabolite repression

- DNA methylation can regulate gene expression

- Posttranscriptional events can regulate gene expression

  Alternative Splicing Pathways for mRNA
  Alternative Splicing and *Drosophila* Sex Determination
  Controlling mRNA Stability

The 1960 discovery of the operon in *Escherichia coli* by Jacob and Monod was the first step in unraveling the mechanisms that regulate gene expression. While some of the principles of regulation found in the *lac* operon are also present in eukaryotes, it is clear that eukaryotes have evolved a more complex system of gene regulation.

In multicellular eukaryotes, differential regulation of gene expression is at the heart of cellular differentiation and function. Cells of the pancreas, for example, do not make retinal pigment, nor do retinal cells make insulin. The question is, how does an organism express a subset of genes in one cell type and a different subset of genes in another cell type? At the cellular level, this regulation is *not* accomplished by eliminating unused genetic information; instead, mechanisms have evolved to activate specific portions of the genome, and to repress the expression of other genes. The activation and repression of selected loci represent a delicate balancing act for an organism; the expression of a gene at the wrong time, in the wrong cell type, or in abnormal amounts can lead to a deleterious phenotype or death, even when the gene itself is normal.

In this chapter we will review the general features of eukaryotic regulation, outline the components needed to control transcription, and discuss how these components interact. We will also consider the role of posttranscriptional mechanisms in regulating gene expression in eukaryotes. When appropriate, we will compare the more complex approaches exhibited by eukaryotes to the less complex counterparts in prokaryotes.

Eukaryotic gene regulation is very different from prokaryotic gene regulation

There are numerous reasons why gene regulation is more complex in eukaryotes than in prokaryotes.

1. Eukaryotic cells contain a much greater amount of genetic information than do prokaryotic cells, and this DNA is complexed with histones and other proteins to form chromatin. As we shall see, the structure of chromatin, whether it is open (decondensed) and available to be transcribed or closed (condensed) and not available, is the major on/off switch for gene regulation (in prokaryotes, remember, the operator is the on/off switch).

2. Genetic information in eukaryotes is carried on many chromosomes (rather than just on one), and these chromosomes are enclosed within a nuclear membrane.

3. In eukaryotes, transcription is spatially and temporally separated from translation—transcription occurs in the nucleus and translation occurs later in the cytoplasm. Because of this, prokaryote attenuation control is not possible.

4. The transcripts of eukaryotic genes are processed before transport to the cytoplasm.

5. Eukaryote mRNA has a much longer half-life ($t_{1/2}$) than does prokaryotic mRNA. When prokaryotes

want to stop making a protein, they turn off transcription and the mRNA decays within minutes.

6. Eukaryotes must have translation level control because the mRNA is much more stable.

7. Most eukaryotes are multicellular with differentiated cell types. These different cell types typically use different sets of overlapping genes to make different proteins even though each cell contains a complete set of genes.

All this means that the regulation of eukaryotic gene expression can potentially occur at many levels (Figure 16.1). These include (1) transcriptional control, (2) posttranscription control (i.e., processing of the pre-mRNA), (3) transport to the cytoplasm, (4) stability of the mRNA, (5) translational control (i.e., selecting which mRNAs are translated), and (6) posttranslational modification of the protein product. Since most eukaryotic genes are regulated, in part, at the transcriptional level, in the following sections, we will emphasize transcriptional control, though we will discuss other levels of control as well.

The promoter is the site of assembly of the basal transcription complex

The structure of eukaryotic genes was first introduced in Chapter 13 and 15. Recall that these genes have two types of regulatory sequences that control their transcription, promoters and enhancers. Figure 16.3 shows the structure of the promoter region at the 5′-end of a eukaryotic gene and compares it to a prokaryotic gene.

Promoters consist of nucleotide sequences that serve as the recognition point for RNA polymerase binding, as they do in prokaryotes. They represent the region necessary to *initiate* transcription and are located a fixed distance from the site where transcription is initiated. Promoters are located immediately adjacent to the genes they regulate and are considered to be a part of the gene. Promoter regions, which include the promoter itself, are usually several hundred nucleotides in length.

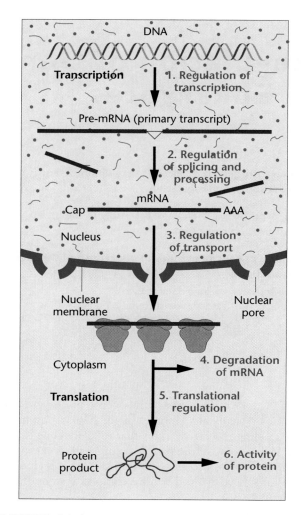

FIGURE 16.1   Various levels of regulation that are possible during the expression of the genetic material.

Eukaryotic promoters require the binding of a number of protein factors to initiate transcription. Promoter regions for genes that are recognized by RNA polymerase II, which transcribes primarily mRNA, consist of short modular DNA sequences usually located within 100 bp upstream (in the

FIGURE 16.2   Organization of typical eukaryotic gene and transcriptional control regions. The promoter region consists of modular DNA sequences, usually located within 100 to 200 bp of the site of transcription initiation. Promoter elements usually consist of modular sequences. The enhancer modulates the activity of the promoter.

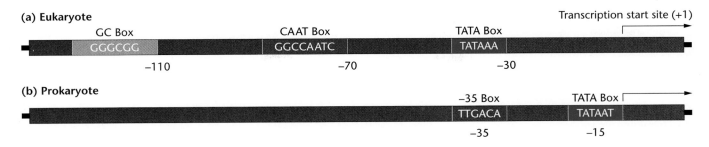

(a) Eukaryote

GC Box   CAAT Box   TATA Box   Transcription start site (+1)

GGGCGG   GGCCAATC   TATAAA

−110   −70   −30

(b) Prokaryote

−35 Box   TATA Box

TTGACA   TATAAT

−35   −15

FIGURE 16.3   (a) The promoter at the 5′-end of eukaryotic genes consists of several modular elements, including the TATA box (−30), the CAAT box (−70), and the GC box (at about −110). (b) The promoter in prokaryotic genes consists of several modular elements including the TATA box (−15) and the −35 box.

5′ direction) of the gene. The promoter region of most genes contains several elements (Figure 16.3a). The promoter itself, where the RNA polymerase will bind, is a sequence called the **TATA box**. Located about 25 to 30 bases upstream from the initial point of transcription (designated as −25 to −30), it consists of an 8-bp consensus sequence (a sequence conserved in most genes studied) composed only of A═T base pairs, often flanked on either side by G≡C rich regions. Mutations in the TATA box reduce transcription, and deletions may alter the initiation point of transcription (Figure 16.4).

Many promoter regions also contain other components; one of the more common, called the **CAAT box**, has the consensus sequence CAAT or CCAAT. The CAAT box frequently appears in the region 70 to 80 bp from the start site, and it can function in a 5′ → 3′ or a 3′ → 5′ orientation. Mutational analysis suggests that the CAAT box and the protein it binds are critical to the promoter's ability to facilitate transcription. Mutations on either side of this element have no effect on transcription, whereas mutations within the CAAT sequence dramatically lower the rate of transcription (Figure 16.4). Another element often seen in some promoter regions, called the **GC box**, has the

consensus sequence GGGCGG and is often found at about position −110. This module is found in either orientation and often occurs in multiple copies. The CAAT and GC elements bind transcription factors and function more like enhancers, which we will cover in the next section.

Compare the structure of the eukaryotic promoter region to the structure of the prokaryotic promoter region [(Figure 16.3(b)]. The eukaryotic promoter has a single element, the TATA box, which is set further back from the start of transcription than the prokaryotic promoter. The prokaryotic promoter has two elements, the TATA box at −10, and a −35 box.

Sequences that make up the upstream regulatory elements in eukaryotic genes vary in location and organization (Figure 16.5). Note that in different genes there are significant differences in the number and orientation of promoter elements and the distance between them. In addition, eukaryotic genes use three RNA polymerases for transcription (prokaryotic genes are all transcribed by a single RNA polymerase). These three are classified as type I (ribosomal RNAs), type II (mRNAs and snRNAs), and type III (tRNAs, 5S rRNA, and several other small-cellular RNAs). The promoter for each type of poly-

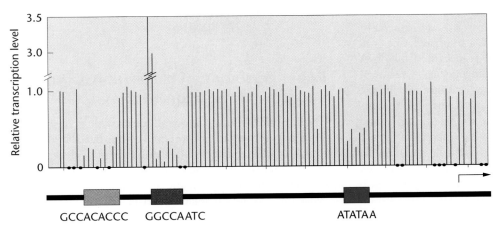

FIGURE 16.4 Summary of the effects of point mutations in the promoter region on transcription of the β-globin gene. Each line represents the level of transcription produced by a single nucleotide mutation (relative to wild type) in a separate experiment. Dots represent nucleotides where no mutation was obtained. Note that mutations within the specific elements of the promoter have the greatest effect on the level of transcription.

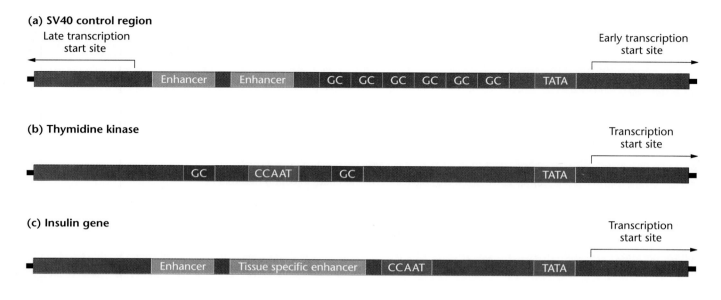

FIGURE 16.5 Upstream organization of several eukaryotic genes, illustrating the variable nature, number, and arrangement of controlling elements.

merase has a different sequence and binds different transcription factors.

There are two essential points to remember when comparing the ways eukaryotic and prokaryotic polymerases bind to the promoter. First of all, the prokaryote chromosome is essentially naked DNA. Its RNA polymerase has no difficulty finding promoter sequences. Eukaryotic chromosomes are chromatin with the vast majority of promoters hidden within the nucleosomes and not available to be bound by RNA polymerase. The chromatin structure must be altered and opened up, as we will discuss in the next section. Second, prokaryotic RNA polymerase can read the DNA sequence with its sigma factor, and find and bind to promoters directly. Eukaryotic RNA polymerases cannot. A set of transcription factors must be preassembled on the promoter to serve as a platform for RNA polymerase to bind.

We now have some insight into the transcriptional apparatus of type II genes, those transcribed by RNA polymerase II, mainly mRNA coding genes. A series of **transcription factors**, proteins that are not part of the RNA polymerase molecule itself, but are needed to control the initiation of transcription are assembled at the promoter in a specific order to provide a platform for RNA polymerase to recognize and bind (Figure 16.6). To initiate formation of the apparatus, the TFIID complex binds to the TATA box via its TBP subunit (Figure 16.7). TFIID is the TBP plus a set of additional factors called TAFs (*T*ATA *A*ssociated *F*actors). About 20 base pairs of

DNA are involved in binding the TBP and the other subunits in the TFIID transcription factor. This has been described as the *commitment* stage. TFIID responds to contact with activator proteins by conformational changes that expedite the binding of additional factors, such as TFIIA and TFIIB. RNA polymerase II and some additional factors such as TFIIF then bind, followed by TFIIE, TFIIH, and TFIIJ. The final stage is *promoter clearance* where the RNA polymerase leaves the TATA box and the transcription of the DNA downstream ensues at a minimal *basal* level. The *induced* state, where transcription is stimulated above the basal level, is not yet well defined, but involves other areas of the promoter region, enhancers, and numerous transcription factors, which control the assembly of the transcription complex and the rate of promoter clearance.

Enhancers control chromatin structure and the rate of transcription

Transcription of most, if not all, eukaryotic genes is regulated not only by the promoter region, but also by additional DNA sequences called **enhancers**. Enhancers can be on either side of a gene, at some distance from the gene, or even within the gene, and are called *cis* regulators because they are found adjacent to the structural genes they regulate, as opposed to *trans* regulators (e.g., binding pro-

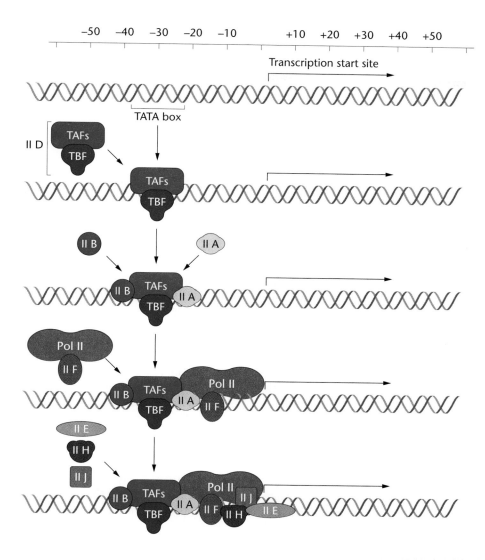

FIGURE 16.6 The assembly of transcription factors required for the initiation of transcription by RNA polymerase II.

teins), which can regulate a gene on any chromosome. Enhancers typically interact with multiple regulatory proteins, transcription factors, and can increase the efficiency of transcription initiation or activate the promoter. Within enhancer sites binding sites are often found for positive as well as negative gene regulators. Thus, there is some degree of analogy between enhancers and operator regions in prokaryotes. However, enhancers appear to be much more complex in both structure and function. Enhancers greatly stimulate the transcriptional activity of a promoter and can be distinguished from promoters by several characteristics.

1. The position of the enhancer need not be fixed; it can be upstream, downstream, or within the gene it regulates.

2. Its orientation can be inverted without significant effect on its action.

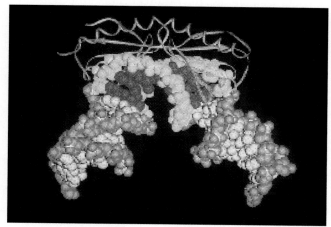

FIGURE 16.7 Molecular complex formed between the TATA-box binding protein (green) and the 8-bp TATA box of DNA (red), itself part of DNA (yellow).

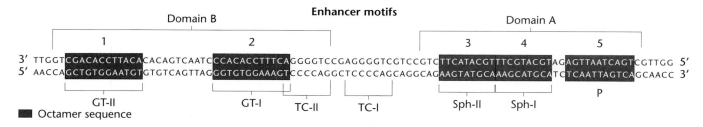

Enhancer motifs

Domain B

Domain A

1   2   3   4   5

3′ TTGGT CGACACCTTACA CACAGTCAATC CCACACCTTTCA GGGGTCCGAGGGGTCGTCCGTC TTCATACGTTTCGTACGTA GAGTTAATCAGT CGTTGG 5′
5′ AACCA GCTGTGGAATGT GTGTCAGTTAG GGTGTGGAAAGT CCCCAGGCTCCCCAGCAGGCAG AAGTATGCAAAGCATGCAT CTCAATTAGTCA GCAACC 3′

GT-II   GT-I   TC-II   TC-I   Sph-II   Sph-I   P

■ Octamer sequence

FIGURE 16.8 DNA sequence for the SV40 enhancer. Screened sequences are those required for maximum enhancer effect. The brackets below the sequence are the various sequence motifs within this region. The two domains of the enhancer (A and B) are indicated.

3. If an enhancer is experimentally moved to another location in the genome, or if an unrelated gene is placed near an enhancer, the transcription of the adjacent gene is enhanced.

An example of an enhancer located *within* the gene it regulates is found in the immunoglobulin heavy-chain gene where an enhancer is located in an intron between two coding regions. This enhancer is also active only in the cells in which the immunoglobulin genes are being expressed, indicating that tissue-specific gene expression can be modulated through enhancers. Internal enhancers have been discovered in other eukaryotic genes, including the immunoglobulin light-chain gene.

Downstream enhancers are found in the human β-globin gene and the chicken thymidine kinase gene. In chickens, an enhancer located between the β-globin and the ε-globin genes apparently works in one direction to control the ε-globin gene during embryonic development and in the opposite direction to regulate the β-globin gene during adult life. In yeast, regulatory sequences similar to enhancers, called **upstream activator sequences (UAS)**, can function upstream at variable distances and in either orientation. They differ from enhancers in that

they cannot function downstream of the transcription start point.

Like the promoter region, enhancers contain modules composed of short DNA sequences. The enhancer of the SV40 virus has a complex structure consisting of two adjacent 72-bp sequences located some 200 bp upstream from a transcriptional start point (Figure 16.8). Each of the two 72-bp regions contains five sequences that contribute to the maximum rates of transcription. If one or the other of these regions is deleted, there is no effect on transcription; but if both are deleted, *in vivo* transcription is greatly reduced.

Enhancers differ from promoters in a number of significant ways. While promoter sequences are essential for basal-level transcription, enhancers are necessary for the full level of transcription. In addition, enhancers are responsible for time- and tissue-specific gene expression. Since enhancers are able to stimulate levels of transcription at a distance, an intriguing question is how they are able to do this. As we will see they perform at least two functions. First, transcription factors bind to enhancers and alter the configuration of chromatin and, second, by bending or looping the DNA they bring distant enhancers and their promoters into direct contact in order to form complexes with transcription factors and polymerases (Figure 16.9). In the new

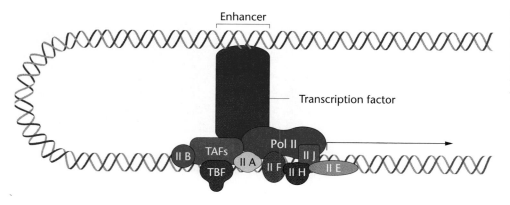

Enhancer

Transcription factor

Pol II

II B   TAFs   II J
TBF   II A   II F   II H   II E

FIGURE 16.9 DNA looping allows factors that bind to enhancers at a distance from the promoter to interact with regulatory proteins in the transcription complex and to maximize transcription.

configuration, transcription is stimulated to a higher level, increasing the overall rate of RNA synthesis.

As in prokaryotes, transcriptional control in eukaryotes involves the interaction between DNA sequences adjacent to genes and DNA-binding proteins. As outlined in Chapter 15, the regulatory sequences adjacent to a wide range of genes have been identified, mapped, and sequenced. The development of DNA-protection and gel-retardation assays have allowed the detection of DNA-binding proteins in cell extracts, and new affinity chromatography techniques have been used to isolate transcriptional factors present in very low concentrations. The result has been a virtual explosion of information about eukaryotic transcription factors. This section will review some of this new information, including the upstream organization of eukaryotic genes, the characteristics of transcription factors, their interaction with DNA sequences, other regulatory factors, and the initiation of transcription by RNA polymerase.

We have already discussed one class of transcription factors, those that work at the promoter to form a platform for the RNA polymerase to bind. We will now discuss a second class, those that bind to enhancer sites. These may act as *positive factors* to increase transcription or as *negative factors* to decrease transcription. These proteins control where and when genes are expressed, and the rate of expression. The positive class of transcription factors can also be divided into two categories, true activators and antirepressors. True activators function by contacting elements of the transcription complex at the promoter and activating transcription. Antirepressors function by altering the chromatin structure to allow other factors to bind.

True Activator Transcription Factors

True activators are modular proteins with at least two functional domains (clusters of amino acids that carry out a specific function). One binds to DNA sequences present in the enhancer (**DNA-binding domain**), and another activates transcription via protein–protein interaction (***trans*-activating domain**). The *trans*-activating domain binds to RNA polymerase or to other transcription factors at the promoter. The existence of separate domains in eukaryotic transcription factors was first demonstrated by the genetic analysis of mutants in yeast.

The domains of eukaryotic transcription factors take on several forms. The DNA-binding domains have distinctive three-dimensional structural patterns or **motifs**. There are a number of major types of these structural motifs, including helix-turn-helix (HTH), zinc finger, and basic leucine zipper (bZIP). This classification is not exhaustive, and other new groups will undoubtedly be established as new factors are characterized.

The first DNA-binding domain to be discovered was the helix-turn-helix (HTH) motif. In prokaryotes, HTH motifs have been identified in the *cro* repressor (in the DNA-binding domain), *lac* repressor, *trp* repressor, and other proteins. Studies indicate that the HTH motif is present in many prokaryotic DNA-binding proteins. This motif is characterized by its geometric conformation rather than a distinctive amino acid sequence (Figure 16.10). Two adjacent α helices separated by a "turn" of several amino acids enables the protein to bind to DNA (hence the name of the motif). In the *cro* repressor, one of the helices binds within a major groove containing six bases, while the other helix stabilizes the interaction by binding to the DNA backbone. Unlike several of the other DNA-binding motifs, the HTH pattern cannot fold or function alone, but is always part of a larger DNA-binding domain. Amino acid residues outside the HTH motif are important in regulating DNA recognition and binding.

The potential for forming helix-turn-helix geometry has been recognized in distinct regions of a large number of eukaryotic genes known to regulate developmental processes. Present almost universally in eukaryotic organisms is the **homeobox**, a stretch of 180 bp specifying a 60-amino-acid **homeodomain** sequence that can form a helix-turn-helix structure. Of the 60 amino acids, many are basic (arginine and lysine), and a conserved sequence is found among these genes. We shall discuss homeobox-containing genes in Chapter 22 because of their significance to animal developmental processes.

Zinc fingers, one of the major structural families of eukaryotic transcription factors, are involved in many aspects of gene regulation. The zinc finger motif was originally discovered in the *Xenopus* transcription factor TFIIIA. This structural motif has now been identified in proto-oncogenes (Chapter 23), in genes that regulate development in *Drosophila* (the *Krüppel* gene, see Chapter 22), in

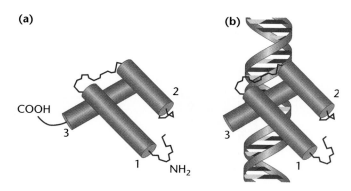

(a)   (b)

FIGURE 16.10   A *helix-turn-helix* or *homeodomain* where (a) three planes of the α helix of the protein are established. (b) These domains bind in the grooves of the DNA molecule.

(a)

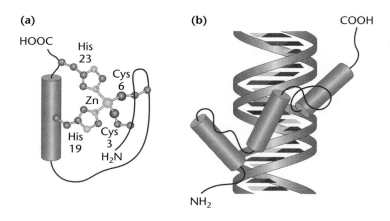

(b)

COOH

NH₂

FIGURE 16.11   (a) A zinc finger where cysteine and histidine residues bind to a Zn⁺⁺ atom, (b) This loops the amino acid chain out into a fingerlike configuration.

proteins whose synthesis is induced by growth factors and differentiation signals, and in transcription factors. There are several types of zinc finger proteins, each with a distinctive structural pattern.

One zinc finger protein contains clusters of two cysteine and two histidine residues at repeating intervals (Figure 16.11). The consensus amino acid repeat is $Cys-N_{2-4}-Cys-N_{12-14}-His-N_3-His$. The interspersed cysteine and histidine residues covalently bind zinc atoms, folding the amino acids into loops known as zinc fingers. Each finger consists of approximately 23 amino acids, with a loop of 12 to 14 amino acids between the Cys and His residues, and a linker between loops consisting of 7 or 8 amino acids. The amino acids in the loop interact with and bind to specific DNA sequences. Studies have shown that zinc fingers bind in the major groove of the DNA helix and wrap at least partway around the DNA. Within the major groove, the zinc finger makes contact with a set of DNA bases and may form hydrogen bonds with the bases, especially in G-rich strands. The number of fingers in a zinc finger transcription factor ranges from 2 to 13, as does the length of the DNA-binding sequence.

A third type of domain, a DNA-binding domain, is represented by the **basic leucine zipper (bZIP)**. This DNA-binding domain is adjacent to the **leucine zipper**, a second domain which allows protein–protein dimerization. The leucine zipper, first seen as a stretch of 35 amino acids in a nuclear protein in rat liver, contains four leucine residues spaced seven amino acids apart and flanked by basic amino acids. The leucine-rich regions form a helix with leucine residues protruding at every other turn. When two such molecules dimerize, the leucine residues "zip" together (Figure 16.12). The dimer contains two basic α-helical regions adjacent to the zipper that bind to phosphate residues and specific bases in DNA, making the dimer look like a pair of scissors.

Transcription factors also contain domains that activate transcription. These regions, which are distinct from the DNA-binding domains, can occupy from 30 to 100 amino acids. The stretches of amino acids interact with other transcription factors (such as those that bind to the promoter) or directly with the RNA polymerase. In addition, many transcription factors also contain domains that

(a)

HOOC   COOH

–L L–
–L L–
–L L–
–L L–

NH₂   NH₂

(b)

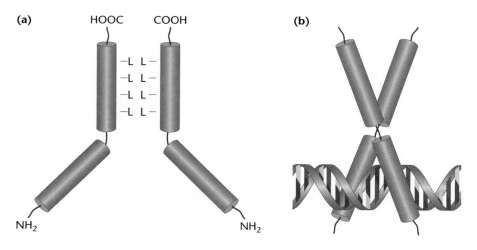

FIGURE 16.12   (a) A leucine zipper when dimers result from leucine residues at every other turn of the α helix in facing stretches of two polypeptide chains. (b) When the α-helical regions form a leucine zipper, the regions beyond the zipper form a Y-shaped region that grips the DNA in a scissors-like configuration.

bind coactivators such as hormones or small metabolites that regulate their activity.

Overall, the picture of transcriptional regulation in eukaryotes is complex, but several important generalizations can be drawn. The regulation of transcription by transcription factors is largely positive, although transcription repressors are now being recognized as important components of gene regulation. The structural organization of chromatin and alterations in chromatin structure to allow the binding of transcription factors are considered to be the primary levels of regulation.

In addition, different transcription factors may compete for binding to a given DNA sequence or to two overlapping sequences, or the same site may bind different factors in two different tissues. Transcription factor concentration and the strength with which each factor binds to the DNA will dictate which factor binds. Finally, because of the variability in the location and modular nature of enhancers, and the variety of transcription factors, an initiation complex at the promoter can be contacted by multiple enhancers, with many individual proteins contacting the basal-transcription complex at the promoter in a number of ways, all involving protein–protein interaction.

Antirepressor Transcription Factors

The second class of transcription factors, the **antirepressors**, work in a totally different fashion; they *remodel the chromatin*. Chromatin is not merely DNA wound around nucleosomes; it is a dynamic structure. Histones make up only half of the chromosomal proteins. The chromatin structure changes as the cell goes from interphase into mitosis and then back to interphase. Euchromatin has a different structure from heterochromatin. Most important for our discussion, chromatin remodeling must occur when genes are turned on or turned off (Figure 16.13). Investigators have shown that genes in chromatin that are actively transcribed or can potentially be transcribed are sensitive or even hypersensitive to *in vitro* digestion by the enzyme DNase I, while inactive or repressed genes are relatively resistant to DNase I digestion. The basic principle is that open chromatin is more easily digested and more DNase sensitive than closed chromatin, which is not easily digested and is DNase insensitive. Chromatin structure is therefore the on/off switch for gene activity.

Chromatin is altered by two very different processes. The first process is catalyzed by an ATP hydrolysis-dependent remodeling complex. One of the first remodeling complexes to be characterized is **SWI/SNF**, a large 11-subunit complex first described in yeast and subsequently found widely distributed in eukaryotes. Remodeling complexes are targeted to specific gene regions by

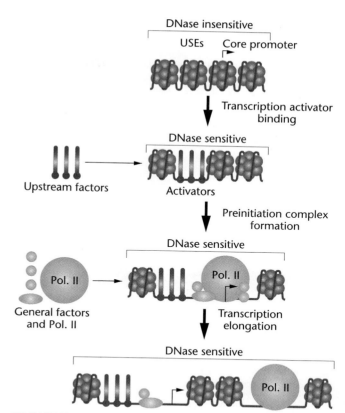

FIGURE 16.13 Nucleosomes can inhibit multiple steps required for gene transcription. The binding of enhancer transcription factors requires accessing nucleosomal DNA and may result in displacement of the nucleosome. Similarly the formation of the basal transcription complex at the TATA box and transcription start site is also suppressed by nucleosomes. Finally, RNA polymerase II is slowed by the presence of nucleosomes.

transcription factors, different transcription factors associating with different remodeling complexes. Some transcription factors can bind to closed chromatin and initiate the opening of the chromatin. They may alter nucleosome structure by several different mechanisms. As seen in Figure 16.14(a), this may entail altering the contacts between the DNA and the nucleosome histone proteins. Alternatively, the path of the DNA around the nucleosome core particle may be altered, as seen in Figure 16.14(b). Another possibility is that the structure of the nucleosome core particle itself may be altered, as seen in Figure 16.14(c).

The second chromatin alteration process is histone modification catalyzed by one of many histone acetyltransferase enzymes (**HAT**). When an acetate group is added to basic amino acids on the histone tails, the attraction between the basic histone protein and acidic DNA is lessened (Figure 16.15). As in the remodeling complex-

(a) Alteration of DNA protein contacts

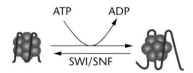

(b) Alteration of the DNA path

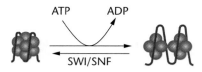

(c) Remodeling of nucleosome core particle

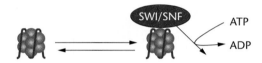

FIGURE 16.14 Three mechanisms that might be used to alter nucleosome structure by the ATP hydrolysis-dependent remodeling complex SWI/SNF. (a) DNA–histone contacts may be loosened. (b) The path of the DNA around an unaltered nucleosome core particle may be altered. (c) The conformation of the nucleosome core particle may be altered.

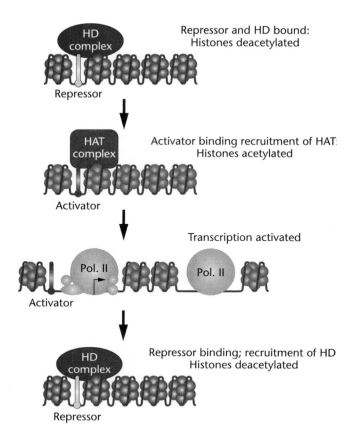

FIGURE 16.15 Proposed model of the action of HAT and HD complexes. Transcription factors recruit the complex to the gene, which either adds or removes acetyl groups, aiding in either opening or closing the chromatin structure.

es, HATs are targeted to genes by specific transcription factors. The remodeling of a gene region would entail the cooperative interaction between a remodeling complex and a HAT. Of course, what can be opened can also be closed. In this case **deacetylases** (**HD** complexes) can be targeted to remove acetate groups (Figure 16.15, last line). Typically, from the focal point of an enhancer, remodeling spreads toward the promoter of the gene and then into the transcription unit. **Insulator elements**, short DNA sequences that bind specific proteins, can act as barriers to prevent the spread into neighboring genes.

Yeast *gal* genes are an example of positive inducibility and catabolite repression

One of the first model systems used to study eukaryotic genetic regulation was the set of genes in yeast that encode the enzymes essential for galactose breakdown, or catabolism. These *gal* genes thus serve a function similar to the *lac* operon and *ara* operon in *E. coli*. Expression of the *gal* genes is **inducible**; that is, they are regulated by the presence or absence of their substrate, galactose. In

the absence of galactose, the genes are not transcribed. If galactose is added to the growth medium, transcription begins immediately, and the mRNA concentration of the transcripts increases by a thousandfold. However, transcription is activated only if the concentration of glucose is low. So like the *lac* and *ara* operons, the genes are under a second level of control, **catabolite repression** (see Chapter 15). A mutation in the regulator of the *gal* genes, *GAL4*, prevents activation, indicating that transcription is under **positive control**, that is, the regulator must be present to turn on gene transcription. We will examine how two of the *gal* genes, *GAL1* and *GAL10* are regulated (Figure 16.16).

The transcription of these two genes is controlled by a single central control region, called **UAS$_G$** (upstream activating sequence), of approximately 170 bp. Recall that UASs are functionally similar to enhancers in higher eukaryotes. The chromatin structure of the UAS is constitutively open, or **DNase hypersensitive**, that is, free of

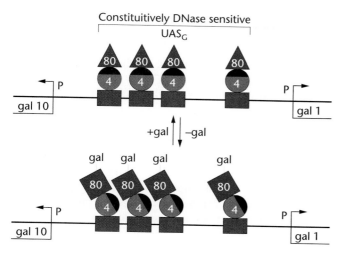

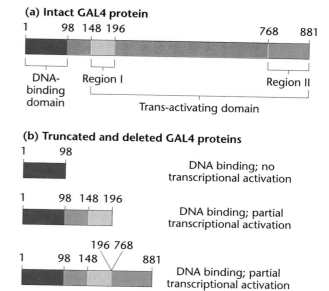

FIGURE 16.16 Model of *GAL1* and *GAL10* UAS structures showing the binding sites for the Gal4p positive regulator and their interaction with the Gal80p negative regulator. Induction is shown as the structure of the Gal80p becomes altered, exposing the activation domain of Gal4p.

FIGURE 16.17 The Gal4p protein has three domains that participate in the activation of the *GAL1* and *GAL10* loci. Amino acids 1–98 or 1–147 bind to DNA-recognition sites in UAS$_G$. Amino acids 148–196 and 768–881 are required for transcriptional activation. Gal80p binding activity resides in the amino acid region 851–881.

nucleosomes. This open chromatin structure requires the SWI/SNF remodeling complex. Within the UAS are four binding sites for the Gal4 protein (Gal4p). These sites are permanently occupied by Gal4p, whether or not the genes have been activated. Gal4p is in turn negatively regulated by Gal80p, another *gal* gene regulator. Gal80p is always bound to Gal4p, covering its activation domain, shown in Figure 16.16 as a dark patch. Induction occurs when the inducer, a phosphorylated galactose, binds to Gal80p and/or Gal4p and alters their structures, exposing the Gal4p activation domain. As in *E. coli*, these genes are under a second level of control—catabolite repression—and thus cannot be turned on in the presence of glucose. In this case, however, the catabolite here does not operate through cAMP but rather through a complex system involving a protein kinase. Activation involves the recruitment of additional factors to further remodel the chromatin, encompassing the two promoters exposing the TATA boxes to the TBP.

Recombinant DNA techniques have revealed that Gal4p is a protein of 881 amino acids. It includes a DNA-binding domain that recognizes and binds to sequences in the UAS$_G$, and a *trans*-activating domain that activates transcription [Figure 16.17(a)]. These functional domains were identified by cloning and expressing truncated *GAL4* genes and assaying the gene products for their ability to bind DNA and to activate transcription. Because the protein contains two functional domains, it is possible to delete one of these regions while the other remains intact and functional.

A region at the N-terminus of the protein was found to be the DNA-binding domain, which is responsible for recognition of and binding to the nucleotide sequences in UAS$_G$. Proteins with deletions extending from amino acid 98 to the C-terminus (amino acid 881) retained the ability to bind to the UAS$_G$ region. Similar experiments also identified two regions of the protein involved in the activation of transcription, region I, consisting of amino acids 148–196, and region II, encompassing amino acids 768–881. Constructs that contain the DNA-binding domain and region I (amino acids 1–96 and 149–196) or the DNA-binding domain and region II (amino acids 1–96 and 768–881) have reduced transcriptional activity [Figure 16.17(b)].

The nature of the activating regions was explored by starting with a protein that contains the DNA-binding region and activating region I (*GAL4* 1–238) and isolating point mutations in region I that result in increased activation. Most single mutations that increase activation produce amino acid substitutions that increase the negative charge of region I, and most cause a threefold increase of transcription. A multiple mutant that contains four more negative charges than the starting protein causes a ninefold increase in transcriptional activation, and has almost 80 percent of the activity of the complete Gal4p. However, negative charge alone is not responsible for the activating function; some mutants with decreased activity do not have fewer negative charges than the parental *GAL4* 1–238 molecule.

These results strongly suggest that activation involves direct contact between the activating domain of the transcription factor and other proteins. How does this protein–protein contact activate transcription? There are several possibilities, including chromatin remodeling, stabilizing the binding between the promoter DNA and RNA polymerase, increasing the rate at which the double-stranded DNA within the transcribed region is unwound, accelerating the release of the RNA polymerase from the promoter, or attracting and stabilizing other factors that bind to the promoter or to the RNA polymerase.

An attractive candidate for an activator target is TFIID, which consists of the TATA binding protein (TBP), and about 10 other TAFs (TBP-associated factors). The idea that TFIID is a target for activation is supported by the finding that some *GAL4* constructs stimulate transcription in mammalian-cell extracts and change the conformation of DNA-bound TFIID. In addition, other studies show that although TBP alone can bind to the promoter, TAFs must be present for the complex to respond to activators.

DNA methylation can regulate gene expression

The alteration of the chromatin conformation is one way in which gene expression can be regulated or modulated. In this section we will consider another type of change that plays a role in gene regulation, a chemical modification of DNA that involves adding or removing methyl groups from the DNA bases.

The DNA of most eukaryotic organisms is modified after replication by the enzyme-mediated addition of methyl groups to bases and sugars. Base methylation most often involves cytosine and in the genome of any given eukaryotic species approximately 5 percent of the cytosine residues are methylated. However, the extent of methylation can be tissue-specific and can vary from less than 2 percent to over 7 percent.

The ability of base methylation to alter gene expression is known from studies on the *lac* operon in *E. coli*. The methylation of DNA in the operator region, even at a single cytosine residue, can cause a marked change in the affinity of the repressor for the operator. Methylation of cytosine occurs at the 5' position, causing the methyl group to protrude into the major groove of the DNA helix where it can alter the binding of proteins to the DNA.

Methylation occurs most often in the cytosine of CG doublets in DNA, usually in both strands:

$$5' - {}^mCpG - 3'$$
$$3' - GpC^m - 5'$$

The state of DNA methylation can be determined by restriction enzyme analysis. The enzyme *Hpa*II cleaves at the recognition sequence CCGG; however, if the second cytosine is methylated, the enzyme will not cut the DNA. The enzyme *Msp*I cuts at the same CCGG site whether or not the second cytosine is methylated. If a segment of DNA is unmethylated, both enzymes produce the same restriction pattern of bands. As shown in Figure 16.18, if one site is methylated, digestion with *Hpa*II produces an altered pattern of fragments. Using this method in eukaryotes to analyze the methylation of a given gene in different tissues shows that, in general, if a gene is expressed, it is not methylated or it has a low level of methylation.

The evidence for the role of methylation as a factor in the regulation of eukaryotic gene expression is somewhat indirect and is based on a number of observations. First, an

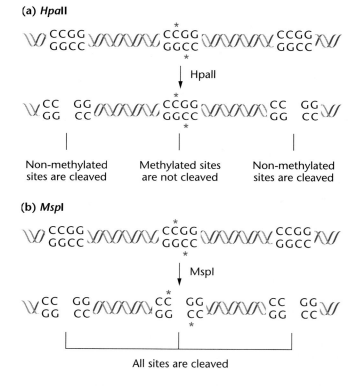

(a) HpaII

Non-methylated sites are cleaved   Methylated sites are not cleaved   Non-methylated sites are cleaved

(b) MspI

All sites are cleaved

FIGURE 16.18 The restriction enzymes *Hpa*II and *Msp*I recognize and cut at CCGG sequences. (a) If the second cytosine is methylated (indicated by an asterisk), *Hpa*II will not cut. (b) The enzyme *Msp*I cuts at all CCGG sites, whether or not the second cytosine is methylated. Thus the state of methylation of a given gene in a given tissue can be determined by cutting DNA extracted from that tissue with *Hpa*II and *Msp*I.

inverse relationship exists between the degree of methylation and the degree of expression. That is, low amounts of methylation are associated with high levels of gene expression, and high levels of methylation are associated with low levels of gene expression. In mammalian females, the inactivated X chromosome, which is almost totally inactive in gene expression, has a higher level of methylation than does the active X chromosome. Within the inactive X, those regions that escape inactivation have much lower levels of methylation than those seen in adjacent inactive regions.

Second, methylation patterns are tissue-specific and, once established, are heritable for all cells of that tissue. Perhaps the strongest evidence for the role of methylation in gene expression comes from studies using base analogues. The nucleotide 5'-azacytidine is incorporated into DNA in place of cytidine and cannot be methylated (Figure 16.19), causing the undermethylation of the sites where it is incorporated. The incorporation of 5'-azacytidine causes changes in the pattern of gene expression and can then stimulate expression of alleles on inactivated X chromosomes.

Recently, 5'-azacytidine has been used in clinical trials for the treatment of sickle-cell anemia. In this autosomal recessive condition, a mutant hemoglobin protein with a defect in the β subunit causes changes in shape of red blood cells that, in turn, generate a cascade of clinical symptoms. During embryogenesis, the epsilon ε- and gamma γ-globin genes are expressed, but normally become transcriptionally inactive after birth, when β globin synthesis begins. Treatment of affected individuals with 5'-azacytidine causes a reduction in the amount of methylation in the ε-globin and γ-globin genes, and initiates the reexpression of these embryonic and fetal genes. The epsilon and gamma proteins replace beta-globin in hemo-globin molecules, bringing about a reduction in the amount of sickling in the red blood cells.

Though the available evidence indicates that the absence of methyl groups in DNA is related to increases in gene expression, methylation cannot be regarded as a general mechanism for gene regulation because methylation is not a general phenomenon in eukaryotes. In *Drosophila*, for example, there is no methylation of DNA. Thus, methylation may represent only one of a number of ways in which gene expression can be regulated by genomic changes, and it seems likely that similar mechanisms remain to be discovered.

How might methylation affect gene regulation? One possibility comes from the observation that there are proteins that bind to 5'-methyl cytosine without regard to the DNA sequence. These proteins could recruit co-repressors and/or histone deacetylases to remodel the chromatin, changing it from an open structure to a closed structure.

Posttranscriptional events can regulate gene expression

As we have seen, the regulation of genetic expression can occur at many points along the pathway from DNA to protein. Although transcriptional control is perhaps the most obvious and widely used mode of regulation in eukaryotes, **posttranscriptional modes** of regulation also occur in many organisms. Eukaryotic nuclear RNA transcripts are modified prior to translation, noncoding introns are removed, the remaining exons are precisely spliced together, and the mRNA is modified by the addition of a cap at the 5'-end and a poly-A tail at the 3'-end. The message is then complexed with proteins and exported to the cytoplasm. Each of these processing steps offers several possibilities for regulation. We will examine two that are especially important in eukaryotes, alternative splicing of a single mRNA transcript to give multiple mRNAs and regulation of the stability of mRNA itself.

Alternative Splicing Pathways for mRNA

Alternative splicing can generate different forms of a protein, so that expression of one gene can give rise to a family of related proteins. Figure 16.20 illustrates an example where the polypeptide products derived from a single type of pre-mRNA are distinct from one another. The initial bovine pre-mRNA transcript is processed into one or two **preprotachykinin mRNAs (PPT mRNAs)**. This precursor mRNA molecule potentially includes the genetic information specifying two neuropeptides, called **P** and **K**. These two peptides, which are members of the family

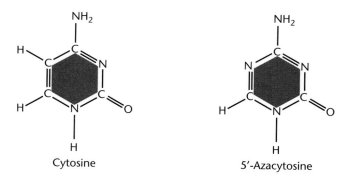

FIGURE 16.19 The base 5'-azacytosine, which has a nitrogen at the 5' position. The base can be incorporated into DNA in place of cytidine. The base 5'-azacytidine cannot be methylated, causing the undermethylation of the CpG dinucleotide wherever it has been incorporated.

Initial PPT RNA transcript

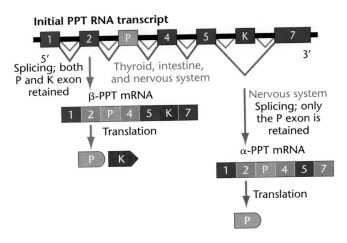

FIGURE 16.20 Alternative splicing of the initial RNA transcript of the preprotachykinin gene (PPT). Exons are either numbered or designated by the letters P or K. When the K exon is excluded, α-PPT mRNA is produced, and only the P neuropeptide is synthesized. The inclusion of P and K exons leads to β-PPT mRNA, which, upon translation, yields both the P and K tachykinin neuropeptides.

of sensory neurotransmitters referred to as **tachykinins**, are believed to play different physiological roles. While the P neuropeptide is largely restricted to tissues of the nervous system, the K neuropeptide is found more predominantly in the intestine and thyroid.

The RNA sequences for both neuropeptides are derived from the same gene. However, the processing of the initial RNA transcript can occur in two ways. In one case (Figure 16.20), exclusion of the K exon during processing results in the α-PPT mRNA, which upon translation yields neuropeptide P but not K. Conversely, processing that includes both the P and K exons yields β-PPT mRNA, which upon translation results in the synthesis of both the P and K neuropeptides. The analysis of the relative levels of the two types of RNA has demonstrated striking differences between tissues. In nervous-system tissues, α-PPT mRNA predominates by as much as a threefold factor, while β-PPT mRNA is the predominant type in the thyroid and intestine.

Given the existence of alternate splicing, how many different polypeptides can be derived from the same pre-mRNA? Work on α-tropomyosin, an accessory protein that regulates muscle contraction, has provided a partial answer to this question. In rats, the α-tropomyosin gene contains a total of 14 exons, six of which make up three pairs that are alternatively spliced. Only one member of each pair ends up in the finished mRNA, never both. Alternative splicing of this pre-mRNA results in 10 different forms of α-tropomyosin, many of which are expressed in

a tissue-specific manner. Another gene, for the muscle form of troponin T, a protein that regulates the calcium needed for contraction, produces 64 known forms of the protein from a single pre-mRNA by alternative splicing.

Alternative Splicing and Drosophila Sex Determination

Alternative splicing not only has the potential to produce many different forms of a protein from a single gene, but can also affect the outcome of developmental processes. Perhaps the most striking example involves sex determination in *Drosophila*, where splicing involving a single exon eventually determines whether the embryo will develop as a male or a female.

As outlined in Chapter 9, sex in *Drosophila* is determined by the ratio of X chromosomes to sets of autosomes (X:A). When the ratio is 0.5 (1X:2A), males are produced, even when no Y chromosome is present; when the ratio is 1.0 (2X:2A), females are produced. Intermediate ratios (2X:3A) produces intersexes. The chromosomal ratios are interpreted by a small number of genes that initiate a cascade of developmental events resulting in the production of male or female somatic cells and the corresponding male or female phenotypes. The major genes in this pathway are *Sex-lethal* (*Sxl*), *transformer* (*tra*), and *doublesex* (*dsx*).

Figure 16.21 illustrates this process. In females, the chromosomal ratio activates transcription of the *Sxl* gene and results in the production of Sxl protein (the mechanism is not completely understood). In males, the chromosomal ratio does not result in production of Sxl protein. When the Sxl protein is present, it binds to and controls the splicing of the pre-mRNA for the next gene in the pathway, *transformer* (*tra*).

The pre-mRNA of *tra* includes a termination codon in exon 2. The female-specific Sxl protein binds to the *tra* pre-mRNA and directs the splicing to remove exon 2 from the mature mRNA. In males, where no Sxl protein is present, splicing proceeds along the ground state or default pathway, and the stop codon is incorporated into the mature mRNA. In males, the translation of *tra* mRNA is prematurely terminated by the stop codon, resulting in an inactive gene product, whereas in females, translation produces a functional gene product. The result is a female-specific tra protein that leads to the female phenotype.

The *dsx* gene is a critical control point in the development of sexual phenotype, for it produces a functional mRNA and protein in both females and males. However, the pre-mRNA is processed in a sex-specific manner to yield different transcripts. In females, the tra protein binds to the *dsx* pre-mRNA and directs its processing in an alternative, female-specific manner. In males, default splicing of the *dsx* pre-mRNA results in a male-specific mRNA

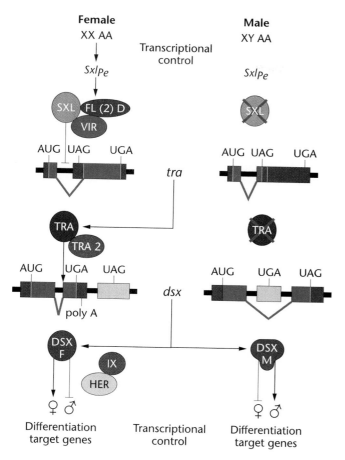

FIGURE 16.21 Hierarchy of gene regulation for sex determination in *Drosophila*. In females, the X:A ratio activates transcription of the *Sxl* gene. The product of this gene binds to pre-mRNA of the *tra* gene and directs its splicing in a female-specific fashion. The female-specific tra protein, in combination with the tra-2 protein, directs female-specific splicing of *dsx* pre-mRNA, resulting in a female-specific dsx protein. This protein, in combination with the ix protein, suppresses the pathway of male sexual development and activates the female pathway. In males, the X:A ratio does not activate *Sxl*. The result is a male-specific processing of *tra* pre-mRNA, resulting in no functional tra protein. This in turn leads to male-specific processing of *dsx* transcripts, resulting in a male-specific dsx protein that activates the pathway for male sexual development.

and protein product. The female dsx protein causes female sexual differentiation in somatic tissues by repressing male sexual development through coordinated action with the *ix* gene product. The male dsx protein brings about male sexual development by suppressing the female pathway. In the absence of *dsx* function, flies develop into phenotypic intersexes.

In summary, the *Sxl* gene acts as a switch gene that selects the pathway of sexual development by eventually controlling the processing of the *dsx* transcript in a fe-

male-specific fashion. The Sxl protein is produced only in embryos with an X:A ratio of 1. Failure to control the splicing of the *dsx* transcript in a female mode results in the default splicing of the transcript in a male mode, leading to the production of a male phenotype.

Controlling mRNA Stability

After mRNA precursors are processed and transported, they enter the population of cytoplasmic mRNA molecules, from which messages are recruited for translation. All mRNA molecules have a characteristic life span (called a half-life, $t_{1/2}$); they are degraded some time after they are synthesized, usually in the cytoplasm. The lifetimes of different mRNA molecules vary widely. Some are degraded within minutes after synthesis, whereas others last hours, or even months and years (in the case of mRNAs stored in oocytes). The mRNA stability, and thus its *turnover rate*, is intrinsic to the sequence of the mRNA.

An mRNA is more than simply an open reading frame that codes for a protein. It also includes 5′ and 3′ untranslated sequences that contain regulatory information. This information can include **stability sequences** as well as an **address sequence** to locate the mRNA in a particular component of the cell. In addition, there are **instability elements**, which can be general (for all cells) or cell-type specific. A typical instability sequence, AUUUA, is found in the mRNA of many oncogenes, such as *fos*.

Although the regulation of gene expression at the level of translation may seem inefficient, a growing body of evidence indicates that altering the stability of mRNA engaged in translation may be a significant point for controlling eukaryotic gene expression. Why would such an inefficient mechanism exist? Clearly, factors that influence the half-life of particular mRNAs have an important effect on the number of protein molecules that can be produced from a single mRNA molecule. Longer-lived mRNAs can produce more protein than short-lived ones. Altering the half-life of an mRNA molecule already engaged in translation is one way a cell can respond to rapidly changing conditions inside or outside the cell.

Another way to control mRNA stability is through **translation level** control; that is, the use of the message controls its stability. One of the best-studied examples of this type of regulation is the synthesis of α- and β-tubulins, the subunit components of eukaryotic microtubules. Treatment of a cell with the drug colchicine leads to a rapid disassembly of microtubules, and an increase in the concentration of the α and β subunits. Under these conditions, the synthesis of α- and β-tubulins drops dramatically. However, when cells are treated with vinblastine, a drug that also causes microtubule disassembly, the synthesis of tubulins is increased. Though the two drugs both cause microtubule disassembly, vinblastine precipitates the subunits, lowering the concentration of free α and β subunits. At low concentrations, the synthesis of tubulins is stimulated, whereas at higher concentrations, synthesis is inhibited. This type of translational regulation is known as **autoregulation**.

Work by Don Cleveland and his colleagues has described the conditions under which the stability of tubulin mRNA is regulated. Fusion of tubulin gene fragments to a cloned thymidine kinase gene shows that the first 13 nucleotides at the 5′-end of the mRNA following the transcription start site (encoding the amino terminal region of the protein) cause the hybrid tubulin–thymidine kinase mRNA to be regulated as efficiently as intact tubulin mRNA. These first 13 nucleotides encode the amino acids Met-Arg-Glu-Ile (abbreviated as MREI). Deletions, translocations, or point mutations in this 13-nucleotide segment abolish regulation. The examination of the cytoplasmic distribution of mRNAs demonstrates that only tubulin mRNAs in polysomes are depleted by drug treatment. Copies of tubulin mRNAs not bound to polysomes were protected from degradation. Also, translation must proceed to at least codon 41 of the mRNA in order for regulation to occur.

A model, shown in Figure 16.22, has been proposed to explain these observations. In this model, regulation occurs after the process of translation has begun. The first four amino acids (Met-Arg-Glu-Ile, MREI) of the tubulin gene product (in this case, β-tubulin) constitute a recognition element to which the regulatory factors bind. The concentration of α- and β-tubulin subunits in the cytoplasm may serve this regulatory function. This protein–protein interaction invokes the action of an RNase that may be a ribosomal component or a nonspecific cytoplasmic RNase. Action by the RNase degrades the tubulin mRNA in the act of translation, shutting down tubulin biosynthesis. As an alternative, the binding of the regulatory element may cause the ribosome to stall in its translocation along the mRNA, leaving it exposed to the action of RNase. Translation must proceed to at least codon 41 because the first 30 to 40 amino acids translated occupy a tunnel within the large ribosomal subunit, and only the translation of this number of amino acids will make the MREI recognition sequence accessible for binding.

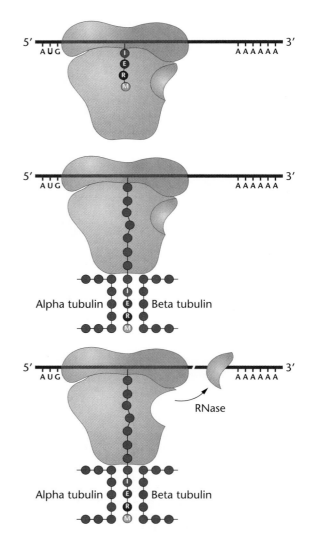

FIGURE 16.22 Model of posttranslational regulation of tubulin synthesis.

Translation-coupled mRNA turnover has been proposed as a regulatory mechanism for other genes, including histones, some transcription factors, lymphokines, and cytokines. This suggests the existence of a regulatory mechanism that acts on mRNAs in the act of translation, probably by protein–protein interaction between a regulatory element and the nascent polypeptide chain, resulting in the activation of RNases.

CHAPTER SUMMARY

1. Mechanisms controlling gene regulation in eukaryotes are governed by the properties of expanded genome size, the spatial and temporal separation between transcription and translation, and, in most organisms, multicellularity.

2. The roles played by chromatin structure, transcription factors, and signal transduction, as well as the expression of specific gene sets, are studied as models for gene regulation in higher organisms.

3. Transcription in eukaryotes is controlled by the interaction between promoters and enhancers. Other regulatory sequences, such as the CCAAT box, are elements of promoter regions found near the promoter itself. Enhancer elements, which appear to control the degree of transcription, can be located before, after, or within the gene expressed.

4. Transcription factors are proteins that bind to DNA-recognition sequences within the promoters and enhancers and activate transcription through protein–protein interaction.

5. A gene may have a different methylation pattern in different tissues. This is often correlated with different regulatory patterns.

6. Several types of posttranscriptional control of gene expression are possible in eukaryotes. One such mechanism is alternative processing of a single class of pre-mRNAs to generate different mRNA species. Another mechanism controls mRNA stability, either through intrinsic sequence elements or through a feedback system in response to the cytoplasmic concentration of the protein.

INSIGHTS AND SOLUTIONS

1. Regulatory sites for eukaryotic genes are usually located within a few hundred nucleotides of the transcriptional starting site, but they can be located up to several kilobases away. DNA sequence-specific binding assays have been used to detect and isolate protein factors present at low concentrations in nuclear extracts. In these experiments, DNA-binding sequences are bound to material on a column, and nuclear extracts are passed over the column. The idea is that if proteins that specifically bind to the sequence represented on the column are present, they will bind to the DNA and they can be recovered from the column after all other nonbinding material has been washed away. Once a DNA-binding protein has been identified and isolated, the problem is to devise a general method for screening cloned libraries for the genes encoding these and other DNA-binding factors. Determining the amino acid sequence of the protein and constructing synthetic oligonucleotide probes is time-consuming and useful only for one factor at a time. Knowing the strong affinity for binding between the protein and its DNA-recognition sequence, how would you screen for genes encoding binding factors?

Solution: Several general strategies have been developed, and one of the most promising has been devised by Steve McKnight's laboratory at the Fred Hutchinson Cancer Center. The cDNA isolated from cells expressing the binding factor is cloned into the lambda vector, gt11. Plaques of this library, containing proteins derived from expression of cDNA inserts, are adsorbed onto nitrocellulose filters and probed with double-stranded radioactive DNA corresponding to the binding site. If a fusion protein corresponding to the binding factor is present, it will bind to the DNA probe. After the unbound probe is washed off, the filter is subjected to autoradiography and the plaques corresponding to the DNA-binding signals can be identified. An added advantage of this strategy is filter recycling by washing the bound DNA from the filters for reuse. This ingenious procedure is similar to the colony-hybridization and plaque-hybridization procedures described for library screening in Chapter 18, and it provides a general method for isolating genes encoding DNA-binding factors.

PROBLEMS AND DISCUSSION QUESTIONS

1. Why is gene regulation assumed to be more complex in a multicellular eukaryote than in a prokaryote? Why is the study of this phenomenon in eukaryotes more difficult?

2. List and define the levels of gene regulation discussed in this chapter.

3. Distinguish between the regulatory elements referred to as promoters and enhancers.

4. Is the binding of a transcription factor to its DNA recognition sequence necessary and sufficient for an initiation of transcription at a regulated gene? What else plays a role in this process?

5. Contrast transcription factors designated as true activators with those designated as antirepressors.

6. Contrast and compare the regulation of the *gal* genes in yeast with the *lac* genes in *E. coli*.

7. Write an essay comparing the control of gene regulation in eukaryotes and prokaryotes at the level of initiation of transcription. How do the regulatory mechanisms work? What are the similarities and differences in these two types of organisms regarding the specific components of the regulatory mechanisms? Address how the differences or similarities relate to the biological context of the control of gene expression.

EXTRA-SPICY PROBLEMS

8. In the autoregulation of tubulin synthesis, two models for the mechanism were proposed: (1) that the tubulin subunits bind to the mRNA and (2) that the subunits interact with the nascent tubulin polypeptides. To distinguish between these two models, Cleveland and colleagues introduced mutations into the 13-base regulatory element of the β-tubulin gene. Some of the mutations resulted in amino acid substitution, while others did not. In addition, they shifted the reading frame of the intact 13-base sequence. Presented here are the results of a mutagenesis study of the mRNA. Which of the two models is supported by the results? What experiments would you do to confirm this?

Wild type	met AUG	arg AGG	glu GAA	lys ATC	Autoregulation
					+
Second codon mutations		UGG			−
		GGG			−
		CGG			+
		AGA			+
		AGC			−
Third codon mutations			GAC		+
			AAC		−
			UAU		−
			UAC		−

9. You are interested in studying transcription factors and have developed an *in vitro* transcription system using a defined segment of DNA that is transcribed under the control of a eukaryotic promoter. The transcription of this DNA occurs when you add purified RNA polymerase II, TFIID (the TATA binding factor), and TFIIB and TFIIE (which bind to RNA polymerase). You perform a series of experiments that compare the efficiency of transcription in this "defined system" with the efficiency of transcription in a crude nuclear extract. You test the two systems with your template DNA and with various deletion templates that you have generated. The results of your study are shown on the next page.

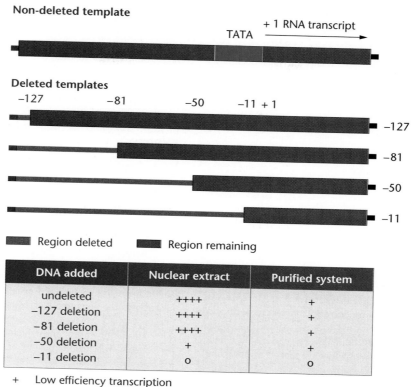

Non-deleted template

TATA · + 1 RNA transcript

Deleted templates

−127 · −81 · −50 · −11 +1

−127

−81

−50

−11

▭ Region deleted · ▬ Region remaining

DNA added	Nuclear extract	Purified system
undeleted	++++	+
−127 deletion	++++	+
−81 deletion	++++	+
−50 deletion	+	+
−11 deletion	o	o

+ Low efficiency transcription
++++ High efficiency transcription
o No transcription

(a) Why is there no transcription from the −11 deletion template?

(b) How do the results for the nuclear extract and the defined system differ from the *undeleted* template? How would you interpret these results?

(c) For the various deleted templates, compare the results with the nuclear extract and the purified system. How would you interpret the results with the deleted templates? Be very specific about what you can conclude from these data.

GENETICS MediaLab

The following resources will help you achieve a better understanding of the concepts presented in this chapter. These resources can be found on the CD packaged with this textbook and on the Companion Website at **http://www.prenhall.com/klug/**.

CD Resources:

Self-grading Chapter Problems

Web Resources:

Web Destinations in Genetics

Self-grading Chapter Problems

Chapter Search Terms

Genetics Newsgroups

Student Bulletin Board

Web Problem 1:

Time for completion = 10 minutes

How is a gene organized in a eukaryote? The nucleotide sequence of a gene includes information about where to begin and end transcription and translation, and how to modify the resulting transcript or protein. Although every gene is a bit different, there are some general features that are shared among many eukaryotic genes. In this exercise, you will review the basic organizational features of the eukaryotic gene. Draw a representative eukaryotic gene, indicating the location of any common features. Indicate on which strand of this double stranded region these elements are located. To complete this exercise, visit Web Problem 1 in Chapter 16 of your Companion Website and select the keyword **TRANSCRIPTIONAL REGULATION.**

Web Problem 2:

Time for completion = 10 minutes

How does a cell know which genes to transcribe? The promoter region of a gene is important for designating when and where to produce a transcript, and how much of a transcript to produce. To understand this process, researchers have focused on understanding what is required for the function of the promoter sequence in *cis* to the gene, and how *trans* factors interact with the promoter region. The work of one researcher in this field is described at the linked website. Discuss what experimental questions have been raised and how they are being tested. To complete this exercise, visit Web Problem 2 in Chapter 16 of your Companion Website and select the keyword **GENE EXPRESSION.**

Web Problem 3:

Time for completion = 15 minutes

How does a cell regulate when and which transcript is made? The tissue and temporal specificity of transcription has been an area of intense work in the last two decades. In particular, much attention has been given to the role of transcriptional regulation in establishing and maintaining the developmental identity of a cell. In this exercise, you will view a database of the genes under tissue-specific transcriptional regulation. Select at least one example from the catalog and create a precis that includes the function of the gene product, when and where the gene is expressed, what transcriptional factors regulate the gene, and where in the promoter these factors bind. To complete this exercise, visit Web Problem 3 in Chapter 16 of your Companion Website and select the keyword **TISSUE SPECIFICITY.**

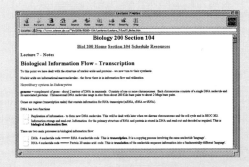

SELECTED READINGS

ASO, T., SHILATIFARD, A., CONAWAY, J.W., and CONAWAY, R.C. 1996. Transcription syndromes and the role of RNA polymerase II general transcription factors in human disease. *J. Clin. Investig.* 97:1561–69.

AYOUBI, T.A., and VAN DE VEN, W.J. 1996. Regulation of transcription by alternate promoters. *FASEB J.* 10:453–60.

BRENNAN, R. 1993. The winged-helix DNA-binding motif: Another helix-turn-helix takeoff. *Cell* 74:773–76.

BURATOWSKI, S. 1995. Mechanisms of gene activation. *Science* 270:1773–74.

BUSCH, S.J., and SASSONE CORSI, P. 1990. Dimers, leucine zippers and DNA-binding domains. *Trends Genet.* 6:36–40.

CLEVELAND, D. 1988. Autoregulated instability of tubulin mRNAs: A novel eukaryotic regulatory mechanism. *Trends Biochem. Sci.* 13:339–43.

CONAWAY, R., and CONAWAY, J. 1993. General initiation factors for RNA polymerase II. *Ann. Rev. Biochem.* 62:161–90.

CLINE, T.W., and MEYER, B.J. 1996. Vive la difference: Males vs females in flies vs worms. *Ann. Rev. Genet.* 30:637–702.

DRAPKIN, R., MERINO, A., and REINBERG, D. 1993. Regulation of RNA polymerase II transcription. *Curr. Opin. Cell Biol.* 5:469–76.

DE LA BROUSSE, F.C., and McKNIGHT, S.L. 1993. Glimpses of allostery in the control of eukaryotic gene expression. *Trends Genet.* 9:151–54.

DYNAN, W.S. 1988. Modularity in promoters and enhancers. *Cell* 58:1–4.

GOODRICH, J.A., CUTLER, G., and TJIAN, R. 1996. Contacts in context: Promoter specificity and macromolecular interaction in transcription. *Cell* 84:825–30.

JACOBSEN, A., and PELTZ, S. 1996. Interrelationships of the pathway of mRNA decay and translation in eukaryotic cells. *Ann. Rev. Biochem.* 65:693–739.

JACOBSON, R.H., and TJIAN, R. 1996. Transcription factor IIA: A structure with multiple functions. *Science* 272:830–36.

KAKIDANI, H., and PTASHNE, M. 1988. GAL4 activates gene expression in mammalian cells. *Cell* 52:161–67.

KASS, S.U., PRUSS, D., and WOLFFE, A.P. 1997. How does DNA methylation repress transcription? *Trends in Genet.* 13:444–49.

LEVINE, M., and MANLEY, J.L. 1989. Transcriptional repression of eukaryotic promoters. *Cell* 59:405–8.

LITTLEWOOD, T.D., and EVANS, G.I. 1994. Transcription factors 2: Helix-loop-helix. *Protein Profile* 1:639–709.

LOPEZ, H.J. 1995. Developmental role of transcription factor isoforms generated by alternate splicing. *Dev. Bio.* 172:396–411.

LOHR, D. 1997. Nucleosome transactions on the promoters of the yeast GAL and PHO genes. *J. Biol. Chem.* 272:26795–98.

MacDOUGALL, C., HARBISON, D., and BOWNES, M. 1995. The developmental consequences of alternate splicing in sex determination and differentiation in *Drosophila. Dev. Biol.* 172:353–76.

McKNIGHT, S.L. 1995. Transcription revisited: A commentary on the 1995 Cold Spring Harbor Meeting, Mechanisms of Eukaryotic Transcription. *Genes Dev.* 10:367–81.

MANIATIS, T., GOODBOURN, S., and FISCHER, J.A. 1987. Regulation of inducible and tissue-specific expression. *Science* 236:1237–45.

MEEHAN, R., LEWIS, J., CROSS, S., NAN, X., JEPPESEN, P., and BIRD, A. 1992. Transcriptional repression by methylation of CpG. *J. Cell Sci. Suppl.* 16:9–14.

MITCHELL, P.J., and TJIAN, R. 1989. Transcriptional regulation in mammalian cells by sequence-specific DNA binding proteins. *Science* 245:371–78.

O'HALLORAN, T. 1993. Transition metals in control of gene expression. *Science* 261:715–25.

PIELER, T., and THEUNISSEN, O. 1993. TFIIIA: Nine fingers—Three hands? *Trends Biochem. Sci.* 18:226–30.

PTASHNE, M., and GANN, A.A.F. 1990. Activators and targets. *Nature* 346:329–31.

SHARP, Z.D., and MORGAN, W.W. 1996. Brain POU-er. *BioEssays* 18:347–50.

SIBLEY, E., KASTELIC, T., KELLY, T.J., and LANE, M.D. 1989. Characterization of the mouse insulin receptor gene promoter. *Proc. Natl. Acad. Sci. USA* 86:9732–36.

STRINGER, K.F., INGLES, C.J., and GREENBLATT, J. 1990. Direct and selective binding of an acidic transcriptional activation domain to the TATA-box factor TFIID. *Nature* 345:783–86.

STRUHL, K. 1993. Yeast transcription factors. *Curr. Opin. Cell Biol.* 5:513–20.

TATE, P., and BIRD, A. 1993. Effects of DNA methylation on DNA-binding proteins and gene expression. *Curr. Opin. Genet. Dev.* 3:226–31.

TJIAN, R. 1995. Molecular machines that control genes. *Sci. Am.* 272:54–61.

WERNER, J.H., GRONENBORN, A.M., and CLORE, G.M. 1996. Intercalation, DNA kinking and the control of transcription. *Science* 271:778–84.

WORKMAN, J.L., and KINGSTON, R.E. 1998. Alteration of nucleosome structure as a mechanism of transcriptional control. *Ann. Rev. Biochem.* 67:545–79.

YEN, T.J., GAY, D.A., PACHTER, J.S., and CLEVELAND, D.W. 1988. Autoregulated changes in stability of polyribosome-bound beta-tubulin mRNAs are specified by the first 13 translated nucleotides. *Mol. Cell Biol.* 8:1224–35.

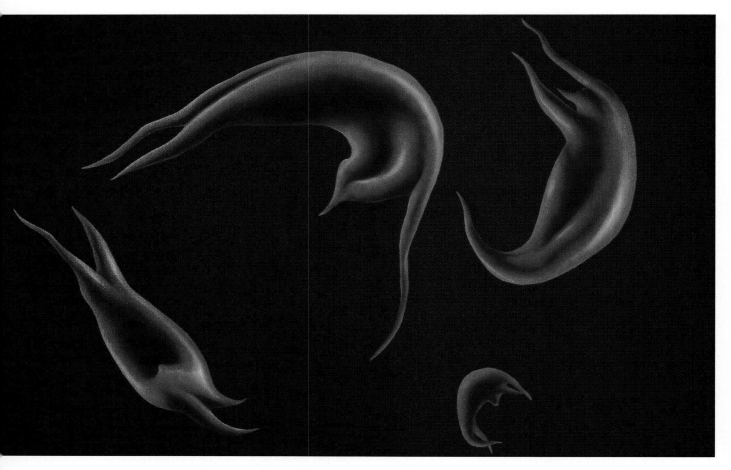

Mutant erythrocytes derived from an individual with sickle-cell anemia.

Gene Mutation, DNA Repair, and Transposable Elements

KEY CONCEPTS

- **Mutations are usually, but not always spontaneous**

  The Luria–Delbruck Fluctuation Test
  Adaptive Mutation in Bacteria

- **Mutations may be classified in various ways**

  Spontaneous vs. Induced Mutations
  Gametic vs. Somatic Mutations
  Other Categories of Mutation

- **Genetic techniques, cell cultures, and pedigree analysis are all used to detect mutations**

  Detection in Bacteria and Fungi
  Detection in *Drosophila*
  Detection in Plants
  Detection in Humans

- **The spontaneous mutation rate varies greatly among organisms**

  Deleterious Mutations in Humans

- **Mutations occur in many forms and arise in different ways**

  Tautomeric Shifts
  Base Analogues
  Alkylating Agents
  Apurinic Sites and Other Lesions
  Ultraviolet Radiation and Thymine Dimers

- **High-energy radiation penetrates cells and induces mutations**

- **Gene sequencing has enhanced understanding of mutations in humans**

  ABO Blood Types
  Muscular Dystrophy

  Trinucleotide Repeats in Fragile-X Syndrome, Myotonic Dystrophy, and Huntington Disease

- **The Ames test is used to assess the mutagenicity of compounds**

- **Organisms can counteract DNA damage and mutations by activating several types of repair systems**

  Photoreactivation Repair: Reversal of UV Damage in Prokaryotes
  Excision Repair in Prokaryotes and Eukaryotes
  Xeroderma pigmentosum and Nucleotide Excision Repair
  Proofreading and Mismatch Repair
  Post-Replication Repair and the SOS Repair System
  Double-Strand Break Repair in Mammals

- **Site-directed mutagenesis allows researchers to investigate specific genes**

  Knockout Genes and Transgenes

- **Transposable genetic elements move within the genome and may disrupt genetic function**

  Insertion Sequences
  Bacterial Transposons
  The *Ac–Ds* System in Maize
  Other Mobile Genetic Elements in Plants: Mendel Revisited
  Copia Elements in *Drosophila*
  P Element Transposons in *Drosophila*
  Transposable Elements in Humans

GenCDX

When you see this icon, there are related animations and exercises on the CD accompanying this text.

In our earlier discussion of DNA as the genetic material (Chapter 11), we defined the four characteristics or functions ascribed to the genetic information: replication, storage, expression, and variation by mutation. In a sense, *mutation* is a failure to store genetic information faithfully. If a change occurs in the stored information, it may be reflected in the expression of that information and will be propagated by replication. Historically, the term mutation includes both chromosomal changes and changes within single genes. We discussed the former alterations in Chapter 10, referring to them collectively as chromosomal mutations or chromosomal aberrations. In this chapter we are concerned with gene mutations. As we will see, a change may be as simple as the substitution, gain, or loss of a single nucleotide or as complex as the addition or deletion of many nucleotides within the normal sequence of DNA.

The term *mutation* was coined in 1901 by Hugo DeVries to describe the variation he observed in crosses involving the evening primrose, *Oenothera lamarckiana*. Most of the variation was due to multiple translocations, but two cases were subsequently shown to be caused by **gene mutations**, actual changes in the chemical composition of the DNA. As the studies of mutations progressed, it soon became clear that gene mutations serve as the source of most new alleles and thus are the origin of much of the genetic variability within populations. As new alleles arise, they constitute the raw material to be tested by the evolutionary process of natural selection, which will determine whether they are detrimental, neutral, or beneficial.

Mutations provide the basis for genetic studies. The resulting phenotypic variability allows geneticists to study the genes that control the traits that have been modified. In genetic investigations, mutations serve as identifying "markers" of genes so that they may be followed during their transmission from parents to offspring. Without the phenotypic variability that mutations provide, genetic analysis would be impossible to conduct. For example, if all pea plants displayed a uniform phenotype, Mendel would have had no basis for his experimentation.

Certain organisms lend themselves to the induction of mutations that can be easily detected and studied throughout reasonably short life cycles. Viruses, bacteria, fungi, fruit flies, certain plants, and mice have been particularly useful in studying mutation and mutagenesis. Because of this, they serve as the foundation for much of our genetic knowledge.

Once we have completed our presentation of mutation, we will turn our attention to two related topics, DNA repair and transposable genetic elements. These topics are logical extensions of our consideration of gene mutation. Repair processes serve to counteract mutation. Transposable elements often disrupt the normal structure of the gene and therefore create mutations.

Mutations are usually, but not always spontaneous

Although it was known well before 1943 that pure cultures of bacteria could give rise to a small number of cells exhibiting heritable variation, particularly with respect to survival under different environmental conditions, the source of the variation was hotly debated. The majority of bacteriologists (now called microbiologists) believed that environmental factors induced changes in certain bacteria that led to their survival or adaptation to the new conditions. For example, strains of *Escherichia coli* are known to be sensitive to infection by the bacteriophage T1; that is, the virus reproduces at the expense of the bacterial cell, which is lysed or destroyed (see Figure 11.5). If a plate of *E. coli* is homogeneously sprayed with T1, almost all the cells are lysed. Rare *E. coli* cells, however, survive the infection and are not lysed. If these cells are isolated and established in pure culture, all their descendants are *resistant* to T1 infection. The **adaptation hypothesis**, put forth to

explain such observations, implied that the interaction of the phage and bacterium was essential to the acquisition of immunity. In other words, the phage had somehow "induced" resistance in the bacteria.

The occurrence of **spontaneous mutations** (sometimes referred to as random mutations) provided an alternative model to explain the origin of T1 resistance in *E. coli*. In 1943, Salvador Luria and Max Delbrück presented the first direct evidence that mutation in bacteria occurs spontaneously within the chromosome; that is, they arise independent of any external force. Thus, there is no way of predicting when or where in the chromosome mutation will occur. This experiment marked the initiation of modern bacterial genetic study.

The Luria–Delbrück Fluctuation Test

In a beautiful example of analytical and theoretical work, Luria and Delbrück carried out experiments with the *E. coli*/T1 systems to differentiate between the adaptation and spontaneous mutation hypotheses. They grew many small individual liquid cultures of phage-sensitive *E. coli* and then added numerous aliquots of each culture to petri dishes of agar medium containing T1 bacteriophages. To obtain precise quantitative data, they also determined the total number of bacteria added to each plate prior to incubation. Following incubation, each plate was scored for the number of phage-resistant bacterial colonies. This was easy to ascertain because only mutant cells were not lysed and thus survived to be counted.

The experimental rationale for distinguishing between the two hypotheses was as follows.

1. **Adaptation.** Every bacterium has a small but constant probability of acquiring resistance as a result of contact with the phages on the petri dish. Therefore, the number of resistant cells will depend only on the number of bacteria and phages added to each plate. The final results should be independent of all other experimental conditions. The adaptation hypothesis predicts, therefore, that if a constant number of bacteria and phages is used for each culture and if the incubation time is constant, there should be little fluctuation in the number of resistant cells from plate to plate and from experiment to experiment.

2. **Spontaneous Mutation.** On the other hand, if resistance is acquired as a result of mutations that occur randomly, resistance will occur at a low rate during the incubation in liquid medium *prior to plating*, that is, before any contact with the phage occurs. When mutations occur *early* during incubation, the subsequent reproduction of the mutant bacteria will produce a relatively large number of resistant cells. When muta-

tions occur *later* during incubation, far fewer resistant cells will be produced. The random mutation hypothesis predicts, therefore, that the number of resistant cells will fluctuate significantly from experiment to experiment, reflecting the varying times when most spontaneous mutations occurred in liquid culture.

Table 17.1 shows a representative set of data from the Luria–Delbrück experiments. The middle column shows the number of mutants recovered from a series of aliquots derived from one large individual liquid culture. In a large culture, experimental differences are evened out because the culture is constantly mixed. As a result, these data serve as a control because the number in each aliquot should be nearly identical. As predicted, little fluctuation is observed. In contrast, the right-hand column shows considerable variation in the number of resistant mutants recovered from each of 10 independent liquid cultures. The amount of fluctuation in the data will support only one of the two alternative hypotheses. For this reason, the experiment has been designated the **fluctuation test**. Fluctuation is measured by the amount of variance, a statistical calculation.

A great fluctuation *was* observed among cultures in the Luria–Delbrück experiment, thus supporting the hypothesis that spontaneous mutations account for inherited variation in bacteria; further support came from other experimentation.

Adaptive Mutation in Bacteria

Although the concept of spontaneous mutation in viruses, bacteria, and even higher organisms has been accepted for some time, the possibility that organisms such as bacteria might also be capable of "selecting" a specific set of mutations occurring as a result of environmental pressures has long intrigued geneticists. Two independent investigations, published in 1988 by John Cairns and Barry Hall and their colleagues, have provided preliminary evidence that this might be the case. Though their findings are still controversial, this general topic is the subject of a current investigation of potentially great significance.

Cairns and his colleagues devised an experimental protocol that improves on the fluctuation test of Luria and Delbrück. The procedures were designed to detect *adaptive mutations* arising in response to factors in the environment in which bacteria are cultured. These are also referred to as *directed* mutations. The results of this work suggest that some bacteria may "select" mutations that are adaptive to the environment.

Instead of using the characteristic of phage T1 resistance, where the only surviving cells are those that have mutated, the Cairns work involved a strain of bacteria that contained a lactose mutation (*lac⁻*). Such bacteria cannot use lactose as a carbon source. Cells first were grown in a rich liquid medium that provided an adequate carbon source other than lactose. Thus, the *lac⁻* cells as well as any spontaneous *lac⁺* mutants were able to grow and reproduce quite well. Aliquots of cells were then plated on petri plates containing minimal medium, to which lactose had been added. The *lac⁻* cells, present in the vast majority, will survive on the plates, but because they lack a carbon source that they can metabolize, they cannot proliferate and form colonies. On the other hand, cells that have mutated to *lac⁺*, whether in the original liquid culture or on the plate, will form colonies and be detected. Those cells that have not mutated to *lac⁺* in liquid medium have the opportunity to do so *after* they are plated, and they may also be detected, since they will then form colonies. This is the major experimental difference between this approach and that of Luria and Delbrück some 45 years earlier.

Using a slightly more sophisticated mathematical analysis than Luria and Delbrück, Cairns attempted to identify the same sort of "fluctuation," or the lack of it, as an indication of either spontaneous or adaptive mutations, respectively. Cairns found both types of data! The distribution was calculated for both cases in which: 1) spontaneous mutations occurred in the liquid cultures at random times (creating "fluctuation"), and 2) directed mutations arose as a response to the presence of lactose on the petri plates. Indeed, such a composite distribution was observed.

But how could it be proved that, in fact, the elevated frequency of mutations occurred in cells growing on the plates in response to lactose? To answer this question, Cairns and colleagues asked whether another mutation, unrelated to the metabolism of lactose, might have also

TABLE 17.1 The Luria–Delbrück Experiment Demonstrating That Spontaneous Mutations are the Source of Phage-Resistant Bacteria

	Number of T1-Resistant Bacteria	
Sample No.	Same Culture (Control)	Different Cultures
1	14	6
2	15	5
3	13	10
4	21	8
5	15	24
6	14	13
7	26	165
8	16	15
9	20	6
10	13	10
Mean	16.7	26.2
Variance	15.0	2178.0

Source: After Luria and Delbrück, (1943).

occurred on the plates. They chose to assay for the production of Val^R mutations, which confer resistance to high concentrations of the amino acid valine. The assay was easily accomplished by overlaying the plates with agar containing valine and glucose. Only Val^R mutants will grow. Cairns and co-workers reasoned that if the lac^+ mutations did not arise specifically in response to the lactose, then Val^R mutations should arise with a similar distribution as earlier observed for the lac mutations. The result was that plates accumulating lac^+ mutants were not, at the same time, accumulating Val^R mutations. The researchers concluded that the higher frequency of lac^+ mutations resulted from the presence of lactose in the medium. This is, of course, an indirect inference.

Cairns's work suggests that the bacterial cell's genetic machinery can respond to its environment by producing adaptive mutations. This conclusion is contrary to the current thinking that mutations are spontaneous events, most of which occur as random errors during replication of DNA. As Cairns has pointed out, these findings, at first glance, seem to support the notion that life is mysterious and that certain observations cannot be explained in simple straightforward terms based on the laws of physics. Whatever the case may be, Cairns's observations are not isolated examples.

The work of Barry Hall and his colleagues demonstrated a similar adaptive response by *E. coli* to growth on the chemical salicin. In this case, a two-step genetic change was required. The first occurred in response to salicin even though that mutation offered no selective advantage without the second genetic alteration, the removal, or deletion, of a segment of nucleotides called an insertion sequence (IS). Further, the changes must occur sequentially. They do so at a much higher frequency than predicted, provided salicin is present in the growth medium. The observed frequency is several orders of magnitude higher.

What can possibly account for such observations? As Barry Hall has pointed out, our failure to explain such phenomena adequately may be attributable to our ignorance of fundamental mechanisms and rates of mutations *in nongrowing cells*, which are much more akin to a natural environment, as opposed to those found with cells grown in rich media under laboratory conditions. One suggestion is that under stressful nutritional conditions (starvation), bacteria may be capable of activating mechanisms that create a hypermutable state in genes that will enhance survival.

Though these ideas are still very much under debate, they highlight the fact that knowledge may often be derived from observations that initially cannot be explained. Such debates—and the often-surrounding controversies—are what constantly maintain intrigue and interest in science.

Mutations may be classified in various ways

We turn now to a consideration of various schemes by which mutations are classified. These organizing schemes are not mutually exclusive; they depend on which aspects of mutation are being described. In this section we describe three sets of distinctions.

Spontaneous vs. Induced Mutations

Putting aside the case of "adaptive" mutations, all mutations are described as either **spontaneous** or **induced**. Although these two categories overlap to some degree, **spontaneous mutations** are those that just happen in nature. No specific agents are associated with their occurrence, and they are generally assumed to be random changes in the nucleotide sequences of genes. Most such mutations are linked to normal chemical processes in the organism that alter the structure of the nitrogenous bases that are part of the existing genes. Most spontaneous mutations are thought to occur during the enzymatic process of DNA replication, an idea that we will discuss later in this chapter. Once an error is present in the genetic code, it may be reflected in the amino acid composition of the specified protein. If the changed amino acid is present in a part of the molecule critical to the structure or biochemical activity, a functional alteration can result.

In contrast to such spontaneous events, those that result from the influence of any artificial factor are considered to be **induced mutations**. It is generally agreed that any natural phenomenon that heightens chemical reactivity in cells will induce mutations. For example, radiation from cosmic and mineral sources and ultraviolet radiation from the sun are energy sources to which most organisms are exposed and, as such, may be factors that cause spontaneous mutations. The earliest demonstration of the artificial induction of mutation occurred in 1927, when Hermann J. Muller reported that X rays could cause mutations in *Drosophila*. In 1928, Lewis J. Stadler reported that X rays had the same effect on barley. In addition to various forms of radiation, a wide spectrum of chemical agents is also known to be mutagenic, as we will see later in this chapter.

Gametic vs. Somatic Mutations

When we consider the effects of mutation in eukaryotic organisms, it is important to distinguish whether the change occurs in somatic cells or in gametes. Mutations arising in somatic cells are not transmitted to future generations. Mutations occurring in somatic cells that create recessive autosomal alleles are rarely of any consequence to the organism. The expression of most such mutations is likely to be masked by the dominant allele. Somatic mutations will have a greater impact if they are dominant or if they are X-linked, since such mutations are most likely to be immediately expressed. Similarly, the impact will be more noticeable if such somatic mutations occur early in development, when undifferentiated cells will give rise to several differentiated tissues or organs. Mutations occurring in adult tissues are often masked by the thousands upon thousands of nonmutant cells performing the normal function.

Mutations in gametes or gamete-forming tissue are of greater significance because, as part of the germ line, they are transmitted to offspring. They have the potential of being expressed in all cells of an offspring. **Dominant autosomal mutations** will be expressed phenotypically in the first generation. **X-linked recessive mutations** arising in the gametes of a homogametic female may be expressed in hemizygous male offspring who receive the affected X chromosome. Because of heterozygosity, the occurrence of an **autosomal recessive mutation** in the gametes of either males or females (even one resulting in a lethal allele) may go unnoticed for many generations until the resultant allele has become widespread in the population. The new allele will become evident only when a chance mating brings two copies of it together in the homozygous condition.

Other Categories of Mutation

In addition to their cause of occurrence or point of origin, mutations may also be classified on the basis of their phenotypic effects. A single mutation may well fall into more than one category. The most obvious mutations are those affecting a **morphological trait**. Such variations are recognized on the basis of their deviation from the normal or wild-type phenotype. All of Mendel's pea characters and many genetic variations encountered in the study of *Drosophila* fit this designation, since they cause obvious changes in the morphology of the organism.

A second broad category of mutations includes those that exhibit **nutritional** or **biochemical variations**. In bacteria and fungi, a typical nutritional mutation is the inability to synthesize an amino acid or vitamin. In humans, hemophilia is an example of a biochemical mutation. While such mutations in these organisms are not visible and do not always affect specific morphological characters, they can have a more general effect on the well-being and survival of the affected individual.

Mutations may also affect an organism's behavior patterns. For example, the mating behavior or circadian rhythms of animals can be altered. The primary target of **behavior mutations** is often difficult to discern. For example, the mating behavior of a fruit fly may be impaired if it cannot beat its wings. However, the defect may be in (1) the flight muscles; (2) the nerves leading to them; or (3) the brain, where the nerve impulses that initiate wing movements originate. The study of behavior and the genetic factors influencing it has benefited immensely from investigations of behavior mutations.

Still another type of mutation may affect the regulation of genes. For example, as we saw in the *lac* operon, discussed in Chapter 15, a regulatory gene may produce a product that controls the transcription of other genes. At the same time, a region of DNA either close to or far away from a gene may also modulate transcription. In either case, **regulatory mutations** can disrupt this process and permanently activate or inactivate a gene. Our knowledge of genetic regulation has been dependent on the study of mutations that disrupt this process.

Another group consists of **lethal mutations**. Various human biochemical disorders, such as Tay-Sachs disease and Huntington disease, are lethal at different points in the life cycle of humans. Nutritional and biochemical mutations can also be lethal. A mutant bacterium that cannot synthesize a specific amino acid it needs will cease to grow in a medium lacking that amino acid.

Finally, any of the above groups can exist as **conditional mutations**. That is, a mutation may be present in the genome of an organism, but it may be evident only under certain conditions. Among the best examples are **temperature-sensitive mutations**, found in a variety of organisms. At certain "permissive" temperatures, a mutant gene product functions normally, but loses its functional capability at a "restrictive" temperature. When the organism is shifted to the restrictive temperature, the impact of the mutation becomes apparent, perhaps even lethal, and is amenable to investigation. The study of conditional mutations has been extremely important in experimental genetics, particularly in understanding the function of genes essential to the viability of organisms.

Genetic techniques, cell cultures, and pedigree analysis are all used to detect mutations

In order to study the mutational process, geneticists must be able to detect the mutations in an organism. The ease and efficiency of detecting mutations has been one of the main criteria in the selection of organisms for genetic studies. In this section we use several examples to illustrate how mutations are detected.

Detection in Bacteria and Fungi

Detection is most efficient in haploid microorganisms, where recessive mutations may be detected without the complications generated by heterozygosity in diploid organisms. In both bacteria and fungi, detection depends on a selection system whereby mutant cells are isolated easily from nonmutant cells. To illustrate, we will look at the fungus *Neurospora crassa*, a mold that normally grows on bread and can be cultured in the laboratory. This eukaryotic mold is haploid in the vegetative phase of its life cycle (see Figure 6.19).

Neurospora shows visible mutant traits such as *albino*, but the full potential of *Neurospora* genetics was realized during the investigation of nutritional mutants. Wild-type *Neurospora* grow on a **minimal culture medium** of glucose, a few inorganic acids and salts, a nitrogen source such as ammonium nitrate, and the vitamin biotin. Induced nutritional mutants will not grow on minimal medium, but will grow on a supplemented or **complete medium** that also contains numerous amino acids, vitamins, and nucleic acid derivatives, among other chemical additives. Microorganisms that are nutritional wild types (requiring

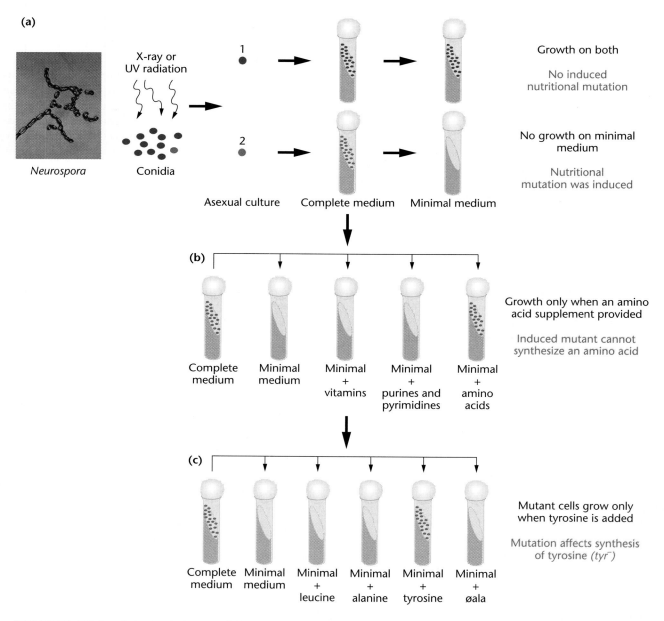

FIGURE 17.1 Induction, isolation, and characterization of a nutritional auxotrophic mutation in *Neurospora*. In (a), conidia are irradiated; most are not affected, but one conidium (shown in red) contains such a mutation, as evidenced by the fact that it will not grow on minimal medium. In (b) and (c), the precise mutation is determined to involve the biosynthesis of tyrosine.

only minimal medium) are called **prototrophs**, whereas those mutants that require a specific supplement to the minimal medium are called **auxotrophs**.

The procedure for detecting and characterizing nutritional mutations in *Neurospora* is illustrated in Figure 17.1. Haploid conidia bearing such a mutation can be detected and isolated by their failure to grow on minimal medium and their ability to grow on complete medium. The mutant cells can no longer synthesize some essential compound absent in minimal medium but present in complete medium [Figure 17.1(a)]. Once a nutritional mutant is detected and isolated, the missing compound can be determined by attempts to grow the mutant strain in a series of tubes, each containing minimal medium supplemented with a single compound [Figure 17.1(b) and (c)]. In this example, mutants can grow if tyrosine is supplied; thus, the mutation is a tyrosine auxotroph (tyr^-). Recall that this was the procedure used by Beadle and Tatum in their pioneering research leading to the "one-gene: one-enzyme" hypothesis, discussed in Chapter 14.

Detection in Drosophila

In his studies demonstrating that X rays are mutagenic, Hermann J. Muller developed a number of systems for detecting and estimating the spontaneous and induced rates of X-linked and autosomal recessive lethal mutations in *Drosophila melanogaster*. We will consider two of these techniques, the *ClB* and the **attached-X procedures**.

The *ClB* system (Figure 17.2) detects the rate of induction of these X-linked recessive lethal mutations: *C*, an inversion that suppresses the recovery of crossover products; *l*, a recessive lethal allele; and *B*, the dominant gene duplication causing *Bar* eye, described in Chapter 10. In this technique, wild-type males are treated with a mutagenic agent—in this case, X rays—and mated to untreated, heterozygous *ClB* females. In the F_1 generation, females expressing *Bar* are selected. They have received one X from their mother (the *ClB* chromosome) and one X from their father. Some of the paternal X chromosomes will bear an induced X-linked lethal mutation. When an F_1 *Bar* female that is heterozygous for such a lethal mutation is then backcrossed to a wild-type male, no male progeny will result. Half of them die because they are hemizygous for the *ClB* chromosome (remember, *l* is lethal) and the other half die because they are hemizygous for the newly induced lethal allele.

Following many such single-pair matings, the percentage of vials bearing *only* females represents the frequency of the induction of X-linked lethal mutations. For example, if 500 such F_2 culture bottles were inspected, and 25 contained only females, the induced rate of such mu-

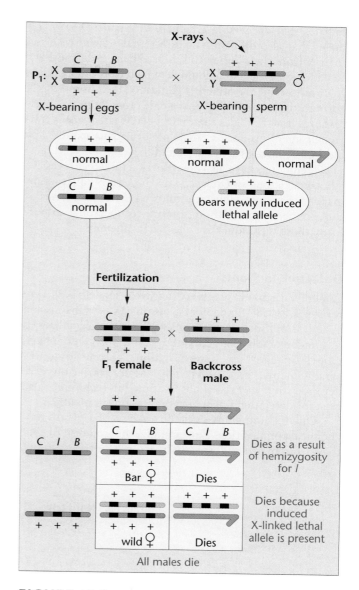

FIGURE 17.2   Muller's classical *ClB* technique for the detection of induced, X-linked recessive lethal mutations in *Drosophila*.

tations would be 5 percent. This method also permits the detection of X-linked morphological mutations. If a mutation of this type has been induced, all surviving F_2 males will show the trait.

The second technique, using females with attached-X chromosomes, is an even simpler way to detect recessive morphological mutations because it requires only one generation. These females have two X chromosomes attached to a single centromere and one Y chromosome, in addition to the normal diploid complement of autosomes. When attached-X females are mated to males with normal sex chromosomes (XY), four types of progeny result: triplo-X females that die; viable attached-X females (X̂XY); YY males that also die; and viable XY males (who received their X from their father and their Y from their mother). In contrast to the *ClB* technique, which tests only a single X chromosome, the attached-X method tests numerous X chromosomes at one time in a single cross. Figure 17.3 illustrates how P_1 males that have been treated with a mutagenic agent produce F_1 male offspring that express any induced X-linked recessive mutation.

Techniques have also been devised to detect recessive autosomal lethals in *Drosophila*. In these techniques, dominant marker mutations are followed through a series of three generations. Although more cumbersome to perform, these techniques are also fairly efficient.

Detection in Plants

Genetic variation in plants is extensive. Mendel's peas were the basis for the fundamental postulates of transmission genetics, and subsequent studies of plants have enhanced our understanding of gene interaction, polygenic inheritance, linkage, sex determination, chromosome rearrangements, and polyploidy. Many variations are detected simply by visual observation, but techniques also exist for detecting biochemical mutations in plants.

One technique involves the analysis of a plant's biochemical composition. For example, the isolation of proteins from maize endosperm, hydrolysis of the proteins,

and determination of the amino acid composition have revealed that the *opaque*-2 mutant strain contains significantly more lysine than do other, nonmutant lines. Because maize protein is usually low in lysine, this mutation significantly improves the nutritional value of the plant. Once this mutant was discovered, plant geneticists and other specialists analyzed the amino acid compositions of various strains of other grain crops, including rice, wheat, barley, and millet. The results of such analyses are useful in combating malnutrition diseases resulting from inadequate protein or the lack of essential amino acids in the diet.

Other detection techniques involve the tissue culture of plant cell lines in defined medium. The plant cells are treated like microorganisms, and their resistance to herbicides or disease toxins can be determined by adding these compounds to the culture medium. The techniques associated with conditional lethal mutants can be used on plant cells in tissue culture and then applied to the genetics of higher plants, providing a means of detection that would not be possible in an intact plant.

Detection in Humans

Because humans are obviously not suitable experimental organisms, techniques developed for the detection of mutations in organisms such as *Drosophila* are not available to human geneticists. To determine the mutational basis for any human characteristic or disorder, geneticists must first analyze a pedigree that traces the family history as far back as possible. Once any trait has been shown to be inherited, it is possible to predict whether the mutant allele is behaving as a dominant or a recessive and whether it is X-linked or autosomal.

Dominant mutations are the simplest to detect. If they are present on the X chromosome, affected fathers pass the phenotypic trait to all their daughters. If dominant mutations are autosomal, approximately 50 percent of the offspring of an affected heterozygous individual are expected to show the trait. Figure 17.4 is a pedigree illustrating the initial occurrence of an autosomal dominant allele for cataracts of the eye. The parents in Generation I were unaffected, but one of their three offspring (Generation II) developed cataracts, so the original mutation presumably occurred in a gamete of one of the parents. The affected female, the proband, produced two children, of which the male child was affected (Generation III). Of his six offspring (Generation IV), four were also affected. Though these observations do not prove an autosomal dominant mode of inheritance, the high percentage of affected offspring in Generation IV favors this conclusion. Provided that the mutant allele is completely penetrant, the presence of an unaffected daughter in this generation

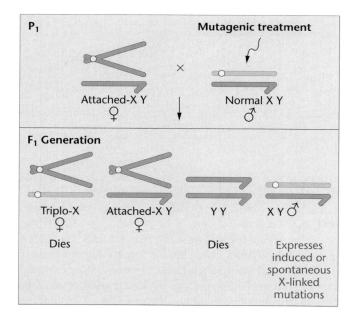

FIGURE 17.3 The attached-X method for detection of induced morphological mutations in *Drosophila*.

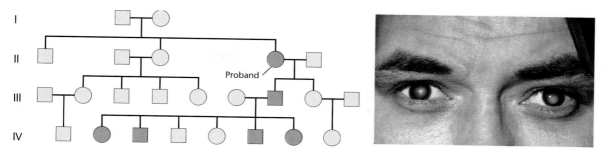

FIGURE 17.4 A hypothetical pedigree of inherited cataract of the eye in humans.

argues against X-linkage because she received her X chromosome from her affected father.

The X-linked recessive mutations can also be detected by pedigree analysis. The most famous example of an X-linked mutation in humans is the **hemophilia** that was

found in the descendants of Britain's Queen Victoria. Inspection of the pedigree in Figure 17.5 leaves little doubt that Victoria was heterozygous (*Hb*) for the trait. Her father was not affected, and there is no reason to believe that her mother was a carrier, as was Victoria. The recessive

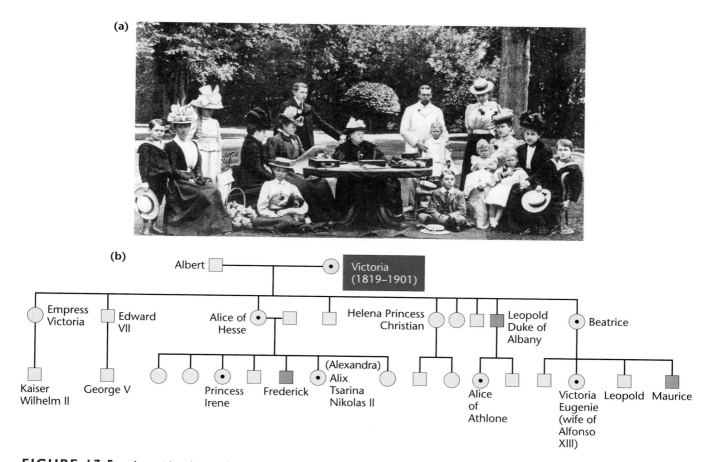

FIGURE 17.5 A partial pedigree of hemophilia in the British royal family descended from Queen Victoria. The pedigree is typical of the transmission of X-linked recessive traits. Circles with a dot in them indicate presumed female carriers heterozygous for the trait. The photograph shows Queen Victoria (seated front center) and some of her immediate family.

TABLE 17.2 Rates of Spontaneous Mutations at Various Loci in Different Organisms

Organism	Character	Gene	Rate	Units
Bacteriophage T2	Lysis inhibition	$r \rightarrow r^+$	1×10^{-8}	Per gene replication
	Host range	$h^+ \rightarrow h$	3×10^{-9}	
	Lactose fermentation	$lac^- \rightarrow lac^+$	2×10^{-7}	
	Lactose fermentation	$lac^+ \rightarrow lac^-$	2×10^{-6}	
	Phage T1 resistance	$Tl\text{-}s \rightarrow Tl\text{-}r$	2×10^{-8}	
	Histidine requirement	$his^+ \rightarrow his^-$	2×10^{-6}	
	Histidine independence	$his^- \rightarrow his^+$	4×10^{-8}	
E. coli	Streptomycin dependence	$str\text{-}s \rightarrow str\text{-}d$	1×10^{-9}	Per cell division
	Streptomycin sensitivity	$str\text{-}d \rightarrow str\text{-}s$	1×10^{-8}	
	Radiation resistance	$rad\text{-}s \rightarrow rad\text{-}r$	1×10^{-5}	
	Leucine independence	$leu^- \rightarrow leu^+$	7×10^{-10}	
	Arginine independence	$arg^- \rightarrow arg^+$	4×10^{-9}	
	Tryptophan independence	$try^- \rightarrow try^+$	6×10^{-8}	
Salmonella typhimurium	Tryptophan independence	$try^- \rightarrow try^+$	5×10^{-8}	Per cell division
Diplococcus pneumoniae	Penicillin resistance	$pen^s \rightarrow pen^r$	1×10^{-7}	Per cell division
Chlamydomonas reinhardi	Streptomycin sensitivity	$str^r \rightarrow str^s$	1×10^{-6}	Per cell division
Neurospora crassa	Inositol requirement	$inos^- \rightarrow inos^+$	8×10^{-8}	Mutant frequency among asexual spores
	Adenine independence	$ade^- \rightarrow ade^+$	4×10^{-8}	
Zea mays	Shrunken seeds	$sh^+ \rightarrow sh^-$	1×10^{-6}	Per gamete per generation
	Purple	$pr^+ \rightarrow pr^-$	1×10^{-5}	
	Colorless	$c^+ \rightarrow c$	2×10^{-6}	
	Sugary	$su^+ \rightarrow su$	2×10^{-6}	
Drosophila melanogaster	Yellow body	$y^+ \rightarrow y$	1.2×10^{-6}	Per gamete per generation
	White eye	$w^+ \rightarrow w$	4×10^{-5}	
	Brown eye	$bw^+ \rightarrow bw$	3×10^{-5}	
	Ebony body	$e^+ \rightarrow e$	2×10^{-5}	
	Eyeless	$ey^+ \rightarrow ey$	6×10^{-5}	
Mus musculus	Piebald coat	$s^+ \rightarrow s$	3×10^{-5}	Per gamete per generation
	Dilute coat color	$d^+ \rightarrow d$	3×10^{-5}	
	Brown coat	$b^+ \rightarrow b$	8.5×10^{-4}	
	Pink eye	$p^+ \rightarrow p$	8.5×10^{-4}	
Homo sapiens	Hemophilia	$h^+ \rightarrow h$	2×10^{-5}	Per gamete per generation
	Huntington disease	$Hu^+ \rightarrow Hu$	5×10^{-6}	
	Retinoblastoma	$R^+ \rightarrow R$	2×10^{-5}	
	Epiloia	$Ep^+ \rightarrow Ep$	1×10^{-5}	
	Aniridia	$An^+ \rightarrow An$	5×10^{-6}	
	Achondroplasia	$A^+ \rightarrow A$	5×10^{-5}	

mutation for hemophilia is not rare in human populations, but the political consequences of the mutation that occurred in the British royal family have been sweeping.*

It also is possible to detect autosomal recessive alleles by pedigree analysis. Because this type of mutation is "hidden" when heterozygous, it is not unusual for the trait to appear only intermittently through a number of generations. A mating between an affected individual and a homozygous normal individual will produce unaffected heterozygous carrier children. Matings between two carriers will produce, on average, one-fourth affected offspring.

In addition to pedigree analysis, human cells may now be routinely cultured *in vitro*. This procedure has allowed

the study of many more mutations than any other form of analysis. Analysis of enzyme activity, protein migration in electrophoretic fields, and direct sequencing of DNA and proteins are among the techniques that have identified mutations and demonstrated the wide genetic variation among individuals in human populations. As we will see later in this chapter, the study of the human disorder xeroderma pigmentosum has relied on the *in vitro* cell culture technique.

The spontaneous mutation rate varies greatly among organisms

The types of detection systems we have been discussing allow geneticists to estimate mutation rates. However, we can only ascertain induced mutation when the induced rate clearly exceeds the spontaneous mutation rate for the or-

---

*Robert Massie's *Nicholas and Alexandra* and Robert and Suzanne Massie's *Journey* (see the list of Selected Readings at the end of this chapter) provide fascinating reading on the topic of hemophilia.

ganism under study. Determining the rate of spontaneous mutation provides the baseline for measuring the rate of experimentally induced mutation.

The examination of this rate in a variety of organisms reveals many interesting points (see Table 17.2). First, the rate of spontaneous mutation is exceedingly low for all the organisms studied. Second, the rate varies considerably among different organisms. Third, even within the same species, the spontaneous mutation rate varies from gene to gene.

Viral and bacterial genes undergo spontaneous mutation on an average of about 1 in 100 million (10^{-8}) cell divisions. *Neurospora* exhibits a similar rate, but maize, *Drosophila*, and humans demonstrate a rate several orders of magnitude higher. The genes studied in these groups average between 1/1,000,000 and 1/100,000 (10^{-6} and 10^{-5}) mutations per gamete formed. Mouse genes are still another order of magnitude higher in their spontaneous mutation rate, 1/100,000 to 1/10,000 (10^{-5} to 10^{-4}). It is not clear why such a large variation occurs in the mutation rate. The variation might reflect the relative efficiency of the enzyme systems, whose function is to repair errors created during replication. We will discuss repair systems later in this chapter.

Deleterious Mutations in Humans

The vast majority of mutations are likely to occur in the large portions of the genome that do not encode genes or in intron regions found within genes. These are considered to be neutral mutations, since they do not usually affect gene products. While this information is of great interest to geneticists, perhaps a more significant piece of information regarding spontaneous mutations is the rate at which those mutations *that are deleterious* occur. Of greatest interest is the rate of occurrence of these mutations in our own species. For decades, this rate has been nearly impossible to determine. However, recent work, using molecular techniques, has allowed an accurate estimate to be made, and it is surprisingly high, at least 1.6 deleterious genetic changes per individual per generation.

The way in which this work was performed and its possible consequences to our species are clearly and insightfully discussed by James Crow in a short article that was published recently (January, 1999) in *Nature*. Dr. Crow is one of the foremost researchers and a true pioneer in mutation research. His article, "The Odds of Losing at Genetic Roulette" is reprinted below. We trust you will

The Odds of Losing at Genetic Roulette

JAMES F. CROW

Department of Genetics, University of Wisconsin, Madison, Wisconsin 53706.

The rate at which deleterious mutations occur in a genome is clearly a quantity of interest—not least when the genome is our own. It becomes particularly important if, as some have argued[1], the rate in some species is high, at several new mutations each generation. Yet the deleterious mutation rate has been notoriously difficult to measure, and no convincing estimates exist for any vertebrate. On page 344 of this issue[2], Eyre-Walker and Keightley give the first such estimates for ourselves, chimpanzees and gorillas.

Getting an estimate of the total mutation rate is relatively simple. Neutral mutations—those which neither enhance nor impair the organism carrying them—accumulate through the generations at a rate equal to the mutation rate[3]. The mutation rate can be determined by the rate of change of presumed neutral regions: areas of the genome, such as introns and pseudogenes, which are not translated into proteins. For mammals, these rates, extrapolated to the whole genome, lead to enormous numbers, on the order of 100 new mutations per individual[1]. Of course these can't all be deleterious, but

no one knows what the proportion is. This proportion could be determined in principle by comparing the slower rate of evolutionary change of the genome as a whole with the faster rate expected under neutral assumptions; but this involves statistical uncertainties and extensive sequencing[4].

Eyre-Walker and Keightley[2] have made the analysis feasible by concentrating on protein-coding regions. They measured the amino-acid changes in 46 proteins in the human ancestral line after its divergence from the chimpanzee. Among 41,471 nucleotides, they found 143 nonsynonymous substitutions—mutations where swapping one DNA base for another changes an amino acid, and therefore the final protein made by that gene. If these had evolved at the neutral rate, 231 would be expected. The difference, 88 (38%), is an estimate of the number of deleterious mutations that have been eliminated by natural selection and have therefore made no contribution to contemporary populations.

Translating these numbers into mutation rates gave a total rate of 4.2 mutations per person per generation, and a deleterious rate of 1.6. The rates of chimpanzees and gorillas were very similar, the deleterious rates being 1.7 and 1.2, respectively. The authors took 60,000 as the gene number and 25 years as the generation length. The number 1.6 is probably an underestimate, for various reasons. For instance, mutations outside the coding region are not counted and some of these regions—such as those controlling gene expression—are expected to be subject to natural selection. The gene number may also be an underestimate. If there have been mutations that increase fitness, they would also cause the number of deleterious mutations to be underestimated. A less conservative, and probably more realistic, estimate doubles the value, giving 3 new deleterious mutations per person per generation.

What's the significance? Every deleterious mutation must eventually be eliminated from the population by premature death or reduced relative reproductive success, a 'genetic death'[5]. That implies three genetic deaths per person! Why aren't we extinct? If harmful mutations were eliminated independently, as in an asexual species, it has been estimated that this would lower population fitness to a fraction e^{-3}, or 5%, of the mutation-free value[6], leading to the inevitable extinction of species with limited reproductive capacity. A way out is for mutations to be eliminated in bunches. This happens if selection operates such that individuals with the most mutations are preferentially eliminated, for example if harmful mutations interact. But such a process can only work in sexual species, where mutations are shuffled each generation by genetic recombination[1,7]. The existence of a high deleterious mutation rate strengthens the argument that a major advantage of sex is that it is an efficient way to eliminate harmful mutations[1]. It also raises again the possibility of fitness decline or even extinction in rare species from too many harmful mutations[8].

Presumably, we humans have profited in the past by sexual reproduction's ability to reduce the effect of a high mutation rate on fitness. In the recent past the intensity of natural selection has been greatly reduced, especially where a high standard of living means that most infants reach reproductive age. From this it would seem that natural selection will weed out mutations more slowly than they accumulate. This effect may be accentuated by trends for males to start or continue reproducing later in life, because the sperm of older men contains more base-substitution mutations[8]. In a time of rapid environmental improvement, how this genetic decline will affect our health can only be guessed at.

Eyre-Walker and Keightley[2] noticed that the proportion of harmful mutations in the 46 genes in their study is greater in humans than in the equivalent genes of rodents. Their preferred explanation is that slightly deleterious mutations have become fixed in the population, by a process known as random genetic drift, during periods of human history when the breeding population size was low—especially during genetic 'bottlenecks.' This would increase whatever effect the accumulated mutations are having on current human welfare. Are some of our headaches, stomach upsets, weak eyesight and other ailments the result of mutation accumulation? Probably, but in our present state of knowledge, we can only speculate.

Reprinted by permission from *Nature* 397, 283, 284. Copyright © 1999 Macmillan Magazines Ltd.

[1]Kondrashov, A.S. *J. Theor. Biol.* **175**, 583–594 (1995).
[2]Eyre-Walker, A. & Keightley, P.D. *Nature* **397**, 344–347 (1999).
[3]Kimura, M. *The Neutral Theory of Molecular Evolution* (Cambridge Univ. Press, 1983).
[4]Kondrashov, A.S. & Crow, J.F. *Hum. Mutat.* **2**, 229–234 (1993).
[5]Muller, H.J. *Am. J. Hum. Genet.* **2**, 111–176 (1950).
[6]Kimura, M. & Maruyama, T. *Genetics* **54**, 1337–1351 (1966).
[7]Crow, J.F. *Proc. Natl Acad. Sci. USA* **94**, 8380–8386 (1997).
[8]Lande, R. *Evolution* **48**, 1460–1469 (1994).

enjoy his perspectives on the general topic and be enlightened by his comments.

Mutations occur in many forms and arise in different ways

Even though we are aware that the gene is a reasonably complex genetic unit, particularly in eukaryotes, we will use a *simplified* definition of a gene in the following description of the molecular basis of mutation. In this context, it is easiest to consider a gene as a linear sequence of nucleotide pairs representing stored chemical information. Because the genetic code is a triplet, each sequence of three nucleotides specifies a single amino acid in the corresponding polypeptide. Any change that dis-

rupts these sequences or the coded information provides sufficient basis for a mutation. The least complex change is the substitution of a single nucleotide. In Figure 17.6, such a change is compared with our own written language, using three-letter words to be consistent with the genetic code. A change of one letter can alter the meaning of the sentence "THE CAT SAW THE DOG," to "THE CAT SAW THE HOG" or "THE BAT SAW THE DOG," creating what is called *missense*. These are analogies to what are most appropriately referred to as **base substitutions** or **point mutations**. The mutation has changed the sense of the information into various forms of missense.

Two other more formal terms are often used to describe nucleotide substitutions. If a pyrimidine replaces a

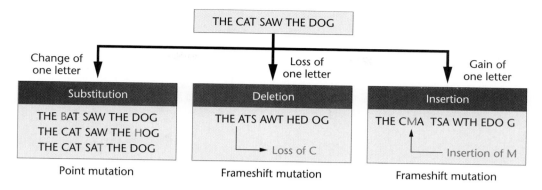

FIGURE 17.6   The impact of the substitution, deletion, and insertion of one letter in a sentence composed of three-letter words as analogies to point and frameshift mutations.   GenCDX

pyrimidine or a purine replaces a purine, a **transition** has occurred. If a purine and a pyrimidine are interchanged, a **transversion** has occurred.

A second type of change is the insertion or deletion of one or more nucleotides at any point within the gene. As illustrated in Figure 17.6, the loss or addition of a single letter causes all the subsequent three-letter words to be changed. These examples are called **frameshift mutations** because the frame of reading is altered. A frameshift mutation will occur when any number of bases are added or deleted except multiples of three, which would reestablish the initial frame of reading. We will return momentarily to a discussion of frameshift mutations as we introduce acridine dyes.

The analogy in Figure 17.6 demonstrates that insertions and deletions have the potential to change all the subsequent triplets in a gene. It is probable that, of the 64 possible triplets, one of the many altered triplets will be UAA, UAG, or UGA, the termination codons. When one of these triplets is encountered during translation, polypeptide synthesis is terminated. Obviously, the results of frameshift mutations can be very severe.

Tautomeric Shifts

In 1953, immediately after they had proposed the molecular structure of DNA, Watson and Crick published a paper in which they discussed the genetic implications of this structure. They recognized that the purines and pyrimidines found in DNA could exist in **tautomeric** forms; that is, a nitrogenous base can exist in alternate chemical forms called structural isomers, each differing by only a single proton shift in the molecule. The biologically important tautomers involve the keto-enol forms of thymine and guanine, and amino-imino forms of cytosine and adenine (Figure 17.7). Because such a shift changes the bonding structure of the molecule, Watson and Crick

suggested that **tautomeric shifts** could result in base-pair changes or mutations.

The most stable tautomers of the nitrogenous bases result in the standard base pairings that serve as the basis of the double-helix model of DNA. The less frequently occurring, transient tautomers are capable of hydrogen bonding with noncomplementary bases. However, the pairing is always between a pyrimidine and a purine. Figure 17.8 compares the normal base-pairing relationships with the rare unorthodox pairings. Anomalous $T \equiv G$ and $C = A$ pairs, among others, may be formed.

The effect leading to mutation occurs during DNA replication when a rare tautomer in the template strand matches with a noncomplementary base. In the next round of replication, the "mismatched" members of the base pair are separated, and each specifies its normal complementary base. The end result is a point mutation (Figure 17.9).

Base Analogues

Molecules that may substitute for purines or pyrimidines during nucleic acid biosynthesis are called **base analogs** and are often mutagenic chemicals. For example, **5-bromouracil (5-BU)*** a derivative of uracil, behaves as a thymine analog and is halogenated at the number 5 position of the pyrimidine ring. Figure 17.10 compares the structure of this analog with that of thymine. The presence of the bromine atom in place of the methyl group increases the probability that a tautomeric shift will occur. If 5-BU is incorporated into DNA in place of thymine and a tautomeric shift to the enol form occurs, 5-BU base-pairs with guanine. After one round of replication, an $A = T$ to $G \equiv C$ transition results (see Figure 17.9). Furthermore, the presence of 5-BU within DNA increases

*If 5-BU is chemically linked to d-ribose, the nucleoside analogue bromodeoxyuridine (BUdR) is formed.

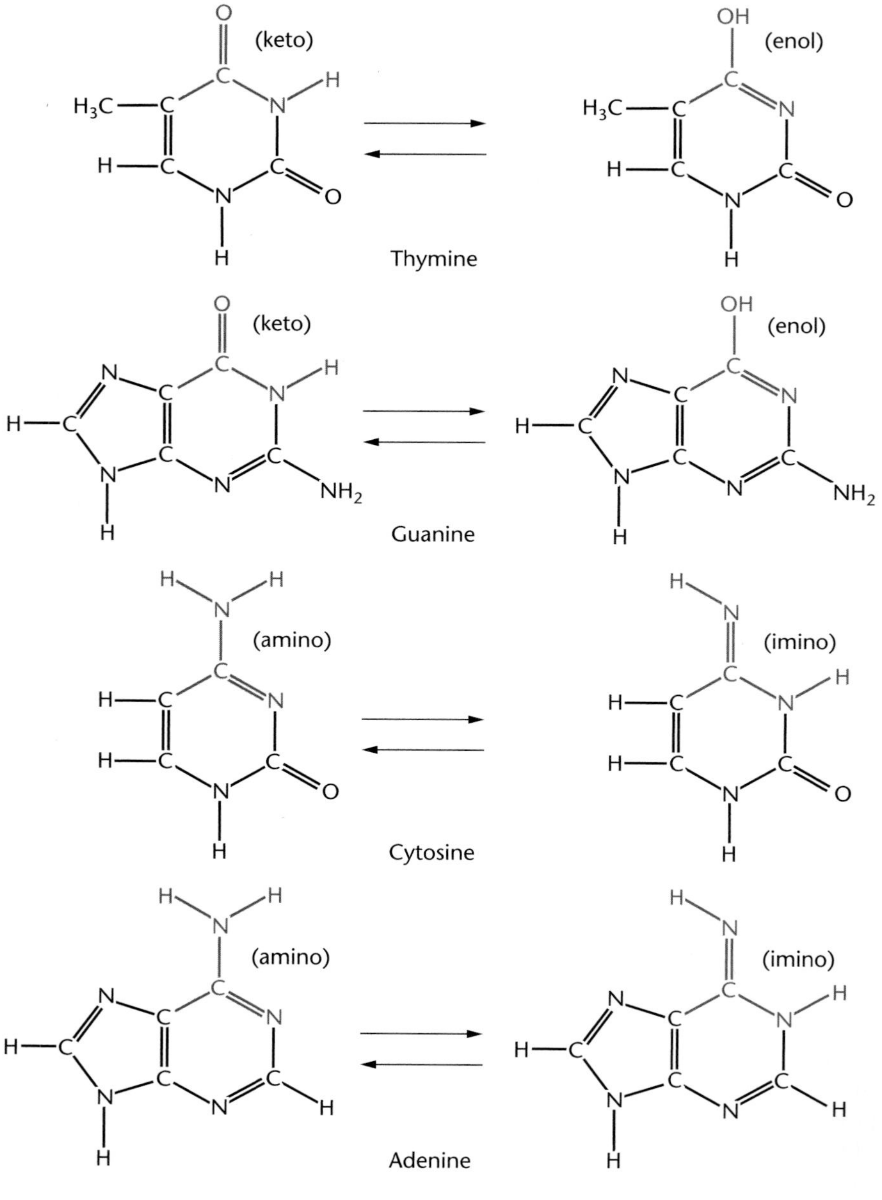

FIGURE 17.7 Rare tautomeric shifts that occur in the chemical structure of the four nitrogenous bases of DNA.

the sensitivity of the molecule to ultraviolet (UV) light, which itself is mutagenic (as discussed in the next section).

There are other base analogues that are mutagenic. For example, **2-amino purine (2-AP)**, can serve successfully as an analogue of adenine. In addition to its base-pairing affinity with thymine, 2-AP can also base-pair with cytosine, leading to possible transitions from $A = T$ to $G \equiv C$ following replication.

Because of the specificity by which base analogues such as 2-AP induce transition mutations, base analogues may also be used to induce reversion to the wild-type nu-

cleotide sequence. This alteration is called **reverse mutation**. The process can also occur spontaneously, although at a much lower rate.

Alkylating Agents

The sulfur-containing **mustard gases**, discovered during World War II, were one of the first groups of chemical mutagens identified in chemical warfare studies. Mustard gases are **alkylating agents**; that is, they donate an alkyl group such as CH_3- or CH_3-CH_2- to amino or keto groups

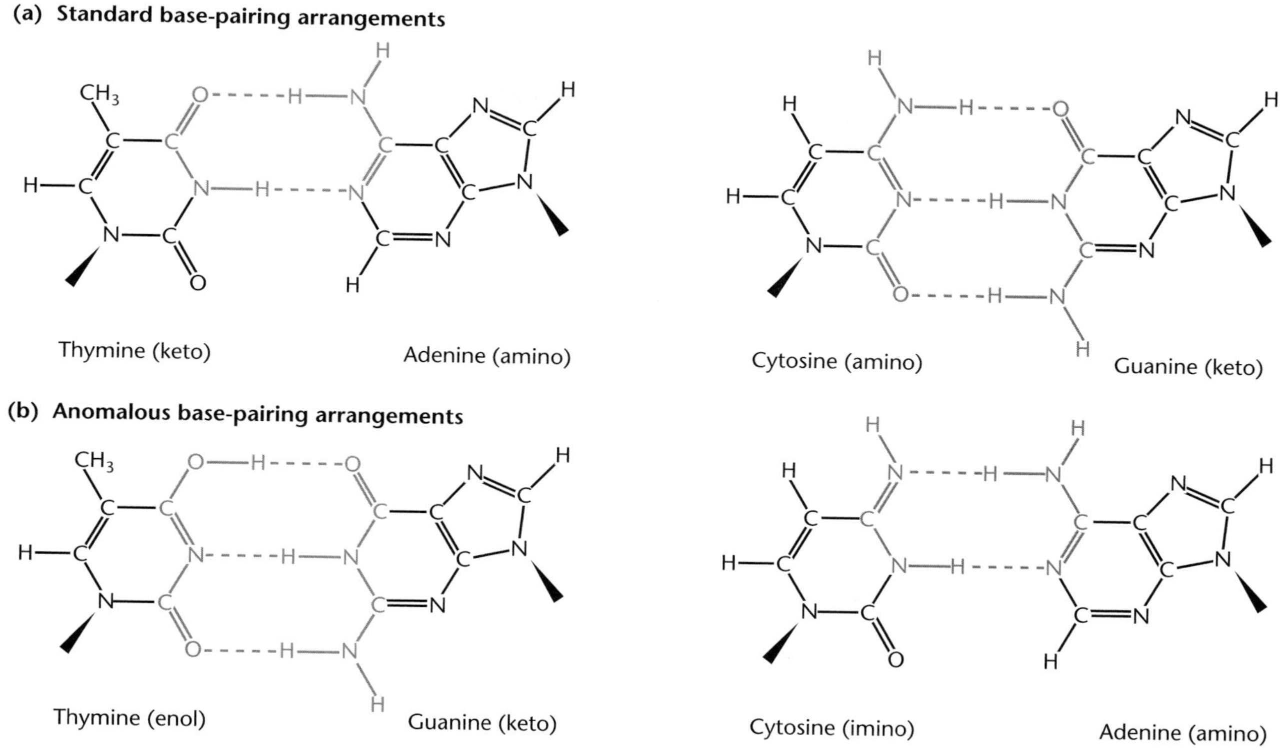

(a) Standard base-pairing arrangements

Thymine (keto)   Adenine (amino)   Cytosine (amino)   Guanine (keto)

(b) Anomalous base-pairing arrangements

Thymine (enol)   Guanine (keto)   Cytosine (imino)   Adenine (amino)

FIGURE 17.8   The standard base-pairing relationships compared with two anomalous arrangements occurring as a result of tautomeric shifts. The dense arrowhead indicates the point of bonding to the pentose sugar.

in nucleotides. **Ethylmethane sulfonate (EMS)**, for example, alkylates the keto groups in the number 6 position of guanine and in the number 4 position of thymine (Figure 17.11). As with base analogues, base-pairing affinities are altered and transition mutations result. In the case of **6-ethyl guanine**, for example, the molecule acts like a base analogue of adenine and pairs with thymine. Table 17.3 lists the chemical names and structures of several frequently used alkylating agents known to be mutagenic.

Chemical mutagens called **acridine dyes** cause frameshift mutations. As illustrated in Figure 17.6 and introduced in the earlier discussion, these mutations result from the addition or removal of one or more base pairs (bp) in the polynucleotide sequence of the gene. Acridine dyes are a group of aromatic organic molecules, illustrated in Figure 17.12 by the widely studied mutagens **proflavin** and **acridine orange**. They are roughly the same dimensions as a nitrogenous base pair and are known to intercalate or wedge between the purines and pyrimidines of intact DNA, inducing contortions in the DNA helix and causing deletions and additions.

One model suggests that the resultant frameshift mutations are generated at gaps produced in DNA during replication, repair, or recombination. During these events, there is the possibility of slippage and improper base-pairing of one strand with the other. The model suggests that the intercalation of the acridine into an improperly base-paired region can extend the existence of these slippage structures. If so, the probability increases that the mispaired configuration will exist when synthesis and rejoining occurs, thereby resulting in an addition or deletion of one or more bases in one of the strands.

Apurinic Sites and Other Lesions

Still another type of mutation involves the spontaneous loss of one of the nitrogenous bases—most frequently, guanine or adenine—in an intact double-helix DNA molecule. These sites, created by the "breaking" of the glycosidic bond linking the 1'-C of d-ribose and the 9-N of the purine ring, are called **apurinic sites (AP sites)**.* It

---

*Apyrimidinic sites, where a pyrimidine has been lost, also occur and are also referred to as AP sites.

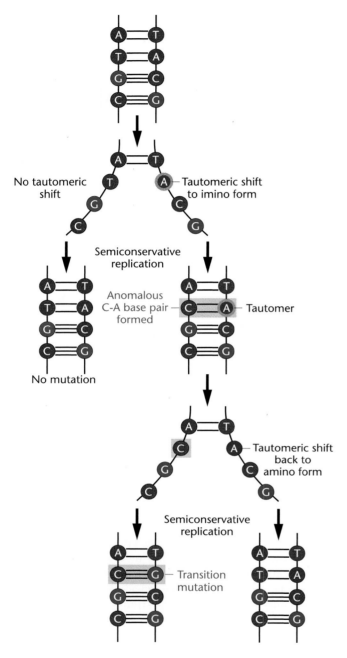

FIGURE 17.9   Formation of a T = A to a C ≡ G transition mutation as a result of a tautomeric shift in adenine.

adequate template and may cause replication to stall. If a nucleotide is inserted, it is frequently incorrect, causing still another mutation. Fortunately, as we will see, cells contain repair systems that often counteract and correct this type of lesion.

Several other types of lesions are known to be the source of some mutations. In the process of **deamination**, an amino group is converted to a keto group in cytosine and adenine (Figure 17.13). In this case, the mutagen is nitrous acid, known to be capable of inducing the deamination of bases in DNA. Cytosine is converted to uracil and adenine is changed to hypoxanthine. The major effect of these changes is to alter the base-pairing specificities of these two molecules during DNA replication. For example, cytosine normally pairs with guanine. Following its conversion to uracil, which pairs with adenine, the original G ≡ C pair is converted to an A = U pair and, following an additional replication, to an A = T pair. When adenine is deaminated, an original A = T pair is converted to a G ≡ C pair because hypoxanthine pairs naturally with cytosine.

Finally, we mention mutational lesions caused by oxidation reactions. Active forms of oxygen radicals, such as hydrogen peroxide (H_2O_2) and superoxides (O_2^-), can cause damage to the DNA bases and result in mispairing during replication.

Ultraviolet Radiation and Thymine Dimers

In Chapter 11 we emphasized the fact that purines and pyrimidines absorb **ultraviolet (UV) radiation** most intensely at a wavelength of about 260 nm, a property that has been useful in the detection and analysis of nucleic acids. In 1934, as a result of studies involving *Drosophila* eggs, it was discovered that UV radiation is mutagenic, and by 1960, several studies concerning the *in vitro* effect on the components of nucleic acids had been completed. The major effect of UV radiation is on pyrimidines, where dimers are induced, particularly between two thymine residues (Figure 17.14). While cytosine–cytosine and thymine–cytosine dimers may also be formed, they are less prevalent. The dimers distort the DNA conformation and inhibit normal replication, which seems to be responsible, at least in part, for the killing effects of UV radiation on microorganisms.

As we will see later, several mechanisms have evolved to correct UV-induced lesions. We will illustrate the essential nature of such "repair" processes in humans by discussing the inherited disorder xeroderma pigmentosum, where mutation has inactivated one UV-repair mechanism. Such individuals have been described as "children of the night," since they cannot risk exposure to the ultraviolet component of sunlight without the risk of epidermal malignancy.

has been estimated that, in the DNA of mammalian cells in culture, thousands of such spontaneous lesions are formed daily.

The absence of a nitrogenous base at an AP site will alter the genetic code if the involved strand is transcribed and translated. If replication occurs, the AP site is an in-

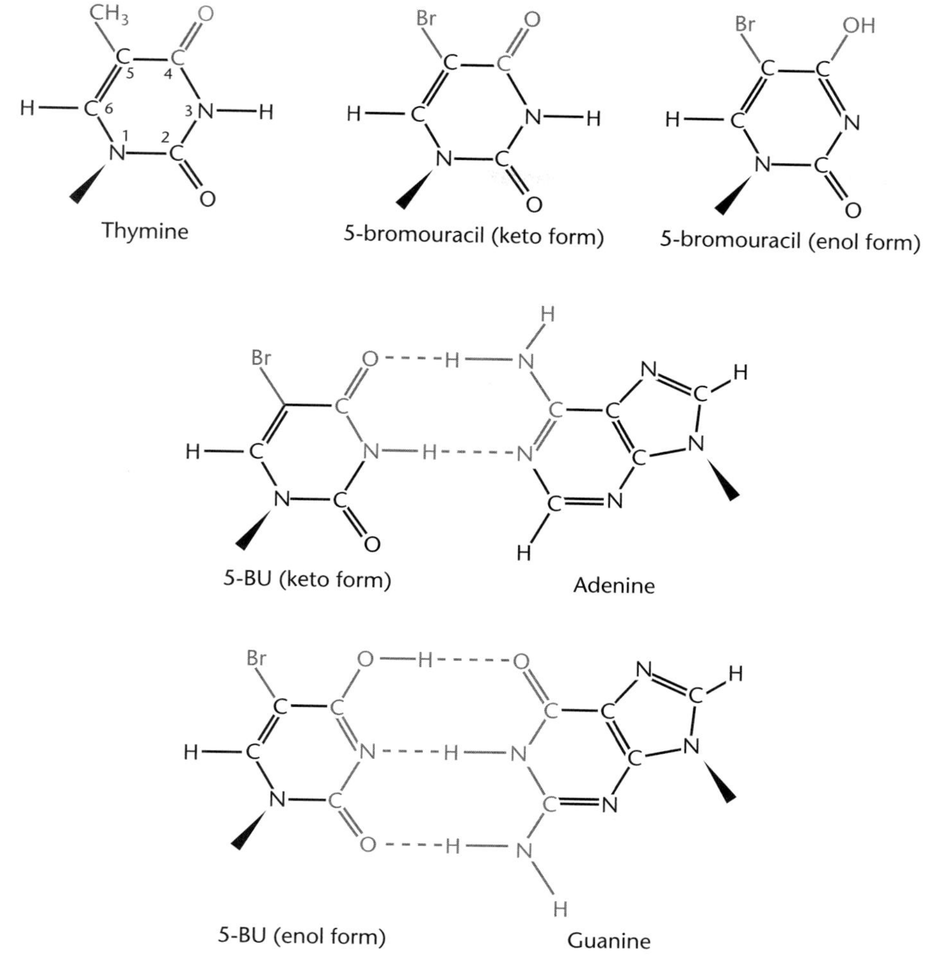

FIGURE 17.10 Similarity of 5-bromouracil (5-BU) structure to thymine structure. In the common keto form, 5-BU pairs normally with adenine, behaving as an analogue. In the rare enol form, it pairs anomalously with guanine.

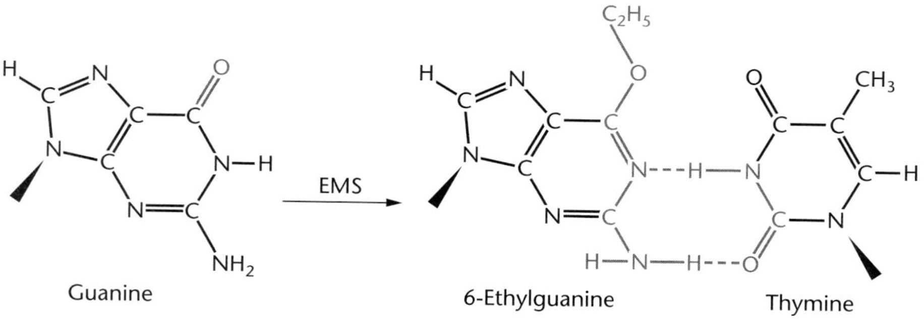

FIGURE 17.11 Conversion of guanine to 6-ethylguanine by the alkylating agent ethylmethane sulfonate (EMS). The 6-ethylguanine base-pairs with thymine.

TABLE 17.3 Alkylating Agents

Common Name or Symbol	Chemical Name	Chemical Structure
Mustard gas (sulfur)	Di-(2-chloroethyl) sulfide	$Cl-CH_2-CH_2-S-CH_2-CH_2-Cl$
EMS	Ethylmethane sulfonate	$CH_3-CH_2-O-\overset{\displaystyle O}{\underset{\displaystyle O}{S}}-CH_3$
EES	Ethylethane sulfonate	$CH_3-CH_2-O-\overset{\displaystyle O}{\underset{\displaystyle O}{S}}-CH_2-CH_3$

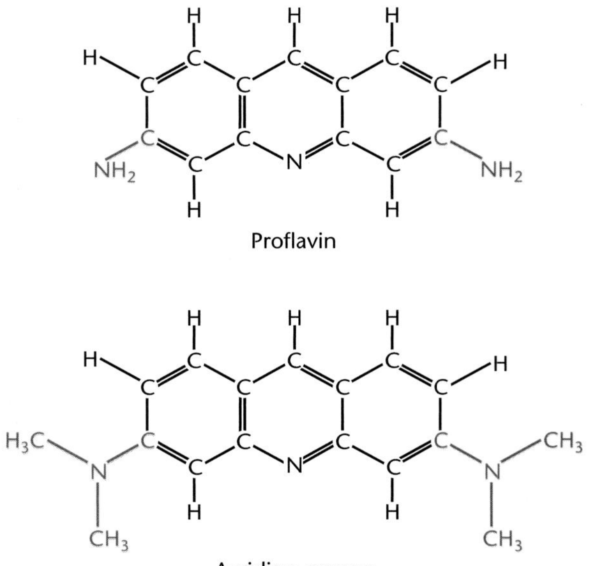

Proflavin

Acridine orange

FIGURE 17.12 Chemical structures of proflavin and acridine orange, which intercalate with DNA and cause frameshift mutations.

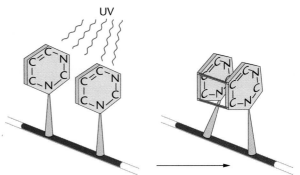

Dimer formed between adjacent thymidine residues along a DNA strand

FIGURE 17.14 Induction of a thymine dimer by UV radiation, leading to distortion of the DNA. The atoms of the pyrimidine ring are shown to illustrate the formation of cross-links. **GenCDX**

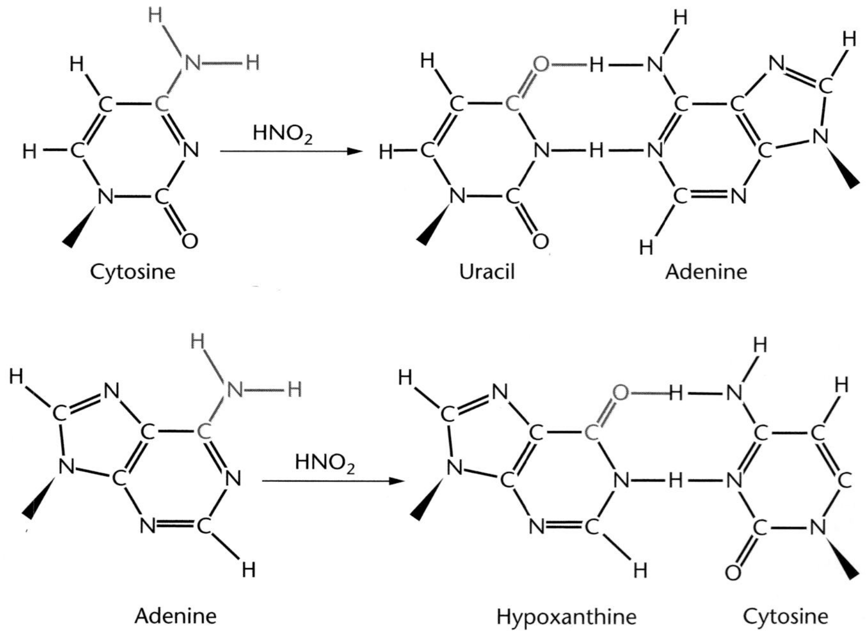

Cytosine → (HNO₂) → Uracil · · · Adenine

Adenine → (HNO₂) → Hypoxanthine · · · Cytosine

FIGURE 17.13 Deaminations caused by nitrous acid (HNO₂) leading to new base-pairing arrangements and mutations. Cytosine may be converted to uracil, which base-pairs with adenine. Adenine may be converted to hypoxanthine, which base-pairs with cytosine.

High-energy radiation penetrates cells and induces mutations

Within the electromagnetic spectrum, energy varies inversely with wavelength. As shown in Figure 17.15, **X rays**, **gamma rays**, and **cosmic rays** have even shorter wavelengths than does UV radiation and are therefore more energetic. As a result, they are strong enough to penetrate deeply into tissues, causing ionization of the molecules encountered along the way. Hermann J. Muller and Lewis J. Stadler established in the 1920s that these sources of **ionizing radiation** are mutagenic. Since that time, the effects

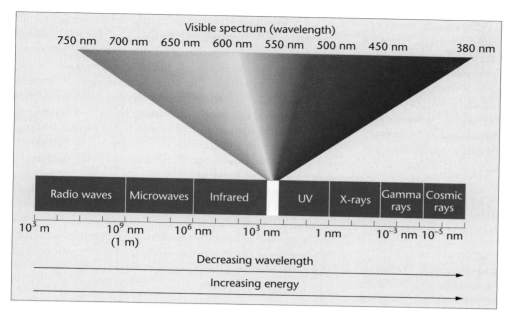

FIGURE 17.15   The components of the electromagnetic spectrum and their associated wavelengths.

of ionizing radiation, particularly X rays, have been studied intensely.

As X rays penetrate cells, electrons are ejected from the atoms of molecules encountered by the radiation. Thus, stable molecules and atoms are transformed into free radicals and reactive ions. The trail of ions left along the path of a high-energy ray can initiate a variety of chemical reactions. These reactions can directly or indirectly affect the genetic material, altering the purines and pyrimidines in DNA and resulting in point mutations. Ionizing radiation is also capable of breaking phosphodiester bonds, disrupting the integrity of chromosomes and producing a variety of aberrations.

Figure 17.16 shows a graph of the percentage of induced X-linked recessive lethal mutations versus the dose of X rays administered. A linear relationship is evident between X-ray dose and the induction of mutation; for each doubling of the dose, twice as many mutations are induced. Because the line intersects near the zero axis, this graph suggests that even very small doses of irradiation are mutagenic.

These observations can be interpreted in the form of the **target theory**, first proposed in 1924 by J. A. Crowther and F. Dessauer. The theory proposes that there are one or more sites, or targets, within cells and that a single event of irradiation at one site will bring about a damaging effect, or mutation. In effect, the target theory suggests that the X rays interact directly with the genetic material.

Two other observations concerning irradiation effects are of particular interest. First, in some organisms studied, the *intensity of the dose* (dose rate) administered seems

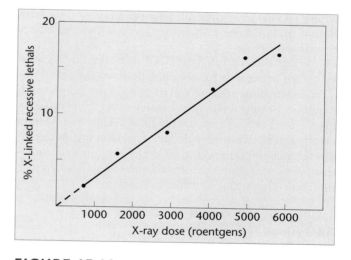

FIGURE 17.16   Plot of the percentage of X-linked recessive mutations induced by increasing doses of X rays. If extrapolated, the graph intersects the zero axis.

to make little difference in the mutagenic effect. Thus, a total exposure of 100 roentgens (a **roentgen** is a measure of energy dose) seems to produce the same overall mutagenic effect whether it is administered in a single acute dose or spread out over time in several smaller chronic doses. *Drosophila* has shown this response. In mammals such as mice and humans, however, this appears not to be the case, suggesting that repair of the damage may occur during the intervals between irradiation.

Second, certain portions of the cell cycle are more susceptible to radiation effects. As we mentioned, X rays can break chromosomes, resulting in terminal or intercalary deletions, translocations, and general chromosome fragmentation. Damage occurs most readily when chromosomes are greatly condensed in mitosis. This property is one of the reasons why radiation is used to treat human malignancy. Because tumorous cells are undergoing division more often than their nonmalignant counterparts, they are more susceptible to the immobilizing effect of radiation.

Gene sequencing has enhanced understanding of mutations in humans

So far we have been discussing the molecular basis of mutation largely in terms of nucleic acid chemistry. We know that various types of nucleotide conversions or frameshift mutations occur, primarily because our analyses of amino acid sequences of many proteins within the populations of a species show substantial diversity. This diversity, which has arisen during evolution, is a reflection of changes in the triplet codons following substitution, insertion, or deletion of one or more nucleotides in the DNA sequences constituting genes.

As our ability to analyze DNA more directly increases, we are better able to look specifically at the actual nucleotide sequence of genes and to gain greater insights into mutation. Several techniques capable of accurate, rapid sequencing of DNA (see Chapter 18 and Appendix A) have greatly extended our knowledge of molecular genetics. In this section we examine the results of a number of studies that have investigated the actual gene sequence of various mutations that seriously affect humans.

ABO Blood Types

The ABO system is based on a series of antigenic determinants found on erythrocytes and other cells, particularly epithelial types. As we discussed in Chapter 4, three alleles of a single gene exist, the product of which modifies the H substance. The modification involves glycosyltransferase activity, converting the H substance to either the A or B antigen, as a result of the product of the I^A or I^B allele, respectively; or failing to modify the H substance, as a result of the I^O allele (see Figure 4.2).

The responsible gene has been sequenced in 14 cases of varying ABO status. When the DNAs of the I^A and I^B alleles were compared, four consistent nucleotide substitutions were found. It is assumed that the resulting changes in the amino acid sequence of the glycosyltransferase gene product lead to the different modifications of the H substance.

The I^O allele situation is unique and interesting. Individuals homozygous for this allele have type O blood, lack glycosyltransferase activity, and fail to modify the H substance. The analysis of the DNA of this allele shows one consistent change that is unique compared to the sequence of the other alleles, the deletion of a single nucleotide early in the coding sequence, causing a frameshift mutation. A complete messenger RNA is transcribed, but at translation, the frame of reading shifts at the point of deletion and continues out of frame for about 100 nucleotides before a stop codon is encountered. At this point, the polypeptide chain terminates prematurely, resulting in a nonfunctional product.

These findings provide a direct molecular explanation of the ABO allele system and the basis for the biosynthesis of the corresponding antigens. The molecular basis for the antigenic phenotypes is clearly the result of structural alterations, or mutations, of the nucleotide sequence of the gene encoding the glycosyltransferase enzyme.

Muscular Dystrophy

Muscular dystrophy is characterized by severe progressive muscular degeneration, or myopathy, resulting in the death of the affected individuals by early adulthood. Because the condition is recessive and X-linked, and because affected males die before they can reproduce, females are rarely affected by the disorder. The incidence of 1/3500 live male births makes muscular dystrophy one of the most common life-shortening hereditary disorders known. Two related forms exist. **Duchenne muscular dystrophy (DMD)** is more common and more severe than the allelic form, **Becker muscular dystrophy (BMD)**.

The region containing the gene, which consists of over 2 million bp, has been analyzed extensively. In normal (unaffected) individuals, transcription results in a messenger RNA containing about 14,000 bases (14 kb) that is translated into the protein **dystrophin**, consisting of 3685 amino acids. This protein can be detected in most cases of the less severe BMD, but is rarely found in DMD (see Figure 2.2). This has led to the hypothesis that most mutations causing BMD do not alter the reading frame, but that most DMD mutations change the reading frame early in the gene, resulting in the premature termination of dystrophin translation. This hypothesis is consistent with the observed differences in severity of these two forms of the disorder.

In an extensive analysis of the DNA of 194 patients (160 DMD and 34 BMD), J. T. Den Dunnen and associates found that 128 of these mutations (65 percent) consisted of substantial deletions or insertions. Of 115 deletions, 17 occurred in BMD cases, and of 13 insertions, 1 occurred in BMD, with the remainder in DMD cases. In most cases, the results were consistent with the "reading frame" hypothesis: With few exceptions, DMD mutations changed the frame of reading, whereas BMD mutations usually did not alter the reading frame.

Perhaps the most noteworthy of Den Dunnen's findings is the high percentage of deleterious mutations that represent the deletion or insertion of nucleotides within the gene. This observation reflects the fact that a mutation caused by a random single-nucleotide substitution within a gene is more likely to be tolerated without the devastating effect of muscular dystrophy than the addition or loss of numerous nucleotides that may alter the frame of reading. There are three reasons for this, as discussed in Chapter 13.

1. A nucleotide substitution may not change the encoded amino acid, since the code is degenerate.

2. If an amino acid substitution does result, the change may not be present at a location within the protein that is critical to its function.

3. Even if the altered amino acid is present at a critical region, it may still have little or no effect on the function of the protein. For example, an amino acid might be changed to another with nearly identical chemical properties or to one with very similar recognition properties, such as shape.

As a result, single-base substitutions sometimes have little or no effect on protein function, or they may simply reduce the efficiency but not eliminate the functional capacity of the gene product. As more mutant genes are analyzed directly, our picture of mutation will become increasingly clear.

Trinucleotide Repeats in Fragile-X Syndrome, Myotonic Dystrophy, and Huntington Disease

Beginning about 1990, the molecular analysis of the DNA representing the genes responsible for a number of inherited human disorders provided a remarkable set of observations. Mutant genes were characterized by an expansion of a simple **trinucleotide repeat sequence**, usually from fewer than 15 copies in normal individuals to a large number in affected individuals. For example, in each of the three separate genes responsible for the autosomal disorders **myotonic dystrophy** and **Huntington disease** and the

X-linked **fragile-X syndrome**, each gene was found to contain a unique trinucleotide DNA sequence repeated many times. While repeated sequences are also present in the non-mutant (normal) allele of each gene, the mutations are characterized by a significant variable increase in the number of times the trinucleotide is repeated. Table 17.4 summarizes the various cases that are discussed below, with particular emphasis on the size of the repeats in normal and diseased individuals.

Recall that we introduced this topic in Chapter 4 when we discussed the onset of the expression of various phenotypes. In several cases, a correlation has been found between the number of repeats and the manifestation of mutant phenotypes. The greater the number of repeats, the earlier disease onset occurs. Further, the number of repeats may increase in each subsequent generation. This general phenomenon, which we called **genetic anticipation** in Chapter 4, represents a unique form of mutation related to an instability of specific regions of the three different genes. Other normal genes in humans, as well as many other species, also contain trinucleotide repeats, however the repeats do not balloon in size, and the genes maintain normal function.

In Chapter 10 we discussed fragile-X syndrome. The responsible gene, *FMR-1*, may have several hundred to several thousand copies of the trinucleotide sequence CGG. Individuals with up to 50 copies are normal and do not display the mental retardation associated with the syndrome. Individuals with 50 to 200 copies are considered "carriers." Although they are normal, their offspring may contain even more copies and express the syndrome.

Myotonic dystrophy, the most common form of adult muscular dystrophy, contains multiple copies of the sequence CTG. Fewer than 35 copies result in normal gene expression. Above this number, symptoms range from mild myopathy and cataracts to severe dystrophy, retardation, and even death in early childhood. Both the severity and onset are directly correlated with the size of the repeated sequence.

More recently, the gene responsible for Huntington disease has been found to demonstrate a similar pattern.

TABLE 17.4 Summary of Trinucleotide-Repeat Disorders

	Trinucleotide Repeat	Number (normal)	Number (affected)	Inheritance Pattern
Huntington disease	CAG	6–35	36–120	Autosomal dominant
Myotonic dystrophy	CTG	5–35	>200	Autosomal dominant
Fragile X syndrome	CGG	6–50	>200	X-linked dominant
Spinobulbar muscular atrophy	CAG	10–30	35–60	X-linked dominant

The trinucleotide CAG repeat, present 10 to 35 times in normal individuals, exists in significantly increased numbers in diseased individuals. Much earlier onset occurs when a very large number of copies are present. Interestingly, in still another disorder, **spinobulbar muscular atrophy** (**Kennedy disease**), the gene also contains repeated copies of the CAG triplet. However, diseased individuals usually have only 35 to 60 copies of the sequence.

The role of such repeated sequences in normal and mutant genes remains a mystery. Their locations within the gene vary in each case. In Huntington and Kennedy diseases, the repeat lies within the coding portion of the gene. This is not the case in the other two disorders, however. In the gene responsible for fragile-X syndrome, the repeat is upstream (5′) in an area that is most often involved in regulating gene expression. In the case of myotonic dystrophy, the repeat is downstream (the 3′-end).

The mechanism by which the repeated sequence expands from generation to generation is of great interest. How such an instability during DNA replication affects only specific areas of certain genes is currently an important research topic. This instability seems to be more prevalent in humans than in many other organisms.

The Ames test is used to assess the mutagenicity of compounds

There is great concern about the possible mutagenic properties of any chemical that enters the human body, whether through the skin, digestive system, or respiratory tract. For example, great attention has been given to residual materials of air and water pollution, food preservatives and additives, artificial sweeteners, herbicides, pesticides, and pharmaceutical products. As we have seen, mutagenicity may be tested in various organisms, including *Drosophila*, mice, and cultured mammalian cells. The most common test, which involves bacteria, was devised by Bruce Ames.

The **Ames test**, (Figure 17.17), uses any of several strains of the bacterium *Salmonella typhimurium* that were selected for sensitivity and specificity for mutagenesis. One strain is used to detect base-pair substitutions, and the

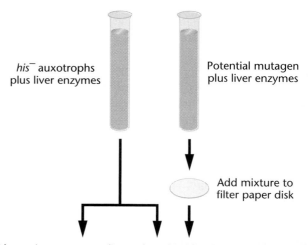

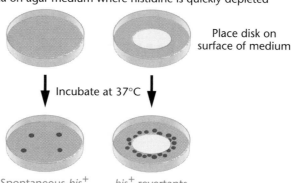

FIGURE 17.17 The Ames test, which screens potential compounds for mutagenicity.

other three detect various frameshift mutations. Each mutant strain is unable to synthesize histidine, and therefore requires histidine for growth (*his⁻*). The assay measures the frequency of reverse mutation, which yields wild-type (*his⁺*) bacteria. Greater sensitivity to mutagens occurs because these strains bear other mutations that eliminate both the DNA excision-repair system (discussed later in this chapter) and the lipopolysaccharide barrier that coats and protects the surface of the bacteria.

It is very interesting to note that many substances entering the human body are relatively innocuous until activated metabolically, usually in the liver, to a more chemically reactive product. Thus, the Ames test includes a step in which the test compound is incubated *in vitro* in the presence of a mammalian liver extract. Or, test compounds may be injected into a mouse, where they are modified by liver enzymes and then recovered.

In the initial use of Ames testing in the 1970s, a large number of known carcinogens were examined and over 80 percent were shown to be strong mutagens! This is not surprising; the transformation of cells to the malignant state undoubtedly occurs as a result of some alteration of DNA. Although a positive response as a mutagen does not prove that a compound is carcinogenic, the Ames test is useful as a preliminary screening device. It is used extensively in conjunction with the industrial and pharmaceutical development of chemical compounds.

Organisms can counteract DNA damage and mutations by activating several types of repair systems

In previous sections of this chapter we established that replicating and nonreplicating DNA molecules are vulnerable to various forms of errors during replication, which induced lesions that lead to gene mutations. Living systems have evolved a variety of very elaborate repair systems that are able to counteract many of the forms of DNA damage that lead to mutation. As we will see, such repair systems are essential to the maintenance of the genetic integrity of organisms, and as such, to the survival of organisms on Earth.

Photoreactivation Repair: Reversal of UV Damage in Prokaryotes

As illustrated in Figure 17.14, UV light is mutagenic as a result of the creation of pyrimidine dimers. The study of the mutagenicity of UV radiation paved the way for the discovery of many forms of natural repair of DNA damage. The first relevant discovery concerning UV repair in bacteria was made in 1949 when Albert Kelner observed

the phenomenon of **photoreactivation repair**. He showed that the UV-induced damage to *E. coli* DNA could be partially reversed if, following irradiation, the cells were exposed briefly to light in the blue range of the visible spectrum. The photoreactivation repair process was subsequently shown to be temperature dependent, suggesting that the light-induced mechanisms involve an enzymatically controlled chemical reaction. Visible light appears to induce the repair process of the DNA damaged by UV radiation.

Further studies of photoreactivation have revealed that the process is dependent on the activity of a protein called the **photoreactivation enzyme** (**PRE**). This molecule can be isolated from extracts of *E. coli* cells. The enzyme's mode of action is to cleave the bonds between thymine dimers, thus reversing the effect of UV radiation on DNA (Figure 17.18). Although the enzyme will associate with a dimer in the dark, it must absorb a photon of light to cleave the dimer. In spite of the potential for diminishing UV-induced mutations, photoreactivation repair is not essential in *E. coli*, since a null mutation in the gene coding for the PRE is not lethal. In addition, the enzyme cannot be detected in humans and other eukaryotes, who must rely on other repair mechanisms to reverse the effects of UV radiation.

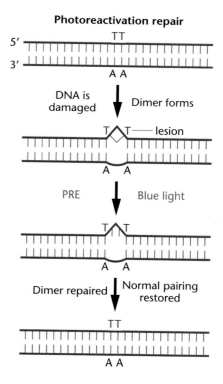

FIGURE 17.18   Damaged DNA repaired by photoreactivation repair. The bond creating the thymine dimer is cleaved by the photoreactivation enzyme (PRE), which must be activated by blue light.

Excision Repair in Prokaryotes and Eukaryotes

Investigations in the early 1960s suggested that, in addition to PRE, a *light-independent* repair system exists in prokaryotes such as *E. coli* that repairs damage to DNA caused by both exogenous and endogenous agents. The basic mechanism of this type of repair, referred to as **excision repair**, has been conserved throughout evolution, occurring in all prokaryotic and eukaryotic organisms. The process consists of three basic steps.

1. The distortion or error is recognized and enzymatically clipped out by a nuclease. This "excision" may affect only a single base, a single nucleotide, or include several nucleotides adjacent to the error as well, leaving a gap in the helix.

2. DNA polymerase I fills this gap by inserting d-ribonucleotides complementary to those on the intact strand. The enzyme adds these bases to the 3'-OH end of the clipped DNA.

3. The joining enzyme DNA ligase seals the final "nick" that remains at the 3'-OH end of the last base inserted, closing the gap.

DNA polymerase I, the enzyme discovered by Arthur Kornberg, was once assumed to be the universal DNA-replication enzyme (see Chapter 12). However, it was the discovery of the *polA1* mutation that demonstrated the role of this enzyme in repairing UV-induced lesions in *E. coli*. Cells carrying the *polA1* mutation lack functional polymerase I, replicate their DNA normally, and are unusually sensitive to UV radiation. Apparently, such cells are unable to fill the gap created by the excision of the thymine dimers.

There are now known to be two types of excision repair: base excision repair and nucleotide excision repair. **Base excision repair (BER)** is involved in correcting damage to nitrogenous bases created by the spontaneous hydrolysis of DNA bases, as well as by agents that chemically alter them. As illustrated in Figure 17.19, the first step of the BER pathway in *E. coli* involves the recognition of the chemically altered base by **DNA glycosylases**, which are specific to different types of DNA damage. For example, the enzyme uracil-DNA glycosylase recognizes the unusual presence of uracil when it is part of a nucleotide in DNA. The enzyme first cuts the glycosidic bond between the base and the sugar, creating an apyrimidinic (AP) site. Such a sugar with a missing base is then recognized by an enzyme called **AP endonuclease**. The endonuclease makes a cut in the sugar backbone at the AP site. This creates a distortion in the DNA helix that is recognized by the excision-repair system, which is then activated, leading ultimately to the correction of the error.

Although much has been learned about glycosylases in *E. coli*, very little is known about the DNA glycosylases in

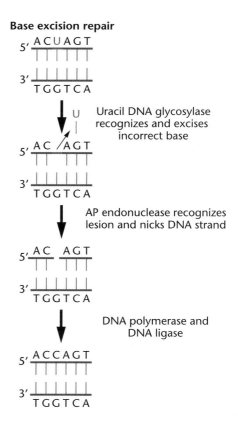

Base excision repair

FIGURE 17.19 Base excision repair (BER) accomplished by uracil DNA glycosylase, AP endonuclease, DNA polymerase, and DNA ligase. Uracil is recognized as a noncomplementary base, excised, and replaced with the complementary base (C). **GenCDX**

eukaryotes. In addition, unlike the nucleotide excision pathway we are going to discuss next, there is no human or animal disease model available that has a defective BER pathway, so it is difficult to assess the importance of BER in eukaryotes.

While base excision repair recognizes and replaces modified bases in DNA, the **nucleotide excision repair** pathway (**NER**) repairs "bulky" lesions in DNA that alter or distort the regular DNA double helix, such as the UV-induced pyrimidine dimers discussed previously. The NER pathway, diagrammed in Figure 17.20, was first discovered in *E. coli* by Paul Howard-Flanders and coworkers who were able to isolate several independent mutants that demonstrated sensitivity to UV radiation. One group of genes was designated *uvr* (ultraviolet repair) and included the *uvrA*, *uvrB*, and *uvrC* mutations. In the NER pathway, the *uvr* gene products are involved in recognizing and clipping out the lesions in the DNA. The repair is then completed by DNA polymerase I and DNA ligase.

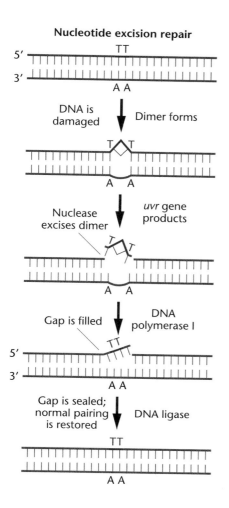

Nucleotide excision repair

FIGURE 17.20 Nucleotide excision repair (NER) of a UV-induced thymine dimer. In actuality, 12 bases are excised in prokaryotes and 28 bases are excised in eukaryotes during repair. **GenCDX**

Xeroderma Pigmentosum and Nucleotide Excision Repair

The mechanism of nucleotide excision repair (NER), which has been elucidated in eukaryotes, is much more complicated than prokaryotic NER because it involves many more proteins. Much of what is known about the system in humans has resulted from detailed studies of individuals with **xeroderma pigmentosum** (**XP**), a rare genetic disorder that predisposes individuals to severe skin abnormalities. These individuals have lost their ability to undergo NER. As a result, individuals suffering from XP who are exposed to the ultraviolet radiation present in sunlight exhibit reactions that range from initial freckling and skin ulceration to the subsequent development of skin cancer. Figure 17.21 contrasts two XP individuals, one of whom has been detected early and protected from sunlight.

The condition is very severe and may be lethal, although early detection and protection from sunlight can arrest it. Because sunlight contains UV radiation, a causal relationship had been predicted between thymine dimer production and XP. Thus, the ability to repair UV-induced lesions was investigated in human fibroblast cultures derived from XP and normal individuals. (Fibroblasts are undifferentiated connective tissue cells.) The results suggested that the XP phenotype may be caused by more than one mutant gene.

In 1968, James Cleaver showed that cells from XP patients were deficient in **unscheduled DNA synthesis** (DNA synthesis other than that occurring during chromosome replication), which is elicited in normal cells by UV radiation. Since this type of synthesis is thought to represent the activity of the excision-repair system, the suggestion is that XP cells are deficient in excision repair.

The link between xeroderma pigmentosum and inadequate excision repair has been strengthened by the use

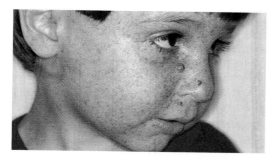
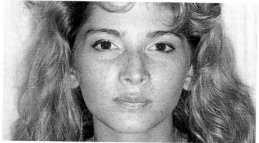

FIGURE 17.21 Two individuals with xeroderma pigmentosum. The 4-year-old boy on the left shows marked skin lesions induced by sunlight. Mottled redness (erythema) and irregular pigment changes that are a response to cellular injury are apparent. Two nodular cancers are present on his nose. The 18-year-old girl on the right has been carefully protected from sunlight since the diagnosis of xeroderma pigmentosum in infancy. Several cancers have been removed and she has worked as a successful model.

of **somatic cell hybridization** studies, a technique that we first introduced in Chapter 6. Cultured fibroblast cells from any two unrelated XP patients may be induced to fuse together, forming a heterokaryon where the two nuclei share a common cytoplasm. After fusion, excision repair, as assayed by unscheduled DNA synthesis, may be reestablished in the heterokaryon. When this occurs, the two variants are said to demonstrate **complementation**. Alone, neither cell type demonstrates excision repair, but together in a heterokaryon the process occurs. In genetic terms, this is strong evidence that in the two patients from whom the cells were derived, different affected genes led to the disease. Complementation occurs because the heterokaryon has at least one normal copy of each gene.

Based on many studies, patients have been divided into seven complementation groups, suggesting that at least seven different genes may be involved in excision repair. These genes or their protein products have now been identified. The genes are found in disparate regions of the genome, and a homologous gene for each has been identified in yeast. Approximately 20 percent of XP patients do not fall into any of the seven groups. They manifest similar symptoms, but their fibroblasts do not demonstrate defective excision repair. There is some evidence that they are less efficient in normal DNA replication.

As a result of the study of the defective genes in xeroderma pigmentosum, a great deal is now known about how NER counteracts DNA damage in normal cells (Figure 17.22). The first step in humans is the recognition of the damage DNA by a specific protein called XPA (*Xe*roderma *P*igmentosum gene *A*). The binding of XPA to the damaged DNA helix causes other proteins of the pathway to be recruited to the site. Interestingly, the transcription factor TFIIH is involved. Once TFIIH is bound to the lesioned site, other proteins join it, creating the repair complex. This repair complex excises a 28-nucleotide-long fragment from the helix, including the lesion. TFIIH is thought to be involved in unwinding the DNA helix around the damaged site, while two proteins, XPF and XPG, are involved in excising the oligonucleotide from the double helix.

Interestingly, in the human NER pathway, the repair of DNA that is actively transcribed occurs before DNA that is not actively transcribed. This preferential repair is called **transcription-coupled repair**. While the exact mechanism is unknown, it is thought that a specific protein (called ERCC6) recognizes RNA polymerase when it has stalled at a lesion and recruits other repair proteins to the damaged DNA sites. A defect in transcription-coupled repair pathway has been linked to **Cockayne syndrome (CS)**, which is characterized by mental retardation, neurological deficiencies, and photosensitivity.

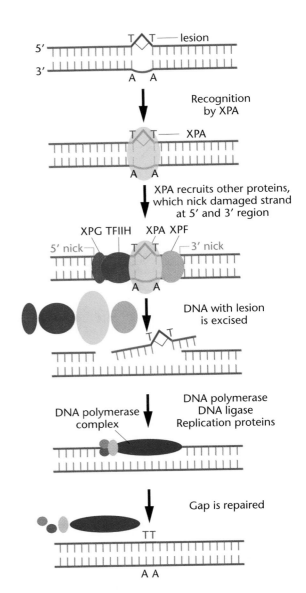

FIGURE 17.22   Nucleotide excision repair of a UV-induced lesion in human DNA, as determined by the study of genes and their products associated with the disorder xeroderma pigmentosum. The initial lesion is recognized by the XPA protein, which then recruits other members of the repair complex, including the transcription factor TFIIH. An oligonucleotide consisting of 28 nucleotides (called a 28mer) is excised. The gap is then filled by DNA polymerase and DNA ligase.

Proofreading and Mismatch Repair

Other types of repair systems have evolved to correct other types of errors in DNA. For example, as we pointed out in our discussion of DNA synthesis in Chapter 12, DNA polymerase III possesses a **proofreading** function. During polymerization, when an incorrect nucleotide is inserted, the enzyme complex has the potential to recognize the error and "reverse" it by cutting out the incorrect nucleotide and replacing it. In bacterial systems, proofreading is thought to increase fidelity during synthesis by two orders of magnitude. If initial mismatches occur in $1/10^5$ nucleotide pairs (a rate of 10^{-5}), proofreading decreases final mismatches to $1/10^7$ (a rate of 10^{-7}).

To cope with those errors that remain after proofreading, still another mechanism, called **mismatch repair**, may be activated. Proposed over 20 years ago by Robin Holliday, the molecular basis of this process is now well established. As in other DNA lesions, the alteration or mismatch must be detected, the incorrect nucleotide(s) must be removed, and the replacement with the correct nucleotide(s) must occur. But a special problem exists with correction of a mismatch. How does the repair system recognize which strand is correct (the template during replication) and which contains the mismatched base (the newly synthesized strand)? How the repair system discriminates and recognizes the "new" nucleotide puzzled geneticists for decades. If the mismatch is recognized, but no discrimination occurred and excision was random, half the time the strand bearing the correct base would be clipped out. The concept of strand discrimination by a repair enzyme is thus a critical step.

At least in some bacteria, including *E. coli*, this process has been elucidated and is based on the process of **DNA methylation**. These bacteria contain an enzyme, **adenine methylase**, that recognizes the DNA sequence

$$5' \ldots GATC \ldots 3'$$
$$3' \ldots CTAG \ldots 5'$$

as a substrate. Upon recognition, a methyl group is added to each of the adenine residues. This modification is stable throughout the cell cycle.

Following a further round of replication, the newly synthesized strands remain temporarily unmethylated. It is at this point that the repair enzyme recognizes the mismatch and preferentially binds to the unmethylated strand. A nick is made by an endonuclease protein, either 5' or 3', to the mismatch on the unmethylated strand. The nicked DNA strand is then unwound, degraded, and replaced until the mismatch is reached and excised. The proteins involved in this process are slightly different depending on which side of the mismatch the nick was made. A series of *E. coli* gene products, MutH, L, S, and U are involved in the discrimination step. Mutations in each result in strains deficient in mismatch repair. While the mechanism described here is based on studies of *E. coli*, similar mechanisms involving homologous proteins are known to exist in yeast and mammals.

Post-Replication Repair and the SOS Repair System

Still other types of repair have been discovered, illustrating the diversity of such mechanisms that have evolved to overcome DNA damage. One system, called **post-replication repair**, was discovered in an excision-defective strain of *E. coli* and first proposed by Miroslav Radman. It is a system that responds *after* damaged DNA has escaped repair and failed to be completely replicated, thus its name. As illustrated in Figure 17.23, when DNA bearing a lesion of some sort (such as a pyrimidine dimer) is being replicated, DNA polymerase at first stalls at the lesion and then skips over it, leaving a gap along the newly synthesized strand. To counteract this, the RecA protein directs a recombinational exchange process whereby this gap is filled as a result of the insertion of a segment initially present on the undamaged complementary strand. This event creates a gap on the "donor" strand, which can then be filled by repair synthesis as replication proceeds. Because a recombinational event is involved, post-replication is also referred to as *homologous recombination repair*, a more general category.

Still another repair pathway in *E. coli* is called the **SOS repair system**, which also responds to damaged DNA, but in a different way. Instead of skipping over a lesion during DNA sysnthesis, as in post-replication repair, the system creates conditions that allow DNA polymerase to continue replication across the lesion. While no gap is created, the fidelity of replication is compromised, leading to the description of the system as being *error prone*. Phil Hanawalt and Paul Howard-Flanders, among others, have established that as many as 20 different genes, including *lexA*, *recA*, and *uvr*, may be induced by DNA damage, so that their products become involved in the SOS response.

Of particular interest is the LexA protein product, which, when produced, serves to repress the transcription of the *recA* and *uvr* genes and other repair genes. However, when a RecA protein binds to single-stranded DNA in the area of a gap, this binding somehow activates the autoproteolytic cleavage of the LexA repressor molecule, disrupting its regulatory capacity. The absence of a functional repressor molecule allows the activation of the *recA* and *uvr* genes, among others, leading to an increased production of the proteins for which they code. These products complete the repair process by converting the lesion into an error-prone site, which enables DNA polymerase to replicate DNA past the lesion.

Double-Strand Break Repair in Mammals

Thus far, we have only discussed repair pathways that deal with damage *within* a particular strand of DNA. We conclude our discussion of DNA repair by considering what

Post replication repair

1. Lesion present in DNA unwound prior to replication

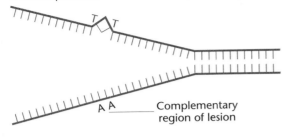

Complementary
region of lesion

2. Replication skips over lesion and continues

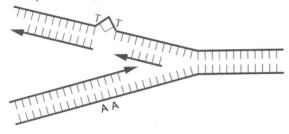

3. Undamaged complementary region of parental strand is recombined

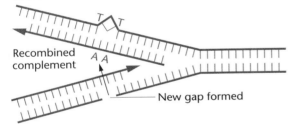

Recombined
complement

New gap formed

4. New gap is filled by DNA polymerase and DNA ligase

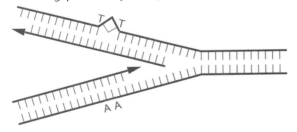

FIGURE 17.23 Post-replication repair that occurs if DNA replication has skipped over a lesion, such as a thymine dimer. Through the process of recombination, the correct complementary sequence is recruited from the parental strand and inserted into the gap opposite the lesion. The new gap created is filled by DNA polymerase and DNA ligase.

happens when both strands of the DNA helix are cleaved, for example, as a result of exposure to ionizing radiation. This causes what are called double-strand breaks. While such damage occurs in bacteria as well, we will focus our discussion on eukaryotic cells. Fortunately, a specialized form of DNA repair, the **DNA double-strand break repair (DSB repair)** pathway, is activated and is responsible for ultimately reannealing the two DNA segments. Recently, interest has grown in DNA DSB repair because defects in this pathway are associated with X-ray hypersensitivity and immune deficiency. Such defects may also underlie familial disposition to breast and ovarian cancer.

Similar to post-replication repair, one pathway involved in double-strand break repair is referred to as **homologous recombinational repair** because damaged DNA is actually recombined and replaced with homologous undamaged DNA. This is necessary because, when both strands are broken, there is no undamaged parental strand available to use as the source of the complementary DNA sequence during repair. Therefore, the genetic information present in the homologous region of either a sister or nonsister homolog is "recruited" to replace the damaged double-stranded break. The undamaged homologous region is actually recombined into the damaged DNA molecule. The process usually occurs during the late S/G2 phase of the cell cycle and depends, in yeast, upon a complex of at least five proteins, called the RAD52 complex. While the exact mechanism is not yet understood, the complex is able to restore the integrity of the broken double helix.

A second pathway, called **nonhomologous recombinational repair**, has also been discovered, which achieves the same type of repair. However, as the name implies, the mechanism does not recruit a homologous region of DNA during repair. Two different protein complexes are involved in nonhomologous repair.

Site-directed mutagenesis allows researchers to investigate specific genes

This section introduces a useful experimental technique, **site-directed mutagenesis**, that allows researchers to introduce a designed mutation at a prescribed site within a gene of interest. The technique relies on the availability of a cloned gene and uses a number of recombinant DNA technology manipulations discussed in Chapter 18. The underlying principles, however, are based on information we have already presented.

The goal of the technique is to specifically alter a gene by changing, deleting, or inserting one or more nucleotides at a predetermined site. For example, if a triplet code is altered, upon transcription and translation the change will cause the insertion of a "mutant" amino acid in the protein encoded by the original gene. Such designed mutations are particularly useful in studying the effects of specific mutations on both gene expression and protein function. Various approaches can be used to accomplish the goal; most are variations on the theme we are about to describe, illustrated in Figure 17.24.

The first step is to determine the nucleotide sequence of the gene being studied. This can be accomplished by DNA-sequencing techniques, or, if the amino acid sequence of the protein is known, it can be predicted by extrapolating from the genetic code.

The next step is to isolate one of the two complementary strands from the DNA of known sequence and decide which nucleotides are to be changed (Step 2). Then, a small piece of DNA, a synthetic oligonucleotide, is chemically synthesized, which is complementary to that region at all points except in the triplet sequence or sequences that are to be altered (Step 3). This sequence includes the triplet encoding the amino acid that will change in the protein.

As illustrated in Step 4, this short piece of DNA is hybridized with the original parent strand, forming a partial duplex because of the complementarity along most of its length. If DNA polymerase and DNA ligase are then added to this hybrid complex, the short sequence is extended so that a duplex of the entire gene is formed. The two strands are perfectly complementary except at the point of alteration.

If this DNA is replicated (Step 5), two types of duplexes are formed; one is like the original, unaltered gene and the other contains the newly designed sequence. Recombinant DNA technology makes it possible not only to complete the above manipulations, but also to allow the altered gene to be expressed so that large amounts of the desired protein are available for study. Or, as we will see in the next section, the altered gene may be introduced into an organism and studied.

Knockout Genes and Transgenes

While a gene bearing a site-directed mutation may be investigated *in vitro*, the insertion of that gene into a living organism allows for the assessment of its *in vivo* function. If the insertion process involves the *replacement* of the comparable gene of the organism from which it originated, the

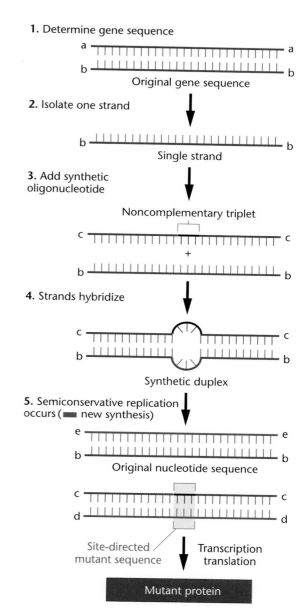

FIGURE 17.24 Flow diagram illustrating the principles of site-directed mutagenesis. A single strand of DNA from a gene of interest is initially isolated. This is hybridized with a synthetic oligonucleotide containing a triplet altered so as to encode an amino acid of choice. Following semiconservative replication, a different complementary base pair is present in one of the new duplexes. Upon transcription and translation, a mutant protein, "designed" in the laboratory, will be produced.

technique is referred to as **gene knockout**. The genetically altered organism is called a **knockout organism** (e.g., a knockout mouse).

Specific gene replacement is difficult to achieve, but it has been accomplished and is now fairly routine in research involving yeast and mice. In mice, the application of gene knockout techniques has been particularly fruitful in studies involving the genetic control of early development and behavior, with a "loss-of-function" mutation most often replacing the normal gene. Furthermore, knockout mice now serve as models for studying human genetic disorders. In such cases, the mouse gene comparable to the gene that causes the human disorder is isolated, subjected to directed mutagenesis, and used to replace the nonmutant mouse gene. Cystic fibrosis and Duchenne muscular dystrophy have been investigated in this way. Applications of this general technology are discussed in more detail in Chapters 21 and 22.

If a gene is inserted into an organism *in addition* to its normal copies, it is called a **transgene**, and the organism is called a **transgenic organism** (e.g., a transgenic tomato). In such cases, the gene may have undergone directed mutagenesis, or it may be a foreign gene isolated from another organism. This technology has been used more extensively than gene knockout, since it is easier to accomplish because specific replacement is not required.

Transgenic organisms, such as *Drosophila*, mice, and a variety of other organisms, are now important in agriculture as well as in gene therapy techniques for our own species.

Transposable genetic elements move within the genome and may disrupt genetic function

We conclude this chapter by discussing the phenomenon of **transposable genetic elements**, sometimes referred to as **transposons**, genetic units that can move or be "transposed" within the genome. It is appropriate to discuss transposable elements in this chapter on mutation because the movement of genetic units from one place in the genome to another often disrupts genetic function and results in phenotypic variation. As such, the impact of the transposition of genetic units often fits into a broad definition of mutation.

Transposable elements were first studied in maize by Barbara McClintock almost 50 years ago. However, even in the 1950s and 1960s the idea that genetic information was *not* fixed within the genome of an organism was slow to find acceptance. Such a notion was quite alien to the classical interpretation of genes on chromosomes. It was not until other transposable elements were discovered, and

their molecular basis revealed, that the phenomenon was found to be nearly universal.

Insertion Sequences

Although the presence of transposable elements in maize had been predicted earlier, the first observation at the molecular level came in the early 1970s. A number of independent researchers, including Peter Starlinger and James Shapiro, visualized a unique class of mutations affecting different genes in various bacterial strains. For example, the expression of a cluster of related genes involving galactose metabolism in *E. coli* was repressed as a result of one such mutation; the phenotypic effect was heritable, but was found not to be caused by a base-pair change characteristic of conventional gene mutations. Instead, it was shown that a short specific DNA segment had been inserted into the bacterial chromosome at the beginning of the galactose gene cluster. When this segment was spontaneously excised from the bacterial chromosome, wild-type function was restored. Such a segment came to be called an **insertion sequence** (**IS**).

It was subsequently revealed that several other distinct DNA segments could behave in a similar fashion, inserting into the chromosome and affecting gene function. These DNA segments are relatively short, not exceeding 2000 bp [2 kilobases (kb)]. The first insertion sequence to be characterized in *E. coli*, IS1, is about 800 bp long; IS2, 3, 4, and 5 are about 1250 to 1400 bp in length.

The analysis of the DNA sequences of most IS units reveals a feature important to their mobility. The nucleotide sequences consist of **inverted terminal repeats** (**ITRs**) of one another; that is, the final sequence at the 5′-end of one strand is the same as the final sequence at the 5′-end of the other strand, except that the sequences run in opposite directions (Figure 17.25). Although the figure shows the terminal-repeating unit to consist of only a few nucleotides, many more are actually involved. For example, in *E. coli* the IS1 termini contain about 20 nucleotide pairs, IS2 and IS3 about 40 pairs, and IS4 about 18 pairs. It seems likely that these terminal sequences are an integral part of the mechanism of the insertion of IS units into DNA. That insertion of IS units is more likely to occur at certain DNA regions than at others suggests that IS termini can recognize certain target sequences in the DNA during the process of insertion.

Careful investigation has revealed IS units in the wild-type *E. coli* chromosome as well as in other autonomous segments of bacterial DNA, called plasmids. Thus, their presence does not always result in mutation. In the *E. coli* chromosome, five or more copies of IS1, IS2, and IS3 are present. The exact number of each varies, depending on the strain examined.

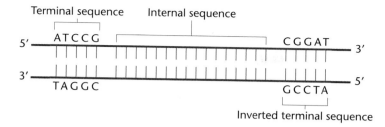

Bacterial Transposons

In addition to their potential mutational effects, IS units play an even more significant role in the formation and movement of the larger **transposon (Tn) elements.** Transposons in bacteria consist of IS units that contain genes whose functions are unrelated to the insertion process. Like IS units, Tn elements are mobile in both bacterial and viral chromosomes and in plasmids, providing a mechanism for the movement of genetic information from place to place both within and between organisms. Transposons were first discovered to move between DNA molecules as a result of observations of antibiotic-resistant bacteria. In the mid-1960s, Susumu Mitsuhashi first suggested that the genes responsible for resistance to several antibiotics were mobile and could move between bacterial plasmids and chromosomes.

Electron microscopic studies may be used to confirm the presence of the terminal-inverted repeat sequences within plasmids harboring transposons. When double-stranded DNA from such a plasmid is separated into single strands and each is allowed to reanneal separately, the inverted repeat units are complementary, forming a heteroduplex. As might be predicted, all areas other than the terminal-repeat units remain single stranded and form loops on either end of the double-stranded stem (Figure 17.26).

Transposons have become the focus of increased interest, particularly because they have been found in organisms other than bacteria. Bacteriophages that demonstrate the ability to insert their genetic material into the host chromosome behave in a similar fashion. The bacteriophage mu, consisting of over 35,000 nucleotides, can insert its DNA at various places in the *E. coli* chromosome. As with IS units, if insertion occurs within a gene, mutant behavior results at that locus. Transposons have also been discovered in higher organisms, including yeast, maize, *Drosophila*, and humans, as we see next.

The Ac–Ds System in Maize

With our knowledge of insertion sequences and transposons in bacteria, it is no surprise that mobile genetic units also exist in eukaryotes. They are, in fact, more widespread, and, like bacteria, they often have the effect of altering the expression of genetic information.

About 20 years before the discovery of transposons in prokaryotic organisms, Barbara McClintock analyzed the genetic behavior of two mutations, *Dissociation* (*Ds*) and *Activator* (*Ac*), in corn plants (maize). She examined the phenotypes of the maize kernels resulting from genes expressed in either the endosperm or aleurone layers (Figure 17.27) and correlated her observations with a cytological examination of the maize chromosomes. Initially, McClintock determined that *Ds* is located on chromosome 9. If *Ac* is also present in the genome, *Ds* induces breakage at a point on the chromosome adjacent to its own location. If breakage occurs in somatic cells during their development, progeny cells often lose part of chromosome 9, causing a variety of phenotypic effects.

Subsequent analysis suggested to McClintock that both the *Ds* and *Ac* genes are sometimes transposed to different chromosomal locations. While *Ds* moves only if *Ac* is also present, *Ac* is capable of autonomous movement. Where *Ds* comes to reside determines its genetic effect; that is, it might cause chromosome breakage or it might inhibit gene expression. In cells where expression is inhibited, *Ds* might move again, releasing this inhibition. In these cases, the *Ds* element is believed to insert into a gene and subsequently to depart from it, causing changes in gene expression.

Figure 17.28 illustrates the sort of movements and effects of the *Ds* and *Ac* elements described. In McClintock's original observation, when the *Ds* element jumped out of chromosome 9, this excision event restored normal gene function. In such cells, pigment synthesis was restored. McClintock concluded that the *Ds* and *Ac* genes are **transposable controlling elements**.

It was not until many years later that anything comparable to the controlling elements in maize was recognized in other organisms. When bacterial insertion sequences and transposons were discovered, many parallels were evident. Transposons and insertion sequences were seen to move into and out of chromosomes, to insert at different positions, and to affect gene expression at the point of insertion.

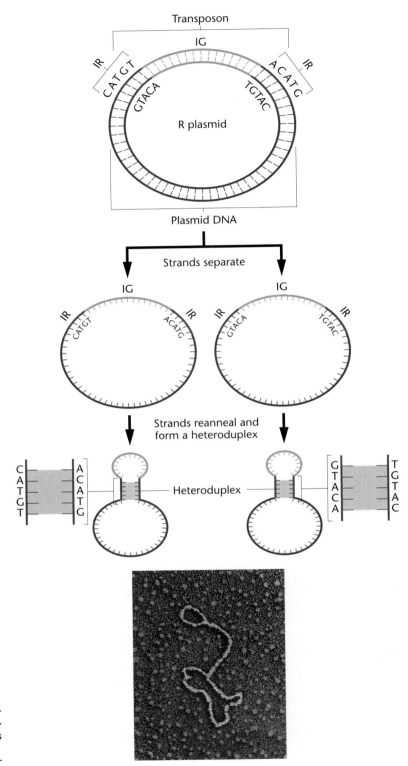

FIGURE 17.26 Heteroduplex formation as a result of inverted repeat sequences within a transposon inserted into a bacterial plasmid. The micrograph illustrates the final heteroduplex.

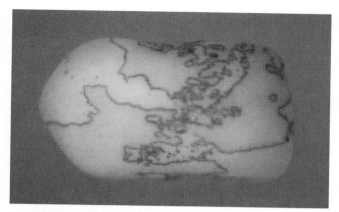

FIGURE 17.27 Corn kernel showing spots of colored aleurone produced by genetic transposition involving the *Ac–Ds* system.

Several *Ac* and *Ds* elements have now been isolated and carefully analyzed, and the relationship between the two elements has been clarified (Figure 17.29). The first *Ac* element sequence is 4563 bases long and strikingly similar to one of the known bacterial transposons. This sequence contains two 11-base-pair imperfect terminal-inverted repeats, two **open reading frames** (**ORFs**), and three noncoding regions. Open reading frames contain initiation and termination sequences and are considered to encode genetic products. The first *Ds* element studied (*Ds-a*) is nearly identical in structure to *Ac* except for a 194-base segment that has been deleted from the largest open reading frame. There is some evidence that this gene encodes a **transposase enzyme**, essential to transposition of both the *Ac* and *Ds* elements. The deletion of part of this gene in the *Ds-a* element explains its dependence on the *Ac* element for transposition. Several other *Ds* elements have also been sequenced, and each reveals an even larger deletion in the same region. In each case, however, the terminal repeats

(a) In absense of *Ac*, *Ds* is not transposable.
Wild type expression of *W* occurs

(b) When *Ac* is present, *Ds* may be transposed

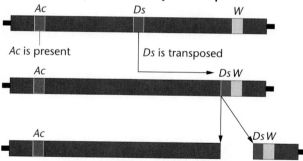

Chromosome breaks and fragment is lost
Expression of *W* ceases, producing mutant effect

(c) *Ds* can move into and out of another gene

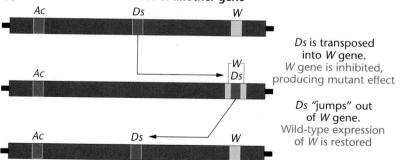

Ds is transposed into *W* gene.
W gene is inhibited, producing mutant effect

Ds "jumps" out of *W* gene.
Wild-type expression of *W* is restored

FIGURE 17.28 Consequences of the influence of the *Activator* (*Ac*) element on the *Dissociation* (*Ds*) element. In (b), *Ds* is transposed to a region adjacent to a theoretical gene *W*. Subsequent chromosomal breakage is induced, the *W*-bearing segment is lost, and mutant gene expression occurs. In (c), *Ds* is transposed to a region within the *W* gene, causing immediate mutant expression. *Ds* may also "jump" out of the *W* gene, with the accompanying restoration of *W* gene activity and its wild-type expression.

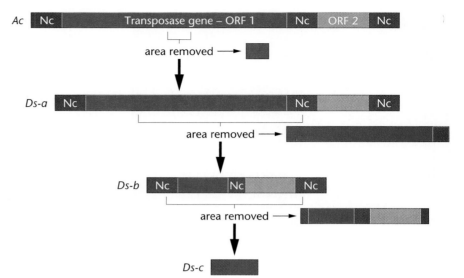

FIGURE 17.29 A comparison of the structure of an *Ac* element with three *Ds* elements, all of which have been isolated and sequenced. The imperfect and inverted repeats are at the ends of the *Ac* element. The transposase gene is in an open reading frame (ORF 1). No function has yet been assigned to ORF 2. Noncoding regions are designated by Nc. As this scheme shows, *Ds-a* appears to be simply an *Ac* element containing a small deletion in the gene encoding the transposase enzyme.

are retained and seem to be essential for transposition, provided that a functional transposase enzyme is supplied by the gene in the *Ac* element.

Although the validity of Barbara McClintock's proposed mobile elements was questioned following her initial observations, molecular analysis has since verified her conclusions. For her work, Barbara McClintock was awarded the Nobel Prize in 1983.

Other Mobile Genetic Elements in Plants: Mendel Revisited

More recent work on transposable elements in plants has led us full circle to the union of Mendel's observations with molecular genetics. Recall that one of the first phenotypes Mendel investigated involved the inheritance of round and wrinkled peas. The two phenotypes are produced by alleles of a single gene, *rugosus*. It is now known that the wrinkled phenotype is associated with the absence of an enzyme, **starch-branching enzyme** (**SBEI**), that controls the formation of branch points in starch molecules. The lack of starch synthesis leads to the accumulation of sucrose and to a higher water content and osmotic pressure in the developing seeds. As the seeds mature, those that are wrinkled (genotype *rr*) lose more water than do the smooth seeds (*RR* or *Rr*), producing the wrinkled phenotype (Figure 17.30).

The structural gene for SBEI has been cloned and characterized in both wild-type and mutant genotypes. In the *rr* genotype, the SBEI protein is nonfunctional, presumably because the SBEI gene is interrupted by a 0.8-kb insertion, resulting in the production of an abnormal RNA transcript. The inserted DNA has 12-bp inverted repeats

FIGURE 17.30 The *wrinkled* trait in garden peas, studied by Gregor Mendel, is caused by the insertion of a transposable element into the structural gene for the starch-branching enzyme. Both *wrinkled* and *smooth* seeds can develop in the same pod.

at each end that are highly homologous to the terminal sequences in the transposable element *Ac* from maize, and to other *Ac*-like elements from snapdragons and parsley. Terminal repeated sequences and the genetic information encoding a transposase enzyme appear to be universal components of transposons in all organisms studied.

Copia *Elements in* Drosophila

In 1975 David Hogness and his colleagues David Finnegan, Gerald Rubin, and Michael Young identified a class of genes in *Drosophila melanogaster* that transcribe "copious" amounts of RNA (thus, their name *copia*). Present in up to 30 copies in the genome of cells, all types of *copia* genes are similar in their nucleotide sequence. Mapping studies show that they are transposable to different chromosomal locations and are dispersed throughout the genome.

Ever since the discovery of *copia*, many other families of transposable elements have been recognized in *Drosophila*, each of which is present in from a few up to 20 to 50 copies in the genome. Some are referred to as *copia*-like. Together, the approximately 30 families constitute about 5 percent of the *Drosophila* genome and over half of the middle repetitive DNA of this organism.

While members of different families vary in DNA sequence, they still share a common structural organization thought to be related to the insertion and excision processes of transposition. For example, each *copia* gene consists of approximately 5000 to 8000 bp of DNA, including a long family-specific **direct terminal repeat (DTR)** sequence of 276 bp at each end. Within each repeat is a short inverted terminal repeat (ITR) of 17 bp. These features are illustrated in Figure 17.31. The shorter ITR sequences are considered universal in *copia* elements. The DTR sequences are found in other transposons in other organisms, but they are not universal.

The insertion of *copia*, as with other transposable elements, appears to be dependent on ITR sequences and seems to occur at specific target sites in the genome. The *copia*-like elements demonstrate regulatory effects at the point of their insertion in the chromosome. Certain mutations, including ones affecting eye color and segment formation, have been found to be due to insertions within genes. For example, the eye color mutation *white-apricot* (*w^a*), an allele of the *white* (*w*) gene, contains a *copia* insertion element within the gene. Removing the transposable element sometimes restores the wild-type allele.

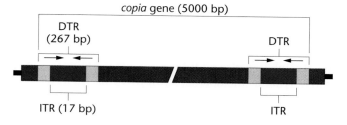

FIGURE 17.31   Structural organization of a *copia* transposable element in *Drosophila melanogaster*, showing the terminal repeats.

P Element Transposons in Drosophila

Still another interesting category of transposable elements in *Drosophila* is the family called **P elements**. These were discovered while studying the phenomenon of **hybrid dysgenesis**, a condition causing sterility, elevated mutation rate, and chromosome rearrangement in the offspring of crosses between certain strains of fruit flies. Hybrid dysgenesis is caused by high rates of P-element transposition in the germ line, where the mobile DNA elements insert themselves into or near genes, thereby causing mutations. P elements are 2.9 kb long, with 31-bp terminal inverted repeats. The elements encode at least two proteins, one of which is the transposase enzyme that is required for transposition. The transposase is expressed only in the germ line, accounting for the tissue specificity of P-element transposition. Strains of flies that contain P elements inserted into their genomes are resistant to further transpositions due to the presence of a repressor protein, also encoded by the P elements.

P elements are useful as insertional mutagens in *Drosophila*. A common method of generating P element-induced mutations is to inject *Drosophila* embryos with a mixture of P elements. One P element lacks the transposase gene and cannot move by itself. The other P element contains the transposase gene, but cannot insert into the genome. The two injected P elements will be taken up by the embryo's cells, transposase will be expressed in the germ line of the fly, and the P element will insert itself into the fly's DNA. The next generation will contain stable copies of P-element DNA, inserted at random locations.

Mutations may arise from several kinds of insertional events. If a P element inserts into the coding region of a gene, it can destroy the normal gene product. If it inserts into the promoter region of a gene, it can affect the level of expression of the gene. Insertions into introns can affect splicing or cause the premature termination of transcription. Researchers screen the flies for mutant phenotypes of interest, isolate the fly's DNA, and clone the gene that was mutated. This technology has been used to identify genes involved in *Drosophila* development, behavior, and regulation of gene expression.

Another equally important P-element technique is germ-line transformation. One of the most powerful ways to analyze a gene's activity and expression pattern is to clone the gene, manipulate its DNA sequences in a test tube, and re-introduce the mutated or modified gene into an organism. P elements provide a vehicle (or "vector") for inserting a cloned gene into a fly's genome. Embryos are injected with a mixture of P elements. One contains the transposase gene (but cannot insert). The other contains the marker gene such as *rosy* (an eye color mutation) and the cloned gene of interest, but lacks a transposase

GENETICS, TECHNOLOGY, AND SOCIETY

Chernobyl's Legacy

On April 26, 1986, Reactor 4 of the Chernobyl Nuclear Power Station overheated, exploded, and ejected massive amounts of radioactive material into the surrounding countryside and ultimately throughout the northern hemisphere. The accident killed 31 emergency workers and caused acute radiation sickness in over 200 others. In the nine days following the initial explosion, as the reactor's temperature approached meltdown, radioactive fission products continued to be discharged into the environment. High levels of radioactive iodine, xenon, strontium, and cesium were released into the atmosphere, along with over 5000 tons of vaporized boron carbide, dolomite, clay, sand, and lead. The plume of fallout traveled in a northwesterly direction through central Europe, reaching Finland and Sweden three days after the initial explosion. By May 5, the radioactive cloud reached the United Kingdom, and remnants of the cloud arrived in North America one week later. Millions of people in the Soviet Union and Europe were exposed to measurable amounts of radioactivity. People living within 30 km of Chernobyl were exposed to high levels of radioactivity prior to their evacuation 36 hours following the accident. Equally high were the exposures of some of the 600,000

military and civilian workers who were sent to Chernobyl to decontaminate the area and to encase the shattered reactor in a sarcophagus.

Chernobyl was the world's largest accidental release of radioactive material. In addition, it was a disaster that inflicted enormous economic and personal burdens on the Soviet people. The question that remains, over a decade after the accident, is whether Chernobyl's radioactive pollution directly threatens the long-term health of millions of people.

More is known about the effects of ionizing radiation on human health than any other agent (except perhaps cigarette smoking). Radiation refers to the energy that is emitted from a source, and includes heat, light, radio waves, microwaves, X rays, and gamma rays from radioactive elements. Ionizing radiation (such as X rays and gamma rays) is a subset of radiation that possesses sufficient energy to eject electrons from atoms. Ionizing radiation can damage any cellular component, and can alter nucleotides and induce strand breaks in DNA. These DNA lesions, if not repaired by the cell, can lead to mutations or chromosomal translocations. It is this DNA-damaging capacity of ionizing radiation that makes it a mutagen.

We know from epidemiological studies that high doses of ionizing radiation increase the risk of developing certain cancers. The survivors of the atomic bomb blasts of Hiroshima and Nagasaki showed an increased incidence of leukemia within two years of the bombing. Leukemias peaked within 10 years and then declined. Breast cancers began to increase 10 years after exposure, as did cancers of the lung, thyroid, colon, ovary, stomach, and nervous system. However, the incidence of other cancers, such as those of the gall bladder, pancreas, uterus, and bone, did not increase due to radiation exposure. As ionizing radiation is known to induce DNA damage, the offspring of A-bomb survivors were expected to show increases in birth defects. However, increases beyond the rates expected in the unexposed population were not detectable. The problem in extrapolating from Hiroshima to Chernobyl is the difference in dose. In A-bomb survivors, the rate of incidence of cancer increased among those exposed to at least 200 mSv (mSv = millisievert, a unit dose of absorbed radiation). It is estimated that people living in the most contaminated areas close to Chernobyl may have received about 50 mSv, and cleanup workers were exposed to approximately 250 mSv. Outside the

---

gene. Flies that have been transformed are selected on the basis of their rosy eye color, and the behavior of the cloned and transposed gene is observed.

Various other elegant techniques, based on P-element transposition, have been developed. These have allowed researchers to study the mechanics of DNA repair and DNA recombination. In addition, researchers are presently developing methods to target P-element insertions to precise single-chromosomal sites. This should increase the precision of germ-line transformation in the analysis of gene activity.

Transposable Elements in Humans

The final class of transposable elements that we shall discuss is represented by the *Alu* family of short interspersed elements (SINEs), which are characteristic of moderately repetitive DNA in mammals. Human DNA contains some 500,000 copies of this 200- to 300-bp sequence in the genome. Originally detected using reassociation kinetic analysis (see Chapter 11), these transposable elements received their name because they contain specific nucleotide sequences that are cleaved by a restriction endonuclease called *Alu*I.

Chernobyl area, radiation doses are estimated at 0.4–0.9 mSv in Germany and Finland, 0.01 mSv in the U.K., and 0.0006 mSv in the United States.

In order to put these numbers in context, the average dose for medical diagnostic procedures (such as chest and dental X rays) is 0.39 mSv per year. A person's radiation exposure from natural background sources (cosmic rays, rocks, and radon gas) is about 2–3 mSv per year. Smokers expose themselves to an average extra dose of about 2.8 mSv per year from the intake of naturally occurring radioactive materials in tobacco smoke. A whole-body X-ray dose of 3000 mSv or more can be fatal.

In the 115,000 people evacuated from Chernobyl, it was estimated that 26 leukemias might appear above the 25 to 30 that would occur spontaneously. It has also been estimated that up to 17,000 additional cancers might occur in Europe, above the 123 million that would occur normally. At present, however, there have been no detectable increases in leukemia in contaminated areas of the Soviet Union, Finland, or Sweden, or in the 600,000 Chernobyl cleanup workers. It is still too early to detect increases in solid-tumor incidence, which should only begin to appear 10 years after the accident.

Despite the lack of detectable increases in leukemias and solid tumors,

one type of cancer does appear to be increasing in the Chernobyl area. The rate of childhood thyroid cancer has reached over 100 cases per million children per year, whereas normal rates would be expected to be between 0.5 and 3 cases per million children per year. The regions with the greatest contamination appear to have the highest rate of thyroid cancer. Although epidemiologists debate whether these increases are due to radiation or to the increased reporting of cases, the scale of the increase, and the fact that radioactive isotopes of iodine made up a significant portion of the Chernobyl fallout, make the link between thyroid cancer and Chernobyl fallout a plausible one.

The most profound effects of Chernobyl have been psychological. A recent study of Chernobyl cleanup workers found no increases in cancers, including leukemias or thyroid cancers. However, there was a 50 percent increase in suicides and a detectable increase in smoking- and alcohol-related disease. This mirrors other studies showing that 45 percent of people living within 300 km of Chernobyl believe that they have a radiation-induced illness. These people may be correct, as health effects such as depression, sleep disturbance, hypertension, and altered perception have been documented. People suffer from the stress of relocation, the

loss of personal and real property, and the feeling of a lack of control over their radiation exposure or their future. Posttraumatic stress may be a greater threat to health than the actual radiation exposure from the accident. Psychological counseling for cleanup workers and for victims living in the contaminated areas is nonexistent. Without solid information about their real risks, people feel that they live in constant danger and are simply awaiting the results of a cancer "lottery." Even if cancer rates and genetic defects do not increase dramatically, the indirect health effects from the Chernobyl Nuclear Power Plant explosion continue to be immense.

REFERENCES
ANSPAUGH, L.R., CATLIN, R.J., and GOLDMAN, M. 1988. The global impact of the Chernobyl reactor accident. *Science* 242:1513–19.
GINZBERG, H.M. 1993. The psychological consequences of the Chernobyl accident—Findings from the International Atomic Energy Agency study. *Public Health Rep.* 108:184–92.
RAHU, M. et al. 1997. The Estonian study of Chernobyl cleanup workers: II. Incidence of cancer and mortality. *Radiation Res.* 147:653–57.
WILLIAMS, D. 1994. Chernobyl, eight years on. *Nature* 371:556.

The *Alu* family is considered to be a significant set of elements in the genome based on the fact that they have been found in the DNA of all the primates and rodents studied. Within the *Alu* elements in mammals a specific 40-bp DNA sequence has been conserved. Finally, *Alu* sequences are represented in some nuclear transcripts.

The *Alu* elements are considered transposable based on several lines of evidence. The most important is the fact that their 200 to 300-bp sequence is flanked on either side by direct repeat sequences consisting of 7 to 20 bp, paralleling bacterial insertion sequences. These flanking

regions are related to the insertion process during transposition. Secondly, clustered regions of *Alu* sequences vary in the DNA of both normal and diseased individuals and in different tissues of the same individual. Additionally, the sequences have been found extrachromosomally.

The potential mobility and mutagenic effects of these and other elements have far-reaching implications, as can be seen in a recent example of a transposon "caught in the act." The case involves a male child with hemophilia. One cause of hemophilia is a defect in blood-clotting factor VIII, the product of an X-linked gene. Haig Kazazian and his colleagues found a transposable element much longer than *Alu* sequences inserted at two points within the gene. Also present elsewhere in the genome, this sequence was shown to be a type of repetitive DNA referred to as a **long interspersed element (LINE)**. Both LINEs and SINEs are discussed in more detail in Chapter 19.

Researchers were very interested in determining if one of the mother's X chromosomes also contained this specific LINE. If so, the unaffected mother would be heterozygous and pass the LINE-containing chromosome to her son. The startling finding was that the LINE sequence is *not* present on either of her *X chromosomes*, but *was* detected on *chromosome 22* of both parents. This suggests that this mobile element may have "moved" from one chromosome to another in the gamete-forming cells of the mother, prior to being transmitted to the son.

Many questions remain concerning this and other transposable elements. What is their origin? Were they once some sort of retrovirus? Exactly how do they move, and what has been their role during evolution? These and other questions will intrigue researchers for many years to come.

CHAPTER SUMMARY

1. The phenomenon of mutation not only provides the basis for most of the inherent variation present in living organisms but it also serves as the working tool of the geneticist in studying and understanding the nature of genes and the mechanisms governing genetic processes.

2. Mutations are distinguished by the tissues affected. Somatic mutations are those that may affect the individual, but are not heritable. Mutations arising in the gametic tissues may produce new alleles that can be passed on to offspring and enter the gene pool.

3. Another classification of mutations is based on their effects. Morphological mutations, for example, can be detected visibly. Other types include biochemical, lethal, conditional, and regulatory mutations, groups that are not mutually exclusive.

4. Organisms in which mutations can be easily induced and detected are most often used in genetic studies. Viruses, bacteria, fungi, *Drosophila*, certain plants, and mice are frequently used because of these properties, as well as the fact that they have short life cycles.

5. Spontaneous mutations may arise naturally as a result of rare chemical rearrangements of atoms, known as tautomeric shifts, and as the result of errors occurring during DNA replication. Spontaneous mutations are very rare.

6. The rate of mutagenesis can be increased experimentally by a variety of mutagenic agents. Agents such as nitrous acid, hydroxylamine, alkylating agents, and base analogues cause chemical changes in nucleotides that alter their base-pairing affinities. As a result, base-substitution mutations arise following DNA replication. The study of mutations caused by these agents has provided considerable insight into the molecular basis of mutations.

7. Frameshift mutations arise when the addition or deletion of one or more nucleotides (but not multiples of three) occurs. Acridine dyes, which intercalate with DNA, are potent inducers of frameshift mutations.

8. The direct analysis of DNA from individuals carrying specific alleles that have arisen through mutations, including the ABO blood types and muscular dystrophy, has been informative. When complete loss of function occurs, as is the case in blood type O and the Duchenne form of muscular dystrophy, deletions or duplications of nucleotides have resulted in a shift in reading frame, eventually causing premature termination of translation of the resulting messenger RNA.

9. Another form of mutation discovered in several human disorders, including fragile-X syndrome and myotonic dystrophy, involves unstable trinucleotide units that are repeated many times within certain genes. The number of repeats may increase in succeeding generations, and when the number of repeats exceeds a critical level, inherited disorders result. An increasing number of repeats often correlates with early onset and increased severity of the disease.

10. Ultraviolet radiation and high-energy radiation from gamma, cosmic, or X-ray sources are other forms of potent mutagenic agents. UV radiation induces the formation of pyrimidine dimers in DNA, whereas high-energy radiation is able to penetrate deeper into tissues, causing the ionization of molecules in its path, including DNA.

11. Various forms of repair of DNA lesions, such as pyrimidine dimers, have been discovered, including photoreactivation, excision repair, mismatch repair, proofreading, and various forms of recombinational repair, including double-stranded break repair. The significance of repair mechanisms is apparent when, in humans, excision repair is lost due to mutation. The result is the severe human disorder xeroderma pigmentosum.

12. Site-directed mutagenesis is a technique allowing researchers to create specific alterations in the nucleotide sequence of the DNA of genes.

13. The insertion sequences in bacteria and other transposable elements in eukaryotes are mobile genetic units characterizing the genomes of all organisms. They have a profound effect on genetic expression, thus serving as a distinct category of mutagenic agent.

INSIGHTS AND SOLUTIONS

1. How could you isolate a mutant strain of bacteria that is resistant to penicillin, an antibiotic that inhibits cell-wall synthesis?

   Solution: Grow a culture of bacterial cells in liquid medium and plate the cells on agar medium to which penicillin has been added. Only penicillin-resistant cells will reproduce and form colonies. Each colony will, in all likelihood, represent a cloned group of cells with the identical mutation. Isolate members of each colony. To enhance the chance of such a mutation arising, you might want to add a mutagen to the liquid culture.

2. The base analogue 2-amino purine (2-AP) substitutes for adenine during DNA replication, but it may base-pair with cytosine. The base analogue 5-bromouracil (5-BU) substitutes for thymidine, but it may base-pair with guanine. Follow the double-stranded trinucleotide sequence shown here through three rounds of replication, assuming that in the first round both analogues are present and become incorporated wherever possible. In the second and third round of replication, they are removed. What final sequences occur?

   Solution:

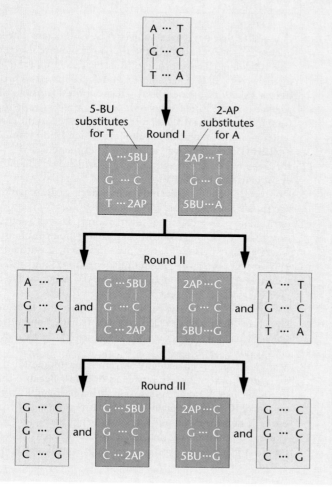

3. A rare dominant mutation expressed at birth was studied in humans. Records showed that six cases were discovered in 40,000 live births. Family histories revealed that in two cases, the mutation was already present in one of the parents. Calculate the spontaneous mutation rate for this mutation. What are some underlying assumptions that may affect our conclusions?

   Solution: Only four cases represent a new mutation. Because each live birth represents two gametes, the sample size is from 80,000 meiotic events. The rate is equal to

   $$\frac{4}{80,000} = \frac{1}{20,000} = 0.5 \times 10^{-4}$$

   There are various reasons why this mutation rate may be inaccurate. We have assumed that the mutant gene is fully penetrant and is expressed in each individual bearing it. If it is not fully penetrant, our calculation may be an underestimate because one or more mutations may have gone undetected. We have also assumed that the screening was 100 percent accurate. One or more mutant individuals may have been "missed," again leading to an underestimate. Finally, we assume that the viability of the mutant and non-mutant individuals is equivalent and that they survive equally *in utero*. Therefore, our assumption is that the number of mutant individuals at birth is equal to the number at conception. If this were not true, our calculation would again be inaccurate.

4. Consider the following estimates:

   (i) There are 4×10^9 humans living on this planet.

   (ii) Each individual has 10^5 genes.

   (iii) The average mutation rate at the 10^5 loci is 10^{-5}.

   How many spontaneous mutations are currently present in the human population? Assuming that these mutations are equally distributed among all genes, how many new alleles have arisen at each locus in the human population?

   Solution: First, since each individual is diploid, there are two copies of each gene per person, each arising from a separate gamete. Therefore, the total number of spontaneous mutations is

   $$\left(\frac{2 \times 10^5 \text{ genes}}{\text{individual}}\right)\left(\frac{4 \times 10^9 \text{ individuals}}{}\right)\left(\frac{10^{-5} \text{ mutations}}{\text{gene}}\right)$$

   $$= (2 \times 10^5)\cdot(4 \times 10^9)\cdot(10^{-5}) \text{ mutations}$$
   $$= 8 \times 10^9 \text{ mutations}$$

   On the average, in the entire human population there are

   $$\frac{8 \times 10^9 \text{ mutations}}{10^5 \text{ genes}} = \frac{8 \times 10^4 \text{ mutations}}{\text{gene}}$$

GENETICS MediaLab

The following resources will help you achieve a better understanding of the concepts presented in this chapter. These resources can be found on the CD packaged with this textbook and on the Companion Website at **http://www.prenhall.com/klug/**.

CD Resources:

Animated Tutorial: *Gene Mutations*

Animated Tutorial: *DNA Repair*

Self-grading Chapter Problems

Web Resources:

Genetics and Society Issue: *Chernobyl's Legacy*

Web Destinations in Genetics

Self-grading Chapter Problems

Chapter Search Terms

Genetics Newsgroups

Student Bulletin Board

Web Problem 1:

Time for completion = 10 minutes

Do mutations arise in response to environmental conditions? The accepted answer for years was no. The work by Luria and Delbrück indicated that mutations occur sponaneously and that selective forces then act upon the variation that is present. Recently, this issue has been reexamined by Cairns and colleagues, who suggest that at times, there may be an adaptive response to some conditions. To understand this controversy, you need to have a clear idea of the experimental conditions used to answer this question. This linked website discusses the basic design of such a fluctuation test. Describe the results you might expect in light of these two hypotheses. To complete this exercise, visit Web Problem 1 in Chapter 17 of your Companion Website and select the keyword **MUTATION.**

Web Problem 2:

Time for completion = 15 minutes

What is the relationship between a mutation and its phenotype? It is useful to have some sense of what types of mutations cause

PROBLEMS AND DISCUSSION QUESTIONS

1. What is the difference between a chromosomal aberration and a gene mutation?

2. Discuss the importance of mutations to the successful study of genetics.

3. Describe the technique for the detection of nutritional mutants in *Neurospora*.

4. Most mutations are thought to be deleterious. Why, then, is it reasonable to state that mutations are essential to the evolutionary process?

5. Why is a random mutation more likely to be deleterious than beneficial?

6. Most mutations in a diploid organism are recessive. Why?

7. What is meant by a conditional lethal mutation?

8. Contrast the concerns about mutation in somatic and gametic tissue.

9. In the *ClB* technique for detection of mutation in *Drosophila*:
   (a) What type of mutation may be detected?
   (b) What is the importance of the *l* gene?
   (c) Why is it necessary to have the crossover suppressor (*C*) in the genome?
   (d) What is *C*?

10. In an experiment with the *ClB* system, a student irradiated wild-type males and subsequently scored 500 F_2 cultures, finding all of them to have both males and females present. What conclusions can you draw about the induced mutation rate?

11. In *Drosophila*, induced mutations on chromosome 2 that are recessive lethals may be detected using a second chromosomal stock, *Curly, Lobe/Plum* (*Cy L/Pm*). These alleles are all dominant and lethal in the homozygous condition. Detection is performed by crossing *Cy L/Pm* females to wild-

type males that have been subjected to a mutagen. Three generations are required. In the F_1, *Cy L* males are selected and individually backcrossed to *Cy L/Pm* females. In the F_2 of each cross, flies expressing *Cy* and *L* are mated to produce a series of F_3 generations. Diagram these crosses and predict how an F_3 culture would vary if a recessive lethal were induced in the original test male compared to the case where no lethal mutation resulted. In which F_3 flies would a recessive morphological mutation be expressed?

12. Describe tautomerism and the way in which this chemical event may lead to mutation.

13. Contrast and compare the mutagenic effects of deaminating agents, alkylating agents, and base analogues.

14. Acridine dyes induce frameshift mutations. Is such a mutation likely to be more detrimental than a point mutation where a single pyrimidine or purine has been substituted?

15. Why are X rays a more potent mutagen than UV radiation?

16. Contrast the induction of mutations by UV radiation and X rays.

17. Contrast the various types of DNA-repair mechanisms known to counteract the effects of UV radiation and other DNA damage. What is the role of visible light in photoreactivation?

18. Mammography is an accurate screening technique for the early detection of breast cancer in humans. Because this technique uses X rays diagnostically, it has been highly controversial. Can you explain why? What reasons justify the use of X rays for this type of medical diagnosis?

19. Compose a short essay that relates the molecular basis of fragile-X syndrome, myotonic dystrophy, and Huntington

a particular phenotype and how the gene is affected. Researchers' increased ability to identify mutations by direct sequencing has led to an explosion of information. To sort the growing volume of information about human diseases in particular, researchers have established databases listing observed mutations. You will find a database of mutations in the androgen receptor gene at the linked website. Severe mutations in this gene disrupt its function and cause androgen insensitivity (AIS) leading to sex reversal in XY individuals. Examine the types of information availiable at this site, including the "selectable options" section. What conclusions can you draw about the effects of mutations in this gene? Discuss alternative ways to present such data or additional types of analyses you think would be useful. To complete this exercise, visit Web Problem 2 in Chapter 17 of your Companion Website and select the keyword **ANDROGEN RECEPTOR.**

Web Problem 3:

Time for completion = 10 minutes
How do transposons move from one place in the genome to another? To answer this question, researchers have devised genetic tests to distinguish among the different molecular steps

required for transpositional movement. This field of research has provided insight into how the genome of an organism may change over time. There are two distinct methods of transposition illustrated at the linked website, replicative and non-replicative. After examining the sections, compare and contrast these two modes of movement. To complete this exercise, visit Web Problem 3 in Chapter 17 of your Companion Website and select the keyword **TRANSPOSITION.**

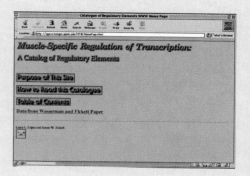

disease to the severity of the disorders, as well as to the phenomenon called genetic anticipation.

20. Describe how the Ames assay screens for potential environmental mutagens. Why is it thought that a compound that tests positively in the Ames assay may also be carcinogenic?

21. Describe the general approach used in site-directed mutagenesis.

22. What genetic defect results in the disorder xeroderma pigmentosum (XP) in humans? How does this defect relate to the phenotype associated with the disorder?

23. Differentiate between point mutations that are transitions and those that are transversions. Using the DNA bases (A, T, C, and G), list the four types of transitions and the eight types of transversions.

24. In a bacterial culture where all cells were unable to synthesize leucine (*leu*⁻) a potent mutagen was added, and the cells were allowed to undergo one round of replication. At that point, samples were taken, a series of dilutions was made, and the cells were plated on either minimal medium or minimal medium to which leucine was added. The first culture condition (minimal medium) allows the detection of mutations from *leu*⁻ to *leu*⁺, while the second culture condition allows the determination of the total cells, since all bacteria can grow. From the results given here, determine the frequency of mutant cells. What is the rate of mutation at the locus involved with leucine biosynthesis?

Culture Condition	Dilution	Colonies
minimal medium	10^{-1}	18
minimal + leucine	10^{-7}	6

25. Contrast the various transposable genetic elements in bacteria, maize, *Drosophila*, and humans. What properties do they share?

26. The initial discovery of IS units in bacteria involved their presence upstream (5′) to three genes controlling galactose metabolism. All three genes were affected despite there being no direct physical interaction with the IS unit. Offer an explanation as to why this might occur.

27. *Ty*, a transposable element in yeast, has been found to contain an open reading frame (ORF) that encodes the enzyme reverse transcriptase. This enzyme synthesizes DNA from an RNA template. Speculate as to the role of this gene product in the transposition of *Ty* within the yeast genome.

 EXTRA-SPICY PROBLEMS

28. Presented here are hypothetical findings from studies of heterokaryons formed from seven human xeroderma pigmentosum cell strains.

	XP1	*XP2*	*XP3*	*XP4*	*XP5*	*XP6*	*XP7*
XP1	0						
XP2	0	0					
XP3	0	0	0				
XP4	+	+	+	0			
XP5	+	+	+	+	0		
XP6	+	+	+	+	0	0	
XP7	+	+	+	+	0	0	0

Note: + = complementing; 0 = noncomplementing.

These data represent the occurrence of unscheduled DNA synthesis in the fused heterokaryon when neither of the strains alone showed synthesis. What does unscheduled DNA synthesis represent? Which strains fall into the same complementation groups? How many different groups are revealed based on these limited data? How does one interpret the presence of these complementation groups?

29. Imagine yourself as one of the team of geneticists who launched the study of the genetic effects of high-energy radiation on the surviving Japanese population immediately following the atomic bomb attacks at Hiroshima and Nagasaki in 1945. Demonstrate your insights into both chromosomal and gene mutation by outlining a comprehensive short-term and long-term study that addresses this topic. Be sure to include strategies for considering the effects on both somatic and germ-line tissues.

30. Cystic fibrosis (CF) is a severe autosomal recessive disorder in humans that results from a chloride ion–channel defect in epithelial cells. Over 500 sequence alterations have been identified in the 24 exons of the responsible gene (*CFTR* for cystic fibrosis conductance regulator), including dozens of different missense mutations and frameshift mutations, as well as numerous splice-site defects. Although all affected CF individuals demonstrate chronic obstructive lung disease, there is variation in pancreatic enzyme insufficiency (PI). Speculate on which types of observed mutations are likely to give rise to less severe symptoms of CF, including only minor PI. Of the 300 sequence alterations, a number of them found within the exon regions of the *CFTR* gene do not give rise to cystic fibrosis. Taking into account your accumulated knowledge of the genetic code, gene expression, protein function, and mutation, explain to a freshman biology major how this might be.

SELECTED READINGS

AMES, B.N., MCCANN, J., and YAMASAKI, E. 1975. Method for detecting carcinogens and mutagens with the *Salmonella*/mammalian microsome mutagenicity test. *Mut. Res.* 31:347–64.

ANDERSON, D.I, SLECHTA, E.S., and ROTH, J. 1998. Evidence that gene amplification underlies adaptive mutability of the bacterial *lac* operon. *Science* 282:1133–35.

AUERBACH, C. 1978. Forty years of mutation research: A pilgrim's progress. *Heredity* 40:177–87.

AUERBACH, C., and KILBEY, B.J. 1971. Mutations in eukaryotes. *Annu. Rev. Genet.* 5:163–218.

BALTIMORE, D. 1981. Somatic mutation gains its place among the generators of diversity. *Cell* 26:295–96.

BATES, G., and LEHARCH, H. 1994. Trinucleotide repeat expansions and human genetic disease. *BioEssays* 16:277–84.

BEADLE, G.W., and TATUM, E.L. 1945. *Neurospora* II. Methods of producing and detecting mutations concerned with nutritional requirements. *Am. J. Bot.* 32:678–86.

BECKER, M.M., and WANG, Z. 1989. Origin of ultraviolet damage in DNA. *J. Mol. Biol.* 210:429–38.

BERG, D., and HOWE, M., eds. 1989. *Mobile DNA*. Washington, DC: American Society of Microbiology.

BRENNER, S., BARNETT, L., CRICK, F.H.C., and ORGEL, L. 1961. The theory of mutagenesis. *J. Mol. Biol.* 3:121–24.

CAIRNS, J., OVERBAUGH, J., and MILLER, S. 1988. The origin of mutants. *Nature* 335:142–45.

CAPECCHI, M.R. 1989. Altering the genome by homologous recombination. *Science* 244:1288–92.

CARTER, P. 1986. Site-directed mutagenesis. *Biochem. J.* 237:1–7.

CLEAVER, J.E. 1968. Defective repair of replication of DNA in xeroderma pigmentosum. *Nature* 218:652–56.

———. 1990. Do we know the cause of xeroderma pigmentosum? *Carcinogenesis* 11:875–82.

CLEAVER, J.E., and KARENTZ, D. 1986. DNA repair in man: Regulation by a multiple gene family and its association with human disease. *Bioessays* 6:122–27.

COHEN, S.N., and SHAPIRO, J.A. 1980. Transposable genetic elements. *Sci. Am.* (Feb.) 242:40–49.

CROW, J.F., and DENNISTON, C. 1985. Mutation in human populations. *Adv. Hum. Genet.* 14:59–216.

CULOTTA, E. 1994. A boost for "adaptive" mutation. *Science* 265:318–19.

DEERING, R.A. 1962. Ultraviolet radiation and nucleic acids. *Sci. Am.* (Dec.) 207:135–44.

DEN DUNNEN, J.T. et al. 1989. Topography of the Duchenne muscular dystrophy (DMD) gene. *Am. J. Hum. Genet.* 45:835–47.

DEVORET, R. 1979. Bacterial tests for potential carcinogens. *Sci. Am.* (Aug.) 241:40–49.

DORING, H.P., and STARLINGER, P. 1984. Barbara McClintock's controlling elements: Now at the DNA level. *Cell* 39:253–59.

DRAKE, J.W. 1991. A constant rate of spontaneous mutation in DNA-based microbes. *Proc. Natl. Acad. Sci. USA* 88:7160–64.

DRAKE, J.W. et al. 1975. Environmental mutagenic hazards. *Science* 187:505–14.

DRAKE, J.W., GLICKMAN, B.W., and RIPLEY, L.S. 1983. Updating the theory of mutation. *Am. Sci.* 71:621–30.

EYRE-WALKER, A., and KEIGHTLEY, P.D. 1999. High genomic deleterious mutation rates in hominids. *Nature* 397:344–47.

FEDOROFF, N.V. 1984. Transposable genetic elements in maize. *Sci. Am.* (June) 250:85–98.

FINNEGAN, D.J. 1985. Transposable elements in eukaryotes. *Int. Rev. Cytol.* 93:281–326.

FRIEDBERG, E.C., WALKER, G.C., and SIEDE, W. 1995. *DNA repair and mutagenesis*. Washington, DC: ASM Press.

HALL, B.G. 1988. Adapative evolution that requires multiple spontaneous mutations: I. Mutations involving an insertion sequence. *Genetics* 120:887–97.

———. 1990. Spontaneous point mutations that occur more often when advantageous than when neutral. *Genetics* 126:5–16.

HANAWALT, P.C., and HAYNES, R.H. 1967. The repair of DNA. *Sci. Am.* (Feb.) 216:36–43.

HASELTINE, W.A. 1983. Ultraviolet light repair and mutagenesis revisited. *Cell* 33:13–17.

HENDRICIKSON, E.A. 1997. Cell-cycle regulation of mammalian DNA double-strand break repair. *Am. J. Hum. Genet.* 61:795–800.

HOWARD-FLANDERS, P. 1981. Inducible repair of DNA. *Sci. Am.* (Nov.) 245:72–80.

JIRICNY, J. 1998. Eukaryotic mismatch repair: An update. *Mutation Research* 409:107–21.

KELNER, A. 1951. Revival by light. *Sci. Am.* (May) 184:22–25.

KNUDSON, A.G. 1979. Our load of mutations and its burden of disease. *Am. J. Hum. Genet.* 31:401–13.

KRAEMER, F.H. et al. 1975. Genetic heterogeneity in xeroderma pigmentosum: Complementation groups and their relationship to DNA repair rates. *Proc. Natl. Acad. Sci. USA* 72:59–63.

LITTLE, J.W., and MOUNT, D.W. 1982. The SOS regulatory system of *E. coli. Cell* 29:11–22.

LURIA, S.E., and DELBRÜCK, M. 1943. Mutations of bacteria from virus sensitivity to virus resistance. *Genetics* 28:491–511.

MASSIE, R. 1967. *Nicholas and Alexandra.* New York: Atheneum.

MASSIE, R., and MASSIE, S. 1975. *Journey.* New York: Knopf.

MCCANN, J., CHOI, E., YAMASAKI, E., and AMES, B. 1975. Detection of carcinogens as mutagens in the *Salmonella*/microsome test: Assay of 300 chemicals. *Proc. Natl. Sci. USA* 72:5135–39.

MCCLINTOCK, B. 1956. Controlling elements and the gene. *Cold Spring Harbor Symp. Quant. Biol.* 21:197–216.

MACDONALD, M.E. et al. 1993. A novel gene containing a trinucleotide repeat that is expanded and unstable in Huntington's disease chromosome. *Cell* 72:971–80.

MCKUSICK, V.A. 1965. The royal hemophilia. *Sci. Am.* (Aug.) 213:88–95.

MIKI, Y. 1998. Retrotransposal integration of mobile genetic elements in human disease. *J. Human Genet.* 43:77–84.

MULLER, H.J. 1927. Artificial transmutation of the gene. *Science* 66:84–87.
—— 1955. Radiation and human mutation. *Sci. Am.* (Nov.) 193:58–68.

NEWCOMBE, H.B. 1971. The genetic effects of ionizing radiation. *Adv. Genet.* 16:239–303.

O'HARE, K. 1985. The mechanism and control of P element transposition in *Drosophila. Trends Genet.* 1:250–54.

OSSANA, N., PETERSON, K.R., and MOUNT, D.W. 1986. Genetics of DNA repair in bacteria. *Trends Genet.* 2:55–58.

RADMAN, M., and WAGNER, R. 1988. The high fidelity of DNA duplication. *Sci. Am.* (Aug.) 259:40–46.

SHERRATT, D.J., ed. 1995. *Mobile genetic elements.* New York: Oxford University Press.

SHORTLE, D., DIMARIO, D., and NATHANS, D. 1981. Directed mutagenesis. *Annu. Rev. Genet.* 15:265–94.

SIGURBJORHSSON, B. 1971. Induced mutations in plants. *Sci. Am.* (Jan.) 224:86–95.

SPRADLING, A.C., STERN, D.M., KISS, I., ROOTE, J., LAVERTY, T., and RUBIN, G.M. 1995. Gene disruptions using P transposable elements: An integral component of the *Drosophila* Genome Project. *Proc. Natl. Acad. Sci. USA* 92:10824–30.

STADLER, L.J. 1928. Mutations in barley induced by X-rays and radium. *Science* 66:84–87.

SUTHERLAND, B.M. 1981. Photoreactivation. *Bioscience* 31:439–44.

TOMLIN, N.V., and APRELIKOVA, O.N. 1989. Uracil DNA glycosylases and DNA uracil repair. *Int. Rev. Cytol.* 114:81–124.

TOPAL, M.D., and FRESCO, J.R. 1976. Complementary base pairing and the origin of substitution mutations. *Nature* 263:285–89.

VOGEL, F. 1992. Risk calculations for hereditary effects of ionizing radiation in humans. *Hum. Genet.* 89:127–46.

WELLS, R.D. 1994. Molecular basis of genetic instability of triplet repeats. *J. Biol. Chem.* 271:2875–78.

WILLS, C. 1970. Genetic load. *Sci. Am.* (March) 222:98–107.

WOLFF, S. 1967. Radiation genetics. *Annu. Rev. Genet.* 1:221–44.

YAMAMOTO, F. et al. 1990. Molecular genetic basis of the histo-blood group ABO system. *Nature* 345:229–33.

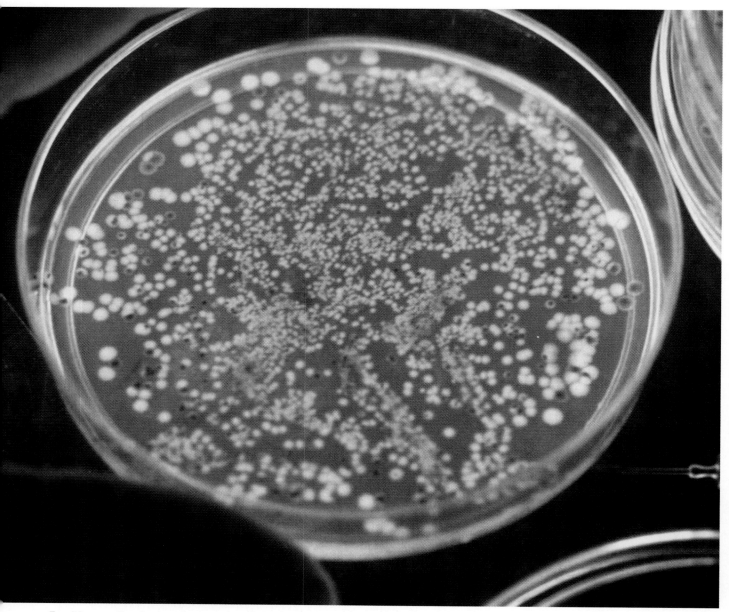

E. coli host cell colonies after uptake of recombinant plasmids. Those that are white have taken up plasmids with inserted DNA fragments.

Construction and Analysis of DNA Clones

KEY CONCEPTS

GenCDX

When you see this icon, there are related animations and exercises on the CD accompanying this text.

In 1982, the U.S. Food and Drug Administration approved the marketing of human insulin produced by genetically engineered bacterial cells. Previously, insulin was chemically extracted from the pancreatic glands of pigs and cows, supplied by slaughterhouses. The production of insulin by recombinant DNA technology was a milestone in the biotechnology revolution, and ensures a steady supply of a purified gene product used to treat diabetes. Today, many human proteins used for the treatment of diseases are produced by biotechnology.

The use of recombinant DNA technology makes it possible to identify and isolate a single gene from the 100,000 or so present in the human genome and to produce large quantities of this gene in the form of cloned DNA molecules. In addition, the gene can be transferred into cells that will synthesize its gene product.

Clones are identical organisms, cells, or molecules descended from a single ancestor. Cloning a gene involves producing many identical copies that can be used for numerous purposes, including research into the structure and organization of a gene or the commercial production of proteins such as insulin. In this chapter, we will review the basic methods of recombinant DNA technology that are used to isolate, replicate, and analyze genes. In Chapter 21, we discuss some of the applications of this technology to research, medicine, the legal system, agriculture, and industry.

Recombinant DNA technology creates artificial combinations of DNA molecules

The term **recombinant DNA** refers to the creation of a new combination of DNA molecules that are not found together naturally. Although processes such as crossing over technically produce recombinant DNA, the term is generally reserved for DNA molecules produced by joining molecules derived from different biological sources.

Recombinant DNA technology uses methods derived from nucleic acid biochemistry coupled with genetic techniques originally developed for the study of bacteria and viruses. As described here, the use of recombinant DNA technology is a powerful tool for the isolation of potentially unlimited quantities of a gene. The basic procedures involve a series of steps:

1. DNA is purified from cells or tissues.

2. DNA fragments are generated using enzymes called **restriction endonucleases** or **restriction enzymes** that recognize and cut DNA molecules at specific nucleotide sequences.

3. The fragments produced by digestion with restriction enzymes are joined to other DNA molecules that serve as **vectors**, or carrier molecules. A vector with an inserted DNA segment is a **recombinant DNA molecule**.

4. The recombinant DNA molecule, consisting of a vector carrying an inserted DNA fragment, is transferred to a host cell. Within the host cell, the recombinant molecule replicates, producing dozens of identical copies known as clones.

5. As host cells replicate, the recombinant DNA molecules within them are passed on to all their progeny cells, creating a population of host cells, each of which carries copies of a cloned DNA sequence.

6. The cloned DNA can be recovered from host cells, purified, and analyzed.

7. Potentially, the cloned DNA can be transcribed, its mRNA translated, and the gene product isolated and used for research, or sold commercially.

Constructing recombinant DNA molecules is a step-wise process

The development of recombinant DNA and gene-cloning technology has provided scientists with methods for the isolation of large quantities of specific genes or other DNA sequences, facilitating studies of gene organization, structure, and expression.

Restriction Enzymes

The cornerstone of recombinant DNA technology is a class of enzymes called restriction endonucleases. These enzymes, isolated from bacteria, received their name because they restrict or prevent viral infection by degrading the invading viral DNA. Restriction enzymes recognize a specific nucleotide sequence and cut both strands of the DNA within that sequence. The 1978 Nobel Prize was awarded to Werner Arber, Hamilton Smith, and Daniel Nathans for their work on restriction enzymes. To date, over 200 restriction enzymes have been identified. Their usefulness in cloning derives from their ability to reproducibly cut DNA into fragments.

One of the first restriction enzymes identified was isolated from *E. coli* and was designated *Eco*RI (pronounced echo-r-one). Its recognition site and points of cleavage are shown in Figure 18.1. Note that the recognition sequence for *Eco*RI reads the same in the 5′-to-3′ direction on one strand as it does in the 5′-to-3′ direction on the complementary strand. Sequences that read the same in both directions are known as **palindromes**.

The DNA fragments produced by *Eco*RI digestion have overhanging single-stranded tails (called "sticky ends") that can reanneal with complementary single-stranded tails on other DNA fragments. If they are mixed together under the proper conditions, DNA fragments from these two sources can form recombinant molecules by hydrogen bonding of their sticky ends. The enzyme **DNA ligase** can be used to covalently link these fragments to form recombinant DNA molecules (Figure 18.2).

Other restriction enzymes such as *Sma*I cleave DNA to produce blunt-end fragments. DNA fragments with blunt ends can also be joined together to create recombinant molecules, after the DNA has been modified (Figure 18.3). The enzyme terminal deoxynucleotidyl transferase is used to create single-stranded ends by the addition of nucleotide

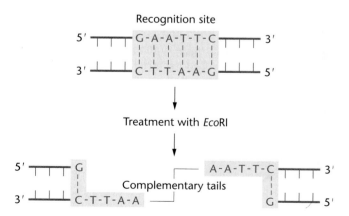

FIGURE 18.1 The restriction enzyme *Eco*RI recognizes and binds to the palindromic nucleotide sequence GAATTC. Cleavage of the DNA at this site produces complementary single-stranded tails. These single-stranded tails can anneal with single-stranded tails from other DNA fragments to form recombinant DNA molecules.

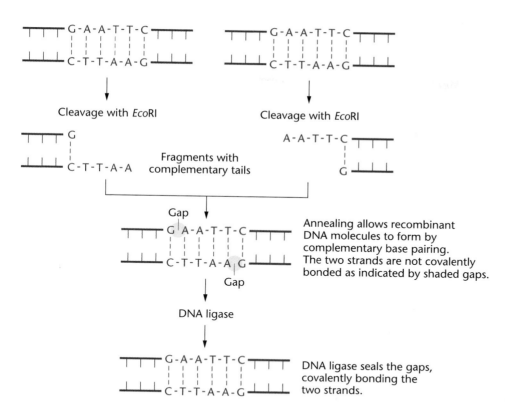

Cleavage with *Eco*RI

Cleavage with *Eco*RI

Fragments with complementary tails

Gap

Gap

Annealing allows recombinant DNA molecules to form by complementary base pairing. The two strands are not covalently bonded as indicated by shaded gaps.

DNA ligase

DNA ligase seals the gaps, covalently bonding the two strands.

FIGURE 18.2 In forming recombinant DNA molecules, DNA from different sources is cleaved with *Eco*RI and mixed together to allow annealing into recombinant molecules. The enzyme DNA ligase is then used to covalently link these annealed fragments into an intact recombinant DNA molecule.

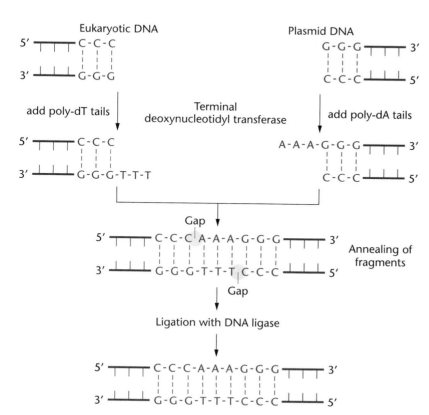

Eukaryotic DNA

Plasmid DNA

add poly-dT tails

Terminal deoxynucleotidyl transferase

add poly-dA tails

Gap

Annealing of fragments

Gap

Ligation with DNA ligase

FIGURE 18.3 Recombinant DNA molecules can be formed from DNA cut with enzymes that leave blunt ends. In this method, the enzyme terminal deoxynucleotidyl transferase is used to create complementary tails by the addition of poly dA and poly dT to the cut fragments. These tails serve to anneal DNA from different sources and to create recombinant DNA molecules that are covalently linked by treatment with DNA ligase.

"tails." If a poly-dA tail is added to DNA fragments from one source, and a poly-dT tail is added to DNA from another source, complementary tails are created, and the fragments can reanneal by hydrogen bond formation. Recombinant molecules can then be created by ligation. Figure 18.4 depicts some common restriction enzymes and their recognition sequences, many of which yield cohesive or "sticky" ends, while others generate blunt ends.

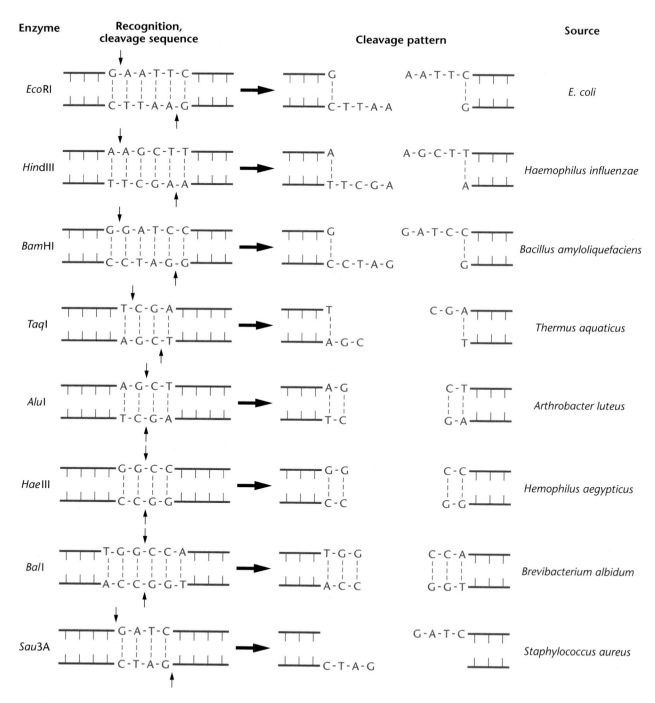

FIGURE 18.4 Some common restriction enzymes, with their recognition and cutting sites, cleavage patterns, and sources.

Vectors are carriers of DNA fragments to be cloned

Fragments of DNA produced by restriction enzyme digestion cannot directly enter bacterial cells for cloning; when a DNA fragment is joined to a vector, however, it can gain entry to a host cell, where it can be replicated or cloned into many copies. Vectors are, in essence, carrier DNA molecules. To serve as a vector, a DNA molecule must have several properties.

1. It must be able to independently replicate itself and the DNA segments it carries.

2. It should contain a number of restriction enzyme cleavage sites that are present only once in the vector. This site is cleaved with a restriction enzyme and used to insert DNA segments cut with the same enzyme.

3. It should carry a selectable marker (usually in the form of antibiotic resistance genes or genes for enzymes missing in the host cell) to distinguish host cells that carry vectors from host cells that do not contain vectors.

4. It should be easy to recover from the host cell.

There are a number of vectors currently in use, which permits the cloning of DNA fragments over a wide size range.

Plasmid Vectors

The first vectors to be developed were genetically modified plasmids, which are versatile and widely used, but which are used to clone only relatively small DNA fragments. Plasmids are naturally occurring, extrachromosomal, double-stranded DNA molecules that replicate autonomously within bacterial cells (Figure 18.5). The genetics of plasmids and their host bacterial cells has been

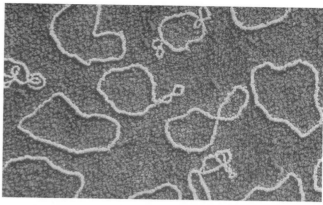

FIGURE 18.5   A color-enhanced electron micrograph of circular plasmid molecules isolated from the bacterium *E. coli*. Genetically engineered plasmids are used as vectors for cloning DNA.

discussed in Chapters 7 and 15. In this section, the emphasis will be on the role of plasmids as vectors. Although only a single plasmid may enter a host cell, many plasmids can increase their copy number in the host cell so that up to 1000 copies are present. When used as a vector, such plasmids allow more copies of cloned DNA to be produced. For use as vectors, plasmids have been modified in the laboratory to contain a limited number of convenient restriction sites and marker genes that can be used to detect the presence of plasmids in host cells.

Many genetically engineered plasmid vectors are now available, and they offer a number of useful features that make it easier to identify host cells carrying a plasmid with an inserted DNA fragment. One such plasmid is pUC18 (Figure 18.6). As a vector, this plasmid has several useful properties.

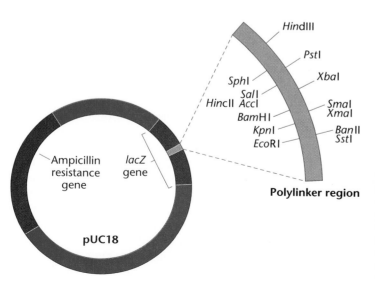

Polylinker region

FIGURE 18.6   The plasmid pUC18 offers several advantages as a vector for cloning. Because of its small size, it can accept relatively large DNA fragments for cloning; it replicates to a high copy number; and it has a large number of restriction sites in the polylinker, located within a *lacZ* gene. Bacteria carrying pUC18 produce blue colonies when grown on a medium containing X-gal. DNA inserted into the polylinker site disrupts the *lacZ* gene, resulting in white colonies, allowing direct identification of colonies carrying cloned DNA inserts.

1. The plasmid is small, allowing it to carry relatively large DNA inserts.

2. In a host cell, pUC18 can replicate to form about 500 copies per cell, producing many clones, or copies, of inserted DNA fragments.

3. A large number of restriction enzyme sites have been engineered into pUC18, and these are conveniently clustered in one region, called a **multiple-cloning polylinker site**.

4. The pUC18 plasmid has a selection system that allows recombinant plasmids to be identified. The pUC18 plasmid carries a fragment of the bacterial *lacZ* gene, and the polylinker is inserted into this fragment. Through action of the *lacZ* gene, bacterial host cells carrying pUC18 will produce blue colonies when grown on medium containing a compound called X-*gal*. When a DNA fragment is inserted into the polylinker site, the *lacZ* gene is inactivated, and a bacterial cell carrying pUC18 with an inserted DNA fragment will form white colonies, making them easy to identify (Figure 18.7).

Lambda and M13 Bacteriophage Vectors

Lambda is a bacteriophage widely used in recombinant DNA work (Figure 18.8). The genes of lambda have all been identified and mapped, and the nucleotide sequence of the entire genome is known. The central third of the lambda chromosome is dispensable, and this

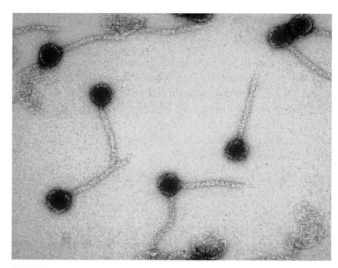

FIGURE 18.8 A colorized electron micrograph showing a cluster of the bacteriophage lambda, widely used as a vector in recombinant DNA work.

region can be replaced with foreign DNA without affecting the ability of the phage to infect cells and form plaques (Figure 18.9). To clone DNA using a lambda-replacement vector, the phage DNA is cut with a restriction enzyme (e.g., *Eco*RI), resulting in a left arm, a right arm, and the dispensable central region. The arms are isolated and combined with DNA fragments generated by cutting genomic DNA with *Eco*RI, followed by ligation with DNA ligase.

The resulting recombinant molecules are transferred into bacterial hosts with a high degree of efficiency by transfection. First, the host cells are made permeable by a chemical treatment, and then they are mixed with the ligated molecules. Vector molecules carrying inserts are taken into the host cells, where they direct the synthesis of the infective phage, each of which carries a DNA insert. Alternatively, the lambda DNA-carrying inserts can be mixed with phage protein components; from this mixture, infective phage particles are formed. Each phage can be amplified by growth on plated bacterial cells to form plaques or by infecting cells in liquid medium and harvesting the lysed cells.

Other bacteriophages are used as vectors, including the single-stranded phage known as M13 (Figure 18.10). When M13 infects a bacterial cell, the single DNA strand (+ strand) replicates to produce a double-stranded molecule, known as the replicative form (RF). RF molecules can be regarded as being similar to plasmids, and foreign DNA can be inserted into the restriction enzyme sites in the phage DNA. When reinserted into bacterial cells, the RF molecules replicate to produce single (+) strands, which include one strand of the inserted DNA. The sin-

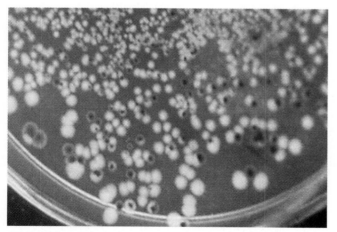

FIGURE 18.7 A petri plate showing the growth of host cells after uptake of recombinant plasmids. The medium on the plate contains a compound called X-gal. DNA inserts in the pUC18 vector disrupt the gene responsible for the formation of blue colonies. Cells in blue colonies do not carry any cloned DNA inserts, whereas white colonies contain vectors carrying DNA inserts.

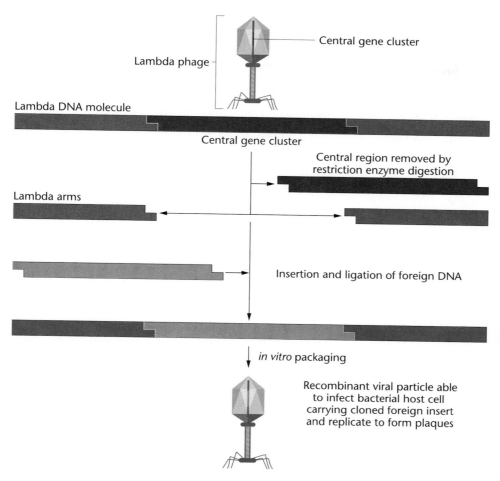

Lambda phage

Central gene cluster

Lambda DNA molecule

Central gene cluster

Central region removed by
restriction enzyme digestion

Lambda arms

Insertion and ligation of foreign DNA

in vitro packaging

Recombinant viral particle able
to infect bacterial host cell
carrying cloned foreign insert
and replicate to form plaques

FIGURE 18.9 Steps in using phage lambda as a vector. DNA is extracted from a preparation of lambda phage, and the central gene cluster is removed by treatment with a restriction enzyme. The DNA to be cloned is cut with the same enzyme and ligated into the arms of the lambda chromosome. The recombinant chromosome is then packaged into phage proteins to form a recombinant virus. This virus is able to infect bacterial cells and replicate its chromosome, including the DNA insert.

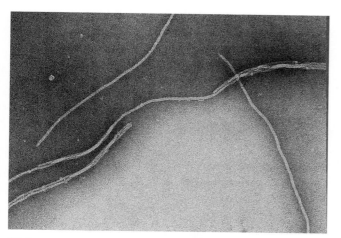

FIGURE 18.10 M13 bacteriophages are long thin filaments containing a single strand of DNA. They infect *E. coli* host cells by attaching to the cellular extensions of the bacterium called *F pili*.

gle-stranded DNA produced by M13 is extruded from the cell, and can be recovered and used directly for DNA sequencing or as a template for mutational alteration of the cloned sequence.

Cosmids and Shuttle Vectors

Cosmids are hybrid vectors constructed using parts of the lambda chromosome and plasmid DNA. Cosmids contain the *cos* sequence of phage lambda, necessary for packing phage DNA into phage protein coats, and plasmid sequences for replication. They also contain an antibiotic resistance gene to identify host cells carrying cosmids (Figure 18.11). Cosmids carrying inserted DNA fragments are packaged into lambda protein heads, forming infective phage particles. After the cosmid infects a bacterial host cell, it replicates as a plasmid. Because almost the entire lambda genome has been deleted, cosmids can carry DNA inserts that are

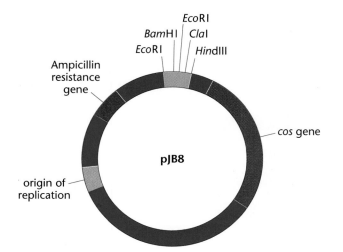

FIGURE 18.11 The cosmid pJB8 contains a bacterial origin of replication (*ori*), a single *cos* site (*cos*), an ampicillin resistance gene (*amp*, for selection of colonies that have taken up the cosmid), and a region containing four restriction sites for cloning (*Bam*HI, *Eco*RI, *Cla*I, and *Hind*III). Because the vector is small (5.4 kb long), it can accept foreign DNA segments between 33 and 46 kb in length. The *cos* site allows cosmids carrying large inserts to be packaged into lambda viral coat proteins as though they were viral chromosomes. The viral coats carrying the cosmid can be used to infect a suitable bacterial host, and the vector, carrying a DNA insert, will be transferred into the host cell. Once inside, the *ori* sequence allows the cosmid to replicate as a bacterial plasmid.

much larger than those carried by lambda vectors. Cosmids can carry almost 50 kb of inserted DNA; phage vectors can accommodate DNA inserts of about 10–15 kb; plasmids are usually limited to inserts 5–10 kb in length.

Other hybrid vectors that have been constructed with origins of replication derived from different sources (e.g., plasmids and animal viruses such as SV40) can replicate in more than one type of host cell. These shuttle vectors usually contain genetic markers that are selectable in both host systems and can be used to shuttle DNA inserts back and forth between *E. coli* and another host cell, such as yeast. Often such vectors are employed in studying gene expression. Other vectors, those used in specific applications, will be described in following sections.

Bacterial Artificial Chromosomes

The mapping and analysis of large complex eukaryotic genomes requires cloning vectors that can accommodate very large DNA fragments. In addition, since some human genes range from 1000 kb to over 2000 kb, vectors with large cloning capacities are useful in studying the organization of these genes. Recently, a number of vectors that use bacterial host cells have been developed.

One of these vectors is based on the fertility plasmid (F factor) of bacteria and is called a **bacterial artificial chromosome** (**BAC**). Recall from Chapter 7 that F factors are independently replicating plasmids that are involved in the transfer of genetic information during bacterial conjugation. Because F factors can carry fragments of the bacterial chromosome up to 1 Mb in length, they have been engineered to act as vectors for eukaryotic DNA and can carry inserts of about 300 kb (Figure 18.12). BAC vectors carry the

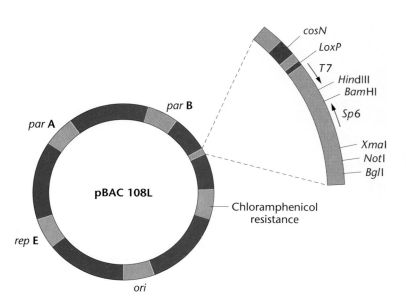

FIGURE 18.12 A bacterial artificial chromosome (BAC). The polylinker carries a number of unique restriction sites for the insertion of foreign DNA. The arrows labeled T7 and Sp6 are promoter regions that allow expression of genes cloned between these regions.

F factor genes for replication and copy number, and incorporate an antibiotic resistance marker and restriction enzyme sites for inserting foreign DNA to be cloned. In addition, the cloning site is flanked by promoter sites that can be used to generate RNA molecules for the expression of the cloned gene, for use as probes in chromosome walking, and for DNA sequencing of the cloned insert.

E. coli can serve as host cells for cloning DNA

As discussed earlier, biotechnology involves not only the construction of recombinant DNA molecules, but also the replication of these molecules to produce many copies, or clones. This is accomplished by transferring the recombinant molecules into host cells where replication takes place. Cell-based cloning was the first cloning method developed, and it has been widely used to prepare copies of

DNA fragments that have been used to study gene organization and function.

A variety of prokaryotic and eukaryotic cells can be used as hosts for the replication of recombinant vectors. One of the most commonly used hosts is a laboratory strain of the bacterium *E. coli* known as K12. *E. coli* strains such as K12 are genetically well characterized, and they can serve as hosts for a wide range of vectors.

Several steps are required to create recombinant DNA molecules (Figure 18.13) in an *E. coli* host cell.

1. The DNA to be cloned is isolated and treated with a restriction enzyme to create fragments ending in a specific sequence.

2. The fragments are ligated to plasmid molecules that have been cut with the same restriction enzyme, creating a recombinant vector molecule.

3. The recombinant vector is transferred into an *E. coli* host cell, where it can replicate to form dozens of copies.

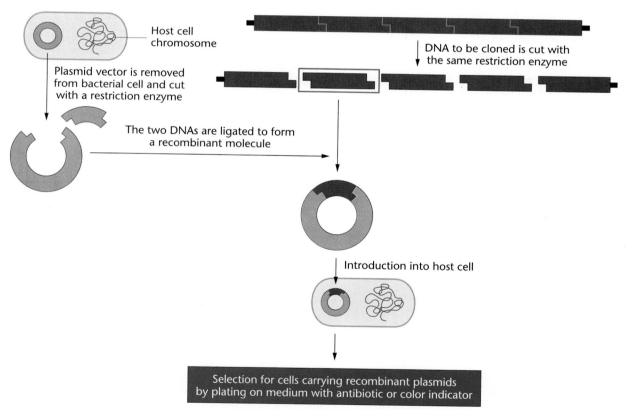

FIGURE 18.13 A summary of steps involved in cloning with a plasmid vector. Plasmid vectors are isolated and cut with a restriction enzyme. The DNA to be cloned is cut with the same restriction enzyme, producing a collection of fragments. These fragments are spliced into the vector and transferred to a bacterial host for replication. Bacterial cells carrying plasmids with DNA inserts can be identified by growth on selective medium and isolated. The cloned DNA can be recovered from the bacterial host for further analysis. ⊙GenCDX

4. The bacteria are grown on a nutrient plate where they form colonies and are screened to identify those that have taken up the recombinant plasmids.

Because the cells in each colony are derived from a single ancestral cell, all the cells in the colony and the plasmids they contain are genetically identical clones.

Similarly, phages containing foreign DNA can be used to infect *E. coli* host cells, and, when plated on solid medium, the resulting plaques each represent a cloned descendant of a single ancestral phage. Cosmids infect bacterial cells like phage, but then replicate as plasmids inside the host cells. Cells carrying cosmids with cloned DNA can be identified and recovered in the same way as can plasmids.

Vectors can be constructed for eukaryotic cells

To study the expression and regulation of eukaryotic genes, it is often convenient and even necessary to use eukaryotic hosts. Several cloning systems using eukaryotic vectors and host cells have been developed. One of these is based on the yeast, *Saccharomyces cerevisiae*.

Vectors Based on Yeast Plasmids

Although yeast is a eukaryotic organism, it can be grown and manipulated in much the same way as bacterial cells. Further, the genetics of yeast has been intensively studied, providing a large catalog of mutations and a highly developed genetic map. In addition, the entire yeast genome has recently been sequenced. Naturally occurring yeast plasmids, such as the 2-micron plasmid, have been used to construct a number of yeast cloning vectors. By combining bacterial plasmid sequences with those of the 2-micron plasmids, vectors for growth in yeast host cells have been produced.

Yeast Artificial Chromosomes

A second type of yeast vector is the **yeast artificial chromosome (YAC)**. In linear form (Figure 18.14), a YAC contains telomeres at each end, an origin of replication (called an autonomously replicating sequence or ARS), and a yeast centromere. The YAC also contains two selectable markers (TRP1 and URA3) and a cluster of restriction sites for DNA inserts.

Segments of DNA 1–2 megabases long (1 Mb = 1 million base pairs) can be inserted into YACs. The ability to clone large pieces of DNA into these vectors has made

them an important tool in the Human Genome Project (discussed in Chapter 21).

Although cloning in yeast vectors and host cells is currently the most advanced eukaryotic system, other systems, including the use of human artificial chromosomes as vectors with tissue culture cells as hosts, are being developed.

Eukaryotic cells can serve as hosts for DNA cloning

Both plant and animal cells can take up DNA from their environment. Moreover, vectors, including YACs, can be used to transfer DNA into eukaryotic cells. When the vec-

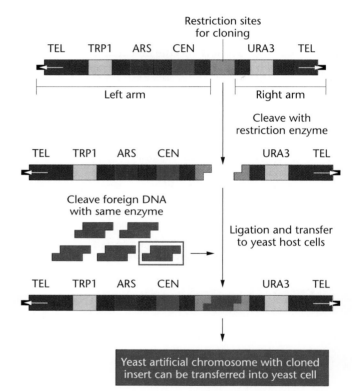

FIGURE 18.14 Cloning into a yeast artificial chromosome. The chromosome contains telomere sequences (TEL) derived from a ciliated protozoan, *Tetrahymena*, a centromere (CEN), and an origin of replication (ARS). These elements give the cloning vector the properties of a chromosome. TRP1 and URA3 are yeast genes that are selectable markers for the left and right arms of the chromosome, respectively. Near the centromere is a region containing several restriction sites. Cleavage with an enzyme in this region breaks the artificial chromosome into two arms. The DNA to be cloned is treated with the same enzyme, producing a collection of fragments. The arms and fragments can be ligated together, and the artificial chromosome can be inserted into yeast host cells. Because yeast chromosomes are large, the artificial chromosome can accept inserts up to several million base pairs.

tor is a plasmid, DNA uptake is referred to as **transformation**. When the vector is a virus, the term **transfection** is used to describe uptake.

Plant Cell Hosts

Cloning using higher plants as hosts has been achieved using a bacterial plasmid. The infectious soil bacterium *Agrobacterium tumifaciens* produces tumors (called plant galls) in many species of plants. Tumor formation is associated with the presence of a tumor-inducing (Ti) plasmid in the bacteria (Figure 18.15). When these bacteria infect plant cells, a segment of the Ti plasmid, known as T-DNA, is transferred into the chromosomal DNA of the host plant cell. The T-DNA controls tumor formation and the synthesis of small molecules known as *opines* that are required for growth of the infecting bacteria. Foreign genes can be inserted into the T-DNA segment of Ti, and the insert-carrying Ti can be transferred into a plant cell by the infecting bacteria. The foreign DNA is inserted into the plant genome when the T-DNA is integrated into a host cell chromosome. Such genetically altered single cells can be induced to grow in tissue culture to form a cell mass known as a callus. By manipulating the culture medium, the callus can be induced to form roots and shoots, and eventual-

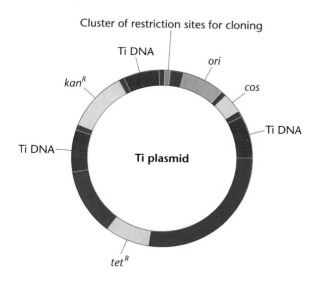

FIGURE 18.15 A Ti plasmid designed for cloning in plants. Segments of Ti DNA, including those necessary for opine synthesis and integration, are combined with bacterial segments that incorporate cloning sites and antibiotic resistance genes (*kan^R* and *tet^R*). The vector also contains an origin of replication (*ori*) and a lambda *cos* site that permits recovery of cloned inserts from the host plant cell.

ly to develop into mature plants carrying a foreign gene. Plants (or animals) carrying a foreign gene are called **transgenic** organisms. In Chapter 21 we will see how gene transfer in plants has been used to genetically alter agriculturally important crop plants.

Mammalian Cell Hosts

DNA can be transferred to mammalian cells by several methods, including endocytosis or encapsulation of DNA into artificial membranes (liposomes) followed by fusion with cell membranes. Also, DNA can be transferred using YACs and vectors based on retroviruses. The DNA introduced into a mammalian cell by any of these methods is usually integrated into the host genome. DNA transfer methods have been used to transfer genes to fertilized eggs, producing transgenic animals. These methods have also been used to study molecular aspects of gene expression and to replicate cloned genes using mammalian cells as hosts.

YACs are used to increase the efficiency of transferring genes into the germ line of mice. Gene transfer is dependent on the introduction of DNA sequences into the nucleus of an appropriate mouse cell, such as a fertilized egg or an embryonic stem cell, followed by the integration of the DNA into a chromosomal site.

YACs are transferred to mice in several ways. The first involves the microinjection of purified YAC DNA into the nucleus of a mouse oocyte (Figure 18.16). The zygotes are then transferred to the uteri of foster mothers for normal development. Other techniques use mouse embryonic stem (ES) cells as hosts. One of these methods fuses a yeast cell carrying a YAC with a stem cell, transferring the YAC and all or a portion of the yeast genome to the stem cell. The transgenic ES cells are then transferred into mouse blastocyst-stage embryos, where they participate in the formation of adult tissues, including those that form germ cells.

The ability to transfer large DNA segments into mice has applications in many areas of research. Some of these are described in Chapter 21.

Vectors for the delivery of functional genes into mammalian cells have been developed using avian and mouse retroviruses. The genome of such retroviruses is a single-stranded RNA that is transcribed by reverse transcriptase into a double-stranded DNA molecule. This DNA is integrated into the host genome and passed on to daughter cells as part of the host chromosome. The retroviral genome can be engineered to remove some viral genes, creating vectors that accept foreign DNA, including human structural genes. These vectors are being used in the treatment of genetic disorders by gene therapy, a topic that is also discussed in Chapter 21.

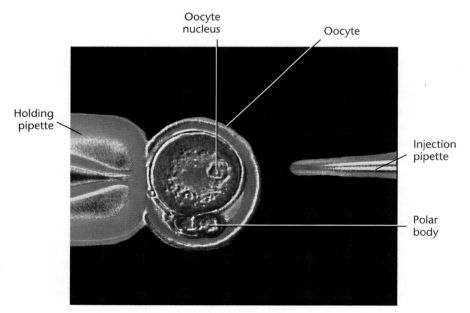

Oocyte
nucleus

Oocyte

Holding
pipette

Injection
pipette

Polar
body

FIGURE 18.16 Cloned DNA can be transferred in mammals by direct injection into the oocytes.

Libraries are collections of cloned DNA sequences

Because each cloned DNA segment is relatively small, many separate clones are required to include even a small portion of an organism's genome. A set of DNA clones derived from a single individual represents a library. Cloned libraries can represent an entire genome, a single chromosome, or a set of genes that are actively transcribed in a single cell type.

Genomic Libraries

Ideally, a genomic library contains at least one copy of all the sequences represented in the genome. Genomic libraries are constructed by extracting DNA from cells or tissues, cutting the DNA with restriction enzymes, and ligating the fragments into vectors. Since each vector can contain only a few thousand bases (kilobases) of foreign DNA, the selection of the proper vector to contain a genome in the smallest number of clones is an important consideration in preparing a genomic library.

The number of clones required to carry all sequences in a genome is dependent on the average size of the cloned inserts carried by the vector and the size of the genome to be cloned. The number of clones in a library can be calculated as:

$$N = \frac{\ln(1 - P)}{\ln(1 - f)}$$

where N is the number of required clones, P is the probability of recovering a given sequence, and f is the fraction of the genome in each clone. Suppose that we wish to prepare a library of the human genome using a lambda phage vector. The human genome contains 3.0×10^6 kb of DNA, and the average size of a cloned insert that can be carried by the vector is 17 kb; therefore, about 8.1×10^5 phages would be required to establish a library housing all sequences. If we had selected a plasmid vector with a capacity of only 5 kb, several million clones would be needed to create the library. Thus, in this example, a phage vector would be the best choice for constructing a library of the human genome.

Chromosome-Specific Libraries

A library made from a subgenomic fraction such as a single chromosome can be of great value in the selection of specific genes and the study of chromosome organization. For example, DNA from a small segment of the X chromosome of *Drosophila* corresponding to a region of about 50 polytene bands was isolated by microdissection. The DNA in this chromosomal fragment was extracted, cut with a restriction endonuclease, and cloned into a lambda vector. This region of the chromosome contains the genes *white*, *zeste*, and *Notch*, as well as the original site of a transposable element that can translocate a chromosomal segment to more than 100 other loci. Although technically difficult, this procedure produces a library that contains only the genes of interest and their adjacent sequences.

Libraries derived from individual human chromosomes have been prepared using a technique known as flow cytometry. In this procedure, chromosomes from mitotic cells are stained with two fluorescent dyes, one that binds to A＝T pairs, the other to G≡C pairs. The stained chromosomes flow past a laser beam that stimulates them to fluoresce, and a photometer sorts and fractionates the chromosomes by differences in dye binding and light scattering (Figure 18.17).

Using this technique, cloned libraries for each of the human chromosomes have been prepared. Having cloned libraries for each human chromosome has greatly facilitated the mapping and analysis of individual chromosomes as part of the Human Genome Project.

A modification of electrophoresis known as pulsed field gel electrophoresis has been used to isolate yeast chromosomes for the construction of chromosome-specific libraries (Figure 18.18). The isolation and construction of a cloned library of yeast chromosome III (315 kb) was the starting point for the formation of a consortium of 35 laboratories to determine the nucleotide sequence of this chromosome. One of the unexpected results of this project was the finding that about half of all the genes on this chromosome were previously unknown. It was difficult for many geneticists to accept the fact that the conventional methods of systematic mutagenesis and mapping of genes used for decades was so inefficient. However, this finding

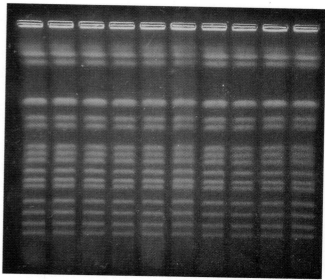

FIGURE 18.18 Intact yeast chromosomes separated using a method of electrophoresis employing a contour-clamped homogeneous electric field (CHEF). In each lane, 15 of the 16 yeast chromosomes are visible, separated by size, with the largest chromosomes at the top.

was confirmed and extended in 1994 with the publication of the nucleotide sequence of yeast chromosome XI (664 kb), and the sequencing of the rest of the yeast genome in 1996. These results indicate that single-chromosome libraries can be used to gain access to genetic loci where conventional methods, such as mutagenesis and genetic analysis, have proven unsuccessful and where other probes, such as mRNA or gene products, are unavailable or unknown. In addition, chromosome-specific libraries provide a means for studying the molecular organization and even the nucleotide sequence in a defined region of the genome.

cDNA Libraries

A library representing the genes active in a specific cell type at a specific time can be constructed using the poly A-containing mRNA molecules from that cell type. Almost all eukaryotic mRNA molecules contain a poly A tail at their 3′-ends. This mRNA can be used to synthesize **complementary DNA (cDNA)** molecules that are subsequently cloned. After the poly A-containing mRNA is isolated, a poly-dT primer is used to pair with the poly-A residues (Figure 18.19). The poly-dT primer serves as the starting point for the synthesis of a complementary DNA strand using the enzyme **reverse transcriptase**. The result is an RNA–DNA double-stranded duplex molecule. The RNA strand is removed, and the remaining single-stranded DNA

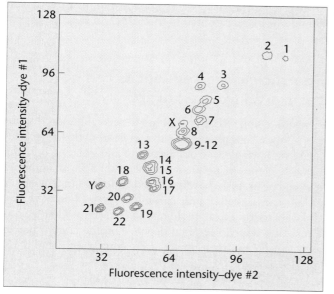

FIGURE 18.17 Each human chromosome has a unique profile resulting from the absorption of two fluorescent dyes. Based on this differential fluorescence, chromosomes can be separated from each other by flow sorting.

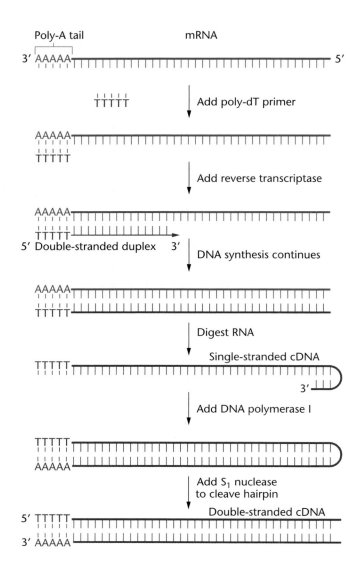

FIGURE 18.19 The production of cDNA from mRNA. Because many eukaryotic mRNAs have a polyadenylated tail of variable length at their 3′-end, a short poly-dT oligonucleotide can be annealed to this tail to serve as a primer for the enzyme reverse transcriptase. Reverse transcriptase uses the mRNA as a template to synthesize a complementary DNA strand (cDNA), forming an mRNA/cDNA double-stranded duplex. The mRNA is digested by alkali treatment, or by the enzyme RNAseH. The 3′-end of the cDNA often folds back to form a hairpin loop. The loop serves as a primer for DNA polymerase, which is used to synthesize the second DNA strand. The enzyme S1 nuclease is used to open the hairpin loop; the result is a double-stranded cDNA molecule that can be cloned into a suitable vector, or used as a probe for library screening.

is used as a template to make the DNA double-stranded, using the enzyme **DNA polymerase I**. The 3′-end of the single-stranded DNA often loops back upon itself and serves as a primer for the synthesis of the second strand. The result is a DNA duplex with the strands joined together at one end. This hairpin loop can be opened using the enzyme S1 nuclease, producing a double-stranded DNA molecule (called **complementary DNA** or **cDNA**), that can be cloned into a plasmid or phage vector.

The cDNA can be cloned by attaching a short double-stranded piece of DNA containing a restriction site to each end. The linkers are cleaved with the appropriate restriction enzyme, producing sticky ends, thus allowing the cDNA to be inserted at a restriction site in a plasmid or phage vector.

A cDNA library is different from a genomic library in that it represents a subset of all the genes in a genome. cDNA libraries use mRNA as a starting point, and thus represent only the expressed sequences in a given cell type, tissue, or stage of embryonic development. The decision whether to construct a genomic or a cDNA library depends on the question at hand. If you are interested in a particular gene, it may be easier to prepare a cDNA library from a tissue that expresses that gene. For example, red blood cells make large amounts of hemoglobin, and most of the mRNA in these cells encodes globin molecules. A cDNA library prepared from red blood cell mRNA is a direct approach to isolating a globin gene. If the regulatory sequences adjacent to the globin gene are of interest, then you need to construct a genomic library, since these regulatory sequences are not present in globin mRNA and would not be represented in the cDNA library.

Specific cloned sequences can be recovered from a library

A genomic library can contain up to several hundred thousand clones. Selecting a gene of interest from this library depends on having a way of identifying and isolating only the clone or clones that contain a gene of interest. In addition, it would be useful to have a method of determining whether a clone contains all or only part of that gene. There are several ways to sort through a library to recover the clones of interest, and the choice often depends on the circumstances and available information about the gene being sought.

DNA Probes

Most methods for sorting through a cloned library use a DNA **probe** to select specific clones. A probe is used to identify complementary nucleic acid sequences present in one or more clones. Probes can be prepared as single-stranded or double-stranded molecules, but are used in a single-stranded form in hybridization reactions (hybridization reactions are described in Chapter 11). Often, probes are radioactive polynucleotides, but other methods use probes that depend on chemical or color reactions to identify a specific clone.

Probes can be derived from a variety of sources; even related genes isolated from other species can serve as probes if enough of the DNA sequence has been conserved. For example, extrachromosomal copies of the ribosomal genes of the clawed frog *Xenopus laevis* can be isolated by centrifugation and cloned into plasmid vectors. Because ribosomal gene sequences have been highly conserved during evolution, clones carrying the human ribosomal genes were recovered from a human genomic library by using cloned *Xenopus* ribosomal DNA fragments as probes.

If the gene to be selected from a genomic library is expressed in certain cell types, a cDNA probe can be prepared. This technique is particularly useful when purified or enriched mRNA for a gene product can be obtained. For example, β-globin mRNA is present in high concentrations in certain stages of red blood cell development. The mRNA purified from these cells can be copied by reverse transcriptase into a cDNA molecule. Such a cDNA was used as a probe to recover the structural gene for human β-globin from a cloned genomic library.

Screening a Library

To screen a plasmid library, clones from a library are grown on nutrient plates, forming hundreds or thousands of colonies (Figure 18.20). A replica of the colonies on the plate is made by gently pressing a nylon or nitrocellulose filter onto the surface of the plate, transferring bacterial cells from the colonies to the filter. The filter is transferred through solutions to lyse the bacteria, denature the double-stranded DNA and convert it to single strands, and bind these strands to the filter.

The colonies on the filter are screened by incubation with a single-stranded nucleic acid probe. If a radioactive DNA probe is to be used, the probe is denatured to form single strands and added to a solution containing the filter. If the DNA sequence of any of the cloned DNA on the filter is complementary to the probe, a DNA–DNA hybrid molecule will form between the probe and the cloned DNA (review the details of hybridization in Chapter 11). After the unbound and excess probe is washed away, the filter is assayed to detect any hybridization that

has taken place. In many types of assays, the filter is overlaid with a piece of photographic film. If a radioactive probe has been used, decay in the probe molecules bound to DNA on the filter will expose the film and produce dark spots on the developed film, representing colonies on the plate containing the cloned gene of interest (Figure 18.20). The corresponding colony can be identified and recovered from the original nutrient plate, and the cloned DNA it contains can be used in further experiments. Nonradioactive probes use a chemical reaction that emits photons of light (chemiluminescence) to expose the photographic film and reveal the location of colonies carrying the gene of interest.

To screen a phage library, a slightly different method, called **plaque hybridization**, is used. A solution of phage carrying DNA inserts is spread over a lawn of bacteria growing on a plate. The phages infect the bacterial cells on the plate and form plaques as they replicate. Each plaque, which appears as a clear spot on the plate, represents the progeny of a single phage and is a clone. The plaques are transferred to a nylon or nitrocellulose membrane, and the phage DNA is denatured into single strands and screened with a labeled probe. Phage plaques are much smaller than plasmid colonies, and many more plaques can be screened on a single filter, making this method more efficient for screening large genomic libraries.

Identification of Nearby Genes: Chromosome Walking

In some cases, when the approximate location of a gene is known, it is possible to clone the gene by first cloning nearby sequences. Often these nearby sequences are identified by linkage analysis and serve as a starting point for **chromosome walking**, which is the isolation of adjacent clones from a library. In a chromosome walk, the end piece of a cloned DNA fragment is subcloned and used as a probe to recover another set of overlapping clones from the genomic library (Figure 18.21). They are analyzed to determine their degree of overlap. A subfragment from one end of the overlapping region is used as a probe to recover another set of overlapping clones, and the analysis is repeated. In this way, it is possible to "walk" along the chromosome, clone by clone. The gene in question can be identified by nucleotide sequencing of the recovered clones and searching for an **open reading frame (ORF)**. An ORF is a stretch of nucleotides that begins with a start codon followed by amino acid-encoding codons, and ends with one or more stop codons. Although laborious and time-consuming, chromosome walking has been used to re-

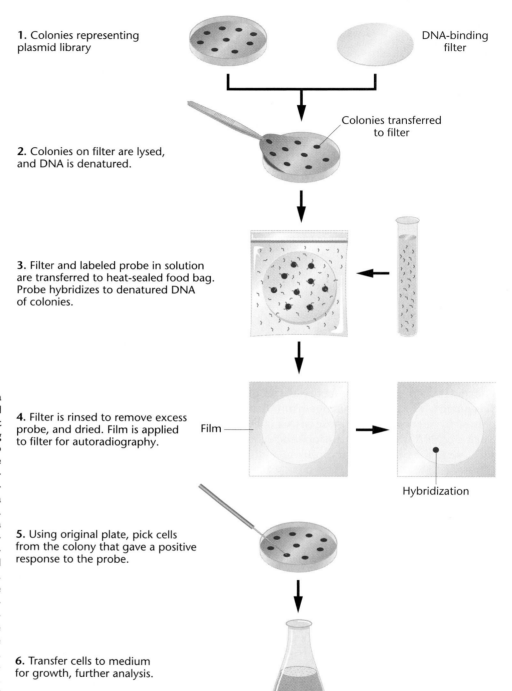

1. Colonies representing plasmid library

DNA-binding filter

Colonies transferred to filter

2. Colonies on filter are lysed, and DNA is denatured.

3. Filter and labeled probe in solution are transferred to heat-sealed food bag. Probe hybridizes to denatured DNA of colonies.

4. Filter is rinsed to remove excess probe, and dried. Film is applied to filter for autoradiography.

Film

Hybridization

5. Using original plate, pick cells from the colony that gave a positive response to the probe.

6. Transfer cells to medium for growth, further analysis.

FIGURE 18.20 Screening a plasmid library to recover a cloned gene. The library, present on nutrient plates, is overlaid with a DNA-binding filter, and colonies are transferred to the filter. Colonies on the filter are lysed, and the DNA is denatured to single strands. The filter is placed in a hybridization bag along with buffer and a labeled single-stranded DNA probe. During incubation, the probe forms a double-stranded hybrid with complementary sequences on the filter. The filter is removed from the bag and washed to remove the excess probe. Hybrids are detected by placing a piece of X-ray film over the filter and exposing it for a short time. The film is developed, and hybridization events are visualized as spots on the film. From the orientation of spots on the film, colonies containing the insert that hybridized to the probe can be identified. Cells are picked from this colony for growth and further analysis.

cover genes for several human genetic disorders, including those for cystic fibrosis and muscular dystrophy.

There are several limitations to chromosome walking in complex eukaryotic genomes, including the human genome. If a probe contains a repetitive se-

quence, such as an *Alu* sequence (see Chapter 11 for a discussion of these and other repetitive sequences), it will hybridize to other clones in the genomic library containing that sequence. Most of these clones will not be adjacent to the clone from which the probe was de-

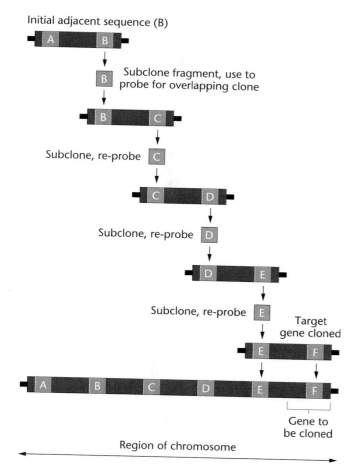

Initial adjacent sequence (B)

Subclone fragment, use to probe for overlapping clone

Subclone, re-probe

Subclone, re-probe

Subclone, re-probe

Target gene cloned

Gene to be cloned

Region of chromosome

FIGURE 18.21   In chromosome walking, the approximate location of a gene to be cloned is known. A subcloned fragment of a linked adjacent sequence is used to recover overlapping clones from a genomic library. This process of subcloning and probing the genomic library is repeated to recover overlapping clones until the gene in question has been reached.

rived, and the walk will be terminated. In some cases, a technique called chromosome jumping is used to skip over the region containing the repetitive sequences and continue the walk.

The polymerase chain reaction (PCR) can replace cell-based cloning

Recombinant DNA techniques were developed in the early 1970s, and in subsequent years they revolutionized the way geneticists and molecular biologists conduct research and gave birth to the booming biotechnology industry. In 1986, another technique, called the **polymerase chain re-**

action (**PCR**), was developed that further accelerated the pace of biological research. The significance of this advance is underscored by the fact that the 1993 Nobel Prize in Chemistry was awarded to Kary Mullis for developing the PCR technique.

PCR is a rapid cell-free method of DNA cloning that has extended the power of recombinant DNA research and, in many cases, has replaced some of the methods of cloning that use host cells. PCR analysis has found applications in a wide range of disciplines, including molecular biology, human genetics, evolution, development, and forensics (the gathering of evidence in criminal cases). Some of these applications are discussed in Chapter 21.

PCR allows a direct amplification of specific target DNA sequences within a population of DNA molecules and can be used on fragments of DNA that are initially present in infinitesimally small quantities. In order to amplify a sequence by PCR, some information about the nucleotide sequence of the target DNA is required. This information is used to synthesize two oligonucleotide primers that are added to a denatured (single-stranded) DNA sample. The primers hybridize to complementary sequences that flank the single-stranded sequence to be amplified. A heat-stable form of DNA polymerase (called *Taq* polymerase) is used to synthesize a second strand of the target DNA (Figure 18.22). There are three basic steps in the PCR reaction, and the amount of amplified DNA produced is limited theoretically only by the number of times these steps are repeated.

1. In the first step, the DNA to be amplified is denatured into single strands. This DNA does not have to be purified and can come from any number of sources, including genomic DNA, forensic samples such as dried blood or semen, dried samples stored as part of medical records, single hairs, mummified remains, and fossils. The double-stranded DNA is denatured by heating (at 90–95°C) until it dissociates into single strands (usually about 5 minutes).

2. The temperature of the reaction is lowered to somewhere between 50 and 70°C, and primers are allowed to bind to the denatured DNA. These primers are synthetic oligonucleotides (15–30 nucleotides long) that bind specifically to sequences flanking the segment to be amplified. These primers are designed to serve as the starting points for the synthesis of new DNA strands complementary to the target DNA.

3. A heat-stable form of DNA polymerase (*Taq* polymerase, an enzyme from a bacterium that lives in hot springs) is added to the reaction mixture and DNA synthesis is carried out at temperatures between 70 and 75°C. The polymerase extends the primers by

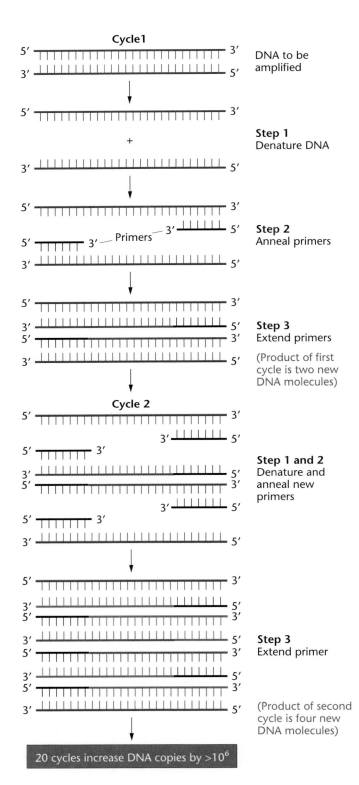

Cycle 1

5′ ||||||||||||||||||||||||||||| 3′   DNA to be amplified
3′ ||||||||||||||||||||||||||||| 5′

5′ ||||||||||||||||||||||||||||| 3′   **Step 1** Denature DNA

+

3′ ||||||||||||||||||||||||||||| 5′

5′ ||||||||||||||||||||||||||||| 3′
               3′ |||||| 5′   **Step 2** Anneal primers
5′ |||||| 3′ — Primers — 3′
3′ ||||||||||||||||||||||||||||| 5′

5′ ||||||||||||||||||||||||||||| 3′
3′ ||||||||||||||||||||||||||||| 5′
5′ ||||||||||||||||||||||||||||| 3′   **Step 3** Extend primers
3′ ||||||||||||||||||||||||||||| 5′

     (Product of first cycle is two new DNA molecules)

Cycle 2

5′ ||||||||||||||||||||||||||||| 3′
             3′ |||||| 5′
5′ |||||| 3′
3′ ||||||||||||||||||||||||||||| 5′
5′ ||||||||||||||||||||||||||||| 3′   **Step 1 and 2** Denature and anneal new primers
             3′ |||||| 5′
5′ |||||| 3′
3′ ||||||||||||||||||||||||||||| 5′

5′ ||||||||||||||||||||||||||||| 3′
3′ ||||||||||||||||||||||||||||| 5′
5′ ||||||||||||||||||||||||||||| 3′
3′ ||||||||||||||||||||||||||||| 5′   **Step 3** Extend primer
5′ ||||||||||||||||||||||||||||| 3′
3′ ||||||||||||||||||||||||||||| 5′
5′ ||||||||||||||||||||||||||||| 3′
3′ ||||||||||||||||||||||||||||| 5′

     (Product of second cycle is four new DNA molecules)

20 cycles increase DNA copies by >10⁶

adding nucleotides in the 5′-to-3′ direction, making a double-stranded copy of the target DNA.

Each set of three steps—**denaturation** of the double-stranded product, **annealing** of primers, and **extension** by polymerase—is referred to as a cycle. PCR is a chain reaction because the number of new DNA strands is doubled in each cycle, and the new strands serve as templates in the next cycle. Each cycle, which takes 4–5 minutes, can be repeated by carrying out each of the steps again. Twenty-five to thirty cycles results in an over 1,000,000-fold increase in the amount of DNA. The process has been automated through the development of machines called *thermocyclers* that can be programmed to carry out a predetermined number of cycles, yielding large amounts of the target DNA, which can be used in other procedures such as cloning, sequencing, clinical diagnosis, and genetic screening.

PCR-based DNA cloning has several advantages over host cell-based cloning. The PCR reaction is fast and can be carried out in a few hours, rather than the weeks required for host cell-based cloning. The design of PCR primers can be done using computer software, and the commercial synthesis of the oligonucleotides is also fast and economical.

The PCR reaction is very sensitive and can be used to amplify target DNA from vanishingly small DNA samples, including the DNA in a single cell. This feature of PCR is valuable in several areas, including genetic testing, forensics, and molecular paleontology. In addition, PCR can be used on DNA samples that have been partly degraded, contaminated with other materials, or embedded in a medium that would make DNA extraction for conventional cloning difficult or impossible.

Limitations of PCR

Although PCR is a major advance, the technique has some limitations. In addition to the requirement that some information about the nucleotide sequence of the target

FIGURE 18.22 PCR amplification. In the polymerase chain reaction (PCR), the DNA to be amplified is denatured into single strands; each strand is then annealed to a short complementary primer. The primers are synthetic oligonucleotides complementary to sequences flanking the region to be amplified. DNA polymerase and nucleotides are added to extend the primers in the 3′ direction, using the single-stranded DNA as a template. The result is a double-stranded DNA molecule with the primers incorporated into the newly synthesized strand. In a second PCR cycle, the products of the first cycle are denatured into single strands, primers are added, and DNA polymerase then extends them. Repeated cycles can amplify the original DNA sequence by more than 1,000,000-fold. *GenCDX*

DNA be known and the relatively short size of the amplified products, even minor contamination of the sample with DNA from other sources can cause difficulties. Cells shed from the skin of a laboratory worker while performing PCR can contaminate the samples, such as those gathered from a crime scene or taken from fossils. This contamination can make it difficult to obtain accurate results. PCR reactions must always be run with carefully designed and appropriate controls.

Other Applications of PCR

Although PCR was developed just over 10 years ago, it has become one of the most widely used techniques in genetics and molecular biology. In addition to its use in cloning, PCR is used in many other fields. It is used to quickly identify restriction-site variants and variations in tandemly repeated DNA sequences that can be used as genetic markers. Using gene-specific primers, PCR is widely used in screening for mutations in genetic disorders, allowing the location and nature of the mutation to be quickly determined. As we discuss in Chapter 21, primers can be designed to distinguish target sequences that differ by only a single nucleotide, making it possible to use allele-specific probes in genetic testing.

The use of random primers allows the indiscriminate amplification of DNA, which is particularly useful in studying samples from single cells, from fossils, or from crime scenes where a single hair or a saliva-licked postage stamp are the sources of DNA. In addition, PCR can be used to explore uncharacterized DNA regions adjacent to known regions and is even used in DNA sequencing. In short, PCR is one of the most widely used and versatile techniques in modern genetics.

Clones can be characterized in several ways

The recovery and identification of genes and other specific DNA sequences by cloning or by PCR is a powerful tool for the analysis of genomic structure and function. In fact, much of the Human Genome Project is based on such techniques. In the following sections, we will consider some of the methods used to answer questions about the organization and function of cloned sequences.

Restriction Mapping

One of the first steps in characterizing a DNA clone is the construction of a **restriction map**. A restriction map is a compilation of the number, order, and distance between restriction enzyme-cutting sites along a cloned segment of DNA. Map units are expressed in **base pairs (bp)** or, for longer distances, **kilobase pairs (kb)**. Restriction maps provide information about the length of a cloned insert and the location of restriction enzyme sites within the cloned DNA that can be used for subcloning fragments of a gene or for comparing its internal organization with that of other cloned sequences. Restriction maps can also be used to compare a gene and its cDNA to identify exons and introns in the genomic copy of the gene.

Fragments generated by cutting DNA with restriction enzymes can be separated by gel electrophoresis, a method that separates fragments by size, with the smallest pieces moving farthest (see Appendix A for a detailed description of electrophoresis). The fragments appear as a series of bands that can be visualized by staining the DNA with ethidium bromide and viewing under ultraviolet illumination (Figure 18.23).

Figure 18.24 shows the steps in constructing a restriction map from a cloned DNA segment. For this map, let us begin with a cloned DNA segment 7.0 kb in length (Figure 18.24). Three samples of the cloned DNA are digested with restriction enzymes; one sample is digested with *Hin*dIII, one with *Sal*I, and one with both *Hin*dIII and *Sal*I. The fragments generated by digestion with the restriction enzymes are separated by electrophoresis, stained with ethidium bromide, and photographed. The sizes of the resulting fragments are calculated by comparison to a set of standards run in adjacent lanes. To construct the map, the fragments generated by the restriction enzymes are analyzed.

1. When the DNA is cut with *Hin*dIII, two fragments are produced, one of 0.8 kb and one of 6.2 kb, confirming that the cloned insert is 7.0 kb in length and

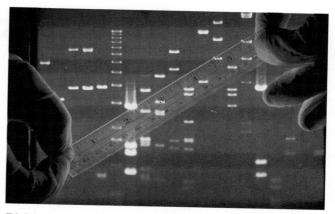

FIGURE 18.23 An agarose gel containing separated DNA fragments stained with a dye (ethidium bromide) and visualized under ultraviolet light. Smaller fragments migrate faster and farther than do larger fragments, resulting in the distributions shown.

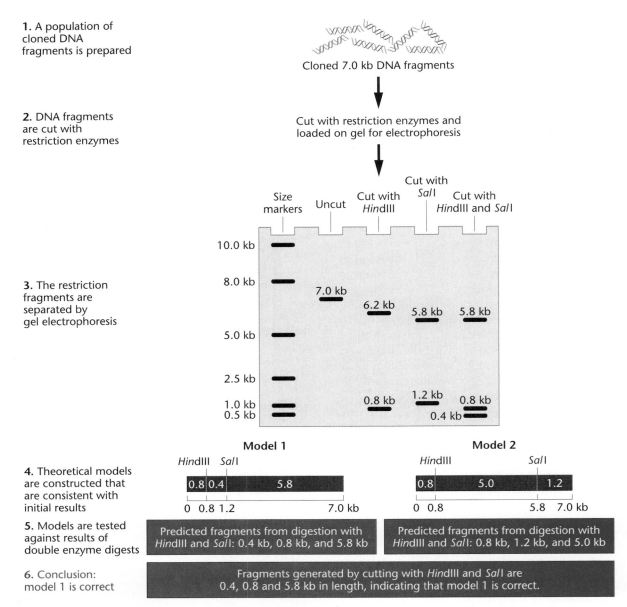

1. A population of cloned DNA fragments is prepared

Cloned 7.0 kb DNA fragments

2. DNA fragments are cut with restriction enzymes

Cut with restriction enzymes and loaded on gel for electrophoresis

3. The restriction fragments are separated by gel electrophoresis

4. Theoretical models are constructed that are consistent with initial results

5. Models are tested against results of double enzyme digests

6. Conclusion: model 1 is correct

Model 1

*Hind*III   *Sal*I

| 0.8 | 0.4 | 5.8 |

0   0.8 1.2   7.0 kb

Predicted fragments from digestion with *Hind*III and *Sal*I: 0.4 kb, 0.8 kb, and 5.8 kb

Model 2

*Hind*III   *Sal*I

| 0.8 | 5.0 | 1.2 |

0   0.8   5.8   7.0 kb

Predicted fragments from digestion with *Hind*III and *Sal*I: 0.8 kb, 1.2 kb, and 5.0 kb

Fragments generated by cutting with *Hind*III and *Sal*I are 0.4, 0.8 and 5.8 kb in length, indicating that model 1 is correct.

FIGURE 18.24   Construction of a restriction map of a 7 kb DNA fragment, as described in the text. One sample is digested with *Hind*III, one with *Sal*I, and one with both *Hind*III and *Sal*I. The resulting fragments are separated by gel electrophoresis. The sizes of the separated fragments are measured by comparison with the molecular-weight standards in an adjacent lane. Models constructed to explain the digestion patterns can be used to predict the fragment sizes generated by digestion with both *Hind*III and *Sal*I. The comparison of the predicted fragments with those observed on the gel indicates that Model 1 is the correct restriction map.

indicating that there is only one cutting site for this enzyme (located 0.8 kb from one end).

2. When the DNA is cut with *Sal*I, two fragments result, one of 1.2 kb and one of 5.8 kb, indicating that there is one cutting site for this enzyme, located 1.2 kb from one end of the insert.

Taken together, the results show that there is one restriction site for each enzyme, but the relationship between the two sites is unknown. From the information available, two different maps are possible. In one map (Model 1), the *Hind*III site is located 0.8 kb from one end, and the *Sal*I site is located 1.2 kb from the same end. In the alternative map (Model 2), the *Hind*III site is located 0.8 kb from one end, and the *Sal*I site is located 1.2 kb from the other end.

The correct model can be selected by considering the results from the sample digested with both *Hind*III and *Sal*I. Model 1 predicts that digestion with both enzymes will generate three fragments: 0.4, 0.8, and 5.8 kb in length; model 2 predicts that digestion with both enzymes will generate three different fragments: 0.8, 5.0, and 1.2 kb in length. The actual fragment pattern observed after cutting with both enzymes indicates that Model 1 is correct (Figure 18.24).

Restriction maps are an important way of characterizing a cloned DNA segment and can be constructed in the absence of any information about the coding capacity or function of the mapped DNA. In conjunction with other techniques, restriction mapping can be used to define the boundaries of a gene, to dissect the molecular organization of a gene and its flanking regions, and to locate mutational sites within genes.

Restriction maps can also be used to refine genetic maps. To a large extent, the accuracy of genetic maps depends on the frequency of recombination between genetic markers and on the number of markers used in construction of the map. In humans, for example, the genome size is large (3.3×10^9 bp) and the number of known genes is small, resulting in long distances between markers on the genetic map. Restriction enzyme-cutting sites are inherited and can be used as genetic markers, reducing the distance between markers, increasing their accuracy, and providing reference points for the correlation of genetic and physical maps.

Restriction sites have played an important role in mapping genes to specific human chromosomes and to defined regions of individual chromosomes. In addition, if a restriction site maps close to a mutant gene, it can be used as a marker in a diagnostic test. These sites, known as **restriction fragment length polymorphisms**, or **RFLPs**, are described in Chapter 21. The use of RFLPs has proven especially useful because the mutant genes underlying many human genetic diseases are poorly characterized at the molecular level, and restriction sites closely linked to mutant genes have been successfully used to identify heterozygotes at risk for having affected children.

Nucleic Acid Blotting

Many of the techniques described in this chapter rely on hybridization between complementary nucleic acid (DNA or RNA) molecules. We will now describe one of the most widely used methods for the detection of such hybrids. This technique, called **Southern blotting** (after Edward Southern, who devised it), involves separating DNA fragments by gel electrophoresis, transferring them to filters, and screening the filters with labeled probes.

To make a Southern blot, cloned DNA is cut into fragments with one or more restriction enzymes, and the fragments are separated by gel electrophoresis (Figure 18.25). The DNA in the gel is denatured into single-stranded fragments by treatment with an alkaline solution, and transferred to a sheet of DNA-binding material, usually nitrocellulose or a nylon derivative. To transfer the fragments, the sheet of membrane is placed on top of the gel, and a buffer solution flows through the gel and the membrane by capillary action (Figure 18.25). As the buffer solution flows through both, the DNA fragments move out of the gel and become immobilized on the membrane.

The DNA fragments bound to the membrane are hybridized with a labeled single-stranded DNA probe. Only the DNA fragments bound to the membrane that are complementary to the nucleotide sequence of the probe will form double-stranded hybrids. The unbound probe is washed away, and the hybridized fragments are visualized on a piece of film (Figure 18.26).

In addition to characterizing cloned DNAs, Southern blots are used for many other purposes, including the mapping of restriction sites within and near a gene, the identification of DNA fragments carrying a single gene from a mixture of many fragments, and the identification of related genes in different species. Southern blots are also used to detect rearrangements and duplications in genes associated with human genetic disorders and cancers.

A related blotting technique can be used to determine whether a cloned gene is transcriptionally active in a given cell or tissue type by probing for the presence of mRNA that is complementary to a cloned gene. This is done by extracting mRNA from one or several cell or tissue types. The RNA is fractionated by gel electrophoresis and the resulting pattern of RNA bands is transferred to a sheet of membrane, as in the Southern blot. The membrane is then hybridized to a single-stranded DNA probe, derived from the cloned gene. If RNA complementary to the DNA probe is present, it will be detected as a band on the film. Because the original procedure using DNA bound to a filter became known as a Southern blot, this inverse procedure using RNA bound to a filter was called a **northern blot**. (Following this somewhat perverse logic, another procedure involving proteins bound to a filter is known as a **western blot**.)

Northern blots provide information about the expression of specific genes and are used to study patterns of gene expression in embryonic and adult tissues. Northern blots can also be used to detect alternatively spliced mRNAs and multiple types of transcripts derived from a single gene. Northern blots also provide other information about transcribed mRNAs. If marker RNAs of known size are run in an adjacent lane, the size of the mRNA of a gene of interest can be calculated. In addition, the amount of transcribed RNA present in the cell or tissue being studied is related to the density of the RNA band on the film. This can be quantified by measuring the density of the band, providing a relative measurement of transcriptional activity. Thus, northern blots can be used to characterize and quantify the transcriptional activity of a given gene in different cells, tissues, or organisms.

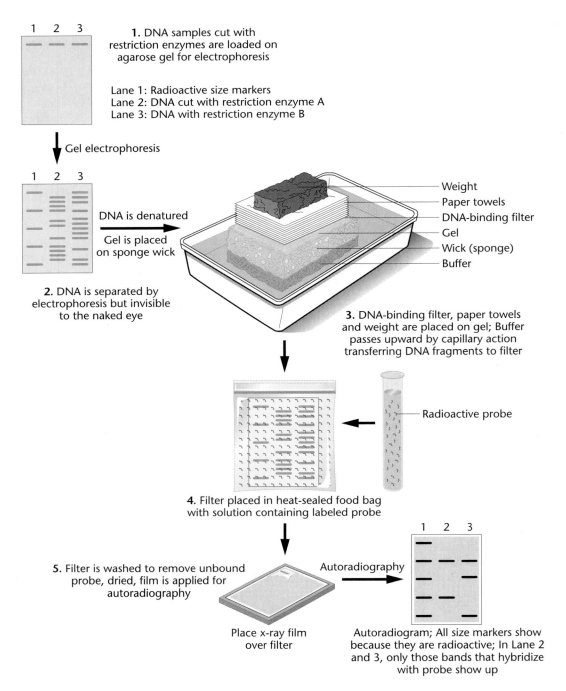

1. DNA samples cut with restriction enzymes are loaded on agarose gel for electrophoresis

Lane 1: Radioactive size markers
Lane 2: DNA cut with restriction enzyme A
Lane 3: DNA with restriction enzyme B

Gel electrophoresis

DNA is denatured

Gel is placed on sponge wick

2. DNA is separated by electrophoresis but invisible to the naked eye

Weight
Paper towels
DNA-binding filter
Gel
Wick (sponge)
Buffer

3. DNA-binding filter, paper towels and weight are placed on gel; Buffer passes upward by capillary action transferring DNA fragments to filter

Radioactive probe

4. Filter placed in heat-sealed food bag with solution containing labeled probe

5. Filter is washed to remove unbound probe, dried, film is applied for autoradiography

Autoradiography

Place x-ray film over filter

Autoradiogram; All size markers show because they are radioactive; In Lane 2 and 3, only those bands that hybridize with probe show up

FIGURE 18.25 The Southern blotting technique. Samples of the DNA to be probed are cut with restriction enzymes and the fragments are separated by gel electrophoresis. The pattern of fragments is visualized and photographed under ultraviolet illumination by staining the gel with ethidium bromide. The gel is then placed on a sponge wick in contact with a buffer solution and covered with a DNA-binding filter. Layers of paper towels or blotting paper are placed on top of the filter and held in place with a weight. Capillary action draws the buffer through the gel, transferring the pattern of DNA fragments from the gel to the filter. The DNA fragments on the binding filter are then denatured into single strands and hybridized with a labeled DNA probe, washed, and overlaid with a piece of X-ray film. The hybridized fragments show up as bands when the X-ray film is developed, as well as the size markers, which are radioactive.

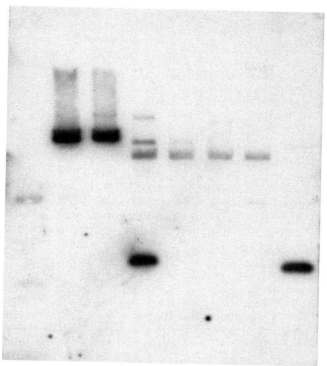

FIGURE 18.26 A Southern blot after hybridization, exposure, and development. The bands show DNA fragments that are complementary to the nucleotide sequence of the probe.

DNA sequencing is the ultimate way to characterize a clone

In a sense, a cloned DNA sequence is completely characterized when its nucleotide sequence is known. The ability to sequence cloned DNA has added immensely to our understanding of gene structure, gene function, and the mechanisms of regulation.

The most widely used method of **DNA sequencing** is based on the 5′-to-3′ elongation of single-stranded DNA templates by the enzyme DNA polymerase. In these reactions, a base-specific analogue, called a **dideoxynucleotide**, is present in addition to the normal deoxynucleotides. A series of four reactions, each in a separate tube, is used to determine the sequence of the nucleotides in a DNA segment. Each tube contains single-stranded template DNA (the DNA to be sequenced), an attached primer, DNA polymerase, the four deoxynucleotides, and a small amount of one of the four dideoxynucleotides. As DNA synthesis takes place, the DNA polymerase occasionally inserts a dideoxynucleotide into a growing DNA strand instead of a deoxynucleotide. Since the analogue lacks a 3′-hydroxyl group, it cannot participate in the formation of a 3′ bond to add another nucleotide to the strand, and DNA synthe-

sis is therefore terminated. Each tube accumulates a series of DNA molecules that differ in length by one nucleotide at their 3′-ends. The fragments from each reaction tube are separated from each other by gel electrophoresis in four adjacent lanes (one for each tube). After visualization, the result is a series of bands forming a ladder-like pattern. The sequence can be read directly from bottom to top, corresponding to the 5′-to-3′ sequence of the DNA strand complementary to the template (Figure 18.27).

G A T C

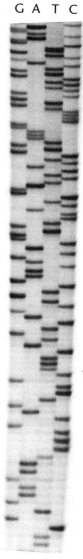

FIGURE 18.27 Photograph of a DNA-sequencing gel showing the separation of bases in the four sequencing reactions (one per lane). The gel is analyzed to reveal the base sequence of the DNA fragment. To do this, the gel is read from the bottom, beginning with the lowest band in any lane, then the next lowest, then the next, and so on. For example, the sequence of the DNA on this gel begins with TTCGTGAAG.

GENETICS, TECHNOLOGY, AND SOCIETY

DNA Fingerprints in Forensics:
The Case of the Telltale Palo Verde

The use of DNA analysis in forensics is making it harder and harder to get away with many crimes. It's now a simple matter to isolate DNA from tissue left at a crime scene, a splattering of blood, some skin left under a victim's fingernails, or even the cells clinging to the base of a hair shaft. A variety of techniques can be used to determine the likelihood that the sample came from a suspect in the case. In recent years, the PCR technique has been increasingly used in forensics, both because it is fast (taking only about a week) and because it requires only a small amount of DNA (about a nanogram). One variation of the PCR method that may be especially valuable is called the random amplified polymorphic DNA (RAPD, pronounced "rapid") procedure. Under the right conditions, the RAPD procedure generates a DNA profile that is unique to each individual and that can thus be used for personal identification.

The RAPD procedure involves the amplification of a set of DNA fragments of unknown sequence. First, an amplification primer is mixed with a sample of DNA. The primer will anneal to all sites in the DNA sample that have a matching base sequence. For primers 10 nucleotides in length, binding will occur at a few thousand sites scattered randomly throughout the genome. When primers happen to bind to DNA sites that are not too far apart (within about 2000 nucleotides), DNA amplification can occur between these sites, eventually producing millions of copies of the DNA sequence lying between the binding sites. When the products of such a reaction are separated by electrophoresis on an agarose gel, each amplified DNA segment will appear as a distinct band. The locations of the primer binding sites in the genome of any two individuals in a population are likely to differ, due to minor DNA sequence differences (e.g., base changes, insertions, and deletions). Since these affect the initial binding to the primers, the number and locations of the DNA bands on the gel will vary from one individual to another. The resulting DNA profile is referred to as a DNA fingerprinting. It represents a characteristic "snapshot" of the genome of an individual, allowing it to be distinguished from those of other individuals in the population.

When a DNA profile from tissue found at a crime scene matches that of a suspect, it does not prove that the tissue belongs to the suspect; instead, it excludes all those who have a different pattern. The strategy, therefore, is to generate five or more DNA profiles from the same sample, using different amplification primers. The more profiles that match between the sample and the suspect, the more unlikely it is that the sample at the crime scene came from someone other than the suspect. The likelihood that a complete match of all the banding patterns occurs simply by chance can be calculated, giving, for example, a 1-in-100,000 to a 1-in-1,000,000 probability of a "random match."

One of the most interesting uses of DNA fingerprinting in a criminal case did not involve the suspect's own DNA, but rather the DNA of plants

For sequencing genomes, large-scale DNA sequencing uses automated DNA sequencers that employ fluorescent dyes instead of radioactive isotopes. Four different-colored dyes are used, producing a colored pattern of peaks that can be read to provide the sequence (Figure 18.28).

DNA sequencing provides information about the organization of genes and the nature and number of mutational events that alter both genes and gene products, confirming the conclusion that genes and proteins are colinear molecules. Sequencing has also been used to study the organization of regulatory regions that flank prokaryotic and eukaryotic genes, and to infer the amino acid sequence of proteins.

Computer analysis of nucleotide sequences may be used to determine whether a cloned segment contains all or just part of a gene. This can be done by searching for exon/intron junctions, or by comparing the DNA sequence

growing at the crime scene. On the night of May 2, 1992, a Phoenix woman was strangled and her body dumped near an abandoned factory in the surrounding desert. Investigators discovered a pager near the body, making its owner a prime suspect. When questioned, this man admitted being with the woman the day of the murder, but claimed he had not been near the factory and suggested that the woman must have stolen his pager from his pickup truck. A search of the pickup provided the crucial clue, placing the suspect at the factory: In its bed were two seed pods from a palo verde tree.

The palo verde tree (*Cercidium floridum*), native to the desert of the Southwest, is a member of the bean family. In an adaptation to the hot, dry climate, it puts out very small leaves. To compensate for this, its stems become green and carry out photosynthesis. This gives the plant its name; palo verde is Spanish for "green stem." The plant may not resemble a bean in some respects, but, like a typical bean, it produces seeds in pods that drop to the ground when mature.

The homicide investigators assigned to the case wondered whether it could be proved that the seed pods found in the bed of the suspect's truck had fallen from a palo verde tree near the place where the body was found. If so, it would be a key piece of evidence placing the suspect at the scene. But how could this be demonstrated? The investigators turned to Dr. Timothy Helentjaris, then at the University of Arizona in nearby Tucson. Helentjaris proceeded to determine whether the DNA profile of the seed pods from the pickup truck matched that of any of the palo verde trees near the scene. He understood that for a match between a seed pod and a tree to have significance, he would first have to show that palo verdes differ genetically from one tree to the next. Otherwise, a pod could never be unambiguously assigned to one particular tree. Fortunately, he found considerable variation in DNA profiles among the palo verde trees.

Helentjaris was then given the two seed pods from the suspect's truck along with pods collected from 12 palo verde trees from the vicinity of the factory. The investigators knew which of the 12 trees was the key one at the crime scene, but they did not tell Helentjaris. He extracted DNA from seeds from each pod and performed RAPD reactions using an amplification primer that he had earlier found to reproducibly yield profiles of 10–15 distinct bands. The result of this "controlled" experiment was unmistakable: The DNA profile from one of the pods found in the truck exactly matched that of only 1 of the 12 trees, the one nearest to where the body was found. In an important additional test, Helentjaris showed that this DNA profile differed from that of pods collected from 18 other trees located at random sites around Phoenix. Collectively, Helentjaris's analysis led to the estimate of the likelihood of a "random match" at a little less than 1 in 1,000,000.

Helentjaris's analysis was admitted as evidence in the trial, placing the suspect at the crime scene. At the completion of the five-week trial, the suspect was found guilty of first-degree murder. His conviction was upheld upon appeal, and he is currently serving a life sentence without parole.

REFERENCES

AYALA, F.J., and BLACK, B. 1993. Science and the courts. *Am. Sci.* 81:230–39.

KRAWCZAK, M., and SCHMIDTKE, J. 1994. *DNA fingerprinting.* Oxford: BIOS Scientific.

MARX, L. 1988. DNA fingerprinting takes the witness stand. *Science* 240: 1616–18.

and the inferred amino acid sequence with the information in various databases that contain the sequence of previously identified genes and proteins. This method was used to scan large cloned segments of DNA to find the gene for cystic fibrosis (CF), an autosomal recessive human genetic disorder that maps to a region on the long arm of chromosome 7. Since the protein product of the gene was unknown, DNA sequencing was used to identify a protein-coding region. The DNA sequence from this region was used to generate a putative amino acid sequence of 1480 amino acids. This amino acid sequence was used to search computer databases containing the sequences of known proteins. The results indicated that the CF protein has an amino acid sequence similar to membrane proteins that play a role in ion transport. Further work confirmed that the product of the cystic fibrosis locus is a membrane protein (called CFTR) that regulates the transport of chloride ions across the plasma membrane. In most cases of CF, the mutant gene has an altered DNA sequence that causes the production of a defective protein.

In addition to identifying DNA defects that cause mutant phenotypes, DNA sequencing is used to study the organization of genes (the number of introns and exons and their boundaries), to provide information about the nature and function of proteins encoded by genes, and to study evolutionary relationships to similar proteins from other organisms.

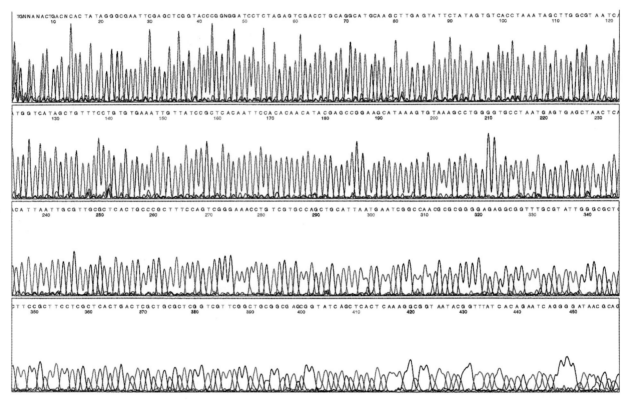

FIGURE 18.28   DNA sequencing has been automated with the use of fluorescent dyes, one for each base; the separated bases are read in order along the axis from left to right.

CHAPTER SUMMARY

1. The cornerstone of recombinant DNA technology is a class of enzymes called restriction endonucleases that cut DNA at specific recognition sites. The fragments produced are joined with DNA vectors to form recombinant DNA molecules.

2. Vectors can replicate autonomously in host cells and facilitate the manipulation of the newly created recombinant DNA molecules. Vectors have been constructed from many sources, including bacterial plasmids and viruses.

3. Recombinant DNA molecules are transferred into a host, and cloned copies are produced during host cell replication. A variety of host cells may be used for replication, including bacteria, yeast, and mammalian cells. Cloned copies of foreign DNA sequences can be recovered, purified, and analyzed.

4. The polymerase chain reaction (PCR) is a method for amplifying a specific DNA sequence that is present in a collection of DNA sequences, such as genomic DNA. The PCR method allows cloning of DNA without the need for host cells and is a rapid sensitive method with wide-ranging applications.

5. Once cloned, DNA sequences can be analyzed using a variety of methods, including restriction mapping and DNA sequencing. Other methods such as blotting hybridization can be used to identify genes and flanking regulatory regions within the cloned sequences.

INSIGHTS AND SOLUTIONS

The recognition site for the restriction enzyme *Sau*3A is GATC (Figure 18.4). The recognition site for the enzyme *Bam*HI is GGATCC, where the four internal bases are identical to the *Sau*3A site. This means that the single-stranded ends produced by the two enzymes are identical. Suppose you have a cloning vector containing a *Bam*HI site and foreign DNA that you have cut with *Sau*3A.

1. Can this DNA be ligated into the *Bam*HI site of the vector? Why?

   Solution: DNA cut with *Sau*3A can be ligated into the *Bam*HI site of the vector because the single-stranded ends generated by the two enzymes are identical.

2. Can the DNA segment cloned into this site be cut from the vector with *Sau*3A? With *Bam*HI? What potential problems do you see with the use of *Bam*HI?

Solution: The DNA can be cut from the vector with *Sau*3A because the recognition site for this enzyme is maintained. Recovering the cloned insert with *Bam*H1 is more problematic. In the ligated vector, the conserved bases are GGATC (left) and GATCC (right). Only about 25 percent of the time will the correct base follow (GGATCC on the left) and only about 25 percent of the time will the correct base lead the conserved sequence (GATCC on the right). Thus, only about 6 percent of the time (0.25 × 0.25) will *Bam*H1 be able to cut the insert from the vector.

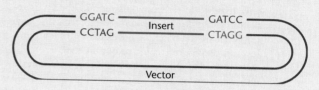

PROBLEMS AND DISCUSSION QUESTIONS

1. In recombinant DNA studies, what is the role of each of the following: restriction endonucleases, terminal transferases, vectors, calcium chloride, and host cells?

2. Why is poly dT an effective primer for reverse transcriptase?

3. An ampicillin-resistant, tetracycline-resistant plasmid is cleaved with *Eco*RI, which cuts within the ampicillin gene. The cut plasmid is ligated with *Eco*RI-digested *Drosophila* DNA to prepare a genomic library. The mixture is used to transform *E. coli* K12.
   (a) Which antibiotic should be added to the medium to select cells that have incorporated a plasmid?
   (b) What antibiotic-resistance pattern should be selected to obtain plasmids containing *Drosophila* inserts?
   (c) How can you explain the presence of colonies that are resistant to both antibiotics?

4. Clones from Problem 3 are found to have an average length of 5 kb. Given that the *Drosophila* genome is 1.5×10^5 kb long, how many clones would be necessary to give a 99 percent probability that this library contains all genomic sequences?

5. Restriction enzymes recognize palindromic sequences in intact DNA molecules and cleave the double-stranded helix at these sites. Inasmuch as the bases are internal in a DNA double helix, how is this recognition accomplished?

6. In a control experiment, a plasmid containing a *Hin*dIII site within a kanamycin-resistant gene is cut with *Hin*dIII, re-ligated, and used to transform *E. coli* K12 cells. Kanamycin-resistant colonies are selected, and plasmid DNA from these colonies is subjected to electrophoresis. Most of the colonies contain plasmids that produce single bands that migrate at the same rate as the original intact plasmid. A few colonies, however, produce two bands, one of original size and one that migrates much higher in the gel. Diagram the origin of this slow band during the religation process.

7. When making cDNA, the single-stranded DNA produced by reverse transcriptase can be made double stranded by treatment with DNA polymerase I. However, no primer is required with the DNA polymerase. Why is this?

8. What facts should you consider in deciding which vector to use in constructing a genomic library of eukaryotic DNA?

9. Using DNA sequencing on a cloned DNA segment, you recover the following nucleotide sequence:

   CAGTATCCTAGGCAT

   Does this segment contain a palindromic recognition site for a restriction enzyme? What is the double-stranded sequence of the palindrome? What enzyme would cut at this site? (Consult Figure 18.4 for a list of restriction enzyme recognition sites.)

10. Figure 18.4 lists restriction enzymes that recognize sequences of four bases and six bases. How frequently should four- and six-base recognition sequences occur in a genome? If the recognition sequence consists of eight bases, how frequently would such sequences be encountered? Under what circumstances would you select such an enzyme for use?

11. You are given a cDNA library of human genes prepared in a bacterial plasmid vector. You are also given the cloned yeast gene that encodes EF-1α, a protein that is highly conserved in protein sequence among eukaryotes. Outline how you would use these resources to identify the human cDNA clone encoding EF-1α.

12. You have recovered a cloned DNA segment of interest and determined that the insert is 1300 bp in length. To characterize this cloned segment, you have isolated the insert and decide to construct a restriction map. Using enzyme I and enzyme II, followed by gel electrophoresis, you determine the number and size of the fragments produced by enzymes I and II alone and in combination as shown.

Enzymes	Restriction Fragment Sizes (bp)
I	350 bp, 950 bp
II	200 bp, 1100 bp
I and II	150 bp, 200 bp, 950 bp

Construct a restriction map from these data, showing the positions of the restriction sites relative to one another and the distance between them in base pairs.

13. Although the potential benefits of cloning in higher plants are obvious, the development of this field has lagged behind cloning in bacteria, yeast, and mammalian cells. Can you think of any reason for this?

14. cDNA can be cloned into vectors to create a cDNA library. In analyzing cDNA clones, it is often difficult to find clones that are full length, that is, that extend to the 5'-end of the mRNA. Why is this so?

15. List the steps involved in screening a genomic library by reverse translation. What needs to be known before starting such a procedure? What are the potential problems with such a procedure? How can they be overcome or minimized?

16. Although the capture and trading of great apes has been banned in 112 countries since 1973, it is estimated that about 1000 chimpanzees are removed annually from Africa and smuggled into Europe, the United States, and Japan. Private owners (such as zoos or circuses) often disguise this illegal trade by simulating births in captivity. Until recently, genetic identity tests to uncover these illegal activities have not been used because of the lack of availability of highly polymorphic markers and the difficulties of obtaining chimp blood samples. Recently a study was reported in which DNA samples were extracted from freshly plucked chimp hair roots and used as templates for the polymerase chain reaction. The primers used in these studies flank highly polymorphic sites in human DNA resulting from variable numbers of tandem nucleotide repeats. Several suspect chimp offspring and their supposed parents were tested to determine if the offspring were "legitimate" or were the "product of the illegal trading" and not the offspring of the putative parents. A sample of the data is shown here. Examine this data carefully and choose the best conclusion.

(a) None of the offspring are legitimate.

(b) Offspring B and C are not the products of these parents and were probably purchased on the illegal market. The data are consistent with offspring A being legitimate.

(c) Offspring A and B are products of the parents shown, but C is not and was therefore probably purchased on the illegal market.

(d) There is not enough data to draw any conclusions. Additional polymorphic sites should be examined.

(e) No conclusion can be drawn because "human" primers were used.

17. Briefly describe the problem that a stretch of repeated sequences would cause in a chromosome walk and name the procedure that is used to overcome this problem.

18. You have obtained a clone of a human gene A that is linked (within 200 kb) to a gene that causes breast cancer. Choose and correctly order six of the items from the following list to outline how you could use the technique of chromosome walking to obtain clones of all genes within 200 kb of gene A.

(1) Partially digest DNA with restriction enzyme *Bam*HI to obtain overlapping fragments of about 20 kb.

(2) Completely digest DNA with restriction enzyme *Eco*RI.

(3) Isolate human genomic DNA.

(4) Isolate *E. coli* genomic DNA.

(5) Probe northern blot with A.

(6) Repeat above step until the desired number of clones is obtained.

(7) Insert fragment mixture into lambda to make a phage bank.

(8) Screen bank (or library) with gene probe A and isolate a hybridizing clone.

(9) Subclone a small fragment from the end of this clone and use it to rescreen the library and isolate a hybridizing clone.

EXTRA-SPICY PROBLEMS

19. The partial restriction map here shows a recombinant plasmid, pBIO220, formed by cloning a piece of *Drosophila* DNA (striped box) including the gene *rosy* into the vector pBR322, which also contains the penicillin-resistance gene, *pen*. The vector part of the plasmid contains only the two E sites shown, and no A or B sites. The gel shows several restriction digests of pBIO220. The enzymes used were:

E-*Eco*RI
A-*Apo*I
B-*Bst*II

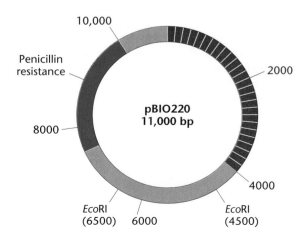

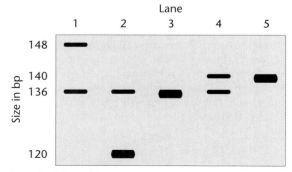

Lane 1: father chimp
Lane 2: mother chimp
Lanes 3-5: putative offspring A, B, C

(a) Use the stained gel pattern to deduce where restriction sites are located in the cloned fragment.

(b) A PCR-amplified copy of the entire 2000-bp *rosy* gene was used to probe a Southern blot of the same gel. Use the Southern-blot results to deduce the locations of *rosy* in the cloned fragment. Redraw the map, showing the location of *rosy*.

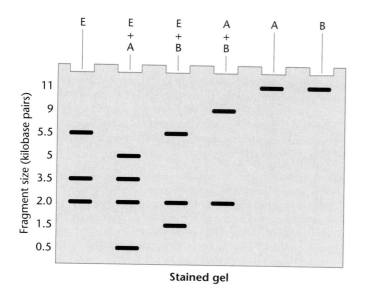

Stained gel

Southern blot

20. One of the restriction maps shown is consistent with the pattern of bands shown here in the gel after digestion with several restriction endonucleases. The enzymes used were:

E-*Eco*RI
N-*Nco*I
A-*Aat*II

(a) From your analysis of the pattern of bands on the gel, select the correct map and explain your reasoning.

(b) In a Southern blot prepared from this gel, the highlighted bands (purple) hybridized with the gene *pep*. Where is the *pep* gene located?

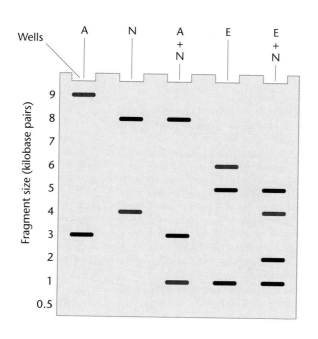

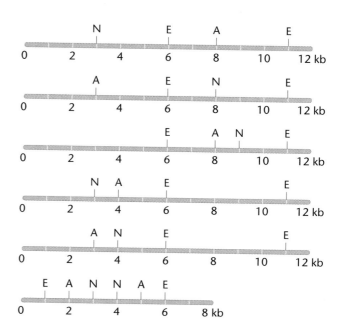

GENETICS MediaLab

The following resources will help you achieve a better understanding of the concepts presented in this chapter. These resources can be found on the CD packaged with this textbook and on the Companion Website at **http://www.prenhall.com/klug/**.

CD Resources:
Animated Tutorial: *Gene Cloning*

Animated Tutorial: *DNA Fingerprinting*

Animated Tutorial: *PCR*

Self-grading Chapter Problems

Web Resources:
Genetics and Society Issue: *DNA Fingerprints in Forensics: The Case of the Telltale Palo Verde*

Web Destinations in Genetics

Self-grading Chapter Problems

Chapter Search Terms

Genetics Newsgroups

Student Bulletin Board

Web Problem 1:
Time for completion = 15 minutes
How can the physical features of a molecule be used to distinguish it from a group of similar molecules? Many of the techniques used in molecular biology use characteristics such as charge, size, shape, and complementarity to separate molecules. Of these features, perhaps the most useful is complementarity between hybridizing molecules. The hybridization of target and probe molecules is a central theme in many molecular techniques. In this exercise, you will review some of these methods. Describe the similarities and differences among Southern, northern, and western blots. For each, list what aspects of the probe or target molecules are essential to perform the technique. To complete this exercise, visit Web Problem 1 in Chapter 18 of your Companion Website and select the keyword **HYBRIDIZATION.**

Web Problem 2:
Time for completion = 5 minutes
PCR has radically changed the field of molecular genetics. The ability to amplify DNA quickly without a cellular host has dramatically increased the pace of molecular genetics. In this exercise, you will view a short animation of the polymerase chain reaction (PCR). After viewing the animation, list the different colored molecules and describe what each colored molecule represents. Indicate how many of each type of colored molecule are present after the three rounds of PCR shown and how many you predict would be present after five rounds of PCR. To complete this exercise, visit Web Problem 2 in Chapter 18 of your Companion Website and select the keyword **PCR.**

Web Problem 3:
Time for completion = 10 minutes
What is a recombinant DNA clone? A DNA clone refers to copies of a small fragment of DNA amplified in a host cell. In contrast, Dolly the cloned sheep was a case of cloning an entire organism. Recombinant DNA methods have allowed researchers to manipulate DNA in a controlled manner in efforts to understand diverse cellular processes. In this exercise, you can test your knowledge of recombinant DNA technology. Feedback will indicate whether your answers are correct and tutorial information is provided to clarify your understanding. To complete this exercise, visit Web Problem 3 in Chapter 18 of your Companion Website and select the keyword **RECOMBINANT DNA.**

SELECTED READINGS

ALWINE, J.C., KEMP, D.J., and STARK, G.R. 1977. Method for detection of specific RNAs in agarose gels by transfer to diazobenzyloxymethyl paper and hybridization with DNA probes. *Proc. Natl. Acad. Sci. USA* 74:5350-54.

ANDERSON, W.F., and DIACUMAKOS, E.G. 1981. Genetic engineering in mammalian cells. *Sci. Am.* (July) 245:106-21.

BOYER, H.W. 1971. DNA restriction and modification mechanisms in bacteria. *Annu. Rev. Microbiol.* 25:153-76.

CAI, L., TAYLOR, J.F., WING, R.A., GALLAGHER, D.S., WOO, S.S., and DAVIS, S.K. 1995. Construction and characterization of a bovine bacterial artificial chromosome library. *Genomics* 29:413-425.

COHEN, S. et al. 1973. Construction of biologically functional bacterial plasmids *in vitro*. *Proc. Natl. Acad. Sci. USA* 70:3240-44.

COLLINS, J., and HOHN, B. 1978. Cosmids: A type of plasmid gene cloning vector that is packageable *in vitro* in bacteriophage heads. *Proc. Natl. Acad. Sci. USA* 75:4242-46.

DAVIES, K.E. et al. 1981. Cloning of a representative genomic library of the human X chromosome after sorting by flow cytometry. *Nature* 293:374-76.

DESALLE, R., and BIRSTEIN, V.J. 1996. PCR identification of black caviar. *Nature* 381:197-98.

FERGUSON-SMITH, M.A.1997. Genetic analysis by chromosome sorting and painting: phylogenetic and diagnostic applications. *Eur. J. Hum. Genet.* 5:253–265.

GLOVER, D.M. 1984. *Gene cloning: The mechanism of DNA manipulation.* London: Chapman and Hall.

GUBLER, U., and HOFFMAN, B.J. 1983. A simple and very effective method for generating cDNA libraries. *Gene* 25:263–69.

HOCHMEISTER, M.N. 1995. DNA technology in forensic applications. *Mol. Aspects Med.* 16:315–437.

KAO, F.T., and TONG, S., SHEN, Y., and YU, J. 1996. Construction and characterization of three region-specific microdissection libraries for human chromosome 18. *Somat. Cell Mol. Genet.* 22:191–199.

KRUMLAUFF, R., JEANPIERRE, M., and YOUNG, B.D. 1982. Construction and characterization of genomic libraries from specific human chromosomes. *Proc. Natl. Acad. Sci. USA* 79:2971–75.

LEBAR, W.D. 1996. Keeping up with new technology: New approaches to diagnosis of *Chlamydia* infection. *Clin. Chem.* 42:809–12.

MESSING, J., and VIERIA, J. 1982. The pUC plasmids, an M13mp7-derived system for insertion mutagenesis and sequencing with synthetic universal primers. *Gene* 19:259–68.

MONACO, A.P. 1994. YACs, BACs, and MACs, artificial chromosomes as research tools. *Trends Biotech.* 12:280–86.

MULLIS, K.B. 1990. The unusual origin of the polymerase chain reaction. *Sci. Am.* (Apr.) 262:56–65.

OLD, R.W., and PRIMROSE, S.B. 1985. *Principles of genetic manipulation: An introduction to genetic engineering.* Palo Alto, CA: Blackwell Scientific.

SAMBROOK, J., FRITSCH, E., and MANIATIS, T. 1992. *Molecular cloning: A laboratory manual*, 2nd ed. Cold Spring Harbor, NY: Cold Spring Harbor Laboratory Press.

SCHLESSINGER, D., and NAGARAJA, R. 1998. Impact and implication of yeast and human artificial chromosomes. *Ann. Med.* 30:186–91.

SHIZUA, H., BIRREN, B., KIM, U-J., MANCINO, V., SLEPAK, T., TACHIIRI, Y., and SIMON, M. 1992. Cloning and stable maintenance of 300 kilo-base-pair fragments of human DNA in *Escherichia coli* using an F-factor-based vector. *Proc. Natl. Acad. Sci. USA* 89:8794–97.

SOUTHERN, E.M. 1975. Detection of specific sequences among DNA fragments separated by gel electrophoresis. *J. Mol. Biol.* 98:503–17.

WATSON, J., GILMAN, M., WITKOWSKI, J., and ZOLLER, M. 1992. *Recombinant DNA*, 2nd ed. New York: Scientific American Books.

Viral and bacterial chromosomes are relatively simple DNA molecules

In comparison with eukaryotes, the chromosomes of viruses and bacteria are much less complicated. They usually consist of a single nucleic acid molecule, largely devoid of associated proteins and containing relatively little genetic information in comparison to the multiple chromosomes comprising the genome of higher forms. These characteristics have greatly simplified analysis, providing a fairly comprehensive view of the structure of viral and bacterial chromosomes.

The chromosomes of viruses consist of a nucleic acid molecule—either DNA or RNA—that can be either single or double stranded. They may exist as circular structures (closed loops), or they may take the form of linear molecules. The single-stranded DNA of the **φX174 bacteriophage** and the double-stranded DNA of the **polyoma virus** is a closed loop housed within the protein coat of the mature virus. The **bacteriophage lambda (λ)**, on the other hand, possesses a linear double-stranded DNA molecule prior to infection, which closes to form a ring upon its infection of the host cell. Still other viruses, such as the **T-even series of bacteriophages,** have linear double-stranded chromosomes of DNA that do not form circles inside the bacterial host. Thus, circularity is not an absolute requirement for replication in some viruses.

Viral nucleic acid molecules have been visualized with the electron microscope. Figure 19.1 shows a mature bacteriophage λ with its double-stranded DNA molecule in the circular configuration. One constant feature shared by viruses, bacteria, and eukaryotic cells is the ability to package an exceedingly long DNA molecule into a relatively small volume. In λ, the DNA is 17 μm long and must fit into the phage head, which is less than 0.1 μm on any side.

Table 19.1 compares the length of the chromosomes of several viruses to the size of their head structure. In each case, a similar packaging feat must be accomplished. The dimensions given for phage T2 may be compared with the

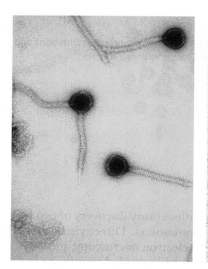
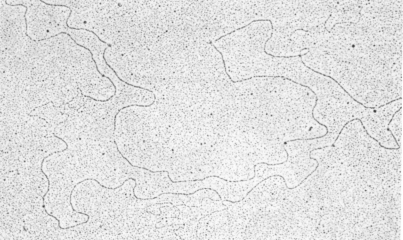

FIGURE 19.1   Electron micrographs of phage λ (left) and the DNA isolated from it. The chromosome is 17 μm long. Note that the phages are magnified about five times more than the DNA (right).

TABLE 19.1   The genetic material of representative viruses and bacteria

| Organism | Nucleic Acid | | | Overall Size of Viral Head or Bacteria (μm) |
	Type	SS or DS*	Length (μm)	
Viruses				
φX174	DNA	SS	2.0	0.025 × 0.025
Tobacco mosaic virus	RNA	SS	3.3	0.30 × 0.02
Lambda phage	DNA	DS	17.0	0.07 × 0.07
T2 phage	DNA	DS	52.0	0.07 × 0.10
Bacteria				
Haemophilus influenzae	DNA	DS	832.0	1.00 × 0.30
Escherichia coli	DNA	DS	1200.0	2.00 × 0.50

*SS = single-stranded; DS = double-stranded.

micrograph of both the DNA and the viral particle shown in Figure 19.2. Seldom does the space available in the head of a virus exceed the chromosome volume by more than a factor of two. In many cases, almost all space is filled, indicating nearly perfect packing. Once packed within the head, the genetic material is functionally inert until it is released into a host cell.

Bacterial chromosomes are also relatively simple in form. They always consist of a double-stranded DNA molecule, compacted into a structure sometimes referred to as the **nucleoid.** *Escherichia coli*, the most extensively studied bacterium, has a large, circular chromosome measuring approximately 1200 μm (1.2 mm) in length. When the cell is gently lysed and the chromosome released, it can be visualized under the electron microscope (Figure 19.3).

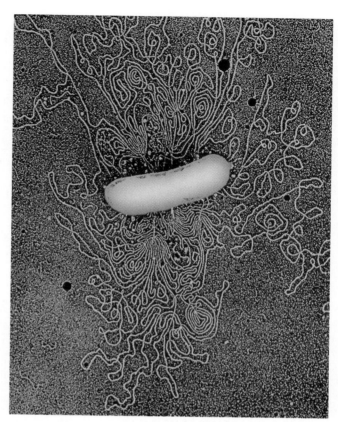

FIGURE 19.3 Electron micrograph of the bacterium *Escherichia coli*, which has had its DNA released by osmotic shock. The chromosome is 1200 μm long.

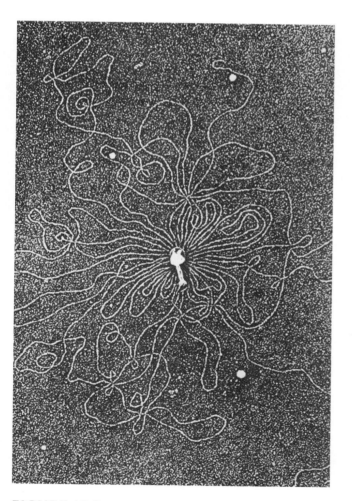

FIGURE 19.2 Electron micrograph of bacteriophage T2, which has had its DNA released by osmotic shock. The chromosome is 52 μm long.

DNA in bacterial chromosomes is found to be associated with several types of **DNA-binding proteins.** Two, called **HU** and **H,** are small but abundant in the cell and contain a high percentage of positively charged amino acids that can bond ionically to the negative charges of the phosphate groups in DNA. These proteins are structurally similar to molecules called **histones** that are associated with eukaryotic DNA. We will discuss histone organization later in this chapter. Unlike the tightly packed chromosome present in the head of a virus, the bacterial chromosome is not functionally inert. Despite its somewhat compacted condition in the bacterial cell, the chromosome can be readily replicated and transcribed.

Supercoiling is common in the DNA of viral and bacterial chromosomes

One major insight into the way in which DNA is organized and packaged in viral and bacterial chromosomes has come from the discovery of **supercoiled DNA,** which is particularly characteristic of closed circular molecules.

Supercoiled DNA was first proposed as a result of a study of double-stranded DNA molecules derived from the polyoma virus, which causes tumors in mice. In 1963, it was observed that when such DNA was subjected to high-speed centrifugation, it was resolved into three distinct components, each of different density and compactness. The one that was least compact, and thus least dense, demonstrated a decreased sedimentation velocity; the other two fractions each showed greater velocities owing to their greater compaction and density. All three were of identical molecular weight.

In 1965 Jerome Vinograd proposed an explanation for these observations. He postulated that the two fractions of greatest sedimentation velocity both consisted of polyoma DNA molecules that are circular, whereas the fraction of lower sedimentation contained polyoma DNA molecules that are linear. Closed circular molecules are more compact and they sediment more rapidly than do linear molecules of the same length and molecular weight.

Vinograd proposed further that the more dense of the two fractions of circular molecules consisted of covalently closed DNA helices that are slightly *underwound* in comparison to the less dense circular molecules. Energetic forces stabilizing the double helix resist this underwinding, causing it to **supercoil** in order to retain normal base pairing. Vinograd proposed that it is the supercoiled shape that causes tighter packing and thus the increase in sedimentation velocity.

The transitions described are illustrated in Figure 19.4. Consider a double-stranded linear molecule that exists in the normal Watson–Crick right-handed helix [Figure 19.4(a)]. This helix contains 20 complete turns, which defines the **linking number** ($L = 20$) of this molecule. If the ends of the molecule are sealed, a closed circle is formed [Figure 19.4(b)], which is described as being *energetically relaxed*. Suppose, however, that the circle were cut open, underwound by several full turns, and resealed [Figure 19.4(c)]. Such a structure, where L is equal to 18, is *energetically "strained,"* and as a result it will exist only temporarily in this form.

In order to assume a more energetically favorable conformation, the molecule can form supercoils in the opposite direction of the underwound helix. In our case [Figure 19.4(d)], two negative supercoils are introduced spontaneously, reestablishing the total number of original "turns" in the helix. The use of the term "negative" refers to the fact that, by definition, the supercoils are left-handed. The end result is the formation of a more compact structure with enhanced physical stability.

In most closed circular DNA molecules in bacteria and their phages, DNA is slightly underwound [as in Figure 19.4(c)]. For example, the virus **SV40** contains 5200

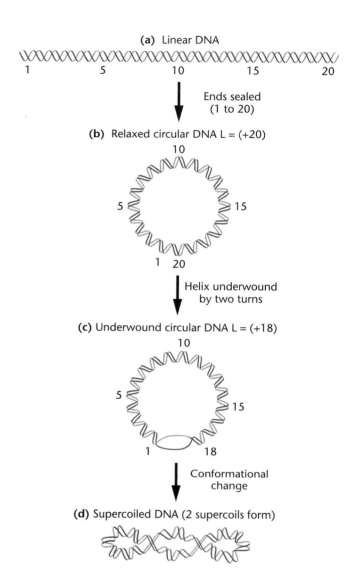

FIGURE 19.4 Depictions of the transformations leading to the supercoiling of circular DNA. *L* equals the linking number.

base pairs, where 10.4 base pairs occupy each complete turn of the helix. The linking number can be calculated as

$$L = 5200/10.4 = 500$$

When circular SV40 DNA is analyzed, it is underwound by 25 turns and L is equal to only 475. Predictably, 25 negative supercoils are observed. In *E. coli*, an even larger number of supercoils is observed, greatly facilitating chromosome condensation in the nucleoid region.

Two otherwise identical molecules that differ only in their linking number are said to be **topoisomers** of one another. But how can a molecule convert from one topoisomer to the other if there are no free ends, as is the case in closed circles of DNA? Biologically, this may be accomplished by any one of a group of enzymes that cut one or both of the strands and wind or unwind the helix before resealing the ends. Appropriately, these enzymes are called **topoisomerases.** First discovered by Martin Gellert and James Wang, these catalytic molecules are known as either type I or type II, depending on whether they cleave one or both strands in the helix, respectively. In *E. coli,* topoisomerase I serves to reduce the number of negative supercoils in a closed-circular DNA molecule. Topoisomerase II introduces negative supercoils into DNA. This latter enzyme is thought to bind to DNA, twist it, cleave both strands, and then pass them through the "loop" that it has created. Once the phosphodiester bonds are reformed, the linking number is decreased and one or more supercoils form spontaneously.

Supercoiled DNA and topoisomerases are also found in eukaryotes. While the chromosomes in these organisms are not usually circular, supercoils can occur when areas of DNA are embedded in a lattice of proteins associated with the chromatin fibers. This association creates "anchored" ends that provide the stability for the maintenance of supercoils once they are introduced by topoisomerases. As occurs in prokaryotes, DNA replication and transcription in eukaryotes creates supercoils downstream as the double helix unwinds and becomes accessible to the appropriate enzyme. These enzymes may well play still other genetic roles involving eukaryotic DNA conformational changes.

Mitochondrial and chloroplast DNA resembles prokaryotic DNA

Numerous studies have demonstrated that both **mitochondria** and **chloroplasts** contain their own genetic information. This was first suggested by the discovery of mutations in yeast, other fungi, and plants that produced altered phenotypes that could be linked to these organelles (see Chapter 8). Furthermore, the transmission of these mutations did not usually demonstrate the *biparental* inheritance patterns characteristic of nuclear genes. Instead, a *uniparental* mode of inheritance was observed, with traits being passed to offspring only from their mother. Because the inheritance of both mitochondria and chloroplasts is usually maternal, these observations suggested that the organelles might house their own DNA that, when mutated, might be responsible for the altered phenotypes.

Thus, geneticists set out to look for more direct evidence of DNA in these organelles. Electron microscopists not only documented the presence of DNA in both organelles, but they also saw DNA in a form quite unlike that seen in the nucleus of the eukaryotic cells that house these organelles. This DNA looked remarkably similar to that seen in viruses and bacteria! This similarity, along with other observations, led to the idea that mitochondria and chloroplasts arose independently more than a billion years ago from free-living, prokaryotic-like organisms that possessed the abililty to undergo aerobic respiration or photosynthesis, respectively. This theory, championed by Lynn Margulis and others and called the **endosymbiotic hypothesis,** proposes that the prokaryotes were engulfed by larger primitive eukaryotic cells, which lacked these bioenergetic functions. A symbiotic relationship developed whereby the prokaryotic organisms eventually lost their ability to function independently, while the eukaryotic host cells gained the ability to either respire aerobically or undergo photosynthesis. Although there are many unanswered questions, the basic tenets of this theory are widely accepted.

Molecular Organization and Function of Mitochondrial DNA

Extensive information is now available about the molecular aspects and gene function of **mitochondrial DNA (mtDNA).** In most eukaryotes, mtDNA exists as a double-stranded closed circle (Figure 19.5) that replicates semiconservatively and is free of the chromosomal proteins characteristic of eukaryotic DNA. In size, mtDNA differs

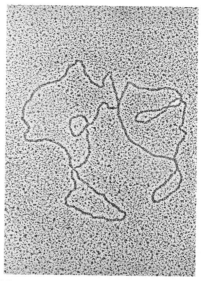

FIGURE 19.5 Electron micrograph of mitochondrial DNA (mtDNA) derived from *Xenopus laevis.*

TABLE 19.2 The Size of mtDNA in Different Organisms

Organism	Size in Kilobases
Human	16.6
Mouse	16.2
Xenopus (frog)	18.4
Drosophila (fruit fly)	18.4
Saccharomyces (yeast)	75.0
Pisum sativum (pea)	110.0
Arabidopsis (mustard plant)	367.0

TABLE 19.3 Sedimentation Coefficients of Mitochondrial Ribosomes

Kingdom	Examples	Sedimentation Coefficient (*S*)
Animals	Vertebrates	55–60
	Insects	60–71
Protists	*Euglena*	71
	Tetrahymena	80
Fungi	*Neurospora*	73–80
	Saccharomyces	72–80
Plants	Maize	77

greatly among organisms, as demonstrated in Table 19.2. In a variety of animals, including humans, mtDNA consists of about 16,000 to 18,000 base pairs (16–18 kb). However, yeast (*Saccharomyces*) contains 75 kb, while up to 387 kb may be present in plant mitochondria such as *Arabidopsis*. In vertebrates, there are 5 to 10 such DNA molecules per organelle, while in plants there may be 20–40 copies per organelle.

We can say several things about mtDNA. With only rare exceptions, there appears to be few or no gene repetitions, and replication is dependent upon enzymes encoded by nuclear DNA. Mitochondrial genes have been identified that code for the ribosomal RNAs, over 20 tRNAs, as well as numerous protein products essential to the cellular respiratory functions of the organelles.

Even so, the protein-synthesizing apparatus and the molecular components for cellular respiration are jointly derived from nuclear and mitochondrial genes. Ribosomes found in the organelle are different from those present in the neighboring cytoplasm. The data in Table 19.3 show

that mitochondrial ribosomes vary considerably in their sedimentation coefficients among different species, ranging from 55*S* to 80*S*. This is another piece of evidence in support of the endosymbiont hypothesis.

Many nuclear-coded gene products are essential to biological activity within mitochondria: DNA and RNA polymerases, initiation and elongation factors essential for translation, ribosomal proteins, aminoacyl tRNA synthetases, and some tRNA species. These imported components are distinct from their cytoplasmic counterparts, even though both sets are coded by nuclear genes. For example, the synthetase enzymes essential for charging mitochondrial tRNA molecules (a process essential to translation) show a distinct affinity for the mitochondrial tRNA species as compared to the cytoplasmic tRNAs. Similar affinity has been shown for the initiation and elongation factors. Furthermore, while bacterial and nuclear RNA polymerases are known to be composed of numerous subunits, the mitochondrial variety consists of only one polypeptide chain.

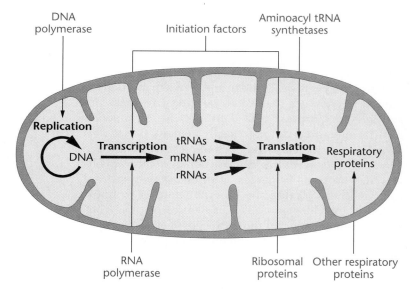

FIGURE 19.6 A comparison of the origin of gene products that are essential to mitochondrial function. Those shown entering the organelle are derived from the cytoplasm and encoded by the nucleus.

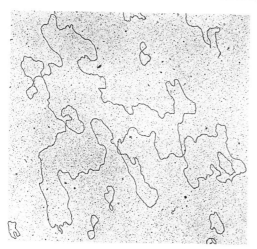

FIGURE 19.7 Electron micrograph of chloroplast DNA derived from lettuce.

Some chloroplast gene products function to synthesize proteins. In a variety of higher plants (beans, lettuce, spinach, maize, and oats), two sets of the genes coding for the ribosomal RNAs—5*S*, 16*S*, and 23*S* rRNA—are present. Additionally, chloroplast DNA codes for at least 25 tRNA species and a number of ribosomal proteins specific to the chloroplast ribosomes. These ribosomes have a sedimentation coefficient slightly less than 70*S*, similar to that of bacteria. Even though chloroplast ribosomal proteins are encoded by both nuclear and chloroplast DNA, most, if not all, such proteins are distinct from their counterparts in cytoplasmic ribosomes.

Still other chloroplast genes have been identified that are specific to the photosynthetic function. Mutations in these genes may have the effect of inactivating photosynthesis in those chloroplasts bearing such a mutation. One of the major photosynthetic enzymes is ribulose-1-5-bisphosphate carboxylase (RuBP). Interestingly, the small subunit of this enzyme is encoded by a nuclear gene, whereas the large subunit is encoded by cpDNA.

This polymerase is generally susceptible to antibiotics that inhibit bacterial RNA synthesis but not to eukaryotic inhibitors. The contributions of nuclear and mitochondrial gene products are contrasted in Figure 19.6.

Molecular Organization and Function of Chloroplast DNA

Chloroplasts, like mitochondria, contain an autonomous genetic system distinct from that found in the nucleus and cytoplasm. This system includes DNA as a source of genetic information and a complete protein-synthesizing apparatus. However, the molecular components of the translation apparatus are jointly derived from both nuclear and chloroplast genetic information. **Chloroplast DNA (cpDNA),** shown in Figure 19.7, is much larger than mitochondrial DNA; nevertheless, it too is similar to that found in prokaryotic cells. It is circular, double stranded, replicated semiconservatively, and free of the associated proteins characteristic of eukaryotic DNA. Compared with nuclear DNA of the same organism, it invariably shows a different buoyant density and base composition.

In the green alga *Chlamydomonas*, there are about 75 copies of the chloroplast DNA molecule per organelle. Each copy consists of a length of DNA that contains 195,000 bases (195 kb). In higher plants such as the sweet pea, multiple copies of the DNA molecule are present in each organelle, but the molecule is considerably smaller than that in *Chlamydomonas*, consisting of 134 kb. Interestingly, genetic recombination between the multiple copies of DNA within chloroplasts has been documented in *Chlamydomonas*.

Specialized chromosomes reveal variations in structure

We now turn to the consideration of two cases of genetic organization that demonstrate the specialized forms that chromosomes may take. Both types, polytene chromosomes and lampbrush chromosomes, are so large that their organization was discerned using light microscopy long before we understood how mitotic chromosomes are formed from interphase chromatin (see Chapter 2). The study of these chromosomes provided many of our initial insights into the arrangement and function of the genetic information.

Polytene Chromosomes

Giant **polytene chromosomes** are found in various tissues (salivary, midgut, rectal, and malpighian excretory tubules) in the larvae of some flies and in several species of protozoans and plants. Such structures were first observed by E. G. Balbiani in 1881. The large amount of information obtained from studies of these genetic structures provided a model system for subsequent investigations of chromosomes. What is particularly intriguing about polytene chromosomes is that they can be seen in the nuclei of interphase cells.

Each polytene chromosome observed under the light microscope reveals a linear series of alternating bands and interbands (Figure 19.8). The banding pattern is distinc-

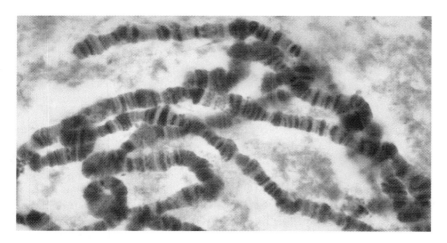

FIGURE 19.8 Polytene chromosomes derived from larval salivary gland cells of *Drosophila*.

tive for each chromosome in any given species. Individual bands are sometimes called **chromomeres,** a more generalized term describing lateral condensations of material along the axis of a chromosome. Each polytene chromosome is 200 to 600 μm long.

Extensive study using electron microscopy and radioactive tracers led to an explanation for the unusual appearance of these chromosomes. First, polytene chromosomes represent paired homologs. This is highly unusual in most organisms, since they are found in somatic cells, where chromosomal material is normally dispersed as chromatin and homologs are not paired. Second, their large size and distinctiveness result from the many DNA strands that compose them. The DNA of these paired homologs undergoes many rounds of replication, but without strand separation or cytoplasmic division. As replication proceeds, chromosomes are created having 1000 to 5000 DNA strands that remain in parallel register with one another. It is apparently the parallel register of so many DNA strands that gives rise to the distinctive band pattern along the axis of the chromosome.

The relationship between the structure of polytene chromosomes and the genes contained within them is intriguing. The presence of bands was initially interpreted as the visible manifestation of individual genes. The discovery that the strands present in bands undergo localized uncoiling during genetic activity further strengthened this view. Each such uncoiling event results in what is called a

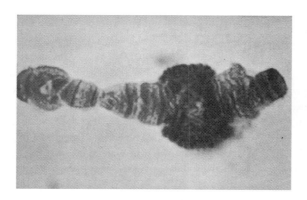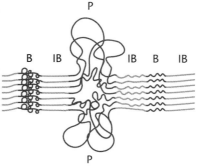

FIGURE 19.9 Photograph of a puff within a polytene chromosome. The schematic representation depicts the uncoiling of strands within a band (B) region to produce a puff (P) in polytene chromosomes. Interband regions (IB) are also labeled.

puff because of its appearance (see Figure 19.9). That puffs are visible manifestations of gene activity (transcription that produces RNA) is evidenced by their high rate of incorporation of radioactively-labeled RNA precursors, as assayed by **autoradiography.** Bands that are not extended into puffs incorporate many fewer radioactive precursors or none at all.

The study of bands during development in insects, such as *Drosophila* and the midge fly *Chironomus*, reveals differential gene activity. A characteristic pattern of band formation that is equated with gene activation is observed as development proceeds. Despite attempts to resolve the issue of the number of genes contained within each band, it is still not clear how many there are. It is known that a band may contain up to 10^7 base pairs of DNA, certainly enough DNA to encode 50–100 average-size genes.

Lampbrush Chromosomes

Another type of specialized chromosome that has provided insights into chromosomal structure is the **lampbrush chromosome,** so named because its appearance is similar to the brushes used to clean kerosene lamp chimneys in the nineteenth century. Lampbrush chromosomes were first discovered in 1892 in the oocytes of sharks and are now known to be characteristic of most vertebrate oocytes as well as the spermatocytes of some insects. Therefore, they are meiotic chromosomes. Most of the experimental work has been done with material taken from amphibian oocytes.

These unique chromosomes are easily isolated from oocytes in the diplotene stage of the first prophase of meiosis, where they are active in directing the metabolic activities of the developing cell. The homologs are seen as synapsed pairs held together by chiasmata. However, instead of condensing, as most meiotic chromosomes do, lampbrush chromosomes are often extended to lengths of 500 to 800 μm. Later in meiosis they revert to their normal length of 15 to 20 μm. Based on these observations, lampbrush chromosomes are interpreted as extended, uncoiled versions of the normal meiotic chromosomes.

The two views of lampbrush chromosomes in Figure 19.10 provide significant insights into their morphology In part (a), the meiotic configuration, as described, is seen under the light microscope. The linear axis of each structure contains a large number of condensed areas, referred to generally as **chromomeres,** that are repeated along the axis. Emanating from each chromomere is a pair of **lateral loops,** which give the chromosome its distinctive appearance. In part

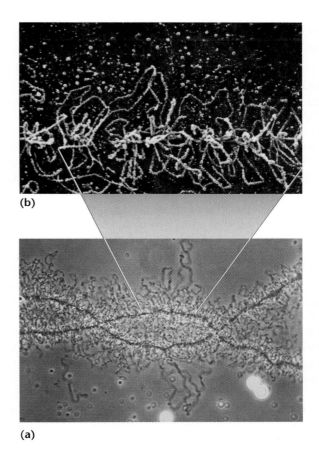

(b)

(a)

FIGURE 19.10 Lampbrush chromosomes derived from amphibian oocytes. Part (a) is a photomicrograph; part (b) is a scanning electron micrograph.

(b), the scanning electron micrograph is an enlargement that reveals adjacent loops present along one of the two axes of the chromosome. As with bands in polytene chromosomes, there is much more DNA present in each loop than is needed to encode a single gene. This SEM provides a clear view of the chromomeres and the chromosomal fibers emanating from them. Each chromosomal loop is thought to be composed of one DNA double helix, while the central axis is made up of two DNA helices. This hypothesis is consistent with the belief that each meiotic chromosome is composed of a pair of sister chromatids. Studies using radioactive RNA precursors reveal that the loops are active in the synthesis of RNA. The lampbrush loops, in a way similar to puffs in polytene chromosomes, represent DNA that has been reeled out from the central chromomere axis during transcription.

DNA is organized into chromatin in eukaryotes

The structure and organization of the genetic material in eukaryotic cells is much more intricate than in viruses or bacteria. This complexity is due to the greater amount of DNA per chromosome and the presence of a large number of proteins associated with eukaryotic DNA. For example, while DNA in the *E. coli* chromosome is 1200 μm long, the DNA in each human chromosome ranges from 19,000 to 73,000 μm in length. In a single human nucleus, all 46 chromosomes contain sufficient DNA to extend almost 2 meters. This genetic material, along with its associated proteins, is contained within a nucleus that usually measures about 5–10 μm in diameter.

Such intricacy parallels the structural and biochemical diversity of the many types of cells present in a multicellular eukaryotic organism. Different cells assume specific functions based upon highly specific biochemical activity. While all cells carry a full genetic complement, different cells activate different sets of genes, so a highly ordered regulatory system governing the readout of this information must exist. Such a system must in some way be imposed on or related to the molecular structure of the genetic material.

Although bacteria can reproduce themselves, they never exhibit a complex process similar to mitosis. As described in Chapter 2, eukaryotic cells exhibit a highly organized cell cycle. During interphase, the genetic material and associated proteins are uncoiled and dispersed throughout the nucleus as **chromatin.** When mitosis begins, the chromatin condenses greatly, and during prophase it is compressed into recognizable chromosomes. This condensation represents a contraction in length of some 10,000 times for each chromatin fiber and is the basis of the **folded-fiber model** of chromosome structure (Figure 2.18). This highly regular cycle of condensation and uncoiling creates special organizational problems in eukaryotic genetic material.

Because of the limitation of light microscopy, early studies of the structure of eukaryotic genetic material concentrated on intact chromosomes, preferably large ones, such as the polytene and lampbrush chromosomes that we discuss in this chapter. Subsequently, new techniques for biochemical analysis, as well as the examination of relatively intact eukaryotic chromatin and mitotic chromosomes under the electron microscope, have greatly enhanced our understanding of chromosome structure.

Chromatin Structure and Nucleosomes

As we have seen, the genetic material of viruses and bacteria consists of strands of DNA or RNA nearly devoid of proteins. In eukaryotic chromatin, a substantial amount of protein is associated with the chromosomal DNA in all phases of the eukaryotic cell cycle. The associated proteins are divided into basic, positively charged **histones** and less positively charged **nonhistones.** Of the proteins associated with DNA, the histones clearly play the most essential structural role. Histones contain large amounts of the positively charged amino acids lysine and arginine, making it possible for them to bond electrostatically to the negatively charged phosphate groups of nucleotides. Recall that a similar interaction has been proposed for several bacterial proteins. There are five main types of histones (Table 19.4).

The general model for chromatin structure is based on the assumption that chromatin fibers, composed of DNA and protein, undergo extensive coiling and folding as they are condensed within the cell nucleus. X-ray diffraction studies confirm that histones play an important role in chromatin structure. Chromatin produces regularly spaced diffraction rings, suggesting that repeating structural units occur along the chromatin axis. If the histone molecules are chemically removed from chromatin, the regularity of this diffraction pattern is disrupted.

A basic model for chromatin structure was worked out in the mid-1970s. Several observations were particularly relevant to the development of this model.

1. Digestion of chromatin by certain endonucleases, such as micrococcal nuclease, yields DNA fragments that are approximately 200 base pairs in length, or multiples thereof. This demonstrates that enzymatic digestion is not random, for if it were, we would expect a wide range of fragment sizes. Thus, chromatin consists of some type of repeating unit, each of which is protected from enzymatic cleavage, except where any two units are joined. It is the area between units that is attacked and cleaved by the nuclease.

2. Electron microscopic observations of chromatin have revealed that chromatin fibers are composed of linear arrays of spherical particles (Figure 19.11). Discovered by Ada and Donald Olins, the particles occur regularly along the axis of a chromatin strand and resemble beads on a string. These particles, initially referred to as *v*-bodies (*v* is the Greek letter nu), are

TABLE 19.4 Categories and Properties of Histone Proteins

Histone Type	Lysine-Arginine Content	Molecular Weight (daltons)
H1	Lysine-rich	23,000
H2A	Slightly lysine-rich	14,000
H2B	Slightly lysine-rich	13,800
H3	Arginine-rich	15,300
H4	Arginine-rich	11,300

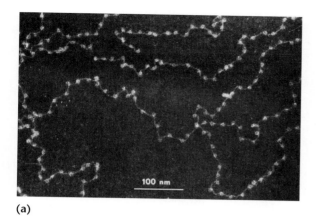

(a)

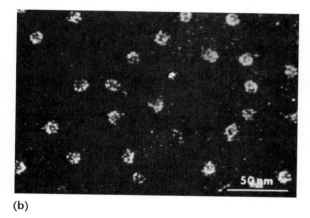

(b)

FIGURE 19.11 (a) Dark-field electron micrograph of nucleosomes present in chromatin derived from a chicken erythrocyte nucleus. (b) Dark-field electron micrograph of nucleosomes produced by micrococcal nuclease digestion.

now called **nucleosomes.** These findings conform nicely to the earlier proposal, which suggests the existence of repeating units.

3. Studies of precise interactions of histone molecules and DNA in the nucleosomes constituting chromatin show that histones H2A, H2B, H3, and H4 occur as two types of tetramers, $(H2A)_2 \bullet (H2B)_2$ and $(H3)_2 \bullet (H4)_2$. Roger Kornberg predicted that each repeating nucleosome unit consists of one of each tetramer in association with about 200 base pairs of DNA. Such a structure is consistent with previous observations and provides the basis for a model that explains the interaction of histones and DNA in chromatin.

4. When nuclease digestion time is extended, some of the 200 base pairs of DNA are removed from the nucleosome, creating what is called a **nucleosome core particle** consisting of 196 base pairs. This number is consistent in all organisms studied. The DNA lost in this prolonged digestion is responsible for linking nucleosomes together. This **linker DNA** is associated with a fifth histone, H1.

5. On the basis of this information, as well as on X-ray and neutron-scattering analyses of crystallized core particles by John T. Finch, Aaron Klug, and others, a detailed model of the nucleosome was put forward in 1984. In this model, the 196-bp DNA core is coiled around an octamer of histones in a left-handed superhelix, which completes about 1.7 turns per nucleosome.

The extensive investigation of nucleosomes now provides the basis for predicting how the the chromatin fiber within the nucleus is formed, and how it coils up into a mitotic chromosome. This model is illustrated in Figure 19.12. The 2-nm DNA molecule is initially coiled into a nucleosome that is about 11 nm in diameter [Figure 19.12(a)], consistent with the longer dimension of the ellipsoidal nucleosome. Significantly, the formation of the nucleosome represents the first level of packing, whereby the DNA helix is reduced to about one-third of its original length.

In the nucleus, the chromatin fiber seldom, if ever, exists in the extended form described here. Instead, the 11-nm chromatin fiber is further packed into a thicker 30-nm fiber, that was initially called a **solenoid** [Figure 19.12(b)]. This larger fiber consists of numerous nucleosomes coiled closely together, creating the second level of packing. The exact details of this structure are not completely clear, but 30-nm chromatin fibers are characteristically seen under the electron microscope.

In the transition to the mitotic chromosome, still another level of packing occurs. The 30-nm fiber forms a series of looped domains that further condense the structure into the chromatin fiber, which is 300 nm in diameter [Figure 19.12(c)]. The fibers are then coiled into the chromosome arms that constitute a chromatid, which is part of the metaphase chromosome [Figure 19.12(d)]. While we show the chromatid arms to be 700 nm in diameter, this value undoubtedly varies among different organisms. At a value of 700 nm, a pair of sister chromatids making up a chromosome measures about 1900 nm.

More recent insights into the structure of the nucleosome were forthcoming in 1997 when Timothy Richmond and members of his research team were able to significantly improve the level of resolution in X-ray diffraction studies of nucleosome crystals (from 7Å in the 1984 studies to 2.8Å in the 1997 studies). One model based on their work is featured on the cover of this text. At this resolution, most atoms are visible, thus revealing the subtle twists and turns of the superhelix of DNA that encircles the histones. Further, details of the location of each

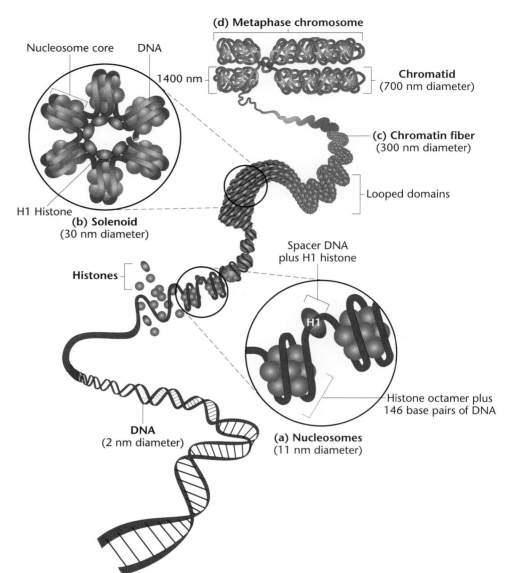

FIGURE 19.12 General model of the association of histones and DNA in the nucleosome, illustrating the way in which the chromatin fiber may be coiled into a more condensed structure, ultimately producing a mitotic chromosome. ⊙GenCDX

histone entity have been revealed. Of particular interest is the observation that there are several unstructured histone tails that are not packed into the folded histone domains within the core of the nucleosome. For example, tails devoid of any secondary structure extending from histones H3 and H2B protrude through the minor groove channels of the DNA helix. The significance of these and other histone tails is that they are seen to interact with neighboring structures, creating nucleosome to nucleosome interactions. These findings add further insight to our knowledge of the structure of chromatin and will help unravel how various genetic functions such as transcription can occur on a DNA molecule that is complexed with histones.

The importance of the organization of DNA into chromatin and chromatin into mitotic chromosomes can

be illustrated by considering a human cell that stores its genetic material in a nucleus that is about 5–10 μm in diameter. The haploid genome contains 3.2×10^9 base pairs of DNA distributed among 23 chromosomes. The diploid cell contains twice that amount. At 0.34 nm per base pair, this amounts to a length of DNA of almost 2 meters! One estimate suggests that about 25×10^6 nucleosomes per nucleus are complexed with this length of DNA.

In the overall transition from a fully extended DNA helix to the extremely condensed status of the mitotic chromosome, a packing ratio (the ratio of DNA length to the length of the structure containing it) of about 500 to 1 must be achieved. In fact, our model accounts for a ratio only of about 50 to 1. Obviously, the larger fiber can be further bent, coiled, and packed as even greater condensation occurs during the formation of a mitotic chromosome.

Heterochromatin

Evidence that the DNA of each eukaryotic chromosome consists of one continuous double-helical fiber along its entire length might suggest that the whole chromosome is structurally uniform. However, in the early part of this century, it was observed that some parts of the chromosome remain condensed and stain deeply during interphase, but most are uncoiled and do not stain. In 1928, the terms **heterochromatin** and **euchromatin** were coined to describe the parts of chromosomes that remain condensed and those that are uncoiled, respectively.

Subsequent investigation revealed a number of characteristics that distinguish heterochromatin from euchromatin. Heterochromatic areas are genetically inactive because they either lack genes or contain genes that are repressed. Also, heterochromatin replicates later during the S phase of the cell cycle than euchromatin does. The discovery of heterochromatin provided the first clues that parts of eukaryotic chromosomes do not always encode proteins. Instead, some chromosome regions are thought to be involved in maintenance of the chromosome's structural integrity and in other functions, such as chromosome movement during cell division.

Heterochromatin is characteristic of the genetic material of eukaryotes. Early cytological studies showed that areas of the centromeres are composed of heterochromatin. The ends of chromosomes, called telomeres, are also heterochromatic. In some cases, in fact, whole chromosomes are heterochromatic. Such is the case with the mammalian Y chromosome, which for the most part is genetically inert. And, as we discussed in Chapter 9, the inactivated X chromosome in mammalian females is condensed into an inert heterochromatic Barr body. In some species, such as mealy bugs, all chromosomes of one entire haploid set are heterochromatic.

When certain heterochromatic areas from one chromosome are translocated to a new site on the same or another nonhomologous chromosome, genetically active areas sometimes become genetically inert if they now lie adjacent to the translocated heterochromatin. As we saw in Chapter 4, this influence on existing euchromatin is one example of what is more generally referred to as a **position effect.** That is, the position of a gene or groups of genes relative to all other genetic material may affect their expression.

Chromosome banding differentiates regions along the mitotic chromosome

Until about 1970, mitotic chromosomes viewed under the light microscope could be distinguished only by their relative sizes and the positions of their centromeres. Even in organisms with a low haploid number, two or more chromosomes are often indistinguishable from one another. However, new cytological procedures made possible differential staining along the longitudinal axis of mitotic chromosomes. Such methods are now called **chromosome banding techniques** because the staining patterns resemble the bands of polytene chromosomes.

One of the first chromosome banding techniques was devised by Mary Lou Pardue and Joe Gall. They found that if chromosome preparations were heat denatured and then treated with Giemsa stain, a unique staining pattern emerged. The centromeric regions of mitotic chromosomes preferentially took up the stain! Thus, this cytological technique stained a specific area of the chromosome composed of heterochromatin. A diagram of the mouse karyotype treated in this way is shown in Figure 19.13. The staining pattern is referred to as **C-banding.**

Other chromosome banding techniques were developed about the same time. A group of Swedish researchers, led by Tobjorn Caspersson, used a technique that provided even greater staining differentiation of metaphase chromosomes. They employed fluorescent dyes that bind to nucleoprotein complexes and produce unique banding patterns. When the chromosomes are treated with the fluorochrome quinacrine mustard and viewed under a fluorescent microscope, precise patterns of differential brightness are seen. Each of the 23 human chromosome pairs can be distinguished by this technique. The bands produced by this method are called **Q-bands.**

Yet another banding technique produces a staining pattern nearly identical to the Q-bands. This method, pro-

FIGURE 19.13 Karyotypes of a male (top) and female (bottom) mouse where chromosome preparations were processed to demonstrate C-banding, where only the centromeres stain.

ducing **G-bands** (Figure 19.14), involves the digestion of the mitotic chromosomes with the proteolytic enzyme trypsin, followed by Giemsa staining. The variety of staining reactions under different conditions reflects the heterogeneity and complexity of chromosome composition. Another technique results in the reverse G-band staining pattern, called an **R-band** pattern.

In 1971, a meeting was held in Paris to establish a uniform nomenclature for human chromosome banding patterns based on G-banding. Figure 19.15 illustrates the application of this nomenclature to the X chromosome. On the left of the chromosome are the various levels of banding that can be identified, while the designations on the right side identify specific regions of the chromosome.

While the molecular mechanisms involved in producing the various banding patterns are still not clear, the significance of banding has been great in cytogenetic analysis, particularly in humans. The pattern of banding of each chromosome is unique, allowing a distinction to be made even between those chromosomes that are identical in size and centromere placement (e.g., human chromosomes 4 and 5 and chromosomes 21 and 22). So precise is the banding pattern of each chromosome that when a segment of one chromosome has been translocated to another chromosome, its origin can be determined with great precision.

Eukaryotic genomes demonstrate complex sequence organization characterized by repetitive DNA

Thus far, we have examined how DNA is organized *into* chromosomes in bacteriophages, bacteria, and eukaryotes.

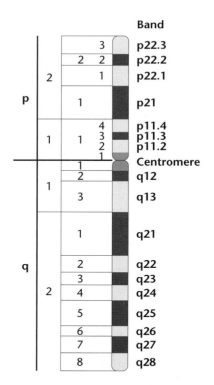

FIGURE 19.15   The regions of the human X chromosome distinguished by its banding pattern. The designations on the right identify specific bands.

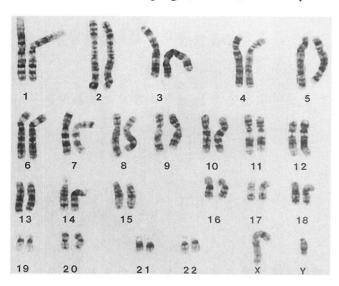

FIGURE 19.14   G-banded karyotype of a normal human male showing approximately 400 bands per haploid set of chromosomes. Chromosomes were derived from cells in metaphase.

We now begin an examination of what we know about the organization of DNA sequences *within* the chromosomes making up an organism's genome, placing our emphasis on eukaryotes. Once we establish the pattern of genome organization, we will turn in the subsequent chapter to a focus on how genes themselves are organized within chromosomes.

We learned in Chapter 11 that, in addition to single copies of unique DNA sequences that make up genes, a great deal of the DNA sequences within chromosomes is repetitive in nature, and that various levels of repetition occur within the genome of organisms. Many studies have now provided insights into repetitive DNA, demonstrating various classes of these sequences and their organization within the genome. Figure 19.16 outlines the various categories of repetitive DNA. As you can see, some functional genes are present in more than one copy and are therefore repetitive in nature. However, the majority of repetitive sequences are nongenic and, in fact, most serve no known function. We will explore three main categories: 1) heterochromatin found associated with centromeres and making up telomeres, 2) tandem repeats of both short and longer DNA sequences, and 3) transposable sequences that are interspersed throughout the genome of eukaryotes.

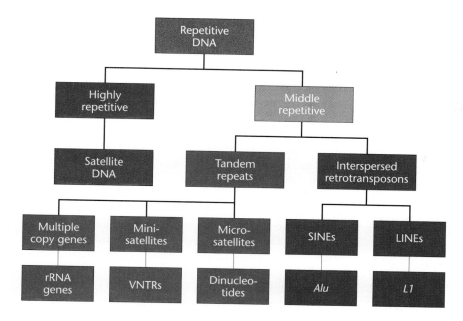

FIGURE 19.16   An overview of the various categories of repetitive DNA.

Repetitive DNA and Satellite DNA

The nucleotide composition of the DNA (e.g., the percentage of G $\equiv$ C versus A $=$ T pairs) of a particular species is reflected in its density, which can be measured with sedimentation equilibrium centrifugation (introduced in Chapter 11). When eukaryotic DNA is analyzed in this way, the majority of it is present as a single main peak or band of fairly uniform density. However, one or more additional peaks represent DNA that differs slightly in density. This component, called **satellite DNA,** represents a variable proportion of the total DNA, depending on the species. For example, a profile of main-band and satellite DNA from the mouse is shown in Figure 19.17. By contrast, prokaryotes contain only main-band DNA.

The significance of satellite DNA remained an enigma until the mid-1960s, when Roy Britten and David Kohne developed a technique for measuring the reassociation kinetics of DNA that had previously been dissociated into single strands (see Chapter 11). They demonstrated that certain portions of DNA reannealed more rapidly than others. They concluded that rapid reannealing was characteristic of multiple DNA fragments composed of identical or nearly identical nucleotide sequences—the basis for the descriptive term **repetitive DNA.**

When satellite DNA was subjected to analysis by reassociation kinetics, it fell into the category of *highly repetitive DNA* and is known to consist of short sequences repeated a large number of times. Further evidence suggested that these sequences are present as tandem repeats clustered in very specific chromosomal areas known to be heterochromatic—the regions flanking *centromeres.* This was discovered in 1969 when several researchers, including

Mary Lou Pardue and Joe Gall, applied the technique of *in situ* **molecular hybridization** to the study of satellite DNA. This technique (see Figure 11.22 and Appendix A) involves the molecular hybridization between an isolated fraction of radioactively labeled DNA or RNA probes and the DNA contained in the chromosomes of a cytological preparation. Following the hybridization procedure, autoradiography is performed to locate the chromosome areas complementary to the fraction of DNA or RNA.

In their work, Pardue and Gall demonstrated that RNA probes made from mouse satellite DNA hybridize

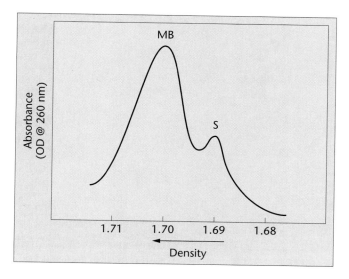

FIGURE 19.17   Separation of main-band (MB) and satellite (S) DNA from the mouse using ultracentrifugation in a CsCl gradient.

with DNA of centromeric regions of mouse mitotic chromosomes (Figure 19.18). Several conclusions can be drawn. Satellite DNA differs from main-band DNA in its molecular composition, as established by buoyant density studies. It is also composed of short repetitive sequences. Finally, satellite DNA is found in the heterochromatic centromeric regions of chromosomes.

Centromere DNA Sequences

Separation of chromatids is essential to the fidelity of chromosome distribution during mitosis and meiosis. Most estimates of infidelity during mitosis are exceedingly low: 1×10^{-5} to 1×10^{-6}, or 1 error per 100,000 to 1 million cell divisions. As a result, it has been generally assumed that the analysis of the DNA sequence of centromeric regions will provide insights into the rather remarkable features of this chromosomal region. This DNA region is designated the **CEN** and its function is now reasonably clear. Its structure within the heterochromatic region binds a platform of proteins forming the centromere, which includes the kinetochore that binds to the spindle fiber during division.

The analysis of the CEN regions of yeast *Saccharomyces cerevisiae* chromosomes provided the basis for a model system first described by John Carbon and Louis Clarke. Because each centromere serves an identical function, it is not surprising that all CENs were found to be remarkably similar in their organization. The CEN region of yeast chromosomes consists of about 225 base pairs, which can be divided into three regions (Figure 19.19). The first and third regions (I and III) are relatively short and highly conserved, consisting of only 8 and 26 base pairs, respectively. Region II, which is larger (80–85 base

pairs) and extremely A-T rich (up to 95%), varies in sequence among different chromosomes.

Mutational analysis suggests that regions I and II are less critical to centromere function than is region III. Mutations in the former regions are often tolerated, but mutations in region III can disrupt centromere function. While the DNA of this region appears to be essential to the eventual binding to the spindle fiber, DNA sequences are not unique to specific chromosomes. They can be exchanged between chromosomes experimentally without altering centromere function.

The amount of DNA associated with the centromeres of multicellular eukaryotes is much more extensive than in yeast. Recall from our discussion that highly repetitive "satellite" DNA is localized in the centromere regions of mice. Such sequences, absent from yeast but characteristic of most multicellular organisms, vary considerably in size. For example, the 10-bp sequence AATAACATAG is tandemly repeated many times in the centromeres of all four chromosomes of *Drosophila*. In humans, one of the most recognized satellite DNA sequences is the **alphoid family.** Found mainly in the centromere regions, alphoid sequences, each about 170 base pairs in length, are present in tandem arrays of up to 1 million base pairs.

The role of this highly repetitive DNA in centromere function remains unclear; it is known that the sequences are not transcribed. There is some sequence variation among members of the alphoid family; the number of repeats is specific to each human chromosome. Perhaps there is a shorter sequence similar to the alphoid family that, like yeast CEN DNA, is critical to centromere function.

Telomeric DNA Sequences

Another important structure component of chromosomes is the **telomere,** found at the ends of linear chromosomes. The function of telomeres is to provide stability to the chromosome by rendering chromosome ends generally inert in interactions with other chromosome ends. In contrast to broken chromosomes, whose ends may rejoin other such ends, telomere regions do not fuse with one another or with broken ends. It is thought that some aspect of the molecular structure of telomeres must be unique compared with most other chromosome regions.

As with centromeres, the analysis of telomeres was first approached by investigating the smaller chromosomes of simple eukaryotes, such as protozoans and yeast. The idea that all telomeres of all chromosomes in a given species might share a common nucleotide sequence has now been borne out.

Two types of telomere sequences have been discovered. The first type, called simply **telomeric DNA sequences,** consists of short tandem repeats. It is this group that contributes to the stability and integrity of the chro-

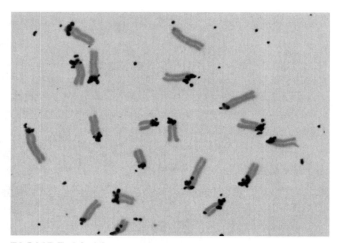

FIGURE 19.18 *In situ* hybridization between RNA transcribed by mouse satellite DNA and mitotic chromosomes. The grains in the autoradiograph localize the chromosome regions (the centromeres) containing satellite DNA sequences.

Centromere regions

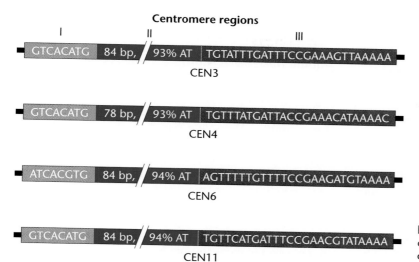

CEN3

CEN4

CEN6

CEN11

FIGURE 19.19 Nucleotide sequence information derived from DNA of the three major centromere regions of chromosomes 3, 4, 6, and 11 of yeast.

mosome. In the ciliate *Tetrahymena*, over 50 tandem repeats of the hexanucleotide sequence GGGGTT occur. In humans, the sequence GGGATT is repeated many times. The analysis of telomeric DNA sequences has shown them to be highly conserved throughout evolution, reflecting the critical role they play in maintaining the integrity of chromosomes.

The second type, **telomere-associated sequences,** is also repetitive and is found both adjacent to and within the telomere. These sequences vary among organisms, and their significance remains unknown.

Interestingly, the telomeres of *Drosophila* are fundamentally different from those found thus far in all other organisms, in that they are made up of transposable elements. The significance of this finding, however, is not yet clear.

As discussed in Chapter 12, replication of the telomere requires a unique RNA-containing enzyme **telomerase.** In its absence, the DNA at the ends of chromosomes becomes shorter during each replication. Because single-celled eukaryotes are immortalized cells, telomerase is critical for the survival of such species. In multicellular organisms, such as humans, telomerase is essential in germ-line cells, but is inactive in somatic cells. Chromosome shortening is considered part of the natural process of cell aging, serving as an internal clock. In human cancer cells, which have become immortalized, the transition to malignancy appears to require the activation of telomerase in order to overcome the normal senescence associated with chromosome shortening.

Middle Repetitive Sequences: VNTRs and Dinucleotide Repeats

A brief review of still another prominent category of repetitive DNA sheds more light on our understanding of the organization of the eukaryotic genome. In addition to highly repetitive DNA, which constitutes about 5 percent

of the human genome (and 10 percent of the mouse genome), a second category, **middle** (or **moderately**) **repetitive DNA,** recognized by C_0t analysis, is fairly well characterized. Because a great deal is being learned about the human genome, we will use our own species to illustrate this category of DNA in genome organization.

Middle repetitive DNA most prominently consists of either tandemly repeated or interspersed sequences. No function has been ascribed to these components of the genome. An example includes those called **variable number tandem repeats (VNTRs).** The repeating DNA sequence or VNTRs may be 15 to 100 base pairs long and may be found within and between genes. Many such clusters are dispersed throughout the genome, and they are often referred to as **minisatellites.**

The number of tandem copies of each specific sequence at each location varies in individuals, creating localized regions of 1000 to 5000 base pairs (1–5 kb) in length. As we will see in Chapter 21, the variation in size (length) of these regions between individuals in humans is the basis for the forensic technique referred to as **DNA fingerprinting.**

Another group of tandemly repeated sequences are dinucleotides, also referred to as **microsatellites.** Like VNTRs, they are dispersed throughout the genome and vary among individuals in the number of repeats present at any site. For example, in humans, the most common microsatellite is the dinucleotide $(CA)_n$, where *n* equals the number of repeats. Most commonly *n* is between 5 and 50. These clusters are also used forensically and, in addition, have served as useful molecular markers during genome analysis.

Repetitive Transposed Sequences: SINEs and LINEs

Still another category of repetitive DNA consists of sequences that are interspersed throughout the genome, rather than being tandemly repeated. They can be either short or long and many have the added distinction of being **transposable sequences,** which are mobile and can move to different locations within the genome. A large portion of eukaryotic genomes are composed of such sequences.

For example, **short interspersed elements,** called **SINES,** are less than 500 base pairs long and may be present 500,000 times or more in the human genome. The best characterized human SINE is a set of closely related sequences called the *Alu* **family** (the name is based on the presence of DNA recognition sequences by the restriction endonuclease *Alu*I). Members of this DNA family, also found in other mammals, are 200 to 300 base pairs long and are dispersed rather uniformly throughout the genome, both between and within genes. In humans, this family encompasses more than 5 percent of the entire genome.

Alu sequences are particularly interesting, although their function, if any, is yet undefined. Members of the *Alu* family are sometimes transcribed. The role of this RNA is not certain, but it is thought to be related to their mobility in the genome. In fact, *Alu* sequences are thought to have arisen from an RNA element whose DNA complement was dispersed throughout the genome as a result of the activity of reverse transcriptase (an enzyme that synthesizes DNA on an RNA template).

The group of **long interspersed elements (LINES)** represents still another category of repetitive transposable DNA sequences. In humans, the most prominent example is a family designated **L1.** Members of this sequence family are about 6400 base pairs long and are present up to 100,000 times, according to one estimate. Their 5′ end is highly variable, and their role within the genome has yet to be defined.

The basis for the transposition of L1 elements is now clear. The L1 DNA sequence is first transcribed into an RNA molecule. The RNA then serves as the template for the synthesis of the DNA complement using the enzyme reverse transcriptase. This enzyme is encoded by a portion of the L1 sequence. The new L1 copy then integrates into the DNA of the chromosome at a new site. Because of the similarity of this mechanism of transposition with that used by retroviruses, LINES are referred to as retrotransposons.

SINES and LINES represent a significant portion of human DNA. Both types of elements share the organizational feature of consisting of a mixture of about 70 percent unique and 30 percent repeating sequences within

the DNA of each entity. Collectively, they constitute about 10 percent of the genome.

Middle Repetitive Multiple Copy Genes

In some cases, middle repetitive DNA includes functional genes present tandemly in multiple copies. For example, many copies exist of the genes encoding ribosomal RNA. For example, *Drosophila* has 120 copies per haploid genome. Single genetic units encode a large precursor molecule that is processed into the 5.8*S*, 18*S*, and 28*S* rRNA components. In humans, multiple copies of this gene are clustered on the p arm of the acrocentric chromosomes 13, 14, 15, 21, and 22. Multiple copies of the genes encoding 5*S* rRNA are transcribed separately from multiple clusters found together on the terminal portion of the p arm of chromosome 1.

The vast majority of a eukaryotic genome does not encode functional genes

Taken together, the various forms of highly repetitive and moderately repetitive DNA comprise up to 40 percent of the human genome. Such an observation is not uncommon in eukaryotes. In addition to repetitive DNA, there is a large amount of single-copy DNA sequences as defined by C_0t analysis that appear to be noncoding. A small portion of them are called **pseudogenes,** which represent evolutionary vestiges of duplicated copies of genes that have undergone sufficient mutations to render them untranscrible. In some cases, multiple copies of such genes exist.

While the proportion of the genome consisting of repetitive DNA varies among organisms, one feature seems to be shared: *Only a very small part of the genome codes for proteins.* For example, the 20,000 to 30,000 genes encoding proteins in sea urchin occupy less than 10 percent of the genome. In *Drosophila,* only 5 to 10 percent of the genome is occupied by genes coding for proteins. In humans, it appears that the estimated 60,000 to 100,000 functional genes occupies less than 5 percent of the genome.

The study of the various forms of repetitive DNA has significantly enhanced our understanding of genome organization. In the next chapter, we will explore the organization of genes within chromosomes.

CHAPTER SUMMARY

1. Knowledge of the organization of the molecular components forming chromosomes is essential to the understanding of the function of the genetic material. Largely devoid of associated proteins, bacteriophage and bacterial chromosomes contain DNA molecules in a form equivalent to the Watson–Crick model.

2. Mitochondria and chloroplasts contain DNA that encodes products essential to their biological function. This DNA is remarkably similar in form and appearance to some bacterial and phage DNA, lending support to the endosymbiotic hypothesis, which suggests that these organelles were once free-living organisms.

3. Polytene and lampbrush chromosomes are examples of specialized structures that have extended our knowledge of genetic organization and function.

4. The eukaryotic chromatin fiber is a nucleoprotein organized into repeating units called nucleosomes. Composed of about 200 base pairs of DNA and an octamer of four types of histones, the nucleosome is important in facilitating the conversion of the extensive chromatin fiber characteristic of interphase into the highly condensed chromosome seen in mitosis.

5. The structural heterogeneity of the chromosome axis has been established as a result of both biochemical and cytological investigation. Heterochromatin, prematurely condensed in interphase, is genetically inert. The centromeric and telomeric regions, the Y chromosome, and the Barr body are examples.

6. DNA analysis has revealed unique nucleotide sequences in both the centromere and telomere regions of chromosomes, which no doubt imparts the characteristic they share as being heterochromatic.

7. Eukaryotic genomes demonstrate complex sequence organization characterized by numerous categories of repetitive DNA.

8. Repetitive DNA consists of either tandem repeats clustered in various regions of the genome or single sequences interspersed uniformly throughout the genome. In the former group, the size of each cluster varies among individuals, providing one form of biochemical identity. The latter group of sequences may be short or long, such as *Alu* and L1, respectively and are transposable elements.

9. The vast majority of a eukaryotic genome does not encode functional genes. In humans, for example, less than 5 percent of the genome is used to encode the 60,000 to 100,000 genes found in our genome.

INSIGHTS AND SOLUTIONS

A previously undiscovered single-celled organism was found living at a great depth on the ocean floor. Its nucleus contained only a single linear chromosome containing 7×10^6 nucleotide pairs of DNA coalesced with three types of histone-like proteins.

1. A short micrococcal nuclease digestion yielded DNA fractions consisting of 700, 1400, and 2100 base pairs. Predict what these fractions represent. What conclusions can be drawn?

   Solution: The chromatin fiber may consist of a variation of nucleosomes containing 700 base pairs of DNA. The 1400- and 2100-bp fractions represent two and three nucleosomes, respectively, linked together. Enzymatic digestion may have been incomplete, leading to the latter two fractions.

2. The analysis of individual nucleosomes revealed that each unit contained one copy of each protein, and that the short linker DNA contained no protein bound to it. If the entire chromosome consists of nucleosomes (discounting any linker DNA), how many are there, and how many total proteins are needed to form them?

   Solution: Since the chromosome contains 7×10^6 base pairs of DNA, the number of nucleosomes, each containing 7×10^2 base pairs, is equal to

   $$7 \times 10^6 / 7 \times 10^2 = 10^4 \text{ nucleosomes}$$

   The chromosome thus contains 10^4 copies of each of the three proteins, for a total of 3×10^4 molecules.

3. Analysis then revealed the organism's DNA to be a double helix similar to the Watson–Crick model, but containing 20 base pairs per complete turn of the right-handed helix. The physical size of the nucleosome was exactly double the volume occupied by that found in all other known eukaryotes, by virtue of increasing the distance along the fiber axis by a factor of two. Compare the degree of compaction of this organism's nucleosome to that found in other eukaryotes.

   Solution: The unique organism compacts a length of DNA consisting of 35 complete turns of the helix (700 base pairs per nucleosome/20 base pairs per turn) into each nucleosome. The normal eukaryote compacts a length of DNA consisting of 20 complete turns of the helix (200 base pairs per nucleosome/10 base pairs per turn) into a nucleosome one-half the volume of that in the unique organism. The degree of compaction is therefore less in the unique organism.

4. No further coiling or compaction of this unique chromosome occurs in the unique organism. Compare this to a eukaryotic chromosome. Do you think an interphase human chromosome 7×10^6 base pairs in length would be a shorter or longer chromatin fiber?

   Solution: The eukaryotic chromosome contains still another level of condensation in the form of "solenoids," which are dependent on the H1 histone molecule associated with linker DNA. Solenoids condense the eukaryotic fiber by still another factor of five.

   The length of the unique chromosome is compacted into 10^4 nucleosomes, each containing an axis length twice that of the eukaryotic fiber. The eukaryotic fiber consists of $7 \times 10^6 / 2 \times 10^2 = 3.5 \times 10^4$ nucleosomes, 3.5 more than the unique organism. However, they are compacted by the factor of five in each solenoid. Therefore, the chromosome of the unique organism is a longer chromatin fiber.

GENETICS MediaLab

The following resources will help you achieve a better understanding of the concepts presented in this chapter. These resources can be found on the CD packaged with this textbook and on the Companion Website at **http://www.prenhall.com/klug/**.

CD Resources:

Animated Tutorial: *Chromatin Structure*

Self-grading Chapter Problems

Web Resources:

Web Destinations in Genetics

Self-grading Chapter Problems

Chapter Search Terms

Genetics Newsgroups

Student Bulletin Board

Web Problem 1:

Time for completion = 10 minutes

How long is DNA? Stretched out end to end, the total length of DNA in the human genome would measure approximately 1.8 meters. Therefore, it is necessary for the cell to package DNA into a more compact chromosome. The linked website illustrates aspects of chromosomal organization. After examining this website, list the levels of chromosomal organization that exist in a eukaryotic cell. Discuss briefly what additional kinds of chromosomal organization exist along the chromosome length. To complete this exercise, visit Web Problem 1 in Chapter 19 of your Companion Website and select the keyword **NUCLEUS.**

Web Problem 2:

Time for completion = 10 minutes

What happens when chromosomes get wound up? Replication and recombination alter the level of supercoiling that occurs within a DNA molecule. Proteins called topoisomerases regulate the level of supercoiling by cutting, winding, or unwinding the DNA helix, and religating the DNA strands. In this exercise, you will examine molecular models describing how toposomerase I of *E. coli*, modulates DNA supercoiling. Directions for downloading the CHIME browser plug-in needed to view this module are found at the top of the linked website. Scroll to the Hall of Recombination and click on the topoisomerase button. In a few words, describe the physical structure of topoisomerase I. Examine the section of DNA cleavage and describe in a few sentences the model of how topoisomerase I allows DNA-strand passage. To complete this exercise, visit Web Problem 2 in Chapter 19 of your Companion Website and select the keyword **TOPOISOMERASE.**

Web Problem 3:

Time for completion = 10 minutes

What is the role of DNA in organelles? Although research has often focused on nuclear DNA, mitochondrial and chloroplast DNA provide significant functions within the eukaryotic cell. The linked website describes the mitochondrial and chloroplast genomes and suggests evidence supporting the endosymbiont hypothesis for the origin of these genomes. Describing the general features, briefly compare and contrast these two organellar genomes to the nuclear genome. Discuss briefly whether you think that the endosymbiosis hypothesis seems reasonable from the data you have available at this website. To complete this exercise, visit Web Problem 3 in Chapter 19 of your Companion Website and select the keyword **ORGANELLE.**

PROBLEMS AND DISCUSSION QUESTIONS

1. Compare and contrast the chemical nature, size, and form assumed by the genetic material of viruses and bacteria.
2. Contrast the DNA associated with mitochondria and chloroplasts.
3. How are giant polytene chromosomes formed?
4. What genetic process is occurring in a "puff" of a polytene chromosome? How do we know?
5. During what genetic process are lampbrush chromosomes present in vertebrates?
6. Why might it be predicted that the organization of eukaryotic genetic material would be more complex than that of viruses or bacteria?

7. Describe the sequence of research findings leading to the model of chromatin structure. What is the molecular composition and arrangement of the nucleosome? Describe the transitions that occur as nucleosomes are coiled and folded, ultimately forming a chromatid.
8. When chloroplasts and mitochondria are isolated, they are found to contain ribosomes that are similar to prokaryotic cells. How does this observation relate to the endosymbiotic hypothesis? What other evidence supports this hypothesis?
9. Provide a comprehensive definition of heterochromatin and list as many examples as you can.

10. Mammals contain a diploid genome consisting of at least 10^9 base pairs. If this amount of DNA is present as chromatin fibers where each group of 200 base pairs of DNA is combined with nine histones into a nucleosome, and each group of five nucleosomes is combined into a solenoid, achieving a final packing ratio of 50, determine:
    (a) The total number of nucleosomes in all fibers.
    (b) The total number of solenoids in all fibers.
    (c) The total number of histone molecules combined with DNA in the diploid genome.
    (d) The combined length of all fibers.

11. Assume that a viral DNA molecule is in the form of a 50-μm-long circular strand of a uniform 20 Å diameter. If this molecule is contained within a viral head that is a sphere with a diameter of 0.08 μm, will the DNA molecule fit into the viral head, assuming complete flexibility of the molecule? Justify your answer mathematically.

12. How many base pairs are in a molecule of phage T2 DNA, which is 52 μm long?

EXTRA-SPICY PROBLEMS

13. If a human nucleus is 10 μm in diameter, and it must hold as much of 2 meters as DNA, which is complexed into nucleosomes that during full extension are 11 nm in diameter, what percentage of the volume of the nucleus is occupied by the genetic material?

SELECTED READINGS

ANGELIER, N. et al. 1984. Scanning electron microscopy of amphibian lampbrush chromosomes. *Chromosoma* 89:243–53.

BAUER, W.R., CRICK, F.H.C., and WHITE, J.H. 1980. Supercoiled DNA. *Sci. Am.* (July) 243:118–33.

BEERMAN, W., and CLEVER, U. 1964. Chromosome puffs. *Sci. Am.* (Apr.) 210:50–58.

CALLAN, H.G. 1986. *Lampbrush chromosomes.* New York: Springer-Verlag.

CARBON, J. 1984. Yeast centromeres: Structure and function. *Cell* 37:352-53.

CHEN, T.R., and RUDDLE, F.H. 1971. Karyotype analysis utilizing differential stained constitutive heterochromatin of human and murine chromosomes. *Chromosoma* 34:51–72.

CORNEO, G. et al. 1968. Isolation and characterization of mouse and guinea pig satellite DNA. *Biochemistry* 7:4373–79.

DUPRAW, E.J. 1970. *DNA and chromosomes.* New York: Holt, Rinehart & Winston.

GALL, J.G. 1963. Kinetics of deoxyribonuclease on chromosomes. *Nature* 198:36–38.

————1981. Chromosome structure and the *C*-value paradox. *J. Cell Biol.* 91:3s–14s.

GREEN, B.R., and BURTON, H. 1970. *Acetabularia* chloroplast DNA: Electron microscopic visualization. *Science* 168:981–82.

HEWISH, D.R., and BURGOYNE, L. 1973. Chromatin sub-structure. The digestion of chromatin DNA at regularly spaced sites by a nuclear deoxyribonuclease. *Biochem. Biophys. Res. Comm.* 52:504–10.

HILL, R.J., and RUDKIN, G.T. 1987. Polytene chromosomes: The status of the band–interband question. *BioEssays* 7:35–40.

HSU, T.C. 1973. Longitudinal differentiation of chromosomes. *Annu. Rev. Genet.* 7:153–77.

JEFFREYS, A.J., WILSON, V., and THEIN, S.L. 1985. Hypervariable minisatellite regions in human DNA. *Nature* 314:66-73.

KORENBERG, J.R. and RYKOWSKI, M.C. 1988. Human genome organization: *Alu*, LINES, and the molecular organization of metaphase chromosome bands. *Cell* 53:391–400.

KORNBERG, R.D. 1975. Chromatin structure: A repeating unit of histones and DNA. *Science* 184:868–71.

KORNBERG, R.D., and KLUG, A. 1981. The nucleosome. *Sci. Am.* (Feb.) 244:52–64.

LORCH, Y., ZHANG, N, and KORNBERG, R.D. 1999. Histone octamer transfer by a chromatin-remodeling complex. *Cell.* 96:389–92.

LUGER, K. et. al. 1997. Crystal structure of the nucleosome core particle at 2.8 Å resolution. *Nature* 389:251–56.

MOYZIS, R.K. 1991. The human telomere. *Sci. Am.* (Aug) 265:48–55.

OLINS, A.L., and OLINS, D.E. 1974. Spheroid chromatin units (*v* bodies). *Science* 183:330–32.

————1978. Nucleosomes: The structural quantum in chromosomes. *Am. Sci.* 66: 704–11.

SINGER, M.F. 1982. SINEs and LINEs: Highly repeated short and long interspersed sequences in mammalian genomes. *Cell* 28:433–34.

VAN HOLDE, K.E. 1989. *Chromatin.* New York: Springer-Verlag.

VERMA, R.S. ed. 1988. *Heterochromatin: Molecular and structural aspects.* Cambridge: Cambridge University Press.

YUNIS, J.J. 1976. High resolution of human chromosomes. *Science* 191:1268–70.

YUNIS, J.J., and PRAKASH, O. 1982. The origin of man: A chromosomal pictorial legacy. *Science* 215:1525–30.

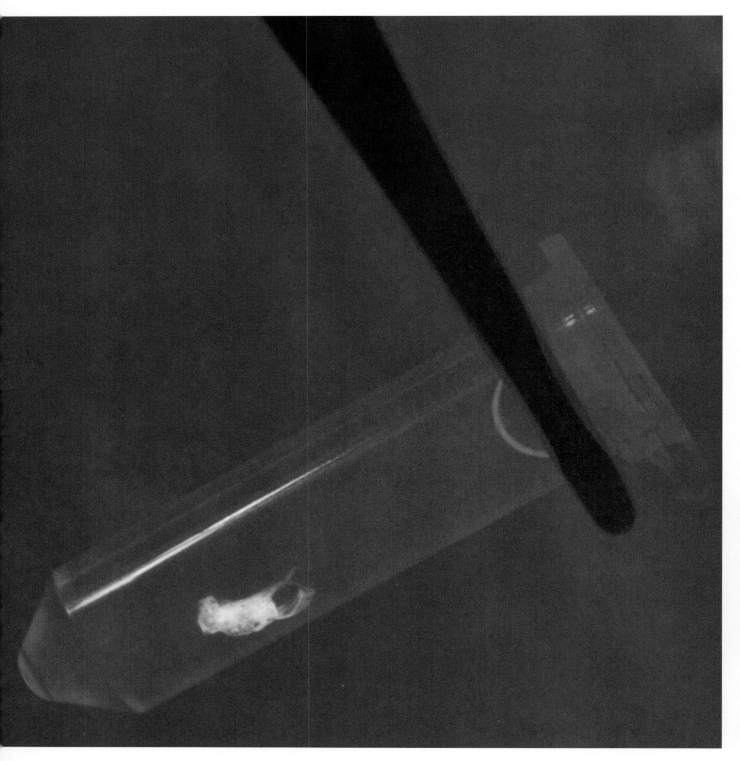

A pellet of DNA prior to molecular analysis.

CHAPTER

Organization of Genes in Chromosomes

20

KEY CONCEPTS

- **Genome analysis provides many insights into genetic organization**

- **Bacterial chromosomes are small, circular, and densely packed with protein-coding genes, mostly organized in polycistronic transcription units without introns**

- **Bacteria show us the minimum genome size necessary for life**

- **Archaea have characteristics of both eubacteria and eukaryotes**

- **Eukaryotic chromosomes are large, linear, and less densely packed with protein-coding genes, mostly organized as single transcription units with introns**

  Eukaryotic Genes Compared to Prokaryotic Genes
  The Difference between Single-Cell Eukaryotes and
    Multicellular Eukaryotes
  The Gene Organization of Higher Plants

- **rRNA and tRNA gene organization is different from mRNA gene organization**

- **Multigene families include the globin genes, immunoglobulin genes, and histone genes**

  The Globin Gene Family
  The Immunoglobulin Family
  The Histone Gene Family

- **How does gene number change with advancing phylogeny?**

- **The Human Genome Project will change our life in the twenty-first century**

GenCDX

When you see this icon, there are related animations and exercises on the CD accompanying this text.

Having established in Chapter 19 how DNA is organized into the chromosomes of bacteriophages, prokaryotes, and eukaryotes, we now turn to the question that is at the forefront of genetic research as we enter the third millennium: *How are genes organized within the chromosomes of organisms?* The answer to this seemingly simple question is, in fact, both complex and far-reaching. It is being derived from the current attempts to analyze the complete **genome,** or the entire haploid set of chromosomes constituting the genetic material, of a variety of diverse organisms from bacteria to humans. Once we have obtained this information, we will be closer then ever to a complete understanding of the genetic basis of life.

In this chapter we will consider how genes are organized as part of chromosomes. As we shall see, this organization is least complex in bacteria, whose typically single chromosome consists largely of an array of tightly packed contiguous genes. The vast majority of the DNA encodes genetic information or consists of control elements involved in the regulation of gene expression.

In contrast, the majority of DNA in most eukaryotes fails to encode genetic information. Nor are genes tightly packed together. As discussed in Chapter 13, a large portion of many eukaryotic genes consists of noncoding DNA sequences called *introns* and flanking sequences that do not become part of the final RNA transcript. Related genes are often arranged and regulated in groups or families. In addition, a large proportion of the total DNA of eukaryotes consists of nongenic regions that often are occupied by repetitive DNA sequences.

Many of the findings over the past several decades related to genomic organization, which we shall discuss in this chapter, represent some of the most exciting and most unexpected discoveries made in the field of genetics. For all geneticists, but particularly those trained prior to the 1970s, these discoveries have served as a constant source of wonderment. We are only now entering the third 20-year phase of molecular genetics. The first phase, classical molecular genetics, began in 1953 with the discovery of the double helix. The second phase, cloning, began in 1974 with the advent of recombinant

DNA. As we embark upon the next millennium, we find ourselves at the beginning of the third phase—genomics. Just as before, we cannot even begin to imagine the questions 10 years from now. It will be you, the students of today, who will develop those questions and determine their answers.

For all of us interested in genetics, it is certain that the next "impossible thing" regarding the organization of the genetic material is just around the corner!

Genome analysis provides many insights into genetic organization

Geneticists have long been interested in studying the genomic organization of organisms, and to do so they have had to develop efficient ways to analyze genes. The methods developed over the past 90 years relied on generating and mapping mutations. The drawback of such methods is that at least one mutation for each of the genes in the genome is required, and this is often a difficult, labor-intensive task.

Geneticists now are using recombinant DNA techniques in a direct approach to genetic analysis. A collection of clones called a genomic library is established, and overlapping clones are assembled to establish genetic and physical maps encompassing the entire genome. The final step involves sequencing the entire genome. The result is a physical map, with all genes in the genome identified by location and their nucleotide sequence. The analysis of the derived amino acid sequence and the generation of gene-specific mutants can be used to study the function of the gene product. A number of these techniques are discussed in detail in Chapter 18. Here we present a brief review of two techniques.

Two basic approaches are used in genome projects. The **bottom-up** approach starts with small overlapping clones and assembles them into larger segments that eventually cover the entire genome. The **top-down** approach isolates very large, randomly cloned DNA segments that cover the entire genome. These are broken down into smaller units for mapping and sequence analysis.

In the bottom-up method, primarily used for organisms with small genomes, a cloned genomic library is prepared by isolating genomic DNA (Figure 20.1, Step 1) and inserting random DNA segments into an appropriate cloning vector, such as a virus. Clones are extensively characterized by restriction mapping (Step 2), and computer analysis of the maps (Step 3) identifies overlapping clones (i.e., clones sharing some sequence) by the presence of shared restriction sites. Clone 1 overlaps with clone 2 and a little with clone 3. The resulting contiguous segment of

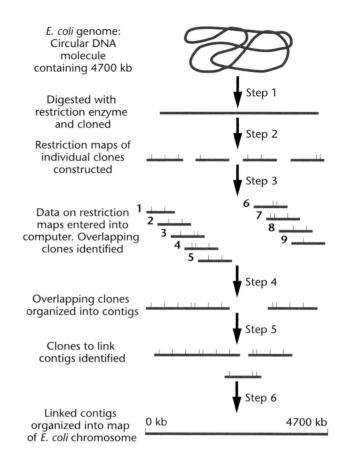

FIGURE 20.1   The bottom-up approach for the *E. coli* genome project. The chromosome is digested with restriction enzymes into small fragments that are cloned into a vector. Individual clones are characterized by restriction mapping and ordered into overlapping clones (numbered 1–9 here) by computer analysis. These are then linked into contigs for DNA sequence analysis. Gaps in the contig map are filled in to produce the chromosome map.

DNA covered by the overlapping clones is called a **contig** (Step 4). Finding clones that span the small gaps between contigs eventually closes the gaps (Step 5). The contigs are arranged in a physical map that covers the entire genome (Step 6). The next stage of the project involves sequencing the clones in the **contig library.**

A variation of the bottom-up approach, the **shotgun method,** employs an alternative technique. In conventional schemes, the genomic library is first organized into a series of contigs. Each of these larger clones is then broken down into smaller clones, which are sequenced and put into the correct order. In the shotgun method, the clones are sequenced randomly and assembled together into contigs by computer analysis. This sequencing method greatly shortens the time required for sequencing a genome, and bypasses the need to prepare physical maps. Investigators are now working to determine whether the shotgun method can be adapted to sequencing larger genomes such as the human genome, which would reduce the estimates for the time and cost of generating this information.

For the top-down approach, a cloned genomic library is constructed using a cloning vector that can hold very large inserts (Figure 20.2). The clones are then mapped to specific regions on the chromosomes by *in situ* hybridization. These clones are then broken down into smaller segments and subcloned into other vectors to be used for DNA sequence analysis.

The **Human Genome Project** is an international coordinated effort whose ultimate goal is to construct a physical map of the 3.3 billion base pairs in the haploid human genome. In the United States, consideration of a genome project began in 1986, and in 1988 the National Institutes of Health and the U.S. Department of Energy created a

joint committee to develop a five-year plan. Called the Human Genome Project, the effort got under way in 1990. Other countries, notably France, Britain, and Japan, began similar projects, which are now coordinated by an international organization, the Human Genome Organization (HUGO).

In addition to new insights into human biology, the genome project has focused attention on the ethical, legal, and social consequences of genomic research. These issues include the application of genetic testing, privacy, and the fair use of genetic information in insurance, employment, and health care. The ELSI (Ethical, Legal, and Social Implications) Program of the Human Genome Project has been set up to address these issues and to involve scientists, health professionals, policy makers, and the public in formulating policy recommendations and legislation.

Within the umbrella of the Human Genome Project, work on numerous other organisms is proceeding. We will discuss the bacterial projects in the section that follows and will then consider eukaryotic organisms.

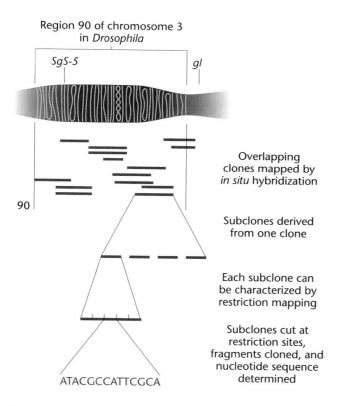

FIGURE 20.2 The top-down approach for the *Drosophila* genome project. A genome library is constructed with very large fragments (~200 kb) in a special vector. The physical location of each is mapped to the polytene chromosomes. Each clone is then broken down into subclones, which are characterized by restriction mapping for DNA sequence analysis.

Bacterial chromosomes are small, circular, and densely packed with protein-coding genes, mostly organized in polycistronic transcription units without introns

Genome projects have been completed for many bacterial species, a few of which are listed in Table 20.1. New genomes are being sequenced at a rate of about one per month, as part of the Human Genome Project. They were selected for several reasons, including their genome size, the availability of detailed genetic maps, and the value of making comparisons about gene number, location, and function across species ranging from bacteria to humans.

TABLE 20.1 Genome Size and Number of Genes in a Variety of Bacterial Species, Each of Whose Genome Has Been Sequenced

Bacteria	Genome Size (Mb)	Total Number of Genes
Escherichia coli	4.64	4397
Bacillus subtilis	4.21	4212
Methanococcus jannaschii	1.66	1813
Haemophilus influenzae	1.83	1791
Aquifex aeolicus	1.55	1552
Rickettsia prowazekii	1.11	834
Mycoplasma pneumoniae	0.82	710
Mycoplasma genitalium	0.58	503

Numerous genome projects in agriculturally important plants and animals have begun. Many of the organisms are pathogens of humans, domestic animals, or plants.

We can make a number of generalizations concerning the organization of protein-coding genes in bacterial chromosomes. Gene density is very high, averaging about one gene per kilobase pair (kb) of DNA. *E. coli*, an example of a bacteria with a large genome, having 4.6 Mb (4.6 million base pairs), has 4288 protein-coding genes, almost one gene per kb. Table 20.1 shows the total number of genes, both protein coding and nonprotein coding. Figure 20.3 shows a small portion of the *E. coli* gene map. The genes are shown as boxes. The bottom line shows distance in kb. The two lines above the genes show how those genes are organized into transcription units with the promoters that locate the start of transcription for each unit. Some of these units are organized as polycistronic **operons,** where one transcription unit codes for multiple proteins (see Chapter 15). The first operon is a three-gene unit containing *thrA*, *thrB*, and *thrC*. Notice also that there are three cases of overlapping genes, where one gene is nested inside another (shown in green). Overlapping genes are common in viruses (where space is at a premium) and uncommon in bacteria; they can also rarely be found in eukaryotes.

M. genitalium, with a genome of 0.6 MB, is an example of a small-genome bacterium. It has 503 genes, also almost one gene per kb. This packing results in a very high proportion of the DNA, (approximately 85–90 percent) serving as coding DNA. The average amount of DNA between genes is only 110–125 bp.

Typically, less than 1 percent of bacterial DNA is noncoding repetitive sequence DNA. As do all organisms, bac-

teria contain transposable elements, or jumping genes, that can move in the genome (discussed in Chapter 17). Some of these are related in structure to eukaryotic transposable elements. Bacteria can also carry cryptic or defective viruses, perhaps remnants of past infections. Polycistronic transcription units, transcription units that contain multiple genes, are characteristic of bacteria. In *E. coli*, 27 percent of the predicted transcription units are polycistronic (as seen in Figure 20.3), while in *Aquifex aeolicus* most of the genes are in polycistronic transcription units. Introns, noncoding sequences that interrupt genes and must be processed out, are extremely rare in bacteria.

In many cases it is possible to predict the function of a gene by knowing the DNA sequence and thus the amino acid sequence. A computer-directed comparison with a database often reveals a similarity to a protein of known function. This is not an easy thing to discern. Typically, a 25 percent match indicates homology. This is confounded by the fact that proteins may be composed of multiple domains, short compact amino acid regions that have a single function, such as binding ATP or binding DNA. It is possible to evolve numerous proteins by mixing different domains. One of the more surprising facts to emerge from bacterial genome studies is that about 40 percent of the genes have no known function; that is, we know the amino acid sequence of the protein, but we cannot predict what the function of the protein is.

Another feature revealed by the studies is the obvious growth in genome size as organisms become more complex. Larger, more complex bacteria such as *E. coli* or *B. subtilis* have larger genomes and more genes than do mycoplasma. This can happen in several ways. Internal gene duplication can be followed by the divergence of gene se-

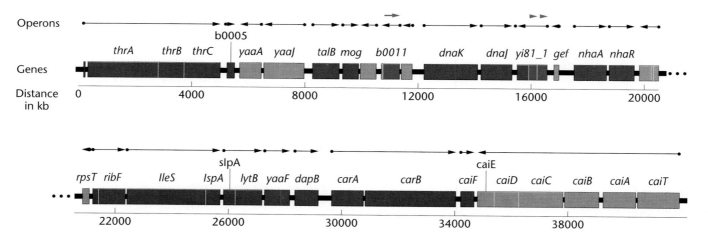

FIGURE 20.3   A section of the *E. coli* genome showing individual genes and the organization of those genes into polycistronic transcription units or operons. The location of the promoter for each operon is shown as a small dot on each arrow line. Light blue and dark blue indicates direction of transcription. Overlapping genes are shown in green.

quences, or organisms can acquire genes from other organisms; in fact both occur. There are 427 known or putative transport proteins in *E. coli* compared to 34 in mycoplasma. Many of these are very similar in sequence. *E. coli* has two completely different sets of genes for the degradation of aromatic compounds. The more recently discovered set of the two resembles the set in the soil bacteria *Pseudomonas*. The analysis of sequence indicates that of the 4288 protein-coding genes, 755 were obtained by transfer from another species in 234 different events. We will return to this topic of genome expansion after we examine the even larger eukaryotic genomes.

Bacteria show us the minimum genome size necessary for life

We now turn to the question of what minimal number of genes is essential for the survival of a free-living organism. To answer this question we shall draw from the complete genomic sequence information now available for two of the smallest bacteria, *Mycoplasma genitalium* and *Mycoplasma pneumoniae*, and then make some comparisons with some larger bacteria. These two mycoplasma are among the simplest self-replicating prokaryotes known. *M. genitalium* has a genome of 0.6 Mb while *M. pneumoniae* is 0.8 Mb. They are members of a group of bacteria that lack a cell wall and invade and often cause disease states in a wide range of hosts, including humans (in the genital and respiratory tracts), insects, and plants.

M. genitalium contains 467 predicted protein-coding genes, while *M. pneumoniae* contains 677. All 467 genes from *M. genitalium* are found within the set of 677 genes in *M. pneumoniae*. In comparison, the more complex *H. influenzae* has a 1.8 Mb genome and 1727 protein-coding genes. Table 20.2 summarizes the biological role ascribed to the genes of two of these bacteria. As we might suspect, genes must exist that code for the enzymes responsible for DNA replication and repair; for transcription and translation, including rRNA, ribosomal proteins, and many tRNAs; for transport proteins; for general cellular processes, including cell division and secretion; for countless biochemical pathways involving energy metabolism; and for biosynthetic pathways, ranging from the synthesis of all nucleic acid components to the synthesis of fatty acids and phospholipids. Genes for all such functions have been identified in *Mycoplasma* and establish the minimal complement necessary to exist as a self-reproducing organism that can thrive independently in nature. In a further comparison, of the 467 genes in *M. genitalium*, 350 are found in *B. subtilis*. This suggests that about 100 genes make the organism uniquely mycoplasma.

TABLE 20.2 A Partial List of Gene Content Classified According to Function in *Mycoplasma genitalium* vs. *Haemophilus influenzae*

Biological Function	Number of Genes	
	M. genitalium	*H. influenzae*
DNA replication	32	87
Transcription	12	27
Translation	101	141
Cell processes	21	53
Transport	34	123
Energy metabolism	31	112
Nucleic acid metabolism	19	53
Amino acid biosynthesis	1	68

Despite having nearly four times as many genes, the physiological capabilities of *Haemophilus* are very similar to *Mycoplasma*. Apparently many more genes are involved in each general category of function. The greatest advance in *Haemophilus* involves its marked increase in biosynthetic capability. For example, this more complex bacterium demonstrates 68 genes that are involved in amino acid biosynthesis, whereas *Mycoplasma* has only one such gene. Without this capability *Mycoplasma* must rely on numerous metabolic products from its host.

Archaea have characteristics of both eubacteria and eukaryotes

Archaea (formerly known as archaebacteria) are one of the three major divisions of living organisms. The other two are the eubacteria (true bacteria, without a membrane-bound nucleus, such as *Haemophilus* and *Mycoplasma*) and eukaryotes (containing a true membrane-bound nucleus). Archaea, like the eubacteria, are prokaryotes; that is, they have no nucleus. They have only recently been recognized as a separate kingdom. The original classification into a third kingdom was proposed by Carl Woese in 1977 based on the sequence from their rRNA. This was later extended by protein sequences and metabolic pathway analysis. Archaea are typically extremophiles; that is, they live in extreme conditions of very high temperature, high salt, high pressure and/or extreme pH. Even though archaea have an overall cellular resemblance to eubacteria, it was recognized early that in many metabolic aspects they more closely resemble eukaryotes. With the completion of several sequences from archaea, we can now examine this relationship more closely.

The sequence of the archaeon, *Methanococcus jannaschii*, was completed in 1996 by a group centered in The Institute for Genome Research (TIGR). It is a medium-size bacteria with a typically circular double-stranded DNA genome of 1.7 Mb, containing 1738 protein-coding genes. It was originally isolated from a high-temperature deep-sea vent at 2600 meters below sea level. It has an optimum growth temperature of 85°C and can survive up to 94°C. The *M. jannaschii* genome consists of a large, major circular chromosome of 1.66 Mb and two mini-circular chromosomes of 58.4 and 16.5 kb. Most of their genes, 58 percent, do not match any other known genes. This is a much higher proportion of unknown genes than in eubacteria. The majority of genes involved in energy production, cell division, and general metabolism most closely resemble those of eubacteria. The organization of genes also resembles eubacteria; that is, they are densely packed and have polycistronic transcription units. They do not possess, introns in their protein-coding genes.

On the other hand, archaea genes have some significant similarities to eukaryotic genes. The genes that are involved in RNA synthesis, protein synthesis, and DNA synthesis more closely resemble those in eukaryotes. Most surprising is the presence of histone chromosomal proteins and evidence that the DNA is organized into chromatin. Although they have no introns in their protein-coding genes, archaea do have introns in their tRNA genes, as do eukaryotes. This information has strengthened the proposal that the eukaryote nuclear apparatus has evolved from the archaea.

Eukaryotic chromosomes are large, linear, and less densely packed with protein-coding genes, and mostly organized as single transcription units with introns

Several eukaryotic genome projects have already reached their goals. Early in 1996, sequencing was completed on the 12 Mb genome of the yeast *Saccharomyces cerevisiae*, and the 100 Mb genome of the worm *Caenorhabditis elegans* was completed in 1998. Work on sequencing the *Drosophila* and human genomes is proceeding, and should be completed by 2000 and 2003, respectively.

Having established the general view of the eukaryotic chromosome in Chapter 19, we turn here to more specific consideration of the organization of genes in the genome of eukaryotic organisms (Table 20.3). We will begin our examination with two of the simplest eukaryotic organisms, the single-cell yeast *Saccharomyces cerevisiae* and the single-cell protozoan, the malaria parasite *Plasmodium falciparum*. The yeast genome was the first complete eukaryotic genome to be sequenced, in 1996. Only one mid-size malaria chromosome has so far been sequenced.

TABLE 20.3 Genome Size and Number of Genes in Various Eukaryotes, Each of Whose Genome Has Been Sequenced

Organism	Genome Size (Mb)	Number of Genes
S. cerevisiae (yeast)	12	6548
P. falciparum (malaria)	30	~6500
C. elegans (worm)	100	>20,000
D. melanogaster (fruit fly)	170	~16,000
H. sapiens (human)	3300	~75,000

Prior to 1970, little was known about eukaryotic gene structure and organization at the molecular level. Nevertheless, many geneticists believed that when the structure of the genes and the organization of the eukaryotic genome were revealed, fundamental differences from bacteria would be evident. Simple single-cell eukaryotes, it was reasoned, are more complex than bacteria due to the fact that their chromosomes are larger and are organized as chromatin and contained within a membrane-bound nucleus. Multicellular eukaryotes, in addition, have several unique requirements distinguishing them from bacteria. For example, cell differentiation, tissue organization, and coordinated development and function depend on genetic expression and its regulation. Do these requirements also depend on different modes of gene structure and organization?

When the technologies of DNA analysis were developed, the answer to this question began to emerge. The eukaryotic gene is far more complex than we could ever have imagined. In this section, we will review many findings related to the organization of the eukaryotic genome and the structure of genes.

Eukaryotic Genes Compared to Prokaryotic Genes

We can make several generalizations concerning the organization of protein-coding genes in eukaryotic chromosomes. Figure 20.4 shows one entire yeast chromosome, chromosome 1, the smallest of 16 chromosomes. On this 230 kb chromosome are 89 protein-coding genes and four tRNA genes. Compared to bacteria, the gene density is much lower. Even in the simple single-cell yeast, the gene density is half that of bacteria—on average, one gene per 2 kb as opposed to one gene per 1 kb. The gene density, however, is not uniform, with the result that about 68 percent of the yeast genome is gene coding as opposed to 85–90 percent of bacterial genomes. Compared to *E. coli*, there is a modest increase in the number of genes, 4397 to 6548, but a relatively large increase in the size of the genome,

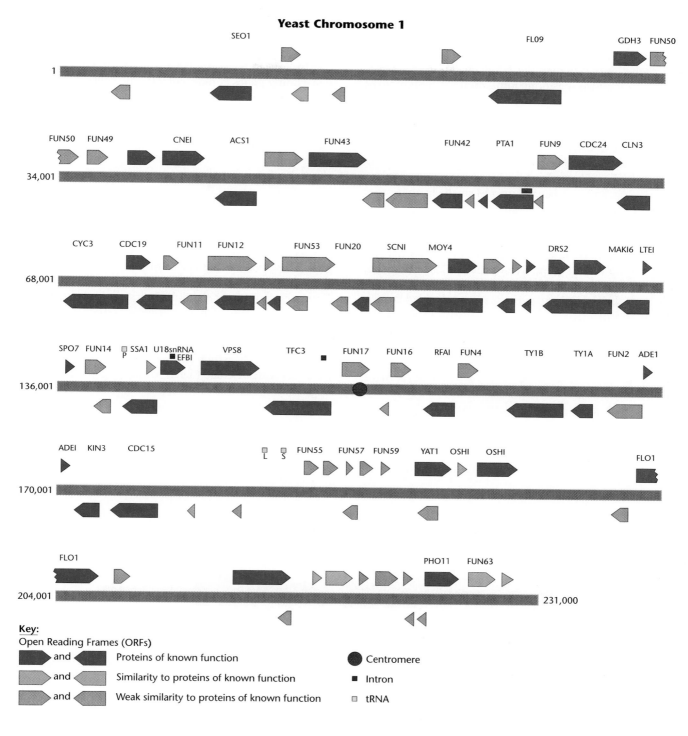

FIGURE 20.4 The entire chromosome 1 of the yeast *Saccharomyces cerevisiae.*

4 Mb to 12 Mb. While *E. coli* has but a single chromosome, the *Saccharomyces* genome has 16 distinct chromosomes ranging in size from 230 kb to 1100 kb. Note that all are smaller in size than the *E. coli* chromosome. Yeast, as do all organisms, has noncoding moderately repetitive DNA in the form of transposable elements or jumping genes and an assortment of small repeats.

The genome of the malaria parasite is similar to yeast, but the gene density is even lower, more comparable to that of multicellular eukaryotes (discussed later), one gene per 4.5 kb. The *Plasmodium* genome is a little larger than yeast, 30 Mb, and is organized into 14 small chromosomes ranging from 650 kb to 3400 kb.

The very nature of the gene is different in eukaryotes [Figure 20.5(a)] than in bacteria [Figure 20.5(b)]. We now briefly review the structure of the eukaryotic gene and, by way of comparison, the prokaryotic gene, a topic first introduced in Chapter 13. An important insight into the eukaryotic gene has been derived from a comparison of mRNA molecules and the DNA of the genes from which they are derived. It has been found that many internal base sequences of genes, referred to as **intervening sequences,** or **introns,** are not represented in the mature mRNAs, the final version of mRNA that is translated into protein. The remaining areas of each gene that are ultimately translated into the amino acid sequence of the encoded protein are called **exons.** The entire gene is transcribed into a large precursor mRNA, but the intron regions of the initial transcript are excised and the exon regions of the transcript are spliced back together prior to translation. As you can see in Figure 20.5(b), a typical bacterial gene has no introns but is polycistronic.

In *Saccharomyces*, approximately 3 percent of the protein-coding genes contain introns and there are no polycistronic transcription units. Yeast introns tend to be small, 50–100 bp, and when they are found in a gene, there is typically only one per gene. As with bacterial genes, a similar proportion of the identified genes have as yet no known function. More of the malaria genes have introns, about 40 percent, than do yeast genes.

The Difference between Single-Cell Eukaryotes and Multicellular Eukaryotes

The second eukaryote to have its complete genome sequenced is the small nematode worm *Caenorhabditis elegans*. This organism has been a major animal model system since the mid-1960s, primarily due to the efforts of Sidney Brenner, who sought a simple multicellular animal model that was less complex than *Drosophila* to study genetics and behavior. The hermaphrodite *C. elegans* has exactly 959 cells in every animal. Moreover, a complete cell lineage map is available, which means that the lineage of every adult cell can be traced back to the single-cell fertilized egg. Thus researchers can follow a combined genetic and developmental approach in studying the worm.

The *C. elegans* genome is 97 Mb and has 6 chromosomes. There is a total of over 20,000 genes, of which 19,099 are protein-coding genes. The gene density is therefore much lower than in yeast—three times more genes

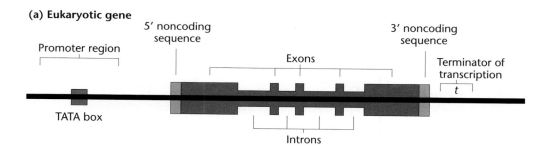

(a) Eukaryotic gene

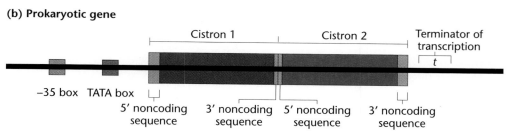

(b) Prokaryotic gene

FIGURE 20.5 Modern concept of the (a) eukaryotic gene and (b) prokaryotic gene.

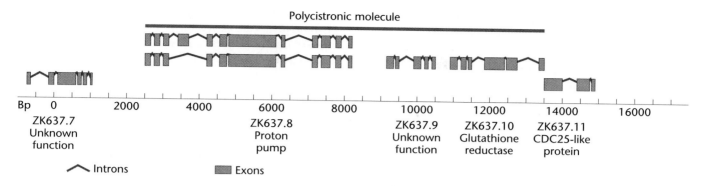

FIGURE 20.6   A small 16-kb section from chromosome III of the worm *C. elegans* showing five genes. Genes ZK637.8, .9, and .10 are transcribed together to produce a single polycistronic message. The ZK673.8 gene can be processed to form two alternative mRNAs.

and eight times more DNA give an average density of about 1 gene per 5 kb, similar to the malaria parasite. Half of the genome is intergenic, that is in-between genes. A significant fraction of the genome is middle repetitive DNA and highly repetitive simple-sequence DNA (sequences such as ATTAT repeated tens of thousands of times).

The genes themselves are surprisingly very different in *C. elegans* than in yeast (very different in fact than most eukaryotes). The biggest surprise came from the finding a few years ago that approximately 25 percent of the *C. elegans* genes are organized into polycistronic transcription units like those in bacteria (Figure 20.6). Compared with yeast, introns are far more prevalent in *C. elegans* genes; the average gene has five introns. In fact, 26 percent of the genome is intronic DNA. Genes are commonly found within introns of other genes (Figure 20.7).

Genome sequencing projects have begun in a number of other eukaryotic organisms, including *Drosophila* and humans. The *Drosophila* genome is somewhat larger than *C. elegans*, 170 Mb, but in terms of gene number and gene organization (although not gene structure) the two are otherwise very similar. *Drosophila* has about 30 percent very highly repetitive, simple-sequence DNA and the moderately repetitive DNA is mostly transposable elements interspersed with genes. Genes are of the standard eukaryotic type, monocistronic with introns.

The human and mouse (and probably most vertebrates) genomes are significantly larger. The human genome is 3300 Mb and contains approximately 75,000 genes. Only approximately 5 percent of the human genome is gene coding. Human genes tend to be larger and to contain more introns, and the introns tend to be larger than the

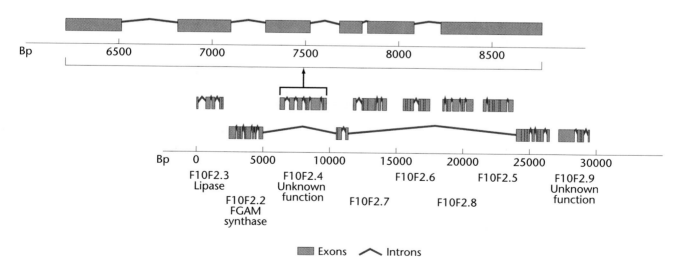

FIGURE 20.7   A small 30-kb section from the genome of the worm *C. elegans* showing genes organized within the intron of another gene. Five genes are contained within the introns of the F10F2.2 gene. The portion containing the gene F10F2.4 has been expanded to show the details of the introns and exons.

TABLE 20.4 Estimates of the Genome Size and Number of Genes in Three Representative Plants

Organism	Genome Size (Mb)	Number of Genes
A. Thaliana	120	~20,000
Rice	430	~20,000
Corn	4500	~20,000

introns in invertebrate genes. The largest gene known so far is the gene encoding dystrophin, which is mutant in Duchenne muscular dystrophy. It is 2.5 Mb, larger than many bacterial chromosomes. Most of the transcription unit is, of course, intronic DNA. It is not uncommon to find 30, 40 or even 50 introns in some human genes.

The Gene Organization of Higher Plants

To geneticists the small flowering plant *Arabidopsis thaliana* is the *Drosophila* of the plant world. Its very small genome, 120 Mb distributed in five chromosomes, is comparable to *C. elegans* and *Drosophila* in size and in gene number. It contains an estimated 20,000 genes with a gene density of one gene per 5 kb, which is also similar to *C. elegans* and *Drosophila*. In fact, at least half of its genes are closely related to genes found in bacteria and humans. Because of the small size and rapid growth rate of *Arabidopsis*, it is possible to perform genetic analyses that are similar to those in *Drosophila*.

Arabidopsis can be used as a model organism for studying other plants that have much larger genomes, such as the economically important grasses rice and corn. Many other plants such as corn have genomes that are more than an order of magnitude larger than *Arabidopsis* but that have about the same number of genes (Table 20.4). The reason

for this is a very different gene organization. These large-genome plants are characterized by having very small islands of unique-sequence DNA containing a gene or two separated from other islands by vast arrays of blocks of transposable elements (TEs). Figure 20.8 shows two corn genes, *adh-1F* and *psg* surrounded by TEs such as *Opie-1*, *Grande-zm1*, and others. These TEs are not arrayed linearly but, rather, TEs are inserted into TEs, which are inserted into TEs.

There is now enough information to summarize the gene organization for protein-coding genes ranging from bacteria through single-cell eukaryotes to multicellular eukaryotes. Figure 20.9 presents the information we have discussed, now in a standard format showing 10-kb segments to more clearly reflect the differences among levels of complexity and organization. Let us focus on gene density first. Notice how tightly packed the bacterial genes are in Figure 20.9(a) compared with even the simple yeast region Figure 20.9(b). Furthermore, as complexity increases in eukaryotes [Figure 20.9(c)], gene density becomes even lower. We cannot properly show a representative human 10-kb region because the gene density is so low.

If we look at the genes themselves, we can see in the bacterial example that a number of genes are found in polycistronic transcription units, called operons. The operon structure is uncommon in eukaryotes (*C. elegans* is a major exception). Also note that introns are not seen in *E. coli* and are found in eukaryotes. Further, as complexity increases, going from yeast to multicellular eukaryotes, the proportion of genes with introns increases, the number of introns per gene increases, and the size of the introns increases. Again, in humans many genes are larger than 10 kb because of introns.

It is important to remember that eukaryotic genes are located in DNA that is more complex than bacterial DNA

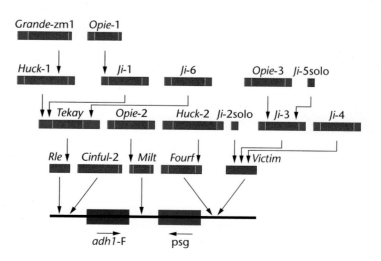

FIGURE 20.8 A small section from the maize genome in the vicinity of the alcohol dehydrogenase (*adh1F*) and *psg* genes showing blocks of transposable elements surrounding islands of unique sequence DNA. Short lines indicate each transposable element (TE) and its arrow indicates how it integrated into the TE below it. Arrows show the direction of transcription of the *adh-1F* and *psg* genes.

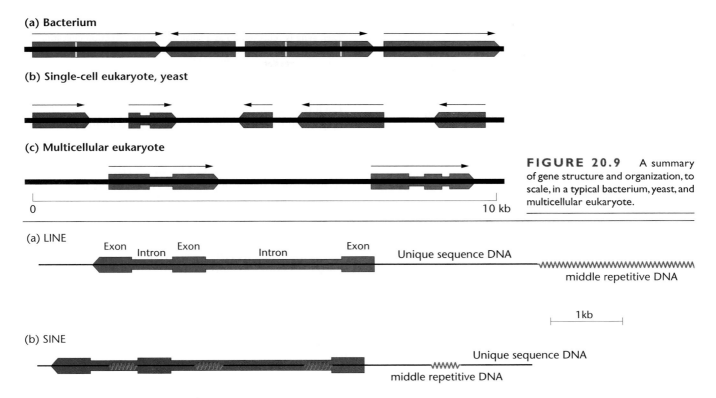

(a) Bacterium

(b) Single-cell eukaryote, yeast

(c) Multicellular eukaryote

0   10 kb

FIGURE 20.9 A summary of gene structure and organization, to scale, in a typical bacterium, yeast, and multicellular eukaryote.

(a) LINE

Exon   Intron   Exon   Intron   Exon   Unique sequence DNA

middle repetitive DNA

1kb

(b) SINE

Unique sequence DNA

middle repetitive DNA

FIGURE 20.10 The gene organization of eukaryotic genes in LINEs and SINEs DNA. The straight line represents unique sequence DNA and the wavy line represents interspersed middle repetitive DNA.

(Chapter 19). The vast majority of eukaryotic single-copy genes are found in euchromatin, which is interspersed middle repetitive and unique-sequence DNA organized into giant blocks of LINEs and SINEs (Figure 20.10). Because the unique-sequence component (the straight line in Figure 20.10) of LINEs DNA is 10 kb on average, genes fit easily into this fraction. However, the unique-sequence component of SINEs DNA is on average only 1 kb, far too short to accommodate most eukaryotic genes, especially from higher eukaryotes. The result is that the coding portion, or exons, are located strictly in the unique sequence and introns contain both unique and interspersed middle repetitive DNA (the zig-zag line in Figure 20.10).

rRNA and tRNA gene organization is different from mRNA gene organization

In addition to the protein-coding genes we have discussed, there are two other major classes of genes, those encoding ribosomal RNA (rRNA) and those encoding transfer RNA (tRNA). These genes do not code for any protein; rather the transcribed RNA is used directly as a component of a ribosome. The organization of these genes is different from typical protein-coding genes because in all but

the smallest bacteria they are multiple-copy genes. Any time there are multiple-copy genes, two models of gene organization are possible (Figure 20.11). One model is that the multiple genes are simply **scattered** in the genome as seen in Figure 20.11(a). The second model is that all or some of the multiple-copy genes are arrayed in **tandem** in one or a few blocks, as seen in Figure 20.11(b). These two arrangements are not necessarily mutually exclusive. We will start with a discussion of rRNA genes, first in bacteria and then in eukaryotes, and finish this section by discussing tRNA genes.

The rRNA genes are in multiple copies even in most bacteria. Larger bacteria have a higher copy number than do smaller bacteria. *E. coli* has seven copies, mycoplasma has only one. In *E. coli* the copies are scattered (Figure 20.12).

(a) Scattered model   **(b) Tandem model**

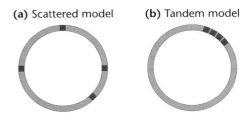

FIGURE 20.11 Two models of how multiple genes can be arranged in a chromosome.

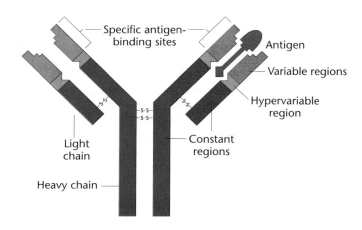

FIGURE 20.16 A typical antibody (immunoglobulin) molecule. The molecule is Y-shaped, and contains four polypeptide chains. The longer arms are heavy chains, and the shorter arms are light chains. Each chain contains a constant region and a variable region. The variable and hypervariable regions form an antibody combining site that interacts with a specific antigen, similar to a lock-and-key mechanism. The chains are joined together by disulfide bonds.

Each **light chain (L)** is composed of 220 amino acids, with the first 110 amino acids making up the variable region (V_L). The rest of the amino acids at the C-terminus make up the constant region (C_L) of the light chain. There are two types of L chains: **kappa chains,** encoded by genes on human chromosome 2, and **lambda chains,** encoded by genes on the long arm of chromosome 22. The variable regions of the heavy and light chains form the **antibody-combining site.** The combining site of each antibody has a unique conformation that allows it to bind to a specific antigen, like a key in a lock.

One of the dominant features of the immune system is the generation of new cells that contain different combinations of antibodies. Because there are billions of such combinations, it is impossible that each combination is coded by a separate gene. There simply is not enough DNA in the human genome to encode tens or hundreds of millions of antibodies. One of the long-standing questions in immunogenetics is how this vast molecular variability in antibodies and antigen receptors is encoded in the genome.

Antibody diversity is due to a number of special features. The key to understanding how tens of millions of antibodies can be produced lies in an understanding of the immunoglobulin gene structure. In humans, the kappa light-chain gene (Figure 20.17) consist of several elements: **L-V (leader-variable)** regions, **J (joining)** regions, and a **C (constant)** region. There are 70 to 100 L-V segments in the kappa light-chain gene region, each with a different nucleotide sequence and each containing a promoter, but no regulator. A significant number of the L-V segments are

pseudogenes. The J region contains six different segments, and there is one C segment. The regulator, but no promoter, is between the last J segment and the C segment. This feature ensures that individual segments cannot be transcribed by themselves.

During maturation in a B cell, one of the 100 L-V regions (L_2-V_2) and its promoter are *randomly* joined by a recombination event to one of the six J regions (J_3) and the C region with its regulator to form a functional light-chain gene as shown in Figure 20.17, line 2. The DNA in between the L-V region selected and the J_3 region is excised and destroyed. This form of recombination is different from the reciprocal recombination occurring in meiosis. Different enzymes are involved in this process. The joining event from the back of the V_2 to the front of J_3 is *imprecise* and can occur over a span of about 6 bp. In addition, the open ends of DNA are subjected to a **break-nibble-add** mechanism, in which a few bases may be randomly removed and a few bases may be randomly added. The newly created gene contains three exons—L, V-J, and C—and two introns, one between L and V-J and the second between J_3 and C. The J_4 through J_6 regions, not selected, become part of the second intron. The light-chain gene is transcribed, processed, and translated to form kappa L-chain proteins that become part of the antibody molecule. The rearranged gene is stable and is passed on to all progeny of the B cell.

Antibody diversity in kappa L-chain production is the result of combining any of the 100 L-V regions with any of the six J regions, generating about 600 different kappa genes. Since each joining reaction can occur over several base pairs, the number of possible genes increases to several thousand. Additional diversity then comes from the nibble-add mechanism. The organization of the lambda L-chain genes is somewhat different from that of the kappa genes, but recombination events also generate a similar number of different lambda genes.

The heavy-chain genes in humans extend over a very large region of DNA and include four types of segments: L-V (leader-variable), D (diversity), J (joining), and C (constant) regions. There are approximately 300 different V regions, 10 to 30 different D regions, and six different J regions. In addition, there are nine C regions for the five classes of immunoglobulins (Figure 20.18). The μ heavy-chain protein is coded by the (μ) region, the δ chain, by the (δ) region, the γ chains by (γ) regions, the ϵ chain by an (ϵ) region, and the α chain by an (α) region. During B-cell maturation, recombination randomly joins an L-V region with one of the D sequences and then to one of the J sequences. The L-V-D-J composite assembled by recombination lies adjacent to the C region of the μ heavy-chain gene, the μ gene. This joining is like that described for the light chain above, break-nibble-add. Processing of this pre-mRNA yields a heavy-chain mRNA molecule.

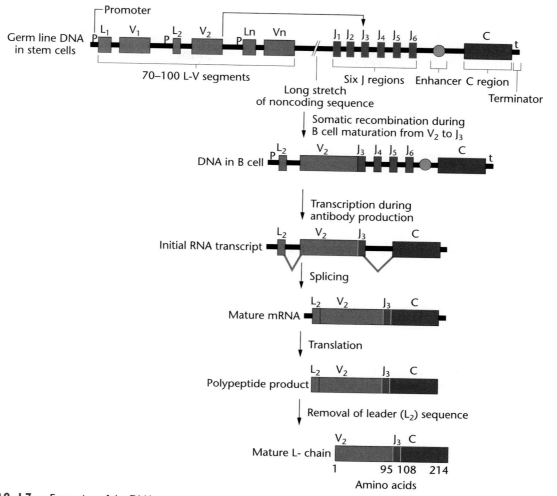

FIGURE 20.17 Formation of the DNA segments encoding a human kappa light chain and the subsequent transcription, mRNA splicing, and translation leading to the final polypeptide chain. In germ-line DNA 70–100 different L-V (leader-variable) segments are present. These are separated from the J regions by a long noncoding sequence. The J regions are separated from a single C gene by an intervening sequence (intron) that must be spliced out of the initial mRNA transcript. Following translation, the amino acid sequence derived from the leader RNA is cleaved off as the mature polypeptide chain passes across the cell membrane.
🌐 GenCDX

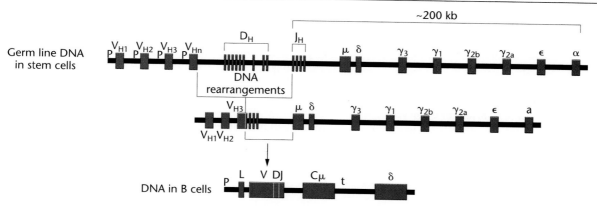

FIGURE 20.18 The variable region of the H chain is assembled by joining three different DNA segments together: L-V, D, and J. In embryonic DNA, these segments occur in clusters separated by long intervals of other DNA sequences adjacent to the μ gene C region, which will produce the μ heavy chain. In a maturing B cell, a random combination of one L-V region with one of the 20 D regions and one of the four J regions produces the μ heavy-chain gene that is transcribed from the promoter (P) to the terminator (t) and translated in the B cell. In other B cells, different combinations of H-chain segments are joined together, producing a large number of different μ heavy chains.
🌐 GenCDX

The random recombination of H-gene components generates a large number of H-chain proteins. If we assume there are 25 D regions, then combining these with any of the 300 V regions and 6 J regions, 45,000 H chains can be generated. Add in imprecise joining and nibble-add for both J to D and V to D-J, and the total climbs markedly. The potential for overall antibody diversity can be estimated by multiplying the combination of all heavy-chain genes with all light-chain genes, resulting in hundreds of millions of possible antibody genes from a few hundred coding sequences.

To summarize, antibody diversity is due to four special features found in the immunoglobulin gene system: (1) multiple numbers of variable regions, (2) multiple numbers of diversity and joining regions, (3) multiple splice locations with break-nibble-add joining, and (4) multiple combination of light chains with heavy chains. As antibody-forming B cells mature, DNA recombination rearranges these genes so that each mature B lymphocyte comes to encode, synthesize, and secrete only one specific type of antibody. Each mature B cell can make one type of light chain (kappa or lambda) and one type of heavy chain. When an antigen is present, it stimulates the B cell, which encodes an antibody against that antigen to divide and differentiate. As a result, populations of differentiated plasma cells are produced, all of which synthesize one type of antibody that interacts with the antigen.

The Histone Gene Family

A variation on the theme established in the globin gene families and immunoglobulin gene families is illustrated by the **histone genes.** Here, a cluster of five related but nonidentical genes, separated from each other by highly divergent nontranscribed spacer sequences, is tandemly repeated many times. Recall from Chapter 19, histones are positively charged (basic) proteins that interact with the negatively charged phosphate groups of DNA to form the

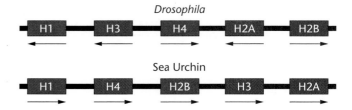

FIGURE 20.19   The histone gene clusters in *Drosophila* and the sea urchin. Arrows indicate the polarity of transcription.

nucleosome characteristic of the chromatin fiber. The need for histone proteins is extremely great because of the large amount of DNA in eukaryotic chromosomes that must be condensed into a small space, the nucleus. In rapidly dividing cells of some organisms, not only is the need great, but the necessary quantity must appear rapidly as one cell cycle runs into the next, time and time again, without pause to catch up with the cell's biochemical needs. To accommodate this need, the entire cluster has been duplicated in order to increase the number of copies of each histone gene.

Many interesting correlations support this contention. Yeast cells, with much less DNA than other eukaryotes, have only two copies of each of four histone genes (they lack histone H1). In contrast, the entire cluster of five histone genes is tandemly repeated from 10 to 800 times in many evolutionarily advanced organisms (Table 20.6). Sea urchins, which display one of the largest number of duplications of the cluster, are noted for their extremely rapid rate of cell division during development.

Beyond its repeat nature, the histone gene family shows several other differences from the other families. First, almost all histone genes lack introns. Second, the individual genes within the cluster of a given species are often oriented with respect to transcription in opposite directions. As shown in Figure 20.19, the arrangement of genes and their polarity of transcription, shown by the direction of the arrow, vary in the sea urchin and *Drosophila*, two organisms where this cluster has been well studied. Polarity differences also exist in other organisms studied. Despite polarity differences, the amino acid sequences of the various histones (reflecting the DNA encoding each one) are remarkably similar in highly divergent organisms. The homology of histone genes is one of the best examples known of sequence conservation throughout evolution.

In mice and humans, members of the histone gene family appear to be clustered, but not present as a tandem array of repeat units. For example, a region of the distal tip of the long arm of the human chromosome 7 contains all the members of the histone gene family, but they may be interspersed with other nonhistone genes. Therefore, no single pattern of organization applies to all organisms.

TABLE 20.6   Number of Tandemly Repeated Histone Gene Clusters Found in a Variety of Eukaryotic Organisms

Organism	Repeats
Yeast	2
Chickens	10
Mammals	20
Xenopus	40
Drosophila	100
Sea urchin	300–600
Newt	600–800

How does gene number change with advancing phylogeny?

In the section on bacterial gene organization we touched briefly on the fact that as organisms become more complex, they undergo an expansion in total genome size and gene number. The amount of DNA contained in the haploid genome of a species is called the **C value.** When such values were determined for a large variety of eukaryotic organisms, several trends were apparent. The most notable trends are that:

1. Eukaryotes contain substantially more DNA in their genomes than do bacteria and exhibit a wide variation among different groups.

2. Development of evolutionary complexity has been accompanied by increased amounts of DNA. In many phylogenetic groups, more evolutionarily complex forms generally contain more DNA than do most less advanced forms.

It might be argued that increasing *C* values are simply the result of a greater need for increased amounts and varieties of gene products in more complex organisms. However, several observations make it unlikely that this is the only reason. First, the total DNA content (in Mb) increases are very dramatic, ranging over 10,000-fold from mycoplasma to rice (0.6 Mb to 4500 Mb) and maize. Gene number increases from 470 in mycoplasma to 75,000 in humans, an increase of less than 200-fold. Second, closely related organisms with the same degree of complexity and the same approximate number of genes can vary 10-fold or more in DNA content, as seen in the comparison between grasses such as maize and *Arabidopsis*. The grasses with fewer genes can even have more DNA than the arguably more complex humans. Therefore, it is doubtful that greater evolutionary complexity alone during evolution can account for the increase in DNA content. This conclusion is the basis of what has been called the **C value paradox.**

An easier question to address is how gene number increases during evolution. As we noted in our discussion of bacterial gene organization, organisms can increase gene number by internal gene duplication followed by divergence of gene sequence, and they can acquire genes from other organisms. Data from the yeast genome project have revealed something that was very unexpected. The yeast genome is a degenerate tetraploid. Sometime in its past history there was a complete genome duplication, followed by an elimination of most of the duplicate material. Only 13 percent was retained in small blocks, which were then rearranged by recombination. Yeast increased its gene content by about 780 genes in one complex event.

Genomes can increase in size by the accumulation of transposable elements as in the grasses, or by the expansion of simple-sequence satellite DNA. But genomes have increased in size also by an increase in single-copy, unique-sequence DNA in between genes as well. More advanced organisms tend to have more and larger introns. The presence of introns may substantially increase the size of the average eukaryotic gene. However, introns represent no more than 10 percent of the genome, and usually much less.

The Human Genome Project will change our life in the twenty-first century

Two areas that will certainly benefit from the knowledge of the complete human genome sequence are medicine and basic research.

Medicine will completely change in the next millennium. Except perhaps for trauma, all of our ills, both physical and mental, have a genetic component. Within 20 years we will be able to predict disease susceptibility far more accurately. That does not mean we will know if you will develop cancer, but we will know what the likelihood is, giving you the opportunity to make various life choices. All of the several thousand genes that cause or predispose us to illnesses will be identified. Treatment for disease will be tailor-made for each individual, according to his or her genotype. Your doctor will either know or be able to obtain a rapid readout of each of your alleles. Drug dosage can be designed to match your tolerance and sensitivity.

One result of the Human Genome Project is that we now know the sequence of each gene for a number of different organisms, both bacterial and eukaryotic. Simply knowing the sequence, however, does not tell us what the gene does or how the gene functions. One research tool used to begin to answer those questions is **genes on a chip.** It is now possible to copy the DNA sequence of every gene in *E. coli* or yeast and fix the DNA spot by spot in a micro-array on a computer chip. It is then possible to use that chip to monitor the gene expression for each gene as conditions change. The response of *E. coli* to starvation or of yeast to mating can be monitored and we can see exactly what each gene is doing. Though we can hardly imagine all the questions that will emanate from the Human Genome Project, we can be sure that the answers and information will be fascinating and rewarding.

CHAPTER SUMMARY

1. Bacterial chromosomes are typically small circular double-stranded DNA. Gene density is very high, averaging one gene per kb of DNA. Typically as much as 90 percent of the chromosome is gene coding. Many genes typically are organized into polycistronic transcription units that do not contain introns.

2. The smallest living cellular organisms are the mycoplasmas. The smallest genome sequenced to date is *M. pneumoniae* at 580 Mb and 467 genes.

3. Archaea is a major prokaryotic kingdom. Their chromosome and gene organization is indistinguishable from the eubacteria. However an examination shows there is a strong similarity of many of their genes to eukaryotic genes.

4. Eukaryotic chromosomes are large, linear double-stranded DNA. The gene density is much lower than the gene density in bacteria. The simplest eukaryote to have its genome sequenced so far has a gene density half that of bacteria, one gene per 2 kb. Genes typically are not organized into polycistronic transcription units, rather each is a separate transcription unit. Genes typically are interrupted with introns.

5. Complex multicellular eukaryotes differ from the less complex yeast by a number of criteria. They have more genes and much more DNA. This results in gene densities falling to one gene per 5 kb and even one gene per 10 to 20 kb or more. A higher proportion of genes have introns, the number of introns per gene increases, and the size of introns increases as complexity increases from yeast to *C. elegans* to humans.

6. Some plants such as *Arabidopsis* have a gene structure and organization that is indistinguishable from animals. Other plants with a gene number comparable to *Arabidopsis*, which have undergone a large amplification in nongenic DNA, have a different organization, with vast blocks of transposable elements separating islands of genes.

7. The ribosomal RNA genes in bacteria are multiple copy and scattered around the genome. Each unit contains all three rRNA coding sequences. The ribosomal RNA genes in eukaryotes are split into two groups, 18S/28S genes and 5S genes. In many eukaryotes, both the 18S/28S genes and the 5S genes provide an example of tandem repeat families, where the same gene is duplicated side by side many times. The transfer RNA genes in both bacteria and eukaryotes are multiple copy and scattered around the genome.

8. Many eukaryotic genes have undergone duplication followed by sequence divergence, leading to multigene families. The globin, immunoglobulin, and histone gene clusters are prime examples of this.

9. As organisms become more complex, gene number and especially total DNA content increases dramatically. The C value paradox, made apparent during the study of the organization of eukaryotic DNA, suggests that eukaryotic organisms contain much more DNA than necessary to encode only the gene products essential to the development and normal functions of an organism.

INSIGHTS AND SOLUTIONS

1. An antibody molecule contains two identical H chains and two identical L chains. This results in antibody specificity, with the antibody binding to a specific antigen. Recall that there are five classes of H-chain genes and two classes of L-chain genes. Because of the high degree of variability in the genes that encode the H and L chains, it is possible (even likely) that an antibody-producing cell contains two different alleles of the H-chain gene and two different alleles of the L-chain gene. Yet the antibodies produced by the plasma cell contain only a single type of H chain and a single type of L chain. How can you account for this, based on the number of classes of H-chain and L-chain genes and the possibility of heterozygosity?

Solution: Although an antibody-producing cell contains different classes of H-chain and L-chain genes, and although it is entirely possible and even likely that a given antibody-producing cell will contain different alleles for the H-chain gene and for the L-chain gene, a phenomenon known as allelic exclusion allows the expression of only one H-chain allele and one L-chain allele in any given plasma cell at a given time. That is not to say that the same antibody is produced over the life span of the antibody-producing cell, however. At first, many plasma cells produce antibodies of the IgM class, and at later times they may produce antibodies of a different class (e.g., IgG). Even in switching between different H-chain genes, allelic exclusion is maintained, with only one allele of an H-chain gene (or L-chain gene) being expressed at a given time.

PROBLEMS AND DISCUSSION QUESTIONS

1. Compare and contrast the chemical nature, size, and form assumed by the genetic material of bacteria and yeast.

2. Why might it be predicted that the organization of eukaryotic genetic material would be more complex than that of viruses or bacteria?

3. Compare the gene organization of bacterial genes to the gene organization of eukaryotic genes.

4. How does the gene organization of eukaryotic genes change as they become more complex?

5. How are eukaryotic genes arranged in the part of euchromatin that is LINEs DNA and SINEs DNA?

6. Describe the rRNA gene family in eukaryotes. What accounts for the major difference in the size of tandem repeats in different organisms?

7. The β-globin gene family consists of 60 kb of DNA, yet only 5 percent of the DNA encodes β-globin gene products. Account for as much of the remaining 95 percent of the DNA as you can.

8. What do the following symbols represent in immunoglobulin structure: V_L, C_H, IgG, J, and D?

9. If germ-line DNA contains 10 V, 30 D, 50 J, and 3 C genes, how many unique DNA sequences can be formed by recombination?

10. If there are 5 V, 10 D, and 20 J genes available to form a heavy-chain gene, and 10 V and 100 J genes available to form a light chain, how many unique antibodies can be formed?

11. Distinguish between an antigen and an antibody.

12. How many polypeptide chains are present in an IgG antibody molecule? How many different polypeptide chains are represented in this molecule? How many gene segments from a germ-line cell were combined to produce the polypeptide chains in this molecule?

13. "In 2000, the C value paradox is not so paradoxical." Agree or disagree with this statement, and support your position.

EXTRA-SPICY PROBLEMS

For these "Extra-Spicy Problems" it is necessary to use the World Wide Web (www). On the web at **http://www.genome.ad.jp/kegg/java/org__list.html** is a master website that lists all the complete genome projects and the links to each one. At the bottom (scroll down) are links to other websites: MAGPIE, TIGR, and "additional sequencing projects." MAGPIE is especially good.

The web sites described here are dynamic sites and are subject to change. You are encouraged to explore links within these sites.

14. At the Kegg site, scroll down to the bacterium *Mycoplasma pneumoniae* line. On the far right column, click on ZMBH, scroll down, then click on genome project, and finally click on gene map. Print (or draw) the circular chromosome to scale, with kb markers, showing any five genes. List their names (not their numbers) and show the direction of transcription for each. To find the names, click on a region for a magnification. The name of each gene is at the bottom under description. Find known genes, that is, genes with four letter names (i.e., *nusA*).

15. From Problem 14 select one gene and find it in *E. coli*. On the line for *E. coli*, click on Wisconsin, then click on NCBI Entrez. Enter the gene you wish to find in the search box and hit enter. Document your answer with the gene location or a printout.

16. At the Kegg site, scroll down to the yeast *Saccharomyces* site and click on the Stanford site (SGD). Under maps, click on "physical and genetic maps" and you will find the key to each of the 16 yeast chromosomes with a map.
    a. Draw chromosome VIII to scale showing the centromere and any four genes with map coordinates.
    b. Approximately how far apart is *spo11* from *msh1* in kb and recombination map units (centimorgans, cM)?
    c. Document your data with a printout of the region.

17. On the web at http:// www.evergreen.edu/user/T4/home.html is the phage T4 genome map.
    a. Draw the chromosome.
    b. Indicate the location of all the tail genes.

18. Using the yeast *Saccharomyces* Stanford web page, click on maps (on the left), click on Gene/Sequence Resources. On option 2, "Pick a Chromosome," select chromosome #1 and click on "submit form" at the bottom. Scroll down to sequence retrieval, click on DNA of Region "GLG", print out the telomere sequence. Underline the terminal repeat sequence.

19. Using the yeast *Saccharomyces* website from Problem 18, click on "Gene/Sequence Resources" and option #1. Scroll down to sequence retrieval, click on DNA of Region "GLG", and print the centromere sequence from chromosome 3 and chromosome 4. Use the gene names, *cen3* and *cen4*.
    a. How do these differ?
    b. Since all centromeres have the same function, how could they have different sequences?

20. Draw the gene organization of the terminal 10 genes from the yeast chromosome III (either end) to scale showing the direction of transcription for each. Compare and draw the first 10 genes from *Mycoplasma pneumoniae* to the same scale.

21. What is the rRNA gene organization in *B. subtilis* and the *Saccharomyces* chromosome XII. Document your answer with a printout. For *B. subtilis* use the BSORF search engine and keyword rrn. For yeast, look in the MIPS database in the Kegg website. Under the yeast genome, click on RNAs, then click on ribosomal RNAs.

GENETICS MediaLab

The following resources will help you achieve a better understanding of the concepts presented in this chapter. These resources can be found on the CD packaged with this textbook and on the Companion Website at **http://www.prenhall.com/klug/**.

CD Resources:

Animated Tutorial: *Immunoglobin Expression*

Self-grading Chapter Problems

Web Resources:

Web Destinations in Genetics

Self-grading Chapter Problems

Chapter Search Terms

Genetics Newsgroups

Student Bulletin Board

Web Problem 1:

Time for completion = 5 minutes

How are members of a gene family related? A gene family consists of genes related by shared sequence homology, including both different genes within an organism or genes found in other species. In an effort to understand the similarities and distinctions in function of the gene products, databases have been established grouping information about the members of gene families. In this exercise you will examine data about one such gene family, the PAX genes, which are thought to function as transcription factors that bind DNA. Discuss whether members of the PAX gene family provide similar or different functions in different organisms. To complete this exercise, visit Web Problem 1 in Chapter 20 of your Companion Website and select the keyword **GENE FAMILY.**

Web Problem 2:

Time for completion = 10 minutes

What kinds of sequences compose a genome? Repetitive DNA accounts for a large portion of the genome of many organisms and contributes to the C-value paradox. Reassociation kinetics measures the time required for denatured fragments of the genome to reanneal as double-stranded DNA. For particular DNA fragments, the time required for reannealling reflects how repetitive the sequence is within the genome. The linked website (keyword 'highly repeated DNA') describes this measure-

ment of reassociation in several organisms for various classes of DNA fragments. What type of sequences are represented by highly repetitive, moderately repetitive and unique DNA? Discuss how the data for *Drosophila* (fruit fly), *Mus* (mouse), *Homo* (human), and *Nicotiana* (tobacco plant) support or refute the C-value paradox. To complete this exercise, visit Web Problem 2 in Chapter 20 of your Companion Website and select the keyword **REASSOCIATION.**

Web Problem 3:

Time for completion = 5 minutes

How similar are humans and mice? humans and pigs? Much of current biomedical research relies on similiarities between humans and model organisms. The relationships between species can be evaluated by examining regions of chromosomes that share sequence homology, called syntenic regions. The linked website contains a database that catalogs pairwise comparisons of chromosomes between species and reports, known homologies. Compare two species by selecting the organisms in the dialog boxes provided and clicking on the retrieve button. The resulting search will show an Oxford grid, comparing all the chromosomes in one species to all the chromosomes in the second species. Each colored square is hypertext-linked to detailed information regarding which genes are found in the orthologous regions. Choose a chromosome in one species and describe its relationship to the chromosomes of the second species. To complete this exercise, visit Web Problem 3 in Chapter 20 of your Companion Website and select the keyword **ORTHOLOGS.**

SELECTED READINGS

BLATTNER, F.R. et al. 1998. The complete genome sequence of *Escherichia coli* K-12. *Science* 277:1453–74.

BULT, C.J. et al. 1996. Complete genome sequence of the methanogenic Archaeon, *Methanococcus jannaschii*. *Science* 273: 1058–72.

FRASER, C.M. et al. 1995. The minimal gene complement of *Mycoplasma genitalium*. *Science* 270:397–403.

GALL, J.G. 1981. Chromosome structure and the *C*-value paradox. *J. Cell Biol.* 91:3s–14s.

GARDNER, M.J. et al. 1998. Chromosome 2 sequence of the human malaria parasite *Plasmodium falciparum*. *Science* 282:1126–32.

GELLERT, M. 1996. A new view of V(D)J recombination. *Genes to Cells* 1:269–75.

GOFFEAU, A. et al. 1996. Life with 6000 genes. *Science* 274:546–67.

HEINTZ, N. 1991. The regulation of histone gene expression during the cell cycle. *Biochim. Biophys. Acta* 1088:327–39.

IYER, V.R. et al. 1999. The transcriptional program in the response of human fibroblasts to serum. *Science* 283:83–87.

MEINKE, D.W. et al. 1998. *Arabidopsis thaliana:* A model plant for genome analysis. *Science* 282:662–82.

MEWES, H.W. et al. 1997. Overview of the yeast genome. *Nature* 387:5–105.

ORKIN, S.H. 1995. Regulation of globin gene expression in erythroid cells. *Eur. J. Biochem.* 231:271–81.

SAN MIGUEL, P. et al. 1998. The paleontology of intergene retrotransposons of maize. *Nature Genet.* 20:43–47

The *C. elegans* Sequencing Consortium 1998. Genome sequence of the nematode *C. elegans:* A platform for investigating biology. *Science* 282:2012–18.

WILLIAMS, S.C. et al. 1996. Sequence and evolution of the human germline V lambda repertoire. *J. Mol. Bio.* 264:200–210.

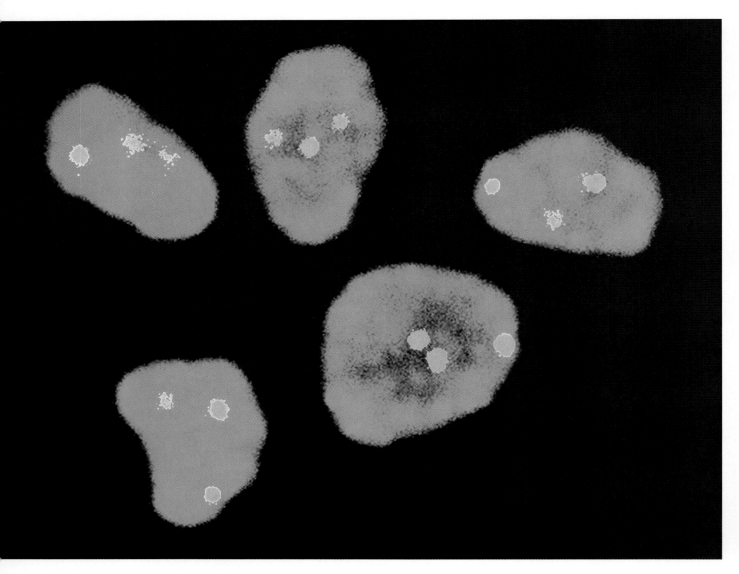

Fluorescence *in situ* hybridization (FISH) used during prenatal diagnosis of Down syndrome (trisomy 21). Each complete nucleus demonstrates three copies of chromosome 21 (pink dots).

Applications and Ethics of Genetic Technology

KEY CONCEPTS

GenCDX

When you see this icon, there are related animations and exercises on the CD accompanying this text.

In 1971, a paper published by Hamilton Smith, Daniel Nathans, and Walter Arber marked the beginning of the recombinant DNA era. The paper described the isolation of an enzyme from a strain of bacteria and the use of this enzyme to cleave viral DNA. It contained the first published photograph of DNA cut with a restriction enzyme. From this modest beginning, recombinant DNA technology has revolutionized all fields of experimental biology and spread beyond the research laboratory. In the intervening years, this technology has expanded to become part of everyday life; recombinant DNA techniques are now being used to help produce medicines, milk, and even the food for the supermarket.

In Chapter 18 we discussed the methods used to create and analyze recombinant DNA molecules. This chapter describes how these tools are being used in research and commercial applications, and the ethical problems posed by the use of this technology. We will consider a cross section of applications illustrating the power of recombinant DNA technology to map and identify human genes, diagnose and treat disease, and generate new plants and animals. We will begin by considering how recombinant DNA

has changed some of the basic methods of genetic mapping. Next, we will examine the impact of recombinant DNA on human genetics and medicine, from the prenatal analysis of genotypes to the treatment of genetic disorders and the mapping of the human genome. Finally, we will discuss some of the products and methods being used in the multibillion-dollar biotechnology industry.

Human genes can be mapped at the molecular level

When the first genes responsible for human genetic disorders were mapped, the technique began with an identified gene product that was defective in affected individuals. With this information the chromosomal locus of the gene was determined. However, in the majority of human genetic disorders the function of the normal gene product is unknown. Therefore, while the genetic basis for a handful of diseases was determined before the recombinant DNA revolution, real progress in the mapping of these genes has only come more recently. Now, with our detailed knowledge of the genome, it is possible to map a gene by first identifying a genetic marker that is linked to the disease phenotype. With this linkage information it is possible to either look in the chromosomal region for potential gene sequences or, with an increasing number of the 50,000 to 100,000 human genes localized to chromosomal sites, check the identified chromosomal area for genes that have already been mapped.

RFLPs as Genetic Markers

Variations in nucleotide sequence occur throughout the human genome (mostly in noncoding regions) with a frequency of about 1 in every 200 nucleotides. These changes in the sequence at a specific site, brought about by either changes in a single nucleotide pair or by deletions or insertions of one or more nucleotide pairs, can create or destroy restriction enzyme cutting sites. If a restriction site created by such variation is present on one chromosome but absent on its homolog, the two chromosomes can be distinguished from one another by their pattern of restriction fragments on a DNA blot (Figure 21.1). The region of chromosome A shown in the figure contains three *Bam*HI sites; its homolog, chromosome B, contains two such sites. When chromosome A is cut with *Bam*HI, 3-kb and 7-kb fragments are generated, whereas only a single 10-kb fragment is generated when chromosome B is cut. Using a probe from this chromosomal region, these fragments can be visualized on a Southern blot (Figure 21.1).

These variations in DNA fragment length generated by cutting with a restriction enzyme are called **restriction fragment-length polymorphisms (RFLPs)**. RFLPs are quite common; several thousand have been identified in the human genome, and many of these have been assigned to individual chromosomes. These variations are inherited as codominant alleles. They can be mapped to specific regions on individual chromosomes and can be used as markers to follow the inheritance of genetic disorders from generation to generation in an affected family.

RFLPs in Linkage Analysis

Determining the chromosomal locus of a genetic disorder using RFLPs involves identifying a mapped RFLP marker that cosegregates with a genetic disorder in multigenerational families (three generations or more). Selecting which RFLP to use in the family being studied is a matter of trial and error. Many different loci are tested in order to find one for which most members of the family are heterozygous, so that each member of a chromosome pair can be identified, with one that shows linkage with the disease phenotype.

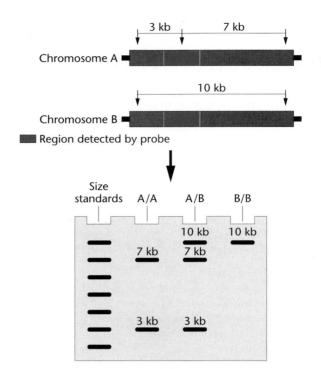

Genotypes	Fragment sizes
Homozygous for chromosome A (A/A)	3 kb, 7 kb
Heterozygous (A/B)	3 kb, 7 kb, 10 kb
Homozygous for chromosome B (B/B)	10 kb

FIGURE 21.1 Restriction fragment-length polymorphisms (RFLPs). The alleles on chromosome A and chromosome B represent DNA segments from homologous chromosomes. The region that hybridizes to a probe is shown in green. Arrows indicate the location of restriction enzyme cutting sites that define the alleles. On chromosome A, three cutting sites generate fragments of 7 kb and 3 kb. On chromosome B, only two cutting sites are present, generating a fragment 10 kb in length. The absence of the cutting site in B could be the result of a single base mutation within the enzyme recognition/cutting site. Because these differences in restriction cutting sites are inherited in a codominant fashion, there are three possible genotypes: AA, AB, and BB. The allele combination carried by any individual can be detected by restriction digestion of genomic DNA (obtained, for example, from a blood sample or skin fibroblasts), followed by gel electrophoresis, transfer to a DNA binding filter, and hybridization to the appropriate probe. The fragment patterns for the three possible genotypes are shown as they would appear on a Southern blot.

To map a genetic disorder, the inheritance of a given RFLP and the disorder are traced through the family (Figure 21.2). If the loci of the RFLP and the disorder are near each other on the same chromosome, they will show linkage. In such studies, most RFLPs will not show linkage; these results serve to identify chromosomes that do *not*

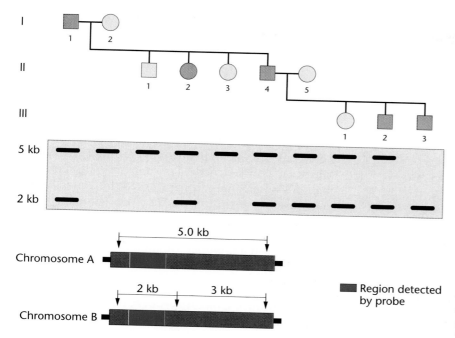

FIGURE 21.2 Establishing linkage between a dominant trait and an RFLP allele. The pedigree shows a family with members affected by a dominant trait (filled symbols). Family members also carry two alleles of an RFLP locus assigned to a specific chromosome. A 5.0 kb allele (Allele *A*) is present on one homolog. The *B* allele on the other homolog consists of two fragments (2.0 kb and 3.0 kb). The probe used in the Southern blot detects the 5 kb *A* allele and the 2.0 kb portion of the *B* allele. The RFLP pattern for each family member is shown below the appropriate pedigree symbol. Individual III-1, who is unaffected, probably received an *A* allele from her father, and a *B* allele from her mother. III-2 is affected and probably received an *A* allele from his mother and a *B* allele from his father. The youngest son (III-3) who is affected, received a *B* allele from each parent. Taken together, the pedigree and the Southern blot suggest that the mutant allele for the dominant trait and the RFLP *B* allele are on the same homolog, and are therefore linked. Assigning a mutant allele to a chromosome by RFLP analysis is the first step in mapping a gene.

■ Region detected by probe

carry the disease locus. Probability analysis is used to determine whether an RFLP marker and a genetic disorder are linked. These probabilities are ultimately expressed as the **logarithm of the odds** or **lod score method** (see Chapter 6). A lod score of 3 means that the odds are 1000 times more likely to have occured if the gene and the RFLP marker are linked than not linked; a lod score of 4 means that the odds are 10,000 times greater in favor of linkage. Lod scores of 3–4 are usually taken as evidence that the gene and the marker are linked.

In mapping human chromosomes, the unit of linkage is the **centimorgan (cM)**, named after the geneticist T. H. Morgan. One centimorgan is equal to a recombination frequency of 1 percent between two loci. The genetic distance between two genes expressed in centimorgans is not directly correlated with the physical distance expressed in nucleotides, but in human chromosomes a distance of 1 cM corresponds roughly to 1–3 million nucleotides of DNA. Many genes have been mapped in humans using techniques such as RFLP analysis, and, by compiling studies from many families, the location of many markers can be determined and genetic maps for human chromosomes can be constructed (Figure 21.3).

Positional Cloning: The Gene for Neurofibromatosis

The use of RFLP analysis in gene mapping can be illustrated by the search for the chromosomal locus of the gene for **type 1 neurofibromatosis (NF1)**. This disorder is in-

herited as an autosomal dominant condition, with an incidence of about 1 in 3000. NF1 is associated with a range of nervous system defects, including benign tumors and an increased incidence of learning disorders.

The mapping of the NF1 gene was accomplished in a series of steps. First, a group of research laboratories compared the inheritance of NF1 and dozens of RFLP markers in multigenerational families. Each RFLP marker represented a specific human chromosome or chromosome region. This produced an **exclusion map**, indicating which RFLPs were not linked to the disease and, therefore, which chromosomes did *not* carry the NF1 locus. This work also produced some evidence of linkage and pointed to chromosomes 5, 10, and 17 as candidates for carrying the NF1 gene. The next stage focused on these chromosomes and produced conclusive evidence that the disorder was closely linked to an RFLP known to reside near the centromere of chromosome 17 (Figure 21.4). Then, more than 30 RFLP markers from this region of chromosome 17 were used to analyze 13,000 individuals from NF1 families, and the gene was mapped to region 17q11.2. Finally, using a collection of genomic clones that spanned a subregion of 17q11.2, the locus for NF1 was identified in 1990 by chromosome walking (see Chapter 18) and DNA sequencing. The identification of the gene was confirmed by finding mutations of the gene in individuals with NF1.

Once the gene was identified, the amino acid sequence of the gene product was reconstructed from the DNA se-

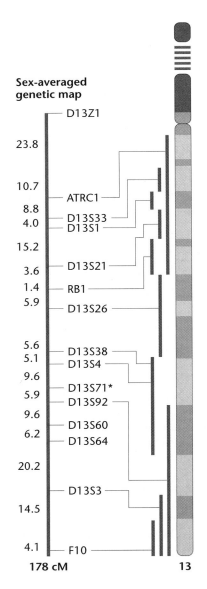

FIGURE 21.3 A genetic and a physical map of human chromosome 13. The genetic map for females is 203 cM, and that for males is 158 cM, reflecting the difference in recombination frequencies between females and males. When the two maps are averaged together, the result is the sex-averaged map of 178 cM shown at left. The location of markers on the physical map is indicated by the brackets adjacent to the chromosome.

tation in the NF1 gene leads to a loss of control of cell growth, causing the production of small tumors that are characteristic of this disorder.

The mapping, cloning, and sequencing of this gene and the identification of the gene product took a little over 3 years, beginning with no direct knowledge of the nature of the gene product or the mutational events that result in the production of the NF1 phenotype. The mapping and cloning of the gene for NF1 described here is an example of **positional cloning**. This recombinant DNA-based method is a departure from previous methods of mapping, which worked from an identified gene product to the gene locus. In positional cloning, the gene can be mapped, isolated, and cloned with no knowledge of the gene product. Using this strategy, an ever-increasing number of human genes are being mapped and isolated.

The Candidate Gene Approach: The Gene for Marfan Syndrome

Another approach to identifying human genes is the investigation of candidate genes. As with positional cloning, the technique begins with the mapping of a chromosomal region by linkage analysis. However, rather than cloning and analyzing all of the surrounding DNA, with the **candidate gene approach** once a chromosomal region is identified it is then scanned for genes that have already been mapped to that region.

The candidate gene approach is most effective when some information about the molecular basis of a genetic disorder is already available. This information provides clues about the function of the defective gene product: Is it likely to be an enzyme, a structural protein, a transport protein, or a hormone? Once the type of protein has been identified, genes that encode similar proteins—and that map to the linked chromosomal region—are identified and examined to see whether mutant alleles are involved in the disorder.

The candidate gene approach was used to map and clone the gene for Marfan syndrome. Marfan syndrome is an autosomal dominant disorder of connective tissue that affects about 1 in 10,000 individuals. It has been speculated that Abraham Lincoln, the 16th president of the United States, had this disorder. Given the nature of the defect, researchers focused on genes encoding structural proteins of connective tissue as the primary candidates. Although this approach seems straightforward, many connective tissue proteins exist, making it difficult to identify the primary gene product that is involved in the disorder.

Fibrillin is a connective tissue protein first identified in 1986 as a component of the eye, aorta, and other

quence and found to include 2485 amino acids. By searching databases containing protein-sequence information, the NF1 gene product was found to be similar to proteins that play a role in signal transduction. Further analysis confirmed that the NF1 protein, **neurofibromin**, is involved in the transduction of intracellular signals and the down-regulation of a gene that controls cell growth. Mu-

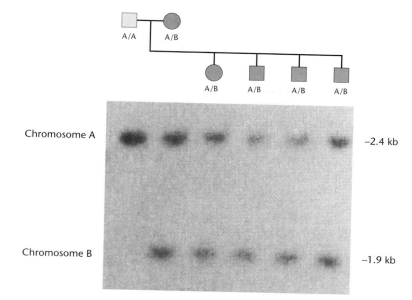

FIGURE 21.4 The segregation of a 2.4-kb RFLP allele with type 1 neurofibromatosis (NF1) in each of four affected offspring and their mother. This RFLP is detected by probe pA10-41, which is known to be a DNA segment near the centromere of human chromosome 17. On the basis of this and results from tests using other probes, the locus for NF1 was assigned to chromosome 17.

elastic tissues, all of which are affected in Marfan syndrome. The gene for fibrillin had been mapped to the long arm of chromosome 15. When RFLP studies using large multigenerational families with Marfan syndrome established a linkage between markers on chromosome 15 and Marfan syndrome, the fibrillin gene became a candidate for Marfan syndrome. In the next step, the fibrillin gene was identified, cloned, and sequenced. Subsequently, a mutation in the fibrillin gene was identified in three unrelated families with Marfan syndrome. In all three families, the mutation was a missense mutation leading to the substitution of a proline for arginine at position 239 in the fibrillin gene product. Unaffected family members carried fibrillin genes with arginine at position 239. Finding a mutant fibrillin gene in affected family members was the final link in establishing that Marfan syndrome was caused by defects in the fibrillin gene.

Fluorescent In Situ Hybridization (FISH) Gene Mapping

Genes can be mapped directly to metaphase chromosomes in a process called *in situ* hybridization (see Chapter 11). Chromosomal preparations on microscope slides are denatured, to convert the double-stranded DNA in the chromosomes into single strands. A labeled probe, containing all or part of a cloned gene, is hybridized to the chromosomal preparation. The probe can be labeled with a radioactive isotope, antibodies or fluorescent dyes. In contrast to the use of gels in Southern blotting, the re-

sulting hybrids in *in situ* hybridization are examined and detected under a microscope.

When fluorescent dyes are used to label the probes, a fluorescence microscope is used to examine the chromosome preparation and to determine the site of hybridization. This technique, known as FISH (fluorescent *in situ* hybridization, see Chapter 11), can be used to directly assign a cloned gene to a chromosomal locus (Figure 21.5).

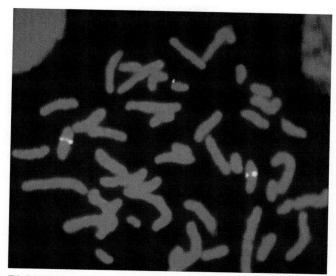

FIGURE 21.5 Localization of a gene by fluorescent *in situ* hybridization (FISH). The sites of hybridization appear as orange dots on the sister chromatids of the chromosome pair.

The technique has a resolution of about 1Mb and is now being used routinely to map human genes.

Genetic disorders can be diagnosed

Often, diagnosis for genetic disease is performed prenatally. The most widely used methods for prenatal diagnosis of genetic disorders are **amniocentesis** and **chorionic villus sampling** (**CVS**). In amniocentesis, a needle is used to withdraw amniotic fluid (Figure 21.6). The fluid and the cells it contains can be analyzed for chromosomal or single-gene disorders. In CVS, a catheter is inserted into the uterus and used to retrieve a small tissue sample of the fetal chorion. This tissue is used for cytogenetic, biochemical, and recombinant DNA-based testing.

Coupled with these methods of recovering samples, recombinant DNA technology has proven to be a highly sensitive and accurate tool for the prenatal detection of genetic disorders. The use of cloned DNA sequences has expanded the range of prenatal testing because it allows direct examination of the fetal genotype, rather than relying on the few tests available for normal or mutant gene products. This is particularly important since, even in cases where tests are available, often the gene product cannot be detected before birth. For example, defects in

the adult β-globin protein cannot be detected prenatally because the β-globin gene is not expressed until a few days after birth.

Genetic Disorders Associated with Mutations in the β-globin Gene

Mutations in the β-globin locus lead to a wide range of genetic disorders. β-thalassemia is an autosomal recessive disorder associated with decreased or absent β-globin production. At the molecular level, this condition can be caused by a variety of deletions or point mutations. One such deletion covers about 600 nucleotides, and includes the third exon of the gene. Restriction enzyme digestion and Southern blot hybridization (see Chapter 18) to a cloned β-globin probe can be used to diagnose the genetic status of a fetus and other members of an affected family (Figure 21.7).

Deletions, however, are relatively rare mutational events. More commonly, mutations are associated with changes in one or a small number of nucleotides (point mutations). However, if changes in a single nucleotide take place in an exon, they can have devastating clinical and phenotypic effects. Therefore more sensitive techniques (discussed later) have been developed to screen for these small-scale changes in sequence.

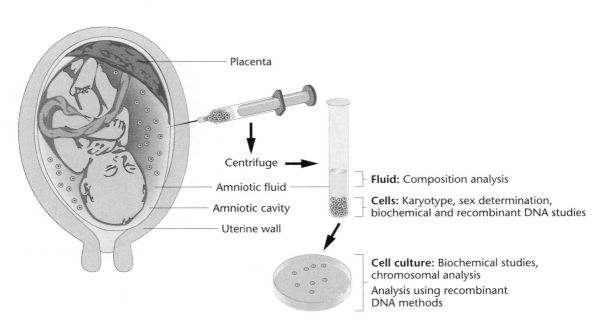

FIGURE 21.6 The technique of amniocentesis. The position of the fetus is first determined by ultrasound, and then a needle is inserted through the abdominal and uterine wall to recover fluid and fetal cells for cytogenetic and/or biochemical analysis.

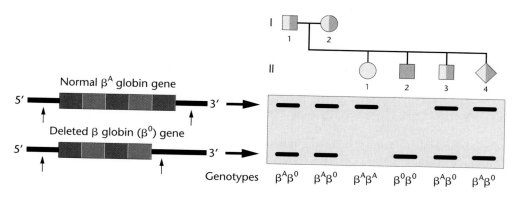

FIGURE 21.7 Diagnosis of β-thalassemia caused by a partial deletion of the β-globin gene. The family pedigree is shown positioned above each individual's genotype on a Southern blot. The normal β-globin gene (β^A) contains three exons and two introns. The deleted β-globin gene (β^0) has the third exon deleted. Arrows indicate the cutting sites for restriction enzymes used in this analysis. The normal gene produces a larger fragment (shown as the top row of fragments on the Southern blot); the smaller fragments produced by the deleted gene are represented at the bottom of the gel. The genotype of each individual in the pedigree can be determined from the pattern of bands on the blot, and these are shown below the blot.

Prenatal Diagnosis of Sickle-Cell Anemia

Sickle-cell anemia is an autosomal recessive condition common in people with family origins in areas of West Africa, the Mediterranean basin, and parts of the Middle East and India (see discussion in Chapter 14). Sickle-cell anemia is also caused by a mutation in the β-globin gene, this time a point mutation that results in an amino acid substitution in the β-globin protein. The single nucleotide change also, coincidentally, eliminates a cutting site for the restriction enzymes *Mst*II and *Cvn*I. As a result, this mutation alters the pattern of restriction fragments seen on Southern blots. These characteristic RFLP patterns can be used for the prenatal diagnosis of sickle-cell anemia and to determine the genotypes of parents and other family members who may be heterozygous carriers of this condition.

For prenatal diagnosis, fetal cells are obtained by amniocentesis or by CVS. DNA is extracted from these cells and digested with a restriction enzyme, such as *Mst*II. This enzyme cuts thrice in the region of the normal β-globin gene, producing two small DNA fragments. In the mutant allele, the second *Mst*II site has been destroyed by the mutation, producing one large restriction fragment (Figure 21.8). The restriction-digested DNA fragments are separated by gel electrophoresis, transferred to a nylon membrane, and visualized by Southern blot hybridization.

In the example shown in Figure 21.8, the parents (I-1 and I-2) are both heterozygous for the mutation. Digestion of DNA from each parent produces a large band (the mutant allele) and two smaller bands (the normal allele). Their first child (II-1) is homozygous normal because she has only the two smaller bands. The second child

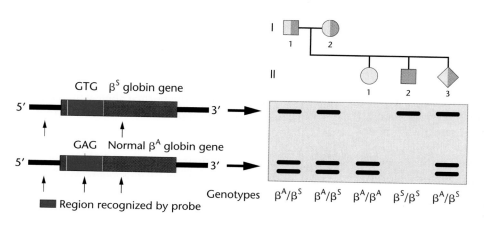

FIGURE 21.8 Southern blot diagnosis of sickle-cell anemia. Arrows represent the location of restriction enzyme cutting sites. In the mutant (β^S) globin gene, a point mutation (GAG → GTG) has destroyed a restriction enzyme cutting site, resulting in a single large fragment on a Southern blot. In the pedigree, the family has one unaffected homozygous normal daughter (II-1), an affected son (II-2), and an unaffected fetus (II-3). The genotype of each family member can be read directly from the blot, and these are shown below the blot.

(II-2) has sickle-cell anemia; he has only one large band and is homozygous for the mutant allele. The fetus (II-3) has a large band and two small bands and is, therefore, heterozygous for sickle-cell anemia. He or she will be unaffected, but will be a carrier.

Only about 5 to 10 percent of all point mutations can be detected by restriction analysis. However, if a mutant gene has been well characterized and the mutated region has been sequenced, it is possible to use synthetic oligonucleotides as probes to detect mutant alleles.

Single Nucleotide Polymorphisms and Genetic Screening

A method for distinguishing between alleles that differ by as little as a single nucleotide involves the use of synthetic probes known as **allele-specific oligonucleotides (ASO)**. In contrast to restriction enzyme analysis, which is restricted to cases where a mutation changes a restriction site, the use of ASOs offers increased resolution and wider application. Under the proper conditions, an ASO will hybridize only with its complementary sequence and not with other sequences that might vary by as little as a single nucleotide. A method using ASOs and PCR analysis is now available to screen for sickle-cell anemia. In this method, DNA from white blood cells is extracted and denatured. A region of the β-globin gene from these cells is amplified by PCR. An aliquot of the amplified DNA is spotted onto filters, and each filter is hybridized to an ASO (Figure 21.9). After visualization, the genotype can be read directly from the filters. Using an ASO for the normal sequence [Figure 21.9(a)], the homozygous normal (AA) genotype produces a dark spot (two copies of the normal allele), and the heterozygous genotype (AS) produces a light spot (one copy of the normal allele). The homozygous recessive sickle-cell genotype will not bind the probe, and no spot will be visualized. Using a probe for the mutant allele [Figure 21.9(b)], the pattern is reversed. This rapid, inexpensive, and highly accurate technique is becoming the method of choice for the diagnosis of a wide range of genetic disorders caused by point mutations.

In cases where the nucleotide sequence of the normal allele is known and where the molecular nature of the mutant allele is known, ASOs can be synthesized directly from the normal and mutant copies of the gene and used to screen for heterozygous carriers of the genetic disorder. For example, in cystic fibrosis a deletion (called Δ508) represents 70 percent of all mutant alleles. Cystic fibrosis is an autosomal recessive disorder associated with a defect in a protein called the **cystic fibrosis transmembrane conductance regulator (CFTR)**, which regulates chloride ion transport across the plasma membrane. To detect het-

DNA extracted from white blood cells

Region covered by ASO probes

DNA is spotted onto binding filters, hybridized with ASO probe

(a) Genotypes   AA   AS   SS

Normal (β^A) ASO: 5′ – CTCCTGAGGAGAAGTCTGC – 3′

(b) Genotypes   AA   AS   SS

(β^S) ASO: 5′ – CTCCTGTGGAGAAGTCTGC – 3′

FIGURE 21.9 Genotype determinations using allele-specific oligonucleotides (ASOs). In this technique, the β-globin gene is amplified by PCR using DNA extracted from blood cells. The amplified DNA is denatured and spotted onto strips of DNA-binding filters. Each strip is hybridized to a specific ASO and visualized on X-ray film after hybridization and exposure. If all three genotypes are hybridized to an ASO from the normal β-globin gene, the pattern in (a) would be observed: AA-homozygous individuals have normal hemoglobin that has two copies of the normal β-globin gene and would show heavy hybridization; AS-heterozygous individuals carry one normal β-globin gene and one mutant gene and would show weaker hybridization; SS homozygous sickle-cell individuals carry no normal copy of the β-globin gene and would show no hybridization to the ASO probe for the normal β-globin gene. (b) The same genotypes hybridized to the probe for the sickle-cell β-globin gene would show the reverse pattern: no hybridization by the AA genotype, weak hybridization by the heterozygote (AS), and strong hybridization by the homozygous sickle-cell genotype (SS).

erozygous carriers for the Δ508 mutation, allele-specific oligonucleotides are made by PCR from cloned samples of the normal allele and the mutant allele. DNA prepared from the white blood cells of the individuals to be tested is applied to a nylon filter and hybridized to each of the ASOs (Figure 21.10) In affected individuals, only the ASO made from the mutant allele will hybridize; in heterozygotes, both ASOs will hybridize; and, in normal homozygotes, only the ASO from the normal allele will hybridize.

Because CF affects approximately 1 in 2000 individuals of northern European descent, screening for CF may be used in certain populations to detect and advise heterozygous carriers of their status for CF. However, since not all of the known mutations for this gene (over 500 mutations have been identified) can be screened, a negative re-

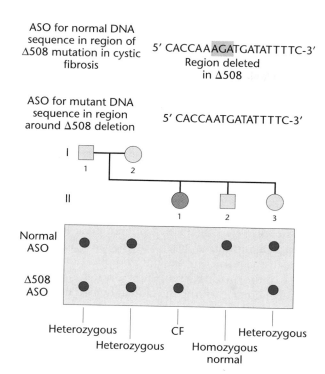

ASO for normal DNA sequence in region of Δ508 mutation in cystic fibrosis

5′ CACCAA AGATGATATTTTC-3′
Region deleted in Δ508

ASO for mutant DNA sequence in region around Δ508 deletion

5′ CACCAATGATATTTTC-3′

FIGURE 21.10 Screening for cystic fibrosis (CF) by allele-specific oligonucleotides (ASOs). ASOs for the region spanning the most common mutation in CF, a three-nucleotide deletion (Δ508), are prepared from normal CF genes and Δ508 CF genes. In screening, the CF gene is amplified by PCR using DNA extracted from blood samples and spotted on a DNA-binding membrane. The membrane is hybridized to a mixture of the two ASOs. The genotype of each family member can be read directly from the filter. DNA from I-1 and I-2 hybridizes to both ASOs, indicating that they carry a normal allele and a mutant allele and are therefore, heterozygous. The DNA from II-1 hybridizes only to the Δ508 ASO, indicating that she is homozygous for the mutation and has cystic fibrosis. The DNA from II-2 hybridizes only to the normal ASO, indicating that he carries two normal alleles. II-3 has two hybridization spots and is heterozygous.

sult does not eliminate someone as a heterozygous carrier, and it is possible that all mutations of the CF gene have not been identified. Such screening will probably become commonplace once tests can cover 98 to 99 percent of all possible CF mutations.

DNA Chips and Genetic Screening

DNA probes similar to allele-specific nucleotides are being coupled with the technology of the semiconductor industry to produce **DNA chips**. The chips themselves are made of glass and are about one-half inch square. The chip is divided into fields (smaller squares), each of which is about half the width of a human hair. Each field contains linker

molecules to which are attached DNA probes about 20 nucleotides in length. Along a row, the sequence of the probe differs by one nucleotide from field to field. Thus, a set of four fields (one for each nucleotide) is needed to test a given position for its nucleotide content. The current generation of chips can hold 280,000 fields, but chips with 1.6 million fields are now being developed. For genetic testing, DNA is extracted from cells and cut with one or more restriction enzymes. The resulting fragments are tagged with a fluorescent dye, melted into single strands, and pumped into the chip. Fragments with a nucleotide sequence that exactly matches the probe sequence will bind, and those with a sequence that does not match are washed off the chip. A laser scanner reads the chip and detects the pattern of fluorescence (Figure 21.11). The fluorescent pattern of hybridization is analyzed by a linked software program, and the data are presented in the form of a nucleotide sequence.

DNA chips are already being used to scan for mutations in the *p53* gene, which is mutated in 60 percent of all cancers, and to screen for mutations in the *BRCA*1 gene, which predispose women to breast cancer.

In addition to testing for mutations in single genes, DNA chips can be made to screen thousands of different genes simultaneously. Chips have been programmed with the 6500 human genes that have already been sequenced, and chips with 50,000 genes are in the planning stages.

This technology will make it possible to analyze someone's DNA for dozens or hundreds of diseases, including those that predispose for heart attacks, diabetes, Alzheimer disease, and other genetically defined disease subtypes.

GeneChip® Expression Analysis

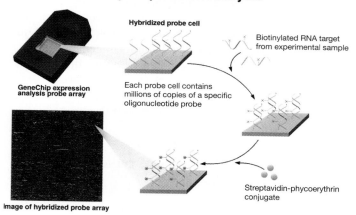

FIGURE 21.11 A DNA chip in its cassette holder (left), the image pattern generated by hybridization between the oligonucleotide probes and the fluorescently labeled DNA target (lower), and a diagram showing the oligonucleotide probe bound to the target DNA (right). This chip contains over 280,000 fields for use in genetic testing.

Genetic testing creates ethical dilemmas

In an earlier section, we described examples of two types of genetic testing, prenatal diagnosis and screening for heterozygous carriers of recessive disorders. Using current technology, genetic testing can also be used to predict risk of disease, identifying those who are presently healthy but are at high risk of contracting a genetic disease in the future, and to establish a prognosis for those with a genetic disorder. In the next few years, DNA chips may be used to test for 50–100 diseases, including many that an individual may not develop for years or decades. These advances will profoundly affect our health, reproductive patterns, and medical care. However, the use of this technology also raises legal, social, and ethical issues that have not been resolved. For example, what should people know before deciding to have a genetic test? How can we protect the information revealed by a genetic test? How can we define and prevent genetic discrimination?

Many of the potential risks and benefits of genetic testing are still unknown. Although we have the ability to test for many genetic diseases, there are almost no effective treatments to cure or improve these disorders. With present technology, the fact that a genetic test gives a negative result does not rule out the future development of the disease, nor does a positive result always mean that one will get the disease.

The development and implementation of public policy and laws about genetic testing have lagged behind the development and application of the technology for testing. The **Ethical, Legal, and Social Implications (ELSI)** Program of the Human Genome Project (described later in this chapter) has set up task forces to identify issues related to genetic testing. This program is charged with developing recommendations for policy makers and lawmakers that will retain the benefits of genetic testing, but reduce or eliminate the potential harm. In addition, other groups made up of scientists, health-care professionals, lawmakers, ethicists, and consumers are debating these issues and formulating policy options.

Gene therapy is a way of treating genetic disorders

Gene *products* such as insulin have been used for decades in therapeutic treatment. Methods for the introduction of specific *genes* into mammalian cells, originally developed as a research tool, are now being used to treat genetic disorders, a process known as **gene therapy**. In theory, gene therapy transfers a normal allele into a somatic cell that carries one or more mutant alleles. The delivery of these

structural genes and their regulatory sequences is accomplished using a vector or gene transfer system.

Several methods can be used to transfer genes and their regulatory sequences into human cells. These include the use of viruses as vectors, the chemically assisted transfer of genes across cell membranes, and the fusion of cells with artificial vesicles containing cloned DNA sequences.

The most common method of gene transfer uses retroviral DNA as a vector. The first retroviral vectors used in a clinical trial were based on a mouse virus called Moloney murine leukemia virus (Figure 21.12). This example illustrates the procedure used to create the recombinant virus. First, a cluster of three genes is removed from the virus and the cloned human gene is inserted. After packaging into a viral protein coat, the recombinant vector can infect cells, but is replication-deficient because of the missing viral genes. Once it is in the cell, the viral genome carrying the cloned human gene moves to the nucleus and integrates into a chromosome. The wild-type gene is now a part of the genome of the infected cell.

Several heritable disorders are currently being treated with gene therapy, such as **severe combined immunodeficiency (SCID)**, **familial hypercholesterolemia**, and **cystic fibrosis**. In the next sections we will describe the first attempt to use gene therapy to treat a young girl with SCID and then give an overview of the trials that are currently underway in several animal model systems using a new generation of viral vectors. Finally, we will discuss the numerous ethical issues raised by the use of gene therapy.

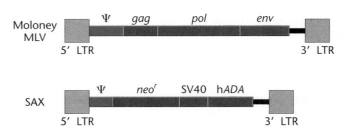

FIGURE 21.12 Retroviral vectors constructed from the Moloney murine leukemia virus (Moloney MLV). The native MLV genome contains a Ψ sequence required for encapsulation and genes that encode viral coat proteins (gag), an RNA-dependent DNA polymerase (pol), and surface glycoproteins (env). At each end, the genome is flanked by long terminal repeat (LTR) sequences that control transcription and integration into the host genome. The SAX vector retains the LTR and Ψ sequences, and it includes a bacterial neomycin resistance (*neo*^{r}) gene that can be used as a selective marker. As shown, the vector carries a cloned human adenosine deaminase (*hADA*) gene, fused to an SV40 early region promoter/enhancer. The SAX construct is typical of retroviral vectors that are used in human gene therapy.

FIGURE 21.13   David, a boy born with severe combined immunodeficiency (SCID), survived by being isolated in a germ-free environment. He died at the age of 12 following a bone marrow transplant undertaken to provide him with a functional immune system.

Severe Combined Immunodeficiency (SCID)

In severe combined immunodeficiency (SCID), affected individuals have no functional immune system and usually die from what would otherwise be minor infections (Figure 21.13). An autosomal form of SCID is caused by a mutation in the gene encoding the enzyme **adenosine deaminase** (**ADA**). Treatment starts with the isolation from the patient of a subpopulation of white blood cells called T cells (Figure 21.14). These cells, which are part of the immune system, are mixed with a genetically modified retrovirus carrying a normal copy of the ADA gene (Figure 21.14). The virus infects the T cells, inserting a functional copy of the ADA gene into the cell's genome. The genetically modified T cells are grown in the laboratory to ensure that the transferred gene is expressed, and the patient is treated by injecting a billion or so of the altered T cells into the bloodstream.

Gene therapy began in 1990, with the treatment of a young girl suffering from SCID. Three years after treatment she had a normal ADA gene in more than 50 percent of her T cells. This level allowed her to lead a normal life, but unfortunately the cells carrying the wild-type gene had a limited life span, and so she has had to have additional treatments. In addition, a second child who was treated a short time later had the normal gene in only 0.1 to 1 percent of her white blood cells after treatment, a level that was not high enough to be effective.

Other gene therapy trials, such as those for cystic fibrosis, have also had poor results, most of which have been traced to inefficient vectors. These first-generation vectors, such as Moloney virus and adenovirus, which causes the common cold, have several drawbacks. First, integration of the retroviral genome (including the cloned human gene) into the host cell genome occurs only if the host cells are replicating their DNA. Under *in vivo* conditions, there is little DNA synthesis in many highly differentiat-

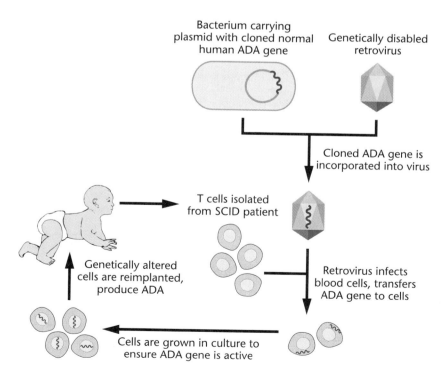

FIGURE 21.14   Gene therapy for treatment of severe combined immunodeficiency (SCID), a fatal disorder of the immune system caused by lack of the enzyme adenosine deaminase (ADA). The cloned human ADA gene is transferred into a viral vector, which is used to infect white blood cells removed from the patient. The transferred ADA gene is incorporated into a chromosome and becomes active. After growth to enhance their numbers, the cells are reimplanted into the patient, where they produce ADA, allowing the development of an immune response.

ed cell types that are suitable target tissues. Second, most of these viruses eventually elicit an immune response in the host. Third, insertion of viral genomes into the host chromosome can inactivate or mutate an indispensable gene. Fourth, retroviruses have a low cloning capacity and cannot carry inserted sequences much larger than 8 kb; many human genes, even without introns, exceed this size. Finally, there is the possibility of producing an infectious virus if a recombination event takes place between the vector and retroviral genomes already present in the host cell. For these reasons, other viral vectors and strategies for targeting cells are being developed.

The Future of Gene Therapy: New Vectors and Target-Cell Strategies

The disappointing results from gene therapy trials that used first-generation vectors led to a widespread crisis of confidence in gene therapy in the early 1990s in both the medical and scientific communities. However, with some promising results in several animal model systems, the tide may finally have turned back. This is thanks primarily to the use of new vectors that both circumvent several of the problems encountered with earlier vectors and have several ingenious design features that allow the levels of gene product to be regulated.

In early 1999 James Wilson from the Institute for Human Gene Therapy at the University of Pennslvania reported the successful production of the blood hormone erythropoeitin, which triggers the production of red blood cells in the bone marrow in rhesus monkeys and mice whose muscle cells had been injected with a viral vector containing the erythropoeitin gene. Furthermore, these animals would only produce the hormone when they were fed a pill of the antibiotic rapamycin. The results are particularly encouraging because the levels of the hormone are quite high, the presence of the foreign DNA has not triggered an immune response, and the gene can be repeatedly activated.

The tremendous success of these trials can be attributed in large part to the new viral vector that Wilson used. Adeno-associated virus (see Table 21.1 for a list of several second-generation vectors) is related to adenovirus, but has several key advantages; it can incorporate into non-replicating DNA, and it is a very small virus, so it does not typically elicit an immune response. Furthermore, the vector used for gene therapy had one additional modification that allowed Wilson to turn the erythropoietin gene on at will—a rapamycin-responsive promoter. Therefore, the gene will only be expressed when the cells are exposed to rapamycin. This therapy is being considered for patients who have low blood counts, such as dialysis patients, who currently have regular injections of erythropoietin itself.

In addition to treating heritable genetic disorders and patients on dialysis, research involving gene therapy is now being directed toward the treatment of skin cancer, breast cancer, brain cancer, and AIDS. In the last few years, at least a dozen biotechnology companies have been founded specifically to develop products for gene therapy, and in the twenty-first century it is possible that it will be a commonplace method of treatment for a large number of genetic disorders.

Ethical Issues and Approaches to Gene Therapy

Gene therapy obviously raises many ethical concerns, and most are still the source of intense debate. As a result, all gene therapy trials currently underway or in the planning stages are restricted to the use of somatic cells as targets for gene transfer. The ethical guidelines for undertaking gene therapy in its present form are well established. This experimental treatment is initiated only after careful review at several levels, and the trials are monitored to protect the interests of the patient. Called **somatic gene therapy**, only one individual is affected, and the therapy is done with the permission and informed consent of the patient.

There are two other potential approaches in gene therapy that have yet to be approved primarily because of the unresolved ethical issues surrounding them. The first is **germ line therapy**, where germ cells (the cells that give rise to the gametes, i.e. the sperm and eggs) or mature gametes are used as targets. In this approach, the corrective DNA will be incorporated into all the cells of the individual produced from the genetically-altered gamete, including his or her own germ cells. This means that many individuals in future generations may also be affected,

TABLE 21.1 Viral Vectors Used for Gene Therapy

	Adenovirus	Adeno-associated Virus	Herpes Virus
Cell targets	Lung cells, cells of respiratory tract	Fibroblasts, T cells	Nerve cells, glial cells
Cloning capacity	7–35 kb	3 kb	Up to 150 kb
Integration into chromosomal DNA	No	Yes	No, but retained in nucleus

without their consent. Is this ethical? Do we have the right to make this decision for future generations? Thus far, the concerns have outweighed the potential benefits, and such research is prohibited.

The second type of gene therapy that raises an even greater ethical dilemma is termed **enhancement gene therapy**, whereby human potential might be enhanced for some desired trait. Such a possible use of gene therapy is extremely controversial and strongly opposed in many quarters. Should genetic technology be used to enhance human potential? For example, if such genes were identified and cloned, should it be permissible to use gene therapy to increase height, to enhance athletic ability, or to extend intellectual potential? Presently, the consensus is that enhancement therapy, like germ line therapy, is an unacceptable use for gene therapy. However, the issues are still being debated, and are unresolved. The outcome of these debates may affect not only the fate of individuals, but that of our species.

DNA fingerprints are based on variability in repetitive sequences

As discussed earlier, restriction sites in the human genome (their presence or absence) can be used as genetic markers. A second type of variation, discovered in the mid-1980s, depends on variation in the length of repetitive DNA sequence clusters. These polymorphisms in human DNA provide a unique molecular pattern for an individual and serve as the basis for a technique called **DNA fingerprinting**. DNA fingerprinting is now being used for a wide variety of applications, from identifying individuals for paternity and forensics cases to establishing degrees of genetic relatedness between individuals in the same or different species.

Minisatellites and VNTRs

One of the most useful forms of restriction fragment-length polymorphisms (RFLPs) arises from variations in the number of tandemly repeated DNA sequences present between two restriction enzyme sites. These sequences are **minisatellites**, or clusters of nucleotides from 2 to 100 nucleotides in length. For example, the base sequence GGAAGGGAAGGGAAGGGAAG is composed of four tandem repeats of the five-nucleotide sequence GGAAG. Clusters of such sequences are widely dispersed in the human genome. Typically, each repeat contains between 14 to 100 nucleotides, and the number of repeats at each locus ranges from 2 to more than 100. These loci are known as **variable-number tandem repeats (VNTRs)**, introduced

initially in Chapter 19 as examples of middle repetitive DNA. The number of repeats at a given locus is variable, and each variation constitutes a VNTR allele. Many loci have dozens of alleles each; as a result, heterozygosity is common.

A pattern of bands is produced when VNTR sequences are cut with restriction enzymes and visualized by Southern blotting. This pattern is one example of what is known as a *DNA fingerprint* (Figure 21.15). These patterns are the equivalent of actual fingerprints because the pattern of bands is always the same for a given individual, no matter what tissue is used as the source of the DNA, but the pattern varies from individual to individual. In fact, there is so much variation in the band pattern from individual to individual that, theoretically, each person's pattern is unique. DNA fingerprint analysis can be performed on very small samples of material [less than 60 microliters (μl) of blood] and on samples that are quite old (VNTR analysis has been performed on Egyptian mummies over 2400 years old), thus increasing its usefulness in a variety of circumstances.

Forensic Applications of DNA Fingerprints

Since 1988, DNA fingerprints have been used as evidence in criminal trials in the United States. The method has also been used in a wide range of other applications, including immigration cases, disputes involving purebred dogs, paternity cases, and in animal conservation studies. At present, just over a dozen different VNTR probes are used in standard forensic testing (Figure 21.16). Most of these are on different chromosomes, and each probe can be used to visualize the VNTR pattern at a particular locus. Standard forensic tests employing four to ten different VNTR probes are used in developing a detailed DNA fingerprint profile.

The results of DNA fingerprints are analyzed and interpreted using statistics, probability, and population genetics. In this interpretation, the frequency of each VNTR allele in the population is calculated, and these are multiplied together to give a combined frequency, which is a calculation of how often the observed DNA fingerprint might occur in the general population. The use of a number of VNTR probes increases the accuracy of the combined frequency. For example, if the frequency of locus 1 is determined to be 1 in 333, and the frequency of locus 2 is 1 in 83, their combined frequency is equal to the product of their individual frequencies, or 1 in 28,000. While this overall frequency may not serve as convincing evidence, if the frequency of a third locus (1 in 100) and a fourth locus (1 in 25) are included in the calculations, the combined frequency becomes 1 in 70 million. This means that the chance than anyone has this combination of al-

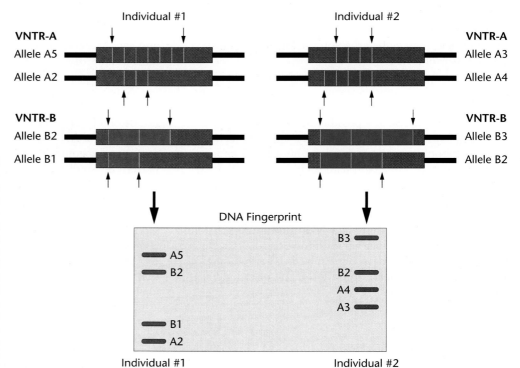

FIGURE 21.15 VNTR loci and DNA fingerprints. VNTR alleles at two loci (A and B) are shown for each individual. Arrows mark restriction cutting sites flanking the VNTRs. Restriction digestion produces a series of fragments that can be detected as bands on a Southern blot (below). Because of differences in the number of repeats at each locus, the overall pattern of bands is distinct for each individual, even though one band is shared (the band representing the B2 allele). Such a pattern is known as a DNA fingerprint. 🌐 GenCDX

leles is about 1 in 70 million (about 4 individuals in the U.S. population). Thus, when a match is made between DNA obtained at a crime scene and the DNA of a suspect, it is highly likely that the suspect was somehow involved at the crime scene. Currently, probabilities that *an individual other than the suspect* could also match the DNA fingerprint is as remote as 1 in 1.5 billion!

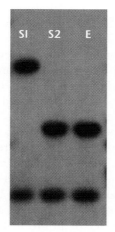

FIGURE 21.16 DNA fingerprinting in a forensic case. The DNA profile of suspect 2 (S2) matches that of the blood sample obtained as evidence E.

Genome projects rely on recombinant DNA technology

Since the 1970s, when the first genes were identified and cloned, a tremendous amount has been learned about the role of genes and their products in biological processes; however, until recently only a small percentage of the genes in any organism were identified. This has all changed with the worldwide effort to characterize the entire genome of a number of organisms, a project which has only recently been feasible, thanks to recombinant DNA techniques. The approach to mapping genes has therefore shifted from a reliance on generating and mapping mutants, which limits the analysis of the genome to regions coding for genes, to a direct analysis of the entire genome of an organism.

Genome analysis begins with the construction of genomic libraries that contain fragments of all of the genomic DNA (genomic clones) of an organism. Overlapping clones are assembled to establish genetic and physical maps encompassing the entire genome. The final step involves sequencing the entire genome. The result is a genomic map, with all regions in the genome identified by location and nucleotide sequence. With the entire genome of an organism characterized, studies on particular processes or diseases can be performed that begin with a DNA sequence, rather than with a gene product. For example, gene sequences can be analyzed and their derived amino acid sequence determined, and gene-specific mutants can be designed to study the function of the gene product.

Genomes of nearly two dozen organisms have been sequenced, and dozens of other genome projects are underway. Several of these projects are being carried out under the auspices of the Human Genome Project. Those nonhuman organisms that are part of the Genome Project were selected for several reasons, including genome size, the availability of detailed genetic maps, and the value of making comparisons about gene number, location, and function across species ranging from bacteria to humans.

The Human Genome Project: An Overview

The **Human Genome Project** is an international, coordinated effort to determine the sequence of the 3.3 billion nucleotide pairs in the haploid human genome. In the United States, the project began in 1990 as a joint program of the National Institutes of Health and the U.S. Department of Energy. Other countries, notably France, Britain, and Japan, began similar projects, which are now coordinated by an international organization, the Human Genome Organization (HUGO).

The task of sequencing the human genome is proceeding in a series of well-defined stages (Figure 21.17). Although the plan is shown as a series of steps, work has progressed more or less simultaneously on all phases of the project in laboratories around the world. The steps in the Project are:

1. The construction of high-resolution genetic maps for each human chromosome, with a 2–5 cM average distance between markers. As you recall, genetic maps are based on recombination frequencies and give the order and distance between different loci along the length of a chromosome. For this phase of the project, maps were assembled using identified genes, RFLPs, and other markers. Genetic maps for each chromosome, incorporating about 15,000 markers, were completed in 1995. These maps have an average distance of 2 million base pairs of DNA between markers.

2. Physical mapping of each chromosome. Physical maps show the order and physical distance between markers expressed in base pairs of DNA. Several methods have been used in making these maps, and the goal has been the creation of a physical map of each chromosome consisting of at least 30,000 markers, spaced at intervals of about 100 kb.

3. Assembling a set of overlapping clones covering each chromosome. This stage of the project uses YACs and other vectors such as BACs (see Chapter 18) to generate sets of overlapping clones that completely cover each chromosome. Generating the clones is only the first step in this stage, however. The real problem is in identifying and ordering a collection of clones that overlap each other and cover the chromosome. The original strategy has been to divide a chromosome into a number of segments and to isolate and map clones that cover each segment. As described later, this has largely been replaced by a new strategy, called shotgun cloning.

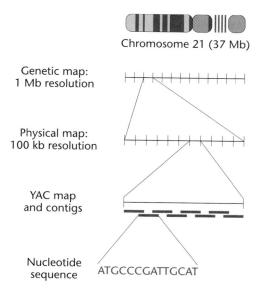

Chromosome 21 (37 Mb)

Genetic map: 1 Mb resolution — Genetic map of markers, such as RFLPs, STSs spaced about 1 Mb apart. This map is derived from recombination studies

Physical map: 100 kb resolution — Physical map with RFLPs, STSs showing order, physical distance of markers. Markers spaced about 100,000 base pairs apart

YAC map and contigs — Set of overlapping ordered clones covering 0.5–1.0 Mb

Nucleotide sequence — ATGCCCGATTGCAT — Each overlapping clone will be sequenced, sequences assembled into genomic sequence of 3.2×10^9 nucleotides, 37 Mb of which will be from chromosome 21

FIGURE 21.17 An overview of the strategy used in the Human Genome Project. The first goal, achieved in 1995, was to have a genetic map of each chromosome, with markers spaced at distances of about 1 Mb (1 million base pairs of DNA). This work was accomplished by finding markers such as RFLPs and STSs and assigning them to chromosomes. Once assigned to chromosomes, the markers' inheritance was observed in heterozygous families to establish the order and distance between them (a genetic map). In the second stage, the goal was to prepare a physical map of each chromosome (our example uses chromosome 21, the smallest chromosome) containing the location of markers spaced about 100,000 base pairs apart. This goal has now been achieved. The third stage involves the construction of a set of overlapping clones, in yeast artificial chromosomes (YACs) or other vectors that cover the length of the chromosome. The last stage will be the sequencing of the entire genome. Sequencing on selected parts of the genome has started.

4. Sequencing the genome. The ultimate goal of the project is to determine the nucleotide sequence of the 3.3 billion nucleotides in the human genome. This task is perhaps the rate-limiting step in the project, and it may require the development of new technology for sequencing DNA, as well as new technology for information storage, analysis, and retrieval. This last stage of the project has started with the establishment of six centers in the United States to improve technologies related to sequencing, and to begin large-scale sequencing of the genome.

The Human Genome Project: A Progress Report

It is clear that the Human Genome Project (HGP) is fulfilling its goals in a timely and cost-efficient manner. By the end of 1998, the HGP had completed or exceeded all its goals for the 1993–1998 period. High-resolution genetic maps are now complete, and although some work is needed to refine the map in certain chromosome regions, this goal was accomplished in 1994 with the publication of a map with 1cM resolution.

The physical maps for each human chromosome are now complete. The original goal was to prepare maps with markers at intervals of 1 million base pairs (1Mb). By 1998, over 52,000 markers were mapped and this phase of the project is now closed.

The HGP has also successfully completed sequencing the genomes of three of the five model organisms, *E. coli* (bacteria), *S. cerevisiae* (yeast), and *C. elegans* (a metazoan roundworm). Sequencing the genome of the fruit fly (*D. melanogaster*) is scheduled to be completed in 2002, and the mouse genome should be sequenced by 2005.

The ultimate goal of the HGP, to sequence the 3.3 billion nucleotides in the human genome, is already underway and about 6.1 percent of the genome has been completely sequenced. Six centers in the U.S. are working to sequence parts of the genome and to develop new technologies related to sequencing. The focus of the project in the next five years will be on sequencing the human genome. It is hoped that advances in sequencing capacity will allow one-third of the genome to be completed by 2001, with about 90 percent of the genome covered by a "working draft" based on mapped clones. Most of the sequencing efforts now use a version of the shotgun method (see Chapter 18), developed by private industry and used with spectacular success to sequence the genomes of several prokaryotic organisms. The new goal is to sequence the entire genome by 2003, two years ahead of schedule.

In addition to the public efforts led by the Human Genome Project, in 1998 J. Craig Venter, the scientist who developed the shotgun method of sequencing, announced the formation of a company to complete the sequencing of the human genome in 2001. In partnership with a corporation that manufactures automated DNA-sequencing machines, Venter plans to use the shotgun method to sequence the human genome faster and at a lower cost than the government-sponsored project. The partnership plans to recover its costs by patenting the genes it discovers during sequencing and by selling subscriptions to its sequence databases.

Whether the sequence is obtained by government sponsorship or by private industry, it is apparent that the knowledge gained will advance our understanding of human genetics, and have a great impact on biomedical research and health care. Applications of the knowledge gained from the project raise ethical, social, and legal issues that need to be identified, debated, and resolved, often in the form of laws or public policy.

The Human Genome Project: The Ethical, Legal, and Social Implications (ELSI) Program

At the time the Human Genome Project (HGP) was being established, scientists raised concerns about how genome information would be used, and how the interests of both individuals and society could be protected. To address these concerns, the Ethical, Legal, and Social Implications (ELSI) Program was established as part of the Human Genome Project. The ELSI program began by considering a number of issues, including the impact of genetic information on individuals, the privacy and confidentiality of genetic information, the impact of genome information and technology on medical practice, genetic counseling, and reproductive decision making. Through research grants, workshops, and public forums, ELSI has started to formulate policy options addressing these issues.

Currently, ELSI is focusing on four areas: (1) privacy and fairness in the use and interpretation of genetic information, (2) the transfer of genetic knowledge from the research laboratory to clinical practice, (3) issues of informed consent for participants in genetic research, and (4) public and professional education.

These steps are being taken to ensure that as the HGP moves from generating information about the genetic basis of disease to the long-range goals of improved treatment, prevention, and cures, that ethical issues of concern are identified and that policies and laws are instituted to deal with them.

After the Genome Projects

As outlined here, several genome projects have already reached their goals. The genomes of two bacteria, *Haemophilus influenzae* and *Mycoplasma genitalium* were both completely sequenced in 1995. The nucleotide sequence of the yeast genome was reported in 1996, and the nucleotide sequence for the genomes of *E. coli* (1997) and *C. elegans* (1998) are now available. Work on sequencing the human genome is proceeding, and by 2001, about one-third of the genome should be completed according to the HGP schedule. As data from these projects become available, it is apparent they are rich sources of information about genome organization, function, and evolution.

From a biological perspective, the *M. genitalium* genome is of interest because it is one of the smallest genomes of any free-living organisms, containing only 482 genes. Ninety of these encode proteins used in translation, and about 30 genes encode products for DNA replication. Because of the small size of the genome, these numbers may be close to the minimum number of genes necessary to complete such tasks. Somewhat surprisingly, 140 of the 482 genes (about 30 percent of the genome) encode membrane-inserted proteins, emphasizing the importance of interactions between the cell and its environment for survival. Further work on this organism and on those with similar-sized genomes may help define the minimum number of genes necessary for life.

Although the flow of information on the human genome is still in its early stages, important insights into human genetic disorders have already been gained. A new mechanism of mutation, the expansion of trinucleotide repeats (see Chapter 17), has been discovered. This form of mutation is responsible for several neurodegenerative disorders, including Huntington disease, myotonic dystrophy, and spinal/bulbar muscular atrophy. Another unexpected finding is that gene dosage can be increased by the duplication of small regions of chromosomes, leading to disorders such as Charcot-Marie-Tooth syndrome. In addition, it is now known that different mutations in a single gene can give rise to different genetic disorders (e.g., multiple endocrine neoplasia, thyroid carcinoma, and Hirschprung disease).

Another significant discovery is that mutations that affect DNA repair can destabilize distant regions of the genome, making the cell carrying the DNA repair mutation more susceptible to cancer (see Chapter 23).

Biotechnology is an outgrowth of recombinant DNA technology

Although recombinant DNA techniques were originally developed to facilitate basic research on gene organization and regulation of expression, scientists have not been blind to the commercial possibilities of this technology. As a result, researchers have participated in the formation of biotechnology companies using recombinant DNA technology to develop products including hormones, clotting factors, herbicide-resistant plants, enzymes for food production, and vaccines. In the last decade, the biotechnology industry has grown into a multibillion-dollar segment of the economy.

Insulin Production by Bacteria

The first human gene product manufactured using recombinant DNA and licensed for therapeutic use was human insulin, which became available in 1982. Insulin is a protein hormone that regulates sugar metabolism, and an inability to produce insulin results in diabetes, a disease that in its more severe form affects more than 2 million individuals in the United States.

Clusters of cells embedded in the pancreas synthesize a precursor peptide known as preproinsulin. As this polypeptide is secreted from the cell, amino acids are cleaved from the end and the middle of the chain to produce the mature insulin molecule, which consists of two polypeptide chains (A and B chains), joined by disulfide bonds.

A look at the method originally used for producing insulin by recombinant DNA methods is instructive, as it shows both the promises and difficulties of this technology. The functional insulin protein consists of two polypeptide chains, A and B. The A polypeptide has 21 amino acids, and the B subunit has 30. Synthetic genes for the A and B subunits were constructed by oligonucleotide synthesis (63 nucleotides for the A polypeptide and 90 nucleotides for the B polypeptide). Each synthetic oligonucleotide was inserted into a vector at a position adjacent to a gene encoding the bacterial form of the enzyme, β-galactosidase. When transferred to a bacterial host, the *β-galactosidase* gene and the synthetic oligonucleotide were transcribed and translated as a unit. The product is a **fusion polypeptide**, consisting of the amino acid sequence for β-galactosidase attached to the amino acid sequence for one of the insulin subunits (Figure 21.18). The fusion proteins were purified from bacterial extracts and treated with cyanogen bromide, to cleave the fusion protein from the β-galactosidase.

Each insulin subunit was produced separately by this method. When mixed together, the two subunits spontaneously unite, forming an intact, active insulin molecule. The purified insulin is then packaged for use by those diabetics who must take insulin injections one or more times daily.

A number of genetically engineered proteins for therapeutic use have been produced by similar methods or are in clinical trials (Table 21.2). In most cases, these proteins are produced by cloning a human gene into a plasmid and inserting the recombinant vector into a bacterial host.

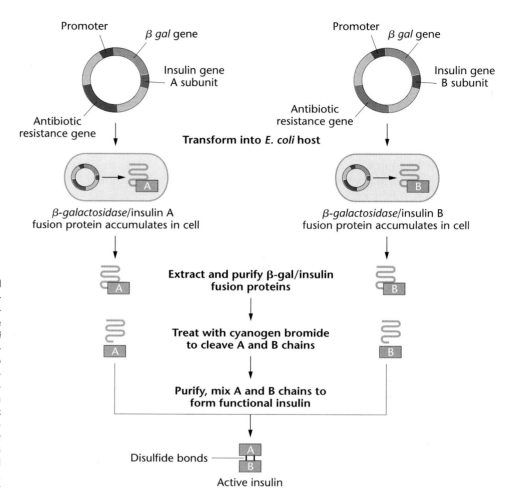

FIGURE 21.18 Method used to synthesize recombinant human insulin. Synthetic oligonucleotides encoding the insulin A and B chains were inserted at the tail end of a cloned *E. coli* β-galactosidase (*β-gal*) gene. These recombinant plasmids were transferred to *E. coli* hosts where the β-gal/insulin fusion protein was synthesized and accumulated in the host cells. Fusion proteins were extracted from the host cells and purified. Insulin chains were released from the β-galactosidase by treatment with cyanogen bromide. The insulin subunits were purified and mixed to produce a functional insulin molecule.

TABLE 21.2 Genetically Engineered Pharmaceutical Products Available or in Clinical Testing

Gene Product	Condition Being Treated
Atrial natriuretic factor	Heart failure, hypertension
Epidermal growth factor	Burns, skin transplants
Erythropoietin	Anemia
Factor VIII	Hemophilia
Gamma interferon	Cancer
Granulocyte colony-stimulating factor	Cancer
Hepatitis B vaccine	Hepatitis
Human growth hormone	Dwarfism
Insulin	Diabetes
Interleukin-2	Cancer
Superoxide dismutase	Transplants
Tissue plasminogen activator	Heart attacks

After ensuring that the transferred gene is expressed, large quantities of the transformed bacteria are produced, and the human protein is recovered and purified.

Transgenic Animal Hosts and Pharmaceutical Products

Bacterial hosts were used to produce the first generation of recombinant proteins, even though there are some disadvantages in using prokaryotic hosts to synthesize eukaryotic proteins. For example, bacterial cells are unable to process and modify eukaryotic proteins. Thus they cannot add sugars and phosphate groups that are often needed for full biological activity. In addition, eukaryotic proteins produced in prokaryotic cells often do not fold into the proper three-dimensional configuration and, as a result, are inactive. To overcome these difficulties and to increase yields, second-generation methods use eukaryotic hosts. Rather than being produced by host cells grown in tissue culture, human proteins such as alpha-1-antitrypsin are being produced in the milk of livestock, as described in the following discussion.

A deficiency of the enzyme alpha-1-antitrypsin is associated with the heritable form of emphysema, a progressive and fatal respiratory disorder common among those of European ancestry. To produce alpha-1-antitrypsin by genetic engineering, the human gene was cloned into a vector at a site adjacent to a sheep promoter sequence that regulates the expression of milk-associated proteins. Any gene placed next to this promoter is expressed only in mammary tissue. This fusion gene was then microinjected into sheep zygotes fertilized *in vitro* (Figure 21.19), which in turn were implanted into foster mothers. The resulting **transgenic** sheep developed normally and, after mating, produced milk that contained high concentrations of functional human alpha-1-antitrypsin. This human protein is present in concentrations up to 35 grams per liter of milk. It is easy to envision that a small herd of lactating sheep could easily provide an adequate supply of this protein and that herds of other transgenic animals, acting as biofactories, might become part of the pharmaceutical industry. In fact, the cloning of Dolly the sheep (see essay) was done in order to facilitate the establishment of a flock of sheep that will produce high levels of human proteins.

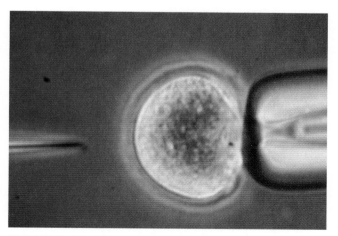

FIGURE 21.19   Injection of cloned DNA. A micropipette is used to transfer cloned genes into the nucleus of a mammalian zygote. The injected zygote will then be transferred to the uterus of a surrogate mother for development.

Transgenic Crop Plants and Herbicide Resistance

Damage from weed infestation destroys about 10 percent of crops worldwide. To combat this problem, more than 100 different herbicides are used, at an annual cost of more than 10 billion dollars. Some of these herbicides also kill crop plants, while others remain in the environment, or are present in run-off and contaminate water supplies. The creation of herbicide-resistant plants using recombinant DNA technology is one way to improve crop yields. Vectors have been used to transfer traits for herbicide resistance to crop plants. As a herbicide, glyphosate is effective at very low concentrations, is not toxic to humans, and is rapidly degraded by soil microorganisms. Glyphosate works by inhibiting the action of a chloroplast enzyme called EPSP synthase. This enzyme is important in amino acid biosynthesis in both bacteria and plants. Without the ability to synthesize these vital amino acids, plants wither and die.

The development of glyphosate-resistant crop plants began with the isolation and cloning of an EPSP synthase gene from a glyphosate-resistant strain of *E. coli*. This gene was cloned into a vector adjacent to plant promoter sequences and upstream from plant transcription-termination sequences. The recombinant vector was transferred into the bacterium *Agrobacterium tumifaciens* (Figure 21.20). Plasmid-carrying bacteria were in turn used to infect cells in discs cut from plant leaves. Calluses formed from these discs were then selected for their ability to grow on glyphosate. Transgenic plants generated from glyphosate-resistant calluses were grown and sprayed with glyphosate at concentrations four times higher than needed to kill wild-type plants. The transgenic plants overproducing the EPSP synthase grew and developed, while the control plants withered and died. Glyphosate-resistant corn and soybeans developed in this way are now on the market.

While use of these genetically engineered crops should reduce costs to the farmer for herbicides, they do present an important ethical dilemma. Is it appropriate to generate crops that encourage the use of herbicides? A less ethically debatable application of genetic engineering to agriculture is the introduction of genes into crop plants that transfer resistance to viruses, bacteria, fungi, and insect infestation. Other work has been directed at improving drought resistance or the nutritional value of crops such as soybeans and corn. Many of these projects are in the developmental stage, and these transgenic products will reach the marketplace over the next few years.

Transgenic Plants and Edible Vaccines

One of the most beneficial applications of recombinant DNA technology may be in the production of vaccines. Vaccines stimulate the immune system to produce anti-

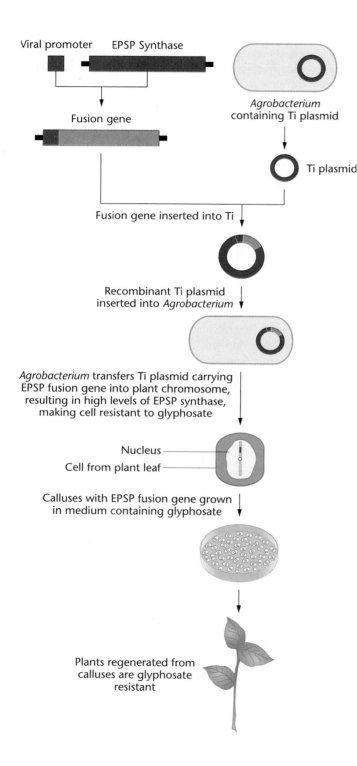

Viral promoter   EPSP Synthase

Fusion gene

Agrobacterium containing Ti plasmid

Ti plasmid

Fusion gene inserted into Ti

Recombinant Ti plasmid inserted into *Agrobacterium*

Agrobacterium transfers Ti plasmid carrying EPSP fusion gene into plant chromosome, resulting in high levels of EPSP synthase, making cell resistant to glyphosate

Nucleus

Cell from plant leaf

Calluses with EPSP fusion gene grown in medium containing glyphosate

Plants regenerated from calluses are glyphosate resistant

bodies against a disease-causing organism and thereby confer immunity against the disease. Two types of vaccines are commonly used, **inactivated vaccines**, prepared from killed samples of the infectious virus or bacteria, and **attenuated vaccines**, which are live viruses or bacteria that can no longer reproduce and cause disease when present in the body.

Using recombinant DNA technology, a new type of vaccine called a **subunit vaccine** is being produced. These vaccines consist of one or more surface proteins of the virus or bacterium. The protein acts as an antigen to stimulate the immune system to make antibodies against the virus or bacterium. One of the first licensed subunit vaccines was for a surface protein of hepatitis B, a virus that causes liver damage and cancer (see Chapter 23 for a discussion of hepatitis B). The gene for the hepatitis B surface protein was cloned into a yeast-expression vector and produced using yeast as a host. The protein is extracted and purified from the host cells and then packaged for use as a vaccine.

Recombinant DNA technology is currently being used to produce the hepatitis B surface protein in edible plants. The idea is to have plant leaves or fruits serve as the source of an oral vaccine. Plant-produced vaccines offer several advantages. They would be inexpensive, as industrial investment for production and extensive purification would be unnecessary. In addition, such vaccines would not require injection and would be useful in developing countries where medical services and delivery of health care are not highly developed.

As a model system, the antigenic subunit of hepatitis B vaccine has been transferred to tobacco plants and expressed in the leaves (Figure 21.21). For use as a source of vaccine, the gene would be inserted into food plants such as grains or vegetables.

Recent clinical tests have used genetically engineered potatoes that carry recombinant bacterial antigens. Human volunteers ate small quantities of these potatoes (50–100 g) and developed antibodies to the antigen, establishing that edible vaccines may be feasible. Similar tests are underway using genetically engineered bananas. If such tests are successful, genetically engineered edible plants will soon be used to vaccinate people against many diseases.

FIGURE 21.20 Gene transfer of glyphosate resistance. The *EPSP* gene is fused to a promoter from cauliflower mosaic virus. This chimeric or fusion gene is then transferred to a Ti plasmid vector, and the recombinant vector is inserted into an *Agrobacterium* host. *Agrobacterium* infection of cultured plant cells transfers the *EPSP* fusion gene into a plant cell chromosome. Cells that acquire the gene are able to synthesize large quantities of EPSP synthase, making it resistant to the herbicide glyphosate. Resistant cells are selected by growth in herbicide-containing medium. Plants regenerated from these cells are herbicide-resistant.

As discussed in Chapter 1, plants and animals were domesticated some 8000 to 10,000 years ago, and by selective breeding we have been modifying these organisms ever since, producing the diversity of domesticated plants and animals present today. Recombinant DNA technology has changed the rate at which new plants and animals can be developed and, by the transfer of genes between species, has altered the types of changes that can be made. Unlike selective breeding programs, biotechnology has generated concerns about the release of genetically modified organisms into the environment and about the safety of eating such products. If biotechnology is to achieve a new green revolution, these concerns need to be addressed through prudent research and education.

FIGURE 21.21   Tobacco leaves expressing the hepatitis B antigen. Transgenic tobacco plants carrying an antigenic subunit of hepatitis B virus were generated. Leaves from transgenic plants were treated with antibodies against hepatitis B antigen, showing that the plants produced the antigen. The central leaf in the photograph is a leaf from a normal tobacco plant and is unstained.

CHAPTER SUMMARY

1. Recombinant DNA technology offers a new approach to genetic analysis. Instead of relying on and being limited to the isolation and mapping of genes defined by mutant phenotypes, it is now possible to analyze any aspect of a genotype directly using molecular markers.

2. The mapping, identification, and isolation of genes responsible for genetic disorders is being approached using either positional cloning or the candidate gene approach, either of which can be used without any knowledge of the nature or function of the gene product.

3. Recombinant DNA techniques are being used in the prenatal diagnosis of human genetic disorders. This method allows the direct examination of the genotype, whereas previous methods relied on gene expression and the identification of the gene product. These methods can also be used to identify carriers of genetic disorders, and they are the basis of proposals to screen the population for a number of genetic disorders, including sickle-cell anemia and cystic fibrosis.

4. The availability of cloned human genes has led to their use in somatic gene therapy. This therapy is carried out by transferring a cloned normal copy of a gene into a vector and using the vector as a means of transferring the gene to a target tissue. The cells of the target tissue express the cloned copy of the gene, which

alters the mutant phenotype. The development of new and more effective vector systems may mean that gene therapy will become a standard method for the treatment of genetic disorders in the near future.

5. Recombinant DNA techniques have been used to create DNA fingerprints. These are the pattern of allelic variants in chromosomal loci, such as RFLPs and variable tandem nucleotide repeats, present in an individual. DNA fingerprinting has found applications in forensics, paternity testing, and a wide range of fields including archeology, conservation biology, and public health.

6. Genome projects are under way for several organisms, including *E. coli*, yeast, *Drosophila*, the mouse, and humans. The goals of these genome projects are to identify and map all genes in the genome and to determine its complete nucleotide sequence.

7. The biotechnology industry is using recombinant DNA methods to produce human gene products in a variety of hosts, ranging from bacteria to farm animals. In addition, gene-transfer techniques are being used to improve crop plants by the transfer of herbicide resistance and resistance to pathogen infestation. In the near future, it may be possible to offer vaccination against infectious disease through food plants, making resistance to infectious agents almost universal.

Beyond Dolly: The Cloning of Humans

The birth of Dolly, a healthy white-faced lamb, took the world by surprise. Until February 1997, the idea that an animal could be cloned from the cells of another adult animal was science fiction—something from *Brave New World* or *The Boys from Brazil*. For many people, the notion that animals could be cloned conjured up scenes of multiple replicated Adolf Hitlers or creations gone astray, as in Shelley's *Frankenstein*. For decades, the public had been comforted by scientists who asserted that it would be impossible to clone a new mammal from cells of an adult. It was thought that cells from adult animals could not be reprogrammed in such a way as to form a new, entire organism. But then Dolly appeared.

Dolly, the first mammal cloned from adult cells, was brought into being by a group of embryologists led by Ian Wilmut and Keith Campbell at the Roslin Institute in Scotland. Their reason for developing methods to clone farm animals was to provide identical transgenic animals that secrete pharmaceutical products, such as blood clotting factors or insulin, into their milk. In this way, animals could be used as bioreactors, to synthesize proteins that are difficult or expensive to synthesize in bacteria or in the test tube.

The cloning method that Wilmut and Campbell used to create Dolly—a procedure called nuclear transfer—was first suggested by embryologist Hans Spemann in 1938. The idea is simply to replace the nucleus of a newly fertilized (or unfertilized) egg with the nucleus from an adult cell, thereby creating a "reconstructed" zygote containing the nucleus from one animal and the egg cytoplasm from another. In theory, the genetic information in the donor nucleus should direct all further embryonic development and the new organism should be a genetic replica of the adult that donated the nucleus. Although this procedure sounds simple in theory, it proved to be extremely difficult in practice. Even when the technical procedures such as collecting eggs, removing nuclei, and transplanting donor nuclei were perfected, the cloning procedure would only work if the donor nucleus came from an embryo. Because adult nuclei come from specialized structures such as the skin, liver, and kidney, they are "differentiated" and express only a small subset of all the genes in the nuclear genome. As all the genes in the genome must be actively expressed in an embryo, adult nuclei proved to be inadequate for the job.

Wilmut and Campbell accomplished the impossible by genetically reprogramming the adult nuclei before transferring them into recipient eggs. They removed donor cells from the udder of an adult ewe and starved the cells so that they became quiescent and entered the G0 phase of the cell cycle. For unknown reasons, this starvation procedure allowed the silent genes within the udder cell nucleus to be turned on after the nucleus was transplanted into the cytoplasm of a sheep's egg. In addition, the researchers passed an electric current through the egg to facilitate nuclear transfer and to stimulate the new "zygote" to begin dividing. To create Dolly, over 200 nuclear transfers were done using udder cell nuclei. Of these, only 29 developed into embryos and 13 were implanted into surrogate mother ewes. One pregnancy resulted, which culminated in the birth of Dolly.

Although Dolly was the first mammal to be cloned from adult cells, the approach has since been used to create cloned mice from adult mouse cells. In addition, transgenic sheep and cattle have been cloned by similar procedures, except that fetal cell nuclei were used as donors, rather than adult nuclei. One of these transgenic sheep ("Polly") bears the human gene for blood-clotting factor IX and secretes factor IX into her milk, paving the way for the production of pharmaceuticals from cloned farm animals. Besides the benefits for drug production, cloning promises other advantages. For example, cloning might allow scientists to preserve and replicate endangered species, cure genetic diseases in farm animals, and provide animal models of human diseases for which there are presently no research models.

Despite the obvious advantages for agriculture and medicine, the cloning of mammals has led to outcry and concern. The idea that humans could be cloned has been termed "immoral," "repugnant," and "ethically wrong." Within days of the announcement about Dolly, bills were introduced into the U.S. Congress to prohibit research into human cloning and worldwide bans were called for. Frightening scenarios were proposed: Rich and powerful people cloning themselves for reasons of vanity, people with serious illnesses cloning replicas to act as organ donors, and legions of human clones suffering loss of autonomy, individuality, and kinship ties. But is it really possible to clone humans? And if so, should human cloning ever be done?

The answer to the first question is simple: The same technology used to create Dolly could be used to clone a human. Most of the necessary technical procedures are being used now for *in vitro* fertilization. And it seems likely that adult human nuclei could be reprogrammed similarly to the adult sheep nuclei that created Dolly.

The answer to the second question is not as simple. Both scientific and ethical issues cloud our judgment about human cloning. Present technology cannot guarantee the health of any cloned individual, human or animal. It is possible that clones could exhibit a higher risk of genetic disease or cancer due to the accumulation of mutations in the donated adult nucleus. Clones might show accelerated aging due to shortened telomeres that are present in the donor's chromosomes. If the donor's nucleus is not completely reprogrammed prior to cloning, the clone could undergo abnormal development. Until the safety issues are resolved, it may be prudent to suspend any attempts to clone humans. The ethical issues are even more problematic. It is possible to foresee positive aspects to human cloning. Infertile couples, or those who suffer from genetic disease on one side of the family, could choose to clone one of the partners in order to raise a child who is biologically related. Cloning a person's cells *in vitro* could provide a source of cells or tissues to treat a number of serious diseases. On the other hand, it is equally possible to foresee negative aspects. Would cloning seriously alter what it means to be a unique human being? What would be the fate of clones created for organ transplantation? Could the technique be misused by the unscrupulous for social or political goals? It is important for society to consider these issues now in order to ensure responsible and beneficial outcomes of this new technology.

REFERENCES:

CIBELLI, J.B. et al. 1998. Cloned transgenic calves produced from nonquiescent fetal fibroblasts. *Science* 280:1256–58.

PENNISI, E. 1998. After Dolly, a pharming frenzy. *Science* 279: 646–48.

ROBERTSON, J.A. 1998. Human cloning and the challenge of regulation. *N. Eng. J. Med.* 339:118–25.

SOLTER, D. 1998. Dolly is a clone—and no longer alone. *Nature* 394: 315–16.

WILMUT, I. 1998. Cloning, for medicine. *Sci. Am.* (Dec.) 58–63.

WILMUT, I. et al. 1997. Viable offspring derived from fetal and adult mammalian cells. *Nature* 385:810–13.

INSIGHTS AND SOLUTIONS

1. DNA fingerprints have been used to test forensic specimens, to identify criminal suspects, and to settle immigration cases and paternity disputes. Probes for DNA fingerprinting can be derived from a single locus or multiple loci. Two multiple-loci probes have been widely employed in both criminal and civil cases and derive from minisatellite loci on chromosome 1 (1cen-q24) and chromosome 7 (7q31.3). These probes, which are widely used because they produce a highly individual fingerprint, have determined paternity in thousands of cases over the last few years.

(a) Shown here are the results of DNA fingerprinting of a mother (M), putative father (F), and child (C) using these probes. As shown, the child has 6 maternal bands, 11 paternal bands, and 5 bands shared between the mother and the alleged father. Based on this fingerprint, can you conclude that the man tested is the father of the child?

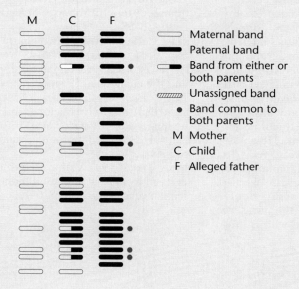

Solution: All bands present in the child can be assigned as coming from either the mother or the father. In other words, all the bands in the child's DNA fingerprint that are not maternal are present in the father. Because the father and the child share 11 bands in common and the child has no unassigned bands, paternity can be assigned with confidence. In fact, the chance that the man tested is not the father is on the order of 10^{-13}.

(b) In a second case, the DNA fingerprints of the mother, the alleged father, and the child are shown.

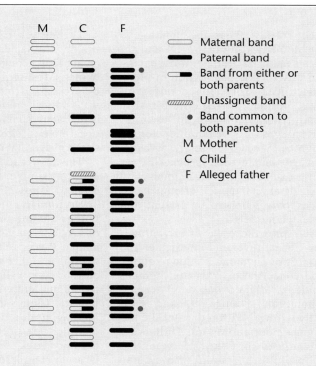

In this case, the child has 8 maternal bands, 15 paternal bands, 6 bands that are common to both the mother and the alleged father, and 1 band that is not present in either the mother or the alleged father. What are the possible explanations for the presence of this band? Based on your analysis of the band pattern, which explanation is most likely?

Solution: In this case, one band in the child cannot be assigned to either parent. Two possible explanations for this finding are that the child is mutant for one band or that the man tested is not the father. In estimating the probability of paternity, the mean number of resolved bands (n) is determined, and the mean probability (x) that a band in individual A matches that in a second unrelated individual, B, is calculated. In this case, because the child and the father share 15 bands in common, the probability that the man tested is not the father is very low (probably 10^{-7} or lower). As a result, the most likely explanation is that the child is a mutant for a single band. In fact, in 1419 cases of genuine paternity resolved by the minisatellite probes on chromosomes 1 and 7, single-mutant bands in the children were recorded in 399 cases, accounting for 28 percent of all cases.

2. Infection by HIV-1 (human immunodeficiency virus) is responsible for the destruction of cells in the immune system and results in the symptoms of AIDS (acquired immunodeficiency syndrome). HIV infects

and kills cells of the immune system that carry a cell-surface receptor known as CD4. An HIV surface protein known as gp120 binds to the CD4 receptor and allows the virus entry into the cell. The gene encoding the CD4 protein has been cloned. How can this clone be used along with recombinant DNA techniques to combat HIV infection?

Solution: Several methods using the CD4 gene are being explored to combat HIV infection. First, because infection depends on an interaction between the viral gp120 protein and the CD4 protein, the cloned CD4 gene has been modified to produce a soluble form of the protein (sCD4). The idea is that HIV can be prevented from infecting cells if the gp120 protein of the virus is bound up with the soluble form of the CD4 protein and thus is unable to bind to CD4 proteins on the surface of immune system cells. Studies in cell-culture systems indicate that the presence of sCD4 effectively prevents HIV infection of tissue culture cells. However, studies in HIV-positive humans has been somewhat disappointing, mainly because the strains of HIV used in the laboratory are different from those found in infected individuals. HIV-infected cells carry the viral gp120 protein on their surface. To kill such cells, the CD4 gene has been fused with those encoding bacterial toxins. The resulting fusion protein contains the CD4 regions that bind to gp120 and the toxin regions that kill the infected cell. In tissue-culture experiments, cells infected with HIV are killed by the fusion protein, whereas noninfected cells survive. It is hoped that the targeted delivery of drugs and toxins can be used in therapeutic applications to treat HIV infection.

PROBLEMS AND DISCUSSION QUESTIONS

1. In attempting to vaccinate against diseases by eating antigens, the antigen (such as the cholera toxin) must be presented to the cells of the small intestine. What are some potential problems in this method? Why don't absorbed food molecules stimulate the immune system and make you allergic to the food you eat?

2. Outline the steps involved in transferring glyphosate resistance to a crop plant. Do you envision that this trait can escape from the crop plant and make weeds glyphosate resistant? Why or why not?

3. What lessons from the *E. coli* genome project can be applied to the Human Genome Project?

4. Gene therapy for human genetic disorders involves transferring a copy of the normal human gene into a vector and using the vector to transfer the cloned human gene into target tissues. Presumably, the gene enters the target tissue, becomes active, and the gene product relieves the symptoms. Although now being used to treat several disorders, there are some unresolved problems with this method.
   (a) Why are disorders such as muscular dystrophy difficult to treat by gene therapy?
   (b) What are the potential problems in using retroviruses as vectors?
   (c) In the long run, should gene therapy involve germ tissue instead of somatic tissue? What are some of the potential ethical problems associated with this approach?

5. In producing physical maps of markers and cloned sequences, what advantage does *Drosophila* offer that other organisms, including humans, do not?

6. Outline the steps in identifying a gene by positional cloning. What steps may cause difficulty in this process? Once a region on a chromosome has been identified as containing a given gene, what kind of mutations would speed the process of identifying the locus?

7. The phenotype of many behavior traits such as manic depression or schizophrenia may be controlled by several genes, each at a different locus. Can positional cloning be used to map and isolate such genes? What if a trait is controlled by six genes, each contributing equally in an additive way to the phenotype? Can positional cloning be used in this case? Why or why not?

8. Suppose that you develop a screening method for cystic fibrosis that allows you to identify the predominant mutation (Δ508) and the next six most prevalent mutations. What do you need to consider before using this method in screening the population at large for this disorder?

9. The DNA sequence surrounding the site of the sickle-cell mutation in the β-globin is shown for normal and mutant genes.

5′ G A C T C C T G A G G A G A A G T 3′

3′ C T G A G G A C T C C T C T T C A 5′

Normal DNA

5′ G A C T C C T G T G G A G A A G T - 3′

3′ C T G A G G A C A C C T C T T C A - 5′

Sickle-cell DNA

Each type of DNA is denatured into single strands and applied to a filter. The paper containing the two spots is hybridized to an ASO of the following sequence: 5′-GACTCCTGAGGAGAAGT-3′. Which (if either) spot will hybridize to this probe? Why?

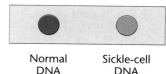

Normal   Sickle-cell
DNA   DNA

10. Dominant mutations can be categorized according to whether they increase or decrease overall activity of a gene or gene product. Although a loss-of-function mutation (that is, a mutation that inactivates the gene product) is usually recessive, for some genes one dose of the gene product is not sufficient to produce a normal phenotype. In this case, a loss-of-function mutation in the gene will be dominant and the gene is said to be haploinsufficient. A second category of dominant mutations are "gain-of-function" mutations that result in an increased activity or expression of the gene or gene product. The phenotype of such a mutation results from too much gene product. The gene therapy technique currently used in clinical trials involves the "addition" to somatic cells of a normal copy of a gene. In other words, a normal copy of the gene is inserted into the genome of the mutant somatic cell, but the mutated copy of the gene is not removed or replaced. Will this strategy work for either of these two types of dominant mutations?

11. One form of hemophilia, an X-linked disorder of blood clotting, is caused by mutation in clotting factor VIII. Many single-nucleotide mutations of this gene have been described, making the detection of mutant genes by Southern blots inefficient. There is, however, an RFLP for the enzyme *Hind*III contained in an intron of the factor VIII gene that can often be used in screening, as shown here.

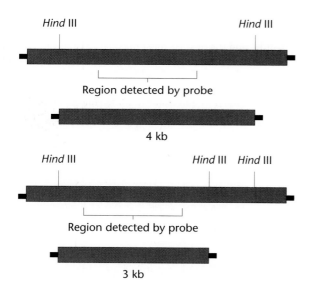

A female whose brother has hemophilia is at 50 percent risk of being a carrier of this disorder. To test her status, DNA is obtained from the white blood cells of family members, and cut with *Hin*dIII, and the fragments probed and visualized by Southern blotting. The results are shown here.

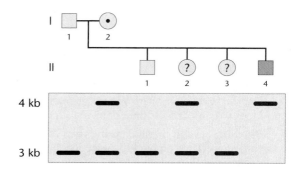

Determine whether any of the females in generation II are carriers for hemophilia.

12. The human insulin gene contains a number of introns. Despite the fact that bacterial cells will not excise introns from mRNA, how can a gene like this be cloned into a bacterial cell and produce insulin?

13. In mice transfected with the rabbit β-globin gene, the rabbit gene is active in a number of tissues including the spleen, brain, and kidney. In addition, some mice suffer from thalassemia, caused by an imbalance in the coordinate production of α- and β-globins. What problems associated with gene therapy are illustrated by these findings?

EXTRA-SPICY PROBLEMS

14. The following is a pedigree that shows the inheritance of a rare disease state.

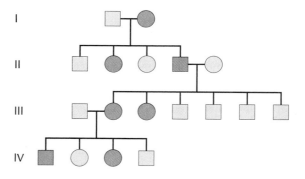

(a) Which mode or modes of inheritance are excluded by or consistent with this pedigree? DNA samples from generations II and III (in the pedigree) are obtained and subjected to RFLP linkage analysis. One RFLP examined is found on chromosome 10 (identified by probe A) and the other is found on chromosome 21 (identified by probe B). The results of the analysis are shown here. In answering the remaining questions, assume that additional data was gathered on this family and that it was consistent with the data shown and statistically significant.

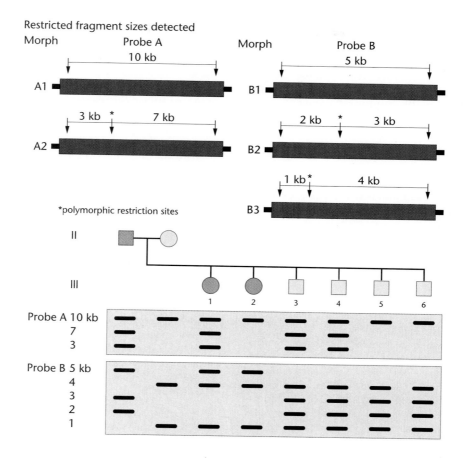

Restricted fragment sizes detected

*polymorphic restriction sites

(b) On which chromosome is the disease gene located?

(c) Individual III-1 is married to a normal man whose RFLP genotype is A1A2 and B1B1. What kind of prenatal diagnostic test can be done to determine if the child they are expecting will be normal? Be sure to describe what you can conclude given this result or that result. Indicate the accuracy of such a test.

(d) Individual III-4 marries a woman of genotype B1B1. They have a child who is B1B1. The father assumes the child is illegitimate, but the mother, who has taken a genetics course, argues that the child could be the result of an event that occurred in the father's germ line, and she cites two possibilities. What are they?

15. You are asked to help in prenatal genetic testing of a couple found to be carriers for a deletion in the β-globin gene that produces β-thalassemia when homozygous. The couple already has one child who is unaffected and is not a carrier. The woman is pregnant, and they wish to know the status of the fetus. You receive DNA samples obtained from the fetus by amniocentesis and from the rest of the family by extraction from white blood cells. Using a probe that detects the deletion, the following blot is obtained.

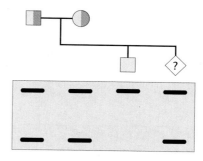

Is the fetus affected? What is its genotype for the β-globin gene?

GENETICS MediaLab

The following resources will help you achieve a better understanding of the concepts presented in this chapter. These resources can be found on the CD packaged with this textbook and on the Companion Website at **http://www.prenhall.com/klug/**.

CD Resources:

Self-grading Chapter Problems

Web Resources:

Genetics and Society Issue: *Beyond Dolly: The Cloning of Humans*

Web Destinations in Genetics

Self-grading Chapter Problems

Chapter Search Terms

Genetics Newsgroups

Student Bulletin Board

Web Problem 1:

Time for completion = 10 minutes
What information will we aquire from the Human Genome Project? The Human Genome Project is designed to map and sequence the entire human genome. The management and analysis of information about the 3 billion nucleotides in the human genome has led to the development of the field of bioinformatics. In this exercise, you will view a linked website that provides an overview of the Human Genome Project. Indicate the different levels of information required to ultimately sequence the human genome. Describe several methods that researchers are using to locate which portions of the DNA sequence encode genes. To complete this exercise, visit Web Problem 1 in Chapter 21 of your Companion Website and select the keyword **GENOME PROJECT**.

Web Problem 2:

Time for completion = 10 minutes
How can the average person deal with the rapidly expanding amount of genetic information available to the public and to health-

care providers? Genetic counselors serve as an interface between this information and the public. They are trained in both genetics and counseling and can convey genetic information to individuals in an accurate and objective manner. The linked website describes a case study of a family with a child diagnosed with neurofibromatosis. Briefly discuss some of the issues that a genetic counselor may encounter in counseling this family. To complete this exercise, visit Web Problem 2 in Chapter 21 of your Companion Website and select the keyword **GENETIC COUNSELING**.

Web Problem 3:

Time for completion = 20 minutes
How is DNA testing used in criminal or paternity cases? The nucleotide sequence of specific alleles provides a unique identifier for each person. Similarly, because a child inherits half of each parent's genetic information, this sequence can be used to determine if two individuals are related. Test your understanding of DNA-testing methods with a series of interactive questions at the linked website. Feedback will indicate whether your answers are correct and tutorial information is provided to clarify your understanding. Discuss whether forensic information is used to prove someone is guilty or to prove someone is innocent. To complete this exercise, visit Web Problem 3 in Chapter 21 of your Companion Website and select the keyword **FORENSICS**.

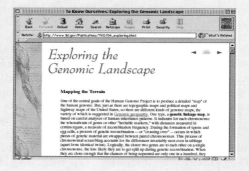

SELECTED READINGS

AMOS, B., and PEMBERTON, J. 1993. DNA fingerprinting in non-human populations. *Curr. Opin. Genet. Dev.* 2:857–60.

BARKER, D. et al. 1987. Gene for von Recklinghausen neurofibromatosis is in the pericentromeric region of chromosome 17. *Science* 236:1001–102.

BEAUDET, A.L. 1999. Making genomic medicine a reality. *Am. J. Hum. Genet.* 64:1–13.

C. elegans Sequencing Consortium 1998. Genome sequence of the nematode *C. elegans*: A platform for investigating biology. *Science* 282:2012–18.

CAWTHON, R.M. et al. 1990. A major segment of the neurofibromatosis type 1 gene: cDNA sequence, genomic structure, and point mutations. *Cell* 62:193–201.

CHANG, J.C., and KAN, Y.W. 1981. Antenatal diagnosis of sickle-cell anemia by direct analysis of the sickle mutation. *Lancet* 2: 1127–29.

COLLINS, F.S. 1995. Ahead of schedule and under budget: The Genome Project passes its fifth birthday. *Proc. Natl. Acad. Sci. USA* 92:10821–23.

CRYSTAL, R.G. 1995. Transfer of genes to humans: Early lessons and obstacles to success. *Science* 270:404–10.

DANNA, K., and NATHANS, D. 1971. Specific cleavage of simian virus 40 DNA by restriction endonuclease of *Hemophilus influenzae*. *Proc. Natl. Acad. Sci. USA* 68:2913–17.

DAVIDSON, B. et al. 1993. A model system for *in vivo* gene transfer into the central nervous system using an adenoviral vector. *Nat. Genet.* 3:219–23.

DIETRICH, W.F. et al. 1995. Mapping the mouse genome: Current status and future prospects. *Proc. Natl. Acad. Sci. USA* 92:10849–53.

DUJON, B. 1996. The yeast genome project: What did we learn? *Trends Genet.* 12:263–70.

ESTRUCH, J.J. 1997. Transgenic plants: An emerging approach to pest control. *Nature Biotech.* 15:137–41.

FLEISCHMANN, R.D. et al. 1995. Whole genome random sequencing and assembly of *Haemophilus influenzae*. *Science* 269:496–512.

FLOTTE, T. 1993. Prospects for virus-based gene therapy for cystic fibrosis. *J. Bioenerg. Biomembr.* 25:37–42.

FRASER, C.M. et al. 1995. The minimal gene complement of *Mycoplasma genitalium*. *Science* 270:397–403.

GOODMAN, H.M., ECKER, J.R., and DEAN, C. 1995. The genome of *Arabadopsis thaliana*. *Proc. Natl. Acad. Sci. USA* 92:10831–35.

GUYER, M. and COLLINS, F.S. 1995. How is the Human Genome Project doing, and what have we learned so far? *Proc. Natl. Acad. Sci. USA* 92:10841–48.

HARTL, D., and LOZOVSKAYA, E. 1992. The *Drosophila* genome project: Current status of the physical map. *Comp. Biochem. Physiol.* [B] 103:1–8.

HUDSON, T.J. et al. 1995. An STS-based map of the human genome. *Science* 270:1945–54.

INBAL, A. et al. 1993. Identification of three candidate mutations causing type IIA von Willebrand disease using a rapid, nonradioactive, allele-specific hybridization method. *Blood* 82:830–36.

KARPATI, G., PARI, G., and MOLNAR, M.J. 1999. Molecular therapy for genetic muscle diseases—status 1999. *Clin. Genet.* 55:1–8.

KMIEC, E.B. 1999. Gene therapy. *Amer. Scient.* 87:240–47.

KRUMLAUFF, R., JEANPIERRE, M., and YOUNG, B.D. 1982. Construction and characterization of genomic libraries from specific human chromosomes. *Proc. Natl. Acad. Sci. USA* 79:2971–75.

LISSENS, W. and SERMON, K. 1997. Preimplantation genetic diagnosis: Current status and new developments. *Human Reproduction* 12:1756–61.

LEWONTIN, R., and HARTL, D. 1991. Population genetics in forensic DNA typing. *Science* 254:1745–50.

MANIATIS, T., FRITSCH, E.F., and SAMBROOK, J. 1988. *Molecular cloning: A laboratory manual*, 2nd ed. Cold Spring Harbor, NY: Cold Spring Harbor Laboratory Press.

MARSHALL, E., and PENNISI, E. 1996. NIH launches the final push to sequence the genome. *Science* 272:188–89.

MARSHALL, E. 1995. Gene therapy's growing pains. *Science* 269:1050–55.

MASON, H., LAM, D., and ARNTZEN, C. 1992. Expression of hepatitis B surface antigen in transgenic plants. *Proc. Natl. Acad. Sci. USA* 89:11745–49.

MOLDOVEANU, Z. et al. 1993. Oral immunization with influenza virus in biodegradable microspheres. *J. Infect. Dis.* 167:84–90.

MULLIS, K.B. 1990. The unusual origin of the polymerase chain reaction. *Sci. Am.* (Apr.) 262:56–65.

NAHREINI, P., WOODY, M., ZHOU, S., and SRIVASTAVA, A. 1993. Versatile adeno-associated virus 2-based vectors for constructing recombinant virions. *Gene* 124:257–62.

REISS, J., and COOPER, D.N. 1990. Application of the polymerase chain reaction to the diagnosis of human genetic disease. *Hum. Genet.* 85:1–8.

SAIKI, R.K. et al. 1986. Analysis of enzymatically amplified beta-globin and HLA-DQ alpha DNA with allele-specific oligonucleotide probes. *Nature* 324:163–66.

SOUTHERN, E.M. 1975. Detection of specific sequences among DNA fragments separated by gel electrophoresis. *J. Mol. Biol.* 98:503–17.

TAYLOR, R. 1993. Food for thought: Seropositive plants may yield cheap oral vaccines. *J. N.I.H. Res.* 5:49–53.

VELANDER, W.H., LUBON, H., and DROHAN, W.N. 1997. Transgenic livestock as drug factories. *Sci. Amer.* (Jan.) 276:70–75.

VILLA-KOMAROFF, L. et al. 1978. A bacterial clone synthesizing proinsulin. *Proc. Natl. Acad. Sci. USA* 75:3727–31.

WATERSON, R., and SULSTON, J. 1995. The genome of *Caenorhabditis elegans*. *Proc. Natl. Acad. Sci. USA* 92:10836–40.

WILLIAMS, J.G.K. et al. 1990. DNA polymorphisms amplified by arbitrary primers are useful as genetic markers. *Nucl. Acids Res.* 18:6531–35.

WU, C.-L., and MELTON, D.W. 1993. Production of a model for Lesch-Nyhan syndrome in hypoxanthine phosphoribosyltransferase-deficient mice. *Nat. Genet.* 3:235–39.

ZUO, J., ROBBINS, C., BAHARLOO, S., COX, D., and MYERS, R. 1993. Construction of cosmid contigs and high-resolution restriction mapping of the Huntington disease region of human chromosome 4. *Hum. Mol. Genet.* 2:889–99.

This unusual four-winged fly has developed an extra set of wings as a result of a homeotic mutation in *Drosophila*.

Developmental Genetics

KEY CONCEPTS

- **Developmental genetics seeks to explain how a differentiated state develops from an organism's genome**

- **Development in eukaryotes is characterized by the selective expression of gene sets in cells with equivalent genomes**
  Genome Equivalence
  Binary Switch Genes

- **Embryonic development in *Drosophila* illustrates the influence of localized cytoplasmic components on gene expression**
  Overview of *Drosophila* Development
  Genetic Analysis of Embryogenesis: Maternal-Effect Genes and Zygotic Genes

- **Maternal-effect genes and zygotic genes interact to determine the basic body plan in *Drosophila***

  The Anterior Genes
  The Posterior Genes
  The Terminal Genes

- **Zygotic genes determine segment formation**
  Gap Genes
  Pair-Rule Genes
  Segment Polarity Genes
  Selector Genes

- **Flower development in *Arabidopsis* is also regulated by the activation of homeotic selector genes**

- **Cell–cell interactions influence developmental fate in *C. elegans***
  Overview of *C. elegans* Development
  Genetic Analysis of Vulva Formation

- **Programmed cell death is genetically regulated and is required for normal development**

In multicellular plants and animals, a fertilized egg, without further stimulus, begins a cycle of developmental events that ultimately give rise to an adult member of the species from which the egg and sperm were derived. Thousands, millions, or even billions of cells are organized into a cohesive and coordinated unit that we perceive as a living organism. The series of events through which organisms attain their final adult forms is studied by developmental biologists. This area of study is perhaps the most intriguing in biology. Not only does the comprehension of developmental processes requires the knowledge of many different biological disciplines—such as molecular, cellular, and organismal biology—but also an understanding of how biological processes change over time and ultimately transform the organism.

In the past hundred years, investigations in embryology, genetics, biochemistry, molecular biology, cell physiology, biophysics, and evolution have all contributed to the study of development. The findings have largely pointed out the tremendous complexity of developmental processes. Unfortunately, a description of *what* happens does not answer the "why" and "how" of development. Over the last

two decades, however, genetic analysis and molecular biology have identified genes that regulate developmental processes, and we are beginning to understand how the action and interaction of these genes control basic developmental processes in a wide range of organisms.

In this chapter the primary emphasis will be on the role of gene action in regulating development. This area of study, called developmental genetics, has contributed tremendously to our understanding of developmental processes because genetic information is both required for the molecular and cellular functions mediating developmental events and contributes to the determination of the final phenotype of the newly formed organism.

Developmental genetics seeks to explain how a differentiated state develops from an organism's genome

The current thrust in developmental genetics is to provide a molecular explanation of developmental processes, including temporal changes in gene expression and cellular differentiation. The goal is to establish a causal relationship between the presence or absence of inducers, receptors, transcriptional events, and cell and tissue interactions, and the observable morphological events that accompany the process of development. These ideas form the theoretical framework that is used to explain how the developmental potential present in a single cell becomes transformed into a recognizable yet individual form that we identify as an adult organism (Figure 22.1).

A useful way to define **development** is to say that it is *the attainment of a differentiated state* by all of the cells in the organism. For example, a cell in a blastula-stage embryo (when the embryo is still just a ball of uniform-looking cells) is undifferentiated, while an erythrocyte active in hemoglobin synthesis is differentiated. In order to accomplish this specialization, different collections of genes must be activated in different cells. Therefore from a genetic perspective, one way of defining the many cell types that develop in multicellular organisms is to define which collection of genes is active in each cell, in other words, their differential transcription patterns. This concept of differentiation, called the **variable gene activity hypothesis** of differentiation, was first entertained by Thomas H. Morgan in 1934 and later proposed by Edgar and Ellen Stedman, as well as Alfred Mirsky, in the 1950s. It has been clearly articulated in more modern form by Eric Davidson in his book *Gene Activity in Early Development*. The theory holds that the differentiated state assumed by any specific cell type is qualitatively determined by those genes that are actively transcribed in that cell. Its underlying assumptions are, first, that each cell contains an entire genome and, second, that differential transcription of selected genes controls the development and differentiation of each cell. While the original theory focused exclusively on gene activity as a means of genetic control, the theory has now been expanded to include additional levels of gene regulation, such as translational and posttranslational control.

This current version of the variable gene activity theory provides an excellent basis for experimental design. The remainder of this chapter examines the validity of this theory and its premises, and provides examples that offer experimental support from several model systems, with particular attention given to *Drosophila* and the roundworm *C. elegans*.

Development in eukaryotes is characterized by the selective expression of gene sets in cells with equivalent genomes

While prokaryotes and simpler eukaryotes, such as the slime mold *Dictyostelium*, have been used to study processes such as cellular differentiation, developmental biologists have focussed their attention primarily on higher eukaryotes, which not only undergo cellular differentiation but also morphogenesis, or the generation of complex forms. In higher eukaryotes, development begins with the formation of the **zygote,** or the cell that is generated by the fusion of the sperm and oocyte. The oocyte is a single cell with a cytoplasm that is heterogeneous and nonuniform in distribution. Following fertilization and early cell divisions, the nuclei of progeny cells find themselves in different environments as the maternal cytoplasm is distributed into the new cells. Evidence suggests that the cytoplasm in different cells exerts influence on the genetic material, causing differential transcription at specific points during development. Therefore **cytoplasmic localization,** or the inheritance of particular cytoplasmic components by individual embryonic cells, plays a major regulatory role during development.

Early gene products synthesized by the zygote further alter the cytoplasm of each cell, producing a still different cellular environment that in turn leads to the activation of other genes, and so on. In addition, as the number of cells increases, they influence one another in a process of **cell–cell interaction.** The total environment acting on the genetic material, therefore, now includes the individ-

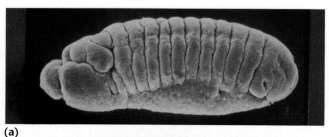

(a)

(b)

FIGURE 22.1   A *Drosophila* embryo and the adult fly that develops from it.

ual cell's cytoplasm as well as the signals sent from other cells. Although early in embryogenesis most cells show no evidence of structural or functional specialization, the combination of their localized cytoplasmic components and their position in the developing embryo seems to determine the ultimate form they will assume. It is as if their fate has been programmed prior to the actual events leading to specialization.

As cells respond to their external and internal environments, which are continually changing, they embark on a pathway comprising a series of developmental stages. The most important stage on this pathway is **determination,** when, in response to the many external and internal cues, a specific developmental fate for a cell becomes fixed. It is now clear that determination precedes the actual events of **differentiation,** the process by which the cell achieves its final form and function.

In the following sections, we will examine how patterns of gene expression relate to the processes of determination, differentiation, and cell–cell interaction during development. We will begin by examining the first assumption of the variable gene hypothesis—namely that each cell of a multicellular organism contains an intact genome. We will then explore the role of differential gene expression in the progressive restriction of developmental options available to cells. This topic will be introduced using the example of binary switch genes to illustrate that differential gene expression leads to different cell types, and then it will be expanded upon as we discuss the two common mechanisms for regulating gene expression, cytoplasmic localization and cell–cell communication. Two model systems, *Drosophila* and *C. elegans,* will be used to illustrate these regulatory mechanisms, as well as the tremendous contribution that genetic analysis has made to our understanding of development in multicellular organisms. Finally, the chapter will conclude with a discussion of the most dramatic restriction of cell fate during development, namely programmed cell death.

Genome Equivalence

The variable gene activity hypothesis is based on the premise of **genome equivalence,** that each somatic cell of a multicellular organism contains a complete set of genetic information. This premise is supported by biochemical analyses showing that the quantity of DNA in each cell within an organism is equivalent and equal to the diploid content in that organism. However, since only a small fraction of the DNA of higher organisms encodes genes, these studies do not address the question of whether different cells have slightly different sets of genes, or genetic information. Therefore other investigations have approached the question differently, by asking whether it is possible

to show that differentiated cells are genetically equivalent; in other words, is a complete set of genes present in each cell?

One way to answer this question is to show that a differentiated nucleus is **totipotent,** or capable of giving rise to a complete organism under the proper conditions. In the early part of the twentieth century, Hans Spemann demonstrated that a nucleus from a cell of the 16-cell stage of a newt embryo was capable of supporting the development of a complete organism. He first constricted a newly fertilized egg prior to the first cell division, forcing the zygote nucleus into one-half of the cell. The half with the nucleus divided and produced a cluster of 16 cells. He then loosened the constriction and allowed one of the 16 nuclei to pass back into the nonnucleated cytoplasm on the other side. Both halves of the embryo, one with 15 nuclei and one with a single nucleus, subsequently produced complete but separate organisms. Therefore, at the 16-cell stage in this organism, genomic equivalence was demonstrated.

Beginning in the 1950s, more sophisticated experiments were performed by Robert Briggs and Thomas King using the grass frog *Rana pipiens,* and by John Gurdon in the 1960s using the African frog *Xenopus laevis.* After inactivating or surgically removing the nucleus from an egg, these investigators were able to test the developmental capacity of somatic nuclei by transplanting those isolated from cells at various stages of differentiation into enucleated oocytes. Nuclei derived from the blastula stage of development were capable of supporting the development of complete and normal adults when transplanted (Figure 22.2). In *Rana,* transplanted nuclei derived from later stages, such as the gastrula and neurula, usually allowed only partial development. In *Xenopus,* however, Gurdon's experiments showed that nuclei from the epithelial gut cells of tadpoles were able to support the development of an adult frog when transplanted into enucleated oocytes. In both cases, it is clear that differentiated cells do carry a complete copy of the genome, and that, under normal developmental conditions, progressive restrictions that prevent the expression of totipotency are placed on differentiating nuclei.

To test whether the nucleus of a highly differentiated adult cell is *irreversibly* specialized or whether it can support the development of a normal embryo, Gurdon used nuclei from adult frog skin cells in serial transplant experiments (Figure 22.3). In these experiments, a donor nucleus was transplanted into an enucleated egg, and the recipient was allowed to develop for a short time, say to the blastula stage. The blastula cells were then dissociated and a nucleus removed from one of them. This nucleus was transplanted into still another enucleated egg and development was allowed to occur. Such serial transfers were repeated a number of times. Subsequently, the blastula was

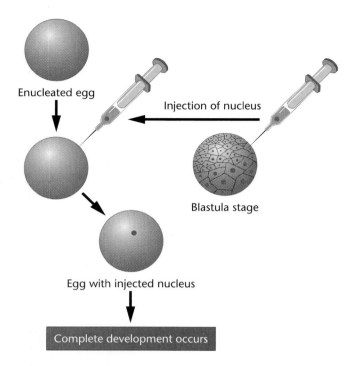

FIGURE 22.2 The process of nuclear transplantation in the frog. The nucleus is removed with a micropipette from an unfertilized egg. A nucleus from a blastula embryo is injected, and development proceeds.

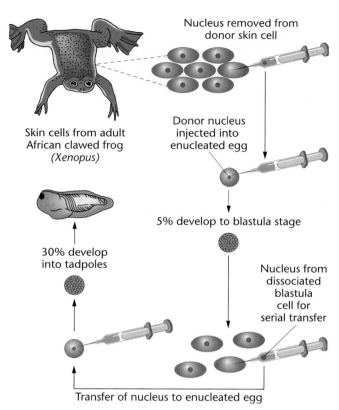

FIGURE 22.3 Serial transplantation experiments using adult frog skin cell nuclei. Serial transplantation dramatically increases the percentage of recipient eggs that undergo complete cleavage.

not dissociated but instead was allowed to continue development. Gurdon found that regardless of the source of the original nucleus, these serial transfers always increased the number of tadpoles that would develop (Table 22.1). Because even nuclei from fully differentiated adult epidermal cells can eventually direct the synthesis of gene products such as myosin, hemoglobin, and crystallin—none of which are normally found in epidermal cells—and promote the organization of cells and tissues into a tadpole, we infer that such cells contain a complete copy of the genome. These experiments show that under the proper circumstances, genes not expressed in more specialized cells can become reactivated.

It took almost half a century until the equivalent procedure was carried out successfully in mammals, both in sheep and mice. In 1996 a sheep was born, named Dolly, that was derived from an enucleated egg that had been fused with a mammary gland cell from an adult female sheep (Figure 22.4). Therefore, with one nuclear transfer, the genes from the fully differentiated somatic cell were reprogrammed so that they could give rise to an entire organism. While this conclusion is dramatic, it is not the reason Dolly became famous. The popular press became interested because in these experiments Dolly is geneti-

cally identical to the sheep from which the somatic cell was taken. In other words, the two are genetic clones. However, since genes are not all that determine the phenotype of an organism, and since the genetic make-up of an individual changes—at least slightly—with every cell division, it is not possible to create truly identical individ-

TABLE 22.1 First Transfer vs. Serial Transplantation in *Xenopus laevis*

Donor Cells	Percentage Reaching Tadpole Stage with Differentiated Cell Types	
	First Transfer	Serial Transfer
Intestinal epithelial cell from tadpoles	1.5	7
Cells from adult skin	0.037	8
Blastula or gastrula	36	57

Source: From Gurdon (1974, p. 24).

FIGURE 22.4 Dolly, the first mammalian genetic clone. This Finn Dorsett sheep was derived from an egg whose nucleus was replaced by that of an adult mammary gland cell.

uals. In spite of this, an important ethical issue is still presented by this technology, since it is now theoretically possible to make human clones using this procedure.

In plants, the reprogramming of the nuclear genome, which required nuclear transfer in animals, can be achieved simply by exposing differentiated cells to specific culture conditions. This totipotency of terminally differentiated, nonmeristematic cells has been demonstrated using the carrot. Frederick Steward observed that when individual phloem cells are explanted into a liquid culture medium, each cell divides and eventually forms a mass called a *callus*. Under appropriate conditions, including the addition of various plant hormones, this cell mass will differentiate into a mature plant. This totipotency has been shown for many other plants as well and illustrates a fundamental difference between plant and animal development. In animals, the adult body structure is essentially completed by the end of embryogenesis. Plants, however, use an *open strategy* of development such that new structures can be added throughout the life of the organism. It is this developmental plasticity that reveals that genetic totipotency is maintained throughout the life of plants.

These studies all convincingly argue that differentiated adult cells have not lost any of the genetic information present in the zygote. Instead, the majority of genes in any given cell type fail to be activated or are repressed, but they can be reprogrammed to direct normal development. We are only beginning to understand the nature of the molecular processes controlling nuclear differentiation during cellular specialization, but the nuclear transplant experiments indicate that the cytoplasm plays a major role in controlling gene expression.

Binary Switch Genes

As part of the progressive restriction of transcriptional capacity that accompanies development, certain genes act as switch points, decreasing the number of alternative developmental pathways. Each decision point is usually binary—that is, there are two alternatives present—and the action of a switch gene programs the cell to follow one of these pathways. Thus, such genes are referred to as **binary switch genes.** We will briefly describe two examples: (1) *myo*D-related genes, which control muscle cell determination and differentiation in both invertebrates and vertebrates; and (2) *eyeless/Small eye,* which is a master regulator of eye formation, again in both invertebrates and vertebrates.

One of the best-characterized classes of switch genes includes those that control muscle cell formation in both invertebrates and vertebrates. These genes are the **myoblast-determining** or *myo*D genes. The nucleotide and amino acid sequences of the four members of this family are very similar, indicating that these genes are highly conserved evolutionarily in vertebrates (Table 22.2). They are less similar to *myo*D-related genes in invertebrates, suggesting a more distant evolutionary relationship. The *myo*D genes are major regulatory genes that normally initiate events leading to skeletal muscle cell differentiation in embryonic cells known as *myoblasts.* Experimentally induced expression of *myo*D genes in nonmuscle-forming cell types, including fibroblasts (connective tissue cells of the skin), adipocytes (fat cells), neurons (nerve cells), and melanocytes (pigment cells), results in the activation of muscle-specific genes in those cells and, in some cases, causes the formation of completely differentiated muscle cells that contain contractile fibers. The ability to switch the phenotype of differentiated cells by the activation of a single gene provides strong evidence for the central role of *myo*D genes in the regulation of muscle cell formation.

Members of the *myo*D gene family encode DNA-binding proteins with a helix-loop-helix (HLH) motif (see Chapter 16 for a review of eukaryotic transcription factors). Members of the *myo*D class of proteins contain an 80-amino-acid basic region (Figure 22.5) that mediates DNA binding and confers specificity for transcriptional activation. These proteins form heterodimers with proteins known as E-box proteins and recognize the sequence

TABLE 22.2 The *myo*D Gene Family

Xenopus	Rodent	Human
XMyoD	*myo*D	*myf* 3
XMyogenin	*myogenin*	*myf* 4
XMyf 5	*myf* 5	*myf* 5
XMRF 4	*MRF* 4	*myf* 6

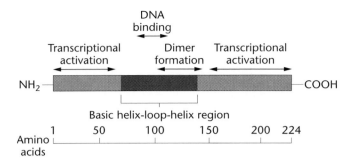

FIGURE 22.5 Structure of myogenin, a 224-amino-acid protein belonging to the *myo*D family. This protein has a characteristic basic region and helix-loop-helix domain near the center. An adjacent region controls dimer formation necessary for activity. Regions near either end control transcriptional activation.

CANNTG (where N is any nucleotide) present in the enhancers and promoters of most skeletal muscle genes.

Results from cell-culture studies on myoblasts indicate that MyoD proteins themselves are under the control of cell growth factors. Myoblasts cultured in the presence of high concentrations of growth factors undergo growth and division, and expressions of *myo*D and *myf*5 are repressed, or are expressed only at very low levels. When growth factors are withdrawn from the culture medium, the myoblasts exit the cell cycle, enter a G0 state, and *myo*D and/or *myf*5 are once again expressed, or derepressed. The action of these transcription factors activates the expression of **myogenin,** which in turn triggers the expression of a cascade of contractile protein genes, causing the myoblast to begin differentiation as a muscle cell. Differences in the timing of expression for the *myo*D-related proteins suggest that they each have separate and distinct functions. According to this model:

1. *myf*5 is the earliest factor to be expressed and is the switch gene involved in *determination.*

2. *myo*D works with *myf*5 to maintain the myoblast identity.

3. All four factors coordinately initiate the process of *terminal differentiation* into muscle.

4. *Myogenin* works to maintain the differentiated adult phenotype.

A second binary switch gene called *eyeless* in *Drosophila* triggers not only the differentiation of a particular cell type, but of an entire organ, the compound eye. The *eyeless* gene has therefore been dubbed the master regulator of eye development. The eyeless protein appears to act very early in development and is expressed in all cells in the embryo that will give rise to the eye. In the absence of this gene, cells that were specified to become part of the eye undergo programmed cell death (discussed at the end of this chapter). Like *myo*D, *eyeless* codes for a transcription factor, although in this case it is a member of the paired box family.

While the ability of this single gene to trigger eye development is remarkable, more recent experiments with *eyeless* and the same gene in mice (called *Small eye/Pax6*) have had a broader impact on biology because they have challenged the widely held assumption that the compound eye of insects and the single-lens eye of vertebrates are not evolutionarily related but, rather, evolved separately. This assumption was based on the fact that both the formation of these organs during development and their final adult structures are completely different. However, in 1995 Walter Gehring and his colleagues generated transgenic flies that had extra copies of either the fly or the mouse gene, which was placed under the control of a strong promoter. The results were striking and led to two inescapable conclusions; first, that this single switch gene was capable of triggering the formation of entire extra eyes on the wings, antennae, and legs of flies (Figure 22.6), and second, that because the mouse gene works in flies, the invertebrate and vertebrate eye *are* in fact homologous at the molecular level. Therefore, the two are related evolutionarily. The downstream targets of these transcription factors also appear to be conserved, indicating that several steps in the molecular regulation of eye development are shared between species that diverged over half a billion years ago. The ultimate impact of this work may be on our understanding of human development because mutations in the

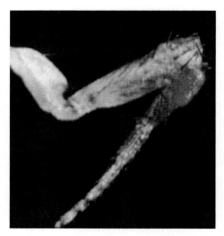

FIGURE 22.6 Ectopic expression of the *eyeless* gene in transgenic flies. Complete eyes are formed at ectopic locations, including the leg as shown in this figure, when the *eyeless* gene is expressed during development.

same gene in humans, called *Aniridia*, cause defects in the iris, lens, cornea, and retina.

Embryonic development in *Drosophila* illustrates the influence of localized cytoplasmic components on gene expression

In the preceding discussion, we examined the mechanisms responsible for making cells different from one another (i.e., differential transcription), and in Chapter 16 we examined the mechanisms that turn specific genes on or off. However, *why* a certain cell turns on or off specific genes during development is a broader and more critical issue and is a central question in developmental biology.

There is no simple answer to this question. However, information derived from the study of many different organisms provides a starting point. One of these examples, embryonic development in *Drosophila*, reflects the direction and scope of our knowledge in this area, and highlights the key regulatory role played by molecular components placed in the oocyte cytoplasm during oogenesis. We will begin with an overview of development and

FIGURE 22.8 Scanning electron micrograph of a newly oviposited *Drosophila* egg. The micropyle is the small white projection at the anterior tip (at the bottom). The dorsal surface (with the two chorionic appendages) is facing up.

then describe how the analysis of mutations has provided insight into the number and action of genes that regulate the processes of determination and differentiation.

Overview of Drosophila Development

Beginning with the fertilized egg, *Drosophila* passes through a pre-adult period of development with five distinct phases: the embryo, three larval stages, and the pupal stage. The adult fly emerges from the pupal case about 10 days after fertilization (Figure 22.7). Externally, the *Drosophila* egg has a number of structures that delineate the anterior, posterior, dorsal, and ventral regions (Figure 22.8). The anterior end of the egg contains the micropyle, a specialized conical structure for the entrance of sperm into the egg, while the posterior end is rounded and marked by a series of aeropyles (openings that allow gas exchange during development). The dorsal side of the egg is flattened and contains the chorionic appendages, while the ventral side is curved.

Internally, the egg cytoplasm is organized into a series of maternally derived molecular gradients. These gradients play a key role in establishing the developmental fates of nuclei that migrate into specific regions of the embryo.

Immediately after fertilization, the zygote nucleus undergoes a series of divisions. After nine rounds of division, most of the approximately 512 nuclei migrate to the egg's outer surface, or cortex, where further divisions take place

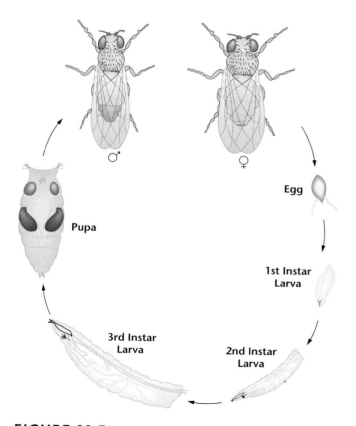

FIGURE 22.7 *Drosophila* life cycle.

Egg

1st Instar Larva

2nd Instar Larva

3rd Instar Larva

Pupa

(a)

Diploid zygote nucleus is produced by
fusion of parental gamete nuclei.

(b)

Nine rounds of nuclear divisions
produce multinucleated syncytium.

(c)

Nuclei migrate to outer surface

(d)

Pole cells form at posterior pole
(precursors to germ cells).

Approximately four further
divisions take place at the cell surface.

(e)

pole cells

Nuclei become enclosed in
membranes, forming a single layer of
cells over embryo surface.

FIGURE 22.9 Early stages of embryonic development in *Drosophila*. (a) Fertilized egg with zygotic nucleus, about 30 minutes after fertilization. (b) Nuclear divisions occur about every 10 minutes, producing a multinucleate cell, the syncytial blastoderm. (c) After approximately nine divisions (512 nuclei), the nuclei migrate to the outer surface or cortex of the egg. (d) At the surface, four additional rounds of nuclear division occur. A small cluster of cells, the pole cells, form at the posterior pole about 2.5 hours after fertilization. These cells will form the germ cells of the adult. (e) About 3 hours after fertilization, the nuclei become enclosed in membranes, forming a single layer of cells over the embryo surface creating the cellular blastoderm.

[Figure 22.9(c) and (d)]. This is the **syncytial blastoderm** stage of embryonic development (a syncytium is any cell with more than one nucleus). The nuclei that migrated are now surrounded by cytoplasm that contains maternally derived transcripts and proteins localized in these regions. Under the influence of these cytoplasmic components a program of gene expression is initiated in the nuclei, leading to the initiation of various developmental programs.

One example that illustrates the regulatory role of these localized cytoplasmic components is the formation of the germ cells at the posterior pole of the embryo [Figure 22.9(d) and (e)]. Experiments have shown that any nucleus that is placed in the posterior cytoplasm (the pole plasm) will generate cells that differentiate into germ cells. Hence the pole plasm contains maternal components that direct the posterior nuclei to give rise to germ cells. The transcriptional programs triggered in migrating nuclei also result in the establishment of a body plan organized along anterior–posterior and dorsal–ventral axes of symmetry.

Following formation of the syncytial blastoderm, cell membranes are deposited around individual nuclei, creating the **cellular blastoderm** [Figure 22.9(e)]. Later, the embryo becomes organized into a series of segments, called parasegments, which are defined by the expression patterns of various embryonic genes (Figure 22.10). These later give rise to the adult segments, though they are shifted back by one-half segment. Cells within the segments first become determined, to form either an anterior or a posterior portion (called a **compartment**) of the segment. In the following stages of development, further restrictions of developmental options occur so that cells in each compartment are eventually restricted to forming a single structure in the adult body.

The body plan of adult *Drosophila* derives its overall organization from the larval body plan and is composed of head, thoracic, and abdominal segments. In addition, many of the discrete body structures of the adult are formed from small sacks of cells, or **imaginal disks,** that form as soon as the larval blastoderm forms. There are 12 bilaterally paired disks—for example, the eye-antennal disks, leg disks, wing disks, and so forth—and one genital disk. Figure 22.11 indicates which imaginal disks will give rise to which adult structures. Cell fates are based on lineage tracing, where individual embryonic cells are marked early in development and then followed until they differentiate. These **fate maps** indicate that in many organisms, such as *Drosophila*, the position of a cell early in development specifies what cell type it will become.

During metamorphosis, most of the larval body parts histolyze, or break down, and the imaginal disks undergo mitosis and differentiate to form the head, thorax, and abdomen of the adult fly.

(a)

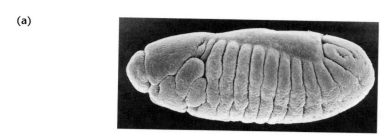

(b)

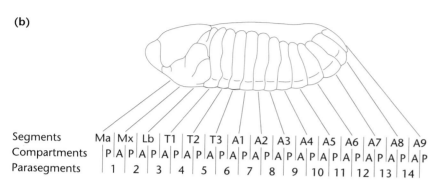

Segments	Ma	Mx	Lb	T1	T2	T3	A1	A2	A3	A4	A5	A6	A7	A8	A9
Compartments	P A	P A	P A	P A	P A	P A	P A	P A	P A	P A	P A	P A	P A	P A	P
Parasegments	1	2	3	4	5	6	7	8	9	10	11	12	13	14	

(c)

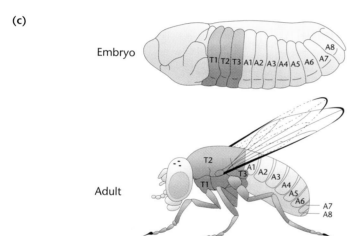

FIGURE 22.10 Segmentation in *Drosophila*. (a) Scanning electron micrograph of a *Drosophila* embryo at about 10 hours after fertilization. By this stage, the segmentation pattern of the body is clearly established. (b) The segments, compartments, and parasegments of the *Drosophila* embryo. Ma, Mx, and Lb represent the segments that will form head structures. T1–T3 are thoracic segments and A1–A9 are abdominal segments. Each segment is divided into anterior (A) and posterior (P) compartments. Parasegments represent an early pattern specification that is later refined into the segmental plan of the body. Note that all the parasegments are shifted forward by one compartment to form the segments. (c) The segmented embryo and the adult structures that will form each segment.

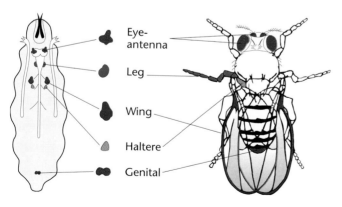

FIGURE 22.11 Imaginal disks of the *Drosophila* larva and the adult structures that are derived from them.

Genetic Analysis of Embryogenesis: Maternal-Effect Genes and Zygotic Genes

Although our knowledge of anatomical development in *Drosophila* has been useful, genetic analysis has provided the most significant information about the events in embryogenesis. Genes that control embryonic development are either **maternal-effect genes** or **zygotic genes**. Maternal-effect genes are those whose gene products (mRNA and proteins) are deposited in the developing egg during oogenesis. Although these products may be distributed evenly throughout the cytoplasm of the egg, they are often distributed in a gradient or concentrated in specific regions of the egg. The phenotype of flies with mutations in maternal-effect genes is expected to be female sterility, since none of the embryos of females homozygous for a recessive mutation would receive wild-type gene products from their mother and, therefore, they would not develop normally. In *Drosophila*, these maternal-effect genes encode transcription factors, receptors, and proteins that regulate translation. During embryonic development, these gene products work to activate or repress the expression of zygotic genes in a temporal and spatial sequence.

Zygotic genes are those of the developing embryo itself, and are therefore transcribed after fertilization. The phenotype of mutations in this class show embryonic lethality; in a cross between two flies heterozygous for a recessive zygotic mutation, one-fourth of the embryos (the homozygotes) fail to develop. In *Drosophila*, many of the zygotic genes are transcribed in patterns that depend on the distribution of maternal-effect proteins.

Much of what is known about the zygotic genes that regulate development came from a mutagenic analysis performed by Christiane Nüsslein-Volhard and Eric Wieschaus, who systematically screened for mutations that affect embryonic development. They examined thousands of dead F₂ offspring of mutagenized flies for recessive embryonic lethal mutations with defects in external structures. The parents were thus identified as carriers of these mutations, which fell into three classes, named *gap*, *pair-rule*, and *segment polarity*. In a paper

published in 1980, they proposed a model in which embryonic development is initiated by gradients of *maternal-effect* gene products. The positional information laid down by these molecular gradients along the anterior–posterior axis of the embryo is interpreted by two sets of *zygotic* genes: (1) those identified in their screen (gap, pair-rule, and segment polarity genes), collectively called the **segmentation genes,** which divide the embryo into a series of stripes or segments and define the number, size, and polarity of each segment; and (2) **selector genes,** which specify the identify or fate of each segment (Figure 22.12).

A diagrammatic view of the model developed by Nüsslein-Volhard and Wieschaus is illustrated in Figure 22.13. The process begins when maternal-effect gene products responsible for the formation of the anterior–posterior axis, which were placed in the egg during

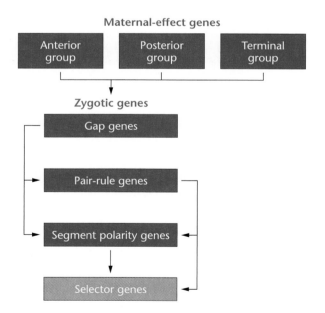

FIGURE 22.12   The hierarchy of genes involved in establishing the segmented body plan in *Drosophila*. Gene products from the maternal genes regulate the expression of the first three groups of zygotic genes (gap, pair-rule, and segment polarity; collectively called the segmentation genes), which in turn control expression of the selector genes.

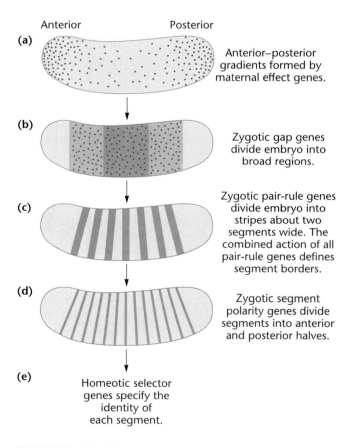

FIGURE 22.13   Overview of progressive restriction of cell fate during development in *Drosophila*. Gradients of maternal proteins are established along the anterior–posterior axis of the embryo (a). The three groups of segmentation genes progressively define the body segments (b), (c), and (d). Individual segments are given identity by the selector genes (e).

oogenesis, are activated immediately after fertilization [Figure 22.13(a)]. Their activity restricts cells to form either anterior or posterior structures. Gene products from this gradient activate the transcription of the gap genes that divide the embryo into a limited number of broad regions [Figure 22.13(b)]. Gap proteins are transcriptional factors that activate pair-rule genes whose products divide the embryo into regions about two segments wide [Figure 22.13(c)]. The combined action of all gap genes defines segment borders. The pair-rule genes in turn activate the segment polarity genes that divide the segments into anterior and posterior compartments [Figure 22.13(d)]. The collective action of the maternal genes that form the anterior–posterior axis and the segmentation genes define the fields of action for the homeotic selector genes that specify the identify of each segment [Figure 22.13(e)].

In a second screening, Wieschaus and Trudi Schüpbach screened thousands of flies for maternal-effect mutations that affected the external structures of the embryo. They estimated that there are about 40 maternal-effect genes and 50–60 zygotic genes that regulate embryonic development. This means that normal embryogenesis is controlled by about 100 genes, a surprisingly low number given the complexity of the structures formed from the fertilized egg. For their work on the genetic control of development in *Drosophila*, Nüsslein-Volhard and Wieschaus, along with Ed Lewis—the geneticist who initially identified and studied many of these genes in the 1970s—were awarded the 1995 Nobel Prize for Physiology or Medicine.

Maternal-effect genes and zygotic genes interact to determine the basic body plan in *Drosophila*

The first group of genes required to form a *Drosophila* embryo are the maternal-effect genes, which are transcribed by the maternal genome during oogenesis and whose products are transported as mRNA or as translated protein into the oocyte for storage. These transcripts and proteins are used early in development to establish the overall bilateral symmetry of the developing embryo, and together they convert the egg into an embryo with anterior–posterior and dorsal–ventral axes. In addition, the maternal-effect genes trigger the expression of the first group of zygotic genes, thus initiating the cascade of gene expression that completes the process of development. The following discussion will emphasize the formation of the anterior–posterior axis as an example of the interaction of maternal-effect gene products and the first group of zygotic genes.

The Anterior Genes

The maternal-effect genes that control the anterior–posterior axis are grouped into three classes: anterior, posterior, and terminal. Of the genes required to form the anterior portion of the embryo, the most important is *bicoid* (Table 22.3). Embryos derived from mothers who were homozygous for a mutation in the *bicoid* gene, and who therefore have no normal bicoid protein, lack head and thoracic structures and also have abnormalities of the first four segments of the abdomen. During oogenesis and the first hours of embryonic development, *bicoid* mRNA is localized to a small region at the anterior pole of the egg [Figure 22.14(a)]. The actions of two genes, *swallow* and *exuperantia*, are involved in localizing the *bicoid* mRNA by docking it to anterior cytoskeletal components. In mutants of these latter genes, the *bicoid* mRNA is distributed to more posterior areas of the egg, and the mutant embryos form enlarged anterior structures.

During early cleavages, *bicoid* mRNA is translated, and the bicoid protein becomes distributed in a gradient along the anterior–posterior axis, with very low levels in the posterior region of the egg [Figure 22.14(b)]. This gradient plays a critical role in determining the developmental pattern of the egg. The bicoid protein is a transcription factor that contains a homeodomain. The bicoid protein binds to the upstream regulatory region of *hunchback*, one of the first zygotic genes to be activated early in development, which is expressed in the nuclei in the anterior half of the embryo. The upstream promoter region of *hunchback* contains five sites, each with the consensus sequence 5′-TC-TAATCCC-3′, to which the bicoid protein binds. Binding of the *bicoid* transcription factor to this consensus sequence activates the expression of the *hunchback* gene, and the degree of the response is dependent on the number of sites occupied by the bicoid protein. In regions of the gradient where there is a higher concentration of bicoid protein, there will be a higher response by the *hunchback* gene, lead-

TABLE 22.3 Maternally Transcribed and Zygotically Transcribed Genes That Control the Anterior–Posterior Axis of the *Drosophila* Embryo

	Anterior	Posterior	Terminal
Maternal	*bicoid* *exuperantia* *swallow*	*staufen* *oskar* *vasa* *valois* *tudor* *mago nashi* *nanos* *pumilio*	*trunk* *fs(1)Nasrat* *fs(1)polehole* *torso* *torso-like* *l(1)polehole*
Zygotic	*hunchback*	*knirps* *giant*	*tailless* *buckebein*

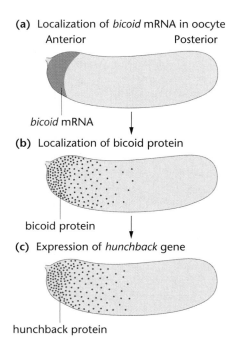

(a) Localization of *bicoid* mRNA in oocyte

(b) Localization of bicoid protein

(c) Expression of *hunchback* gene

FIGURE 22.14 Formation of the anterior–posterior axis of the *Drosophila* egg depends on the action of three independent systems. The anterior portion of the axis is formed by a gradient of *bicoid* mRNA (a) that is translated into a transcription factor that distributes itself in a gradient at the anterior end of the embryo (b). The gradient of bicoid protein triggers expression of the *hunchback* gene, forming a gradient of hunchback protein along the anterior–posterior axis (c).

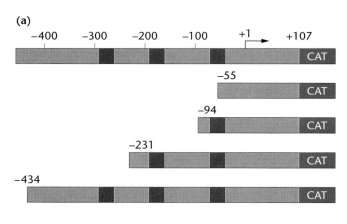

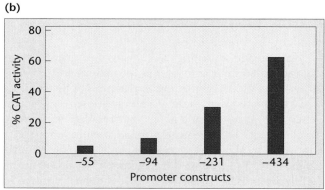

FIGURE 22.15 (a) Fusion genes between a reporter gene for chloramphenicol acetyltransferase (CAT) with various promoter regions upstream from the *hunchback* gene. Each construct was tested by injection into bicoic *bicoid* embryos and wild-type embryos. When injected into *bicoid* embryos, the fusion gene was not transcribed, and no CAT protein was produced. When injected into wild-type embryos (b), the full expression of the CAT protein requires the presence of three of the five *hunchback* promoter sites.

ing to the formation of a gradient of *hunchback* protein along the anterior–posterior axis [Figure 22.14(c)].

This interaction between the bicoid protein and the promoter region of *hunchback* was demonstrated by fusing a reporter gene encoding the enzyme chloramphenicol acetyltransferase (CAT) to promoter sites of the *hunchback* gene (Figure 22.15). When these constructs were injected into *bicoid* mutant embryos, no CAT was produced. When they were injected into wild-type embryos, the CAT constructs were expressed. The full expression of the CAT gene required the presence of three of the five promoter sites. These experiments demonstrated that the bicoid protein determines the spatial pattern of expression of the *hunchback* gene. The gradient of bicoid protein causes transcription of *hunchback* in the anterior half of the embryo, but not in the posterior half. The hunchback protein then stimulates the transcription of the genes forming the head and thorax, and represses the activity of the genes that form the abdomen.

The Posterior Genes

The posterior of the embryo is formed through the action of eight maternal genes (Table 22.3). Mutants of these genes all have a similar phenotype: The posterior portion of the body is dwarfish and abdominal segments are lacking. The names of some of these mutants reflect this phenotype: *oskar* (named after the dwarf in the Günter Grass novel *The Tin Drum*) and *nanos* (derived from the Spanish word for dwarf). The *nanos* gene is transcribed by the maternal genome during oogenesis, and its mRNA is stored at the posterior pole of the egg (Figure 22.16). Other genes involved in posterior formation ensure that the *nanos* mRNA is packaged and stored at the posterior pole (*oskar, staufen, valois,* and *vasa*) and is distributed properly (*pumilio*).

While the nanos protein is produced soon after fertilization and controls the formation of the posterior structures in the developing embryo, it does so in a way different from the bicoid protein. In contrast to the action of the anterior genes, the posterior genes work by

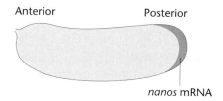

Anterior   Posterior

nanos mRNA

FIGURE 22.16 The posterior end of the anterior–posterior axis is formed by action of the *nanos* gene. The mRNA from *nanos* transcribed during oogenesis is stored at the posterior pole of the oocyte. After fertilization, the mRNA is distributed throughout the posterior region of the embryo by action of the *pumilio* gene. The nanos protein acts to suppress the translation of *hunchback* mRNA, ensuring the formation of abdominal structures in the posterior region.

a negative effect on a maternal-gene transcript. The nanos protein acts to prevent the translation of any *hunchback* mRNA that might be present in the posterior region of the embryo, ensuring the expression of abdomen-forming genes (Figure 22.17). The *hunchback* gene, therefore, is regulated by maternal-effect gene products from both the anterior and posterior parts of the developing embryo.

The Terminal Genes

The *terminal genes* represent a third set of genes controlling the formation of the anterior–posterior axis (Table 22.3). Mutations in these genes result in the loss of the unsegmented structures that develop from the anterior and posterior poles of the embryo. The cloning of two of the terminal genes, *torso* and *fs(1)polehole*, suggests that this group contains genes that encode transmembrane signal receptors associated with the reception and transduction of an extracellular signal. Genetic studies have now demonstrated that this signal comes from the ovarian follicle cells surrounding the developing egg.

Together, the anterior, posterior, and terminal genes establish the anterior–posterior axis of the embryo as a

spatial pattern of gene products. The anterior of the embryo is organized by a gradient of the bicoid protein that acts to activate anterior-specific genes and to repress posterior-specific genes. The posterior of the embryo is formed through the action of the *nanos* protein, which suppresses the formation of the *hunchback* protein, preventing the *hunchback*-mediated inhibition of abdominal-specific gene expression. Critical to the formation of the terminal region of the embryo is the *torso* gene product, which allows the expression of the genes that form the unsegmented terminal structures. Two of these systems function by sequestering a specific mRNA at the poles of the egg (*bicoid* at the anterior, *nanos* at the posterior) and by the formation of a protein gradient. The third system works via a signal transduction system coupled to the anterior–posterior axis system during oogenesis.

The dorsal–ventral axis of the embryo is specified by a set of maternal-effect genes that act essentially completely independently of the anterior, posterior, and terminal genes. Together they create a single gradient of a protein called dorsal along the dorsal–ventral axis. This gradient is unlike the gradients of bicoid and nanos, however, because it is a nuclear protein gradient, with the highest levels of dorsal protein in the ventral nuclei. Thus the determination of the ventral side of the embryo requires the selective transport of dorsal protein into ventral nuclei. This is accomplished immediately after fertilization when dorsal protein, which was distributed evenly throughout the oocyte's cytoplasm, is transported into only those nuclei that have migrated to what will be the future ventral side of the embryo.

Together these maternal-effect gene products regulate gene expression in the embryo along the anterior–posterior and dorsal–ventral axes, and illustrate the critical role of cytoplasmic gradients in determining the fates of cells in different embryonic regions. The next section describes how components of the anterior–posterior axes control the transcription of gene sets that regulate the further development of the embryo.

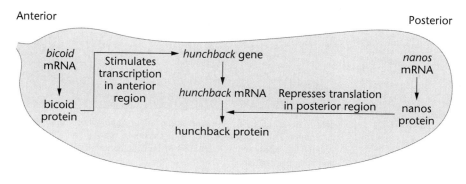

Anterior

Posterior

bicoid mRNA → Stimulates transcription in anterior region → *hunchback* gene

bicoid protein

hunchback mRNA ← Represses translation in posterior region

hunchback protein

nanos mRNA → nanos protein

FIGURE 22.17 Coordinated action of bicoid and nanos proteins produces the anterior–posterior axis of the early embryo. Throughout the anterior end, bicoid protein activates the *hunchback* gene. The hunchback protein activates transcription of head and thorax genes. In the posterior region, the nanos protein inhibits production of hunchback protein, allowing the expression of abdominal genes.

Zygotic genes determine segment formation

The zygotic genes are activated or repressed in a positional gradient by the maternal-effect gene products, and three subsets of these genes (gap, pair-rule, and segment polarity), collectively called the segmentation genes, divide the embryo into a series of segments along the anterior–posterior axis. These segmentation genes are transcribed in the developing embryo, and mutations in these genes are seen as embryonic lethal phenotypes.

The segmentation genes are expressed soon after the *bicoid* gradient has been established [see Figure 22.14(b)]. Over 20 segmentation loci have been identified (Table 22.4). They are classified on the basis of their mutant phenotypes: (1) the gap genes (e.g., *Krüppel* and *knirps*) delete a group of adjacent segments; (2) the pair-rule genes (e.g., *even-skipped* and *fushi-tarazu*) affect every other segment and eliminate a specific part of each affected segment; and (3) the segment polarity genes (e.g., *hedgehog* and *gooseberry*) cause defects in homologous portions of each segment.

Gap Genes

The transcription of the **gap genes** is activated or inactivated by gene products previously expressed along the anterior–posterior axis, including *bicoid* and the other genes of the maternal gradient systems. When mutated, these genes produce large gaps in the segmentation pattern of the embryo. *Hunchback* mutants lose head and thorax structures, *Krüppel* mutants lose thoracic and abdominal structures, and mutations in *knirps* result in the loss of most abdominal structures. The transcription of gap genes divides the embryo into a series of broad regions (roughly, the head, thorax, and abdomen) within which different combinations of gene activity will even-

tually specify both the type of segment that will form and the proper order of segments in the body of the larva, pupa, and adult. To date, all gap genes that have been cloned encode transcription factors with zinc finger DNA-binding motifs. The expression of the gap genes correlates roughly with their mutant phenotypes: *hunchback* at the anterior, *Krüppel* in the middle, and *knirps* at the posterior (Figure 22.18). Gap genes control the transcription of pair-rule genes.

Pair-Rule Genes

The **pair-rule genes** divide the broad regions established by the gap genes into regions about one segment wide. Mutations in pair-rule genes eliminate segment-size regions at every other segment. The pair-rule genes are expressed in narrow bands or stripes of nuclei that extend around the circumference of the embryo. The expression of this gene set first establishes the boundaries of segments and then establishes the developmental fate of the cells within each segment by controlling the segment polarity genes. At least eight pair-rule genes act to divide the embryo into a series of stripes. However, the boundaries of these stripes overlap, meaning that cells in the stripes express different combinations of pair-rule genes in an overlapping fashion (Figure 22.19).

Many pair-rule genes encode transcription factors containing helix-turn-helix homeodomains. The transcription of the pair-rule genes is mediated by the action of gap gene products, but the pattern is resolved into highly delineated stripes by the interaction among the gene products of the pair-rule genes themselves (Figure 22.20).

TABLE 22.4 Segmentation Genes in *Drosophila*

Gap Genes	Pair-Rule Genes	Segment Polarity Genes
Krüppel	*hairy*	*engrailed*
knirps	*even-skipped*	*wingless*
hunchback	*runt*	*cubitis interruptus*[D]
giant	*fushi-tarazu*	*hedgehog*
tailless	*odd-paired*	*fused*
huckebein	*odd-skipped*	*armadillo*
	sloppy-paired	*patched*
		gooseberry
		paired
		naked
		disheveled

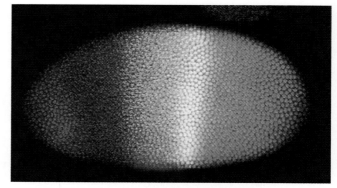

FIGURE 22.18 Expression of several gap genes in a *Drosophila* embryo. The hunchback protein is shown in orange and Krüppel in green. The yellow stripe is created when cells contain both hunchback and Krüppel proteins.

(a)

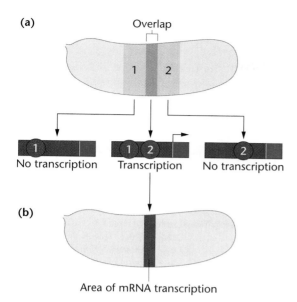

(b)

FIGURE 22.19 New patterns of gene expression can be generated by overlapping regions containing gene products. (a) Transcription factors 1 and 2 are present in an overlapping region of expression. If both transcription factors must bind to their binding sites in a promoter in order to trigger the expression of a target gene, the gene will be active only in cells containing both factors (most likely in the zone of overlap). (b) The expression of the target gene in the restricted region of the embryo.

Segment Polarity Genes

Expression of the **segment polarity genes** is controlled by transcription factors encoded by the pair-rule genes. Each segment polarity gene becomes active in a single band of cells within each segment that extends around the embryo (Figure 22.21). These expression patterns divide the embryo into 14 segments and the gene products control the cellular identity within each segment. Some of the segment polarity genes, including *engrailed*, encode transcription factors. Rather than activating transcription, however, the engrailed protein competitively inhibits activation by other homeodomain proteins, resulting in the establishment of a segment border.

Selector Genes

As segment boundaries are being established by the action of the segmentation genes, the **selector genes** are activated. These are genes whose expression determines the structures to be formed by each body segment, including the antennae, mouth parts, legs, wings, thorax, and abdomen. Mutants of these genes are known as **homeotic mutants** (from the Greek word for "same") because the

(a)

(b)

FIGURE 22.20 Stripe pattern of pair-rule gene expression in *Drosophila* embryo. This embryo is stained to show patterns of expression of the genes *even-skipped* and *fushi-tarazu*; (a) low-power view, and (b) high-power view of the same embryo.

FIGURE 22.21 The fourteen stripes of expression of the segment polarity gene *engrailed* in a *Drosophila* embryo.

(a)

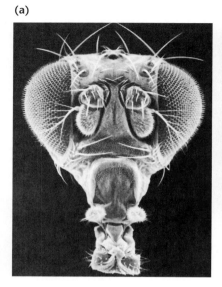

(b)

FIGURE 22.22 *Antennapedia (Antp)* mutation in *Drosophila*. (a) Head from wild-type *Drosophila* showing the structure of the antenna and other head parts. (b) Head from an *Antp* mutant, showing the replacement of normal antenna structures with a leg. This is caused by activation of the *Antp* gene in the head region.

identity of one segment is transformed so that it is the same as the identity of a neighboring segment. For example, the wild-type allele of the *Antennapedia (Antp)* gene is required to specify structures in the second thoracic segment (which carries a leg). Dominant gain-of-function *Antp* mutations lead to the expression of the gene in the head and thorax, and the antennae of mutant flies are transformed and acquire the identity of structures on the thorax or, in other words, legs (Figure 22.22).

The *Drosophila* homeotic selector genes are found in two clusters on chromosome 3 (Table 22.5). The *Antp* complex contains five genes required for the specification of structures in the head and first two thoracic segments (Figure 22.23). The *bithorax (BX-C)* complex contains three genes required for the specification of structures formed by the posterior portion of the second thoracic segment, the entire third thoracic segment, and the abdominal segments (Figure 22.24).

The activation of the selector genes is under the control of the gap genes and the pair-rule genes. Although selector gene expression is confined to certain domains in the embryo, the timing and patterns of expression are rather complex and involve interactions between segmen-

tation and selector genes as well as interactions among the various selector genes. For example, the *Antennapedia* gene has two promoters (P1 and P2) and can be transcribed to produce two different pre-mRNAs (Figure 22.25). Transcription from P1 is stimulated by Krüppel protein (a gap-gene product) and repressed by Ultrabithorax protein (a selector-gene product). Transcription from P2 is activated by hunchback and fushi-tarazu proteins (maternal-effect and gap-gene products) and inhibited by oskar protein (a maternal-effect gene product). Although the protein produced from these two pre-mRNAs is identical, the two

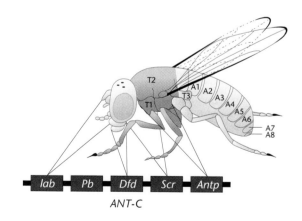

ANT-C

FIGURE 22.23 Genes of the *Antennapedia* complex and the adult structures they specify. The *labial (lab)* and *Deformed (Dfd)* genes control the formation of head segments. The *Sex comb reduced (Scr)* and *Antennapedia (Ant)* genes specify the identity of the first two thoracic segments. The remaining gene in the complex, *Proboscipedia (Pb)*, may not act during embryogenesis, but may be required to maintain the differentiated state in adults. In mutants, the labial palps are transformed into legs.

TABLE 22.5 Homeotic Selector Genes of *Drosophila*

Antennapedia Complex	*Bithorax* Complex
labial	*Ultrabithorax*
Antennapedia	*Abdominal a*
Sex comb reduced	*Abdominal B*
Deformed	
proboscipedia	

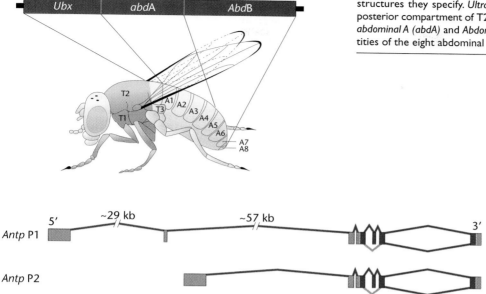

FIGURE 22.24 Genes of the *bithorax* complex and the adult structures they specify. *Ultrabithorax (Ubx)* controls structures in the posterior compartment of T2 and structures in T3. The two other genes, *abdominal A (abdA)* and *Abdominal B (AbdB)*, specify the segmental identities of the eight abdominal segments (A1–A8).

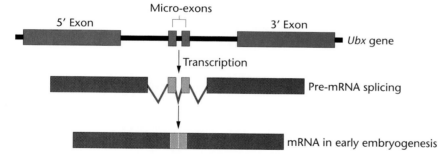

FIGURE 22.25 Transcription and processing of mRNA from the *Antp* gene. Transcription from P1 or P2 in the *Antp* gene results in two different pre-mRNAs. The exons in these pre-mRNAs (indicated by boxes) are then spliced together, with different introns (indicated by bent lines) removed from different pre-mRNAs.

promoters direct the expression of this protein at different times and in different places in the embryo.

The regulation of expression of selector genes is also accomplished by the alternative splicing of pre-mRNAs to yield different transcripts (Figure 22.26). The *Ultra-bithorax* gene transcripts produced during early embryogenesis can be processed at varying 5′ sites in the first exon and will include two "micro-exons," producing a number of different but related proteins. Transcripts from the *Ultrabithorax* gene later in development (during

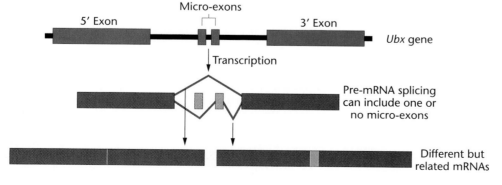

FIGURE 22.26 Alternative processing of *Ubx* pre-mRNA. (a) In early embryogenesis, splicing includes both micro-exons in the mature mRNA. (b) During development of the nervous system, splicing of pre-mRNA includes either no micro-exons or one micro-exon, producing different but related proteins upon translation.

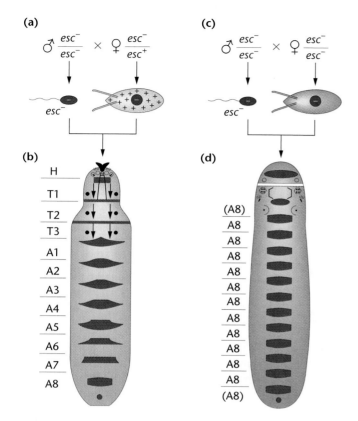

FIGURE 22.27 Action of the *extra sex combs* mutation in *Drosophila*. (a) Heterozygous females form wild-type *esc* gene product and store it in the oocyte. (b) When the *esc⁻* egg formed at meiosis by the heterozygous female is fertilized by *esc⁻* sperm, it produces wild-type larva with a normal segmentation pattern (maternal rescue). Borderlines between the head and thorax and between the thorax and abdomen are marked with arrows. (c) Homozygous *esc⁻* females produce defective eggs, which, when fertilized by *esc⁻* sperm, (d) produce a larva in which most of the segments of the head, thorax, and abdomen are transformed into the eighth abdominal segment. Maternal rescue demonstrates that the *esc⁻* gene product is produced by the maternal genome and is stored in the oocyte for use in the embryo. (H, head; T1–T3, thoracic segments; A1–A8, abdominal segments).

formation of the nervous system) include one or neither of the micro-exons.

A number of additional genes that control the expression of selector genes have been identified, including *extra sex combs (esc), Polycomb (Pc), super sex combs* (sxc),* and *trithorax (trx).* In the second chromosomal mutant *extra sex combs (esc),* some of the head and all of the thoracic and abdominal segments develop as posterior abdominal segments (Figure 22.27), indicating that this gene normally controls the expression of *BX-C* genes in all body seg-

ments. The mutation does not affect either the number or the polarity of the segments. It does affect their developmental fate, indicating that the *esc⁺* gene product that is synthesized and stored in the egg by the maternal genome may be required for the correct interpretation of the information gradient in the egg cortex.

Several of the genes regulating selector gene expression, including the *Polycomb* genes, appear to operate by binding to many sites along chromosomes, rather than just to the promoters and enhancers of individual genes. This observation may begin to explain an intriguing observation about the patterns of expression of the selector genes. When embryos are stained by *in situ* hybridization for each of the mRNAs for the homeotic selector genes, the anterior borders of expression for each gene parallels its position on the chromosome. Genes that are at the 3′-end of the two clusters have borders of expression at the anterior end of the embryo, and as you move toward the 5′-end of the clusters the borders of expression move more and more toward the posterior (Figure 22.28). In addition, as you move more toward the posterior the patterns of expression then overlap such that several genes are expressed in the same segments. These overlapping patterns of expression suggest, first, that the regulation of selector gene activity may be coordinated during development and, second, that the formation of each segment requires the activity of a particular combination of homeotic genes.

E. B. Lewis has proposed that genes in the *bithorax* complex, and perhaps other genes involved in segmentation, arose from a common ancestral gene by tandem duplication and subsequent divergence of structure and function. In fact, each of the homeotic selector genes listed

(a) Expression domains of *HOM-C* genes

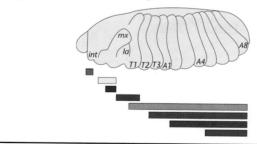

(b) Chromosomal locations of HOM-C genes

FIGURE 22.28 The colinear relationship between the expression domains and chromosomal locations of homeotic genes in *Drosophila*. *Drosophila* embryo and the domains of homeotic gene expression in the embryonic epidermis and central nervous system (a); chromosomal location of homeotic genes (b). Note that the order of genes on the chromosome correlates with the sequential anterior borders of their expression domains.

---

*The sex combs are a set of bristles found only on the legs of males.

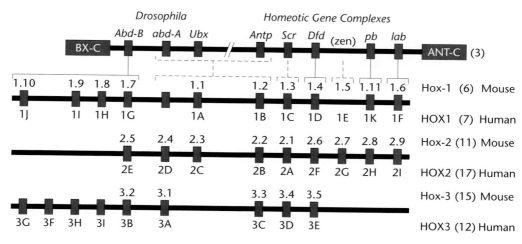

FIGURE 22.29 The *Hox* genes of mammals (human and mouse) and their organizational alignment with the *BX-C* and *ANT-C* complexes of *Drosophila*.

in Table 22.5 encodes a transcription factor that includes a DNA-binding domain encoded by a 180-bp sequence known as a **homeobox**. The homeobox encodes a 60-amino-acid sequence known as a **homeodomain** (see Chapter 16). Similar sequences have been found in the genomes of other eukaryotes with segmented body plans, including *Xenopus*, chicken, mice, and humans. In mammals, the complexes of selector genes are called **Hox** *gene clusters* (Figure 22.29). Homeodomains from all organisms examined to date are very similar in amino acid sequence and encode a protein associated with the transcriptional regulation of a specific gene set. This suggests that the metameric or segmented body plan may have evolved only once.

To summarize, the genes that control development in *Drosophila* act in a temporally and spatially ordered cascade, beginning with the genes that establish the anterior–posterior and dorsal–ventral axes of the egg and early embryo. The gradients of maternal mRNAs and proteins along the anterior–posterior axis activate the gap genes, which subdivide the embryo into broad bands. The gap genes in turn activate the pair-rule genes, which divide the embryo into segments. The final group of segmentation genes, the segment polarity genes, divide each segment into anterior and posterior regions arranged linearly along the anterior–posterior axis. These segments are then given identity by the selector genes. Therefore, this progressive restriction of the developmental potential of the cells in the *Drosophila* embryo, all of which occurs during the first one-third of embryogenesis, involves a cascade of gene regulation, with regulatory proteins acting at both the transcriptional and translational levels.

The tremendous success of the work with *Drosophila* has inspired scientists to take a similar approach in other species. Christiane Nüsslein-Volhard and Wolfgang Driever performed separate large-scale mutagenesis screens in zebrafish, a vertebrate model system, and in 1998 completed the first stage of their work with the identification of nearly 400 different zygotic mutations that affect embryogenesis. Genetic epistasis experiments are now being performed in order to establish the pathways of regulation that mediate the development of a vertebrate embryo. These studies will allow scientists to compare the developmental programs in invertebrates and vertebrates, and, based on the similarity or homology between species, determine when during evolution pathways branched off and created the diverse array of species now present.

Flower development in *Arabidopsis* is also regulated by the activation of homeotic selector genes

The question of the evolutionary conservation of developmental processes and the homology between distantly related species is also being addressed by plant biologists. Elliot Meyerowitz has pioneered the work on flower development using the genetic approach in *Arabidopsis*, a tiny flowering plant in the mustard family that has become a favorite of plant biologists. He and his colleagues have generated a collection of homeotic mutants in flowering in which the identity of one of the four organs (sepals, petals, stamens, and carpels) has been transformed into its neighboring organ type (Figure 22.30). Furthermore, an *Arabidopsis* gene has been cloned, called *CURLY LEAF**, that

---

*By convention gene names in *Arabidopsis* are capitalized.

(a)

(b)

FIGURE 22.30 Flowers in wild-type and mutant *Arabidopsis*. Wild-type flowers have four petals, six stamens, a pistil fused from the two carpels, and sepals, which are hidden in this photo (a). Homeotic *agamous* mutants show extra petals and sepals at the expense of stamens and carpels (b).

regulates the expression of the plant homeotic genes, and it shares significant homology with the *Drosophila Polycomb* genes. Thus both plants and animals use the same mechanism, the activation of unique gene sets in each neighboring region of a particular structure, to determine its identity.

The homeotic selector genes used in *Arabidopsis*, like *Drosophila*, are also transcription factors that are expressed in overlapping patterns in the developing flower; however they are members of a different family of transcription factors, called the MADS box proteins. Therefore, while the same general mechanism is used during development in both multicellular plants and animals, the final task of determining regional identity is performed by a different collection of genes.

Cell–cell interactions influence developmental fate in *C. elegans*

During development in multicellular organisms, cells influence the transcriptional patterns and developmental fate of neighboring cells; these interactions involve the generation and reception of signal molecules. **Cell–cell interaction** is an important process in the development of most eukaryotic organisms, including *Drosophila*, as well as vertebrates such as *Xenopus*, mice, and humans. To study the developmental role of higher-level processes such as cell–cell interactions, as well as the role of individual genes and the genesis of behavior, Sidney Brenner began working with the soil nematode *Caenorhabditis elegans*.

Overview of *C. elegans* Development

The nematode **Caenorhabditis elegans,** which we first introduced in our discussion of sex determination in Chap-

ter 9, is widely used to study the genetic control of development. This organism has two advantages for such studies, the genetics of the organism are well known and adults have a small number of cells that follow a developmental program that is unchanged from individual to individual. The adult *Caenorhabditis elegans*, about 1 mm long, matures from a fertilized egg in about 2 days (Figure 22.31). The life cycle consists of an embryonic stage (about 16 hours), four larval stages (L1 through L4), and the adult stage. The diploid chromosome number is 12 (two X chromosomes and five pairs of autosomes), and it is estimated that the haploid genome consists of about 15,000 genes. Adults are of two sexes—XX self-fertilizing hermaphrodites that can make both eggs and sperm, and XO males. Self-crossing mutagen-treated hermaphrodites quickly results in homozygous stocks of mutant strains, and hundreds of mutants have been generated, cataloged, and mapped.

The adult hermaphrodite consists of 959 somatic cells (and about 2000 germ cells). The exact cell lineage from fertilized egg to adult has been mapped (Figure 22.32) and is in-

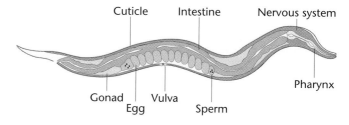

FIGURE 22.31 An adult *Caenorhabditis elegans*. This nematode, about 1 mm in length, consists of 959 cells and has been used to study many aspects of the genetic control of development.

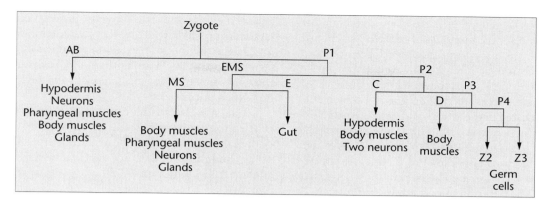

FIGURE 22.32 A truncated cell lineage chart for *C. elegans* showing early divisions and the tissues and organs. Each vertical line represents a cell division, and horizontal lines connect the two cells produced. For example, the first cell division creates two new cells from the zygote, AB and P1. The cells in this chart refer to those present in the first-stage larva L1. During subsequent larval stages, further cell divisions will produce the 959 somatic cells of the adult hermaphrodite worm.

variant from individual to individual. With knowledge of the lineage of each cell, it is easy to follow the events that result either from mutational alterations in cell fate or the killing of cells by laser microbeams or ultraviolet irradiation.

In *C. elegans* hermaphrodites, the fate of cells in the development of the reproductive system is determined by cell–cell interactions and provides some insight into how gene expression and cell–cell interactions are linked in the specification of developmental pathways.

Genetic Analysis of Vulva Formation

Adult hermaphrodites lay eggs through the **vulva,** an opening located about midbody (see Figure 22.31). The vulva is formed in stages during larval development and the process involves several rounds of cell–cell interactions.

During development in *C. elegans*, two neighboring cells, Z1.ppp and Z4.aaa, interact with each other so that one becomes the gonadal anchor cell and the other becomes a precursor to the ventral uterus. The determination of which is which comes during the second larval stage (L2) and is controlled by the *lin*-12 gene, which encodes a cell surface receptor protein. In recessive *lin*-12(0) mutants (a loss-of-function mutant) both cells become anchor cells. The dominant mutant *lin*-12(d) (a gain-of-function mutation) causes both to become uterine precursors. Thus it appears that the *lin*-12 gene causes the selection of the uterine pathway, since in the absence of the lin-12 protein, both cells become anchor cells. However, as shown in Figure 22.33, both cells normally begin to synthesize and secrete a chemical signal (as yet unknown) for uterine differentiation and also synthesize the lin-12 protein, which is a receptor for the signal. At a critical time in L2, the cell that by chance is secreting more of the signal molecule causes its neighbor to increase tran-

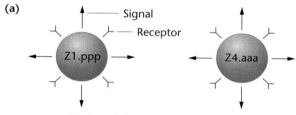

(a)

During L2, both cells begin secreting signal for uterine differentiation

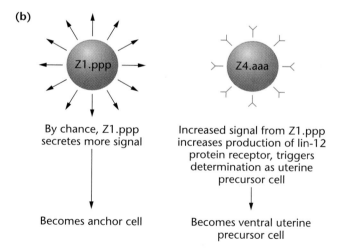

(b)

By chance, Z1.ppp secretes more signal

Becomes anchor cell

Increased signal from Z1.ppp increases production of lin-12 protein receptor, triggers determination as uterine precursor cell

Becomes ventral uterine precursor cell

FIGURE 22.33 Cell–cell interaction in anchor cell determination. (a) During L2, two neighboring cells begin the secretion of chemical signals for the induction of uterine differentiation. (b) By chance, cell Z1.ppp secretes more of these signals, causing cell Z4.aaa to increase production of the receptor for signals. The action of increased signals causes Z4.aaa to become the ventral uterine precursor cell and allows Z1.ppp to become the anchor cell.

scription of the *lin*-12 gene, thus increasing production of the receptor. The cell with more *lin*-12 receptors becomes the ventral uterine precursor, and the other cell becomes the anchor cell. The critical factor in this first round of cell–cell interaction and determination is the *lin*-12 gene product.

A second round of cell–cell interactions in L3 involves the anchor cell, located in the gonad, and six precursors located in the skin (hypodermis) adjacent to the gonad. The precursor cells are named P3.p, P4.p, P5.p, P6.p, P7.p, and P8.p and collectively are called Pn.p cells. The fate of each of the Pn.p cells is specified by its position relative to the anchor cell. The determination pathway is illustrated in Figure 22.34 and described in the following paragraphs.

Sometime in L3, a gene called *lin-3* is activated in the anchor cell and secretes a protein structurally related to the vertebrate epidermal growth factor (EGF). All six of the Pn.p cells express a cell surface receptor encoded by the *let-23* gene, which is homologous to the vertebrate EGF receptor. The binding of the *lin-3* protein to the let-23 receptor triggers an intracellular cascade of events that causes the Pn.p cells to become determined to form either vulval precursor cells or secondary vulvar cells. The let-23

gene is required for establishing the primary and secondary fates of Pn.p cells. Recessive loss-of-function mutations in *let-23* cause all the Pn.p cells to develop as hypodermis (a tertiary fate). In *let-23* mutants, the Pn.p cells act as though they have not received a signal from the anchor cell, and no vulva is formed.

The signal transduction cascade from the anchor cells to the Pn.p cells also involves the *let-60* gene. Recessive mutations in *let-60* cause the Pn.p cells to develop as though they have not received a signal from the anchor cell. Dominant *let-60* alleles have the opposite phenotype: They cause all the Pn.p cells to respond to form vulval cells, forming multiple vulvas. The *let-60* gene has been cloned, and it is the *C. elegans* homolog of the *ras* gene, which in mammals acts downstream of receptor proteins in signal transduction. The *let-60* dominant gain-of-function mutant that causes multiple vulva formation has a Gly-Glu mutation at amino acid 13, the same mutation that converts *c-ras* to an oncogene (see Chapter 23).

The data on the signal transduction cascade initiated by the binding of lin-3 to let-23 suggest that the cell (P6.p)

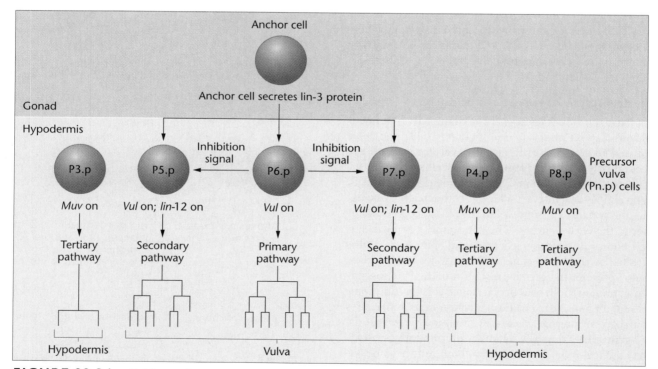

FIGURE 22.34 Cell lineage determination in *C. elegans* vulva formation. A signal from the anchor cell in the form of LIN3 protein is received by three precursor vulval cells (Pn.p cells). The cells closest to the anchor cell become primary vulval precursor cells, and adjacent cells become secondary precursor cells. Primary cells secrete a signal that activates the *lin*-12 gene in secondary cells, preventing them from becoming primary cells. Flanking Pn.p cells, which receive no signal from the anchor cell activity of the *Muv* gene, cause these cells to become hypodermis cells instead of vulval cells.

closest to the anchor cell receives the strongest signal. In P6.p, expression of the *Vulvaless (Vul)* gene (the gene is named for its mutant phenotype) is activated; the primary fate of this cell is to divide three times to produce vulva cells. The two neighboring cells receive a lower amount of signal, which specifies a secondary fate—asymmetric division to form more vulva cells. To reinforce this determination, the primary vulval precursor activates the *lin*-12 gene in the two neighboring cells. This lateral inhibition signal prevents the neighboring secondary cells from adopting the division pattern of the primary cell. In other words, cells in which both *Vul* and *lin*-12 are active cannot become primary vulvar cells. The three remaining Pn.p cells receive no signal from the anchor cell. In these cells (P3.p, P4.p, and P8.p) the *Multivulva (Muv)* gene is expressed, *Muv* represses *Vul*, and the three cells develop as hypodermal (skin) cells.

Thus, three levels of cell–cell interaction are required to specify the developmental pathway leading to vulva formation in *C. elegans*. First, two neighboring cells interact during L2 to establish the identity of the anchor cell. Second, in L3 the anchor cell interacts with three Pn.p cells to establish the primary vulvar precursor cell and two secondary cells. Third, the primary vulvar cell interacts with the secondary cells to suppress their selection of the primary vulvar pathway. Each of these interactions is accompanied by the secretion of molecular signals and by the reception and processing of these signals by neighboring cells.

This example of cell–cell interactions acting in a spatial and temporal cascade to specify the developmental fates of individual cells is a developmental theme repeated over and over in developing organisms from prokaryotes to higher vertebrates, including mice and humans.

Programmed cell death is genetically regulated and is required for normal development

While there is now ample evidence that most cells retain a complete genome, and can therefore theoretically be reprogrammed, there is at least one cellular fate that is irreversible, and that is cell death. **Programmed cell death,** or **apoptosis,** is a normal process whereby cells undergo a characteristic series of events, such as chromosome condensation and blebbing of the plasma membrane, that lead to their death. During development these programmed cell suicides contribute to the shaping and molding of tissues and organs; for example, the formation of the digits of the vertebrate limbs requires the death of the cells in between the digits.

This cellular process is beautifully illustrated in *C. elegans*, for in this organism development not only involves but relies on programmed cell death. The number of cells that die during the worm's development is always the same, 131 of 1090 in hermaphrodites, and 147 of 1178 in males. In addition, the time during development at which a given cell dies is always the same, and the identity of the cells that die is always the same.

Genetic analysis of mutants indicates that although programmed cell death occurs in cells of different developmental origins, all the cells that die use the same genetic pathway. The expression of two genes, *ced-3* and *ced-4*, is necessary for cell death; mutations that inactivate either of these genes result in the survival of cells that normally die. Expression of *ced-3* and *ced-4* is controlled by the gene *ced-9*. Gain-of-function mutations that lead to constitutive expression or overexpression of *ced-9* prevent cell deaths during development. Conversely, loss-of-function mutants that inactivate *ced-9* cause embryonic lethality. Based on these observations, it has been concluded that the *ced-9* gene works by preventing the expression of *ced-3* and *ced-4* in cells that survive (Figure 22.35). This means that *ced-9* acts as a *binary switch gene* to regulate programmed cell death. Cells that express *ced-9* survive and those that do not, die.

The *ced-9* gene of *C. elegans* has a human homolog, the *bcl-2* proto-oncogene, which plays a role in programmed cell death in mammals. The overexpression of *bcl-2* prevents cell death in cells that would normally die. In humans, this overexpression is associated with follicular lymphoma, a form of cancer. The transfer of a cloned *bcl-2* gene into *ced-9* mutant *C. elegans* embryos prevents programmed cell death, indicating that nematodes and mammals share a common pathway for programmed cell death. Thus the homology in molecules and mechanisms among species across the phylogenetic tree, illustrated throughout this chapter, extends to the molecules and mechanisms required for cells to die.

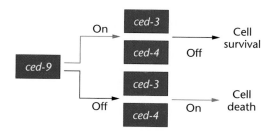

FIGURE 22.35 In the genetic pathway controlling cell death, the gene *ced-9* acts as a binary switch. If *ced-9* is active, it represses the expression of *ced-3* and *ced-4*, and the cell lives. If *ced-9* is inactive, *ced-3* and *ced-4* are expressed, and the cell dies.

CHAPTER SUMMARY

1. The role of genetic information during development and differentiation is one of the major questions in biology and has been studied extensively. Geneticists are exploring this question by isolating developmental mutations and identifying the genes involved in controlling developmental processes.

2. The variable gene activity theory, which applies to higher organisms, assumes that all somatic cells in an organism contain equivalent genetic information, but that the information is differentially expressed. Studies conducted on genomic equivalence have demonstrated that somatic cells undergoing differentiation retain the entire gene set, which can be reprogrammed to direct the development of the entire organism.

3. Determination is the regulatory event whereby cell fate becomes fixed during early development. Determination precedes the actual differentiation or specialization of distinctive cell types.

4. During embryogenesis, specific gene activity appears to be affected by the internal environment of the cell, or localized cytoplasmic components. The regulation of early events is mediated by the maternal cytoplasm, which then influences zygotic gene expression. As development proceeds, both the cell's internal environment and its external environment become further altered by the presence of early gene products and communication with other cells.

5. In *Drosophila*, both genetic and molecular studies have confirmed that the egg contains information that specifies the body plan of the larva and adult, and that interactions of embryonic nuclei with the maternal cytoplasm initiate transcriptional programs characteristic of specific developmental pathways.

6. Extensive genetic analysis of embryonic development in *Drosophila* has led to the identification of maternal-effect genes that lay down the anterior–posterior and dorsal–ventral axes of the embryo. In addition, these maternal-effect genes activate sets of zygotic segmentation genes, initiating a cascade of gene regulation that ends with the determination of segment identity by the selector genes.

7. In *C. elegans*, the stereotyped lineage of all cells allows developmental biologists to study the cell–cell signaling required for organogenesis and to determine which genes are required for the normal process of programmed cell death.

INSIGHTS AND SOLUTIONS

1. In the slime mold *Dictyostelium*, experimental evidence suggests that cyclic AMP (cAMP) plays a central role in the developmental program leading to spore formation. The genes encoding the cAMP cell-surface receptor have been cloned, and the amino acid sequence of the protein components is known. To form reproductive structures, free-living individual cells aggregate together and then differentiate into one of two cell types, prespore cells or prestalk cells. Aggregating cells secrete waves or oscillations of cAMP to foster the aggregation of cells, and then continuously secrete cAMP to activate genes in the aggregated cells at later stages of development. It has been proposed that the role of cAMP lies in cell–cell interaction and gene expression. It is important to test this hypothesis by using several experimental techniques. What different approaches can you devise to test this hypothesis, and what specific experimental systems would you employ to test them?

Solution: Two of the most powerful forms of analysis in biology involve the use of biochemical analogues or inhibitors to block gene transcription or the action of gene products in a predictable way, and the use of mutations to alter the gene and its products. These two approaches can be used to study the role of cAMP in the developmental program of *Dictyostelium*. First, compounds chemically related to cAMP, such as GTP and GDP, can be used to test whether they have any effect on the processes controlled by cAMP. In fact, both GTP and GDP lower the affinity of cell-surface receptors for cAMP, effectively blocking the action of cAMP. To inhibit the synthesis of the cAMP receptor, it is possible to construct a vector that contains a DNA sequence that transcribes an antisense RNA (a molecule that has a base sequence complementary to the mRNA). Antisense RNA forms a double-stranded structure with the mRNA, preventing it from being transcribed. If normal cells are transformed with a vector that expresses antisense RNA, no cAMP receptors will be produced. It is possible to predict that such cells will fail to respond to a gradient of cAMP and, consequently, will not migrate to an aggregation center. In fact, that is what happens. Such cells remain dispersed and nonmigratory in the presence of cAMP. Similarly, it is possible to determine whether this response to cAMP is necessary to trigger changes in the transcriptional program by assaying for the expression of developmentally regulated genes in cells expressing this antisense RNA.

Mutational analysis can be used to dissect components of the cAMP receptor system. One approach is to use transformation with wild-type genes to restore mutant function. Similarly, because the genes for the receptor proteins have been cloned, it is possible to construct mutants with known alterations in the component proteins and transform them into cells to assess their effects.

PROBLEMS AND DISCUSSION QUESTIONS

1. Carefully distinguish between the terms *differentiation* and *determination*. Which phenomenon occurs initially during development?

2. The *Drosophila* mutant *spineless aristapedia (ss^a)* results in the formation of a miniature tarsal structure (normally part of the leg) on the end of the antenna. This is a *homeotic* mutation. From your knowledge of imaginal disks, what insight is provided by *ss^a* concerning the role of genes during determination?

3. In the sea urchin, early development up to gastrulation may occur even in the presence of actinomycin D, which inhibits RNA synthesis. However, if actinomycin D is present early on but removed at the end of bastula formation, gastrulation does not proceed. In fact, if actinomycin D is present only between the 6th and 11th hours of development, gastrulation (normally occurring at the 15th hour) is arrested. What conclusions can be drawn concerning the role of gene transcription between hours 6 and 15?

4. How can you determine whether a particular gene is being transcribed in different cell types?

5. Observing that a particular gene is being transcribed during development, how can you tell whether the expression of this gene is under transcriptional or translational control?

6. Both the *ftz* gene and the *engrailed* gene encode homeobox transcription factors and are capable of eliciting the expression of other genes. Both genes work at about the same time during development and in the same region to specify cell fate in body segments. The question is: Does *ftz* regulate the expression of *engrailed*, or does *engrailed* regulate *ftz?* Or are they both regulated by another gene? To answer these questions, mutant analysis is performed. In *ftz²⁻* embryos *(ftz/ftz)*, engrailed protein is absent; in *engrailed²⁻* embryos *(eng/eng)*, *ftz* expression is normal. What does this tell you about the regulation of these two genes? Does the *engrailed* gene regulate *ftz?* Does the *ftz* gene regulate *engrailed?*

7. Define what is meant by a homeotic mutant. If it were possible to introduce one of the homeotic genes in *Drosophila* into an *Arabidopsis* embryo that was homozygous for a homeotic flowering gene, would you expect any of the *Drosophila* genes to be able to rescue the *Arabidopsis* mutation? Why or why not?

8. Nuclei from almost any source may be injected into *Xenopus* oocytes. Studies have shown that these nuclei remain active in transcription and translation. How can such an experimental system be useful in developmental genetic studies?

9. The concept of epigenesis indicates that an organism develops by forming cells that acquire new structures and functions, which become greater in number and complexity as development proceeds. This theory is in contrast to the preformationist doctrine that miniature adult entities are contained in the egg that must merely unfold and grow to give rise to a mature organism. What sorts of isolated evidence presented in this chapter might have led to the preformation doctrine? Why is the epigenetic theory held as correct today?

10. You are given two snails, a female snail that has a *d/d* genotype and a *d* phenotype, and a male snail that has a *D/D* genotype and a *D* phenotype. What genetic crosses are required to generate a snail that has a *d/d* genotype but a *D* phenotype?

 EXTRA-SPICY PROBLEMS

11. In studying gene action during development, it is desirable to be able to position genes in a hierarchy or pathway of action, to establish which genes are primary and in what order genes act. There are several ways of doing this. One way is to make double mutants and study the outcome. The gene *fushi-tarazu (ftz)* is expressed in early embryos at the seven-stripe stage. All genes involved in forming the anterior–posterior pattern affect the expression of this gene, as do the gap genes. However, expression of segment-polarity genes is affected by *ftz*. What is the location of *ftz* in this hierarchy?

12. (a) The identification and characterization of genes that control sex determination has been another focus of investigators working with *C. elegans*. As with *Drosophila*, sex in this organism is determined by the ratio of X chromosomes to sets of autosomes. A diploid wild-type male has one X chromosome and a diploid wild-type hermaphrodite has two X chromosomes. Many different mutations have been identified that affect sex determination. Loss-of-function mutations in a gene called *her-1* cause an XO animal to develop into a hermaphrodite and have no effect on XX development (that is, XX animals are normal hermaphrodites). In contrast, loss-of-function mutations in a gene called *tra-1* cause an XX animal to develop into a male. From this in-

GENETICS MediaLab

The following resources will help you achieve a better understanding of the concept presented in this chapter. These resources can be found on the CD packaged with this textbook and on the Companion Web site at **http://www.prenhall.com/klug/**.

CD Resources:

Self-grading Chapter Problems

Web Resources:

Web Destinations in Genetics

Self-grading Chapter Problems

Chapter Search Terms

Genetics Newsgroups

Student Bulletin Board

Web Problem 1:

Time for completion = 10 minutes
Does cellular differentiation alter the genetic material within the nucleus such that totipotency is lost? Dolly, the sheep cloned in a Scottish genetics lab, was the first of several successful mammalian cloning experiments that suggest the answer is no. After reading the article "Virtual Embryo" about mammalian cloning, you should be able to answer the following questions. Is the differentiation of a particular cell most likely a result of changes in the nucleus or changes in the cytoplasm? Are cells cloned by injecting a nucleus into an enucleated egg truly clones of the donor cells? What, if any, genetic components of the cell might be different between the clone and the donor? What are some important implications of cloning for human medical treatment? To complete this exercise, visit Web Problem 1 in Chapter 22 of your Companion Website and select the keyword **CLONING**.

Web Problem 2:

Time for completion = 10 minutes
What combination of technology and ingenuity allows certain scientists to make critical breakthroughs in understanding how organisms function? The German scientists Christiane Nüsslein-Volhard and Eric Wieschaus won the 1995 Nobel Prize in Physiology or Medicine for discovering the role of genetics in the development of the *Drosophila* embryo. While reading and thinking about Dr. Nüsslein-Volhard's training and achievements, pay particular attention to the critical factors that led her to the discovery. Did the discovery of these genes depend on the development of sophisticated molecular techniques or on the

ingenuity of the investigators? How did Nüsslein-Volhard decide to study *Drosophila* embryogenesis; were the events leading her to the field carefully planned or serendipitous? Were Nüsslein-Volhard's major collaborators in the first part of this research her professors or fellow students? This article presents two views of the particular challenges facing women in science; what do you think of Dr. Nüsslein-Volhard's approach? To complete this exercise, visit Web Problem 2 in Chapter 22 of your Companion Website and select the keyword **NÜSSLEIN-VOLHARD**.

Web Problem 3:

Time for completion = 15 minutes
How can expression of a single homeotic gene cause the differentiation of different appendages in different body segments? The homeotic genes discovered in *Drosophila* have homologs in many other taxa, including plants and mammals, and appear to regulate the differentiation of various body segments. For example, the *Ultrabithorax* (*Ubx*) locus regulates the expression of thoracic appendages in *Drosophila* and other insects. In this exercise, you will read an article on *Ubx* and use the information in Chapter 22 to answer the following questions. What are the functions of the proteins encoded by *Ubx*? Flies (order Diptera), like *Drosophila,* have one pair of wings, whereas other insect orders such as butterflies and dragonflies have two pair of wings. Is the suppression of wing development on the third thoracic segment in Diptera associated with suppression of expression of *Ubx*? How does the expression of this gene differ between flies and butterflies? A single RNA transcript of *Ubx* appears to serve different roles in different tissues; how is the transcript modified in different tissues? To complete this exercise, visit Web Problem 3 in Chapter 22 of your Companion Website and select the keyword **HOMEOTIC**.

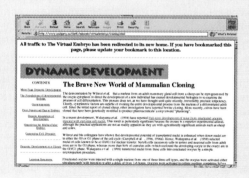

formation, deduce the roles of these genes in wild-type sex determination.

(b) Based on the phenotypes of single and double mutant strains, a model for sex determination in *C. elegans* has been generated. This model proposes that the *her-1* gene controls sex determination by determining the level of activity

of the *tra-1* gene, which in turn controls the expression of genes involved in generating the various sexually dimorphic tissues. Given this information, does the *her-1* gene product have a negative or a positive effect on the activity of the *tra-1* gene? What would be the phenotype of a *tra-1, her-1* double mutant?

SELECTED READINGS

BONCINELLI, E., and MALLAMACI, A. 1995. Homeobox genes in vertebrate gastrulation. *Curr. Opin. Genet. Dev.* 5:619–27.

BRIGGS, R., and KING, T. 1952. Transplantation of living nuclei from blastula cells into enucleated frog eggs. *Proc. Natl. Acad. Sci. USA* 38:455–63.

DAVIDSON, E. 1976. *Gene activity in early development.* New York: Academic Press.

DESCHAMPS, J., and MEIJLINK, F. 1992. Mammalian homeobox genes in normal development and neoplasia. *Crit. Rev. Oncog.* 3:117–73.

DRIEVER, W., THOMA, G., and NÜSSLEIN-VOLHARD, C. 1989. Determination of spatial domains of zygotic gene expression in the *Drosophila* embryo by the affinity of binding sites for the bicoid morphogen. *Nature* 340:363–67.

DUKE, R.C., OJCIUS, D.M., and YOUNG, J.D. 1996. Cell suicide in health and disease. *Sci. Am.* (Dec.) 275:80–87.

EDMONDSON, D., and OLSON, E. 1993. Helix-loop-helix proteins as regulators of muscle-specific transcription. *J. Biol. Chem.* 268:755–58.

GAUNT, S. 1991. Expression patterns of mouse *Hox* genes: Clues to an understanding of development and evolutionary strategies. *BioEssays* 13:505–13.

GEHRING, W. 1968. The stability of the differentiated state in cultures of imaginal disks in *Drosophila*. In *The stability of the differentiated state*, ed. H. Unsprung. New York: Springer-Verlag.

GILBERT, S. 1997. *Developmental biology*, 5th ed. Sunderland, MA: Sinauer Associates.

GOODRICH, J., et al. 1997. A Polycomb-group gene regulates homeotic gene expression in *Arabidopsis*. *Nature* 386:44–51.

GURDON, J.B. 1968. Transplanted nuclei and cell differentiation. *Sci. Am.* (Dec.) 219:24–35.

———. 1974. *The control of gene expression in animal development.* Cambridge, MA: Harvard U. Press.

GURDON, J.B., LASKEY, R.A., and REEVES, O.R. 1975. The developmental capacity of nuclei transplanted from keratinized skin cells of adult frogs. *J. Embryol. Exp. Morphol.* 34:93–112.

GURDON, J.B., MITCHELL, A., and MAHONY, D. 1995. Direct and continuous assessment by cells of their position in a morphogen gradient. *Nature* 376:520–21.

HALDER, G., CALLAERTS, P., and GEHRING, W.J. 1995. Induction of ectopic eyes by targeted expression of the *eyeless* gene in *Drosophila*. *Science* 267:1788–92.

HENGARTNER, M.O. 1995. Life and death decisions: *ced-9* and programmed cell death in *Caenorhabditis elegans*. *Science* 270:931.

HENGARTNER, M.O., and HORVITZ, H.R. 1994. The ins and outs of programmed cell death during *C. elegans* development. *Phil. Trans. R. Soc. London* B 345:243–46.

HOLLAND, P.W., and GARCIA-FERNANDEZ, J. 1996. *Hox* genes and chordate evolution. *Dev. Biol.* 173:382–95.

JÄCKLE, H., et al. 1992. Transcriptional control by *Drosophila* gap genes. *J. Cell Sci. Suppl.* 16:39–51.

KAPPEN, C., SCHUGHART, K., and RUDDLE, F.H. 1989. Organization and expression of homeobox genes in mouse and man. *Ann. NY Acad. Sci.* 567:243–52.

KAUFMANN, T., SEEGER, M., and OLSON, G. 1990. Molecular organization of the *Antennapedia* gene cluster of *Drosophila melanogaster. Adv. Genet.* 27:309–62.

KENNISON, J., and TAMKUN, J. 1992. *Trans*-regulation of homeotic genes in *Drosophila*. *New Biol.* 4:91–96.

KING, T.J. 1966. Nuclear transplantation in amphibia. In *Methods in cell physiology*, ed. D. Prescott, Vol. 2. Orlando, FL: Academic Press.

KRAUSE, M. 1995. *MyoD* and myogenesis in *C. elegans*. *BioEssays* 17:219–28.

KRUMLAUFF, R., and GOULD, A. 1992. Homeobox cooperativity. *Trends Genet.* 8:297–300.

KYRAICOU, C. 1992. Sex variations. *Trends Genet.* 8:261–67.

LAWRENCE, P. 1992. *The making of a fly: The genetics of animal design*. Oxford: Blackwell Scientific.

LEVINE, M., RUBIN, G., and TJIAN, R. 1984. Human DNA sequences homologous to a protein coding region conserved between homeotic genes of *Drosophila*. *Cell* 38:667–73.

LEWIS, E.B. 1976. A gene complex controlling segmentation in *Drosophila*. *Nature* 276:565–70.

———. 1994. Homeosis: The first 100 years. *Trends Genet.* 10:341–43.

LOCKSHIN, R.A., ZAKERI, A., and TILLY, J. 1998. *When cells die.* New York: John Wiley.

MANSEAU, L.J., and SCHÜPBACH, T. 1989. The egg came first, of course! Anterior-posterior pattern formation in *Drosophila* embryogenesis and oogenesis. *Trends Genet.* 5:400–5.

MEYEROWITZ, E. 1997. Plants and the logic of development. *Genetics* 145:5–9.

NÜSSLEIN-VOLHARD, C. 1994. Of flies and fishes. *Science* 266:572–74.

NÜSSLEIN-VOLHARD, and WEISCHAUS, E. 1980. Mutations affecting segment number and polarity in *Drosophila*. *Nature* 287:795–801.

OLSON, E. 1990. *MyoD* family: A paradigm for development? *Genes Dev.* 4:1454–61.

RUDDLE, F.H., HART, C.P., and McGINNIS, W. 1985. Structural and functional aspects of the mammalian homeobox sequences. *Trends Genet.* 1:46–50.

RUDNICKI, M.A., and JAENISCH, R. 1995. The MyoD family of transcription factors and skeletal myogenesis. *BioEssays* 17:203–9.

SANCHEZ-HERRERO, E., and AKAM, M. 1989. Spatially ordered transcription of regulatory DNA in the *bithorax* complex of *Drosophila*. *Development* 107:321–29.

SATHE, S.S., and HARTE, P.J. 1995. *Drosophila* extra sex combs protein contains WD motifs essential for its function as a repressor of homeotic genes. *Mech. Dev.* 52:77–87.

SMALL, S., and LEVINE, M. 1991. The initiation of pair-rule stripes in the *Drosophila* blastoderm. *Curr. Opin. Genet. Dev.* 1:255–60.

ST. JOHNSTON, D., and NÜSSLEIN-VOLHARD, C. 1992. The origin of pattern and polarity in the *Drosophila* embryo. *Cell* 68:201–19.

STRUHL, G. 1981. A gene product required for correct initiation of segmental determination in *Drosophila*. *Nature* 293:36–41.

TAUTZ, D. 1996. Selector genes, polymorphisms, and evolution. *Science* 271:160–61.

WEIGEL, D., and MEYEROWITZ, E. 1994. The ABCs of floral homeotic genes. *Cell* 78:203–9.

WEINTRAUB, J., et al. 1991. The *myoD* gene family: Nodal point during specification of the muscle cell lineage. *Science* 251:761–66.

WIESCHAUS, E. 1996. Embryonic transcription and the control of developmental pathways. *Genetics* 142:5–10.

WILMUT, I., et al. 1997. Viable offspring from fetal and adult mammalian cells. *Nature* 385:810–13.

WRIGHT, W. 1992. Muscle basic helix-loop-helix proteins and the regulation of myogenesis. *Curr. Opin. Genet. Dev.* 2:243–48.

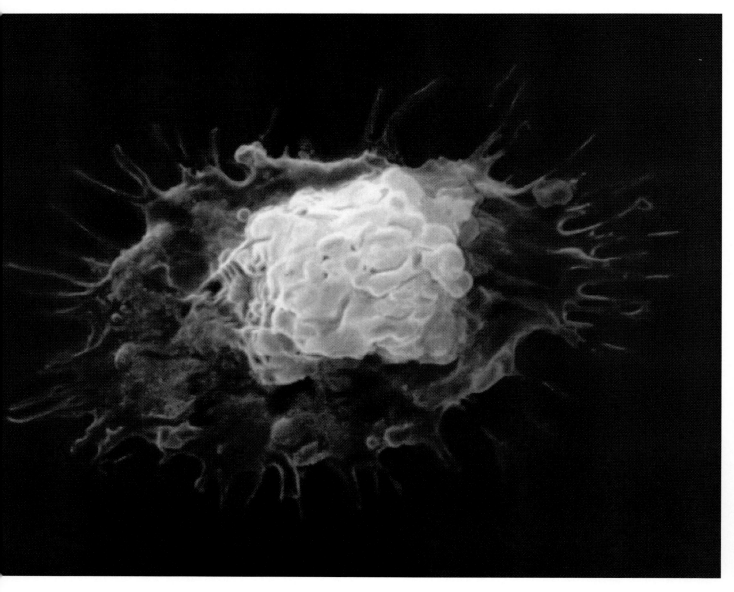

A human breast cancer cell.

Genetics and Cancer

KEY CONCEPTS

Although often viewed as a single disease, cancer is actually a complex group of diseases affecting a wide range of cells and tissues. A genetic link to cancer was first proposed early in the twentieth century, and this idea has served as one of the foundations of cancer research. Mutations that alter gene expression are now regarded as a common feature of all cancers. In most cancer cases, mutations arise in somatic cells and are not passed on to future generations through the germ cells. However, in about 1 percent of all cancer cases, germ-line mutations of various genes are transmitted to offspring and are responsible for susceptibility to cancer. Although the frequency of these cases is low, studies of these mutations have provided many insights into the origins of cancer. Sometimes for cancer to occur, the inherited mutation is itself insufficient and must be accompanied by an additional somatic mutation at the homologous locus, creating homozygosity. Whichever the case, *cancer is now considered a genetic disorder at the cellular level.*

Genomic alterations associated with cancer can involve small-scale changes, such as a single nucleotide substitution, or large-scale events, including chromosome rearrangement, chromosome gain or loss, or even the integration of viral genomes into chromosomal sites. Large-scale genomic alterations are a common feature of cancer; the majority of human tumors are characterized by visible chromosomal changes. Some of these chromosomal changes, particularly in leukemia, are so characteristic they can be used to diagnose the disorder and to make accurate predictions about the severity and course of the disease.

That cancer runs in families has been known for over 200 years. Most often, no clear-cut pattern of inheritance can be discerned for these familial forms. This is primarily because these patients inherit only one mutant allele of a cancer-causing gene, which predisposes them to cancer. The likelihood that an individual will ultimately develop cancer varies depending on the particular mutant allele, mutations in other genes, and environmental factors. These variables may influence the age of onset and the severity of the disease. Thus, we can identify a class of genes called **cancer susceptibility genes** that increase the risk of cancer. Variant alleles of these cancer susceptibility genes may have an important role in sporadic cancers as well as familial forms of cancer.

TABLE 23.2 Number of Mutations Associated with Some Cancers

Cancer	Chromosome Sites	Minimum Number of Mutations Required
Retinoblastoma	13q	2
Wilms tumor	11p	2
Colon cancer	5p, 12p, 17p, 18q	4–5
Small-cell lung cancer	3p, 11p, 13q, 17p	10–15

disposed to develop retinoblastoma, as only one additional mutational event is required to cause tumor formation. This does not happen in all cases. About 10 percent of those inheriting a mutant *RB* allele do not develop cancer. Presumably, the wild-type allele of the *RB* gene does not mutate in any retinal cells.

In the *non-familial sporadic cases* [Figure 21.(4b)], independent mutations in both normal *RB* genes must occur in the same retinal cell for a tumor to develop. As might be expected, these events are far less frequent and occur at a much later age. As predicted by Knudson's model, such sporadic forms of retinoblastoma are more likely to occur in a single eye.

Similar studies on the predisposition to other cancers have led to the conclusion that the number of mutations necessary for the development of cancer ranges from 2 to perhaps as many as 20 (Table 23.2). Thus, retinoblastoma, while serving as an excellent model to investigate cancer, may be the exception rather than the rule.

Tumor-suppressor genes normally suppress cell division

In general, mitosis can be regulated in two ways: (1) by genes that normally function to suppress cell division and (2) by genes that normally function to promote cell division. The first class, called **tumor-suppressor genes,** inactivates or represses passage through the cell cycle and the resulting cell division. These genes and/or their gene products must be absent or inactive for cell division to take place. If these genes become permanently inactivated or lost through mutation, control over cell division is lost, and the cell begins to proliferate in an uncontrollable fashion. As we will see later, in the case of tumor-suppressor genes both alleles of the gene must be mutated in order for cancer to develop. In this section, we shall consider some examples of how mutations in tumor-suppressor genes can lead to a loss of control over cell division and the development of cancer. In the following section, we will consider the genes that normally promote cell division and the results of their mutation.

The Retinoblastoma (RB) Gene

The first tumor-suppressor gene we will examine is the gene for retinoblastoma. The **RB gene,** located on chromosome 13, encodes a protein that is designated **pRb.** It is 928 amino acids long. The pRb protein is present in the nuclei of all cell and tissue types examined so far, and is found in both resting (G0) cells and actively dividing cells. Moreover, pRb is present at all stages of the cell cycle. Because the normal gene is expressed ubiquitously in the body, regulation occurs by modifying pRb—specifically, by adding or removing phosphate groups. The pRb is phosphorylated in the S and the G2/M stages of the cell cycle, but is not phosphorylated in the G0 and G1 stages of the cycle.

The observation that pRb alternately has phosphate groups added and removed, and that this pattern of modification occurs in synchrony with phase of the cell cycle, suggests that pRb may be part of a control point in G1. In resting cells (G0 or G1), when pRb is not phosphorylated, the protein is active and cells do not progress through the cell cycle. Just after the cell enters S phase, pRb is modified by the addition of phosphate groups and becomes inactive. When pRb activity is suppressed, cells pass through the S phase and G2 and undergo mitosis.

Direct evidence for the role of pRb in regulating the cell cycle comes from several types of experiments. Cultured osteosarcoma (bone cancer) cells carry *RB* mutations, and these cells do not produce pRb. When osteosarcoma cells are injected into a cancer-prone strain of mice, tumors are formed. If a normal *RB* gene is transferred to the cancer cells, pRb is produced and no tumors are formed when these genetically modified cells are injected into mice.

In a separate two-step experiment, a normal *RB* gene was transferred into osteosarcoma cells grown in culture. These cells produced pRb and stopped cell division. When D or E cyclins were added to these pRb-blocked cells, cell division resumed. The analysis of pRb after the addition of cyclins indicates that the cyclin causes phosphorylation of pRb, presumably by activating CDK1 or another kinase. These results demonstrate that the presence of active pRb stops cell division, that pRb is a target of a G1 cyclin/CDK

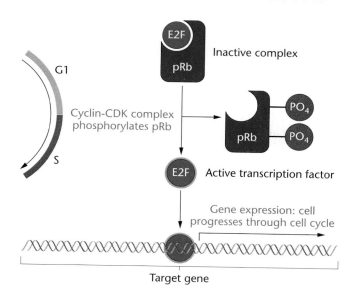

FIGURE 23.5 In the nucleus during G1, pRb, the product of the retinoblastoma gene, interacts with and inactivates transcription factor E2F. As the cell moves from G1 to S, a CDK/cyclin complex forms and adds phosphate groups to pRb. As pRb becomes hyperphosphorylated, E2F is released and becomes transcriptionally active, allowing the cell to pass through S phase. Phosphorylation of pRb is transitory; as cyclin is degraded, phosphorylation declines. The identity of the G1 cyclin is uncertain.

complex, and that inactivation of pRb allows passage through the cell cycle and subsequent division, confirming the role of pRb in G1 control.

Recent studies at the molecular level suggest that wild-type pRb normally prevents the proliferation of retinal cells by interacting with the transcription factor E2F (Figure 23.5). Transcription factors are involved in binding to specific regions of DNA and regulate transcription. The transcription factor in this case, E2F, is responsible for the regulation of the expression of many genes required for cell cycle progression. When bound to pRb, E2F is inactive, and the cell stops at the G1/S checkpoint. In cancerous cells, both copies of the *RB* allele are defective and no pRb binding occurs. As a result, E2F activates genes required for progression through the the G1/S checkpoint in the cell cycle. This important checkpoint is continually overridden, resulting in uncontrolled cell growth.

The Wilms Tumor Gene

Wilms tumor (WT) is a cancer of the kidney found primarily in children. It occurs with a frequency of about 1 in 10,000 births, and, like retinoblastoma, it is found in two forms, a noninherited sporadic form and a familial form conferring a predisposition to WT. The familial predisposition is inherited as an autosomal dominant trait. According to the model developed by Alfred Knudson and colleagues described earlier, familial cases inherit one mutant allele through the germ line and develop a second mutation in the remaining normal allele in a somatic cell. Sporadic cases, on the other hand, require two independent mutations of the *WT* gene within the same cell, and they more often develop tumors involving only one kidney. Because the mutation of the gene and/or the loss of its functional gene product are both associated with the development of tumors, the normal gene is regarded as a tumor suppressor.

The *WT* gene has been mapped to the short arm of chromosome 11. The gene product encoded by the *WT* gene contains four contiguous zinc finger domains, motifs characteristic of DNA-binding proteins that regulate transcription. Further, the amino acid sequence upstream from the zinc fingers is similar to other proteins that are known transcription factors.

The *WT* gene has a very restricted pattern of expression; it is normally active only in the mesenchymal cells of the fetal kidney, and only during the brief time when the nephron (the basic filtration unit of the kidney) is being formed. The only other cells to express the *WT* gene are the tumorous nephroblastomas.

The structure of the *WT*-gene product, its pattern and time of expression, and the resultant mutant phenotype have led to the hypothesis that the *WT* gene may encode a nuclear protein that functions to turn off genes that sustain cell proliferation. Alternatively, the gene product may switch on genes that begin the process of differentiation of the mesenchymal cells into kidney structures. In either case, in tumor cells the mutant gene is switched on at the appropriate time in the appropriate cells, but the altered gene product is unable to regulate its target genes, resulting in continued proliferation of the mesenchymal cells, aberrant differentiation, and tumor formation (Figure 23.6).

Although both the *RB* and *WT* genes encode proteins restricted to the nucleus and normally involved in the suppression of tumor formation, conspicuous differences exist in their properties and modes of action. The pRb protein does not bind to DNA; instead, it interacts with E2F, as described in the previous section. The WT protein has all the characteristics of acting directly as a transcription factor. pRb is expressed in all dividing cells, whereas the expression of WT occurs during a restricted period of specific prenatal development (of the kidney). pRb is a general regulator of cell division; WT is a cell- or tissue-specific regulator of gene activity during fetal or neonatal development.

FIGURE 23.6 The proposed role of the WT protein in regulating cell division. In normal cells (at left), the WT1 protein is produced and acts on a set of target genes to switch off cell division or switch on cell differentiation. Any mutation that causes the loss or inactivation of the WT1 protein (at right) will result in a failure to regulate the target genes, allowing cell division and tumor formation to occur.

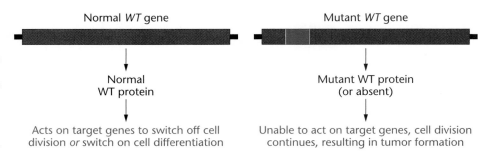

Normal *WT* gene

Normal WT protein

Acts on target genes to switch off cell division *or* switch on cell differentiation

Mutant *WT* gene

Mutant WT protein (or absent)

Unable to act on target genes, cell division continues, resulting in tumor formation

Breast Cancer Genes

A small proportion of all cases of breast cancer, especially those in younger women, is related to genes that confer a dominantly inherited predisposition to the disease. Mutations in *BRCA1*, a gene that maps to the long arm of chromosome 17, are associated with a predisposition to breast cancer, and this predisposition is inherited as an autosomal dominant trait. About 90 percent of women with a mutant *BRCA1* gene will develop breast cancer; these women also have an increased risk of ovarian cancer.

A second breast cancer gene, *BRCA2*, which maps to the long arm of chromosome 13, is also inherited as an autosomal dominant predisposition to breast cancer, but is not associated with an increased risk of ovarian cancer. Together, these genes account for a large majority of all inherited cases of breast cancer, but have little or no role in sporadic cases of breast and ovarian cancer, which make up 80 percent of all cases.

Both the *BRCA1* and the *BRCA2* genes are expressed in rapidly dividing cells, with expression at its highest at the G1/S boundary in the cell cycle. This common pattern of expression, the similar phenotype of the mutant alleles (breast cancer), and experimental evidence from knockout mice indicate that both genes may work in the same pathway. Both the BRCA1 and BRCA2 proteins bind to another protein, RAD51, which is involved in the repair of breaks in double-stranded DNA, which may explain their normal function as tumor suppressor genes.

The role of BRCA1 and BRCA2 proteins in DNA repair may also explain why they are not associated with sporadic cases of breast cancer. In sporadic cases, at least three mutations would have to accumulate in a breast cell for it to become cancerous: Both copies of the *BRCA1* or *BRCA2* genes would have to be mutant, and at least one other mutation in a cell-cycle regulation gene would have to occur. On the other hand, those who inherit a mutant copy of either the *BRCA1* or *BRCA2* gene already have one mutation and need only two more to trigger breast cancer.

The p53 Tumor-Suppressor Gene

Another highly significant tumor-suppressor gene, *p53*, has been implicated in a wide range of human cancers. The gene encodes a nuclear protein that acts as a transcription factor. Normally, *p53*, as a tumor-suppressor gene, controls passage of the cell from G1 into S phase of the cell cycle. Mutations of *p53* are found in a wide range of cancers, including breast, lung, bladder, and colon cancers. It is estimated that over half of all cancers are associated with mutations in the *p53* gene, suggesting that *p53* controls a key event in cell proliferation and is not involved in a cell- or tissue-specific form of regulation. Inherited mutations in the *p53* gene are associated with the Li-Fraumeni syndrome, an autosomal dominant trait associated with a predisposition to develop cancers in several tissues at a high frequency.

Studies have shown that normal cells contain low levels of p53 protein, but these levels rise dramatically following irradiation of cells by ultraviolet light (UV), which damages DNA. The p53 protein is involved in initiating two possible responses: 1) it can arrest the cell cycle, thereby promoting DNA repair; or 2) it can initiate a sequence of events leading to **apoptosis,** whereby genetically programmed cell death results. These tasks are accomplished as a result of the activation of target genes by p53 through its function as a transcription factor.

For example, in order to arrest the cell cycle, p53 stimulates the transcription of a gene encoding a CDK inhibitor protein called p21. The p21 protein then binds to cyclin/CDK complexes, which, in the absence of DNA damage and p21, are normally responsible for the cell's progression into the S phase. The binding blocks their action and the cell cycle is arrested. Cells lacking functional p53 protein are unable to arrest in G1 following irradiation, and they move immediately from G1 into S. These cells therefore do not repair the DNA damage and, as a result, have a high rate of mutation. p53 is thus often referred to as the "guardian of the genome."

Recently, *p53* has been linked to apoptosis. It is clearly an integral component in the regulation of this program, in which cells initiate a programmed series of steps leading to their death. This genetic program should be familiar to anyone who has suffered a severe sunburn, followed by peeling skin a few days later. The dead cells have undergone apoptosis induced by exposure to the ultraviolet rays of the sun. In cells lacking a functional *p53* gene, UV irradiation is not followed by apoptosis. Such cells may survive, escape cell-cycle control, and undergo further mutational events leading to the formation of a cancerous lesion on the skin.

The central role of the *p53* gene in controlling the cell cycle emphasizes both the relationship between cancer and the cell cycle and the relationship between genes that regulate cell growth and cancer.

Oncogenes induce uncontrolled cell division

We turn now to the second category of genes involved in the regulation of the cell cycle: those that normally function to *promote* cell division, called **proto-oncogenes.** When they are expressed, these genes promote cell division. To regulate cell division, these genes and/or their gene products must be inactivated. If proto-oncogenes become permanently switched on, then uncontrolled cell division occurs, leading to tumor formation. When this occurs as a result of the mutation of proto-oncogenes, they become known as **oncogenes,** because they induce or maintain uncontrolled cellular proliferation associated with cancer. Unlike tumor-suppressor genes, where mutations in both alleles of a gene are necessary to promote the development of cancer, only one of the two copies of a proto-oncogene needs to be mutated to induce malignancy.

Rous Sarcoma Virus and Oncogenes

The existence of specific genes associated with the transformation of normal cells into cancerous cells was first inferred by Francis Peyton Rous in 1910. Rous studied a kind of tumor known as a **sarcoma** (a tumor of connective tissue), injecting cell-free extracts from such tumors in chickens into healthy chickens and inducing the formation of sarcomas. He postulated the existence of an agent that was responsible for transmitting the disease, which decades later was shown by other investigators to be a virus. It is now known as the **Rous sarcoma virus (RSV).** Rous received the Nobel Prize in 1966 for his pioneering work establishing a relationship between viruses and cancer.

The Rous sarcoma virus belongs to a group of oncogenic viruses in which single-stranded RNA molecules serve as the genetic material. These viruses use an enzyme called **reverse transcriptase** to convert the RNA genome into a single-stranded DNA molecule. The enzyme then uses the single-stranded DNA as a template to synthesize the complementary strand, creating a double-stranded DNA molecule. The DNA can be integrated into the genome of the infected cell, forming a **provirus.** At a later time, the DNA can be transcribed into RNA, which can be translated into viral proteins. The packaging of the RNA molecules into these proteins forms new virus particles. Because the life cycle of these viruses "reverse" the flow of genetic information, they are called **retroviruses.**

In RSV, the tumor-forming ability results from a single gene, called the *src* gene, present in the viral genome. This gene, responsible for the induction of tumor formation in chicken cells, is designated as an oncogene. Retroviruses that carry oncogenes are known as **acute transforming viruses.** Other retroviruses that do not carry oncogenes but that induce the activity of cellular genes that bring about tumor formation are designated as **nonacute** or **nondefective viruses**.

Oncogenes (*onc*) carried by acute transforming viruses are acquired from the host's genome during infection, when a portion of the viral genome is exchanged for a cellular proto-oncogene (Figure 23.7). Retroviruses carry oncogenes designated *v-onc*; the cellular version of the gene is designated a *c-onc* gene or proto-oncogene. Retroviruses that carry a *v-onc* gene are able to infect and transform a specific type of host cell into a tumor cell. For RSV, the

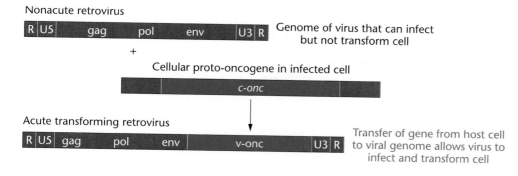

FIGURE 23.7 A transforming retrovirus has acquired a copy of a gene from the host genome, converting it from a *c-onc* or proto-oncogene into an oncogene that confers on the virus the ability to transform a specific type of host cell into a cancerous cell.

T A B L E   2 3 . 3   The Origin and Function of Representative Oncogenes

c-onc	Origin	Species	Cellular Function
src	Rous sarcoma virus	Chicken	Tyrosine kinase signal protein
fos	FBJ osteosarcoma virus	Mouse	Transcription factor
sis	Simian sarcoma virus	Monkey	Platelet-derived growth factor
abl	Abelson murine leukemia virus	Mouse	Tyrosine kinase
myc	Avian myelocytomatosis virus	Chicken	Transcription factor
erbB	Avian erythroblastosis virus	Chicken	EGF receptor
N-ras	Neuroblastoma, leukemia	Human	GTP-binding protein

T A B L E   2 3 . 4   Structural Alterations in Oncogenes

Gene	Number of Codons	Number of Changed Amino Acids from c-onc gene	Region Missing in v-onc Protein
H-ras	189	3	None
K-ras	189	7	None
mos	369	11	None
myc	417	2	None
erbA	408	22	Deletion at N terminus
src	533	16	Deletion at C terminus
erbB	1210	49	Deletion at N terminus and C terminus

oncogene captured from the chicken genome, *v-src*, confers the ability to transform chicken cells into sarcomas. The cellular version of the same gene, found in the chicken genome, is called *c-src*. More than 20 oncogenes have been identified by their presence in retroviral genomes, and over 50 oncogenes have been identified. Some of these are listed in Table 23.3.

Origins and Characterizations of Oncogenes

Several key questions about oncogenes come to mind:

1. What differences exist between corresponding *v-onc* and *c-onc* versions of a gene?

2. How do such differences arise?

3. How do oncogenes bring about cellular transformation and tumor formation?

Let's address the first question. In some cases, comparison of *v-onc* DNA sequences (such as *v-ras* and *v-mos*) with the corresponding *c-onc* sequences shows only minimal differences, probably generated by point mutations. In other cases, segments of the *c-onc* sequence have been replaced or lost. In *v-src*, a string of 19 amino acids of the *c-onc* sequence has been replaced with a sequence of 12 different amino acids. In another oncogene, *v-erb*, the N-terminus portion of the gene has been lost (see Table 23.4). Mutations, therefore, play a role in differentiating some *v-onc* se-

quences from their *c-onc* counterparts. However, not all oncogenes are mutant versions of cellular genes carried by retroviruses. We must, therefore, consider a broad range of mechanisms in oncogenetic events.

At least three mechanisms can explain how proto-oncogenes are converted into oncogenes: **point mutations, translocations,** and **overexpression** (see Table 23.5). Although some of these are mediated by viruses, others are generated solely by intracellular events in the absence of retroviruses.

The ***ras* gene family,** which encodes a 189-amino acid protein involved in the transduction of signals across the plasma membrane, illustrates how single nucleotide changes can convert a *c-onc* sequence into the oncogenic version of the gene. The normal ras protein functions as a molecular switch, alternating between an on-and-off state

T A B L E   2 3 . 5   Conversion of Proto-Oncogenes to Oncogenes

Mechanism	c-onc
Point mutation	ras
Translocation	abl
Overexpression of gene product	
New promoter by viral insertion	mos, myb
New enhancer by viral insertion	myc
Amplification of proto-oncogene	myc

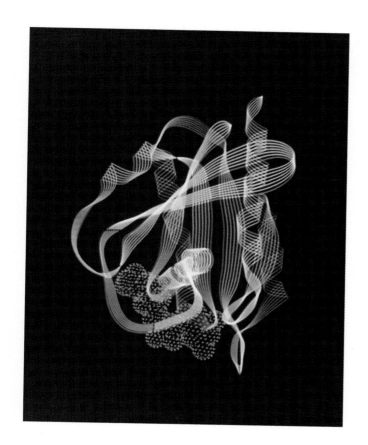

FIGURE 23.8 A three-dimensional computer-generated image of ras proteins in two different conformations. Normal ras proteins act as molecular switches controlling cell growth and differentiation. The switch is "on" (conformation shown in blue) when GTP binds to the protein, and "off" (conformation shown in yellow) when the GTP is hydrolyzed to GDP. Switching the protein between states alters the conformation of the protein in the two regions (blue and yellow). Oncogenic mutations of *ras* are stuck in the "on" state, continuously signaling for cell growth.

(Figure 23.8). In some cases, the mutant ras protein becomes stuck in the "on" position, stimulating cell growth. In some tumors, this event is the result of a somatic mutation. The comparison of the amino acid sequence of ras proteins from a number of different human carcinomas (tumors of epithelial tissue) reveals that *ras* mutations involve single amino acid substitutions at either position 12 or 61 (Figure 23.9).

One of the well-characterized examples of oncogene activation by translocation is that of *c-abl*, an oncogene associated with chronic myelogenous leukemia (CML). In this case, described in detail in a later section, the translocation results in altered gene activity that causes tumor formation.

At least three separate mechanisms of proto-oncogene activation are associated with overexpression. First, the *c-onc* may acquire a new promoter, causing an increase in the level of transcript production or causing a silent locus to become activated. This is the case in avian leukosis, where strong viral promoters are inserted adjacent to a proto-oncogene, causing an increase in mRNA produc-

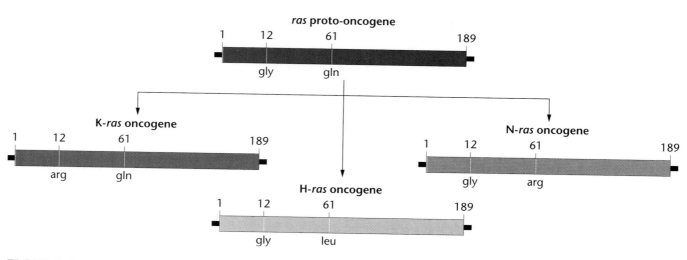

FIGURE 23.9 The *ras* proto-oncogene encodes a protein of 189 amino acids. In the normal protein, glycine is encoded at position 12, and glutamine at position 61. Analysis of *ras* oncogene proteins from several tumors shows a single amino acid substitution at one of these positions, which can convert a proto-oncogene into a tumor-promoting oncogene.

TABLE 23.6 Cellular Location of c-onc and v-onc Proteins

Gene	Location of c-onc Protein	Location of v-onc Protein
src	Membranes	Membranes
ras	Membranes	Membranes
myc	Nucleus	Nucleus
fps	Cytoplasm	Cytoplasm and membranes
abl	Nucleus	Cytoplasm
erbB	Plasma membrane	Plasma membrane and Golgi

tion and the amount of the gene product. A second mechanism of overexpression involves the acquisition of new upstream regulatory sequences, including enhancers. The third mechanism involves the amplification of the proto-oncogene. In human tumors, members of the *myc* family of oncogenes are frequently amplified. The *c-myc* proto-oncogene is found amplified up to several hundred copies in some human tumors.

It is also of interest to know where in the cell these genes function. The protein products of proto-oncogenes are found in the plasma membrane, cytoplasm, and nucleus (Table 23.6). In spite of their wide range of location within the cell, which suggests varying functions, all the products characterized to date alter gene expression in a direct or an indirect manner.

A final note involves the link between oncogenes and apoptosis. As we described earlier, apoptosis is associated with a *p53*-induced pathway that responds to DNA damage. More recent evidence has shown that oncogenes and their corresponding oncoproteins can also induce apoptosis by sensitizing the cell to apoptotic triggers (such as DNA damage). It is currently believed that the processes mediating growth and apoptosis are coupled. In a current model, the activation of cell proliferation (e.g., proto-oncogene conversion to an oncogene) primes the cellular apoptotic program. As a result, many cells containing oncogenes do not proliferate but instead apoptose. This model helps us to understand why cancer does not develop every time a proto-oncogene changes into an oncogene.

Metastasis is genetically controlled

Cancer cells have lost the ability to regulate their own growth and division. As a result, they can develop into malignant tumors that may potentially invade neighboring tissue. Sometimes, cancer cells detach from the primary tumor and settle elsewhere in the body, where they grow and divide, producing secondary tumors. This process, called **metastasis**, is often the cause of death in cancer patients.

A metastatic cancer cell can spread from a primary tumor by entering the blood or lymphatic circulatory system. These cells are carried in the circulation until they become lodged in a capillary bed. Normally, more than 99 percent of the cells die. The surviving cells invade tissue adjacent to the capillary bed and begin dividing to form a secondary tumor. To reach a new site, the tumor cells pass through the layer of epithelial cells lining the interior wall of the capillary (or lymph vessel) and penetrate the surrounding extracellular matrices.

The extracellular matrix is a meshwork of proteins and carbohydrate molecules separating tissues; it acts as a scaffold for tissue growth and inhibits the migration of cells. To establish a secondary tumor, metastatic cells secrete enzymes that digest proteins in the basement membrane, creating holes through which they can move. The cell "tunnels" through the matrix, enters new tissue, and establishes a secondary tumor.

The invasive ability of tumor cells is also a property of some normal cell types. The implantation of the embryo in the uterine wall during pregnancy, for example, requires cell migration across the extracellular matrix. White blood cells normally reach the site of infection by penetrating capillary walls. The mechanism of invasion is probably the same in such normal cells as it is in cancer cells. The difference is that in normal cells, the invasive ability is controlled and tightly regulated, while in tumor cells, regulation has been lost.

Metastasis may be regulated by genes that encode or regulate matrix-cutting enzymes. Recent studies have shown that protein-cutting enzymes called **metalloproteinases** are necessary for cell invasion. Tumor cell lines with high metastatic ability produce greater amounts of certain metalloproteinases than do tumor cell lines with low invasive properties. Activated metalloproteinases are inhibited by the presence of a tissue inhibitor of metalloproteinase (TIMP). Normal cells produce TIMP and thereby suppress inappropriate activity of metalloproteinase. In tumor cells, TIMP inhibits metalloproteinase only when the number of TIMP molecules is greater than the number of metalloproteinase molecules. On the basis of its action and regulation, TIMP can be considered as a protein that suppresses metastasis.

Colon cancer results from a series of mutations

Although cancer has been studied in great detail in only a small number of cases, the development of the disease is clearly a multistep process resulting from a number of specific genetic alterations. The studies of tumors such as retinoblastoma and Wilms tumor have been useful in establishing that, in some cases, only a limited number of steps are required to transform a normal cell into a malignant one. However, these tumors are of limited value in establishing the sequence of the genetic events leading to tumor formation and metastasis.

For several reasons, the study of colorectal cancer has provided more detailed information and greater insights into how malignancy occurs. First, the malignant tumors develop from preexisting benign tumors. Second, several discrete precancerous phases occur that may be isolated and studied. Furthermore, both hereditary and sporadic forms of colorectal cancer exist. As a result, colorectal cancer has served as a useful model in studying the interaction of genetic and environmental factors in the genesis of tumors.

Hereditary Colon Cancer and DNA-Repair Defects

Two forms of genetic predisposition to colon cancer are known, a genetically complex trait, known as **hereditary nonpolyposis colorectal cancer (HNPCC)**, and an au-

tosomal dominant trait, known as **familial adenomatous polyposis (FAP)**. HNPCC is an example of a cancer that develops because of a defective gene in the DNA-repair pathway. Two genes, *MSH2* and *MLH1*, have been identified in this form of cancer and have been shown to be an essential part of the mismatch repair pathway (see Chapter 17 for a description of mismatch repair). A defect in one or more enzymes involved in DNA repair leads to a higher rate of mutation and therefore an increased likelihood of cancer-causing genetic lesions.

The Multistep Nature of Colon Cancer

Through an analysis of mutations in many tumors, the number and nature of the genetic steps involved in changing normal intestinal epithelial cells into tumor cells has been defined. Such analysis has led to the genetic model for colon cancer shown in Figure 23.10. The first feature of this model is that mutations in four or five specific genes are needed to bring about malignant growth. If fewer changes take place, benign growths or intermediate stages of tumor formation result. Second, the order of mutations usually follow a sequence indicated in the figure. Ultimately, however, the accumulation of specific mutations is more important than the order in which they occur.

The first mutation in the sequence occurs in a normal epithelial cell. In cases of FAP, a first mutation is inherited and results in the development of dozens or hundreds of benign adenomas in the colon and rectum. In spontaneous cases, evidence suggests that the initial mutational event takes place in a single cell and that the resulting adenoma consists of a clone of cells, all of which carry the mutation. This first mutation takes place in a gene called *APC*, located on the long arm of chromosome 5. The loss of the corresponding allele on the homologous copy of chromo-

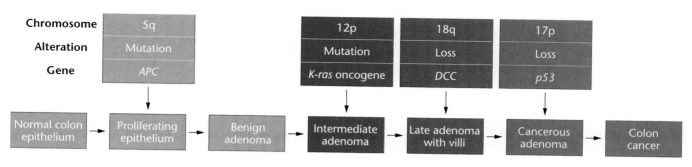

FIGURE 23.10　A model for the multistep production of colon cancer. The first step is the loss or inactivation of both alleles of the *APC* gene on chromosome 5. In familial cases, one mutation of the *APC* gene is inherited. Loss of both alleles leads to the formation of benign adenomas. Subsequent mutations involving genes on chromosomes 12, 17, and 18 in cells of the benign adenomas can lead to a malignant transformation resulting in colon cancer. Although the mutations in chromosomes 12, 17, and 18 usually occur at a later stage than those involving chromosome 5, the sum of the changes is more important than the order in which they occur.

some 5 is *not* necessary for proliferation and adenoma formation. The relative order of subsequent mutations is shown in Figure 23.10. Mutations in the *ras* oncogene may precede or follow the loss of a segment of chromosome 18p. In either case, the accumulation of these two mutations in adenoma cells with a preexisting mutation on chromosome 5 causes the adenoma to grow larger and develop a number of fingerlike villous outgrowths. Finally, a mutation on 17p, involving the loss or inactivation of *p53*, causes the transition to a cancerous cell. As discussed earlier, mutations in the *p53* gene are pivotal to the development of a number of cancers, including lung, brain, and breast cancers as well as colon cancer. Recall that the normal allele of *p53* is a tumor-suppressor gene that regulates the passage of cells from late G1 to S phase. Metastasis occurs after the formation of colon cancer, and it involves an unknown number of mutational steps.

In sum, the genetic model for colon cancer involves sequential mutations in oncogenes and tumor-suppressor genes, and the disruption of the cell cycle at a specific transition point, although the nature and function of the normal and mutant gene products of the *p53* gene have not yet been identified with certainty. In cases of an inherited predisposition to colon cancer, the first mutation is genetically transmitted; the rest are believed to occur by action of environmental agents, as we will discuss shortly. This multistep model has applications to other forms of cancer, but many questions remain unanswered, including the normal functions of the genes involved and the molecular mechanisms by which the mutated genes and/or their gene products bring about tumor formation.

Translocation of chromosomes is often associated with cancer

Changes in chromosome structure or number are often associated with various forms of cancer. In selected cases, the relationship between the chromosome aberration and the development and/or maintenance of the cancerous state is clear. Such a connection is most clearly seen in leukemias, where the presence of a specific translocation of a chromosome is well defined (Table 23.7). One of the best-studied examples is the translocation between chromosomes 9 and 22 that is associated with **chronic myelogenous leukemia (CML)** (Figure 23.11). Originally, this translocation was described as an abnormal chromosome 21 and called the **Philadelphia chromosome** (because it was discovered in that city). Later, Janet Rowley showed that the Philadelphia chromosome results from the exchange of genetic material between chromosomes 9 and

TABLE 23.7 Specific Chromosome Aberration and Cancer

Cancer	Chromosome Alteration*
Chronic myelogenous leukemia	t(9;22)
Acute promyelocytic leukemia	t(15;17)
Acute lymphocytic leukemia	t(4;11)
Acute myelogenous leukemia	t(8;21)
Prostate cancer	del(10q)
Synovial sarcoma	t(X;18)
Testicular cancer	inv(12p)
Retinoblastoma	del(13q)
Wilms tumor	del(11p)

*t = translocation, del = deletion, inv = inversion.

22. This translocation is never seen in a normal cell and is found only in the white blood cells involved in CML. These observations indicate that the translocation is a primary and causal event in the generation of CML, and that this form of cancer may originate from a single cell bearing this translocated chromosome.

By examining a large number of cases involving the Philadelphia chromosome, researchers were able to determine the exact location of the break points on chromosomes 9 and 22. Genetic mapping studies using recombinant DNA techniques established that a proto-oncogene *c-abl* maps to the breakpoint region on chro-

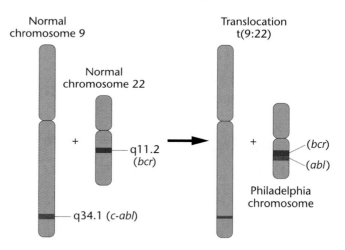

FIGURE 23.11 The Philadelphia chromosome. A reciprocal translocation involving the long arms of chromosomes 9 and 22 results in the production of a characteristic chromosome, the Philadelphia chromosome, which is associated with cases of chronic myelogenous leukemia (CML). The t(9;22) translocation results in the fusion of the *c-abl* oncogene on chromosome 9 with the *bcr* gene on chromosome 22. The fusion protein is a powerful hybrid molecule that allows cells to escape control of the cell cycle, resulting in leukemia.

mosome 9, and that the gene *bcr* maps near the breakpoint on chromosome 22. The normal c-abl protein is a kinase and the normal bcr protein activates a phosphorylation reaction. In the translocation event, all or most of the *c-abl* gene is translocated to a region within the *bcr* gene, generating a hybrid *bcr/c-abl* oncogene that is transcriptionally active. The abnormal hybrid 200-kDa protein product of the fused genes has been implicated in the generation of CML.

Both cytogenetic and molecular approaches have been applied to the cytogenetic events in translocations involving chromosome 8 [these include t(8;14), t(8;22), and t(2;8)] in a disease related to leukemia, known as **lymphoma**. The breakpoint on chromosome 8 in all these tranlocations is the same, and the proto-oncogene *c-myc* has been mapped to this locus. The loci at the breakpoint on the other chromosomes involved in this series of translocations all have immunoglobulin genes at the breakpoints. The movement of the *c-myc* gene to a position near these immunoglobulin genes leads to the overexpression of the *c-myc* gene, resulting in the transformation of the lymphoid cells into leukemic cells.

Other genes at translocation breakpoints have also been isolated and characterized. In these cases, as in CML, the translocation results in the formation of an abnormal gene product that causes the cell to undergo a malignant transformation even though a second and presumably normal copy of the gene is present and active in synthesizing a normal gene product. It is hoped that, if the abnormal gene products at other translocation loci associated with leukemia can be identified, therapeutic strategies can be developed to administer high levels of the normal gene product, or that the abnormal gene product can be inactivated.

Genomic instability can lead to cancer

The term *genomic instability* describes events resulting in genomic alterations characteristic of cancer cells. At least three classes of genetic defects can lead to genomic instability: defects in DNA repair and replication, aberrant chromosome segregation, and defects in cell-cycle control. Cultured cells from individuals with Fanconi anemia, ataxia telangiectasia (AT), and Werner syndrome all show an increase in chromosome breaks and translocations, indicating that genomic instability is an heritable trait. The main defect in these syndromes appears to be in DNA repair and/or faulty control of DNA replication.

The role of genetic instability in cancer can be seen in familial adenomatous polyposis (FAP), where a number of mutations are distributed throughout the genome.

A more dramatic relationship between genomic instability and another form of colon cancer, not associated with polyp formation, has recently been discovered. This form, **hereditary nonpolyposis colon cancer, HNPCC**, accounts for up to 15 percent of all cases of colon cancer. Loci for HNPCC have been mapped to 2p16 and 3p21.

Malignant cells from affected individuals show changes in short, repetitive DNA sequences (called microsatellite DNA or variable nucleotide tandem repeats; see Chapter 19) scattered throughout the genome. These large-scale genomic alterations, representing perhaps thousands of mutations, indicate that the loci on chromosomes 2 and 3 may affect the accuracy of DNA replication or DNA repair. It is estimated that the mutant gene on chromosome 2 may be carried by 1 in 200 individuals in the Western world, making it one of the most common genetic disorders known.

The results of studies on tumor development *in vivo* have established that, in many cases, genomic instability precedes the formation of tumors. Individuals with a chronic regurgitation of stomach acid into the esophagus have a high risk of esophageal cancer. Cytogenetic studies on esophageal cells of affected individuals show that the first detectable change is a progressive alteration in cell-cycle control in small clones of cells (descended from a single ancestral cell), followed by the development of aneuploidy, and finally by the appearance of cancer. In addition, preliminary studies on the HNPCC loci indicate that microsatellite instability precedes the formation of colon tumors and may be an early event in cancer development.

Various environmental agents, including viruses, are directly or indirectly linked to cancer

The relationship between environmental agents and the genesis of cancer is often elusive, and in the early stages of an investigation this relationship is based on indirect evidence. Such studies often begin with an epidemiological survey, comparing the cancer death rates among different geographic populations. When differences are found in the death rate for a given type of cancer, a given age group, or a cluster of related occupations such as chemical workers, further work is necessary to identify one or more environmental factors that correlate with these cancer deaths. These correlations are not conclusive, but they serve to identify factors that, upon additional investigation, may turn out to be directly related to the development of cancer. Finally, extensive laboratory investigations may establish the mechanism by which an environmental agent generates cancer. In other cases, such as for certain viruses, the relationship between cancer and the environment is more straightforward, as we shall see next.

Hepatitis B and Cancer

Epidemiological surveys have shown that individuals who develop a form of liver cancer known as **hepatocellular carcinoma (HCC)** are infected with the **hepatitis B virus (HBV)** (Figure 23.12). In fact, for those carrying HBV, the risk of cancer is increased by a factor of 100. Aside from the risk of cancer, HBV infection is a public health risk affecting some 300 million people worldwide; it is most prevalent in Asia and tropical regions of Africa. The infection produces a wide range of responses, from a chronic, self-limiting infection with few symptoms, to active hepatitis and cirrhosis, to fatal liver disorders and hepatocellular carcinoma. In the last few years, efforts have centered on understanding the mechanism by which HBV replicates and its role in causing liver cancer.

The genome of HBV is a mostly double-stranded DNA molecule of 3200 nucleotides. After cellular infection, the DNA moves to the nucleus, where it inserts into a chromosome. The viral DNA is transcribed into an RNA molecule and packaged into a viral capsid. The capsid containing the RNA pregenome moves to the cytoplasm, where it is reverse-transcribed into a DNA strand, which in turn is made double-stranded. The copied DNA genome is repackaged into a new capsid for release from the cell or returns to the nucleus for another round of replication.

The key to the role of HBV in carcinogenesis apparently resides in its entry into the nucleus. While in the nucleus, the HBV genome can insert itself at many different sites into human chromosomes (Figure 23.13). Insertion often results in cytogenetic alterations of the host genome that include translocations, deletions, or amplifications of adjacent regions. As outlined earlier, most forms of cancer are associated with chromosomal rearrangements of these types, and it is possible that such rearrangements trigger the development of a cancerous transformation. The HBV genome has been shown to integrate into the cyclin A gene, and abnormal expression of this gene can disrupt the cell cycle and normal control of proliferation. Similarly,

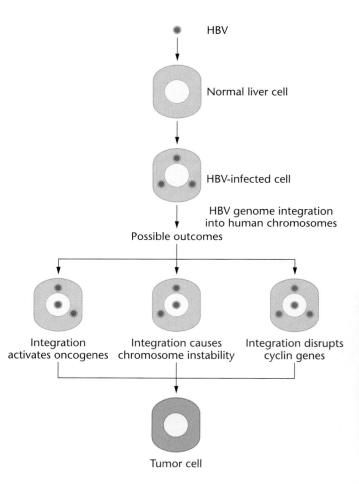

FIGURE 23.13 The possible outcomes following cellular infection with hepatitis B virus (HBV). Following the integration of the viral chromosome into the human genome, oncogenes can be activated, resulting in tumor formation. Integration of the HBV genome can also cause chromosome instability, including the production of translocations and deletions that trigger a malignant cyclin A gene, triggering abnormal regulation of the cell cycle and cellular proliferation, resulting in cancer formation.

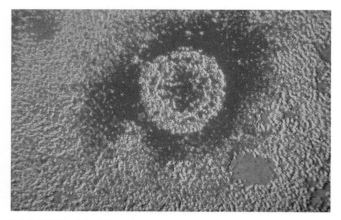

FIGURE 23.12 A false-color transmission electron micrograph of a hepatitis B virus. Infection with this virus often results in cancer of the liver. Worldwide, viral infections of all kinds are thought to be responsible for about 15 percent of the total number of cancers.

HBV integration has resulted in activation of proto-onco-gene loci, and altered expression of such loci has been clearly implicated in the development of malignant cell growth. Thus, a strong link exists between the HBV virus as an environmental agent and the development of cancer in infected individuals.

Environmental Agents

Many surveys of cancer incidence (Table 23.8) have pointed to the role of the physical environment and personal behavior as factors contributing to the development of cancer. The precise role of the environment in the genesis of cancer can be difficult to ascertain. Obviously, the induction of cancer can involve an interaction between the genotype and environmental agents. The differential expression of a specific genotype in various environments is often neglected in making calculations about the role of the environment in cancer, but it has been estimated that at least 50 percent of all cancers are environmentally induced. Environmental agents responsible for cancer include background levels of radiation, occupational exposure to physical and chemical agents, exposure to sunlight, and personal behavior such as diet and the use of tobacco. To show how environmental factors are identified, let us consider a recent study on the risk of colon cancer.

Epidemiological surveys reveal that cancer of the colon occurs with a much higher frequency in North America and Western Europe than in Asia, Africa, or other parts of the world. When people migrate from low-risk areas to high-risk areas, their risk of colon cancer increases to match that of the native residents. This finding suggests that environmental factors, including diet, may play a role in the incidence of colon cancer. In a long-range health study of more than 88,000 registered nurses across the United States, dietary habits have been monitored by questionnaires since 1980. In that population, 150 cases of colon cancer were observed through 1986. Detailed analysis indicated that nurses who had daily meals of pork, beef, or lamb had a 2.5-fold higher risk of colon cancer than did those who ate such meals less than once a month. A more detailed analysis of diets strongly suggested that animal fat in the diet is the environmental risk factor for colon cancer. At the same time, a negative correlation existed between eating skinless chicken meat and the incidence of colon cancer. Laboratory studies have been undertaken to identify the mechanisms by which fat brings about a cancerous transformation of the intestinal epithelium. However, even in the absence of information about the mechanism of action, it would seem prudent to reduce one's intake of animal fat in order to reduce the risk of colon cancer.

The accumulated results of research on the role of external factors as a cause of cancer indicate that neither the environment in general nor pollution in particular is responsible for a large fraction of cancer cases. Rather, diet, tobacco, and other agents, including medical and dental X rays, viruses, drugs, and ultraviolet light, play significant roles in cancer development. It is estimated that 50 percent of all cases of cancer are caused by these agents and personal choices (e.g., the composition of food in the diet, tobacco use, and suntanning) play an important part. Education and more judicious choices on the part of individuals could lead to the prevention of a high percentage of all human cancers.

TABLE 23.8 Epidemiology of Cancer in Various Countries

Country	Death Rate per 100,000	
	Male	Female
Australia	212	125
Canada	214	136
Dominican Republic	54	48
Egypt	39	18
England	248	156
Greece	188	103
Israel	174	145
Japan	190	109
Nicaragua	22	35
Portugal	180	108
Singapore	249	130
United States	216	137
Venezuela	135	128

GENETICS, TECHNOLOGY, AND SOCIETY

The Double-Edged Sword of Genetic Testing: The Case of Breast Cancer

As a result of the Human Genome Project, the sequence of all 75,000 or so genes residing on each set of 23 chromosomes will be known, among which will be thousands that potentially contribute to disease. This is expected to revolutionize medicine, as it will open the door to treating genetic diseases at their source by replacing or even correcting defective genes. However, the lag between the identification of a gene and the development of a new treatment is likely to be quite lengthy, measured in decades rather than years. In the meantime, it is already becoming possible to test people to see whether they carry mutant alleles that impart a high risk of developing a genetic disease. In some cases, those with a genetic predisposition to disease then have the opportunity to take preventive measures to lower their risk. Although genetic testing has the potential to save or extend lives, at the same time it may raise difficult questions for which the answers are not always clear. To illustrate how genetic testing may create troubling choices, consider recent advances in the genetics of breast cancer.

Breast cancer is the most common type of cancer among women and the second leading cause of cancer deaths (after lung cancer). Each year, more than 180,000 new cases are diagnosed and about 43,000 women die of breast cancer. The likelihood of developing breast cancer is low before age 35, but the risk rises after that. If a woman lives a long life, her risk of breast cancer is about 12 percent.

For many years it has been known that 5 to 10 percent of the cases of breast cancer are familial, meaning that they are clustered in families in which generation after generation of women are stricken. The familial form of breast cancer also tends to develop earlier than spontaneous occurrences, often in a woman's 30s or 40s, and is more often bilateral (involving both breasts) and multifocal (involving several separate tumors).

The clustering of breast cancer in certain families suggested that genetic mutations are involved in the development of these cases of breast cancer. The study of such families subsequently led to the identification of two genes that, when mutated, impart an increased susceptibility to breast cancer. These genes, named *BRCA1* and *BRCA2*, were isolated in 1994 and 1995, respectively. Since then, several other genes have been found that affect the risk of breast cancer to a lesser degree.

Mutations in *BRCA1* and *BRCA2* result in a greatly increased likelihood of developing breast cancer over the lifetime risk in the general population. The exact risk is somewhere between 56 and 85 percent, depending on the specific mutation and the extent of the family history of breast cancer. Mutations in *BRCA1* and *BRCA2* are responsible for about two-thirds of all familial breast cancer. In addition, *BRCA1* mutations result in a 10-fold increased risk of ovarian cancer.

The role of *BRCA1* and *BRCA2* in the origin of cancer is indirect. The protein products of both normal genes appear to be involved in the repair of damaged DNA. They thus help prevent the accumulation of mutations in somatic cells, including those that cause the cell to lose control over the cell cycle and become cancerous. *BRCA1* also behaves as a tumor-suppressor gene. Why mutations in these

CHAPTER SUMMARY

1. Cancer is a genetic disorder at the cellular level that can result from the mutation of a given subset of genes or from alterations in the timing and amount of gene expression. Although some forms of cancer show familial patterns of inheritance, few show clear evidence for Mendelian inheritance.

2. Cancer results from the uncontrolled proliferation of cells and from the ability of such cells to metastasize or migrate to other sites and form secondary tumors. Mutant forms of genes involved in the regulation of the

cell cycle are obvious candidates for cancer-causing genes. The cell cycle is apparently regulated at several sites, with two types of gene products primarily involved, cyclins and kinases. The actions of these genes and the genes they control are important in regulating cell division, and links between these genes and the process of tumor formation are being discovered.

3. Although Mendelian patterns of cancer development are not clear-cut, there are genes that predispose to cancer. The study of genes such as that for retinoblastoma

two genes increase the likelihood of cancer specifically in the breast (and in the case of *BRCA1* in the ovary) is not yet clearly understood, but it may be related to the physiological responses of these tissues to hormones.

The identification of breast cancer-susceptibility genes was hailed as a breakthrough in the understanding of this disease at the molecular level, an understanding that would pave the way for new therapies and better preventive measures. But unraveling the exact chain of events from a mutated *BRCA* gene to runaway cell division will take many years. In the short term, women in high-risk families now have the option of being tested to see whether they carry a mutant allele of *BRCA1* or *BRCA2*. If so, they face a lifetime risk of breast cancer as high as 85 percent; if not, they still have a 12 percent risk.

There are some very good reasons why a woman in a high-risk family would want to be tested. Foremost among them is that a negative test result gives tremendous peace of mind, allowing a woman to live her life out from under a cloud of fear. She will also be secure in the knowledge that she will not pass along the mutant allele to her children. It's the consequences of a positive test result that makes the decision whether to be tested so difficult. For women with *BRCA1* or *BRCA2* muta-

tions, there are two alternative courses of action, neither very appealing. The conservative approach is close medical surveillance, including frequent breast self-exams, clinical exams, and mammographies, with the goal of identifying a tumor at its earliest stages, when it is most treatable. The more radical strategy is the surgical removal of the breasts (bilateral prophylactic mastectomy) and, in the case of a mutation in *BRCA1*, possibly of the ovaries. Unfortunately, there currently seems to be no middle ground.

Recent studies have suggested that women in high-risk families who undergo a bilateral prophylactic mastectomy reduce their breast cancer risk by about 90 percent. Even so, it is unclear whether the personal and psychological costs of such extreme surgery are outweighed by the increased life expectancy. There are other fears, as well. A positive test result, some believe, may lead to discrimination in employment and to higher insurance premiums or loss of coverage altogether.

Issues raised in genetic testing for breast cancer susceptibility are likely to be played out again and again with other genes that predispose individuals to other cancers. Already, genes that increase the risk of cancers of the prostate, colon, skin, and lung are in hand, and more will soon follow. In ad-

dition, several genes have been identified that increase the risk of other late-onset disorders, including Alzheimer disease, Parkinson disease, and heart disease. This list is sure to grow with each passing year.

We are now on the verge of having the ability to predict diseases based on genotype many years before they might occur. From our experience with testing for breast cancer susceptibility, we must learn how to extend the use of this new technology while gaining the wisdom to know when to test and when not to test.

REFERENCES

KAHN, P. 1996. Coming to grips with genes and risk. *Science* 274:496–98.

MARX, J. 1997. Possible function found for breast cancer genes. *Science* 276:531–32.

MIKI, Y. et al. 1994. A strong candidate for the breast and ovarian cancer susceptibility gene *BRCA1*. *Science* 266:66–71.

SCHRAG, D., KUNTZ, K.M., GARBER, J.E., and WEEKS, J.C. 1997. Decision analysis—Effects of prophylactic mastectomy and oophorectomy on life expectancy among women with *BRCA1* or *BRCA2* mutations. *New Engl. J. Med.* 336:1465–71.

WOPSTER, R. et al. 1995. Identification of the breast cancer susceptibility gene *BRCA2*. *Nature* 378:789–92.

has provided insight into the number and sequence of mutations that result in the development of tumors.

4. Tumor-suppressor genes normally act to suppress cell division. When these genes are mutated or have altered expression, control over cell division is lost. Molecular analysis of two such genes, *retinoblastoma (RB)* and *Wilms tumor (WT)*, indicates that these genes work to control gene expression in different ways, either by affecting the activity of transcription factors or by themselves acting as transcription factors. Mutations in both alleles of a tumor-suppressor gene are necessary to promote the development of cancer.

5. Oncogenes are genes that normally function to maintain cell division, and these genes must be mutated or inactivated to halt cell division. If these genes escape control and become permanently switched on, cell division occurs in an uncontrolled fashion. In contrast to tumor-suppressor genes, a mutation in only one copy of an oncogene can promote the development of cancer.

6. In most cases of cancer, a series of mutations is necessary to cause the malignant state. Colon cancer is a useful model illustrating the multistep nature of cancer.

7. The cells of most tumors have visible chromosomal alterations, and the study of these aberrations has provided some insight into the steps involved in the development of cancer. This relationship between cancer and chromosome alterations has been best studied in leukemias, where the formation of hybrid genes or substitution of regulatory sequences is associated with the transformation of normal cells into malignant tumors.

8. Although cancer is the result of genetic alterations, the environment appears to be an important factor in cancer induction. Environmental agents include occupational exposure to physical or chemical substances, viruses, diet, and other personal choices such as the use of tobacco and suntanning.

INSIGHTS AND SOLUTIONS

1. In disorders such as retinoblastoma, a mutation in one allele of the *retinoblastoma* (*RB*) gene can be inherited from the germ line, causing an autosomal dominant predisposition to the development of eye tumors. To develop tumors, a somatic mutation in the second copy of the *RB* gene is necessary, indicating that the mutation itself acts as a recessive trait. In sporadic cases, two independent mutational events, involving both *RB* alleles, are necessary for tumor formation. Given that the first mutation can be inherited, what are the ways in which a second mutational event can occur?

   Solution: In considering how this second mutation can arise, several levels of mutational events need to be considered, including changes in nucleotide sequence and events involving whole chromosomes or chromosome parts. Retinoblastoma results when both copies of the *RB* locus have been lost or inactivated. With this in mind, perhaps the best way to proceed is to prepare a list of the phenomena that can result in a mutational loss or inactivation of a gene.

   One way for the second *RB* mutation to occur is by a nucleotide alteration that converts the remaining normal *RB* allele to a mutant form of the gene. This alteration may occur through a nucleotide substitution or by a frameshift mutation caused by the insertion or deletion of one or more nucleotides. A second mechanism of mutation can involve the loss of the chromosome carrying the normal allele. This event would take place during mitosis, resulting in chromosome 13 monosomy, leaving the mutant copy of the gene as the only *RB* allele. This mechanism does not necessarily involve loss of the entire chromosome; deletion of the long arm (*RB* is on 13q) or an interstitial deletion involving the *RB* locus and some surrounding material would have the same result.

   As an alternative, a chromosome aberration involving loss of the normal copy of the *RB* gene might be followed by a duplication of the chromosome carrying the mutant allele. Two copies of chromosome 13 would be restored to the cell, but no normal allele of the *RB* gene would be present. Finally, a recombination event followed by chromosome segregation could produce a homozygous combination of mutant *RB* alleles.

   More can be discovered about the mechanisms involved in RB by analyzing the cells from tumors using a combination of cytogenetic and molecular techniques (such as RFLP analysis and hybridizations to look for deletions) to see which mechanisms are actually found in tumors and to what extent they play a role in generating the second mutation. Although such analysis is still in the preliminary stages, all the mechanisms proposed have been found in tumors, indicating that a variety of spontaneous events can bring about the second mutation that triggers retinoblastoma.

PROBLEMS AND DISCUSSION QUESTIONS

1. As a genetic counselor, you are asked to assess the risk for a couple who plans to have children, but where there is a family history of retinoblastoma. In this case, both the husband and wife are phenotypically normal, but the husband has a sister with familial retinoblastoma in both eyes. What is the probability that this couple will have a child with retinoblastoma? Are there any tests that you could recommend to help in this assessment?

2. Review the stages of the cell cycle. What events occur in each of these stages? Which stage is most variable in length?

3. Where are the major regulatory points in the cell cycle?

4. Progression through the cell cycle depends on the interaction of two types of regulatory proteins, kinases and cyclins. List the functions of each and describe how they interact with each other to cause cells to move through the cell cycle.

5. What is the difference between saying that cancer is inherited and saying that predisposition to cancer is inherited?

6. Define tumor-suppressor genes. Why are most tumor suppressor genes expected to be recessive?

7. Review the differences among transcriptional activity, tissue distribution, and function of the tumor-suppressor genes associated with retinoblastoma and Wilms tumor. What properties do they have in common? Which are most different?

8. Distinguish between oncogenes and proto-oncogenes. In what ways can proto-oncogenes be converted to oncogenes?

9. How do translocations such as the Philadelphia chromosome lead to oncogenesis?

10. Given that 50 percent of all cancers are environmentally induced, and most of these are caused by actions such as smoking, suntanning, and diet choices, how much of the money spent on cancer research do you think should be devoted to research and education on the prevention of cancer rather than on finding a cure?

11. The compound benzo[a]pyrene is found in cigarette smoke. This compound chemically modifies guanine bases in DNA. Such abnormal bases are typically removed by an enzyme, which hydrolyzes the base, leaving an apurinic site. If such a site is left unrepaired, an adenine is preferentially inserted *across from* the apurinic site. In a study of lung cancer patients, tumor cells from 15 out of 25 patients had a G to T transversion in the *p53* gene, which has a known role in cancer formation. Imagine that you are asked to testify as an expert witness in the following court case. A widow of a man who died of lung cancer is suing R. J. Reynolds for selling tobacco products that killed her husband (who was a lifelong smoker). What do you tell the jury? (Science only; no personal expositions on lawyers or the legal system.) [*Reference: Nature* 350:377–78 (1991).]

EXTRA-SPICY PROBLEMS

12. About 5 to 10 percent of breast cancer cases are caused by an inherited susceptibility. An inherited mutation in a gene called *BRCA1* is thought to account for approximately 80 percent of families with a high incidence of both early-onset breast and ovarian cancer. Table 1 summarizes some of the data that has been collected on *BRCA1* mutations in such families. Table 2 shows neutral polymorphisms found in control families (not showing an increased frequency of breast and ovarian cancer). Study these data and then answer the questions listed below.

T A B L E 1 Predisposing Mutations in *BRCA1*

		Mutation		Frequency in Control Chromosomes
Kindred	Codon	Nucleotide Change	Coding Effect	
1901	24	−11bp	Frameshift or splice	0/180
2082	1313		Gln → Stop	0/170
1910	1756	Extra C	Frameshift	0/162
2099	1775	T → G	Met → Arg	0/120
2035	NA*	?	Loss of transcript	NA*

*NA indicates not applicable, as the regulatory mutation is inferred and the position has not been identified.

T A B L E 2 Neutral Polymorphisms in *BRCA1*

			Frequency in Control Chromosomes*			
Name	Codon Location	Base in Codon†	A	C	G	T
PM1	317	2	152	0	10	0
PM6	878	2	0	55	0	100
PM7	1190	2	109	0	53	0
PM2	1443	3	0	115	0	58
PM3	1619	1	116	0	52	0

*The number of chromosomes with a particular base at the indicated polymorphic site (A, C, G, or T) is shown.
†That is, position 1, 2, or 3 of the codon.

GENETICS MediaLab

The following resources will help you achieve a better understanding of the concepts presented in this chapter. These resources can be found either on the CD packaged with this textbook or the Companion Website found at **http://www.prenhall.com/klug/**.

CD Resources:

Self-grading Chapter Problems

Web Resources:

Genetics and Society Issue: *The Double-Edged Sword of Genetic Testing: The Case of Breast Cancer*

Web Destinations in Genetics

Self-grading Chapter Problems

Chapter Search Terms

Genetics Newsgroups

Student Bulletin Board

Web Problem 1:

Time for completion = 10 minutes
Do the telomeres of chromosomes hold a key to detecting and perhaps preventing many forms of cancer? The American Cancer Society's article on cloning the human telomerase gene, along with information in Chapter 23, will help you answer the following questions. How are the telomeres implicated in the aging and death of normal cells? What is the significance of telomerase in cancer cells? Why did cloning the telomerase gene in a yeast and a protozoan help in isolating and cloning the human gene? To complete this exercise, visit Web Problem 1 in Chapter 23 of your Companion Website and select the keyword **TELOMERASE**.

Web Problem 2:

Time for completion = 10 minutes
What role does heredity play in susceptibility to cancer? The article on the characteristics of inherited forms of breast cancer, along with information in Chapter 23, will help you answer the following questions. Individuals normally inherit a predisposition to cancer, not the cancer itself; explain how the predisposition is transmitted. Individuals who carry *BRCA1* and *BRCA2* muta-

tions have an increased risk of breast cancer; is the risk associated with a specific protein product of these genes or the lack of a functional product? What is the function of these gene products in unaffected individuals? Are individuals with *BRCA1* and *BRCA2* mutations at higher than average risk for cancers other than breast cancer? To complete this exercise, visit Web Problem 2 in Chapter 23 of your Companion Website and select the keyword **BREAST CANCER**.

Web Problem 3:

Time for completion = 15 minutes
Where are the genes associated with cancer in the human genome? Dr. Philip McClean has produced a map of the human genome with the names and locations of oncogenes and tumor-suppressor genes hot-linked to the Online Mendelian Inheritance in Man database. Are the genes associated with human cancers clustered together or distributed throughout the genome? Pick a chromosome with four or more cancer genes and examine the Online Mendelian Inheritance description of those linked genes. Do closely linked genes seem to cause similar or different cancers? Are most of the mutations you examined oncogenes or tumor-suppressor genes? Do any of the oncogenes you found have an apparent viral origin? How did the virus's genes get into the human DNA? To complete this exercise, visit Web Problem 3 in Chapter 23 of your Companion Website and select the keyword **CANCER MAP**.

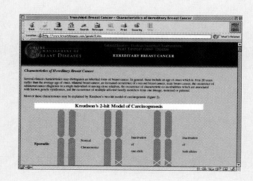

12. (a) Note the coding effect of the mutation found in kindred group 2082 in Table 1. This results from a single base pair substitution. Starting with a drawing of the normal double-stranded DNA sequence for this codon (with the 59 and 39 ends labeled), show the sequence of events that generated this mutation, assuming that it resulted from an uncorrected mismatch event during DNA replication.
   (b) Examine the types of mutations that are listed in Table 1. Is the *BRCA1* gene likely to be a tumor-suppressor gene or an oncogene?
   (c) Although the mutations in Table 1 are clearly deleterious, causing breast cancer in women at very young ages, each of the kindred groups examined had at least one woman who carried the mutation but lived until age 80 without developing cancer. Name at least two different

mechanisms (or variables) that could underlie variation in the expression of a mutant phenotype. Then, in the context of the current model for cancer formation, propose an explanation for the incomplete penetrance of this mutation. How do these variables relate to this explanation? [*Reference: Science* 266:66–71 (1994).]

13. (a) Examine Table 2. What is meant by a neutral polymorphism? What is the significance of this table in the context of examining a family or population for *BRCA1* mutations that predispose an individual to cancer?
   (b) Examine Table 2. Is the polymorphism PM2 likely to result in a neutral missense mutation or a silent mutation?
   (c) Answer question (b) for the polymorphism PM3. [*Reference: Science* 266:66–71 (1994).]

SELECTED READINGS

AALTONEN, L.A. 1993. Clues to the pathogenesis of familial colorectal cancer. *Science* 260: 812–16.

AMES, B., MAGRAW, R., and GOLD, L. 1990. Ranking possible cancer hazards. *Science* 236:71–80.

AMES, B., PROFET, M., and GOLD, L. 1990. Dietary pesticides (99.99% all natural). *Proc. Natl. Acad. Sci. USA* 87:7777–81.

BEIJERSBERGEN, R.L., and BERNARDS, R. 1996. Cell cycle regulation by the retinoblastoma family of growth inhibitory proteins. *Biochim. Biophys. Acta* 1287:103–20.

BENEDICT, W., XU, H., HU, S., and TAKAHASHI, R. 1990. Role of the retinoblastoma gene in the initiation and progression of a human cancer. *J. Clin. Invest.* 85:988–93.

BIECHE, I., and LIDEREAU, R. 1995. Genetic alterations in breast cancer. *Genes Chromosomes Cancer* 14:227–51.

BIRRER, M., and MINNA, J. 1989. Genetic changes in the pathogenesis of lung cancer. *Annu. Rev. Med.* 40:305–17.

BROWN, M.A. 1997. Tumor suppressor genes and human cancer. *Adv. Genet.* 36:45–135.

CAVENEE, W.K., and WHITE, R.L. 1995. The genetic basis of cancer. *Sci. Am.* (Mar.) 272:72–79.

CHEN, J. *et al.* 1998. Stable interaction between the products of the *BRCA1* and *BRCA2* tumor suppressor genes in mitotic and meiotic cells. *Molec. Cell* 2:317–28.

CORNELIS, J.F. et al. 1998. Metastasis. *Am. Scient.* 86:130–41.

COTTER, F. 1993. Molecular pathology of lymphomas. *Cancer Surv.* 16:157–74.

CROCE, C., and KLEIN, G. 1985. Chromosome translocations and human cancer. *Sci. Am.* (Mar.) 252:54–60.

DAMM, K. 1993. ErbA: Tumor suppressor turned oncogene? *FASEB J.* 7:904–09.

EASTON, D. et al. 1993. Genetic linkage analysis in familial breast and ovarian cancer: Results from 214 families. *Am. J. Hum. Genet.* 52:678–701.

ELLEDGE, S.J. 1996. Cell cycle checkpoints: Preventing an identity crisis. *Science* 274:1664–72.

EVAN, G., and LITTLEWOOD, T. 1998. A matter of life and cell death. *Science* 281:1317–22.

FEARON, E.R. 1997. Human cancer syndromes: Clues to the origin and nature of cancer. *Science* 278:1043–50.

FELMAN, M., and EISENBACH, L. 1988. What makes a tumor cell metastatic? *Sci. Am.* (Nov.) 259:60–85.

FEUNTEUN, J., and LENOIR, G.M. 1996. *BRCA1*, a gene involved in inherited predisposition to breast cancer. *Biochim. Biophys. Acta* 1242:177–180.

GIACCONE, G. 1996. Oncogenes and antioncogenes in lung tumorigenesis. *Chest* 109 (Suppl. 5): 130S–35S.

GRIGNANI, F. et al. 1993. The molecular genetics of acute promyelocytic leukemia. *Blood Rev.* 7:87–93.

HABER, D., and BUCKLER, A. 1992. *WT1*: A novel tumor suppressor gene inactivated in Wilms tumor. *New Biol.* 4:97–106.

HASTIE, N. 1993. Wilms' tumour gene and function. *Curr. Opin. Genet. Dev.* 3:408–13.

HATKEYAMA M. et al. 1994. The cancer cell and the cell cycle clock. *Cold Spring Harbor Symp. Quant. Biol.* 59:1–10.

HOROWITZ, J.M. 1993. Regulation of transcription by the retinoblastoma protein. *Genes Chromosomes Cancer* 6:124–31.

HUAN, B., and SIDDIQUE, A. 1993. Regulation of hepatitis B virus gene expression. *J. Hepatol.* 17(Suppl. 3): 520–S23.

KINZLER, K.W., and VOGELSTEIN, B. 1996. Lessons from hereditary colorectal cancer. *Cell* 87:156–70.

———1997. Gatekeepers and caretakers. *Nature* 386: 761–63.

LEAKE, R. 1996. The cell cycle and regulation of cancer cell growth. *Ann. NY Acad. Sci.* 784: 252–62.

LENGAUER, C., KINZLER, K.W., and VOGELSTEIN, B. 1997. Genetic instability in colorectal cancer. *Nature* 386: 623–27.

LOTEM, J., and SACHS, L. 1996. Control of apoptosis in hematopoiesis and leukemia by cytokines, tumor suppressor and oncogenes. *Leukemia* 10:925–31.

NURSE, P. 1997. Checkpoint pathways come of age. *Cell* 91:865–67.

OLSSON, H., and BORG, A. 1996. Genetic predisposition to breast cancer. *Acta Oncol.* 35:1–8.

PELTOMAKI, P., et al. 1993. Genetic mapping of a locus predisposing to human colorectal cancer. *Science* 260:810–12.

PINES, L. 1996. Cyclin from sea urchin to HeLas: Making the human cell cycle. *Biochem. Soc. Trans.* 24:15–33.

RAFF, M. 1998. Cell suicide for beginners. *Nature* 396:119–22.

RUSTGI, A., and PODOLSKY, D. 1992. The molecular basis of colon cancer. *Annu. Rev. Med.* 43:61–68.

SHERR, C.J. 1996. Cancer cell cycles. *Science* 274:1672–77.

SMITH, M.L., and FORNACE, A.J., Jr. 1996. The two faces of tumor suppressor p53. *Am. J. Pathol.* 148:1019–22.

SOLOMON, E. et al. 1987. Chromosome 5 allele loss in human colorectal carcinomas. *Nature* 328:616–29.

STAHL, A. et al. 1994. The genetics of retinoblastoma. *Ann. Genet.* 37:172–178.

STANBRIDGE, E. 1992. Functional evidence for human tumor suppressor genes: Chromosome and molecular genetic studies. *Cancer Summ.* 12:43–57.

TIMAR, J. et al. 1995. Interaction of tumor cells with elastin and the metastatic phenotype. *Ciba Found. Symp.* 192:321–35.

VON LINDERN, M. et al., 1992. Translocation t (6;9) in acute non-lymphocytic leukaemia results in the formation of a DEK-CAN fusion gene. *Baillieres Clin. Haematol.* 5:857–79.

WEINBERG, R.A. 1995. The molecular basis of oncogenes and tumor suppressor genes. *Ann. NY Acad. Sci.* 758:331–338.

———1995. The retinoblastoma protein and cell cycle control. *Cell* 81:323–30.

———1996. How cancer arises. *Sci. Am.* (Sept.) 275:62–70.

———1996. E2F and cell proliferation: A world turned upside down. *Cell* 85:457–59.

WHITE, R. 1992. Inherited cancer genes. *Curr. Opin. Genet. Dev.* 2:53–57.

ZANG, H., TOMBLINE, G., and WEBER, B.L. 1998. *BRCA1, BRCA2*, and DNA damage response: Collision or collusion? *Cell* 92:433–36.

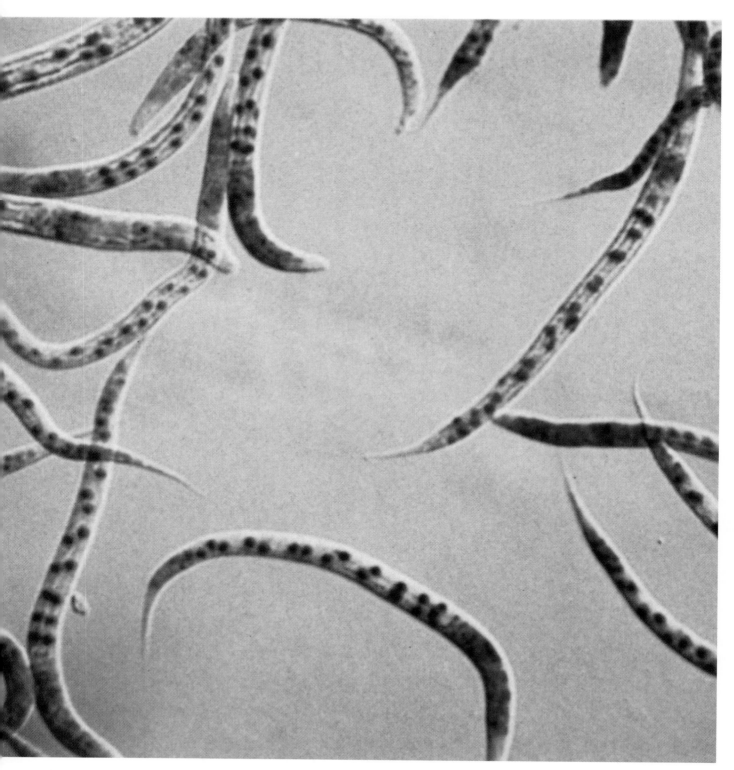

Genetically engineered *Caenorhabditis elegans* roundworms that have turned blue in response to an environmental stress, such as toxins or heat.

Genetics and Behavior

KEY CONCEPTS

- **Behavioral differences in related organisms can be shown to be genetically influenced**

  Alcohol Preference in Mice
  Open-Field Behavior in Mice

- **Modified behavior may be artificially selected from heterogeneous populations**

  Maze Learning in Rats
  Geotaxis in *Drosophila*

- **The effect of single genes that control behavior can be studied**

  Nest-Cleaning in Bees
  Molecular Motors in Bacteria
  Locomotor Behavior in Mice

- ***C. elegans* has served as a model organism in the study of behavior genetics**

- ***Drosophila* provides a unique approach in the study of behavior genetics**

  Mosaic Flies
  Neurogenetics
  Learning in *Drosophila*

- **Behavioral genetics has been studied in humans**

  Single Genes
  Multifactorial Traits

☉ GenCDX

When you see this icon, there are related animations and exercises on the CD accompanying this text.

Behavior is generally defined as a reaction to stimuli or environment. In broad terms, every action, re-action, and response represents a type of behavior. Animals run, remain still, or counterattack in the presence of a predator; birds build complex and distinctive nests; fruit flies execute intricate courtship rituals; plants bend toward light; and humans reflexively avoid painful stimuli as well as "behave" in a variety of ways as guided by their intellect, emotions, and culture.

Even though clear-cut cases of genetic influence on behavior were known in the early 1900s, the study of be-havior was of greater interest to psychologists, who were concerned with learning and conditioning. Although some traits were recognized as innate or instinctive, behavior that could be modified by prior experience received the most attention. Such traits or patterns of behavior were thought to reflect the previous environmental setting to the exclusion of the organism's genotype. This philoso-phy served as the basis of the behaviorist school.

Such thinking provided a somewhat distorted view of the nature of behavioral patterns. It is logical that a geno-type can be expressed in a hierarchy, each level with a unique environmental setting (e.g., cell, tissue, organ, organism, population, surrounding environment) and that a behav-ioral pattern must rely on the expression of the individual genotype for its execution. Nevertheless, the nature–nur-ture controversy flourished well into the 1950s. By that time it became clear that while certain behavioral patterns, par-ticularly in less complex animals, seemed to be innate, oth-ers were the result of environmental modifications limited by genetic influences. The latter condition is particularly true in organisms with more complicated nervous systems.

Since about 1950, studies of the genetic component of behavioral patterns have intensified and support for the importance of genetics in understanding behavior has in-creased. The prevailing view is that all behavior patterns are influenced both genetically and environmentally. The genotype provides the physical basis and/or mental abili-ty essential to execute the behavior and further determines the limitations of environmental influences.

Behavior genetics has blossomed into a distinct spe-cialty within genetics as more and more behaviors have been found to be under genetic control. As we will see, several approaches have been developed that document the contribution genetics makes to specific behaviors. This chapter examines many examples illustrating how behav-iors can be shown to be influenced by genes. There seems to be no question that this topic will be one of the most ex-citing areas of genetics in the years to come.

Behavioral differences in related organisms can be shown to be genetically influenced

Several methodologies have been used to ascertain whether a behavioral trait is influenced genetically. The first ap-proach involves the identification of behavioral differences between genetic strains of the same species or between closely related species. If closely related organisms exist in similar environments and their survival needs are identical, any observed behavioral differences may be due to genotypic differences. This comparative approach, in which different strains are closely scrutinized, will be illustrated with several examples.

Alcohol Preference in Mice

Studies of alcohol preference in mice have compared preference or aversion to ethanol in different strains. In a typical study, the alcohol consumption rates in four strains of inbred mice were measured over a period of 3 weeks. Each strain was presented with seven vessels containing either pure water or alcohol varying in strength from 2.5 to 15.0 percent. The daily consumption was measured. Table 24.1 shows the proportion of alcohol to total liquid consumed on a weekly basis. The examination of these data shows that the C57BL and C3H/2 strains exhibit a preference for alcohol, while BALB/c and A/3 demonstrate an aversion toward it. As the mice have been raised over many generations in a constant environment, the differences in alcohol preference are attributed to genotypic differences between the strains. Presumably, these strains vary only by the fixation of different alleles at particular loci.

Despite the observed differences in alcohol consumption among mouse strains, the underlying mechanisms remain unclear. It has been suggested that differences in alcohol preference, metabolism, and severity of withdrawal symptoms seen in various mouse strains are related to differences in the major enzymes of alcohol metabolism, particularly alcohol dehydrogenase (ADH) and acetaldehyde dehydrogenase (AHD). The major enzymes and their isozymes have been isolated and characterized with respect to biochemical and kinetic properties, and their distribution and activities in different mouse strains have been cataloged. Genetic variants of these enzymes present in different strains of inbred mice have been used to try to establish an association between a given form of the enzyme and a form of alcohol-related behavior. To date, no correlation between alcohol preference or metabolism and these biochemical markers has been established. Other genetic markers, including the neuropharmacological effects of alcohol, may be needed to establish a link between specific genes and alcohol-related behavior in mice.

Open-Field Behavior in Mice

Open-field tests are used to study exploratory and emotional behavior in mice. When placed in a new environment, mice normally explore their surroundings, but they are a bit cautious or "nervous" about the new setting. The latter response is evidenced by their elevated rate of defecation and urination. To study this behavior in the laboratory, an enclosed, brightly illuminated box with the floor marked into squares is used. Exploration is measured by counting movements into different squares, and emotion is measured by counting the number of defecations.

As in the study of alcohol preference, different inbred strains of mice vary significantly in their response to the open-field setting. John C. DeFries and his associates have centered their work on two strains, BALB/cJ and C57BL/6j. The BALB strain is homozygous for a coat color allele, c, and is albino, whereas the C57 strain has normal pigmentation (CC). BALB mice have low exploratory activity and are very emotional, whereas C57 mice are active in exploration and relatively unemotional.

To test for genotypic differences, the two strains were crossed and then interbred for several generations. Each generation beyond the F_1 contained albino and nonalbino mice, and these were tested for open-field behavior. In all cases, pigmented mice behaved as strain C57, whereas albino mice behaved as BALB. The general conclusion is that the c allele behaves pleiotropically, affecting both coat color and behavior.

Heritability analysis (see Chapter 5) has been used to assess the input of the c gene to these behavioral patterns. Analysis reveals that that this locus accounts for 12 percent of the variance in open-field activity and 26 percent of the variance in defecation-related emotion. The heritability values indicate that these behaviors are polygenically controlled.

TABLE 24.1 Alcohol Consumption in Mice

Strain	Week	Proportion of Absolute Alcohol to Total Liquids	$\bar{x}$
C57BL	1	0.085	
	2	0.093	9.4% alcohol
	3	0.104	
C3H/2	1	0.065	
	2	0.066	6.9% alcohol
	3	0.075	
BALB/c	1	0.024	
	2	0.019	2.0% alcohol
	3	0.018	
A/3	1	0.021	
	2	0.016	1.7% alcohol
	3	0.015	

Source: Modified from Rogers and McClearn, 1962. Reprinted by permission from *Quarterly Journal of Studies on Alcohol*, Vol. 23, pp. 26–33, 1962. Copyrighted by Journal of Studies on Alcohol, Inc. New Brunswick, NJ 08903.

To assess the relationship between albinism and behavior, albino and nonalbino mice were tested for open-field behavior under white light and red light. Red light provides little visual stimulation to mice. The behavioral differences between mice with the two types of coat pigmentation disappeared under red light, indicating that the open-field responses of albino mice are visually mediated. This is not surprising; albino mice are photophobic, and the lack of pigmentation in albinos extends to the iris as well as to the coat.

Modified behavior may be artificially selected from heterogeneous populations

In a second major approach used in behavior genetics, organisms are artificially selected from a genetically heterogeneous population that contains members which exhibit a behavioral trait. If genetic strains can be established where this behavior is uniformly expressed, and then this trait can be transferred by genetic crosses to another strain that initially does not exhibit the behavior under study, the positive influence of the genotype is established. The study of maze learning in rats and geotaxis in *Drosophila* will be used to illustrate this approach.

Maze Learning in Rats

The first experiment of this kind was reported by E. C. Tolman in 1924. He began with 82 white rats of heterozygous ancestry and measured their ability to "learn" to obtain food at the end of a multiple-T maze (Figure 24.1) by recording the number of errors and trials. When first exposed to the maze, a rat explores all alleys and eventually arrives at the end, to be rewarded with food. In succeeding trials, fewer and fewer mistakes are made as the rat learns the correct route. Eventually, a hungry rat may proceed to the food with no errors.

From the initial 82 rats, nine pairs of the "brightest" and "dullest" rats were selected and mated to produce two lines. In each generation, selection was continued. Even in the first generation, Tolman demonstrated that he could select and breed rats whose offspring performed more efficiently in the maze. Subsequently, his approach was pursued by others, notably R. C. Tryon, who in 1942 published results of 18 generations of selection.

As shown in Figure 24.2, two lines with significant differences in maze-learning ability were established. There is some variation around the mean, but by the eighth generation there was no overlap between lines. That is, the dullest of the bright rats were superior to the brightest of the dull rats by the eighth generation.

These two lines were also used to study other behavior traits, with varying results. Bright rats were found to be better in solving hunger-motivation problems but inferior in escape-from-water tests. Bright rats were also found to be more emotional in open-field experiments. These results show that selection of genetic strains superior in certain traits is possible, but care must be taken not to generalize such studies to overall intelligence, which is composed of many learning parameters.

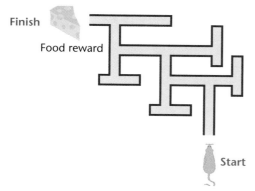

FIGURE 24.1 A multiple T-maze used in learning studies with rats.

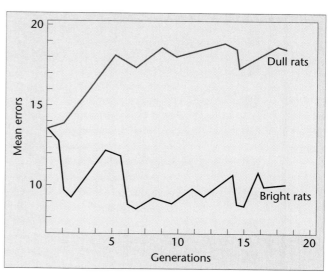

FIGURE 24.2 Selection for the ability and inability of rats in learning to negotiate a maze.

Geotaxis in Drosophila

A taxis is the movement toward or away from an external stimulus. The response may be positive or negative, and the sources of stimulation may include chemicals (chemotaxis), gravity (geotaxis), light (phototaxis), and so on.

To investigate geotaxis in *Drosophila*, Jerry Hirsch and his colleagues designed a mass-screening device that tests about 200 flies per trial, as shown in Figure 24.3. In this test, flies are added to a vertical maze. Flies that turn up at each intersection will arrive at the top of the maze; those that always turn down will arrive at the bottom; and those making both "up" and "down" decisions will end up somewhere in between. Flies can be selected for both positive and negative geotaxis, establishing a genetic influence on this behavioral response.

As shown in Figure 24.4, mean scores can vary from about +4 to −6, corresponding to the number of T-junctions the fly encounters. These data show that selection for negative geotaxis is stronger than for positive geotaxis. The two lines have now undergone selection for almost 30 years, encompassing over 500 generations, and the testing of more than 80,000 flies. Throughout the experiment, clear-cut but fluctuating differences have been observed. These results indicate that geotaxis in *Drosophila* is controlled in a polygenic fashion.

Hirsch and his colleagues have analyzed the relative contribution of loci on different chromosomes to geot-

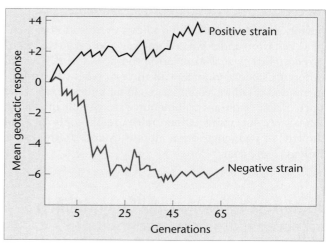

FIGURE 24.4 Selection for positive and negative geotaxis in *Drosophila* over many generations.

axis. Loci on chromosomes 2 and 3 and the X were identified in an ingenious way. Crosses were performed to produce flies that were either heterozygous or homozygous for a given chromosome. Figure 24.5 shows how this is accomplished. From a selected line, a male is crossed to a "tester" female whose chromosomes are each marked with a dominant mutation. Each marked chromosome carries

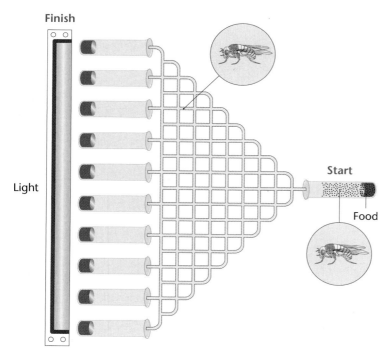

FIGURE 24.3 Schematic drawing of a maze used to study geotaxis in *Drosophila*.

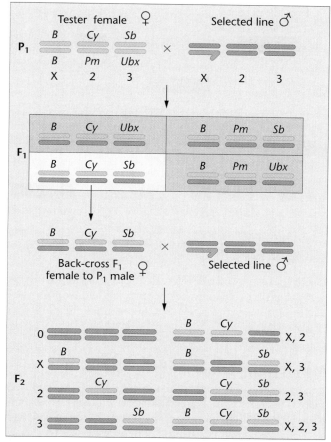

FIGURE 24.5 In this mating scheme in *Drosophila*, the effect of genes located on specific chromosomes that contribute to geotaxis can be assessed. The progeny produced by backcrossing the F$_1$ female contains all combinations of chromosomes. Examining the phenotypes of these flies makes it possible to determine which chromosomes from the selected strain are present. Subsequent testing for geotaxis is then performed (B = Bar eyes; Cy = Curly wing; Sb = Stubble bristles; Ubx = ultrabithorax). The designations alongside each genotype (e.g., X, 2, and 3) indicate which chromosomes are heterozygous.

maze, it is possible to assess the behavioral influence of genes on any given chromosome.

For negatively geotaxic flies (the ones that go up in the maze), the genes on the second chromosome make the largest contribution to the phenotype, followed by loci on chromosome 3 and the X (a gradient of chromosomes 2 > 3 > X). For the positive line (flies that go down in the maze), the reverse arrangement (a gradient of chromosomes X > 3 > 2) is true. The overall results indicate that geotaxis is under polygenic control and that the loci controlling this trait are distributed on all three major chromosomes of *Drosophila*.

Further genetic testing in which each chromosome from a selected line has been isolated in homozygous form in an unselected background has been used to estimate the number of genes controlling the geotaxic response in *Drosophila*. This work indicates that a small number of genes, perhaps two to four loci, are responsible for this behavior.

The effect of single genes that control behavior can be studied

The two approaches discussed above identify behavioral patterns as being under genetic control. Genetic crosses can then be used to establish inheritance patterns. In many cases, such studies have revealed that the behavior under investigation does not exhibit a simple Mendelian pattern of inheritance. Instead, many behavioral traits appear to be polygenic (i.e., where more than one gene influences the behavioral trait). A third approach used in behavior genetics, used extensively today, attempts to isolate and study the effects of single genes on behavior.

Using inbred strains of organisms with relatively constant genotypes, spontaneous or induced mutants with behavioral deviations are collected and analyzed. Although a behavioral trait may be controlled by many genes, a mutation in a single gene may alter that behavior. Genetic analysis of the mutant allele can provide insight into the genetic control of single components of a behavioral trait. This approach offers the most definitive information concerning the role of genes in behavior. In selected organisms, defining a gene-controlling behavior is a prelude to the isolation, cloning, and molecular characterization of such genes.

Nest-Cleaning in Bees

Honeybee nests are frequently infected with *Bacillus* larvae, a bacterium causing foulbrood disease. The disease may be controlled by hygienic behavior on the part of worker bees. To combat infection, worker bees open the cells containing infected larvae and remove them from the hive.

an inversion to suppress the recovery of any crossover products. An F$_1$ female heterozygous for each chromosome is backcrossed to a male from the original line. The resulting female offspring contain all combinations of the chromosomes. The dominant mutations make it possible to recognize which chromosomes from the selected lines are present in homozygous or heterozygous configurations.

Combinations O and X (Figure 24.5) are homozygous and heterozygous, respectively, for the X chromosome. Similar combinations exist for chromosome 2 (0 and 2) and chromosome 3 (0 and 3). By testing these flies in the

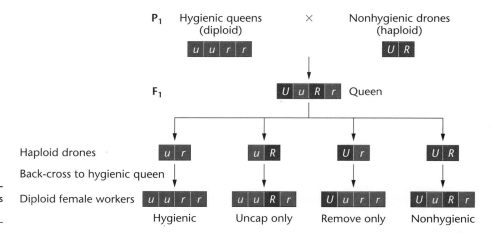

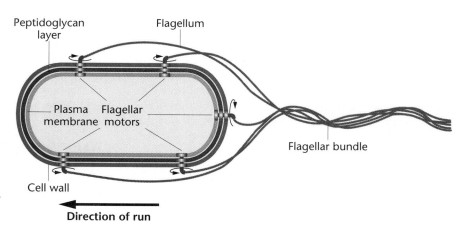

FIGURE 24.6 Results of a honey-bee cross between hygienic diploid females and nonhygienic haploid males.

Hives in which the workers exhibit hygienic behavior are resistant to infection, whereas hives containing workers that do not display this removal behavior are susceptible to the disease.

In 1964, Walter Rothenbuhler published the results of his cross between a hygienic (Brown) line with a non-hygienic (Van Scoy) line. This work strongly favors the idea that either a gene complex or two independently assorting recessive alleles (*u* and *r*) are responsible for hygienic behavior.

The F_1 offspring of a cross between hygienic and nonhygienic bees were all nonhygienic. However, when F_1 males were backcrossed to hygienic females, four phenotypes were produced in roughly equal proportions, as shown in Figure 24.6. While one phenotypic class was hygienic and one was nonhygienic, the other two classes had intermediate phenotypes. One could uncap cells but not remove infected larvae. The other class was able to remove larvae if the cells were artificially uncapped. It appears that one gene pair (*u/u*) controls the uncapping behavior, and a second gene (*r/r*) determines the ability to remove larvae.

Molecular Motors in Bacteria

Behavioral responses also exist in single-celled organisms such as bacteria, which lack an organized nervous system. Bacteria exhibit chemotaxis and are attracted to or repelled by a variety of chemical stimuli. These responses have now been analyzed in some detail, and mutants that disrupt the normal behavior have been isolated and characterized.

Motile bacteria such as *Escherichia coli* and *Salmonella* respond to gradients of chemicals by moving along the gradient. This movement is controlled by flagellar action (Figure 24.7). As they move in response to a chemical gradient, the cells exhibit two forms of motion. The

FIGURE 24.7 When the flagellar motors rotate in a counterclockwise (CCW) direction (as shown), the flagella form a bundle that works to push the cell along a mostly linear path, called a run. When the flagellar motors rotate in a clockwise direction (CW) (not shown), the flagella operate independently, causing the cell to tumble.

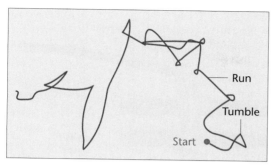

FIGURE 24.8 Runs and tumbles executed by an *E. coli* cell in a medium where no chemical gradient is present. Tracking of the cell began at the large dot at lower right. The cell moves via a series of runs interspersed with shorter periods of tumbling.

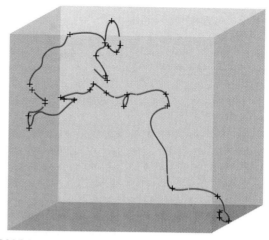

FIGURE 24.9 Runs and tumbles executed by a single *E. coli* cell in an exponential gradient of attractant (highest concentration at top). Movement up the gradient occurred in 35 runs and 34 tumbles. Two extended runs (indicated by the solid lines) represent most of the vertical distance traveled.

first type of movement consists of periods of smooth swimming, called runs, generated by counterclockwise (CCW) rotation of flagella, which act like propellers, driving the bacteria through the medium. During runs, the flagella coil together to form a cohesive bundle. The movement of this bundle drives the cell in a single direction. Also during runs, which have an average duration of 1 second, the bacteria swim about 10 to 20 body lengths. Runs alternate with the second form of movement, called tumbles. During tumbles, the flagella switch direction and rotate clockwise (CW). In clockwise rotation, the flagella become dispersed, and each acts independently. As a result, during tumbles, the cell covers little or no distance (Figure 24.8). Tumbles, which last only about 0.1 second, serve to change the direction of the cell. In the presence of a chemical attractant, runs that carry the cell up the gradient are extended, whereas those that move the cell down the gradient are not (Figure 24.9). In the presence of a chemical repellent, runs that carry the cell down the gradient are extended.

The molecular details of the chemotactic response are becoming known, with the major components of the system identified and the role of several gene products described. The response of bacteria to chemical stimuli serves as a model for the molecular mechanisms by which cells process and transduce sensory signals. In this case, chemical signals are converted into the kinetic action of flagella, moving the cell toward or away from a stimulus. *E. coli* has at least four types of membrane-spanning receptors that sense chemicals in the environment and initiate intracellular responses (Figure 24.10). These receptor proteins are the products of a gene family called transducers. Each protein in the family has a unique receptor domain extending into the periplasmic space and a conserved domain that extends into the cytoplasm. The

cytoplasmic surface of the chemoreceptor protein forms a complex with two cytoplasmic proteins, CheA and CheW. CheA and the chemoreceptor interact directly, and CheW enhances this association. CheA integrates the response of the cell to chemical stimuli. The CheA mutants are nonchemotactic.

A change in the conformation of the chemoreceptor caused by the binding or release of a sensory molecule generates an intracellular signal. This signal is transmitted by changes in the phosphorylation of a group of cytoplasmic proteins. These proteins act as a molecular switch to change the direction (clockwise or counterclockwise) of flagellar rotation. In the presence of a repellent molecule, the rate of CheA phosphorylation is increased (Figure 24.10). Phosphorylated CheA, in turn, leads to the phosphorylation of CheY. When CheY is phosphorylated, it binds to the base of the flagellar motor and favors clockwise rotation of the flagellum. Recall that clockwise rotation causes the cell to tumble, during which the cell switches direction. Another protein, CheZ, turns off the action of CheY by accelerating the dephosphorylation of CheY.

In the presence of an attractant, the cycle of phosphorylation is reversed, decreasing the rate of CheA phosphorylation, which in turn slows the phosphorylation of CheY. Dephosphorylated CheY favors counterclockwise rotation of the flagella, moving the cell on a run toward the attractant. The result of this behavior is the detection of

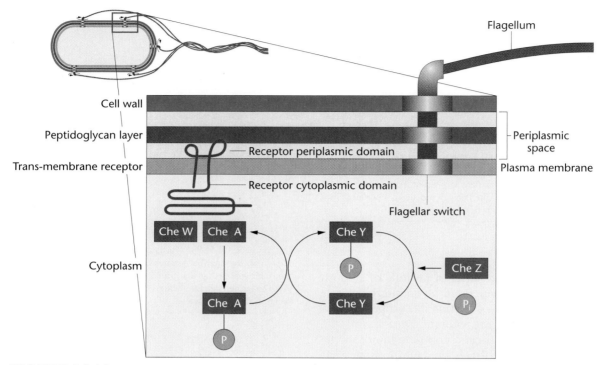

FIGURE 24.10 Sensory transduction during bacterial chemotaxis. Stimulation of the transmembrane receptor leads to the phosphorylation of CheA with the aid of CheW. Activated CheA transfers a phosphate group to CheY. Phosphorylated CheY interacts with the flagellar switch to alter the direction of flagellar rotation. CheZ inactivates CheY by removing the phosphate group.

chemical gradients, and movement toward attractants (food) and away from repellents.

The chemoreceptors are aggregated into clusters, located at the poles of the cell. They are not clustered near the base of the flagellar motors. As expected, CheA and CheW proteins are also localized in the cytoplasm at the poles, with the CheA protein associated with the inner membrane of the cell. In mutants lacking all four receptors, CheA and CheW proteins are randomly distributed throughout the cytoplasm, indicating that the polar localization of these gene products depends on the presence of receptors. Studies of CheW mutants indicate that this protein is also required for the aggregation and polar localization of the receptors. In CheW deletion mutants, the CheA receptor complexes are randomly distributed. These results suggest that signal transduction is not initiated by the formation of the complex, but rather through conformational changes in the complex brought about by sensory molecules.

The chemotactic behavior of *E. coli* involves a number of gene products that receive and process signals and generate a response through a metabolic network that involves protein phosphorylation. At the present time, chemotaxis in bacteria represents the best-known example of the relationship between a behavior and its underlying molecular mechanism.

Locomotor Behavior in Mice

One of the oldest recorded behavior mutants is the waltzer mutation in the mouse, recorded in 80 B.C. in China. This mutant was described as a mouse "found dancing with its tail in its mouth." Waltzing mice run in a tight circle and demonstrate both horizontal and vertical head shaking, as well as hyperirritability. Some waltzers are also deaf.

Genetic crosses between mutant and normal house mice (*Mus musculus*) reveal a simple recessive inheritance pattern characteristic of Mendel's monohybrid matings. Investigation of the inner ear of mutants has revealed degeneration of both the cochlea and semicircular canals, accounting for the deafness and circling behavior. This is an

TABLE 24.2 Inherited Neurological Defects in Mice

Group	Name
Waltzer-shaker	*Waltzer, shaker, pirouette, jerker, fidget, twirler, zig-zag*
Convulsive	*Trembler, tottering, spastic*
Incoordination	*Jumping, quaking, reeler, staggerer, leaner, Purkinje cell degeneration, cerebral degeneration*

example of a mutation causing a structural anomaly that, in turn, alters behavior.

Many other single-gene behavior effects are known in the mouse, an organism particularly well characterized genetically. Of over 300 mutants discovered representing about 250 loci, over 90 are neurological in nature and alter behavior. These are classified into three groups of syndromes, as shown in Table 24.2. The neurological bases of many of these have now been determined. The incoordination mutants' quaking and jumping are due to faulty myelination of nervous tissue. Quaking is an autosomal recessive mutation, and jumping is inherited as an X-linked recessive. The former cannot synthesize adequate myelin, while the latter demonstrates a degenerative process involving myelin. Cerebral degeneration is a mutation exhibiting progressive deterioration of behavior. As its name implies, part of the brain deteriorates.

The study of such mutants clearly establishes the genetic basis of normal neurological development.

C. elegans has served as a model organism in the study of behavior genetics

In 1968, Sydney Brenner began an investigation of behavior genetics in the nematode *Caenorhabditis elegans*. In his earlier work, Brenner made important contributions to the understanding of DNA replication, F factors, mRNA, and the genetic code. When he turned to the study of behavior, he hoped that it would be possible to dissect genetically the nervous system of *C. elegans* using techniques previously applied successfully to other organisms.

He chose this nematode because it has a relatively simple nervous system. Adult worms are about 1 mm long and contain only 959 somatic cells, about 350 of which are neurons. As a result, it is possible to cut serial sections of an embedded organism and reconstruct the entire organism and its nervous system in the form of a three-dimensional model. Brenner then hoped to induce a large

number of behavioral mutations and to correlate aberrant behavior with structural and biochemical alterations in the nervous system. Some progress has been made toward these goals, particularly in isolating large numbers of mutations. In early experiments, three general types of behavioral mutants were characterized. Worms are positively chemotactic to a variety of stimuli (cyclic AMP and GMP; anions such as Cl^-, Br^-, and I^-, and cations such as Na^+, Li^+, K^+, and Mg^{++}). As shown in Figure 24.11, positive attraction can be tracked in gradients on agar plates. The study of chemotactic mutants has shown that sensory receptors in the head alone mediate the orientation responses to attractants.

A second class of behavior studied involves thermotaxis. Cryophilic mutants move toward cooler temperatures, and thermophilic mutants move toward warmer temperatures. However, this behavior has not yet been correlated with the responsible component of the nervous system.

A third class of behavior involves generalized movement on the surface of an agar plate. Of 300 induced mutations, 77 affected the movement of the animal. Whereas wild-type worms move with a smooth, sinuous pattern, mutants are either uncoordinated (*unc*) or rollers (*rol*). Those that are uncoordinated vary from the display of par-

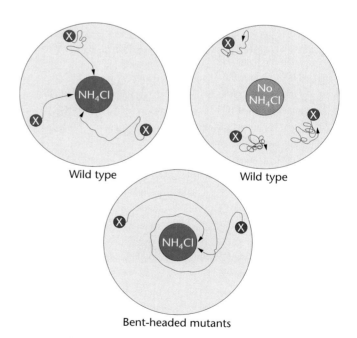

FIGURE 24.11 Chemotactic response to ammonium chloride (NH_4Cl) of wild-type and mutant *Caenorhabditis*.

tial paralysis to small aberrations of movement, including twitching. Rollers move by rotating along their long axis, creating circular tracks on an agar surface. Many of these mutants have been correlated with defects in the dorsal or ventral nerve cord or in the body musculature. Mutants are distributed on six linkage groups, corresponding to the haploid number of chromosomes characteristic of this organism. Numerous *unc* mutants are found on each of the six chromosomes, indicating extensive genetic control of nervous system development.

As the genetics of *C. elegans* has developed, the analyses of the links among genes, the nervous system, and behavior have become more refined, beginning with a complex behavioral repertoire and isolating genes that control this behavior. The recent sequencing of the *C. elegans* genome will accelerate the identification and isolation of the genes that control behavior.

As an example of another genetic approach, feeding behavior in the worm can be used to study behavior. In *C. elegans*, the pharynx is a self-contained feeding pump composed of three parts: (1) the corpus, which ingests bacteria, (2) the isthmus, which conducts bacteria to the terminal bulb, and (3) the terminal bulb, where bacteria are ground up and passed into the intestine (Figure 24.12). The pharynx is surrounded by a basement membrane and contains a total of 60 cells (60 out of the 959 in the organism), 20 of which are neurons making up the pharyngeal nervous system.

The pharyngeal nervous system is somewhat self-contained, with only one pair of nerves connecting it to the rest of the nervous system; this connection can be severed without impairing the action of the pharyngeal nerves. Only 1 of the 20 neurons, M4, which innervates the posterior isthmus, is essential for life. When M4 is missing, the isthmus remains closed, bacteria are not transported for grinding and digestion, and starvation ensues. Worms with an intact M4 neuron, but missing the 19 others, are viable, although these neurons are necessary for normal patterns of feeding.

In an ongoing study of the genetics of feeding, Leon Avery has set out to isolate and characterize the genes that control the presence or absence, developmental fate, patterns of innervation, and function of the 20 neurons responsible for feeding behavior. Almost 38,000 progeny of mutagenized worms were screened for feeding-defective mutants. By linkage and mapping studies, 52 mutations were assigned to 35 genes located on all six chromosomes. It is estimated that at least 60 genes are involved in feeding behavior in *C. elegans*, meaning that roughly half the genes remain to be identified.

The 52 mutations recovered to date fall into three broad phenotypic classes. The *eat* mutants affect the motion of the pharyngeal muscles; in addition, some *eat* mutants also affect the function of body-wall muscles. The *pha* mutants have misshapen pharynxes that prevent normal feeding behavior. The third class of mutants, *phm* mutants, all have weak and irregular pharyngeal muscle contractions. The distribution of mutants is shown in Table 24.3. Preliminary work indicates that many of the *eat* mutants may affect the function of the nervous system or muscle functions that control contraction. In addition, *phm* mutants may have defects in muscle contraction. The *pha* mutants may represent defects in morphogenesis or the specification of cell fate.

Work is now proceeding on the isolation, cloning, and characterization of the genes affecting muscle excitability and neuron function. Many of the genes cloned to date encode a diverse array of ion channels, or components of signal transduction systems. Results from this study are providing insight

(a)

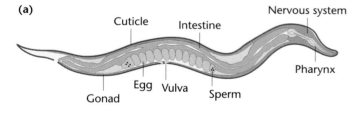

(b)

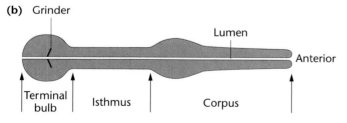

FIGURE 24.12 The pharynx of *C. elegans*. (a) Body plan of intact animal, showing location and arrangement of internal organs, including the pharynx. (b) Pharynx, showing three regional divisions and associated structures.

TABLE 24.3 Distribution of Pharynx Mutants in *C. elegans* by Phenotypic Class

Mutations/Gene	*eat*	*pha*	*phm*
1	16	2	5
2	7	0	2
3	1	0	0
4	2	0	0
Total genes:	26	2	7
Total mutations:	41	2	9

Source: Modified from Avery (1993).

into molecular events through which behavior is mediated by both the nervous system and the muscular system.

Drosophila provides a unique approach in the study of behavior genetics

We have already discussed geotaxis in *Drosophila* as an example of the selection methodology, but much more extensive information on the behavior genetics of this organism is available. This is not at all surprising because of our extensive knowledge of its genetics and the ease with which *Drosophila* can be manipulated experimentally.

As early as 1915, Alfred Sturtevant observed that the X-linked recessive gene *yellow* affects mating preference in *Drosophila* females, in addition to its effects on body pigmentation. Sturtevant found that both wild-type and *yellow* females, when given the choice of wild-type or *yellow* males, prefer to mate with wild types. Both wild-type and *yellow* males prefer to mate with *yellow* females. These conclusions were based on measurements of mating success in all combinations of *yellow* and wild-type (gray-bodied) males and females.

In 1956, Margaret Bastock extended these observations by investigating which, if any, component of courtship behavior was affected by the *yellow* mutant gene. Courtship in *Drosophila* is a complex ritual. The male first shows orientation. The male follows the female, often circling her, and then orients at a right angle to her body and taps her on the abdomen. Once he has her attention, the male begins wing display. He raises the wing closest to the female and vibrates this wing rapidly for several seconds. He then moves behind her, and makes contact with her genitalia. If she signals acceptance by remaining in place, he mounts her and copulation occurs.

Bastock compared the courtship ritual in wild-type and *yellow* males. The *yellow* males prolong orientation but spend much less time in the vibrating and genital contact phases. Her observations indicate that the *yellow* mutation alters the pattern of male courtship, making these males less successful in mating.

Mosaic Flies

In 1967, Seymour Benzer and his colleagues initiated a study of behavior genetics in *Drosophila*. Benzer's approach uses mutant alleles of behavior to "dissect" a complex biological phenomenon into its simpler components. This approach has several goals.

1. Isolate mutants that disrupt normal behavior.
2. Identify the mutant genes by chromosome localization and mapping.

3. Determine the structural component at which gene expression influences the behavioral response.
4. Establish a causal link between the mutant allele and its behavioral phenotype.

All four steps can be illustrated in a discussion of phototaxis, one of the first behaviors studied by Benzer. Normal flies are positively phototactic; that is, they move toward a light source. Mutations were induced by feeding male flies sugar-water containing EMS (ethylmethanesulfonate, a potent mutagen) and mating them to attached-X virgin females. As shown in Figure 24.13, the F_1 males receive their X chromosome from their fathers. Because they are hemizygous, induced X-linked recessive mutations are phenotypically expressed. The

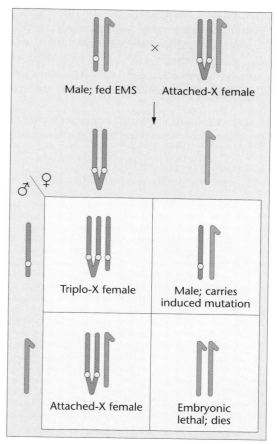

FIGURE 24.13 Genetic cross in *Drosophila* that facilitates the recovery of X-linked induced mutations. The female parent contains two X chromosomes that are attached, in addition to a Y chromosome. In a cross between this female and a normal male that has been fed the mutagen ethylmethanesulfonate, all surviving males receive their X chromosomes from their father and express all mutations induced on that chromosome.

F_1 males were tested for phototaxic responses, and those with abnormal behavior were isolated. Benzer found runner mutants, which move quickly to and from light; negative phototactic mutants, which move away from light; and nonphototactic mutants, which show no preference for light or darkness. The genetic basis of these behavioral changes was confirmed by mating these F_1 males to attached-X virgin females. The male progeny of this cross also showed the abnormal phototactic responses, indicating that these behavioral alterations are the result of X-linked recessive mutations.

Nonphototactic mutants walk normally in the dark, but show no phototactic response to light. Benzer and Yoshiki Hotta tested electrical activity at the surface of mutant eyes in response to a flash of light. The pattern of electrical activity was recorded as an electroretinogram. Various types of abnormal responses were detected in different nonphototactic mutant strains. None of the mutants had a normal pattern of electrical activity. When these mutations were mapped, they were not all allelic; instead, they were shown to occupy several loci on the X chromosome. This analysis indicates that several gene products contribute to the phototactic response.

Where, within the fly, must gene expression occur to produce a normal electroretinogram? In an ingenious ap-

proach to answer this question, Benzer turned to the use of mosaics. In mosaic flies, some tissues are mutant and others are wild type. If it can be ascertained which part must be mutant in order to yield the abnormal behavior, the primary focus of the genetic alteration can be determined.

To produce mosaic flies, Benzer used a *Drosophila* strain that carries one of its X chromosomes in an unstable ring shape. When the ring-X is present in a zygote undergoing cell division, it is frequently lost by nondisjunction. If the zygote is female and has two X chromosomes (one normal and one ring-X), loss of the ring-X at the first mitotic division will result in two cells—one with a single X (normal X) and one with two X chromosomes (one normal and one ring-X). The former cell goes on to produce male tissue (XO) and expresses all alleles on the remaining X, whereas the latter produces female tissue and does not express heterozygous recessive X-linked genes. This is illustrated in Figure 24.14. The loss of the ring-X results in mosaic flies with male and female parts that differ in the expression of X-linked recessive genes.

When and where the ring-X is lost in development determines the pattern of mosaicism. The loss usually occurs early in development, before the cells migrate to the embryo surface. As shown in Figure 24.15, different types of mosaics are produced, depending on the orien-

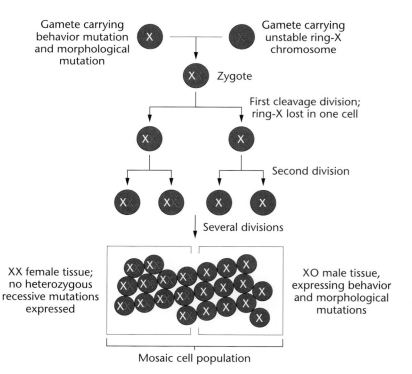

FIGURE 24.14 Production of a mosaic *Drosophila* as a result of fertilization by a gamete carrying an unstable ring-X chromosome (shown in color). If this chromosome is lost in one of the two cells following the first mitotic division, the body of the fly will consist of one side that is male (XO) and the other side that is female (XX). The male side will express all mutations contained on the X chromosome.

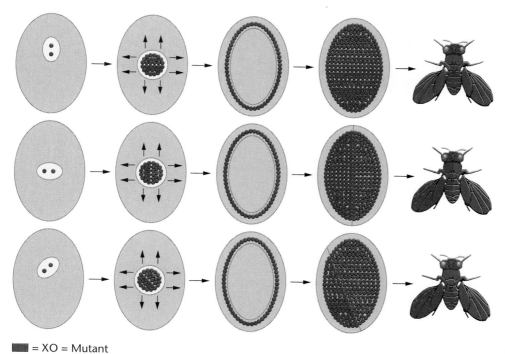

= XO = Mutant

= XX = Wild type

FIGURE 24.15 The effect of spindle orientation on the production of mosaic flies as shown in Figure 24.14.

tation of the spindle when loss of the ring-X takes place. If the normal X chromosome carries the behavior mutation and an obvious mutant allele (*yellow*, for example), the pattern of mosaicism will be easy to distinguish. By examining the distribution of body color, flies with a combination of normal and mutant structures can be identified, such as a fly with a mutant head on a wild-type body, or a normal head on a mutant body, or a fly with one normal and one mutant eye on a normal or mutant body, and so on.

When nonphototactic mosaics were studied, the focus of the genetic defect was found to be in the eye itself. In mosaics where every part of the fly except the eye was normal, abnormal behavior was still detected. When one eye was mutant, and the other normal, the fly had an unusual behavior. Instead of crawling straight up toward light as the normal fly does, the mosaic fly with one mutant eye crawls upward to light in a spiral pattern. In the dark, this fly moves in a straight line.

Using a combination of genetics, physiology, and biochemistry, Benzer and other workers in the field have identified and analyzed a large number of genes affecting behavior in *Drosophila*. As shown in Table 24.4, mutations that affect locomotion, response to stress, circadian rhythm, sexual behavior, visual behavior, and even learning have been isolated. In keeping with the long-standing tradition of naming mutants in *Drosophila* genetics, some

of these mutations have received very descriptive and often humorous names.

Many mutations have been analyzed with the mosaic technique to localize the focus of gene expression. While it was easy to predict that the focus of the nonphototactic mutant would be in the eye, other mutations are not so predictable. For example, the focus of mutations affecting circadian rhythms has been located in the head, presumably in the brain. A mutant with a wings-up phenotype might have a defect in the wings, muscle attachment sites in the thorax, abnormal muscle formation, or have a neuromuscular defect. The analysis of mosaics established that the defect is in flight muscles of the thorax. Cytological studies have confirmed this finding, showing a complete lack of myofibrils in these muscles.

The mosaic technique has also been used to determine which regions of the brain are associated with sex-specific aspects of courtship and mating behavior. Jeffrey Hall and his associates have shown that mosaics with male cells in the most dorsal region of the brain, the protocerebrum, exhibit the initial stages of male courtship toward females. Later stages of male sexual behavior, including wing vibrations and attempted copulations, require male cells in the thoracic ganglion. Similar studies of female-specific sexual behavior have shown that the ability of a mosaic to induce courtship by a male depends on female cells in the posterior thorax or abdominal region. A region of the brain

TABLE 24.4 Some Behavioral Mutants of *Drosophilia*

Class	Name	Characteristics
Locomotor	*sluggish*	Moves slowly
	hyperkinetic	High consumption of O_2; shaking of legs; early death
	wings-up	Wings perpendicular to body
	flightless	Does not fly well, although wings are well developed
	uncoordinated	Lacks coordinated movements
	nonclimbing	Fails to climb
	easily shocked	Mechanical shock induces "coma"
	stoned	Stagger induced by mechanical shock
	Shaker	Vibrates all legs while etherized
	freaked out	Grotesque, random gyrations under the influence of ether
Response to stress	*paralyzed*	Collapses above a critical temperature
	parched	Dies quickly in low humidity conditions
	tko	Epileptic-like response
	comatose	Paralyzed by cool temperature
	out-cold	Similar to comatose
Circadian rhythm	*period^o*	Eclosion at any time; locomotor activity spread randomly over the day
	period^s	19-hour cycle rather than 24-hour cycle
	period^l	28-hour cycle rather than 24-hour cycle
Sexual	*savoir-faire*	Males unsuccessful in courtship
	fruitless	Males pursue each other
	stuck	Male is often unable to withdraw after copulation
	coitus interruptus	Males disengage in about half the normal time
Visual	*nonphototactic*	Blind
	negatively phototactic	Moves away from light
Learning	*dunce*	Fails to learn conditioned response

within the protocerebrum must be female for receptivity to copulation. Anatomical studies have confirmed that there are fine structural differences in the brains of male and female *Drosophila*, indicating that some forms of behavior may be dependent on the development and maturation of specific parts of the nervous system.

Neurogenetics

The analysis of behavior mutants in *Drosophila* has led to an understanding of fundamental mechanisms in the animal nervous system. Nerve impulses are generated at one end of a neuron and move along the cell to the opposite end (Figure 24.16). During this process, sodium

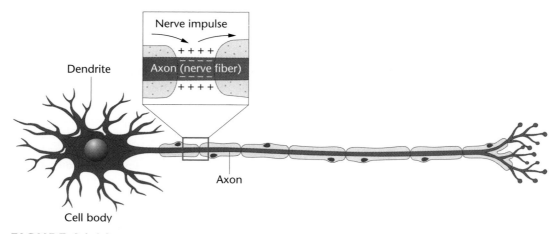

FIGURE 24.16 A nerve cell (neuron) has a cell body and extensions called dendrites that carry impulses toward the cell body, and one or more axons that carry impulses away from the cell body. Electrodes placed on either side of the plasma membrane can record the electric potential across the membrane. At rest, there is more sodium outside the cell, and more potassium inside the cell. When a nerve impulse is generated, sodium moves into the cell, and potassium moves out. As the impulse moves away, ions are pumped across the membrane to restore the electric potential.

and potassium ions move across the plasma membrane. The movement of these electrically charged ions can be monitored by measuring changes in the electrical potential of the neuron's membrane. To screen for genes that control the generation and transmission of nerve impulses, Barry Ganetzky and his colleagues screened behavioral mutants of *Drosophila* to identify those with electrophysiological abnormalities in the generation and propagation of nerve impulses. Two general classes of such mutants have been isolated, those with defects in the movement of sodium and those with defects in potassium transport (Table 24.5).

One mutant identified in this screening procedure is a temperature-sensitive allele called *paralytic*. Flies homozygous for this mutant allele become paralyzed when exposed to temperatures at or above 29°C, but they recover rapidly when the temperature is lowered to 25°C. Mosaic studies revealed that both the brain and thoracic ganglia are the focus of this abnormal behavior. Electrophysiological studies showed that mutant flies have defective sodium transport associated with the conduction of nerve impulses. Subsequently, Ganetzky and his colleagues mapped, isolated, and cloned the *paralytic* gene. This locus encodes a protein, called the sodium channel, that controls the movement of sodium across the membrane of nerve cells.

A second mutant, called *Shaker*, originally isolated over 40 years ago as a behavioral mutant, encodes a potassium channel gene that has also been cloned and characterized. Because the mechanism of nerve impulse conduction has been highly conserved during animal evolution, the cloned *Drosophila* genes were used as probes to isolate the equivalent human ion channel genes. The cloned human genes are being used to provide new insights into the molecular basis of neuronal activity. One of the human ion channel genes first identified in *Drosophila* is defective in a heritable form of cardiac arrhythmia. The identification and cloning of the human gene now makes it possible to screen for family members at risk for this potentially fatal condition.

Learning in *Drosophila*

To study the genetics of a complex behavior such as learning, it would be advantageous to use an organism such as *Drosophila*. However, the first question is: Can *Drosophila* learn? Work from a number of laboratories has shown that *Drosophila* can learn. To demonstrate this, flies are presented with a pair of olfactory cues, one of which is associated with an electric shock. Flies quickly learn to avoid the odor associated with the shock. For several reasons, this response is thought to be learned. First, performance is associated with the pairing of a stimulus/response with a reinforcement and, in addition, the response is reversible. Second, flies can be trained to select an odor they previously avoided. Third, flies exhibit short-term memory for the training they have received.

The demonstration that *Drosophila* can learn opens the way to selecting mutants that are defective in learning and memory. To accomplish this, males from an inbred wild-type strain are mutagenized and mated to females from the same strain. Their progeny are recovered and mated to produce populations of flies, all of which carry a mutagenized X chromosome. Mutations that affect learning are selected by testing for responses in the olfactory/shock apparatus. A number of learning-deficient mutants including *dunce*, *turnip*, *rutabaga*, and *cabbage* have been recovered. In addition, a memory-deficient mutant, *amnesia*, which learns normally but forgets four times faster than normal, has been recovered. Each of these mutations represents a single gene defect that affects a specific form of behavior. Because of the method used to recover them, all the mutants found so far are X-linked genes. Presumably similar genes controlling behavior are also located on the autosomes.

Since in many cases, mutation results in the alteration or abolition of a single protein, the nervous system mu-

TABLE 24.5 Behavioral Mutants of *Drosophila* Affecting Nerve Impulse Transmission

Mutation	Map Location	Ion Channel Affected	Phenotype
nap[ts]	2–56.2	Sodium	Adults, larvae paralyzed at 37.5°C; reversible at 25°C
para[ts]	1–53.9	Sodium	Adults paralyzed at 29°C, larvae at 37°C; reversible at 25°C
tip-E	3–13.5	Sodium	Adults, larvae paralyzed at 39–40°C; reversible at 25°C
sei[ts]	2–106	Sodium	Adults paralyzed at 38°C, larvae unaffected; adults recover at 25°C
Sh	1–57.7	Potassium	Aberrant leg shaking in adults exposed to ether
eag	1–50.0	Potassium	Aberrant leg shaking in adults exposed to ether
Hk	1–30.0	Potassium	Ether-induced leg shaking
slo	3–85.0	Potassium	At 22°C, adults are weak fliers; at 38°C, adults are weak, uncoordinated

tants described here can provide a link between behavior and molecular biology. One group of *Drosophila* learning mutants involves defects in a cellular signal transduction system.

In cells of the nervous system, and in many other cell types, cyclic AMP (cAMP) is produced from ATP in the cytoplasm in response to signals received at the cell's surface. Cyclic AMP activates protein kinases, which, in turn, phosphorylate proteins and initiate a cascade of metabolic effects that control gene expression. Behavioral mutants of *Drosophila* were among the first to show the link between cyclic AMP and learning. The *rutabaga* locus encodes one form of adenylyl cyclase, the enzyme that synthesizes cyclic AMP from ATP. In the mutant allele of *rutabaga*, a missense mutation destroys the catalytic activity of adenylyl cyclase and is associated with learning deficiency in homozygous flies. The *dunce* locus encodes the structural gene for the enzyme cyclic AMP phosphodiesterase, which degrades cAMP. The *turnip* mutation occurs in a gene encoding a G protein, a class of molecules that bind GTP and, in turn, activate adenylyl cyclase. The clustering of these independently derived mutations in a signal-transduction pathway involving cyclic nucleotides supports the idea that these molecules play a role in learning and provide the basis for further studies of the molecular impetus of learning and memory in flies, as well as in humans.

Behavioral genetics has been studied in humans

The genetic control of behavior in humans has proven more difficult to characterize than that of other organisms. Not only are humans unavailable as experimental subjects in genetic investigations, but the types of responses considered to be the most interesting forms of behavior, including some aspects of **intelligence**, **language**, **personality**, and **emotion**, are difficult to study. Two problems arise in studying such behaviors. First, all are difficult to define objectively and to measure quantitatively. Second, they are affected by environmental factors. In each case, the environment is extremely important in shaping, limiting, or facilitating the final phenotype for each trait.

Historically, the study of human behavior genetics has been hampered by other factors. Many studies of human behavior were performed by psychologists without adequate input from geneticists. Second, traits involving intelligence, personality, and emotion have the greatest social and political significance. As such, these traits are more

likely to be the subject of sensationalism when reported to the lay public. Because the study of these traits comes closest to infringing upon individual liberties, such as the right to privacy, they are the basis of the most controversial investigations.

In lamenting the gulf between psychology and genetics in explaining human behavior genetics, C. C. Darlington wrote in 1963, "Human behavior has thus become a happy hunting ground for literary amateurs. And the reason is that psychology and genetics, whose business it is to explain behavior, have failed to face the task together." Since 1963, some progress has been made in bridging this gap, but the genetics of human behavior remains a controversial area.

Single Genes

Many genetic disorders in humans result in a behavioral abnormality. One of the most prominent examples is **Huntington disease** (**HD**). Inherited as an autosomal dominant disorder, HD affects the nervous system, including the brain. Symptoms usually appear in the fifth decade of life with a gradual loss of motor function and coordination. Structural degeneration of the nervous system is progressive, and personality changes occur. Most victims die within 10 to 15 years after the onset of the disease. Because HD usually appears after a family has been started, all the children of an affected person must live with the knowledge that they face a 50 percent probability of developing the disorder (affected individuals are usually heterozygotes). HD is associated with elevated brain levels of quinolinic acid, a naturally occurring neurotoxin. The HD gene has been identified and cloned. It encodes a large protein (348 kd) that is unrelated to any known gene product. The mutant form of the gene is associated with the presence of extra CAG trinucleotide repeats near the 5′-end (see Chapter 17). Normal individuals have 11 to 24 repeats, but those affected by HD carry 42 to 86 CAG repeats. The number of CAG repeats in HD is inversely related to the age of onset in a large number of individuals, but there seems to be no correlation between the number of repeats and the severity of the behavioral and psychiatric symptoms. The expansion of a trinucleotide repeat in HD is similar to expansions of other repeats in several disorders that affect the brain and nervous system, indicating that such "stutter" mutations may be a common form of mutation in neurobehavioral traits.

Monoamine oxidases are enzymes that degrade chemical signals (called neurotransmitters) in the nervous system. Recently, studies have established a link between mutation in an X-linked gene encoding monoamine oxidase A (MAOA) and a syndrome that involves mild mental retardation and aggressive behavior. The pedigree of one affected family is shown in Figure 24.17. Six males are affected, and all showed a characteristic pattern of aggressive behavior and lack of impulse control. Using a variety of RFLP markers, the locus for this behavior was mapped to the short arm of the X chromosome near the MAOA locus. The analysis of the structural gene for MAOA in five of the six affected males showed that they carry a single nucleotide substitution that introduces a stop codon that causes the premature termination of translation, resulting in a lack of MAOA activity. This mutation is carried by heterozygous females, but not by normal males. Other studies show that mutations and polymorphisms in the MAOA gene are not common, strengthening the link between a mutation in this gene and an abnormal form of behavior.

Other metabolic disorders are also known to affect mental function. These include **Lesch-Nyhan syndrome**, an X-linked recessive disorder. The onset is within the first year, and the disease is most often fatal early in childhood. The disorder involves a defect in purine biosynthesis. Affected individuals lack an enzyme, hypoxanthine-guanine phosphoribosyltransferase (HGPRT), and accumulate high levels of uric acid. Mental and developmental retardation occurs, and many of these individuals demonstrate uncontrolled self-mutilation. **Tay-Sachs disease**, an autosomal recessive disorder that maps to chromosome 15, involves severe mental retardation. The disease is apparent soon after birth and is fatal. The autosomal recessive disease **phenylketonuria**, unless detected and treated early, results in mental retardation. All of these disorders alter the normal biochemistry of the affected individuals and are inherited in a Mendelian fashion.

Chromosome abnormalities also produce syndromes with behavioral components. **Down syndrome** (trisomy 21) results in varying degrees of mental retardation. Although there is a wide range, the mean IQ of affected individuals is estimated to be 25 to 50. The onset of walking and talking is often delayed until 4 to 5 years of age.

Multifactorial Traits

Other aspects of human behavior, notably **schizophrenia** and **manic-depressive illness**, have been the subject of extensive investigations. Twin studies and studies on adopted and natural siblings have supported the idea that manic depression has a genetic component. Mapping studies originally identified loci on the X chromosome and chromosome 11 as being associated with manic depression, but these studies were later invalidated. Several laboratories are using RFLP markers to screen large pedigrees in which manic depression is segregating in an attempt to find the loci for this disorder, which affects about 1 percent of the U.S. population.

Schizophrenia is a mental disorder characterized by withdrawal and bizarre and sometimes delusional behavior. Those affected by the disease are unable to lead organized lives and are periodically disabled by the condition. It is clearly a familial disorder, with relatives of schizophrenics having a much higher incidence of this order than the general population. Furthermore, the closer the relationship to the index case or proband, the greater is the probability of the disorder occurring.

The concordance of schizophrenia in monozygotic and dizygotic twins has been the subject of many studies. In almost every investigation, concordance has been higher in monozygotic twins than in dizygotic twins reared together. Although these results suggest that a genetic component exists, they do not reveal the precise genetic basis of schizophrenia. Simple monohybrid inheritance, dihybrid inheritance, and multiple gene control have been proposed for schizophrenia. It seems unlikely that only one or two loci are involved, nor is it likely that the control is strictly quantitative, as in polygenic inheritance. As with manic depression, a large-scale collaborative effort is using DNA markers to identify the genomic regions that may contain loci controlling this behavioral disorder. Once identified, these regions will be studied in detail to identify, isolate, and clone genes for schizophrenia.

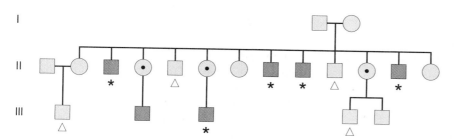

FIGURE 24.17 Partial pedigree of a family in which mental retardation and aggressive behavior segregates with a mutation in the monoamine oxidase A gene (MAOA). Those males marked with an asterisk are hemizygous for a mutant allele of MAOA. Those marked with a triangle have the normal allele. Female heterozygotes have one normal and one mutant allele.

GENETICS, TECHNOLOGY, AND SOCIETY

Coming to Terms with the Heritability of IQ

Few topics in human heredity are as contentious as the genetics of intelligence. The debate concerning "genetic" vs. "environmental" influences over intelligence can have far-reaching societal ramifications, and may even affect the formulation of educational policy. What does genetic analysis tell us about the hereditary basis of intelligence?

To answer this question, the first challenge is to define and assess intelligence. From a genetic standpoint, is it a single phenotypic entity, and, if so, can it be measured? Obviously, intelligence is not a physical characteristic like height or weight, which are easy to see and easy to measure. However, performance on IQ tests has been taken as a measure of intelligence since Alfred Binet's original test was introduced in the United States in 1916. Despite many objections, this practice is unlikely to change. The debate about the nature of intelligence, whether or not it is an innate mental process that can be reliably assessed, is certain to continue for some time.

What about the genetic basis for IQ scores? For a complex trait such as intelligence, there is no readily dis-

cernible chain of cause-and-effect events leading from genotype to phenotype, as there is for, say, albinism or hemophilia. The genetic analysis of intelligence therefore requires the assumption that it is a quantitative trait, whereby the final phenotype is influenced by the additive contributions of many genes, as well as a variety of environmental factors. Statistical techniques can then be employed that were first developed for the analysis of other quantitative traits in plants and animals. The most frequently used statistic is heritability.

Heritability is the fraction of the total variance in the phenotypic expression of a trait within a specific population that is due to genetic factors. The heritability of many traits in plants and animals can be estimated from the results of breeding experiments. Several methods of estimating heritabilities have been used in humans.

One method involves examining sets of identical twins and fraternal twins. Identical (monozygotic) twins develop from a single zygote and thus are genetically identical, whereas fraternal (dizygotic) twins are no more alike genetically than any other two

siblings, sharing, on average, 50 percent of their genes. If the trait in question has a genetic basis, then, given common environments, identical twins should resemble each other more closely than fraternal twins do. A still better way is to study identical twins who have been separated very early in life and reared apart. In theory, this allows the effects of a common genetic inheritance to be distinguished from the effects of different environments.

Reflecting the uncertainties of such approaches, these and similar methods yield a range of heritability values from a low of around 0.30 to a high of around 0.80. A commonly accepted heritability value for IQ score is 0.60. So let's consider what a heritability of 0.60 for IQ score really means, and, perhaps more importantly, what it does *not* mean. Strictly speaking, a heritability of 0.60 for IQ score means that *within the population under study,* 60 percent of the variance in IQ scores can be attributed to different genotypes within the population.

It should be understood, first of all, that heritability is a statistic pertaining to variation within a population and has no meaning when applied to an in-

CHAPTER SUMMARY

1. Behavioral genetics has emerged as an important specialty within genetics because both the genotype and environment have been found to have an impact in determining an organism's behavioral response.

2. Historically, three approaches have been used in studying the genetic control of behavior: the use of closely related organisms from similar environments whose survival needs appear to be identical; the artificial selection of traits that can subsequently be trans-

ferred from one strain to another; and the identification and isolation of single genes that have an effect on behavior patterns.

3. Studies done on alcohol preference and open-field behavior in mice illustrate behaviors strongly influenced by the genotype.

4. Studies of maze learning in rats and geotaxis in *Drosophila* have successfully established both bright and dull lines of rats and positively and negatively ge-

dividual. It is therefore not correct to say that 60 percent of a person's IQ is determined by genes and the other 40 percent by environment.

Second, heritability estimates pertain to variation only *within* the specific population that was measured and not to variation *between* populations. A heritability value of 0.60 *does not* mean that 60 percent of the difference in IQ scores between two separate populations (or groups) is due to different genes.

Third, the heritability of a trait is not a permanent characteristic of a population. In fact, a heritability value, whether for IQ score or anything else, pertains to that particular population *at a specific time*, under the conditions then prevailing. Heritability values may change as environmental factors change and may thus vary from one year to the next. Perhaps most important, "heritable" does not therefore equal "inevitable." A high heritability value, such as 0.60, does not mean that the trait cannot be modified substantially by environmental factors. The genotype may determine an upper and lower limit within a broad range of possible phenotypes, but exactly what phenotype occurs in an individual depends on interactions with the environment.

Attempts to elucidate the genetic basis of human intelligence will remain controversial, and for good reason. The results of such studies, especially when misinterpreted, can be taken as a biological justification for social policies that may be viewed as discriminatory. For example, some people prefer to believe that the problems of the disadvantaged in our society are not the result of social forces beyond their control, but are based on their genes. Recently, heritability estimates of IQ scores have been used to bolster claims that intelligence is largely genetic and not significantly influenced by the child's environment. On the basis of such faulty reasoning, it has been concluded that efforts to boost IQ through better educational and social interventions are doomed to failure.

It should be clear that the heritability of intelligence, even an estimate as high as 0.60, is an inadequate foundation on which to base such claims and construct social policy. This is not to say that intelligence, however it is defined and measured, is not influenced to a substantial degree by the genes each of us has inherited, or that differences in IQ score will not be found in populations that are studied. However, we cannot allow the heritability of IQ scores to serve as the rationale for making policies that discriminate against anyone in our society.

REFERENCES

HERRNSTEIN, R.J., and MURRAY, C. 1994. *The bell curve: Intelligence and class structure in American life.* New York: The Free Press.

JACOBY, R., and GLAUBERMAN, N., eds. 1995. *The bell curve debate.* New York: Times Books.

TERMAN, L.M., and MERRILL, M.A. 1973. *Stanford-Binet intelligence scale: 1972 norms edition.* Boston: Houghton Mifflin.

otaxic *Drosophila.* These results have led researchers to ascribe the relative contributions of genes on a specific chromosome to these behaviors.

5. By isolating mutations that cause deviations from normal behavior, the role of a corresponding wild-type allele in the response is established. Numerous examples have been examined, including honeybee hygienic behavior; chemotaxis, thermotaxis, and general movement in nematodes; neuro-

logical mutations in mice; and a variety of behaviors in *Drosophila.*

6. Any aspect of human behavior is difficult to study because the individual's environment makes an important contribution to trait development. In humans, studies using twins have shown that, while a family may have a predisposition to schizophrenia, the expression of this disorder may be modified by the environment.

INSIGHTS AND SOLUTIONS

1. Manic depression is an affective disorder associated with recurring mood changes. It is estimated that one in four individuals will suffer from some form of affective disorder at least once in his or her lifetime. Genetic studies indicate that manic depression is familial and that single genes may play a major role in controlling this behavioral disorder. In 1987, two separate studies using RFLP analysis and other genetic markers reported a linkage between manic depression and markers on the X chromosome and to the short arm of chromosome 11. At the time, these reports were hailed as landmark discoveries, opening the way to the isolation and characterization of genes that control specific forms of behavior and to the development of therapeutic strategies based on knowledge of the nature of the gene product and its action. However, further work on the same populations (reported in 1989 and 1990) demonstrated that the original results were invalid, and it was concluded that no linkage to markers on the X chromosome and to markers on chromosome 11 could be established. These findings do not exclude the role of major genes on the X chromosome and autosomes in manic depression, but they do exclude linkage to the markers used in the original reports.

   This setback not only caused embarrassment and confusion but it also forced a reexamination of the validity of the methods used in mapping human genes, and of the analysis of data from linkage studies involving complex behavioral traits. Several factors have been proposed to explain the flawed conclusions reported in the original studies. What do you suppose some of these factors to be?

   Solution: While some of the criticisms were directed at the choice of markers, most of the factors at work in this situation appear to be related to the phenotype of manic depression. At least three confounding elements have been identified. One is age of onset. There is a positive correlation between age and the appearance of manic depression. Therefore, at the time of a pedigree study, younger individuals who will be affected later in life may not show any signs of manic depression. Another confusing factor is that, in the populations studied, there may in fact be more than one major X-linked gene and more than one major autosomal gene that trigger manic depression. A third factor relates to the diagnosis of manic depression itself. The phenotype is complex and not as easily quantified as height or weight. In addition, mood swings are a universal part of everyday life, and it is not always easy to distinguish transient mood alterations and the role of environmental factors from affective disorders having a biological and/or genetic basis.

   These factors point up the difficulty in researching the genetic basis of complex behavioral traits. Individually or in combinations, the factors described might skew the results enough so that the guidelines for proof that are adequate for other traits are not stringent enough for behavioral traits with complex phenotypes and complex underlying causes.

PROBLEMS AND DISCUSSION QUESTIONS

1. Contrast the methodologies used in studying behavior genetics. What are the advantages of each method with respect to the type of information gained?
2. Contrast the advantages of using *Drosophila* vs. *Caenorhabditis* for studying behavior genetics.
3. If a haploid honeybee drone of the genotype *uR* is mated to a queen of genotype *UuRR*, what ratio of behavioral phenotypes will be observed in the hive in the offspring with respect to American foulbrood disease? For *Ur × UURr*? For *uR × Uurr*?
4. Assume that you discovered a fruit fly that walked with a limp and was continually off balance as it moved. Describe how you would determine whether this behavior is due to an injury (induced in the environment) or is an inherited trait. Assuming that it is inherited, what are the various possibilities for the focus of gene expression causing the imbalance? Describe how you would locate the focus experimentally if it were X-linked.
5. In humans, the chemical phenylthiocarbamide (PTC) is either tasted or not. When the offspring of various combinations of taster and nontaster parents are examined, the data given here are obtained.

Parents	Offspring
Both tasters	All tasters
Both tasters	1/2 tasters
	1/2 nontasters
Both tasters	3/4 tasters
	1/4 nontasters
One taster	
One nontaster	All tasters
One taster	1/2 tasters
One nontaster	1/2 nontasters
Both nontasters	All nontasters

Based on these data, how is PTC tasting behavior inherited?

GENETICS MediaLab

The following resources will help you achieve a better understanding of the concept presented in this chapter. These resources can be found on the CD packaged with this textbook and on the Companion Website at **http://www.prenhall.com/klug/**.

CD Resources:

Self-grading Chapter Problems

Web Resources:

Genetics and Society Issue: *Coming to Terms with the Heritability of IQ*

Web Destinations in Genetics

Self-grading Chapter Problems

Chapter Search Terms

Genetics Newsgroups

Student Bulletin Board

Web Problem 1:

Time for completion = 10 minutes

Do single genes control complex behaviors? Dr. Marta Sokolowski has studied alleles at the *foraging* locus in *Drosophila* that influence feeding behavior. After reading the article on Dr. Sokolowski's research, you should be able to answer the following questions. Roving in search of food is energetically costly; when would this behavior be advantageous? What effect did the experimental manipulation of larval density have on feeding behavior? Were the changes that evolved in the response to variation in larval density associated with alleles at the *foraging* locus? To complete this exercise, visit Web Problem 1 in Chapter 24 of your Companion Website and select the keyword **FORAGING**.

Web Problem 2:

Time for completion = 10 minutes

Is alcoholism a genetic disease? After reading the paper on "The Genetics of Alcoholism," you should be able to answer the following questions. What methods are used to estimate the genetic contribution to alcoholic behavior in humans? Is the evidence of a genetic contribution similar between the different types of studies? Does the inherited component of alcoholism behave like a simple Mendelian locus or as a quantitative trait? Are any known genes linked to alcoholic behavior? To complete this exercise, visit Web Problem 2 in Chapter 24 of your Companion Website and select the keyword **ALCOHOLISM**.

Web Problem 3:

Time for completion = 10 minutes

Mammalian MHC loci are involved in discriminating "self" from "non-self", but could they also be used by potential mates to size up a potential partner? Read the article "Scent of a Man" and use it and information from Chapter 24 to answer the following questions. Do young mice prefer MHC odors that are similar or different to their own? Does this preference change when they reach maturity? Why might mice use MHC odors to select potential mates? Can human females also detect differences in MHC loci by odor? To complete this exercise, visit Web Problem 3 in Chapter 24 of your Companion Website and select the keyword **MATE CHOICE**.

6. Discuss why the study of human behavior genetics has lagged behind that of other organisms.

7. J. P. Scott and J. L. Fuller studied 50 traits in five pure breeds of dogs. Almost all the traits varied significantly in the five breeds, but very few bred true in crosses. What can you conclude with respect to the genetic control of these behavioral traits?

EXTRA-SPICY PROBLEMS

8. This chapter introduces *C. elegans* as a model system for studying the genetic specification of the nervous system. The life cycle of this organism makes it an excellent choice for the genetic dissection of many biological processes. This species has two natural sexes, hermaphrodite and male. The hermaphrodite is essentially a female that can generate sperm as well as oocytes, so reproduction can occur by hermaphrodite self-fertilization or hermaphrodite—male mating. In the context of studying mutations in the nervous system, what is the advantage of hermaphrodite self-fertilization with respect to the identification of recessive mutations and the propagation of mutant strains?

9. Theoretical data concerning the genetic effect of *Drosophila* chromosomes on geotaxis are shown here. What conclusions can you draw?

	Chromosome		
	X	2	3
Positive geotaxis	+0.2	+0.1	+3.0
Unselected	+0.1	−0.2	+1.0
Negative geotaxis	−0.1	−2.6	+0.1

SELECTED READINGS

AVERY, L. 1993. The genetics of feeding in *Caenorhabditis elegans*. *Genetics* 133:897–917.

BASTOCK, M. 1956. A gene which changes a behavior pattern. *Evolution* 10:421–39.

BENZER, S. 1973. Genetic dissection of behavior. *Sci. Am.* (Dec.) 229:24–37.

BERG, H.C. 1988. A physicist looks at bacterial chemotaxis. *Cold Spring Harbor Symp. Quant. Biol.* 53 (1):1–9.

BRENNER, S. 1974. The genetics of *Caenorhabditis elegans*. *Genetics* 77:71–94.

BUCK, K.J., 1995. Strategies for mapping and identifying quantitative trait loci specifying behavioral responses to alcohol. *Alcohol Clin. Exp. Res.* 19:795–801.

BYERS, D., DAVID, R.L., and KIGER, J.A., Jr. 1981. Defect in the cyclic AMP phosphodiesterase due to the *dunce* mutation of learning in *Drosophila*. *Nature* 289:79–81.

DEVOR, E.J., and CLONINGER, C.R. 1989. Genetics of alcoholism. *Annu. Rev. Genet.* 23:19–36.

EISENBACH, M., and CAPLAN, S.R. 1998. Bacterial chemotaxis: unsolved mystery of the flagellar switch. *Curr. Biol.* 8:R444–46.

FULLER, J.C., and THOMPSON, W.R. 1978. *Foundations of behavior genetics*. St. Louis: C. V. Mosby.

GAILEY, D.A., HALL, J.C., and SIEGEL, R.W. 1985. Reduced reproductive success for a conditioning mutant in experimental populations of *Drosophila melanogaster*. *Genetics* 111:795–804.

GANETZKY, B. 1994. Cysteine strings, calcium channels and synaptic transmission. *BioEssays* 16:461–63.

GANETZKY, B., WARMKE, J.W., ROBERTSON, G., and PALLANCK, L. 1995. New potassium channel gene families in flies and mammals: From mutants to molecules. *Soc. Gen. Physiol Ser.* 50:29–39.

GERHARD, D.S., LaBUDA, M.C., BLAND, S.D., ALLEN, C., EGELAND, J.A., and PAULS, D.L. 1995. Initial report of a genome search for the affective disorder predisposition gene in the old order Amish pedigrees: Chromosomes 1 and 11. *Am. J. Med. Genet.* 54:398–404.

GERSHON, E.S., BADNER, J.A. et al. 1998. Closing in on genes for manic-depressive illness and schizophrenia. *Neuropsychopharmacology* 18:233–42.

HAZELBANER, G.L., BERG, H.C., and MATSUMURA, P. 1993. Bacterial motility and signal transduction. *Cell* 73:15–22.

HOTTA, Y., and BENZER, S. 1972. Mapping behavior of *Drosophila* mosaics. *Nature* 240:527–35.

HUNTINGTON'S DISEASE COLLABORATIVE GROUP. 1993. A novel gene containing a trinucleotide repeat that is expanded and unstable on Huntington's disease chromosomes. *Cell* 72:971–83.

KIDD, K.K., and CAVALLI-SFORZA, L.L. 1973. An analysis of the genetics of schizophrenia. *Social Biol.* 20:254–65.

LEE, R.Y., and SAWIN, E.R. 1999. EAT-4, a homolog of a mammalian sodium-dependent inorganic phosphate cotransporter, is necessary for glutamatergic neurotransmission in *Caenorhabditis elegans*. *J. Neurosci.* 19:159–67.

LIVINGSTONE, K.S. 1985. Genetic dissection of *Drosophila* adenylate cyclase. *Proc. Natl. Acad. Sci. USA* 82:5795–99.

McINNES, L.A., and FREIMER, N.B. 1995. Mapping genes for psychiatric disorders and behavioral traits. *Curr. Opin. Genet. Dev.* 5:376–81.

MADDOCK, J.R., and SHAPIRO, L. 1993. Polar location of the chemoreceptor complex in the *E. coli* cell. *Science* 259:1717–23.

MANSON, M.D., ARMITAGE, J.D., HOCH, J.A., and McNAB, R.M. 1998. Bacterial locomotion and signal transduction. *J. Bact.* 180:1009–22.

MELO, J.A., SHENDURE, J., POCIASK, K., and SILVER, L.M. 1996. Identification of sex-specific quantitative trait loci controlling alcohol preference in C57BL/6 mice. *Nature Genet.* 13:147–53.

PARKINSON, J.S. 1993. Signal transduction schemes of bacteria. *Cell* 73:857–71.

PARKINSON, J.S., and BLAIR, D. 1993. Does *E. coli* have a nose? *Science* 259:1701–2.

PETRONIS, A., SHERRINGTON, R.P., PATERSON, A.D., and KENNEDY, J.L. 1995. Genetic anticipation in schizophrenia: Pro and con. *Clin. Neurosci.* 3:76–80.

PLOMIN, R. 1997. *Behavioral genetics.* 3rd ed. New York: W.H. Freeman.

PLOMIN, R., DeFRIES, J.C., and McCLEARN, G.E. 1990. *Behavioral genetics: A primer*, 2nd ed. New York: W. H. Freeman.

QUINN, W.B., and GREENSPAN, R.J. 1984. Learning and courtship in *Drosophila*: Two stories with mutants. *Annu. Rev. Neurosci.* 7:67–93.

RAIZEN, D.M., LEE, R.Y., and AVERY, L. 1995. Interacting genes required for pharyngeal excitation by motor neuron MC in *Caenorhabditis elegans. Genetics* 141:1365–82.

RICKER, J.P., and HIRSCH, J. 1988(a). Genetic changes occurring over 500 generations in lines of *Drosophila melanogaster* selected divergently for geotaxis. *Behav. Genet.* 18:13–24.

———. 1988(b). Reversal of genetic homeostasis in laboratory populations of *Drosophila melanogaster* under long-term selection for geotaxis and estimates of gene correlates: Evolution of behavior-genetic systems. *J. Comp. Psychol.* 102:203–14.

RIDDLE, D.L. 1978. The genetics of development and behavior in *Caenorhabditis elegans. J. Nematol.* 10:1–15.

ROTHENBUHLER, W.C., KULINCEVIC, J.M., and KERR, W.E. 1968. Bee genetics. *Annu. Rev. Genet.* 2:413–38.

SIEGEL, R.W., HALL, J.C., GAILEY, D.A., and KYRIACOU, C.P. 1984. Genetic elements of courtship in *Drosophila*: Mosaics and learning mutants. *Behav. Genet.* 14:383–410.

STOLTENBERG, S.F., HIRSCH, J., and BERLOCHER, S.H. 1995. Analyzing correlations of three types in selected lines of *Drosophila melanogaster* that have evolved stable extreme geotactic performance. *J. Comp. Psychol.* 105: 85–94.

TRYON, R.C. 1942. Individual differences. In *Comparative psychology*, 2nd ed., ed. F.A. Moss. Englewood Cliffs, NJ: Prentice-Hall.

This sockeye salmon run constitutes a population, a group of individuals that actually or potentially interbreed.

Population Genetics

KEY CONCEPTS

- **Populations and gene pools are the focus of study in population genetics**

- **The first step in studying the genetic structure of a population is to calculate the frequencies of alleles**

- **The Hardy-Weinberg law shows that under simple assumptions allele frequencies do not change from one generation to the next**
  Demonstration of the Hardy-Weinberg Law
  Testing for Equilibrium

- **The Hardy-Weinberg law can be extended to include more than two alleles**

- **The Hardy-Weinberg law provides a method for calculating the frequency of heterozygotes in a population**

- **Natural selection is a potent force of evolution**
  Natural Selection

Fitness and Selection
Selection in Natural Populations
Natural Selection Affecting Quantitative Traits

- **Mutation is the source of new alleles, but a weak force of change in allele frequencies**

- **Migration homogenizes allele frequencies across populations**

- **Genetic drift is random change in allele frequencies and is especially important in small populations**

- **Nonrandom mating changes genotype frequencies, but not allele frequencies**
  Genetic Effects of Inbreeding

🌀 GenCDX

When you see this icon, there are related animations and exercises on the CD accompanying this text.

Charles Darwin, in work he co-authored with Alfred Russel Wallace and in his 1859 book *On the Origin of Species*, identified natural selection as the mechanism of adaptive evolution. Darwin's theory of evolution by natural selection consists of four claims about populations of organisms, plus a consequence that must logically follow if all four claims are true. The four claims are: (1) there are differences among individuals within populations; (2) these differences are passed from parents to offspring; (3) more offspring are born than will survive and reproduce; and (4) some variants are more successful at surviving and/or reproducing than others. The consequence for a population in which all four claims are true is that the relative abundance of the different forms within the population changes across generations. In other words, the population evolves.

Although Darwin lacked an accurate model of the mechanism responsible for variation and inheritance, he realized that a complete understanding of evolution by natural selection would require such knowledge. Gregor Mendel published his theory of genetics in 1865, but his work was universally ignored. When Mendel's paper was rediscovered in 1900, biologists immediately recognized its importance and began a 30-year effort to synthesize Mendel's concept of genes and alleles with the theory of evolution. A key insight in this effort was the realization that changes in the relative abundance of different phenotypic traits in a population can be tied to changes in the relative abundance of the alleles that influence the traits. The discipline within evolutionary biology that studies the changes in the frequencies of alleles is known as **population genetics.**

Early in the twentieth century a number of workers, including G. Udny Yule, William Castle, Godfrey Hardy, and Wilhelm Weinberg, formulated the basic principles of population genetics. For many years the emphasis was on the development of mathematical models to describe the genetic structure of populations. Prominent among the theoreticians who developed these models were Sewall Wright, Ronald Fisher, and J. B. S. Haldane. Following their work, experimentalists and field workers have used biochemical and molecular techniques to directly measure variation at the protein and DNA levels in order to test the mathematical models. Allele frequencies and the forces that alter these frequencies, such as selection, mutation, migration, and random genetic drift, are examined. In this chapter, we shall consider some general aspects of population genetics and also discuss other areas of genetics relating to evolution.

Populations and gene pools are the focus of study in population genetics

The key to understanding genetic evolution is to focus not on individual organisms but on populations. A **population** is a group of individuals, all the same species and all living in the same geographic area, that actually or potentially interbreed with each other. If we consider a single genetic locus in the genome of the species we are studying, we may find that different individuals within a population have different genotypes. The fraction of the population that has a particular genotype is called the **frequency** of that genotype. We will be concerned with whether the frequencies of various genotypes change from one generation to the next.

When considering changes from one generation to the next, we will often rely on a useful conceptual device called the gene pool. The **gene pool** consists of all the gametes made by all the breeding members of a population in one particular generation. These gametes will combine to form the zygotes that will become the population's next generation. The eggs and sperm that constitute the gene pool are haploid and thus contain only a single allele for each genetic locus. If we consider a single locus, we may find that different gametes carry different alleles. The fraction of the gametes in the gene pool that carry a particular allele is the frequency of the allele. As with genotypes in the population, we will be concerned with whether the frequencies of various alleles in the gene pool change over time.

Populations are dynamic; they may grow and expand or diminish and contract through changes in birth or death rates, by migration, or by merging with other populations. The dynamic nature of populations has important consequences and, over time, can lead to changes in the genetic structure of the population.

The first step in studying the genetic structure of a population is to calculate the frequencies of alleles

The first step in most population genetic studies is to measure the allele frequencies at the locus of interest. Measuring allele frequencies requires determining the genotypes of a large number of individuals. In some cases researchers can infer genotypes directly from phenotypes. In other cases the direct analysis of proteins or DNA sequences is required. Let us consider an example of calculating allele frequencies.

Our example comes from research into the genetic factors that influence an individual's susceptibility to infection by HIV-1, the virus responsible for the bulk of the global AIDS epidemic. Of particular interest to AIDS researchers are the small number of individuals who have lived high-risk lifestyles, involving unprotected sex with HIV-positive partners, yet remain uninfected. In 1996 Rong Liu and colleagues discovered that two of the exposed-but-uninfected individuals they were studying were homozygous for a mutant allele of a gene called *CC-CKR-5*.

The *CC-CKR-5* gene, located on chromosome 3, encodes a cell-surface protein called C-C chemokine receptor-5, often abbreviated CCR5. Chemokines are signalling molecules used by the immune system. When the immune system cells that carry CCR5 on their surface receive signals in the form of chemokines binding to their receptors, the cells move into inflamed tissues to help fight an infection.

The CCR5 protein is also exploited as a co-receptor by strains of HIV-1 that are transmitted by sexual contact and that tend to dominate the viral population in newly infected hosts. To gain entry into the host's cells, HIV-1 virions use a protein called Env (short for envelope protein). Env's mode of action appears to be as follows. Env first binds to a protein on the surface of the host cell called CD4. Binding to CD4 induces a conformational change in Env, opening a second binding site. This second site binds to CCR5, which in turn initates the fusion of the viral envelope with the host cell membrane. The merging of the viral envelope with the cell membrane transports the HIV viral core into the host cell's cytoplasm.

The mutant allele of the *CCR5* gene that Liu discovered contains a 32-bp deletion in one of its coding regions. As a result of this deletion the protein encoded by the gene is severely shortened and is nonfunctional. The protein never makes it to the cell membrane, so the HIV-1 virions cannot bind. We will call the normal allele of the gene *CCR5 +* (or just +) and the allele with the 32-bp deletion

TABLE 25.1 *CCR5* Genotypes and Phenotypes

Genotype	Phenotype
+/+	Susceptible to sexually transmitted strains of HIV-1
+/Δ32	Susceptible, but may progress to AIDS slowly
Δ32/Δ32	Resistant to most sexually transmitted strains of HIV-1

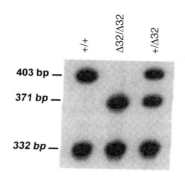

FIGURE 25.1 Allelic variation in the *CCR5* gene. Michel Samson and colleagues used PCR to amplify a part of the *CCR5* gene containing the site of the 32-bp deletion, cut the resulting DNA fragments with a restriction enzyme, and ran the fragments on an electrophoresis gel. Each lane reveals the genotype of a single individual. The + allele produces a 332-bp fragment and a 403-bp fragment; the *Δ32* allele produces a 332-bp fragment and a 371-bp fragment. Heterozygotes produce three bands. (From Samson et al. 1996.)

CCR5-Δ32 (or just *Δ32*). The two individuals described by Liu both had the genotype *Δ32/Δ32*. As a result, they had no CCR5 on the surface of their cells, and were immune to infection by the strains of HIV-1 that require CCR5 as a co-receptor.

At least three *Δ32/Δ32* individuals who are infected with HIV-1 have been found. However, most *Δ32/Δ32* individuals are resistant to most sexually transmitted strains of the virus. Curiously, researchers have been unable to discover any adverse effects associated with this genotype. Heterozygotes, with genotype *+/Δ32*, are susceptible to

HIV-1 infection, but there is evidence that they progress more slowly to full-blown AIDS. Table 25.1 summarizes the genotypes possible at the *CCR5* locus, and the phenotypes associated with each.

Upon learning of the *CCR5-Δ32* allele, and the fact that it provides some protection against AIDS, the obvious questions are: Which human populations harbor the *Δ32* allele, and how common is it? To answer these questions, several teams of researchers have surveyed a large number of people from a variety of populations. The researchers determined each individual's genotype by direct analysis of DNA (Figure 25.1). Consider a sample population of 100 French individuals from Brittany surveyed by Frédérick Libert and colleagues. Among this group, 79 individuals had genotype +/+, 20 had genotype *+/Δ32*, and 1 had genotype *Δ32/Δ32*. To calculate the allele frequencies, imagine a gene pool to which each of the 100 individuals in the sample contributes two gametes, each carrying only one of the individual's two alleles. There are 200 gametes in this gene pool. Counting up the gametes carrying the *CCR5* + allele, there are 158 produced by +/+ individuals, plus 20 produced by *+/Δ32* individuals, for a total of 178. The frequency of the *CCR5* + allele is thus 178/200 = 0.89 = 89 percent. Counting up the *CCR5-Δ32* alleles, there are 20 produced by *+/Δ32* individuals, plus 2 produced by the *Δ32/Δ32* individual, for a total of 22. The frequency of the *CCR5-Δ32* allele is thus 22/200 = 0.11 = 11 percent. Notice that 0.89 + 0.11 = 1.00, which confirms that we have accounted for the entire gene pool. Table 25.2 illustrates two methods for computing the frequencies of the + and *Δ32* alleles in the Brittany sample.

TABLE 25.2 Methods of Determining Allele Frequencies from Data on Genotypes

A. Counting Alleles				
Genotype	*+/+*	*+/Δ32*	*Δ32/Δ32*	Total
Number of individuals	79	20	1	100
Number of + alleles	158	20	0	178
Number of Δ32 alleles	0	20	2	22
Total number of alleles	158	40	2	200

Frequency of *CCR5* + in sample: 178/200 = 0.89 = 89%
Frequency of *CCR5-Δ32* in sample: 22/200 = 0.11 = 11%

B. From Genotype Frequencies				
Genotype	*+/+*	*+/Δ32*	*Δ32/Δ32*	Total
Number of individuals	79	20	1	100
Genotype Frequency	79/100 = 0.79	20/100 = 0.20	1/100 = 0.01	1.00

Frequency of *CCR* + in sample: 0.79 + (1/2) 0.20 = 0.89
Frequency of *CCR5-Δ32* in sample: (1/2) 0.20 + 0.01 = 0.11

Figure 25.2 shows the frequency of the *CCR5-Δ32* allele in the 18 European populations that Libert surveyed. Table 25.3 gives the results of a larger study by Jeremy Martinson and colleagues. Together, the studies show that the populations with the highest frequencies of the *Δ32* allele, 20% or more, are in northern Europe. The frequency of the allele declines as we move away from northern Europe to the south and to the east. In populations without European ancestry the *Δ32* allele is essentially absent. The highly patterned global distribution of the *Δ32* allele presents an evolutionary puzzle that we will return to later in the chapter.

The Hardy-Weinberg law shows that under certain assumptions allele frequencies do not change from one generation to the next

The large variation among populations in the frequency of the *CCR5-Δ32* allele raises a number of questions. Among them is whether the allele can be expected to become more common in populations in which it is currently rare. Population genetics provides a set of tools for exploring such questions. These tools are based on a mathematical model developed independently by the British mathematician

Godfrey H. Hardy and the German physician Wilhelm Weinberg. This mathematical model, called the **Hardy-Weinberg law,** shows what will happen to genotype and allele frequencies in a population under a set of simple assumptions. The assumptions specify an ideal population that is free of many of the complications that affect real populations.

1. Individuals of all genotypes have equal rates of survival and equal reproductive success. That is, there is no selection.

2. There are no mutations that create new alleles or convert one allele into another.

3. There is no migration of individuals into or out of the population.

4. The population is infinitely large, which in practical terms means that the population is large enough that sampling errors and other random effects are negligible.

5. The individuals in the population mate at random.

The Hardy-Weinberg law demonstrates that a population conforming to these assumptions has the following properties.

1. The allele frequencies in the population do not change from generation to generation; in other words, the population does not evolve.

2. After one generation of random mating, the genotype frequencies can be predicted from the allele frequencies.

A model in which allele frequencies do not change from one generation to the next may not seem like a promising tool for the study of evolution. What makes the Hardy-Weinberg law useful, however, is the explicit assumptions on which it is based. By specifying the ideal conditions under which allele frequencies will not change, the Hardy-Weinberg law identifies the real-world forces that can cause allele frequencies to change. In other words, the Hardy-Weinberg law identifies the forces of evolution.

Demonstration of the Hardy-Weinberg Law

To demonstrate the Hardy-Weinberg law, we will first work a numerical example, and then develop the general case. In both versions we will focus on a single locus with two alleles, *A* and *a*.

Imagine the gene pool for a population in which the frequency of allele *A*, in both eggs and sperm, is 0.7, and the frequency of allele *a* is 0.3. Note that $0.7 + 0.3 = 1$, accounting for all the alleles in the gene pool. We are assuming that individuals mate at random, which we can visualize as follows. We take all the gametes in the gene pool,

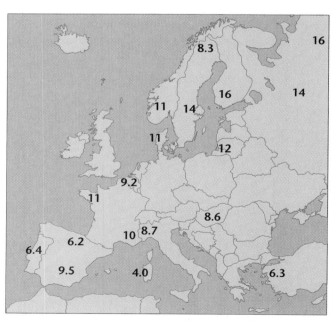

FIGURE 25.2 The frequency of the *CCR5-Δ32* allele in 18 European populations. (From F. Leibert, P. Cochaux et al. The CCR5 mutation conferring protection against HIV-1 in Caucasian populations has a single and recent origin in Northeastern Europe. *Hum. Molec. Genet.* 7:399–406 by permission of Oxford University Press.)

TABLE 25.3 Frequencies of *CCR5+* and *CCR5-Δ32* Alleles in Various Populations

Population	Number of People Tested	Number with Each Genotype			Allele Frequency (%)	
		+/+	*+/Δ32*	*Δ32/Δ32*	*CCR5+*	*CCR5+Δ32*
Europe						
Ashkenazi	43	26	16	1	79.07	20.93
Iceland	102	75	24	3	85.29	14.71
Britain	283	223	57	3	88.87	11.13
Italy	91	81	10	0	94.51	5.49
Greece	63	60	3	0	97.62	2.38
Middle East						
Caucasus: Daghestan	110	96	14	0	93.64	6.36
Saudi Arabia	241	231	10	0	97.93	2.07
Yemen	34	34	0	0	0	0
Asia						
Russia: Udmurtia	46	38	7	1	91.22	9.78
Pakistan	34	32	2	0	97.06	2.94
Punjab	34	33	1	0	98.53	1.47
Bengal	25	25	0	0	100	0
Taiwan	83	83	0	0	100	0
Filipino	26	26	0	0	100	0
Mongolia	59	59	0	0	100	0
Thailand	101	100	1	0	99.50	0.50
Borneo	151	151	0	0	100	0
Africa						
Nigeria	111	110	1	0	99.55	0.45
Gambia	56	56	0	0	100	0
Central African Republic	52	52	0	0	100	0
Kenya	80	80	0	0	100	0
Zambia	96	96	0	0	100	0
Kalahari San	36	36	0	0	100	0
Oceania						
New Guinea Coast	96	96	0	0	100	0
French Polynesia	94	94	0	0	100	0
Aboriginal Australian	98	96	2	0	98.98	1.02
Guam	59	58	1	0	99.15	0.85
Fiji	17	17	0	0	100	0
Americas						
Nuu-Chah-Nulth	38	37	1	0	98.68	1.32
Mexican (Huicholes)	52	52	0	0	100	0
Brazilian Amerindian	98	98	0	0	100	0

Excerpted from: Martinson, et al. 1997

place them in a barrel, and stir. We then draw eggs and sperm from the barrel at random and pair them to make zygotes. What are the genotype frequencies among the zygotes? For any one zygote, the probability that the egg will contain *A* is 0.7, and the probability that the sperm will contain *A* is 0.7. Thus the probability that both the egg *and* the sperm will contain *A* is 0.7 × 0.7 = 0.49. In other words, the frequency of genotype *AA* among the zygotes will be 49 percent. The probability that a zygote will be formed from an egg carrying *A* and a sperm carrying *a* is 0.7 × 0.3 = 0.21, and the probability that a zygote will be formed from an egg carrying *a* and a sperm carrying *A* is

0.3 × 0.7 = 0.21. Thus the frequency of genotype *Aa* among the zygotes is 0.21 + 0.21 = 0.42 = 42 percent. Finally, the probability that a zygote will be formed from an egg carrying *a* and a sperm carrying *a* is 0.3 × 0.3 = 0.09, meaning that the frequency of genotype *aa* among the zygotes will be 9 percent. As a check on our calculation, note that 0.49 + 0.42 + 0.09 = 1.0, which confirms that we have accounted for all of the zygotes. Figure 25.3 shows these calculations in graphical form.

We started with the allele frequencies in a gene pool and calculated the genotype frequencies among the zygotes produced from it. When the zygotes develop into adults

Sperm

	fr(A) = 0.7	fr(a) = 0.3
fr(A) = 0.7	fr(AA) = 0.7 × 0.7 = 0.49	fr(Aa) = 0.7 × 0.3 = 0.21
fr(a) = 0.3	fr(aA) = 0.3 × 0.7 = 0.21	fr(aa) = 0.3 × 0.3 = 0.09

Eggs

FIGURE 25.3 Calculating genotype frequencies from allele frequencies. Gametes represent withdrawals from the gene pool to form the genotypes of the next generation. In this population, the frequency of the A allele is 0.7, and the frequency of the a allele is 0.3. The frequencies of the genotypes in the next generation can be calculated: 0.49 for AA, 0.42 for Aa, and 0.09 for aa. Under the Hardy-Weinberg law, the frequencies of A and a remain constant from generation to generation. **GenCDX**

and reproduce, what will be the allele frequencies in the new gene pool? Recall that we are assuming that all genotypes have equal rates of survival and reproduction. This means that all genotypes contribute equally to the gene pool. *AA* individuals constitute 49 percent of the population, so the gametes they produce will constitute 49 percent of the gene pool. All of these gametes carry allele *A*. *Aa* individuals constitute 42 percent of the population, so their gametes will constitute 42 percent of the gene pool. Half of these gametes carry allele *A*. Thus the frequency of allele *A* in the gene pool is 0.49 + (1/2) 0.42 = 0.7. The other half of the gametes produced by *Aa* individuals carry allele *a*. And *aa* individuals constitute 9 percent of the population, so their gametes constitute 9 percent of the gene pool. All of these gametes carry allele *a*. Thus the frequency of allele *a* in the gene pool is (1/2) 0.42 + 0.09 = 0.3. As a check on our calculation, note that 0.7 + 0.3 = 1.0, accounting for all of the gametes in the gene pool.

We have arrived right back where we began, with a gene pool in which the frequency of allele *A* is 0.7 and the frequency of allele *a* is 0.3. We have demonstrated the Hardy-Weinberg law: The allele frequencies in our population remain unchanged from one generation to the next, and after just one generation of random mating the genotype frequencies can be predicted from the allele frequencies. The population does not evolve.

For the general case of the Hardy-Weinberg law, we replace the specific numerical values for the allele frequencies with variables. Imagine a gene pool in which the frequency of allele *A* is p and the frequency of allele *a* is q, such that $p + q = 1$. If we draw an egg and a sperm at random from the gene pool and pair them to make a zygote, the probability that both the egg and the sperm will carry

allele *A* is $p \times p$. Thus the frequency of genotype *AA* among the zygotes is p^2. The probability that the egg will carry *A* and the sperm will carry *a* is $p \times q$, and the probability that the egg will carry *a* and the sperm will carry *A* is $q \times p$. Thus the frequency of genotype *Aa* among the zygotes is $2pq$. Finally, the probability that both the egg and the sperm will carry *a* is $q \times q$. Thus the frequency of genotype *aa* among the zygotes will be q^2. The distribution of genotypes among the zygotes can be expressed as:

$$p^2 + 2pq + q^2 = 1 \qquad (25.1)$$

Figure 25.4 shows these calculations in graphical form.

When the zygotes develop into adults and reproduce, what will be the allele frequencies in the new gene pool? All of the gametes made by *AA* individuals carry allele *A*, as do half of the gametes made by *Aa* individuals. Thus the frequency of allele *A* is

$$p^2 + (\tfrac{1}{2}) 2pq = p^2 + pq \qquad (25.2)$$

Recall that $p + q = 1$. We can therefore substitute $(1 - p)$ for q in Equation 25.2, which gives

$$p^2 + p(1 - p) = p^2 + p - p^2 = p \qquad (25.3)$$

Likewise, half of the gametes made by *Aa* individuals carry allele *a*, as do all of the gametes made by *aa* individuals. Thus the frequency of allele *a* is

$$(\tfrac{1}{2}) 2pq + q^2 = pq + q^2 \qquad (25.4)$$

We can substitute $(1 - q)$ for p in Equation 25.4, which gives

$$(1 - q)q + q^2 = q - q^2 + q^2 = q \qquad (25.5)$$

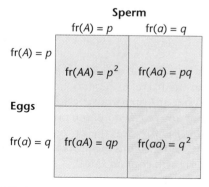

FIGURE 25.4 The general case of allele and genotype frequencies under Hardy-Weinberg assumptions. The frequency of allele *A* is p and the frequency of allele *a* is q. After mating, the three genotypes *AA*, *Aa*, and *aa* have the frequencies p^2, $2pq$, and q^2, respectively.

We have again arrived back where we began, with a gene pool in which the frequency of allele A is p and the frequency of allele a is q. The Hardy-Weinberg law is true for any frequencies of A and a, so long as the frequencies sum to 1 and the five assumptions hold. A population in which the allele frequencies remain constant from generation to generation and in which the genotype frequencies can be predicted from the allele frequencies is said to be in a state of **Hardy-Weinberg equilibrium.**

The Hardy-Weinberg law has several important consequences. First, the law shows that dominant traits do not automatically tend to increase in frequency from one generation to the next. Second, the law demonstrates that **genetic variability** can be maintained in a population, since, once established in a population, allele frequencies remain unchanged. Third, if we assume Hardy-Weinberg conditions, then knowing the frequency of just one genotype enables us to calculate the frequencies of other genotypes. This relationship is particularly useful in human genetics because it enables us to calculate the frequency of heterozygous carriers for a recessive genetic disorder even when all we know is the frequency of affected individuals (see later in the chapter).

We began this section by asking whether the frequency of the $CCR5$-$\Delta 32$ allele is expected to become more common in populations in which it is currently rare. We can now give a partial answer to this question. If individuals of all genotypes have equal rates of survival and reproduction, and if there is no mutation, and if no one migrates into or out of the population, and if the population is extremely large, and if individuals in the population choose their mates at random, then the frequency of the $\Delta 32$ allele will not change. Of course, for any real population, few if any of these assumptions are likely to be exactly true.

This observation illustrates the most important role of the Hardy-Weinberg law: It is the foundation upon which the rest of population genetics is built. By showing that under a short list of explicit assumptions populations do not evolve, the Hardy-Weinberg law identifies the forces that can cause evolution. As we will see later in the chapter, when the assumptions of the Hardy-Weinberg law are broken, because of natural selection, mutation, migration, and random sampling errors (also known as genetic drift), the allele frequencies in a population may change from one generation to the next. Nonrandom mating does not, by itself, alter allele frequencies, but by altering genotype frequencies it can indirectly affect the course of evolution. The Hardy-Weinberg law tells both theoreticans and empiricists where to look to find the causes of evolution in real populations. We will return to the $CCR5$-$\Delta 32$ allele later in the chapter to see how a population genetic perspective has generated fruitful hypotheses for further research.

Testing for Equilibrium

One way to establish whether one or more of the assumptions of the Hardy-Weinberg law is being violated in a population is to determine whether genotypes in the population are in equilibrium. To do this, researchers must first determine the frequencies of the genotypes, either directly from the phenotypes if heterozygotes are recognizable or by the analysis of proteins or DNA sequences. Researchers then calculate the allele frequencies from the genotype frequencies, as demonstrated earlier. Finally, they use the allele frequencies to predict the genotype frequencies they were calculated from. According to the Hardy-Weinberg law (Equation 25.1), the genotype frequencies are predicted to fit a $p^2 + 2pq + q^2 = 1$ relationship. If they do not, then one or more of the assumptions is invalid for the population.

The $CCR5$ genotypes of individuals in Britain (Table 25.3) provide an example of this application of the Hardy-Weinberg law. The sample includes 283 individuals, of which 223 had genotype $+/+$, 57 had genotype $+/\Delta 32$, and 3 had genotype $\Delta 32/\Delta 32$. These numbers represent genotype frequencies of $223/283 = 0.788$, $57/283 = 0.201$, and $3/283 = 0.011$, respectively. From the genotype frequencies we can calculate that the frequency of the $CCR5+$ allele is 0.89 and that the frequency of the $CCR5$-$\Delta 32$ allele is 0.11. From these allele frequencies, the Hardy-Weinberg law predicts the genotype frequencies as follows:

$$\text{Expected frequency of genotype } +/+ = p^2 = (0.89)^2 = 0.792$$

$$\text{Expected frequency of genotype } +/\Delta 32 = 2pq$$
$$= 2(0.89)(0.11) = 0.196$$

$$\text{Expected frequency of genotype } \Delta 32/\Delta 32 = q^2 =$$
$$(0.11)^2 = 0.012$$

These **expected frequencies** are nearly identical to the **observed frequencies** we just calculated from the data in Table 25.3. Our test of this population has failed to provide evidence that any of the assumptions are being violated. This conclusion is confirmed by a χ^2 analysis (see Chapter 3). The χ^2 value in this case is tiny: 0.00023. To be statistically significant at even the most generous, accepted level, $p = 0.05$, the χ^2 value would have to be 3.84. (In a test for Hardy-Weinberg equilibrium, the degrees of freedom are given by $k - 1 - m$, where k is the number of genotypes and m is the number of independent allele frequencies estimated from the data. Here, $k = 3$ and $m = 1$, since calculating only one allele frequency allows us to determine the other by subtraction. Thus there is $3 - 1 - 1 = 1$ degree of freedom.)

On the other hand, if the Hardy-Weinberg test had demonstrated that the population is not in equilibrium, it would have indicated that one or more of the assumptions is not being met. Imagine two hypothetical populations, one living on East Island, the other living on West Island. Everyone living on East Island has genotype $+/+$, so the frequency of the $CCR5+$ allele is 100 percent. Everyone living on West Island has genotype $\Delta32/\Delta32$, so the frequency of the $CCR5-\Delta32$ allele is 100 percent. Both of these populations are in Hardy-Weinberg equilibrium.

Now imagine that 500 people from each population move to the previously uninhabited Central Island. The allele and genotype frequencies for the source populations and the new population appear in Table 25.4. Both the $+$ allele and the $\Delta32$ allele are at frequencies of 0.5 in the new population, so the Hardy-Weinberg law predicts that the genotype frequencies would be $0.25 + 0.50 + 0.25 = 1$. These expected frequencies are obviously different from the observed frequencies; there are no heterozygotes at all on Central Island. The new population is not in Hardy-Weinberg equilibrium (and a χ^2 test would confirm this).

Two assumptions are being violated in the new population. First, everyone living on Central Island is a migrant. Second, the individuals on Central Island are not the products of random mating; everyone has either two parents from East Island or two parents from West Island. Note, however, that it would take only one generation of random mating on Central Island to bring the new population to the expected allele frequencies, as shown in Figure 25.5.

An example of a real human population that is not in Hardy-Weinberg equilibrium was documented by Therese Markow and colleagues. These researchers studied 122 Havasupai people, a population of Native Americans in Arizona. The biologists determined the genotype of each Havasupai individual at two loci in the major histocompatibility complex (MHC). These genes, *HLA-A* and *HLA-B*, encode proteins that are involved in the immune system's discrimination between self versus non-self. The immune systems of individuals that are heterozygous at MHC loci may recognize a greater diversity of foreign invaders and thus may be better able to fight disease. Markow and colleagues observed significantly more individuals heterozygous at both loci, and significantly fewer

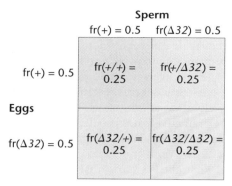

FIGURE 25.5 Hardy-Weinberg equilibrium. If mating is random in a population that also meets the other assumptions of the Hardy-Weinberg law, an equilibrium will be reached in one generation. Here the genotype frequencies after one generation of random mating are 0.25, 0.5, and 0.25; compare these with the genotype frequencies of the new population in Table 25.4.

homozygous individuals, than would be expected under the Hardy-Weinberg law. Violation of either or both of two assumptions could explain the excess of heterozygotes among the Havasupai. First, Havasupai fetuses, children, or adults who are heterozygous for *HLA-A* and *HLA-B* may have higher rates of survival than individuals who are homozygous. Second, rather than choosing their mates at random, Havasupai people may prefer mates whose MHC genotypes are different than their own.

The Hardy-Weinberg law can be extended to include more than two alleles

It is common to find several alleles of a single locus in a population. The ABO blood group in humans (discussed in Chapter 4) is such an example. The locus I (isoagglutinin) has three alleles (I^A, I^B, and I^O), yielding six possible genotypic combinations (I^AI^A, I^BI^B, I^OI^O, I^AI^B, I^AI^O, I^BI^O). Recall that in this case A and B are codominant alleles, and

TABLE 25.4 Frequencies of *CCR5+* and *CCR5-Δ32* Alleles in a Hypothetical Population Composed of 500 *+/+* Individuals and 500 *Δ32/Δ32* Individuals

Population	Allele Frequencies		Genotype Frequencies		
	CCR5+	*CCR5-Δ32*	*+/+*	*+/Δ32*	*Δ32/Δ32*
Source populations					
East Island	1.0	0	1.0	0	0
West Island	0	1.0	0	0	1.0
New population					
Observed	$p = 0.5$	$q = 0.5$	0.5	0	0.5
Expected			$p^2 = 0.25$	$2pq = 0.5$	$q^2 = 0.25$

both of these are dominant to *O*. The result is that homozygous *AA* and heterozygous *AO* individuals are phenotypically identical, as are *BB* and *BO* individuals, so we can distinguish only four phenotypic combinations.

By adding another variable to the Hardy-Weinberg equation, we can calculate both the genotype and allele frequencies for the situation involving three alleles. Let *p*, *q*, and *r* represent the frequencies of alleles *A*, *B*, and *O*, respectively. Note that because there are only three alleles

$$p + q + r = 1 \qquad (25.6)$$

Under Hardy-Weinberg assumptions, the frequencies of the genotypes will be given by

$$(p + q + r)^2 = p^2 + q^2 + r^2 + 2pq + 2pr + 2qr = 1 \quad (25.7)$$

If we know the frequencies of blood types for a population, we can then estimate the frequencies for the three alleles of the ABO system. For example, in one population sampled, the following blood type frequencies were observed: A = 0.53, B = 0.13, AB = 0.08, and O = 0.26. Because the *O* allele is recessive, the frequency of type O blood in the population is equal to the frequency of the recessive genotype r^2. Thus,

$$r^2 = 0.26$$
$$r = \sqrt{0.26}$$
$$= 0.51$$

By using the value estimated for *r*, we can estimate the allele frequencies for the *A* and *B* alleles. The *A* allele is present in two genotypes, *AA* and *AO*. The frequency of the *AA* genotype is represented by p^2, and the *AO* genotype by $2pr$. Therefore, the combined frequency of type A blood and type O blood is given by

$$p^2 + 2pr + r^2 = 0.53 + 0.26$$

If we factor the left side of the equation and compute the sum of the terms on the right we get

$$(p + r)^2 = 0.79$$
$$p + r = \sqrt{0.79}$$
$$p = 0.89 - r$$
$$p = 0.89 - 0.51$$
$$= 0.38$$

Having estimated *p* and *r*, the frequencies of allele *A* and allele *O*, the frequency for the *B* allele can be estimated from Equation 25.6:

$$p + q + r = 1$$
$$q = 1 - p - r$$
$$= 1 - 0.38 - 0.51$$
$$= 0.11$$

The phenotypic frequencies and genotypic frequencies for this population are summarized in Table 25.5.

The Hardy-Weinberg law provides a method for calculating the frequency of heterozygotes in a population

One of the practical applications of the Hardy-Weinberg law is that it allows us to estimate the frequency of heterozygotes in a population. The frequency of a recessive phenotype usually can be determined by counting such individuals in a sample of the population. With this information and the Hardy-Weinberg law, we can then calculate the allele and genotype frequencies.

Cystic fibrosis, an autosomal recessive trait, has an incidence of about $1/2500 = 0.0004$ in people of northern European ancestry. Individuals with cystic fibrosis are easily distinguished from the population at large by a suite of

TABLE 25.5 Calculating Genotype Frequencies for Multiple Alleles Where the Frequency of Allele *A* = 0.38, Allele *B* = 0.11, and Allele *O* = 0.51

Genotype	Genotype Frequency	Phenotype	Phenotype Frequency
AA	$p^2 = (0.38)^2 = 0.14$	A	0.53
AO	$2pr = 2(0.38)(0.51) = 0.39$		
BB	$q^2 = (0.11)^2 = 0.01$	B	0.12
BO	$2qr = 2(0.11)(0.51) = 0.11$		
AB	$2pq = 2(0.38)(0.11) = 0.084$	AB	0.08
OO	$r^2 = (0.51)^2 = 0.26$	O	0.26

symptoms, including salty skin, production of an excess of thick mucus in the lungs, and susceptibility to bacterial infections. Because this is a recessive trait, individuals with cystic fibrosis must be homozygous. Their frequency in a population is represented by q^2, provided that mating has been at random and all Hardy-Weinberg conditions have been met in the previous generation.

The frequency of the recessive allele therefore is

$$q = \sqrt{q^2} = \sqrt{0.0004} \qquad (25.8)$$
$$= 0.02$$

Since $p + q = 1$, then the frequency of p is

$$p = 1 - q \qquad (25.9)$$
$$= 1 - 0.02$$
$$= 0.98$$

In the Hardy-Weinberg equation, the frequency of heterozygotes is given as $2pq$. Therefore, we have

$$\text{Heterozygote frequency} = 2pq \qquad (25.10)$$
$$= 2(0.98)(0.02)$$
$$= 0.04, \text{ or } 4 \text{ percent, or } 1/25$$

Thus, heterozygotes for cystic fibrosis are rather common in the population (4 percent), even though the incidence of homozygous recessives is only 1/2500, or 0.04 percent.

In general, the frequencies of all three genotypes can be estimated once the frequency of either allele is known and Hardy-Weinberg conditions are assumed. The relationship between genotype frequency and allele frequency is shown in Figure 25.6. It is important to note how fast heterozygotes increase in a population as the values of p and q move away from zero or one. This observation confirms our conclusion that when a recessive trait such as cystic fibrosis is rare, the majority of those carrying the allele are heterozygotes. In populations in which the frequencies of p and q are between 0.33 and 0.67, heterozygotes are at higher frequency than is either homozygote.

Natural selection is a potent force of evolution

We have said that the Hardy-Weinberg law establishes a set of ideal conditions that allows us to estimate allele frequencies and genotype frequencies in populations in which the initial assumptions about random mating, absence of selection and mutation, and equal viability and fecundity

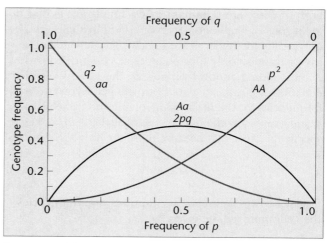

FIGURE 25.6 Relationship between genotype frequencies and allele frequencies derived from the Hardy-Weinberg equation.

hold true. Obviously, it is difficult to find natural populations in which all these conditions are met. In nature, populations are dynamic, and changes in size and structure are common. This situation allows us to use the Hardy-Weinberg law to investigate exceptions to the initial assumptions. In this section, and the sections that follow, we will discuss factors that prevent populations from reaching equilibrium, or that drive populations toward a different equilibrium, and the relative contribution of these factors to evolutionary change.

Natural Selection

The first assumption of the Hardy-Weinberg law is that individuals of all genotypes have equal rates of survival and equal reproductive success. If this assumption is violated, allele frequencies may change from one generation to the next. To see why, imagine a population of 100 individuals in which the frequency of allele A is 0.5 and the frequency of allele a is 0.5. Assuming the previous generation mated at random, the genotype frequencies in the present generation are $(0.5)^2 = 0.25$ for AA, $2(0.5)(0.5) = 0.5$ for Aa, and $(0.5)^2 = 0.25$ for aa. Because our population contains 100 individuals, we have 25 AA individuals, 50 Aa individuals, and 25 aa individuals.

Now imagine that individuals with different genotypes have different rates of survival: All 25 AA individuals survive to reproduce, 90 percent or 45 of the Aa individuals survive to reproduce, and 80 percent or 20 of the aa individuals survive to reproduce. When the survivors reproduce, imagine that each contributes two gametes to the gene pool, giving us $2(25) + 2(45) + 2(20) = 180$ gametes. What are the frequencies of the two alleles? We have 50

A gametes from *AA* individuals, plus 45 *A* gametes from *Aa* individuals, so the frequency of allele *A* is (50 + 45)/180 = 0.53. We have 45 *a* gametes from *Aa* individuals, plus 40 *a* gametes from *aa* individuals, so the frequency of allele *a* is (45 + 40)/180 = 0.47. These are different from the frequencies we started with, the frequency of *A* has increased, while the frequency of *a* has declined.

A difference among individuals in the rate of survival and/or reproduction is called **natural selection.** Natural selection is the principal force that shifts allele frequencies within large populations and is one of the most important factors in evolutionary change.

Fitness and Selection

Selection occurs whenever individuals with a particular genotype enjoy an advantage in survival or reproduction over other genotypes. However, selection may be weak or strong, depending on the magnitude of the advantage. In the numerical example we worked above, selection was strong. Weak selection might involve just a fraction of a percent difference in the survival rates of different genotypes. Advantages in survival and reproduction both ultimately translate into an increased genetic contribution to future generations. An individual's genetic contribution to future generations is called its **fitness.** Thus, genotypes that are associated with high rates of survival and/or high reproductive success are said to have high fitness, whereas genotypes associated with low rates of survival and/or low reproductive success are said to have low fitness.

We can incorporate differences in fitness into the Hardy-Weinberg analysis by using variables to represent the relative fitness of different genotypes. By convention, population geneticists use the letter w to represent fitness. Thus we can use w_{AA} to represent the relative fitness of genotype *AA*, w_{Aa} to represent the relative fitness of genotype *Aa*, and w_{aa} to represent the relative fitness of genotype *aa*. Assigning the values $w_{AA} = 1$, $w_{Aa} = 0.9$, and $w_{aa} = 0.8$ could mean, for example, that all *AA* individuals survive, 90 percent of the *Aa* individuals survive, and 80 percent of the *aa* individuals survive, as in the example above.

Let us consider selection against deleterious alleles. Fitness values $w_{AA} = 1$, $w_{Aa} = 1$, and $w_{aa} = 0$ describe a situation in which allele *a* is a lethal recessive. As homozygous recessive individuals die without leaving offspring, the frequency of allele *a* will decline. The decline in the frequency of allele *a* is described by the equation

$$q_g = q_0 / (1 + gq_0) \qquad (25.11)$$

where q_g is the frequency of allele *a* in generation *g*, q_0 is the starting frequency of *a* (that is, the frequency of *a* in

generation zero), and *g* is the number of generations that have passed.

Figure 25.7 shows what happens to a lethal recessive allele with an initial frequency of 0.5. At first, because of the high percentage of *aa* genotypes, the frequency of allele *a* declines rapidly. The frequency of *a* is halved in only two generations. By the sixth generation, the frequency is halved again. By now, however, the majority of *a* alleles are carried by heterozygotes. Because *a* is recessive, these heterozygotes are not selected against. This means that as time continues to pass, the frequency of allele *a* declines ever more slowly. As long as heterozygotes continue to mate, it is difficult for selection to completely eliminate a recessive allele from a population.

Of course, a deleterious allele may not be completely recessive. A great many other scenarios are possible. A deleterious allele may be codominant, so that heterozygotes have intermediate fitness. Or an allele may be dele-

Generation	p	q	p²	2pq	q²
0	0.50	0.50	0.25	0.50	0.25
1	0.67	0.33	0.44	0.44	0.12
2	0.75	0.25	0.56	0.38	0.06
3	0.80	0.20	0.64	0.32	0.04
4	0.83	0.17	0.69	0.28	0.03
5	0.86	0.14	0.73	0.25	0.02
6	0.88	0.12	0.77	0.21	0.01
10	0.91	0.09	0.84	0.15	0.01
20	0.95	0.05	0.91	0.09	<0.01
40	0.98	0.02	0.95	0.05	<0.01
70	0.99	0.01	0.98	0.02	<0.01
100	0.99	0.01	0.98	0.02	<0.01

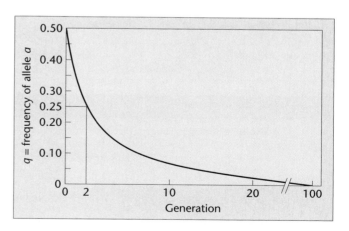

FIGURE 25.7 Change in the frequency of a lethal recessive allele, *a*. The frequency of *a* is halved in two generations, and halved again by the sixth generation. Subsequent reductions occur slowly because the majority of *a* alleles are carried by heterozygotes.

terious in the homozygous state, but beneficial in the heterozygous state, like the allele for sickle-cell anemia. The following formula calculates the frequency of allele a in the next generation for any selection scenario, given the frequencies of a and A in this generation and the fitnesses of all three genotypes:

$$q_{g+1} = [\mathrm{w}_{Aa}p_g q_g + \mathrm{w}_{aa}q_g^2] / [\mathrm{w}_{AA}p_g^2 + \mathrm{w}_{Aa}2p_g q_g + \mathrm{w}_{aa}q_g^2]$$

$$(25.12)$$

where q_{g+1} is the frequency of allele a in the next generation, q_g is the frequency of allele a in this generation, p_g is the frequency of allele A in this generation, and w_{AA}, w_{Aa}, and w_{aa} are the fitnesses of genotype AA, Aa, and aa, respectively.

Figure 25.8 shows a variety of examples in which a deleterious allele is codominant, but the intensity of selection varies from strong to weak. In each case the frequency of the deleterious allele, a, starts at 0.99 and declines over time. However, the rate of decline depends heavily on the strength of selection. When only 90 percent of the heterozygotes and 80 percent of the aa homozygotes survive (red curve), the frequency of allele a drops from 0.99 to less than 0.01 in 85 generations. However, when 99.8 percent of the heterozygotes and 99.6 percent

of the aa homozygotes survive (blue curve), it takes 1000 generations for the frequency of allele a to drop from 0.99 to 0.93. There are two important messages in this graph. First, given thousands of generations, even weak selection can cause substantial changes in allele frequencies; because evolution occurs over vast spans of time, selection is a powerful force of evolution. Second, for selection to produce rapid changes in allele frequencies, changes measurable within a human life span, the differences in fitness among genotypes must be large.

Our analysis of how selection affects allele frequencies allows us to make some inferences about the *CCR5-Δ32* allele that we discussed earlier. Because individuals with genotype *Δ32/Δ32* are immune to most sexually transmitted strains of HIV-1, while individuals with genotypes *+/+* and *+/Δ32* are susceptible, we might expect that the AIDS epidemic will cause the frequency of the *Δ32* allele to increase over time. Indeed it probably will, but the increase in frequency is likely to be slow in human terms. We can do a rough calculation as follows. Imagine a population in which the current frequency of the *Δ32* allele is 0.10. Under Hardy-Weinberg conditions, the genotype frequencies in this population are 0.81 for *+/+*, 0.18 for *+/Δ32*, and 0.01 for *Δ32/Δ32*. Imagine also that 1 percent of the *+/+* and *+/Δ32* individuals in this population will contract HIV and die of AIDS. (This figure is high for North American and European populations; the Joint United Nations Programme on HIV/AIDS estimates that as of December 1998, 0.56 percent of the adult population of North America and 0.25 percent of the adult population of western Europe were infected with HIV.)

Based on our assumptions, we can assign fitnesses to the genotypes as follows: $\mathrm{w}_{+/+} = 0.99$; $\mathrm{w}_{+/Δ32} = 0.99$; $\mathrm{w}_{Δ32/Δ32} = 1.0$. Given these fitnesses, Equation 25.12 predicts that the frequency of the *CCR5-Δ32* allele in the next generation will be 0.100091. In fact, it will take about 100 generations for the frequency of the *Δ32* allele to reach just 0.11 (Figure 25.9). In other words, the frequency of the *Δ32* allele will probably not change much over the next few generations in most populations that currently harbor it.

This conclusion depends heavily on our assumptions about the rate of infection and the intitial frequency of the allele. If the infection rate were higher in our hypothetical population (UNAIDS estimates that in Botswana and Zimbabwe 25 percent of the adult population is infected with HIV) and if the initial frequency of the *Δ32* allele were higher (so that the frequency of heterozygotes were higher), Equation 25.12 predicts that the frequency of the *Δ32* allele would change at a rate of a few percentage points per generation (Figure 25.9).

A population genetic perspective sheds light on the *CCR5-Δ32* story in other ways as well. Frédérick Libert and colleagues, and J. Claiborne Stephens and colleagues,

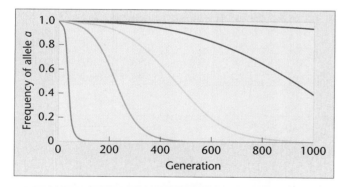

Selection against allele *a*				
Strong ◄─────────────────────────────► Weak				
	───	───	───	───	───
w_{AA}	1.0	1.0	1.0	1.0	1.0
w_{Aa}	0.90	0.98	0.99	0.995	0.998
w_{aa}	0.80	0.96	0.98	0.99	0.996

FIGURE 25.8 The effect of selection on allele frequency. The rate at which a deleterious allele is removed from a population depends heavily on the strength of selection.

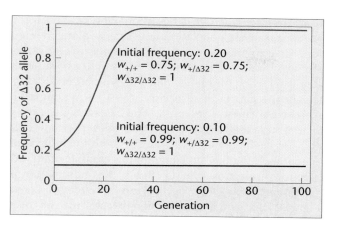

FIGURE 25.9 The rate at which the frequency of the *CCR5-Δ32* allele changes in hypothetical populations with different initial frequencies and different fitnesses.

have analyzed genetic variation at marker loci closely linked to the *CCR5* gene. Both groups concluded that most, if not all extant copies of the *Δ32* allele are descended from a single ancestral copy that appeared in northeastern Europe at most a few thousand years ago. In fact, Stephens estimates that the common ancestor of all *Δ32* alleles existed just 700 years ago. How could a new allele rise from a frequency of virtually zero to as high as 20 percent in roughly 30 generations?

It seems that there must have been strong selection in favor of the *Δ32* allele. The agent of selection cannot have been HIV-1, because HIV-1 moved from chimpanzees to humans too recently. Stephens suggests that the agent of selection was instead bubonic plague. During the Black Death of 1346 to 1352, between a quarter and a third of all Europeans died from plague. Bubonic plague is caused by the bacterium *Yersinia pestis*. This bacterium manufactures a protein that can kill some kinds of white blood cells. Stephens hypothesizes that the process through which the bacterial protein kills white cells involves the product of the *CCR5* gene. If true, this hypothesis would indicate a mechanism that would make individuals lacking the CCR5 protein more likely to survive plague epidemics. As of this writing, the bubonic plague hypothesis has not been tested.

If past epidemics of some infectious disease are responsible for the high frequency of the *CCR5-Δ32* allele in European populations, then the virtual absence of the allele in non-European populations is puzzling. Perhaps the disease responsible has been more common in Europe than elsewhere. Alternatively, perhaps other populations have different alleles of the *CCR5* gene that also confer protection against the disease. Three teams of researchers looking for other alleles of the *CCR5* gene in various pop-

ulations have found a total of 20 mutant alleles, including *Δ32*. Sixteen of these alleles encode proteins different in structure from that encoded by the *CCR5+* allele. Some, like *Δ32*, are loss-of-function alleles. Some of the alleles appear confined to Asian populations, others to African populations. Their frequencies are as high as 3 to 4 percent. Together, these discoveries are consistent with the hypothesis that alteration or loss of the CCR5 protein protects against an as-yet-unidentified infectious disease.

Selection in Natural Populations

Evolutionary geneticists have done a great deal of research on the effect of natural selection on allele frequencies in both laboratory and natural populations. Among the most detailed studies of natural populations are those that involve insects exposed to pesticides. Christine Chevillon and colleagues, for example, studied the effect of the insecticide chlorpyrifos on allele frequencies in populations of house mosquitoes. Chlorpyrifos kills mosquitoes by interfering with the function of the enzyme acetylcholinesterase (ACE), which under normal circumstances breaks down the neurotransmitter acetylcholine. An allele of the gene encoding ACE called *Ace^R*, encodes a slightly altered version of ACE that is immune to interference by chlorpyrifos.

Chevillon measured the frequency of the *Ace^R* allele in nine populations located geographically along a line. In the first four locations along the line, chlorpyrifos had been used to control mosquitoes for 22 years; in the last five locations, chlorpyrifos had never been used. Chevillon predicted that the frequency of *Ace^R* would be higher in the exposed than in the unexposed populations. The researchers also predicted that the frequencies of alleles for enzymes unrelated to the physiological effects of chlorpyrifos would show no such pattern. Among the control enzymes the researchers studied was aspartate amino transferase 1.

The results appear in Figure 25.10. As the researchers predicted, the frequency of the *Ace^R* allele was significantly higher in the populations on the exposed end of the geographic line than in the populations on the unexposed end. Also as predicted, the frequencies of the most common alleles of the control enzyme genes showed no geographic trends. The explanation is that during the past 22 years mosquitoes in the exposed populations had higher rates of survival if they carried the *Ace^R* allele. In other words, the *Ace^R* allele had been favored by natural selection.

Natural Selection Affecting Quantitative Traits

Most phenotypic traits are controlled not by alleles at a single locus but instead by the combined influence of the individual's genotype at many different loci and the envi-

House mosquito, *Culex pipiens*

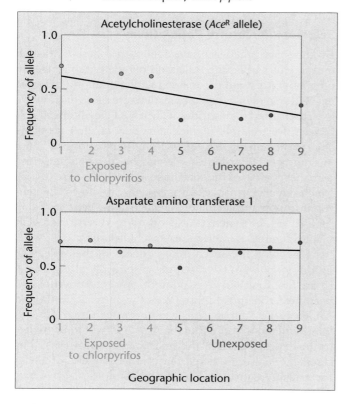

FIGURE 25.10 The effect of selection on allele frequencies in natural populations. (a) The frequency of the *Ace*^R^ allele, which confers resistance to the insecticide chlorpyrifos, is higher in house mosquito populations exposed to chlorpyrifos than in unexposed populations. (b) The frequency of an allele for an enzyme unrelated to chlorpyrifos metabolism (aspartate amino transferase 1) shows no such pattern.

ronment. To the extent that such quantitative traits are heritable, they respond to selection. Quantitative traits, including adult body height and weight in humans, often demonstrate a continuously varying distribution resembling a bell-shaped curve. Selection on such traits can be classified as directional, stabilizing, or disruptive.

In **directional selection,** important to plant and animal breeders, desirable traits, often representing phenotypic extremes, are selected. If the trait is polygenic, the

most extreme phenotypes that the genotype can express will appear in the population only after prolonged selection. An example of directional selection is the long-running experiment at the State Agricultural Laboratory in Illinois selecting for high and low oil content in corn kernels. At the start of the experiment in 1896, a population of 163 corn ears was surveyed. The 24 ears highest in oil content were used as the founders of a new population. In each succeeding generation, the ears with the highest oil content were used for breeding. In this example of upward directional selection, the oil content was raised after 50 generations from about 4 percent to just over 16 percent (Figure 25.11), and there is no sign that a plateau has been reached yet. A second population was founded with the 12 ears with the lowest oil content and downward directional selection from this founding group has lowered the oil content from 4 percent to less than 1 percent over the period of 50 generations (Figure 25.11). In this case, the high and low lines have diverged to the point where the distribution of oil content in each line does not overlap with that of the other line.

In the upwardly selected population, the alleles for high oil content increase in frequency and replace the alleles for lower oil content. At some point, all individuals in the population will have the genotype for the highest oil content, and all will express the most extreme phenotype for high oil content. In the downwardly selected population, the opposite situation will occur, eventually producing a population with a genotype and phenotype for the lowest oil content. When the lines fail to respond to further selection, there will be no remaining genetic variance for oil content, heritability will be zero, and each line will have a homozygous genotype for oil content.

In nature, directional selection can occur when one of the phenotypic extremes becomes selected for or

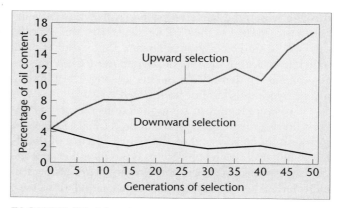

FIGURE 25.11 Long-term selection to alter the oil content of corn kernels.

against, usually as a result of changes in the environment. A meticulously documented example comes from research by Peter R. Grant and colleagues on the medium ground finches (*Geospiza fortis*) of Daphne Major island in the Galápagos archipelago. The beak size of these birds varies enormously. In 1976, for example, some birds in the population had beaks less than 7 mm deep, while others had beaks more than 12 mm deep. Beak size is heritable, which means that large-beaked parents tend to have large-beaked offspring, and small-beaked parents tend to have small-beaked offspring. In 1977 a severe drought on Daphne killed some 80 percent of the finches. Big-beaked birds survived at higher rates than small-beaked birds because when food became scarce the big-beaked birds were able to eat a greater variety of seeds. When the drought ended in 1978 and the survivors paired off and bred, the chicks inherited their parents' big beaks. Between 1976 and 1978, the beak depth of the average finch in the Daphne Major population increased by just over 0.5 mm.

Stabilizing selection, in contrast, tends to favor intermediate types, with both extreme phenotypes being selected against. One of the clearest demonstrations of stabilizing selection is provided by the data of Mary Karn and Sheldon Penrose on human birth weight and survival for 13,730 children born over an 11-year period. Figure 25.12 shows the distribution of birth weight and the percentage of mortality at 4 weeks of age. Infant mortality increases on either side of the optimal birth weight of 7.5 pounds, and quite dramatically so at the low end. At the genetic level, stabilizing selection acts to keep a population well adapted to its environment. In this situation, individuals closer to the average for a given trait will have higher fitness.

Disruptive selection is selection against intermediates and for both phenotypic extremes, within a single pop-

ulation. It may be viewed as the opposite of stabilizing selection because the intermediate types are selected against. In one set of experiments, John Thoday applied disruptive selection to a population of *Drosophila* based on bristle number. In every generation he allowed only the flies with high or low bristle numbers to breed. After several generations, most of the flies could be easily placed in a low or high bristle category (Figure 25.13). In natural populations, such a situation might exist for a population in a heterogeneous environment.

The types and effects of selection are summarized in Figure 25.14.

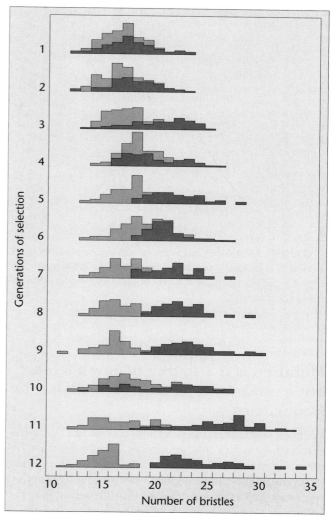

FIGURE 25.13 The effect of disruptive selection on bristle number in *Drosophila*. When individuals with the highest and lowest bristle number were selected, the population showed a nonoverlapping divergence in only 12 generations.

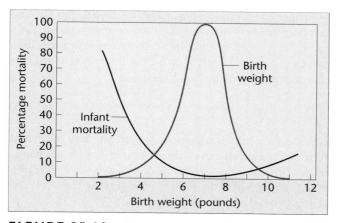

FIGURE 25.12 Relationship between birth weight and mortality in humans.

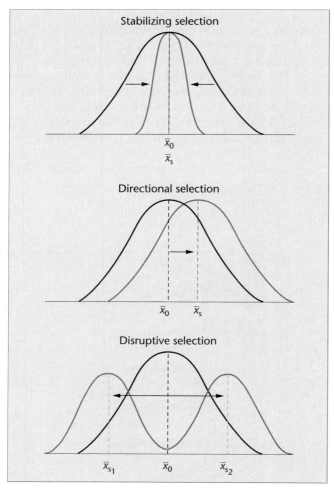

FIGURE 25.14 Comparison of the impact of stabilizing, directional, and disruptive selection. In each case, $\overline{X}_0$ is the mean of an original population (blue), and $\overline{X}_S$ is the mean of the population following selection (red).

Mutation is the source of new alleles, but a weak force of change in allele frequencies

Within a population, the gene pool is reshuffled each generation to produce new genotypes in the offspring. Because the number of possible multilocus genotypes is so large, the members of a population alive at any given time represent only a fraction of all the possible genotypes. The enormous genetic reserve present in the gene pool allows Mendelian assortment and recombination to continuously produce new multilocus genotypes. But assortment and recombination do not produce new alleles. **Mutation** alone acts to create new alleles. It is important to keep in

mind that mutational events occur at random—that is, without regard for any possible benefit or disadvantage to the organism. In this section we consider whether mutation is, by itself, a significant factor in causing allele frequencies to change.

To know whether mutation is a significant force in changing allele frequencies, we must measure the rate at which mutations are produced. As most mutations are recessive, it is difficult to observe mutation rates directly in diploid organisms. Indirect methods using probability and statistics or large-scale screening programs must be employed. For certain dominant mutations, however, a direct method of measurement can be used. To ensure accuracy, several conditions must be met:

1. The trait must produce a distinctive phenotype that can be distinguished from similar traits produced by recessive alleles.

2. The trait must be fully expressed or completely penetrant so that mutant individuals can be identified.

3. An identical phenotype must never be produced by nongenetic agents such as drugs or chemicals.

Mutation rates can be expressed as the number of new mutant alleles per given number of gametes. Suppose that for a given gene that undergoes mutation to a dominant allele, 2 out of 100,000 births exhibit a mutant phenotype. In these two cases, the parents are phenotypically normal. Because the zygotes that produced these births each carry two copies of the gene, we have actually surveyed 200,000 copies of the gene (or 200,000 gametes). If we assume that the affected births are each heterozygous, we have uncovered 2 mutant alleles out of 200,000. Thus the mutation rate is 2/200,000 or 1/100,000, which in scientific notation would be written as 1×10^{-5}.

In humans, a dominant form of dwarfism known as **achondroplasia** fulfills the requirements for the measurement of mutation rates. Individuals with this skeletal disorder have an enlarged skull, short arms and legs, and can be diagnosed by radiologic examination at birth. In a survey of almost 250,000 births, the mutation rate (μ) for achondroplasia has been calculated as

$$1.4 \times 10^{-5} \pm 0.5 \times 10^{-5} \qquad (25.13)$$

Knowing the rate of mutation, we can estimate the extent to which mutation can cause allele frequencies to change from one generation to the next. We will use d to represent the normal allele, and D to represent the allele for achondroplasia.

Imagine a population of 500,000 individuals in which everyone has genotype dd. The initial frequency of d is one, and the initial frequency of D is zero. If each individual contributes 2 gametes to the gene pool, the gene pool will contain 1,000,000 gametes, all carrying allele d. While the gametes are in the gene pool, 1.4 of every 100,000 d alleles is converted by mutation into a D allele. The frequency of allele d is now $(1,000,000 - 14)/1,000,000 = 0.999986$, and the frequency of D is $14/1,000,000 = 0.000014$. It is going to be long time before mutation, by itself, causes any appreciable change in the allele frequencies in this population.

More generally, if we have two alleles, A with frequency p and a with frequency q, and if μ represents the rate of mutations converting copies of A into copies of a, then the frequencies of the alleles in the next generation are given by:

$$p_{g+1} = p_g - \mu p_g \qquad (25.14)$$

and

$$q_{g+1} = q_g + \mu p_g \qquad (25.15)$$

where p_{g+1} and q_{g+1} represent the allele frequencies in the next generation, and p_g and q_g represent the allele frequencies in this generation.

Figure 25.15 shows the change over time in the frequency of allele A for a population in which the initial frequency of A is one, and the rate of mutations converting A into a is 1.0×10^{-5}. It takes about 70,000 generations to reduce the frequency of A to 0.5. Even if the rate of mutation were to increase through exposure to higher levels of radioactivity or chemical mutagens, the impact of mutation on allele frequencies would be extremely weak. Mutation is the ultimate source of the genetic variability that provides the raw material for evolution, but *by itself* mutation plays a relatively insignificant role in changing allele

frequencies. Instead, the fate of alleles created by mutation is more likely to be determined by natural selection (discussed previously) and genetic drift (discussed later).

An evolutionary perspective on mutation can lead to medical discoveries. The autosomal recessive disease cystic fibrosis, which we discussed previously in relation to calculating heterozygote frequencies, is caused by loss-of-function mutations in the gene for a cell-surface protein called the cystic fibrosis transmembrane conductance regulator (CFTR). The overall frequency of mutant alleles is about 2 percent in European populations. Until recently, most individuals with two mutant alleles died before reproducing, meaning that selection against homozygous recessives was rather strong. This creates a puzzle: In the face of selection against them, what has maintained the mutant alleles at a total frequency of 2 percent?

One hypothesis, the **mutation-selection balance hypothesis**, posits that mutation is constantly creating new alleles to replace the ones being eliminated by selection. However, for this scenario to work, the rate of mutations creating new cystic fibrosis alleles would have to be very high, on the order of 5×10^{-4}, to counteract the effects of selection. Many evolutionary geneticists instead prefer an alternative explanation, the heterozygote superiority hypothesis. According to the **heterozygote superiority hypothesis**, selection against homozygous mutant individuals is counterbalanced by selection in favor of heterozygotes. The most popular agent of selection in favor of heterozygotes was resistance to an as-yet-unidentified disease.

Recent work by Gerald Pier and colleagues suggests that CFTR heterozygotes may, in fact, have enhanced resistance to typhoid fever. Typhoid fever is caused by the bacterium *Salmonella typhi*. Typhoid fever bacteria launch an infection by infiltrating cells of the intestinal lining. In laboratory studies, Pier exposed mouse intestinal cells to *S. typhi* bacteria. Mouse cells that were heterozygous for *CFTR-ΔF508*, the analogue of the most common cystic fibrosis mutation in humans, acquired 86 percent fewer bacteria than did cells homozygous for the wild-type allele. Whether humans heterozygous for *CFTR-Δ508* also enjoy resistance to typhoid fever remains to be definitively established. If they do, then cystic fibrosis will join sickle-cell anemia as an example of heterozygote superiority.

Migration homogenizes allele frequencies across populations

Frequently, a species of plant or animal becomes divided into subpopulations that are to some extent geographically separated. Various evolutionary forces, including selection, can establish different allele frequencies in the

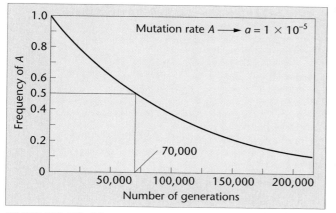

FIGURE 25.15 Rate of replacement of an allele by mutation alone, assuming an average mutation rate of 1.0×10^{-5}.

subpopulations. **Migration** occurs when individuals move between the subpopulations. Imagine a population in which a single locus has two alleles, A and a. Let p_m represent the frequency of A in the mainland subpopulation, and p_i represent the frequency of A in the subpopulation on an island to which a number of new individuals have just migrated. The island frequency of A in the next generation (p_i') is given by

$$p_i' = (1 - m)p_i + mp_m \qquad (26.16)$$

where m represents the frequency of migrants in the island population.

For example, assume that $p_i = 0.4$ and $p_m = 0.6$, and that 10 percent of the parents giving rise to the next generation on the island are immigrants, so that $m = 0.1$. In the next generation, the frequency of A on the island will increase as follows

$$p_i' = [(1 - 0.1) \times 0.4] + (0.1 \times 0.6) \qquad (25.17)$$
$$= 0.36 + 0.06$$
$$= 0.42$$

If either m is large or if p_m is very different from p_i, then a rather large change in the frequency of A can occur in a single generation. If migration is the only force acting to change the allele frequency on the island, then an equilibrium will be attained only when $p_i = p_m$. These calculations reveal that the change in allele frequency attributable to migration is proportional to the differences in allele frequency between the donor and recipient populations and to the rate of migration. As m can have a wide range of values, the effect of migration can substantially alter allele frequencies in populations, as shown for the ABO blood group in Figure 25.16. Although migration can be somewhat difficult to quantify, it can often be estimated.

Migration can also be regarded as the flow of genes between populations that were once, but are no longer, geographically isolated. Esteban Parra and colleagues measured allele frequencies for several different DNA sequence polymorphisms in African-American and European-American populations, and in African and European populations representative of the ancestral populations from which African-Americans and European-Americans are descended. One locus they studied, a restriction site polymorphism called *FY-NULL*, has two alleles, *FY-NULL*1* and *FY-NULL*2*. Figure 25.17 shows the frequency of *FY-NULL*1* in each population. The frequency of this allele is 0 in the three African populations, and 1.0 in the three European populations, but lies between these extremes in all African-American and Euro-

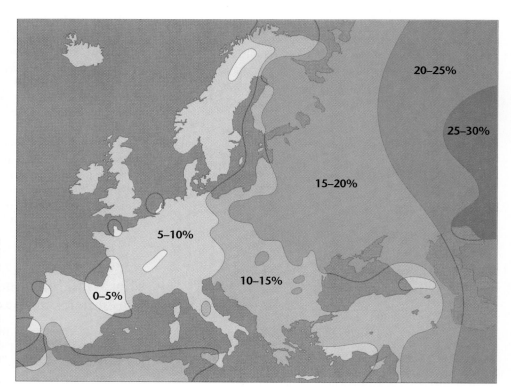

FIGURE 25.16 Migration as a force of evolution. The frequency of the *B* allele of the *ABO* locus is present in a gradient from east to west. This allele has its highest frequency in Central Asia, and declines to its lowest point in northeastern Spain. The gradient parallels the waves of Mongol migration into Europe following the fall of the Roman Empire and is a genetic relic of human history.

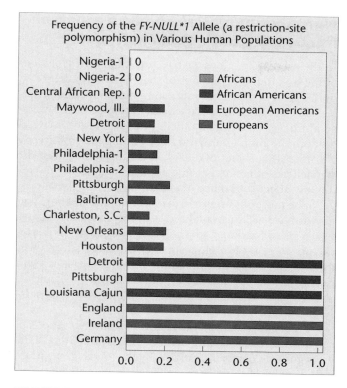

FIGURE 25.17 Frequencies of a restriction site polymorphism allele in several populations. These and other data demonstrate that African-American and European-American populations are of mixed ancestry.

pean-American populations. The simplest explanation for these data is that there has been admixture of genes between American populations with predominantly African ancestry and American populations with predominantly European ancestry. Based on *FY-NULL* and several other loci, Parra and colleagues estimate that African-American populations derive between 11.6 and 22.5 percent of their ancestry from Europeans, and that European-American populations derive between 0.5 and 1.2 percent of their ancestry from Africans.

Genetic drift is random change in allele frequencies and is especially important in small populations

In laboratory crosses, one condition that is essential to realizing theoretical genetic ratios (e.g., 3:1, 1:2:1, 9:3:3:1) is a fairly large sample size. A large sample size is also important to the study of population genetics as allele and genotype frequencies are examined or predicted. For ex-

ample, if a population consists of 1000 randomly mating heterozygotes (*Aa*), the next generation will consist of approximately 25 percent *AA*, 50 percent *Aa*, and 25 percent *aa* genotypes. Provided that the initial and subsequent populations are large, there will be at most only minor deviations from this mathematical ratio, and the frequencies of *A* and *a* will remain about equal (at 0.50).

However, if a population consists of only one set of heterozygous parents and they produce only two offspring, the allele frequency can change drastically. In this case, genotypes and allele frequencies in the offspring can be predicted, as shown in Table 25.6. As shown, 10 of 16 times such a cross is made, the allele frequencies would be altered. In 2 of 16 times, either the *A* or *a* allele would be eliminated in a single generation.

Founding a population with only one set of heterozygous parents that produce two offspring is an extreme example, but it illustrates the point that large interbreeding populations are essential to the maintenance Hardy-Weinberg equilibrium. In small populations, significant random fluctuations in allelic frequencies are possible by chance deviation. The degree of fluctuation will increase as the population size decreases, a situation known as **genetic drift**. In the extreme case, genetic drift may lead to the chance fixation of one allele to the exclusion of another allele.

To study genetic drift in laboratory populations of *Drosophila melanogaster*, Warwick Kerr and Sewall Wright set up over 100 lines, with four males and four females as the parents for each line. Within each line, the frequency of the sex-linked bristle mutant *forked* (*f*) and its wild-type allele (*f*⁺) was 0.5. In each generation, four males and four females were chosen at random to be parents of the next generation. After 16 generations, the complete loss of one allele and the fixation of the other had occurred in 70 lines—29 in which only the *forked* allele was present and 41 in which the wild-type allele had become fixed. The rest of the lines were still segregating the two alleles or had gone extinct. If fixation had occurred randomly,

TABLE 25.6 All Possible Pairs of Offspring Produced by Two Heterozygous Parents (*Aa* × *Aa*)

Possible Genotype of Two Offspring	Probability	Allele Frequency	
		A	*a*
AA and *AA*	(1/4)(1/4) = 1/16	1.00	0.00
aa and *aa*	(1/4)(1/4) = 1/16	0.00	1.00
AA and *aa*	2(1/4)(1/4) = 2/16	0.50	0.50
Aa and *Aa*	(2/4)(2/4) = 4/16	0.50	0.50
AA and *Aa*	2(1/4)(2/4) = 4/16	0.75	0.25
aa and *Aa*	2(1/4)(2/4) = 4/16	0.25	0.75

then an equal number of lines should have become fixed for each allele. In fact, the experimental results do not differ statistically from the expected ratio of 35:35, demonstrating that alleles can spread through a population and eliminate other alleles by chance alone.

How are small populations created in nature? In one instance, a large group of organisms may be split by some natural event, creating a small isolated subpopulation. A natural disaster such as an epidemic might also occur, leaving a small number of survivors to constitute the breeding population. Third, a small group might emigrate from the larger population, as founders, in a new environment, such as a volcanically created island.

Allelic frequencies in certain isolated human populations demonstrate the role of drift as an evolutionary force in natural populations. The Pingelap atoll in the western Pacific Ocean (lat. 6° N, long. 160° E) has in the past been devastated by typhoons and famine, and around 1780 there were only about nine surviving males. Today there are fewer than 2000 inhabitants, all of whose ancestry can be traced to the typhoon survivors. From 4–10 percent of the current population is blind from infancy. These people are affected by an autosomal recessive disorder, **achromatopsia,** which causes ocular disturbances, a form of color blindness, and cataract formation. The disorder is extremely rare in the human population as a whole. However, the mutant allele is present at a relatively high frequency in the Pingelap population. From genealogical reconstruction, it was found that one of the original survivors (about 30 individuals) was heterozygous for the condition. If we assume that he was the only carrier in the founding population, the initial gene frequency was 1/60, or 0.016. On average, about 7 percent of the current population is affected (a homozygous recessive genotype), so the frequency of the allele in the present population has increased to 0.26.

Another example of genetic drift involves the Dunkers, a small isolated religious community who emigrated from the German Rhineland to Pennsylvania. Because their religious beliefs do not permit marriage with outsiders, the population has grown only through marriage within the group. When the frequencies of the ABO and MN blood group alleles are compared, significant differences are found among the Dunkers, the German population, and the U.S. population. The frequency of blood group A in the Dunkers is about 60 percent, in contrast to 45 percent in the U.S. and German populations. The I^B allele is nearly absent in the Dunkers. Type M blood is found in about 45 percent of the Dunkers, compared with about 30 percent in the U.S. and German populations. As there is no evidence for a selective advantage of these alleles, it is apparent that the observed frequencies are the result of chance events in a relatively small isolated population.

Nonrandom mating changes genotype frequencies, but not allele frequencies

In the previous sections we have explored how violations of the first four assumptions of the Hardy-Weinberg law, in the form of selection, mutation, migration, and genetic drift, can cause allele frequencies to change. The fifth assumption is that the members of a population mate at random. Nonrandom mating does not, by itself, directly alter the frequencies of alleles. It can, however, alter the frequencies of genotypes in a population and thereby indirectly affect the course of evolution.

The most important form of nonrandom mating, and the form we will focus on here, is **inbreeding,** mating between relatives. For a given allele, inbreeding increases the proportion of homozygotes in the population. Over time, with complete inbreeding, only homozygotes will remain in the population. To illustrate this concept, we shall consider the most extreme form of inbreeding, **self-fertilization.**

Figure 25.18 shows the results of four generations of self-fertilization starting with a single individual heterozygous for one pair of alleles. By the fourth generation, only about 6 percent of the individuals are still heterozygous, and 94 percent of the population is homozygous. Note, however, that the frequencies of alleles A and a still remain at 50 percent.

In humans, inbreeding (called **consanguineous marriage**) is related to population size, mobility, and social customs governing marriages among relatives. To describe the amount of inbreeding in a population, Sewall Wright devised the coefficient of inbreeding. Expressed as F, the

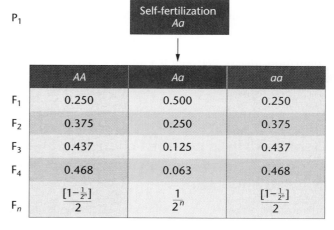

	AA	Aa	aa
F_1	0.250	0.500	0.250
F_2	0.375	0.250	0.375
F_3	0.437	0.125	0.437
F_4	0.468	0.063	0.468
F_n	$\dfrac{[1-\frac{1}{2^n}]}{2}$	$\dfrac{1}{2^n}$	$\dfrac{[1-\frac{1}{2^n}]}{2}$

P_1 — Self-fertilization Aa

FIGURE 25.18 Reduction in heterozygote frequency brought about by self-fertilization. After n generations, the frequencies of the genotypes can be calculated according to the formulas in the bottom row.

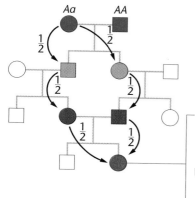

The chance that this female will inherit 2 copies of her great-grandmother's *a* allele is:

$$\frac{1}{2} \times \frac{1}{2} \times \frac{1}{2} \times \frac{1}{2} \times \frac{1}{2} \times \frac{1}{2} = \frac{1}{64}$$

Because the female's 2 alleles could be identical by descent from any of four different alleles,

$$F = 4 \times \frac{1}{64} = \frac{1}{16}$$

FIGURE 25.19 Calculating the coefficient of inbreeding (*F*) for the offspring of a first-cousin marriage.

coefficient of inbreeding can be defined as the probability that the two alleles of a given gene in an individual are identical *because they are descended from the same single copy of the allele in an ancestor.* If $F = 1$, all individuals are homozygous and both alleles in every individual are derived from the same ancestral copy. If $F = 0$, no individual has two alleles derived from a common ancestral copy.

Figure 25.19 shows a pedigree of a first-cousin marriage. The fourth-generation female (shaded pink) is the daughter of first cousins (purple). Imagine that her great-grandmother (green) was a carrier of a recessive lethal allele, *a*. What is the probability that the fourth-generation female will inherit two copies of her great-grandmother's lethal allele? For this to happen, (1) the great-grandmother had to pass a copy of the allele to her son, (2) her son had to pass it to his daughter, and (3) his daughter has to pass it to her daughter (the pink female). And also, (4) the great-grandmother had to pass a copy of the allele to her daughter, (5) her daughter had to pass it to her son, and (6) her son has to pass it to his daughter (the pink female). Each of the six necessary events has an individual probability of 1/2, and they all have to happen together, so the overall probability that the pink female will inherit two copies of her great-grandmother's lethal allele is $(1/2)^6 = 1/64$. This takes us most of the way toward calculating F for a child of a first-cousin marriage. We need only note that the fourth-generation female could also inherit two copies of any of the other three alleles present in her great-grandparents. Because any of four possibilities would give the pink female two alleles identical by descent from an ancestral copy, $F = 4 \times (1/64) = 1/16$.

Genetic Effects of Inbreeding

Inbreeding results in the production of individuals homozygous for recessive alleles that were previously concealed in heterozygotes. Because many recessive alleles are deleterious when homozygous, one of the consequences of inbreeding is an increase in the chance that an individual will be homozygous for a recessive deleterious allele. Inbred populations often have a lowered mean fitness. **Inbreeding depression** is a measure of the loss of fitness caused by inbreeding. In domesticated plants and animals, inbreeding and selection have been used for thousands of years, and these organisms already have a high degree of homozygosity at many loci. Further inbreeding will usually produce only a small loss of fitness. However, inbreeding among individuals from large, randomly mating populations can produce high levels of inbreeding depression. This effect can be seen by examining the mortality rates in offspring of inbred animals in zoo populations (Table 25.7). To counteract inbreeding depression, many zoos use DNA fingerprinting to estimate the relatedness of animals used in breeding programs and choose the most unrelated animals as parents.

As natural populations of endangered species become smaller, the concern about inbreeding is a factor in de-

TABLE 25.7 Mortality in Offspring of Inbred Zoo Animals

Species	N		Non-inbred	Inbred	Inbreeding Coefficient
Zebra	32	Lived:	20	3	0.250
		Died:	7	2	
Eld's deer	24	Lived:	13	0	0.250
		Died:	4	7	
Giraffe	19	Lived:	11	2	0.250
		Died:	3	3	
Oryx	42	Lived:	35	0	0.250
		Died:	2	5	
Dorcas gazelle	92	Lived:	36	17	0.269
		Died:	14	25	

GENETICS, TECHNOLOGY, AND SOCIETY

Gene Pools and Endangered Species: The Plight of the Florida Panther

In the last 400 years, more than 700 of Earth's animal and plant species have become extinct. In the U.S. alone, at least 30 species have suffered extinction in the last decade. In addition, hundreds of genetically distinct plant and animal species are now endangered. The rates of species and ecosystem losses are accelerating rapidly and are predicted to continue unabated into the next millennium. This dramatic loss of biological diversity is a direct consequence of human activity. As we humans increase our numbers and spread over Earth's surface, we harness more of Earth's resources for our use—clearing the forests, polluting the water, and inexorably and permanently altering Earth's natural balance.

Although the destruction continues, we are beginning to understand the implications of our actions and to make efforts to save some of Earth's more endangered life forms. However, despite our best efforts to save threatened species, we often intercede after population numbers have suffered se-vere declines. This means that much of the genetic diversity that existed in the species is lost, and the population must be rebuilt from a reduced gene pool. The resulting genetic uniformity can reduce fitness, as well as exposing genetic diseases. This reduced fitness may further deplete the numbers of the threatened plant or animal, and the population spirals downward to extinction.

The story of the Florida panther provides a poignant example of how an animal can be brought to the verge of extinction, and how the efforts of scientists and the general public might restore this creature to a healthy place in the ecosystem. The Florida panther, *Felis concolor coryi*, is one of 30 subspecies of cougar (puma or mountain lion), and is one of the most endangered mammals in the world. Florida panthers once roamed the southeastern corner of North America, from South Carolina and Arkansas to the southern tip of Florida. As people settled the southern and eastern states, panthers were killed as potential threats to livestock and humans. In the 19th and 20th centuries, panthers were killed by hunting, highway collisions, poisoning, and loss of habitat. Today, only 30–50 Florida panthers remain, isolated in southern Florida in the Big Cypress Swamp and Everglades National Park regions. Population projections indicate that, unless drastic measures are taken, there is an 85 percent likelihood that the Florida panther will be extinct in 25 years.

The geographical isolation and inbreeding of the few remaining Florida panthers have resulted in the loss of genetic variability and declining health. The panther has been separated from other cougar subspecies for 15 to 25 generations. As a result, Florida panthers have the lowest levels of genetic heterozygosity of any subspecies of cougar. This loss of genetic diversity has manifest itself in the appearance of several severe genetic defects. Almost 80 percent of panther males born after 1989 in the Big Cypress region show a

signing programs to restore these species. One example is a project in Scandanavian zoos established to preserve the Fennoscandic wolf. Conservationists established a captive-bred wolf population with four founding individuals. Later, two Russian wolves were added to the population to reduce the level of inbreeding, a strategy some conservationists later came to regret because it introduced foreign alleles into the Fennoscandic gene pool. Because the captive wolf population was founded by such a small number of individuals, it is severely inbred, and individuals show inbreeding depression in the form of smaller body weight, reduced reproductive success, and reduced longevity. Furthermore, a number of wolves in the population are blind.

Linda Laikre, Nils Ryman, and Elizabeth Thompson analyzed the pedigree for the captive-bred wolf population with three objectives. First, the researchers wanted to determine the mode of inheritance of blindness; second, they wanted to identify the individuals most likely to be the carriers of the gene or genes for blindness; and third, they wanted to determine whether removing the likely carriers from the breeding population would severely reduce the standing genetic variation in this already genetically impoverished population.

The researchers concluded that blindness in the captive wolf population is caused by an autosomal recessive allele at a single locus. The bad news from their analysis is that all the remaining pure-bred Fennoscandic wolves (that is, individuals with no genes from the two Russian wolves) have at least a 6 percent chance of being carriers for blindness. Some pure-bred individuals have a 67 percent chance of being carriers. The good news is that all individuals with a greater than 30 percent chance of being carriers can be removed from the breeding population without reducing

rare heritable (autosomal dominant or sex-linked recessive) condition known as cryptorchidism, manifested as the failure of one or both testicles to descend. This defect is associated with lower testosterone levels and reduced sperm count. Florida panther males in general have the poorest seminal quality of any cat species or any cougar subspecies, with approximately 93 percent abnormal sperm. Life-threatening congenital heart defects are also appearing in the Florida panther population, possibly due to an autosomal dominant gene defect. In addition, some immune deficiencies are appearing, and these may have a genetic component. Reduced immunity could leave the small panther population susceptible to diseases that could wipe out the population. Other less serious genetic features have appeared, such as a kink in the tail and a whorl of fur on the back.

Over the last two decades, a faint glimmer of hope has appeared for the Florida panther. Federal and state agencies, as well as private individuals, are implementing a "Florida Panther Recovery Program." The goal of the plan is to have 130 breeding animals (wild and captive) by the year 2000,

and 500 by the year 2010. If successful, the plan would grant the panther a 95 percent probability of survival while retaining 90 percent of its genetic diversity. The plan is multifaceted and includes a captive breeding program, strict protection, increasing and improving the panther habitat, and educating the public and private landowners. Wildlife underpasses have been constructed on highways in panther territory, and these have significantly reduced panther highway fatalities (which account for half of panther deaths). In 1995, a genetic restoration program began. In order to introduce genetic diversity into the Florida panther population and to retard the detrimental effects of inbreeding, eight female wild Texas panthers (a related subspecies from western Texas) were released into Florida panther territory. Two of the females have been killed—one from an automobile collision and one from gunshot wounds. However, the remaining Texas females have given birth to 12 healthy kittens (as of the summer of 1998), and one of these F_1 offspring has produced three kittens of her own. None of the kittens appear to have the

"kinked" tail of the inbred Florida panther.

The survival of the Florida panther is far from certain. Its recovery will require years of monitoring and frequent intervention. In addition, people must be willing to share their land with wild creatures that do not directly further their self-interests. However, as public support for the return of the Florida panther has been strong, there may be hope for this unique, impressive animal.

REFERENCES

FERGUS, C. 1991. The Florida panther verges on extinction. *Science* 251: 1178–80.

HEDRICK, P.W. 1995. Gene flow and genetic restoration: The Florida panther as a case study. *Conservation Biology* 9: 996–1007.

O'BRIEN, S.J. 1994. A role for molecular genetics in biological conservation. *Proc. Natl. Acad. Sci. USA* 91: 5748–55.

WEBSITE:

The Florida Panther Society **http://members.atlantic.net/~oldfla/panther**

the remaining genetic variation in the population by more than 10 percent. Removing the likely carriers would reduce the frequency of the blindness allele from 14 percent to 7 percent, and would improve the long-term prospects of maintaining a viable Fennoscandic wolf population.

In humans, inbreeding increases the risk of spontaneous abortions, neonatal deaths, and congenital deformities and recessive genetic disorders. Although less common than in the past, inbreeding occurs in many regions of the world where social customs favor marriage between first cousins. Alan Bittles and James Neel analyzed data from numerous studies on a variety of cultures. They found that the rate of child mortality (that is, death in the first several years of life) varies dramatically from culture to culture. No matter what the baseline mortality rate for children of unrelated parents, however, children of first cousins virtually always have a higher death rate—

typically by about 4.5 percentage points. Over many generations, inbreeding should eventually reduce the frequency of deleterious recessive alleles (if homozygotes die before reproducing). However, some studies indicate that parents who are first cousins tend to have more children to compensate for those lost to genetic disorders. On average, two-thirds of the surviving offspring are heterozygous carriers of the deleterious allele.

It is important to note that inbreeding is not always harmful. Indeed, inbreeding has long been recognized as a useful tool for breeders of domesticated plants and animals. When an inbreeding program is initiated, homozygosity increases, and some breeding stocks become fixed for favorable alleles and others for unfavorable alleles. By selecting the more viable and vigorous plants or animals, the proportion of individuals carrying desirable traits can be increased.

If members of two favorable inbred lines are mated, hybrid offspring are often more vigorous in desirable traits than is either of the parental lines. This phenomenon is called **hybrid vigor.** When such an approach was used in breeding programs established for maize, crop yields increased tremendously. Unfortunately, the hybrid vigor extends only through the first generation. Many hybrid lines are sterile, and those that are fertile show subsequent declines in yield. Consequently, the hybrids must be regenerated each time by crossing the original inbred parental lines.

Hybrid vigor has been explained in two ways. The first theory, the **dominance hypothesis,** incorporates the obvious reversal of inbreeding depression, which inevitably must occur in out-crossing. Consider a cross between two strains of maize with the following genotypes:

$$\text{Strain A} \qquad \text{Strain B}$$

$$aaBBCCddee \times AAbbccDDEE \longrightarrow AaBbCcDdEe$$

The F_1 hybrids are heterozygotes at all loci shown. The deleterious recessive alleles present in the homozygous form in the parents will be masked by the more favorable dominant alleles in the hybrids. Such masking is thought to cause hybrid vigor.

The second theory, **overdominance,** holds that in many cases the heterozygote is superior to either homozygote. This may be related to the fact that in the heterozygote two forms of a gene product may be present, providing a form of biochemical diversity. Thus, the cumulative effect of heterozygosity at many loci accounts for the hybrid vigor. Most likely, hybrid vigor results from a combination of phenomena explained by both hypotheses.

We have seen that nonrandom mating can drive the genotype frequencies in a population away from their expected values under the Hardy-Weinberg law. This can indirectly affect the course of evolution. As in stocks purposely inbred by animal and plant breeders, inbreeding in a natural population may increase the frequency of homozygotes for a deleterious recessive allele. As in domestic stocks, this increases the efficiency with which selection removes the deleterious allele from the population.

CHAPTER SUMMARY

1. Individuals do not evolve; populations evolve. A key insight in understanding evolution was the recognition that changes in the relative abundance of different phenotypes can be tied to changes in the relative abundance of the alleles that influence the phenotypes. This insight led to the development of population genetics, which studies the factors that cause allele frequencies to change.

2. The Hardy-Weinberg law specifies what will happen to allele frequencies in a population under a set of simple assumptions. If there is no selection, no mutation, and no migration, if the population is large, and if individuals mate at random, then allele frequencies will not change from one generation to the next. The population will not evolve.

3. The Hardy-Weinberg formula can be used to investigate whether or not a population is in evolutionary equilibrium, and to estimate the frequency of heterozygotes in a population from the frequency of homozygous recessives.

4. By specifying the conditions under which allele frequencies will not change, the Hardy-Weinberg law identifies the forces that can cause evolution. Selection, mutation, migration, and genetic drift can cause allele frequencies to change and are thus forces of evolution. Nonrandom mating does not alter the frequencies of alleles, but by altering the frequencies of genotypes, nonrandom mating can indirectly affect the course of evolution.

5. Natural selection is the most powerful of the forces that can alter the frequencies of alleles in a population. The rate of evolution under natural selection depends on the initial frequencies of the alleles and on the relative fitness of different genotypes. In small populations, random genetic drift can also be an important force altering the frequencies of alleles.

6. Mutation and migration introduce new alleles into a population, but by themselves mutation and migration usually have little effect on allele frequencies. Instead, the retention or loss of introduced alleles depends on the fitness they confer and the action of selection.

7. Genetic drift is change in allele frequencies as a result of chance events; it is an important force of evolution in small populations.

8. The most important form of non-random mating is inbreeding, or mating between relatives. Inbreeding increases the frequency of homozygotes in a population, and decreases the frequency of heterozygotes.

INSIGHTS AND SOLUTIONS

1. Tay-Sachs disease is caused by loss-of-function mutations in a gene on chromosome 15 that encodes a lysosomal enzyme. Tay-Sachs is inherited as an autosomal recessive condition. Among Ashkenazi Jews of central European ancestry, about 1 in 3600 children is born with the disease. What fraction of the individuals in this population are carriers?

   Solution: If we let p represent the frequency of the wild-type enzyme allele, and q the total frequency of recessive loss-of-function alleles, and if we assume that the population is in Hardy-Weinberg equilibrium, then the frequencies of the genotypes are given by p^2 for homozygous wild-type, $2pq$ for carriers, and q^2 for individuals with Tay-Sachs. The frequency of Tay-Sachs alleles is thus

   $$q = \sqrt{q^2} = \sqrt{(1/3600)} = 0.017$$

   Since $p + q = 1$, we have

   $$p = 1 - q = 1 - 0.017 = 0.983$$

   Therefore, we can estimate that the frequency of carriers is

   $$2pq = 2(0.983)(0.017) = 0.033$$

   or 1 in 30.

2. Eugenics is the term employed for the selective breeding of humans to bring about improvements in populations. As a eugenic measure, it has been suggested that individuals suffering from serious genetic disorders should be prevented (sometimes by force of law) from reproducing (by sterilization, if necessary) in order to reduce the frequency of the disorder in future generations. Suppose that such a recessive trait were present in the population at a frequency of 1 in 40,000, and that affected individuals did not reproduce. In 10 generations, or about 250 years, what would be the frequency of the condition? Are the eugenic measures effective in this case?

   Solution: Let q represent the frequency of the recessive allele responsible for the disorder. Because the disorder is recessive, we can estimate that

   $$q = \sqrt{(1/40,000)} = 0.005$$

   If all affected individuals are prevented from reproducing, then in an evolutionary sense the disorder is lethal: Affected individuals have zero fitness. That means we can predict the frequency of the recessive allele 10 generations in the future using Equation 25.11:

   $$q_g = q_0 / (1 + gq_0)$$

   In this case, $q_0 = 0.005$, and $g = 10$, so we have

   $$q_{10} = (0.005) / [1 + (10 \times 0.005)]$$
   $$= 0.0048$$

   If $q_{10} = 0.0048$ then the frequency of homozygous recessive individuals will be roughly

   $$q_{10}^2 = 0.0048^2 = 0.000023 = 1/43,500$$

   The frequency of the genetic disorder has been reduced from 1 in 40,000 to 1 in 43,500 in 10 generations, indicating that this eugenic measure has limited effectiveness.

PROBLEMS AND DISCUSSION QUESTIONS

1. The ability to taste the compound PTC is controlled by a dominant allele T, while individuals homozygous for the recessive allele t are unable to taste this compound. In a genetics class of 125 students, 88 were able to taste PTC, and 37 could not. Calculate the frequency of the T and t alleles in this population, and the frequency of the genotypes.

2. Calculate the frequencies of the AA, Aa, and aa genotypes after one generation if the initial population consists of 0.2 AA, 0.6 Aa, and 0.2 aa genotypes. What genotype frequencies will occur after a second generation?

3. Consider rare disorders in a population caused by an autosomal recessive mutation. From the frequencies of the disorder in the population given, calculate the percentage of heterozygous carriers:
   (a) 0.0064
   (b) 0.000081
   (c) 0.09
   (d) 0.01
   (e) 0.10

4. What must be assumed in order to validate the answers in Problem 3?

5. In a population where only the total number of individuals with the dominant phenotype is known, how can you calculate the percentage of carriers and homozygous recessives?

6. For the following two sets of data, determine whether they represent populations that are in Hardy-Weinberg equilibrium. Use χ^2 analysis if necessary.
   (a) CCR5 genotypes: $+/+$, 60 percent; $+/\Delta32$, 35.1 percent; $\Delta32/\Delta32$, 4.9 percent
   (b) Sickle-cell hemoglobin: *AA*, 75.6 percent; *AS*, 24.2 percent; *SS*, 0.2 percent

7. If 4 percent of the population in equilibrium expresses a recessive trait, what is the probability that the offspring of two individuals who do not express the trait will express it?

8. Consider a population in which the frequency of allele $A = p = 0.7$ and the frequency of allele $a = q = 0.3$, and where the alleles are co-dominant. What will be the allele frequencies after one generation if
   (a) $w_{AA} = 1$, $w_{Aa} = 0.9$, and $w_{aa} = 0.8$?
   (b) $w_{AA} = 1$, $w_{Aa} = 0.95$, and $w_{aa} = 0.9$?
   (c) $w_{AA} = 1$, $w_{Aa} = 0.99$, and $w_{aa} = 0.98$?
   (d) $w_{AA} = 0.8$, $w_{Aa} = 1$, and $w_{aa} = 0.8$?

9. If the initial allele frequencies are $p = 0.5$ and $q = 0.5$, and allele a is a lethal recessive, what will be the frequencies after 1, 5, 10, 25, 100, and 1000 generations?

10. Determine the frequency of allele A in an island population after one generation of migration from the mainland under the following conditions:
    (a) $p_i = 0.6$; $p_m = 0.1$; $m = 0.2$
    (b) $p_i = 0.2$; $p_m = 0.7$; $m = 0.3$
    (c) $p_i = 0.1$; $p_m = 0.2$; $m = 0.1$

11. Assume that a recessive autosomal disorder occurs in one of 10,000 individuals (0.0001) in the general population. Assume that in this population about 2 percent (0.02) of the individuals are carriers for the disorder. Estimate the probability of this disorder occurring in the offspring of a marriage between first cousins. Compare this probability to the population at large.

12. What is the basis of inbreeding depression?

13. Describe how inbreeding may be used in the domestication of plants and animals. Discuss the theories underlying these techniques.

14. Evaluate the following statement: Inbreeding increases the frequency of recessive alleles in a population.

15. In a breeding program to improve crop plants, which of the following mating systems should be employed to produce a homozygous line in the shortest possible time?
    (a) Self-fertilization
    (b) Brother–sister matings
    (c) First-cousin matings
    (d) Random matings
    Illustrate your choice with pedigree diagrams.

16. If the recessive trait albinism (*a*) is present in 1/10,000 individuals, calculate the frequency of

(a) The recessive mutant allele.
(b) The normal dominant allele.
(c) Heterozygotes in the population.
(d) Matings between heterozygotes.

17. One of the first Mendelian traits identified in humans was a dominant condition known as *brachydactyly*. This gene causes an abnormal shortening of the fingers and/or toes. At the time, it was thought by some that this dominant trait would spread until 75 percent of the population would be affected (since the phenotypic ratio of dominant to recessive is 3:1). Show that this line of reasoning is incorrect.

18. Achondroplasia is a dominant trait that causes a characteristic form of dwarfism. In a survey of 50,000 births, five infants with achondroplasia were identified. Three of the affected infants had affected parents, while the rest had normal parents. Calculate the mutation rate for achondroplasia and express the rate as the number of mutant genes per given number of gametes.

19. A prospective groom, who is normal, has a sister with cystic fibrosis, an autosomal recessive disease state. Both of his parents are normal. He plans to marry a woman who has no history of cystic fibrosis (CF) in her family. What is the probability that they will produce a CF child? They are both Caucasian and the overall frequency of CF in the Caucasian population is 1/2500; that is, 1 affected child per 2500. (Assume the population meets the Hardy-Weinberg conditions.)

EXTRA-SPICY PROBLEMS

20. A form of dwarfism known as Ellis-van Creveld syndrome was first discovered in the late 1930s, when Richard Ellis and Simon van Creveld shared a train compartment on the way to a pediatrics meeting. In the course of conversation, they discovered they each had a patient with this syndrome. They published a description of the syndrome in 1940. Affected individuals have a short-limbed form of dwarfism and often have defects of the lips and teeth, and polydactyly (extra fingers). The largest pedigree for this condition was reported in an Old Order Amish population in eastern Pennsylvania by Victor McKusick and his colleagues (1964). In this community, about 5 per 1000 births are affected, and in the population of 8000, the observed frequency is 2 per 1000. All affected individuals have unaffected parents, and all affected cases can trace their ancestry to Samuel King and his wife, who arrived in the area in 1774. It is known that neither King nor his wife were affected with this disorder. There are no cases of this disorder in other Amish communities, such as those in Ohio or Indiana.
    (a) From the information provided, derive the most likely mode of inheritance of this disorder. Using the Hardy-Weinberg law, calculate the frequency of the mutant allele in the population and the frequency of heterozygotes, assuming Hardy-Weinberg conditions.

GENETICS MediaLab

The following resources will help you achieve a better understanding of the concept presented in this chapter. These resources can be found on the CD packaged with this textbook and on the Companion Website at **http://www.prenhall.com/klug/**.

CD Resources:

Animated Tutorial: *Population Genetics*

Self-grading Chapter Problems

Web Resources:

Genetics and Society Issue: *Gene Pools and Endangered Species: The Plight of the Florida Panther*

Web Destinations in Genetics

Self-grading Chapter Problems

Chapter Search Terms

Genetics Newsgroups

Student Bulletin Board

Web Problem 1:

Time for completion = 10 minutes

How does a population remain in Hardy-Weinberg Equilibrium (HWE)? Study the page on "The Hardy-Weinberg Equilibrium Model" and Chapter 25 of your text to help you answer the following questions. If a population with allele frequencies of $p = q = 0.5$ actually met all the assumptions of HWE, what would the allele frequencies be after 100 generations? Which of the assumptions of HWE is violated in all natural populations? If you knew only the proportion of non-albinos in a population, could you determine the frequency of the allele for albinism? To complete this exercise, visit Web Problem 1 in Chapter 25 of your Companion Website and select the keyword **HARDY-WEINBERG.**

Web Problem 2:

Time for completion = 10 minutes

How does population size affect allele frequencies over time? Use the "Hardy-Weinberg Simulator" and Chapter 25 of your text to investigate how genetic drift caused by finite population size affects evolution. First, set the "Number of Runs" to 10 and the "Population" to 10 individuals. Leave all the other settings at the defaults and run the simulation. After it is completed, take a look at your results before hitting "Return to Simulator." How many of the populations went to fixation for either the "+" or the "−" allele? Click the "Append" box (so that your second set

of experiments will be plotted along with the first) and change the "Population" to 1000 individuals. Run the simulation. For these runs, how many populations went to fixation? Is there more genetic variation *within* populations of large size or small size? Is there more genetic variation *between* populations of large size or small size? To complete this exercise, visit Web Problem 2 in Chapter 25 of your Companion Website and select the keyword **GENETIC DRIFT.**

Web Problem 3:

Time for completion = 15 minutes

How does natural selection interact with dominance to affect allele frequencies over time? You will use the "Hardy-Weinberg Simulator" we introduced in the exercise 2, along with Chapter 25 of your text to explore these problems. First, consider selection against a deleterious recessive allele. Reset the simulator to the defaults, then set "Population" to 10000, "−/− Survival Rate" at 50, and "Initial + Percentage" at 95. Do you think the "−" allele will be eliminated within 100 generations? Run the simulation and find out. After examining your results, hit "Return to Simulator" and make the deleterious allele dominant by setting both "+/− Survival Rate" and "−/− Survival Rate" to 50. Leave all the other settings as in the previous run. Do you think the "−" allele will be eliminated this time? Run the simulation and find out. Why did changing the fitness of the heterozygotes make such a big difference? Would your results be the same with a smaller population size? Why or why not? Do these examples illustrate stabilizing, directional, or disruptive selection? To complete this exercise, visit Web Problem 3 in Chapter 25 of your Companion Website and select the keyword **DOMINANCE SELECTION.**

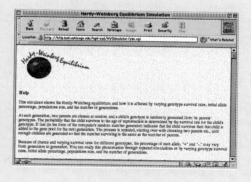

(b) What is the most likely explanation for the high frequency of this disorder in the Pennsylvania Amish community, and its absence in other Amish communities?

21. The following graph shows the variation in the frequency of a particular allele (*A*) that occurred over time in two relatively small, independent populations exposed to very similar environmental conditions. A student has analyzed the graph and concluded that the best explanation for these data is that the selection for allele *A* is occurring in Population 1. Is the student's conclusion correct? Explain.

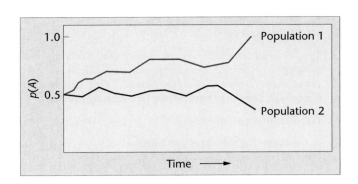

SELECTED READINGS

ANSARI-LARI, M.A., LIU, X.-M. et al. 1997. The extent of genetic variation in the CCR5 gene. *Nature Genetics.* 16: 221–22.

BALLOU, J., and RALLS, K. 1982. Inbreeding and juvenile mortality in small populations of ungulates: A detailed analysis. *Biol. Conserv.* 24:239–72.

BARBUJANI, G., MAGAGNI, A., MINCH, E., and CAVALLI-SFORZA, L.L. 1997. An apportionment of human DNA diversity. *Proc. Natl. Acad. Sci. USA* 94: 4516–19.

BARTON, N.H., and TURELLI, M. 1989. Evolutionary quantitative genetics: How little do we know? *Annu. Rev. Genet.* 23: 337–70.

BITTLES, A.H., and NEEL, J.V. 1994. The costs of human inbreeding and their implications for variations at the DNA level. *Nature Genet.* 8: 117–21.

BODMER, W.F., and CAVALLI-SFORZA, L.L. 1976. *Genetics, evolution and man.* New York: W. H. Freeman.

CARRINGTON, M., and KISSNER, T. et al. 1997. Novel alleles of the chemokine-receptor gene CCR5. *Am. J. Hum. Genet.* 61: 1261–67.

CHEVILLON, C., PASTEUR, N., MARQUINE, M., HEYSE, D., and RAYMOND, M. 1995. Population structure and dynamics of selected genes in the mosquito *Culex pipiens. Evolution* 49: 997–1007.

CROW, J.F. 1986. *Basic concepts in population, quantitative and evolutionary genetics.* New York: W. H. Freeman.

CROW, J.F., and KIMURA, M. 1970. *An introduction to population genetic theory.* New York: Harper & Row.

DAVID, J.R., and CAPY, P. 1988. Genetic variation of *Drosophila melanogaster* natural populations. *Trends Genet.* 4:106–11.

DUDLEY, J.W. 1977. 76 generations of selection for oil and protein percentage in maize. Ed. E. Pollack et al. In *Proceedings of the International Conference on Quantitative Genetics,* pp. 459–73. Ames, IA: Iowa State University Press.

EDWARDS, J.H. 1989. Familiarity, recessivity and germline mosaicism. *Ann. Hum. Genet.* 53:33–47.

FISHER, R.A. 1930. *The genetical theory of natural selection.* Oxford: Clarendon Press. (Reprinted Dover Press, 1958.)

FREEMAN, S., and HERRON, J.C. 1998. *Evolutionary analysis.* Upper Saddle River, NJ: Prentice Hall.

FREIRE-MAIA, N. 1990. Five landmarks in inbreeding studies. *Am. J. Med. Genet.* 35:118–20.

GAYLE, J.S. 1990. *Theoretical population genetics.* Boston: Unwin-Hyman.

GRANT, P.R. 1986. *Ecology and evolution of Darwin's finches.* Princeton, NJ: Princeton University Press.

HARTL, D.L. 1988. *A primer of population genetics,* 2nd ed. Sunderland, MA: Sinauer Associates.

JONES, J.S. 1981. How different are human races? *Nature* 293:188–90.

KARN, M.N., and PENROSE, L.S. 1951. Birth weight and gestation time in relation to maternal age, parity and infant survival. *Ann. Eugen.* 16:147–64.

KERR, W.E., and WRIGHT, S. 1954. Experimental studies of the distribution of gene frequencies in very small populations of *Drosophila melanogaster.* I. *Forked. Evolution* 8: 172–77.

KEVLES, D.J. 1985. *In the name of eugenics: Genetics and the uses of human heredity.* Berkeley: University of California Press.

KHOURY, M.J., and FLANDERS, W.D. 1989. On the measurement of susceptibility to genetic factors. *Genet. Epidemiol.* 6: 699–701.

LAIKRE, L., RYMAN, N., and THOMPSON, E.A. 1993. Hereditary blindness in a captive wolf (*Canis lupus*) population: Frequency reduction of a deleterious allele in relation to gene conservation. *Conservation Biol.* 7: 592–601.

LI, C.C. 1977. *First course in population genetics.* Pacific Grove, CA: Boxwood Press.

LIBERT, F., COCHAUX, P. et al. 1998. The ΔCCR5 mutation conferring protection against HIV-1 in Caucasian populations has a single and recent origin in northeastern Europe. *Hum. Molec. Genet.* 7: 399–406.

LIU, R., PAXTON, W.A. et al. 1996. Homozygous defect in HIV-1 coreceptor accounts for resistance in some multiply-exposed individuals to HIV-1 infection. *Cell* 86: 367–77.

LUCOTTE, G., and MERCIER, G. 1998. Distribution of the CCR5 gene 32-bp deletion in Europe. *J. Acquired Immune Deficiency Syndrome and Human Retrovirology* 19: 174–77.

MARKOW, T., HEDRICK, P.W. et al. 1993. HLA polymorphism in the Havasupai: Evidence for balancing selection. *Am. J. Hum. Genet.* 53: 943–52.

MARTINSON, J.J., and CHAPMAN, N.H. et al. 1997. Global distribution of the CCR5 gene 32-basepair deletion. *Nature Genet.* 16: 100–103.

METTLER, L.E., GREGG, T., and SCHAFFER, H. E. 1988. *Population genetics and evolution,* 2nd ed. Englewood Cliffs, NJ: Prentice-Hall.

PARRA, E.J., MARCINI, A. et al. 1998. Estimating African American admixture proportions by use of population-specific alleles. *Am. J. Hum. Genet.* 63:1839–51.

PIER, G.B., and GROUT, M. et al. 1998. *Salmonella typhi* uses CFTR to enter intestinal epithelial cells. *Nature* 393: 79–82.

QUILLENT, C., and OBERLIN, E. et al. 1998. HIV-1-resistance pheno-type conferred by combination of two separate inherited mutations of CCR5 gene. *The Lancet* 351:14–18.

RENFREW, C. 1994. World linguistic diversity. *Sci. Am.* (Jan.) 270:116–23.

ROBERTS, D.F. 1988. Migration and genetic change. *Hum. Biol.* 60:521–39.

SAMSON, M. and LIBERT, F. et al. 1996. Resistance to HIV-1 infection in Caucasian individuals bearing mutant alleles of the CCR-5 chemokine receptor gene. *Nature* 382: 722–25.

SPIESS, E.B. 1989. *Genes in populations*, 2nd ed. New York: John Wiley and Sons.

STEPHENS, J.C., and REICH, D.E. et al. 1998. Dating the origin of the *CCR5-Δ32* AIDS-resistance allele by the coalescence of haplotypes. *Am. J. Hum. Genet.* 62: 1507–15.

UNAIDS JOINT UNITED NATIONS PROGRAMME ON HIV/AIDS. 1998. *AIDS epidemic update: December 1998*. Geneva UNAIDS/World Health Organization.

WALLACE, B. 1989. One selectionist's perspective. *Quart. Rev. Biol.* 64:127–45.

WOODWORTH, C.M., LENG, E.R., and JUGENHEIMER, R.W. 1952. Fifty generations of selection for protein and oil in corn. *Agron.* J. 44:60–66.

YUDIN, N.S., VINOGRADOV, S.V. et al. 1998. Distribution of *CCR5-Δ32* gene deletion across the Russian part of Eurasia. *Hum. Genet.* 102: 695–98.

ZHU, J., NESTOR, K.E., and MORITSU, Y. 1996. Relationship between band sharing levels of DNA fingerprints and inbreeding coefficients and estimation of true inbreeding in turkey lines. *Poult. Sci.* 75:25–28.

These lady-bird beetles, from the Chiricahua Mountains in Arizona, show considerable phenotypic variation.

Genetics and Evolution

KEY CONCEPTS

- **Evolutionary history includes the transformation and divergence of lineages**

- **Most populations and species harbor considerable genetic variation**
  Protein Polymorphisms
  Variations in Nucleotide Sequence
  Explaining the High Level of Genetic Variation in Populations

- **The genetic structure of populations changes across space and time**

- **A reduction in gene flow between populations, accompanied by divergent selection and/or genetic drift, can lead to speciation**

Examples of Speciation
The Minimum Genetic Divergence Required for Speciation
In at Least Some Instances, Speciation is Rapid

- **We can use genetic differences among populations or species to reconstruct evolutionary history**
  A Method for Estimating Evolutionary Trees from Genetic Data
  Molecular Clocks

- **Reconstructing evolutionary history allows us to answer a variety of questions**
  Transmission of HIV from a Dentist to His Patients
  The Relationship of Neanderthals to Modern Humans
  The Origin of Mitochondria

In late 1986 a Florida dentist tested positive for HIV. Several months later he was diagnosed with AIDS. He continued to practice general dentistry for two more years, until one of his patients, a young woman with no obvious risk factors, discovered that she, too, was infected with HIV. When the dentist publicly urged his other patients to have themselves tested, several more were found to be HIV-positive. Did this dentist transmit HIV to his patients, or did the patients become infected by some other means?

Paleontological evidence indicates that the Neanderthals, *Homo neanderthalensis*, lived in Europe and western Asia from some 300,000 to 30,000 years ago. During their later years, the Neanderthals coexisted with anatomically modern humans. Did Neanderthals and modern humans interbreed, so that descendants of the Neanderthals are alive today, or did the Neanderthals die off without issue, so that today their lineage is extinct?

The mitochondria that inhabit our cells have their own DNA. The organization and function of the mitochondrial genome bears many similarities to the organization and function of bacterial genomes. Did our mitochondria originate as free-living bacteria that took up residence inside other cells, or is the mitochondrial genome ultimately derived from nuclear chromosomes?

The preceding paragraphs pose three apparently disparate questions. What these questions have in common is that they all can be addressed by using genetic data to reconstruct evolutionary history. The use of genetic data to reconstruct evolutionary history is the focus of this chapter.

Evolutionary history involves the transformation and splitting of lineages to produce a diversity of populations and species. In Chapter 25 we described the evolution of populations as changes in allele frequencies and outlined the forces that can cause allele frequencies to change. Mutation, migration, selection, and drift, individually and collectively, alter allele frequencies and bring about evolutionary divergence and species formation. This process depends not only upon genetic divergence but also upon environmental or ecological diversity. If a population is spread over a geographic range encompassing a number of subenvironments or **niches,** the populations occupying these niches will adapt and become genetically differentiated. Differentiated populations are dynamic: They may remain in existence, become extinct, merge with the parental population, or continue to diverge from the parental population until they become reproductively isolated and form new species.

It is difficult to define the exact moment of transition when a new species forms. A **species** is defined as a group of interbreeding or potentially interbreeding populations that are reproductively isolated in nature from all other such groups. In sexually reproducing organisms, speciation divides a single gene pool into two or more separate gene pools. Changes in morphology, physiology, and adaptation to an ecological niche may also occur, but are not necessary components of the speciation event. Speciation can take place gradually or within a few generations.

Because the divergence of populations and species is accompanied by genetic differentiation, we can use the pattern of genetic differences among populations or species to reconstruct evolutionary history. After exploring the genetic structure of populations, the divergence of populations across space and time, and the process of speciation, we will make good on our claim that genetic data can be used to answer the questions we posed at the beginning of the chapter.

Evolutionary history includes the transformation and divergence of lineages

Figure 26.1 shows an evolutionary tree, or **phylogeny,** that describes the history of a group of hypothetical lizard species. Movement along the horizontal axis represents the passage of time; movement along the vertical axis represents change in the form, or phenotype, of the lizards. The heavy lines on the graph trace the phenotype of the average lizard in each species over time.

The history of these lizards began with species 1. For some time species 1 experienced evolutionary **stasis;** that is, it did not change. Species 1 then underwent a period of steady transformation, also called **phyletic evolution,** or

anagenesis, and became species 2. Species 2 then underwent **speciation,** also called **cladogenesis;** it gave rise to two distinct and independent daughter species. Further transformation of these daughter species produced the extant kinds, species 3 and species 4.

In his 1859 book, *On the Origin of Species*, Charles Darwin amassed evidence that all species were produced, by transformation and speciation, from a single common ancestor:

> ". . .all living things have much in common, in their chemical composition, their germinal vesicles, their cellular structure, and their laws of growth and reproduction. . . . Therefore I should infer . . . that probably all the organic beings which have ever lived on this earth have descended from some one primordial form. . . ."

Everything biologists have learned since Darwin's time supports his conclusion that there is just one tree of life. To understand evolution, then, we must understand the mechanisms responsible for both the transformation of lineages and speciation.

Chapter 25 covered the mechanisms responsible for the transformation of lineages. Chief among them is natural selection, discovered independently by Darwin and by Alfred Russel Wallace.

1. Among individuals of a species, variations in phenotype exist. These can include differences in size, agility, coloration, or ability to obtain food.

2. Many of these variations, even small and seemingly insignificant ones, are heritable and passed on to offspring.

3. Organisms tend to reproduce in an exponential fashion. More offspring are produced than can survive. This causes members of a species to engage in a struggle for survival, competing with other members of the species for scarce resources.

4. In the struggle for survival, individuals with particular phenotypes will be more successful than individuals with others, allowing them to survive and reproduce at higher rates.

As a consequence of items 1–4, species change over time. The phenotypes that confer improved ability to survive and reproduce become more common, and the phenotypes that confer poor prospects for survival and reproduction disappear.

Although Wallace and Darwin proposed that natural selection can explain how evolution occurs, they could not explain the origin of the variations that provided the raw material for evolution nor how such variation is passed from parents to offspring. In the twentieth century, as the principles of Mendelian genetics were applied to populations, the source of variation (mutation) and the mecha-

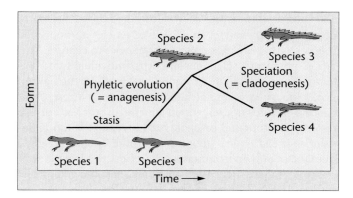

FIGURE 26.1 In phyletic evolution, or anagenesis, one species is transformed over time into another species. At all times, only one species exists. In speciation or cladogenesis, one species splits into two or more species.

nism of inheritance (segregation of alleles) were both explained, and evolution was seen as changes in allele frequencies in populations over time. This union of population genetics with natural selection generated a new view of the evolutionary process, called neo-Darwinism.

In this chapter we will consider the mechanisms responsible for speciation. As with natural selection, a modern understanding of speciation rests on an insight about genetics. That insight is the subject of the next section.

Most populations and species harbor considerable genetic variation

We might assume that the members of a well-adapted population are genetically homogeneous because the most favorable allele at each locus has become fixed. Certainly, an examination of most populations of plants and animals reveals many phenotypic similarities among individuals. However, considerable evidence indicates that most populations contain a high degree of heterozygosity. This built-in genetic diversity is concealed, so to speak, because it is not necessarily apparent in the phenotype. The detection of this concealed genetic variation is not simple. Nevertheless, such investigation has been successful using the techniques discussed below.

Protein Polymorphisms

Gel electrophoresis (see Appendix A) is a technique that can be used to separate protein molecules on the basis of differences in size and electrical charge. If a nucleotide variation in a structural gene results in the substitution of a charged amino acid, such as glutamic acid, for an uncharged amino acid, such as glycine, the net electrical charge on the protein will be altered. This difference in

charge can be detected as a change in the rate at which proteins migrate through an electrical field. In the mid-1960s, John Hubby and Richard Lewontin used gel electrophoresis to measure protein variation in natural populations of *Drosophila*. Since then, genetic variation has been studied using gel electrophoresis in a wide range of organisms (Table 26.1).

The electrophoretically distinct forms of a protein produced by different alleles are called allozymes. As shown in Table 26.1, a surprisingly large percentage of loci examined from diverse species produce distinct allozymes. Of the populations shown in Table 26.1, approximately 30 loci per species were examined, and about 30 percent of the loci were polymorphic. An approximate average of 10 percent allozyme heterozygosity per diploid genome was revealed.

These values apply only to genetic variation detectable by altered protein migration in an electric field. Electrophoresis probably detects only about 30 percent of the actual variation that is due to amino acid substitutions because many substitutions do not change the net electric charge on the molecule. Richard Lewontin has estimated that about two-thirds of all loci are polymorphic in a population and that, in any individual within that group, about one-third of the loci exhibit genetic variation in the form of heterozygosity.

There is some controversy over the significance of genetic variation as detected by electrophoresis. Some argue that allozymes are biochemically equivalent and therefore do not play any role in adaptive evolution. We shall address this argument later in this section.

Variations in Nucleotide Sequence

The most direct way to estimate genetic variation is to compare the nucleotide sequences of genes carried by individuals in a population. With the development of tech-

TABLE 26.1 Heterozygosity at the Molecular Level

Species	Number of Populations Studied	Number of Loci Examined	Polymorphic Loci per Population (%)	Hetero-zygotes per Locus (%)
Homo sapiens (humans)	1	71	28	6.7
Mus musculus (mouse)	4	41	29	9.1
Drosophila pseudoobscura (fruit fly)	10	24	43	12.8
Limulus polyphemus (horseshoe crab)	4	25	25	6.1

Note: A polymorphic locus is one for which a population harbors more than one allele.
Source: From Lewontin, (1974, p. 117).

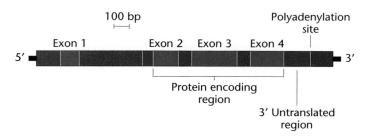

FIGURE 26.2 The *Adh* locus of *Drosophila melanogaster*.

niques for cloning and sequencing DNA, nucleotide sequence variations have been cataloged for an increasing number of gene systems. Using restriction endonucleases to detect polymorphisms, Alec Jeffreys surveyed 60 unrelated individuals to estimate the total number of DNA-sequence variants in humans. His results show that within the genes of the β-globin cluster, 1 in 100 base pairs shows polymorphic variation. If this region is representative of the genome, this indicates that at least 3×10^7 nucleotide variants per genome are possible.

In another study, Martin Kreitman investigated the *alcohol dehydrogenase* locus (*Adh*) in *Drosophila melanogaster* (Figure 26.2). This locus encodes two allozymic variants, the *Adh-f* and the *Adh-s* alleles. These differ by a single amino acid (*thr* versus *lys* at codon 192). To determine whether the amount of genetic variation detectable at the protein level (one amino acid difference) corresponds to the variation at the nucleotide level, Kreitman cloned and sequenced *Adh* loci from five natural populations of *Drosophila*.

The 11 cloned loci contained 43 nucleotide variations from the consensus *Adh* sequence of 2721 base pairs. These variations are distributed throughout the locus: 14 in the exon coding regions, 18 in the introns, and 11 in

the nontranslated and flanking regions. Of the 14 in the coding regions, only one leads to an amino acid replacement, accounting for the two observed electrophoretic variants. The other 13 nucleotide substitutions do not lead to amino acid replacements. Are the differences in the number of allozyme and nucleotide variants the result of natural selection? If so, of what significance is this? These issues will be discussed below.

Among the most intensively studied loci to date is the locus encoding the cystic fibrosis transmembrane conductance regulator (*CFTR*). Recessive loss-of-function mutations in the *CFTR* locus cause cystic fibrosis, a disease with symptoms including salty skin, production of an excess of thick mucus in the lungs, and susceptibility to bacterial infections. Geneticists have examined the *CFTR* locus in some 30,000 chromosomes from individuals with cystic fibrosis and have found over 500 different mutations that can cause the disease. These include missense mutations, amino acid deletions, nonsense mutations, frameshifts, and splice defects.

Figure 26.3 shows a map of the 27 exons in the *CFTR* locus, with most exons identified by function. The histogram above the map shows the locations of the known disease-causing mutations and the number of copies of each

FIGURE 26.3 The locations of disease-causing mutations in the cystic fibrosis gene. The histogram shows the number of copies of each mutation geneticists have found (note that the vertical axis is on a logarithmic scale). The genetic map below the histogram shows the locations and relative sizes of the 27 exons of the *CFTR* locus. The boxes at the bottom indicate the functions of different domains of the CFTR protein. Reprinted from *Trends Genet.* 8: 392–98, L. Tsui, The spectrum of cystic fibrosis mutations, 1992, with permission from Elsevier Science.

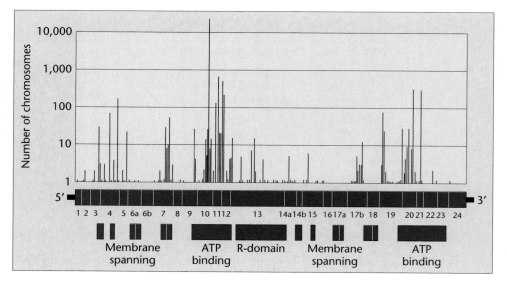

that have been found. A single mutation, a 3-bp deletion in exon 10 called *ΔF508*, accounts for 67 percent of the cystic fibrosis allele copies found, but several other mutations have been found in at least 100 chromosomes. In populations of European ancestry between 1 in 44 and 1 in 20 individuals are heterozygous carriers of cystic fibrosis alleles. Note that Figure 26.3 includes only the sequence variants that alter the function of the CFTR protein. There are undoubtedly many more *CFTR* alleles with silent sequence variants that do not change the structure of the protein and do not affect its function. The *CFTR* locus, by itself, shows considerable genetic variation.

Studies of other organisms, including the rat and the mouse, have produced similar estimates of nucleotide diversity. These studies indicate that there is an enormous reservoir of genetic variability within most populations and that, at the level of DNA, most and perhaps all genes exhibit diversity from individual to individual.

Explaining the High Level of Genetic Variation in Populations

As we mentioned above, the first demonstrations that populations harbor considerable genetic diversity at the level of nucleotide and amino acid sequences came as a surprise to many evolutionary biologists. They had expected that selection would favor a single optimal allele at each locus, and that as a result populations would be genetically homogenous. This expectation was obviously wrong, prompting considerable research and argument concerning the forces that maintain genetic variation.

One point of view argues that the existence of a high degree of genetic variation primarily reflects the action of mutation and genetic drift. The **neutral theory** of molecular evolution, proposed by Motoo Kimura, argues that mutations leading to amino acid substitutions are rarely favorable. They are sometimes detrimental, but most often are neutral or genetically equivalent to the allele that is replaced. Those few polymorphisms that are favorable or detrimental are either preserved or removed from the population, respectively, by natural selection. However, the vast majority of mutations are neutral genetic changes that are not affected by selection. Their frequency will be determined by the rate of mutation and the principles of random genetic drift. Some neutral mutations will drift to fixation in the population; other neutral mutations will be lost. At any given time, the population may contain several neutral alleles at any particular locus. The diversity of alleles at most polymorphic loci does not, however, reflect the action of natural selection.

Opposed to the neutral theory are the **selectionists.** These geneticists point to examples where enzyme or protein polymorphisms are associated with adaptation to certain environmental conditions. The well-known advantage of sickle-cell anemia heterozygotes in infection by malarial parasites is such an example. We discussed another example in Chapter 25, when we considered evidence that polymorphism at the *CFTR* locus may reflect the superior fitness of heterozygotes in areas where typhoid fever is common.

Selectionists also stress that enzyme polymorphisms may often appear to offer no advantage, but exist in such a frequency that it is impossible to explain as a random occurrence. Thus, even though no currently available analytical technique can detect any physiological difference, some slight advantage associated with any given amino acid substitution may exist. There may, in fact, be an advantage to having two forms of a given protein, allowing optimum performance under a wider range of cellular conditions.

The neutralist and selectionist perspectives should not be viewed as mutually incompatible positions. Instead, they are best seen as the endpoints of a spectrum of possibilities. The neutralists do not discount natural selection as a guiding force in evolution. Rather, they suggest that some features of organisms' genotypes are nonadaptive, fluctuate randomly, and may have been fixed by genetic drift. On the other hand, the selectionists certainly do not deny that genetic drift is an important factor in establishing differences in allele frequencies. It is difficult to argue against the notion that some genetic variation must be neutral. The difference between the two theories is in the degree of neutrality that exists.

Current data are insufficient to determine what fraction of molecular genetic variation is neutral vs. naturally selected. The neutral theory nonetheless serves a crucial function. By pointing out that some genetic variation is expected simply as a result of mutation and drift, the neutral theory provides a null hypothesis for studies of molecular evolution, a hypothesis that should be accepted until it can be definitively refuted. In other words, biologists must find positive evidence that selection is acting on the allele frequencies at a particular locus before they can reject the simpler assumption that only mutation and drift are at work.

The genetic structure of populations changes across space and time

As population geneticists were discovering that most populations harbor considerable genetic diversity, they also found that the genetic structure of populations within most species varies across space and time. To illustrate, we shall consider studies on *Drosophila pseudoobscura* conducted by Theodosius Dobzhansky and his colleagues. This species is found over a wide range of environmental habitats, in-

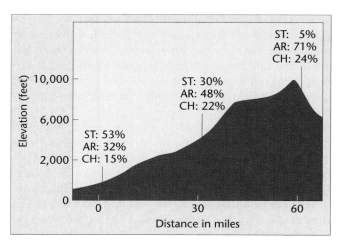

FIGURE 26.4 Inversions in chromosome 3 of *D. pseudoobscura* at different elevations in the Sierra Nevada range near Yosemite National Park.

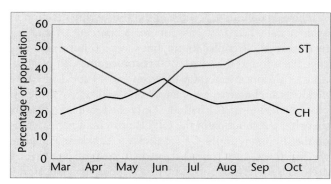

FIGURE 26.5 Changes in the ST and CH arrangements in *D. pseudoobscura* throughout the year.

cluding the western and southwestern United States. Although the flies throughout this range are morphologically similar, Dobzhansky discovered that populations from different locations vary in the arrangement of genes on chromosome 3. He found several different inversions in this chromosome that can be detected by loop formations in larval polytene chromosomes. Each inversion sequence was named after the locale in which it was first discovered (e.g., AR 5 Arrowhead, British Columbia; and CH 5 Chiricahua Mountains). The inversion sequences were compared with one standard sequence, designated ST.

Figure 26.4 shows a comparison of three arrangements found in populations at three elevations in the Sierra Nevada chain in California. The ST arrangement is most common at low elevations; at 8000 feet, AR is the most common and ST least common. In these populations, the frequency of the CH arrangement gradually increases with elevation. The gradual change in inversion frequencies is probably the result of natural selection and parallels the gradual environmental changes occurring at ascending elevations. As the populations along this gradient show a continuous and gradual change in inversion frequencies, it is difficult to classify a fly as a member of one racial group.

Dobzhansky also found that if populations were collected at a single site throughout the year, inversion frequencies also changed. Through the seasons, cyclic variation in chromosome arrangements occurred, as shown in Figure 26.5. Such variation was consistently observed over a period of several years. The frequency of ST always declined during the spring, with a concomitant increase in CH during the same period.

To test the hypothesis that this cyclic change is a response to natural selection, Dobzhansky devised the fol-

lowing laboratory experiment. He constructed a large population cage from which samples of *D. pseudoobscura* could be periodically removed and studied. He began with a population of a known inversion frequency, 88 percent CH and 12 percent ST. He reared it at 25°C and sampled it over a 1-year period. As shown in Figure 26.6, the frequency of ST increased gradually until it was present at a level of 70 percent. At this point, an equilibrium between ST and CH was reached. When the same experiment was performed at 16°C, no change in inversion frequency occurred. It can be concluded that the equilibrium reached at 25°C was in response to the elevated temperature, the only variable in the experiment.

The results are evidence that a balance in the frequency of the two inversions and their respective gene arrangements in a population is superior to either inversion by itself. The equilibrium attained presumably represents the

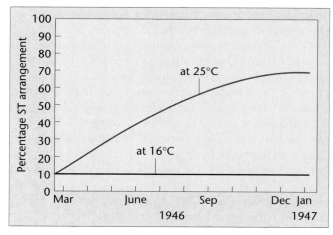

FIGURE 26.6 Increase in the ST arrangement of *D. pseudoobscura* in population cages under laboratory conditions.

highest mean fitness in the population under controlled laboratory conditions. This interpretation of the experiment suggests that natural selection is the driving force maintaining the diversity in chromosome 3 inversions.

In a more extensive study, Dobzhansky and his colleagues sampled *D. pseudoobscura* populations over a broader geographic range. Twenty-two different chromosome arrangements were found in populations from 12 locations. In Figure 26.7, the frequencies of five of these inversions are shown according to geographic location. The differences are largely quantitative, with most populations differing only in the relative frequencies of inversions.

Collectively, Dobzhansky's data show that the genetic structure of *D. pseudoobscura* populations changes from place to place and from one time to another. At least some of this variation in population genetic structure is the result of natural selection.

Research by Dennis A. Powers and colleagues on the mummichog (*Fundulus heteroclitus*) provides another example of a species in which genetic structure varies among populations. The mummichog is a small fish (5 to 10 cm long) that lives in inlets, bays, and estuaries along the Atlantic coast of North America from Florida to Newfoundland. Powers studied allele frequencies at the locus encoding the protein lactate dehydrogenase-B (LDH-B). LDH-B is an enzyme made in the liver, heart, and red skeletal muscle that converts lactate to pyruvate, and it is thus pivotal in both the manufacture of glucose and aerobic metabolism. There are two electrophoretically distinguishable allozymes of LDH-B; they differ at two amino acid positions. The alleles encoding the allozymes are called *Ldh-B^a* and *Ldh-B^b*.

The frequencies of the *Ldh-B* alleles vary dramatically among mummichog populations [Figure 26.8(a)]. In

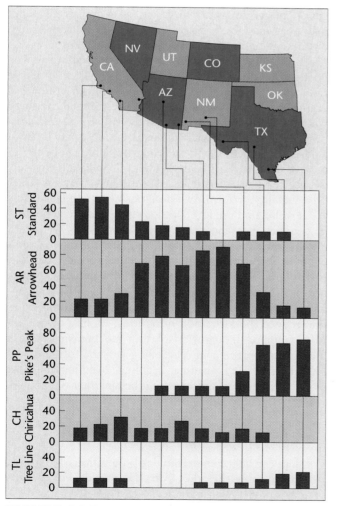

FIGURE 26.7 Frequencies of five chromosomal inversions in *D. pseudoobscura* in different geographic regions.

(a)

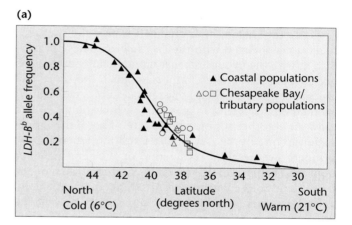

(b)

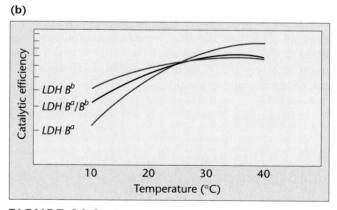

FIGURE 26.8 Variation in genetic structure among mummichog populations. (a) Frequencies of the *Ldh-B^b* allele in populations along the Atlantic coast of North America. (b) Catalytic efficiency of LDH-B allozymes as a function of temperature. From D.A. Powers and P.M. Schulte. Evolutionary adaptations of gene structure and expression in natural populations in relation to a changing environment, *The Journal of Experimental Zoology*, ©1998. Reprinted by permission of Wiley-Liss, Inc., a subsidiary of John Wiley & Sons, Inc. and with permission from *Annu. Rev. Genet.* 2S, © 1991 by Annual Reviews.

northern populations, where the mean water temperature is about 6°C, *Ldh-B^b* predominates. In southern populations, where the mean water temperature is about 21°C, *Ldh-B^a* predominates. Between the geographic extremes, allele frequencies are intermediate.

To determine whether the geographic variation in *Ldh-B* allele frequencies is due to natural selection, Powers studied the biochemical properties of the LDH-B allozymes. He found that the enzyme encoded by *Ldh-B^b* shows higher catalytic efficiency at low temperatures, whereas the gene product of *Ldh-B^a* is more efficient at high temperatures [Figure 26.8(b)]. A mixture of the two forms has intermediate efficiency at all temperatures.

In addition to the functional differences between the LDH-B allozymes, the *Ldh-B* alleles have sequence differences in their regulatory regions. The *Ldh-B^b* allele is transcribed at over twice the rate of the *Ldh-B^a* allele. As a result, northern fish have higher concentrations of the LDH-B enzyme in their cells.

The differences in catalytic efficiency and transcription rate associated with the two *Ldh-B* alleles are consistent with the hypothesis that mummichog populations are adapted to the temperatures at which they live. The fish are ectotherms; their body temperature is determined by their environment. In general, low body temperatures slow an ectotherm's metabolic rate. The higher transcription rate of the *Ldh-B^b* allele and the superior low-temperature catalytic efficiency of its gene product appear to help northern fish compensate for the tendency of their cold environment to reduce their metabolic rate. The superior high-temperature catalytic efficiency and lower transcription rate associated with the *Ldh-B^a* allele appear to allow southern fish to economize on the resources devoted to the machinery for glucose production and aerobic metabolism. Based on this and other evidence, Powers suggests that the differences among mummichog populations in *Ldh-B* allele frequencies are the result of natural selection.

A reduction in gene flow between populations, accompanied by divergent selection and/or genetic drift, can lead to speciation

In the preceding sections we have seen that most populations harbor considerable genetic variation, and that different populations within a species may have different alleles and/or allele frequencies at a variety of loci. The divergence of populations within a species can be caused by natural selection, genetic drift, or both. In Chapter 25 we saw that the migration of individuals between populations, and the gene flow that accompanies this migration, tends to homogenize allele frequencies among populations. In other words, migration counteracts the tendency of populations to diverge.

When gene flow between populations is reduced or absent, the populations may diverge to the point that members of one population are no longer able to successfully interbreed with members of the other. When populations reach the point that they are reproductively isolated from one another, they have become different species.

The barriers that prevent interbreeding between populations can be physiological, behavioral, or mechanical. The biological and behavioral properties of organisms that prevent or reduce interbreeding are called **reproductive isolating mechanisms**. These mechanisms are classified in Table 26.2.

Prezygotic isolating mechanisms prevent individuals from mating in the first place. Individuals from different populations may not find each other at the right time, may not recognize each other as suitable mates, or may try to mate but find that they are unable to do so.

Postzygotic isolating mechanisms create reproductive isolation even when the members of two populations are willing and able to mate with each other. For example, genetic divergence may have reached the stage where the viability or fertility of hybrids is reduced. Hybrid zygotes may be formed, but all or most may be inviable. Alternatively, the hybrids may be viable but be sterile or have reduced fertility. In another possibility, the hybrids themselves may be fertile, but their progeny may have lowered viability or fertility. These postzygotic mechanisms act at or beyond the level of the zygote and are generated by genetic divergence.

In some models of speciation, postzygotic isolation evolves first, and is followed by prezygotic isolation. Postzygotic isolating mechanisms waste gametes and zygotes and lower the reproductive fitness of hybrid survivors. Selection will therefore favor the spread of alleles that reduce the formation of hybrids, leading to the development of prezygotic isolating mechanisms. These mechanisms prevent interbreeding and the formation of hybrid zygotes and offspring. In animal evolution, the most effective prezygotic mechanism is behavioral isolation, involving courtship behavior.

Examples of Speciation

We shall consider two examples of speciation, one from a laboratory study, the other from a field study. For the laboratory study, biologist Diane Dodd took advantage of experimental populations of *Drosophila pseudoobscura* that had already been established by two of her colleagues. Dodd's

TABLE 26.2   Reproductive Isolating Mechanisms

Prezygotic Mechanisms (prevent fertilization and zygote formation)

1. Geographic or ecological: The populations live in the same regions but occupy different habitats
2. Seasonal or temporal: The populations live in the same regions but are sexually mature at different times
3. Behavioral (only in animals): The populations are isolated by different and incompatible behavior before mating
4. Mechanical: Cross-fertilization is prevented or restricted by differences in reproductive structures (genitalia in animals, flowers in plants)
5. Physiological: Gametes fail to survive in alien reproductive tracts

Postzygotic Mechanisms (fertilization takes place and hybrid zygotes are formed, but these are nonviable or give rise to weak or sterile hybrids)

1. Hybrid nonviability or weakness
2. Developmental hybrid sterility: Hybrids are sterile because gonads develop abnormally or meiosis breaks down before completion
3. Segregational hybrid sterility: Hybrids are sterile because of abnormal segregation to the gametes of whole chromosomes, chromosome segments, or combinations of genes
4. F_2 breakdown: F_1 hybrids are normal, vigorous, and fertile, but F_2 contains many weak or sterile individuals

Source: From G. Ledyard Stebbins, *Processes of organic evolution*, 3rd edition, copyright 1977, p. 143. Reprinted by permission of Prentice-Hall, Upper Saddle River, NJ.

colleagues were interested in the evolution of digestive physiology. They collected flies from a wild population and used them to establish several separate laboratory populations. Some of the laboratory populations received starch-based medium, others received maltose-based medium. Both food sources were stressful for the flies. It was only after several months of evolution by natural selection that the populations became adapted to their artificial diets and began to thrive.

Dodd wanted to know whether the starch-adapted populations and the maltose-adapted populations, which had been diverging under strong selection and in the absence of gene flow, had become different species. Roughly a year after the populations were established, Dodd performed a series of mating trials. For each trial she placed 48 flies together in a bottle: 12 males and 12 females from a starch-adapted population, and 12 males and 12 females from a maltose-adapted population. Then she simply watched which flies mated with which.

Dodd predicted that if the populations adapted to different media had speciated, then the flies would prefer to mate with members of their own population. If the populations had not speciated, then the flies would mate at random. The results appear in Table 26.3. Roughly 600 of the 900 matings Dodd observed were between males and females from the same population. In other words, the differently adapted fly populations showed partial pre-mating isolation. Dodd concluded that the populations had begun to speciate, but had not yet completed the process.

TABLE 26.3   Incipient Speciation in Laboratory Populations of *Drosophila pseudoobscura* (Number of Matings Involving Each Kind of Male–Female Pair)

		Female	
		Starch-adapted	Maltose-adapted
Male	Starch-adapted	290	153
	Maltose-adapted	149	312

Source: Compiled from Dodd, D.M.B. 1989. Reproductive isolation as a consequence of adaptive divergence in *Drosophila pseudoobscura*. *Evolution* 43:1308–11.

For the field study of speciation, Nancy Knowlton and colleagues took advantage of a natural experiment that the formation of the Isthmus of Panama had performed on several species of snapping shrimps (Figure 26.9). The Isthmus of Panama, which created a land bridge connecting North and South America and simultaneously separated the Carribbean Sea from the Pacific Ocean, formed roughly 3 million years ago. Knowlton identified seven Caribbean species of snapping shrimp, each of which formed a natural pair with a Pacific species. The members of each species pair were more similar to each other in structure and appearance than either was to any other species in its own ocean. Analysis of allozyme allele frequencies and mitochondrial DNA sequences confirmed that the members of each pair were one another's closest genetic relatives. Knowlton's interpretation of these data is that, prior to the formation of the Isthmus, the ancestors of each species pair were a single species. When the Isthmus closed, the seven ancestral species were each divided into two separate populations, one in the Caribbean, the other in the Pacific.

If they met in a dish in Knowlton's lab for the first time in 3 million years, would Caribbean and Pacific members of a species pair recognize each other as suitable mates? Knowlton placed males and females together and noted their behavior toward each other. She then calculated the relative inclination of Caribbean/Pacific couples to mate vs. that of Caribbean/Caribbean or Pacific/Pacific couples. For three of the seven species pairs, transoceanic couples refused to mate altogether. For the other four species pairs, transoceanic couples were 33, 45, 67, and 86 percent as likely to mate with each other as were same-ocean pairs. Of the same-ocean couples that mated, 60 percent produced viable clutches of eggs. Of the transoceanic couples that mated, only 1 percent

produced viable clutches. We can conclude from these results that 3 million years of separation has resulted in complete or nearly complete speciation, involving strong pre- and postzygotic isolating mechanisms, for all seven species pairs.

The Minimum Genetic Divergence Required for Speciation

How much genetic divergence is required between two populations before they become different species? We shall consider two examples, one from an insect, the other from a plant, which demonstrate that in some cases the answer is not very much.

Drosophila heteroneura and *D. silvestris*, found only on the island of Hawaii, are estimated to have diverged from a common ancestral species only about 300,000 years ago. They are thought to be descended from *D. planitibia* colonists from the older island of Maui (Figure 26.10). The two species are clearly separated from each other by different and incompatible courtship and mating behaviors (a prezygotic isolating mechanism), by morphology, and by pigmentation of the body and wings (Figure 26.11).

In spite of their morphological and behavioral divergence, it is difficult to demonstrate significant differences between these species in chromosomal inversion patterns or protein polymorphisms. When DNA-hybridization studies are carried out on these species, the sequence diversity between the two is only about 0.55

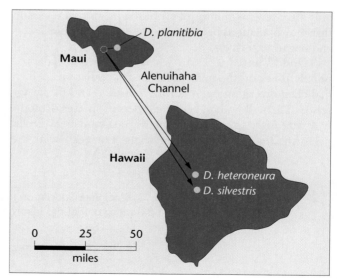

FIGURE 26.10 Proposed pathway of colonization of Hawaii by members of the *D. planitibia* species complex. The open circle represents a population ancestral to the three present-day species.

FIGURE 26.9 A snapping shrimp (genus *Alpheus*).

(a)

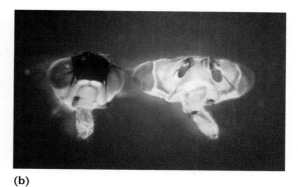

(b)

FIGURE 26.11 (a) Differences in pigmentation patterns in *D. sylvestris* (left) and *D. heteroneura* (right). (b) Head morphology in *D. sylvestris* (left) and *D. heteroneura* (right).

percent (Figure 26.12). Thus, nucleotide sequence diversity may precede the development of protein or chromosomal polymorphisms.

Genetic evidence suggests that the differences between *D. heteroneura* and *D. silvestris* are controlled by a relatively small number of genes. For example, as few as 15 to 19 major loci may be responsible for the morphological differences between the species, demonstrating that the process of speciation need only involve a small number of genes.

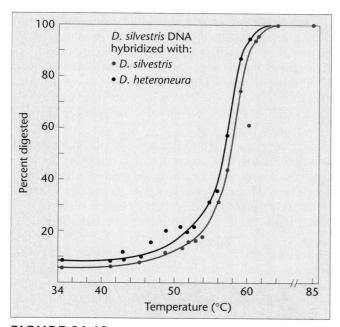

FIGURE 26.12 Nucleotide sequence diversity in the *Drosophila planitibia* species complex. The degree of shift to the left by the heterologous hybrids is an indication of the degree of nucleotide sequence divergence.

More recent studies using two closely related species of monkey flowers, a plant that grows in the Rocky Mountains and areas west, confirm that species can be separated by only a few genes. One species, *Mimulus cardinalis* is fertilized by hummingbirds and does not interbreed with *Mimulus lewisii*, which is fertilized by bumblebees. Toby Bradshaw and his colleagues studied genetic differences related to reproduction in the two species: flower shape, size, and color; and nectar production. For each trait, a difference in a single gene provided at least 25 percent of the variation observed among laboratory-created *M. cardinalis* × *lewisii* hybrids. *M. cardinalis* makes 80 times more nectar than does *M. lewisii*, and a single gene is responsible for at least half the difference. A single gene also controls a large part of the differences in flower color between the two species (Figure 26.13). In this case, as in the Hawaiian *Drosophila*, species differences can be traced to a relatively small number of genes.

In at Least Some Instances, Speciation is Rapid

How much time is required for speciation? In many cases, speciation may take place slowly over a long period of time. In other cases, however, speciation can be surprisingly rapid.

The Rift lakes of East Africa support hundreds of species of cichlid fish. Lake Victoria (Figure 26.14), for example, has over 400 species. These species have some morphological diversity, but, more dramatically, they are highly specialized for different niches (Figure 26.15). Some eat algae floating on the water's surface, others are bottom feeders, insect feeders, mollusc eaters, and predators on other fish species. Lake Tanganyika and Lake Malawi have a similar array of species. Remarkably, however, genetic analyses indicate that Lake Victoria's species are all more

(a)

(b)

FIGURE 26.13 The flowers of two closely related species of monkey flowers, (a) *Mimulus cardinalis* and (b) *Mimulus lewisii*.

FIGURE 26.14 Lake Victoria in the Rift Valley of East Africa is home to more than 400 species of cichlids.

closely related to each other than any of them is to any species from another lake. The implication is that most or all of Lake Victoria's 400 species are descended from a single common ancestor, and that they evolved within Lake Victoria itself.

Lake Victoria is between 250,000 and 750,000 years old, and there is evidence that the lake may have dried out nearly completely less than 14,000 year ago. If further research confirms that Lake Victoria's 400 cichlids evolved in less than 14,000 years, they will represent the most rapid evolutionary radiation ever documented in vertebrates.

Even faster speciation is possible through the special mechanism of polyploidy. The formation of animal species by polyploidy is rare, but polyploidy is an important factor in the evolution of plants. It is estimated that one-half of all flowering plants have evolved by this mechanism. One form of polyploidy is allopolyploidy (see Chapter 10), produced by doubling the chromosome number in an interspecific hybrid.

If two species of related plants have the genetic constitution SS and TT (where S and T represent the haploid set of chromosomes in each species), then the F_1 hybrid would have the chromosome constitution ST. Normally such a plant would be sterile because there are few or no homologous chromosome pairs and aberrations would arise during meiosis. However, if the hybrid undergoes a spontaneous doubling of chromosome number, a tetraploid SSTT plant would be produced. This chromosomal aberration might occur during mitosis in somatic tissue, giving rise to a partially tetraploid plant that would produce some tetraploid flowers. Alternatively, aberrant meiotic events may produce ST gametes, which when fertilized would yield SSTT zygotes. The SSTT plants would be fertile because they would possess homologous chromosomes producing viable ST gametes. This new, true-

FIGURE 26.15 Cichlids occupy a diverse array of niches, and each species is specialized for a distinct food source.

breeding tetraploid would have a combination of characters derived from the parental species and would be reproductively isolated from them because F₁ hybrids would be triploids and consequently sterile.

The tobacco plant *Nicotiana tabacum* ($2n = 48$) is the result of the doubling of the chromosome number in the hybrid between *N. otophora* ($2n = 24$) and *N. silvestris* ($2n = 24$) (Figure 26.16). The origin of *N. tabacum* is an example of virtually instantaneous speciation.

FIGURE 26.16 The cultivated tobacco plant *Nicotiana tabacum* is the result of hybridization between two other species.

We can use genetic differences among populations or species to reconstruct evolutionary history

Early in this chapter we pointed out that evolution involves the transformation and splitting of lineages. The examples we have reviewed have demonstrated that the transformation of lineages is associated with changes in genetic structure and that speciation is associated with genetic divergence. It follows that we should be able to use the genetic differences among species to reconstruct their evolutionary history.

An important early example of phylogeny reconstruction will serve to introduce this section. In the 1960s W. M. Fitch and E. Margoliash assembled data on the amino acid sequence for cytochrome c in a variety of organisms. **Cytochrome c** is a respiratory pigment found in the mitochondria of eukaryotes. The amino acid sequence of cytochrome c has changed very slowly during evolution. The amino acid sequence in humans and chimpanzees is identical; between humans and rhesus monkeys only one amino acid is different. This is remarkable considering that the fossil record indicates that the lines leading to humans and monkeys diverged from a common ancestral species approximately 20 million years ago.

Column (a) of Table 26.4 shows the number of amino acid differences between cytochrome c in humans and a variety of other organisms. The table is broadly consistent with our intuitions about how closely related we are to these other species. For example, we are more closely related to other mammals than we are to insects, and we are more closely related to insects than we are to fungi. Likewise, our cytochrome c differs in 13 amino acids from that of dogs, in 36 amino acids from that of moths, and in 56 amino acids from that of yeast.

TABLE 26.4   A Comparison of the Number of Amino Acid Differences and the Minimal Mutational Distance Between Cytochrome c in Humans and Other Organisms

Organism	(a) Number of Amino Acid Differences	(b) Minimal Mutational Distance
Human	0	0
Chimpanzee	0	0
Rhesus monkey	1	1
Rabbit	9	12
Pig	10	13
Dog	10	13
Horse	12	17
Penguin	11	18
Moth	24	36
Yeast	38	56

Source: From W. M. Fitch and E. Margoliash, Construction of phylogenetic trees, *Science* 155:279–84, 20 January 1967. Copyright 1967 by the American Association for the Advancement of Science.

More than one nucleotide change may be required to change a given amino acid. When the nucleotide changes necessary for all amino acid differences observed in a protein are totaled, the **minimal mutational distance** between the genes of any two species is established. Column (b) in Table 26.4 shows such an analysis of the genes encoding cytochrome c. As expected, these values are larger than the corresponding number of amino acids separating humans from the other nine organisms.

Fitch used data on the minimal mutational distances between the cytochrome c genes of 19 organisms to reconstruct their evolutionary history. The result is an estimate of the evolutionary tree, or phylogeny, that unites the species (Figure 26.17). The blue dots on the tips of the twigs represent the extant species. These are connected to the inferred common ancestors, represented by red dots, that speciated and diverged to produce the extant organisms. The common ancestors are connected to still earlier common ancestors, culminating in a single common ancestor for all the species on the tree, represented by the red dot on the extreme left.

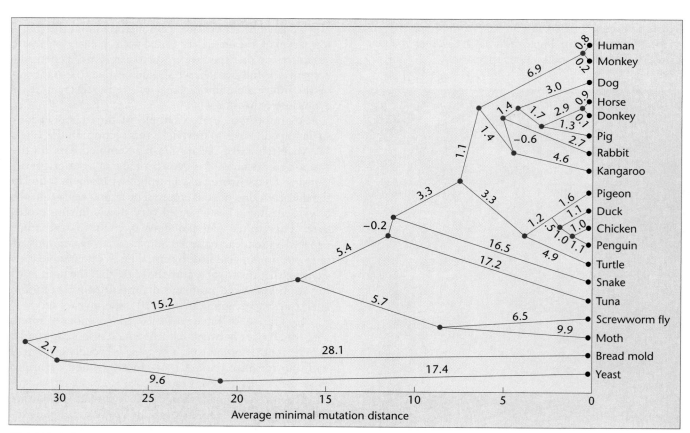

FIGURE 26.17   Phylogeny constructed by comparison of homologies in cytochrome c amino acid sequences.

A Method for Estimating Evolutionary Trees from Genetic Data

There are a great many methods for using data on genetic differences to estimate phylogenies. It is beyond the scope of this chapter to review all of them. Instead we shall present just one method. It is called the unweighted pair group method using arithmetic averages, or UPGMA. UPGMA is not the most powerful method for estimating phylogenies from genetic data, but it is, in spite of its name, intuitively straightforward. Furthermore, UPGMA works reasonably well under many circumstances.

The starting point for UPGMA is a table of pairwise genetic distances among a group of species [Figure 26.18(a)]. The data we shall use to illustrate the method come from DNA-hybridization studies by Charles Sibley and Jon Alquist, although nowadays researchers are far more likely to use data derived from comparisons of amino acid or nucleotide sequences. The subset of Sibley's data that we shall use provide genetic distances among humans and four species of apes: the common chimpanzee, the gorilla, the siamang, and the common gibbon.

The assumption that drives UPGMA is that the least genetically distant species are the most closely related. The steps of the method are as follows:

1. Search the table for the smallest genetic distance between any pair of species. In our table, this is 1.628, the distance between human and chimpanzee [Figure 26.18(a)]. Place these species on neighboring branches of an evolutionary tree [Figure 26.18(b)]. Define the lengths of these two branches to be half the genetic distance between the species, so that moving down from one tip and back out to the other is a trip equal to the total genetic distance. The length of the branches connecting human and chimpanzee to their common ancestor is about 0.81.

(a)

	Human	Chimp	Gorilla	Siamang	Gibbon
Human	–				
Chimp	**1.628**	–			
Gorilla	2.267	2.21	–		
Siamang	4.7	5.133	4.543	–	
Gibbon	4.779	4.76	4.753	1.95	–

(c)

	Hu-Ch	Gorilla	Siamang	Gibbon
Hu-Ch	–			
Gorilla	2.2385	–		
Siamang	4.9165	4.543	–	
Gibbon	4.7695	4.753	**1.95**	–

(e)

	Hu-Ch	Gorilla	Si-Gi
Hu-Ch	–		
Gorilla	**2.239**	–	
Si-Gi	4.843	4.648	–

(g)

	Hu-Ch-Go	Si-Gi
Hu-Ch-Go	–	
Si-Gi	4.778	–

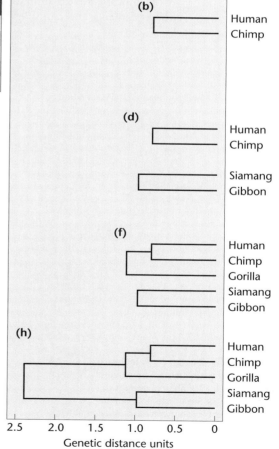

FIGURE 26.18   An example of phylogeny reconstruction by UPGMA. From Tab. 1, p. 126 from *Journal of Molecular Evolution* 26:99–122.

2. Reduce the table of genetic distances, defining the species just joined in step 1 as a cluster [Figure 26.18(c)]. Recalculate the genetic distance between the cluster and the other species as the average of the distances between each cluster member and the other species. For example, the genetic distance between the human–chimp cluster and gorilla is the average of the human–gorilla distance and the chimp–gorilla distance. This is $(2.267 + 2.21)/2 = 2.2385$.

3. Repeat steps 1 and 2 until all the species have been added to the tree.

The smallest distance in the reduced table is 1.95, the distance between siamang and gibbon [Figure 26.18(c)]. We therefore place siamang and gibbon on neighboring branches of our evolutionary tree, with the branch lengths connecting each to their common ancestor equal to 0.98 [Figure 26.18(d)].

Reducing the table again, we have two clusters plus gorilla [Figure 26.18(e)]. The genetic distance between the human–chimp cluster and the siamang–gibbon cluster is the average of four distances: human–siamang, human–gibbon, chimp–siamang, and chimp–gibbon.

Now the smallest genetic distance is 2.239, the distance between the human–chimp cluster and gorilla. We therefore add a branch to the tree for gorilla and connect it to the common ancestor of the human–chimp cluster [Figure 26.18(f)]. We arrange the branch lengths so that the total horizontal distance between the tips of any two branches in the human–chimp–gorilla cluster is 2.239.

Reducing the table for the final time [Figure 26.18(g)], we calculate the genetic distance between the human–chimp–gorilla cluster and the siamang–gibbon cluster as the average of six genetic distances: human–siamang, human–gibbon, chimp–siamang, chimp–gibbon, gorilla–siamang, and gorilla–gibbon. This distance is 4.778, which allows us to complete our evolutionary tree [Figure 26.18(h)].

The evolutionary tree we have constructed indicates that humans and chimpanzees are one another's closest relatives. That is not to say that humans evolved from chimpanzees. Rather, humans and chimpanzees share a more recent common ancestor than either shares with any of the other species on the tree.

The chief shortcoming of UPGMA is that it provides no means of determining how well the tree it produces fits the data, compared to other possible trees. For example,

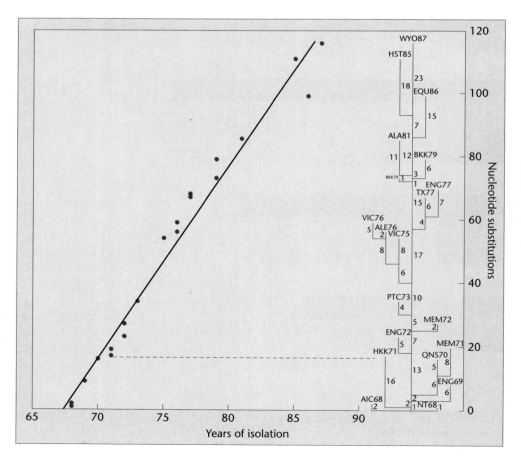

FIGURE 26.19 A molecular clock in the influenza A hemagglutinin gene. (a) Number of nucleotide differences between the first isolate and each subsequent isolate as a function of year of isolation. (b) An estimate of the phylogeny of the isolates. From Fitch et al. 1991. Positive Darwinian evolution in human influenza A viruses. *PNAS* 88:4270–73.

how much better does the tree we just created fit our data than a tree in which chimpanzees and gorillas are one another's closest relatives? Evolutionary geneticists have developed a variety of methods for searching the set of all possible trees connecting a group of species, and calculating the relative performance of each. **Parsimony** methods compare trees based on the minimum number of evolutionary changes each requires, and select the simplest possible tree. **Maximum likelihood** methods start with a model of the evolutionary process, calculate how likely it is that evolution will produce each possible tree under the model, and select the most likely tree.

It is important to keep in mind that a phylogeny produced from data on genetic distances is not the truth. Instead, it is an estimate of the truth. How do we know whether our methods produce reasonable estimates? First, biologists like David M. Hillis and his colleagues have tested the performance of various methods in cases where the true phylogeny is known. They generate known phylogenies in two ways—by using computers to simulate the evolutionary divergence of species, or by dividing laboratory populations of organisms, such as viruses, and letting them evolve in isolation. Most methods currently being used for phylogeny reconstruction have been tested and found to work reasonably well, provided they are used correctly and are appropriate for the data at hand. Second, phylogenies estimated on the basis of genetic data generally agree with phylogenies estimated on the basis of other kinds of data, such as morphology. Third, phlogenies estimated on the basis of different genetic data sets can be compared against each other. Phylogenies based on combined data sets from several different proteins or genetic loci inspire greater confidence than phylogenies based on a single molecule.

Molecular Clocks

In many cases we would like to estimate not only which of a set of species are most closely related, but also when their common ancestors lived. Sometimes we can do so, thanks to molecular clocks. **Molecular clocks** are amino acid sequences or nucleotide sequences in which evolutionary changes accumulate at a constant rate over time.

Research by Walter M. Fitch and colleagues on the influenza A virus provides an example. Fitch sequenced a portion of the hemagglutinin gene from flu viruses that had been isolated at different times over a 20-year period and stored in a freezer. He calculated the number of nucleotide differences among the different isolates and estimated their evolutionary tree [Figure 26.19(b)]. Most of the isolated strains appear to have gone extinct; they do not have descendents among the more recently isolated viruses. Fitch then plotted the number of nucleotide sub-

stitutions between the first isolate and each subsequent isolate as a function of the year of isolation [Figure 26.19(a)]. The data points all fall remarkably close to a straight line, indicating that nucleotide substitutions have accumulated in this gene at a steady rate. The hemagglutinin gene thus serves as a molecular clock. If we compare the sequences of two new isolates, we can use this graph to estimate the amount of time that has passed since each diverged from their common ancestor.

Molecular clocks must be carefully calibrated and used with caution. For example, Fitch's data indicate that strains of influenza A that have jumped from birds to humans have evolved much more rapidly than strains that have remained in birds. Thus a molecular clock calibrated from human strains of the virus would be highly misleading if applied to bird strains.

Figure 26.20 shows a phylogeny based on Sibley's complete data set of genetic differences among humans and apes. The horizontal axis gives estimated ages for the common ancestors, based on a molecular clock calibrated from the fossil record. This tree suggests that the most recent common ancestor of humans and chimps lived between 5 and 10 million years ago.

Reconstructing evolutionary history allows us to answer a variety of questions

We are now in a position to make good on our claim that the questions we asked at the beginning of the chapter can be answered by using genetic data to reconstruct evolutionary history.

Transmission of HIV from a Dentist to His Patients

In late 1986 a Florida dentist tested positive for HIV. Several months later he was diagnosed with AIDS. He continued to practice general dentistry for two more years, until one of his patients, a young woman with no obvious risk factors, discovered that she, too, was infected with HIV. When the dentist publicly urged his other patients to have themselves tested, several more were found to be HIV-positive. Did this dentist transmit HIV to his patients, or did the patients become infected by some other means?

The movement of a virus from one individual to another is like the founding of a new island population by a small number of migrants. Gene flow between the ancestral population and the new population is non-existent. The populations are free to diverge, as a result of genetic drift, adaptation to different environments, or both.

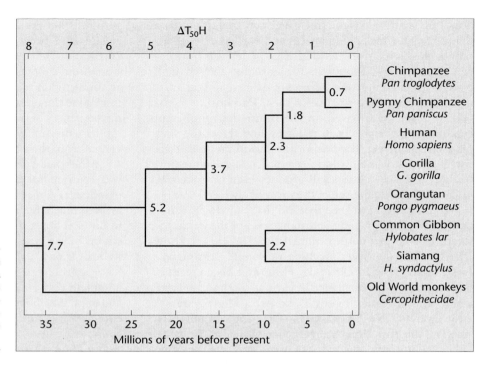

FIGURE 26.20 Phylogenetic sequence of the hominoid primates and the Old World monkeys as estimated by DNA hybridization. The numbers at the branch points are $\Delta T_{50}H$ measurements. The evolutionary branch points are dated by reference to the fossil record and nucleotide divergence.

If the dentist passed his HIV infection to his patients, then the viral strains isolated from these individuals should be more closely related to each other and to the dentist's strain than to strains from any other individuals living in the same area. Chin-Yih Ou and colleagues sequenced portions of the gene for the HIV envelope protein from viruses collected from the dentist, 10 of his patients, and several other HIV-infected individuals from the area who served as local controls.

An evolutionary tree for the HIV isolates, produced by David Hillis and colleagues in a reanalysis of Ou's sequence data, appears in Figure 26.21. The tree is wrapped in a circle to make it fit more easily on the page. The viruses isolated from various individuals appear at the tips of the branches around the outside of the tree; the inferred common ancestors are toward the center. The HIV strains taken from patients A, B, C, E, G, and I all share a more recent common ancestor with each other and the dentist's strains than with HIV from any other local individual. It appears that the dentist did, indeed, transmit his infection to these patients. In contrast, the viruses taken from patients D, F, H, and J are all more closely related to strains from local controls (LC) than they are to the dentist's strains. These patients appear to have gotten their infection from someone other than the dentist. Patient J, in fact, seems to have acquired his infection from two different sources.

The Relationship of Neanderthals to Modern Humans

Paleontological evidence indicates that the Neanderthals, *Homo neanderthalensis*, lived in Europe and Western Asia from some 300,000 to 30,000 years ago. During their later years, the Neanderthals coexisted with anatomically modern humans. Did Neanderthals and modern humans interbreed, so that descendants of the Neanderthals are alive today, or did the Neanderthals die off without issue, so that today their lineage is extinct?

Svante Pääbo, Matthias Krings, and their colleagues extracted fragments of mitochondrial DNA from a Neanderthal skeleton. After completing a variety of careful analyses to confirm that they had indeed isolated Neanderthal gene fragments, the researchers placed the Neanderthal sequences on a phylogeny with over 2000 modern humans (Figure 26.22). The Neanderthal appears to be a distant relative. Using a molecular clock calibrated with chimpanzees and humans, the researchers calculated that the last common ancestor between Neanderthals and modern humans lived roughly 600,000 years ago, four times as long ago as the last common ancestor of all modern humans.

Considerable caution is required when drawing conclusions based on a single Neanderthal specimen. However, Pääbo's data suggest that the Neanderthals were a distinct species that did not freely interbreed with modern humans.

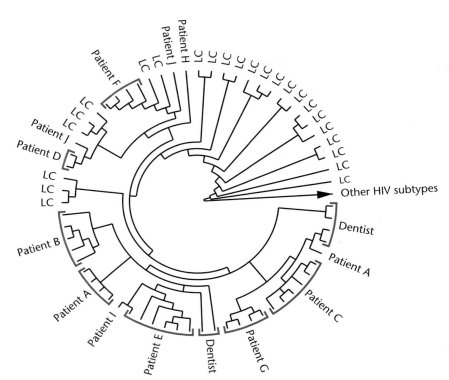

FIGURE 26.21 A phylogeny of HIV strains taken from a dentist, his patients, and several local controls (LC). From Hillis (1998), *Current Biology* 7:R129–R131.

If this interpretation is correct, then when the Neanderthals disappeared, their lineage went extinct with them.

The Origin of Mitochondria

The mitochondria that inhabit our cells have their own DNA. The organization and function of the mitochondrial genome bears many similarities to the organization and function of bacterial genomes. Did our mitochondria orig-

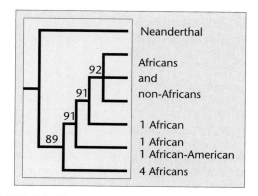

FIGURE 26.22 A simplified diagram depicting a phylogeny estimated from mitochondrial DNA sequences of one Neanderthal and over 2000 modern humans. From M. Krings, A. Stone, R.W. Schmitz, H. Krainitzki, M. Stoneking, and S. Pääbo. 1997. ©1997 Cell Press. *Cell* 90:19–30. Fig. 7A, p. 26.

inate as free-living bacteria that took up residence inside other cells, or is the mitochondrial genome ultimately derived from nuclear chromosomes?

Accurate phylogeny reconstruction demands that researchers choose a genetic sequence appropriate for the question at hand. If researchers want to reconstruct recent speciation events, they must use rapidly evolving sequences, or else there will be no differences among the species they want to study. Conversely, if researchers want to reconstruct ancient speciation events, they must use slowly evolving sequences, so that the species under study still show recognizable similarities.

The origin of mitochondria was an extremely ancient event. Researchers investigating this question have used nucleotide sequences in the genes for small-subunit ribosomal RNAs. These genes are present in mitochondria, chloroplasts, and every cellular organism known. Furthermore, their sequences are under strong functional constraint, and so are highly conserved.

Stephen Giovannoni and colleagues reconstructed a phylogeny based on rRNA sequences from a variety of eukaryotes, bacteria, and archea. The species used represented the full diversity of life on Earth. The researchers also included the rRNA sequence from the maize mitochondrion. If mitochondria are descended from free-living bacteria, then the maize mitochondrion's rRNA sequence should be most closely related to the sequences

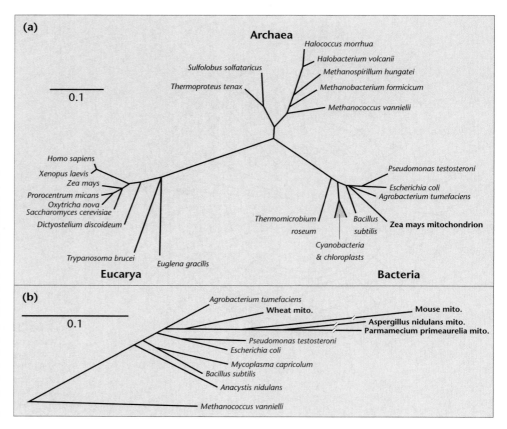

FIGURE 26.23 The evolutionary relationships of mitochondria. (a) A phylogeny estimated from nucleotide sequences of the small-subunit rRNA genes of representatives of all three domains in the tree of life. The *Zea mays* mitochondrion is closely related to bacteria. The scale bar represents the number of point mutations fixed per nucleotide. (b) A phylogeny estimated from nucleotide sequences of the small-subunit rRNA genes of several bacteria and four mitochondria. The mitochondria are closely related to each other and to the purple bacteria, including *E. coli*. The scale bar represents the estimated number of mutations per nucleotide. From *Journal of Bacteriology.* 170:3584–92, Fig. 2.

of the bacteria. If, on the other hand, mitochondrial genomes are derived from nuclear genes, then the mitochondrion's sequence should be most closely related to the sequences of eukaryotes. Giovannoni's phylogeny appears in Figure 26.23(a). The mitochondrial gene branches from within the bacteria.

In a similar study, D. Yang and colleagues reconstructed a phylogeny of rRNA sequences from several bacteria and the mitochondria of wheat, mouse, *Aspergillus* (a fungus), and *Paramecium* [Figure 26.23(b)]. The mitochondria are one another's closest relatives,

and they branch from within the bacteria. Their closest free-living relatives are the purple bacteria, which include *E. coli*.

Both studies point to the same conclusion. They strongly support the hypothesis that mitochondria are descended from free-living bacteria.

Giovannoni's phylogeny also included several chloroplasts, which like mitochondria have their own genome. The chloroplasts all branch from within the cyanobacteria. In an evolutionary sense mitochondria and chloroplasts are both bacteria.

CHAPTER SUMMARY

1. Today's organisms are the products of an evolutionary history that included the transformation, splitting, and divergence of lineages. Alfred Russel Wallace and Charles Darwin formulated the theory of natural selection, which provides a mechanism for the transformation of lineages. The genetic basis of evolution and the role of natural selection in changing allele frequencies were discovered in the twentieth century.

2. When population geneticists began studying the genetic structure of populations, they discovered that most populations harbor considerable genetic diversity. This diversity is apparent in electrophoretic studies of proteins and at the level of DNA sequences. Whether the genetic diversity of populations is maintained primarily by mutation plus genetic drift or by natural selection is a matter of some debate.

3. The geographic ranges of most species encompass a degree of environmental diversity. As a result of both adaptation to different environments and genetic drift, different populations within a species may have different alleles and/or allele frequencies at many loci.

4. Gene flow among populations tends to homogenize their genetic composition. When gene flow is reduced, genetic drift and adaptation to different environments can cause populations to diverge. Eventually populations may become so different that the individuals in one population either will not or cannot mate with the individuals in the other. At this point the divergent populations have become different species.

5. Because speciation is associated with genetic divergence, we can use the genetic differences among species to infer their evolutionary history. By comparing amino acid or nucleotide sequences we can determine the genetic distances among species. We can then use the genetic distances to reconstruct evolutionary trees. The simplest methods for reconstructing phylogenies are based on the assumption that the least divergent species are one another's closest relatives.

6. The reconstruction of evolutionary trees is a key technique in answering a diversity of interesting questions. Examples include tracing the route of transmission in an epidemic, deciphering recent events in human evolution, and determining the evolutionary relationships among all organisms.

INSIGHTS AND SOLUTIONS

1. Sequence analysis of DNA can be accomplished by a number of techniques (see Appendix A for a detailed description). Protein sequencing, on the other hand, is made more complex by the fact that 20 different subunits need to be unambiguously identified and enumerated, rather than the four nucleotides of DNA. Because of their unique properties, the N-terminal and C-terminal amino acids in a protein are easy to identify, but the array in between offer a difficult challenge, since many proteins contain hundreds of amino acids. How is it that protein sequencing is accomplished?

   Solution: The strategy for protein sequencing is the same as for DNA sequencing—divide and conquer. To accomplish this, specific enzymes are used that reproducibly cleave proteins between certain amino acids. The use of different enzymes produces overlapping fragments. Each fragment is isolated and its amino acid sequence is determined by chemical means. Sequences from overlapping fragments are then assembled to give sequence for the entire protein. Alternatively, researchers can first sequence the gene that encodes the protein, and then use the genetic code to infer the protein's amino acid sequence.

2. A single plant twice the size of others in the same population suddenly appears. Normally, plants of this species reproduce by self-fertilization and by cross-fertilization. Is this new giant plant simply a variant, or could it be a new species? How would you determine this?

   Solution: One of the most widespread mechanisms of speciation in higher plants is polyploidy, the multiplication of entire chromosome sets. The result of polyploidy is usually a larger plant with larger flowers and seeds. There are two ways of testing this new variant to determine whether it is a new species. First, the giant plant should be crossed with a normal-size plant to see whether it produces viable, fertile offspring. If it does not, the two different types of plants would appear to be reproductively isolated. Second, the giant plant should be cytogenetically screened to examine its chromosome complement. If it has twice the number of its normal-size neighbors, it is a tetraploid that may have arisen spontaneously. If the chromosome number differs by a factor of two and the new plant is reproductively isolated from its normal-size neighbors, it is a new species.

PROBLEMS AND DISCUSSION QUESTIONS

1. Discuss the rationale behind the statement that inversions in chromosome 3 of *Drosophila pseudoobscura* represent genetic variation.
2. Describe how populations with substantial genetic differences can form. What is the role of natural selection?
3. What types of nucleotide substitutions will not be detected by electrophoretic studies of a gene's protein product?
4. Shown here are two homologous lengths of the alpha and beta chains of human hemoglobin. Consult the genetic code dictionary (Figure 13.6) and determine how many amino acid substitutions may have occurred as a result of a single nucleotide substitution. For any that cannot occur as the result of a single change, determine the minimal mutational distance.

Alpha:	Ala	Val	Ala	His	Val	Asp	Asp	Met	Pro
Beta:	Gly	Leu	Ala	His	Leu	Asp	Asn	Leu	Lys

5. Determine the minimal mutational distances between these amino acid sequences of cytochrome c from various organisms. Compare the distance between humans and each organism.

Human:	Lys	Glu	Glu	Arg	Ala	Asp
Horse:	Lys	Thr	Glu	Arg	Glu	Asp
Pig:	Lys	Gly	Glu	Arg	Glu	Asp
Dog:	Thr	Gly	Glu	Arg	Glu	Asp
Chicken:	Lys	Ser	Glu	Arg	Val	Asp
Bullfrog:	Lys	Gly	Glu	Arg	Glu	Asp
Fungus:	Ala	Lys	Asp	Arg	Asn	Asp

6. The data here are exerpted from a paper by Heui-Soo Kim and Osamu Takenaka. They are short DNA sequences from five species: human, chimpanzee, gorilla, orangutan, and baboon. The sequences are each 50 bases long. They represent a short piece of a gene for testis-specific protein Y, which is located on the Y chromosome. The complete human sequence is given. The sequences for the other four species are shown only in the spots where they differ from the human sequence. Calculate the genetic difference between each pair of species. For example, chimps and gorillas differ in 2 out of 50 bases, or 4 percent. Thus the genetic difference between chimps and gorillas is 0.04. Then use UPGMA to reconstruct the phylogeny for these five species. Is your phylogeny consistent with the ones shown in Figure 26.18 and Figure 26.20?
7. The genetic difference between two species of *Drosophila*, *D. heteroneura* and *D. sylvestris*, as measured by nucleotide di-

versity, is about 1.8 percent. The difference between chimpanzees (*Pan troglodytes*) and humans (*Homo sapiens*) is about the same, yet the latter species are classified in different genera. In your opinion, is this valid? If so, why; if not, why not?

8. As an extension of Problem 7, consider the following: In sorting out the complex taxonomic relationships among birds, species with $\Delta T_{50}H$ values of 4.0 are placed in the same genus, even by traditional taxonomy based on morphology. Using the data in Figure 26.20, construct a classification that obeys this rule, using any of the appropriate genus names (*Pongo*, *Pan*, *Homo*), or creating new ones.
9. The use of nucleotide-sequence data to measure genetic variability is complicated by the fact that the genes of higher eukaryotes are complex in organization and contain 5′ and 3′ flanking regions as well as introns. Slightom and colleagues have compared the nucleotide sequence of two cloned alleles of the γ-globin gene from a single individual and found a variation of 1 percent. Those differences include 13 substitutions of one nucleotide for another, and three short DNA segments that have been inserted in one allele or deleted in the other. None of the changes take place in the exons (coding regions) of the gene. Why do you think this is so, and should it change the concept of genetic variation?
10. Discuss the arguments supporting the neutralist hypothesis. What counterarguments are proposed by the selectionists?
11. Of what value to our understanding of genetic variation and evolution is the debate concerning the neutralist hypothesis?

EXTRA-SPICY PROBLEMS

12. Native tribes of both North and South America are descendants of one or a few groups of Mongoloid peoples who immigrated from the Asian land mass sometime between 11,000 and 40,000 years ago.

The HLA genes in humans code for the histocompatibility antigens found on the surface of most cells. These genes are the most polymorphic genes known in humans. For example, over 40 alleles of the HLA B gene have been identified. In contrast to Old World populations, however, native Americans do not show this phenomenal allelic diversity. Instead, a limited set of alleles of all of the HLA genes are found in varying frequency in all tribes, regardless of location of the tribe. What are the various forces that determine allele frequencies? What is the best explanation for these observations?

Human	AGAGGTTTTTCAGTGAATGAAGCTATTTTTAAGGGAGTGTGATTGCTGCC
Chimpanzee	C
Gorilla	T C C
Orangutan	T GT C C C C
Baboon	C C G G T C G GC C C G

GENETICS MediaLab

The following resources will help you achieve a better understanding of the concept presented in this chapter. These resources can be found on the CD packaged with this textbook and on the Companion Website at **http://www.prenhall.com/klug/**.

CD Resources:
Self-grading Chapter Problems

Web Resources:
Web Destinations in Genetics

Self-grading Chapter Problems

Chapter Search Terms

Genetics Newsgroups

Student Bulletin Board

Web Problem 1:
Time for completion = 10 minutes
What are the genetic changes that take place during speciation? John Werren studies speciation in wasps of the genus *Nasonia*. After reading about the genetics of speciation among these parasitoids, you should be able to answer the following questions. Which species are allopatric and which are sympatric? What are the isolating mechanisms between *Nasonia vitripennis* and *Nasonia giraulti*? Why does haplodiploidy facilitate the study of differences among these species? Are there interactions between nuclear and/or cytoplasmic factors that reduce the fitness of hybrids? To complete this exercise, visit Web Problem 1 in Chapter 26 of your Companion Website and select the keyword **SPECIATION**.

Web Problem 2:
Time for completion = 10 minutes
How does natural selection lead to divergence among populations? Use the "EvolveIt" simulation to examine how divergent natural selection of the finches of two Galápagos Islands can lead to dramatic differences in morphology. The default setting specifies identical phenotypic distributions of beak size on the two islands. Go to the "Precipitation" parameter and adjust the precipitation level on Wallace Island to 80 percent (so that soft seeds represent 64 percent of the resource base); leave the other island at the default. Set the duration of the experiment to 300 years and run the simulation; then look at the plots of the results and answer the following questions. On which island did most of the evolution occur? Describe how the environment selected for birds with larger beaks. What happened to the size of the population under the strongest selection? Would increasing the variance in beak size increase or decrease the rate of evolution? To complete this exercise, visit Web Problem 2 in Chapter 26 of your Companion Website and select the keyword **NATURAL SELECTION**.

Web Problem 3:
Time for completion = 15 minutes
How can we determine the pattern of evolutionary relationships among a group of species? The study of phylogenetics has benefited from the increased availability of molecular sequence data. Follow the tutorial on phylogeny reconstruction from molecular sequence data and then answer the following questions. Why do all phylogenetic trees eventually converge on a common ancestor? What is the relationship between mutation and time of divergence? Are all mutations equally useful in estimating phylogenies? What processes might cause different rates of evolution for "essential" vs. "irrelevant" mutations? To complete this exercise, visit Web Problem 3 in Chapter 26 of your Companion Website and select the keyword **PHYLOGENY**.

Another interesting observation about the HLA alleles of native Americans is that South American tribes have alleles of the HLA B gene that are not found in the present-day Asian population. Provide two possible explanations for this observation.

13. Nauru is a remote Pacific atoll occupied by 5000 Micronesians. Colonization by Britain, Australia, and New Zealand, as well as income from phosphate mining, has changed the lifestyle of the Nauruans. Although these people formerly depended on fishing and subsistence farming and had an active lifestyle, food is now imported, and high in energy content and obesity occurs at a high frequency in the population. The multifactorial disease NIDDM (non-insulin-dependent diabetes mellitus) results from a combination of genetic and environmental influences. This disease used to be nonexistent on this island, but after 1950 the prevalence of the disease increased from 1 percent in the 1950s to 21 percent in the mid-1970s and then dropped to 9 percent in 1987. At one point in the 1970s, a severe form striking many young adults reached epidemic proportions in this population. Diabetic women had more stillbirths and less than half as many live births as non-diabetic women. Although the NIDDM epidemic on Nauru has passed its peak, this outcome cannot be attributed to a decline in environmental risk factors, since their lifestyle has not changed since the 1970s.

Consider these data carefully and then propose an explanation for the increase and then subsequent decrease in the NIDDM incidence in this population.

SELECTED READINGS

ANDERSON, W. et al. 1975. Genetics of natural populations: XLII. Three decades of genetic change in *Drosophila pseudoobscura*. *Evolution* 29:24–36.

ARMOUR, J.A., ANTTINEN, T., MAY, C.A., VEGA, E.E., SAJANTILA, A., KIDD, J. R., KIDD, K.K., BERTRANPETIT, J., PÄÄBO, S., and JEFFREYS, A.J. 1996. Minisatellite diversity supports a recent African origin for modern humans. *Nat. Genet.* 13:154–60.

AVISE, J.C. 1990. Flocks of African fishes. *Nature* 347:512–13.

AYALA, F.J. 1976. *Molecular evolution.* Sunderland, MA: Sinauer Associates.

———. 1984. Molecular polymorphism: How much is there, and why is there so much? *Dev. Genet.* 4:379–91.

BARTON, N.H., and HEWITT, G.M. 1989. Adaptation, speciation and hybrid zones. *Nature* 341:497–503.

BOWCOCK, A.M., RUIZ-LINARES, A., TOMFOHRDE, J., MINCH, E., KIDD, J.R., and CAVALLI-SFORZA, L.L. 1994. High resolution of human evolutionary trees with polymorphic microsatellites. *Nature* 368:455–57.

BULT, C., WHITE, O., OLSEN, G.J., ZHOU, L., FLEISCHMANN, R.D., SUTTON, G.G., BLAKE, J.A., FITZGERALD, L.M. et al. 1996. Complete genome sequence of the methanogenic Archeon, *Methanococcus jannaschii*. *Science* 273:1058–73.

CARSON, H. 1970. Chromosome tracers of the origin of species. *Science* 168:1414–18.

———. 1975. The genetics of speciation at the diploid level. *Am. Natural.* 109:83–92.

COYNE, J.A. 1992. Genetics and speciation. *Nature* 355:511–15.

DAYHOFF, M.O. 1969. Computer analysis of protein evolution. *Sci. Am.* (July) 221:86–95.

DIAMOND, J. 1992. *The third chimpanzee: The evolution and future of the human animal.* New York: HarperCollins.

DOBZHANSKY, T. 1947. Adaptive changes induced by natural selection in wild populations of *Drosophila*. *Evolution* 1:1–16.

———. 1948. Genetics of natural populations, XVI. Altitudinal and seasonal changes produced by natural selection in certain populations of *Drosophila pseudoobscura* and *Drosophila persimilis*. *Genetics* 33:158–76.

———. 1955. *Genetics of the evolutionary process.* New York: Columbia University Press.

DOBZHANSKY, T. et al. 1966. Genetics of natural populations: XXXVII. Continuity and change in populations of *Drosophila pseudoobscura* in western United States. *Evolution* 20:418–27.

DODD, D.M.B. 1989. Reproductive isolation as a consequence of adaptive divergence in *Drosophila pseudoobscura*. *Evolution* 43:1308–11.

ELDREDGE, N. 1985. *Time frames: The evolution of punctuated equilibria.* Princeton, NJ: Princeton University Press.

FELSENSTEIN, J. 1987. Estimation of Hominoid phylogeny from a DNA hybridization data set. *J. Mol. Evol.* 26:123–31.

FITCH, W.M. 1973. Aspects of molecular evolution. *Annu. Rev. Genet.* 7:343–80.

———. 1996. The variety of human virus evolution. *Mol. Phylog. Evol.* 5: 247–258.

FITCH, W.M., and MARGOLIASH, E. 1967. Construction of phylogenetic trees. *Science* 155:279–84.

———. 1970. The usefulness of amino acid and nucleotide sequences in evolutionary studies. *Evol. Biol.* 4:67–109.

FREEMAN, S. , and HERRON, J.C. 1998. *Evolutionary analysis.* Upper Saddle River, NJ: Prentice Hall.

FRYER, G. 1997. Biological implications of a suggested Late Pleistocene desiccation of Lake Victoria. *Hydrobiologia* 354:177–82.

GILLESPIE, J.H. 1992. *The causes of molecular evolution.* New York: Oxford University Press.

GIOVANNONI, S.J., TURNER, S., OLSEN, G.J., BARNS, S., LANE, D.J., and PACE, N.R. 1988. Evolutionary relationships among cyanobacteria and green chloroplasts. *J. Bacteriol.* 170:3584–92.

GOULD, S.J. 1982. Darwinism and the expansion of evolutionary theory. *Science* 216:380–87.

HARDISON, R. 1999. The evolution of hemoglobin. *Amer. Scient.* 87:126–37.

HILLIS, D.M. 1998. Phylogenetic analysis. *Curr. Biol.* 7:R129–31.

HILLIS, D.M., BULL, J.J., WHITE, M.E., BADGETT, M.R., and MOLINEUX, I.J. 1992. Experimental phylogenetics: Generation of a known phylogeny. *Science* 255:589–92.

HILLIS, D.M., HUELSENBECK, J.P., and CUNNINGHAM, C.W., 1994. Application and accuracy of molecular phylogenies. *Science* 264:671–77.

HILLIS, D.M., and MORITZ, C., eds. 1990. *Molecular systematics.* Sunderland, MA: Sinauer Associates.

HUNT, J. et al. 1981. Evolution distance in Hawaiian *Drosophila*. *J. Mol. Evol.* 17:361–67.

JOHNSON, T.C., SCHOLZ, C.A. et al. 1996. Late pleistocene desiccation of Lake Victoria and rapid evolution of cichlid fishes. *Science* 273: 1091–93.

KIM, H.-S., and TAKENAKA, O. 1996. A comparison of TSPY genes from Y-chromosomal DNA of the great apes and humans: Sequence, evolution, and phylogeny. *Am. J. Phys. Anthropol.* 100:301–9.

KIMURA, M. 1979a. Model of effectively neutral mutations in which selective constraint is incorporated. *Proc. Natl. Acad. Sci. USA* 76:3440–44.

———. 1979b. The neutral theory of molecular evolution. *Sci. Am.* (Nov.) 241:98–126.

———. 1989. The neutral theory of molecular evolution and the world view of the neutralists. *Genome* 31:24–31.

KING, M.C., and WILSON, A.C. 1975. Evolution at two levels: Molecular similarities and biological differences between humans and chimpanzees. *Science* 188:107–16.

KNOWLTON, N., WEIGT, L.A., SOLÓRZANO, L.A., MILLS, D.K., and BERMINGHAM, E. 1993. Divergence in proteins, mitochondrial DNA, and reproductive compatibility across the Isthmus of Panama. *Science* 260: 1629–32.

KREITMAN, M. 1983. Nucleotide polymorphism at the alcohol dehydrogenase locus of *Drosophila melanogaster*. *Nature* 304:412–17.

KRINGS, M., STONE, A., SCHMITZ, R.W., KRAINITZKI, H., STONEKING, M., and PÄÄBO, S. 1997. Neandertal DNA sequences and the origin of modern humans. *Cell* 90:19–30.

LANDE, R. 1989. Fisherian and Wrightian theories of speciation. *Genome* 31:221–27.

LEWIN, R. 1993. *Human evolution*, 3rd ed. Cambridge, MA: Blackwell Scientific.

LEWONTIN, R.C., and HUBBY, J.L. 1966. A molecular approach to the study of genic heterozygosity in natural populations: II. Amount of variation and degree of heterozygosity in natural populations of *Drosophila pseudoobscura*. *Genetics* 54:595–609.

MAYR, E. 1963. *Animal species and evolution.* Cambridge, MA: Harvard University Press.

MEYER, A., KOCHER, T.D., BASASIBWAKI, P., and WILSON, A.C. 1990. Monophyletic origin of Lake Victoria cichlid fishes suggested by mitochondrial DNA sequences. *Nature* 347:550–53.

OU, C.-Y., CIESIELSKI, C.A. et al. 1992. Molecular epidemiology of HIV transmission in a dental practice. *Science* 256:1165–71.

PÄÄBO, S. 1993. Ancient DNA. *Sci. Am.* (Nov.) 269:86–92.

PAGEL, M.D., and HARVEY, P.H. 1989. Comparative methods for examining adaptation depend on evolutionary models. *Folia Primatol.* 53:203–20.

POWERS, D.A., and SCHULTE, P.M. 1998. Evolutionary adaptations of gene structure and expression in natural populations in relation to a changing environment: A multidisciplinary approach to address the million-year saga of a small fish. *J. Exper. Zool.* 282:71–94.

RICE, W.R., and HOSTERT, E.E. 1993. Laboratory experiments on speciation: What have we learned in 40 years? *Evolution* 47:1637–53.

SIBLEY, C., and AHLQUIST, J. 1984. The phylogeny of the hominoid primates, as indicated by DNA-DNA hybridization. *J. Mol. Evol.* 20:2–15.

SIBLEY, C.G., COMSTOCK, J.A., and AHLQUIST, J.E. 1990. DNA evidence of hominoid phylogeny: A re-analysis of the data. *J. Mol. Evol.* 30:202–36.

SMITHIES, O., BLECHL, A.E., SHEN, S., SLIGHTOM, J.L., and VANIN, E.F. 1981. Co-evolution and control of globin genes. In *Levels of genetic control in development*, ed. S. Subtelny and U. Abbot, pp. 185–200. New York: Alan R. Liss.

STIASSNY, M.L.J., and MEYER, A. 1999. Cichlids of the Rift Lakes. *Sci. Am.* (Feb.) 280:64–69.

STONEKING, M. 1995. Ancient DNA: How do you know when you have it and what can you do with it? *Am. J. Hum. Genet.* 57:1259–62.

TEMPLETON, A.R. 1985. Phylogeny of the hominoid primates: A statistical analysis of the DNA-RNA hybridization data. *Mol. Biol. Evol.* 2:420–33.

———. 1994. The role of molecular genetics in speciation studies. *EXS* 69:455–57.

THORNE, A.G., and WOLPOFF, M.H. 1992. The multiregional evolution of humans. *Sci. Am.* (Apr.) 266:76–83.

TISHKOFF, S.A., DIETZSCH, E., SPEED, W., PAKSTIS, A.J., KIDD, J.R., CHEUNG, K., BONNE-TAMIR, B., SANTACHIARA-BENERECETTI, A.S. et al. 1996. Global patterns of linkage disequilibrium at the CD4 locus and modern human origins. *Science* 271:1380–87.

TSUI, L.-C. 1992. The spectrum of cystic fibrosis mutations. *Trends in Genet.* 8:392–98.

VAL, F.C. 1977. Genetic analysis of the morphological differences between two interfertile species of Hawaiian *Drosophila*. *Evolution* 31:611–29.

WHITE, M.J.D. 1977. *Modes of speciation.* New York: W. H. Freeman.

WILSON, A.C., and CANN R.L. 1992. The recent African genesis of humans. *Sci. Am.* (Apr.) 266:68–73.

YANG, D., OYAIZU, Y., OYAIZU, H., OLSEN, G.J., and WOESE, C.R. 1985. Mitochondrial origins. *Proc. Nat. Acad. Sci. USA* 82: 4443–47.

YUNIS, J.J., and PRAKASH, O. 1982. The origin of man: A chromosomal pictorial legacy. *Science* 215:1525–30.

EXPERIMENTAL METHODS

In addition to the techniques of genetic analysis, physical and chemical techniques for the separation and analysis of macromolecular components of the cell nucleus and cytoplasm have been instrumental in advancing our understanding of genetics at the molecular level. In this appendix we will describe the background and theoretical basis of some techniques that have been important in molecular genetics.

Isotopes

Isotopes are forms of an element that have the same number of protons and electrons but differ in the number of neutrons contained in the atomic nucleus. For example, the most common form of carbon has an atomic number of 6 (the number of protons in the nucleus) and an atomic weight of 12 (the sum of the protons and neutrons in the nucleus). In a very small percentage of carbon atoms, a seventh neutron is present, producing an atom with an atomic weight of 13. This is an example of a so-called **heavy isotope.** Since the number of protons and electrons, and thus the net charge, has not changed, the atom has the same chemical properties as **carbon-12 (^{12}C)** and differs only in mass. **Carbon-13 (^{13}C)** is thus a stable, heavy isotope of carbon.

Although the addition of neutrons does not alter the chemical properties of an atom, it can produce instabilities in the atomic nucleus. If another neutron is added to a carbon-13 atom, the isotope **carbon-14 (^{14}C)** results. However, the presence of eight neutrons and six protons is an unstable condition, and the atom undergoes a nuclear reaction in which radiation is emitted during the transition to a more stable condition. Therefore, carbon-14 is a **radioactive isotope** of carbon.

The type of radiation emitted and the rate at which these nuclear events take place are characteristic of the element. Table A.1 lists types of radioactivity. The rate at which a radioactive isotope emits radiation is expressed as its **half-life,** which is the time required for a given amount of a radioactive substance to lose one-half of its radioactivity. Table A.2 lists some of the isotopes available for use in research.

Detection of Isotopes

The choice of which isotope to use as a tracer in biological experiments depends on a combination of its physical and chemical properties, which enable the investigator to quantitate the amount of radioactivity or measure the ratio of heavy to light isotopes. For the detection of heavy isotopes, two methods are commonly employed: **mass spectrometry** and **equilibrium density gradient centrifugation** (discussed in the following section). Although the use of heavy isotopes has been more restricted than that of radioisotopes, they have been instrumental in several basic advances in molecular biology [e.g., demonstrating the semiconservative nature of DNA replication and the existence of messenger RNA (mRNA)].

There are a number of methods to detect radioisotopes, the foremost being **liquid scintillation spectrometry,** which provides quantitative information about the amount of radioactive isotope present in a sample, and **autoradiography,** which is used to demonstrate the cytological distribution and localization of radioactively labeled molecules.

In recording radioactivity by liquid scintillation counting, a small sample of the material to be counted is solubilized and immersed in a solution containing a **phosphor,**

TABLE A.1 Properties of Ionizing Radiation

Type	Relative Penetration	Relative Ionization	Range in Biological Tissue
Alpha particle (2 protons + 2 neutrons)	1	10,000	Microns
Beta particle (electron)	100	100	Microns–mm
Gamma ray	>1000	<1	∞

TABLE A.2 Some Isotopes Used in Research

Element	Isotope	Half-life	Radiation
H	^2H	—	Stable
	^3H	12.3 years	β
C	^{13}C	—	Stable
	^{14}C	5700 years	β
N	^{15}N	—	Stable
O	^{18}O	—	Stable
P	^{32}P	14 days	β
S	^{35}S	87 days	β
K	^{40}K	1.2×10^9 years	β, gamma
Fe	^{59}Fe	45 days	β, gamma
I	^{125}I	60 days	β, gamma
	^{131}I	8 days	β, gamma

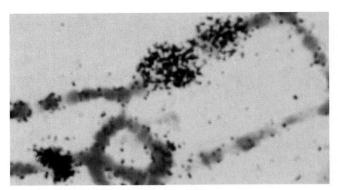

FIGURE A.2 Autoradiogram of RNA synthesis in salivary glands of *Drosophila* larva. Silver grains are deposited over sites of RNA synthesis at chromosome puffs.

an organic compound that emits a flash of light after it absorbs energy released by decay of the radioactive compound. The counting chamber of the liquid scintillation spectrometer is equipped with very sensitive photomultiplier tubes that record the light flashes emitted by the phosphor. The data are recorded as counts of radioactivity per unit time and are displayed on a printout or can be fed into a computer for storage.

In autoradiography, a gel, chromatogram, or plant or animal part is placed against a sheet of photographic film. Radioactive decay from the incorporated isotope behaves just as light energy does and reduces silver grains in the emulsion. After exposure, the sheet or film is developed and fixed, revealing a deposit of silver grains corresponding to the location of the radioactive substance (Fig. A.1).

Alternatively, to record the subcellular localization of an incorporated labeled isotope, cells or chromosomal preparations that have been incubated with radioactively labeled compounds are affixed to microscope slides and covered with a thin layer of liquid photographic emulsion. After they are exposed, the slides are processed to develop and fix the reduced silver grains in the emulsion. After staining, microscopic examination reveals the location and extent of labeled isotope incorporation (Fig. A.2).

Centrifugation Techniques

The centrifugation of biological macromolecules is widely employed to provide information about their physical characteristics (e.g., size, shape, density, and molecular weight) and to purify and concentrate cells, organelles, and their molecular components.

Differential centrifugation is commonly used to separate materials such as cell homogenates according to size. Initially, the homogenate is distributed uniformly in the centrifuge tube. After a period of centrifugation, the pellet obtained is enriched for the largest and most dense particles, such as nuclei, in the homogenate. After each step, the supernatant can be recentrifuged at higher speeds to pellet the next heavier component. A typical fractionation scheme for cell homogenates is shown in Fig. A.3. Further fractionation using density gradient techniques can be used to purify any of the fractions obtained by differential centrifugation.

Rate zonal centrifugation is used to separate particles on the basis of differences in their sedimentation rates. It may be used to separate mixtures of macromolecules such as proteins or nucleic acids and cellular organelles such as mitochondria. In this technique, which employs a

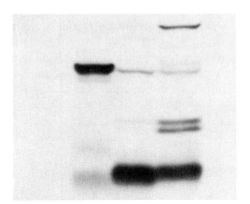

FIGURE A.1 Autoradiogram of radioactive bacterial proteins synthesized in a minicell system.

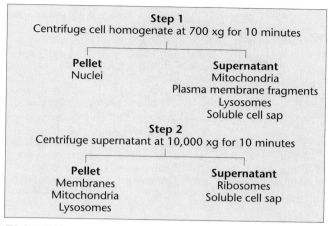

FIGURE A.3 Fractionation scheme for cell homogenates.

in a centrifugal field, the molecule will eventually sediment toward the bottom at a constant velocity. At this point, the molecular weight (M) can be calculated as

$$M = f \times v/\omega^2 r$$

where f is the frictional coefficient of the solvent system (which has been calculated from other measurements) and $v/\omega^2 r$ is the rate of sedimentation per unit applied centrifugal field. The latter value is given the symbol S, or sedimentation coefficient. The S value for most proteins is between 1×10^{-13} sec and 2×10^{-11} sec. A sedimentation coefficient of 1×10^{-13} is defined as one **Svedberg unit (S)**; this unit is named for The Svedberg, a pioneer in the field of centrifugation. Thus, a protein with a sedimentation value of 2×10^{-11} sec would have a value of $200S$. Fig. A.4 shows the S values of selected molecules and particles.

Isopycnic centrifugation, or **equilibrium density gradient centrifugation**, is one of the most widely used techniques in genetics and molecular biology. In this technique, the solvent varies in density from one end of the tube to the other. A mixture of molecules layered on top and centrifuged through this gradient will migrate toward the bottom of the tube until each particle reaches its isopycnic point—that is, the place in the gradient where the density of the solvent equals the buoyant density of the particle.

medium of increasing density, the rate at which particles sediment depends on size, shape, density, and the frictional resistance of the solvent.

In addition to the preparation and purification of macromolecules and cellular components, rate zonal centrifugation can be used to determine the **sedimentation coefficients** and **molecular weights** of biological macromolecules. If a purified molecule such as a protein is spun

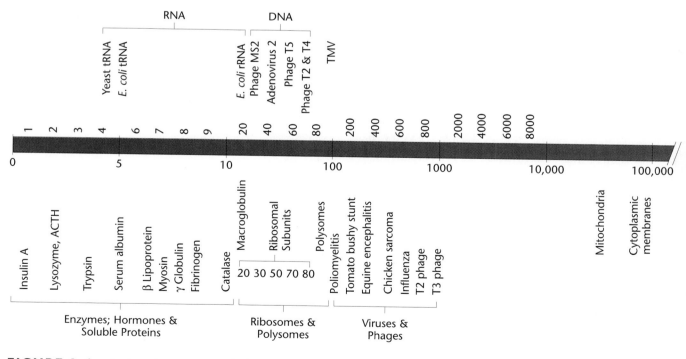

FIGURE A.4 S values of some common biological molecules and particles.

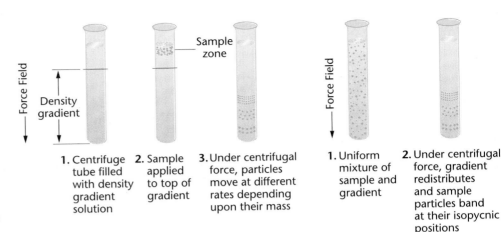

1. Centrifuge tube filled with density gradient solution
2. Sample applied to top of gradient
3. Under centrifugal force, particles move at different rates depending upon their mass

1. Uniform mixture of sample and gradient
2. Under centrifugal force, gradient redistributes and sample particles band at their isopycnic positions

FIGURE A.5 Two methods of forming density gradients.

When each molecular species in the mixture migrates to its own characteristic isopycnic point, it no longer moves and is at equilibrium no matter how much longer the centrifugal field is applied. After separation by this method, components may be recovered by puncturing the bottom of the tube and collecting fractions. Two methods of forming density gradients are commonly employed (Fig. A.5), one using preformed gradients of sucrose, or soluble salts of heavy metals such as cesium chloride or cesium sulfate. In the second method, the gradient is formed by the action of the centrifugal field on the salt solution.

Renaturation and Hybridization of Nucleic Acids

The ability of separated complementary strands of nucleic acids to unite and form stable, double-stranded molecules has been used to measure the relatedness of nucleic acids from different parts of the cell, different organs, and even different species. If both strands are DNA, the process is known as **renaturation** or **reassociation**. If one strand is RNA and the other DNA, it is known as **hybridization**. Renaturation and hybridization involve two steps: (1) a rate-limiting step, in which collision or nucleation between two homologous strands initiates base pairing, and (2) a rapid pairing of complementary bases, or "zippering" of the strands, to form a double-stranded molecule. For DNA renaturation, the formation of double-stranded molecules can be assayed at any time during an experiment. A sample is passed over a column of **hydroxyapatite**, which selectively binds double-stranded DNA but allows single-stranded molecules to pass through. The double-stranded molecules can be released from the column by raising the temperature or salt concentration.

The process of renaturation follows second-order kinetics according to the equation

$$\frac{C}{C_0} = \frac{1}{1 + kC_0 t}$$

where C is the single-stranded DNA concentration at time t, C_0 is the total DNA concentration, and k is a second-order rate constant. It is usually convenient to express the data from renaturation experiments as the fraction of single-stranded DNA at any time t versus the product of total DNA concentration and time, as shown in Fig. A.6.

In the process of renaturation, the DNA is initially single-stranded and is renatured to double-stranded structures in the final state. The time necessary to reassociate half of the DNA in a sample at a given concentration should be proportional to the number of different pieces of DNA present. Consequently, the half-reassociation time should be proportional to the DNA content of the genome, with smaller genomes having shorter half-re-

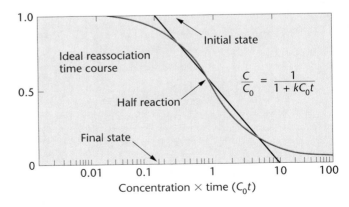

FIGURE A.6 Idealized $C_0 t$ curve.

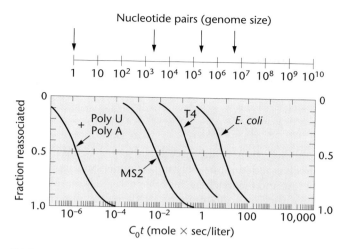

FIGURE A.7 Changes in C_0t value as genome size increases.

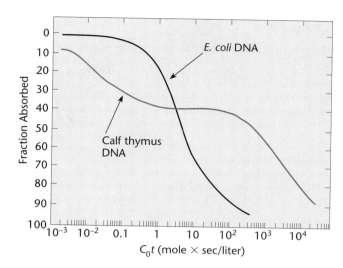

FIGURE A.8 Renaturation curve for *E. coli* DNA, containing no repetitive DNA sequences, and for calf thymus DNA, containing several families of repetitive DNA.

naturation rates. Fig. A.7 confirms this expectation and shows increases in C_0t values as the size of the genome increases. This proportional relationship between C_0t and genome size is valid only in cases where repetitive DNA sequences are absent from the DNA being studied. DNA from calf thymus (and many other eukaryotic sources) exhibits a complex pattern of reassociation, indicating that bovine DNA contains some sequences (in this case, 40% of the total DNA) that reassociate rapidly and others that reassociate more slowly. The rapidly reassociating fraction must therefore contain sequences that are present in many copies. The more slowly renaturing DNA, however, contains sequences present in only one copy per genome. The *E. coli* DNA shown in Fig. A.8 renatures with a pattern close to that of an ideal second-order reaction, indicating the absence of significant amounts of repeated DNA sequences.

In the case of hybridization between DNA and RNA, two approaches can be used—either the RNA or the DNA can be in excess. RNA-excess hybridization is usually preferred because most double-stranded molecules that are formed are RNA:DNA hybrids. Since DNA is present in low concentration, and since single-stranded RNA molecules lack complementary RNA strands in the mixture, the number of RNA:RNA and DNA:DNA hybrids is negligible.

The extent of hybridization can be followed by using hydroxyapatite columns or by using radioactively labeled RNA or DNA. In practice, one of the components of the hybridization reaction is usually immobilized on a substrate such as nitrocellulose paper. For example, DNA may be sheared to a uniform size, denatured to single strands, immobilized on nitrocellulose filters, and hybridized with an excess of labeled RNA. After hybridization, the unbound

RNA is removed by washing, and the hybrids assayed by liquid scintillation counting. DNA:RNA hybridization can also be performed using cytological preparations, a technique called *in situ* **hybridization**. DNA, which is a part of an intact chromosome preparation fixed to a slide, can be denatured and hybridized to radioactive RNA. Hybrid formation is detected using autoradiography (Fig. A.9).

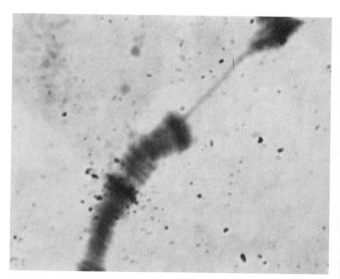

FIGURE A.9 Light micrograph of *in situ* hybridization of radioactive 5S RNA to a single band of polytene chromosome in the salivary gland of a *Drosophila* larva.

Electrophoresis

Electrophoresis is a technique that measures the rate of migration of charged molecules in a liquid-containing medium when an electrical field is applied to the liquid. Negatively charged molecules (anions) will migrate toward the positive electrode (anode) and positively charged molecules (cations) will migrate toward the negative electrode (cathode). Several factors affect the rate of migration, including the strength of the electrical field and the molecular sieving action of the medium (paper, starch, or gel) in which migration takes place. Since proteins and nucleic acids are electrically charged, electrophoresis has been used extensively to provide information about the size, conformation, and net charge of these macromolecules. On a larger scale, electrophoresis provides a method of fractionation that can be used to isolate individual components in mixtures of proteins or nucleic acids.

More recently, the analytical separation of proteins has been enhanced with the development of two-dimensional electrophoresis techniques, in which separation in the first dimension is by net charge, and in the second dimension by size and molecular weight.

In practice, electrophoresis employs a buffer system; a medium (paper, cellulose acetate, starch gel, agarose gel, or polyacrylamide gel); and a source of direct current.

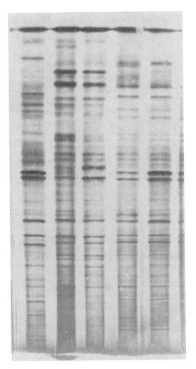

FIGURE A.10   Coomassie blue-stained protein slab gel.

Samples are applied, and current is passed through the system for an appropriate time. Following migration of the molecules, the gel or paper may be treated with selective stains to reveal the location of the separated components (Fig. A.10).

DNA Sequencing

The ability to determine directly the sequence of DNA segments has added immensely to our understanding of gene structure and the mechanisms of gene regulation. Although techniques were available in the 1940s to determine the base composition of DNA, it was not until the 1960s that methods providing direct chemical analysis of nucleotide sequence were developed and used. These first methods were based on those used in determining the amino acid sequence of proteins, and were time-consuming and laborious. For example, in 1965 Robert Holley determined the sequence of a tRNA molecule consisting of 74 nucleotides, a task that required almost a year of concentrated effort.

In the 1970s more efficient and direct methods of analysis of nucleotide sequencing were developed. These methods developed in parallel with recombinant DNA techniques that allow the isolation of large quantities of purified DNA segments from any organism. A chemical method, developed by Allan Maxam and Walter Gilbert, cleaves DNA at specific bases. A second method, developed by Fred Sanger and colleagues, synthesizes a stretch of DNA that terminates at a given base. In both methods, the DNA to be sequenced is subjected to four individual reactions (one for each base). The products of the four reactions are a series of DNA fragments that differ in length by only one nucleotide. These reaction products are electrophoretically separated in four adjacent lanes on a gel. Each band on the gel corresponds to a base, and the sequence of the DNA segment can be read from the bands on the gel.

In the Maxam-Gilbert method, the complementary strands of the DNA fragment to be sequenced are separated and recovered. The strand to be sequenced is labeled at its 5'-end with radioactive ^{32}P using the enzyme polynucleotide kinase. This provides a means of identifying specific DNA fragments after gel electrophoresis. In the next step, aliquots of the DNA are subjected to each of four chemical treatments that cleave the strand at a specific nucleotide. The reaction is carried out for a limited time so that any given molecule is cleaved at only a small number of the target nucleotides. The result is a collection of fragments, all labeled at the 5'-end, but differing in length, depending on the point of cleavage.

Four different reactions cleave DNA at guanine (G > A), adenine (A > G), cytosine alone (C), or cytosine and thymine (C + T). In these reactions, purines are cleaved by using dimethylsulfate. This reagent methylates guanine far more efficiently than adenine, and when heat is applied, the strand is broken at the methylated site, producing a DNA fragment most frequently cleaved at a G residue (G > A). This can be reversed by cleaving the strand in acid, producing fragments most often cleaved at an A residue (A > G). The reaction for pyrimidines uses hydrazine, which cleaves both cytosine and thymine; however, in high salt (2M NaCl), only cytosine reacts. Thus, in the two reactions, one represents only cytosine (C), and the other represents cytosine and thymine (C + T).

For each set of reactions, the fragments produced are then subjected to gel electrophoresis in adjacent lanes on a gel. Each reaction contains a series of fragments in which strand breakage has occurred at one of the target bases, resulting in a series of fragments that differ in length. Under the influence of the electric field during electrophoresis, the different-sized fragments separate from each other, with the smallest fragments migrating the farthest. Since the DNA fragments contain a radioactive 5′-end, the gel is subjected to autoradiography by placing a sheet of X-ray film over the gel for an appropriate exposure time. The DNA sequence is then analyzed by reading the bands on the gel from the bottom up, reading across all four lanes. In the example shown in Fig. A.11, the first bases on the gel read GCGG.

In the Sanger method for DNA sequencing, which relies on an initial enzymatic treatment, the DNA strand to be sequenced is used as a template for the synthesis of a new DNA strand catalyzed by DNA polymerase I. This method employs a modified dideoxynucleoside triphosphate to generate a series of DNA fragments. The dideoxynucleoside triphosphates lack a 3′-OH group. This allows them to be added to a DNA strand undergoing synthesis, but since they lack a 3′-OH group, no nucleotide can be added to them, causing termination of strand synthesis, producing a DNA fragment. In use, four separate reaction mixtures are set up, each with a template DNA strand, a primer, all four radioactive nucleoside triphosphates, and a small amount of a single dideoxyribonucleoside triphosphate. Each of the four reactions contains a different dideoxynucleoside triphosphate that acts as a chain terminator. Because only a small amount of the modified nucleoside is used in each reaction, the newly synthesized strands are randomly terminated, producing a collection of fragments.

After synthesis, the radioactive fragments are electrophoresed in four adjacent lanes, one corresponding to each of the reactions. The fragments are visualized by autoradiography, and the gel is analyzed by reading the sequence from the bottom, as in the Maxam-Gilbert reaction. In the example shown in Fig. A.12, the first band is in the lane with ddT, so it is a T residue, and the next few bands have the sequence GCAATCG.

DNA sequencing methods have generated a great deal of information about the structure and organization of many genes in a wide range of organisms. In the case of some viruses, the sequence of the entire genome is known, and the genomes of other organisms, including *E. coli* and yeast have been sequenced. To date, only a very small portion of the human genome has been sequenced, but the U.S. Department of Energy and the National Institutes of Health are coordinating efforts to develop the technology to sequence the more than 3 billion bases that constitute the haploid human genome.

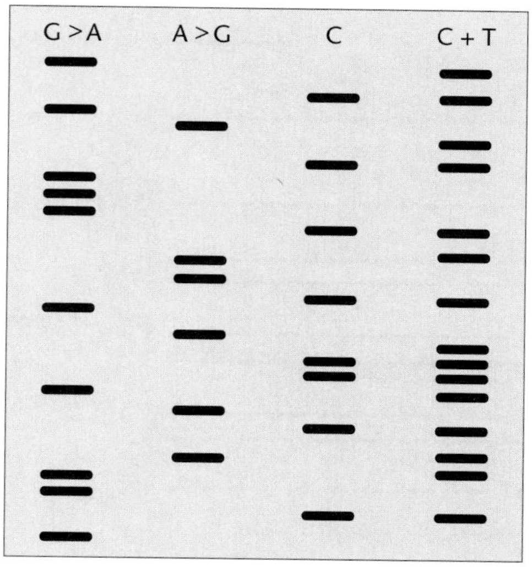

FIGURE A.11   Nucleotide sequence derived by the Maxam-Gilbert method.

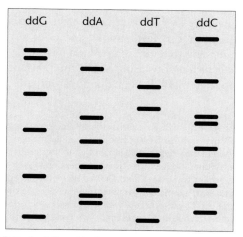

FIGURE A.12   Nucleotide sequence derived by the Sanger method.

The Polymerase Chain Reaction

The polymerase chain reaction (PCR) technique, described in Chapter 18, relies on the fact that the enzyme DNA polymerase requires a double-stranded primer section of DNA in order to initiate DNA synthesis. In the PCR technique, that primer is supplied in the form of a synthetic oligonucleotide, allowing the replication of a specific region of DNA. In other words, DNA polymerase can be directed to replicate only one specific region from an entire genome. Because these same primer sites are copied onto the newly synthesized strands of DNA, after each round of replication they can be used as primers. By repeating the replication process, a region of DNA can be amplified millions or billions of times, producing enough copies of the DNA for experimental purposes without the need for cloning (Fig. A.13).

Instead of a single technique, the PCR reaction has become a versatile tool used in combination with other methods and has become the basis of a technical revolution in molecular genetics. The PCR method has been used to detect the presence of mutations, produce *in vitro* mutations, diagnose genetic disorders, prepare DNA for sequencing, identify viruses and bacteria in infectious diseases, amplify DNA from fossils, analyze genetic defects in gametes and single cells from human embryos, and a host of other applications.

Mutation Detection

A PCR-based method called single-strand conformation polymorphism (SSCP) has been used to screen for mutations caused by single base substitutions. Under certain conditions, single-stranded DNA fragments fold into nucleotide-sequence-dependent conformations. These conformations affect the mobility of the fragment during electrophoresis. Single base substitutions change the conformation of the DNA, altering its electrophoretic mobility. The method is fast, simple, and does not require DNA sequencing to detect single base changes.

To detect mutations using SSCP, the DNA of interest is first amplified by PCR. The double-stranded amplified

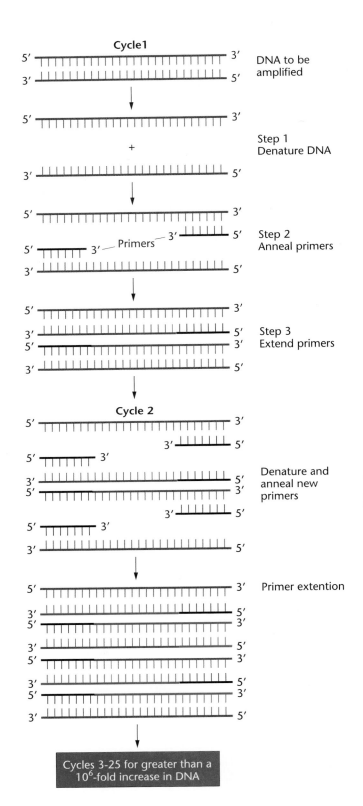

FIGURE A.13 PCR amplification. In the polymerase chain reaction (PCR) method, the DNA to be amplified is first denatured into single strands, and then each strand is annealed to a different primer. The primers are synthetic oligonucleotides complementary to sequences flanking the region to be amplified. DNA polymerase is added along with nucleotides to extend the primers in the 3′ direction, resulting in a double-stranded DNA molecule with the primers incorporated into the newly synthesized strand. In a second PCR cycle, the products of the first cycle are denatured into single strands, primers are added, and they are annealed and extended by DNA polymerase. Repeated cycles can amplify the original DNA sequence by more than a million times.

DNA is then converted to a single-stranded form by boiling and is loaded onto a gel for electrophoresis. DNA from a normal gene is similarly amplified, denatured, and loaded in adjacent lanes. After electrophoresis, any variations in base sequence in the DNA being tested can be detected as a shifted band. The method gives best results when the DNA fragments being tested are around 200 base pairs, so screening an entire gene requires the use of several to many fragments, each about 200 base pairs in length.

In the development of new PCR-based techniques, one technique is often piggy-backed onto another. While SSCP is a rapid and simple technique, its accuracy is somewhat limited. DNA sequencing is a more accurate technique for detecting single base changes, but it is relatively slow and laborious. Recently, these two techniques have been combined into a method called dideoxy fingerprinting (ddF). The technique begins by amplifying a DNA segment by PCR. The amplified segment is used as a template for one of the Sanger dideoxy sequencing reactions, and the gel is run under conditions where the position of the band reflects both size (like DNA sequencing) and conformation (like SSCP). This combined technique is three times faster than sequencing alone and although more complex than SSCP, has a degree of accuracy approaching 100 percent.

PCR Cycle Sequencing

Conventional DNA sequencing requires subcloning of fragments into vectors such as a plasmid, followed by growth of a host colony and extraction and purification of the cloned DNA. This DNA is then used as a template in the four reactions for DNA sequencing, followed by gel electrophoresis. The technique of cycle sequencing incorporates parts of the PCR reaction with parts of the standard reaction for the dideoxy method of DNA sequencing. The double-stranded DNA segment to be sequenced is first heated to form single strands that serve as templates. Primers are added, and DNA polymerase then begins the process of DNA replication. Extension of the primer continues until a labeled dideoxynucleotide is incorporated. The labeled strand is separated from the template (by heating) and is then used in another round of replication. The amplified extension products are then analyzed by gel electrophoresis. This adaptation of DNA sequencing is faster than conventional sequencing, requires only a very small amount of template, and can be used to directly sequence plasmid or phage clones. To further simplify the process, commercial kits are available that provide all materials and solutions for this procedure.

Versatility: The Advantage of PCR

The examples above illustrate the versatility of the polymerase chain reaction and some of its applications in molecular genetics. This remarkable technique has so many adaptations and variations that several journals are devoted to reports on PCR.

GLOSSARY

A-DNA An alternative form of the right-handed double-helical structure of DNA in which the helix is more tightly coiled, with 11 base pairs per full turn of the helix. In this form, the bases in the helix are displaced laterally and tilted in relation to the longitudinal axis. It is not yet clear whether this form has biological significance.

abortive transduction An event in which transducing DNA fails to be incorporated into the recipient chromosome. (See *transduction*.)

acentric chromosome Chromosome or chromosome fragment with no centromere.

acquired immunodeficiency syndrome (AIDS) An infectious disease caused by a retrovirus designated as human immunodeficiency virus (HIV). The disease is characterized by a gradual depletion of T lymphocytes, recurring fever, weight loss, multiple opportunistic infections, and rare forms of pneumonia and cancer associated with collapse of the immune system.

acridine dyes A class of organic compounds that bind to DNA and intercalate into the double-stranded structure, producing local disruptions of base pairing. These disruptions result in additions or deletions in the next round of replication.

acrocentric chromosome Chromosome with the centromere located very close to one end. Human chromosomes 13, 14, 15, 21, and 22 are acrocentric.

active immunity Immunity gained by direct exposure to antigens followed by antibody production.

active site That portion of a protein, usually an enzyme, whose structural integrity is required for function (e.g., the substrate binding site of an enzyme).

adaptation A heritable component of the phenotype that confers an advantage in survival and reproductive success. The process by which organisms adapt to the current environmental conditions.

additive genes See *polygenic inheritance*.

additive variance Genetic variance that is attributed to the substitution of one allele for another at a given locus. This variance can be used to predict the rate of response to phenotypic selection in quantitative traits.

albinism A condition caused by the lack of melanin production in the iris, hair, and skin. In humans, most often inherited as an autosomal recessive.

aleurone layer In seeds, the outer layer of the endosperm.

alkaptonuria An autosomal recessive condition in humans caused by the lack of an enzyme, homogentisic acid oxidase. Urine of homozygous individuals turns dark upon standing due to oxidation of excreted homogentisic acid. The cartilage of homozygous adults blackens from deposition of a pigment derived from homogentisic acid. Such individuals often develop arthritic conditions.

allele One of the possible mutational states of a gene, distinguished from other alleles by phenotypic effects.

allele frequency Measurement of the proportion of individuals in a population carrying a particular allele.

allele-specific nucleotide (ASO) Synthetic nucleotides, usually 15 to 20 bp in length that by hybridization under carefully controlled conditions will hybridize only to a complementary sequence with a perfect match. Under these same conditions, ASOs with a one-nucleotide mismatch will not hybridize.

allelic exclusion In plasma cell heterozygous for an immunoglobulin gene, the selective action of only one allele.

allelism test See *complementation test*.

allolactose A lactose derivative that acts as the inducer for the *lac* operon.

allopatric speciation Process of speciation associated with geographic isolation.

allopolyploid Polyploid condition formed by the union of two or more distinct chromosome sets with a subsequent doubling of chromosome number.

allosteric effect Conformational change in the active site of a protein brought about by interaction with an effector molecule.

allotetraploid Diploid for two genomes derived from different species.

allozyme An allelic form of a protein that can be distinguished from other forms by electrophoresis.

alpha fetoprotein (AFP) A 70-kd glycoprotein synthesized in embryonic development by the yolk sac. High levels of this protein in the amniotic fluid are associated with neural tube defects such as spina bifida. Lower than normal levels may be associated with Down syndrome.

***Alu* sequence** An interspersed DNA sequence of approximately 300 bp found in the genome of primates that is cleaved by the restriction enzyme *Alu* I. *Alu* sequences are composed of a head-to-tail dimer, with the first monomer approximately 140 bp and the second approximately 170 bp. In humans, they are dispersed throughout the genome and are present in 300,000 to 600,000 copies, constituting some 3 to 6 percent of the genome. See *SINEs*.

amber codon   The codon UAG, which does not code for an amino acid but for chain termination.

Ames test   An assay developed by Bruce Ames to detect mutagenic and carcinogenic compounds using reversion to histidine independence in the bacterium *Salmonella typhimurium*.

amino acid   Any of the subunit building blocks that are covalently linked to form proteins.

aminoacyl tRNA   Covalently linked combination of an amino acid and a tRNA molecule.

amniocentesis   A procedure used to test for fetal defects in which fluid and fetal cells are withdrawn from the amniotic layer surrounding the fetus.

amphidiploid   See *allotetraploid*.

anabolism   The metabolic synthesis of complex molecules from less complex precursors.

analogue   A chemical compound structurally similar to another, but differing by a single functional group (e.g., 5-bromodeoxyuridine is an analogue of thymidine).

anaphase   Stage of cell division in which chromosomes begin moving to opposite poles of the cell.

anaphase I   The stage in the first meiotic division during which members of homologous pairs of chromosomes separate from one another.

aneuploidy   A condition in which the chromosome number is not an exact multiple of the haploid set.

angstrom   Unit of length equal to 10^{-10} meter. Abbreviated Å.

antibody   Protein (immunoglobulin) produced in response to an antigenic stimulus with the capacity to bind specifically to the antigen.

anticipation   A phenomenon first observed in myotonic dystrophy, where the severity of the symptoms increases from generation to generation and the age of onset decreases from generation to generation. This phenomenon is caused by the expansion of trinucleotide repeats within or near a gene.

anticodon   The nucleotide triplet in a tRNA molecule that is complementary to, and binds to, the codon triplet in an mRNA molecule.

antigen   A molecule, often a cell surface protein, that is capable of eliciting the formation of antibodies.

antiparallel   Describing molecules in parallel alignment, but running in opposite directions. Most commonly used to describe the opposite orientations of the two strands of a DNA molecule.

apoenzyme   The protein portion of an enzyme that requires a cofactor or prosthetic group to be functional.

apoptosis   A genetically controlled program of cell death, activated as part of normal development, or as a result of cell damage.

ascospore   A meiotic spore produced in certain fungi.

ascus   In fungi, the sac enclosing the four or eight ascospores.

asexual reproduction   Production of offspring in the absence of any sexual process.

assortative mating   Nonrandom mating between males and females of a species. Selection of mates with the same genotype is positive; selection of mates with opposite genotypes is negative.

ATP   Adenosine triphosphate.

attached-X chromosome   Two conjoined X chromosomes that share a single centromere.

attenuator   A nucleotide sequence between the promoter and the structural gene of some operons that can act to regulate the transit of RNA polymerase, diminishing transcription of the related structural gene.

autogamy   A process of self-fertilization resulting in homozygosis.

autoimmune disease   The production of antibodies that results from an immune response to one's own molecules, cells, or tissues. Such a response results from the inability of the immune system to distinguish self from nonself. Diseases such as arthritis, scleroderma, systemic lupus erythematosus, and perhaps diabetes are considered to be autoimmune diseases.

autopolyploidy   Polyploid condition resulting from the replication of one diploid set of chromosomes.

autoradiography   Production of a photographic image by radioactive decay. Used to localize radioactively labeled compounds within cells and tissues.

autosomes   Chromosomes other than the sex chromosomes. In humans, there are 22 pairs of autosomes.

autotetraploid   An autopolyploid condition composed of four similar genomes. In this situation, genes with two alleles (*A* and *a*) can have five genotypic classes: *AAAA* (quadraplex), *AAAa* (triplex), *Aaaa* (duplex), *Aaaa* (simplex), and *aaaa* (nulliplex).

auxotroph   A mutant microorganism or cell line which requires a substance for growth that can be synthesized by wild-type strains.

back-cross A cross involving an F_1 heterozygote and one of the P_1 parents (or an organism with a genotype identical to one of the parents).

bacteriophage A virus that infects bacteria (synonym is *phage*).

bacteriophage mu A group of phages whose genetic material behaves as an insertion sequence that can cause inactivation of host genes and rearrangement of host chromosomes.

bacteriostatic A compound that inhibits the growth of bacteria, but does not kill them.

balanced lethals Recessive, nonallelic lethal genes, each carried on different homologous chromosomes. When organisms carrying balanced lethal genes are interbred, only organisms with genotypes identical to the parents (heterozygotes) survive.

balanced polymorphism Genetic polymorphism maintained in a population by natural selection.

Barr body Densely staining nuclear mass seen in the somatic nuclei of mammalian females. Discovered by Murray Barr, this body is thought to represent an inactivated X chromosome.

base analogue See *analogue*.

base substitution A single base change in a DNA molecule that produces a mutation. There are two types of substitutions: *transitions*, in which a purine is substituted for a purine or a pyrimidine for a pyrimidine; and *transversions*, in which a purine is substituted for a pyrimidine, or vice versa.

β-galactosidase A bacterial enzyme encoded by the *lacZ* gene that converts lactose into galactose and glucose.

bidirectional replication A mechanism of DNA replication in which two replication forks move in opposite directions from a common origin of replication.

biometry The application of statistics and statistical methods to biological problems.

biotechnology Commercial and/or industrial processes that utilize biological organisms or products.

bivalents Synapsed homologous chromosomes in the first prophase of meiosis.

Bombay phenotype A rare variant of the ABO system in which affected individuals do not have A or B antigens, and thus appear as blood type O, even though their genotype may carry unexpressed alleles for the A and/or B antigens.

bottleneck effect Fluctuation in allele frequency when a population undergoes a temporary reduction in size.

BrdU (5-bromodeoxyuridine) A mutagenically active analogue of thymidine in which the methyl group at the $5'$ position in thymine is replaced by bromine.

buoyant density A property of particles (and molecules) that depends upon their actual density, as determined by partial specific volume and degree of hydration. Provides the basis for density gradient separation of molecules or particles.

CAAT box A highly conserved DNA sequence found about 75 base pairs $5'$ to the site of transcription in eukaryotic genes.

cAMP Cyclic adenosine monophosphate. An important regulatory molecule in both prokaryotic and eukaryotic organisms.

canonical sequence See *consensus sequence*.

CAP Catabolite activator protein. A protein that binds cAMP and regulates the activation of inducible operons.

carcinogen A physical or chemical agent that causes cancer.

carrier An individual heterozygous for a recessive trait.

cassette model First proposed to explain mating type interconversion in yeast, this model proposes that both genes for mating types, *a* and *alpha* are present as silent or unexpressed genes in transposable DNA segments (cassettes) that are activated (played) by transposition to the mating type locus.

catabolism A metabolic reaction in which complex molecules are broken down into simpler forms, often accompanied by the release of energy.

catabolite activator protein See *CAP*.

catabolite repression The selective inactivation of an operon by a metabolic product of the enzymes encoded by the operon.

cdc mutation A class of mutations in yeast that affect the timing and progression through the cell cycle.

cDNA DNA synthesized from an RNA template by the enzyme reverse transcriptase.

cDNA library A collection of cloned cDNA sequences.

cell cycle Sum of the phases of growth of an individual cell type; divided into G1 (gap 1), S (DNA synthesis), G2 (gap 2), and M (mitosis).

cell-free extract A preparation of the soluble fraction of cells, made by lysing cells and removing the particulate matter, such as nuclei, membranes, and organelles. Often used to carry out the synthesis of proteins by the addition of specific, exogenous mRNA molecules.

CEN In yeast, fragments of chromosomal DNA, about 120 bp in length, that when inserted into plasmids confer the ability to segregate during mitosis. These segments contain at least three types of sequence elements associated with centromere function.

centimeter A unit of length equal to 10^{-2} meter. Abbreviated cm.

centimorgan A unit of distance between genes on chromosomes. One centimorgan represents a value of 1 percent crossing over between two genes.

central dogma The concept that information flow progresses from DNA to RNA to proteins. Although exceptions are known, this idea is central to an understanding of gene function.

centric fusion See *Robertsonian translocation.*

centriole A cytoplasmic organelle composed of nine groups of microtubules, generally arranged in triplets. Centrioles function in the generation of cilia and flagella and serve as foci for the spindles in cell division.

centromere Specialized region of a chromosome to which the spindle fibers attach during cell division. Location of the centromere determines the shape of the chromosome during the anaphase portion of cell division. Also known as the primary constriction.

centrosome Region of the cytoplasm containing the centriole.

character An observable phenotypic attribute of an organism.

charon phages A group of genetically modified lambda phages designed to be used as vectors for cloning foreign DNA. Named after the ferryman in Greek mythology who carried the souls of the dead across the River Styx.

chemotaxis Negative or positive response to a chemical gradient.

chiasma (pl., **chiasmata**) The crossed strands of non-sister chromatids seen in diplotene of the first meiotic division. Regarded as the cytological evidence for exchange of chromosomal material, or crossing over.

chiasmatype theory The theory that crossing over between nonsister chromatids is the cause of chiasma formation.

chi-square (χ^2) analysis Statistical test to determine if an observed set of data fits a theoretical expectation.

chloroplast A cytoplasmic self-replicating organelle containing chlorophyll. The site of photosynthesis.

chorionic villus sampling (CVS) A technique of prenatal diagnosis that intravaginally retrieves fetal cells from the chorion and uses them to detect cytogenetic and biochemical defects in the embryo.

chromatid One of the longitudinal subunits of a replicated chromosome, joined to its sister chromatid at the centromere.

chromatin Term used to describe the complex of DNA, RNA, histones, and nonhistone proteins that make up uncoiled chromosomes characteristic of the eukaryotic interphase nucleus.

chromatography Technique for the separation of a mixture of solubilized molecules by their differential migration over a substrate.

chromocenter An aggregation of centromeres and heterochromatic elements of polytene chromosomes.

chromomere A coiled, beadlike region of a chromosome most easily visible during cell division. The aligned chromomeres of polytene chromosomes are responsible for their distinctive banding pattern.

chromosomal aberration Any change resulting in the duplication, deletion, or rearrangement of chromosomal material.

chromosomal mutation See *chromosomal aberration.*

chromosomal polymorphism Alternative structures or arrangements of a chromosome that are carried by members of a population.

chromosome In prokaryotes, an intact DNA molecule containing the genome; in eukaryotes, a DNA molecule complexed with RNA and proteins to form a threadlike structure containing genetic information arranged in a linear sequence and visible during mitosis and meiosis.

chromosome banding Technique for the differential staining of mitotic or meiotic chromosomes to produce a characteristic banding pattern or selective staining of certain chromosomal regions such as centromeres, the nucleolus organizer regions, and GC- or AT-rich regions. Not to be confused with the banding pattern present in polytene chromosomes, which is produced by the alignment of chromomeres.

chromosome map A diagram showing the location of genes on chromosomes.

chromosome puff A localized uncoiling and swelling in a polytene chromosome, usually regarded as a sign of active transcription.

chromosome theory of inheritance The idea put forward by Walter Sutton and Theodore Boveri that chromosomes are the carriers of genes and the basis for the Mendelian mechanisms of segregation and independent assortment.

chromosome walking A method for analyzing long stretches of DNA, in which the end of a cloned segment of DNA is subcloned and used as a probe to identify other clones that overlap the first clone.

***cis* configuration** The arrangement of two mutant sites within a gene on the same homolog, such as

$$\frac{a^1 \; a^2}{+ \; +}$$

Contrasts with a *trans* arrangement, where the mutant alleles are located on opposite homologs.

***cis-trans* test**   A genetic test to determine whether two mutations are located within the same cistron.

cistron   That portion of a DNA molecule that codes for a single polypeptide chain; defined by a genetic test as a region within which two mutations cannot complement each other.

cline   A gradient of genotype or phenotype distributed over a geographic range.

clonal selection   Theory of the immune system that proposes that antibody diversity precedes exposure to the antigen, and that the antigen functions to select the cells containing its specific antibody to undergo proliferation.

clone   Genetically identical cells or organisms all derived from a single ancestor by asexual or parasexual methods. For example, a DNA segment that has been enzymatically inserted into a plasmid or chromosome of a phage or a bacterium and replicated to form many copies.

cloned library   A collection of cloned DNA molecules representing all or part of an individual's genome.

code   See *genetic code*.

codominance   Condition in which the phenotypic effects of a gene's alleles are fully and simultaneously expressed in the heterozygote.

codon   A triplet of bases in a DNA or RNA molecule that specifies or encodes the information for a single amino acid.

coefficient of coincidence   A ratio of the observed number of double-crossovers divided by the expected number of such crossovers.

coefficient of inbreeding   The probability that two alleles present in a zygote are descended from a common ancestor.

coefficient of selection   A measurement of the reproductive disadvantage of a given genotype in a population. If for genotype *aa*, only 99 of 100 individuals reproduce, then the selection coefficient (*s*) is 0.1.

colchicine   An alkaloid compound that inhibits spindle formation during cell division. Used in the preparation of karyotypes to collect a large population of cells inhibited at the metaphase stage of mitosis.

colicin   A bacteriocidal protein produced by certain strains of *E. coli* and other closely related bacterial species.

colinearity   The linear relationship between the nucleotide sequence in a gene (or the RNA transcribed from it) and the order of amino acids in the polypeptide chain specified by the gene.

competence   In bacteria, the transient state or condition during which the cell can bind and internalize exogenous DNA molecules, making transformation possible.

complementarity   Chemical affinity between nitrogenous bases as a result of hydrogen bonding. Responsible for the base pairing between the strands of the DNA double helix.

complementation test   A genetic test to determine whether two mutations occur within the same gene. If two mutations are introduced into a cell simultaneously and produce a wild-type phenotype (i.e., they complement each other), they are often nonallelic. If a mutant phenotype is produced, the mutations are noncomplementing and are often allelic.

complete linkage   A condition in which two genes are located so close to each other that no recombination occurs between them.

complexity   The total number of nucleotides or nucleotide pairs in a population of nucleic acid molecules as determined by reassociation kinetics.

complex locus   A gene within which a set of functionally related pseudoalleles can be identified by recombinational analysis (e.g., the *bithorax* locus in *Drosophila*).

concatemer   A chain or linear series of subunits linked together. The process of forming a concatemer is called concatenation (e.g., multiple units of a phage genome produced during replication).

concordance   Pairs or groups of individuals identical in their phenotype. In twin studies, a condition in which both twins exhibit or fail to exhibit a trait under investigation.

conditional mutation   A mutation that expresses a wild-type phenotype under certain (permissive) conditions and a mutant phenotype under other (restrictive) conditions.

conjugation   Temporary fusion of two single-celled organisms for the sexual transfer of genetic material.

consanguineous   Related by a common ancestor within the previous few generations.

consensus sequence   A basically common, although not necessarily identical, sequence of nucleotides in DNA or amino acids in proteins.

continuous variation   Phenotype variation exhibited by quantitative traits distributed from one phenotypic extreme to another in an overlapping or continuous fashion.

cosmid   A vector designed to allow cloning of large segments of foreign DNA. Cosmids are hybrids composed of the cos sites of lambda inserted into a plasmid. In cloning, the recombinant DNA molecules are packaged into phage protein coats, and after infection of bacterial cells, the recombinant molecule replicates and can be maintained as a plasmid.

coupling conformation   See *cis configuration*.

covalent bond A nonionic chemical bond formed by the sharing of electrons.

cri-du-chat syndrome A clinical syndrome in humans produced by a deletion of a portion of the short arm of chromosome 5. Afflicted infants have a distinctive cry which sounds like that of a cat.

crossing over The exchange of chromosomal material (parts of chromosomal arms) between homologous chromosomes by breakage and reunion. The exchange of material between nonsister chromatids during meiosis is the basis of genetic recombination.

cross-reacting material (CRM) Nonfunctional form of an enzyme, produced by a mutant gene, which is recognized by antibodies made against the normal enzyme.

C-terminal amino acid The terminal amino acid in a peptide chain which carries a free carboxyl group.

C terminus The end of a polypeptide that carries a free carboxyl group of the last amino acid. By convention, the structural formula of polypeptides is written with the C terminus at the right.

C **value** The haploid amount of DNA present in a genome.

C **value paradox** The apparent paradox that there is no relationship between the size of the genome and the evolutionary complexity of species. For example, the C value (haploid genome size) of amphibians varies by a factor of 100.

cyclins A class of proteins found in eukaryotic cells that are synthesized and degraded in synchrony with the cell cycle, and regulate passage through stages of the cycle.

cytogenetics A branch of biology in which the techniques of both cytology and genetics are used to study heredity.

cytokinesis The division or separation of the cytoplasm during mitosis or meiosis.

cytological map A diagram showing the location of genes at particular chromosomal sites.

cytoplasmic inheritance Non-Mendelian form of inheritance involving genetic information transmitted by self-replicating cytoplasmic organelles such as mitochondria, chloroplasts, etc.

cytoskeleton An internal array of microtubules, microfilaments, and intermediate filaments that confers shape and the ability to move on a eukaryotic cell.

dalton A unit of mass equal to that of the hydrogen atom, which is 1.67×10^{-24} gram. A unit used in designating molecular weights.

Darwinian fitness See *fitness.*

deficiency (deletion) A chromosomal mutation involving the loss or deletion of chromosomal material.

degenerate code Term used to describe the genetic code, in which a given amino acid may be represented by more than one codon.

deletion See *deficiency.*

deme A local interbreeding population.

denatured DNA DNA molecules that have been separated into single strands.

de novo Newly arising; synthesized from less complex precursors rather than having been produced by modification of an existing molecule.

density gradient centrifugation A method of separating macromolecular mixtures by the use of centrifugal force and solvents of varying density. In sedimentation velocity centrifugation, macromolecules are separated by the velocity of sedimentation through a preformed gradient such as sucrose. In density gradient equilibrium centrifugation, macromolecules in a cesium salt solution are centrifuged until the cesium solution establishes a gradient under the influence of the centrifugal field, and the macromolecules sediment until the density of the solvent equals their own.

deoxyribonuclease A class of enzymes that breaks down DNA into oligonucleotide fragments by introducing single-stranded breaks into the double helix.

deoxyribonucleic acid (DNA) A macromolecule usually consisting of antiparallel polynucleotide chains held together by hydrogen bonds, in which the sugar residues are deoxyribose. The primary carrier of genetic information.

deoxyribose The five-carbon sugar associated with the deoxyribonucleotides found in DNA.

dermatoglyphics The study of the surface ridges of the skin, especially of the hands and feet.

determination A regulatory event that establishes a specific pattern of gene activity and developmental fate for a given cell.

diakinesis The final stage of meiotic prophase I in which the chromosomes become tightly coiled and compacted and move toward the periphery of the nucleus.

dicentric chromosome A chromosome having two centromeres.

differentiation The process of complex changes by which cells and tissues attain their adult structure and functional capacity.

dihybrid cross A genetic cross involving two characters in which the parents possess different forms of each character (e.g., tall, round × short, wrinkled peas).

diploid A condition in which each chromosome exists in pairs; having two of each chromosome.

diplotene A stage of meiotic prophase I immediately after pachytene. In diplotene, one pair of sister chromatids begins separating from the other, and chiasmata become visible. These overlaps move laterally toward the ends of the chromatids (terminalization).

directional selection   A selective force that changes the frequency of an allele in a given direction, either toward fixation or toward elimination.

discontinuous replication of DNA   The synthesis of DNA in discontinuous fragments on the lagging strand of the replication fork. The fragments, known as Okazaki fragments, are joined by DNA ligase to form a continuous strand.

discontinuous variation   Phenotypic data that fall into two or more distinct, non-overlapping classes.

discordance   In twin studies, a situation where one twin expresses a trait but the other does not.

disjunction   The separation of chromosomes at the anaphase stage of cell division.

disruptive selection   Simultaneous selection for phenotypic extremes in a population, usually resulting in the production of two discontinuous strains.

dizygotic twins   Twins produced from separate fertilization events; two ova fertilized independently. Also known as fraternal twins.

DNA   See *deoxyribonucleic acid.*

DNA footprinting   See *footprinting.*

DNA gyrase   One of the DNA topoisomerases that functions during DNA replication to reduce molecular tension caused by supercoiling. DNA gyrase produces, then seals, double-stranded breaks.

DNA ligase   An enzyme that forms a covalent bond between the 5′-end of one polynucleotide chain and the 3′-end of another polynucleotide chain. Also called polynucleotide-joining enzyme.

DNA polymerase   An enzyme that catalyzes the synthesis of DNA from deoxyribonucleotides and a template DNA molecule.

DNase   Deoxyribonucleosidase, an enzyme that degrades or breaks down DNA into fragments or constitutive nucleotides.

dominance   The expression of a trait in the heterozygous condition.

dominant suppression   A form of epistasis in which a dominant allele at one locus suppresses the effect of a dominant allele at another locus, resulting in a 13:3 phenotypic ratio.

dosage compensation   A genetic mechanism that regulates the levels of gene products at certain autosomal loci; this results in homozygous dominants and heterozygotes having the same amount of a gene product. In mammals, random inactivation of one X chromosome in females leads to equal levels of X chromosome-coded gene products in males and females.

double-crossover   Two separate events of chromosome breakage and exchange occurring within the same tetrad.

double helix   The model for DNA structure proposed by James Watson and Francis Crick, involving two antiparallel, hydrogen-bonded polynucleotide chains wound into a right-handed helical configuration, with 10 base pairs per full turn of the double helix. Often called B-DNA.

Duchenne muscular dystrophy   An X-linked recessive genetic disorder caused by a mutation in the gene for dystrophin, a protein found in muscle cells.

duplication   A chromosomal aberration in which a segment of the chromosome is repeated.

dyad   The products of tetrad separation or disjunction at the first meiotic prophase. Consists of two sister chromatids joined at the centromere.

dystrophin   See *Duchenne muscular dystrophy.*

effector molecule   Small, biologically active molecule that acts to regulate the activity of a protein by binding to a specific receptor site on the protein.

electrophoresis   A technique used to separate a mixture of molecules by their differential migration through a stationary phase (such as a gel) in an electrical field.

endocytosis   The uptake by a cell of fluids, macromolecules, or particles by pinocytosis, phagocytosis, or receptor-mediated endocytosis.

endogenote   The segment of the chromosome in a partially diploid bacterial cell (merozygote) that is homologous to the chromosome transmitted by the donor cell.

endomitosis   Chromosomal replication that is not accompanied by either nuclear or cytoplasmic division.

endonuclease   An enzyme that hydrolyzes internal phosphodiester bonds in a polynucleotide chain or nucleic acid molecule.

endoplasmic reticulum   A membranous organelle system in the cytoplasm of eukaryotic cells. The outer surface of the membranes may be ribosome-studded (rough ER) or smooth ER.

endopolyploidy   The increase in chromosome sets that results from endomitotic replication within somatic nuclei.

endosymbiont theory   The proposal that self-replicating cellular organelles such as mitochondria and chloroplasts were originally free-living organisms that entered into a symbiotic relationship with nucleated cells.

enhancer   Originally, a 72-bp sequence in the genome of the virus, SV40, that increases the transcriptional activity of nearby structural genes. Similar sequences that enhance transcription have been identified in the genomes of eukaryotic cells. Enhancers can act over a distance of thousands of base pairs and can be located 5′, 3′ or internal to the gene they affect, and thus are different from promoters.

enhansor   The DNA sequence that represents the core sequence of an enhancer.

environment   The complex of geographic, climatic, and biotic factors within which an organism lives.

enzyme   A protein or complex of proteins that catalyzes a specific biochemical reaction.

epigenesis   The idea that an organism develops by the appearance and growth of new structures. Opposed to preformationism, which holds that development is the growth of structures already present in the egg.

episome   A circular genetic element in bacterial cells that can replicate independently of the bacterial chromosome or integrate and replicate as part of the chromosome.

epistasis   Nonreciprocal interaction between genes such that one gene interferes with or prevents the expression of another gene. For example, in *Drosophila*, the recessive gene *eyeless*, when homozygous, prevents the expression of eye color genes.

epitope   That portion of a macromolecule or cell that acts to elicit an antibody response; an antigenic determinant. A complex molecule or cell can contain several such sites.

equational division   A division of each chromosome into longitudinal halves that are distributed into two daughter nuclei. Chromosome division in mitosis is an example of equational division.

equatorial plate   See *metaphase plate*.

euchromatin   Chromatin or chromosomal regions that are lightly staining and are relatively uncoiled during the interphase portion of the cell cycle. The region of the chromosomes thought to contain most of the structural genes.

eugenics   The improvement of the human species by selective breeding. Positive eugenics refers to the promotion of breeding of those with favorable genes, and negative eugenics refers to the discouragement of breeding among those with undesirable traits.

eukaryotes   Those organisms having true nuclei and membranous organelles and whose cells demonstrate mitosis and meiosis.

euphenics   Medical or genetic intervention to reduce the impact of defective genotypes.

euploid   Polyploid with a chromosome number that is an exact multiple of a basic chromosome set.

evolution   The origin of plants and animals from preexisting types. Descent with modifications.

excision repair   Repair of DNA lesions by removal of a polynucleotide segment and its replacement with a newly synthesized, corrected segment.

exogenote   In merozygotes, the segment of the bacterial chromosome contributed by the donor cell.

exon (extron)   The DNA segment(s) of a gene that are transcribed and translated into protein.

exonuclease   An enzyme that breaks down nucleic acid molecules by breaking the phosphodiester bonds at the 3′ or 5′ terminal nucleotides.

expressed sequence tags (ESTs)   The nucleotide sequence of cDNA clones. Used as markers in construction of genetic maps.

expression vector   Plasmids or phages carrying promoter regions designed to cause expression of inserted DNA sequences.

expressivity   The degree or range in which a phenotype for a given trait is expressed.

extranuclear inheritance   Transmission of traits by genetic information contained in cytoplasmic organelles such as mitochondria and chloroplasts.

F⁺ cell   A bacterial cell having a fertility (F) factor. Acts as a donor in bacterial conjugation.

F⁻ cell   A bacterial cell that does not contain a fertility (F) factor. Acts as a recipient in bacterial conjugation.

F factor   An episome in bacterial cells that confers the ability to act as a donor in conjugation.

F′ factor   A fertility (F) factor that contains a portion of the bacterial chromosome.

F_1 generation   First filial generation; the progeny resulting from the first cross in a series.

F_2 generation   Second filial generation; the progeny resulting from a cross of the F_1 generation.

F pilus   See *pilus*.

facultative heterochromatin   Regions of, or whole chromosomes that may or may not become heterochromatic such as the mammalian X chromosomes.

familial trait   A trait transmitted through and expressed by members of a family.

fate map   A diagram or "map" of an embryo showing the location of cells whose development fate is known.

fertility (F) factor   See *F factor*.

filial generations   See *F_1, F_2 generations*.

fingerprint   The pattern of ridges and whorls on the tip of a finger. The pattern obtained by enzymatically cleaving a protein or nucleic acid and subjecting the digest to two-dimensional chromatography or electrophoresis.

FISH   See *fluorescence in situ hybridization*.

fitness A measure of the relative survival and reproductive success of a given individual or genotype.

fixation In population genetics, a condition in which all members of a population are homozygous for a given allele.

fixity of species The idea that members of a species can give rise only to other members of the species, thus implying that all species are independently created.

fluctuation test A statistical test developed by Salvadore Luria and Max Delbrück to determine whether bacterial mutations arise spontaneously or are produced in response to selective agents.

fluorescence *in situ* hybridization (FISH) A method of *in situ* hybridization that utilizes probes labeled with a fluorescent tag, causing the site of hybridization to fluoresce when viewed in ultraviolet light under a microscope.

flush-crash cycle A period of rapid population growth followed by a drastic reduction in population size.

fmet See *formylmethionine.*

folded-fiber model A model of eukaryotic chromosome organization in which each sister chromatid consists of a single fiber, composed of double-stranded DNA and protein, which is wound up like a tightly coiled skein of yarn.

footprinting A technique for identifying a DNA sequence that binds to a particular protein, based on the idea that the phosphodiester bonds in the region covered by the protein are protected from digestion by deoxyribonucleases.

formylmethionine (fmet) A molecule derived from the amino acid methionine by attachment of a formyl group to its terminal amino group. This is the first amino acid inserted in all bacterial polypeptides. Also known as *N*-formyl methionine.

founder effect A form of genetic drift. The establishment of a population by a small number of individuals whose genotypes carry only a fraction of the different kinds of alleles in the parental population.

fragile site A heritable gap or nonstaining region of a chromosome that can be induced to generate chromosome breaks.

fragile X syndrome A genetic disorder caused by the expansion of a CGG trinucleotide repeat and a fragile site at Xq27.3, within the *FMR-1* gene.

frameshift mutation A mutational event leading to the insertion of one or more base pairs in a gene, shifting the codon reading frame in all codons following the mutational site.

fraternal twins See *dizygotic twins.*

G1 checkpoint A point in the G1 phase of the cell cycle when a cell becomes committed to initiate DNA synthesis and continue the cycle or withdraw into the G0 resting stage.

G0 The G-zero stage of the cell cycle is a point in G1 where cells withdraw from the cell cycle and enter a nondividing but metabolically active state.

gamete A specialized reproductive cell with a haploid number of chromosomes.

gap genes Genes expressed in contiguous domains along the anterior-posterior axis of the *Drosophila* embryo, which regulate the process of segmentation in each domain.

gene The fundamental physical unit of heredity whose existence can be confirmed by allelic variants and which occupies a specific chromosomal locus. A DNA sequence coding for a single polypeptide.

gene amplification The process by which gene sequences are selected for differential replication either extrachromosomally or intrachromosomally.

gene conversion The process by which one allele in a heterozygote is converted into the corresponding allele.

gene duplication An event in replication leading to the production of a tandem repeat of a gene sequence.

gene flow The gradual exchange of genes between two populations, brought about by the dispersal of gametes or the migration of individuals.

gene frequency The percentage of alleles of a given type in a population.

gene interaction Production of novel phenotypes by the interaction of alleles of different genes.

gene mutation See *point mutation.*

gene pool The total of all alleles possessed by reproductive members of a population.

generalized transduction The transduction of any gene in the bacterial genome by a phage.

genetic anticipation The phenomenon of a progressively earlier age of onset and increasing severity of symptoms for a genetic disorder in successive generations.

genetic background All genes carried in the genome other than the one being studied.

genetic burden Average number of recessive lethal genes carried in the heterozygous condition by an individual in a population. Also called genetic load.

genetic code The nucleotide triplets that code for the 20 amino acids or for chain initiation or termination.

genetic counseling Analysis of risk for genetic defects in a family and the presentation of options available to avoid or ameliorate possible risks.

genetic drift Random variation in gene frequency from generation to generation. Most often observed in small populations.

genetic engineering The technique of altering the genetic constitution of cells or individuals by the selective removal, insertion, or modification of individual genes or gene sets.

genetic equilibrium Maintenance of allele frequencies at the same value in successive generations. A condition in which allele frequencies are neither increasing nor decreasing.

genetic fine structure Intragenic recombinational analysis that provides mapping information at the level of individual nucleotides.

genetic load See *genetic burden*.

genetic polymorphism The stable coexistence of two or more discontinuous genotypes in a population. When the frequencies of two alleles are carried to an equilibrium, the condition is called balanced polymorphism.

genetics The branch of biology that deals with heredity and the expression of inherited traits.

genome The array of genes carried by an individual.

genomic imprinting A condition where the expression of a trait depends on whether the trait has been inherited from a male or a female parent.

genotype The specific allelic or genetic constitution of an organism; often, the allelic composition of one or a limited number of genes under investigation.

germplasm Hereditary material transmitted from generation to generation.

Goldberg-Hogness box A short nucleotide sequence 20 to 30 bp 5′ to the initiation site of eukaryotic genes to which RNA polymerase II binds. The consensus sequence is TATAAAA. Also known as TATA box.

graft versus host disease (GVHD) In transplants, reaction by immunologically competent cells of the donor against the antigens present on the cells of the host. In human bone marrow transplants, often a fatal condition.

gynandromorph An individual composed of cells with both male and female genotypes.

gyrase One of a class of enzymes known as topoisomerases. Gyrase converts closed circular DNA to a negatively supercoiled form prior to replication, transcription, or recombination.

H substance The carbohydrate group present on the surface of red blood cells. When unmodified, it results in blood type O; when modified by the addition of monosaccharides, it results in type A, B, and AB.

haploid A cell or organism having a single set of unpaired chromosomes. The gametic chromosome number.

haplotype The set of alleles from closely linked loci carried by an individual and usually inherited as a unit.

Hardy-Weinberg law The principle that both gene and genotype frequencies will remain in equilibrium in an infinitely large population in the absence of mutation, migration, selection, and nonrandom mating.

heat shock A transient response following exposure of cells or organisms to elevated temperatures. The response involves activation of a small number of loci, inactivation of previously active loci, and selective translation of heat shock mRNA. Appears to be a nearly universal phenomenon observed in organisms ranging from bacteria to humans.

helicase An enzyme that participates in DNA replication by unwinding the double helix near the replication fork.

helix-turn-helix motif The structure of a region of DNA-binding proteins in which a turn of four amino acids holds two alpha helices at right angles to each other.

hemizygous Conditions where a gene is present in a single dose. Usually applied to genes on the X chromosome in heterogametic males.

hemoglobin (Hb) An iron-containing, oxygen-carrying protein occurring chiefly in the red blood cells of vertebrates.

hemophilia An X-linked trait in humans associated with defective blood-clotting mechanisms.

heredity Transmission of traits from one generation to another.

heritability A measure of the degree to which observed phenotypic differences for a trait are genetic.

heterochromatin The heavily staining, late replicating regions of chromosomes that are condensed in interphase. Thought to be devoid of structural genes.

heteroduplex A double-stranded nucleic acid molecule in which each polynucleotide chain has a different origin. These structures may be produced as intermediates in a recombinational event, or by the *in vitro* reannealing of single-stranded, complementary molecules.

heterogametic sex The sex that produces gametes containing unlike sex chromosomes.

heterogeneous nuclear RNA (hnRNA) The collection of RNA transcripts in the nucleus, representing precursors and processing intermediates to rRNA, mRNA, and tRNA. Also represents RNA transcripts that will not be transported to the cytoplasm, such as snRNA (small nuclear RNA).

heterokaryon A somatic cell containing nuclei from two different sources.

heterosis The superiority of a heterozygote over either homozygote for a given trait.

heterozygote An individual with different alleles at one or more loci. Such individuals will produce unlike gametes and therefore will not breed true.

Hfr A strain of bacteria exhibiting a high frequency of recombination. These strains have a chromosomally integrated F factor that is able to mobilize and transfer part of the chromosome to a recipient F⁻ cell.

histocompatibility antigens See *HLA*.

histones Proteins complexed with DNA in the nucleus. They are rich in the basic amino acids arginine and lysine and function in the coiling of DNA to form nucleosomes.

HLA Cell surface proteins, produced by histocompatibility loci, which are involved in the acceptance or rejection of tissue and organ grafts and transplants.

hnRNA See *heterogeneous nuclear RNA*.

holandric A trait transmitted from males to males. In humans, genes on the Y chromosome are holandric.

Holliday structure An intermediate in bidirectional DNA replication seen in the transmission electron microscope as an X-shaped structure showing four single-stranded DNA regions.

homeobox A sequence of about 180 nucleotides that encodes a 60-amino-acid sequence called a homeodomain, which is a DNA-binding protein that acts as a transcription factor.

homeotic mutation A mutation that causes a tissue normally determined to form a specific organ or body part to alter its differentiation and form another structure. Alternatively spelled: homoeotic.

homogametic sex The sex that produces gametes that do not differ with respect to sex chromosome content; in mammals, the female is homogametic.

homogeneously staining regions (hsr) Segments of mammalian chromosomes that stain lightly with Giemsa following exposure of cells to a selective agent. These regions arise in conjunction with gene amplification and are regarded as the structural locus for the amplified gene.

homogenote A bacterial merozygote in which the donor (exogenote) chromosome carries the same alleles as the chromosome of the recipient (endogenote). A homozygous merozygote.

homologous chromosomes Chromosomes that synapse or pair during meiosis. Chromosomes that are identical with respect to their genetic loci and centromere placement.

homozygote An individual with identical alleles at one or more loci. Such individuals will produce identical gametes and will therefore breed true.

homunculus The miniature individual imagined by preformationists to be contained within the sperm or egg.

human immunodeficiency virus (HIV) A human retrovirus associated with the onset and progression of acquired immunodeficiency syndrome (AIDS).

hybrid An individual produced by crossing two parents of different genotypes.

hybridoma A somatic cell hybrid produced by the fusion of an antibody-producing cell and a cancer cell, specifically, a myeloma. The cancer cell contributes the ability to divide indefinitely, and the antibody cell confers the ability to synthesize large amounts of a single antibody.

hybrid vigor See *heterosis*.

hydrogen bond An electrostatic attraction between a hydrogen atom bonded to a strongly electronegative atom such as oxygen or nitrogen and another atom that is electronegative or contains an unshared electron pair.

hypervariable regions The regions of antibody molecules that attach to antigens. These regions have a high degree of diversity in amino acid content.

identical twins See *monozygotic twins*.

Ig See *immunoglobulin*.

imaginal disk Discrete groups of cells set aside during embryogenesis in holometabolous insects which are determined to form the external body parts of the adult.

immunoglobulin The class of serum proteins having the properties of antibodies.

inborn error of metabolism A biochemical disorder that is genetically controlled; usually an enzyme defect that produces a clinical syndrome.

inbreeding Mating between closely related organisms.

inbreeding depression A decrease in viability, vigor or growth in progeny following several rounds of inbreeding

incomplete dominance Expression of heterozygous phenotype which is distinct from, and often intermediate to, that of either parent.

incomplete linkage Occasional separation of two genes on the same chromosome by a recombinational event.

independent assortment The independent behavior of each pair of homologous chromosomes during their segregation in meiosis I. The random distribution of maternal and paternal homologues into gametes.

inducer An effector molecule that activates transcription.

inducible enzyme system An enzyme system under the control of a regulatory molecule, or inducer, which acts to block a repressor and allow transcription.

initiation codon The triplet of nucleotides (AUG) in an mRNA molecule that codes for the insertion of the amino acid methionine as the first amino acid in a polypeptide chain.

insertion sequence See *IS element.*

***in situ* hybridization** A technique for the cytological localization of DNA sequences complementary to a given nucleic acid or polynucleotide.

intercalary deletion A form of chromosome deletion where material is lost from within the chromosome. Deletions that involve the end of the chromosome are called terminal deletions.

intercalating agent A compound that inserts between bases in a DNA molecule, disrupting the alignments and pairing of bases in the complementary strands (e.g., acridine dyes).

interference A measure of the degree to which one crossover affects the incidence of another crossover in an adjacent region of the same chromatid. Negative interference increases the chances of another crossover; positive interference reduces the probability of a second crossover event.

interferon One of a family of proteins that acts to inhibit viral replication in higher organisms. Some interferons may have anticancer properties.

interphase That portion of the cell cycle between divisions.

intervening sequence See *intron.*

intron A portion of DNA between coding regions in a gene which is transcribed, but which does not appear in the mRNA product.

inversion A chromosomal aberration in which the order of a chromosomal segment has been reversed.

inversion loop The chromosomal configuration resulting from the synapsis of homologous chromosomes, one of which carries an inversion.

in vitro Literally, in glass; outside the living organism; occurring in an artificial environment.

in vivo Literally, in the living; occurring within the living body of an organism.

IS element A mobile DNA segment that is transposable to any of a number of sites in the genome.

isoagglutinogen An antigenic factor or substance present on the surface of cells that is capable of inducing the formation of an antibody.

isochromosome An aberrant chromosome with two identical arms and homologous loci.

isolating mechanism Any barrier to the exchange of genes between different populations of a group of organisms. In general, isolation can be classified as spatial, environmental, or reproductive.

isotopes Forms of a chemical element that have the same number of protons and electrons, but differ in the number of neutrons contained in the atomic nucleus.

isozyme Any of two or more distinct forms of an enzyme that have identical or nearly identical chemical properties, but differ in some property such as net electrical charge, pH optima, number and type of subunits, or substrate concentration.

kappa particles DNA-containing cytoplasmic particles found in certain strains of *Paramecium aurelia.* When these self-reproducing particles are transferred into the growth medium, they release a toxin, paramecin, which kills other sensitive strains. A nuclear gene, *K*, is responsible for the maintenance of kappa particles in the cytoplasm.

karyokinesis The process of nuclear division.

karyotype The chromosome complement of a cell or an individual. Often used to refer to the arrangement of metaphase chromosomes in a sequence according to length and position of the centromere.

kilobase A unit of length consisting of 1000 nucleotides. Abbreviated kb.

kinetochore A fibrous structure with a size of about 400 nm, located within the centromere. It appears to be the location to which microtubules attach.

Klenow fragment A part of bacterial DNA polymerase that lacks exonuclease activity, but retains polymerase activity. It is produced by enzymatic digestion of the intact enzyme.

Klinefelter syndrome A genetic disorder in human males caused by the presence of an extra X chromosome. Klinefelter males are XXY instead of XY. This syndrome is associated with enlarged breasts, small testes, sterility, and, occasionally, mental retardation.

knockout mice in producing knockout mice, a cloned normal gene is inactivated by the insertion of a marker, such as an antibiotic resistance gene. The altered gene is transferred to embryonic stem cells, where the altered gene will replace the normal gene (in some cells). These cells are injected into a blastomere embryo, producing a mouse that is bred to yield mice that are homozygous for the mutated gene.

***lac* repressor protein** A protein that binds to the operator in the *lac* operon and blocks transcription.

lagging strand In DNA replication, the strand synthesized in a discontinuous fashion, 5′ to 3′ away from the replication fork. Each short piece of DNA synthesized in this fashion is called an Okazaki fragment.

lampbrush chromosomes Meiotic chromosomes characterized by extended lateral loops, which reach maximum extension during diplotene. Although most intensively studied in amphibians, these structures occur in meiotic cells of organisms ranging from insects through humans.

lariat structure A structure formed by an intron via a $5' \rightarrow 2'$ bond during processing and removal of that intron from mRNA.

leader sequence That portion of an mRNA molecule from the 5'-end to the beginning codon; may contain regulatory or ribosome binding sites.

leading strand During DNA replication, the strand synthesized continuously 5' to 3' toward the replication fork.

lectins Carbohydrate-binding proteins found in the seeds of leguminous plants such as soybeans. Although their physiological role in plants is unclear, they are useful as probes for cell surface carbohydrates and glycoproteins in animal cells. Proteins with similar properties have also been isolated from organisms such as snails and eels.

leptotene The initial stage of meiotic prophase I, during which the chromosomes become visible and are often arranged in a bouquet configuration, with one or both ends of the chromosomes gathered at one spot on the inner nuclear membrane.

lethal gene A gene whose expression results in death.

leucine zipper A structural motif in a DNA-binding protein that is characterized by a stretch of leucine residues spaced at every seventh amino acid residue, with adjacent regions of positively charged amino acids. Leucine zippers on two polypeptides may interact to form a dimer that binds to DNA.

LINEs Long interspersed elements are repetitive sequences found in the genomes of higher organisms, such as the 6-kb *Kpnl* sequences found in primate genomes.

linkage Condition in which two or more nonallelic genes tend to be inherited together. Linked genes have their loci along the same chromosome, do not assort independently, but can be separated by crossing over.

linkage group A group of genes that have their loci on the same chromosome.

linking number The number of times that two strands of a closed, circular DNA duplex cross over each other.

locus The site or place on a chromosome where a particular gene is located.

lod score A statistical method used to determine whether two loci are linked or unlinked. A lod (log of the odds) score of +3 by convention indicates linkage.

long period interspersion Pattern of genome organization in which long stretches of single copy DNA are in-

terspersed with long segments of repetitive DNA. This pattern of genome organization is found in *Drosophila* and the honeybee.

long terminal repeat (LTR) Sequence of several hundred base pairs found at the ends of retroviral DNAs.

Lutheran blood group One of a number of blood group systems inherited independently of the ABO, MN, and Rh systems. Alleles of this group determine the presence or absence of antigens on the surface of red blood cells. Gene is on human chromosome 19.

Lyon hypothesis The random inactivation of the maternal or paternal X chromosome in somatic cells of mammalian females early in developement. All daughter cells will have the same X chromosome inactivated, producing a mosaic pattern of expression of genes on the X chromosome.

lysis The disintegration of a cell brought about by the rupture of its membrane.

lysogenic bacterium A bacterial cell carrying the DNA of a temperate bacteriophage integrated into its chromosome.

lysogeny The process by which the DNA of an infecting phage becomes repressed and integrated into the chromosome of the bacterial cell it infects.

lytic phase The condition in which a temperate bacteriophage loses its integrated status in the host chromosome (becomes induced), replicates, and lyses the bacterial cell.

major histocompatibility loci See *MHC*.

map unit A measure of the genetic distance between two genes, corresponding to a recombination frequency of 1 percent. See *centimorgan*.

mapping functions Map distance estimates from recombination when the recombination frequency in a region exceeds 15–20 percent, and double crossovers are undetectable.

maternal effect Phenotypic effects on the offspring produced by the maternal genome. Factors transmitted through the egg cytoplasm that produce a phenotypic effect in the progeny.

maternal influence See *maternal effect*.

maternal inheritance The transmission of traits via cytoplasmic genetic factors such as mitochondria or chloroplasts.

mean The arithmetic average.

median The value in a group of numbers below and above which there is an equal number of data points or measurements.

meiosis The process in gametogenesis or sporogenesis during which one replication of the chromosomes is followed by two nuclear divisions to produce four haploid cells.

melting profile   See T_m.

merozygote   A partially diploid bacterial cell containing, in addition to its own chromosome, a chromosome fragment introduced into the cell by transformation, transduction, or conjugation.

messenger RNA   See *mRNA*.

metabolism   The sum of chemical changes in living organisms by which energy is generated and used.

metacentric chromosome   A chromosome with a centrally located centromere, producing chromosome arms of equal lengths.

metafemale   In *Drosophila*, a poorly developed female of low viability in which the ratio of X chromosomes to sets of autosomes exceeds 1.0. Previously called a superfemale.

metamale   In *Drosophila*, a poorly developed male of low viability in which the ratio of X chromosomes to sets of autosomes is less than 0.5. Previously called a supermale.

metaphase   The stage of cell division in which the condensed chromosomes lie in a central plane between the two poles of the cell, and in which the chromosomes become attached to the spindle fibers.

metaphase plate   The arrangement of mitotic or meiotic chromosomes at the equator of the cell during metaphase.

MHC   Major histocompatibility loci. In humans, the HLA complex; and in mice, the H2 complex.

micrometer   A unit of length equal to 1×10^{-6} meter. Previously called a micron. Abbreviated μm.

micron   See *micrometer*.

migration coefficient   An expression of the proportion of migrant genes entering the population per generation.

millimeter   A unit of length equal to 1×10^{-3} meter. Abbreviated mm.

minimal medium   A medium containing only those nutrients that will support the growth and reproduction of wild-type strains of an organism.

mismatch repair   DNA repair by a mechanism known as cut and patch. In this repair, the defective single-stranded region is excised, followed by the synthesis of a new segment, using the complementary strand as a template.

missense mutation   A mutation that alters a codon to that of another amino acid, causing an altered translation product to be made.

mitochondrion   Found in the cells of eukaryotes, a cytoplasmic, self-reproducing organelle that is the site of ATP synthesis.

mitogen   A substance that stimulates mitosis in nondividing cells (e.g., phytohemagglutinin).

mitosis   A form of cell division resulting in the production of two cells, each with the same chromosome and genetic complement as the parent cell.

mode   In a set of data, the value occurring in the greatest frequency.

monohybrid cross   A genetic cross between two individuals involving only one character (e.g., $AA \times aa$).

monosomic   An aneuploid condition in which one member of a chromosome pair is missing; having a chromosome number of $2n - 1$.

monozygotic twins   Twins produced from a single fertilization event; the first division of the zygote produces two cells, each of which develops into an embryo. Also known as identical twins.

mRNA   An RNA molecule transcribed from DNA and translated into the amino acid sequence of a polypeptide.

mtDNA   Mitochondrial DNA.

multigene family   A gene set descended from a common ancestor by duplication and subsequent divergence from a common ancestor. The globin genes represent a multigene family.

multiple alleles   Three or more alleles of the same gene.

multiple-factor inheritance   See *polygenic inheritance*.

multiple infection   Simultaneous infection of a bacterial cell by more than one bacteriophage, often of different genotypes.

mu phage   A phage group in which the genetic material behaves like an insertion sequence, capable of insertion, excision, transposition, inactivation of host genes, and induction of chromosomal rearrangements.

mutagen   Any agent that causes an increase in the rate of mutation.

mutant   A cell or organism carrying an altered or mutant gene.

mutation   The process that produces an alteration in DNA or chromosome structure; the source of most alleles.

mutation rate   The frequency with which mutations take place at a given locus or in a population.

muton   The smallest unit of mutation in a gene, corresponding to a single base change.

nanometer   A unit of length equal to 1×10^{-9} meter. Abbreviated nm.

natural selection   Differential reproduction of some members of a species resulting from variable fitness conferred by genotypic differences.

nearest-neighbor analysis   A molecular technique used to determine the frequency with which nucleotides are adjacent to each other in polynucleotide chains.

neutral mutation   A mutation with no immediate adaptive significance or phenotypic effect.

nonautonomous transposon   A transposable element that lacks a functional transposase gene.

noncrossover gamete   A gamete that contains no chromosomes that have undergone genetic recombination.

nondisjunction   An error during cell division in which the homologous chromosomes (in meiosis) or the sister chromatids (in mitosis) fail to separate and migrate to opposite poles; responsible for defects such as monosomy and trisomy.

nonsense codon   The nucleotide triplet in an mRNA molecule that signals the termination of translation. Three such codons are known: UGA, UAG, and UAA.

nonsense mutation   A mutation that alters a codon to one which encodes no amino acid, that is, UAG (amber codon), UAA (ochre codon), or UGA (opal codon). Leads to premature termination during the translation of mRNA.

NOR   See *nucleolar organizer region.*

normal distribution   A probability function that approximates the distribution of random variables. The normal curve, also known as a Gaussian or bell-shaped curve, is the graphic display of the normal distribution.

np   See *nucleotide pair.*

N-terminal amino acid   The terminal amino acid in a peptide chain that carries a free amino group.

N terminus   The end of a polypeptide that carries a free amino group of the first amino acid. By convention, the structural formula of polypeptides is written with the N terminus at the left.

nu body   See *nucleosome.*

nuclease   An enzyme that breaks bonds in nucleic acid molecules.

nucleoid   The DNA-containing region within the cytoplasm in prokaryotic cells.

nucleolar organizer region (NOR)   A chromosomal region containing the genes for rRNA; most often found in physical association with the nucleolus.

nucleolus   A nuclear organelle that is the site of ribosome biosynthesis; usually associated with or formed in association with the NOR.

nucleoside   A purine or pyrimidine base covalently linked to a ribose or deoxyribose sugar molecule.

nucleosome   A complex of four histone molecules, each present in duplicate, wrapped by two turns of a DNA molecule. One of the basic units of eukaryotic chromosome structure. Also known as a nu body.

nucleotide   A nucleoside covalently linked to a phosphate group. Nucleotides are the basic building blocks of nucleic acids. The nucleotides commonly found in DNA are deoxyadenylic acid, deoxycytidylic acid, deoxyguanylic acid, and deoxythymidylic acid. The nucleotides in RNA are adenylic acid, cytidylic acid, guanylic acid, and uridylic acid.

nucleotide pair   The pair of nucleotides (A and T, or G and C) in opposite strands of the DNA molecule that are hydrogen-bonded to each other.

nucleus   The membrane-bounded cytoplasmic organelle of eukaryotic cells that contains the chromosomes and nucleolus.

null hypothesis   Used in statistical tests, it states the there is no difference between the observed and expected data sets. Statistical methods are used to test the probability to this hypothesis.

nullisomic   Describes an individual with a chromosomal aberration in which both members of a chromosome pair are missing.

ochre codon   A codon that does not code for the insertion of an amino acid into a polypeptide chain, but signals chain termination. The ochre codon is UAA.

Okazaki fragment   The small, discontinuous strands of DNA produced during DNA synthesis on the lagging strand.

oligonucleotide   A linear sequence of nucleotides (about 10–20) connected by 5′ to 3′ phosphodiester bonds.

oncogene   A gene whose activity promotes uncontrolled proliferation in eukaryotic cells.

open reading frame (ORF)   The interval between the start and stop codon that encodes amino acids for insertion into polypeptide chains.

operator region   A region of a DNA molecule that interacts with a specific repressor protein to control the expression of an adjacent gene or gene set.

operon   A genetic unit that consists of one or more structural genes (that code for polypeptides) and an adjacent operator gene that controls the transcriptional activity of the structural gene or genes.

origin of replication (ori)   Sites along the length of the chromosome where DNA replication begins.

overdominance   The phenomenon where heterozygotes have a phenotype that is more extreme than either homozygous genotype.

overlapping code   A genetic code first proposed by George Gamow in which any given nucleotide is shared by three adjacent codons.

pachytene   The stage in meiotic prophase I when the synapsed homologous chromosomes split longitudinally (except at the centromere), producing a group of four chromatids called a tetrad.

pair-rule genes   Genes expressed as stripes around the blastoderm embryo during development of the *Drosophila* embryo.

palindrome   A word, number, verse, or sentence that reads the same backward or forward (e.g., *able was I ere I saw elba*). In nucleic acids, a sequence in which the base pairs read the same on complementary strands ($5' \rightarrow 3'$). For example: $5'$GAATTC$3'$, $3'$CTTAAG$5'$. These often occur as sites for restriction endonuclease recognition and cutting.

pangenesis   A discarded theory of development that postulated the existence of pangenes, small particles from all parts of the body that concentrated in the gametes, passing traits from generation to generation, blending the traits of the parents in the offspring.

paracentric inversion   A chromosomal inversion that does not include the centromere.

parasexual   Condition describing recombination of genes from different individuals that does not involve meiosis, gamete formation, or zygote production. The formation of somatic cell hybrids is an example.

parental gamete   See *noncrossover gamete*.

parthenogenesis   Development of an egg without fertilization.

partial diploids   See *merozygote*.

partial dominance   See *incomplete dominance*.

passive immunity   A form of immunity produced by receipt of antibodies synthesized by another individual.

patroclinous inheritance   A form of genetic transmission in which the offspring have the phenotype of the father.

pedigree   In human genetics, a diagram showing the ancestral relationships and transmission of genetic traits over several generations in a family.

P element   Transposable DNA elements found in *Drosophila* that are responsible for hybrid dysgenesis.

penetrance   The frequency (expressed as a percentage) with which individuals of a given genotype manifest at least some degree of a specific mutant phenotype associated with a trait.

peptide bond   The covalent bond between the amino group of one amino acid and the carboxyl group of another amino acid.

pericentric inversion   A chromosomal inversion that involves both arms of the chromosome and thus involves the centromere.

permissive condition   Environmental conditions under which a conditional mutation (such as temperature-sensitive mutant) expresses the wild-type phenotype.

phage   See *bacteriophage*.

phenocopy   An environmentally induced phenotype (nonheritable) that closely resembles the phenotype produced by a known gene.

phenotype   The observable properties of an organism that are genetically controlled.

phenylketonuria (PKU)   A hereditary condition in humans associated with the inability to metabolize the amino acid phenylalanine. The most common form is caused by the lack of the enzyme phenylalanine hydroxylase.

Philadelphia chromosome   The product of a reciprocal translocation that contains the short arm of chromosome 9 carrying the *c-abl* oncogene and the long arm of chromosome 22 carrying *bcr*.

phosphodiester bond   In nucleic acids, the covalent bond between a phosphate group and adjacent nucleotides, extending from the $5'$ carbon of one pentose (ribose or deoxyribose) to the $3'$ carbon of the pentose in the neighboring nucleotide. Phosphodiester bonds form the backbone of nucleic acid molecules.

photoreactivation enzyme (PRE)   An exonuclease that catalyzes the light-activated excision of ultraviolet-induced thymine dimers from DNA.

photoreactivation repair   Light-induced repair of damage caused by exposure to ultraviolet light. Associated with an intracellular enzyme system.

phyletic evolution   The gradual transformation of one species into another over time; vertical evolution.

pilus   A filamentlike projection from the surface of a bacterial cell. Often associated with cells possessing F factors.

plaque   A clear area on an otherwise opaque bacterial lawn caused by the growth and reproduction of phages.

plasmid   An extrachromosomal, circular DNA molecule (often carrying genetic information) that replicates independently of the host chromosome.

platysome   Term originally used in electron and X-ray diffraction studies to describe the flattened appearance of the DNA in the nucleosome core.

pleiotropy   Condition in which a single mutation simultaneously affects several characters.

ploidy   Term referring to the basic chromosome set or to multiples of that set.

point mutation   A mutation that can be mapped to a single locus. At the molecular level, a mutation that results in the substitution of one nucleotide for another.

polar body   A cell produced at either the first or second meiotic division in females that contains almost no cytoplasm as a result of an unequal cytokinesis.

polycistronic mRNA   A messenger RNA molecule that encodes the amino acid sequence of two or more polypeptide chains in adjacent structural genes.

polygenic inheritance   The transmission of a phenotypic trait whose expression depends on the additive effect of a number of genes.

polylinker   A segment of DNA that has been engineered to contain multiple sites for restriction enzyme digestion. Polylinkers are usually found in engineered vectors such as plasmids.

polymerase chain reaction (PCR)   A method for amplifying DNA segments that uses cycles of denaturation, annealing to primers, and DNA polymerase-directed DNA synthesis.

polymerases   The enzymes that catalyze the formation of DNA and RNA from deoxynucleotides and ribonucleotides, respectively.

polymorphism   The existence of two or more discontinuous, segregating phenotypes in a population.

polynucleotide   A linear sequence of more than 20 nucleotides, joined by 5′-to-3′ phosphodiester bonds. See *oligonucleotide.*

polypeptide   A molecule made up of amino acids joined by covalent peptide bonds. This term is used to denote the amino acid chain before it assumes its functional three-dimensional configuration.

polyploid   A cell or individual having more than two sets of chromosomes.

polyribosome   See *polysome.*

polysome   A structure composed of two or more ribosomes associated with mRNA, engaged in translation. Formerly called polyribosome.

polytene chromosome   A chromosome that has undergone several rounds of DNA replication without separation of the replicated chromosomes, forming a giant, thick chromosome with aligned chromomeres producing a characteristic banding pattern.

population   A local group of individuals belonging to the same species, which are actually or potentially interbreeding.

position effect   Change in expression of a gene associated with a change in the gene's location within the genome.

postzygotic isolation mechanism   Factors that prevent or reduce inbreeding by acting after fertilization to produce nonviable, sterile hybrids or hybrids of lowered fitness.

preadaptive mutation   A mutational event that later becomes of adaptive significance.

preformationism   The discredited idea that an organism develops by growth of structures already present in the egg or sperm.

prezygotic isolation mechanism   Factors that reduce inbreeding by preventing courtship, mating, or fertilization.

Pribnow box   A 6-bp sequence 5′ to the beginning of transcription in prokaryotic genes, to which the sigma sub-unit of RNA polymerase binds. The consensus sequence for this box is TATAAT.

primary protein structure   Refers to the sequence of amino acids in a polypeptide chain.

primary sex ratio   Ratio of males to females at fertilization.

primer   In nucleic acids, a short length of RNA or single-stranded DNA that is necessary for the functioning of polymerases.

prion   An infectious pathogenic agent devoid of nucleic acid and composed mainly of a protein, PrP, with a molecular weight of 27,000 to 30,000 daltons. Prions are known to cause scrapie, a degenerative neurological disease in sheep, and are thought to cause similar diseases in humans, such as Kuru and Creutzfeldt-Jakob disease.

probability   Ratio of the frequency of a given event to the frequency of all possible events.

proband   An individual in whom a genetically determined trait of interest is first detected. Also known as a propositus.

probe   A macromolecule such as DNA or RNA that has been labeled and can be detected by an assay such as autoradiography or fluorescence microscopy. Probes are used to identify target molecules, genes, or gene products.

product law   The law that holds that the probability of two independent events occurring simultaneously is the product of their independent probabilities.

proflavin   An acridine dye that acts as a mutagen. See *acridine dyes.*

progeny   The offspring produced from a mating.

prokaryotes   Organisms lacking nuclear membranes, meiosis, and mitosis. Bacteria and blue-green algae are examples of prokaryotic organisms.

promoter   Region having a regulatory function and to which RNA polymerase binds prior to the initiation of transcription.

proofreading   A molecular mechanism for correcting errors in replication, transcription, or translation. Also known as editing.

prophage   A phage genome integrated into a bacterial chromosome. Bacterial cells carrying prophages are said to be lysogenic.

propositus (female, **proposita**)   See *proband.*

protein   A molecule composed of one or more polypeptides, each composed of amino acids covalently linked together.

proto-oncogene   A cellular gene that normally functions to control cellular proliferation. Proto-oncogenes can be converted to oncogenes by alterations in structure or expression.

protoplast   A bacterial or plant cell with the cell wall removed. Sometimes called a spheroplast.

prototroph   A strain (usually microorganisms) that is capable of growth on a defined, minimal medium. Wild-type strains are usually regarded as prototrophs.

pseudoalleles   Genes that behave as alleles to one another by complementation, but that can be separated from one another by recombination.

pseudoautosomal inheritance   Inheritance of alleles located within the regions of the Y chromosome that are homologous to the X chromosome. Because these alleles are located on both the X and Y chromosome, their pattern of inheritance is indistinguishable from that of autosomal inheritance.

pseudodominance   The expression of a recessive allele on one homolog caused by the deletion of the dominant allele on the other homolog.

pseudogene   A nonfunctional gene with sequence homology to a known structural gene present elsewhere in the genome. They differ from their functional relatives by insertions or deletions and by the presence of flanking direct repeat sequences of 10 to 20 nucleotides.

puff   See *chromosome puff.*

punctuated equilibrium   A pattern in the fossil record of brief periods of species divergence punctuated with long periods of species stability.

quantitative inheritance   See *polygenic inheritance.*

quantitative trait loci (QTL)   Two or more genes that act on a single polygenic trait.

quantum speciation   Formation of a new species within a single or a few generations by a combination of selection and drift.

quaternary protein structure   Types and modes of interaction between two or more polypeptide chains within a protein molecule.

R point   The point (also known as the restriction point) during the G1 stage of the cell cycle when either a commitment is made to DNA synthesis and another cell cycle, or the cell withdraws from the cycle and becomes quiescent.

race   A genotypically or geographically distinct subgroup within a species.

rad   A unit of absorbed dose of radiation with an energy equal to 100 ergs per gram of irradiated tissue.

radioactive isotope   One of the forms of an element, differing in atomic weight and possessing an unstable nucleus that emits ionizing radiation during decay.

random amplified polymorphic DNA   A PCR method that uses random primers about 10 nucleotides in length to amplify unknown DNA sequences.

random mating   Mating between individuals without regard to genotype.

reading frame   Linear sequence of codons (groups of three nucleotides) in a nucleic acid.

reannealing   Formation of double-stranded DNA molecules from dissociated single strands.

recessive   Term describing an allele that is not expressed in the heterozygous condition.

reciprocal cross   A paired cross in which the genotype of the female in the first cross is present as the genotype of the male in the second cross, and vice versa.

reciprocal translocation   A chromosomal aberration in which nonhomologous chromosomes exchange parts.

recombinant DNA   A DNA molecule formed by the joining of two heterologous molecules. Usually applied to DNA molecules produced by *in vitro* ligation of DNA from two different organisms.

recombinant gamete   A gamete containing a new combination of genes produced by crossing over during meiosis.

recombination   The process that leads to the formation of new gene combinations on chromosomes.

recon   A term used by Seymour Benzer to denote the smallest genetic units between which recombination can occur.

reductional division   The chromosome division that halves the diploid chromosome number. The first division of meiosis is a reductional division.

redundant genes   Gene sequences present in more than one copy per haploid genome (e.g., ribosomal genes).

regulatory site   A DNA sequence that is involved in the control of expression of other genes, usually involving an interaction with another molecule.

rem   Radiation equivalent in man; the dosage of radiation that will cause the same biological effect as one roentgen of X-rays.

renaturation   The process by which a denatured protein or nucleic acid returns to its normal three-dimensional structure.

repetitive DNA sequences   DNA sequences present in many copies in the haploid genome.

replicating form (RF)   Double-stranded nucleic acid molecules present as an intermediate during the reproduction of certain viruses.

replication   The process of DNA synthesis.

replication fork   The Y-shaped region of a chromosome associated with the site of replication.

replicon   A chromosomal region or free genetic element containing the DNA sequences necessary for the initiation of DNA replication.

replisome   The term used to describe the complex of proteins, including DNA polymerase, that assembles at the bacterial replication fork to synthesize DNA.

repressible enzyme system   An enzyme or group of enzymes whose synthesis is regulated by the intracellular concentration of certain metabolites.

repressor   A protein that binds to a regulatory sequence adjacent to a gene and blocks transcription of the gene.

reproductive isolation   Absence of interbreeding between populations, subspecies, or species. Reproductive isolation can be brought about by extrinsic factors, such as behavior, and intrinsic barriers, such as hybrid inviability.

resistance transfer factor (RTF)   A component of R plasmids that confers the ability for cell-to-cell transfer of the R plasmid by conjugation.

resolution   In an optical system, the shortest distance between two points or lines at which they can be perceived to be two points or lines.

restriction endonuclease   Nuclease that recognizes specific nucleotide sequences in a DNA molecule, and cleaves or nicks the DNA at that site. Derived from a variety of microorganisms, those enzymes that cleave both strands of the DNA are used in the construction of recombinant DNA molecules.

restriction fragment length polymorphism (RFLP) Variation in the length of DNA fragments generated by a restriction endonuclease. These variations are caused by mutations that create or abolish cutting sites for restriction enzymes. RFLPs are inherited in a codominant fashion and can be used as genetic markers.

restrictive condition   Environmental conditions under which a conditional mutation (such as a temperature-sensitive mutant) expresses the mutant phenotype.

restrictive transduction   See *specialized transduction.*

retrovirus   Viruses with RNA as genetic material that utilize the enzyme reverse transcriptase during their life cycle.

reverse transcriptase   A polymerase that uses RNA as a template to transcribe a single-stranded DNA molecule as a product.

reversion   A mutation that restores the wild-type phenotype.

R factor (R plasmid)   Bacterial plasmids that carry antibiotic resistance genes. Most R plasmids have two components: an r-determinant, which carries the antibiotic resistance genes, and the resistance transfer factor (RTF).

RFLP   See *restriction fragment length polymorphism.*

Rh factor   An antigenic system first described in the rhesus monkey. Recessive *r/r* individuals produce no Rh antigens and are Rh negative, while *R/R* and *R/r* individuals have Rh antigens on the surface of their red blood cells and are classified as Rh positive.

ribonucleic acid   A nucleic acid characterized by the sugar ribose and the pyrimidine uracil, usually a single-stranded polynucleotide. Several forms are recognized, including ribosomal RNA, messenger RNA, transfer RNA, and heterogeneous nuclear RNA.

ribose   The five-carbon sugar associated with the ribonucleotides found in RNA.

ribosomal RNA   See *rRNA.*

ribosome   A ribonucleoprotein organelle consisting of two subunits, each containing RNA and protein. Ribosomes are the site of translation of mRNA codons into the amino acid sequence of a polypeptide chain.

RNA   See *ribonucleic acid.*

RNA editing   Alteration of the nucleotide sequence of an mRNA molecule after transcription and before translation. There are two main types of editing: substitution editing, which changes individual nucleotides, and insertion/deletion editing, where individual nucleotides are added or deleted.

RNA polymerase   An enzyme that catalyzes the formation of an RNA polynucleotide strand using the base sequence of a DNA molecule as a template.

RNase   A class of enzymes that hydrolyze RNA molecules.

Robertsonian translocation   A form of chromosomal aberration that involves the fusion of long arms of acrocentric chromosomes at the centromere.

roentgen   A unit of measure of the amount of radiation corresponding to the generation of 2.083×10^9 ion pairs in one cubic centimeter of air at $0°C$ at an atmospheric pressure of 760 mm of mercury. Abbreviated R.

rolling circle model   A model of DNA replication in which the growing point or replication fork rolls around a circular template strand; in each pass around the circle, the newly synthesized strand displaces the strand from the previous replication, producing a series of contiguous copies of the template strand.

rRNA   The RNA molecules that are the structural components of the ribosomal subunits. In prokaryotes, these are the 16*S*, 23*S*, and 5*S* molecules; and in eukaryotes, they are the 18*S*, 28*S*, and 5*S* molecules.

RTF   See *resistance transfer factor.*

S₁ nuclease   A deoxyribonuclease that cuts and degrades single-stranded molecules of DNA.

satellite DNA   DNA that forms a minor band when genomic DNA is centrifuged in a cesium salt gradient. This DNA usually consists of short sequences repeated many times in the genome.

SCE   See *sister chromatid exchange.*

secondary protein structure The alpha helical or pleated-sheet form of a protein molecule brought about by the formation of hydrogen bonds between amino acids.

secondary sex ratio The ratio of males to females at birth.

secretor An individual having soluble forms of the blood group antigens A and/or B present in saliva and other body fluids. This condition is caused by a dominant, autosomal gene unlinked to the *ABO* locus (*I* locus).

sedimentation coefficient See *Svedberg coefficient unit.*

segment polarity genes Genes that regulate the spatial pattern of differentiation within each segment of the developing *Drosophila* embryo.

segregation The separation of homologous chromosomes into different gametes during meiosis.

selection The force that brings about changes in the frequency of alleles and genotypes in populations through differential reproduction.

selection coefficient (*s*) A quantitative measure of the relative fitness of one genotype compared with another.

selfing In plant genetics, the fertilization of ovules of a plant by pollen produced by the same plant. Reproduction by self-fertilization.

semiconservative replication A model of DNA replication in which a double-stranded molecule replicates in such a way that the daughter molecules are composed of one parental (old) and one newly synthesized strand.

semisterility A condition in which a proportion of all zygotes are inviable.

sex chromatin body See *Barr body.*

sex chromosome A chromosome, such as the X or Y in humans, which is involved in sex determination.

sexduction Transmission of chromosomal genes from a donor bacterium to a recipient cell by the F factor.

sex-influenced inheritance Phenotypic expression that is conditioned by the sex of the individual. A heterozygote may express one phenotype in one sex and the alternate phenotype in the other sex.

sex-limited inheritance A trait that is expressed in only one sex even though the trait may not be X-linked.

sex ratio See *primary* and *secondary sex ratio.*

sexual reproduction Reproduction through the fusion of gametes, which are the haploid products of meiosis.

Shine-Dalgarno sequence The nucleotides AGGAGG present in the leader sequence of prokaryotic genes that serve as a ribosome binding site. The 16*S* RNA of the small ribosomal subunit contains a complementary sequence to which the mRNA binds.

short period interspersion Pattern of genome organization in which stretches of single-copy DNA (about 1000 bp) are interspersed with short segments of repetitive DNA (300 bp). This pattern is found in *Xenopus*, humans, and the majority of organisms examined to date.

shotgun experiment The cloning of random fragments of genomic DNA into a vehicle such as a plasmid or phage, usually to produce a bank or library of clones from which clones of specific interest will be selected.

sibling species Species that are morphologically almost identical, but which are reproductively isolated from one another.

sickle-cell anemia A genetic disease in humans caused by an autosomal recessive gene, fatal in the homozygous condition if untreated. Caused by an alteration in the amino acid sequence of the beta chain of globin.

sickle-cell trait The phenotype exhibited by individuals heterozygous for the sickle-cell gene.

sigma factor A polypeptide subunit of the RNA polymerase that recognizes the binding site for the initiation of transcription.

SINEs Short interspersed elements are repetitive sequences found in the genomes of higher organisms, such as the 300-bp *Alu* sequence.

single stranded binding proteins (SSBs) In DNA replication, proteins that bind to and stabilize single stranded regions of DNA that result from the action of unwinding proteins.

sister chromatid exchange (SCE) A crossing over event that can occur in meiotic and mitotic cells; involves the reciprocal exchange of chromosomal material between sister chromatids (joined by a common centromere). Such exchanges can be detected cytologically after BrdU incorporation into the replicating chromosomes.

site-directed mutagenesis A process that uses a synthetic oligonucleotide containing a mutant base or sequence as a primer for inducing a mutation at a specific site in a cloned gene.

small nuclear RNA (snRNA) Species of RNA molecules ranging in size from 90 to 400 nucleotides. The abundant snRNAs are present in 1×10^4 to 1×10^6 copies per cell. snRNAs are associated with proteins and form RNP particles known as snRNPs or snurps. Six uridine-rich snRNAs known as U1–U6 are located in the nucleoplasm, and the complete nucleotide sequence of these is known. snRNAs have been implicated in the processing of pre-mRNA and may have a range of cleavage and ligation functions.

snurps See *small nuclear RNA (snRNA).*

solenoid structure A level of eukaryotic chromosome structure generated by the supercoiling of nucleosomes.

somatic cell genetics The use of cultured somatic cells to investigate genetic phenomena by parasexual techniques.

somatic cells All cells other than the germ cells or gametes in an organism.

somatic mutation A mutational event occurring in a somatic cell. In other words, such mutations are not heritable.

somatic pairing The pairing of homologous chromosomes in somatic cells.

SOS response The induction of enzymes to repair damaged DNA in *E. coli*. The response involves activation of an enzyme that cleaves a repressor, activating a series of genes involved in DNA repair.

spacer DNA DNA sequences found between genes, usually repetitive DNA segments.

special creation An idea that each species originated through a special act of creation by a divine force.

specialized transduction Genetic transfer of only specific host genes by transducing phages.

speciation The process by which new species of plants and animals arise.

species A group of actually or potentially interbreeding individuals that is reproductively isolated from other such groups.

spheroplast See *protoplast*.

spindle fibers Cytoplasmic fibrils formed during cell division that are involved with the separation of chromatids at anaphase and their movement toward opposite poles in the cell.

spliceosome The nuclear macromolecule complex within which splicing reactions occur to remove introns from pre-mRNAs.

spontaneous generation The origin of living systems from nonliving matter.

spontaneous mutation A mutation that is not induced by a mutagenic agent.

spore A unicellular body or cell encased in a protective coat that is produced by some bacteria, plants, and invertebrates; is capable of survival in unfavorable environmental conditions; and can give rise to a new individual upon germination. In plants, spores are the haploid products of meiosis.

SRY A gene (sex-determining region of the Y) found near the pseudoautosomal boundary of the Y chromosome. Accumulated evidence indicates that this gene is the testis-determining factor (TDF).

stabilizing selection Preferential reproduction of those individuals having genotypes close to the mean for the population. A selective elimination of genotypes at both extremes.

standard deviation A quantitative measure of the amount of variation in a sample of measurements from a population.

standard error A quantitative measure of the amount of variation in a sample of measurements from a population.

sterility The condition of being unable to reproduce; free from contaminating microorganisms.

strain A group with common ancestry that has physiological or morphological characteristics of interest for genetic study or domestication.

structural gene A gene that encodes the amino acid sequence of a polypeptide chain.

sublethal gene A mutation causing lowered viability, with death before maturity in less than 50 percent of the individuals carrying the gene.

submetacentric chromosome A chromosome with the centromere placed so that one arm of the chromosome is slightly longer than the other.

subspecies A morphologically or geographically distinct interbreeding population of a species.

sum law The law that holds that the probability of one or the other of two mutually exclusive events occurring is the sum of their individual probabilities.

supercoiled DNA A form of DNA structure in which the helix is coiled upon itself. Such structures can exist in stable forms only when the ends of the DNA are not free, as in a covalently closed circular DNA molecule.

superfemale See *metafemale*.

supermale See *metamale*.

suppressor mutation A mutation that acts to restore (completely or partially) the function lost by a previous mutation at another site.

Svedberg coefficient unit A unit of measure for the rate at which particles (molecules) sediment in a centrifugal field. This unit is a function of several physico-chemical properties, including size and shape. A sedimentation value of 1×10^{-13} sec is defined as one Svedberg coefficient (*S*) unit.

symbiont An organism coexisting in a mutually beneficial relationship with another organism.

sympatric speciation Process of speciation involving populations that inhabit, at least in part, the same geographic range.

synapsis The pairing of homologous chromosomes at meiosis.

synaptonemal complex (SC)   An organelle consisting of a tripartite nucleoprotein ribbon that forms between the paired homologous chromosomes in the pachytene stage of the first meiotic division.

syndrome   A group of signs or symptoms that occur together and characterize a disease or abnormality.

synkaryon   The nucleus of a zygote that results from the fusion of two gametic nuclei. Also used in somatic cell genetics to describe the product of nuclear fusion.

syntenic test   In somatic cell genetics, a method for determining whether or not two genes are on the same chromosome.

T_m   The temperature at which a population of double-stranded nucleic acid molecules is half-dissociated into single strands. This is taken to be the melting temperature for that species of nucleic acid.

target theory   In radiation biology, a theory stating that damage and death from radiation is caused by the inactivation of specific targets within the organism.

TATA box   See *Goldberg–Hogness box.*

tautomeric shift   A reversible isomerization in a molecule brought about by a shift in the localization of a hydrogen atom. In nucleic acids, tautomeric shifts in the bases of nucleotides can cause changes in other bases at replication and are a source of mutations.

TDF (testis-determining factor)   The product of a gene on the Y chromosome that controls the developmental switch point for the development of the indifferent gonad into a testis. See *SRY.*

telocentric chromosome   A chromosome in which the centromere is located at the end of the chromosome.

telomerase   The enzyme that adds short, tandemly repeated DNA sequences to the ends of eukaryotic chromosomes.

telomere   The terminal chromomere of a chromosome.

telophase   The stage of cell division in which the daughter chromosomes reach the opposite poles of the cell and reform nuclei. Telophase ends with the completion of cytokinesis.

telophase I   In the first meiotic division, when duplicated chromosomes reach the poles of the dividing cell.

temperate phage   A bacteriophage that can become a prophage and confer lysogeny upon the host bacterial cell.

temperature-sensitive mutation   A conditional mutation that produces a mutant phenotype at one temperature range and a wild-type phenotype at another temperature range.

template   The single-stranded DNA or RNA molecule that specifies the nucleotide sequence of a strand synthesized by a polymerase molecule.

teratocarcinoma   Embryonal tumors that arise in the yolk sac or gonads and are able to undergo differentiation into a wide variety of cell types. These tumors are used to investigate the regulatory mechanisms underlying development.

terminalization   The movement of chiasmata toward the ends of chromosomes during the diplotene stage of the first meiotic division.

tertiary protein structure   The three-dimensional structure of a polypeptide chain brought about by folding upon itself.

test cross   A cross between an individual whose genotype at one or more loci may be unknown and an individual who is homozygous recessive for the genes in question.

tetrad   The four chromatids that make up paired homologs in the prophase of the first meiotic division. The four haploid cells produced by a single meiotic division.

tetrad analysis   Method for the analysis of gene linkage and recombination using the four haploid cells produced in a single meiotic division.

tetranucleotide hypothesis   An early theory of DNA structure proposing that the molecule was composed of repeating units, each consisting of the four nucleotides containing adenine, thymine, cytosine, and guanine.

tetraparental mouse   A mouse produced from an embryo that was derived by the fusion of two separate blastulas.

theta structure   An intermediate in the bidirectional replication of circular DNA molecules. At about midway through the cycle of replication, the intermediate resembles the Greek letter theta.

thymine dimer   A pair of adjacent thymine bases in a single polynucleotide strand between which chemical bonds have formed. This lesion, usually the result of damage caused by exposure to ultraviolet light, inhibits DNA replication unless repaired by the appropriate enzyme system.

topoisomerase   A class of enzymes that convert DNA from one topological form to another. During DNA replication, these enzymes facilitate the unwinding of the double-helical structure of DNA.

totipotent   The ability of a cell or embryo part to give rise to all adult structures. This capacity is usually progressively restricted during development.

trailer sequence   A transcribed but nontranslated region of a gene or its mRNA that follows the termination signal.

trait   Any detectable phenotypic variation of a particular inherited character.

trans **configuration**   The arrangement of two mutant sites on opposite homologs, such as

$$\frac{a^1 \; +}{+ \; a^2}$$

Contrasts with a *cis* arrangement, where they are located on the same homolog.

transcription Transfer of genetic information from DNA by the synthesis of an RNA molecule copied from a DNA template.

transdetermination Change in developmental fate of a cell or group of cells.

transduction Virally mediated genes transfer from one bacterium to another, or the transfer of eukaryotic genes mediated by retrovirus.

transfer RNA See *tRNA*.

transformation Heritable change in a cell or an organism brought about by exogenous DNA.

transgenic organism An organism whose genome has been modified by the introduction of external DNA sequences into the germ line.

transition A mutational event in which one purine is replaced by another, or one pyrimidine is replaced by another.

translation The derivation of the amino acid sequence of a polypeptide from the base sequence of an mRNA molecule in association with a ribosome.

translocation A chromosomal mutation associated with the transfer of a chromosomal segment from one chromosome to another. Also used to denote the movement of mRNA through the ribosome during translation.

transmission genetics The field of genetics concerned with the mechanisms by which genes are transferred from parent to offspring.

transposable element A defined length of DNA that translocates to other sites in the genome, essentially independent of sequence homology. Usually such elements are flanked by short, inverted repeats of 20 to 40 base pairs at each end. Insertion into a structural gene can produce a mutant phenotype. Insertion and excision of transposable elements depends on two enzymes, transposase and resolvase. Such elements have been identified in both prokaryotes and eukaryotes.

transversion A mutational event in which a purine is replaced by a pyrimidine, or a pyrimidine is replaced by a purine.

trinucleotide repeat A tandemly repeated cluster of three nucleotides (such as CTG) in or near a gene, that undergo an expansion in copy number, resulting in a disease phenotype.

triploidy The condition in which a cell or organism possesses three haploid sets of chromosomes.

trisomy The condition in which a cell or organism possesses two copies of each chromosome, except for one, which is present in three copies. The general form for trisomy is therefore $2n + 1$.

tritium (^{3}H) A radioactive isotope of hydrogen, with a half-life of 12.36 years.

trivalent An association between three homologous chromosomes.

tRNA Transfer RNA; a small ribonucleic acid molecule that contains a three-base segment (anticodon) that recognizes a codon in mRNA, a binding site for a specific amino acid, and recognition sites for interaction with the ribosomes and the enzyme that links it to its specific amino acid.

tumor suppressor gene A gene that encodes a gene product that normally functions to suppress cell division. Mutations in tumor suppressor genes result in the activation of cell division and tumor formation.

Turner syndrome A genetic condition in human females caused by a 45,X genotype (XO). Such individuals are phenotypically female but are sterile because of undeveloped ovaries.

unequal crossing over A crossover between two improperly aligned homologs, producing one homolog with three copies of a region and the other with one copy of that region.

unique DNA DNA sequences that are present only once per genome. Single copy DNA.

universal code The assumption that the genetic code is used by all life forms. In general, this is true; some exceptions are found in mitochondria, ciliates, and mycoplasmas.

unwinding proteins Nuclear proteins that act during DNA replication to destabilize and unwind the DNA helix ahead of the replicating fork.

up promoter A promoter sequence, often mutant, that increases the rate of transcription initiation. Also known as strong promoter.

variable number tandem repeats (VNTRs) Short repeated DNA sequences (2–20 nucleotides) present as tandem repeats between two restriction enzyme sites. Variations in the number of repeats creates DNA fragments of differing lengths following restriction enzyme digestion.

variable region Portion of an immunoglobulin molecule that exhibits many amino acid sequence differences between antibodies of differing specificities.

variance A statistical measure of the variation of values from a central value, calculated as the square of the standard deviation.

variegation Patches of differing phenotypes, such as color, in a tissue.

vector In recombinant DNA, an agent such as a phage or plasmid into which a foreign DNA segment will be inserted.

viability The measure of the number of individuals in a given phenotypic class that survive, relative to another class (usually wild type).

virulent phage A bacteriophage that infects and lyses the host bacterial cell.

VNTR See *variable number tandem repeats.*

W, Z chromosomes Sex chromosomes in species where the female is the heterogametic sex (WZ).

western blot A technique in which proteins are separated by gel electrophoresis and transferred by capillary action to a nylon membrane or nitrocellulose sheet. A specific protein can be identified through hybridization to a labeled antibody.

wild type The most commonly observed phenotype or genotype, designated as the norm or standard.

wobble hypothesis An idea proposed by Francis Crick stating that the third base in an anticodon can align in several ways to allow it to recognize more than one base in the codons of mRNA.

writhing number The number of times that the axis of a DNA duplex crosses itself by supercoiling.

X inactivation In mammalian females, the random cessation of transcriptional activity of one X chromosome. This event, which occurs early in development, is a mechanism of dosage compensation. Molecular basis of inactivation is unknown, but involves a region called the X-inactivation center (XIC) on the proximal end of the p arm. Some loci on the tip of the short arm of the X can escape inactivation. See *Barr body, Lyon hypothesis.*

XIST A locus in the X-chromosome inactivation center that may control inactivation of the X chromosome in mammalian females.

X linkage The pattern of inheritance resulting from genes located on the X chromosome.

X-ray crystallography A technique to determine the three-dimensional structure of molecules through diffraction patterns produced by X-ray scattering by crystals of the molecule under study.

YAC A cloning vector in the form of a yeast artificial chromosome, constructed using chromosomal elements including telomeres (from a ciliate), centromeres, origin of replication, and marker genes from yeast. YACs are used to clone long stretches of eukaryotic DNA.

Y chromosome Sex chromosome in species where the male is heterogametic (XY).

Y linkage Mode of inheritance shown by genes located on the Y chromosome.

Z-DNA An alternative structure of DNA in which the two antiparallel polynucleotide chains form a left-handed double helix. Z-DNA has been shown to be present along with B-DNA in chromosomes and may have a role in regulation of gene expression.

zein Principal storage protein of corn endosperm, consisting of two major proteins, with molecular weights of 19,000 and 21,000 daltons.

zinc finger A DNA-binding domain of a protein that has a characteristic pattern of cysteine and histidine residues that complex with zinc ions, throwing intermediate amino acid residues into a series of loops or fingers.

zygote The diploid cell produced by the fusion of haploid gametic nuclei.

zygotene A stage of meiotic prophase I in which the homologous chromosomes synapse and pair along their entire length, forming bivalents. The synaptonemal complex forms at this stage.

Solutions to Problems and Discussion Questions

Chapter 1

2. *Pangenesis* refers to a theory that various parts of the body contain "humors" which bear the hereditary traits and gather in the reproductive organs. *Epigenesis* refers to the theory that organisms are derived from the assembly and reorganization of substances in the egg which eventually lead to the development of the adult. *Preformationism* is a seventeenth century theory which states that the sex cells (eggs or sperm) contain miniature adults, called homunculi.

4. Their theory of natural selection proposed that more offspring are produced than can survive, and that in the competition for survival, those with favorable variations survive. Darwin did not understand the nature of heredity and variation which led him to lean toward older theories of pangenesis and inheritance of acquired characteristics.

8. Norman Borlaug applied Mendelian principles of hybridization and trait selection to the development of superior varieties of wheat. Such varieties are now grown in many countries, including Mexico, and have helped maintain the world supply of food. This change in worldwide agricultural food production has been called the "Green Revolution."

10. In the last 40 years, human transmission, cytological and molecular genetics have provided an understanding of many aspects of both plant and animal biology including development of pest resistant crops and identification of hazardous organisms in our food (*E. coli*, for example). In addition, much has been learned about many human diseases. There is promise that a certain amount of human suffering will be minimized by the application of genetics to crop production (disease resistance, protein content, growth conditions) and medicine. Major medical areas of activity include genetic counseling, gene mapping and identification, disease diagnosis, and genetic engineering.

Chapter 2

2. Chromosomes which are homologous share many properties including *overall length, position of the centromere, banding patterns, type and location of genes, autoradiographic pattern.*

6. Refer to K/C figures for an explanation. Notice the different anaphase shapes of chromosomes as they are moving to the poles.

8. Regulatory checkpoints include those in G1/S, G2/M, and mitosis. *Autoradiography* is a technique which can be used to determine that there is DNA synthesis during the interphase.

10. Meiosis provides for a reduction in chromosome number and an opportunity for exchange of genetic material from homologous chromosomes.

12. Sister chromatids are genetically identical, except where mutations may have occurred during DNA replication. Nonsister chromatids are genetically similar if on homologous chromosomes or genetically dissimilar if on nonhomologous chromosomes. If crossing over occurs, then chromatids attached to the same centromere will no longer be identical.

14. (a) 8 tetrads **(b)** 8 dyads **(c)** 8 monads migrating to *each* pole **(d)** $(1/2)^8$

16. Genetic variation occurs through independent assortment of chromosomes at anaphase I of meiosis and crossing over.

18. One half of each tetrad will have a maternal homologue: $(1/2)^{10}$.

20. In angiosperms, meiosis results in the formation of microspores (male) and megaspores (female) which give rise to the haploid male and female gametophyte stage. Micro- and megagametophytes produce the pollen and the ovules respectively. Following fertilization, the sporophyte is formed.

22. The transition occurs at the end of interphase and the beginning of mitosis (prophase) when the chromosomes are in the condensation process.

24. *cdc* mutations are those involved in regulating the *c*ell *d*ivision *c*ycle.

26. (a) Duplicated chromosomes A^m, A^p, B^m, B^p, C^m, and C^p will align at metaphase, with the centromeres dividing and sister chromatids going to opposite poles at anaphase.

(b) Side-by-side alignment of A^mA^p, B^mB^p, C^mC^p will occur in various arrangements.

(c,d) Eight possible combinations of meiotic II products will occur.

(e) After meiosis I, the two product cells would be as follows: A^m or A^p, B^m or B^p, C^m *and* C^p; A^m or A^p, B^m or B^p, no C^m or C^p.

(f) Two products will be trisomic for chromosome C and disomic for chromosomes A and B. Two products will be monosomic for chromosome C and disomic for chromosomes A and B.

Chapter 3

2. (a) The parents must both be heterozygous (*Aa*).

(b) The normal male could have either the *AA* or *Aa* genotype. The female must be *aa*. Since all the children are normal one would consider the male to be *AA* instead of *Aa*. However, the male *could* be *Aa*. Under that circumstance, the likelihood of having six children, all normal is 1/64.

(c) The normal male could have either the *AA* or *Aa* genotype. The female must be *aa*. The *Aa* male is mated to the *aa* female.

(d)

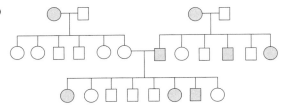

6.

	P_1 Cross	Checkered	Plain
(a)	$PP \times PP$	PP	
(b)	$PP \times pp$	Pp	
(c)	$pp \times pp$		pp
(d)	$PP \times pp$	Pp	
(e)	$Pp \times pp$	Pp	pp
(f)	$Pp \times Pp$	PP, Pp	pp
(g)	$PP \times Pp$	PP, Pp	

8. $WWgg = 1/16$

10. (d)

14. Homozygosity refers to a condition where both genes of a pair are the same (i.e. *AA* or *GG* or *bb*) whereas heterozygosity refers to the condition where members of a gene pair are different (i.e. *Aa* or *Gg* or *Bb*).

16. The general formula for determining the number of kinds of gametes produced by an organism is 2^n where n = number of *heterozygous* gene pairs.

- (a) *4: AB, Ab, aB, ab*
- (b) *2: AB, aB*
- (c) *8: ABC, ABc, AbC, Abc, aBC, aBc, abC, abc*
- (d) *2: ABc, aBc*
- (e) *4: ABc, Abc, aBc, abc*
- (f) $2^5 = 32$

18.

	Phenotypes	Genotypes
P_1:	Yellow × green	$GG \times gg$
F_1:	all yellow	Gg
F_2:	6022 yellow	1/4 GG; 2/4 Gg
	2001 green	1/4 gg

20. 1:1:1:1

22. (a) $\chi^2 = .064$, (*P*) value between 0.9 and 0.5. **(b)** $\chi^2 = 0.39$, *P* value in the table for 1 degree of freedom is still between 0.9 and 0.5.

24. For the test of a 3:1 ratio, the χ^2 value is 33.3 with an associated P value of less than 0.01 for 1 degree of freedom. For the test of a 1:1 ratio, the χ^2 value is 25.0 again with an associated P value of less than 0.01 for 1 degree of freedom. Based on these probability values, both null hypotheses should be rejected.

26. I-1 (*Aa*), I-2 (*aa*), I-3 (*Aa*), I-4(*Aa*)

II-1 (*aa*), II-2 (*Aa*), II-3 (*aa*), II-4 (*Aa*), II-5 (*Aa*), II-6 (*aa*), II-7 (*AA* or *Aa*), II-8 (*AA* or *Aa*)

III-1 (*AA* or *Aa*), III-2 (*AA* or *Aa*), III-3 (*AA* or *Aa*), III-4 (*aa*), III-5 (probably *AA*), III-6 (*aa*)

IV-1 through IV-7 all *Aa*.

28. (a) There are two possibilities. Either the trait is dominant, in which case I-1 is heterozygous as are II-2 and II-3, or the trait is recessive and I-1 is homozygous and I-2 is heterozygous recessive. Under the condition of recessiveness, both II-1 and II-4 would be heterozygous, II-2 and II-3 homozygous.

(b) Recessive: Parents *Aa, Aa*

(c) Recessive: Parents *Aa, Aa*

(d) Recessive: Parents *AA* (probably), *aa*; Second pedigree? Recessive or dominant, not sex-linked, if recessive, parents *Aa, aa*

(e) See initial explanation in this problem. It is identical to the first pedigree.

30. zero

32. (a) 1/6 **(b)** 1/36 **(c)** 1/18 **(d)** 1/3

34. $2/3 \times 2/3 \times 1/4 = 1/9$

36. $P = \dfrac{8!}{6!2!}{}^{(3/4)^6(1/4)^2}$

38. (a) (one of several possibilities):

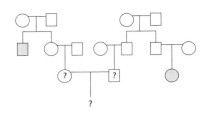

(b) 1/12
(c) 6/12
(d) 5/12

40. Because there are three possibilities within the group, it probably represents a case of incomplete dominance or codominance.

Chapter 4

2. *Incomplete dominance* can be viewed more as a quantitative phenomenon where the heterozygote is intermediate (approximately) between the limits set by the homozygotes. *Codominance* can be viewed in a more qualitative manner where both of the alleles in the heterozygote are expressed.

4.   Cross 1:

    Ss × ss →

    1/2 *Ss* (short), 1/2 *ss* (long)

    Cross 2:

    Ss × Ss →

    1/4 *SS* (lethal), 2/4 *Ss* (short), 1/4 *ss* (long)

6. Male Parent: $I^B I^O$ Female Parent: $I^A I^O$
The ratio in offspring would be
1(A):1(B):1(AB):1(O).

8. (a) 3/8 A, secretor
1/8 A, nonsecretor
3/8 B, secretor
1/8 B, nonsecretor

(b) 1/4 of all individuals will have blood type O.

10.(a) $c^a c^a \times c^{cb} c^a \rightarrow$ 1/2 chinchilla; 1/2 albino

(b) $c^a c^a \times Cc^a \rightarrow$ 1/2 full color; 1/2 albino

(c) $c^b c^a \times c^b c^a \rightarrow$ 3/4 Himalayan; 1/4 albino

12. (a) all

(b) 18/64

14. (a) $C^{cb}C^{cb}$ = chestnut, $C^c C^c$ = cremello, $C^{cb}C^c$ = palomino

(b) The F_1 resulting from matings between cremello and chestnut horses would be expected to be all palomino. The F_2 would be expected to fall in a 1:2:1 ratio as in the third cross in part (a) above. The final ratio is 9:3:4.

16. (a) 9:3:4 final ratio agouti, black, albino.

(b) Results of crosses of female agouti
(1) $AACc$ (2) $AaCC$ (3) $AaCc$

18. (a) $AaBbCc \rightarrow$ gray (C allows pigment)

(b) $A\_B\_Cc \rightarrow$ gray (C allows pigment)

(c) 16/32 albino;
9/32 gray;
3/32 yellow;
3/32 black;
1/32 cream

(d) 9/16 (gray);
3/16 (black);
4/16 (albino)

(e) 3/8 (grey);
1/8 (yellow);
4/8 (albino)

20. (a) Cross A:

P_1: $AABB \times aaBB$
F_1: $AaBB$
F_2: 3/4 A_BB: 1/4 $aaBB$

Cross B:

P_1: $AABB \times AAbb$
F_1: $AABb$
F_2: 3/4 $AAB\_$: 1/4 $AAbb$

Cross C:

P_1: $aaBB \times AAbb$
F_1: $AaBb$
F_2: 9/16 $A\_B\_$: 3/16 A_bb:
3/16 $aaB\_$: 1/16 $aabb$

(b) The genotype of the unknown P_1 individual would be $AAbb$ (brown) while the F_1 would be $AaBb$ (green).

22. (a)

$A\_B\_$ = 9/16 (yellow)
A_bb = 3/16 (blue)
$aaB\_$ = 3/16 (red)
$aabb$ = 1/16 (mauve)

(b) $aaBb \times aaBb \rightarrow$

3/4 $aaB\_$ (red): 1/4 $aabb$ (mauve)

24. (a) 1/4

(b) 1/2

(c) 1/4

(d) zero

26. Symbolism:
Normal wing margins = sd^+;
scalloped = sd

(a) P1: $X^{sd}X^{sd} \times X^+/Y \rightarrow$
F_1:
1/2 X^+X^{sd} (female, normal)
1/2 X^{sd}/Y (male, scalloped)
F_2:
1/4 X^+X^{sd} (female, normal)
1/4 $X^{sd}X^{sd}$ (female, scalloped)
1/4 X^+/Y (male, normal)
1/4 X^{sd}/Y (male, scalloped)

(b) P1: $X^+/X^+ \times X^{sd}/Y \rightarrow$
F_1:
1/2 X^+X^{sd} (female, normal)
1/2 X^+/Y (male, normal)
F_2:
1/4 X^+X^+ (female, normal)
1/4 X^+X^{sd} (female, normal)
1/4 X^+/Y (male, normal)
1/4 X^{sd}/Y (male, scalloped)

If the *scalloped* gene were not X-linked, then all of the F_1 offspring would be wild (phenotypically) and a 3:1 ratio of normal to scalloped would occur in the F_2.

28. Set up the symbolism and the cross in the following manner:

P_1: $X^+X^+; su\text{-}v/su\text{-}v \times X^v/Y; su\text{-}v^+/su\text{-}v^+ \rightarrow$
F_1:
1/2 $X^+ X^v; su\text{-}v^+/su\text{-}v$ (female, normal)
1/2 $X^+/Y; su\text{-}v^+/su\text{-}v$ (male, normal)

F_2:
8/16 wild type females;
5/16 wild type males;
3/16 vermilion males;

30. (a) $w/w; se^+/se^+ \times w^+/Y; se/se \rightarrow$
F_1: $w^+/w; se^+/se$ = wild females
$w/Y; se^+/se$ = white-eyed males

F₂:

    3/16 males wild

    4/16 males white

    1/16 males sepia

    3/16 females wild

    4/16 females white

    1/16 females sepia

(b) $w^+/w^+;se/se \times w/Y;se^+/se^+ \rightarrow$

F₁: $w^+/w;se^+/se$ = wild females

       $w^+/Y;se^+/se$ = wild males

F₂:

    3/16 males wild

    4/16 males white

    1/16 males sepia

    6/16 females wild

    2/16 females sepia

    2/16 females sepia

32. P₁: female: *RR* (red) × male: *rr* (mahogany)

F₁:

    Rr = females red; males mahogany

    1/2 females (red)

    1/2 males (mahogany)

	1/2 females	1/2 males
1/4 *RR*	1/8 red	1/8 red
2/4 *Rr*	2/8 red	2/8 mahogany
1/4 *rr*	1/8 mahogany	1/8 mahogany

34. The fact that the father in couple #2 has hemophilia would not predispose his son to hemophilia. The #1 couple has no valid claim.

36. *Penetrance* refers to the percentage of individuals which express the mutant phenotype while *expressivity* refers to the range of expression of a given phenotype.

38. (a) *AAB- × aaBB* (other configurations possible but each must give all offspring with *A* and *B* dominant alleles)

(b) *AaB- × aaBB* (other configurations are possible but no *bb* types can be produced)

(c) *AABb × aaBb*

(d) *AABB × aabb*

(e) *AaBb × Aabb*

(f) *AaBb × aabb*

(g) *aaBb × aaBb*

(h) *AaBb × AaBb*

Those genotypes which will breed true will be as follows:

    black = *AABB*

    golden = all genotypes which are *bb*

    brown = *aaBB*

40. A 27(purple):37(white) ratio fits well. To test this hypothesis, one might take the purple F₁'s and cross them to the pure breeding (*aabbcc*) white type. Such a cross should give a 1(purple):7(white).

42. (a) Because the denominator in the ratio is 64 one would begin to consider that there are three independently assorting gene pairs operating in this problem. Because there are only two characteristics (eye color and croaking) however, one might hypothesize that two gene pairs are involved in the inheritance of one trait while one gene pair is involved in the other.

(b) Because of these relationships one would conclude that croaking is due to one gene pair while eye color is due to two gene pairs. Because there is a (9:4:3) ratio regarding eye color, some gene interaction (epistasis) is indicated.

(c,d) Symbolism:

Croaking: *R_* = rib-it; *rr* = knee-deep

Eye color: Since the most frequent phenotype is blue eye, let *A_B_* represent the genotypes. For the purple class, "a 3/16 group" use the *A_bb* genotypes. The "4/16" class (green) would be the *aaB_* and the *aabb* groups.

(e)   *AAB_R_* = blue-eyed, rib-it

      AAB_rr = blue-eyed, knee-deep

      *AAbbR_* = purple-eyed, rib-it

      AAbbrr = purple-eyed, knee-deep

(f) The following genotypes can define the green phenotype: *aaBB, aaBb, aabb*

(g) Both parents would be *AabbRr*.

44. Given that both parents are true-breeding and the sort of gene interaction described is occurring, one can come up with the following symbols:

(a) P₁: *YYBB × yWbb*

F₁: *YyBb* and *YWBb*

Crossing these F₁'s gives the observed ratios in the F₂.

(b) A blue male with the genotype *yyBb* and a green female with the genotype *YWBb*.

Chapter 5

4. (a) It *is possible* that two parents of moderate height can produce offspring that are much taller or shorter than either parent because segregation can produce a variety of gametes. Therefore offspring as illustrated below:

$$\frac{rrSsTtuu}{(\text{moderate})} \times \frac{RrSsTtUu}{(\text{moderate})}X$$

Offspring from this cross can range from very tall *RrSSTTUu* (12 "tall" units) to very short *rrssttuu* (8 "small" units).

(b) If the individual with a minimum height, *rrssttuu*, is married to an individual of intermediate height *RrSsTtUu*, the offspring can be no taller than the height of the tallest parent.

6. (a) Because the extreme phenotypes (6cm and 30cm) each represent 1/64 of the total, it is likely that there are three gene pairs in this cross. The genotypes of the parents would be combinations of alleles which would produce a 6cm (*aabbcc*) tail and a

30cm (*AABBCC*) tail while the 18cm offspring would have a genotype of *AaBbCc*.

(b) 1:3:3:1

10. (a) 140 cm, **(b)** 374.18, **(c)** 19.34, **(d)** 0.70

The plot approximates a normal distribution. Variation is continuous.

12. 0.53.

14. h² = (7.5 − 8.5/6.0 − 8.5) = 0.4

Selection will have little relative influence on olfactory learning in *Drosophila*.

16. Chromosome 2 seems to confer considerable resistance to the insecticide, somewhat in the heterozygous state and more so in the homozygous state. Thus, some partial dominance is occurring.

Chapter 6

2. It is likely that the side-by-side pairing which occurs during synapsis is the earliest time during the cell cycle that chromosomes achieve that necessary proximity. Chiasmata are visible during prophase I of meiosis and it is likely that these structures are intimately associated with the genetic event of crossing over.

4. Because crossing over occurs at the four-strand stage of the cell cycle (that is, after S phase) notice that each single crossover involves only two of the four chromatids.

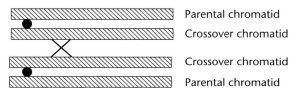

Parental chromatid
Crossover chromatid
Crossover chromatid
Parental chromatid

6. Interference is often explained by a physical rigidity of chromatids such that they are unlikely to make sufficiently sharp bends to allow crossovers to be close together.

8. Since the distance between *dp* and *ap* is greatest, they must be on the "outside" and *cl* must be in the middle. The genetic map would be as follows:

dp—dp———————ap
3mu.   39mu.

10. *Coupled RY/ry × ry/ry* with the two dominant genes on one chromosome and the two recessives on the homologue. The map distance would be 10 map units (20/200 × 100 to convert to percentages) between the *R* and *Y* loci.

12. The original parental arrangement in the test cross was *PZ/pz × pz/pz*, with 14 map units between the two genes.

14.

	female A:	female B:	Frequency:
NCO	3, 4	7, 8	first
SCO	1, 2	3, 4	second
SCO	7, 8	5, 6	third
DCO	6, 5	1, 2	fourth

16.

(a) *y w +/+ + ct × y w +/Y*

(b)

y —————— w ————————————— ct
0.0   1.5   20.0

(c) There were .185 × .015 × 1000 = 2.775 double crossovers expected.

(d) It would not be possible to unequivocally determine the genotypes from the F₂ phenotypes for all classes.

18.

0.20	wild type
0.05	ebony
0.05	pink
0.20	pink, ebony
0.20	dumpy
0.05	dumpy, ebony
0.05	dumpy, pink
0.20	dumpy, pink, ebony

For the reciprocal cross there would be no crossover classes.

.25	wild type
.25	pink, ebony
.25	dumpy
.25	dumpy, pink, ebony

The results would change as a result of no crossing over in males.

20. (a,b) *a − b* = 7 map units

b − c = 2 map units

(c) The progeny phenotypes that are missing are + + *c* and *a b*+, which, of 1000 offspring, 1.4 (.07 × .02 ×1000) would be expected.

22. Because sister chromatids are genetically identical (with the exception of rare new mutations) crossing over between sisters provides no increase in genetic variability. Individual genetic variability could be generated by somatic crossing over because certain patches on the individual would be genetically different from other regions. This variability would be of only minor consequence in all likelihood. Somatic crossing over would have no influence on the offspring produced.

24. (a) 2ⁿ = 8 genotypic and phenotypic classes and they would occur in a 1:1:1:1:1:1:1:1 ratio.

(b) There would be two classes and they would occur in a 1:1 ratio.

(c) There are 20 map units between the *A* and *B* loci and locus *C* assorts independently from both *A* and *B* loci.

26. Progeny A: *Ro/rO × rroo* = 10 map units

Progeny B: *RO/ro × rroo* = 10 map units

28. 10 map units

30. For Cross 1, 50 map units. Because there are 50 map units between genes *a* and *b*, they are not linked. For Cross 2, 12 map units. Because genes *a* and *b* are not linked they could be on non-homologous chromosomes or far apart (50 map units or more) on the same chromosome. Because genes *c* and *b* are linked and therefore on the same chromosome, it is also possible that genes *a* and *c* are on different chromosome pairs. Under that condition, the NP and P (parental ditypes) would be equal; however, there is a possibility that the following arrangement occurs and that genes *a* and *c* are linked.

32. (a)

Tetrad in Problem	Class
1	NP
2	T
3	P
4	NP
5	T
6	P
7	T

(b) P = 44 and NP = 2, therefore the genes are linked.
(c) For the centromere to *c* distance: 7.2 map units
For the centromere to *d* distance: 15.9 map units
The map would be, according to these figures:

$$C \underset{<7.2>}{—c—} \underset{<15.9>}{————} d$$

or

$$c \underset{<7.2>}{—C—} \underset{<15.9>}{————} d$$

Since the likelihood of multiple crossovers decreases as the number of crossovers increases, it would seem reasonable that the genes are on the same side of the centromere.
(d) 20 map units
(e)

$$C \underset{<7.9>}{—c—} \underset{<16.4>}{————} d$$

The discrepancy between the two mapping systems is caused by the manner in which first and second division segregation products are scored. For instance, in tetrad arrangement #1 there are actually two crossovers between the *d* gene and the centromere, but it is still scored as a first division segregation. In tetrad arrangement #4, three crossovers occur between the *d* gene and the centromere but they are scored as one. If one draws out all the crossovers needed to produce the tetrad arrangements in this problem it would become clear that there are many crossovers between the *d* gene and the centromere which go undetected in the scoring of the arrangements of the *d* gene itself. This will cause one to underestimate the distance and give the discrepan-

cy noted. One could account for these additional crossover classes to make the map more accurate.
(f)

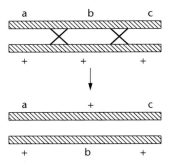

34. *ENO1:* chromosome 1
MDH1: chromosome 2
PEPS: chromosome 4
PMG1: chromosome 1

36. What you will have is the following gametes from the females (left) and males (right):

bw+ *st*+ 1/4	*bw*+ *st*+ 1/2
bw+ *st* 1/4	*bw st* 1/2
bw st+ 1/4	
bw st 1/4	

Combining these gametes will give the ratio presented in the table of results.

Chapter 7

2. (a) The requirement for physical contact between bacterial cells during conjugation was established by placing a filter in a U-tube such that the medium can be exchanged but the bacteria can not come in contact.
(b) By treating cells with streptomycin, an antibiotic, it was shown that recombination would not occur if one of the two bacterial strains was inactivated. However, if the other was similarly treated, recombination would occur. Thus, directionality was suggested, with one strain being a donor strain and the other being the recipient.
(c) An F+ bacterium contains a circular, double-stranded, structurally independent, DNA molecule which can direct recombination.
4. The length of the chromosome being transferred is contingent on the duration of conjugation, thus mapping of genes is based on time.
6. The F+ element can enter the host bacterial chromosome and upon returning to its independent state, it may pick up a piece of a bacterial chromosome. When combined with a bacterium with a complete chromosome, a partial diploid, or merozygote, is formed.

8. One would consider no linkage between these two loci.

10. In their experiment a filter was placed between the two auxotrophic strains which would not allow contact. F-mediated conjugation requires contact and without that contact, such conjugation can not occur. The treatment with DNase showed that the filterable agent was not naked DNA.

12. In *generalized transduction* virtually any genetic element from a host strain may be included in the phage coat and thereby be transduced. In *specialized (restricted) transduction* only those genetic elements of the host which are closely linked to the insertion point of the phage can be transduced.

Because only certain genetic elements are involved in specialized transduction, it is not useful in determining linkage relationships. Cotransduction of genes in generalized transduction allows linkage relationships to be determined.

14. Viral recombination occurs when there is a sufficiently high number of infecting viruses so that there is a high likelihood that more than one type of phage will infect a given bacterium.

16. (a) The concentration of phage is greater than 10^4.

(b) The concentration of phage is around 1.4×10^6.

(c) The concentration of phage is less than 10^6.

18. *Group A: 1,4,5   Group B: 2,3*

(a) 2×3 = no lysis;

2×4 = lysis;

3×4 = lysis.

(b) Because mutant 5 failed to complement with mutations in either cistron, it probably represents a major alteration in the gene such that both cistrons are altered. A deletion which overlaps both cistrons could cause such a major alteration.

(c) 5×10^{-4}

(d) Because mutant 6 complemented mutations 2 and 3, it is likely to be in the cistron with mutants 1, 4, and 5. A lack of recombinants with mutant 4 indicates that mutant 6 is a deletion which overlaps mutation 4. Recombinants with 1 and 5 indicate that the deletion does not overlap these mutations.

20. (a)

Combination	Complementation
1, 2	−
1, 3	+
2, 4	+
4, 5	−

(b) The recombination frequency would be 4×10^{-5}.

(c) The dilution would be 10^{-3} and the colony number would be 8×10^3.

(b) Mutant 7 might well be a deletion spanning parts of both A and B cistrons.

22. (a) Rifampicin eliminates the donor strain which is *rif^s*.

(b) *b a   c   F*

(c) To determine the location of the *rif* gene one could use a donor strain which was *rif^r* but sensitive to another antibiotic (ampicillan for example). The interrupted mating experiment is conducted as usual on an ampicillin-containing medium but the recombinants must be replated on a rifampicin medium to determine which ones are sensitive.

Chapter 8

2. In the reciprocal cross, with the *mt$^+$* strain being streptomycin sensitive, the nuclear genome of half of the offspring will contain a streptomycin resistance gene and therefore be resistant.

4. (a) neutral

(b) segregational (nuclear mutations)

(c) suppressive

6. One would expect, in the diploid zygote, the *segregational* allele to be "covered" by normal alleles from the neutral strain. As the nuclear genes are again "exposed" in the haploid state of the ascospores, one would expect a 1:1 ratio of normals to petites.

8. (a) It is likely that mitochondria and chloroplasts evolved from bacteria in a symbiotic relationship, therefore it is not surprising that certain antibiotics which influence bacteria will also influence all mitochondria and chloroplasts.

(b) The *mt$^+$* strain is the donor of the cpDNA.

10. Parents: $Dd \times dd$

Offspring (F$_1$):   1/2 *Dd*, 1/2 *dd*

(all dextral because of the maternal genotype)

Progeny (F$_2$):

All those from *Dd* parents will be dextral while all those from *dd* parents will be sinistral.

12. Since there is no evidence for segregation patterns typical of chromosomal genes and Mendelian traits, some form of extranuclear inheritance seems possible. If the *lethargic* gene is dominant then a maternal effect may be involved. In that case, some of the F2 progeny would be hyperactive because maternal effects are only temporary, affecting only the immediate progeny. If the lethargic condition is caused by some infective agent, then perhaps injection experiments could be used. If caused by a mitochondrial defect, then the condition would persist in all offspring of lethargic mothers, through more than one generation.

14. (a) The presence of *bcd$^-$/bcd$^-$* males can be explained by the maternal effect: mothers were *bcd$^+$/bcd$^+$*.

(b) The cross

female *bcd$^+$/bcd$^-$* $\times$ male *bcd$^-$/bcd$^-$*

will produce an F1 with normal embryogenesis because of the maternal effect. In the F2, any cross having *bcd$^+$/bcd$^-$* mothers will have phenotypically normal embryos. Any cross involving homozygous *bcd$^-$/bcd$^-$* mothers will have problems with embryogenesis.

16. (a) A locus, Segregation Distortion *(SD)*, is present on the wild type chromosome.

(b) One could use this *SD* chromosome in a variety of crosses and determine that the abnormal segregation is based on a particular chromosomal element. One could even map the *SD* locus on the second chromosome (as has been done).

(c) Segregation Distortion describes a condition in which typical Mendelian segregation is distorted from the 50:50.

Chapter 9

2. The *Protenor* form of sex determination involves the XX/XO condition while the *Lygaeus* mode involves the XX/XY condition.

4. In *primary* nondisjunction half of the gametes contain two X chromosomes while the complementary gametes contain no X chromosomes. Fertilization, by a Y-bearing sperm cell, of those female gametes with two X chromosomes would produce the XXY Klinefelter syndrome. Fertilization of the "no-X" female gamete with a normal X-bearing sperm will produce the Turner syndrome.

6. Males and females share a common placenta and therefore hormonal factors carried in blood. Hormones and other molecular species (transcription factors perhaps) triggered by the presence of a Y chromosome lead to a cascade of developmental events which both suppress female organ development and enhance masculinization.

8. It would be detrimental to sex-determining mechanisms to have sex-determining loci on the Y chromosome transferred, through crossing over, to the X chromosome.

10.

Klinefelter syndrome (XXY)	= 1
Turner syndrome (XO)	= 0
47, XYY	= 0
47, XXX	= 2
48, XXXX	= 3

12. Females will display mosaic retinas with patches of defective color perception.

14. Many organisms have evolved over millions of years under the fine balance of numerous gene products. Many genes required for normal cellular and organismic function in *both* males and females are located on the X chromosome. These gene products have nothing to do with sex determination or sex differentiation.

16. One could account for the significant departures from a 1:1 ratio of males to females by suggesting that at anaphase I of meiosis, the Y chromosome more often goes to the pole which produces the more viable sperm cells. One could also speculate that the Y-bearing sperm has a higher likelihood of surviving in the female reproductive tract, or that the egg surface is more receptive to Y-bearing sperm. At this time the mechanism is unclear.

18. Because of the homology between the red and green genes, there exists the possibility for an irregular synapsis which, following crossing over, would give a chromosome with only one (green) of the duplicated genes.

20. (a) Something is missing from the male-determining system of sex determination either at the level of the genes, gene products, or receptors, etc.

(b) The *SOX9* gene or its product is probably involved in male development. Perhaps it is activated by *SRY*.

(c) There is probably some evolutionary relationship between the *SOX9* gene and *SRY*. There is considerable evidence that many other genes and pseudogenes are also homologous to *SRY*.

(d) Normal female sexual development does not require the *SOX9* gene or gene product(s).

(e) *SRY* may activate *SOX9*, which is also required for normal skeletal development. A good review of *SRY* and its relatives among a variety of vertebrates is found in Jeyasuria and Pace. 1998. *Journal of Experimental Zoology* 281:428–449.

Chapter 10

4. There is a significant maternal age effect associated with Down syndrome and certain genetic and cytogenetic marker data indicate the influence of female nondisjunction.

6. Because an allotetraploid has a possibility of producing bivalents at meiosis I, it would be considered the most fertile of the three. Having an even number of chromosomes to match up at the metaphase I plate, autotetraploids would be considered to be more fertile than autotriploids.

8. It is likely that an interspecific hybridization occurred followed by chromosome doubling. These events probably produced a fertile amphidiploid (allotetraploid).

10. The appearance that crossing over is suppressed in inversion "heterozygotes" extends from the fact that the crossover chromatids end up being abnormal in genetic content. As such they fail to produce viable (or competitive) gametes or lead to zygotic or embryonic death.

12. Females: N^+/N × males: B/Y

1/4	N^+/B	females; Bar
1/4	N/B	females; Notch, Bar
1/4	N^+/Y	males; wild
1/4	N/Y	lethal

The final phenotypic ratio would be 1:1:1 for the phenotypes shown above.

14. By having the genes in an inversion, crossover chromatids are not recovered and therefore are not passed on to future generations. Translocations offer an opportunity for new gene combinations by associations of genes from nonhomologous chromosomes. Under certain conditions such new combinations may be of selective advantage and meiotic conditions have evolved so that segregation of translocated chromosomes yields a relatively uniform set of gametes.

16. The primrose, *Primula kewensis*, with its 36 chromosomes, is likely to have formed from the hybridization and subsequent chromosome doubling of a cross between the two other species, each with 18 chromosomes.

18. The rare double crossovers in the boundaries of a paracentric or pericentric inversion produce only minor departures from the standard chromosomal arrangement as long as the crossovers involve the same two chromatids. With two-strand double crossovers, the second crossover negates the first. However, three-strand and four-strand double crossovers have consequences which lead to anaphase bridges as well as a high degree of genetically unbalanced gametes.

20. In the trisomic, segregation will be "2 × 1" as illustrated below:

P.:

$$b/b/b \times b^+/b^+$$

F.:

$$b^+/b^1/b^2 \qquad \times \qquad b^+/b$$

(normal bristles)   (normal bristles)

F.:

b^+b^+	= normal bristles
$b^+b^1b^2$	= normal bristles
$b^+b^+b^1$	= normal bristles
$b^+b^+b^2$	= normal bristles
b^+b^2	= normal bristles
b^+b^1	= normal bristles
b^+b	= normal bristles
bb^1b^2	= bent bristles
b^+bb^1	= normal bristles
b^+bb^2	= normal bristles
bb^2	= bent bristles
bb^1	= bent bristles

22.

$(35/36)^2$ $W—A—$

$35/(36)^2$ $W—aaaa$

$35/(36)^2$ $wwwwA—$

$1/(36)^2$ $wwwwaaaa$

24.

26. (a) The father must have contributed the abnormal X-linked gene.
(b) Thus non-disjunction must have occurred during meiosis I.
(c) This son's mosaic phenotype is caused by X-chromosome inactivation, a form of dosage compensation in mammals.

Chapter 11

2. Proteins are composed of as many as twenty different subunits (amino acids) thereby providing ample structural and functional variation for the multiple tasks which must be accomplished by the genetic material. The tetranucleotide hypothesis (structure) provided insufficient variability to account for the diverse roles of the genetic material.

4. Specific degradative enzymes, proteases, RNase, and DNase were used to selectively eliminate components of the extract and,

if transformation is concomitantly eliminated, then the eliminated fraction is the transforming principle. DNase eliminates DNA and transformation, therefore DNA must be the transforming principle.

6. The T2 phage, in its mature state, contains very little if any RNA, therefore DNA would be interpreted as being the genetic material in T2 phage.

8. Transformation in bacteria, DNA content in various cell types (sperm and somatic cells), the *action* and *absorption* spectra of ultraviolet light were correlated.

12.

Guanine:	2-amino-6-oxypurine
Cytosine:	2-oxy-4-aminopyrimidine
Thymine:	2,4-dioxy-5-methylpyrimidine
Uracil:	2,4-dioxypyrimidine

16. Because in double-stranded DNA, A$=$T and G$\equiv$C (within limits of experimental error), the data presented would have indicated a lack of pairing of these bases in favor of a single-stranded structure or some other nonhydrogen-bonded structure.

18. Three main differences between RNA and DNA are the following: (1) uracil in RNA replaces thymine in DNA, (2) ribose in RNA replaces deoxyribose in DNA, and (3) RNA often occurs as both single- and double-stranded forms whereas DNA most often occurs in a double-stranded form.

20. The nitrogenous bases of nucleic acids (nucleosides, nucleotides, and single- and double-stranded polynucleotides), absorb UV light maximally at wavelengths 254 to 260 nm. Since proteins absorb UV light maximally at 280 nm, this is a relatively simple way of dealing with mixtures of biologically important molecules. UV absorption is greater in single-stranded molecules (hyperchromic shift) as compared to double-stranded structures and A$=$T rich DNA denatures more readily than G$\equiv$C rich DNA,. Therefore one can estimate base content by denaturation kinetics.

22. Because G$\equiv$C base pairs are more compact, they are more dense than A$=$T pairs. The percentage of G$\equiv$C pairs in DNA is proportional to the buoyant density of the molecule.

24. For curve A in the problem, there is evidence for a rapidly renaturing species (repetitive) and a slowly renaturing species (unique). Fraction B contains mostly unique, relatively complex DNA.

26. It takes more energy (higher temperature) to separate G$\equiv$C pairs.

28. In one sentence of Watson and Crick's paper in *Nature*, they state, "It has not escaped our notice that the specific pairing we have postulated immediately suggests a possible copying mechanism for the genetic material."

30. MS-2 = 200 base pairs

E. coli = 2×10^6 base pairs

32. Left side (a) = left, right side (b) = right.
34. thymine

36. (a) Heat application would yield a hyprochromic if the DNA is double-stranded. One could also get a rough estimation of the GC content from the kinetics of denaturation and the degree of sequence complexity from comparative renaturation studies.

(b) Determination of base content by hydrolysis and chromatography could be used for comparative purposes and could also provide evidence as to the strandedness of the DNA.

(c) Antibodies for Z-DNA could be used to determine the degree of left-handed structures, if present.

(d) Sequencing the DNA from both viruses would indicate sequence homology.

Chapter 12

2. By labeling the pool of nitrogenous bases of the DNA of *E. coli* with a heavy isotope of ^{15}N, it would be possible to "follow" the "old" DNA.

4. (a) Under a conservative scheme all of the newly labeled DNA will go to one sister chromatid, while the other sister chromatid will remain unlabeled.

(b) Under a dispersive scheme all of the newly labeled DNA will be interspersed with unlabeled DNA.

6. DNA template, a divalent cation (Mg^{++}), and all four of the deoxyribonucleoside triphosphates: dATP, dCTP, dTTP, and dGTP.

8. *base composition* and *nearest neighbor frequencies*

10. The *in vitro* rate of DNA synthesis using DNA polymerase I is slow and it is capable of degrading as well as synthesizing DNA. In addition, DeLucia and Cairns discovered a strain of *E. coli* (*polA1*) which still replicated its DNA but was deficient in DNA polymerase I activity.

12. DNA that is capable of supporting typical metabolic activities of the cell or organism and is capable of faithful reproduction.

18. Given a stretch of double-stranded DNA, one could initiate synthesis at a given point and either replicate strands in one direction only (unidirectional) or in both directions (bidirectional). The synthesis of complementary strands occurs in a *continuous* 5′ > 3′ mode on the leading strand in the direction of the replication fork, and in a *discontinuous* 5′ > 3′ mode on the lagging strand opposite the direction of the replication fork.

22. Eukaryotic DNA is replicated in a manner which is very similar to that of *E. coli*. Synthesis is bidirectional, continuous on one strand and discontinuous on the other, and the requirements of synthesis (four deoxyribonucleoside triphosphates, divalent cation, template, and primer) are the same. Okazaki fragments of eukaryotes are about one-tenth the size of those in bacteria. DNA replication is slower and there are multiple initiation sites for replication in eukaryotes.

24. (a) 4,000,000bp.

(b) 1.3mm

26. Gene conversion is now considered a result of heteroduplex formation which is accompanied by mismatched bases. When these mismatches are corrected, the "conversion" occurs.

28. about 8,800 replication sites

30. If replication is conservative, the first autoradiographs would have label distributed only on one side (chromatid) of the metaphase chromosome as shown below.

Conservative Replication

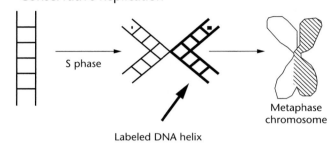

Labeled DNA helix

Metaphase chromosome

Chapter 13

2. No

4. Assume that you have introduced a copolymer (ACACACAC...) to a cell free protein synthesizing system. There are two possibilities for establishing the reading frames: ACA, if one starts at the first base and CAC if one starts at the second base. These would code for two different amino acids (ACA = threonine; CAC = histidine) and would produce repeating polypeptides which would alternate *thr-his-thr-his*... or *his-thr-his-thr*. ...

Because of a triplet code, a trinucleotide sequence will, once initiated, remain in the same reading frame and produce the same code all along the sequence regardless of the initiation site.

Given the sequence CUACUACUACUA, the different reading frames produce three different sequences each containing the same amino acid.

6. Threonine would have the codon ACA.

8. The translation initiation complex is trapped in the filter whereas the components by themselves are not trapped.

10. Apply the most conservative pathway of change.

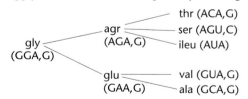

12. Because Poly U is complementary to Poly A, double-stranded structures will be formed.

14. (b) TCCGCGGCTGAGATGA (use complementary bases, substituting T for U), **(c)** GCU, **(d)** Assuming that the AGG... is the 5′ end of the mRNA, then the sequence would be

arg-arg-arg-leu-tyr

16. (a) Starting from the 5′ end and locating the AUG triplets one finds two initiation sites leading to the following two sequences:

met-his-thr-tyr-glu-thr-leu-gly
met-arg-pro-leu-gly

(b) In the shorter of the two reading sequences (the one using the internal AUG triplet), a UGA triplet was introduced at the second codon. While not in the reading frames of the longer polypeptide (using the first AUG codon) the UGA triplet eliminates the product starting at the second initiation codon.

18. DNA produces, through transcription, RNA, which is "decoded" (during translation) to produce proteins.

20. RNA polymerase is composed of subunits (α, β, β', σ) in the proportion $\alpha 2$, β, β', σ for the holoenzyme. The β subunit provides catalytic function while the sigma (σ) subunit is involved in recognition of specific promoters.

22. While some folding (from complementary base pairing) may occur with mRNA molecules, they generally exist as single-stranded structures which are quite labile. Eukaryotic mRNAs are generally processed such that the 5′ end is "capped" and the 3′ end has a considerable string of adenine bases. It is thought that these features protect the mRNAs from degradation. Such stability of eukaryotic mRNAs probably evolved with the differentiation of nuclear and cytoplasmic functions.

24.

Proline:	C_3, and one of the C_2A triplets
Histidine:	one of the C_2A triplets
Threonine:	one C_2A triplet, and one A_2C triplet
Glutamine:	one of the A_2C triplets
Asparagine:	one of the A_2C triplets
Lysine:	A_3

Chapter 14

2. Transfer RNAs are "adaptor" molecules in that they provide a way for amino acids to interact with sequences of bases in nucleic acids. Messenger RNA contains a copy of the triplet codes which are stored in DNA. The sequences of bases in mRNA interact, three at a time, with the anticodons of tRNAs. Enzymes involved in transcription include the following: RNA polymerase (*E. coli*), and RNA polymerase I, II, III (eukaryotes). Those involved in translation include the following: aminoacyl tRNA synthetases, peptidyl transferase, and GTP-dependent release factors.

4. The sequence of base triplets in mRNA constitutes the sequence of codons. A three-base portion of the tRNA constitutes the anticodon.

6. An amino acid in the presence of ATP, Mg^{++}, and a specific aminoacyl synthetase produces an amino acid-AMP enzyme complex (+ PP$_i$). This complex interacts with a specific tRNA to produce the aminoacyl tRNA.

8. Phenylalanine is an amino acid which, like other amino acids, is required for protein synthesis.

10. Tyrosine is a precursor to melanin, skin pigment. Individuals with PKU fail to convert phenylalanine to tyrosine.

12.

trp-8	trp-2	trp-3	trp-1

X -----> AA -----> IGP ----->I ----->TRY

14. Some proteins are made up of subunits, each different type of subunit (polypeptide chain) being under the control of a different gene. Many functions of cells and organisms are controlled by stretches of DNA which either produce no protein product (operator and promoter regions, for example) or have more than one function as in the case of overlapping genes and differential mRNA splicing.

16.

Hemoglobin	Polypeptide chains	
HbA	$2\alpha 2\beta$	(alpha, beta)
HbA₂	$2\alpha 2\delta$	(alpha, delta)
HbF	$2\alpha 2\gamma$	(alpha, gamma)
Gower 1	$2\zeta 2\epsilon$	(zeta, epsilon)

18. In the late 1940's Pauling demonstrated a difference in the electrophoretic mobility of HbA and HbS (sickle-cell hemoglobin) and concluded that the difference had a chemical basis. Ingram determined that the chemical change occurs in the primary structure of the globin portion of the molecule using the fingerprinting technique. He found a change in the 6th amino acid in the β chain.

20. *Colinearity* refers to the sequential arrangement of subunits, amino acids and nitrogenous bases in proteins and DNA, respectively.

22. *Primary:* the sequence of amino acids. This sequence determines the higher level structures. *Secondary:* α-helix and β-pleated-sheet structures generated by hydrogen bonds between components of the peptide bond. *Tertiary:* folding which occurs as a result of interactions of the amino acid side chains. *Quaternary:* the association of two (dimer) or more polypeptide chains. Called *oligomeric*, such a protein is made up of more than one *promoter.*

26. All of the substitutions involve one base change.

28.

(a) F₁: *AABbCC*   =   speckled
      F₂: 3   *AAB_CC*   =   speckled
           1   *AAbbCC*   =   yellow

(b) F₁: *AABbCc*   =   speckled
      F₂: 9   *AAB_C_*   =   speckled
           3   *AAB_cc*   =   green
           3   *AAbbC_*   =   yellow
           1   *AAbbcc*   =   yellow

(c) F₁: *AaBBCc*   =   speckled
      F₂: 9   *A_BBC_*   =   speckled
           3   *A_BBcc*   =   green
           3   *aaBBC_*   =   colorless
           1   *aaBBcc*   =   colorless

30. (a) Outcomes from the last cross with its 9:4:3 ratio suggest two gene pairs.

(b) orange = $Y\_R\_$
yellow = $yyrr$; $yyR\_$
red = Y_rr

(c) white —-> yellow —y-> red —r-> orange

Chapter 15

2. Under *negative* control, the regulatory molecule interferes with transcription while in *positive* control, the regulatory molecule stimulates transcription.

4. The *lac* I^q mutation causes a 10X increase in repressor protein production, thus facilitating its isolation. With the use of dialysis against a radioactive gratuitous inducer (IPTG), Gilbert and Muller-Hill were able to identify the repressor protein in certain extracts of *lac* I^q cells. The material which bound the labeled IPTG was purified and shown to be heat labile and have other characteristics of protein. Extracts of *lac* I^- cells did not bind the labeled IPTG. The IPTG-binding protein was labeled with sulfur-containing amino acids and mixed with DNA from λ phage which contained the *lac* O^+ section of DNA. By glycerol gradient centrifugation it was shown that the labeled repressor protein binds only to DNA which contained the *lac* O^+ region, thus indicating a *specific* binding to DNA.

6.

$I^+ O^+ Z^+ =$ **Inducible**
$I^- O^+ Z^+ =$ **Constitutive**
$I^+ O^c Z^+ =$ **Constitutive**
$I^- O^+ Z^+ /$ F$'$ $I^+ =$ **Inducible**
$I^+ O^c Z^+ /$ F$'$ $O^+ =$ **Constitutive**
$I^s O^+ Z^+ =$ **Repressed**
$I^s O^+ Z^+ /$ F$'$ $I^+ =$ **Repressed**

8. (a) With no lactose and no glucose, the operon is off because the *lac* repressor is bound to the operator and although CAP is bound to its binding site, it will not override the action of the repressor. **(b)** With lactose added to the medium, the *lac* repressor is inactivated and the operon is transcribing the structural genes. With no glucose, the CAP is bound to its binding site, thus enhancing transcription.
(c) With no lactose present in the medium, the *lac* repressor is bound to the operator region, and since glucose inhibits adenyl cyclase, the CAP protein will not interact with its binding site. The operon is therefore "off."
(d) With lactose present, the *lac* repressor is inactivated, however since glucose is also present, CAP will not interact with its binding site. Under this condition transcription is severely diminished and the operon can be considered to be "off."

10. *C* codes for the structural gene. *B* must be the promoter. The mutant *A* gene is dominant to its wild type allele, whereas the mutant *d* allele is recessive (behaving as wild type in the first row). Therefore, the *A* locus is the operator and the *D* locus is the repressor gene.

12. The different results in strains #2 and #4 suggest a *cis*- acting system. Because the operon by itself (when mutant as in strain #3) gives constitutive synthesis of the structural genes, *cis*-acting system is supported. The *cis*-acting element is most likely part of the operon.

14. You will need to identify the complementary regions. To get started, find the CACUUCC sequence. It pairs, with one mismatch with a second region. *Hint:* The third region is composed of seven bases and starts with an AG.

Chapter 16

2. Organization of DNA, Gene amplification, Transcription, Processing and Transport Translation: Posttranslation

4. Transcription factors are proteins which are *necessary* for the initiation of transcription. However, they are not *sufficient* for the initiation of transcription. Transcription factors contain at least two functional domains: one binds to the DNA sequences of promoters and/or enhancers, the other interacts with RNA polymerase or other transcription factors.

6. Both the *lac* and *gal* systems are influenced by catabolite repression, however, the *lac* system is under negative control whereas the *gal* system is under positive control. Both systems are inducible.

8. The model which depicts binding of factors to the nascent polypeptide chain is supported. One might stabilize the proposed MREI-protein complex with "crosslinkers," treat with RNAse to digest mRNA and to break up polysomes, then isolate individual ribosomes. One may use some specific antibody or other method to determine whether tubulin subunits contaminate the ribosome population.

Chapter 17

2. Mutations are the "windows" through which geneticists look at the normal function of genes, cells, and organisms. When a mutation occurs it allows the investigator to formulate questions as to the function of the normal allele of that mutation.

4. *All* mutations may not be deleterious. Those few, rare variations which are beneficial will provide a basis for possible differential propagation of the variation.

6. A diploid organism possesses at least two copies of each gene (except for "hemizygous" genes) and in most cases, the amount of product from one gene of each pair is sufficient for production of a normal phenotype.

8. Mutations constantly occur in both somatic and gametic tissues. When in somatic tissue, the mutation will not be passed to the next generation, however the physiological or structural role of the mutant cell may be compromised, but of little or no impact. If a somatic mutation occurs early in development and if numerous progeny cells are produced from the mutant cell, then a significant tissue mass may have abnormal function. In addition, mutation can alter the normal regulatory aspects of a cell cycle. When such regulation is compromised, then abnormal rates of cell proliferation, cancer, may result. Mutations in the gametic tissue may pose problems for future generations.

10. The rate of induced mutation for each essential X-linked gene is less than 1/500.

12. Important tautomers involve keto-enol pairs for thymine and guanine, and amino-imino pairs for cytosine and adenine.

14. As the reading frame is shifted, new codons are generated. In addition, there is the possibility that a nonsense triplet could be introduced.

16. In contrast to UV light, X-rays penetrate surface layers of cells and thus can affect gamete-forming tissues in multicellular organisms. In addition, X-rays break chromosomes and a variety of chromosomal aberrations can result. Ions and free radicals are formed in the paths of X-rays and these interact with components of DNA to cause mutations. UV light generates pyrimidine dimers, primarily thymine, which distort the normal conformation of DNA and inhibit normal function.

18. Because mammography involves the use of X-rays and X-rays are known to be mutagenic, it has been suggested that frequent mammograms may do harm.

20. In the *Ames assay* the compound to be tested is incubated with a mammalian liver extract to simulate an *in vivo* environment. This solution is then placed on culture plates with an indicator microorganism, *Salmonella typhimurium*, which is defective in its normal repair processes. The frequency of mutations in the tester strains is an indication of the mutagenicity of the compound.

22. *Xeroderma pigmentosum* is a form of human skin cancer caused by perhaps several rare autosomal genes which interfere with the repair of damaged DNA. Studies with heterokaryons provided evidence for complementation, indicating that there may be as many as seven different genes involved. The photoreactivation repair enzyme appears to be involved.

24. 1.5×10^{-6}

26. It is probable that the IS occupied or interrupted normal function of a controlling region related to the galactose genes.

28. Completing complementation pairings allows one to determine the following groupings:

XP1	XP4	XP5
XP2		XP6
XP3		XP7

The groupings (complementation groups) indicate that there are at least three "genes" which form products necessary for unscheduled DNA synthesis. All of the cell lines which are in the same complementation group are defective in the same product.

30. The cystic fibrosis gene produces a complex membrane transport protein which contains several major domains: a highly conserved ATP binding domain, two hydrophobic domains, and a

large cytoplasmic domain which probably serves in a regulatory capacity. The protein is like many ATP-dependent transport systems, some of which have been well studied. When a mutation causes clinical symptoms, fluid secretion is decreased and dehydrated mucus accumulates in the lungs and air passages. Mutations which radically alter the structure of the protein (frameshift, splicing, nonsense, deletions, duplications, *etc.*) would probably have more influence on protein function than those which cause relatively minor amino acid substitutions, although this generalization does not always hold true. A protein with multiple functional domains would be expected to react to mutational insult in a variety of ways.

Chapter 18

2. Eukaryotic mRNAs typically have a 3′ polyA tail. The poly dT segment provides a double-stranded section which serves to prime the production of the complementary strand.

4. 1.38×10^5

6. Given that there is only one site for the action of *Hind*III, then the following will occur. Cuts will be made such that a four base single-stranded set of sticky ends will be produced. For the antibiotic resistance to be present, the ligation will reform the plasmid into its original form. However, two of the plasmids can join to form a dimer which will migrate higher in the gel.

8. All other factors being equal (appropriate cloning sites and selectable markers), it is important to consider the size of the foreign DNA which can be cloned into the vector. Generally for large genomes, it is best to use a vector which will accept relatively large fragments.

10. Assuming a random distribution of all four bases, the four-base sequence would occur (on average) every 256 base pairs (4^4), the six-base sequence every 4096 base pairs (4^6), and the eight-base sequence every 65,536 base pairs (4^8). One might use an eight-base restriction enzyme to produce a relatively few large fragments. If one wanted to construct a eukaryotic genomic library, such large fragments would have to be cloned into special vectors, such as yeast artificial chromosomes.

12.

14. There may be several factors contributing to the lack of representation of the 5′ end of the mRNA. One has to deal with the possibility that the reverse transcriptase may not completely synthesize the DNA from the RNA template. The other reason may be that the 3′ end of the copied DNA tends to fold back on itself thus providing a primer for the DNA polymerase. Additional preparation of the cDNA requires some digestion at the folded region. Since this folded region corresponds to the 5′ end of the mRNA, some of the message is often lost.

16. Option (b) fits the expectation because both bands in the off-spring are found in both parents. Option (d) would be a possibility, however since the primers were stated as being highly polymorphic in humans, it is likely that they are also highly polymorphic in other primates. Thus, the likelihood of such a match is expected to be low in the general population. Additional polymorphic sites would add credibility to the claim in option (b).

18. An appropriate sequence would be the following: 3, 1, 7, 8, 9, 6. Note that (2) may be used to make restriction maps for detemining the direction of the walk and the relationships of overlapping clones.

20. (a) Pattern #5 is the likely choice. (b) The only place of consistent overlap to the probe is the 1kb fragment between A and N.

Chapter 19

2.

mtDNA	cpDNA
circular	circular
double-stranded	double-stranded
semiconservative repl.	semiconservative repl.
animal (16 to 18 kb)	>100 kb
plant (>100 kb)	genes(rRNAs, tRNAs, etc.)
genes (rRNAs, tRNAs, etc.)	
diverse (introns in some)	
variations in genetic code	

4. Puffs represent active genes as evidenced by staining and uptake of labeled RNA precursors as assayed by autoradiography.

6. Eukaryotic cells have a greater variety of cellular structures and are involved in a greater number of functions.

8. In addition to the size (sedimentation properties) of ribosomes, the ribosomes of mitochondria and chloroplasts show antibiotic sensitivity which is similar to present-day bacteria, thus supporting the endosymbiont theory.

10. (a) 5×10^6 nucleosomes
(b) 1×10^6 solenoids
(c) 4.5×10^7 histone molecules
(d) 6.8×10^7 Å

12. 152,941 base pairs

Chapter 20

2. It is likely that a much higher *percentage* of the genome of a prokaryote is actually involved in phenotype production than in a eukaryote. Eukaryotes have evolved the capacity to obtain and maintain what appears to be large amounts of "extra" perhaps "junk" DNA. Prokaryotes are extremely efficient in their accumulation and use of their genome. Given the larger amount of DNA per cell and the requirement that the DNA be partitioned in an orderly fashion to daughter cells during cell division, certain mechanisms and structures (mitosis, nucleosomes, centromeres, etc.) have evolved for *packaging* the DNA. In addition, the genome is divided into separate entities (chromosomes) to perhaps facilitate the partitioning process in mitosis and meiosis.

4. As complexity increases, going from yeast to multicellular eukaryotes, the proportion of genes with introns increases, the number of introns per gene increases, and the size of the introns increases.

6. The *rRNA* gene family consists of tandem repeats of three molecules in the following order: 18S, 5.8S, and 28S rRNA. The initial transcript is processed similar to the removal of introns. Between each three-gene unit in the cluster is a non-transcribed spacer DNA sequence.

8. V_L = variable region of the light chain, C_H = constant region of the heavy chain, IgG = an immunoglobin class which represents approximately 80% of the antibodies in the blood. J = genes that specify a portion of the V region which includes a portion of the hypervariable region. D = a region between V and J in the heavy immunoglobin chain. The *recombination theory* is supported most heavily by experimental evidence.

10. 10^6

12. There are four polypeptide chains in each IgG antibody molecule. There are three different types of polypeptide chains represented in the IgG molecule: κ, λ, γ. Each light chain is encoded by at least three gene segments (C, V, and J) while each heavy chain is encoded by at least four gene segments (C, V, D, and J).

14. Follow the instructions as given. Click directly on the map. The top left will be 0bp, bottom 816,394bp. Click on a given gene, then scroll down to locate it by bp and number in the list.

16. By clicking on any gene you will be taken (a) to its location and information about the gene. (b) They are approximately 287,000bp apart and about 93cM.

18. The telomere will have $C_{1-3}A$ repeats at the start region.

20. Follow the links from the KEGG site. For *Mycoplasma* the direction of transcription is given by the arrows, however for *Saccharomyces*, click MIPS rather than SGD. From there click on Physical Map, III, and the arrows will indicate the direction of transcription.

Chapter 21

2. To generate glyphosate resistance in crop plants a fusion gene was created which introduced a viral promoter to control the EPSP synthetase gene. The fusion product was placed into the Ti vector and transferred to *A. tumifaciens* which was used to infect crop cells. Calluses were selected on the basis of their resistance to glyphosate. Resistant calluses were later developed into transgenic plants. It seems more likely that the trait will not "escape" from the plant; rather that the engineered *A. tumifaciens*

may escape, infect and transfer glyphosate resistance to pest species.

4. (a,b) Several of the problems involving the use of retroviral vectors are the following. (1) Integration into the host must be cell specific so as not to damage non-target cells. (2) Retroviral integration into host cell genomes only occurs if the host cell is replicating. (3) Insertion of the viral genome might influence non-target but essential genes. (4) Retroviral genomes have a low cloning capacity and can not carry large inserted sequences as are many human genes. (5) There is a possibility that recombination with host viruses will produce an infectious virus which may do harm.

(c) The question posed here plays on the practical versus the ethical. It would certainly be more efficient (although perhaps more difficult technically) to engineer germ tissue, for once it is done in a family, the disease would be eliminated. However, there are considerable ethical problems associated with germ plasm therapy. It recalls previous attempts of the eugenics movements of past decades which involved the use of selective breeding to purify the human stock. Some present-day biologists have said publically that germ line gene therapy will *not* be conducted.

6. Difficulties in positional cloning relate to the availability of sufficient RFLP markers in the region of the gene and sufficient kindreds to do the actual genetic association of the gene in question to the RFLP markers. Any mutation or alteration which can serve as a molecular landmark would be useful: VNTRs, alterations which change the banding patterns of chromsomes (deletions, duplication, inversions. translocations), or those which cause sufficient base changes which would alter probe hybridization.

8. The specific probes (or allele-specific oligonucleotides) that have been developed will not necessarily be useful for screening all mutant genes. In addition, the cost-effectiveness of such a screening proposal would need to be considered.

10. In the case of haplo-insufficient mutations, gene therapy holds promise; however in "gain-of-function" mutations in all probability, the mutant gene's activity or product must be compromised. Addition of a normal gene probably will not help.

12. Use the amino acid sequence of the protein to produce the gene synthetically or, if mRNA can be obtained, it can be used to make DNA through the use of reverse transcriptase.

14. (a) Y-linked excluded, X-linked recessive excluded, autosomal recessive possible but unlikely, X-linked dominant possible if heterozygous, autosomal dominant possible. **(b)** Chromosome 21 with the B1 marker probably contains the mutation. **(c)** The disease gene is segregating with some certainty with the B1 RFLP marker in the family. Since the mother also has the B3 marker, the offspring could be tested. If the child carries the B3 marker, then he/she does not carry the B1 marker which has been segregating with the defective gene. However, this prediction is not completely accurate because a crossover in the mother could put the undesirable gene with the B3 marker. **(d)** A crossover between the restriction sites in the father giving a B1 chromosome, or a mutation eliminating either the B2 or B3 restriction site.

Chapter 22

2. Because the replacement of the arista (end of the antenna) can occur by a mutation in a single gene, one would consider that one "selector" gene distinguishes aristal from tarsal structures.

4. If protein products of a given gene are present in different cell types, it can be assumed that the responsible gene is being transcribed. If one is able to actually observe, microscopically, gene activity, as is the case in some specialized chromosomes (polytene chromosomes), gene activity can be inferred by the presence of localized chromosomal puffs. If a labeled probe can be obtained which contains base sequences that are complementary to the transcribed RNA, then such probes will hybridize to that RNA if present in different tissues.

6. The *ftz* gene product regulates, either directly or indirectly, *eng*.

10.

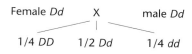

(all phenotypically "D")

12. (a) One may hypothesize that the *her−1*⁺ gene produces a product which suppresses hermaphrodite development, while the *tra−1*⁺ gene product is needed for hermaphrodite development. Information provided in part (b) supports this hypothesis. **(b)** If the *her−1*⁺ product acts as a negative regulator, then when the gene is mutant, suppression over *tra−1*⁺ is lost and hermaphroditism would be the result. This hypothesis fits the information provided. The double mutant should be male because even though there is no suppression from her-1⁻, there is no tra-1⁺ product to support hermaphrodite development.

Chapter 23

2. Review Chapter 2 in the text. Most cell cycle time variation is caused by changes in the duration of G_1.

4. Kinases regulate other proteins by adding phosphate groups. Cyclins bind to the kinases, switching them on and off. Several cyclins, including D and E, can move cells from G_1 to S. At the G_2/mitosis border a CDK1 (cyclin dependent kinase) combines with another cyclin (cyclin B). Phosphorylation occurs bringing about a series of changes in the nuclear membrane, cytoskeleton, and histone 1.

6. A tumor suppressor gene is a gene that normally functions to suppress cell division. Since tumors and cancers represent a significant threat to survival and therefore Darwinian fitness, strong evolutionary forces would favor a variety of co-evolved and perhaps complex conditions in which mutations in these suppressor genes would be recessive.

8. Oncogenes are genes that induce or maintain uncontrolled cellular proliferation associated with cancer. They are mutant forms of proto-oncogenes which normally function to regulate cell division.

12. (a) If the G mutated to an A (transition), then the DNA strand would be $3'-ATC(T)-5'$ which would cause a UAG(A) triplet to be produced and this would cause the stop.

(b) Tumor suppression because loss-of-function causes predisposition to cancer.

(c) Some women may carry genes (perhaps mutant) which "spare" for the *BRAC1* gene product. Some women may have immune systems which recognize and destroy precancerous cells or they may have mutations in breast signal transduction genes so that cell division suppression occurs in the absence of *BRAC1*.

Chapter 24

2. The advantages of *Drosophila* in behavioral studies include the following: an immense repertoire of chromosomal and single-gene alterations and a fairly elaborate set of behaviors (reproductive, locomotor, taxic, *etc.*) which allows one to generate and isolate a variety of abnormal behaviors upon which genetic studies depend. Because of the relative developmental simplicity and developmental properties of *Caenorhabditis*, most scientists view these organisms as eventually contributing greatly to our understanding of such complex processes as development and behavior.

4. One of the easiest ways to determine whether a genetic basis exists for a given abnormality is to cross the abnormal fly to a normal fly. If the trait is determined by a dominant gene, the trait should appear in the offspring, probably half of them if the gene was in the heterozygous state. If the gene is recessive and homozygous, then one may not see expression in the offspring of the first cross. However, if one crosses the F_1 one should see the trait appearing in approximately 1/4 of the offspring. Modifications of these patterns would be expected if the mode of inheritance is X-linked or shows other modifications of typical Mendelian ratios.

One might hypothesize that the focus is in the nervous or muscular system. Mapping the primary focus of the gene could be accomplished with patience and the use of the unstable ring-X chromosome to generate gynandromorphs. Given that the gene is X-linked, one would use classical recombination methods to place a recessive X-linked marker, such as *singed bristles*, on the X chromosome with the gene causing the limp. This would help one identify the male/female boundaries. One would then cross homozygous (for the trait and markers) females to ring-X males, or the reciprocal, then examine the offspring for gynandromorphs (and singed mosaics). If one obtained a pool of gy-

nandromorphs, one could then assess the phenotype (limp or normal) with respect to exposure of the recessive gene in the male tissue. Correlating such would allow one to provide an educated guess as to the primary focus of the gene causing the limp.

6. 1. With a relatively small number of offspring produced per mating, standard genetic methods of analysis are difficult. 2. Records on family illnesses are difficult to obtain. 3. The long generation time. 4. The scientist can not direct matings. 5. The scientists can't always subject humans to the same types of experimental treatments as with other organisms. 6. Traits which are of interest to study are often extremely complex and difficult to quantify.

8. Self-fertilization, the ultimate form of inbreeding, greatly enhances the likelihood that recessive genes will become homozygous.

Chapter 25

2. p = frequency of A
      $q = 1 - p = 0.5$

Frequency of AA = .25 or 25%
Frequency of Aa = .5 or 50%
Frequency of aa = .25 or 25%

After one generation of mating under the Hardy-Weinberg assumptions, the population is in equilibrium and will continue to be so (and not change) until one or more of the Hardy-Weinberg assumptions is not met.

4. In order for the Hardy-Weinberg equations to apply, the population must be in equilibrium.

6. (a)

Frequency of +/+ = .6014 or 60.14%
Frequency of +/$\Delta 32$ = .3482 or 34.82%
Frequency of $\Delta 32$/$\Delta 32$ = .0504 or 5.04%

Comparing these equilibrium values with the observed values strongly suggests that the observed values are drawn from a population in equilibrium.

(b)

Frequency of AA = .7691 or 76.91%
Frequency of AS = .2157 or 21.57%
Frequency of SS = .0151 or 1.51%

$X^2 = 1.47$

In calculating degrees of freedom in a test of gene frequencies, the "free variables" are reduced by an additional degree of freedom because one estimated a parameter (p or q) used in determining the expected values. Therefore, there is one degree of freedom even though there are three classes. Checking the X^2 table with 1 degree of freedom gives a value of 3.84 at the 0.05 probability level. Since the X^2 value calculated here is smaller, the null hypothesis (the observed values fluctuate from the equilibrium values by chance and chance alone) should not be rejected. Thus the frequencies of *AA, AS, SS* sampled a population which is in equilibrium.

8. $q_1 = (q_o - sq_o^2)/(1 - sq_o^2)$,
with $p = .7$ and $q = .3$
and *s* being the selection coefficient,

(a) $q_1 = .23$
      $p_1 = .77$
(b) $q_1 = .267$
      $p_1 = .733$
(c) $q_1 = .293$
      $p_1 = .707$
(d) $q_1 = .299$
      $p_1 = .701$

10. For this question, apply the equations
(a) $p_1 = 0.6 + 0.2(0.1 - 0.6) = 0.5$
(b) $p_1 = 0.2 + 0.3(0.7 - 0.2) = 0.35$
(c) $p_1 = 0.1 + 0.1(0.2 - 0.1) = 0.11$

12. *Inbreeding depression* refers to the reduction in fitness observed in populations which are inbred. With inbreeding comes an increase in the number of homozygous individuals and a decrease in genetic variability. Genetic variability is necessary for a genetic response to environmental change. As deleterious genes become homozygous, more individuals are less fit in the population.

14. While inbreeding increases the frequency of homozygous individuals in a population, it does not change the *gene* frequencies. There will be fewer heterozygotes in the population to compensate for the additional homozygotes.

16. (a) q is 0.01
(b) $p = 1 - q$ or .99
(c) $2pq = 0.0198$ (or about 1/50)
(d) $2pq \times 2pq = 0.000392$ or about 1/255

18. There are 50,000 births, therefore 100,000 genes involved. The frequency of mutation is therefore given as follows:

     2/100,000
     or 2×10^{-5}

20. (a) The gene is most likely recessive because all affected individuals have unaffected parents and the condition clearly runs in families. For the population, since $q^2 = .002$, then $q = .045$, $p = .955$, and $2(pq) = 0.086$. For the community, since $q^2 = .005$, $q = .07$, $p = .93$, and $2(pq) = 0.13$.
(b) The "founder effect" is probably operating here.

Chapter 26

2. As stated in Chapter 25, factors such as selection, migration, genetic drift, or even mutation, may be important in divergent population formation. One would certainly include geographic isolation as a major barrier to gene flow and thus an important process in such formation. *Natural selection* occurs when there is non-random elimination of individuals from a population. Since such selection is a strong force in changing gene

frequencies, it should also be considered as a significant factor in subspecies formation.

4. All of the amino acid substitutions
(Ala -> Gly, Val -> Leu, Asp -> Asn, Met -> Leu)
require only one nucleotide change. The last change from
Pro (CC-) -> Lys (AAA,G)
requires two changes (the minimal mutational distance).

6.

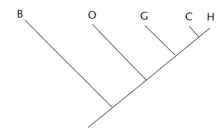

8. In looking at the figure, notice that the $\Delta T_{50}H$ value of 4.0 on the right could be used as a decision point such that any group which diverged above that line would be considered in the same genus while any group below would be in a different genus. Under this rule, one would have the chimpanzee, pygmy chimpanzee, human, gorilla and orangutan in the same genus. If one assumed that 3.7 is close enough to be considered above 4.0, given considerable experimental error, one could provide a scheme where the orangutan is not included with the chimpanzee, pygmy chimpanzee, human, and gorilla.

10. The text lists several cornerstones of the *neutral mutation theory:* (a) the relatively uniform rate of amino acid substitution, (b) there is no particular pattern to the substitutions, (c) the rate of mutation is relatively high and has remained relatively constant for millions of years, (d) certain regions of molecules and certain functions of those molecules should logically be less likely to have amino acid substitutions influence the phenotype, (e) the rate of amino acid substitution in some proteins is much too high to have been produced by selection. The *selectionists* suggest that even though amino acid substitutions *appear* to be neutral, it is more likely that their influence has just not been determined. In addition, they point out that many polymorphisms are clearly maintained in the population *by* selection.

12. Given the small range of *HLA* diversity, one might conclude that the Native Americans descended from a relatively small population either because of a small number who originally arrived or because of significant population crashes over time. Additionally, either the South American groups received an influx of genes from other populations or only small bands of North Americans survived.

Credits

7.14(a) M. Wurtz/Biozentrum, University of Basel/Science Photo Library/Photo Researchers, Inc.

7.16(b) Bruce Iverson

7.20 from Hershey & Chase, 1951

MediaLab screen William Sofer

CO8 Bob Newmann/Visuals Unlimited

8.1(b) Professor Peter A. Peterson, Iowa State University

8.2(b) Eric Grave/Phototake NYC

8.3 John D. Cunningham/Visuals Unlimited

8.4 Dr. Ronald A. Butow, Dept. of Molecular Biology and Oncology, University of Texas Southwestern Medical Center

8.5 Douglas C. Wallace, "Mitochondral Diseases in Man and Mouse." Reprinted with permission from *Science*, 283:1482, Fig. 1, parts A and C, March 5, 1999. © 1999 American Association for the Advancement of Science

8.6(b) John D. Cunningham/Visuals Unlimited

8.9(b) Robert & Linda Mitchell Photography

Table 8.1 Park Seed Company

MediaLab screen Josh Bond, Georgia Perimeter College

CO9 James King-Holmes/Science Photo Library/Photo Researchers, Inc.

9.1(b) Biophoto Assoc./Photo Researchers, Inc.

9.3(b) Bill Beatty/Visuals Unlimited

9.4(a) Dr. Maria Gallegos, University of California, San Francisco

9.6 Courtesy of the Greenwood Genetic Center, Greenwood, SC

9.7 Catherine B. Palmer, Dept. of Medical Genetics/Indiana University-Indianapolis

9.9 Stuart Kenter Associates

9.11(a) Reed/Williams/Animals Animals/Earth Scenes

9.11(b) W. Layer/Okapia/Photo Researchers, Inc.

MediaLab screen Kenneth R. Miller

CO10 Evelin Schrock, Stan du Manoir, and Tom Reid, National Institutes of Health

10.2(a) Courtesy of the Greenwood Genetic Center, Greenwood, SC

10.2(b) Five P Minus Society

10.3(b) Richard Shiell/Animals Animals/Earth Scenes

10.5(a) Courtesy of the Greenwood Genetic Center, Greenwood, SC

10.5(b) William McCoy/Rainbow

10.7(a) David D. Weaver, M.D.

10.8(a) David D. Weaver, M.D.

10.12 Ken Wagner/Phototake NYC

10.13(b) Pfizer, Inc./Phototake NYC

10.18 Mary Lilly/Carnegie Institution of Washington

10.25(b) Dr. Jorge Yunis

10.26 Visuals Unlimited

MediaLab screen Peter V. Sengbusch

CO11 Richard Megna/Fundamental Photographs

11.3 Bruce Iverson

11.5 Olver Meckes/MPI-Tubingen/Photo Researchers, Inc.

11.8(b) Runk/Schoenberger/Grant Heilman Photography, Inc.

11.8(c) Biology Media/Photo Researchers, Inc.

11.13 Science Source/Photo Researchers, Inc.

11.17(a) Ken Eward/Science Source/Photo Researchers, Inc.

11.22 Oncor, Inc.

11.27(b) Dr. William S. Klug

MediaLab screen Nucleic Acid Database Project, Rutgers University

CO12 Dr. Gopal Murti/Science Photo Library/Photo Researchers, Inc.

12.5(a) Walter H. Hodge/Peter Arnold, Inc.

12.5(c) "Molecular Genetics," Pt. 1, p. 74075, J.H. Taylor, ed. Reprinted by permission of Academic Press, Inc. (1963).

12.5(d) "Molecular Genetics," Pt. 1, p. 74075, J.H. Taylor, ed. Reprinted by permission of Academic Press, Inc. (1963).

12.6(b) Sundin and Varshavsky, *Cell* 25:659 (1981). Courtesy of A. Vashavsky.

12.15 H. J. Kreigstein and D.S. Hogness, "Proceedings National Academy of Sciences," 71, 136, 1974. (Fig. 2, p. 137).

12.17 Dr. Harold Weintraub, Howard Hughes Medical Institute, Fred Hutchinson Cancer Center/Essential Molecular Biology" 2e, Freifelder & Malachinski, Jones & Bartlett, Fig. 7-24, p. 141.

12.20(b) David Dressler, Oxford University, England

MediaLab screen David Marcey

CO13 Prof. Oscar L. Miller/Science Photo Library/Photo Researchers, Inc.

13.11(a) Courtesy of Bert W. O'Malley, M.D.

13.11(b) O.L. Miller, Jr., Barbara A. Hamkalo, C.A. Thomas, Jr., *Science*, 169:392-395, 1970 by the American Association for the Advancement of Science

13.15(c) O.L. Miller, Jr., Barbara A. Hamkalo, C.A. Thomas, Jr., *Science*, 169:392-395, 1970 by the American Association for the Advancement of Science

13.15(d) O.L. Miller, Jr., B.R. Beatty, *Journal of Cellular Physiology*, vol. 74 (1969). Reprinted by permission of Wiley-Liss, Inc., a subsidiary of John Wiley & Sons, Inc.

MediaLab screen Steven Hahn, Howard Hughes Medical Institute and Fred Hutchinson Cancer Center

CO14 J. Gross/Science Photo Library/Photo Researchers, Inc.

14.9(a) from Rich et al., 1963. Reproduced by permission of the Cold Spring Harbor Laboratory Press Symp. 38 (1963) lfg 4D, p. 273, © 1964

14.9(b) E.V. Kiseleva

14.13(a) Dennis Kunkel/Phototake NYC

14.13(b) Francis Leroy/Biocosmos/Science Photo Library/Photo Researchers, Inc.

14.19

MediaLab screen Courtesy of WGBH Educational Foundation

CO15 Thomas A. Steitz, Yale University, Department of Molecular Biophysics and Biochemistry

15.11(a) Science Lewis et al./Johnson Research Foundation

15.11(b) Science Lewis et al. 271 pp. 1247–1254/Johnson Research Foundation

15.11(c) Science Lewis et al./Johnson Research Foundation

MediaLab screen University of Tennessee

CO16 Yale University Medical Center and the Howard Hughes Medical Institute/Dr. Paul Sigler

16.7 Dr. Paul Sigler

MediaLab screen University of British Columbia

CO17 Francis Leroy/Biocosmos/Science Photo Library/Photo Researchers, Inc.

17.1(a) John D. Cunningham/Visuals Unlimited

17.4(b) Sue Ford/Science Photo Library/Photo Researchers, Inc.

Index

Bold page number indicates a figure. *Italic* page number indicates a table.

System Requirements:

Macintosh
PowerPC
System 7.0 or above
10 MB RAM
4X CD-ROM drive
Internet browser 4.0 or above with Flash plugin
(included on CD-ROM)
256-Color monitor with 640x480 pixel resolution

Windows
Pentium processor
Windows 95/98/NT 4.0x
10 MB RAM
4X CD-ROM drive
Internet browser 4.0 or above with Flash plugin
(included on CD-ROM)
256-Color monitor with 640x480 pixel resolution

To Begin:

The Concepts of Genetics 6/e CD-ROM requires an Internet browser 4.0 or above with the Flash Player plugin to run properly. If this is already your default configuration, simply open the CD-ROM directory and double-click index.html.

For more information, consult the README file available on the CD-ROM.